MOLECULAR PRINCIPLES OF
FUNGAL PATHOGENESIS

MOLECULAR PRINCIPLES OF
FUNGAL PATHOGENESIS

EDITED BY

JOSEPH HEITMAN
Duke University Medical School
Durham, North Carolina

SCOTT G. FILLER
Harbor-UCLA Medical Center
Torrance, California

JOHN E. EDWARDS, Jr.
Harbor-UCLA Medical Center
Torrance, California

AND

AARON P. MITCHELL
Columbia University College of Physicians & Surgeons
New York, New York

ASM PRESS

Washington, DC

Top left: Bud cell of *Cryptococcus neoformans* separating from parent cell in human cerebrospinal fluid. India ink preparation, differential interference contrast microscopy. Photo by Wiley A. Schell.

Top right: Conidiophore of *Aspergillus flavus* var. *columnaris*. In vitro sporulation, differential interference contrast microscopy. Photo by Wiley A. Schell.

Lower left: Depth view of a *Candida albicans* biofilm, stained with Alexa-ConA, with pseudocolor added to represent depth. Photo provided by Clarissa J. Nobile.

Lower right: Top view of the coils in the cleistothecia produced by mating an American strain and an African strain of *Histoplasma capsulatum*. Photo provided by K. J. Kwon-Chung.

Copyright © 2006 ASM Press
American Society for Microbiology
1752 N Street, N.W.
Washington, DC 20036-2804

Library of Congress Cataloging-in-Publication Data

Molecular principles of fungal pathogenesis / edited by Joseph Heitman . . . [et al.].
   p. ; cm.
Includes bibliographical references and index.
ISBN-13: 978-1-55581-368-0
ISBN-10: 1-55581-368-2
   1. Pathogenic fungi—Molecular aspects.   2. Histology, Pathological.
   [DNLM: 1. Fungi—pathogenicity. 2. Fungi—genetics. 3. Fungi—immunology.   4. Genome, Fungal.
5. Mycoses.   QW 180 M7186 2006] I. Heitman, Joseph.

QR245.M65 2006
616.9′6901—dc22

                                                                                2006006924

All Rights Reserved
Printed in the United States of America

10 9 8 7 6 5 4 3 2 1

Address editorial correspondence to: ASM Press, 1752 N St., N.W., Washington, DC 20036-2904, U.S.A.

Send orders to: ASM Press, P.O. Box 605, Herndon, VA 20172, U.S.A.
Phone: 800-546-2416; 703-661-1593
Fax: 703-661-1501
Email: Books@asmusa.org
Online: estore.asm.org

Myra Kurtz was trained in Bacterial Molecular Genetics at Harvard University (Ph.D. 1971) and soon afterwards developed an independent project on *Aspergillus nidulans* development in the laboratory of Sewell Champe at Rutgers. Her main impact on our field was through her career of over 20 years in the pharmaceutical industry, first with Squibb and then with Merck. She developed the first molecular genetic methods for *Candida albicans* and spearheaded the development of caspofungin. She had numerous interactions with the academic community through collaborations and extensive service as a reviewer and organizer. We remember her pronounced enthusiasm for new ideas and rigorous science, which serves as an inspiration to us all.
(Photo courtesy of Jennifer Nielsen Kahn)

CONTENTS

III.   Specific Pathogens

IV.   The Host

V. Future Directions

CONTRIBUTORS

Alex Andrianopoulos
Dept. of Genetics, University of Melbourne, Melbourne, Victoria, 3010, Australia

Frederick M. Ausubel
Dept. of Molecular Biology, Wellman 10, Massachusetts General Hospital, Boston, MA 02114

Catherine Bachewich
Biology Dept., Concordia University, 7141 Sherbrooke St. West, Montreal, Quebec H4B 1R6, Canada

Charles Boone
Banting and Best Dept. of Medical Research, University of Toronto, Toronto, Ontario M5G 1L6, Canada

Priscila Borra
South Texas Center for Emerging Infectious Diseases, Dept. of Biology, University of Texas at San Antonio, San Antonio, TX 78249

Ricardo Borra
South Texas Center for Emerging Infectious Diseases, Dept. of Biology, University of Texas at San Antonio, San Antonio, TX 78249

Axel A. Brakhage
Dept. of Molecular and Applied Microbiology, Leibniz Institute for Natural Products, Research and Infection Biology, Hans Knoell Institute, D-07745 Jena, Germany

Alistair J. P. Brown
Aberdeen Fungal Group, School of Medical Sciences, Institute of Medical Sciences, University of Aberdeen, Foresterhill, Aberdeen AB25 2ZD, United Kingdom

Howard Bussey
Dept. of Biology, McGill University, Montreal, Quebec H2A 1B1, Canada

Stephen B. Calderwood
Division of Infectious Diseases, Massachusetts General Hospital, 55 Fruit St., Gray 5, GRJ-504, Boston, MA 02114

Arturo Casadevall
Depts. of Medicine, Microbiology and Immunology, Albert Einstein College of Medicine, Bronx, NY 10461

Irene Castaño
Instituto Potosino de Investigacion Cientifica y Tecnologica, Division de Biologia Molecular, Camino a la Presa San Jose #2055, 78216 San Luis Potosi, San Luis Potosi, Mexico

Garry T. Cole
South Texas Center for Emerging Infectious Diseases, Dept. of Biology, University of Texas at San Antonio, San Antonio, TX 78249

Brendan P. Cormack
Dept. of Molecular Biology and Genetics, Johns Hopkins University School of Medicine, Baltimore, MD 21205

Karoll Cortez
Immunocompromised Host Section, Pediatric Oncology Branch, National Cancer Institute, Bethesda, MD 20892

Melanie T. Cushion
Dept. of Internal Medicine, Division of Infectious Diseases, University of Cincinnati College of Medicine, 231 Albert Sabin Way, Cincinnati, OH 45267-0560

Alejandro De Las Peñas
Instituto Potosino de Investigacion Cientifica y Tecnologica, Division de Biologia Molecular, Camino a la Presa San Jose #2055, 78216 San Luis Potosi, San Luis Potosi, Mexico

George S. Deepe, Jr.
Dept. of Medicine, University of Cincinnati College
of Medicine, 231 Bethesda Ave. [ML560],
Cincinnati, OH 45267

Fred S. Dietrich
Dept. of Molecular Genetics and Microbiology,
287 CARL Bldg., Duke University Medical Center,
Durham, NC 27710

John E. Edwards, Jr.
Los Angeles Biomedical Research Institute at
Harbor-UCLA Medical Center, Torrance, CA
90502, and The David Geffen School of Medicine
at UCLA, Los Angeles, California

Paul L. Fidel, Jr.
Dept. of Microbiology, Immunology, and
Parasitology, Louisiana State University Health
Sciences Center, New Orleans, LA 70112

Scott G. Filler
Los Angeles Biomedical Research Institute at
Harbor-UCLA Medical Center, Torrance, CA
90502, and The David Geffen School of Medicine
at UCLA, Los Angeles, California

Beatriz Finkel-Jimenez
Dept. of Microbiology, Immunology, and
Parasitology, Louisiana State University Health
Sciences Center, New Orleans, LA 70112

James A. Fraser
Dept. of Molecular Genetics and Microbiology, Duke
University Medical Center, Durham, NC 27710

Mahmoud A. Ghannoum
Center for Medical Mycology, Dept. of Dermatology,
University Hospitals of Cleveland and Case Western
Reserve University, Cleveland, OH 44106-5028

Neil A. R. Gow
Aberdeen Fungal Group, School of Medical Sciences,
Institute of Medical Sciences, University of Aberdeen,
Foresterhill, Aberdeen AB25 2ZD, United Kingdom

Joseph Heitman
Dept. of Molecular Genetics and Microbiology, Duke
University Medical Center, Durham, NC 27710

Deborah A. Hogan
Dept. of Microbiology and Immunology,
Dartmouth Medical School, Hanover, NH 03755

Gary B. Huffnagle
Dept. of Internal Medicine, Division of Pulmonary
& Critical Care Medicine, University of Michigan,
Ann Arbor, MI 48109

Christina M. Hull
Dept. of Biomolecular Chemistry and Dept. of
Medical Microbiology & Immunology, University
of Wisconsin, Madison, WI 53706

Chiung-Yu Hung
South Texas Center for Emerging Infectious
Diseases, Dept. of Biology, University of Texas at
San Antonio, San Antonio, TX 78249

Lena Hwang
Dept. of Microbiology and Immunology, University
of California, San Francisco, 513 Parnassus, Box
0414, San Francisco, CA 94143-0414

Ashraf S. Ibrahim
Division of Infectious Diseases, David Geffen
School of Medicine at the University of California,
Los Angeles, and Los Angeles Biomedical Research
Institute at Harbor-UCLA Medical Center, 1124 W.
Carson St., Torrance, CA 90502

Timothy Y. James
Dept. of Biology, Duke University, Durham, NC
27708

Alexander Johnson
Dept. of Microbiology and Immunology, University
of California, San Francisco, San Francisco, CA
94158

Nancy Keller
Dept. of Plant Pathology, University of Wisconsin,
Madison, WI 53706

Bruce Klein
Dept. of Pediatrics, Internal Medicine, and Medical
Microbiology and Immunology, and the University
of Wisconsin Comprehensive Cancer Center,
University of Wisconsin, Madison, WI 53792

Roberto Kolter
Dept. of Microbiology and Molecular Genetics,
Harvard Medical School, Boston, MA 02115

Thomas R. Kozel
Dept. of Microbiology and Immunology,
University of Nevada School of Medicine, Reno,
NV 59557

Kyung J. Kwon-Chung
Laboratory of Clinical Infectious Diseases,
National Institute of Allergy and Infectious
Diseases, National Institutes of Health, Bethesda,
MD 20892

Stuart M. Levitz
Depts. of Medicine and Microbiology, Boston
University, Boston, MA 02118

Dennis M. Lindell
Dept. of Internal Medicine, Division of Pulmonary &
Critical Care Medicine, University of Michigan,
Ann Arbor, MI 48109

Jennifer K. Lodge
Edward A. Doisy Dept. of Biochemistry and
Molecular Biology, Saint Louis University School of
Medicine, 1402 S. Grand Blvd., St. Louis, MO 63104

José L. López-Ribot
Dept. of Biology, The University of Texas at San
Antonio, San Antonio, TX 78249

Michael C. Lorenz
Dept. of Microbiology and Molecular Genetics, The
University of Texas Health Science Center, 6431
Fannin, Houston, TX 77030

Caron Lyman
Immunocompromised Host Section, Pediatric
Oncology Branch, National Cancer Institute,
Bethesda, MD 20892

John H. McCusker
Dept. of Molecular Genetics & Microbiology, 3020,
Duke University Medical Center, Durham, NC 27710

Mathew Miller
Dept. of Microbiology and Immunology, University
of California, San Francisco, San Francisco,
CA 94158

Aaron P. Mitchell
Dept. of Microbiology, Columbia University, New
York, NY 10032

Eleftherios Mylonakis
Division of Infectious Diseases, Massachusetts
General Hospital, 55 Fruit St., Gray 5, GRJ-504,
Boston, MA 02114

Simon L. Newman
Dept. of Medicine, Division of Infectious Diseases,
University of Cincinnati College of Medicine,
Cincinnati, OH 45267

Clarissa J. Nobile
Dept. of Microbiology, Columbia University, New
York, NY 10032, and Biological Sciences Program,
Dept. of Biological Sciences, Columbia University,
New York, NY 10027

Joshua D. Nosanchuk
Dept. of Medicine, Albert Einstein College of
Medicine, Bronx, NY 10461

Mairi C. Noverr
Dept. of Internal Medicine, Division of Pulmonary &
Critical Care Medicine, University of Michigan,
Ann Arbor, MI 48109

Frank C. Odds
Aberdeen Fungal Group, School of Medical
Sciences, Institute of Medical Sciences, University of
Aberdeen, Foresterhill, Aberdeen AB25 2ZD,
United Kingdom

Ainslie B. Parsons
Banting and Best Dept. of Medical Research, University
of Toronto, Toronto, Ontario M5G 1L6, Canada

John R. Perfect
Dept. of Medicine, Duke University Medical Center,
Durham, NC 27710

Liise-anne Pirofski
Depts. of Medicine, Microbiology and Immunology,
Albert Einstein College of Medicine, Bronx, NY
10461

Gordon Ramage
Dept. of Biological and Biomedical Sciences,
Glasgow Caledonian University, Glasgow G4 0BA,
United Kingdom

Judith C. Rhodes
Dept. of Pathology and Laboratory Medicine,
University of Cincinnati College of Medicine,
Cincinnati, OH 45267-0529

Emmanuel Roilides
Immunocompromised Host Section, Pediatric
Oncology Branch, National Cancer Institute,
Bethesda, MD 20892

Luigina Romani
Dept. of Experimental Medicine and Biochemical
Science, University of Perugia, 06122 Perugia, Italy

Dominique Sanglard
Institute of Microbiology, University Hospital
Lausanne, Rue de Bugnon 44, CH-1011 Lausanne,
Switzerland

Ruth Schaller
South Texas Center for Emerging Infectious
Diseases, Dept. of Biology, University of Texas at
San Antonio, San Antonio, TX 78249

Kalpathi Seshan
South Texas Center for Emerging Infectious
Diseases, Dept. of Biology, University of Texas at
San Antonio, San Antonio, TX 78249

Donald C. Sheppard
Depts. of Microbiology and Immunology, Medicine,
McGill University, Montreal, Quebec, Canada

Anita Sil
Dept. of Microbiology and Immunology, University
of California, San Francisco, 513 Parnassus, Box
0414, San Francisco, CA 94143-0414

A. George Smulian
University of Cincinnati College of Medicine,
VAMC, 3200 Vine St., Cincinnati, OH 45220

David R. Soll
Dept. of Biological Sciences, The University of
Iowa, Iowa City, IA 52242

Brad Spellberg
Division of Infectious Diseases, David Geffen
School of Medicine at the University of California,
Los Angeles, and Los Angeles Biomedical Research
Institute at Harbor-UCLA Medical Center, 1124
W. Carson St., Torrance, CA 90502

Jason E. Stajich
University Program in Genetics & Genomics, Dept.
of Molecular Genetics and Microbiology, Institute
for Genome Science & Policy, Duke University,
Durham, NC 27710

Paula Sundstrom
Microbiology and Immunology, Dartmouth Medical
School, Hanover, NH 03755

Paul J. Szaniszlo
Section of Molecular Genetics and Microbiology
and Institute of Cell and Molecular Biology,
The University of Texas at Austin, Austin,
TX 78712-0162

Eric Tarcha
South Texas Center for Emerging Infectious
Diseases, Dept. of Biology, University of Texas at
San Antonio, San Antonio, TX 78249

John W. Taylor
Dept. of Plant and Microbial Biology, University of
California, Berkeley, Berkeley, CA 94720-3102

Galen B. Toews
Dept. of Internal Medicine, Division of Pulmonary &
Critical Care Medicine, University of Michigan,
Ann Arbor, MI 48109

Rytas Vilgalys
Dept. of Biology, Duke University, Durham, NC
27708

Thomas J. Walsh
Immunocompromised Host Section, Pediatric
Oncology Branch, National Cancer Institute,
Bethesda, MD 20892

Theodore C. White
Dept. of Pathobiology, School of Public Health and
Community Medicine, University of Washington, and
Seattle Biomedical Research Institute, 307 Westlake
Ave., N., Suite 500, Seattle, WA 98109-5219

Malcolm Whiteway
Health Sector, Biotechnology Research Institute,
National Research Council of Canada, 6100
Royalmount Ave., Montreal, Quebec H4P 2R2,
Canada

Brian L. Wickes
Dept. of Microbiology and Immunology, The
University of Texas Health Center at San Antonio,
San Antonio, TX 78284

Jon P. Woods
Dept. of Medical Microbiology and Immunology,
University of Wisconsin, Madison, WI 53706

Jianmin Xue
South Texas Center for Emerging Infectious
Diseases, Dept. of Biology, University of Texas at
San Antonio, San Antonio, TX 78249

Lauren E. Yauch
Depts. of Medicine and Microbiology, Boston
University, Boston, MA 02118

Jieh-Juen Yu
South Texas Center for Emerging Infectious
Diseases, Dept. of Biology, University of Texas at
San Antonio, San Antonio, TX 78249

Sophie Zuber
Quality and Safety Assurance Dept., Nestle
Research Center, Vers-chez-les-Blanc, Box 44,
CH-1000 Lausanne 26, Switzerland

PREFACE

The study of infectious diseases is an ancient one, arising out of a desire to understand the dramatic maladies that can befall individuals. We now appreciate that infectious agents occur as bacteria, parasites, viruses, prions, and fungi. Yet only two of these classes are eukaryotic cells like our own: the fungi and the parasites. As such, these organisms can serve as models for the conserved machinery present in all eukaryotes; they also represent a special challenge in the development of therapies, as fewer pathogen-specific targets are available. Historically, our understanding of the parasites and pathogenic fungi has lagged behind advances in bacteriology and virology, yet, with the advent of improved genetics approaches and the impact of a multitude of genome projects, our understanding of these pathogenic agents now rivals our understanding of bacteria and viruses. Given the current impact of pathogenic fungi as agents of infection in humans, it seemed timely to bring together experts in the area to review the current state of knowledge.

The conception of this book arose out of our shared experiences in teaching the Molecular Mycology course at the Marine Biological Laboratories in Woods Hole, Massachusetts. Beginning in 1997, each of us has made an annual sojourn to Woods Hole to teach and, joined by 18 students, four teaching assistants, and a cadre of invited faculty, focus on the basic principles and approaches to the study of pathogenic fungi. This course includes state-of-the-art molecular and genetic approaches wedded with classical mycology and the use of both cell lines and animal models to probe the host-pathogen interface.

Increasingly, however, we found ourselves in dire need of a definitive text, not only for the students' use but also to ensure that we ourselves were keeping abreast of rapid developments in our own field. Here we have worked to bring together leading experts in the field to compose a series of chapters spanning a cross section of the field. It is our hope and intent that the product of these efforts, *Molecular Principles of Fungal Pathogenesis*, will serve as a valued resource to the entire community. We hope that this text not only encapsulates all that has been accomplished in the field over the past several intense decades, but will also serve to illuminate areas about which our current state of knowledge is insufficient and which will serve to drive research in the field. As well, on the eve of the upcoming 10th anniversary of the course in August 2006, it also seemed a propitious time to review the state of the art in the field and set the stage for the next decade of teaching, research, and discovery.

The book is organized into five sections covering general principles, model host systems, specific pathogens, the host, and future directions in the areas of genomics and proteomics. Each chapter is self-contained enough such that one should be able to read any individual chapter in isolation. Alternatively, the entire book could be read as one continuous text. We have worked to include areas of current excitement, including model heterologous hosts and the impact of genomics. Because much of the field is divided into individuals focusing on specific organisms, we have organized some chapters by this thematic device, but we also include a large number of chapters that span multiple organisms. The deliberate intent here is to bring a common voice to the field and to ensure that common principles shared among diverse pathogens are not overlooked. Wherever possible we have also tried to ensure that general principles of relevance beyond the specific confines of mycology are considered. Finally, because the nature of pathogenesis involves a dialog between an infectious agent and the host, a critical feature of the book is a section on the second component of this interaction. We are cognizant that the immune system is a formidable topic, and thus it has not been our intent to review it in all its glory, but rather to focus on specific aspects relevant in the realm of fungal pathogenesis.

Any text is ultimately judged by how well it stands the test of time, but another measure of success is when a text stimulates and advances a field so well it is ultimately outdated. As all good scientific models are ultimately replaced by more refined models, it is our hope that this book will stimulate the field, but that a decade hence we will have progressed to the point where it will be time to refresh our thinking once again. We invite you to peruse and study the current state of knowledge here, and hope that these musings will spur you to further advances in the field. We invite you to communicate to us your experiences with the book.

We thank the numerous individuals who have contributed both to the success of the Woods Hole Mycology course and, by extension, to the inception and realization of this text. Finally, we dedicate this effort to the students of the Mycology course, past and future, for stimulating this further teaching experiment, and also to our families, without whose forbearance and tolerance this book would not have been possible.

Joseph Heitman, *Duke University*
Scott G. Filler, *Harbor-UCLA Medical Center*
John E. Edwards, Jr., *Harbor-UCLA Medical Center*
Aaron P. Mitchell, *Columbia University*

I. GENERAL PRINCIPLES

Molecular Principles of Fungal Pathogenesis
Edited by Joseph Heitman et al.
©2006 ASM Press, Washington, D.C.

Chapter 1

Fungal Molecular Pathogenesis:
What Can It Do and Why Do We Need It?

JOHN R. PERFECT AND ARTURO CASADEVALL

Human fungal diseases are largely a 20th- and 21st-century phenomenon. Most systemic fungal diseases were described in the early 1900s at a time when the pathogenic potential of certain endemic fungi was first recognized. Two medical developments in the mid-1900s resulted in a major increase in the number of fungal diseases: the clinical use of corticosteroids and the discovery of antibacterial drugs. However, it has been in the last two decades that fungal diseases have come to the forefront of medical practice. As potent antibacterial agents were widely administered in empirical fashion and a variety of foreign body devices were inserted into critically ill patients, these patients became susceptible to fungal diseases. In fact, with medical advances in the management of cancers with chemotherapy and antibodies and the management of organ transplantations with potent immunosuppressive agents, an unintended consequence has been a dramatic rise in the number of individuals with impaired immunity at risk for yeast- and mold-related diseases. Furthermore, advances in surgery and the increasing use of prosthetic devices inside the host are also associated with increased risk for fungal diseases. These nosocomial developments in invasive mycoses were paralleled over the last two decades by the human immunodeficiency virus (HIV) pandemic, which has resulted in an even larger number of patients at risk for fungal diseases. There is simply no location in the world that has not seen a dramatic recent rise in invasive mycoses. This in turn has resulted in a terrible toll of morbidity, mortality, and costs taken on the human population. Fortunately, several highly effective antifungal agents, which provide options for the therapy of fungal diseases, have been recently introduced. It is reassuring that clinicians in advanced medical systems have access to three major classes of antifungal agents for systemic use, but there remains significant clinical resistance to invasive mycoses and solutions for their management continue to evolve.

During the last two decades, when the incidence of invasive mycoses has dramatically risen, there has also been a significant revolution in the molecular biology and genetics of medically important fungi. The platform for progress in the molecular biology of fungi has been substantially built on the foundation of knowledge generated from studies of *Saccharomyces cerevisiae*. A significant part of our general understanding of eukaryotic molecular biology, genetics, and biochemistry has come from studies with this model organism. However, the last decade has witnessed a significant change in research emphasis as investigators have now begun to carry out direct molecular studies of the specific fungal pathogen. There are presently numerous scientists focused on molecular, biological, and immunological studies of *Candida albicans*, *Candida glabrata*, *Cryptococcus neoformans*, *Aspergillus fumigatus*, the dimorphic fungi (*Coccidioides*, *Histoplasma*, and *Blastomyces*), and the less common black molds (*Wangiella*). These early molecular biological strategies have now paved the road for a renaissance in pathogenesis studies from the fungal side of the host-parasite equation. With the completion of the genome sequences of many of the major fungal pathogens (*Candida albicans*, *A. fumigatus*, *Cryptococcus neoformans*) (28, 34) and substantial progress in the determination of the genome sequences of some agents of endemic mycoses (*Coccidioides* and *Histoplasma*), our ability to understand the microbial variables that contribute to virulence should be significantly enhanced in future years (36). Over 1,500,000 fungal species are present in the Earth's biosphere, but only several dozen fungal species can consistently produce disease in

John R. Perfect • Division of Infectious Diseases, Department of Medicine, Duke University Medical Center, Durham, NC 27710.
Arturo Casadevall • Department of Medicine, Albert Einstein Medical College, Bronx, NY 10461.

humans. Fungal pathogens either originate from the endogenous host flora (*Candida*) or are acquired from the host's immediate environment. Diseases associated with the commensal flora are usually associated with immune suppression or a breakdown in our protective integument. On the other hand, those acquired from the environment have the capacity through their natural niche selection of characteristics to establish themselves in mammalian hosts, often in the setting of an apparent intact immunity.

As molecular studies are applied to study the pathogenesis of invasive mycoses, it is essential that we describe the parameters of the host-parasite interactions. Casadevall and Pirofski have carefully defined host-parasite interactions into a "damage response" framework (8). When considering fungi with pathogenic potential, the damage response framework describes infection as the acquisition of a fungus on or within a host, and evidence of infection only documents the presence of a fungus in a host but does not mean that it is necessarily a cause of disease. Infection results in disease when the host-fungus interactions produce sufficient damage to disrupt normal homeostasis and/or become clinically apparent. This definition and the concept it supports are important as we attempt to fully understand molecular fungal pathogenesis. Many fungus-host interactions are carefully balanced between the two parties toward a final end point of the interactions. Hence, the outcome of infection can be clearance, asymptomatic infection, latency, or disease, and this outcome is a function of the interaction between the microbe and the host. From the microbial perspective, all the genes and their respective null mutants or knockout strains must be eventually related to disease and their outcome in various animal models must be carefully correlated with this goal. From the host perspective, the genetic correlates of susceptibility and resistance need to be understood and correlated with the outcome of infection. Fortunately, the molecular age has reached the mammalian genome, and the ability to manipulate the host through studies with transgenic mice or the ability to perform transcriptional or translational profiling of the host response allows rapid assessment of how the host responds to a fungal challenge (17). Although such studies are in their infancy with regard to mycotic diseases, evidence from other fields suggests that both genetics and the study of gene polymorphisms will allow specific identification of host susceptibility or resistance genes. The host-parasite interaction is intricately molded together, and scientific advances mean that both microbe-centric and host-centric investigators now have the molecular tools to dissect the molecular characteristics that define the outcome of the interaction.

With the recent developing power of molecular genetics in medical mycology, will this progress be used to simply satisfy intellectual curiosity in pathobiology or can it be directed toward translational studies to help in the management of invasive mycoses? In fact, molecular pathogenesis research is likely to do both. A strong argument can be made that molecular biology studies will be the primary weapon to combat clinical drug resistance which now occurs in a substantial number of human mycoses. For instance, the attributable mortality continues to be listed at 30 to 40% for cases of candidemia or invasive aspergillosis (21, 48). The issue of clinical antifungal drug resistance and the impact of molecular studies on these invasive mycoses can be discussed under several categories: (i) diagnosis, (ii) immunomodulation, (iii) drug prescription, (iv) surgery, (v) prevention, (vi) new antifungal agents, and (vii) drug combinations. The following discussion includes an illustration of how molecular studies of medically relevant fungi have begun to find uses in the clinical arena and discusses their potential to make a major impact on the outcome of human mycoses.

DIAGNOSIS

There are two major issues in clinical diagnosis in which molecular biology is likely to lead to significant advances. First, despite the availability of several new diagnostic tests for invasive candidiasis and aspergillosis, which include serologic tests for measurement of fungal products and the frequent use of imaging techniques such as X rays, computed tomography, and magnetic resonance imaging, there remain, from a clinical standpoint, opportunities to make an earlier specific fungal diagnosis or prevent the use of empirical antifungal therapies. The use of PCR techniques continues to be evaluated, and in most cases PCR competes favorably with other testing options. In fact, there are situations in which molecular biology-based tests outperform standard tests. For instance, when used with dermatological specimens from patients with skin diseases, a LightCycler PCR protocol provided new fungal information from over 20% of specimens that was not available from the use of standard methods (22). When used with the nonsterile bronchoalveolar lavage fluids in hematopoietic stem cell transplant patients, a quantitative PCR test showed excellent specificity for aspergillosis and appeared to correlate with the burden of organisms in the host (42). Furthermore, the use of real-time panfungal PCR in blood has performed reasonably well in high-risk patients (29). With the recent use of nucleic acid sequence-based amplification that amplifies mRNA, the detection system suggests that the pathogen is active rather than dead or latent (60). Despite these results, the use of fungal PCR technology remains primarily a research tool. Some of this inertia against its more widespread use is a result of the cost of converting clinical laboratories to these methods. However, there are also issues of sensitivity, and this is particularly true with the frequent use of empirical and prophylactic antifungal strategies in the clinics. Furthermore, there are concerns about the specificity of molecular tests. For instance, sampling of nonsterile sites such as the human sinuses can give discordant results between fungal cultures and amplification of fungal DNA (50). As these PCR tests increasingly become a part of the diagnostic lexicon, it will be necessary to rigorously validate them by careful clinical research and treatment algorithm designs. For example, what does it mean for a clinical condition to have fungal DNA or RNA present in a clinical specimen and what will be the therapeutic response? Despite these issues, it is clear that molecular biology-based diagnostics have great potential to rapidly, accurately, and quantitatively establish the presence or absence of a fungus in the host. The challenges for these testing methods are (i) to make them clinically usable in a routine clinical laboratory with direct specimens and avoidance of contamination; (ii) to improve the speed of analysis compared to culture and serologic testing; (iii) to optimize the sensitivity of detection in the setting of a low burden of organisms without compromising specificity; (iv) to validate the clinical relevance of quantitative data on organism burden and outcome of infection; and (v) to produce an economically viable strategy within a health cost-challenged environment.

The second area in which molecular biology is poised to make a major diagnostic contribution in the clinical mycology laboratory is in strain or isolate identification. Currently, we are witnessing the unfortunate decline of classical or botanical mycology since the training of new personnel with fungal expertise for the clinical laboratories is lagging behind current needs. It is not helpful for clinicians to have delayed or inaccurate species identification, and with limited morphological examination expertise, the use of molecular PCR targeting of sequences in the ribosomal DNA (rDNA) or ribosomal internal transcribed spacer regions and the use of large databases can make the identification of fungal strains more accurate and rapid (14). Furthermore, the discovery that fungi previously assigned to a species are in fact genetically distinct despite having similar morphology has raised the possibility of biological differences that contribute to differences in disease outcome. Here, molecular biology provides a new technology for strain identification and classification

that is not based on morphological or biochemical characteristics. Correlation of genetic information with clinical outcome could result in new insights into the pathogenic potential of certain fungal subspecies. Given the increasing prevalence of fungal diseases and the emergence of new species as pathogens, the development of these molecular tests will be necessary. Molecular biology-based studies for fungal identification are likely to become more standard in the clinical laboratory once costs are lowered and the technological problems with existing platforms are solved. Furthermore, with the use of a series of global molecular screening devices such as randomly amplified polymorphic DNAs, amplification fragment length polymorphism, PCR fingerprinting, and a genotype chip for sequenced species, it is and will be possible to identify and monitor individual strains in disease outbreak populations (10, 23). The ability to genotype strains has significant implications for infection control practices and prevention strategies.

IMMUNOMODULATION

It is apparent that the vast majority of invasive mycoses in humans are associated with an identifiable immunosuppressive event such as neutropenia, corticosteroid use, graft-versus-host disease, and HIV infection. The remarkable resistance of the human host may be related to the combination of a high basal body temperature and sophisticated innate and acquired immune mechanisms. In this aspect of immunity, advances in the molecular biology age have brought us new therapies that have been associated with a variety of infectious disease complications including fungal diseases. For instance, the introduction of recombinant monoclonal antibodies such as inflixamab or alemtuzumab to treat certain underlying diseases has had the unfortunate side effect of creating new groups of immunosuppressed patients at high risk for the development of invasive mycoses. On the other hand, advances in molecular biology have also produced numerous compounds that provide clinicians with the ability to clinically intervene with the goal of enhancing immune function. In this regard, recombinant antibodies and proteins have begun to come to the forefront of immune reconstitution strategies. For instance, the recombinant cytokines, granulocyte colony-stimulating factor (G-CSF), granulocyte-macrophage CSF (GM-CSF), and M-CSF have made some impact on the management of mycoses during neutropenia and its consequences (43). Recombinant gamma interferon is occasionally used in case studies. Furthermore, it has been used successfully in a comparative study to aid in rapidly reducing the cerebrospinal fluid yeast burden during cryptococcal meningitis (45) and in patients with chronic granulomatous disease and fungal infections. In over a decade of studies of antibodies used for management of cryptococcal meningitis (7), a recombinant monoclonal antibody used in humans has begun to be the subject of clinical evaluation as a potential adjunctive modality in the management of life-threatening cryptococcal meningitis (32). In addition, a recombinant antibody fragment to the C. albicans heat shock 90 protein is currently in advanced clinical trials as an adjunct to antifungal chemotherapy with amphotericin B (40).

It is important to emphasize in this era of molecular biology-based studies and recombinant technology for immune enhancement that we have just begun to realize its tremendous potential. To fully exploit these new agents and strategies, we need to understand how to administer pharmacological doses of immunomodulators, especially when one considers that their physiological function is likely to occur at much lower dosages. Furthermore, it is also possible that immunomodulators are harmful agents, since overstimulation of the immune system during treatment of infection can lead to host-mediated inflammatory damage. In this regard, it is noteworthy that there is increased awareness of an immune reconstitution syndrome (IRS) that can develop when immune function returns during therapies in hosts who retain significant microbial burdens (54). As proposed in the damage response framework, host damage can result from either the microbe, the host, or both, and most damage tends to occur at the extremes of the immune response. In certain situations the immune response can become the host's own worst enemy as it attempts to restore homeostasis. Further precise studies of the molecular immunomodulation of the host with an invasive mycosis are possible, given that many biological compounds are now available and critically necessary to improve outcome of fungal diseases. On the other hand, it is important to recognize that the manipulation of the immune system represents a double-edged sword: The host cannot live without it, but too much of a good thing can also be bad.

DRUG PRESCRIPTION

The use of molecular biology and genetics in antifungal drug administration is a new clinical direction that is starting to gain momentum. The field of pharmacogenomics is beginning to yield insights into the fact that there is tremendous individual and group variation to drug responses (59). A good example of how molecular biology can explain issues related to antifungal dosing is in the use of voriconazole. The hepatic cytochrome P450 isoenzyme CYP2C19 plays

a significant role in voriconazole metabolism. The CYP2C19 gene exhibits genetic polymorphisms that have a consequence for the drug's metabolism. For instance, individuals in some ethnic groups such as the Japanese frequently manifest certain polymorphisms in CYP2C19 and consequently metabolize voriconazole poorly, resulting in increased plasma drug levels which might have therapeutic consequences (26).

It is also becoming apparent that antifungal drugs can have variable impact on the host, which can translate into immune modulation and exhibit variable host toxicities. For instance, liposomal amphotericin B appears to divert signaling from Toll-like receptor 2 (TLR2) to TLR4, which can activate neutrophils while attenuating the proinflammatory effects of amphotericin B deoxycholate (2). It is possible that polymorphisms in these innate Toll-like receptors might explain some of the variation in patient susceptibility to amphotericin B infusion-related toxicities. It has also been shown that when human peripheral blood mononuclear cells were exposed to amphotericin B and then their transcriptome was assessed by using cDNA microarray analysis, inflammatory genes like the one encoding interleukin-1α were induced and variability was observed between individual donors (51). These preliminary studies provide an excellent foundation for formulating hypotheses that can then be validated by further performing genomic analyses of host gene responses. The discoveries of correlations between certain gene polymorphisms in drug metabolisms and/or host immunomodulating genes with clinical antifungal drug efficacy or toxicity suggest that many other similar relationships exist that are ready to be uncovered.

Increased antifungal drug resistance in some fungal strains has been encountered as more antifungal drug prescriptions have been written (49). The problem is most acute for Candida spp., where the commensal population is exposed to selection for resistant strains by frequent or chronic antifungal drug use. The use of molecular biological studies has elegantly identified the mechanisms of antifungal drug resistance from target alterations to induction of genes encoding drug efflux pumps (1, 39, 53, 56). The ability to document resistance mechanisms and their frequencies will be a great resource as the design of new drugs or the more creative use of old drugs matures.

SURGERY

Surgical intervention in the management of fungal infections is generally a clinical bedside decision in which it becomes necessary to remove necrotic or abscessed tissue since antifungal agents cannot clear damaged host sites or there is simply no antifungal agent available to eliminate the infection. The direct impact of molecular studies on surgery at this time is not clear. However, it is conceivable that molecular typing of isolates may identify the strains that are most resistant to medical therapy and are likely to require surgical management. Furthermore, an indirect benefit of better diagnostics for early identification of infection or development of more potent antifungal agents through molecular biological studies may make future surgeries an unlikely clinical event as management starts earlier and medical treatment is more effective.

PREVENTION

Numerous clinical studies have attempted to characterize risk factors for fungal infections in order to identify those individuals at high risk for disease. In certain patients at very high risk for fungal diseases, such as those undergoing intensive chemotherapy for hematological malignancy or bone marrow transplant recipients, the risk is often great enough that the use of prophylactic antifungal agents is of benefit (18, 57). However, even carefully focused antimicrobial drug prescription strategies still expose many patients to unnecessary antifungal drugs. Therefore, it is fair to ask whether molecular biological studies can make "risk" more precise. The answer awaits further studies, but the field of immunogenetics offers great promise to help answer this question (19). For instance, a recent study showed a correlation between a common haplotype of the interleukin-4 promoter and the development of chronic disseminated candidiasis in patients with acute leukemia (9). This is a potential landmark study because it suggests a new paradigm in that variants of certain "immune genes" exist that can predispose individuals to fungal diseases and they can be identified (52). With careful population genetic studies backed by basic molecular pathogenesis investigations to identify target genes, this genetic variability may be used to assess individual risk and then carefully target very high-risk patients for fungal infections with more aggressive prevention strategies. The old adage has never been truer: we are what our genes say we are. We now have the ability to harness human genomic information in order to help predict individual susceptibility to a fungal infection. It is not a leap of faith to believe that if 20 of 100 patients treated similarly with an allogenic bone marrow transplant for the same disease, all living in a similar "sea of aspergillus spores" in their environment, present with aspergillosis, there must be some genetic perturbation of immunity which

selectively predisposes certain patients to disease. Whether it is polymorphisms in innate Toll-like receptors or other unknown factors, we have the genomic and molecular tools to understand or more precisely predict the potential risks. The next decade represents a very fertile area for the application of immunogenetics in the management and study of invasive mycoses.

NEW ANTIFUNGAL AGENTS

In the arena of drug target identification, molecular fungal pathogenesis has made great strides. The ability to identify and validate potential drug targets directly in the fungus has matured for several of the major fungal pathogens (5, 31, 47). For instance, the ability to perform global genetic screens with microarrays (24, 30, 35) and signature-tagged libraries of mutants has been combined with the creation of specific gene (knockout or null) mutants and determination of their impact in robust animal models of infection. Comparative genomics can be used to explore the uniqueness of the target or pathway among various fungi and confirm its presence or absence in the mammalian host. Finally, with the creation of a recombinant protein for the gene target such as an enzyme, the screening of large chemical libraries for specific inhibitors can be performed.

The most difficult choice today is not in finding potential molecular targets for antifungal drug discovery but in choosing the best target for drug development. The first basic issue that needs to be addressed is selection. For instance, can a virulence gene target be relevant to drug development or does it have to be an essential gene for basic survival and therefore a fungicide target? Most antimicrobial drugs function by blocking a basic cellular function and not by interfering with the function of a specific virulence phenotype. Although focus on essential, unique genes and their targets remains a very attractive strategy, work on classic virulence genes has its advantages. For instance, the study of the fungal virulence phenotype has led to the identification of potential antifungal compounds. The focus on the phospholipase gene is a good example of how virulence identification of several genes can be exploited to identify antifungal drugs. In both *C. neoformans* and *C. albicans*, a phospholipase gene (*PLB1*) was found to be part of the molecular virulence network (12, 41). With this validation of the enzyme's importance to the pathobiology of the yeast, a search for a phospholipase inhibitor was conducted and compounds with antiphospholipase activity were found. Several of these compounds also demonstrated potent antifungal in vitro activity against *C. neoformans* (20). In another virulence-directed strategy, a direct study of melanin formation led to identification of the herbicide glysophosphate (44). This compound had little direct anticryptococal activity, but when it was used in the treatment of murine cryptococcal infection, it delayed melanization of yeast cells in vivo and prolonged mouse survival. Similarly, the identification of fungal virulence factors can lead to the development of immunological strategies for the prevention and treatment of fungal diseases. For example, studies showing that the polysaccharide capsule of *C. neoformans* is an essential virulence factor have led to the development of vaccines that elicit protective antibodies (16) and inspired the development of passive antibody therapy as an adjunct to antifungal therapy (32).

It is clear that both gene targets of the virulence composite and genes for essential functions in pathogenic fungi can be validated and used for compound library screens and examination of structure-function relationships with these antifungal compounds. The molecular studies of fungal pathogenesis have matured and can routinely be used to define and validate specific antifungal targets. However, recent molecular biology-based studies have also been used to clone fungal genes which express immunoreactive proteins, and thus the ability to create the first protective fungal vaccines has been predicted (3, 4, 11, 15, 25, 33, 37, 58). It may also be possible to use attenuated null mutants to create protective immunity. With the paucity of antifungal drug classes for invasive mycoses, it is encouraging that there is now a robust platform of molecular biology for many of the major fungal pathogens, which may be used for antifungal drug discovery and vaccine development.

DRUG COMBINATIONS

The use of antifungal combinations to treat invasive mycoses has been a controversial topic because there are few evidence-based studies that support their widespread clinical use (27). In fact, despite substantial in vitro and animal studies to support combination drug use, only the combination of amphotericin B plus flucytosine for induction therapy of cryptococcal meningitis has received widespread clinical acceptance (6, 55). However, the use of combination therapy for synergistic antifungal purposes, reduction of drug resistance development, and broader antifungal spectrum remains a very attractive option for the clinician. Therefore, in the study of fungal pathogenesis, molecular targets that might positively interact with one another could provide important clues about which drug combinations are likely to be effective. A recent example of synergistic behavior between antifungal targets was the positive interaction between the azole block in ergosterol membrane formation with fluconazole and the

stress pathway block at the serine-threonine phosphatase, calcineurin (13). For instance, fluconazole, which is a fungistatic agent, becomes fungicidal when there is also a block at the calcineurin locus either by mutation or through specific use of inhibitors such as cyclosporine or tacrolimus (38). Other molecular biology-based drug studies have shown how the cell wall can be interrupted by both glucan and chitin synthase inhibitors in a synergistic manner (46). With the use of genomics combined with careful biochemistry, scientists are in a better position to attack the pathogenic fungus at several weak points in its biology either together or in sequence.

Unfortunately, it is very unlikely that the epidemic of invasive mycoses that has plagued humanity over the last two or three decades will abate any time soon. We live in an environment intimately associated with fungi at every level. Given that both natural and acquired immunosuppressive events remain common, it is likely that we will continue to see a distressingly high prevalence of the major fungal diseases. Furthermore, since there are over 1.5 million fungal species and many fungi possess some attributes that can affect their virulence, we can anticipate that each year will see the description of new fungal pathogens. In this environment of constant fungal threat, it is both exciting and helpful that fungal molecular pathogenesis is maturing into a science with practical applications. It is urgently needed and is now available. One always shudders to think what may have happened if the molecular biology infrastructure for retrovirology was not available prior to the HIV pandemic. Similarly, the molecular biology of pathogenic fungi has now developed to meet the challenges of invasive human mycoses today and tomorrow.

REFERENCES

1. **Balashov, S. V., R. Gardiner, S. Park, and D. S. Perlin.** 2005. Rapid, high-throughput, multiplex, real-time PCR for identification of mutations in the *cyp51A* gene of *Aspergillus fumigatus* that confer resistance to itraconazole. *J. Clin. Microbiol.* **43:**214–222.
2. **Bellocchio, S., R. Gaziano, S. Bozza, G. Rossi, C. Montagnoli, K. Perruccio, M. Calvitti, L. Pitzurra, and L. Romani.** 2005. Liposomal amphotericin B activates antifungal resistance with reduced toxicity by diverting Toll-like receptor signalling from TLR-2 to TLR-4. *J. Antimicrob. Chemother.* **55:**214–222.
3. **Biondo, C., C. Beninati, D. Delfino, M. Oggioni, G. Mancuso, A. Midiri, M. Bombaci, G. Tomaselli, and G. Teti.** 2002. Identification and cloning of a cryptococcal deacetylase that produces protective immune responses. *Infect. Immun.* **70:**2383–2391.
4. **Borges, C. L., M. Pereira, M. S. Felipe, F. P. de Faria, F. J. Gomez, G. S. Deepe, Jr., and C. M. Soares.** 2005. The antigenic and catalytically active formamidase of

Paracoccidioides brasiliensis: protein characterization, cDNA and gene cloning, heterologous expression and functional analysis of the recombinant protein. *Microbes Infect.* 7:66–77.

5. Brakhage, A. A., and K. Langfelder. 2002. Menacing mold: the molecular biology of *Aspergillus fumigatus*. *Annu. Rev. Microbiol.* 56:433–455.

6. Brouwer, A. E., A. Rajanuwong, W. Chierakul, G. E. Griffin, R. A. Larsen, N. J. White, and T. S. Harrison. 2004. Combination antifungal therapies for HIV-associated cryptococcal meningitis: feasibility and power of quantitative CSF cultures to determine fungicidal activity. *Lancet* 363:1764–1767.

7. Casadevall, A. 1999. Passive antibody therapies: progress and continuing challenges. *Clin. Immunol.* 93:5–15.

8. Casadevall, A., and L. A. Pirofski. 2003. The damage-response framework of microbial pathogenesis. *Nat. Rev. Microbiol.* 1:17–24.

9. Choi, E. H., C. B. Foster, J. G. Taylor, H. C. Erichsen, R. A. Chen, T. J. Walsh, V. J. Anttila, T. Ruutu, A. Palotie, and S. J. Chanock. 2003. Association between chronic disseminated candidiasis in adult acute leukemia and common IL4 promoter haplotypes. *J. Infect. Dis.* 187:1153–1156.

10. Clark, T. A., S. A. Slavinski, J. Morgan, T. Lott, B. A. Arthington-Skaggs, M. E. Brandt, R. M. Webb, M. Currier, R. H. Flowers, S. K. Fridkin, and R. A. Hajjeh. 2004. Epidemiologic and molecular characterization of an outbreak of *Candida parapsilosis* bloodstream infections in a community hospital. *J. Clin. Microbiol.* 42:4468–4472.

11. Cole, G. T., J. M. Xue, C. N. Okeke, E. J. Tarcha, V. Basrur, R. A. Schaller, R. A. Herr, J. J. Yu, and C. Y. Hung. 2004. A vaccine against coccidioidomycosis is justified and attainable. *Med. Mycol.* 42:189–216.

12. Cox, G. M., H. C. McDade, S. C. Chen, S. C. Tucker, M. Gottfredsson, L. C. Wright, T. C. Sorrell, S. D. Leidich, A. Casadevall, M. A. Ghannoum, and J. R. Perfect. 2001. Extracellular phospholipase activity is a virulence factor for *Cryptococcus neoformans*. *Mol. Microbiol.* 39:166–175.

13. Cruz, M. C., A. L. Goldstein, J. R. Blankenship, M. Del Poeta, D. Davis, M. E. Cardenas, J. R. Perfect, J. H. McCusker, and J. Heitman. 2002. Calcineurin is essential for survival during membrane stress in *Candida albicans*. *EMBO J.* 21:546–559.

14. de Aguirre, L., S. F. Hurst, J. S. Choi, J. H. Shin, H. P. Hinrikson, and C. J. Morrison. 2004. Rapid differentiation of *Aspergillus* species from other medically important opportunistic molds and yeasts by PCR-enzyme immunoassay. *J. Clin. Microbiol.* 42:3495–3504.

15. Deepe, G. S., Jr., and R. S. Gibbons. 2002. Cellular and molecular regulation of vaccination with heat shock protein 60 from *Histoplasma capsulatum*. *Infect. Immun.* 70:3759–3767.

16. Devi, S. J. N. 1996. Preclinical efficacy of a glucuronoxylomannan-tetanus toxoid conjugate vaccine of *Cryptococcus neoformans* in a murine model. *Vaccine* 14:841–842.

17. Diehn, M., and D. A. Relman. 2001. Comparing functional genomic datasets: lessons from DNA microarray analyses of host-pathogen interactions. *Curr. Opin. Microbiol.* 4:95–101.

18. Diekema, D. J., and M. A. Pfaller. 2004. Nosocomial candidemia: an ounce of prevention is better than a pound of cure. *Infect. Control Hosp. Epidemiol.* 25:624–626.

19. Fierer, J., L. Walls, F. Wright, and T. N. Kirkland. 1999. Genes influencing resistance to *Coccidioides immitis* and the interleukin-10 response map to chromosomes 4 and 6 in mice. *Infect. Immun.* 67:2916–2919.

20. Ganendren, R., F. Widmer, V. Singhal, C. Wilson, T. Sorrell, and L. Wright. 2004. In vitro antifungal activities of inhibitors of phospholipases from the fungal pathogen *Cryptococcus neoformans*. *Antimicrob. Agents Chemother.* 48:1561–1569.

21. Gudlaugsson, O., S. Gillespie, K. Lee, B. J. Vande, J. Hu, S. Messer, L. Herwaldt, M. Pfaller, and D. Diekema. 2003. Attributable mortality of nosocomial candidemia, revisited. *Clin. Infect. Dis.* 37:1172–1177.

22. Gutzmer, R., S. Mommert, U. Kuttler, T. Werfel, and A. Kapp. 2004. Rapid identification and differentiation of fungal DNA in dermatological specimens by LightCycler PCR. *J. Med. Microbiol.* 53:1207–1214.

23. Huang, Y. C., L. H. Su, T. L. Wu, and T. Y. Lin. 2004. Genotyping analysis of colonizing candidal isolates from very-low-birthweight infants in a neonatal intensive care unit. *J. Hosp. Infect.* 58:200–203.

24. Hwang, L., D. Hocking-Murray, A. K. Bahrami, M. Andersson, J. Rine, and A. Sil. 2003. Identifying phase-specific genes in the fungal pathogen *Histoplasma capsulatum* using a genomic shotgun microarray. *Mol. Biol. Cell* 14:2314–2326.

25. Ibrahim, A. S., B. J. Spellberg, V. Avenissian, Y. Fu, S. G. Filler, and J. E. Edwards, Jr. 2005. Vaccination with recombinant N-terminal domain of Als1p improves survival during murine disseminated candidiasis by enhancing cell-mediated, not humoral, immunity. *Infect. Immun.* 73:999–1005.

26. Ikeda, Y., K. Umemura, K. Kondo, K. Sekiguchi, S. Miyoshi, and M. Nakashima. 2004. Pharmacokinetics of voriconazole and cytochrome P450 2C19 genetic status. *Clin. Pharmacol. Ther.* 75:587–588.

27. Johnson, M. D., C. MacDougall, L. Ostrosky-Zeichner, J. R. Perfect, and J. H. Rex. 2004. Combination antifungal therapy. *Antimicrob. Agents Chemother.* 48:693–715.

28. Jones, T., N. A. Federspiel, H. Chibana, J. Dungan, S. Kalman, B. B. Magee, G. Newport, Y. R. Thorstenson, N. Agabian, P. T. Magee, R. W. Davis, and S. Scherer. 2004. The diploid genome sequence of *Candida albicans*. *Proc. Natl. Acad. Sci. USA* 101:7329–7334.

29. Jordanides, N. E., E. K. Allan, L. A. McLintock, M. Copland, M. Devaney, K. Stewart, A. N. Parker, P. R. Johnson, T. L. Holyoake, and B. L. Jones. 2005. A prospective study of real-time panfungal PCR for the early diagnosis of invasive fungal infection in haemato-oncology patients. *Bone Marrow Transplant.* 35:389–395.

30. Kaur, R., I. Castano, and B. P. Cormack. 2004. Functional genomic analysis of fluconazole susceptibility in the pathogenic yeast *Candida glabrata*: roles of calcium signaling and mitochondria. *Antimicrob. Agents Chemother.* 48:1600–1613.

31. Larriba, G., J. J. Rubio Coque, A. Ciudad, and E. Andaluz. 2000. *Candida albicans* molecular biology reaches its maturity. *Int. Microbiol.* 3:247–252.

32. Larsen, R. A., P. G. Pappas, J. Perfect, J. A. Aberg, A. Casadevall, G. A. Cloud, R. James, S. Filler, and W. E. Dismukes. 2005. Phase I evaluation of the safety and pharmacokinetics of murine-derived anticryptococcal antibody 18B7 in subjects with treated cryptococcal meningitis. *Antimicrob. Agents Chemother.* 49:952–958.

33. Levitz, S. M., S. Nong, M. K. Mansour, C. Huang, and C. Specht. 2001. Molecular characterization of a mannoprotein with homology to chitin deacetylases that stimulates T-cell responses to *Cryptococcus neoformans*. *Proc. Natl. Acad. Sci. USA* **98:**10422–10427.

34. Loftus, B. J., E. Fung, P. Roncaglia, D. Rowley, P. Amedeo, D. Bruno, J. Vamathevan, M. Miranda, I. J. Anderson, J. A. Fraser, J. E. Allen, I. E. Bosdet, M. R. Brent, R. Chiu, T. L. Doering, M. J. Donlin, C. A. D'Souza, D. S. Fox, V. Grinberg, J. Fu, M. Fukushima, B. J. Haas, J. C. Huang, G. Janbon, S. J. Jones, H. L. Koo, M. I. Krzywinski, J. K. Kwon-Chung, K. B. Lengeler, R. Maiti, M. A. Marra, R. E. Marra, C. A. Mathewson, T. G. Mitchell, M. Pertea, F. R. Riggs, S. L. Salzberg, J. E. Schein, A. Shvartsbeyn, H. Shin, M. Shumway, C. A. Specht, B. B. Suh, A. Tenney, T. R. Utterback, B. L. Wickes, J. R. Wortman, N. H. Wye, J. W. Kronstad, J. K. Lodge, J. Heitman, R. W. Davis, C. M. Fraser, and R. W. Hyman. 2005. The genome of the basidiomycetous yeast and human pathogen *Cryptococcus neoformans*. *Science* **307:**1321–1324.

35. Lorenz, M. C., J. A. Bender, and G. R. Fink. 2004. Transcriptional response of *Candida albicans* upon internalization by macrophages. *Eukaryot. Cell* **3:**1076–1087.

36. Magee, P. T., C. Gale, J. Berman, and D. Davis. 2003. Molecular genetic and genomic approaches to the study of medically important fungi. *Infect. Immun.* **71:**2299–2309.

37. Mandel, M. A., G. G. Grace, K. I. Osborn, F. Schafer, J. W. Murphy, M. J. Orbach, and J. N. Galgiani. 2002. The *Cryptococcus neoformans* gene *DHA1* encodes an antigen that elicits a delayed-type hypersensitivity reaction in immune mice. *Infect. Immun.* **68:**6196–6201.

38. Marchetti, O., P. Moreillon, M. P. Glauser, J. Bille, and D. Sanglard. 2000. Potent synergism of the combination of fluconazole and cyclosporine in *Candida albicans*. *Antimicrob. Agents Chemother.* **44:**2373–2381.

39. Markovich, S., A. Yekutiel, I. Shalit, Y. Shadkchan, and N. Osherov. 2004. Genomic approach to identification of mutations affecting caspofungin susceptibility in *Saccharomyces cerevisiae*. *Antimicrob. Agents Chemother.* **48:**3871–3876.

40. Matthews, R. C., G. Rigg, S. Hodgetts, T. Carter, C. Chapman, C. Gregory, C. Illidge, and J. Burnie. 2003. Preclinical assessment of the efficacy of mycograb, a human recombinant antibody against fungal HSP90. *Antimicrob. Agents Chemother.* **47:**2208–2216.

41. Mukherjee, P. K., K. R. Seshan, S. D. Leidich, J. Chandra, G. T. Cole, and M. A. Ghannoum. 2001. Reintroduction of the *PLB1* gene into *Candida albicans* restores virulence in vivo. *Microbiology* **147:**2585–2597.

42. Musher, B., D. Fredricks, W. Leisenring, S. A. Balajee, C. Smith, and K. A. Marr. 2004. *Aspergillus* galactomannan enzyme immunoassay and quantitative PCR for diagnosis of invasive aspergillosis with bronchoalveolar lavage fluid. *J. Clin. Microbiol.* **42:**5517–5522.

43. Nemunaitis, J., J. D. Meyers, C. D. Buckner, K. Shannondorcy, H. Shulman, J. A. Bianco, C. S. Higano, E. Groves, R. Storb, J. Hansen, F. R. Applebaum, and J. W. Singer. 1991. Phase I trial of recombinant human macrophage colony-stimulating factor in patients with invasive fungal infections. *Blood* **78:**907–913.

44. Nosanchuk, J. D., R. Ovalle, and A. Casadevall. 2001. Glyphosate inhibits melanization of *Cryptococcus neoformans* and prolongs survival of mice after systemic infection. *J. Infect. Dis.* **183:**1093–1099.

45. Pappas, P. G., B. Bustamante, E. Ticona, R. J. Hamill, P. C. Johnson, A. Reboli, J. Aberg, R. Hasbun, and H. H. Hsu. 2004. Recombinant interferon-gamma 1b as adjunctive therapy for AIDS-related acute cryptococcal meningitis. *J. Infect. Dis.* **189:**2185–2191.

46. Perfect, J. R., K. A. Wright, and R. F. Hector. 1993. Synergistic interaction of nikkomycin and cilofungin against diverse fungi, p. 369–380. *In* H. Yamaguchi, G. S. Kobayashi, and H. Takahashi (ed.), *Recent Progress in Antifungal Chemotherapy*. Marcel Dekker, Inc., New York, N.Y.

47. Perfect, J. R. 1996. Fungal virulence genes as targets for antifungal chemotherapy. *Antimicrob. Agents Chemother.* **40:**1577–1583.

48. Perfect, J. R., G. M. Cox, J. Y. Lee, C. A. Kauffman, L. de Repentigny, S. W. Chapman, V. A. Morrison, P. Pappas, J. W. Hiemenz, and D. A. Stevens. 2001. The impact of culture isolation of *Aspergillus* species: a hospital-based survey of aspergillosis. *Clin. Infect. Dis.* **33:**1824–1833.

49. Pfaller, M. A., and D. J. Diekema. 2004. Rare and emerging opportunistic fungal pathogens: concern for resistance beyond *Candida albicans* and *Aspergillus fumigatus*. *J. Clin. Microbiol.* **42:**4419–4431.

50. Polzehl, D., M. Weschta, A. Podbielski, H. Riechelmann, and D. Rimek. 2005. Fungus culture and PCR in nasal lavage samples of patients with chronic rhinosinusitis. *J. Med. Microbiol.* **54:**31–37.

51. Rogers, P. D., M. M. Pearson, J. D. Cleary, D. C. Sullivan, and S. W. Chapman. 2002. Differential expression of genes encoding immunomodulatory proteins in response to amphotericin B in human mononuclear cells identified by cDNA microarray analysis. *J. Antimicrob. Chemother.* **50:**811–817.

52. Schroder, N. W., and R. R. Schumann. 2005. Single nucleotide polymorphisms of Toll-like receptors and susceptibility to infectious disease. *Lancet Infect. Dis.* **5:**156–164.

53. Silva Ferreira, M. E., J. L. Capellaro, M. E. dos Reis, I. Malavazi, D. Perlin, S. Park, J. B. Anderson, A. L. Colombo, B. A. Arthington-Skaggs, M. H. Goldman, and G. H. Goldman. 2004. In vitro evolution of itraconazole resistance in *Aspergillus fumigatus* involves multiple mechanisms of resistance. *Antimicrob. Agents Chemother.* **48:**4405–4413.

54. Stoll, M., and R. E. Schmidt. 2004. Adverse events of desirable gain in immunocompetence: the immune restoration inflammatory syndromes. *Autoimmun. Rev.* **3:**243–249.

55. van der Horst, C. M., M. S. Saag, G. A. Cloud, and National Institute of Allergy and Infectious Diseases Mycoses Study Group and AIDS Clinical Trials Group. 1997. Treatment of cryptococcal meningitis associated with the acquired immunodeficiency syndrome. *N. Engl. J. Med.* **337:**15–21.

56. White, T. C., S. Holleman, F. Dy, L. F. Mirels, and D. A. Stevens. 2002. Resistance mechanisms in clinical isolates of *Candida albicans*. *Antimicrob. Agents Chemother.* **46:**1704–1713.

57. Winston, D. J., C. Emmanouilides, K. Bartoni, G. J. Schiller, R. Paquette, and M. C. Territo. 2004. Elimination of *Aspergillus* infection in allogeneic stem cell transplant recipients with long-term itraconazole prophylaxis: prevention is better than treatment. *Blood* **104:**1581.

58. Wuthrich, M., H. I. Filutowicz, T. Warner, and B. S. Klein. 2002. Requisite elements in vaccine immunity

to *Blastomyces dermatitidis*: plasticity uncovers vaccine potential in immune-deficient hosts. *J. Immunol.* **169:**6969–6976.

59. **Xie, H. G., R. B. Kim, A. J. Wood, and C. M. Stein.** 2001. Molecular basis of ethnic differences in drug disposition and response. *Annu. Rev. Pharmacol. Toxicol.* **41:**815–850.

60. **Yoo, J. H., J. H. Choi, S. M. Choi, D. G. Lee, W. S. Shin, W. S. Min, and C. C. Kim.** 2005. Application of nucleic acid sequence-based amplification for diagnosis of and monitoring the clinical course of invasive aspergillosis in patients with hematologic diseases. *Clin. Infect. Dis.* **40:**392–398.

Molecular Principles of Fungal Pathogenesis
Edited by Joseph Heitman et al.
©2006 ASM Press, Washington, D.C.

Chapter 2

Sex, *MAT*, and the Evolution of Fungal Virulence

JAMES A. FRASER AND JOSEPH HEITMAN

The theory of evolution provides a foundation to our understanding of how species arise and either adapt to novel challenges in the environment or become extinct. Selection for mutations that increase fitness is the force that drives evolution. While mutations and their corresponding phenotypes that confer a selective advantage can be transmitted via clonal propagation, a broader repertoire of genetic diversity is produced via sexual recombination, which yields recombinant offspring and helps purge deleterious transposons from the genome (2).

The vast majority of organisms undergo sexual reproduction, and the mechanisms underlying this process are diverse and fascinating. While many fungi have a characterized sexual cycle, the pathogenic fungi represent a special example in which sexuality is uncommon and sexual mechanisms are unusual or cryptic. Given their evolutionary relationship to model organisms such as *Saccharomyces cerevisiae* (in which sexual reproduction and recombination is common and readily observed), it is striking that many species of pathogenic fungi have never been observed to mate in the laboratory despite population genetic evidence for recombining population structures (*Coccidioides immitis, Aspergillus fumigatus*). Conversely, other pathogenic fungi such as *Cryptococcus neoformans* readily mate in the laboratory and yet mating structures have never been observed in the wild, while others such as *Candida albicans* and *Candida glabrata* retain all of the machinery required for sex and yet rarely (*C. albicans*) or possibly never (*C. glabrata*) mate and have population structures that are largely if not entirely clonal. It would therefore appear that while the machinery for sex has been retained in the pathogenic fungi, this capacity is utilized only infrequently if at all.

Recent discoveries have forced us to reevaluate hypotheses on how sexual recombination impacts the evolution and virulence of pathogenic fungi. The genomic revolution has drastically changed the way we think about fungal sex, enabling the identification of genomic sex-determining structures irrespective of whether these fungi have been observed to mate. *C. albicans*, which was thought to be strictly asexual for more than a century, has now been shown to undergo mating and a parasexual cycle (50), and the sexual cycle of the predominant pathogenic form of *C. neoformans* (var. *grubii*, serotype A) has been defined (53, 93). Prior to these discoveries, it had been thought that most pathogenic fungi might be evolving to be asexual.

One prevailing hypothesis posited that, if virulence is a polygenic trait, mating might lead to the reassortment and loss of unique polygenic traits required for pathogenesis. And yet *C. albicans*, *C. neoformans*, and many other pathogenic fungi have retained both the machinery and the ability to undergo at least some form of a sexual cycle. Therefore, the opposite can also be argued—an actively recombining population may also give rise to genotypes that are more virulent, in addition to contributing to adaptation to environmental challenges outside of the infection cycle. It would appear that both clonal and sexual life cycles confer advantages and costs, and pathogens might primarily maintain a clonal lifestyle once adapted to a host and only rarely deploy their sexual cycles when confronted by novel conditions. Sex might therefore punctuate the evolution of pathogens by enabling them to expand their environmental range, infect new hosts, evade the immune system, or thwart antimicrobial therapy. The study of the role of sexual cycles of pathogenic fungi therefore takes center stage, promising to reveal much about how these pathogenic microbes evolve, enabling infection of humans and continued adaptation to unique challenges, such as the advent of new therapeutic interventions.

James A. Fraser and Joseph Heitman • Department of Molecular Genetics and Microbiology, Duke University Medical Center, Durham, NC 27710.

This chapter provides a summary of our current knowledge of the role of mating in pathogenic fungi and discusses future directions in this field.

CRYPTOCOCCUS NEOFORMANS: A ROBUST AND COMPLETE LABORATORY-DEFINED SEXUAL CYCLE WHOSE ROLE IN NATURE REMAINS MYSTERIOUS

While a laboratory-defined complete sexual cycle for *C. neoformans* has been known for three decades, the vast majority of clinical and environmental isolates are of a single mating type (α), and the organism has never been directly observed to mate in nature (60, 65). *C. neoformans* grows as a budding yeast in the environment and the infected host but can also undergo a dimorphic transition and grow as a filamentous fungus during the sexual cycle and monokaryotic fruiting (46). The organism is haploid, and mating occurs between cells of the **a** and α mating types.

Infection is acquired by inhalation. It is most frequently asymptomatic and can be cleared or enter a dormant, latent state in the hilar lymph nodes, involving facultative intracellular growth and survival in granulomas. In response to immunosuppression, the organism can be reactivated later in life and disseminate to infect most prominently the central nervous system. Since spores and desiccated yeast cells are small enough to fit into the alveoli of the lungs, they are the suspected infectious propagules (11). However, given that α strains predominate over **a** strains by a ratio of 50:1 to 1,000:1 in nature, it has been a mystery if and where *C. neoformans* might mate in the environment. The dramatically mating-type-skewed population, combined with early studies documenting a virulence difference between cells of the two mating types in certain varieties and genetic backgrounds (68), has focused interest on the sexual cycle. Together, these observations suggest that the cellular machinery involved in establishing cell identity and enabling mating might be linked to virulence. We review here recent studies that shed light on the role of sex and the evolution of virulence in *C. neoformans*.

Based on capsular antigens, *C. neoformans* has been classified into three divergent varieties that may represent a species cluster of human fungal pathogens. These divergent forms are represented by four serotypes: variety *neoformans* (serotype D), variety *grubii* (serotype A), and variety *gattii* (serotypes B and C). Serotypes A (the most clinically prevalent) and D are opportunistic pathogens that infect immunocompromised individuals worldwide, whereas the divergent serotype B and C isolates are primary pathogens that can infect immunocompetent hosts (11). This primary pathogenic form of the organism has recently been recognized as an independent species, *Cryptococcus gattii* (67). In contrast to serotypes A and D, which are cosmopolitan and most frequently associated with pigeon guano, serotype B and C isolates are usually restricted to tropical and subtropical regions of the world, where they are found in association with the woody debris of trees.

In fungi, a specialized region of the genome, known as the mating-type locus, governs the establishment of cell type identity and differs in DNA sequence between cells of opposite mating types. *Cryptococcus* belongs to the phylum of basidiomycetes, which usually have tetrapolar mating systems composed of two unlinked loci, and strains must differ at both loci for mating to succeed (12, 31, 57). One locus encodes pheromones and pheromone receptors, and the other encodes homeodomain transcription factors. It is this second locus that resembles the single mating-type locus that is common in ascomyceteous fungi such as *S. cerevisiae*. However, landmark studies of the least pathogenic form, *C. neoformans* var. *neoformans* (serotype D), by Kwon-Chung revealed that *Cryptococcus* has a bipolar mating system, unlike many other basidiomycetous fungi (61). For mating to occur, strains must differ at only a single locus, and these haploid strains are designated either **a** or α. Subsequent studies have revealed multiple environmental signals that stimulate the sexual cycle, including desiccation, nutrient limitation, low-temperature growth (25°C), and the absence of light (46). A related process has been characterized by which cells can also respond to nitrogen limitation, desiccation, and pheromone signals and differentiate by a process known as monokaryotic fruiting, which also involves filamentation and sporulation and which was thought until recently to be haploid and strictly mitotic (78, 121, 133, 135).

In response to appropriate growth conditions, the vegetative budding-yeast cells undergo different developmental programs dependent on their mating type. When nutrient limited, **a** and α cells secrete peptide pheromones that trigger conjugation tube formation and cell-cell fusion (88). In contrast to the model budding yeast *S. cerevisiae*, nuclear fusion is delayed and the zygote forms dikaryotic hyphae with fused clamp connections. The filament tips ultimately form terminal basidia, wherein the paired nuclei undergo karyogamy and meiosis, producing four chains of rough, ovoid basidiospores through sequential rounds of basipetal budding. In each basidiospore, mitochondria are inherited uniparentally from only the **a** parent (141, 142).

The importance of the mating process in the infectious cycle is significant. Unlike *Candida*, *Cryptococcus* is not a part of the normal human flora. Rather, the route of infection is via inhalation. The initial route of entry is a major barrier to the organism, with the ability to infect restricted by alveolar size. Vegetative cells of *Cryptococcus* are normally 4 to 10 μm in diameter and are therefore too large to enter the alveoli. In contrast, the basidiospores are smaller, at 1 to 3 μm, and are therefore ideally sized to penetrate the lung and infect the host. Furthermore, in a murine model, spores have been shown to be up to 100 times more infectious than vegetative cells (118). An understanding of the sexual cycle of *Cryptococcus* is therefore of central importance, since it represents a route by which the suspected infectious propagules are produced.

Based on the identification of the sexual cycle and robustly mating strains, Kwon-Chung et al. employed a backcrossing approach to create a congenic strain pair suitable for genetic and molecular studies (68). First, a pair of serotype D strains of opposite mating type was isolated from a meiotic cross of the type strains NIH433 (**a**) and NIH12 (α). An α progeny of the two strains was then selected and backcrossed to its **a** parent nine times, producing the JEC20 (**a**) and JEC21 (α) congenic strain pair (39, 68). Mathematically, these strains can be calculated to be more than 99.95% identical, with one major exception: one strain bears the **a** allele at the mating-type locus, whereas the other has inherited the α allele. The finding that α strains can be more virulent than congenic **a** strains, along with the vast preponderance of α strains in environmental and clinical isolates, has focused interest on what makes α cells unique in this fungal pathogen (68).

Cell type identity is established by the mating-type locus. Insight into this genomic region was first provided by Moore and Edman via a differential cloning approach, which identified a 35-kb region of the JEC21 *MAT*α genome that was not present in the genome of JEC20, its *MAT***a** counterpart (90). Subsequent analysis of this *MAT*-specific region identified a subclone capable of inducing mating structures in **a** isolates under mating conditions. Sequence analysis revealed a gene predicted to encode a lipid-modified pheromone similar to those identified in other fungi. This hypothesis was later confirmed by molecular and genetic analysis of the pheromone genes (107) and synthesis of active MFα pheromone (18).

The original identification of a *MAT* locus structure over 35 kb in size defined *Cryptococcus* as an exceptional system. In other fungi, the *MAT* loci are normally restricted in size; in *S. cerevisiae* the *MAT* locus is less than 1 kb in length. Over the ensuing years, several laboratories contributed to the ongoing story of the discovery of this unique genomic structure. In an overexpression screen, Wickes et al. identified

an α-specific *STE12* homolog that induced filamentation when overexpressed; remarkably, this *MAT*-specific gene was not part of the originally defined *MAT* locus (134). A second gene, *STE20*α, was also identified that shared these features—it was *MAT* specific but was not part of the known *MAT* region (76, 132). Together, these findings revealed that the bipolar *Cryptococcus MAT* locus was even larger than originally estimated, spanning ~50 kb encompassing the *STE12*α and *STE20*α genes and the original 35-kb region (51).

Hull et al. deleted the known 50-kb *MAT*α region in an **a**/α diploid background (a deletion that is lethal in haploid cells) and found that the resulting **a**/Δ strain still exhibited the self-filamentous phenotype characteristic of **a**/α cells (44). Thus, additional α-specific gene information remained to be defined. Importantly, a central component of *MAT* that had not been identified was the transcription factors normally associated with this locus in fungi with both bipolar and tetrapolar mating systems. A search of the *C. neoformans* genome revealed six potential homeodomain factors related to the *C. albicans MAT* locus homeobox gene α2, only one of which was associated exclusively with one mating type, α (43). Subsequent deletion analysis revealed that the encoded Sxi1α homeodomain protein plays a crucial role in establishing **a**/α cell identity, and disruption of this gene prevents proper completion of the sexual cycle. Importantly, the *SXI1*α gene did not lie within the previously defined 50-kb *MAT*α locus.

To define the complete *MAT* locus, Lengeler et al. isolated and sequenced multiple overlapping large-insert bacterial artificial chromosome clones from both **a** and α serotype A and D strains (75). This analysis revealed that the bipolar *MAT* locus of *Cryptococcus* is remarkably large in both mating types (over 100 kb long) and consists of divergent alleles of a cohort of more than 20 genes (Fig. 1). While this gene set contained all of those previously identified to reside in the mating-type-specific region of the genome, additional genes were present that would not be predicted to play a role in the sexual cycle.

Surprisingly, a number of genes that are either in the *MAT* locus or involved in the mating process have been implicated in virulence. In serotype D, both *ste12***a** and *ste12*α mutants are attenuated (13, 14, 19), yet mutations in the mitogen-activated protein kinase (MAPK) cascade that regulates Ste12 function have no change in virulence (19). Further confounding this analysis, mutation of the RGS protein-encoding *CRG1* implicated in deactivating the heterotrimeric G-protein that regulates this cascade are hypervirulent (131), and a pheromoneless *mf*α*1*, *mf*α*2*, *mf*α*3* triple mutant is modestly reduced in virulence (107). Together,

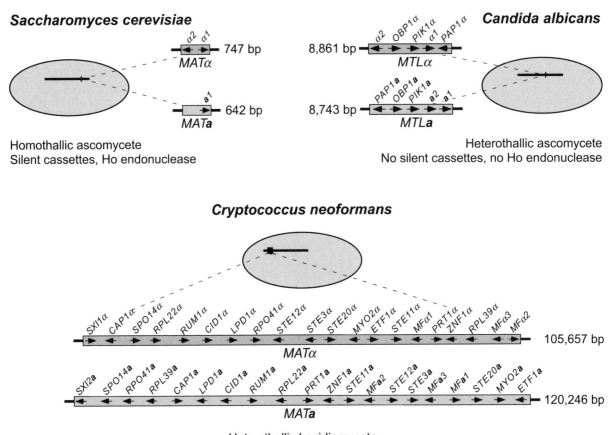

Figure 1. Fungal *MAT* locus paradigms. Selection of mating partners during the sexual cycle is directed by bipolar mating systems in a wide variety of fungi. The budding yeast *S. cerevisiae* is an ascomycete with a *MAT* locus of only 642 or 747 bp, encoding only one or two transcription factors. The related pathogenic ascomycete *C. albicans* contains a *MAT* locus over 10 times as large and, along with the components present in *S. cerevisiae*, has an additional three or four genes. In contrast, the bipolar *MAT* alleles of the pathogenic basidiomycete *C. neoformans* are over 140 times the size of their *S. cerevisiae* counterparts, containing more than 20 genes, many of which contribute to the sexual cycle including homeodomain transcription factors, pheromones and pheromone receptors, and other elements of the pheromone-activated MAPK cascade. The drawing is not to scale.

these data indicate that some genes involved in mating play a role in virulence, but they do not yet explain how or why *MAT*α isolates are more common in the clinical setting.

The identification of *MAT*-specific sequences provided a unique tool to analyze the population structure of this pathogenic basidiomycete. First, analysis of **a** isolates enabled the identification of the *STE20***a** gene, which showed a high degree of nucleotide divergence from its α counterpart and provided the first sequence information about the heretofore unknown **a** allele of *MAT* (76). However, the power to discriminate between **a** and α strains on a purely molecular level allowed an even greater potential: these sequences could be used to search for the elusive phenomenon of mating in the more pathogenic and clinically significant serotype A (var. *grubii*).

While earlier mating studies had identified multiple **a** and α isolates capable of undergoing sexual reproduction from the less virulent serotype D, only α isolates of serotype A had ever been identified, despite large-scale efforts by several research groups.

The *STE20***a** and *STE20*α gene sequences provided an indispensable tool to identify var. *grubii* *MAT***a** strains, irrespective of their ability to mate. In a large-scale screen of >1,000 serotype A isolates, a single *MAT***a** clinical isolate designated 125.91 was found, originating in cerebrospinal fluid isolated from an AIDS patient with cryptococcal meningitis in Tanzania (76). However, this strain was unable to mate with the serotype A and D reference strains H99 and JEC21.

Evidence to support the existence of fertile *MAT***a** serotype A strains was found from an unexpected source. Artificial diploid isolates created in the laboratory had revealed thermal dimorphism: at 37°C, serotype D a/α diploid isolates could be generated that grow stably in the yeast form (108). On transfer to mating conditions (25°C, growth on V8), these strains undergo a dimorphic transition from budding yeast to the hyphal form and complete the sexual cycle. Similar observations held with clinical AD hybrid isolates (74). While some displayed exclusively vegetative growth, others exhibited different degrees of thermal dimorphism analogous to the laboratory-isolated serotype D a/α diploids, but few if any fertile haploid basidiospores could be recovered.

The hybrid serotype AD isolates are therefore the products of an incomplete sexual cycle between isolates of the two divergent varieties or sibling species. Molecular analysis proved this hypothesis to be correct with respect to the serotype AD hybrids, with most strains containing both an a and an α allele at *MAT* (17, 74). In each case, one *MAT* allele originated from serotype A and the other originated from serotype D, giving rise to two distinct classes of serotype AD strains that have been designated aADα and αADa to reflect their distinct ancestries. Furthermore, multiple isolates contain a alleles that clearly originate from serotype A, not only indicating the environmental presence of these elusive isolates but also providing evidence of successful mating with serotype D. This suggests that the AD hybrids resulted from intervarietal mating between two parents of opposite mating type which are trapped in the diploid state because genomic differences and nucleotide polymorphisms prevent the proper completion of meiosis. Analogous findings have previously been reported for hybrids produced via interspecies mating in the sensu stricto group of *Saccharomyces* species. Population genetic studies indicate that the origin of the unusual AD hybrids was within the past several million years (5, 139, 140).

These discoveries were rapidly followed by the identification of more serotype aA isolates. First a fertile environmental isolate was discovered in Italy (IUM96-2828), and later a fertile Hungarian clinical a isolate (IUM99-3617) was reported, supporting the earlier hypothesis that fertile serotype aA isolates would exist based on the analysis of AD hybrids (128, 129). Further study of clinical isolates from sub-Saharan isolates was performed by Mitchell and colleagues, yielding the exciting presence of multiple a isolates in a unique population of isolates from the cerebrospinal fluid of AIDS patients in Botswana (79). These isolates are robustly fertile in the laboratory when crossed with the serotype A reference strain H99, yielding viable fertile, haploid recombinant progeny. Furthermore, these studies revealed

evidence of a recombining population structure, indicating that mating may still occur in the environment, a heretofore unproven step in the environmental life cycle of this pathogen. Together, these data suggest that while a isolates of the most successful pathogenic clade (serotype A) exist, current sampling indicates that they appear to be largely geographically restricted to sub-Saharan Africa (79).

Given the discovery of both a and α isolates of serotypes A and D, the existence of AD hybrids produced by mating, and evidence of recombination in a geographically restricted region, it has become apparent that the sexual cycle is still playing a role in the evolution of this important pathogen. Furthermore, this lends support to the hypothesis that sexual spores exist in the environment and may be an infectious propagule.

Importantly, the serotype A *MATa* allele from the sterile isolate 125.91 appeared intact and contained the entire cohort of *MAT* genes (75). If the serotype A *MATa* isolate 125.91 had an apparently intact *MAT* locus, why was it sterile? In the search for mating of this novel a isolate, Nielsen et al. performed pairwise mating comparisons against not only the commonly used clinical α isolate H99 but also an additional 150 serotype A strains of various origins (93). Three isolates were identified with which strain 125.91 could mate to produce different levels of filamentation and basidiospores. However, the vast majority of serotype A studies are performed with strain H99 and its genome is being sequenced, so mating was therefore desirable in this background. Fortunately, when progeny from a cross with the most robust mating partner (8-1) were analyzed for fertility, one a offspring (KNA14) now exhibited the capacity to mate with strain H99 and robustly mated with a pheromone-hypersensitive H99 *crg1* mutant strain. In the same manner as had been performed for serotype D, this novel strain was backcrossed to H99 nine times, introducing the 125.91 a allele into the H99 genetic background and creating the serotype A congenic strain pair KN99a and KN99α (93). These congenic serotype A strains are proving invaluable as a foundation for genetic analysis.

The failure of the serotype A *MATa* strain 125.91 to mate with most α isolates may reflect cryptic speciation occurring or other barriers to mating, possibly similar to vegetative incompatability in other fungi. In fact, ~50% of clinical and environmental serotype A isolates are sterile in genetic crosses. Further studies are warranted to define the population and genomic structures of sterile and fertile isolates.

Why are serotype A *MATa* strains so uncommon? One possibility is that they are less virulent than their α counterparts, and therefore do not appear in the clinical setting. Studies with various serotype D isolates and three congenic strain pairs showed that in this variety the α mating type is associated with increased virulence in a murine tail vein injection model in some genetic backgrounds (68, 94) (Fig. 2). Supporting this observation with serotype A, neither 125.91 nor the Italian isolate IUM96-2828 was as virulent as H99 in a virulence model (3, 93, 128); however, these strains are not congenic, confounding the analysis. Nielsen et al. (92, 93) showed that the congenic KN99**a** and KN99α strains have equivalent virulence to each other in a murine inhalation assay, indicating that the difference in virulence of the original **a** isolates is not attributable to their mating-type locus allele (Fig. 2). However, when genetically marked derivatives of the serotype A congenic pair were coinoculated, the α strains predominated in the central nervous system (92). Furthermore, deletion of the only α-specific gene *SXI1α* does not affect virulence in either serotype (43). These results therefore indicate that the role of mating type in virulence may differ between the laboratory-adapted, less virulent serotype D strains and a clinical isolate of the most common pathogenic variety serotype A.

The presence of AD hybrid strains in both clinical and environmental samples provides evidence that mating can occur in nature, and the detailed analysis of these unusual hybrid strains provides a clue to the present unequal distribution of mating types. A striking feature of the AD hybrid population is that there is a near-equal distribution of the two diploid classes, the αAD**a** and the **a**ADα hybrids, even though **a**A strains are extremely uncommon (17, 74). We hypothesize that the AD hybrids arose at a time when **a**A strains were more abundant and that a more recent event has then led to global colonization beginning with an unusual αA isolate. We hypothesize that this could have been the emergence of isolates that show a unique predilection to colonize pigeon guano and that, as pigeons followed humans in colonizing the world, a recombinant or mutant α clone swept the globe. The now clonal α isolate would then be stranded in an environment in the absence of compatible **a** partners, inhibiting access to meiosis during the life cycle.

If this model holds true, does sex still play a role in the evolution of this pathogen outside of sub-Saharan Africa? Recent advances by Lin et al. have revealed that the process of monokaryotic fruiting is in fact a mating event, with a level of meiotic recombination occurring equivalent to that seen in the classic **a**/α sexual cycle (78). Diploidization can occur by either endoduplication or cell-cell fusion between cells of the same mating type and is therefore an example of an organism that has evolved the ability to utilize the sexual cycle in the absence of a partner of the opposite mating type, an ability particularly useful in a fungus where one of the mating types is geographically restricted (78).

The finding that a pathogenic fungus retains the ability to engage in infrequent sexual reproduction that could impact its virulence shares several features with recent models proposed for the role of sexual recombination in parasites. In that case, unusual recombinants resulted in efficient oral transmission of the human parasite *Toxoplasma gondii*, an event that is

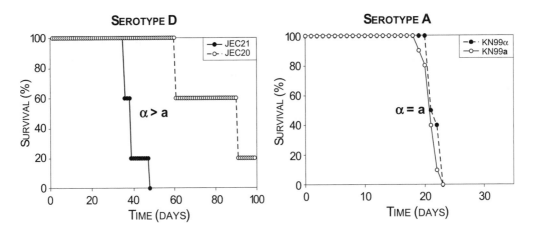

Figure 2. Association of mating type with virulence in *C. neoformans*. Why are *MAT*α isolates so common in the clinical setting? For serotype D, studies with various isolates and the JEC20-JEC21 congenic strain pair showed that in this variety the α mating type is associated with increased virulence (68). However, in the more clinically relevant serotype A, the KN99**a**-KN99α congenic strain pair have equivalent virulence to each other in a murine inhalation assay (93). The role of mating type in virulence may therefore differ between the less virulent serotype D strains and the most common pathogenic variety, serotype A.

hypothesized to have occurred ~10,000 years ago (37, 117). Because serotype A *MAT**a*** strains are common in sub-Saharan Africa, retain sexual fecundity and belong to a population that shows genetic features of recombination (79), we hypothesize that *Cryptococcus* originated on this continent. However, since pigeons are uncommon in Botswana and Tanzania, we postulate that there is another unique environmental reservoir, from which a recombinant arose with altered growth properties, allowing worldwide distribution with its new environmental niche. This model makes several testable predictions, including the prediction that there will be genetic determinants for growth on pigeon guano and that unique environmental hosts or niches remain to be discovered. Finally, we note that this model shares features with those proposed for the potato blight *Phytophthora infestans*, an oomycete fungus-like stramenophile which is also thought to undergo geographically restricted sexual recombination in central Mexico, which then gives rise to unique variants that sweep the globe to cause pandemics, such as the Irish Potato Famine of the 1800s (35, 102).

Why is it that so many isolates of *Cryptococcus* are infertile? Laboratory passage experiments with serotype D have revealed that certain strains of *Cryptococcus* relatively easily lose the ability to undergo sex (138). Furthermore, the serotype A strain H99 has been observed to lose mating ability during laboratory culture but regains it following passage through the murine central nervous system (T. A. Missall and J. K. Lodge, personal communication). One potential explanation is that this observed loss of mating is not via mutation but via an epigenetic "switch," a model supported by the observation that mating ability can be regained in a strain from which it has been lost.

Why would such a mechanism exist? The recent completion of the genome sequence for two sibling strains of *C. neoformans* var. *neoformans* revealed that one of the first steps in the creation of the JEC20-JEC21 congenic strain pair produced a chromosomal translocation and segmental duplication event, possibly during meiosis (32). Meiotic mapping, chromoblots, and sequence analysis demonstrate that two chromosomes underwent a telomere-telomere fusion, generating a dicentric chromosome that broke to form two novel chromosomes that now share a large segmental duplication of 62,872 identical nucleotides encompassing 22 predicted genes. If the translocation and segmental duplication events indeed occurred during meiosis, this discovery highlights the potential hazards of sexual recombination to an organism which has a high genomic repeat content; such a danger could lead to the avoidance of sex

while maintaining the machinery of the sexual cycle for uncommon use, as in *C. albicans* (79).

The story of the role of sex in virulence of *Cryptococcus* is therefore far from completed. There are still questions to be answered, both in the laboratory and in the environment. How is it that the ability to undergo sex can be lost and then regained? Do gross genome rearrangements occur frequently, or is the serotype D congenic strain pair an exceptional case? Beyond the dissection of mating as a basic biological trait of the organism, recent findings in multiple clinical situations have stressed the importance of discovering its role in the environment. In sub-Saharan Africa, does coinfection occur with spores of both mating types, and if so does this alter the course of infection or result in selection of recombinant, more infectious spores? Alternatively, is the major role of mating to provide the opportunity to rearrange allele combinations in order to allow altered environmental niche range, such as may have led to the spread of serotype A *MATα*, and is this the selective pressure that has led to the evolution of same-sex mating? Or is it a mutation closely linked to the *MATα* allele that has enabled the spread of this mating type throughout the world? Questions such as these will continue to drive research on the role of mating and virulence in *Cryptococcus* for years to come.

CRYPTOCOCCUS GATTII—SEX IN A PRIMARY PATHOGEN

At the time that mating was first observed in *C. neoformans*, a similar observation of a bipolar mating system was made in the divergent sibling species *C. gattii* (62). Unlike the lemon-shaped spores of *C. neoformans*, *C. gattii* mating is distinguished by the formation of spores that are more elongated, giving a bacillus rod-shaped structure (33, 62). Moreover, the mechanics of clamp cell fusion differ between the two, with a prominent peg cell being formed in *C. gattii* mating filaments that fuses with the clamp cell from an adjacent filament cell (33). While *C. gattii* is represented by two serotypes, phylogenetic analysis has revealed that these are still interbreeding (5). However, as in *C. neoformans*, although population studies revealed isolates of both mating types in equal ratios inhabiting common environmental niches in Australia, no evidence for mating in these isolates in these locations was found (38).

In contrast to serotypes A and D, which are cosmopolitan and most frequently associated with pigeon guano, the *C. gattii* serotypes B and C are usually restricted to tropical and subtropical regions of the world and are found in association with woody debris of trees (114). However, interest in *C. gattii* has increased due to an outbreak on Vancouver Island

(33, 42, 116). Since 1999 more than 134 human cases with at least four fatalities have been detected in otherwise healthy individuals and are attributable to infection with serotype B isolates. An equally large increase in the rate of infection of animals, including dogs, cats, and porpoises, has also been observed (77). Unlike previous identification of environmental sources of this variety, on Vancouver Island C. *gattii* has been found in association with a number of native tree species (Douglas fir, alder, maple and Garry oak) rather than eucalyptus trees, and also with the air and soil surrounding them. This expansion of the serotype B habitat to include nontropical areas and different host trees increases the risk of acquiring cryptococcosis and makes it paramount to understand the mechanisms by which C. *gattii* sexually reproduces.

While the majority of worldwide C. *gattii* strains are sterile under most laboratory conditions, almost all isolates from the Vancouver Island outbreak are fertile, and all are of the α mating type (33) (Fig. 3). Furthermore, 95% of the clinical and 100% of the environmental isolates are molecular type VGII,

suggesting that this outbreak is caused by a clonal population of C. *gattii*, with patients occasionally succumbing to latent infections acquired elsewhere (22, 54) (Fig. 3). One potential model is therefore that the Vancouver Island outbreak may be attributable to a recent recombination event in an unusual, fertile clade of C. *gattii* with increased basidiospore production or altered environmental range. This therefore parallels the model of C. *neoformans* worldwide expansion through the meiotic creation of progeny with altered growth traits. This model is supported by observations that rare isolates of the VGII molecular type from Australia, where this fungus is endemic, are fertile and display evidence of a recombining population structure (9, 10). Although the particles of C. *neoformans* detected in the contaminated air on Vancouver Island are small enough to be spores, no **a** isolates have been identified thus far and monokaryotic fruiting has only been reported once in this species; therefore, the way in which spores might be produced remains a mystery.

The C. *gattii* species represented an ideal system with which to perform a comparative analysis of the

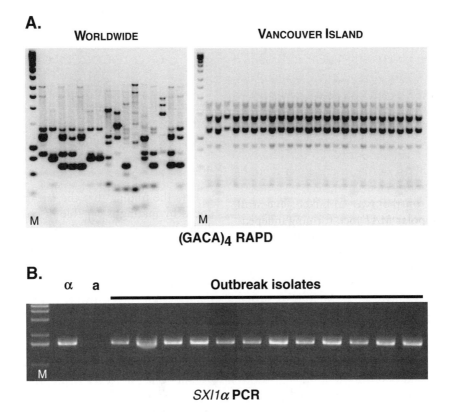

Figure 3. The Vancouver Island isolates of C. *gattii* are clonal. (A) Comparison of C. *gattii* isolates using the (GACA)₄ random amplification of polymorphic DNA (RAPD) fingerprinting method shows that isolates from around the world exhibit a high degree of variation. In contrast, clinical, veterinary, and environmental isolates from the Vancouver Island outbreak all appear to be clonal. (B) PCR of the *MAT*α-specific gene *SXI1*α reveals that the isolates from the Vancouver Island outbreak are exclusively of the α mating type. M, marker.

formation of the *C. neoformans MAT* locus (29). Sequencing of candidate **a** and α alleles cloned from Australian environmental isolates revealed that the major features of *MAT* were conserved. As in *C. neoformans*, *C. gattii MAT* is over 100 kb long, and contains the same cohort of genes (29). In short, the most common ancestor of serotypes A, B, and D probably contained a *MAT* locus that, while genetically functioning as a bipolar system (as in ascomycetes), actually contained not only all of the components of typical basidiomycete *MAT* loci but also numerous additional genes.

How did this system evolve to be so different from other basidiomycetes that are tetrapolar? By employing analysis of synteny, phylogeny, nucleotide identity, and synonymous substitution rate, it was observed that *MAT* consists of four major classes of genes, which represent the relative evolutionary time that the alleles of these components have been diverging in the nonrecombining *MAT* locus (29). The most ancient class represents the genes contained within this recombinationally suppressed region for the longest time and includes those encoding pheromones, pheromone receptors, and elements of the pheromone-sensing MAPK pathway. The second and third classes represent intermediate genes that have been within this recombinationally suppressed region for shorter periods, with progressively higher nucleotide identity between the **a** and α alleles and with less discrete although still *MAT*-specific phylogenetic patterns. These three classes represent genes within the ancestral *MAT* structure. Finally, the last group contains the five most recently acquired genes that exhibit a species-specific but not mating-type-specific phylogenetic pattern, similar to genes outside *MAT*.

Together, these data indicate that *MAT* evolved via four main steps (29). First, additional genes were acquired by each of the two unlinked ancestral basidiomycete tetrapolar *MAT* loci, forming independent gene clusters involved in pheromone sensing/production and meiosis/karyogamy. Second, these gene clusters were fused via chromosomal translocation to give a transitional tripolar intermediate state where the most recent class of genes were captured between the two tetrapolar gene clusters. Third, the opposite mating type converted to an equivalent bipolar structure via gene conversion or recombination between the linked and unlinked sex-determining regions. Finally, *MAT* was subjected to ongoing intra- and interallelic gene conversion, and inversions that suppress recombination that may be driven by the high transposon content of the locus. These events resemble those that have shaped the genomic sex-determining structures in the animal (71, 109) and plant (80, 91) kingdoms, suggesting that the *Cryptococcus MAT*

locus represents an early stage in the evolution of sex chromosomes (29, 30).

Our understanding of the closely related *C. neoformans* and *C. gattii* sexual cycles and their role in virulence is therefore advancing concurrently. The *C. gattii* Vancouver outbreak represents a unique opportunity to study a potential link between fertility and the evolution of fungal virulence. The evolutionary mechanisms behind this outbreak, which is due exclusively to one mating type of a clonal isolate (33, 54), are therefore of paramount interest and represent a question which invites further investigation.

CANDIDA ALBICANS—THE RELUCTANTLY SEXUAL DIPLOID YIELDS TO TEMPTATION

C. albicans is the most common fungal human pathogen, and as part of our normal mucosal microbial flora is uniquely poised to cause diseases that range from superficial infections of mucosal surfaces to systemic life-threatening disease in neutropenic hosts. Yet studies of this important pathogen have been impeded by the absence of a genetic system. *C. albicans* was thought to be strictly asexual for more than a century; however, over the past 5 years we have witnessed a renaissance in our understanding of the mating system of *C. albicans* (50). These breakthroughs resulted from the following:

1. the discovery of the mating-type-like locus (*MTL*), which revealed that most isolates are **a**/α diploids (47)
2. the finding that strains engineered to have only one or the other mating-type allele are capable of mating with each other (48, 84)
3. the discovery that the white-to-opaque switch enables high-efficiency mating (89)
4. the finding that the tetraploid cell type produced by mating is capable of collapsing again to a diploid state by concerted chromosome loss (4)

Thus, we now appreciate that *C. albicans* can complete a parasexual cycle, but so far no meiosis or sporulation has been observed. It is exciting to discover, then, that this parasexual cycle might contribute to evolution and even pathogenesis in the human host, and we may capitalize on these discoveries to conduct genetic studies in the laboratory. In addition, these breakthroughs further suggest that continued studies may yet reveal elusive conditions or strains in which *C. albicans* may undergo meiosis, with important implications for the evolution of virulence.

The discovery of the *C. albicans* mating-type locus was the result of a search for genes encoding homeodomain proteins that were encoded by only one allele in the genome. The genome project revealed a sequence trace that encoded a portion of a homeo-domain protein related to the a1 protein of *S. cerevisiae*. This sequence trace served as the entry point to clone the entire *MTL*a allele and then, using the flanking homologous regions of the genome, to clone the *MTL*α allele (47). The *MTL* locus was found to be ~10 times larger than the *S. cerevisiae MAT* alleles (8,861 and 8,742 bp versus 747 and 642 bp) and to encode both unique cell identity factors similar to those found in *S. cerevisiae* (a1, α1, and α2) as well as three additional genes that are encoded as divergent alleles by the two mating-type locus alleles (Fig. 1). These are the *PIK1* gene, encoding a phosphatidylinositol-4-kinase homolog, a gene encoding a poly(A) polymerase homolog (*PAP1*), and a gene encoding an oxysterol binding protein (*OBP1*). The functions of these additional *MTL* genes are, as yet, unknown. An additional *MTL* gene was recently discovered (a2) and found to encode an HMG box factor that is conserved and plays a role in fertility in some ascomycetes (e.g., *Neurospora crassa*) but not in others (*S. cerevisiae*) (8, 122). This factor is absent from *S. cerevisiae*. In short, these sequences indicated that *Candida* may have a cryptic mating system of a bipolar nature, analogous to that observed in other ascomycetes. *C. albicans* is a diploid, and the vast majority of isolates are a/α heterozygotes. This arrangement at *MAT* is analogous to that of *S. cerevisiae* a/α diploid cells, a cell type that is specialized to undergo meiosis and sporulation and is unable to mate. Thus, this discovery revealed a unifying molecular basis to explain why *C. albicans* had not been observed to mate.

The discovery of the *C. albicans MTL* locus suggested a facile route to construct strains capable of mating: strains engineered to be hemizygous or homozygous at *MTL* (*MTL*a/Δ or *MTL*a/*MTL*a) might fuse with cells of the opposite mating type. In fact, two groups independently succeeded in these efforts. First, Hull et al. demonstrated that deletion of one or the other *MTL* allele resulted in *MTL* hemizygous strains that can fuse to produce tetraploid isolates (48). Cell-cell fusion requires the presence of only one mating type at the *MTL* locus and thus mimics the cell fusion that occurs during mating of a and α isolates of *S. cerevisiae*, or of diploid *S. cerevisiae* strains in which one or the other *MAT* allele has been deleted (40, 41). Mating in *C. albicans* was not observed initially in vitro but was found to occur at low efficiency during experimental infection in a murine model.

In parallel with these studies, Magee and Magee achieved mating of *C. albicans* by performing a cunning genetic trick (84). Growth on sorbose was known to result in loss of one copy of chromosome

5, to result in an intermediate $2n - 1$ aneuploid state (49). Following the release of cells from selection for growth on sorbose, the one remaining copy of chromosome 5 is reduplicated to restore the cell to a full $2n$ diploid state. The finding that the *MTL* locus resides on chromosome 5 provided the impetus to apply this growth selection, resulting in the isolation of a/a and α/α homozygous isolates. *MTL* homozygous strains were found to undergo mating to produce a/a/α/α tetraploid cells under in vitro growth conditions of rich medium and growth at 24°C. Together, these studies revealed that mating could be achieved with engineered strains, but at very low efficiency. The findings that mating requires specific *MTL* configurations in this diploid fungus and occurred only at extremely low efficiency in these engineered strains led some to suggest that this might simply be a laboratory artifact rather than being representative of true sexual events that occur in nature.

Prior to the discovery of the *MTL* locus and mating of engineered strains, there were two hints that mating might in fact occur in *C. albicans*. In *S. cerevisiae* the Gpa1 Gα protein homolog is required for pheromone response. Sadhu et al. discovered that the *C. albicans* ortholog is regulated via a native a1/α2 consensus binding site when heterologously expressed in *S. cerevisiae* and can complement an *S. cerevisiae* *gpa1* mutation (106). The observation that at least one component of the pheromone-responsive mating machinery was conserved, and could be regulated in a cell-type-specific fashion, suggested that other mating machinery might be similarly conserved for a functional role in an as yet undiscovered mating cycle. In accord with this theory, both comparative studies and genetic analysis of mutants later revealed that most of the mating elements are indeed present in the *C. albicans* genome (15, 83, 124). The second clue that a sexual cycle might occur was based on population genetics studies indicating that, while the vast majority of the *C. albicans* population worldwide is clonal, some degree of genetic exchange can be detected (36). Such studies cannot discern the mechanisms of exchange but have been taken as evidence that cryptic mating might occur naturally in the population. Subsequent studies have revisited this question by examining the inheritance of the mitochondrial genome and have drawn a different conclusion: namely, that while evidence for clonality is clearly apparent, evidence for genetic exchange could not be definitively established, since clonal loss of markers via mitotic recombination might also mimic the apparent exchange observed (1).

These conflicting observations were therefore inconclusive about the presence of a sexual cycle.

However, insight was provided from an unexpected source: the white-to-opaque phenotypic switch (113). The white-to-opaque switch is a change in colony and cell morphology that can occur at high frequency under certain conditions and was discovered by Soll and colleagues two decades ago (110, 112). The fact that the white-to-opaque switch had been observed to occur only in a subset of isolates suggested to some that this phenomenon might not be of general importance to the biology or life cycle of *C. albicans*. However, the findings that gene expression patterns are globally altered, that temperature regulates the switch, and that both colony and cell morphologies are dramatically altered all suggested that this dimorphic transition could play critical roles in virulence or the life cycle.

Johnson and colleagues discovered that *MTL* locus hemizygous strains (*MTLa/Δ* or *MTLΔ/α*) undergo the white-to-opaque switch at high frequency, whereas the parental strain bearing the heterozygous a/α configuration does not (45, 89). That this represents a true white-to-opaque switch was demonstrated by colony morphology and staining, by gene expression patterns, and by electron microscopy to observe the surface protrusions (or "pimples") known to adorn cells of the opaque form. Heterozygosity of *MTL* therefore prevents the white-to-opaque switch, and this developmental pathway is negatively regulated by the a1/α2 complex. Furthermore, cells that had undergone the white-to-opaque switch were found to be dramatically enhanced for mating. Mating efficiency is increased 10^3-fold in crosses involving one white and one opaque partner and increased over 10^6-fold when both mating partners are in the opaque phase. Thus, the *MTL* locus controls the white-to-opaque switch, which in turn controls mating efficiency. It has been suggested that the protrusions that decorate the surface of opaque cells might represent specialized structures for pheromone production, sensing, or the sites of cell-cell fusion during mating (50).

Following the discovery that the white-to-opaque switch governs mating efficiency, previously identified switching-competent strains in the population were reinvestigated and shown to represent naturally occurring a/a or α/α isolates (82). Furthermore, ~3% to 7% of clinical isolates were also found to be *MTL* homozygotes (73, 82). *MTL* homozygous strains have been identified that fail to undergo the white-to-opaque switch, possibly as a consequence of unknown mutations that preclude the switch or differences in genetic background that alter the developmental cascade (82). In short, while all switching strains are *MTL* homozygotes, not all *MTL* homozygotes switch.

That *MTL* homozygotes exist in nature, undergo the white-to-opaque switch, and mate efficiently in the laboratory suggest that mating might not be restricted to engineered strains. The finding that *MTL* homozygous strains occur naturally in the population could reflect an environmental selection, similar to sorbose selection in the laboratory, that results in chromosome loss and regain. Alternatively, these isolates could be the consequence of homozygosis of the *MTL* locus as a result of DNA damage. Similar events are readily apparent in mating-type tests of a/α isolates of *S. cerevisiae* in which a/a and α/α isolates are observed at a low frequency (41, 103).

An exciting possibility is that the *MTL* homozygous *C. albicans* isolates are not exclusively the precursors of mating (i.e., a/a or α/α strains produced from an a/α diploid by some host condition that mimics in vitro sorbose selection) but are in fact the products produced by the parasexual reduction of the a/a/α/α tetraploid. The tetraploid sexual cycle that has been characterized in *S. cerevisiae* involves mating of a/a and α/α strains to produce a/a/α/α isolates that undergo meiosis and produce a mixture of diploid progeny, including not only a/α isolates that can undergo another round of meiotic reduction but also a/a and α/α isolates that are mating competent (99, 104). In a similar fashion, reduction of the *C. albicans* a/a/α/α isolates produced by mating could potentially lead to some or all of the *MTL* homozygous strains that are apparent in nature. If so, then examining this unique population of a/a and α/α isolates for signatures of genetic exchange might provide a more sensitive approach to address whether a sexual or parasexual cycle indeed occurs in nature.

The discovery that the white-to-opaque switch generates high-efficiency mating in *C. albicans* created a conundrum, since this developmental transition is inhibited at 37°C. Do special microenvironments exist in the mammalian host that support the opaque cell type? In fact, Soll and colleagues discovered that *C. albicans* mating readily occurs on skin, which is several degrees cooler (31.5 to 32°C) than physiological temperature and low enough to support the opaque cell type (70). First, their studies examined the cell biology of mating, revealing the production of dramatic conjugation tubes by both partners that fuse at their tips to produce the zygote (81). Similar conjugation tubes have recently been found to occur in response to synthetic mating pheromones (4, 81, 96). Second, these events occur robustly on the skin of mice, with conjugation tubes following the ridges of the skin and resulting in up to 50% zygote formation. These findings reveal at least one opaque-cell-permissive physiological niche in which mating of *C. albicans* might plausibly occur in nature.

While *C. albicans* is capable of mating to form a tetraploid, how does it return to the diploid state? Bennett and Johnson reported that prespore medium induces chromosome loss of the tetraploid to produce a diploid state (4). In contrast to sporulation medium, which is nutrient limiting, prespore medium is a nutrient-rich medium used to drive cells to rapid proliferation prior to nutrient limitation; in *S. cerevisiae*, this feast and then famine enhances sporulation efficiency. There has been no hint yet of meiosis in *C. albicans*, and intermediate aneuploid states between tetraploid and diploid can be observed. At this stage, these studies document at least the presence of a parasexual cycle, but no studies have yet been conducted to measure the frequency of recombination or to examine whether synaptonemal complexes might be present. One hypothesis might be that meiosis is occurring but that the chromosome reduction step is inefficient, leading to partial aneuploidy during the process.

Does *C. albicans* mate in nature, perhaps during colonization or infection of the human host? There are several hints that the laboratory-observed parasexual cycle might occur in nature. First, a variety of studies have revealed that the genomic loci can be triploid rather than diploid in *C. albicans* (16). While this could result from spontaneous chromosome nondisjunction or other genetic events, the incomplete reduction of the tetraploid state produced by mating could also yield strains that are still aneuploid, such as $2n + 1$ diploids. Second, studies by White and colleagues reveal that *MTL* homozygosity is common among azole-resistant isolates (105) whereas other studies reveal that *MTL* homozygous strains not exposed to azoles are not drug resistant (100). It has been noted that both the *ERG11* gene as well as a regulator of ABC transporter genes (*TAC1*) are linked to the *MTL* locus, and thus selection for azole resistance could lead to homozygosis of these loci and the linked *MTL* locus (85). We propose an alternative model: namely, that mating and parasexual reduction have occurred, leading to homozygosis of *MTL* and linked markers via chromosome assortment. Third, a recent study by Magee and colleagues reveals that karyotype heterogeneity and aneuploidy $(2n + 1)$ is commonly associated with *MTL* homozygosity, and these investigators posited that a common process promoting mitotic hyperrecombination might lead to homozygosis at *MTL* and chromosomal rearrangements (73). Again, we hypothesize that these isolates might represent the products, rather than the precursors, of either the parasexual cycle or an as yet unknown meiotic cycle that occurs during infection. In this model, one role of mating and reduction could be to reassort, and

also rearrange, the genome and thereby provide genetic diversity.

The past 5 years has revolutionized our understanding of the role of sexual recombination in *C. albicans*. From an organism known to be an obligate diploid that was thought to be strictly asexual, *C. albicans* is now known to retain the machinery for at least a parasexual, if not sexual, cycle. Our understanding was enhanced by the discovery of the *MTL* locus and of mating of engineered strains, the remarkable finding that the white-to-opaque switch yields mating-specialized cells, and finally the revelation that under appropriate conditions the tetraploid state can be reduced efficiently to the diploid state. When these laboratory studies are combined with the discoveries of naturally occurring fertile isolates and conditions that enable efficient mating on the host, we now appreciate that mating might play significant roles in both the evolution and pathogenicity of the most common human fungal pathogen. However, conditions have not yet been discovered that allow meiosis or sporulation. These may yet be discovered, or may have been lost during evolution, or may be restricted to only some isolates of the population. Under certain conditions *C. albicans* can produce chlamydospores, but these are not thought to be meiotic products. Some serotype B strains have been reported to form unusual "multibudded" structures on cornmeal agar that could represent abortive attempts at meiosis and sporulation (66), but these have not been pursued further.

Given that *C. albicans* is specialized to survive on our mucosal surfaces and must continually run the gauntlet of the human immune response, generating asci and spores might not be an appropriate strategy. Moreover, the main benefits of meiosis, ploidy reduction and recombination, may be fulfilled during the parasexual cycle of *C. albicans* through tetraploidization, reduction, and mitotic recombination. Potentially, an increase in mitotic recombination could occur during the parasexual cycle of *C. albicans* and thereby confer many of the benefits of meiosis. Finally, there is ample evidence for abundant genomic polymorphisms in *C. albicans*, and reduction of the tetraploid to the diploid state would generate diversity at these loci to produce isolates that are homozygous for either of the two alleles. Thus, our understanding of the role of sex in this important microbial pathogen has begun, but the final chapter has not yet been written.

In contrast to *C. albicans*, which retains the ability to mate but has not yet been observed to undergo meiosis or sporulate, several other *Candida* species, including *Candida guilliermondii*, *Candida krusei*, and *Candida lusitaniae*, retain the ability to undergo complete sexual cycles, including meiosis and sporulation (28, 143). These species may therefore provide a unique resource to compare and contrast the components required for meiosis to ascertain what *C. albicans* may be missing and which may then be reintroduced to enable more sophisticated genetic approaches to be applied.

ASPERGILLUS FUMIGATUS—ASEXUAL IN A CLADE OF FERTILITY?

In contrast to *Cryptococcus*, to which our exposure is limited to occasional environmental encounters with spores or yeast cells, the infectious particle of *A. fumigatus* is abundant worldwide and we are all continually exposed on a daily basis to conidia, the infectious agent (72). Produced via a developmental cascade well studied in its saprophytic cousin *A. nidulans*, this opportunistic pathogenic mold rapidly generates masses of asexually produced conidia that are readily airborne and universally dispersed (120). However, unlike the homothallic fungus *A. nidulans*, *A. fumigatus* has never been observed to undergo a sexual cycle.

Analysis of the *A. nidulans* haploid genome revealed that both alleles of the *MAT* loci corresponding to a traditional ascomycete bipolar system are present—one encoding an α box domain protein, and the other encoding an HMG domain transcription factor, but unlinked in the genome (20, 24) (Fig. 4). Together these enable a homothallic lifestyle that allows both inbreeding and outcrossing. Analysis of the surrounding gene organization provides evidence that this could have arisen via a chromosomal translocation event, producing a homothallic organism from a heterothallic ancestor.

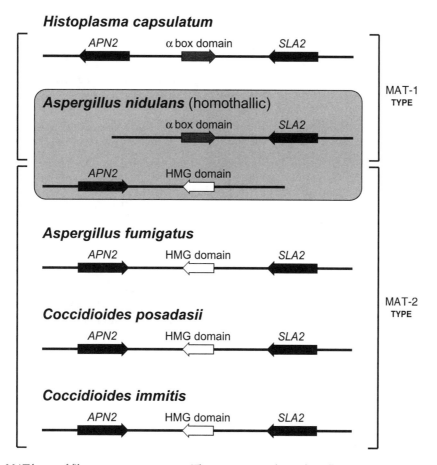

Figure 4. The *MAT* locus of filamentous ascomycetes. The genomic revolution has allowed the identification of the *MAT* locus from a large variety of ascomycetous fungi based on homology and synteny, even in the absence of a strain of the opposite mating type. Based on the nomenclature system proposed by Turgeon and Yoder (123), most of these can be classified into one of two types. *MAT-1* idiomorphs contain an α-box-domain-encoding homeobox gene, as is seen in the genome sequence of *H. capsulatum*. *MAT-2* idiomorphs contain an HMG-domain-encoding gene, as seen in the genome sequences of *A. fumigatus*, *C. posadasii*, and *C. immitis*. Homothallism can arise from a wide variety of events—in the saprophyte *A. nidulans*, this appears to have arisen by a translocation event leading to the presence of both idiomorphs in the haploid genome, as suggested by synteny with the adjacent orthologs of *SLA2* (encoding a protein involved in cytoskeleton assembly) and *APN2* (encoding DNA lyase). The drawing is not to scale.

Indeed, the incomplete *A. fumigatus* genome reveals just such a heterothallic system with only one *MAT* locus allele present, in this example the allele of the locus that encodes an HMG transcription factor (98, 125) (Fig. 4). Furthermore, a genomic survey has revealed that both the components required for the production of both peptide and lipid-modified pheromones, and also several components of the meiotic machinery, are conserved, suggesting that *A. fumigatus* has maintained the machinery required for a sexual cycle (97, 98, 125).

Is the *MAT* locus of *A. fumigatus* functional, and is the sexual cycle of this important pathogen awaiting discovery? Mating is readily observed not only in the homothallic organism *A. nidulans* but also in the more closely related homothallic fungus *Neosartorya fischeri* (34, 127). Additionally, molecular population studies provide evidence of sexual recombination in *A. fumigatus*, suggesting that appropriate mating strains and conditions may remain to be defined (126).

What might be the role of this cryptic mating process? One important question that has recently been answered is whether the opposite mating type still exists in the environment and, if so, whether it is as ubiquitous or virulent as the currently identified mating type. Studies of *A. fumigatus* isolates from around the world identified the α box domain-encoding allele and revealed that both mating types occur in similar proportions, further supporting the model that this species is actively undergoing a sexual cycle in the environment (97). This is in direct contrast to *C. neoformans*, in which one mating type is abundant and ubiquitous and the other is uncommon and geographically restricted. This discovery now sets the stage for studies to discover an elusive cryptic sexual cycle of this important pathogen, a discovery that would be invaluable in the laboratory, creating a tractable genetic system in this difficult-to-transform organism.

COCCIDIOIDES IMMITIS AND C. POSADASII—CRYPTIC SEX

The infectious particle of the primary pathogen *C. immitis* is also an asexually produced structure, the arthroconidium (95). Under appropriate conditions, the soil-borne saprobic multicellular hyphal form of this fungus septates into single-celled arthroconidia that can become airborne and be inhaled to produce life-threatening infections in otherwise healthy individuals. The role of this airborne structure during infection is highlighted by the increased incidence of coccidioidomycosis following dust storms.

While little is known about this fungus in the environment, much has been learned from studies of its population structure. Unlike most fungal pathogens, *Coccidioides* is restricted to the Western hemisphere (101), and although a sexual cycle has never been described, molecular phylogenetic analysis of isolates from throughout the Americas has revealed evidence supporting sexual reproduction (7, 56). Surprisingly, these studies indicate that *Coccidioides* actually represents two nearly identical yet independent species. The two species (*C. immitis* and *C. posadasii*) are distinct from each other and represent two discrete populations undergoing independent sexual recombination (27, 55, 56). Furthermore, these studies of population structure provide insight into the recent evolutionary history of this pathogen, revealing that the expansion of *C. posadasii* into South America probably corresponds to movement of humans onto this continent (26).

Analogous to that observed for *A. fumigatus*, analysis of the incomplete *C. immitis* and *C. posadasii* genomes reveals the presence of the components of the pheromone production system, in addition to a *MAT* locus with a single open reading frame (ORF) predicted to encode an HMG domain transcription factor (J. A. Fraser and J. Heitman, unpublished results) (Fig. 4). Comparison between the two sibling species reveals that the encoded proteins are almost identical, with only three changes across the predicted 339-amino-acid protein. The structure of this *MAT* locus is consistent with a classic ascomycete bipolar system in which the presumed second mating type encoding an α box domain protein has not yet been identified.

The identification of the second mating type of this organism could provide further evidence for the biogeographic range expansion into South America. The South American *Coccidioides* isolates appear largely clonal, and this is predicted to be the result of a founder effect (26). If this expansion is indeed due to a single founding isolate, we predict that only one mating type will be present in this population but that both would exist in the North American population from which the founder presumably arose.

Further characterization of a sexual cycle in this pathogen could provide an opportunity to derive classical genetic data to support the observation that coccidioidomycosis is caused by two closely related species (56). Classic genetic dogma dictates that these reproductively isolated taxa should be unable to interbreed to complete the sexual cycle or produce fertile offspring.

HISTOPLASMA CAPSULATUM—THE RAPID LOSS OF FECUNDITY

Histoplasma capsulatum is a primary pathogen that is distributed worldwide in association with soil enriched with bird and bat guano, and it is thought to represent multiple cryptic species (52). In the soil, the mycelial phase asexually produces microconidia that are inhaled into the lungs, causing histoplasmosis.

The sexual cycle of *H. capsulatum* was first described by Kwon-Chung in 1972 (59, 63), and represents one of the first cases of identification of the sexual structures of a pathogenic fungus. These genetic studies revealed that *H. capsulatum* has a heterothallic bipolar mating system, with strains designated either + or −. Furthermore, it has been reported that fertility is lost rapidly during laboratory passage, implying that selective pressures may serve to maintain fecundity in the environment (69).

The ongoing *H. capsulatum* genome project reveals a single *MAT* locus homolog encoding an α box domain protein, in contrast to the HMG-type locus observed in the genome projects of *Coccidioides* and *A. fumigatus* (Fraser and Heitman, unpublished) (Fig. 4). Whether the *MAT* locus allele identified thus far represents the + or − allele is unknown, since this strain is of unknown mating type. Given that the infectious particle can be produced asexually, does the *MAT* locus play any role in virulence? The original studies of the *Histoplasma* sexual cycle revealed an unusual bias; although both mating types were identified, all clinical specimens were of the − mating type. In contrast, environmental soil samples were either + or − (64). Subsequent analysis of *H. capsulatum* from more widespread clinical and environmental locations throughout the United States revealed a surprising correlation with this study. While environmental samples exhibit a 1:1 ratio of the two mating types, the incidence of the + mating type was disproportionately low in clinical samples, yielding a −/+ ratio of 7:1 (69). However, despite the clinical prevalence of the − mating type, mouse infection experiments revealed equivalent virulence between + and − isolates (58). *H. capsulatum* + and − isolates may therefore differ in the ability to produce the infectious propagule. Kwon-Chung et al. observed that in soil isolates, − strains displayed an increased ability to convert from the hyphal to the yeast form, the cell type that is more commonly the cause of infection (69). Mating type may therefore regulate this dimorphic transition and increase virulence.

A similar role of mating type in virulence has also been reported for *C. neoformans* (68). In this case, the role of the locus in virulence was addressed via the creation of a congenic strain pair, and a similar approach could address the seeming disparity between clinical observations and mouse experiments with *Histoplasma*.

CANDIDA GLABRATA—RETAINS MACHINERY FOR SEX BUT APPEARS CLONAL

C. glabrata shares several important features with *C. albicans* in that it is a human fungal pathogen of increasing importance, given its rapid propensity to develop azole resistance. While the two often occupy similar niches, *C. glabrata* is more closely related to the model budding yeast, *S. cerevisiae* (23). For example, *C. glabrata* is haploid, in contrast to the obligate diploid nature of *C. albicans*. *C. glabrata* also employs CTG codons to encode leucine rather than serine, similar to *S. cerevisiae* and in contrast to *C. albicans*, and thus is not part of the clade of *Candida* spp. that underwent the CTG codon capture event some 170 million years ago (86). *C. glabrata* and *S. cerevisiae* share the genomic signatures of a relatively recent whole-genome duplication event that occurred ~100 million years ago in the progenitor to both of these organisms but not in the lineage leading to *C. albicans* (23).

Remarkably, an early partial genome sequencing effort revealed the presence of many genes encoding components involved in mating, including the *MAT* locus (Fig. 5) and mating pheromones (136), and essentially the entire repertoire has been revealed now by the sequencing of the complete genome (23). Based on these findings, both haploid mating types have been discovered to be common in clinical isolates (115). In accordance with the close evolutionary link that *C. glabrata* shares with *S. cerevisiae*, the *C. glabrata* genome contains copies of the budding yeast silent mating-type cassettes. Moreover, synteny between the entire left arm of *S. cerevisiae* chromosome III has been maintained from *HML* (*MTL3* in *C. glabrata*) to the presumed active *MAT* locus (*MTL1*), while a single chromosomal translocation has transposed *HMR* (*MTL2*) to another chromosome. Although both **a** and α cassettes exist at *MTL1*, *MTL2*, and *MTL3*, giving eight potential genotypes, most isolates have an *MTL2***a**/*MTL3*α configuration at the silent cassettes. Irrespective of the genotypes at the silent cassettes, molecular mating type is bipolar since it is dependent on the allele present at *MTL1*. An Ho endonuclease homolog is also present, and evidence has been presented that mating-type switching can occur in the population (6, 8, 25).

These findings, combined with additional genomic information for other organisms, allowed the recent proposal of a detailed evolutionary model for the hemiascomycete mating-type system involving an ancestral bipolar mating-type system with an active *MAT* cassette shared by *C. albicans*, *C. glabrata*, and *S. cerevisiae* (8). Subsequently, the silent cassettes were gained by two sequential duplications

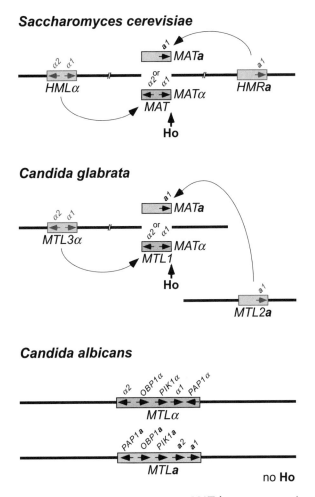

Figure 5. Variations on a common *MAT* locus structure in related ascomycetes. The model budding yeast *S. cerevisiae* is a homothallic ascomycete with a bipolar mating system. Cells have the ability to undergo mating-type switching in response to cleavage by the Ho endonuclease, allowing the active *MAT* cassette to be replaced with the silent cassette of the opposite mating type. The closely related pathogenic ascomycetous yeast *C. glabrata* bears a similar system; however, one of the silent cassettes resides on another chromosome. While the silent cassettes usually contain the *MAT* alleles shown here, they have been observed to encode either **a** or α at lower frequencies. The more distantly related pathogen *C. albicans* also has a bipolar system but lacks silent cassettes or the Ho endonuclease and does not undergo mating-type switching. The drawing is not to scale.

in one lineage (progenitor to *C. glabrata* and *S. cerevisiae*) but not the other (*C. albicans*). This gave rise to an inefficient mating-type switching system via mitotic recombination that is retained in some yeasts, such as *Kluveryomyces lactis*. Finally, the Ho endonuclease was acquired by capture of an intein from a mobile element, giving rise to high-efficiency switching in *S. cerevisiae*, *C. glabrata*, and their close relatives.

Thus genomic studies of *C. glabrata* reveal in elegant detail the retention of virtually all elements of

the mating-type locus, an elaborate mating-type switching system, and even the machinery for mating. Yet no diploids have been recovered, the population appears to be largely clonal with evidence for rare recombination events which could be ancestral or ongoing yet infrequent (21), and no one has as yet described mating in the laboratory despite considerable effort by several groups (D. R. Soll, personal communication; B. Dujon, personal communication; J. L. Reedy and J. Heitman, unpublished results). Thus, it remains a mystery why this close cousin of the sexually fertile model budding yeast retains not only the mating-type locus but also silent mating-type cassettes and numerous components of the mating machinery. Specialized conditions could be required to support mating, as has recently been found in *C. albicans* and *C. neoformans* var. *grubii*. *C. glabrata* thus joins the ranks of several pathogenic fungi, including *A. fumigatus* and *C. immitis*, that appear as if they should mate but are reluctant to reveal their secrets of nature in the laboratory.

PNEUMOCYSTIS—SEX OF AN OBLIGATE PATHOGEN IN THE HOST?

Pneumocystis jiroveci is a common fungal pathogen of humans that is often acquired early in life (119). It is transmitted via the respiratory route among individuals but causes symptomatic infection only in severely immunocompromised hosts. Since it is an obligate intracellular pathogen, the organism cannot be independently cultured, posing a formidable experimental challenge. Thus, many studies of this system have focused on interactions with the host pulmonary system and on the structure and nature of the genome. However, several features of the genome and life cycle of *P. jiroveci* suggest that an atypical sexual cycle might be an integral component of its virulence cycle.

The ongoing genome project from the closely related rat Form 1 *P. carinii* has revealed a unique region resembling the *C. neoformans MAT* locus, which also encodes homologs of the Ste3 pheromone receptor, the Ste20 kinase, the Ste12 transcription factor, and a farnesylated mating pheromone (111). This is surprising, given that *Pneumocystis* is an ascomycete and yet appears to have a *MAT* locus-like region more closely related to those of basidiomycetous fungi. Beyond the presence of what appears to be a *MAT* locus in the genome, additional lines of evidence suggest that an active sexual cycle may exist in this fungus. First, recent expression studies revealed that the *Pneumocystis* pheromone receptor homolog Ste3 is expressed in the trophic but not the cyst form of the organism (130). Second, evidence has suggested that a ploidy shift from haploid to diploid might occur in the trophic stage, but this conclusion remains controversial (119, 137). Finally, the cystic form can contain two, four, or eight nuclei, and a single publication has reported the identification of synaptonemal complexes (although this has not yet been independently verified) (87), suggesting that meiosis may occur in the cystic form. Only one allele of the *Pneumocystis MAT* locus-like region and no silent cassettes or Ho endonuclease have as yet been identified. These studies are intriguing and suggest that some form of unusual sexual cycle might occur in the lungs of infected patients. In *Cryptococcus*, α cells undergo sexual reproduction in the absence of a partner of the opposite mating type, either by endoduplication or cell-cell fusion between cells of the same mating type (78). Taken together, these two divergent human fungal pathogens, *Pneumocystis* and *C. neoformans*, which enter the human host via inhalation may similarly deploy a modified sexual cycle that plays central roles in infection and virulence and involves endoreplication or fusion between cells of the same mating type (78).

SUMMARY AND FUTURE DIRECTIONS

The mechanisms by which organisms recombine their genomes to produce recombinant offspring are central to our understanding of how organisms evolve and adapt to new environments. How do environmental saprophytes evolve into a pathogenic form? It is becoming apparent that pathogenic fungi share much of their mating machinery with related nonpathogenic fungi, and yet the pathogens have evolved to apparently rarely deploy the sexual cycle, perhaps to gain the benefits of clonal expansion while retaining the ability to undergo sex in response to extreme challenge. It is likely that these features will be shared with other pathogenic microbes, such as the parasites and even bacteria, to reveal general principles involved in the host-pathogen interface. Genomics has provided a window on the sexual behavior of many additional fungi, and it remains now to be established if and when they engage in sexual reproduction and what the consequences of mating are for their interactions, both with the environment and with the host.

REFERENCES

1. Anderson, J. B., C. Wickens, M. Khan, L. E. Cowen, N. Federspiel, T. Jones, and L. M. Kohn. 2001. Infrequent genetic exchange and recombination in the mitochondrial genome of *Candida albicans*. *J. Bacteriol.* **183:**865–872.
2. Arkhipova, I., and M. Meselson. 2005. Deleterious transposable elements and the extinction of asexuals. *Bioessays* **27:**76–85.
3. Barchiesi, F., M. Cogliati, M. C. Esposto, E. Spreghini, A. M. Schimizzi, B. L. Wickes, G. Scalise, and M. A. Viviani. 2005. Comparative analysis of pathogenicity of *Cryptococcus neoformans* serotypes A, D and AD in murine cryptococcosis. *J. Infect.* **51:**10–16.
4. Bennett, R. J., and A. D. Johnson. 2003. Completion of a parasexual cycle in *Candida albicans* by induced chromosome loss in tetraploid strains. *EMBO J.* **22:**2505–2515.
5. Boekhout, T., B. Theelen, M. Diaz, J. W. Fell, W. C. Hop, E. C. Abeln, F. Dromer, and W. Meyer. 2001. Hybrid genotypes in the pathogenic yeast *Cryptococcus neoformans*. *Microbiology* **147:**891–907.
6. Brockert, P. J., S. A. Lachke, T. Srikantha, C. Pujol, R. Galask, and D. R. Soll. 2003. Phenotypic switching and mating type switching of *Candida glabrata* at sites of colonization. *Infect. Immun.* **71:**7109–7118.
7. Burt, A., D. A. Carter, G. L. Koenig, T. J. White, and J. W. Taylor. 1996. Molecular markers reveal cryptic sex in the human pathogen *Coccidioides immitis*. *Proc. Natl. Acad. Sci. USA* **93:**770–773.
8. Butler, G., C. Kenny, A. Fagan, C. Kurischko, C. Gaillardin, and K. H. Wolfe. 2004. Evolution of the *MAT* locus and its Ho endonuclease in yeast species. *Proc. Natl. Acad. Sci. USA* **101:**1632–1637.
9. Campbell, L., B. J. Currie, M. Krockenberger, R. Malik, W. Meyer, J. Heitman, and D. A. Carter. 2005. Clonality and recombination in genetically differentiated subgroups of *Cryptococcus gattii*. *Eukaryot. Cell* **4:**1410–1419.
10. Campbell, L., J. A. Fraser, C. B. Nichols, F. Dietrich, D. A. Carter, and J. Heitman. 2005. Clinical and environmental isolates of *Cryptococcus gattii* from Australia that retain sexual fecundity. *Eukaryot. Cell* **4:**1410–1419.
11. Casadevall, A., and J. R. Perfect. 1998. *Cryptococcus neoformans*. ASM Press, Washington, D.C.
12. Casselton, L. A., and N. S. Olesnicky. 1998. Molecular genetics of mating recognition in basidiomycete fungi. *Microbiol. Mol. Biol. Rev.* **62:**55–70.
13. Chang, Y. C., L. A. Penoyer, and K. J. Kwon-Chung. 2001. The second *STE12* homologue of *Cryptococcus neoformans* is MATa-specific and plays an important role in virulence. *Proc. Natl. Acad. Sci. USA* **98:**3258–3263.
14. Chang, Y. C., B. L. Wickes, G. F. Miller, L. A. Penoyer, and K. J. Kwon-Chung. 2000. *Cryptococcus neoformans STE12α* regulates virulence but is not essential for mating. *J. Exp. Med.* **191:**871–882.
15. Chen, J., J. Chen, S. Lane, and H. Liu. 2002. A conserved mitogen-activated protein kinase pathway is required for mating in *Candida albicans*. *Mol. Microbiol.* **46:**1335–1344.
16. Chen, X., B. B. Magee, D. Dawson, P. T. Magee, and C. A. Kumamoto. 2004. Chromosome 1 trisomy compromises the virulence of *Candida albicans*. *Mol. Microbiol.* **51:**551–565.
17. Cogliati, M., M. C. Esposto, D. L. Clarke, B. L. Wickes, and M. A. Viviani. 2001. Origin of *Cryptococcus neoformans* var. *neoformans* diploid strains. *J. Clin. Microbiol.* **39:**3889–3894.
18. Davidson, R. C., T. D. Moore, A. R. Odom, and J. Heitman. 2000. Characterization of the MFα pheromone of the human fungal pathogen *Cryptococcus neoformans*. *Mol. Microbiol.* **38:**1017–1026.
19. Davidson, R. C., C. B. Nichols, G. M. Cox, J. R. Perfect, and J. Heitman. 2003. A MAP kinase cascade composed of cell type specific and non-specific elements controls mating and differentiation of the fungal pathogen *Cryptococcus neoformans*. *Mol. Microbiol.* **49:**469–485.

20. Denning, D. W., M. J. Anderson, G. Turner, J. P. Latge, and J. W. Bennett. 2002. Sequencing the *Aspergillus fumigatus* genome. *Lancet Infect. Dis.* 2:251–253.

21. Dodgson, A. R., C. Pujol, M. A. Pfaller, D. W. Denning, and D. R. Soll. 2005. Evidence for recombination in *Candida glabrata. Fungal Genet. Biol.* 42:233–243.

22. Dromer, F., O. Ronin, and B. Dupont. 1992. Isolation of *Cryptococcus neoformans* var. *gattii* from an Asian patient in France: evidence for dormant infection in healthy subjects. *J. Med. Vet. Mycol.* 30:395–397.

23. Dujon, B., D. Sherman, G. Fischer, P. Durrens, S. Casaregola, I. Lafontaine, J. De Montigny, C. Marck, C. Neuveglise, E. Talla, N. Goffard, L. Frangeul, M. Aigle, V. Anthouard, A. Babour, V. Barbe, S. Barnay, S. Blanchin, J. M. Beckerich, E. Beyne, C. Bleykasten, A. Boisrame, J. Boyer, L. Cattolico, F. Confanioleri, A. De Daruvar, L. Despons, E. Fabre, C. Fairhead, H. Ferry-Dumazet, A. Groppi, F. Hantraye, C. Hennequin, N. Jauniaux, P. Joyet, R. Kachouri, A. Kerrest, R. Koszul, M. Lemaire, I. Lesur, L. Ma, H. Muller, J. M. Nicaud, M. Nikolski, S. Oztas, O. Ozier-Kalogeropoulos, S. Pellenz, S. Potier, G. F. Richard, M. L. Straub, A. Suleau, D. Swennen, F. Tekaia, M. Wesolowski-Louvel, E. Westhof, B. Wirth, M. Zeniou-Meyer, I. Zivanovic, M. Bolotin-Fukuhara, A. Thierry, C. Bouchier, B. Caudron, C. Scarpelli, C. Gaillardin, J. Weissenbach, P. Wincker, and J. L. Souciet. 2004. Genome evolution in yeasts. *Nature* 430:35–44.

24. Dyer, P. S., M. Paoletti, and D. B. Archer. 2003. Identification of a mating-type gene in the homothallic fungus *Aspergillus nidulans. Fungal Genet. Newsl.* 50:145.

25. Fabre, E., H. Muller, P. Therizols, I. Lafontaine, B. Dujon, and C. Fairhead. 2005. Comparative genomics in hemiascomycete yeasts: evolution of sex, silencing, and subtelomeres. *Mol. Biol. Evol.* 22:856–873.

26. Fisher, M. C., G. L. Koenig, T. J. White, G. San-Blas, R. Negroni, I. G. Alvarez, B. Wanke, and J. W. Taylor. 2001. Biogeographic range expansion into South America by *Coccidioides immitis* mirrors New World patterns of human migration. *Proc. Natl. Acad. Sci. USA* 98:4558–4562.

27. Fisher, M. C., G. L. Koenig, T. J. White, and J. W. Taylor. 2001. Molecular and phenotypic description of *Coccidioides posadasii* sp. nov., previously recognised as the non-Californian population of *Coccidioides immitis. Mycologia* 94:73–84.

28. Francois, F., T. Noel, R. Pepin, A. Brulfert, C. Chastin, A. Favel, and J. Villard. 2001. Alternative identification test relying upon sexual reproductive abilities of *Candida lusitaniae* strains isolated from hospitalized patients. *J. Clin. Microbiol.* 39:3906–3914.

29. Fraser, J. A., S. Diezmann, R. L. Subaran, A. Allen, K. B. Lengeler, F. S. Dietrich, and J. Heitman. 2004. Convergent evolution of chromosomal sex-determining regions in the animal and fungal kingdoms. *PLoS Biol.* 2:2243–2255.

30. Fraser, J. A., and J. Heitman. 2004. Evolution of fungal sex chromosomes. *Mol. Microbiol.* 51:299–306.

31. Fraser, J. A., and J. Heitman. 2003. Fungal mating-type loci. *Curr. Biol.* 13:R792–R795.

32. Fraser, J. A., J. C. Huang, R. Pukkila-Worley, J. A. Alspaugh, T. G. Mitchell, and J. Heitman. 2005. Chromosomal translocation and segmental duplication in *Cryptococcus neoformans. Eukaryot. Cell* 4:401–406.

33. Fraser, J. A., R. L. Subaran, C. B. Nichols, and J. Heitman. 2003. Recapitulation of the sexual cycle of the primary fungal pathogen *Cryptococcus neoformans* var. *gattii*: implications for an outbreak on Vancouver Island, Canada. *Eukaryot. Cell* 2:1036–1045.

34. Geiser, D. M., J. C. Frisvad, and J. W. Taylor. 1998. Evolutionary relationships in *Aspergillus* section *Fumigati* inferred from partial β-tubulin and hydrophobin sequences. *Mycologia* 90:831–845.

35. Goodwin, S. B., B. A. Cohen, and W. E. Fry. 1994. Panglobal distribution of a single clonal lineage of the Irish potato famine fungus. *Proc. Natl. Acad. Sci. USA* 91:11591–11595.

36. Graser, Y., M. Volovsek, J. Arrington, G. Schonian, W. Presber, T. G. Mitchell, and R. Vilgalys. 1996. Molecular markers reveal that population structure of the human pathogen *Candida albicans* exhibits both clonality and recombination. *Proc. Natl. Acad. Sci. USA* 93:12473–12477.

37. Grigg, M. E., S. Bonnefoy, A. B. Hehl, Y. Suzuki, and J. C. Boothroyd. 2001. Success and virulence in *Toxoplasma* as the result of sexual recombination between two distinct ancestries. *Science* 294:161–165.

38. Halliday, C. L., T. Bui, M. Krockenberger, R. Malik, D. H. Ellis, and D. A. Carter. 1999. Presence of α and a mating types in environmental and clinical collections of *Cryptococcus neoformans* var. *gattii* strains from Australia. *J. Clin. Microbiol.* 37:2920–2926.

39. Heitman, J., B. Allen, J. A. Alspaugh, and K. J. Kwon-Chung. 1999. On the origins of congenic *MAT*α and *MAT*a strains of the pathogenic yeast *Cryptococcus neoformans. Fungal Genet. Biol.* 28:1–5.

40. Herskowitz, I., and Y. Oshima. 1981. Control of cell type in *Saccharomyces cerevisiae*: mating type and mating-type interconversion, p. 181–210. *In* J. N. Strathern, E. W. Jones, and J. R. Broach (ed.), *The Molecular Biology of the Yeast Saccharomyces: Life Cycle and Inheritance.* Cold Spring Harbor Laboratory Press, Cold Spring Harbor, N.Y.

41. Herskowitz, I., J. Rine, and J. N. Strathern. 1992. Mating-type determination and mating-type interconversion, p. 583–656. *In* E. W. Jones, J. R. Pringle, and J. R. Broach (ed.), *The Molecular and Cellular Biology of the Yeast Saccharomyces: Gene Expression*, vol. 2. Cold Spring Harbor Laboratory Press, Cold Spring Harbor, N.Y.

42. Hoang, L. M., J. A. Maguire, P. Doyle, M. Fyfe, and D. L. Roscoe. 2004. *Cryptococcus neoformans* infections at Vancouver Hospital and Health Sciences Centre (1997–2002): epidemiology, microbiology and histopathology. *J. Med. Microbiol.* 53:935–940.

43. Hull, C. M., G. M. Cox, and J. Heitman. 2004. The α-specific cell identity factor Sxi1α is not required for virulence of *Cryptococcus neoformans. Infect. Immun.* 72:3643–3645.

44. Hull, C. M., R. C. Davidson, and J. Heitman. 2002. Cell identity and sexual development in *Cryptococcus neoformans* are controlled by the mating-type-specific homeodomain protein Sxi1α. *Genes Dev.* 16:3046–3060.

45. Hull, C. M., and J. Heitman. 2002. Fungal mating: *Candida albicans* flips a switch to get in the mood. *Curr. Biol.* 12:R782–R784.

46. Hull, C. M., and J. Heitman. 2002. Genetics of *Cryptococcus neoformans. Annu. Rev. Genet.* 36:557–615.

47. Hull, C. M., and A. D. Johnson. 1999. Identification of a mating type-like locus in the asexual pathogenic yeast *Candida albicans. Science* 285:1271–1275.

48. Hull, C. M., R. M. Raisner, and A. D. Johnson. 2000. Evidence for mating of the "asexual" yeast *Candida albicans* in a mammalian host. *Science* 289:307–310.

49. Janbon, G., F. Sherman, and E. Rustchenko. 1998. Monosomy of a specific chromosome determines L-sorbose utilization: a novel regulatory mechanism in *Candida albicans*. *Proc. Natl. Acad. Sci. USA* **95:** 5150–5155.

50. Johnson, A. 2003. The biology of mating in *Candida albicans*. *Nat. Rev. Microbiol.* **1:**106–116.

51. Karos, M., Y. C. Chang, C. M. McClelland, D. L. Clarke, J. Fu, B. L. Wickes, and K. J. Kwon-Chung. 2000. Mapping of the *Cryptococcus neoformans MAT*α locus: presence of mating type-specific mitogen-activated protein kinase cascade homologs. *J. Bacteriol.* **182:**6222–6227.

52. Kasuga, T., J. W. Taylor, and T. J. White. 1999. Phylogenetic relationships of varieties and geographical groups of the human pathogenic fungus *Histoplasma capsulatum* Darling. *J. Clin. Microbiol.* **37:**653–663.

53. Keller, S. M., M. A. Viviani, M. C. Esposto, M. Cogliati, and B. L. Wickes. 2003. Molecular and genetic characterization of a serotype A *MAT*a *Cryptococcus neoformans* isolate. *Microbiology* **149:**131–142.

54. Kidd, S. E., F. Hagen, R. L. Tscharke, M. Huynh, K. H. Bartlett, M. Fyfe, L. Macdougall, T. Boekhout, K. J. Kwon-Chung, and W. Meyer. 2004. A rare genotype of *Cryptococcus gattii* caused the cryptococcosis outbreak on Vancouver Island (British Columbia, Canada). *Proc. Natl. Acad. Sci. USA* **101:**17258–17263.

55. Koufopanou, V., A. Burt, T. Szaro, and J. W. Taylor. 2001. Gene genealogies, cryptic species, and molecular evolution in the human pathogen *Coccidioides immitis* and relatives (Ascomycota, Onygenales). *Mol. Biol. Evol.* **18:**1246–1258.

56. Koufopanou, V., A. Burt, and J. W. Taylor. 1997. Concordance of gene genealogies reveals reproductive isolation in the pathogenic fungus *Coccidioides immitis*. *Proc. Natl. Acad. Sci. USA* **94:**5478–5482.

57. Kronstad, J. W., and C. Staben. 1997. Mating type in filamentous fungi. *Annu. Rev. Genet.* **31:**245–276.

58. Kwon-Chung, J. 1981. Virulence of the two mating types of *Emmonsiella capsulata* and the mating experiments with *Emmonsiella capsulata* var. *duboisii*. *In* C. de Vroey and R. Vanbreuseghem (ed.), *Sexuality and Pathogenicity of Fungi*. Masson, Paris, France.

59. Kwon-Chung, K. J. 1972. *Emmonsiella capsulata*: perfect state of *Histoplasma capsulatum*. *Science* **177:**368–369.

60. Kwon-Chung, K. J. 1976. Morphogenesis of *Filobasidiella neoformans*, the sexual state of *Cryptococcus neoformans*. *Mycologia* **68:**821–833.

61. Kwon-Chung, K. J. 1975. A new genus, *Filobasidiella*, the perfect state of *Cryptococcus neoformans*. *Mycologia* **67:**1197–1200.

62. Kwon-Chung, K. J. 1976. A new species of *Filobasidiella*, the sexual state of *Cryptococcus neoformans* B and C serotypes. *Mycologia* **68:**943–946.

63. Kwon-Chung, K. J. 1972. Sexual stage of *Histoplasma capsulatum*. *Science* **175:**326.

64. Kwon-Chung, K. J. 1973. Studies on *Emmonsiella capsulata*. I. Heterothallism and development of the ascocarp. *Mycologia* **65:**109–121.

65. Kwon-Chung, K. J., and J. E. Bennett. 1978. Distribution of α and **a** mating types of *Cryptococcus neoformans* among natural and clinical isolates. *Am. J. Epidemiol.* **108:**337–340.

66. Kwon-Chung, K. J., and J. E. Bennett. 1992. *Medical Mycology*. Lea & Febiger, Philadelphia, Pa.

67. Kwon-Chung, K. J., T. Boekhout, J. W. Fell, and M. Diaz. 2002. Proposal to conserve the name *Cryptococcus gattii* against *C. hondurianus* and *C. bacillisporus* (Basidiomycota, Hymenomycetes, Tremellomycetiadae). *Taxon* **51:**804–806.

68. Kwon-Chung, K. J., J. C. Edman, and B. L. Wickes. 1992. Genetic association of mating types and virulence in *Cryptococcus neoformans*. *Infect. Immun.* **60:**602–605.

69. Kwon-Chung, K. J., R. J. Weeks, and H. W. Larsh. 1974. Studies on *Emmonsiella capsulata* (*Histoplasma capsulatum*). II. Distribution of the two mating types in 13 endemic states of the United States. *Am. J. Epidemiol.* **99:**44–49.

70. Lachke, S. A., S. R. Lockhart, K. J. Daniels, and D. R. Soll. 2003. Skin facilitates *Candida albicans* mating. *Infect. Immun.* **71:**4970–4976.

71. Lahn, B. T., and D. C. Page. 1999. Four evolutionary strata on the human X chromosome. *Science* **286:**964–967.

72. Latge, J. P. 2001. The pathobiology of *Aspergillus fumigatus*. *Trends Microbiol.* **9:**382–389.

73. Legrand, M., P. Lephart, A. Forche, F. M. Mueller, T. Walsh, P. T. Magee, and B. B. Magee. 2004. Homozygosity at the *MTL* locus in clinical strains of *Candida albicans*: karyotypic rearrangements and tetraploid formation. *Mol. Microbiol.* **52:**1451–1462.

74. Lengeler, K. B., G. M. Cox, and J. Heitman. 2001. Serotype AD strains of *Cryptococcus neoformans* are diploid or aneuploid and are heterozygous at the mating-type locus. *Infect. Immun.* **69:**115–122.

75. Lengeler, K. B., D. S. Fox, J. A. Fraser, A. Allen, K. Forrester, F. S. Dietrich, and J. Heitman. 2002. Mating-type locus of *Cryptococcus neoformans*: a step in the evolution of sex chromosomes. *Eukaryot. Cell* **1:**704–718.

76. Lengeler, K. B., P. Wang, G. M. Cox, J. R. Perfect, and J. Heitman. 2000. Identification of the *MAT*a mating-type locus of *Cryptococcus neoformans* reveals a serotype A *MAT*a strain thought to have been extinct. *Proc. Natl. Acad. Sci. USA* **97:**14455–14460.

77. Lester, S. J., N. J. Kowalewich, K. H. Bartlett, M. B. Krockenberger, T. M. Fairfax, and R. Malik. 2004. Clinicopathologic features of an unusual outbreak of cryptococcosis in dogs, cats, ferrets, and a bird: 38 cases (January to July 2003). *J. Am. Vet. Med. Assoc.* **225:**1716–1722.

78. Lin, X., C. M. Hull, and J. Heitman. 2005. Sexual reproduction between partners of the same mating type in *Cryptococcus neoformans*. *Nature* **434:**1017–1021.

79. Litvintseva, A. P., R. E. Marra, K. Nielsen, J. Heitman, R. Vilgalys, and T. G. Mitchell. 2003. Evidence of sexual recombination among *Cryptococcus neoformans* serotype A isolates in sub-Saharan Africa. *Eukaryot. Cell* **2:**1162–1168.

80. Liu, Z., P. H. Moore, H. Ma, C. M. Ackerman, M. Ragiba, Q. Yu, H. M. Pearl, M. S. Kim, J. W. Charlton, J. I. Stiles, F. T. Zee, A. H. Paterson, and R. Ming. 2004. A primitive Y chromosome in papaya marks incipient sex chromosome evolution. *Nature* **427:**348–352.

81. Lockhart, S. R., K. J. Daniels, R. Zhao, D. Wessels, and D. R. Soll. 2003. Cell biology of mating in *Candida albicans*. *Eukaryot. Cell* **2:**49–61.

82. Lockhart, S. R., C. Pujol, K. J. Daniels, M. G. Miller, A. D. Johnson, M. A. Pfaller, and D. R. Soll. 2002. In *Candida albicans*, white-opaque switchers are homozygous for mating type. *Genetics* **162:**737–745.

83. Magee, B. B., M. Legrand, A. M. Alarco, M. Raymond, and P. T. Magee. 2002. Many of the genes required for

mating in *Saccharomyces cerevisiae* are also required for mating in *Candida albicans*. *Mol. Microbiol.* 46:1345–1351.

84. Magee, B. B., and P. T. Magee. 2000. Induction of mating in *Candida albicans* by construction of *MTL*α and *MTL*a strains. *Science* 289:310–313.

85. Magee, P. T., and B. B. Magee. 2004. Through a glass opaquely: the biological significance of mating in *Candida albicans*. *Curr. Opin. Microbiol.* 7:661–665.

86. Massey, S. E., G. Moura, P. Beltrao, R. Almeida, J. R. Garey, M. F. Tuite, and M. A. Santos. 2003. Comparative evolutionary genomics unveils the molecular mechanism of reassignment of the CTG codon in *Candida* spp. *Genome Res.* 13:544–557.

87. Matsumoto, Y., and Y. Yoshida. 1984. Sporogony in *Pneumocystis carinii*: synaptonemal complexes and meiotic nuclear divisions observed in precysts. *J. Protozool.* 31:420–428.

88. McClelland, C. M., J. Fu, G. L. Woodlee, T. S. Seymour, and B. L. Wickes. 2002. Isolation and characterization of the *Cryptococcus neoformans MATa* pheromone gene. *Genetics* 160:935–947.

89. Miller, M. G., and A. D. Johnson. 2002. White-opaque switching in *Candida albicans* is controlled by mating-type locus homeodomain proteins and allows efficient mating. *Cell* 110:293–302.

90. Moore, T. D., and J. C. Edman. 1993. The α-mating type locus of *Cryptococcus neoformans* contains a peptide pheromone gene. *Mol. Cell. Biol.* 13:1962–1970.

91. Nicolas, M., G. Marais, V. Hykelova, B. Janousek, V. Laporte, B. Vyskot, D. Mouchiroud, I. Negrutiu, D. Charlesworth, and F. Moneger. 2004. A gradual process of recombination restriction in the evolutionary history of the sex chromosomes in dioecious plants. *PLoS Biol.* 3:47–56.

92. Nielsen, K., G. M. Cox, A. P. Litvintseva, E. Mylonakis, S. D. Malliaris, D. K. Benjamin, S. S. Giles, T. G. Mitchell, A. Casadevall, J. R. Perfect, and J. Heitman. 2005. *Cryptococcus neoformans* α strains preferentially disseminate to the central nervous system during coinfection. *Infect. Immun.* 73:4922–4933.

93. Nielsen, K., G. M. Cox, P. Wang, D. L. Toffaletti, J. R. Perfect, and J. Heitman. 2003. Sexual cycle of *Cryptococcus neoformans* var. *grubii* and virulence of congenic a and α isolates. *Infect. Immun.* 71: 4831–4841.

94. Nielsen, K., R. E. Marra, F. Hagen, T. Boekhout, T. G. Mitchell, G. Cox, and J. Heitman. 2005. Interaction between genetic background and the mating type locus in *Cryptococcus neoformans* virulence potential. *Genetics* 171:975–983.

95. Nosanchuk, J. D., J. Snedeker, and J. S. Nosanchuk. 1998. Arthroconidia in coccidioidoma: case report and literature review. *Int. J. Infect. Dis.* 3:32–35.

96. Panwar, S. L., M. Legrand, D. Dignard, M. Whiteway, and P. T. Magee. 2003. MFα1, the gene encoding the a mating pheromone of *Candida albicans*. *Eukaryot. Cell* 2:1350–1360.

97. Paoletti, M., C. Rydholm, E. U. Schwier, M. J. Anderson, G. Szakacs, F. Lutzoni, J. P. Debeaupuis, J. P. Latge, D. W. Denning, and P. S. Dyer. 2005. Evidence for sexuality in the opportunistic fungal pathogen *Aspergillus fumigatus*. *Curr. Biol.* 15:1242–1248.

98. Poggeler, S. 2002. Genomic evidence for mating abilities in the asexual pathogen *Aspergillus fumigatus*. *Curr. Genet.* 42:153–160.

99. Pomper, S., K. M. Daniels, and D. W. McKee. 1954. Genetic analysis of polyploid yeast. *Genetics* 39:343–355.

100. Pujol, C., S. A. Messer, M. Pfaller, and D. R. Soll. 2003. Drug resistance is not directly affected by mating type locus zygosity in *Candida albicans*. *Antimicrob. Agents Chemother.* 47:1207–1212.

101. Rippon, J. W. 1988. *Medical Mycology*. The W. B. Saunders Co., Philadelphia, Pa.

102. Ristaino, J. B., C. T. Groves, and G. R. Parra. 2001. PCR amplification of the Irish potato famine pathogen from historic specimens. *Nature* 411:695–697.

103. Roman, H., D. C. Hawthorne, and H. C. Douglas. 1951. Polyploidy in yeast and its bearing on the occurrence of irregular genetic ratios. *Proc. Natl. Acad. Sci. USA* 37:79–84.

104. Roman, H., M. M. Phillips, and S. M. Sands. 1955. Studies of polyploid *Saccharomyces*. I. Tetraploid segregation. *Genetics* 40:546–561.

105. Rustad, T. R., D. A. Stevens, M. A. Pfaller, and T. C. White. 2002. Homozygosity at the *Candida albicans MTL* locus associated with azole resistance. *Microbiology* 148:1061–1072.

106. Sadhu, C., D. Hoekstra, M. J. McEachern, S. I. Reed, and J. B. Hicks. 1992. A G-protein α subunit from asexual *Candida albicans* functions in the mating signal transduction pathway of *Saccharomyces cerevisiae* and is regulated by the a1-α2 repressor. *Mol. Cell. Biol.* 12:1977–1985.

107. Shen, W. C., R. C. Davidson, G. M. Cox, and J. Heitman. 2002. Pheromones stimulate mating and differentiation via paracrine and autocrine signaling in *Cryptococcus neoformans*. *Eukaryot. Cell* 1:366–377.

108. Sia, R. A., K. B. Lengeler, and J. Heitman. 2000. Diploid strains of the pathogenic basidiomycete *Cryptococcus neoformans* are thermally dimorphic. *Fungal Genet. Biol.* 29:153–163.

109. Skaletsky, H., T. Kuroda-Kawaguchi, P. J. Minx, H. S. Cordum, L. Hillier, L. G. Brown, S. Repping, T. Pyntikova, J. Ali, T. Bieri, A. Chinwalla, A. Delehaunty, K. Delehaunty, H. Du, G. Fewell, L. Fulton, R. Fulton, T. Graves, S. F. Hou, P. Latrielle, S. Leonard, E. Mardis, R. Maupin, J. McPherson, T. Miner, W. Nash, C. Nguyen, P. Ozersky, K. Pepin, S. Rock, T. Rohlfing, K. Scott, B. Schultz, C. Strong, A. Tin-Wollam, S. P. Yang, R. H. Waterston, R. K. Wilson, S. Rozen, and D. C. Page. 2003. The male-specific region of the human Y chromosome is a mosaic of discrete sequence classes. *Nature* 423:825–837.

110. Slutsky, B., J. Buffo, and D. R. Soll. 1985. High-frequency switching of colony morphology in *Candida albicans*. *Science* 230:666–669.

111. Smulian, A. G., T. Sesterhenn, R. Tanaka, and M. T. Cushion. 2001. The *ste3* pheromone receptor gene of *Pneumocystis carinii* is surrounded by a cluster of signal transduction genes. *Genetics* 157:991–1002.

112. Soll, D. R. 1992. High-frequency switching in *Candida albicans*. *Clin. Microbiol. Rev.* 5:183–203.

113. Soll, D. R., S. R. Lockhart, and R. Zhao. 2003. Relationship between switching and mating in *Candida albicans*. *Eukaryot. Cell* 2:390–397.

114. Sorrell, T. C., A. G. Brownlee, P. Ruma, R. Malik, T. J. Pfeiffer, and D. H. Ellis. 1996. Natural environmental sources of *Cryptococcus neoformans* var. *gattii*. *J. Clin. Microbiol.* 34:1261–1263.

115. Srikantha, T., S. A. Lachke, and D. R. Soll. 2003. Three mating type-like loci in *Candida glabrata*. *Eukaryot. Cell* 2:328–340.

116. Stephen, C., S. Lester, W. Black, M. Fyfe, and S. Raverty. 2002. Multispecies outbreak of cryptococcosis on southern Vancouver Island, British Columbia. *Can. Vet. J.* **43:**792–794.

117. Su, C., D. Evans, R. H. Cole, J. C. Kissinger, J. W. Ajioka, and L. D. Sibley. 2003. Recent expansion of *Toxoplasma* through enhanced oral transmission. *Science* **299:**414–416.

118. Sukroongreung, S., K. Kitiniyom, C. Nilakul, and S. Tantimavanich. 1998. Pathogenicity of basidiospores of *Filobasidiella neoformans* var. *neoformans*. *Med. Mycol.* **36:**419–424.

119. Thomas, C. F., Jr., and A. H. Limper. 2004. *Pneumocystis* pneumonia. *N. Engl. J. Med.* **350:**2487–2498.

120. Timberlake, W. E. 1991. Temporal and spatial controls of *Aspergillus* development. *Curr. Opin. Genet. Dev.* **1:**351–357.

121. Tscharke, R. L., M. Lazera, Y. C. Chang, B. L. Wickes, and K. J. Kwon-Chung. 2003. Haploid fruiting in *Cryptococcus neoformans* is not mating type α-specific. *Fungal Genet. Biol.* **39:**230–237.

122. Tsong, A. E., M. G. Miller, R. M. Raisner, and A. D. Johnson. 2003. Evolution of a combinatorial transcriptional circuit: a case study in yeasts. *Cell* **115:**389–399.

123. Turgeon, B. G., and O. C. Yoder. 2000. Proposed nomenclature for mating type genes of filamentous ascomycetes. *Fungal Genet. Biol.* **31:**1–5.

124. Tzung, K. W., R. M. Williams, S. Scherer, N. Federspiel, T. Jones, N. Hansen, V. Bivolarevic, L. Huizar, C. Komp, R. Surzycki, R. Tamse, R. W. Davis, and N. Agabian. 2001. Genomic evidence for a complete sexual cycle in *Candida albicans*. *Proc. Natl. Acad. Sci. USA* **98:**3249–3253.

125. Varga, J. 2003. Mating type gene homologues in *Aspergillus fumigatus*. *Microbiology* **149:**816–819.

126. Varga, J., and B. Toth. 2003. Genetic variability and reproductive mode of *Aspergillus fumigatus*. *Infect. Genet. Evol.* **3:**3–17.

127. Varga, J., Z. Vida, B. Toth, F. Debets, and Y. Horie. 2000. Phylogenetic analysis of newly described *Neosartorya* species. *Antonie Leeuwenhoek* **77:**235–239.

128. Viviani, M. A., M. C. Esposto, M. Cogliati, M. T. Montagna, and B. L. Wickes. 2001. Isolation of a *Cryptococcus neoformans* serotype A MATa strain from the Italian environment. *Med. Mycol.* **39:**383–386.

129. Viviani, M. A., R. Nikolova, M. C. Esposto, G. Prinz, and M. Cogliati. 2003. First European case of serotype A MATa *Cryptococcus neoformans* infection. *Emerg. Infect. Dis.* **9:**1179–1180.

130. Vohra, P. K., J. G. Park, B. Sanyal, and C. F. Thomas, Jr. 2004. Expression analysis of PCSTE3, a putative pheromone receptor from the lung pathogenic fungus *Pneumocystis carinii. Biochem. Biophys. Res. Commun.* **319:**193–199.

131. Wang, P., J. Cutler, J. King, and D. Palmer. 2004. Mutation of the regulator of G protein signaling Crg1 increases virulence in *Cryptococcus neoformans*. *Eukaryot. Cell* **3:**1028–1035.

132. Wang, P., C. B. Nichols, K. B. Lengeler, M. E. Cardenas, G. M. Cox, J. R. Perfect, and J. Heitman. 2002. Mating-type-specific and nonspecific PAK kinases play shared and divergent roles in *Cryptococcus neoformans*. *Eukaryot. Cell* **1:**257–272.

133. Wang, P., J. R. Perfect, and J. Heitman. 2000. The G-protein β subunit GPB1 is required for mating and haploid fruiting in *Cryptococcus neoformans*. *Mol. Cell. Biol.* **20:**352–362.

134. Wickes, B. L., U. Edman, and J. C. Edman. 1997. The *Cryptococcus neoformans* STE12α gene: a putative *Saccharomyces cerevisiae* STE12 homologue that is mating type specific. *Mol. Microbiol.* **26:**951–960.

135. Wickes, B. L., M. E. Mayorga, U. Edman, and J. C. Edman. 1996. Dimorphism and haploid fruiting in *Cryptococcus neoformans*: association with the α-mating type. *Proc. Natl. Acad. Sci. USA* **93:**7327–7331.

136. Wong, S., M. A. Fares, W. Zimmermann, G. Butler, and K. H. Wolfe. 2003. Evidence from comparative genomics for a complete sexual cycle in the 'asexual' pathogenic yeast *Candida glabrata*. *Genome Biol.* **4:**R10.

137. Wyder, M. A., E. M. Rasch, and E. S. Kaneshiro. 1998. Quantitation of absolute *Pneumocystis carinii* nuclear DNA content. Trophic and cystic forms isolated from infected rat lungs are haploid organisms. *J. Eukaryot. Microbiol.* **45:**233–239.

138. Xu, J. 2002. Estimating the spontaneous mutation rate of loss of sex in the human pathogenic fungus *Cryptococcus neoformans*. *Genetics* **162:**1157–1167.

139. Xu, J., G. Luo, R. J. Vilgalys, M. E. Brandt, and T. G. Mitchell. 2002. Multiple origins of hybrid strains of *Cryptococcus neoformans* with serotype AD. *Microbiology* **148:**203–212.

140. Xu, J., R. Vilgalys, and T. G. Mitchell. 2000. Multiple gene genealogies reveal recent dispersion and hybridization in the human pathogenic fungus *Cryptococcus neoformans*. *Mol. Ecol.* **9:**1471–1481.

141. Yan, Z., C. M. Hull, J. Heitman, S. Sun, and J. Xu. 2004. SXI1α controls uniparental mitochondrial inheritance in *Cryptococcus neoformans*. *Curr. Biol.* **14:**R743–R744.

142. Yan, Z., and J. Xu. 2003. Mitochondria are inherited from the MATa parent in crosses of the basidiomycete fungus *Cryptococcus neoformans*. *Genetics* **163:**1315–1325.

143. Young, L. Y., M. C. Lorenz, and J. Heitman. 2000. A STE12 homolog is required for mating but dispensable for filamentation in *Candida lusitaniae*. *Genetics* **155:**17–29.

Molecular Principles of Fungal Pathogenesis
Edited by Joseph Heitman et al.
©2006 ASM Press, Washington, D.C.

Chapter 3

Postgenomic Strategies for Genetic Analysis: Insight from *Saccharomyces cerevisiae* and *Candida albicans*

CLARISSA J. NOBILE AND AARON P. MITCHELL

The fungal pathogens are a diverse group of organisms. Their habitats range from environmental to commensal; they initiate infection in the lungs, gastrointestinal tract, or genitourinary tract; some infect as yeast, whereas some infect as hyphae; they include primary and opportunistic pathogens; there are ascomycetes, basidiomycetes, and a genus grouped among fungi only through rRNA sequence comparison. Viewed in this context, it is remarkable that scientists who study these varied pathogens ever have anything to talk about with one another. Yet we do: we study the same antifungal drugs and related resistance mechanisms; we are always contemplating new targets that might distinguish between the eukaryotic pathogen and eukaryotic host; we seek to understand the host defenses that limit susceptibility to infection. But most importantly, we are driven to understand the unique biology of these organisms and how their biological attributes contribute to disease. One of the fundamental tools for analysis of biological attributes—for translating a phenomenon into a mechanism—is genetics. And that gives us a lot to talk about.

Much of what we say is "good," in the sense that it inspires new ideas or pulls together unifying themes. (Chapter 10 provides many important examples.) But some of what we say should probably be reserved for a psychiatrist's couch—if only there were a psychiatrist who could help us! That is because our organisms make us jump through more flaming hoops than a circus pony. We routinely have to circumvent problems that include inefficient targeting of DNA, incomplete sexual cycles, being stuck as a haploid, being stuck as a diploid, or even being stuck without genetic manipulation at all. All of us have already entered or will soon enter the postgenomic era (as discussed in chapter 40), and this helps us because we can choose DNA sequences for manipulation that may be most informative. And

yet, this may cause us all to return to the psychiatrist's couch, muttering to unseen companions about which sequences to choose.

Our goal in this chapter is to try to save you $300, and, if you are a quick reader, about 30 minutes of that hour with the psychiatrist, by discussing the choice of genes for study in this postgenomic era. We do not propose a solution—in that way we are pretty much like the psychiatrist. Instead, we attempt to provide a framework for thought and discussion . . . with those companions of yours!

FROM GENOME TO GENE FUNCTION

On sequencing of any new genome, the most pressing need is to describe genes through annotation—to define open reading frames (ORFs) and their sequence homologs, which gives us a spectrum of possible biochemical activities, inferred functional relationships, and a sense of the number of unique genes. From that point on (or, if you are clever, before you even get a sequence [18]), a very inviting strategy is to describe genes through their expression patterns. Expression profiling allows us to focus on genes that are expressed at their highest levels when an organism responds to a potentially significant signal—interaction with human tissue, attack by host defense systems, or a change in a critical environmental parameter (see chapter 40). Interpreting gene function clearly is always a challenging feat. Annotation and expression profiling together can provide a sturdy framework to begin to elucidate gene function, and then we can look toward the prospect of testing those hypotheses by defining the set of genes that clearly function to support infection, pathogenicity, persistence, drug resistance, and the many other detailed traits that combine to characterize the organism's unique biological attributes.

Clarissa J. Nobile • Department of Microbiology and Biological Sciences Program, Department of Biological Sciences, Columbia University, New York, NY 10027.   **Aaron P. Mitchell** • Department of Microbiology, Columbia University, New York, NY 10032.

As the postgenomic era progresses, approaches to address the gene function question have expanded and certainly evolved. Perhaps the most universally applicable approach is that of heterologous expression, in which a pathogen's gene or cDNA is expressed in a neutral host. The engineered host is then tested for the trait that is thought to be governed by the pathogen gene. One might look for complementation of a specific defect in a mutant host strain or acquisition of a novel trait related to pathogenicity by a wild-type host strain. (For some recent examples, see references 12, 21, 23, and 37.) One major limitation of using a host organism as a model for the heterologous expression of a gene of interest in another organism is differences in genetic coding between the two organisms. This problem, for example, exists when using *Saccharomyces cerevisiae* as a host for *Candida albicans* genes, where the codon CUG specifies leucine in *S. cerevisiae* but serine in *C. albicans*; thus, *C. albicans* proteins may be nonfunctional when translated in *S. cerevisiae*. With this in mind, one could argue that the ultimate test of pathogen gene function comes from manipulation of a gene in the pathogen itself. For an increasing number of pathogenic fungi, it is possible to reduce or abolish the function of specific genes through gene disruption or RNA interference strategies. The phenotypic impact of a loss of gene function has served as the cornerstone of any argument concerning gene function, since the logic that connects mutant, wild type, and their phenotypic differences is straightforward.

Now, how do we decide which genes in a fungal pathogen are worth manipulating? Which genes provide the most useful insight into the diverse traits that together promote pathogenicity? The simplest answer is "all of them." It is intuitive that a comprehensive panel of disruption mutant strains provides an ideal resource for analysis of critical traits and pathways. For haploid organisms with suitable transposons or transformation systems, or for more recalcitrant genetic systems in which expense is no object, it is feasible to create such a set of strains. However, for many fungal pathogens, the resources required for such an endeavor are prohibitive. How might one strategically and feasibly choose a subset of pathogen genes for further analysis, optimizing the likelihood that one could identify functionally significant genes and, ultimately, connect gene function to biochemical and biological mechanisms without any further information?

Strategies for gene discovery—the identification of functionally significant genes—may follow individual beliefs rather than defined strategies. Rather than simply to preach, though, we will turn to a large-scale analysis of gene function from the budding yeast *S. cerevisiae* (17) to develop a rational basis for evaluating the functional range of gene subsets.

FUNCTIONAL PROFILING IN *S. CEREVISIAE*

The large-scale study by Giaever et al. (17) sought to identify *S. cerevisiae* genes that participate in various environmental interactions, based solely on deletion mutant phenotypes. Each nonessential *S. cerevisiae* ORF was represented by an individual haploid deletion mutant strain, and each deletion was marked by a unique sequence or "bar code" for identification purposes. The 4,811 viable deletion mutants were pooled, and the pool was grown under various different environmental conditions. The pools were analyzed after growth by hybridization of the bar code sequences to a corresponding detection microarray. Fitness of each strain under a specific growth condition was expressed as a ratio of the strain's bar code signal strength under that particular condition to its reference signal strength. In essence, this is signature-tagged mutagenesis conducted on a comprehensive scale.

The specific growth conditions examined, as shown in Table 1, were not intended to represent pathogenicity in any way, although some of these traits (such as adaptation to high salt levels or alkaline pH) probably correlate with pathogenicity in many organisms. What do we find if we examine subsets of mutants that represent various classes of genes? We have chosen classes of genes that might be reasonable choices for further analysis in a newly sequenced genome, using gene ontology descriptions and search tools at the *Saccharomyces* Genome Database (http://www.yeastgenome.org/). These classifications are listed in Table 1. Our gene subsets include genes encoding transcription factors (TxnF), cell wall proteins (CWP), plasma membrane proteins (MbP), and protein kinases (PrK), as well as genes annotated to be involved in a stress response (StrR). In addition, we wanted a group of random genes (in terms of function), so we chose the 395 genes on chromosome 7 (Chr7) to represent a large random group, as well as the 113 genes on chromosome 3 (Chr3) to represent a small random group. Finally, we wanted to examine a group of genes about which little is known, and so we chose genes annotated as "function unknown, biological process unknown" (UKUK). These genes may represent the unique set of genes that is found in any newly sequenced genome. These groups are simple enough to allow us to decipher general expectations and to make extrapolations that may be applicable to other less well understood genomes.

What can we learn from these *S. cerevisiae* genes along with the corresponding properties of their mutants? One point is that there are different numbers of genes in each group (Table 1). (In any category, essential genes contribute to the total number but

Table 1. Distribution of *S. cerevisiae* growth-defective mutants under various selection conditions

Gene class[a]	Total no.[b]	No. of severe growth-defective mutants under selection condition[c]:					
		NaCl	YPG	Sorb	Min	Nys	pH 8
All	4,811	325	24	56	92	33	200
Chr7	395	15	0	8	5	1	6
StrR	339	15	2	7	1	3	13
TxnF	319	16	2	7	6	1	15
MbP	218	23	3	2	3	1	13
UKUK	176	8	0	1	0	2	3
PrK	118	16	0	5	2	0	13
Chr3	113	9	0	0	1	2	3
CWP	84	5	0	0	0	0	4

[a]This column lists gene classes, which are defined as follows. All, all viable mutants; Chr7, verified ORFs on chromosome 7; StrR, verified ORFs with GO biological process "response to stress"; TxnF, verified ORFs with GO molecular function "transcription regulator activity"; MbP, verified ORFs with GO cellular component "plasma membrane"; UKUK, verified ORFs with GO biological process "unknown" and molecular function "unknown"; PrK, verified ORFs with GO molecular function "protein kinase activity"; Chr3, verified ORFs on chromosome 3; CWP, verified ORFs with GO cellular component "cell wall."

[b]This column lists the number of ORFs in the respective gene class, including both essential and nonessential genes.

[c]These columns list the number of mutants with severe growth defects, defined as a fitness defect of 100 or greater (17), under each growth condition. Only experiments with a duration of 15 generations in series A of Giaever et al. (17) are analyzed here. Selection conditions were as follows: NaCl, 1 M sodium chloride; YPG, YEP glycerol medium; Sorb, 1 M sorbitol; Min, minimal medium; Nys, nystatin; pH 8, growth medium adjusted to pH 8. Details of growth conditions are described in reference 17.

corresponding mutants were not analyzed, of course.) The larger groups include the Chr7, StrR, and TxnF genes, with 319 to 395 members represented. The smaller groups are CWP, PrK, and Chr3 genes, with 84 to 118 members. The UKUK and MbP genes are in the middle, with 176 to 218 members. So, if you had the time and materials to create 300 mutant strains, you could, for example, choose between all transcription factor genes or all PrK plus all MbP genes.

In terms of function, we want to know whether one group of genes is more likely than the others to yield clear mutant phenotypes in a variety of different screens. The idea of considering several different screens is an extrapolation from the well-accepted idea that many different phenotypes contribute to the virulence of a pathogen (6, 26). So first, which groups yield severe mutant phenotypes in all screens? The answer is the StrR, TxnF, and MbP genes—in other words, most of the larger groups. The exception among large groups is Chr7 genes, which failed to yield mutants severely defective in glycerol growth. Second, among the small groups, the PrK genes consistently yielded more severely defective mutants in most screens than did any other group of genes. So there is no single group of genes among these that is much more likely to yield phenotypically defective mutants than all the others; it looks simply like "more mutants is better."

On the other hand, there is an advantage to analyzing genes chosen with an eye to biochemical function rather than those chosen randomly (Fig. 1). If we consider mutant frequencies among all genes that could be identified through biochemically informative sequence motifs—TxnF, MbP, PrK, and CWP genes—the frequency of severe mutants is consistently higher than for the randomly chosen genes, i.e., Chr3 and Chr7 genes. This increased likelihood of severe mutant phenotypes is not simply a literature bias hidden in the categories, because even the UKUK group has a yield of severe mutants that is as great as or greater than that of the random mutants (Fig. 1); the UKUK group is arguably the least extensively studied category. Indeed, the severe mutant frequency among this function-based gene group (TxnF, MbP, PrK, and CWP genes) is consistently greater than that among all 4,811 genes (Fig. 1). So, if you want to find mutants for further study, this analysis argues that you should choose genes whose products have sequence motifs indicative of known biochemical functions.

One reason that mutants may fail to be recognized is that the selection or screening criteria are too restrictive. We can ask, for any of these screens, what would happen if our criteria were two- or fourfold less stringent? The NaCl sensitivity screen yielded several mutants in all gene groups (Table 1). For NaCl sensitivity (Table 2), we see that decreasing the

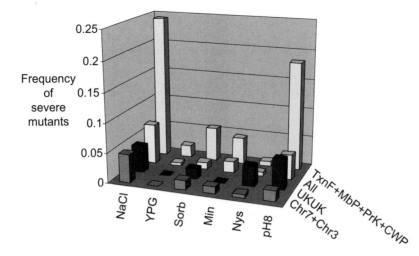

Figure 1. Frequencies of *S. cerevisiae* mutants with severe growth defects under various selection conditions. Mutant growth properties (17) are grouped by gene categories. Selection conditions and the severity of defects are as described in Table 1. Gene categories include a representative random group (Chr7+Chr3), a group about which little is known (UKUK), the entire set of nonessential genes (All), and those with biochemically informative sequence motifs (TxnF+MbP+PrK+CWP). Abbreviations are described in Table 1.

stringency of the screen two- or fourfold yields a corresponding overall increase in the number of mutants in most gene classes. The glycerol growth screen yielded among the fewest mutants in all gene classes (Table 1). Decreasing the stringency of the screen yields detectable mutants in more of the groups (Table 2), but does not necessarily increase the number of mutants in every group (e.g., TxnF and CWP). These different behaviors are expected from statistical considerations of large versus small numbers. Nonetheless, if you went to the trouble to make a set of mutant strains and then had difficulty finding a mutant with the phenotypic alteration you expect, you might do better to try a different screen rather than to make more mutants or to decrease the

stringency of your screen. This advice goes against the grain of many biologists, who typically have a specific process that they want to study. Nonetheless, we suggest that you pick a biochemically defined group of genes to disrupt, like TxnF or PrK genes, and screen them for as many phenotypes as possible (including, but not limited to, your favorite phenotypic process). As a real-life example of our gene disruption choices, and how we benefited from the genes we chose to disrupt, we will recount to you our experiences with the pathogenic fungus *C. albicans*.

OUR WORK WITH *C. ALBICANS* MUTANTS

The sequencing of the *C. albicans* genome indicates that there are approximately 7,110 ORFs in

Table 2. Effect of reduced selection stringency on mutant detection

Gene class[a]	Total no.[b]	No. of severe growth-defective mutants with fitness defects indicated under selection condition[c]:					
		NaCl, >100	NaCl, >50	NaCl, >25	YPG, >100	YPG, >50	YPG, >25
All	4,811	325	1,150	2,287	24	42	90
Chr7	395	15	54	117	0	1	3
StrR	339	15	59	121	2	2	6
TxnF	319	16	51	101	2	2	2
MbP	218	23	60	98	3	3	3
UKUK	176	8	30	78	0	1	1
PrK	118	16	32	53	0	0	1
Chr3	113	9	20	42	0	0	1
CWP	84	5	18	48	0	0	0

[a]Gene classes are defined in Table 1 footnote *a*.

[b]Total number is defined in Table 1 footnote *b*.

[c]These columns list the number of mutants with fitness defects (17) of 100 or greater, 50 or greater, and 25 or greater under the NaCl and YPG growth conditions described in Table 1.

the haploid complement of its genome (http://www-sequence.stanford.edu/group/candida/). Genetic manipulation of *C. albicans*, unfortunately, is unwieldy and labor-intensive for several reasons. The major reason is that the organism is naturally diploid, so that loss-of-function mutations have to become homozygous in order to yield a pronounced phenotype. The main strategy that yields homozygotes in the premier genetically tractable diploid organisms (e.g., fruit flies, worms, and zebrafish) is a genetic cross. However, this approach is not an option with *C. albicans* because it lacks a complete sexual cycle (see chapters 6 and 7). The organism has two huge assets, though: it is transformable, and it has a very active homologous recombination pathway. Thus, it is fairly straightforward to create defined genetic changes such as insertions or deletions, and this is generally done through two successive transformations (3, 11).

Our philosophy has been that having a large set of defined mutant strains that can be subjected to phenotypic screens may be almost as useful as having a truly facile genetic system. (One may look no further than the model yeast *S. cerevisiae* to see that this is true [17].) Such a set of *C. albicans* strains would permit a definition of new genes that govern unique *C. albicans* traits. Then both genetic and nongenetic tools could be used to define the mechanistic relationships between the gene products and the trait.

Our initial efforts were aimed at streamlining gene disruption tools to accelerate the use of two successive transformations (44, 45). Homology of the gene disruption cassette to the genome was created with long PCR primers, making it feasible to create large numbers of mutant strains by ordering primers instead of cloning genes. However, the use of two sequential transformations to disrupt a gene meant that bookkeeping of strains and primers was a significant consideration in the creation of large quantities of mutants. In addition, reliance on long primers made costs significant, as well as increasing the chances of primer synthesis errors, which often necessitated reordering and sometimes redesigning primers. It is notable, though, that some of our more talented and wealthier colleagues have been able to use long primers to disrupt many *C. albicans* genes (35).

These considerations prompted us to develop a method for the creation of homozygous mutants after a single transformation (15). The method employs a gene disruption cassette called *UAU1* (*ura3-ARG4-ura3*) that has three segments: *ura3Δ3'*, a nonfunctional 3'-deleted copy of *URA3*; *ARG4*; and *ura3Δ5'*, a nonfunctional 5'-deleted copy of *URA3* (Fig. 2A). *UAU1* expresses *ARG4* but not a functional *URA3*. Homology between the *ura3* segments permits recombination to yield an intact

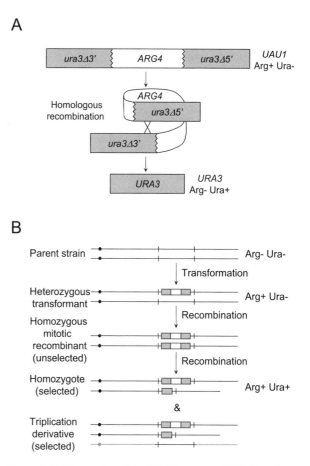

Figure 2. (A) Properties of the *UAU1* cassette. *UAU1* is a functional *ARG4* gene flanked by nonfunctional *ura3Δ3'* and *ura3Δ5'* deletion alleles. The *ura3* alleles are in direct repeat orientation and share 500 bp of homology, so they can undergo recombination. Recombination between the *ura3* alleles yields a functional *URA3* gene and excises *ARG4*. (B) Double-disruption selection. Transformation of *C. albicans* strain BWP17 with *UAU1* results in expression of *ARG4* and confers an Arg+ phenotype. If the insertion becomes homozygous through mitotic recombination or gene conversion, then a second recombination event within one cassette yields an Arg+ Ura+ segregant. These Arg+ Ura+ segregants are of two classes. One class carries a *UAU1* insertion allele and a *URA3* insertion allele. We refer to these segregants as homozygotes. The other class carries three copies of the gene: a *UAU1* insertion allele, a *URA3* insertion allele, and a wild-type allele. We refer to these segregants as allelic triplication derivatives. A trisomic genetic structure is diagrammed for simplicity, but their precise genetic structure has not been determined.

URA3 gene and excision of *ARG4*. The recombination product expresses *URA3* but not *ARG4*. We reasoned that the cassette would permit the identification of homozygous mutants (Fig. 2B). A heterozygous *UAU1* insertion, introduced by transformation, might occasionally become homozygous through gene conversion or mitotic recombination. The homozygous mutant would be uniquely capable of yielding segregants that express both *ARG4* and

URA3. Thus, homozygous mutants should be found among Arg+ Ura+ segregants from a strain that is heterozygous for a *UAU1* insertion allele.

We tested this prediction with insertions at three genetic loci that are not essential for viability: *ADE2*, *YGR189*, and *RIM20* (15). In each case, we transformed in a linear disruption fragment (e.g., *ade2::UAU1*) to create a heterozygous insertion-deletion mutant. Then, for each heterozygote, 30 independent Arg+ Ura+ segregants were isolated and screened by PCR for presence of the initial *UAU1*-disrupted allele, a *URA3*-disrupted allele, and a wild-type allele. For each mutation, approximately one-third of the Arg+ Ura+ segregants were homozygous mutants (e.g., *ade2::UAU1/ade2::URA3*). The remaining two-thirds of the Arg+ Ura+ segregants had three copies of the gene in which the insertion was targeted (e.g., *ade2::UAU1/ade2::URA3/ADE2*). We refer to these strains as allelic triplication derivatives. These results indicated that an insertion of the *UAU1* cassette into a nonessential gene can permit selection of homozygous insertion mutants with roughly 30% efficiency (15).

As might be expected, *UAU1* insertions in essential genes fail to yield homozygous segregants, but they do yield triplication derivatives at frequencies similar to nonessential gene insertions (15). This method has been useful for gathering initial indications that genes are essential (10, 14, 38, 43) and for synthetic lethality tests (5, 39).

For construction of sets of mutants, we incorporated the *UAU1* cassette into the GPS Tn7 transposon derivative (New England BioLabs). We then used the Tn7-*UAU1* cassette to create insertions in vitro. Our overall approach has been to use *C. albicans* target DNA cloned into an ampicillin-resistant vector, transpose in the Tn7-*UAU1* kanamycin resistance cassette, and select insertion-bearing clones as those doubly resistant to ampicillin and kanamycin. Although Tn7 naturally displays extreme target site specificity, the Tn7 transposase we used (New England BioLabs) is modified so that insertions occur at a very broad range of sites.

A significant concern with homozygous mutants created via *UAU1* is that the same allelic recombination event that creates a homozygous insertion may also create homozygosity for linked mutations. About 30% of *C. albicans* ORFs have alleles with nonidentical sequences (19), so this concern is reasonable. When we find an insertion mutant with a phenotype of interest, we verify that the insertion itself (rather than an unidentified linked mutation) is the cause of the phenotype by three approaches (10, 24, 27, 46). First, we test multiple independent insertion mutant isolates, all of which should have the same phenotype. Second, we complement the insertion mutation with a clone of the complete ORF and flanking sequences, which should restore a wild-type phenotype. Third, we create engineered *orf∆/orf∆* deletion strains with two successive transformations rather than an allelic recombination step; we expect that the deletion mutant will have a phenotype similar to the initial insertion mutant. In practical terms, our experience has been that whenever multiple mutant isolates have the same phenotype, we are able to complement the defect and to create a deletion mutant that behaves similarly.

Our first approach to large-scale mutagenesis was to make random insertions into *C. albicans* library DNA (10). We tried a few different strategies, but what worked the best was to isolate individual insertion-bearing clones, sequence them from a Tn7 primer to read out into *C. albicans* DNA, and choose insertions that were in ORFs. Thus, we eliminated insertions in vector sequences or in noncoding regions, whose effects may be subtle, and focused on insertions that were likely to reduce or abolish the function of the affected gene. Each *orf::Tn7-UAU1* insertion DNA was released from vector sequences by restriction enzyme digestion and then transformed into *C. albicans*. We picked 12 to 20 independent transformants for each DNA, grew them as patches on rich YPD medium, and replica-plated them to selective medium to obtain Arg+ Ura+ segregants. We then genotyped one segregant per transformant (ensuring the independence of the segregants analyzed) through a PCR test with flanking primers and an internal *ARG4* primer (10), yielding resolvable PCR products for each of the alleles (*orf::Tn7-UAU1*, *orf::Tn7-URA3*, and *ORF*) that may be present (see examples in references 10 and 15). This approach permitted the construction of 217 random insertion mutants representing 197 genes. From screening this library of random insertion mutants, we identified new genes that govern pH responses (10, 24), chlamydospore formation (27), and azole drug susceptibility (5). These studies defined a new pH response regulator, identified a member of the known Rim101p pH-response pathway, implicated protein kinase CK2 in azole resistance, and defined several genes required for chlamydospore formation. The chlamydospore mutants, in particular, defined several different phenotypes, so there is the promise that genetic studies may contribute to dissection of the developmental pathway. In a broader sense, the studies provided an important demonstration that the overall insertional mutagenesis strategy would be workable—we were not swamped by secondary mutations, triplication derivatives, and other nightmarish scenarios.

Our second approach was to create insertions in a defined group of genes for further study (46). Since we are interested in pH regulation, we picked a group of *C. albicans* genes to disrupt that are homologous to some of the pH response regulators in *S. cerevisiae* in order to find new genes involved in *C. albicans* pH regulation. There we cloned individual ORFs as PCR products, transposed Tn7-*UAU1* into each one, and chose insertions near the middle of each ORF (which we reasoned would reduce or abolish function for most genes). Mutants were created and analyzed as described above. Using this approach, we were able to show that the same subset of vesicle-trafficking "ESCRT" pathway genes is required for pH response regulation in *S. cerevisiae* and in *C. albicans* (46).

APPLICATIONS TO A NOVEL
C. ALBICANS TRAIT

The process of biofilm formation is of particular interest because of its medical relevance (9, 13). Biofilm formation on implanted medical devices has severe consequences for human health by providing a portal of entry as well as a sanctuary for invading bacterial and fungal pathogens within the human host (13, 20). Facile genetics has provided insight into the regulatory pathways and adherence molecules that govern bacterial biofilm properties (30, 41), but our understanding of these functions in *C. albicans* biofilms has been more limited.

Three general kinds of approaches have proven useful for dissection of gene function in *C. albicans* biofilm formation (see chapter 12). One is to use known mutants to test general hypotheses. For example, hyphae are a major component of biofilms, and mutants defective in the transcription factor Efg1p, a major activator of hyphal development, have a severe defect in biofilm formation (33). A second is to use gene expression comparisons of biofilm cells and free-living (planktonic) cells. This approach has implicated *ALS* adhesin genes in biofilm adherence (7), several azole drug efflux genes (*CDR* and *MDR* gene family members) in biofilm drug resistance (25, 32), and amino acid biosynthesis genes in biofilm biomass accumulation (16). The third strategy has been to develop a more genetically tractable model with the budding yeast *S. cerevisiae*. *S. cerevisiae* forms a thin layer of adherent cells on semisolid agar (34) or on polystyrene (22). The validity of the model may be limited (7), but it has nonetheless implicated the surface adhesin Flo11p in biofilm formation. Since Flo11p is structurally related to the *C. albicans* ALS proteins, the model clearly can provide insight into *C. albicans* biology.

We first screened the library of random insertion mutants for those defective in biofilm formation (34a). The screen and genetic follow-up studies revealed four new genes that are required for normal biofilm development. The four mutants were defective in hyphal formation under several conditions, and so the work reinforced the idea that biofilm formation depends on hyphal formation. The study also revealed two different kinds of phenotypic defects, thus suggesting (as did the chlamydospore study) that genetic analysis might contribute to dissection of a biofilm developmental pathway. Unfortunately, though, there was no single unifying theme among the functions of the new biofilm regulatory genes, based on their *S. cerevisiae* homologs. In addition, there was no clear relationship between biochemical functions of these gene products and the mechanisms through which they might promote biofilm formation. These limitations underscored the importance of choosing a gene set whose biochemical functions would be useful for understanding biofilm regulation.

Therefore, we sought to disrupt a set of genes that would have a strong potential to yield an altered phenotype in a variety of screens and to yield mechanistic insight into the process under study. Again, there are many gene sets that meet these criteria (as discussed for *S. cerevisiae* above), and we settled on transcription factors for three main reasons. First, the number of genes is manageable. We identified 313 unique genes in the *C. albicans* genome sequence whose products have site-specific DNA binding motifs (e.g., zinc finger), using BLAST searches with known DNA binding proteins and relevant motif sequences. (We count both alleles as one unique gene, using sequence contig assignments to guide the definition of alleles.) Second, the likelihood of a mutant phenotype is good, as discussed earlier in this chapter. Third, the likelihood of making a mechanistic connection to the process under study is good. That is, genes whose expression is altered in the transcription factor mutant are candidates whose products may participate directly in the process under study. Among many examples are studies of iron (31, 47), phosphate (29), and meiotic (8) regulation in *S. cerevisiae*.

We cloned PCR products corresponding to each of the 313 *C. albicans* transcription factor genes. Thus far, 99 genes have been subjected to Tn7-*UAU1* in vitro mutagenesis, and insertions in the middle of each gene have been transformed into *C. albicans* (28). Among 99 such genes analyzed so far, 16 failed to yield homozygous mutants, suggesting that these transcription factors may be essential under our selection conditions. Viable homozygous mutants were created with insertions in the remaining 83 transcription factor genes, and these transcription factor mutants were tested for biofilm formation ability (28). Among the insertion mutants, we found two strains that were unable to form biofilms when compared to a reference strain (Color Plate 1). One carried an insertion in *TEC1* (orf19.5908 [36]), and the other carried an insertion in the new gene we call *BCR1* (orf19.8342; biofilm and cell wall regulator).

Tec1p is a TEA/ATTS transcription factor that is required for hyphal formation in vitro (36). The *tec1/tec1* insertion mutant grew largely as free-living cells in our biofilm formation assay. It created a thin rudimentary biofilm composed exclusively of yeast cells (Color Plate 1C and D). The mutant defects were complemented by introduction of a copy of the wild-type *TEC1* gene, thus indicating that the *tec1* insertion mutation is the cause of the biofilm formation defect. The biofilm defects of the *tec1/tec1* mutant are expected for a strain that is defective in hyphal development (2, 33, 36).

Bcr1p is a C_2H_2 zinc finger transcription factor. A *bcr1/bcr1* insertion mutant produced a thin rudimentary biofilm that, unlike the *tec1/tec1* mutant, included yeast cells interspersed with a few hyphal filaments (Color Plate 1E and F). The biofilm formation defect of the *bcr1/bcr1* insertion mutant was reversed by introduction of a wild-type copy of *BCR1*, and a *bcr1Δ/bcr1Δ* mutant had the same phenotype as the initial insertion mutant (28).Thus, Bcr1p is required for biofilm formation.

So, is *BCR1* yet another gene that is required for hyphal development and, as a consequence, biofilm formation? Three lines of evidence argue against that interpretation (28). First, there were some hyphae in the mutant's rudimentary biofilm, although one might argue that it has a severe quantitative defect in hyphal formation. Second, the mutant forms hyphae under a variety of planktonic inducing conditions, including the medium surrounding the rudimentary biofilm. And third, when a green fluorescent protein (GFP)-expressing *bcr1/bcr1* mutant is mixed with a wild-type strain to form a mixed biofilm, the biofilm includes many GFP-expressing hyphae (Color Plate 1G). This final observation proves that the *bcr1/bcr1* mutant cells form hyphae efficiently in the context of a biofilm. Thus, Bcr1p is not required for hyphal formation under any tested condition, including a condition most directly relevant to biofilm development.

Since Bcr1p is a transcription factor, we decided to elucidate the nature of the *bcr1/bcr1* defect though microarray gene expression comparisons of mutant (*bcr1/bcr1*) and complemented (*bcr1/bcr1*+p*BCR1*) strains. What we found, interestingly, was that among the 20 genes whose expression is most severely altered in the *bcr1/bcr1* mutant, 11 specify either cell surface proteins or cell wall modification enzymes. In particular, we noted the altered expression of several glycosylphosphatidylinositol (GPI)-linked cell wall protein genes in the *bcr1/bcr1* mutant, including *HYR1*,

ECE1, *RBT5*, *ECM331*, and adhesins *HWP1* and *ALS3*. These proteins are predicted (or, in some cases, shown) to exist at the cell surface by virtue of a covalent linkage between the GPI anchor remnant and β-glucan (42). Therefore, Bcr1p is required for normal expression levels of several genes that govern cell surface properties. The idea that Bcr1p is required for adherence is completely consistent with its role in biofilm formation, and the hypothesis that the mutant adherence defect is the cause of its biofilm defect is now being tested carefully.

There are several important general points from the Bcr1p story. First, mutant analysis has defined a new step in biofilm formation. Of course, everyone would expect adherence to be required for biofilm formation, but we now have direct evidence as well as a model to understand, for example, whether biofilm maturation steps are dependent on adherence. Second, the fact that Bcr1p is a transcription factor means that we can use microarray analysis to define the mechanistic basis for its defect. We can do that with any mutant of course, but it is less certain that the gene expression defect associated with a defect in a plasma membrane protein, for example, is the cause rather than the effect of the mutant's biological defect. And finally, we note that we found pretty much the same frequency of biofilm-defective mutants among transcription factor genes (2 of 83 or 2 of 99, as you prefer) as we did among random genes (4 of 197), although the numbers are small. So much for the *S. cerevisiae* model and Giaever et al.? We think not: perhaps the biofilm formation screen is fairly specific, yielding small numbers of mutants among any group of genes. The value of the transcription factor screen in this case was that it invited a mechanistic interpretation from results obtained with existing methods. After all, what's the point of having the cake if we can't eat it too?

REFERENCES

1. (Reference deleted.)
2. **Baillie, G. S., and L. J. Douglas.** 1999. Role of dimorphism in the development of *Candida albicans* biofilms. *J. Med. Microbiol.* **48:**671–679.
3. **Berman, J., and P. E. Sudbery.** 2002. *Candida albicans*: a molecular revolution built on lessons from budding yeast. *Nat. Rev. Genet.* **3:**918–930.
4. (Reference deleted.)
5. **Bruno, V. M., and A. P. Mitchell.** 2005. Regulation of azole drug susceptibility by *Candida albicans* protein kinase CK2. *Mol. Microbiol.* **56:**559–573.
6. **Calderone, R. A., and W. A. Fonzi.** 2001. Virulence factors of *Candida albicans*. *Trends Microbiol.* **9:** 327–335.
7. **Chandra, J., D. M. Kuhn, P. K. Mukherjee, L. L. Hoyer, T. McCormick, and M. A. Ghannoum.** 2001. Biofilm formation by the fungal pathogen *Candida albicans*: development, architecture, and drug resistance. *J. Bacteriol.* **183:**5385–5394.

8. Chu, S., J. DeRisi, M. Eisen, J. Mulholland, D. Botstein, P. O. Brown, and I. Herskowitz. 1998. The transcriptional program of sporulation in budding yeast. *Science* 282:699–705.

9. Costerton, J. W., P. S. Stewart, and E. P. Greenberg. 1999. Bacterial biofilms: a common cause of persistent infections. *Science* 284:1318–1322.

10. Davis, D. A., V. M. Bruno, L. Loza, S. G. Filler, and A. P. Mitchell. 2002. *Candida albicans* Mds3p, a conserved regulator of pH responses and virulence identified through insertional mutagenesis. *Genetics* 162:1573–1581.

11. De Backer, M. D., P. T. Magee, and J. Pla. 2000. Recent developments in molecular genetics of *Candida albicans*. *Annu. Rev. Microbiol.* 54:463–498.

12. De Smet, K., I. Eberhardt, R. Reekmans, and R. Contreras. 2004. Bax-induced cell death in *Candida albicans*. *Yeast* 21:1325–1334.

13. Donlan, R. M., and J. W. Costerton. 2002. Biofilms: survival mechanisms of clinically relevant microorganisms. *Clin. Microbiol. Rev.* 15:167–193.

14. Dunyak, D. S., D. S. Everdeen, J. G. Albanese, and C. L. Quinn. 2002. Deletion of individual mRNA capping genes is unexpectedly not lethal to *Candida albicans* and results in modified mRNA cap structures. *Eukaryot. Cell* 1:1010–1020.

15. Enloe, B., A. Diamond, and A. P. Mitchell. 2000. A single-transformation gene function test in diploid *Candida albicans*. *J. Bacteriol.* 182:5730–5736.

16. Garcia-Sanchez, S., S. Aubert, I. Iraqui, G. Janbon, J. M. Ghigo, and C. d'Enfert. 2004. Candida albicans biofilms: a developmental state associated with specific and stable gene expression patterns. *Eukaryot. Cell* 3:536–545.

17. Giaever, G., A. M. Chu, L. Ni, C. Connelly, L. Riles, S. Veronneau, S. Dow, A. Lucau-Danila, K. Anderson, B. Andre, A. P. Arkin, A. Astromoff, M. El-Bakkoury, R. Bangham, R. Benito, S. Brachat, S. Campanaro, M. Curtiss, K. Davis, A. Deutschbauer, K. D. Entian, P. Flaherty, F. Foury, D. J. Garfinkel, M. Gerstein, D. Gotte, U. Guldener, J. H. Hegemann, S. Hempel, Z. Herman, D. F. Jaramillo, D. E. Kelly, S. L. Kelly, P. Kotter, D. LaBonte, D. C. Lamb, N. Lan, H. Liang, H. Liao, L. Liu, C. Luo, M. Lussier, R. Mao, P. Menard, S. L. Ooi, J. L. Revuelta, C. J. Roberts, M. Rose, P. Ross-Macdonald, B. Scherens, G. Schimmack, B. Shafer, D. D. Shoemaker, S. Sookhai-Mahadeo, R. K. Storms, J. N. Strathern, G. Valle, M. Voet, G. Volckaert, C. Y. Wang, T. R. Ward, J. Wilhelmy, E. A. Winzeler, Y. Yang, G. Yen, E. Youngman, K. Yu, H. Bussey, J. D. Boeke, M. Snyder, P. Philippsen, R. W. Davis, and M. Johnston. 2002. Functional profiling of the *Saccharomyces cerevisiae* genome. *Nature* 418:387–391.

18. Hwang, L., D. Hocking-Murray, A. K. Bahrami, M. Andersson, J. Rine, and A. Sil. 2003. Identifying phase-specific genes in the fungal pathogen *Histoplasma capsulatum* using a genomic shotgun microarray. *Mol. Biol. Cell* 14:2314–2326.

19. Jones, T., N. A. Federspiel, H. Chibana, J. Dungan, S. Kalman, B. B. Magee, G. Newport, Y. R. Thorstenson, N. Agabian, P. T. Magee, R. W. Davis, and S. Scherer. 2004. The diploid genome sequence of *Candida albicans*. *Proc. Natl. Acad. Sci. USA* 101:7329–7334.

20. Kojic, E. M., and R. O. Darouiche. 2004. *Candida* infections of medical devices. *Clin. Microbiol. Rev.* 17:255–267.

21. Kottom, T. J., C. F. Thomas, Jr., and A. H. Limper. 2001. Characterization of *Pneumocystis carinii PHR1*, a pH-regulated gene important for cell wall integrity. *J. Bacteriol.* 183:6740–6745.

22. Kuchin, S., V. K. Vyas, and M. Carlson. 2002. Snf1 protein kinase and the repressors Nrg1 and Nrg2 regulate FLO11, haploid invasive growth, and diploid pseudohyphal differentiation. *Mol. Cell. Biol.* 22:3994–4000.

23. Li, F., and S. P. Palecek. 2003. *EAP1*, a *Candida albicans* gene involved in binding human epithelial cells. *Eukaryot. Cell* 2:1266–1273.

24. Li, M., S. J. Martin, V. M. Bruno, A. P. Mitchell, and D. A. Davis. 2004. *Candida albicans* Rim13p, a protease required for Rim101p processing at acidic and alkaline pHs. *Eukaryot. Cell* 3:741–751.

25. Mukherjee, P. K., J. Chandra, D. M. Kuhn, and M. A. Ghannoum. 2003. Mechanism of fluconazole resistance in *Candida albicans* biofilms: phase-specific role of efflux pumps and membrane sterols. *Infect. Immun.* 71:4333–4340.

26. Navarro-Garcia, F., M. Sanchez, C. Nombela, and J. Pla. 2001. Virulence genes in the pathogenic yeast *Candida albicans*. *FEMS Microbiol. Rev.* 25:245–268.

27. Nobile, C. J., V. M. Bruno, M. L. Richard, D. A. Davis, and A. P. Mitchell. 2003. Genetic control of chlamydospore formation in *Candida albicans*. *Microbiology* 149:3629–3637.

28. Nobile, C. J., and A. P. Mitchell. 2005. Regulation of cell-surface genes and biofilm formation by the *C. albicans* transcription factor Bcr1p. *Curr. Biol.* 15:1150–1155.

29. Ogawa, N., J. DeRisi, and P. O. Brown. 2000. New components of a system for phosphate accumulation and polyphosphate metabolism in *Saccharomyces cerevisiae* revealed by genomic expression analysis. *Mol. Biol. Cell* 11:4309–4321.

30. O'Toole, G., H. B. Kaplan, and R. Kolter. 2000. Biofilm formation as microbial development. *Annu. Rev. Microbiol.* 54:49–79.

31. Protchenko, O., T. Ferea, J. Rashford, J. Tiedeman, P. O. Brown, D. Botstein, and C. C. Philpott. 2001. Three cell wall mannoproteins facilitate the uptake of iron in *Saccharomyces cerevisiae*. *J. Biol. Chem.* 276:49244–49250.

32. Ramage, G., S. Bachmann, T. F. Patterson, B. L. Wickes, and J. L. Lopez-Ribot. 2002. Investigation of multidrug efflux pumps in relation to fluconazole resistance in *Candida albicans* biofilms. *J. Antimicrob. Chemother.* 49:973–980.

33. Ramage, G., K. VandeWalle, J. L. Lopez-Ribot, and B. L. Wickes. 2002. The filamentation pathway controlled by the Efg1 regulator protein is required for normal biofilm formation and development in *Candida albicans*. *FEMS Microbiol. Lett.* 214:95–100.

34. Reynolds, T. B., and G. R. Fink. 2001. Bakers' yeast, a model for fungal biofilm formation. *Science* 291:878–881.

34a. Richard, M. L., C. J. Nobile, V. M. Bruno, and A. P. Mitchell. 2005. *Candida albicans* biofilm-defective mutants. *Eukaryot. Cell* 4:1493–1502.

35. Roemer, T., B. Jiang, J. Davison, T. Ketela, K. Veillette, A. Breton, F. Tandia, A. Linteau, S. Sillaots, C. Marta, N. Martel, S. Veronneau, S. Lemieux, S. Kauffman, J. Becker, R. Storms, C. Boone, and H. Bussey. 2003. Large-scale essential gene identification in *Candida albicans* and applications to antifungal drug discovery. *Mol. Microbiol.* 50:167–181.

36. Schweizer, A., S. Rupp, B. N. Taylor, M. Rollinghoff, and K. Schroppel. 2000. The TEA/ATTS transcription factor

CaTec1p regulates hyphal development and virulence in *Candida albicans. Mol. Microbiol.* **38:**435–445.

37. **Smith, D. G., M. D. Garcia-Pedrajas, S. E. Gold, and M. H. Perlin.** 2003. Isolation and characterization from pathogenic fungi of genes encoding ammonium permeases and their roles in dimorphism. *Mol. Microbiol.* **50:**259–275.

38. **Song, J. L., and T. C. White.** 2003. RAM2: an essential gene in the prenylation pathway of *Candida albicans. Microbiology* **149:**249–259.

39. **Spreghini, E., D. A. Davis, R. Subaran, M. Kim, and A. P. Mitchell.** 2003. Roles of *Candida albicans* Dfg5p and Dcw1p cell surface proteins in growth and hypha formation. *Eukaryot. Cell* **2:**746–755.

40. (Reference deleted.)

41. **Stanley, N. R., and B. A. Lazazzera.** 2004. Environmental signals and regulatory pathways that influence biofilm formation. *Mol. Microbiol.* **52:**917–924.

42. **Sundstrom, P.** 2002. Adhesion in *Candida* spp. *Cell Microbiol.* **4:**461–469.

43. **Warenda, A. J., and J. B. Konopka.** 2002. Septin function in *Candida albicans* morphogenesis. *Mol. Biol. Cell* **13:**2732–2746.

44. **Wilson, R. B., D. Davis, B. M. Enloe, and A. P. Mitchell.** 2000. A recyclable *Candida albicans URA3* cassette for PCR product-directed gene disruptions. *Yeast* **16:**65–70.

45. **Wilson, R. B., D. Davis, and A. P. Mitchell.** 1999. Rapid hypothesis testing with *Candida albicans* through gene disruption with short homology regions. *J. Bacteriol.* **181:**1868–1874.

46. **Xu, W., F. J. Smith, Jr., R. Subaran, and A. P. Mitchell.** 2004. Multivesicular body-ESCRT components function in pH response regulation in *Saccharomyces cerevisiae* and *Candida albicans. Mol. Biol. Cell* **15:**5528–5537.

47. **Yun, C. W., T. Ferea, J. Rashford, O. Ardon, P. O. Brown, D. Botstein, J. Kaplan, and C. C. Philpott.** 2000. Desferrioxamine-mediated iron uptake in *Saccharomyces cerevisiae.* Evidence for two pathways of iron uptake. *J. Biol. Chem.* **275:**10709–10715.

Molecular Principles of Fungal Pathogenesis
Edited by Joseph Heitman et al.
©2006 ASM Press, Washington, D.C.

Chapter 4

Candida albicans Hypha Formation and Virulence

PAULA SUNDSTROM

The role of hypha formation in virulence of the yeast *Candida albicans* continues to be a regular source of dialogue and debate among scientists interested in how this fungus damages human hosts (8, 10). The presence of hyphal forms amid mixtures of yeasts and elongated yeasts termed pseudohyphae in host tissues prompts a natural curiosity about the reasons for the presence of multiple growth forms and their individual and collective contributions to pathogenicity. Growth as yeasts in minimal media under most laboratory conditions, coupled with the rapid induction of hyphae in serum at physiological temperature, hints at a predilection for hyphae in host tissue, although hyphal forms are easily grown in minimal media in vitro as well. Less is known about cells in intermediate pseudohyphal form because of the lack of standardized conditions leading to uniform growth as pseudohypae without buds or true hyphae. Variability among strains in their propensity to grow as pseudohyphae, along with the wide variety of growth conditions leading to pseudohyphal growth in mixtures of yeast and hyphae, has contributed to the lack of understanding of pseudohyphal growth forms. Recent progress has been made in defining culture conditions supporting high proportions of pseudohyphae (9), which will advance our understanding of this growth form. Whereas laboratory conditions allowing mass conversion from yeast to hyphal forms are well defined and permit analyses of the various growth form properties, the factors contributing to morphogenesis in the host and the factors driving the relative proportions of hyphae, pseudohyphae, and yeast in any given host are unknown. Since the presence and severity of candidiasis is dictated by the immune status of the host, with commensalism being the norm for immunocompetent hosts, the forms taken by *C. albicans* in tissues are likely to be intimately coupled to host susceptibility. Fungal attributes necessary for tissue invasion may differ depending on the particular set of risk factors present in an individual host. Current research shows that major changes in gene expression accompany morphogenesis and suggest that fungal attributes necessary for proliferation and survival in the host are present in both yeast and hyphal growth forms. Expression of an optimal set of genes by fungi in response to specific host environments reflects the assortment of hyphae, pseudohyphae, and yeast that are observed in clinical specimens. The following paragraphs focus on recent evidence for the possible mechanisms by which hyphal forms contribute to tissue damage and are not intended to diminish the importance of yeast forms for growth in the host, which have been previously discussed (8).

INSIGHTS INTO HYPHA-MEDIATED TISSUE DAMAGE FROM EPITHELIAL PATHOLOGY

The clinical manifestations of candidiasis are broadly categorized into superficial invasion of mucosal surfaces and systemic infections with deep-organ involvement. Because of the existence of *C. albicans* in the gastrointestinal tract, the most common infections occur superficially on mucosal surfaces. The invasive properties of hyphae are suggested by their ability to penetrate several cell layers deeper than yeast forms into the epithelium in superficial candidiasis. Histological examination of lingual tissue from an AIDS patient with candidiasis shows yeast forms within vacuolated regions of tissue whereas hyphae appear to be embedded in tissue laterally adjacent to yeast forms and are also present in deeper layers (Color Plate 2A). In superficial candidiasis, *C. albicans* generally impinges on but does not penetrate into the prickle cell layer (Color Plate 2B). These findings suggest that hyphae are more invasive than yeasts and that hyphae increase the level of damage to host tissue by extending the region of tissue susceptibility to invasion. Breakdown of tissue by hyphal forms may facilitate

Paula Sundstrom • Microbiology and Immunology, Dartmouth Medical School, Hanover, NH 03755.

budding growth from hyphal septa, thereby extending the area of tissue populated by yeast. Further evidence for the role of hyphae in superficial candidiasis is suggested by the appearance of material taken from raised, opaque lesions of pseudomembranous candidiasis (Color Plate 3). Epithelial cells from the mucosal surface appear to be stitched together by fungal hyphae. The adherent mixtures of *C. albicans* hyphae and host epithelial cells are not dispersible by treatment with sodium dodecyl sulfate and heat, reflecting the tight attachments formed between fungus and host involving mammalian transglutaminase and the *C. albicans* transglutaminase substrate Hwp1 (17). The histological picture of mucosal candidiasis supports a role for hyphae in lesion formation and in tissue penetration.

INFERENCES FROM ANIMAL MODELS

Host Effects on Fungal Morphology

Manipulation of host-fungus interactions in animal models has provided insight into the role of hypha formation in virulence. The idea that host-mediated inhibition of germ tube formation influences the spectrum of candidal morphologies seen in tissue is suggested by a study of murine oroesophageal candidiasis (7) in which a decreased ratio of germ tube to yeast forms was correlated with clearance of the organism from host tissue. The influence of the host on fungal morphology is further supported by the finding that genetically defined strains of immunodeficient mice exhibit differing profiles of morphologies following inoculation with the same strain of *C. albicans* (18). Evidence for tissue factors that promote hyphal formation in susceptible hosts comes from studies showing that some *C. albicans* mutants that are defective in hyphal formation under most in vitro conditions overcome these defects when placed in susceptible hosts (14). These studies support the hypothesis that *C. albicans* morphologies can be controlled by host factors and that hyphal formation favors the fungus and is disadvantageous to the host.

Attenuated Virulence of Morphological Mutants

The use of genetic manipulation has led to the discovery that transcription factors and signal transduction pathways are critical for the proper regulation of morphogenic conversions in *C. albicans*. Mutants derived from these studies have provided seminal insights into the role of candidal morphogenesis in virulence. Mutant strains of *C. albicans* that are defective in hypha production in host tissue have reduced virulence in murine models of systemic candidiasis (2, 11, 16). Reduced virulence is also a property of mutants that lack transcriptional repressors and grow

primarily as pseudohyphae (3, 4, 12) or have an unregulated cyclic AMP signaling pathway that accelerates germ tube formation and inhibits the production of yeasts from septa and tips of hyphal forms (1). These studies support the hypothesis that a complete spectrum of morphologies is required for eliciting a wild-type extent of host damage and that an unhindered ability to make the transition between growth forms is important for virulence. In a recent innovative study to manipulate bud-hypha transitions within the host by placing a negative regulator of hyphal growth under the control of a tetracycline-regulatable promoter, it was shown that hyphal growth was necessary for mortality but not for hematogenous dissemination to internal organs (15). This finding is consistent with basic principles of mycology which state that, in general, budding-yeast growth promotes dispersal in liquid films (5). The studies of morphological mutants do not diminish the importance of either growth form in pathogenesis but instead predict that critical determinants mediating tissue damage and survival in the host reside in both yeast and hyphal morphologies.

Attributes of Hyphae, not Hyphae Per Se, Are Required for Optimal Damage to the Host

A caveat associated with the use of morphogenic mutants in attempts to understand the role of hypha formation in virulence is that the roles of apical and budding modes of growth cannot be separated from morphology-specific gene expression. Analysis of mutants that do not affect morphology but do affect virulence has led to insights into the properties of hyphae that contribute to virulence. For example, deletion of the hypha-specific adhesin Hwp1 does not affect hypha formation but *HWP1* knockout strains are attenuated in virulence in both systemic and mucosal models of candidiasis (17–19). In another example, *C. albicans* mutants lacking the phospholipase D1 gene are able to form hyphae in vivo; however, the hyphae are defective in their ability to cause oroesophageal and gastric candidiasis and to kill *C. albicans*-colonized mice (6). Studies of mutants displaying defects in hypha-specific gene expression but not in hypha production support the principle that the role of hyphae in virulence is more closely related to the changes in gene expression profiles that accompany hyphal growth than to hyphal growth itself.

PUTATIVE VIRULENCE FACTOR GENES IN YEAST AND HYPHAE

Recent research using genome-wide transcriptional profiling (13) has considerably extended our knowledge of differentially induced genes. In addition to *HWP1*, *PLD1*, and other potential virulence factors, genes that could potentially regulate virulence factor gene expression during hypha production, such as those encoding transcription factors, signal transduction components, and proteins that function in secretion, as well as others, were identified. Similarly, a different collection of genes was identified that were down-regulated in hyphae compared to yeast. Based on the importance of yeast in pathogenesis (8), it is likely that some of the identified genes will eventually be shown to encode pathogenically important virulence factors and regulators that are specific for yeast growth. The dichotomy in gene expression between yeast and hyphal growth forms suggests an advantageous status for pseudohyphae, if this intermediate but distinct morphological growth form were to display intermediate levels of both hypha- and yeast-specific genes in the same cell. Combinations of yeast- and hypha-specific genes could be advantageous for invasion of and survival in host tissues. In conclusion, whereas the enhanced adhesive and invasive properties of hyphae have been established, survival of *C. albicans* in the host is inextricably connected to morphological transitions to yeast and pseudohyphal forms that are accompanied by gene expression profiles optimized for growth and invasion of the host in candidiasis.

Acknowledgments. I thank Carl Allen for photographs of oral candidiasis. Edward Balish and Carl Allen provided helpful comments on the manuscript.

REFERENCES

1. **Bahn, Y. S., J. Staab, and P. Sundstrom.** 2003. Increased high-affinity phosphodiesterase *PDE2* gene expression in germ tubes counteracts *CAP1*-dependent synthesis of cyclic AMP, limits hypha production and promotes virulence of *Candida albicans. Mol. Microbiol.* 50:391–409.
2. **Bahn, Y. S., and P. Sundstrom.** 2001. *CAP1*, an adenylate cyclase-associated protein gene, regulates bud-hypha transitions, filamentous growth, and cyclic AMP levels and is required for virulence of *Candida albicans. J. Bacteriol.* 183:3211–3223.
3. **Braun, B. R., and A. D. Johnson.** 1997. Control of filament formation in *Candida albicans* by the transcriptional repressor *TUP1. Science* 277:105–109.
4. **Braun, B. R., D. Kadosh, and A. D. Johnson.** 2001. *NRG1*, a repressor of filamentous growth in *C. albicans*, is down-regulated during filament induction. *EMBO J.* 20:4753–4761.
5. **Deacon, J.** 1997. *Modern Mycology*, 3rd ed. Blackwell Science, Oxford, United Kingdom.
6. **Dolan, J. W., A. C. Bell, B. Hube, M. Schaller, T. F. Warner, and E. Balish.** 2004. *Candida albicans PLD I* activity is required for full virulence. *Med. Mycol.* 42:439–447.
7. **Elahi, S., G. Pang, R. Clancy, and R. B. Ashman.** 2000. Cellular and cytokine correlates of mucosal protection in murine model of oral candidiasis. *Infect. Immun.* 68:5771–5777.
8. **Gow, N. A., A. J. Brown, and F. C. Odds.** 2002. Fungal morphogenesis and host invasion. *Curr. Opin. Microbiol.* 5:366–371.
9. **Hornby, J. M., R. Dumitru, and K. W. Nickerson.** 2004. High phosphate (up to 600 mM) induces pseudohyphal development in five wild type *Candida albicans. J. Microbiol. Methods* 56:119–124.
10. **Hube, B.** 2004. From commensal to pathogen: stage- and tissue-specific gene expression of *Candida albicans. Curr. Opin. Microbiol.* 7:336–341.
11. **Lo, H. J., J. R. Kohler, B. DiDomenico, D. Loebenberg, A. Cacciapuoti, and G. R. Fink.** 1997. Nonfilamentous *C. albicans* mutants are avirulent. *Cell* 90:939–949.
12. **Murad, A. M., C. d'Enfert, C. Gaillardin, H. Tournu, F. Tekaia, D. Talibi, D. Marechal, V. Marchais, J. Cottin, and A. J. Brown.** 2001. Transcript profiling in *Candida albicans* reveals new cellular functions for the transcriptional repressors CaTup1, CaMig1 and CaNrg1. *Mol. Microbiol.* 42:981–993.
13. **Nantel, A., D. Dignard, C. Bachewich, D. Harcus, A. Marcil, A. P. Bouin, C. W. Sensen, H. Hogues, M. van het Hoog, P. Gordon, T. Rigby, F. Benoit, D. C. Tessier, D. Y. Thomas, and M. Whiteway.** 2002. Transcription profiling of *Candida albicans* cells undergoing the yeast-to-hyphal transition. *Mol. Biol. Cell* 13:3452–3465.
14. **Riggle, P. J., K. A. Andrutis, X. Chen, S. R. Tzipori, and C. A. Kumamoto.** 1999. Invasive lesions containing filamentous forms produced by a *Candida albicans* mutant that is defective in filamentous growth in culture. *Infect. Immun.* 67:3649–3652.
15. **Saville, S. P., A. L. Lazzell, C. Monteagudo, and J. L. Lopez-Ribot.** 2003. Engineered control of cell morphology in vivo reveals distinct roles for yeast and filamentous forms of *Candida albicans* during infection. *Eukaryot. Cell* 2:1053–1060.
16. **Sobel, J. D., G. Muller, and H. R. Buckley.** 1984. Critical role of germ tube formation in the pathogenesis of candidal vaginitis. *Infect. Immun.* 44:576–580.
17. **Staab, J. F., S. D. Bradway, P. L. Fidel, and P. Sundstrom.** 1999. Adhesive and mammalian transglutaminase substrate properties of *Candida albicans* Hwp1. *Science* 283:1535–1538.
18. **Sundstrom, P., E. Balish, and C. M. Allen.** 2002. Essential role of the *Candida albicans* transglutaminase substrate, hyphal wall protein 1, in lethal oroesophageal candidiasis in immunodeficient mice. *J. Infect. Dis.* 185:521–530.
19. **Sundstrom, P., J. E. Cutler, and J. F. Staab.** 2002. Reevaluation of the role of *HWP1* in systemic candidiasis by use of *Candida albicans* strains with selectable marker *URA3* targeted to the *ENO1* locus. *Infect. Immun.* 70:3281–3283.

Molecular Principles of Fungal Pathogenesis
Edited by Joseph Heitman et al.
©2006 ASM Press, Washington, D.C.

Chapter 5

The Conversion from Classical Studies in Fungal Pathogenesis to the Molecular Era

KYUNG J. KWON-CHUNG AND BRIAN L. WICKES

One of the most important challenges of research in fungal pathogenesis is to dissect the interplay between the genetic makeup of pathogenic fungi and the environmental conditions that influence the patterns of the diseases they cause. This challenge requires the identification of the genes associated with virulence in order to design adequate drugs and vaccines for therapeutic intervention. Identification of species-specific genes should contribute to the development of rapid diagnostic tools, especially for slow-growing pathogens.

The conversion from classical approaches used in the past for studies of fungal pathobiology to the molecular era took nearly 3 decades, which badly lagged the field of bacterial pathogenesis in spite of the head start in medical mycology. In fact, the history and development of medical mycology preceded the formal study of bacteriology by several decades. For example, Koch's postulates for determining whether a specific bacterium was the causative agent of a specific disease were applied to dermatophytes by a German dermatologist, David Gruby, 40 years before Koch formulated them (52). However, the study of pathogenic fungi declined for a considerable time period as bacteriology began to overshadow medical mycology, beginning with the towering works of Koch and Pasteur.

This trend has been reversed during the past 3 decades. The world has witnessed a sharp surge in fungal diseases as immunocompromised patient populations have increased. The drastic increases in these populations have been caused in large part by increasingly aggressive cytotoxic therapies for malignancies and organ transplantations as well as other aggressive medical procedures. The AIDS epidemic worsened the situation by increasing both the frequency and scope

of fungi that were able to successfully infect humans (81). As the occurrence rate of fungal infections was escalating, so was the interest in pathogenic mycology. As a result, basic research in pathogenic fungi and clinical mycology became rejuvenated and followed bacteriology into the molecular era. The increases in workshops and conferences on specific mycotic agents along with funding and the resulting literature on pathogenic fungi generated in the past 20 years attest to this rejuvenation (29; D. Dixon, 2004, personal communication). Conferences specific for candidiasis, cryptococcosis, coccidioidomycosis, and paracoccidioidomycosis have been held periodically and have fostered interest and enthusiasm in research in their respective fields. The first aspergillosis-specific conference (Advances against Aspergillosis) was organized in the fall of 2004 and close to 370 investigators from 28 countries participated.

While the incidences of mycoses have increased, considerable progress has been made during the past three decades in the development of new technologies applicable for fungal research starting with model organisms such as *Saccharomyces cerevisiae*, *Aspergillus nidulans*, and *Neurospora crassa* (Fig. 1 shows the major biotechnological milestones). As a result, mycologists can now exploit a wide range of techniques and methods that allow a sophisticated analysis of fungi at the molecular level, which was previously unthinkable when classical methods were the only ones available. Research into the pathobiology of fungi, especially fungi that cause systemic infection, has made significant progress that is comparable to any other field of microbiology. As we transition from the era of genome sequencing of model organisms into the era of sequencing genomes

Kyung J. Kwon-Chung • Laboratory of Clinical Infectious Diseases, National Institute of Allergy and Infectious Diseases, National Institutes of Health, Bethesda, MD 20892.   **Brian L. Wickes** • Department of Microbiology and Immunology, The University of Texas Health Center at San Antonio, San Antonio, TX 78284.

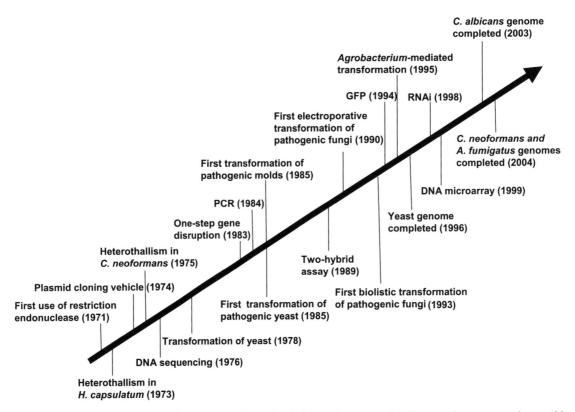

Figure 1. Mycology milestones. The transition from classical mycology to molecular mycology was made possible through a number of biological and technological milestones that were achieved over the last 25 years.

of pathogens from all the major taxonomic groups (http://www.broad.mit.edu/annotation/fungi/fgi/), vital information will become available, enabling us to address the key pathobiological questions for each pathogenic species.

The classical methods and approaches used for differentiation of fungal pathogens, which have relied primarily on morphology and physiology at the species level, are now assisted by molecular biology-based tools that employ conserved nucleotide sequences of either housekeeping genes or the subunits and the spacer regions of ribosomal DNA (rDNA) clusters. This conversion to molecular typing methods has resulted in the widening of the boundaries of the Kingdom Fungi. For example, the Kingdom Fungi now includes organisms such as *Pneumocystis* (32) and *Microsporidia* (65), which had been previously classified as protozoa. Sequencing of particular regions of the rDNA cluster has also revealed extraordinary and unsuspected degrees of genetic heterogeneity (41) among strains of fungi classified in the same species based on their morphology and/or physiology. Delineation of monophyletic clusters based on sequence data offered a new classification system. As a result, the number of fungi being recognized as new species is multiplying rapidly. For example, *Cryptococcus laurentii*, an infrequently isolated basidiomycetous

yeast from clinical specimens, is now known to be composed of three different species based on molecular data (139). Molecular studies have revealed that the etiologic agents of coccidioidomycosis are composed of at least two genetically distinguishable species, *Coccidioides immitis* and *C. posadasii* (43), instead of one species. Although heterogeneity of *Cryptococcus neoformans*, the etiologic agent of cryptococcosis, was known by the existence of two morphologically distinct teleomorphs (78, 79) among four serotypes (A to D), the extent of their genetic diversity was not appreciated until molecular strain typing became available. Multigene sequence analysis of serotype A clinical isolates from sub-Saharan Africa alone revealed two distinct subpopulations (91).

While we have gained an understanding of the genetic disparity between each fungal pathogen by using molecular approaches, this trend has also resulted in diagnostic confusion between research mycologists and the clinical laboratories that rely solely on automated diagnostic kits that are often attached to limited databases. Consequently, we are frequently reminded of the need to retain and apply classical mycological skills to molecular advances. In fact, in spite of the tremendous power and creativity that molecular biology brings to the field of medical mycology, most molecular mycologists would agree that the accuracy of classical mycology in establishing order in fungal systematics, as confirmed by recent applications of molecular or other methods, has been striking.

The purpose of this chapter is not to present a chronology of the conversion from classical to molecular studies of fungal pathobiology or to present all the available molecular tools for the study of fungal pathogenesis. Instead, it focuses on the pioneering works that have laid the foundation for the conversion from the classical study of fungal pathobiology to the molecular era. The most widely used and important molecular biology-based technologies that are now available and in use for the identification of pathogenic fungi and for study of fungal pathobiology are also discussed.

CONVERSION OF CLASSICAL STUDIES OF VIRULENCE FACTORS TO THE MOLECULAR ERA

A complex and intriguing aspect of pathogenicity in most fungal species that cause infection in humans is that their pathogenicity is neither essential nor beneficial for their survival. Except for commensal species such as *Candida albicans*, certain species of ringworm fungi that evolved to become anthropophilic or zoophilic such as *Microsporum audouinii* or *M. canis*, respectively (81), and *Pneumocystis* species, whose ecological niche is unknown, fungal pathogens are environmental organisms that cause infection only as a result of accidental encounters with a susceptible host (81). Regardless of whether the approach is classical or molecular, the objective of studies in fungal pathogenesis is to search for distinct genetic properties that provide fungi with a significantly greater ability to colonize the host, ward off host immune system assault, and multiply.

Classical approaches to the identification of virulence genes from the 1960s to the early 1980s were based on forward genetics. Investigations started with random mutagenesis using various chemical or radiation sources, which allowed mutants to be selected based on a detectable phenotype. Whenever possible, sexual or parasexual recombinational analysis was then attempted to confirm cosegregation of low virulence and the mutant phenotype. Examples of the classical identification of virulence factors using mutagenesis and recombinational analysis can be seen in the study of melanin formation (Color Plate 4) (84) and polysaccharide capsule production in *C. neoformans* (62, 85). Protease in *C. albicans* (83, 94) and conidial pigment in *Aspergillus fumigatus* (63) as virulence factors are additional examples of classical identification of virulence factors based on mutant phenotypes. However, for most pathogenic fungi, classic recombinational analysis is either impossible or impractical.

The transition from the classical era to the molecular era necessitated the development of models, techniques, and strategies which would facilitate the definition of the genetic and molecular basis of pathogenicity according to the working hypothesis in the molecular Koch's postulates formulated by Falkow (37, 38). In short, Falkow's postulates for linking a microbial gene to pathogenicity required that (i) the phenotype or the property under investigation be associated with pathogenic members of a genus or pathogenic strains of a species, (ii) specific inactivation of the gene(s) associated with the suspected virulence trait should lead to a measurable loss in pathogenicity or virulence, and (iii) restoration of pathogenicity should accompany the reintroduction of the wild-type gene(s). At present, Falkow's postulates have been fulfilled for various virulence factors in nearly all the important systemic mycotic agents, starting with the capsule genes in *C. neoformans* (15). The most important molecular technique that enabled investigators to fulfill Koch's molecular postulates was the development and application of DNA transformation and gene disruption.

Transformed Species

Classical approaches to the characterization of fungal virulence genes using mutant strains obtained by chemical or radiation mutagenesis fall short in defining whether the gene is an essential component of the ability of the fungus to cause diseases in humans for a number of reasons. First, the mutant may harbor multiple undefined mutations that cannot always be removed by backcrossing, such as in asexual pathogens or even in sexual pathogens such as *Histoplasma capsulatum* and *Blastomyces dermatitidis*, where recombinational analysis is impractical. The genetic consequences of these mutations cannot be assessed and may result in an erroneous interpretation of the relationship observed between phenotypes and pathogenicity. Second, even if the mutant carries a defect in a single gene, precise characterization of the type of genetic lesion and the role of mutation in giving rise to the phenotype is unattainable.

Functional analysis of genes became possible with the advent of reverse genetic approaches based on homologous recombination with transforming DNA, which results in precise disruption as well as reconstitution of the wild-type genes. In classical studies, in lieu of reconstituting the mutated genes, revertants were sought to confirm the relationship between the genes/phenotypes and virulence. Examples are found in the studies of protease in *C. albicans* (83), melanin in *C. neoformans* (119), and conidial pigment production by *A. fumigatus* (63), in which animal experiments were performed using a set of three strains: the wild type, a mutant, and a revertant. Molecular studies that utilized transformation eventually confirmed the association of proteases (58, 123), melanin (122), and conidial pigment (144) with virulence in the three respective species.

The development of DNA-mediated transformation systems generally consists of (i) insertion of exogenous DNA into recipient cells, (ii) expression of genes present on the incoming DNA, and (iii) stable maintenance and replication of the transformed DNA leading to expression of the expected phenotype. Transformation, at this writing, has been accomplished in over 20 species of pathogenic fungi starting from the mid-1980s (Table 1). Transformation not

Table 1. Pioneering transformation protocols used in different pathogenic fungi

Disease and fungus	Method	Marker	Frequency[a]	Episomes[b]	Date	Reference
Aspergillosis						
A. nidulans	Spheroplast	ArgB	500	−	1985	64
A. flavus	Spheroplast	URA3	20	−	1989	154
A. fumigatus	Spheroplast	hph	Up to 100	−	1992	140
A. niger	Spheroplast	ArgB	4	−	1985	13
Blastomycosis						
B. dermatitidis	Electroporation	hph	45	−	1997	57
Candidiasis						
C. guilliermondii	Spheroplast	ARG4	1,500	−	1985	72
C. albicans	Spheroplast	ADE2	0.5–5	−	1986	73
C. tropicalis	Spheroplast	URA3		−	1990	53
C. glabrata	Lithium acetate	LEU2	>10^4	+[c]	1992	101
C. lusitaniae	Biolistic	URA3		−	2000	165
C. parapsilosis	Lithium acetate	GAL1	1,200–2,800	−	2002	110
Coccidioidomycosis						
C. immitis	Biolistic	hph	0.5–6	−	1998	166
Cryptococcosis						
C. neoformans	Electroporation	URA5	50–100	+	1990	33
C. gattii	Biolistic	hph	1	NA[d]	1996	20
Dermatomycosis						
Trichophyton mentagrophytes	Spheroplast	hph	0.004–6	−	1989	50
Fusariosis						
F. oxisporum	Spheroplast	hph	1	−	1988	70
F. solani	Spheroplast	hph	10^4	−	1992	21
Histoplasmosis						
H. capsulatum	Electroporation	URA5	30	+	1990	161
Zygomycosis						
Rhizomucor pusillus	Spheroplast	LeuA	20	−	1996	147
Rhizopus oryzae	Biolistic	URA3	10–50	+	1998	Skory and Bothast, abstract
Paracoccidioidomycosis						
P. brasiliensis	A. tumefaciens mediated	hph	3.4/10^6 yeast	−	2004	88
Penicilliosis						
P. marneffeii	Spheroplast	URA3	1,400	−	2001	9
Pheohyphomycosis						
Wangiella dermatitidis	Electroporation	URA5, hph	10–203	−	1995	113
Trichosporonosis						
T. cutaneum	Spheroplast	hph, ble	10–500	−	1989	49

[a]Per microgram of DNA except for *P. brasiliensis,* in which 3.4 transformants per 10^6 yeast cells were obtained by ATMT.
[b]Presence (+) and absence (−) of episomal transforming plasmids (without an autonomously replicating sequence present in the vector).
[c]The pioneering transformation was accomplished with a plasmid that contained the *MT-11a* gene harboring an *ARS* element.
[d]NA, information not available.

only offered a molecular biology-based tool for gene cloning and functional analysis by disruption but also became the most widely applied tool for random mutagenesis. As we are in the genomic era, transformation-mediated random mutagenesis is a powerful tool for forward genetic analysis (see below). In this section, the pioneering approaches of transformation in representative pathogens of yeast, mold, and dimorphic species are reviewed.

Candida albicans

Transformation within the genus *Candida* was first developed in *C. guilliermondii* in 1985 (72), using *ARG4* as a selectable marker and spheroplast formation as the transformation method, which was developed earlier for *S. cerevisiae* (7, 56) and *N. crassa* (14). A year later, *C. albicans* was transformed using the *ADE2* gene as a selectable marker since auxotrophic, adenine-requiring, red-pigmented mutants used for classical genetic studies were available as the recipient strains (82, 115). The *C. albicans ADE2* gene was first cloned by complementing the *S. cerevisiae ade2* mutants with a *C. albicans* genomic library constructed in a shuttle vector, Yep13. The cloned *C. albicans ADE2* gene was then used to transform the corresponding *ade2* mutant of *C. albicans*. Transformed DNA was found to be integrated into the genome, which not only opened the way for molecular analysis by reverse genetics but also helped confirm the diploid status of the species suggested by classical genetic studies (115, 124, 151). The transformation protocol for *Candida* species was soon improved to obtain higher transformation frequencies by using less time-consuming procedures. The use of plasmids carrying an autonomously replicating sequence (CARS) resulted in a 200-fold increase in transformation frequency (74). The utility of the CARS-carrying plasmid, however, has been limited due to its unpredictable behavior. The successful application of both lithium acetate (149) and electroporation (141) protocols, methods developed for other organisms, provided alternate transformation methods that were less time-consuming and more efficient.

One of the most significant technical developments following transformation of *C. albicans* was targeted gene disruption as a means of direct mutagenesis (66). The disruption procedure was the same as that described for *S. cerevisiae* (121), which is based on the recombinogenic ability of free DNA ends with homologous sequences in the genome. *C. albicans* is an ascomycetous yeast in which homologous integration readily occurs. This tendency has been exploited to get around the problem of having to inactivate both homologs of a particular gene using the *URA3* gene. One *URA3* homolog was disrupted by introducing

a mutated *ura3* allele by transformation. Subsequently, the resulting heterozygous (*URA3/ura3*) strain was subjected to UV irradiation to induce mitotic recombination and segregation of a homozygous *ura3* strain (66). To circumvent the UV exposure step, which might introduce additional silent mutations and augment the lack of dominant selection markers in *C. albicans*, two sequential transformation methods, in which the mutant could be recycled and then transformed with the same marker, were developed (44, 51). Of these, the Ura-blaster cassette developed for repeated gene disruption in *S. cerevisiae* (2) and adapted for use in *C. albicans* by Fonzi and Irwin (44) became the standard method of gene disruption. The disruption construct contains the gene of interest (*GOI*) disrupted by the Ura-blaster cassette in which the *C. albicans URA3* gene (encoding OMP decarboxylase) is flanked by direct repeats of the *Salmonella enterica* serovar Typhimurium *hisG* sequence. A *ura3Δ* strain, most often CAI4 derived from a clinical strain (SC5314), is transformed with the disruption construct, and *URA3* colonies are selected. When the transforming DNA is integrated into the genome by allele replacement, the strain becomes *ura3Δ/ura3ΔGOI/goiΔ::hisG-URA3-hisG*. This heterozygote can undergo spontaneous recombination between the flanking direct repeats of the *hisG* sequence, which results in the excision of the *URA3* gene along with one copy of the *hisG* repeats. The other copy of the *hisG* sequence is left behind, keeping one allele of the *GOI* disrupted. The resulting Ura− clone can be selected on 5-fluoroorotic acid-containing medium and used for a second round of gene disruption with the same cassette. Numerous genes of pathogenic importance were characterized using this molecular tool. The drawback of this method is the limited utility of *ura3* laboratory strains (due to the uracil auxotrophy) and potential positional effects of *URA3* on virulence, since the Ura-blaster technique results in an ectopic placement of *URA3* within the locus of the gene of interest. Depending on the genomic position of the *URA3*, variable levels of OMP decarboxylase production have been demonstrated in vitro (10, 87). Survival curves of mice infected with various *HWP1/hwp1* or *hwp1/hwp1* strains with different locations of *URA3* confirmed that the location of the *URA3* selectable marker influences the virulence of *C. albicans* (138).

Molecular and genetic analysis of clinical isolates, which have prototrophic phenotypes, has been hampered by the lack of dominant selectable markers. *C. albicans* strains are innately resistant to the antibiotics commonly used as dominant selectable markers for other fungi, such as hygromycin, phleomycin, or benomyl (6). Moreover, noncanonical codon usage of *C. albicans* (CTG is decoded as serine and not leucine) probably affects the function of potential heterologous resistance genes and/or reporter genes unless codons are optimized for *C. albicans*. Such an example is seen with the green fluorescent protein of *Aequorea victoria*, which became functional in *C. albicans* only after CTG was changed to leucine codon TTG (18). Recently, *C. albicans* was found to be resistant to nourseothricin and the *SAT1* flipper cassette containing a nourseothricin resistance gene has been shown to be useful for targeted gene disruption (118).

Cryptococcus neoformans

Unlike ascomycetous yeasts, spheroplast or lithium acetate methods fail to give rise to *C. neoformans* transformants. *C. neoformans* was first transformed in serotype D strains (33) by an electroporation method first used to transfer genes into mouse lyoma cells (108) and later developed for the transformation of *S. cerevisiae* (5). Subsequently, a transformation system based on the biolistic method (3) was applied to *C. neoformans* serotype A strains (143) as well as serotype D and serotype B strains. Both systems yielded high transformation frequencies in serotype A and D, while the frequency was significantly lower in *C. gattii* (serotype B and C) (20, 145). When the transforming plasmids were linear, the transformation frequency was always higher (10- to 300-fold), regardless of the serotype (33, 146). The selectable markers most widely used are nutritional markers such as *URA5* (33) or *ADE2* (143), or dominant markers that confer resistance to aminoglycoside antibiotics such as hygromycin B (146) or nourseothricin (99). Circular plasmids not only transformed poorly but were always linearized and/or rearranged following entry into *C. neoformans* cells (33, 146).

DNA molecules delivered into the cell by electroporation are maintained as autonomously replicating linear plasmids with telomeres (33), as genomic integrants, or infrequently as a minichromosome that is readily detectable by pulsed-field gel electrophoresis (75). Genomic integration is mostly ectopic, and homologous integration is infrequent, especially in serotype D. Linear plasmid DNA in transformed cells was found to be stabilized by the addition of telomeric repeats at the termini (31). Once the termini of the episomes were protected from degradation by the addition of telomeric repeats, not only was further modification prevented but also transformation efficiency increased drastically (up to 90,000 transformants/μg of DNA). A library constructed in the plasmid CnTel, which contains telomeric sequences, has served as an efficient tool for cloning cryptococcal

genes. The first four capsule genes (*CAP59*, *CAP64*, *CAP10*, and *CAP60*) shown to be essential for virulence were cloned by complementation of acapsular mutants with the CnTel genomic library. Since the complementing plasmids remained episomal, these genes could be readily rescued in *E. coli* (15, 16).

Since homologous integration in serotype D is a rare event (1/1,000 to 1/10,000) after electroporative transformation, an enrichment strategy employing positive-negative selection using *ADE2* and *URA5* markers (15) was first applied for recovery of disrupted capsule genes. A double selection method using a similar concept was subsequently applied to *H. capsulatum*, another pathogenic fungus in which homologous integration is also rare (159). Transformation of serotype A *C. neoformans* by biolistics (143), however, has resulted in gene disruption at a frequency of between 2 and 50% (114). Significantly higher frequencies (1 to 4%) of gene disruption by homologous recombination were also achieved in serotype D strains by biolistics (22), indicating that homologous recombination frequency is dependent on the DNA delivery system rather than the serotype of the strains.

As an alternative method to gene disruption, RNA interference (RNAi), first developed to down regulate gene expression in *Caenorhabditis elegans* (42), was successfully used for targeted inactivation in *C. neoformans* (92). Unlike gene disruption, which depends on the homologous recombination machinery, RNAi can silence native genes by forming double-stranded RNA molecules that are homologous to the target. The double-stranded RNA molecules are cleaved into shorter species-specific 21- to 28-nucleotide small interfering RNA (siRNA) by the RNAse III-like Dicer enzyme. The siRNA then guides specific degradation of corresponding mRNA (reviewed in references 55 and 142). RNAi has been successfully used to inactivate several *C. neoformans* genes and offers some advantages over gene disruption. The requirement for complementation of a mutant allele in gene disruption to demonstrate specificity of phenotype for the target locus can be omitted in RNAi since a mutant phenotype will revert on loss of the RNAi plasmid. Other advantages include the need for only a short segment of the gene, and so identification of only the open reading frame is sufficient to make an interfering construct. Additional advantages include tandem interference with multiple genes and, perhaps most importantly, the ability to study essential genes. It appears to be a reliable and rapid method for genome-wide screening for functional analysis in *C. neoformans* since RNAi-inactivated strains have shown the same phenotype as the gene-deleted strains (92). The application of RNAi, however, may be limited to in vitro studies since the plasmids used are likely to be unstable in vivo. Other disadvantages include phenotypic heterogeneity among transformants resulting from different degrees of interference.

Filamentous fungi

In classical studies of pathobiology with filamentous fungi, parasexual genetics has been explored in *A. fumigatus* by using conidial pigment mutants (8). The molecular aspect of pigment synthesis and its significance in pathobiology was not understood until a transformation system was established for the species. The spheroplast method used to transform model fungi such as *N. crassa* (14) and *A. nidulans* (4) was adapted for transformation of the agents of aspergillosis (140), fusariosis, dermatomycosis, and mucormycosis during the mid-1980s to late 1990s (Table 1). Hygromycin resistance (116) or nutritional markers were used for selection of transformants. To reduce time and labor, as well as variability in efficiency of spheroplast formation, electroporation of germinating conidia was carried out in certain filamentous pathogens. Electroporation of germinating *A. fumigatus* conidia yielded a 2- to 10-fold increase in transformation frequency but with a lower frequency of homologous integration (reviewed in reference 24). Biolistic transformation has also been used in filamentous fungi. Transformation of *Rhizopus oryzae*, the most prevalent cause of zygomycosis, was first achieved by the biolistic method (C. D. Skory and R. J. Bothast, Soc. Ind. Microbiol. Annu. Meet., abst. P43, p. 86, 1998). Recently, *Agrobacterium tumefaciens*-mediated transformation (ATMT) has been exploited for use in several filamentous pathogens such as *A. fumigatus* (136), *A. niger* (23), and *Fusarium oxysporum* (106). ATMT was found to be superior to the conventional method of transformation of *A. fumigatus* (136), being less laborious with a higher yield of transformants and fewer multicopy insertions.

Gene disruption in filamentous pathogenic fungi follows the same principles as those used for unicellular fungi. For most pathogenic molds, the homologous integration rate using DNA-mediated transformation has been reasonably high, except for *R. oryzae*, where homologous integration frequency is lower than 10% (130). In *A. fumigatus*, *Agrobacterium*-mediated transformation resulted in a higher frequency (up to 60%) of gene replacement by double crossover (136). Gene inactivation by RNAi has also been tested with *A. fumigatus* and showed its applicability for genome-wide screens of gene function in this species (105).

Dimorphic (biphasic) pathogens

Pathogenic fungi with well-characterized dimorphism include *Blastomyces dermatitidis*, *Coccidioides immitis* (*C. posadasii*), *Exophiala dermatitidis* (*Wangiella dermatitidis*), *Histoplasma capsulatum*, *Paracoccidioides brasiliensis*, and *Sporothrix schenckii*. Only two of these species, *B. dermatitidis* (100) and *H. capsulatum* (76), are heterothallic, and classical genetic studies of the two species have been limited to the identification of the number of mating alleles (77, 80). Classical genetic studies of these two species were hampered by the instability of the meiotic life cycle in laboratory strains in addition to long incubation periods for the formation of meiotic spores. Transformation systems that enabled the conversion from classical to molecular studies have been established in all the dimorphic pathogens listed above except for *S. schenckii*.

P. brasiliensis is the latest of the dimorphic pathogens in which transformation has been developed (88). *B. dermatitidis* and *H. capsulatum* are both thermally dimorphic fungi with yeast and mycelial phases and were transformed by electroporation during the 1990s (57, 161). Transformation of *H. capsulatum* using the lithium acetate method yielded a substantially lower transformation frequency than did electroporation (162). Although *B. dermatitidis* and *H. capsulatum* are closely related and belong to the same teleomorphic genus, *Ajellomyces* (81), the two fungi differ in the number of nuclei in their yeast phase and the way they handle transforming DNA. *B. dermatitidis* yeast cells are multinucleate, and the transforming DNA integrates primarily ectopically into the chromosome with single or multiple insertion events (57). Yeast cells of *H. capsulatum*, on the other hand, are uninucleate, and the transforming DNA integrates randomly into the genome but also frequently exists as an episomal plasmid (158, 159). Hygromycin resistance (encoded by the *Escherichia coli hph* gene) has proven to be an efficient selectable marker for both species, and in *H. capsulatum*, a heterologous *URA5* gene (from *Podospora anserine*) also functions well as a selectable marker (161). The *H. capsulatum ura5* strain transformed with the *URA5* gene of *Podospora anserina*, an ascomycetous mold, resulted largely in random integration into the genome with extremely rare homologous integration.

Although the species are phylogenetically remote, there are several similarities between *C. neoformans* and *H. capsulatum* transformation besides the low rate of homologous integration. By electroporation, transforming plasmids are frequently maintained extrachromosomally and always found to have acquired telomeric repeats at both ends of linearized plasmids (158). As was the case in *C. neoformans*, maintenance of episomal telomeric plasmids by *H. capsulatum* offered a means of constructing shuttle vectors that can replicate both in *E. coli* and *H. capsulatum*. Furthermore, shuttle vectors with telomeric repeats increased the transformation frequency 10- to 100-fold compared to nontelomeric plasmids (159).

Recently, the system of ATMT was established for both *B. dermatitidis* and *H. capsulatum* (137). ATMT yielded a higher frequency of transformation with a substantially reduced frequency of multiple integration in both species. Unlike electroporation, transformants of *H. capsulatum* obtained by ATMT showed no evidence of episomal plasmids (137). Interestingly, progeny of *B. dermatitidis* transformants showed that a single round of colony growth under hygromycin selection generated homokaryotic yeast cells, in contrast to the requirement of several rounds of colony growth under selection pressure to generate homokaryons after electroporative transformation. These results indicate that ATMT is a superior molecular biology-based tool for cloning and analysis of insertion alleles compared to electroporative transformation.

C. immitis (*C. posadasii*), unlike *B. dermatitidis* or *H. capsulatum*, is a dimorphic fungus with mycelium and spherule phases. The species was first transformed in 1999 by the biolistic method, using the *hph* gene as the selection marker (166). The frequency of transformation was extremely low (one or two transformants per microgram of DNA), and the transforming DNA was integrated into the chromosome as tandem repeats. The transformation frequency of *C. immitis* increased to 30 to 50 transformants/10^7 arthroconidia when the spheroplast method was used (117). Further improvement in transformation frequency was obtained with ATMT, in which 1 in 10^5 arthroconidia were found to be transformed (1).

Gene disruption in the above dimorphic pathogens has been a challenging task since the integration of incoming DNA is primarily ectopic. Gene disruption in *B. dermatitidis* was first achieved with allelic replacement of a virulence gene WI-1, which encodes an adhesin (11, 71). The disruption frequency ranged from 1 to 2% when the disruption construct had two features: a long stretch of *hph* flanking sequence homologous to the 5′ (and 3′) region of the WI-1 open reading frame and a short stretch of WI-1 minipromoter that served to express *hph*. The pattern of crossing over at the 5′ end suggested the effect of "promoter trapping," where the expression of selection genes is dependent on transcriptional start signals from *B. dermatitidis* DNA (11).

Gene disruption in *H. capsulatum* was accomplished by an enrichment strategy using the positive (*hph*)-negative (*URA5*) selection procedure. The relative frequency of targeted *URA5* disruption was approximately 1 in 10^3 transformants (160). The homologous integration rate was drastically increased, however, when the positive-negative selection procedure

was carried out in two steps. A telomeric plasmid containing *URA5* at the distal end next to the telomeric sequence and the gene of interest disrupted with *hph* was transformed into the *ura5* mutant of *H. capsulatum* and (i) selected for prototrophy, promoting maintenance of the telomeric plasmid (this step presumably enhances the chance of a desired double crossover) and (ii) selected for homologous recombinants by using positive-negative selection with hygromycin and 5-fluoroorotic acid. This method enriches the recombinant clones derived from the double crossover between the disruption construct and the gene of interest. One-third of clones grown on 5-fluoroorotic acid medium containing hygromycin were found to be the desired homologous recombinants (128). The drawbacks of this procedure are the time requirement of weeks to even months to enrich the desired recombinant population.

More recently, RNAi has been applied to gene inactivation in *H. capsulatum* as an alternative to allelic replacement. RNAi inactivation of the *H. capsulatum* AGS1 gene encoding α-(1,3)-glucan resulted in the same phenotype in vitro and in vivo with the *ags1*Δ strain.

Gene disruption in *C. immitis* was first accomplished with the *CTS1* gene encoding chitinase 1 (117). The gene-targeting strategy included no unique features, except that the transformation was carried out by spheroplasting, which yielded a higher frequency of transformation compared to electroporation.

Molecular Biology-Based Tools for Mutational Analysis in the Genomic Era

In classical studies, the genes that are associated with virulence were identified by forward genetics: mutants with recognizable phenotype were isolated, and the mutant was than crossed with the wild-type strain of opposite mating type to analyze the segregation pattern of mutant phenotype and virulence. In fungi, where meiotic analysis is not feasible, revertants were sought in order to observe coreversion between the mutant phenotype and virulence. As molecular manipulation became possible with the development of DNA-mediated transformation, directed mutagenesis approaches (reverse genetic approach) have been used to identify virulence genes. Genes of interest were cloned, and disruption constructs were made and transformed into wild-type strains to isolate mutants resulting from homologous integration. This approach is disadvantageous since it is a biased search for genes associated with pathogenesis and does not permit large-scale screening in a timely manner. Recently, forward genetic approaches have taken place based on insertional mutagenesis. Insertional mutagenesis is a powerful tool for identification of genes and their

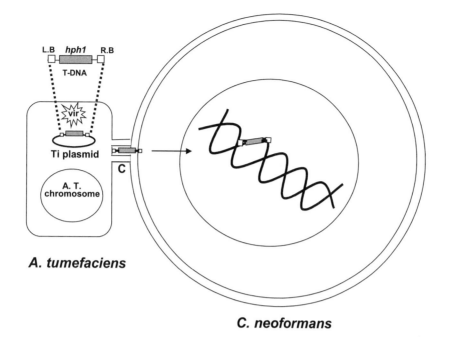

A. tumefaciens

C. neoformans

Figure 2. ATMT. Transformation is mediated by selection on hygromycin. The hygromycin B phosphotransferase gene is transferred by *A. tumafaciens* to the *C. neoformans* host cell, where it integrates into the genome. L.B, left border; *hph*, hygromycin B phosphotransferase gene; R.B., right border, T-DNA, transfer DNA present in Ti plasmid; Vir, virulence proteins that lead to production of the single-stranded T-DNA molecule as well as formation of the channel; C, channel through which T-DNA is exported into host cells (*C. neoformans*).

functions on a large scale. This approach permits genome-wide screening of pools of insertional mutants with tags that serve for the quick identification of the mutated region. Based on insertional mutagenesis of various techniques, many genes related to pathogenicity have been identified in *Candida* species (19, 145), *A. fumigatus* (136), and *C. neoformans* (35, 60, 107).

Of the various insertional mutagenesis techniques used, such as signature-tagged mutagenesis, restriction enzyme-mediated integration, and ATMT, ATMT (Fig. 2) appears to yield the highest frequency of simple insertional mutations (60, 136). *A. tumefaciens* was routinely used for the genetic modification of a wide range of plant species (148) before it was found to be useful for insertional mutagenesis in yeast (12) and filamentous fungi (23). In *A. fumigatus*, single copy insertion of T-DNA occurred in 95% of transformants with the *hph* gene as a selectable marker. ATMT was also found to be an efficient tool for targeted disruption in *A. fumigatus* (136). In *C. neoformans*, ATMT yielded a transformation frequency comparable to that of electroporation in serotypes A and D but yielded at least a 10-fold higher frequency in strains of serotypes B and C. A total of 80% to 90% of transformants contained ectopically inserted T-DNA, and, unlike the case for electroporative transformation, no episomal plasmid was detected (98a).

FUNGAL DIAGNOSIS AND EPIDEMIOLOGY: CLASSICAL VERSUS MOLECULAR APPROACHES APPLIED TO THE IDENTIFICATION OF FUNGAL PATHOGENS

Of the 72,000 different species of fungi that have been identified, more than 300 are known to be pathogenic for humans (17). One of the most important tools for treating fungal infections is the ability to diagnose them faster, more accurately, and with more sensitive assays. A major factor in proper diagnosis is the number of fungi that have to be considered as the causative agent of a particular infection. Unfortunately, classical diagnostic methods rely heavily on morphology, staining properties, and/or growth characteristics under different conditions. These strategies are not sufficient to meet the increased spectrum and frequency of fungal infections encountered in modern clinical laboratories.

Current Methodology

Current fungal identification methods can be divided into two broad categories, culture and nonculture. Culture-based identification of fungi can be done using a number of methods; however, all methods have the common characteristic of relying on whole organisms for identification. Organisms can be identified biochemically or morphologically after growth on the appropriate media or histologically from tissue specimens. Biochemical identification, through either carbohydrate assimilation patterns (API test strips, etc.) or chromogenic properties on primary differential media (Chromagar, etc.), may be the easiest and most frequently employed identification method in the clinical laboratory. Other types of biochemical-based assays, such as the Vitek system (Vitek, Inc.), ID32 C strip (bioMérieux, Inc.), or Biolog YT MicroPlate (Biolog, Inc.), have even been automated for higher throughput and rely on a battery of assays, using assimilation, that are coded based on a growth (+) or no-growth (−) output. Biochemical identification and other assays (i.e., cellular fatty acid analysis) are usually only applicable to yeasts; consequently, culture-based identification of the majority of filamentous fungi must still be done by morphological examination or nonculture methods (45, 104, 111).

By and large, culture-based identification methods have been the cornerstone of diagnostic mycology and arguably are the standards by which all other identification methods are measured. Unfortunately, the major drawbacks of culture-based methods are low sensitivity, variable species-specific growth rates, and requirement of mycological expertise (needed mainly for mold identification). To compensate for these problems, alternate nonculture methods of identification have been pursued over the years.

Non-culture-based methods can loosely be placed into different categories, which would include imaging, serodiagnostic assays, metabolite detection, and molecular identification. The major advantage of nonculture methods is that evidence of infection does not require isolation or visual demonstration of whole organisms. Consequently, these methods can be less invasive, faster, and more sensitive. They also have the advantage of not requiring specific mycological expertise, in addition to being useful for monitoring the course of infection to determine whether it responds to treatment. Computed tomography, for example, has become an extremely valuable tool for diagnosing invasive aspergillosis in suspected patients because of the hallmark "halo sign." However, the halo sign also appears in mucormycosis, tumor cell infiltration, pulmonary candidiasis, granulomatous lung diseases, and tuberculoma (34). Fungal metabolites were first identified as infection markers almost 20 years ago and were of interest because they were not detectable in the uninfected host (68). They have been associated with Candida (D-arabinitol), C. neoformans (mannitol), and Aspergillus infections (mannitol) (120, 156, 157). Part of the attraction of using metabolites as markers is that some studies have shown that the amount is somewhat proportional to fungal load (155). However, analytical and artifactual problems have hindered the development of metabolites as valid markers of infection. Some

of the serodiagnostic assays, on the other hand, are widely employed due to their ease of use. Perhaps the best example is the exoantigen test used for the identification of Histoplasma spp. (134). The utility of the assay is that it is a rapid method for diagnosing a potentially slow-growing fungus, which can display morphologic characteristics similar to other fungi (28, 134). Continued research in this area has identified some promising assays. For example, very recent modifications of Aspergillus detection by using a double-direct-sandwich enzyme-linked immunosorbent assay technique (Platelia Aspergillus) have made the assay one of the most sensitive for fungal galactomannan detection (135). This strategy has also been applied with some success for Candida detection (129). Among the simplest and fastest assays are the latex agglutination tests. A number of commercially available assays are presently in use, including kits for the identification of C. neoformans, H. capsulatum, C. immitis, and S. schenckii. However, potential problems with serodiagnostic techniques that need to be studied further include, but are not limited to, cross reactivity (36, 164), decreased sensitivity in debilitated patients (98), transient antigen presence (25), and high false-positive frequency (129). Serodiagnosis, imaging, and metabolite detection, although promising in some areas, all suffer from a common drawback: they are narrow-spectrum techniques that are directed largely to a few different, albeit important mycoses. Molecular methods, in contrast, have the potential to be easily adaptable to a broad spectrum of organisms, which can be identified from a wide range of specimens recovered from a wide variety of patient conditions.

Molecular Biology-Based Diagnostic Assays

There are a number of different approaches to the molecular identification of fungi. Some of these approaches require an a priori choice of assay based on suspicion of the etiologic agent, which is a drawback associated with some of the non-molecular biology-based methods. However, most molecular methods offer a number of advantages over traditional methods of fungal diagnosis. In many assays, the results are unambiguous and not subject to interpretation. Importantly, for some techniques no prior information regarding the suspected identity of the organism is required before running the test. Additionally, in contrast to many of the other techniques, the sensitivity of some of the molecular techniques can be increased prior to analysis.

Nucleic acid-based molecular assays can be divided into two classes: PCR based and non-PCR. The major PCR-based assays for fungal diagnosis consist of three different strategies, each of which has a standard *Taq* polymerase amplification of template DNA as a central component. In its simplest form, traditional PCR has been employed using primers designed to specific sequences to yield a product that usually displays a predicted size based on extensive pilot testing. Discrimination can be through the presence or absence of a band, bands that differ in size or number, or digestion of PCR products to yield multiple bands. Amplification products are usually just visualized on an ethidium bromide gel, which contains an appropriately sized marker. The advantages of this approach are its simplicity and sensitivity and the ability of the reaction to easily distinguish and amplify target sequences from mixed specimens. In fact, an extension of these characteristics is the uncomplicated sample preparation—in some cases, template can be prepared simply by boiling cells or even just by adding whole cells to a PCR mixture directly. The single most important requirement for employing this strategy as a diagnostic tool is the need to have the sequence of the target site in hand before primers can be accurately designed. Consequently, the strategy has little or no use for specimens that contain DNA from fungi for which no sequence information is available. A second potential drawback is that for most applications, only a single primer pair can be used for each reaction. This drawback obviously limits the number of PCR to the number of templates that can be recovered from each specimen. In the case of live cultures, this limitation is not a concern; however, for specimens in which viable cells are not present, only a finite number of reactions can be performed. Additional problems include an inability to accurately and easily quantify fungal burdens as well as the danger of amplifying other DNA that contaminates the specimen. The desire to more accurately quantify starting material led to the development of real-time PCR (RT-PCR).

RT-PCR uses fluorogenic probes with reporter and quencher dyes in various strategies to extend the power of traditional PCR. Because reaction mixtures contain both a reporter and a quencher, amplifications can be visualized live, in real time, using an integrated system to both detect and quantitate the presence of DNA. Cycle times are much shorter because amplicons are much smaller, by design, than traditional PCR amplicons. Additionally, results of the amplification are available earlier because they can be visualized with each cycle instead of relying on a single end point.

This ability to observe the entire amplification profile greatly expands the quantification range. Integration of the system with a computer eliminates the need to physically separate products on a gel, and because of the microtiter-based format, the system is easily adapted to a high throughput that can handle hundreds of samples. Automation of plate loading combined with additional instrumentation can further increase the throughput. The major advantage of RT-PCR, however, is its quantitative nature. This advantage translates into the ability to estimate fungal loads from individual specimens or the ability to monitor infections over time, which can be useful for early detection or monitoring responses to treatment. Application of the strategy for estimating fungal burdens is particularly important since the standard for quantifying tissue loads has been colony counts. While useful, estimating fungal burdens by using colony counts can be problematic for filamentous fungi due to decreased viability caused by hyphal fracture during specimen preparation. RT-PCR can overcome this problem by quantifying the nuclei present in a given specimen instead of relying on outgrowth of organisms from an aliquot of specimen. This approach also compensates for sampling errors caused by nonviable or nongrowing cells or by isolates that grow in clumps. Importantly, nonviable cells are encountered during the analysis of pathology specimens, which are normally fixed. Unfortunately, the fixation process leads to cross-linking of DNA, which makes the use of amplification products longer than a few hundred base pairs extremely difficult. Since RT-PCR utilizes short amplicons (in the 100-bp range), it is well suited for histological specimens. When combined with other techniques such as laser capture microdissection, RT-PCR becomes an extremely powerful diagnostic strategy.

The quantitative nature of the technology easily lends itself to the study of gene expression levels, which is particularly useful for genes that may play a role in virulence or that are known targets of specific inhibitors. This capability, combined with the increasing amount of molecular data from a wide variety of fungi, has enabled the technique to become highly integrated into mycology research laboratories. The major disadvantages of RT-PCR are high costs. There is also a requirement for advanced skills during data interpretation, which can put the technology out of reach for basic clinical mycology laboratories. Lastly, because of the fluorogenic nature of the reaction, primer costs can be high and, in cases where they will only be used a few times at most, may be hard to justify. Nonetheless, because of its power, new applications are continually being found for this technology in mycology.

Both standard PCR and RT-PCR require that something be known about the organism (specifically, the target sequence) prior to development of a diagnostic system. Consequently, these techniques have reduced applicability for the diagnosis of unknown organisms based solely on the presence or absence of the amplification product. This requirement is particularly important for RT-PCR because the sequence of the probe-binding site must be known for proper binding. There are, however, PCR applications where only a partial sequence (primer-binding sites) of the region to be amplified needs to be known. If the priming sites are conserved and well designed and the distance between them is not too large (precluding amplification), useful diagnostic assays can still be employed. One such assay is sequence analysis of selected regions within the genes encoding the ribosomal DNA (rDNA) complex.

The rDNA complex contains the 18S and 26S rDNA sequences, which are separated by two internal transcribed spacer regions, designated ITS1 and ITS2, and an intervening 5.8S coding region that lies between ITS1 and ITS2 (see http://www.biology. duke.edu/fungi/mycolab/primers.htm and reference 61 for reviews). These three genes are flanked on one side by the gene encoding the 5S rDNA, which lies between two nontranscribed spacer regions (IGS1 and IGS1) and is transcribed independently from the 18S, 26S, and 5.8S genes. The entire set is repeated many times in fungi (Fig. 3). The organization of the rDNA cluster provides a number of useful characteristics for fungal identification. The ribosomal genes normally are highly conserved; consequently, sequence from within specific genes generally does not display enough variation to distinguish fungi at the species level. The D1-D2 region within the 26S subunit, however, is sufficiently variable to allow for species identification in most cases. Even more variable is the region separating the 18S and 26S rDNA genes, which encompasses the ITS1-5.8S-ITS2 region. This span contains the variable ITS1 and ITS2 regions, as well as the conserved 5.8S gene. The relatively short length of this sequence eliminates the need to skip the uninformative 5.8S sequence since the entire length is usually ~600 to 700 bp. This length is extremely convenient since sequencing from both ends generally falls within the range of highly accurate data.

Ribosomal sequencing as a method for fungal identification may represent the major transition point from classical mycology to molecular mycology in the area of fungal identification and taxonomy. The utility of the ITS1-5.8S-ITS2 sequence for identification purposes derives from a number of factors. Because the rDNA gene cluster contains both highly conserved regions (the ribosomal genes) interspersed with variable to highly variable regions (spacer/intergenic regions), it is much easier to obtain sequence from unknown fungi than when sequencing of conserved structural genes, such as the actin gene, is used. Second, the conserved ribosomal genes offer universal priming sites that can be used to PCR across previously uncharacterized regions (47, 152). As a result, rDNA sequences represent the most frequent GenBank entries for fungi, with the number of entries in the tens of thousands. In fact, for many fungi, rDNA sequence represents the sole GenBank entry. This situation largely reflects the utility of rDNA sequences for taxonomic classification. The highly conserved regions of the rDNA genes provide near-universal priming sites, greatly simplifying primer selection for unknown organisms. The absence of a requirement to know everything about the diagnostic target (in this case the complete sequence) is arguably the greatest asset of rDNA sequencing over most diagnostic methods. Finally, the repetitive nature of the ribosomal genes, often approaching 100 copies per cell, adds an additional layer of sensitivity to the PCR-based detection method, which is generally unrivaled for other PCR targets that could be used for detection and diagnostic purposes.

Variations of this approach have focused on simplifying the strategy such that identification can be performed using only basic laboratory skills. For example, a rapid nonisotopic hybrid protection assay (Accuprobe; Gen-Probe, Inc., San Diego, Calif.) that utilizes a chemiluminescent acridinium ester DNA probe directed to the rDNA region has proven to be highly specific and effective (54, 59, 112). Another variation includes instrumentation based on flow cytometry (Luminex 100 IS system; Luminex Corp., Austin, Tex.) that captures and identifies microspheres coupled to species-specific probes in a multiplex format

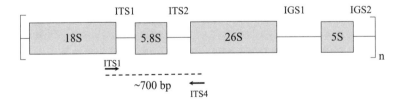

Figure 3. rDNA gene cluster. Ribosomal subunits are shown as a single cluster within brackets. The four subunits are identified (18S, 5.8S, 26S, and 5.8S) as boxes. The intervening regions are ITS1 and ITS2 (internal transcribed spacer) and IGS1 and IGS2 (intergenic region). Primers for amplifying the ITS1-ITS2 region are shown by arrows (ITS1 and ITS4), with the general size of the amplicon being ~700 bp.

(26). The significance of these approaches is that they couple classical mycology, molecular mycology, and instrumentation in a format that potentially fits accurate fungal identification into the day-to-day requirements of a clinical laboratory. The combination of more molecular data, increasingly sophisticated instrumentation, and novel strategies arising from traditional basic research argues for future improvements in fungal diagnosis and identification that should move these fields out of the realm of mycology into the more generalized fields of clinical microbiology and medical technology.

Molecular Epidemiology

The transition of medical mycology into the molecular era has probably been more important for fungal epidemiology than for any other area inside the field of clinical mycology. While the molecular era greatly improved diagnostic mycology, it essentially created the field of fungal epidemiology. Previous methods of determining whether two strains were related relied on extensions of many of the techniques used for diagnostic mycology, with a goal of making these types of assay more sensitive and thus enhancing their discriminatory power to the level of intraspecies distinction. These methods relied largely on a battery of phenotypic methods that allowed a comparison of numerous outputs from these assays. Some of the major methods included morphologic typing, serotyping, antibiogram typing, resistogram typing, biotyping, killer toxin typing, and isozyme analysis (for a review, see reference 102), although most were geared toward discriminating within the pathogenic species of *Candida*. It is worth noting that virtually all of the most important fungal molecular typing methods were adapted from techniques employed for other organisms. One exception is electrophoretic karyotyping. This technique has some important applications in mycology which even include strain typing; however, because of the need for specialized expensive equipment, laborious nature, long assay times, and limited sample analysis, it is not a mainstream technique. Consequently, most of the major applications of karyotyping involve basic research. The following techniques are included because of their widespread use at one time or another in fungal epidemiology.

Restriction fragment length polymorphism

The impact of molecular biology on fungal epidemiology first appeared with the development of restriction fragment length polymorphism (RFLP) detection, which proved successful in discriminating isolates of *Candida* spp. (95). This technique is among the simplest of all methods for the molecular analysis of strain relatedness. The technique works by preparing high-molecular-weight, total DNA from a set of isolates and then digesting these DNAs with a restriction enzyme. The digests are then run on a gel, which typically yields a smeared pattern after ethidium bromide staining. However, in areas typically representative of repetitive DNA, a large number of fragments appear as an intense band against a background of diffusely stained DNA (Fig. 4). The discriminatory power is assisted by increasing the numbers of bands and is dependent on the nature of the repetitive DNA. As a consequence, the more variable the sequence recognized by the restriction enzyme, the more powerful the technique. The limits are, of course, the ability to visualize the bands on a gel, which is affected by DNA quality and the variability of the target sequence. The strength of this technique is the simplicity and the necessity to know very little about the target site that the enzyme cleaves. In fact, the best enzyme can be determined by randomly picking frequently used enzymes and testing each one individually. RFLPs are rarely used for epidemiological studies since other techniques have been developed that have proven to be more sensitive. However, RFLPs provided the foundation for what arguably is one of the gold standards of fungal typing methods, DNA fingerprinting using a specific hybridization probe.

DNA fingerprinting

As molecular biology-based techniques began to be increasingly applied to fungal epidemiology, the

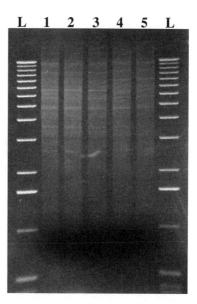

Figure 4. RFLP. Genomic DNA from five isolates (lanes 1 to 5) of *C. albicans* digested with EcoRI is shown. DNA appears as a smear except when the restriction enzyme yields fragments identical in size. These fragments appear as more intensely staining bands on the gel against the background of sheared DNA because they migrate to the same place on the gel. The number and pattern can be used to discriminate strains, depending on the enzyme used for digestion. L, size marker (1-kb ladder).

development of DNA probes that could be useful for DNA fingerprinting probably did more than any other technique to advance the field of fungal epidemiology. While "fingerprinting" is a term that is applied to a number of related yet distinct methods for discriminating different banding patterns of DNA and even RNA, depending on the assay, in this case we are using the term to describe the hybridization of a probe to Southern blots of restriction enzyme-digested DNA. In this type of assay, the key component is the probe target. It is almost always a repeated sequence that is dispersed throughout the genome. Sources of these probes have included repeated sequences isolated by probing plaque lifts from a genomic library with whole DNA (131), transposon sequences identified by probing phage with fungal genomic DNA (109), and functional sequences, such as telomeres, known to be present in higher copy number (31), to name a few. Because the target is dispersed, restriction digestion with the appropriate enzyme releases target-containing fragments of different sizes. As the number of copies of the target increases, the number of informative bands also increases. Finally, although it is important for the target sequence to be both numerous and variable, it should be somewhat stable, such that in a clonal lineage the patterns should be nearly identical.

A number of effective probes have developed for some of the most important human fungal pathogens. They include Ca3, 27A, and CARE-1 for *C. albicans* (86, 125, 133), CNRE-1, Cntel, and UT-4p for *C. neoformans* (30, 46, 133), and *Afut1*, 3.11, and 3.19 for *A. fumigatus* (48, 109). In addition to these fungi, fingerprinting probes exist for a number of other organisms, although the majority are primarily *Candida* spp. Reviews of these probes and how they were developed, as well fungi that can be fingerprinted, were recently published by Lockhart et al. and Soll (93, 131).

One of the major attractions of using fingerprinting probes as a means of typing strains is that the outputs are largely qualitative. After hybridization, data are interpreted from autoradiographs; depending on the band pattern, this allows investigators to make qualitative judgments about the relatedness of two strains based solely on visual inspection. The widespread use of this strategy has led to more sophistication in data interpretation, including the development of software for data analysis (127). However, the ease of interpreting results still makes this technique one of the most popular typing systems in use today.

Random amplification of polymorphic DNA

When it comes to ease of use, there is probably no more convenient and easy method than random amplification of polymorphic DNA (RAPD) analysis

as an epidemiologic tool for determining whether two isolates are related. The technique is based on PCR amplification of genomic DNA (150, 153) but differs from standard PCR in that nothing needs to be known about the target primer-annealing sequence. Among the earliest applications of RAPD analysis for human fungal pathogens were the discrimination of isolates of *H. capsulatum*, *A. fumigatus*, and *C. albicans* in 1992 (67, 89). The discriminating aspect of the technique relies on multiple binding sites of the primer (typically one, but as many as three primers in a reaction), which are close enough together to permit amplification (usually 3.0 kb). Primer design is typically done arbitrarily and in fact can be as simple as using a single M13 primer. The primers are shorter than standard PCR primers (9- to 10-mers), which increases the probability of finding multiple priming sites, and hence primers have a relatively low annealing temperature. Amplification products are visualized after separation on an agarose gel and staining with ethidium bromide (Fig. 5).

RAPD analysis for fungal epidemiology is severely criticized for its poor reproducibility and wide variation among laboratories. In fact, virtually

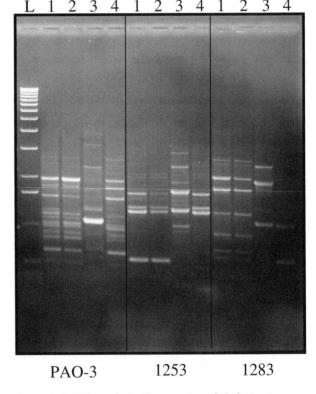

Figure 5. RAPD analysis. Four strains of *C. lusitaniae* were analyzed by RAPD analysis. Strains 1 and 2 are clinical isolates, and strains 3 and 4 are unrelated controls. RAPD primers are listed under each panel and are from references 69 and 163. RAPD patterns suggest that strains 1 and 2 are related.

any variable that is associated with PCR, such as the *Taq* polymerase source, the thermocycler, or the DNA quality, can alter the pattern. However, for ease of use, there are few techniques that compare to RAPD. Since DNA from crude preparations is generally sufficient as template, this allows even physical DNA methods to be used for DNA preparation and eliminates the need for laborious spheroplasting protocols. High-molecular-weight template DNA is also an option. These factors combine to make RAPD analysis a preferred epidemiological method when little is known about the strains of interest or when a short turnaround time is needed. While it may not yield useful data for all species of fungi, RAPD analysis is a starting point for the epidemiological analysis of fungal outbreaks due to virtually any species.

Microsatellite typing

As more is being learned about the sequences of specific genes and genomes of medically important fungi, it is becoming possible to study the sequences in more detail, including comparing multiple loci of multiple strains. Microsatellite typing is increasingly common for studying epidemiologic relationships of fungal pathogens. Microsatellites, or simple sequence repeats, are tandemly repeated stretches of one or more nucleotides (usually one to six) that are distinguished based on repeat number (90, 126). Simple sequence repeats are dispersed throughout the genome and are highly polymorphic, which makes them an excellent typing tool (27). Microsatellite typing requires sequence knowledge of the organism in order to design the primer pair flanking the locus. However, data are portable, highly reproducible among laboratories, quantitative, and easily amenable to high-throughput analysis.

Microsatellite markers are analyzed by amplifying specific loci with fluorescent primers and then separating the products with a capillary sequencer, which discriminates the amplified loci based on the sizes of each amplified fragment. Size resolution is typically within a base. Variations of the general strategy exist where single primers designed with various combinations of specific microsatellite repeats, i.e., $(GTG)_5$ and $(GACA)_4$ (103), are informative when amplified under RAPD-like conditions. Since the major drawback of microsatellite typing is the need to determine the flanking sequences to design primers, time and effort must be invested into each organism of interest to identify the best markers. However, with increasing sequence data available for medically important fungi, including whole genomes, microsatellites can be identified by using bioinformatics to identify candidate markers. Consequently, this typing method may become predominant for many fungi in the future.

Multilocus sequence typing

Multilocus sequence typing (MLST) was first described by Maiden et al., who used it to type *Neisseria meningitidis* strains (96, 97). The technique is based on multilocus enzyme electrophoresis (MLEE); however, allele numbers are assigned directly and unambiguously instead of indirectly as is done for MLEE, using gel migration distance. MLST utilizes the sequences of six to eight housekeeping genes by assigning an allele number to each sequence, with each nucleotide difference generating a new allele number for that specific gene. Numbers are unweighted, regardless of the source of allele difference (i.e., point mutation versus deletion). Therefore, each isolate of a species will have a specific allele designation for each gene. The distinct alleles form a profile or sequence type of the specific isolate. Housekeeping genes are used for sequence analysis because each isolate of a species is likely to possess all the genes used for typing due to the essential nature of this class of genes, yet variability in the sequence allows differences to be detected. Since housekeeping genes are conserved, universal primers can be used for each gene, and by using 450 to 500 bp of sequence and seven genes, there is the potential for massive discriminatory power.

Statistical analysis is performed using the eBURST algorithm (39, 40), which differs from other clustering algorithms by allowing the analysis of larger samples and identification of the ancestral strains that give rise to a particular outbreak cluster. The technique is especially useful for identifying lineages containing hypervirulent strains. Since sequencing data from housekeeping genes are unambiguous regardless of laboratory, the data can be maintained in a central database for direct comparison of new isolates. A database currently exists at http://www.mlst.net/ for the analysis of numerous species of bacterial pathogens but has recently started adding sequence data for fungal pathogens (*C. albicans* and *C. glabrata*). Since the technique is PCR driven, it is possible to analyze small specimens or even nonculturable isolates, and the sequence-driven data are especially amenable to high throughput using automated capillary sequencers.

CONCLUSION

We have reviewed the pioneering works that contributed to the conversion from classical studies of medically important fungi to the molecular era by focusing primarily on five systemic pathogens. These major pathogens, however, are only a fraction of the 150 species that are regularly involved in human and animal mycoses (81), in addition to a few hundred more species of opportunistic pathogens that are occasionally detected in the expanding populations of immunocompromised patients. Just as model, nonpathogenic fungi formed the basis for understanding the major systemic pathogens, there is a large enough body of information on the major fungal pathogens to allow researchers to expand their studies into fungi that are not as frequently encountered. Consequently, the field stands to greatly increase its depth with regard to what is known about the common pathogens, as well as to expand the breadth about what is known about even the rarely encountered pathogens.

In the 1950s, medical mycology was a small field that was concerned primarily with fungal taxonomy (which was based mainly on the phenotypic species concept), environmental reservoirs of the well-known pathogens, infections caused by unusual fungi, and development of laboratory criteria for the diagnosis of fungal diseases. When the epidemic of fungal infections began in the mid-1960s, almost no information was available on the epidemiology (except for skin testing results of *H. capsulatum*), genetics, and pathogenesis of medically relevant species, especially for those that caused systemic infections. During the late 1960s to mid-1970s, avenues to classical genetic studies were opened for three important systemic pathogens: *B. dermatitidis*, *H. capsulatum*, and *C. neoformans*. Classical genetic studies of the pathobiology of these species, however, made progress only with *C. neoformans* because recombinational analysis of the other two fungi was impractical. Classical genetic approaches using parasexual cycles allowed limited progress with *A. fumigatus* and later with *C. albicans*.

The conversion from classical studies of pathogenic fungi to molecular studies had to await the development of molecular biological techniques and reagents that were being used for studies of model fungi such as *S. cerevisiae*, *A. nidulans*, and *N. crassa* during the late 1970s to mid-1980s. Molecular biology-based technologies developed for these model fungi were quickly adapted for the study of pathogenic fungi, starting with *C. albicans* in the mid-1980s. Soon thereafter, the fulfillment of Koch's molecular postulates for fungal virulence factors was first accomplished with the capsule genes in *C. neoformans* at the beginning of the 1990s. Mutant strains isolated by classical studies served as the springboard for transformation and cloning by complementation in these fungi. In addition to the availability of technology, the influx of young investigators trained in the molecular biology of model fungi into the field of pathogenic mycology during the 1990s hastened this conversion. This influx was especially noteworthy in the field of pathogenic yeasts, where many people migrated from *S. cerevisiae* research. The progress in molecular pathogenic mycology attained the level of model fungi by the late 1990s to early 2000. Much of the information

about virulence-associated traits found by classical studies or classification of pathogens based on the phenotypic species concept has been confirmed by molecular studies. Since phenotype is determined by the genes, it is not surprising how accurate classical mycology has been in diagnosis as well as identification of virulence traits. However, in spite of the talent and information available from the study of model organisms, two other distinct but related factors probably did more than anything else to drive the field of medical mycology into the molecular era: AIDS and the biotechnology industry.

The AIDS epidemic predisposed patients to a tremendously broad spectrum of infections from increasingly diverse species of fungi, which occurred with unprecedented frequency. These infections revealed a woefully inadequate capability for managing AIDS patients and other immunosuppressed patients due to the lack of effective antifungal drugs. Consequently, industry responded with massive interest that culminated in the entry of the highly successful antifungal, fluconazole, into the marketplace. As a result, AIDS and industry arguably can be considered to have provided the final push of medical mycology into the mainstream of pathogenic microbiology. The recently recognized importance of fungal pathogens clearly primed the field for other advances, including leading-edge research such as genome-sequencing projects.

Genomic sequencing of all the major human fungal pathogens is either being completed or close to completion (Table 2). In the near future, genome sequences of all the important species from every phylum of the Kingdom Fungi will become available, due in large part to the Broad Institute fungal genome initiative. Currently, construction of deletion libraries is under way, which will allow functional genomics for at least several pathogenic yeasts: *C. albicans*, *C. glabrata*, and *C. neoformans*. At the same time, insertional libraries, which offer a means of forward genetic analysis, are also being developed. These approaches, combined with technology derived from genome sequences, including transcriptome analysis via microarrays, proteomics, and genome interrogation using bioinformatics, promise to accelerate what is known about pathogenic fungi and the fungal kingdom in general. Predictably, the basic research enabled by the molecular era has reached into the most important area of clinical mycology: diagnosis.

The diagnosis of mycotic agents no longer depends solely on phenotypic criteria since a variety of methods are presently available and new ones are becoming available. The availability of species-specific DNA sequences allows rapid, accurate identification of many pathogens, especially those that require long

Table 2. Genome sequence status of human fungal pathogens

Fungus	Site(s)[a]	NCBI BLAST[a]
C. neoformans serotype D	TIGR Stanford Univ.	Yes
C. neoformans serotype A	Duke Univ. MIT	Yes
C. neoformans serotype B	Univ. of British Columbia	No
P. carinii	Univ. of Kentucky Univ. of Cincinnati Univ. of Georgia	No
C. albicans	Stanford	Yes
C. glabrata	Pasteur Institute	Yes
H. capsulatum	Washington Univ.	No
C. immitis	MIT	Yes
A. fumigatus	TIGR Sanger Institute	Yes

[a]Web address for BLAST searches of the deposited fungal databases at NCBI: http://www.ncbi.nlm.nih.gov/sutils/genom_tree.cgi. Web addresses for the sequencing sites: TIGR (The Institute for Genomic Research), http://www.tigr.org/tdb/e2k1/cna1/ and http://www.tigr.org/tdb/e2k1/afu1/; Duke University, http://cneo.genetics.duke.edu/; Massachusetts Institute of Technology (MIT), http://www.broad.mit.edu/annotation/fungi/cryptococcus_neoformans/ and http://www.broad.mit.edu/annotation/fungi/coccidioides_immitis/; University of British Columbia, http://www.bcgsc.ca/gc/cryptococcus/; Washington University, http://www.genome.wustl.edu/projects/hcapsulatum/; Stanford University, http://www-sequence.stanford.edu/group/C.neoformans/ and http://www-sequence.stanford.edu/group/candida/search.html; Sanger Institute http://www.sanger.ac.uk/Projects/A_fumigatus/; Pasteur Institute, http://genopole.pasteur.fr/glabrata/.

incubation periods to exhibit distinguishing morphologies or phenotypes. Molecular strain typing has revealed extraordinary and unsuspected degrees of genetic diversity within the species. Knowledge thus obtained offers a new classification system based on the phylogenetic species concept. It is still a matter of debate whether and how species boundaries should be established by the phylogenetic species concept alone. However, because of the application of modern technology to what was previously a field of classical science governed by descriptive studies, important taxonomic questions can now be asked and thoroughly debated at the molecular level.

The most challenging tasks ahead consist of how to apply our knowledge of pathobiology and new identification systems obtained in the recent molecular era to the prevention of and therapeutic intervention for fungal diseases and management of patients. The discovery of effective drugs and the development of vaccines based on knowledge of molecular biology will indeed contribute to the prevention of and therapy for these infections and better management of patients. An equally important issue is the keen awareness of the importance of fungal diseases by the public and academia, as well as the pharmaceutical industry. Education in medical mycology, therefore, must be viewed on par with that of other infectious diseases in order to realize the full benefit of the evolution from the classical to molecular era of studies on pathogenic fungi.

REFERENCES

1. **Abuodeh, R. O., M. J. Orbach, M. A. Mandel, A. Das, and J. N. Galgiani.** 2000. Genetic transformation of *Coccidioides immitis* facilitated by *Agrobacterium tumefaciens. J. Infect. Dis.* **181:**2106–2110.

2. **Alani, E., and N. Kleckner.** 1987. A method for gene disruption that allows repeated use of *URA3* selection in the construction of multiply disrupted yeast strains. *Genetics* **116:**541–545.

3. **Armaleo, D., G. N. Ye, T. M. Klein, K. B. Shark, J. C. Sanford, and S. A. Johnston.** 1990. Biolistic nuclear transformation of *Saccharomyces cerevisiae* and other fungi. *Curr. Genet.* **17:**97–103.

4. **Ballance, D. J., and G. Turner.** 1985. Development of a high frequency transforming vector for *Aspergillus nidulans. Gene* **36:**321–331.

5. **Becker, D. M., and L. Guarente.** 1991. High-efficiency transformation of yeast by electroporation. *Methods Enzymol.* **194:**182–187.

6. **Beckerman, J., H. Chibana, J. Turner, and P. T. Magee.** 2001. Single-copy *IMH3* allele is sufficient to confer resistance to mycophenolic acid in *Candida albicans* and to mediate transformation of clinical *Candida* species. *Infect. Immun.* **69:**108–114.

7. **Beggs, J. D.** 1978. Transformation of yeast by a replicating hybrid plasmid. *Nature* **275:**104–109.

8. **Berg, C. M., and E. D. Garber.** 1962. A genetic analysis of color mutants of *Aspergillus fumigatus. Genetics* **47:**1139–1146.

9. **Borneman, A. R., M. J. Hynes, and A. Andrianopoulos.** 2001. A *STE12* homolog from the asexual, dimorphic fungus *Penicillium marneffei* complements the defect in sexual development of an *Aspergillus nidulans steA* mutant. *Genetics* **157:**1003–1014.

10. **Brand, A., D. M. MacCallum, A. J. Brown, N. A. Gow, and F. C. Odds.** 2004. Ectopic expression of *URA3* can influence the virulence phenotypes and proteome of *Candida albicans* but can be overcome by targeted reintegration of *URA3* at the *RPS10* locus. *Eukaryot. Cell* **3:**900–909.

11. **Brandhorst, T. T., M. Wuthrich, T. Warner, and B. Klein.** 1999. Targeted gene disruption reveals an adhesin indispensable for pathogenicity of *Blastomyces dermatitidis. J. Exp. Med.* **189:**1207–1216.

12. **Bundock, P., and P. J. J. Hooykaas.** 1996. Integration of *Agrobacterium tumefaciens* T-DNA in the *Saccharomyces cerevisiae* genome by illegitimate recombination. *Proc. Natl. Acad. Sci.* **93:**15272–15275.

13. **Buxton, F. P., D. I. Gwynne, and R. W. Davies.** 1985. Transformation of *Aspergillus niger* using the *argB* gene of *Aspergillus nidulans. Gene* **37:**207–214.

14. **Case, M. E., M. Schweizer, S. R. Kushner, and N. H. Giles.** 1980. Efficient transformation of *Neurospora crassa* utilizing hybrid plasmid DNA. *Proc. Natl. Acad. Sci. USA* **77:**5259–5263.

15. **Chang, Y. C., and K. J. Kwon-Chung.** 1994. Complementation of a capsule-deficient mutation of *Cryptococcus*

neoformans restores its virulence. *Mol. Cell. Biol.* **14:**4912–4918.

16. Chang, Y. C., L. A. Penoyer, and K. J. Kwon-Chung. 1996. The second capsule gene of *Cryptococcus neoformans, CAP64,* is essential for virulence. *Infect. Immun.* **64:**1977–1983.

17. Collier, L., A. Balows, and M. Sussman (ed.). 1998. *Microbiology and Microbial Infections,* 9th ed, vol. 4. Arnold, London, United Kingdom.

18. Cormack, B. P., G. Bertram, M. Egerton, N. A. Gow, S. Falkow, and A. J. Brown. 1997. Yeast-enhanced green fluorescent protein (yEGFP) a reporter of gene expression in *Candida albicans. Microbiology* **143:**303–311.

19. Cormack, B. P., N. Ghori, and S. Falkow. 1999. An adhesin of the yeast pathogen *Candida glabrata* mediating adherence to human epithelial cells. *Science* **285:**578–582.

20. Cox, G. M., D. L. Toffaletti, and J. R. Perfect. 1996. Dominant selection system for use in *Cryptococcus neoformans. J. Med. Vet. Mycol.* **34:**385–391.

21. Crowhurst, R. N., J. Rees-George, E. H. Rikkerink, and M. D. Templeton. 1992. High efficiency transformation of *Fusarium solani* f. sp. *cucurbitae* race 2 (mating population V). *Curr. Genet.* **21:**463–469.

22. Davidson, R. C., M. C. Cruz, R. A. L. Sia, B. Allen, J. A. Alspaugh, and J. Heitman. 2000. Gene disruption by biolistic transformation in serotype D strains of *Cryptococcus neoformans. Fungal Genet. Biol.* **29:**38–48.

23. De Groot, M. J. A., P. Bundock, P. J. J. Hooykaas, and A. G. M. Beijersbergen. 1998. *Agrobacterium tumefaciens*-mediated transformation of filamentous fungi. *Nat. Biotechnol.* **16:**839–842.

24. d'Enfert, C., G. Weidner, P. C. Mol, and A. A. Brakhage. 1999. Transformation systems of *Aspergillus fumigatus.* New tools to investigate fungal virulence. *Contrib. Microbiol.* **2:**149–166.

25. de Repentigny, L. 1992. Serodiagnosis of candidiasis, aspergillosis, and cryptococcosis. *Clin. Infect. Dis.* **14:**S11–S22.

26. Diaz, M. R., and J. W. Fell. 2004. High-throughput detection of pathogenic yeasts of the genus *Trichosporon. J. Clin. Microbiol.* **42:**3696–3706.

27. Dib, C., S. Faure, C. Fizames, D. Samson, N. Drouot, A. Vignal, P. Millasseau, S. Marc, J. Hazan, E. Seboun, M. Lathrop, G. Gyapay, J. Morissette, and J. Weissenbach. 1996. A comprehensive genetic map of the human genome based on 5,264 microsatellites. *Nature* **380:**152–154.

28. DiSalvo, A. F., A. S. Sekhon, G. A. Land, and W. H. Fleming. 1980. Evaluation of the exoantigen test for identification of *Histoplasma* species and *Coccidioides immitis* cultures. *J. Clin. Microbiol.* **11:**238–241.

29. Dixon, D. 2001. US-Japan workshops in medical mycology: past, present and future. *Jpn. J. Med. Mycol.* **42:**75–80.

30. Dromer, F., A. Varma, O. Ronin, S. Mathoulin, and B. Dupont. 1994. Molecular typing of *Cryptococcus neoformans* serotype D clinical isolates. *J. Clin. Microbiol.* **32:**2364–2371.

31. Edman, J. C. 1992. Isolation of telomerelike sequences from *Cryptococcus neoformans* and their use in high-efficiency transformation. *Mol. Cell. Biol.* **12:**2777–2783.

32. Edman, J. C., J. A. Kovacs, H. Masur, D. V. Santi, H. J. Elwood, and M. L. Sogin. 1988. Ribosomal RNA sequence shows *Pneumocystis carinii* to be a member of the fungi. *Nature* **334:**519–522.

33. Edman, J. C., and K. J. Kwon-Chung. 1990. Isolation of the *URA5* gene from *Cryptococcus neoformans* var.

neoformans and its use as a selective marker for transformation. *Mol. Cell. Biol.* **10:**4538–4544.

34. Ellis, M. 2002. Invasive fungal infections: evolving challenges for diagnosis and therapeutics. *Mol. Immunol.* **38:**947–957.

35. Erickson, T., L. Liu, A. Gueyikian, X. Zhu, J. Gibbons, and P. R. Williamson. 2001. Multiple virulence factors of *Cryptococcus neoformans* are dependent on *VPH1. Mol. Microbiol.* **42:**1121–1131.

36. Erjavec, Z., and P. E. Verweij. 2002. Recent progress in the diagnosis of fungal infections in the immunocompromised host. *Drug. Resist. Update* **5:**3–10.

37. Falkow, S. 2004. Molecular Koch's postulates applied to bacterial pathogenicity—a personal recollection 15 years later. *Nat. Rev. Microbiol.* **2:**67–72.

38. Falkow, S. 1988. Molecular Koch's postulates applied to microbial pathogenicity. *Rev. Infect. Dis.* **10**(Suppl. 2): S274–S276.

39. Feil, E. J., and M. C. Enright. 2004. Analyses of clonality and the evolution of bacterial pathogens. *Curr. Opin. Microbiol.* **7:**308–313.

40. Feil, E. J., B. C. Li, D. M. Aanensen, W. P. Hanage, and B. G. Spratt. 2004. eBURST: inferring patterns of evolutionary descent among clusters of related bacterial genotypes from multilocus sequence typing data. *J. Bacteriol.* **186:**1518–1530.

41. Fell, J. W., T. Boekhout, A. Fonseca, G. Scorzetti, and A. Statzell-Tallman. 2000. Biodiversity and systematics of basidiomycetous yeasts as determined by large-subunit rDNA D1/D2 domain sequence analysis. *Int. J. Syst. Evol. Microbiol.* **50:**1351–1371.

42. Fire, A., S. Xu, M. K. Montgomery, S. A. Kostas, S. E. Driver, and C. C. Mello. 1998. Potent and specific genetic interference by double-stranded RNA in *Caenorhabditis elegans. Nature* **391:**806–811.

43. Fisher, M. C., G. L. Koenig, T. J. White, and J. T. Taylor. 2002. Molecular and phenotypic description of *Coccidioides posadasii* sp. nov., previously recognized as the non California population of *Coccidioides immitis. Mycologia* **94:**73–84.

44. Fonzi, W. A., and M. Y. Irwin. 1993. Isogenic strain construction and gene mapping in *Candida albicans. Genetics* **134:**717–728.

45. Freydiere, A. M., R. Guinet, and P. Boiron. 2001. Yeast identification in the clinical microbiology laboratory: phenotypical methods. *Med. Mycol.* **39:**9–33.

46. Garcia-Hermoso, D., F. Dromer, S. Mathoulin-Pelissier, and G. Janbon. 2001. Are two *Cryptococcus neoformans* strains epidemiologically linked? *J. Clin. Microbiol.* **39:**1402–1406.

47. Gardes, M., and T. D. Bruns. 1993. ITS primers with enhanced specificity for basidiomycetes—application to the identification of mycorrhizae and rusts. *Mol. Ecol.* **2:**113–118.

48. Girardin, H., J. P. Latge, T. Srikantha, B. Morrow, and D. R. Soll. 1993. Development of DNA probes for fingerprinting *Aspergillus fumigatus. J. Clin. Microbiol.* **31:**1547–1554.

49. Glumoff, V., O. Kappeli, A. Fiechter, and J. Reiser. 1989. Genetic transformation of the filamentous yeast, *Trichosporon cutaneum,* using dominant selection marker. *Gene* **84:**311–318.

50. Gonzalez, R., S. Ferrer, J. Buesa, and D. Ramon. 1989. Transformation of the dermatophyte *Trichophyton mentagrophytes* to hygromycin B resistance. *Infect. Immun.* **57:**2923–2925.

51. Gorman, J. A., W. Chan, and J. W. Gorman. 1991. Repeated use of *GAL1* for gene disruption in *Candida albicans*. *Genetics* **129**:19–24.

52. Gruby, D. 1841. Sur les mycodermes que constituent la teigne faveus. *C. R. Acad. Sci.* (Paris) **13**:309–312.

53. Haas, L. O., J. M. Cregg, and M. A. Gleeson. 1990. Development of an integrative DNA transformation system for the yeast *Candida tropicalis*. *J. Bacteriol.* **172**:4571–4577.

54. Hall, G. S., K. Pratt-Rippin, and J. A. Washington. 1992. Evaluation of a chemiluminescent probe assay for identification of *Histoplasma capsulatum* isolates. *J. Clin. Microbiol.* **30**:3003–3004.

55. Hammond, S. M., A. A. Caudy, and G. J. Hannon. 2001. Post-transcriptional gene silencing by double-stranded RNA. *Nat. Rev. Genet.* **2**:110–119.

56. Hinnen, A. J., B. Hickes, and G. R. Fink. 1978. Transformation of yeast. *Proc. Natl. Acad. Sci. USA* **75**:1929–1933.

57. Hogan, L. H., and B. S. Klein. 1997. Transforming DNA integrates at multiple sites in the dimorphic fungal pathogen *Blastomyces dermatitidis*. *Gene* **186**:219–226.

58. Hube, B., D. Sanglard, F. C. Odds, D. Hess, M. Monod, W. Schafer, A. J. Brown, and N. A. Gow. 1997. Disruption of each of the secreted aspartyl proteinase genes *SAP1*, *SAP2*, and *SAP3* of *Candida albicans* attenuates virulence. *Infect. Immun.* **65**:3529–3538.

59. Huffnagle, K. E., and R. M. Gander. 1993. Evaluation of Gen-Probe's *Histoplasma capsulatum* and *Cryptococcus neoformans* AccuProbes. *J. Clin. Microbiol.* **31**:419–421.

60. Idnurm, A., J. L. Reedy, J. C. Nussbaum, and J. Heitman. 2004. *Cryptococcus neoformans* virulence gene discovery through insertional mutagenesis. *Eukaryot. Cell* **3**:420–429.

61. Iwen, P. C., S. H. Hinrichs, and M. E. Rupp. 2002. Utilization of the internal transcribed spacer regions as molecular targets to detect and identify human fungal pathogens. *Med. Mycol.* **40**:87–109.

62. Jacobson, E. S., D. J. Ayers, A. C. Harrell, and C. C. Nicholas. 1982. Genetic and phenotypic characterization of capsule mutants of *Cryptococcus neoformans*. *J. Bacteriol.* **156**:460–462.

63. Jahn, B., A. Koch, A. W. Schmidt, G. H. Gehringer, S. Bhakdi, and A. A. Brakhage. 1997. Isolation and characterization of a pigment-conidium mutant of *Aspergillus fumigatus* with altered conidial surface and reduced virulence. *Infect. Immun.* **65**:5110–5117.

64. Johnstone, I. L., S. G. Hughes, and A. J. Clutterbuck. 1985. Cloning an *Aspergillus nidulans* developmental gene by transformation. *EMBO J.* **4**:1307–1311.

65. Keeling, P. J., and N. M. Fast. 2002. *Microsporidia*: biology and evolution of highly reduced intracellular parasites. *Annu. Rev. Microbiol.* **56**:93–116.

66. Kelly, R., S. M. Miller, M. B. Kurtz, and D. R. Kirsh. 1987. Direct mutagenesis in *Candida albicans*: one-step gene disruption to isolate *ura3* mutants. *Mol. Cell. Biol.* **7**:199–207.

67. Kersulyte, D., J. P. Woods, E. J. Keath, W. E. Goldman, and D. E. Berg. 1992. Diversity among clinical isolates of *Histoplasma capsulatum* detected by polymerase chain reaction with arbitrary primers. *J. Bacteriol.* **174**:7075–7079.

68. Kiehn, T. E., E. M. Bernard, J. W. Gold, and D. Armstrong. 1979. Candidiasis: detection by gas-liquid chromatography of D-arabinitol, a fungal metabolite, in human serum. *Science* **206**:577–580.

69. King, D., J. Rhine-Chalberg, M. A. Pfaller, S. A. Moser, and W. G. Merz. 1995. Comparison of four DNA-based methods for strain delineation of *Candida lusitaniae*. *J. Clin. Microbiol.* **33**:1467–1470.

70. Kistler, H., and U. Benny. 1988. Genetic transformation of the fungal plant wilt pathogen, *Fusarium oxysporum*. *Curr. Genet.* **13**:145–149.

71. Klein, B. S., S. Chaturvedi, L. H. Hogan, J. M. Jones, and S. L. Newman. 1994. Altered expression of surface protein WI-1 in genetically related strains of *Blastomyces dermatitidis* that differ in virulence regulates recognition of yeasts by human macrophages. *Infect. Immun.* **62**:3536–3542.

72. Kunze, G., C. Petzoldt, R. Bode, I. Samsonova, M. Hecker, and D. Birnbaum. 1985. Transformation of *Candida maltosa* and *Pichia guilliermondii* by a plasmid containing *Saccharomyces cerevisiae ARG4* DNA. *Curr. Genet.* **9**:205–209.

73. Kurtz, M. B., M. W. Cortelyou, and D. R. Kirsch. 1986. Integrative transformation of *Candida albicans*, using a cloned *Candida ADE2* gene. *Mol. Cell. Biol.* **6**:142–149.

74. Kurtz, M. B., M. W. Cortelyou, S. M. Miller, M. Lai, and D. R. Kirsch. 1987. Development of autonomously replicating plasmids for *Candida albicans*. *Mol. Cell. Biol.* **7**:209–217.

75. Kwon-Chung, K. 1998. Gene disruption to evaluate the role of fungal candidate virulence genes. *Curr. Opin. Microbiol.* **1**:381–389.

76. Kwon-Chung, K. J. 1972. *Emmonsiella capsulata*: perfect state of *Histoplasma capsulatum*. *Science* **177**:368–369.

77. Kwon-Chung, K. J. 1971. Genetic analysis on the incompatibility system of *Ajellomyces dermatitidis*. *Sabouraudia* **9**:231–238.

78. Kwon-Chung, K. J. 1975. A new genus, *Filobasidiella*, the perfect state of *Cryptococcus neoformans*. *Mycologia* **67**:1197–1200.

79. Kwon-Chung, K. J. 1976. A new species of *Filobasidiella*, the sexual state of *Cryptococcus neoformans* B and C serotypes. *Mycologia* **68**:943–946.

80. Kwon-Chung, K. J. 1973. Studies on *Emmonsiella capsulata*. I. Heterothallism and development of the ascocarp. *Mycologia* **65**:109–121.

81. Kwon-Chung, K. J., and J. E. Bennett. 1992. *Medical Mycology*. Lea & Febiger, Philadelphia, Pa.

82. Kwon-Chung, K. J., and W. B. Hill. 1970. Studies on the pink adenine-deficient strains of *Candida albicans*. I. Cultural and morphological characteristics. *Sabouraudia* **8**:48–59.

83. Kwon-Chung, K. J., D. Lehman, C. Good, and P. T. Magee. 1985. Genetic evidence for role of extracellular proteinase in virulence. *Infect. Immun.* **49**:571–575.

84. Kwon-Chung, K. J., I. Polacheck, and T. J. Popkin. 1982. Melanin-lacking mutants of *Cryptococcus neoformans* and their virulence for mice. *J. Bacteriol.* **150**:1414–1421.

85. Kwon-Chung, K. J., and J. C. Rhodes. 1986. Encapsulation and melanin formation as indicators of virulence in *Cryptococcus neoformans*. *Infect Immun.* **51**:218–223.

86. Lasker, B. A., L. S. Page, T. J. Lott, G. S. Kobayashi, and G. Medoff. 1991. Characterization of CARE-1: *Candida albicans* repetitive element-1. *Gene* **102**:45–50.

87. Lay, J., L. K. Henry, J. Clifford, Y. Koltin, C. E. Bulawa, and J. M. Becker. 1998. Altered expression of selectable marker *URA3* in gene-disruptred *Candida albicans* strains complicates interpretation of virulence studies. *Infect. Immun.* **66**:5301–5306.

88. Leal, C. V., B. A. Montes, A. C. Mesa, A. L. Rus, M. Corredor, A. Restrepo, and J. G. McEwen. 2004. *Agrobacterium tumefaciens*-mediated transformation of *Paracoccidioides brasiliensis*. *Med. Mycol.* 42:391–395.

89. Lehmann, P. F., D. Lin, and B. A. Lasker. 1992. Genotypic identification and characterization of species and strains within the genus *Candida* by using random amplified polymorphic DNA. *J. Clin. Microbiol.* 30:3249–3254.

90. Litt, M., and J. A. Luty. 1989. A hypervariable microsatellite revealed by in vitro amplification of a dinucleotide repeat within the cardiac muscle actin gene. *Am. J. Hum. Genet.* 44:397–401.

91. Litvintseva, A. P., R. E. Marra, K. Nielsen, J. Heitman, R. Vilgalys, and T. G. Mitchell. 2003. Evidence of sexual recombination among *Cryptococcus neoformans* serotype A isolates in sub-Saharan Africa. *Eukaryot. Cell* 2:1162–1168.

92. Liu, H., T. R. Cottrell, L. M. Pierini, W. E. Goldman, and T. L. Doering. 2002. RNA interference in the pathogenic fungus *Cryptococcus neoformans*. *Genetics* 160:463–470.

93. Lockhart, S. R., C. Pujol, S. Joly, and D. R. Soll. 2001. Development and use of complex probes for DNA fingerprinting the infectious fungi. *Med. Mycol.* 39:1–8.

94. Macdonald, F., and F. C. Odds. 1983. Virulence for mice of a proteinase-secreting strain of *Candida albicans* and proteinase-deficient mutant. *J. Gen. Microbiol.* 129:431–438.

95. Magee, B. B., T. M. D'Souza, and P. T. Magee. 1987. Strain and species identification by restriction fragment length polymorphisms in the ribosomal DNA repeat of *Candida* species. *J. Bacteriol.* 169:1639–1643.

96. Maiden, M. C. 1998. Horizontal genetic exchange, evolution, and spread of antibiotic resistance in bacteria. *Clin. Infect. Dis.* 27(Suppl. 1):S12–S20.

97. Maiden, M. C., J. A. Bygraves, E. Feil, G. Morelli, J. E. Russell, R. Urwin, Q. Zhang, J. Zhou, K. Zurth, D. A. Caugant, I. M. Feavers, M. Achtman, and B. G. Spratt. 1998. Multilocus sequence typing: a portable approach to the identification of clones within populations of pathogenic microorganisms. *Proc. Natl. Acad. Sci. USA* 95:3140–3145.

98. Martino, P., and C. Girmenia. 2000. Making the diagnosis of fungal infection: when to start treatment. *Int. J. Antimicrob. Agents* 16:323–329.

98a. McClelland, C. M., Y. C. Chang, and K. J. Kwon-Chung. 2005. High frequency transformation of *Cryptococcus neoformans* and *Cryptococcus gattii* by *Agrobacterium tumefaciens*. *Fungal Genet. Biol.* 42:904–913.

99. Mcdade, H. C., and G. M. Cox. 2001. A new dominant selectable marker for use in *Cryptococcus neoformans*. *Med. Mycol.* 39:151–154.

100. McDonough, E. S., and A. L. Lewis. 1968. The ascigerous stage of *Blastomyces dermatitidis*. *Mycologia* 60:76–83.

101. Mehra, R. K., J. L. Thorvaldsen, I. G. Macreadie, and D. R. Winge. 1992. Cloning system for *Candida glabrata* using elements from the metallothionein-IIa-encoding gene that confer autonomous replication. *Gene* 113:119–124.

102. Merz, W. G. 1990. *Candida albicans* strain delineation. *Clin. Microbiol. Rev.* 3:321–334.

103. Meyer, W., T. G. Mitchell, E. Z. Freedman, and R. Vilgalys. 1993. Hybridization probes for conventional DNA fingerprinting used as single primers in the polymerase chain reaction to distinguish strains of *Cryptococcus neoformans*. *J. Clin. Microbiol.* 31:2274–2280.

104. Miller, R. E., Jr., and L. P. Lu. 1976. Evaluation of a multitest microtechnique for yeast identification. *Am. J. Med. Technol.* 42:238–242.

105. Mouyna, I., C. Henry, T. L. Doering, and J. P. Latge. 2004. Gene silencing with RNA interference in the human pathogenic fungus *Aspergillus fumigatus*. *FEMS Microbiol. Lett.* 237:317–324.

106. Mullins, E. D., X. Chen, P. Romaine, R. Raina, D. M. Geiser, and S. Kang. 2001. *Agrobacterium*-mediated transformation of *Fusarium oxysporum*: an efficient tool for insertional mutagenesis and gene transfer. *Phytopathology* 91:173–180.

107. Nelson, R. T., J. Hua, B. Pryor, and J. K. Lodge. 2001. Identification of virulence mutants of the fungal pathogen *Cryptococcus neoformans* using signature-tagged mutagenesis. *Genetics* 157:935–947.

108. Neumann, E., M. Schaefer-Ridder, Y. Wang, and P. H. Hofscheider. 1982. Gene transfer into mouse lyoma cells by electroporation in high electric fields. *EMBO J.* 1:841–845.

109. Neuveglise, C., J. Sarfati, J. P. Latge, and S. Paris. 1996. *Afut1*, a retrotransposon-like element from *Aspergillus fumigatus*. *Nucleic Acids Res.* 24:1428–1434.

110. Nosek, J., L. Adamikova, J. Zemanova, L. Tomaska, R. Zufferey, and C. B. Mamoun. 2002. Genetic manipulation of the pathogenic yeast *Candida parapsilosis*. *Curr. Genet.* 42:27–35.

111. Odds, F. C. 1987. *Candida* infections: an overview. *Crit. Rev. Microbiol.* 15:1–5.

112. Padhye, A. A., G. Smith, D. McLaughlin, P. G. Standard, and L. Kaufman. 1992. Comparative evaluation of a chemiluminescent DNA probe and an exoantigen test for rapid identification of *Histoplasma capsulatum*. *J. Clin. Microbiol.* 30:3108–3111.

113. Peng, M., C. R. Cooper, Jr., and P. J. Szaniszlo. 1995. Genetic transformation of the pathogenic fungus *Wangiella dermatitidis*. *Appl. Microbiol. Biotechnol.* 44:444–450.

114. Perfect, J. R., D. L. Toffaletti, and T. H. Rude. 1993. The gene encoding phosphoribosylaminoimidazole carboxylase (*ADE2*) is essential for growth of *Cryptococcus neoformans* in cerebrospinal fluid. *Infect. Immun.* 61:4446–4451.

115. Poulter, R. T. M., V. Hanrahan, K. Jeffery, D. Markie, M. G. Shepherd, and P. A. Sullivan. 1982. Recombination analysis of naturally diploid *Candida albicans*. *J. Bacteriol.* 152:969–975.

116. Punt, P. J., R. P. Oliver, M. A. Dingemanse, P. H. Pouwels, and C. A. van den Hondel. 1987. Transformation of *Aspergillus* based on the hygromycin B resistance marker from *Escherichia coli*. *Gene* 56:117–124.

117. Reichard, U., C.-Y. Hung, P. W. Thomas, and G. T. Cole. 2000. Disruption of the gene which encodes a serodiagnostic antigen and chitinase of the human fungal pathogen *Coccidioides immitis*. *Infect. Immun.* 68:5830–5838.

118. Reuss, O., A. Vik, R. Kolter, and J. Morschhauser. 2004. The *SAT1* flipper, an optimized tool for gene disruption in *Candida albicans*. *Gene* 341:119–127.

119. Rhodes, J. C., I. Polacheck, and K. J. Kwon-Chung. 1982. Phenoloxidase activity and virulence in isogenic strains of *Cryptococcus neoformans*. *Infect. Immun.* 36:1175–1184.

120. Roboz, J., R. Suzuki, and J. F. Holland. 1980. Quantification of arabinitol in serum by selected ion monitoring as a diagnostic technique in invasive candidiasis. *J. Clin. Microbiol.* 12:594–601.

121. Rothstein, R. S. 1983. One-step gene disruption in yeast. *Methods Enzymol. C* **101:**202–211.

122. Salas, S. D., J. E. Bennett, K. J. Kwon-Chung, J. R. Perfect, and P. R. Williamson. 1996. Effect of the laccase gene *CNLAC1*, on virulence of *Cryptococcus neoformans. J. Exp. Med.* **184:**377–386.

123. Sanglard, D., B. Hube, M. Monod, F. C. Odds, and N. A. Gow. 1997. A triple deletion of the secreted aspartyl proteinase genes *SAP4*, *SAP5*, and *SAP6* of *Candida albicans* causes attenuated virulence. *Infect. Immun.* **65:**3539–3546.

124. Saracheck, A., and D. A. Weber. 1984. Temperature-dependent internuclear transfer of genetic material in heterokaryons of *Candida albicans. Curr. Genet.* **8:**181–187.

125. Scherer, S., and D. A. Stevens. 1988. A *Candida albicans* dispersed, repeated gene family and its epidemiologic applications. *Proc. Natl. Acad. Sci. USA* **85:**1452–1456.

126. Schlotterer, C. 2000. Evolutionary dynamics of microsatellite DNA. *Chromosoma* **109:**365–371.

127. Schmid, J., M. Rotman, B. Reed, C. L. Pierson, and D. R. Soll. 1993. Genetic similarity of *Candida albicans* strains from vaginitis patients and their partners. *J. Clin. Microbiol.* **31:**39–46.

128. Sebghati, T. S., J. T. Engle, and W. E. Goldman. 2000. Intracellular parasitism by *Histoplasma capsulatum*: fungal virulence and calcium dependence. *Science* **290:**1368–1372.

129. Sendid, B., M. Tabouret, J. L. Poirot, D. Mathieu, J. Fruit, and D. Poulain. 1999. New enzyme immunoassays for sensitive detection of circulating *Candida albicans* mannan and antimannan antibodies: useful combined test for diagnosis of systemic candidiasis. *J. Clin. Microbiol.* **37:**1510–1517.

130. Skory, C. D. 2002. Homologous recombination and double-strand break repair in the transformation of *Rhizopus oryzae. Mol. Genet. Genomics* **268:**397–406.

131. Soll, D. R. 2000. The ins and outs of DNA fingerprinting the infectious fungi. *Clin. Microbiol. Rev.* **13:**332–370.

132. Soll, D. R., M. Staebell, C. Langtimm, M. Pfaller, J. Hicks, and T. V. Rao. 1988. Multiple *Candida* strains in the course of a single systemic infection. *J. Clin. Microbiol.* **26:**1448–1459.

133. Spitzer, S. G., and E. D. Spitzer. 1994. Characterization of the CNRE-1 family of repetitive DNA elements in *Cryptococcus neoformans. Gene* **144:**103–106.

134. Standard, P. G., and L. Kaufman. 1976. Specific immunological test for the rapid identification of members of the genus *Histoplasma. J. Clin. Microbiol.* **3:**191–199.

135. Stynen, D., A. Goris, J. Sarfati, and J. P. Latge. 1995. A new sensitive sandwich enzyme-linked immunosorbent assay to detect galactofuran in patients with invasive aspergillosis. *J. Clin. Microbiol.* **33:**497–500.

136. Sugui, J., Y. C. Chang, and K. J. Kwon-Chung. 2005. *Agrobacterium tumefaciens*-mediated transformation of *Aspergillus fumigatus*: efficient tool for insertional mutagenesis and targeted gene disruption. *Appl. Environ. Microbiol.* **71:**1798–1802.

137. Sullivan, T. D., P. J. Rooney, and B. S. Klein. 2002. *Agrobacterium tumefaciens* integrates transfer DNA into single chromosomal sites of dimorphic fungi and yields homokaryotic progeny from multinucleate yeast. *Eukaryot. Cell* **1:**895–905.

138. Sundstrom, P., J. E. Cutler, and J. F. Staab. 2002. Reevaluation of the role of *HWP1* in systemic candidiasis by use of *Candida albicans* strains with selectable marker *URA3* targeted to the *ENO1* locus. *Infect. Immun.* **70:**3281–3283.

139. Takashima, M., T. Sugita, T. Shinoda, and T. Nakase. 2003. Three new combinations from the *Cryptococcus laurentii* complex: *Cryptococcus aureus*, *Cryptococcus carnescens* and *Cryptococcus peneaus. Int. J. Syst. Evol. Microbiol.* **53:**1187–1194.

140. Tang, C. M., J. Cohen, and D. W. Holden. 1992. An *Aspergillus fumigatus* alkaline protease mutant constructed by gene disruption is deficient in extracellular elastase activity. *Mol. Microbiol.* **6:**1663–1671.

141. Thompson, J. R., E. Register, J. Curotto, M. B. Kurtz, and R. Kelly. 1998. An improved protocol for the preparation of yeast cells for transformation by electroporation. *Yeast* **14:**565–571.

142. Tijsterman, M., R. F. Ketting, and R. H. Plasterk. 2002. The genetics of RNA silencing. *Annu. Rev. Genet.* **36:**489–519.

143. Toffaletti, D. L., T. H. Rude, S. A. Johnston, D. T. Durack, and J. R. Perfect. 1993. Gene transfer in *Cryptococcus neoformans* by use of biolistic delivery of DNA. *J. Bacteriol.* **175:**1405–1411.

144. Tsai, H. F., Y. C. Chang, R. G. Washburn, M. H. Wheeler, and K. J. Kwon-Chung. 1998. The developmentally regulated *alb1* gene of *Aspergillus fumigatus*: its role in modulation of conidial morphology and virulence. *J. Bacteriol.* **180:**3031–3038.

145. Uhl, M. A., M. Biery, N. Craig, and A. D. Johnson. 2003. Haplosufficiency-based large-scale forward genetic analysis of filamentous growth in the diploid human fungal pathogen *Candida albicans. EMBO J.* **22:**2668–2678.

146. Varma, A., and K. J. Kwon-Chung. 1999. Characterization of the glyceraldehyde-3-phosphate dehydrogenase gene [correction of glyceraldehyde-3-phosphate gene] and the use of its promoter for heterologous expression in *Cryptococcus neoformans*, a human pathogen. *Gene* **232:**155–163.

147. Wada, M., T. Beppu, and S. Horinouchi. 1996. Integrative transformation of the zygomycete *Rhizomucor pusillus* by homologous recombination. *Appl. Microbiol. Biotechnol.* **45:**652–657.

148. Walden, R., B. Reiss, C. Koncz, and J. Schell. 1997. The impact of Ti-plasmid-derived gene vectors on the study of the mechanism of action of phytohormones. *Annu. Rev. Phytopathol.* **35:**45–66.

149. Walther, A. W. J. 2003. An improved transformation protocol for the human fungal pathogen *Candida albicans. Curr. Genet.* **42:**339–343.

150. Welsh, J., N. Rampino, M. McClelland, and M. Perucho. 1995. Nucleic acid fingerprinting by PCR-based methods: applications to problems in aging and mutagenesis. *Mutat. Res.* **338:**215–229.

151. Whelan, W. L., R. M. Partridge, and P. T. Magee. 1980. Heterozygosity and segregation in *Candida albicans. Genetics* **180:**107–113.

152. White, T. J., T. D. Bruns, S. B. Lee, and J. W. Taylor. 1990. Amplification and direct sequencing of fungal ribosomal RNA genes for phylogenetics, p. 315–322. *In* M. A. Innis, D. H. Gelfand, J. J. Sninsky, and T. J. White (ed.), *PCR Protocols. A Guide to Methods and Applications.* Academic Press, Inc., San Diego, Calif.

153. Williams, J. G., A. R. Kubelik, K. J. Livak, J. A. Rafalski, and S. V. Tingey. 1990. DNA polymorphisms amplified by arbitrary primers are useful as genetic markers. *Nucleic Acids Res.* **18:**6531–6535.

154. Woloshuk, C. P., E. R. Seip, G. A. Payne, and C. R. Adkins. 1989. Genetic transformation system for the aflatoxin-producing fungus *Aspergillus flavus*. *Appl. Environ. Microbiol.* **55:**86–90.

155. Wong, B., E. M. Bernard, J. W. Gold, D. Fong, A. Silber, and D. Armstrong. 1982. Increased arabinitol levels in experimental candidiasis in rats: arabinitol appearance rates, arabinitol/creatinine ratios, and severity of infection. *J. Infect. Dis.* **146:**346–352.

156. Wong, B., K. L. Brauer, R. R. Tsai, and K. Jayasimhulu. 1989. Increased amounts of the *Aspergillus* metabolite D-mannitol in tissue and serum of rats with experimental aspergillosis. *J. Infect. Dis.* **160:**95–103.

157. Wong, B., J. R. Perfect, S. Beggs, and K. A. Wright. 1990. Production of the hexitol D-mannitol by *Cryptococcus neoformans* in vitro and in rabbits with experimental meningitis. *Infect. Immun.* **58:**1664–1670.

158. Woods, J. P., and W. E. Goldman. 1992. *In vivo* generation of linear plasmids with addition of telomeric sequences by *Histoplasma capsulatum*. *Mol. Microbiol.* **6:**3603–3610.

159. Woods, J. P., E. L. Heinecke, and W. E. Goldman. 1998. Electrotransformation and expression of bacterial genes encoding hygromycin phosphotransferase and β-galactosidase in the pathogenic fungus *Histoplasma capsulatum*. *Infect. Immun.* **66:**1697–1707.

160. Woods, J. P., D. M. Retallack, E. L. Heinecke, and W. E. Goldman. 1998. Rare homologous gene targeting in *Histoplasma capsulatum*: disruption of the *URA5Hc* gene by allelic replacement. *J. Bacteriol.* **180:** 5135–5143.

161. Worsham, P. L., and W. E. Goldman. 1990. Development of a genetic transformation system for *Histoplasma capsulatum*: complementation of uracil auxotrophy. *Mol. Gen. Genet.* **221:**358–362.

162. Worsham, P. L., and W. E. Goldman. 1988. Selection and characterization of *ura5* mutants of *Histoplasma capsulatum*. *Mol. Gen. Genet.* **214:**348–352.

163. Xu, J., C. M. Boyd, E. Livingston, W. Meyer, J. F. Madden, and T. G. Mitchell. 1999. Species and genotypic diversities and similarities of pathogenic yeasts colonizing women. *J. Clin. Microbiol.* **37:**3835–3843.

164. Yeo, S. F., and B. Wong. 2002. Current status of nonculture methods for diagnosis of invasive fungal infections. *Clin. Microbiol. Rev.* **15:**465–484.

165. Young, L. Y., M. C. Lorenz, and J. Heitman. 2000. A *STE12* homolog is required for mating but dispensable for filamentation in *Candida lusitaniae*. *Genetics* **155:**17–29.

166. Yu, J. J., and G. T. Cole. 1998. Biolistic transformation of the human pathogenic fungus *Coccidioides immitis*. *J. Microbiol. Methods* **33:**129–141.

Molecular Principles of Fungal Pathogenesis
Edited by Joseph Heitman et al.
©2006 ASM Press, Washington, D.C.

Chapter 6

Mating and Parasexual Genetics in *Candida albicans*

MATHEW MILLER AND ALEXANDER JOHNSON

Candida spp. are the fourth most common cause of hospital-acquired infections (15), and the species *Candida albicans* is the primary fungal pathogen of humans (10). It normally exists as a commensal organism but can cause life-threatening systemic infections in immunocompromised individuals (38). Until recently, *C. albicans* was considered to be completely asexual, a conclusion based largely on the apparent absence of sexual forms (59). Moreover, the population structure of clinical isolates exhibits a high degree of clonality (81), suggesting that sexual reproduction and recombination are rare compared to an asexual mode of reproduction.

Recent discoveries in *C. albicans* have challenged the idea that *C. albicans* is purely asexual. In 1999, the discovery of a mating-type locus in *C. albicans* (28) permitted new experimental approaches to the study of mating in this "asexual" organism. This chapter outlines the discovery that *C. albicans* can be coaxed to mate in the laboratory and the recognition that a phenotypic switching system—white-opaque switching—is intimately involved in this process. Other subjects will include the regulation of cell type determination, the cell biology of the pheromone response, nuclear fusion following mating (karyogamy), and a concerted process of chromosome loss that can complete a parasexual cycle in *C. albicans*. We also attempt to reconcile the largely clonal population structure of *C. albicans* with the existence of a complex and highly conserved mating apparatus.

Many features of mating in *C. albicans* relate to prior knowledge of the mating machinery of the model organism *Saccharomyces cerevisiae*. Of particular interest are the departures from the *S. cerevisiae* "paradigm" which serve to underscore the unique features of the *C. albicans* mating apparatus. For these reasons, we begin by briefly comparing the two organisms and reviewing the mating cycle in *S. cerevisiae*.

COMPARISON OF *S. CEREVISIAE* AND *C. ALBICANS*

S. cerevisiae and *C. albicans* diverged approximately 200 million to 800 million years ago (22, 26) and have since adapted to very different environments. *S. cerevisiae* is found naturally on fruit and in soil and commercially in bakeries, breweries, and scientific laboratories. In the laboratory, *S. cerevisiae* undergoes a well-described sexual cycle: it exists in mating-competent haploid forms and a diploid form that is capable of meiosis and sporulation (39, 55). Most laboratory strains of *S. cerevisiae* have been extensively selected for their ease of genetic manipulation (70). On the other hand, *C. albicans* is a successful opportunistic pathogen of warm-blooded animals (61). It is capable of a benign commensal lifestyle or, in an immunocompromised host, of invading a large variety of tissues (18). *C. albicans* is a diploid organism for which no haploid forms have been observed (60). The diploid genome of *C. albicans* contains many examples of allelic polymorphism, both inside and outside of coding regions (34, 53, 69). Most strains of *C. albicans* used in laboratory studies are clinical isolates or very closely related derivatives of these isolates. For example, a set of auxotrophically marked laboratory strains have been derived from the clinical isolate SC5314 (19). Despite these differences, *S. cerevisiae* and *C. albicans* have a number of similarities that allow for useful comparisons. Roughly two-thirds of the ~6,500 genes in the *C. albicans* genome have clear orthologs in *S. cerevisiae* (34), from which biochemical or physiological function can be inferred.

MATING IN *S. CEREVISIAE*

The sexual cycle of *S. cerevisiae* is controlled by two alleles of a genetic locus termed the mating-type (*MAT*) locus (27). In *S. cerevisiae*, the two alleles are known as *MAT***a** and *MAT*α. These two alleles

Mathew Miller and Alexander Johnson • Department of Microbiology and Immunology, University of California San Francisco, San Francisco, CA 94158.

encode different sets of transcriptional regulators that work together and in combination with other DNA-binding proteins to regulate the transcription of genes involved in the mating process (27, 33).

Cells containing only the *MAT*a or only the *MAT*α allele are typically haploid and are known as **a** and α cells, respectively (Fig. 1). Cells that carry both the *MAT*a and *MAT*α alleles are **a**/α cells and are typically diploid and therefore heterozygous for the *MAT* locus. *MAT*a encodes a single protein, a1, a homeodomain protein that apparently has no function in **a** cells even though it is expressed. The **a**-specific genes—the genes required for **a** cells to mate—are expressed constitutively. *MAT*α encodes two sequence-specific DNA-binding proteins, the alpha-domain protein α1 and the homeodomain protein α2. In α cells, α1 activates the transcription of

α-specific genes and α2 represses the transcription of **a**-specific genes. The product of successful mating between **a** and α haploids is an **a**/α diploid cell, which expresses both alleles of the *MAT* locus. The a1 and α2 proteins form a heterodimer that binds DNA and represses the transcription of a large set of genes, the so-called haploid-specific genes. Thus, **a**/α cells are unable to mate but are able to undergo meiosis and sporulation in response to environmental signals. Sporulation of **a**/α diploids produces **a** and α haploid cells, thus completing the sexual cycle (39) (Fig. 1).

EARLY EVIDENCE FOR A SEXUAL CYCLE IN *C. ALBICANS*

Although *C. albicans* had been classified as asexual for nearly a century, evidence began to emerge in the 1990s indicating that *C. albicans*

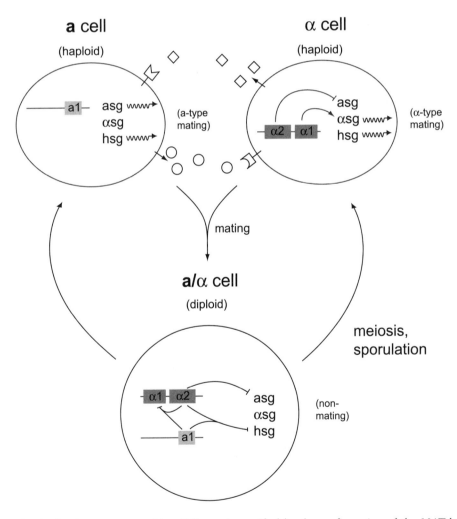

Figure 1. Mating in *Saccharomyces cerevisiae*. Cell type is specified by the configuration of the *MAT* locus, which encodes regulators of mating gene expression. **a** cells express a-specific genes (asg), including the a-factor pheromone (circles) and the α-factor receptor. α cells encode α2, which represses the a-specific genes, and α1, which activates the α-specific genes (αsg), including α-factor mating pheromone (squares) and the a-factor pheromone receptor. **a**/α cells contain both a1 and α2, which heterodimerize and together repress the haploid-specific genes (hsg) that are required for mating in both cell types. The a1-α2 heterodimer also represses α1, thereby repressing the α-specific genes.

contained genes homologous to those required for mating in *S. cerevisiae*. Two such genes encoded the transcriptional regulator Cph1 (Ste12) (47) and the kinase Hst20 (Ste20) (44). The *S. cerevisiae* homologs of these genes were known to participate in both the pheromone signaling and filamentous growth pathways, and it seemed possible that the *C. albicans* homologs might have been co-opted solely for the purpose of regulating filamentous growth, a prominent and much-studied feature of *C. albicans* associated with pathogenesis. However, homologs of other *S. cerevisiae* proteins whose only known roles were in mating were also present in *C. albicans*, including the pheromone transporter Hst6 (Ste6) (63) and the mating-specific trimeric G-protein signaling G_β component Cag1 (Gpa1) (67). The completed genome sequence of *C. albicans* would eventually reveal the conservation of many mating-gene homologs (34, 78). However, the discovery of mating in *C. albicans* ultimately depended on the discovery and subsequent manipulation of a mating-type locus.

THE *C. ALBICANS* MATING-TYPE LOCUS

The experimental demonstration that *C. albicans* could mate derived from the identification of a mating-type locus in the genome of *C. albicans*. Hull and Johnson (28) used a sequence trace from the Stanford *C. albicans* Sequencing Project to describe two alleles of a mating-type-like (*MTL*) locus via chromosome walking through a lambda library of *C. albicans* genomic DNA fragments. The *MTL*

locus resembles the *MAT* locus of *S. cerevisiae* in many respects (Fig. 2). It encodes homologs of all three transcriptional regulators (a1, α1, and α2) discussed above. These orthologs show 26 to 30% amino acid identity throughout their open reading frames, with even higher conservation within the DNA-binding regions of the regulators. In addition, the position of the genes and their relative orientations within the loci are conserved, despite a general lack of conserved synteny with *S. cerevisiae* (24, 25). Finally, the locations of the introns within the genes *MTLa1* and *MTLα2* are similar to the placement of introns within *MATa1* (28). All of these observations indicated that the *S. cerevisiae MAT* locus and the *C. albicans MTL* locus descended from a common ancestral locus.

Despite the similarities in sequence and overall structure, there are also important differences between the two mating-type loci. While *S. cerevisiae* maintains silenced copies of both *MAT* alleles at separate genomic loci (*HMR* and *HML*) (27), *C. albicans* contains only single copies of the information in the *MTLa* and *MTLα* alleles at the *MAT* locus (28), which is located on chromosome 5 (52). The *MTL* locus (8.8 kb) is considerably larger that the *MAT* locus (0.7 kb), and it contains several genes in addition to those encoding the transcriptional regulatory proteins (Fig. 2). Each allele of the *MTL* locus encodes genes for a poly(A) polymerase (*PAP*), a phosphatidylinositol kinase (*PIK*), and an oxysterol-binding protein-like protein (*OBP*). On average, the

A *S. cerevisiae MAT locus:*

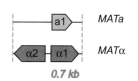

B *C. albicans MTL locus:*

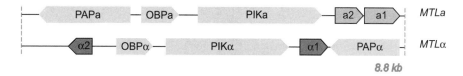

Figure 2. A comparison of the *S. cerevisiae MAT* locus and the *C. albicans MTL* locus. (A) The *S. cerevisiae MAT* locus contains genes for three transcriptional regulatory genes: **a1**, α1, and α2. DNA sequence inside of the dashed lines is nonidentical between homologs. (B) The homologous *C. albicans MTL* locus contains three transcriptional regulatory genes (outlined): **a1**, α1, α2, and an additional regulator, **a2**. In addition, the *MTL* locus contains two alleles each of genes homologous to a poly(A) polymerase (*PAP*), an oxysterol binding protein (*OBP*), and a phosphatidylinositol kinase (*PIK*) with no apparent function in mating. These genes are omitted in subsequent figures for clarity.

sequences of the proteins encoded by each allelic pair are only 60% identical, far less than the typical pair of alleles in the diploid genome (34). This observation raises the possibility that allele-specific functions could exist for the nonregulatory *MTL* genes, although none have so far been described. To date, these genes have not been demonstrated to function in the mating process of *C. albicans*; moreover, homologs of these genes are not found in the mating-type loci of other fungi. Thus, it is possible that they play no role in mating. The presence of "nonmating" genes in mating loci was subsequently described for *Cryptococcus neoformans*: the 100-kb *MAT* locus of this fungus contains at least 14 genes with no apparent function in mating regulation (46).

Although it was not immediately appreciated, the *C. albicans* MTL**a** allele also encodes a transcriptional regulator, **a**2, with similarity to HMG box proteins present at the mating-type loci of other fungal species, including *Cryphonectria parasitica*, *Kluyveromyces lactis*, *Neuorospora crassa*, and *Podospora anserina* (4, 12). This gene is missing in the *S. cerevisiae* MAT locus, and this absence has important implications for the evolution and regulation of cell type determination discussed later in this review. It is worth emphasizing that the *MTL***a**2 gene is not homologous to the apparently nonfunctional *MAT***a**2 transcript produced by the *MAT* locus of *S. cerevisiae* (33).

In addition to the sequence and organizational similarities of the *MAT* and *MTL* loci, the original description of the *C. albicans* *MTL* locus included an experiment showing the functional similarity of two of the regulators encoded by the mating-type loci. Multimerized **a**1-α2 DNA-binding sites from either *S. cerevisiae* or the *CAG1* promoter of *C. albicans* directed repression in heterologous reporter constructs when integrated into the *C. albicans* genome (28). The fact that this repression was dependent on the *C. albicans* **a**1 gene strongly suggested that *C. albicans* produced an **a**1-α2 activity (probably a heterodimer) that could recognize the **a**1-α2 DNA site for *S. cerevisiae*.

The common laboratory strain of *C. albicans*, CAI4 (19), is an **a**/α diploid since it contains one copy of *MTL***a** and one copy of *MTL*α (28). By analogy to *S. cerevisiae*, an **a**/α cell would not be expected to mate. This observation provided one explanation for why mating had not been observed in the laboratory. The presence of both alleles at the mating-type locus in CAI4 also ruled out the possibility that *C. albicans* had lost the ability to mate as a consequence of losing one allele of the mating-type locus. Furthermore, these observations suggested a new approach to examining mating in *C. albicans*:

"**a**" and "α" strains could be constructed genetically and their behavior could be studied in the laboratory.

THE DISCOVERY OF MATING

Two independent approaches were taken to generate putative mating-competent strains from *CAI4*. Hull et al. (29) employed targeted deletion of the entire *MTL***a** or *MTL*α locus and, alternatively, deletion of the individual *MTL* genes encoding the transcriptional regulatory proteins (Fig. 3). A second approach by Magee and Magee exploited the unstable inheritance of chromosome 5, which carries the *MTL* locus (52). Strains were passaged on sorbose, which induces the loss of one homolog of chromosome 5, including its *MTL* locus (31, 52). Subsequent passaging on rich medium results in the duplication of the remaining homolog, producing an **a**/**a** or α/α strain (Fig. 3). In both sets of strains, mating was monitored by selecting for progeny with complementary auxotrophic markers from each parental strain.

Cell-type-specific mating was observed using both of these approaches: **a** cells could mate with α cells, but no mating was observed between cells of the same mating type or with **a**/α diploid cells (29, 52). The mating products contained genetic and physical markers from both parents, and they had doubled DNA content (~4*n* by flow cytometry) compared to the parental strains. They also had single nuclei (29), as would be expected for cell fusion followed by nuclear fusion. Mating was observed under two conditions: after injecting the parental strains into the tail vein of a mouse and recovering the mating products in the kidney (29) and after mating on standard laboratory medium at 25°C (52). However, mating under both conditions was inefficient. Quantitative mating assays on laboratory medium later established that mating efficiency under these conditions was 10^{-7} (57). One reason for this inefficiency would become apparent with the discovery of a relationship between mating and a phenotypic switching phenomenon known as white-opaque switching.

WHITE-OPAQUE SWITCHING

White-opaque switching, first described by David Soll and colleagues (71), is a reversible switch between two heritable states known as white and opaque. The phenomenon has been most extensively studied in the strain WO-1, named for its ability to undergo white-opaque switching (71). Colonies of white and opaque cells are easily distinguished by their morphologies: opaque colonies and opaque sectors are flat and have a darker appearance than the typical domed white colonies of *C. albicans* that

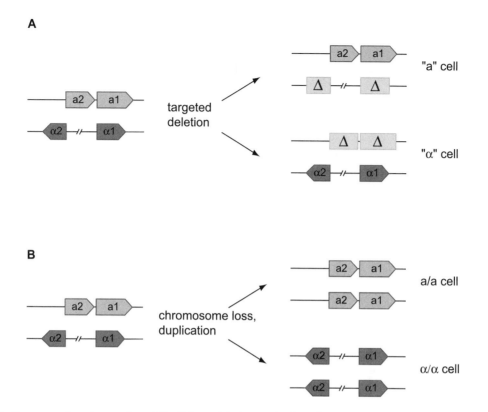

Figure 3. Constructed mating strains of *C. albicans*. Mating strains have been constructed from a/α diploids by two methods. (A) Targeted deletion of both regulatory proteins in one allele of the *MTL* locus results in a and α mating strains. (B) When grown on sorbose, *C. albicans* diploids are induced to lose one homolog of chromosome 5 (on which the *MTL* locus resides), resulting in a hemizygous strain. When cultured on rich medium (YEPD), the single remaining homolog of chromosome 5 duplicates, producing a homozygous a/a or α/α strain.

grow on standard laboratory media (Fig. 4). When a colony of white WO-1 cells is resuspended and plated on standard laboratory medium, the majority of the resulting colonies are white. However, a few colonies are opaque, and some of the colonies are white with opaque sectors. Likewise, when an opaque colony is resuspended and plated, the majority of the resulting colonies are opaque but a few are completely white (71). Opaque cells are larger and more elongated than white cells, and their surface appears to be speckled with structures known as pimples when visualized by scanning electron microscopy (3) (Fig. 4C to F). The mechanism that controls white-opaque switching is not understood in detail, but studies with the deacetylase inhibitor trichostatin-A (35) and strains carrying mutations in the *HDA1* and *RPD3* histone deacetylase genes (35, 74) have implicated chromatin structure as playing a role. White-opaque switching was studied largely in the context of virulence, where it was found that opaque cells are unstable (they reverted to white) and relatively avirulent in systemic infections but more persistent and virulent on an epithelial surface (40, 41). This observation raised the possibility that white and

opaque cells may be specialized for different niches in the host. However, one curious feature of white-opaque switching was that only a few *C. albicans* clinical isolates could undergo the transition (49). The majority of clinical isolates do not switch and remain locked in the white phase. The reason for this difference became clear with the discovery of the relationship between white-opaque switching and mating.

WHITE-OPAQUE SWITCHING AND MATING

The relationships between white-opaque switching, the *MTL* locus, and mating were discovered through a fortuitous observation. The common laboratory strain CAI4, which does not undergo white-opaque switching, was genetically manipulated to create a- and α-type mating strains through genetic deletions within the *MTL* locus. Unlike their parent strain, these mating strains appeared to undergo a phenotypic switch that resembled the "classic" white-opaque switching described for strain WO-1 (57). It was found that two transcriptional regulatory proteins encoded by the *MTL* locus, the homeodomain proteins a1 and α2, work together to

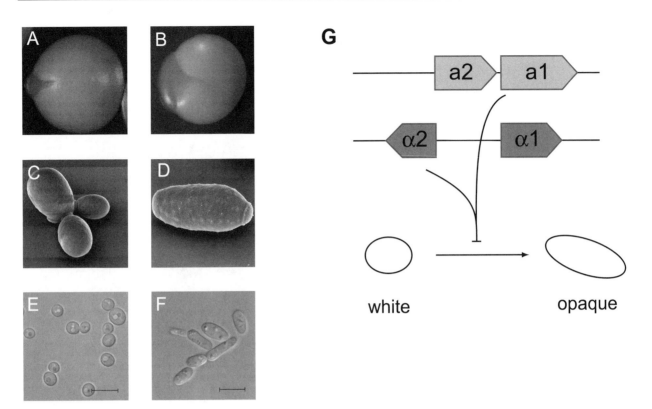

Figure 4. White-opaque switching. (A, B) White colonies with opaque sectors. When grown on medium containing the dye phloxine B, opaque sectors are darkly stained. (C) White cells of *C. albicans* are ellipsoidal when grown under rich conditions. Here, they are visualized by scanning electron microscopy. Preparation and visualization were done with the assistance of the Electron Microscopy Laboratory, University of California, Berkeley. (D) Opaque cells of *C. albicans* are larger and more elongated than white cells. In addition, their surface is covered with "pimple" structures of unknown function. (E) White cells are visualized by DIC microscopy. (F) Opaque cells are visualized by DIC microscopy. (G) The **a**1 and α2 transcriptional regulators together repress white-opaque switching in **a**/α *C. albicans* cells. Reprinted from *Cell* (57) and *Nature Reviews in Microbiology* (32) with permission of the publishers.

repress white-opaque switching (57) (Fig. 4G). This idea was consistent with a large body of biochemical and genetic analysis of the **a**1 and α2 proteins of *S. cerevisiae*, which form a heterodimer and together bind to DNA to repress the expression of many genes.

Because they contain both alleles of the *MTL* locus, *C. albicans* **a**/α strains (such as CAI4) are unable to undergo white-opaque switching and, unless they become an **a** or α cell, remain locked in the white phase. However, natural isolates of *C. albicans* that lack either the *MTLa1* or *MTLα2* genes would be expected to undergo white-opaque switching. Consistent with this prediction, the classic white-opaque switching strain WO-1 was found to lack *MTLa1* (57). This finding has now been generalized to many clinical isolates of *C. albicans*; two independent studies have shown that about 6% of clinical isolates are homozygous **a**/**a** or α/α at the *MTL* locus and most of these can undergo white-opaque switching (45, 49). In contrast, none of the stable **a**/α heterozygotes have been observed to form opaque colonies (45, 49). These studies show that

a- and α-type strains exist in nature and that, like the laboratory strain CAI4, their switching behavior is controlled by the *MTL* locus. Moreover, many of the natural **a**/**a** and α/α isolates are able to mate (45, 48).

The coregulation of mating and white-opaque switching by the *MTL* locus suggested that these two seemingly disparate phenomena might be related. Indeed, quantitative mating experiments showed that although mating between populations of white cells on laboratory medium is extremely inefficient (10^{-7}), mating between **a** and α populations of opaque cells occurs readily (40% of cells mated in the most efficient experiments) (57). Given the 10^6-fold advantage of opaque cells over white cells in mating efficiency, it is possible that the early mating experiments carried out with white cells resulted from mating between opaque cells that spontaneously arose in the parental populations. It is also consistent with the earlier observation that mating was most efficient at 25°C (52), which is the temperature at which opaque cells are stabilized. Taking into account the specialization of the opaque cell for mating, the mating efficiency of *C. albicans* approaches that of *S. cerevisiae*.

PHEROMONE RESPONSE

Mating in *C. albicans* is dependent on signaling between cells of opposite mating types. When **a** and α opaque cells are mixed in culture, they respond by forming long mating projections that resemble, in principle if not in detailed morphology, the cell elongations observed in *S. cerevisiae* known as shmoos (48, 57) (Fig. 4A). In both organisms, each cell type secretes a mating pheromone and expresses a cell surface receptor that recognizes pheromone of the opposite mating type. The altered cell morphologies in mating mixtures result from polarized chemotropic growth toward pheromone secreted by mating partners.

The *C. albicans* gene encoding the α-factor mating pheromone (*MFα*) was identified based on its resemblance to the *S. cerevisiae* α-factor gene: the *C. albicans* genome was searched for short open reading frames with repeated peptide motifs separated by processing sites for the Kex2 protease (8, 50, 62). The full-length *C. albicans* protein is probably processed into two 13-mer peptides and one 14-mer peptide (50, 62). Synthetic α-factor peptides (either the 13-mer or the 14-mer) are sufficient to induce the majority of opaque **a** cells to form long mating projections (over two cell diameters in length) within 4 h of treatment in liquid culture (8, 50, 62). Opaque cells from many **a/a** clinical isolates form mating projections from both their tips and their sides, and in some cases more than one mating projection is visible from a single opaque cell (50). When spotted onto a petri plate of opaque **a** cells, α-factor induces a faint halo of growth inhibition indicative of at least partial cell cycle arrest (62). The observation that this effect is only partial may be due to hydrolysis of the pheromone by a secreted protease. Indeed, *C. albicans* has a large family of genes coding for secreted aspartyl proteases (58), some of which (*SAP4*, *SAP5*, and *SAP6*) are highly upregulated upon pheromone induction (8). It is also possible that the cell cycle machinery in *C. albicans* is not as tightly controlled by α-factor as it is in *S. cerevisiae*.

Mating in *C. albicans* depends on a pheromone response pathway that is largely conserved with that of *S. cerevisiae* (5) (Fig. 5B). In *S. cerevisiae*, cell-type-specific pheromone receptors (Ste2 and Ste3) bind to pheromone produced by cells of the opposite mating type (α-factor and **a**-factor). A heterotrimeric G-protein (Gpa1, Ste4, and Ste18) then transduces the pheromone signal from the receptor to a cascade of protein kinases (Ste20, Ste11, Ste7, and Fus3). The mitogen-activated protein (MAP) kinase Fus3 ultimately phosphorylates a transcriptional regulator

(Ste12) that induces the expression of many genes. In *C. albicans*, homologs of several of these signaling components are required for mating. These components are shaded in gray in Fig. 5B. (In the text, *S. cerevisiae* homologs are indicated in parentheses when named differently.) The *C. albicans* α-factor is required for mating in α cells (8, 62), and its cognate α-factor receptor, *Ste2*, is required for mating in **a** cells (8). In addition, the **a**-factor transporter Hst6 (Ste6) and the Kex2 protease that processes α-factor are required for mating in **a** and α cells, respectively (51). Components of the heterotrimeric G-protein (Cag1 [Gpa1], Ste4, and Ste18) are present in *C. albicans* and expressed (50, 77) but have not yet been tested for their requirement in mating. The protein kinase Hst7 (Ste7) is required for mating (11, 51), and the protein kinase Cst20 (Ste20) is required for efficient mating (11). The Cek2 (Fus3) and Cek1 kinases appear to be partially redundant, because individually the knockout mutants can mate inefficiently whereas the cek1/cek1 cek2/cek2 double mutant is completely sterile (11). Finally, the transcription factor Cph1 (Ste12) is required for mating (11, 51).

The pheromone signals of *S. cerevisiae* and *C. albicans* are transduced from the plasma membrane to the nucleus, where transcriptional regulators direct changes in gene expression required for mating. When *S. cerevisiae* **a** cells are treated with synthetic α-factor peptide, many genes that are required for mating are induced while many others are repressed as a consequence of cell cycle arrest in G_1 (65). When opaque **a** cells of *C. albicans* are treated with α-factor, approximately 62 genes are induced while very few are repressed (8), again suggesting the absence of a tight cell cycle arrest under the conditions tested. Of the *C. albicans* genes that are pheromone induced, only 15 are shared with the *S. cerevisiae* pheromone response, indicating a major evolutionary diversification of this response. Most of the 15 shared genes are involved in some aspect of mating. The remaining *C. albicans* genes are made up mostly of proteins with no close relatives in any sequenced genome and conserved proteins with no apparent role in mating. Among these, the most surprising group of genes are cell surface and secreted proteins that had previously been implicated in *C. albicans* virulence in a murine model of infection (*HWP1* [75], *SAP4*, *SAP5*, *SAP6* [17], *RBT1* [9], and *RBT4* [9]) (8, 48). These genes could, in principle, participate in some aspect of mating (for example, cell-cell adhesion or pheromone degradation) and could also be used for virulence, perhaps to interact with host cells or to degrade host proteins. Alternatively, these genes could be induced as a consequence of polarized

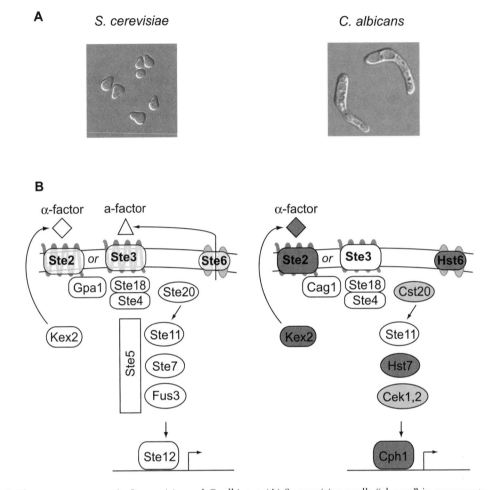

Figure 5. Pheromone response in *S. cerevisiae* and *C. albicans*. (A) *S. cerevisiae* **a** cells "shmoo" in response to α-factor pheromone. (Image courtesy of Emma McCullagh, University of California, San Francisco.) *C. albicans* **a** opaque cells undergo a similar polarized growth response in the presence of *C. albicans* α-factor. (B) Pheromone signaling in *S. cerevisiae* requires many signaling components, including a pheromone receptor (Ste2), a heterotrimeric G-protein (Gpa1, Ste4, Ste18), a protein kinase signaling cascade (Ste20, Ste11, Ste7, Fus3), and a transcriptional regulator (Ste12). *C. albicans* encodes homologs of several pheromone signaling components. *C. albicans* signaling components that are required for mating are shaded in dark gray, and components that are only partially required are shaded in light gray. The Cek1 and Cek2 MAP kinases are partially redundant in mating. Unshaded components have not been tested.

growth during formation of mating projections, which resemble the polarized growth required for production of hyphae. Indeed, the Hwp1 protein is localized to the cell wall of both hyphae and mating projections (13). The apparent interplay between virulence and the pheromone response in *C. albicans* is intriguing, and it is safe to say that this subject has not yet been thoroughly explored.

CELL TYPE DETERMINATION

This section addresses how the *C. albicans* mating-type loci regulate gene expression to give rise to three distinctive cell types, **a**, α, and **a**/α. Since diverging from their common ancestor 200 million to 800 million years ago (22, 26), *S. cerevisiae* and *C. albicans* have maintained some common features of

cell type control but have diversified in others. A crucial discovery in the work on cell type determination of *C. albicans* was the presence of a gene encoding the **a**2 mating-type regulator at the *C. albicans MTL* locus (77). Homologs of this HMG domain DNA-binding protein are found in the mating-type loci of many fungi (4, 12) but are absent from the *MAT* locus of *S. cerevisiae* and closely related yeasts (73).

To establish the role of the *MTL* transcriptional regulatory proteins in cell type determination, precise deletions of the four transcriptional regulators (**a**1, **a**2, α1, and α2) were constructed in each of 16 possible permutations of single, double, triple, and quadruple mutants. Each of these mutants was tested for its ability to undergo white-opaque switching and to mate with both **a** and α tester strains (77). To identify the target genes of each *MTL*-encoded transcriptional regulator, microarray-based gene expression profiles were determined for each of the 16 *MTL* mutants in the white phase and (where applicable) the opaque phase (77). The gene expression data for each of the mutant strains were consistent with mating behavior and identify three broad classes of genes based on their expression profiles: a1-α2 repressed genes, a-specific genes, and α-specific genes. Because white-opaque switching is also under cell type control, an additional and very large set of genes—the opaque-specific genes—are under indirect control of the *MTL* locus.

As can be seen in Fig. 6, there are many similarities and many differences between the cell type determination circuitry in *C. albicans* and *S. cerevisiae*. Among the differences, three are most significant.

(i) In *S. cerevisiae*, the a-specific genes are expressed constitutively; that is, they require no input from the mating-type locus. In *C. albicans*, expression of the a-specific genes requires the activator protein a2, which is encoded at the a mating-type locus. (ii) In *S. cerevisiae*, the a-specific genes are repressed in α cells by the α2 protein, encoded at the α mating-type locus. In *C. albicans*, α2 lacks this activity. There is a simple rationale for this: because the *C. albicans* a-specific genes require activation by a2 and since this protein is not expressed in α cells, there is no need to actively turn off these genes. (iii) In both *C. albicans* and *S. cerevisiae*, a1 and α2 together repress a set of genes. In *S. cerevisiae*, they repress a relatively small number of haploid-specific genes (20). In *C. albicans*, white-opaque switching (with its nearly 400 changes in gene expression) is also controlled by

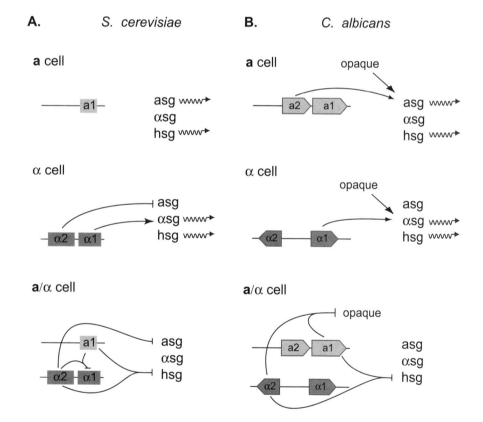

Figure 6. Cell-type determination in *S. cerevisiae* and *C. albicans*. (A) In *S. cerevisiae* a cells, the a-specific genes (asg) and haploid-specific genes (hsg) are transcribed constitutively. Expression is indicated by an arrow or block next to each class of genes. In α cells, the a-specific genes are repressed by the transcriptional regulator α2, and the α-specific genes (αsg) are activated by the α1 transcriptional regulator. As in a cells, the haploid-specific genes are constitutively expressed. In a/α cells, the a1-α2 heterodimer directly represses expression of haploid specific genes (hsg) and the α1 regulator required for α-specific gene expression. Additionally, a2 represses expression of the a-specific genes. (B) In *C. albicans* a cells, expression of the a-specific genes depends on the positive transcriptional regulator a2. The opaque state is also required for the expression of a-specific genes. In α cells, the a2 regulator is absent, and the a-specific genes are not expressed. Instead, α1 and the opaque state activate expression of the α-specific genes. In a/α cells, the a1-α2 heterodimer represses the switch to opaque, thereby preventing expression of the a- and α-specific genes. Additionally, the a1-α2 regulator represses genes that participate in mating in both cell types, the equivalents of "haploid-specific genes" in the diploid organism.

a1-α2. Opaque-specific genes include genes involved in metabolic specialization, cell adhesion, cell surface composition, and virulence (43, 77). Thus, by indirect means, the mating-type locus in *C. albicans* controls aspects of pathogenesis that are lacking in *S. cerevisiae*.

We now briefly discuss the identities of the genes that make up the different classes of cell-type-specific genes in *C. albicans*: a1-α2-repressed, **a**-specific, and α-specific genes. Each class of genes is organized according to its regulator in Fig. 7.

The first class of genes is repressed by the a1-α2 regulator, and many of these genes are implicated in the response to mating pheromone (Fig. 7). In *C. albicans*, a1-α2 represses genes through two layers of regulation. Six a1-α2-repressed genes (*STE2*,

CAG1 [*GPA1*], *CEK2* [*FUS3*], *FAR1*, *ORF 19.1616*, and *TOS9*) are derepressed in white cells lacking a1 or α2 (Fig. 7, first column) (77). Additionally, three genes (*CEK2*, *STE4*, and *TOS9*) require a switch to opaque for their full expression (77). The opaque state itself is repressed by a1-α2; thus, these genes are indirectly controlled by a1-α2 through this second layer of regulation. Many of the *C. albicans* genes repressed by a1-α2 (*CAG1* [*GPA1*], *STE4*, *CEK2* [*FUS3*], and *FAR1*) have homologs in *S. cerevisiae* that are directly repressed by a1-α2 (20). The subtle differences between *S. cerevisiae* and *C. albicans* mating-gene regulation preserve the output of gene expression in *C. albicans* mating strains but may be instrumental in restricting mating to opaque-phase cells. The finding that the gene encoding the *C. albicans*

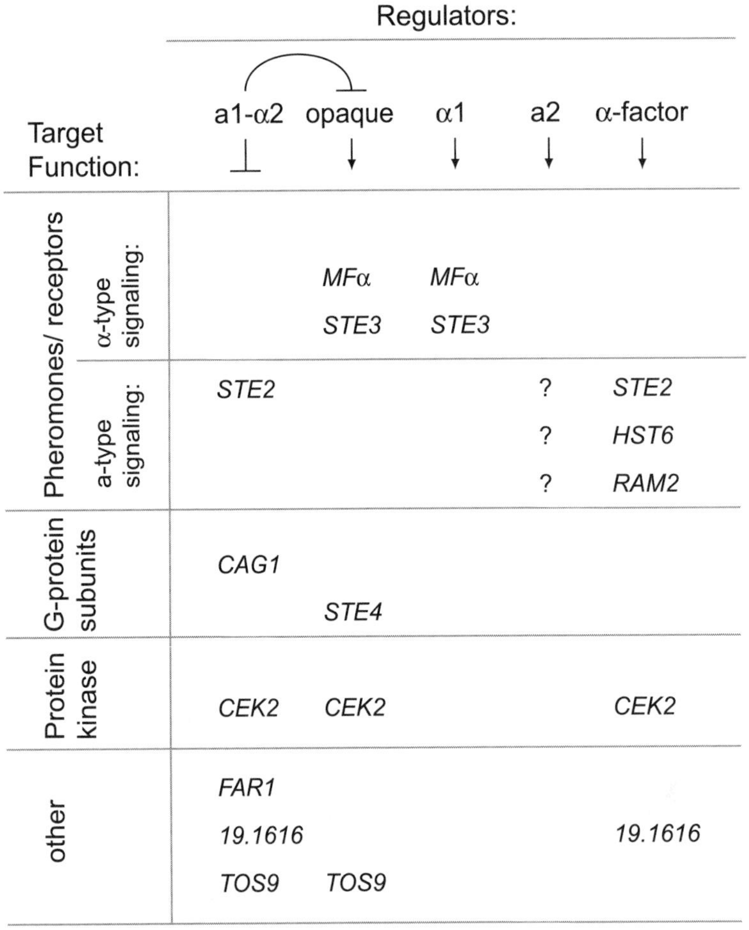

Target Function:		a1-α2 ⊥	opaque ↓	α1 ↓	a2 ↓	α-factor ↓
Pheromones/ receptors	α-type signaling:		MFα STE3	MFα STE3		
	a-type signaling:	STE2			? ? ?	STE2 HST6 RAM2
G-protein subunits		CAG1	STE4			
Protein kinase		CEK2	CEK2			CEK2
other		FAR1 19.1616 TOS9	TOS9			19.1616

Figure 7. Mating gene regulation in *C. albicans*. A number of *C. albicans* genes are controlled by a1-α2, α1, and possibly by a2, in the SC5314 background. Genes that are controlled by a particular regulator (top row) are entered into each column below the regulator. In the leftmost column, these target genes are classified according to their function in mating. Six genes are regulated by a1-α2 in white-phase *C. albicans* cells. Genes that are upregulated in the opaque phase are indirectly controlled by a1-α2 and include members from all categories of genes on the left. While the α-specific genes (controlled by α1) have been identified, the **a**-specific genes (depending on a2 for their expression) have not been rigorously identified. Three candidate **a**-specific genes are induced upon exposure to α-factor and are likely to be **a**-specific genes, based on their roles in *S. cerevisiae* mating. Genes induced in **a** cells upon exposure to α-factor mating pheromone are also included above.

α-factor receptor *STE2* is regulated by a1-α2 (77) was unexpected because its *S. cerevisiae* homolog is strictly a specific.

The second cell-type-specific gene class is the α-specific genes, which require α1 for their expression. The α-factor pheromone, *MFα*, and the a-factor pheromone receptor, *STE3*, are α-specific genes in both *C. albicans* and *S. cerevisiae* (77). However, in *C. albicans*, they require a switch to opaque in order to be fully expressed (43, 77).

The final class of cell-type-specific genes contains the a-specific genes. Three putative a-specific genes are expressed in opaque a cells only on pheromone induction, which is itself dependent on a2. Thus, they have not been conclusively demonstrated to depend on a2 for their expression. They code for an a-factor transporter (*HST6* [*STE6*]), a farnesyl transferase subunit required for a-factor production (*RAM2*), and the α-factor pheromone receptor (*STE2*) (77). All three are known to play specific roles in a cells of *S. cerevisiae* (68, 80), although *RAM2* expression in *S. cerevisiae* is not cell type regulated. Additionally, *C. albicans STE2* is likely to be regulated by a2 because it is expressed more highly in an a/a clinical isolate than in an a/α clinical isolate (50), and it is required for mating as an a cell but not as an α cell (8).

Cell type regulation of mating genes may vary among clinical isolates of *C. albicans*. *STE4* is expressed more highly in white cells in the a/a strain P37005 than in the nonisogenic a/α strain 3153A, suggesting that *STE4* is regulated by a1/α2 in the 3153A background (50). Also in strain P37005, *STE4* is not upregulated on the switch to opaque but is induced on exposure to pheromone (50). Each of these observations is in contrast to the regulation observed in the SC5314 background (outlined in Fig. 7), where *STE4* is not directly regulated by a1-α2 in white cells (77), *STE4* is upregulated in opaque cells (77), and α-factor pheromone does not induce its expression (8). In both strain backgrounds, however, *STE4* is expressed where it presumably functions, i.e., in opaque cells exposed to pheromone.

Perhaps the most interesting change in the regulatory circuit for cell type determination is the replacement of a positive circuit in *C. albicans* (activation of the a-specific genes by a2) with a negative circuit in *S. cerevisiae* (repression of the α-specific genes by α2). Because homologs of the a2 protein are found in the mating-type loci of many ascomycetes, it appears as though a2 has been lost in the *S. cerevisiae* lineage. To regain effective cell type control, this loss would require that (i) the a-specific genes bypass the requirement for a2 in order to be expressed and (ii) the homeodomain protein α2

acquire the ability to repress the a-specific genes and thereby prevent their expression in α cells. The fact that such dramatic differences in circuitry can underlie what is, in essence, the same basic scheme (a mating locus determines cell type) testifies to the "evolvability" of transcription.

FUNCTIONS OF WHITE-OPAQUE SWITCHING

In addition to the upregulation of mating genes, opaque cells have several characteristics that may enhance their ability to mate. The unique surface structures of opaque cells, known as pimples, may play an important part in the process of membrane fusion. Transmission electron microscopy of sectioned opaque cells reveals that the cell wall tapers in thickness near the center of the pimple, and a membrane bilayer can be observed at the center of the structure (1). Membrane fusion is an essential step in mating, and fungi must degrade their cell wall for the membranes of participating cells to fuse (21). While *S. cerevisiae* degrades the cell wall between mating partners only after the two cells have come into contact, it is possible that opaque cells of *C. albicans* are primed for mating through exposure of the membrane prior to cell contact.

Mating is probably one of many specialized functions of opaque cells. Nearly 400 genes are differentially regulated between the white and opaque phases (43, 77), and although it is possible that some of the uncharacterized opaque-specific genes play roles in mating, it seems unlikely that most do. Instead, many of the differences between white and opaque cells seem to be in the specialization of metabolic enzymes and regulators (43). This specialization may serve to direct opaque cells (and therefore mating) to specific compartments within the host. For example, Soll and colleagues have demonstrated that opaque cells proliferate on skin more efficiently than do white cells (40); moreover, opaque cells mate efficiently on the skin of newborn mice (42). As described above, opaque cells are more stable at lower temperatures, and skin temperature is perhaps an ideal environment for mating. Furthermore, *C. albicans* cells on the skin surface may be more likely to come into contact with mating partners from different genetic backgrounds.

Some of the specializations required for mating such as secretion of pheromones and their detection by pheromone receptors may be incompatible with many host environments. For example, opaque cells are more sensitive to oxidative stress and neutrophils than white cells are (37). Thus, the white-opaque switch may be a strategy to generate a small number of sensitive but mating-competent cells within a much larger population of white cells that can withstand the rigors of life in a mammalian host.

ZYGOTE FORMATION AND KARYOGAMY

It is thought that, for most fungi, the principal function of mating is to allow recombination of genetic material. Following the fusion of two cells to form a zygote, an important step on the path to recombination is the process of nuclear fusion, or karyogamy. Karyogamy combines the chromosomes of both parents into the same nuclear compartment to allow for various types of genetic exchange.

The earliest observations of mating in *C. albicans* (monitored through genetic experiments) suggested that karyogamy could occur, but the efficiency of this process was not determined. In these experiments, mating and karyogamy were detected by selecting for progeny which contained complementary auxotrophic markers from each parent strain (29, 52). Several pieces of evidence indicated that these products had undergone karyogamy. All progeny observed had a single nucleus, visualized by 4′,6-diamidino-2-phenylindole (DAPI) staining (29). In addition, flow cytometry analysis indicated that the progeny had tetraploid DNA content (29, 52). Finally, the progeny were shown to contain four different alleles at one genetic locus (the *MTL* locus); this situation could arise only from the combination of the four alleles, two from each parent (29). These observations established that karyogamy was possible for *C. albicans*, but they did not allow an assessment of its frequency because both mating and karyogamy were scored many generations after the initial mating event.

The first real-time microscopic studies of the mating process of *C. albicans* suggested that karyogamy did not readily occur. In these studies, microscopy of living cells was used to observe mating between several clinical isolates of **a**/**a** and α/α strains that were mixed and cultured under nonselective conditions (48). In these mixed cultures, **a**/**a** and α/α opaque cells formed long mating projections (or conjugation tubes) up to five times the diameter of the mother cell. If no partners of the opposite mating-type were nearby, the projections reverted to normal budding growth after several hours; they formed constrictions and septated. In the presence of a mating partner, projections on cells of opposite mating-types grew toward one another. Concurrently, the nuclei of mating cells migrated into the conjugation tubes and, on fusion of the mating projections at their apices, the two parental nuclei were juxtaposed in the cytoplasm of the newly formed zygote. In these experiments, the nuclei did not appear to fuse, but, rather, a vacuole expanded within the conjugation bridge and pushed apart the two nuclei. The nuclei subsequently migrated back into the mating projections without having fused (48).

The resolution of the apparent inconsistency between the cell-biological and genetic experiments came from experiments that studied karyogamy both genetically and microscopically in strains of several genetic backgrounds, under the same culture conditions, and in the absence of genetic selection. Karyogamy between genetically matched SC5314-derived strains in these studies was very efficient (>90% of zygotes underwent nuclear fusion), whereas karyogamy between the clinical strains used in the prior cytological studies (WO-1 and P37005) was relatively inefficient (~10%). In addition, karyogamy was found to be highly dependent on the culture medium, with high karyogamy rates observed in some media and much lower rates in others. In contrast, the efficiency of zygote formation (the initial cell fusion event) depended much less on media conditions. Taken together, these results explain the discrepancy of karyogamy observations between earlier genetic and cytological experiments. These results also suggest that the capacity for karyogamy may be highly variable among *C. albicans* clinical isolates. Indeed, in a set of crosses involving strains from nine genetic backgrounds, the efficiency of karyogamy ranged from over 90% to less than 4% under the same conditions (7). It has been proposed that *C. albicans* clinical isolates can be divided into at least five clades based on genetic similarities (72). For the strains tested so far, the efficiency of nuclear fusion seems unrelated to whether the mating partners belong to the same clades or to different clades (7).

For *C. albicans* strains that are competent for nuclear fusion, karyogamy resembles, at least in outline, that observed for *S. cerevisiae* (66). **a** and α strains constructed in the karyogamy-competent SC5314 background and marked with yellow fluorescent protein- and cyan fluorescent protein-labeled nuclei were observed during mating by videomicroscopy (7). Following zygote formation, the yellow and blue nuclei migrated toward one another, and their signals colocalized on nuclear fusion. The fluorescent signals did not separate on evagination at the site of the nascent bud; rather, the single nucleus divided and a daughter nucleus containing signal from both parents migrated into the nascent bud formed near the site of cell fusion (Fig. 8).

Further evidence that nuclear fusion is an active process in *C. albicans* is provided by genetic and cytological evidence that it requires the Kar3 protein, a microtubule-based kinesin motor protein that, in *S. cerevisiae*, is required for nuclear migration prior to nuclear fusion during mating (56, 66). In *C. albicans*, Kar3 is required for both efficient karyogamy and efficient mating, as scored genetically (7). Kar3 is not

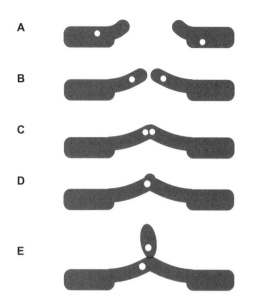

Figure 8. A schematic for karyogamy in *C. albicans*. (A) *C. albicans* **a** and α opaque cells respond to secreted pheromones by forming polarized growth projections. (B) Mating projections grow toward one another, and nuclei migrate into the projections. (C) Cell fusion between two cells forms a zygote and combines the two parent nuclei into the same cytoplasmic compartment. (D) The two nuclei undergo karyogamy (nuclear fusion). A nascent bud begins to emerge near the site of cell fusion. (E) The single remaining nucleus divides, and the daughter nucleus migrates into the daughter bud prior to septation. The daughter cell is tetraploid and contains information from both parental nuclei.

required for zygote formation or any of the mating activities preceding this event.

The apparent regulation of karyogamy in *C. albicans* by environmental signals is a topic that deserves further study. This phenomenon may relate to the wide range of karyogamy efficiencies observed for different clinical isolates. It is possible, for example, that differences in efficiencies of nuclear fusion among clinical isolates reflect differences in the environmental regulation of this process rather than differences in the karyogamy machinery itself.

PHENOTYPES OF MATING PRODUCTS, CHROMOSOME LOSS, AND A PARASEXUAL CYCLE

Crosses between opaque **a** and α cells form tetraploid cells, most of which rapidly revert to the white phase (57). The small fraction of mating products that remain in the opaque phase reveal an important feature of this reversion. In the rare cases where opaque cells persist after mating, the mating products have lost either *MTL***a***1* or *MTL*α*2* (57). These observations suggest that the **a**1-α2 transcriptional repressor actively destabilizes the opaque state when its two components are combined into the

zygote during mating, thereby returning the products of a successful mating to the white phase.

Tetraploid progeny formed from mating of **a** and α diploid cells can be stably propagated in the laboratory, although tetraploid *C. albicans* cells are apparently rare in nature (60). In principle, there are two processes that could return tetraploids to the diploid state: a true sexual cycle involving meiosis, or a parasexual cycle accomplished through reductional mitotic divisions. To date, meiosis has not been observed for *C. albicans*. The *C. albicans* genome encodes several homologs of genes that are responsible for meiosis and sporulation in *S. cerevisiae*, including *SPO11*, *DMC1*, and *NDT80* (14, 78); however, other *S. cerevisiae* "meiosis" genes, including *IME1*, *ZIP2*, and *SPO13*, are apparently absent from the *C. albicans* genome (34, 78). At this stage of analysis, it is difficult to convincingly argue one way or another for the existence of meiosis in *C. albicans*.

While the presence of meiosis remains unresolved, there is evidence of another pathway that *C. albicans* tetraploids can use to return to diploidy. Tetraploid **a**/α cells obtained through mating can undergo rapid chromosome loss on certain types of media that are nutritionally poor (6). Under these conditions, chromosome loss by tetraploids appears to be random but cooperative. Loss of one chromosome predisposes cells to lose others until the diploid state (within the resolution of flow cytometry) is reached. The exact mechanism of this chromosome loss is unknown, but it does not have the characteristics of a true meiosis, nor is it dependent on several of the "meiosis" genes discussed above (6).

The loss of chromosomes by tetraploids probably reflects a more general property of *C. albicans*, namely, the plasticity of its genome. In *C. albicans*, chromosome loss and genomic rearrangements appear fairly common, particularly in response to stress (53). One example is the loss of one homolog of chromosome 5 when cells are grown on media containing sorbose, discussed above (30, 31). Chromosome loss in tetraploids resembles this phenomenon in that the stress of low-nutrient conditions (which causes death in the majority of tetraploid cells) can be overcome by chromosome loss. It is possible that a regulated, active mechanism for genomic instability operates both in the loss of chromosome 5 from diploids and in the loss of a whole complement of chromosomes from tetraploids.

Chromosome loss can reduce tetraploid cells to diploidy, thus completing a parasexual cycle that can be carried out in the laboratory (Fig. 9). The cycle begins when **a** and α strains in the white phase switch to opaque and become competent for mating. Opaque cells mate to form an **a**/α tetraploid, and the

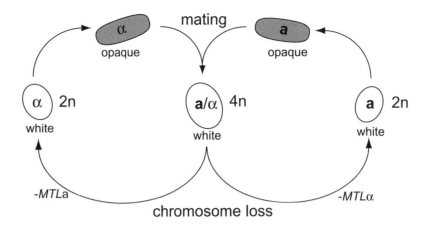

Figure 9. A parasexual cycle for *C. albicans*. *C. albicans* diploid **a/a** and α/α opaque cells must switch to the opaque phase in order to mate. The mating products are **a**/α tetraploids, and they quickly revert to the white phase. Tetraploids are stable under standard laboratory conditions, but can be induced to undergo rapid, concerted chromosome loss on certain growth media. Chromosomes are lost randomly until a diploid state is reached (within the resolution of flow cytometry). Because chromosome loss is random, all three possible combinations of *MTL* alleles are generated during chromosome loss: **a**/α, **a/a**, and α/α. The resulting diploid **a/a** and α/α products are competent to switch to opaque and mate with cells of the opposite type, thereby completing a parasexual cycle.

tetraploid quickly reverts to the white phase. Under the appropriate environmental conditions, these tetraploids randomly and cooperatively lose chromosomes until the diploid state is restored. Because chromosome loss from tetraploids is random, three types of diploids are generated, **a**/α, **a/a**, and α/α. Importantly, the **a/a** and α/α cells generated through chromosome loss remain competent for white-opaque switching and mating, and therefore the parasexual cycle can be perpetuated for multiple generations (6).

The lack of a tractable sexual cycle has made classical genetic analysis of *C. albicans* difficult. Although limited compared to the sexual cycles of "model" fungi, the parasexual cycle of *C. albicans* can be employed for some simple genetic tests. One application is for linking a phenotype to a genotype. To demonstrate that a specific phenotype is caused by a particular mutation and not a second, unlinked mutation, heterozygous null mutants of a gene of interest can be crossed and their tetraploid progeny can be induced to lose chromosomes and revert to diploid. If the phenotype is caused by the candidate gene, only diploid products of chromosome loss that have two null alleles of the intended mutation should display the phenotype (6). A second application of the parasexual cycle is in demonstrating that a mutation occurs in an essential gene. In this case, mating of two heterozygous null strains followed by chromosome loss should produce no diploid colonies that are homozygous null for the essential gene of interest (6). A third application is in creating new

mutant strains. In principle, homozygous null mutants could be constructed by mating two heterozygous mutants followed by chromosome loss. This strategy could bypass the sometimes difficult step of deleting the second allele of a gene of interest. Alternatively, new combinations of (unlinked) homozygous null mutations could be created through the same process (6).

POPULATION GENETICS AND MATING

This section discusses the apparent paradox that, although *C. albicans* has a robust parasexual cycle that can be readily observed in the laboratory, population studies indicate that the *C. albicans* population is largely clonal (76, 81). Notably, however, population studies have also detected evidence for a low level of genetic exchange (23, 79). Nonetheless, recombination between genetically unrelated isolates of *C. albicans* appears to be a rare event. Below, we discuss three possible hypotheses that might reconcile the existence of an elaborate and conserved mating mechanism, seen in both clinical isolates and laboratory strains, with the observation of a largely clonal population structure in *C. albicans*.

According to the first hypothesis, the mating apparatus is highly conserved but successful mating and recombination in the wild is simply very rare. The majority of *C. albicans* clinical isolates are **a**/α strains; to mate, they must undergo *MTL* gene conversion, deletion, or chromosome loss to generate mating-competent **a/a** or α/α strains. White-opaque switching itself could present a second barrier to mating because switching is infrequent (64, 71), and opaque cells appear to be unstable in many niches within the host (41). A third possible block to successful recombination is the inability of some strains to efficiently undergo nuclear fusion (7, 48). Even in strains that do efficiently undergo this process in the laboratory, environmental conditions in the wild may downregulate this process. All of these barriers to mating and karyogamy have been overcome in the laboratory environment, but they may severely limit recombination in the wild. Nevertheless, conservation of a large number of mating genes would argue for a selective advantage conferred by the capacity to mate. According to this scenario, the main purpose of mating would be to generate a very low level of mating and recombination in the wild.

A second hypothesis is that mating could be conserved for a function unrelated to recombination. For example, the plant pathogen *Ustilago maydis* mates to form a stable dikaryon, which is the form responsible for pathogenesis (16). According to this idea, mating—but not recombination—is required for some aspect of *C. albicans* pathogenesis. It has been proposed, for example, that one purpose of mating is to regulate white-opaque switching, which in turn regulates virulence (54). Another potential benefit of mating without karyogamy or recombination is the transfer of cytoplasmic elements such as prions, viruses, or mitochondria. There is evidence that mitochondrial populations exhibit clonality (2), but to our knowledge, prions and viruses have not been observed in *C. albicans*.

The third scenario that remains consistent with the primarily clonal population structure is that mating and physical recombination are actually frequent but occur most often between identical or closely related strains. Recombination between such strains would be difficult or impossible to detect in population studies. In support of this idea, multiple isolates of *C. albicans* sampled from the same host are much more closely related (by Ca3 fingerprinting) than samples from multiple patients in the same city or from the population at large; however, they are not identical (36).

What would be the purpose of mating and recombination between highly similar isolates? Mating between related isolates followed by chromosome loss or meiosis could lead to a limited amount of genetic diversity within a commensal or infectious population. *C. albicans*, being diploid, has two copies of each gene, and many (if not most) show allelic differences (34, 53, 69). In principle, these alleles could be shuffled by a sexual or parasexual cycle within the host, after commensalism is established, to create a population of limited diversity. These variant strains may not be transmissible or become widespread within a given commensal population, but they could theoretically perform specific functions for virulence or persistence within a specific host niche, according to their fitness.

CONCLUSIONS

Recent discoveries in the mating process of *C. albicans* have revealed a complex, conserved mating apparatus that remained hidden from investigators for many years. The critical role of white-opaque switching in mating seems unique to *C. albicans* and may reflect a long evolutionary relationship between mating and life inside a mammalian host. Chromosome loss, a general propensity of *C. albicans*, provides a means of completing a parasexual cycle that in principle can be used to generate diversity. Although mating and a parasexual cycle have been well documented in the laboratory, we are only beginning to understand how they pertain to the success of *C. albicans* as a commensal organism and opportunistic pathogen.

REFERENCES

1. **Anderson, J., R. Mihalik, and D. R. Soll.** 1990. Ultrastructure and antigenicity of the unique cell wall pimple of the *Candida* opaque phenotype. *J. Bacteriol.* **172:**224–235.
2. **Anderson, J. B., C. Wickens, M. Khan, L. E. Cowen, N. Federspiel, T. Jones, and L. M. Kohn.** 2001. Infrequent genetic exchange and recombination in the mitochondrial genome of *Candida albicans*. *J. Bacteriol.* **183:**865–872.
3. **Anderson, J. M., and D. R. Soll.** 1987. Unique phenotype of opaque cells in the white-opaque transition of *Candida albicans*. *J. Bacteriol.* **169:**5579–5588.
4. **Astrom, S. U., A. Kegel, J. O. Sjostrand, and J. Rine.** 2000. *Kluyveromyces lactis* Sir2p regulates cation sensitivity and maintains a specialized chromatin structure at the cryptic alpha-locus. *Genetics* **156:**81–91.
5. **Bardwell, L.** 2005. A walk-through of the yeast mating pheromone response pathway. *Peptides* **26:**339–350.
6. **Bennett, R. J., and A. D. Johnson.** 2003. Completion of a parasexual cycle in *Candida albicans* by induced chromosome loss in tetraploid strains. *EMBO J.* **22:**2505–2515.
7. **Bennett, R. J., M. G. Miller, P. R. Chua, M. E. Maxon, and A. D. Johnson.** 2005. Nuclear fusion occurs during mating in *Candida albicans* and is dependent on the *KAR3* gene. *Mol. Microbiol.* **55:**1046–1059.
8. **Bennett, R. J., M. A. Uhl, M. G. Miller, and A. D. Johnson.** 2003. Identification and characterization of a *Candida albicans* mating pheromone. *Mol. Cell. Biol.* **23:**8189–8201.
9. **Braun, B. R., W. S. Head, M. X. Wang, and A. D. Johnson.** 2000. Identification and characterization of TUP1-regulated genes in *Candida albicans*. *Genetics* **156:**31–44.
10. **Calderone, R. A.** 2002. Taxonomy and biology of *Candida*, p. 15–27. *In* R. A. Calderone (ed.), *Candida and Candidiasis*. ASM Press, Washington, D.C.
11. **Chen, J., S. Lane, and H. Liu.** 2002. A conserved mitogen-activated protein kinase pathway is required for mating in *Candida albicans*. *Mol. Microbiol.* **46:**1335–1344.

12. Coppin, E., R. Debuchy, S. Arnaise, and M. Picard. 1997. Mating-types and sexual development in filamentous ascomycetes. *Microbiol. Mol. Biol. Rev.* **61:**411–428.

13. Daniels, K. J., S. R. Lockhart, J. F. Staab, P. Sundstrom, and D. R. Soll. 2003. The adhesin Hwp1 and the first daughter cell localize to the a/a portion of the conjugation bridge during *Candida albicans* mating. *Mol. Biol. Cell* **14:**4920–4930.

14. Diener, A. C., and G. R. Fink. 1996. *DLH1* is a functional *Candida albicans* homologue of the meiosis-specific gene *DMC1*. *Genetics* **143:**769–776.

15. Edmond, M. B., S. E. Wallace, D. K. McClish, M. A. Pfaller, R. N. Jones, and R. P. Wenzel. 1999. Nosocomial bloodstream infections in United States hospitals: a three-year analysis. *Clin. Infect. Dis.* **29:**239–244.

16. Feldbrugge, M., J. Kamper, G. Steinberg, and R. Kahmann. 2004. Regulation of mating and pathogenic development in *Ustilago maydis*. *Curr. Opin. Microbiol.* **7:**666–672.

17. Felk, A., M. Kretschmar, A. Albrecht, M. Schaller, S. Beinhauer, T. Nichterlein, D. Sanglard, H. C. Korting, W. Schafer, and B. Hube. 2002. *Candida albicans* hyphal formation and the expression of the Efg1-regulated proteinases Sap4 to Sap6 are required for the invasion of parenchymal organs. *Infect. Immun.* **70:**3689–3700.

18. Filler, S. G., and B. J. Kullberg. 2002. Deep-seated candidal infections, p. 341–348. *In* R. A. Calderone (ed.), *Candida and Candidiasis*. ASM Press, Washington, D.C.

19. Fonzi, W. A., and M. Y. Irwin. 1993. Isogenic strain construction and gene mapping in *Candida albicans*. *Genetics* **134:**717–728.

20. Galgoczy, D. J., A. Cassidy-Stone, M. Llinas, S. M. O'Rourke, I. Herskowitz, J. L. DeRisi, and A. D. Johnson. 2004. Genomic dissection of the cell-type-specification circuit in *Saccharomyces cerevisiae*. *Proc. Natl. Acad. Sci. USA* **101:**18069–18074.

21. Gammie, A. E., V. Brizzio, and M. D. Rose. 1998. Distinct morphological phenotypes of cell fusion mutants. *Mol. Biol. Cell* **9:**1395–1410.

22. Gargas, A., P. T. DePriest, M. Grube, and A. Tehler. 1995. Multiple origins of lichen symbioses in fungi suggested by SSU rDNA phylogeny. *Science* **268:**1492–1495.

23. Graser, Y., M. Volovsek, J. Arrington, G. Schonian, W. Presber, T. G. Mitchell, and R. Vilgalys. 1996. Molecular markers reveal that population structure of the human pathogen *Candida albicans* exhibits both clonality and recombination. *Proc. Natl. Acad. Sci. USA* **93:**12473–12477.

24. Hamer, L., H. Pan, K. Adachi, M. J. Orbach, A. Page, L. Ramamurthy, and J. P. Woessner. 2001. Regions of microsynteny in *Magnaporthe grisea* and *Neurospora crassa*. *Fungal Genet. Biol.* **33:**137–143.

25. Hartung, K., D. Frishman, A. Hinnen, and S. Wolfl. 1998. Single-read sequence tags of a limited number of genomic DNA fragments provide an inexpensive tool for comparative genome analysis. *Yeast* **14:**1327–1332.

26. Hedges, S. B. 2002. The origin and evolution of model organisms. *Nat. Rev. Genet.* **3:**838–849.

27. Herskowitz, I., J. Rine, and J. Strathern. 1992. Mating-type determination and mating-type interconversion in *Saccharomyces cerevisiae.*, p. 583–657. *In* E. W. Jones, J. R. Pringle, and J. R. Broach (ed.), *The Molecular and Cellular Biology of the Yeast Saccharomyces*, vol. 2. Cold Spring Harbor Laboratory Press, Plainview, N.Y.

28. Hull, C. M., and A. D. Johnson. 1999. Identification of a mating type-like locus in the asexual pathogenic yeast *Candida albicans*. *Science* **285:**1271–1275.

29. Hull, C. M., R. M. Raisner, and A. D. Johnson. 2000. Evidence for mating of the "asexual" yeast *Candida albicans* in a mammalian host. *Science* **289:**307–310.

30. Janbon, G., F. Sherman, and E. Rustchenko. 1999. Appearance and properties of L-sorbose-utilizing mutants of *Candida albicans* obtained on a selective plate. *Genetics* **153:**653–664.

31. Janbon, G., F. Sherman, and E. Rustchenko. 1998. Monosomy of a specific chromosome determines L-sorbose utilization: a novel regulatory mechanism in *Candida albicans*. *Proc. Natl. Acad. Sci. USA* **95:**5150–5155.

32. Johnson, A. 2003. The biology of mating in *Candida albicans*. *Nat. Rev. Microbiol.* **1:**106–116.

33. Johnson, A. D. 1995. Molecular mechanisms of cell-type determination in budding yeast. *Curr. Opin. Genet. Dev.* **5:**552–558.

34. Jones, T., N. A. Federspiel, H. Chibana, J. Dungan, S. Kalman, B. B. Magee, G. Newport, Y. R. Thorstenson, N. Agabian, P. T. Magee, R. W. Davis, and S. Scherer. 2004. The diploid genome sequence of *Candida albicans*. *Proc. Natl. Acad. Sci. USA* **101:**7329–7334.

35. Klar, A. J., T. Srikantha, and D. R. Soll. 2001. A histone deacetylation inhibitor and mutant promote colony-type switching of the human pathogen *Candida albicans*. *Genetics* **158:**919–924.

36. Kleinegger, C. L., S. R. Lockhart, K. Vargas, and D. R. Soll. 1996. Frequency, intensity, species, and strains of oral *Candida* vary as a function of host age. *J. Clin. Microbiol.* **34:**2246–2254.

37. Kolotila, M. P., and R. D. Diamond. 1990. Effects of neutrophils and in vitro oxidants on survival and phenotypic switching of *Candida albicans* WO-1. *Infect. Immun.* **58:**1174–1179.

38. Kullberg, B. J., and S. G. Filler. 2002. Candidemia, p. 327–340. *In* R. A. Calderone (ed.), *Candida and Candidiasis*. ASM Press, Washington, D.C.

39. Kupiec, M., B. Byers, R. E. Esposito, and A. P. Mitchell. 1997. Meiosis and sporulation in *Saccharomyces cerevisiae*, p. 889–1036. *In* E. W. Jones, J. R. Pringle, and J. R. Broach (ed.), *The Molecular and Cellular Biology of the Yeast Saccharomyces*, vol. 3. Cold Spring Harbor Laboratory Press, Plainview, N.Y.

40. Kvaal, C., S. A. Lachke, T. Srikantha, K. Daniels, J. McCoy, and D. R. Soll. 1999. Misexpression of the opaque-phase-specific gene PEP1 (SAP1) in the white phase of *Candida albicans* confers increased virulence in a mouse model of cutaneous infection. *Infect. Immun.* **67:**6652–6662.

41. Kvaal, C. A., T. Srikantha, and D. R. Soll. 1997. Misexpression of the white-phase-specific gene WH11 in the opaque phase of *Candida albicans* affects switching and virulence. *Infect. Immun.* **65:**4468–4475.

42. Lachke, S. A., S. R. Lockhart, K. J. Daniels, and D. R. Soll. 2003. Skin facilitates *Candida albicans* mating. *Infect. Immun.* **71:**4970–4976.

43. Lan, C. Y., G. Newport, L. A. Murillo, T. Jones, S. Scherer, R. W. Davis, and N. Agabian. 2002. Metabolic specialization associated with phenotypic switching in *Candida albicans*. *Proc. Natl. Acad. Sci. USA* **99:**14907–14912.

44. Leberer, E., D. Harcus, I. D. Broadbent, K. L. Clark, D. Dignard, K. Ziegelbauer, A. Schmidt, N. A. Gow, A. J. Brown, and D. Y. Thomas. 1996. Signal transduction through homologs of the Ste20p and Ste7p protein kinases can trigger hyphal formation in the pathogenic fungus *Candida albicans*. *Proc. Natl. Acad. Sci. USA* **93:**13217–13222.

45. Legrand, M., P. Lephart, A. Forche, F. M. Mueller, T. Walsh, P. T. Magee, and B. B. Magee. 2004. Homozygosity at the *MTL* locus in clinical strains of *Candida albicans*: karyotypic rearrangements and tetraploid formation. *Mol. Microbiol.* 52:1451–1462.

46. Lengeler, K. B., D. S. Fox, J. A. Fraser, A. Allen, K. Forrester, F. S. Dietrich, and J. Heitman. 2002. Mating-type locus of *Cryptococcus neoformans*: a step in the evolution of sex chromosomes. *Eukaryot. Cell.* 1:704–718.

47. Liu, H., J. Kohler, and G. R. Fink. 1994. Suppression of hyphal formation in *Candida albicans* by mutation of a STE12 homolog. *Science* 266:1723–1726.

48. Lockhart, S. R., K. J. Daniels, R. Zhao, D. Wessels, and D. R. Soll. 2003. Cell biology of mating in *Candida albicans. Eukaryot. Cell* 2:49–61.

49. Lockhart, S. R., C. Pujol, K. J. Daniels, M. G. Miller, A. D. Johnson, M. A. Pfaller, and D. R. Soll. 2002. In *Candida albicans*, white-opaque switchers are homozygous for mating-type. *Genetics* 162:737–745.

50. Lockhart, S. R., R. Zhao, K. J. Daniels, and D. R. Soll. 2003. Alpha-pheromone-induced "shmooing" and gene regulation require white-opaque switching during *Candida albicans* mating. *Eukaryot. Cell* 2:847–855.

51. Magee, B. B., M. Legrand, A. M. Alarco, M. Raymond, and P. T. Magee. 2002. Many of the genes required for mating in *Saccharomyces cerevisiae* are also required for mating in *Candida albicans. Mol. Microbiol.* 46:1345–1351.

52. Magee, B. B., and P. T. Magee. 2000. Induction of mating in *Candida albicans* by construction of MTLa and MTLalpha strains. *Science* 289:310–313.

53. Magee, P. T., and H. Chibana. 2002. The genomes of *Candida albicans* and other *Candida* species, p. 293–306. *In* R. A. Calderone (ed.), *Candida and Candidiasis.* ASM Press, Washington, D.C.

54. Magee, P. T., and B. B. Magee. 2004. Through a glass opaquely: the biological significance of mating in *Candida albicans. Curr. Opin. Microbiol.* 7:661–665.

55. Marsh, L., and M. D. Rose. 1997. The pathway of cell and nuclear fusion during mating in *S. cerevisiae*, p. 827–888. *In* E. W. Jones, J. R. Pringle, and J. R. Broach (ed.), *The Molecular and Cellular Biology of the Yeast Saccharomyces*, vol. 3. Cold Spring Harbor Laboratory Press, Plainview, N.Y.

56. Meluh, P. B., and M. D. Rose. 1990. KAR3, a kinesin-related gene required for yeast nuclear fusion. *Cell* 60:1029–1041.

57. Miller, M. G., and A. D. Johnson. 2002. White-opaque switching in *Candida albicans* is controlled by mating-type locus homeodomain proteins and allows efficient mating. *Cell* 110:293–302.

58. Naglik, J. R., S. J. Challacombe, and B. Hube. 2003. *Candida albicans* secreted aspartyl proteinases in virulence and pathogenesis. *Microbiol. Mol. Biol. Rev.* 67:400–428.

59. Odds, F. C. 1998. Biological aspects of pathogenic *Candida* species, p. 7–15. *In Candida and Candidosis*, 2nd ed. Bailliere Tindall, London, United Kingdom.

60. Odds, F. C. 1988. *Candida* genetics, p. 16–21. *In Candida and Candidosis*, 2nd ed. Bailliere Tindall, London, United Kingdom.

61. Odds, F. C. 1988. Ecology of *Candida* and epidemiology of candidosis, p. 68–92. *In Candida and Candidosis*, 2nd ed. Bailliere Tindall, London, United Kingdom.

62. Panwar, S. L., M. Legrand, D. Dignard, M. Whiteway, and P. T. Magee. 2003. MFalpha1, the gene encoding the alpha mating pheromone of *Candida albicans. Eukaryot. Cell* 2:1350–1360.

63. Raymond, M., D. Dignard, A. M. Alarco, N. Mainville, B. B. Magee, and D. Y. Thomas. 1998. A Ste6p/P-glycoprotein homologue from the asexual yeast *Candida albicans* transports the a-factor mating pheromone in *Saccharomyces cerevisiae. Mol. Microbiol.* 27:587–598.

64. Rikkerink, E. H., B. B. Magee, and P. T. Magee. 1988. Opaque-white phenotype transition: a programmed morphological transition in *Candida albicans. J. Bacteriol.* 170:895–899.

65. Roberts, C. J., B. Nelson, M. J. Marton, R. Stoughton, M. R. Meyer, H. A. Bennett, Y. D. He, H. Dai, W. L. Walker, T. R. Hughes, M. Tyers, C. Boone, and S. H. Friend. 2000. Signaling and circuitry of multiple MAPK pathways revealed by a matrix of global gene expression profiles. *Science* 287:873–880.

66. Rose, M. D. 1996. Nuclear fusion in the yeast *Saccharomyces cerevisiae. Annu. Rev. Cell Dev. Biol.* 12:663–695.

67. Sadhu, C., D. Hoekstra, M. J. McEachern, S. I. Reed, and J. B. Hicks. 1992. A G-protein alpha subunit from asexual *Candida albicans* functions in the mating signal transduction pathway of *Saccharomyces cerevisiae* and is regulated by the a1-α2 repressor. *Mol. Cell Biol.* 12:1977–1985.

68. Schafer, W. R., C. E. Trueblood, C. C. Yang, M. P. Mayer, S. Rosenberg, C. D. Poulter, S. H. Kim, and J. Rine. 1990. Enzymatic coupling of cholesterol intermediates to a mating pheromone precursor and to the ras protein. *Science* 249:1133–1139.

69. Scherer, S. 2002. Gene discovery and comparative genomics: progress and prospects, p. 259–266. *In* R. A. Calderone (ed.), *Candida and Candidiasis.* ASM Press, Washington, D.C.

70. Sherman, F. 1991. Getting started with yeast, p. 3–20. *In* C. Guthrie and G. R. Fink (ed.), *Guide to Yeast Genetics and Molecular Biology*, vol. 194. Academic Press, Inc., San Diego, Calif.

71. Slutsky, B., M. Staebell, J. Anderson, L. Risen, M. Pfaller, and D. R. Soll. 1987. "White-opaque transition": a second high-frequency switching system in *Candida albicans. J. Bacteriol.* 169:189–197.

72. Soll, D. R., and C. Pujol. 2003. *Candida albicans* clades. *FEMS Immunol. Med. Microbiol.* 39:1–7.

73. Srikantha, T., S. A. Lachke, and D. R. Soll. 2003. Three mating-type-like loci in *Candida glabrata. Eukaryot. Cell* 2:328–340.

74. Srikantha, T., L. Tsai, K. Daniels, A. J. Klar, and D. R. Soll. 2001. The histone deacetylase genes *HDA1* and *RPD3* play distinct roles in regulation of high-frequency phenotypic switching in *Candida albicans. J. Bacteriol.* 183:4614–4625.

75. Staab, J. F., S. D. Bradway, P. L. Fidel, and P. Sundstrom. 1999. Adhesive and mammalian transglutaminase substrate properties of *Candida albicans* Hwp1. *Science* 283:1535–1538.

76. Tibayrenc, M. 1997. Are *Candida albicans* natural populations subdivided? *Trends Microbiol.* 5:253–254; discussion 254–257.

77. Tsong, A. E., M. G. Miller, R. M. Raisner, and A. D. Johnson. 2003. Evolution of a combinatorial transcriptional circuit: a case study in yeasts. *Cell* 115:389–399.

78. Tzung, K. W., R. M. Williams, S. Scherer, N. Federspiel, T. Jones, N. Hansen, V. Bivolarevic, L. Huizar, C. Komp, R. Surzycki, R. Tamse, R. W. Davis, and N. Agabian.

2001. Genomic evidence for a complete sexual cycle in *Candida albicans*. *Proc. Natl. Acad. Sci. USA* **98:** 3249–3253.

79. **Vilgalys, R., Y. Graser, G. Schonian, and W. Presber.** 1997. Response: Is *Candida albicans* clonal or recombining? *Trends Microbiol.* **5:**254–257.

80. **Wilson, K. L., and I. Herskowitz.** 1984. Negative regulation of *STE6* gene expression by the alpha 2 product of *Saccharomyces cerevisiae. Mol. Cell. Biol.* **4:**2420–2427.

81. **Xu, J., and T. G. Mitchell.** 2002. Population genetic analyses of candida, p. 55–66. *In* R. A. Calderone (ed.), *Candida and Candidiasis.* ASM Press, Washington, D.C.

Molecular Principles of Fungal Pathogenesis
Edited by Joseph Heitman et al.
©2006 ASM Press, Washington, D.C.

Chapter 7

The Mating-Type Locus and Mating
of *Candida albicans* and *Candida glabrata*

David R. Soll

The majority of *Candida* infections can be attributed to two species, *C. albicans* and *C. glabrata* (33, 38, 86, 91). These two species also represent the two most frequent human commensals, inhabiting a variety of body locations of the healthy human host without causing disease (68, 86). Both are therefore opportunistic, overgrowing and invading body locations in response to a compromising change in the host. *C. albicans* is clearly the more predominant of the two. However, it has received disproportionately more attention by researchers because it has been wrongly assumed that most of what is learned about its biology and pathogenesis will be applicable to *C. glabrata*. Unfortunately, although *C. albicans* and *C. glabrata* share several pathogenic characteristics, they are at opposite ends of the evolutionary tree of the hemiascomycetes (17) and exhibit fundamental differences in filamentation (25, 56), switching (54, 56), and mating (15, 116, 134). For these reasons, the mating systems of these two *Candida* species will be dealt with separately in this review.

DISCOVERY OF THE *MTL* LOCUS
IN *C. ALBICANS*

Through the mid-1990s, it was thought that *C. albicans* was asexual, lacking mating-type genes. However, there were hints that at least a cryptic mating system might exist. First, population studies suggested that there might be low levels of recombination that could be attributed to mating (35, 94, 138), although other mechanisms for genetic recombination were also considered (93). A more provocative observation, however, was made by Sadhu et al. in 1992 (104). They identified the gene *CAG1*, the ortholog to the *Saccharomyces cerevisiae* gene *GPA1*, which encodes a G-protein α subunit involved in the pheromone response. The *C. albicans CAG1* gene complemented an *S. cerevisiae GPA1* null mutant for mating (104). Leberer et al. (59) subsequently

identified *CST20*, the homolog to *S. cerevisiae STE20*, a p21-activated kinase (PAK) that functions early in the pheromone pathway, and Raymond et al. (99) identified the homolog to *S. cerevisiae STE6*, a mating-factor transporter in *S. cerevisiae*. What were these orthologs of genes involved in *S. cerevisiae* mating doing in *C. albicans* if there was no mating system?

In 1999, Hull and Johnson (45) answered this question by identifying in the emerging *C. albicans* genome database a mating-type-like (*MTL*) locus. The configuration of this putative mating-type locus contrasted markedly with those of *S. cerevisiae*. Haploid *S. cerevisiae* possesses three loci containing mating-type genes on chromosome 3 (Fig. 1A), two silent and one expressed, in a remarkable cassette system for mating-type switching (36, 43, 136). In a haploid cell, while one silent locus, *HML*, contains α genes and the other silent locus, *HMR*, contains **a** genes, the expressed locus *MAT* contains either **a** or α genes and dictates mating type. The organization of *MAT* genes is presented for *MAT***a** and *MAT*α in Fig. 1B. *C. albicans*, which is diploid, possesses a single mating-type locus located on chromosome 5, which, in the laboratory strain analyzed by Hull and Johnson (45), contained **a** genes on one homolog and α genes on the other (Fig. 1E). It was immediately evident from this configuration that *C. albicans* would have to use a mechanism quite different from that of *S. cerevisiae* to express a mating type. Indeed, *S. cerevisiae*, which is generally diploid and heterozygous at the *MAT* locus in nature (22, 76, 84), undergoes meiosis to express a mating type (Fig. 2A). Switching of mating type, from **a** to α or from α to **a**, is facilitated on haploidization by up-regulation of the *HO* gene, which encodes an endonuclease involved in the gene conversion event (42, 85). Switching is accomplished by recombination at the expressed *MAT* locus with a DNA copy of the silent

David R. Soll • Department of Biological Sciences, The University of Iowa, Iowa City, IA 52242.

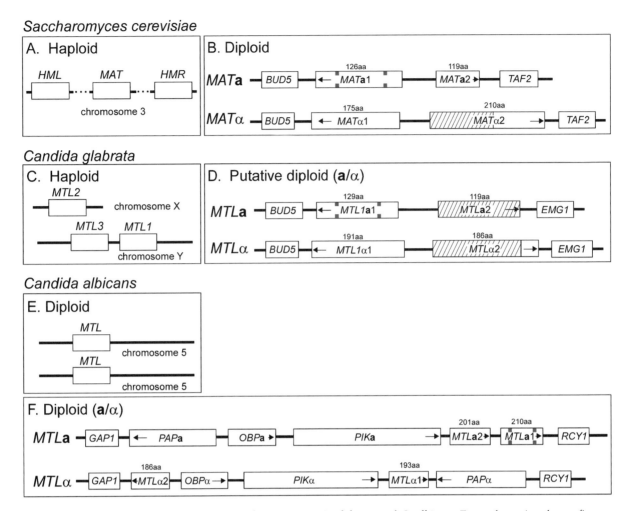

Figure 1. A comparison of mating-type loci of *S. cerevisiae*, *C. glabrata*, and *C. albicans*. For each species, the configuration of loci in the genome and the structure of genes in each locus are presented. The gray boxes within genes represent introns. Hatched regions represent identical protein sequences. Genes bordering the mating-type loci are presented. Arrows represent the directions of open reading frames. A putative diploid of *C. glabrata* is presented in panel D, although **a**/α strains have not been identified in nature. Note that the *MATa2* and *MTLa2* sequences of *S. cerevisiae* and *C. glabrata*, respectively, are not related to *MATa2* of *C. albicans*. The information for *C. glabrata* was obtained from references 116 and 134. The information for *C. albicans* is from references 17, 45, and 127.

locus harboring alternative mating-type genes (Fig. 2A). *C. albicans*, on the other hand, must undergo homozygosis at the *MTL* locus in order to express a mating type (Fig. 2C). Hence, while *S. cerevisiae* retains alternative mating-type information at a silent locus when expressing a mating type, *C. albicans* loses alternative mating-type information.

The *MTL* locus of an **a**/α strain of *C. albicans* (17, 45, 127) is diagrammed in Fig. 1F. The *MTLa* locus on one chromosome 5 homolog contains the *MTLa1* and *MTLa2* genes, as well as alleles of the genes encoding a phosphotidylinositol kinase (*PIK*), an oxysterol-binding protein-like protein (*OBP*), and a poly(A) polymerase (*PAP*). The locus on the other chromosome 5 homolog contains the *MTLα1* and *MTLα2* genes and alleles of the genes *PIK*, *OBP*, and

PAP. *MTLa1* and *MTLα2* encode homeobox proteins, and *MTLα1* encodes an α domain protein, all transcription factors that bind to specific DNA sequences. These genes are orthologs of *S. cerevisiae MATa1*, *MATα2*, and *MATα1*, respectively. The *MTLa1* protein is 30% identical and 56% similar to *S. cerevisiae MATa1*, the *MTLα1* protein is 26% identical and 49% similar to *S. cerevisiae MATα1*, and the *MTLα2* protein is 28% identical and 58% similar to *S. cerevisiae MATα2* (45). *MTLa2*, on the other hand, contains an HMG box (127) and encodes a transcription factor with homology to the MAT1-1-3 protein of the chestnut blight fungus (78). There is no ortholog to this gene in the *S. cerevisiae MATa* locus or in the *S. cerevisiae* genome. In *S. cerevisiae*, the gene products of *MATa1* and *MATα2*, the

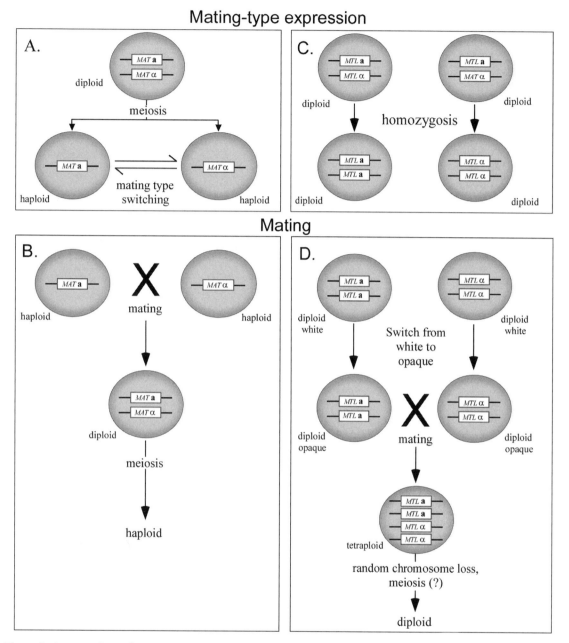

Figure 2. A comparison of mating-type expression and the general steps in the mating process between *S. cerevisiae* (A, B) and *C. albicans* (C, D).

homologs of *C. albicans MTLa1* and *MTLα2*, encode proteins which form a heterodimer repressor complex, Mata1p-Matα2p, which suppresses Matα1p and hence α-specific genes, including the **a**-pheromone receptor gene and the α-pheromone gene (37). Together with Tup1p and Ssn6p, the Mata1p-Matα2 complex also suppresses haploid-specific genes by binding to haploid-specific gene (hsg) operator sequences (34). In *S. cerevisiae*, the gene product

of *MATα2*, the homolog of *C. albicans MTLα2*, when complexed with Mcm1p, suppresses **a**-specific genes, including the α-pheromone receptor gene and **a**-pheromone gene (49). To test whether the Mtla1p-Mtlα2p complex played a role as a repressor in *C. albicans*, as does the Mata1p-Matα2p complex of *S. cerevisiae*, Hull and Johnson (45) constructed in *C. albicans* green fluorescent protein reporters driven by the *C. albicans ADH1* promoter with the addition of *S. cerevisiae* consensus hsg operator sequences, mutated consensus operator sequences, or consensus operator sequences from the *C. albicans CAG1* gene (104). Hull and Johnson (45) also placed green fluorescent protein under the control of the *ADH1* promoter plus mutated *CAG1* consensus operator sequences so that it was not recognized by the Mata1p-Matα2p complex. The results of Hull and Johnson (45) demonstrated that the presence of hsg operators in the *ADH1* promoter depressed fluorescence, strongly indicating that Mtla1p-Mtlα2p functioned as a repressor in *C. albicans* a/α cells. With this discovery, it seemed at the time that the regulatory circuitry controlling mating would be similar, if not the same, in *C. albicans* and *S. cerevisiae*.

DEMONSTRATION OF *C. ALBICANS* MATING IN VIVO AND IN VITRO

After identification of the mating locus, the next logical step was to demonstrate mating. In 2000, two independent laboratories simultaneously demonstrated that mating-type-dependent fusion occurred in *C. albicans*, Hull et al. (46) in a mouse model of systemic infection and Magee and Magee (70) in vitro. In both studies, fusion was identified by complementation between a and α strains, each carrying a different auxotrophic marker, Ade^- and Ura^-, respectively. Hull et al. (46) generated two a strains by deleting either the entire *MTLα* locus or only the α1 and α2 genes, and two α strains by deleting either the entire *MTLa* locus or only the a1 gene. a and α cells were mixed and injected into mice. After 24 h, the mice were sacrificed and yeast from macerates of their kidneys were plated on selection agar lacking uradine and adenine. Yeast clones that grew proved to be a/α, to possess increased levels of DNA, and to contain a single nucleus, indicating that they had undergone mating-type-dependent fusion and karyogamy, resulting in tetraploid cells. In the parallel report by Magee and Magee (70), advantage was taken of an earlier observation by Janbon et al. (47) that culturing cells on medium containing sorbose leads to loss of a chromosome 5 homolog. Again, auxotrophs of a and α cells were used so that fusants could be selected for by complementation. In this case, however, a and α strains were cross-streaked on selection agar on which only fusants could grow. Again, putative fusants proved to be a/α and to contain increased DNA.

Three years after the initial demonstration of mating by Hull et al. (46) and Magee and Magee (70), Lockhart et al. (64) took advantage of culture conditions that facilitated fusion to describe for the first time the cell biology of mating between natural *C. albicans* a/a and α/α strains. Using the microscopically observed fusion event as an assay, they demonstrated that mating occurred only between natural a/a and α/α strains in vitro. Furthermore, mating occurred readily between members of different *C. albicans* clades. Under optimum conditions, up to 25% of the cells in mating mixtures fused.

MTL GENOTYPE AND MATING COMPETENCE OF *C. ALBICANS*

In *S. cerevisiae*, haploid a and α cells generated by meiosis are immediately mating competent and spatially in close association. In the absence of the Mata1p-Matα2p repressor complex, the endonuclease gene *HO*, which is required for mating-type switching at the *MAT* locus, is up-regulated. Switching ensues by gene conversion. Interestingly, the majority of natural *C. albicans* strains, like *S. cerevisiae* strains, are also heterozygous at the mating-type locus (a/α). In an analysis of natural isolates from around the world, Lockhart et al. (63) found that approximately 97% were *MTL*-heterozygous and approximately 3% were *MTL*-homozygous. In a similar analysis, Legrand et al. (61) found that 90% were a/α while 10% were either a/a or α/α. To generate either an a or an α cell, a/α cells do not undergo meiosis like *S. cerevisiae* (Fig. 2A). Rather, *C. albicans* undergoes *MTL* homozygosis to either a/a or α/α, with no change in ploidy (Fig. 2C). Lockhart et al. (63) found that approximately 4% of a/α strains underwent *MTL* homozygosis at increased frequencies. The mechanisms of homozygosis were assessed by Wu et al. (135), who identified intergenic and gene polymorphisms between homologs of chromosome 5 on either side of the *MTL* locus in *MTL*-heterozygous parent strains and then tested for the retention or loss of these polymorphisms in *MTL*-homozygous offspring. Wu et al. (135) found that homozygosis was achieved in vitro primarily through the loss of one of the two chromosome 5 homologs followed by duplication of the retained homolog. It also was achieved by spontaneous mitotic recombination along chromosome 5, but this mechanism appeared to be far less common, at least in vitro (135). By analyzing polymorphisms on chromosomes other than chromosome 5, meiosis was ruled out by Wu et al. (135) as a mechanism for *MTL* homozygosis in vitro.

In their analysis of polymorphisms, Wu et al. (135) discovered that specific alleles of genes other than those located at the *MTL* locus along chromosome 5 were linked to either *MTLa* or *MTLα*. The same allelic linkages held true across clade boundaries. Hence, the chromosome 5 homolog harboring the *MTLa* locus also harbors *MTLa*-associated alleles of other genes and, likewise, the chromosome 5 homolog harboring the *MTLα* locus also harbors *MTLα*-associated alleles of these other genes. *MTL* homozygosis, therefore, may lead to homolog-specific phenotypic changes that are associated with genes other than those of mating-type genes. In addition, the **a**-associated and α-associated alleles of the three non-*MTL* genes harbored by the *MTL* locus, *PAP*, *OBP*, and *PIK*, are far more dissimilar than the **a**- and α-associated alleles of other chromosome 5 genes outside of the *MTL* locus (C. Pujol and D. R. Soll, unpublished observations). Caution must therefore be applied in interpreting phenotypes resulting from *MTL* homozygosis as being due solely to changes in the mating-type genes, unless one generates specific gene mutations, as did Hull and Johnson (45), or isogenic strains in which a single gene is added to a neutral site, as did Lockhart et al. (67).

Mtla1p-Mtlα2p SUPPRESSES WHITE-OPAQUE SWITCHING IN C. ALBICANS

Although Hull et al. (46) and Magee and Magee (70) demonstrated mating-type-dependent fusion in vivo and in vitro, respectively, neither study provided direct cytological evidence of fusion, primarily because it represented a rare event that had to be identified after the fact by genetic complementation of auxotrophic markers. In 2002, however, Miller and Johnson (80) made a quite extraordinary observation. They noticed that when plated on agar containing phloxine B, which preferentially stains opaque cells of the white-opaque switching system (107) red (3), engineered **a** and α strains formed colonies with red sectors but colonies of the parent **a**/α strain did not. Cells from these sectors, when examined microscopically, were found to exhibit the morphological characteristics of opaque cells, including an elongate, bean-shaped cell morphology and pimples. Cells from these putative opaque sectors expressed opaque-specific genes and down-regulated white-specific genes (80). Miller and Johnson (80) further demonstrated that while fewer than 0.03% of colonies formed by *MTL***a**/*MTLα* cells contained opaque cells (no opaque cells were detected in 3×10^3 cells examined), 22% of colonies of strain WO-1, which is α/α, 10% of colonies of strain CHY420a, which is *MTL***a**/*mtlα1Δmtlα2Δ*, 4.5% colonies of strain CHY247α, which is *mtla1Δ*/*MTLα*, <0.1% of colonies of strain CHY317 (no opaque cells were detected in 10^3 cells examined), which is *MTL***a**/*mtlα1Δ*, and 4.4% of colonies of strain CHY405, which is *MTL***a**/*mtlα2Δ*, contained opaque cells. Hence, opaque cells were present only when a cell lacked either a functional *MTL***a**1 or a *MTLα*2 gene. These results indicated that an Mtla1p-Mtlα2p

complex homologous to the Mata1p-Matα2p repressor complex of *S. cerevisiae* repressed white-opaque switching in the **a**/α laboratory strain CAI4, which had been shown to undergo 3153A-like switching but not white-opaque switching (115). Furthermore, it suggested that the reason why most natural strains of *C. albicans* did not undergo white-opaque switching was that they may be **a**/α.

Lockhart et al. (63) tested this hypothesis by analyzing the *MTL* genotypes and the switching capabilities of natural strains of *C. albicans* from around the world. As noted, of 220 strains analyzed, 97% were **a**/α while only 3% were either **a**/**a** or α/α. Of 20 randomly selected **a**/α strains, only 2 underwent switching. These two, however, underwent *MTL* homozygosis before switching (63). Of the seven identified *MTL*-homozygous natural strains, five underwent white-opaque switching (63). Furthermore, seven strains identified earlier in a larger genetic analysis of population structure as undergoing white-opaque switching (Pujol and Soll, unpublished) were demonstrated to be *MTL* homozygous. The results of Lockhart et al. (63) therefore generalized the discovery of Miller and Johnson (80) that a cell had to undergo homozygosis at the *MTL* locus in order to switch and further supported the conclusion of the latter that the Mtla1p-Mtlα2p complex suppressed white-opaque switching.

THE WHITE-OPAQUE TRANSITION IS ESSENTIAL FOR MATING

The experiments by Miller and Johnson (80), which suggested that *MTL* homozygosis derepressed white-opaque switching, further revealed that when **a** and α cells expressed the opaque phenotype, they mated at frequencies many orders of magnitude higher (10^5 to 10^6) than did cells expressing the white phenotype. This was truly an extraordinary discovery since it indicated that the transition from the white to the opaque phenotype might be a requisite for mating (Fig. 2D) and that the opaque phenotype was in fact the mating-competent phenotype of *C. albicans*. The reversible, complex phenotypic transitions between white and opaque phenotypes, first discovered in 1987 by Slutsky et al. (107), affects virtually every morphological parameter of the cell (3, 107, 110, 111) and involves the regulation of hundreds of genes (57), many of them white or opaque specific (82, 83, 111, 112). It also affects virulence. While white cells of the natural α/α strain WO-1 were found to be virulent in the mouse model of systemic infection, opaque cells were found to be relatively avirulent (53; S. Lockhart and D. R. Soll, unpublished observations). In marked contrast, opaque cells were found to be far more virulent than white cells in a mouse model of skin colonization (52). Remarkably, there is no correlate to the white-opaque transition in the *S. cerevisiae* mating process. One must therefore ask why *C. albicans* mating includes this complex developmental transition while *S. cerevisiae* mates quite well without it. Could it have something to do with the intimate relationship *C. albicans* has evolved with its host, given its pathogenic lifestyle (72)?

Lockhart et al. (64) tested whether expressing the opaque phenotype was necessary for mating between natural **a**/**a** and α/α strains of *C. albicans*. They performed 12 mating reactions between **a**/**a** and α/α opaque cells from different natural strains and observed fusions microscopically for every mating mixture. They observed no fusions in mating mixtures of white and opaque or white and white cells of opposite mating type. Their results generalized the observation of Miller and Johnson (80), who had made their original observations on engineered isogenic **a**/− and α/− cells of the laboratory strain Sc5314.

One additional early observation was also relevant to the requisite that a cell switch to opaque to mate. Magee and Magee (70) reported that mating-type-dependent fusion, assessed in this case by genetic complementation, occurred far less frequently at physiological temperature than at lower temperatures. This was consistent with the conclusion that cells had to express the opaque phenotype in order to mate, since it had previously been demonstrated that opaque cells mass-converted to the white phenotype when the temperature was raised from 25 to 35°C or greater (100, 107, 111, 112). Presumably, the mating events in the in vitro crosses performed by Magee and Magee (70) resulted from fusions between opaque cells generated by switching in the white cell populations, and increased temperature presumably drove newly formed opaque cells back to white, possibly decreasing the frequency of fusion. Indeed, the low fusion rates between white **a** and α cells in vitro at low temperature were estimated by Miller and Johnson (80) to be in the range of 10^{-6} to 10^{-7}. Presumably the mating events that occurred in vivo in the experiments performed by Hull et al. (46) occurred either between opaque cells that formed at high temperatures, during the two-division window during which a cell remained opaque at 37°C before reverting to white (111, 112), or between opaque cells in the populations just before injection into mice. These results raise questions whether mating can or does occur within a host and whether host-specific factors regulate mating.

CELL BIOLOGY OF
C. ALBICANS MATING

Taking advantage of the high efficiency of mating observed when opaque **a**/**a** and opaque α/α cells derived from saturation-phase cultures are mixed at high density in suspension, Lockhart et al. (64) were able to describe for the first time the mating process at the cellular level, employing videomicroscopy, computer-assisted three-dimensional reconstruction software for living cells (39, 131), phase-contrast microscopy, and fluorescence microscopy. The sequence of cytological events that they documented in the mating process proved to be surprisingly similar to that of *S. cerevisiae* but with at least a few interesting differences. For that reason, it is worth reviewing the cell biology of *S. cerevisiae* mating for comparison. Meiosis of an **a**/α diploid *S. cerevisiae* cell results in two **a** and two α haploid spores. Because each haploid spore undergoes mating-type interconversion, the opposite mating types that are generated are spatially juxtaposed (40). Let us consider an *S. cerevisiae* α cell generated by sporulation. After the first division, at the two-cell stage, the mother cell switches to **a**. After the second cell division, its second daughter is **a** as well. The first daughter cell and its offspring, however, remain α. Hence, at the four-cell stage, two **a** and two α cells are spatially juxtaposed and can readily mate. Further cell multiplication juxtaposes additional **a** and α cells. In the mating process, an **a** and an α cell in close proximity each extends a single evagination toward the other in response to pheromones released by the alternative cell types (21, 41, 105). An **a** cell responds to the α-pheromone released by an α cell, and an α cell responds to the **a**-pheromone released by an **a** cell. These pheromones form opposing gradients. An evagination forms in a polarized fashion toward the opposite mating type through receptor-mediated assessment of the increasing pheromone gradient, in a process referred to as "chemotropism." The evagination that forms from each mating type is wide, with no constriction at the cell-evagination junction. The shape of the early evaginating cell looks very much like a shmoo, the cartoon creature in Li'l Abner. Hence a cell that has evaginated in response to pheromone has been aptly referred to as a "shmoo," and the process has been referred to as "shmooing" (69). When the apices of the evaginations of an **a** and α cell contact one another, they fuse. The fused cellular extension is referred to as a "conjugation bridge." The **a** and α cell nuclei migrate into the bridge and fuse, generating an **a**/α diploid nucleus (24, 102). Near the site of fusion, a daughter bud forms from the bridge. The diploid nucleus then divides, with one of the two diploid nuclei entering the daughter bud.

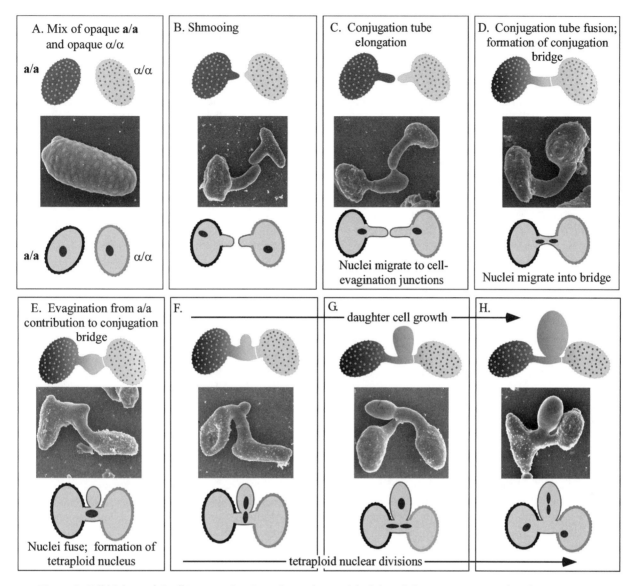

Figure 3. Cell biology of *C. albicans* mating. In each panel, a model of the cellular stage is presented at the top, a scanning electron micrograph of the stage is presented in the middle, and a diagram of the nuclear event is presented at the bottom. In conjugation bridges, a thin white line represents the point of tube fusion.

The cytology of *C. albicans* mating, diagrammed in Fig. 3, is superficially similar to that of *S. cerevisiae*, but there are interesting dissimilarities that may be related to differences in how the two species generate mating-competent **a** and α cells and in lifestyles, in particular the unique relationship *C. albicans* appears to have with its host. Unlike *S. cerevisiae*, *C. albicans* generates **a** and α cells through *MTL* homozygosis. Rather than generating spatially juxtaposed, mating-competent **a** and α cells, *C. albicans* generates an **a**/**a** or an α/α cell that must then switch to opaque. Hence, when *C. albicans* generates an **a**/**a** or α/α cell, it is not immediately juxtaposed to an α/α or **a**/**a** cell, respectively. Furthermore, the requirement to switch spontaneously to opaque may

mean that mating-competent **a**/**a** and α/α cells are even rarer in populations. Hence, fusion may be a rare event, and opaque **a**/**a** and α/α cells may have to work much harder to find each other than **a** and α cells of *S. cerevisiae* do. When equal concentrations of *C. albicans* **a**/**a** and α/α cells are mixed in high-density shaking cultures, they agglutinate, forming large, tight clumps, presumably in response to pheromones (64, 95). One way in which rare *C. albicans* opaque **a**/**a** and α/α cells could therefore find one another in a predominantly white population is through selective adhesion.

The clumps that form in suspension cultures of opaque **a**/**a** and α/α cells are quite large and dense and appear to create microenvironments that protect pheromone gradients and facilitate chemotropism (64). Just as in the case of *S. cerevisiae*, *C. albicans* **a/a** and α/α cells shmoo (Fig. 3B) in response to pheromone released from opposite mating types (Fig. 3). However, in contrast to *S. cerevisiae*, the evaginations can and do grow quite long within clumps, looking superficially like mycelia (64), which supports the suggestion that they must travel greater distances than *S. cerevisiae* in order to fuse. These structures have been referred to as "conjugation tubes" (64). Conjugation tubes do not grow indefinitely if they do not find a conjugation tube of opposite mating type with which to fuse. These "unfulfilled" tubes eventually revert apically to the budding growth form (26, 64).

Several groups have demonstrated that chemically synthesized α-pheromone induces shmooing in engineered **a** cells and natural **a/a** cells (11, 65, 69). Under the right conditions, the evaginations that are formed will grow into long tubes and then revert apically to the budding growth form (65), just as is the case in mating mixtures (64). One would assume that apical reversion results from the rapid digestion of pheromones. However, since apical reversion occurs in mating mixtures in which pheromone production is continuous, it is likely that there may be a limit to the length of conjugation tubes before apical reversion proceeds.

Conjugation tubes have an uncanny capacity to find each other (64), traveling circuitous routes within clumps (Fig. 3C). On contact, they fuse end-to-end to form the conjugation bridge (Fig. 3D), which can be quite long, depending on the lengths of the **a/a**- and α/α-contributed tubes. On fusion, the portion of the bridge contributed by the **a/a** cell may selectively swell (Fig. 3E) (26). After fusion, an evagination forms from the bridge (Fig. 3E). Employing indirect immunofluorescence, Daniels et al. (26) demonstrated that the hypha-associated adhesin Hwp1p (117, 118) is expressed selectively on conjugation tubes formed by **a/a** but not α/α cells and that the first daughter bud invariably forms from the stained **a/a** contribution to the bridge. Hence, **a/a** cells serve as the "female" in the mating process, a distinction that has not yet been tested in *S. cerevisiae*.

Bennett et al. (12) recently demonstrated that isogenic **a** and α strains, when mated, undergo nuclear fusion. They also demonstrated that the efficiency of nuclear fusion is both strain and medium dependent. They performed elegant live imaging of individual cells that expressed fluorescently tagged histone H2B. They found that nuclear fusion preceded evagination and that the tetraploid nucleus then divided, resulting in a nucleus in the daughter cell and a nucleus in the conjugation bridge (12). Subsequent progeny cells were demonstrated by

fluorescence-activated cell sorter analysis to retain tetraploid nuclei. Bennett et al. (12) further demonstrated that the reason why Lockhart et al. (64) did not observe nuclear fusion in their original study of the cytology of mating was due to the modified Lee's medium and the two clinical strains that were employed. Finally, Bennett et al. (12) demonstrated that, as is the case in *S. cerevisiae* (79), nuclear fusion is dependent on *KAR3*, which encodes a microtubule-based motor molecule. In another recent study of nuclear dynamics during fusion of natural **a/a** and α/α cells (K. Daniels, S. Lockhart, and D. R. Soll, unpublished data), similar to that performed by Pujol et al. (95) for *C. dubliniensis*, it was similarly demonstrated that at the onset of the fusion process, each cell contains a single diploid nucleus positioned randomly. On polarization and growth of the conjugation tubes, each nucleus moved toward the junction of mother cell and evagination (Fig. 3B and C). At the time of tube fusion, the nuclei had migrated into the newly formed conjugation bridge. After tube fusion and prior to evagination from the conjugation bridge of the zygote, the nuclei fused, generating a tetraploid nucleus (Fig. 3E). After growth of the daughter cell to at least half the volume of the parent cells, the single tetraploid nucleus divided, with one nucleus entering the daughter cell and the other remaining in the bridge.

G1 IS REQUIRED FOR *C. ALBICANS* MATING

When **a** cells of *S. cerevisiae* derived from an exponential-phase growth culture are treated with α-pheromone, cells cycle to G1 and arrest, then evaginate in the process of shmooing (16, 19, 106, 133). Bennett et al. (11) found that in contrast to *S. cerevisiae*, *C. albicans* did not undergo "whole-scale" cell cycle arrest when treated with α-pheromone but suggested that this did not preclude the possibility that cells individually arrested at G1 of the cell cycle during shmooing. Interestingly, to analyze the cell biology of *C. albicans* mating, Lockhart et al. (64) exploited the observations that mixtures of natural opaque-phase **a/a** and α/α cells derived from the saturation phase of a liquid growth culture formed shmoos and underwent fusion at higher frequencies than cells derived from exponential phase growth cultures. Zhao et al. (137a) hypothesized that the cessation of cell multiplication, a major characteristic of cells in saturation phase, may, therefore, be a requisite for mating. However, in testing this hypothesis, they found that mating mixtures of exponential-phase cells formed shmoos and fused at lower frequencies for two reasons. First, they produced limiting amounts of pheromone. Second, only a fraction of cells were unbudded and in G1 and therefore responsive. Zhao et al. (137a) indeed found that G1 alone was the requisite for mating competency. Saturation-phase cells were mating competent simply because they accumulated in G1. Exponential-phase cells in G1 were fully competent to shmoo when treated with chemically synthesized α-pheromone (137a). In an exponential-phase growth culture, this represents over 30% of the entire cell population. Furthermore, pheromone was found to block cells in G1 through shmooing and conjugation tube growth, as it does in *S. cerevisiae* cells (133).

SIMILARITIES AND DIFFERENCES IN THE ROLE OF *S. CEREVISIAE MAT* AND *C. ALBICANS MTL* GENES IN MATING

C. albicans possesses homologs of the three mating-type genes that control the general mating process in *S. cerevisiae* (45). However, it possesses a fourth gene, *MTLa2*, which has no homolog in *S. cerevisiae*. The *MATa2* gene at the mating locus of *S. cerevisiae* is not

a homolog of *MTLa2* and has no known function (5, 126). It also has no homology to *MTLa2* in *C. glabrata* (116). *MTLa2* plays a unique role in *C. albicans* mating, with no analogy in the *S. cerevisiae* mating process. As noted by Tsong et al. (127), K. Wolfe (Dublin University) brought to their attention the existence and uniqueness of the *MTLa2* open reading frame in the *C. albicans MTL* locus. The protein encoded by *MTLa2* is similar to HMG box proteins, which are found in the mating-type loci of a number of other fungi, including *Neurospora crassa* (23, 32, 119), *Kluyveromyces lactis* (6, 17), and *Podospora anserina* (27).

Tsong et al. (127) demonstrated through an analysis of mating between various mutants of the four *MTL* genes that although three of the four players are the same in *C. albicans* and *S. cerevisiae*, the way in which these genes determine mating type differs significantly. First, in *S. cerevisiae* **a**/α cells, the Mata1p and Matα2p heterodimer suppresses **a** and α cell mating, and Matα2p suppresses **a** cell mating (Fig. 4A). In *C. albicans* **a**/α cells, while Mtla1p

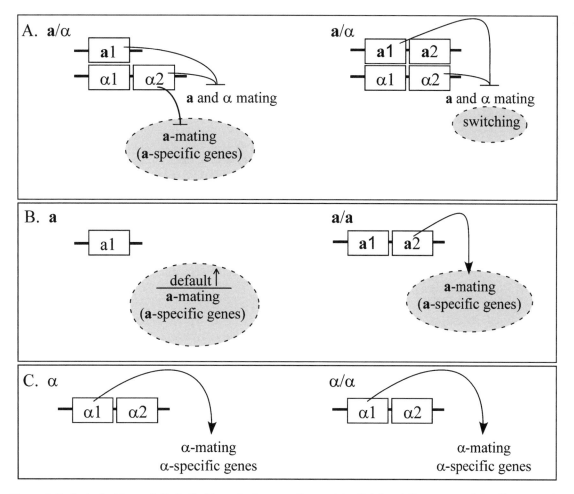

Figure 4. Both similarities and dissimilarities exist between *S. cerevisiae* (left-hand diagrams) and *C. albicans* (right-hand diagrams) in the roles played by mating-type genes in the regulation of mating. The dashed ellipsoids highlight differences between the two species. See reference 11 for details.

and Mtlα2p similarly suppress **a** and α cell mating, Mtlα2p dissimilarly does not suppress mating of **a** cells (Fig. 4A). Second, in *S. cerevisiae* **a** cells, **a** cell mating occurs by default without any input from a *MAT* gene (Fig. 4B). On the other hand, in *C. albicans* **a** cells, Mtla2p plays the role of a mating activator (Fig. 4B). In both *S. cerevisiae* and *C. albicans*, however, Matα1p and Mtlα1p, respectively, activate α mating in a similar fashion (Fig. 4C). Transcriptional profiling studies of null mutants of the four mating-type genes, individually and in combination, supported these conclusions in terms of the regulation of **a**-specific and α-specific gene expression (127). Tsong et al. (127) further argued that the circuit of *C. albicans* was ancestral to that of *S. cerevisiae*, which underwent a sequence of evolutionary changes that included loss of the *MTLa2* equivalent, bypass of the a2 requirement for activation of **a**-specific genes, and development of α2 control over (i.e., suppression of) **a** cell-specific genes. Tsong et al. (127) argued that *C. albicans* uniquely acquired the white-opaque transition in the evolution of its mating system.

Butler et al. (17) undertook a comparison of *MAT* and mating-type-like (*MTL*) loci of nine yeast species in the development of a hemiascomycete phylogenetic tree. They developed an evolutionary tree with three landmark events related to the mating system. The earliest event was the development of silent cassettes, the next was the acquisition of the *HO* endonuclease gene, and the last was the loss of the *MATa2* gene and genomic duplication. In this phylogenetic tree, *Yarrowia lipolytica* and *C. albicans* branched off early and did not undergo the three landmark events. In contrast, both *S. cerevisiae* and *C. glabrata* underwent all three landmark events. The evolutionary sequence of landmark events proposed by Tsong et al. (127) and that proposed by Butler et al. (17) suggest that what we have referred to as "unique" characteristics of the *C. albicans* mating system may in part simply reflect ancestral characteristics retained by *C. albicans* and lost by *S. cerevisiae*.

HEMOGLOBIN, SWITCHING, AND THE *C. ALBICANS* MATING-TYPE LOCUS

It has been repeatedly noted throughout this review, and most recently in a review by Magee and Magee (72), that the increased complexity of the developmental programs associated with *C. albicans* mating may very well be due in part to the intimate relationship *C. albicans* has as a commensal and pathogen with its host. Pendrak et al. (90) recently added more ammunition to this conclusion. The gene *HBR1* (for "hemoglobin response gene 1") was demonstrated to

be up-regulated severalfold in *C. albicans* by hemoglobin in growth cultures. When one of the two *HBR1* alleles was disrupted, the generated HBR1:hbr1 strain formed colonies containing a mixed population of white and bona fide opaque cells, even though the *MTL* genotype was **a**/α. The opaque cells were demonstrated to possess wall pimples and a bean-shaped morphology, to express differentially the opaque-specific genes *OP4* and *SAP1*, and to down-regulate the white-specific gene *WH11*. Pendrak et al. (90) concluded that Hbr1p was a "haplo-insufficient" repressor of white-opaque switching in strains containing all of the *MTL* genes (i.e., strains that were genetically **a**/α). In other words, cells with a single copy of *HBR1* exhibited dosage-dependent derepression of white-opaque switching. They could not delete both alleles of *HBR1* and concluded that possession of one allele was essential. They did not, however, consider the possibility that the two *HBR1* alleles are different and confer synergistic activities, as has been demonstrated, for instance, for *ALS3* alleles (88). This type of interallelic interaction, termed "transvection," has been observed in other fungi (4). Pendrak et al. (90) further demonstrated by reverse transcription-PCR that *HBR1* controls switching by up-regulating *MTLα* genes. Indeed, they demonstrated that the *HBR1* heterozygote functions as a mating-competent **a** cell. Hence, they concluded that *HBR1* expression in normal **a**/α cells is stimulated by growth in general and is further influenced by hemoglobin, presumably through a hemoglobin receptor. They presented evidence that *MTLα* expression is down-regulated in **a**/α cells when a growth culture enters saturation phase but can be maintained by the addition of hemoglobin (90). The cues and receptor(s) that function during exponential phase have not yet been identified.

SIMILARITIES IN THE PHEROMONE-ACTIVATED PATHWAYS OF *C. ALBICANS* AND *S. CEREVISIAE*

As noted above, three research groups (10, 65, 89) simultaneously reported that opaque **a**/− or **a**/**a** cells responded to chemically synthesized α-pheromone, either a 13-mer or 14-mer, by forming shmoos. Bennett et al. (10) and Panwar et al. (89) demonstrated that deletion of the *MFα* gene, which encodes α-pheromone, resulted in the loss of **a** cell mating. Bennett et al. (10) further demonstrated that deletion of the *STE2* gene, which encodes the α-pheromone receptor, results also in the loss of **a** cell mating. Hence, as is the case in *S. cerevisiae*, one mating type responds to the pheromone secreted by the opposite mating type. It therefore seemed logical to expect that the regulatory pathways emanating from the pheromone receptor that induced shmooing in *C. albicans* would be similar to those in *S. cerevisiae*.

Deletion and overexpression studies have been performed and have demonstrated that the mitogen-activated pathway central to the α-pheromone response in *S. cerevisiae* was also essential for the α-pheromone response in *C. albicans* opaque **a** cells. Magee et al. (71) and Chen et al. (20) demonstrated that genes related to the mitogen-activated protein (MAP) kinase pathway, which play a central role in the pheromone response in *S. cerevisiae* (73), also play a central role in the pheromone response in *C. albicans*. By mutational analyses, both groups demonstrated that *CPH1*, the homolog of *S. cerevisiae STE12*, which is downstream of the MAP kinase cascade and the key transcription factor regulating pheromone-activated genes, is essential for *C. albicans* mating (20, 71) and that *HST7*, the homolog of *S. cerevisiae STE7*, a MAP kinase in that cascade, is also essential. While Magee et al. (71) found that *CST20*, the homolog of *STE20*, which initiates the MAP kinase pathway, was not necessary for mating (i.e., the deletion mutant of *CST20* exhibited no decrease in mating efficiency), Chen et al. (20) found a decrease in efficiency, as reported for *STE20* in *S. cerevisiae* (58). Hence, while *CST20* may be involved, it is not essential. Chen et al. (20) also demonstrated that *CEK1* and *CEK2*, the *S. cerevisiae* homologs of which are MAP kinases, serve overlapping functions essential for *C. albicans* mating. Finally, Magee et al. (71) demonstrated that *KEX2*, whose *S. cerevisiae* homolog encodes a proteinase that processes α-pheromone through a *KEX2* proteolytic cleavage site and is essential for α cell mating (62), and *HST6*, whose *S. cerevisiae* homolog encodes an ABC transporter of **a**-pheromone (77), were both essential for α and **a** cell mating, respectively, in *C. albicans*. Together, these studies revealed a high degree of similarity between *C. albicans* and *S. cerevisiae* in the signal transduction pathway activated by pheromone.

GENES REGULATED BY PHEROMONE IN *C. ALBICANS*

To explore further the similarities between the mating programs of *C. albicans* and *S. cerevisiae*, several studies were performed to develop transcription profiles for α-pheromone-induced **a** cells and mating mixtures of **a/a** and α/α cells, employing microarray and Northern analyses. Bennett et al. (11), using a PCR product-based microarray analysis, identified 62 genes that were up-regulated and 4 that were down-regulated in cultures of opaque **a** cells treated with α-pheromone, in which α1 and α2 had been disrupted (80). Samples were analyzed through 4 h of incubation with α-pheromone. Lockhart et al. (65), using Northern analysis, identified five genes that were up-regulated in opaque cells of the natural **a/a** strain P37005 treated with chemically synthesized α-pheromone for 2 h. Finally, Zhao et al. (137a), using an oligonucleotide-based microarray, identified 50 genes up-regulated and 30 genes down-regulated in mating mixtures of the natural **a/a** strain P37005 and the natural α/α strain WO-1. The most interesting aspects of pheromone-regulated genes in *C. albicans* were not the similarities but, rather, the dissimilarities to that of pheromone-treated *S. cerevisiae*.

Genes Associated with Mating

In Fig. 5A, a synopsis is presented of the data from the three studies (11, 65, 137a) of pheromone-induced expression of *C. albicans* genes associated with the mating process, and in Fig. 5B, a comparison is made between the regulation of these genes in *C. albicans* and in *S. cerevisiae* (101). Twenty mating-associated genes have so far been demonstrated to be up-regulated either in **a** cells treated with α-pheromone or in mating mixtures of **a/a** and α/α cells. This represents an incomplete list for two reasons. First, the fact that the microarray analysis by Bennett et al. (11) and that by Zhao et al. (137a) contained both overlapping and nonoverlapping up-regulated and down-regulated genes demonstrated that neither provided a complete list of pheromone-regulated genes. Second, in Northern analyses performed to verify microarray results, Zhao et al. (137a) found that while all genes identified as up-regulated or down-regulated by their microarray analyses proved to be similarly regulated by Northern analyses, several genes not identified by the former were found to be up-regulated by Northern analysis.

Mating-associated genes identified as up-regulated (Fig. 5) included the following: *STE2* and *STE3*, which encode the **a**-pheromone and α-pheromone receptors, respectively; *CAG1* and *STE4*, which encode the G-protein α and β subunits; *CEK2* and *CEK1*, which encode mitogen-activated kinases; *CPH1*, which encodes the transcription factor downstream of the kinase cascade that is responsible for activating pheromone-induced genes; *MFα1*, which encodes the α-pheromone; and a number of genes involved in membrane-related functions, pheromone adaptation, maturation and transport of pheromone, reentry into the mitotic cell cycle, and karyogamy. Two of these genes, *FIG1* and *FUS1*, were demonstrated by Chen et al. (20) to be up-regulated in cells overexpressing *CPH1*, which is consistent with its

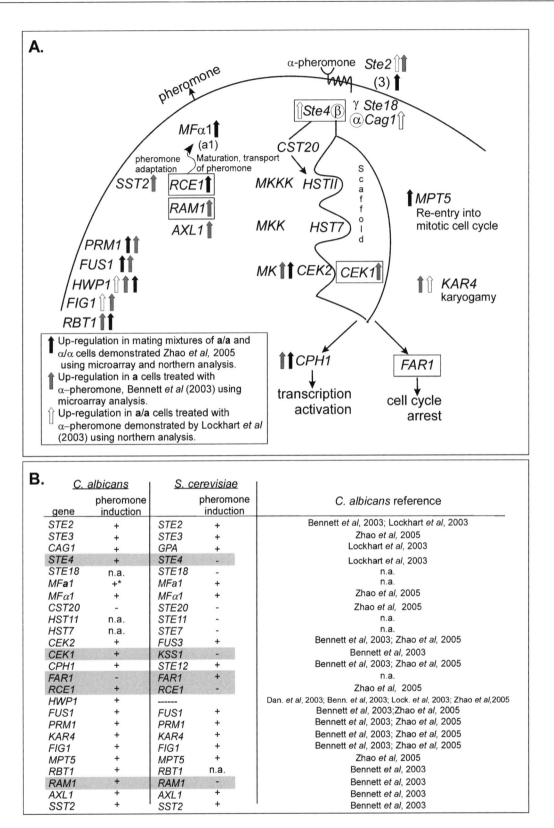

Figure 5. Transcription-profiling studies indicate that while a majority of genes associated with the mating process are regulated similarly by pheromone in *C. albicans* and *S. cerevisiae*, a minority of genes are regulated dissimilarly. (A) A model incorporating the transcription profiling data from references 11, 65, and 137a. Vertical arrows indicate up-regulation demonstrated by the three studies. Boxed genes represent dissimilar regulation between *C. albicans* and *S. cerevisiae*. (B) Comparison of the combined profiling data for *C. albicans* with that of Roberts et al. (101) for *S. cerevisiae*. Boxed genes represent dissimilar regulation between *C. albicans* and *S. cerevisiae*.

role as a transcription activator of mating-associated genes. If one compares the *C. albicans* transcription profile with that of *S. cerevisiae*, it is interesting that while 15 genes are similarly regulated, 5 are dissimilarly regulated (Fig. 5B). Hence, in general the circuit has been conserved through evolution from presumably an ancestral organism to *C. albicans* and *S. cerevisiae*. The five genes dissimilarly regulated were distributed throughout the program and included genes encoding the G-protein β subunit (*STE4*), a MAP kinase (*CEK1*), the kinase responsible for cell cycle arrest (*FAR1*), and proteins involved in pheromone maturation (*RCE1* and *RAM1*). While four of these genes are up-regulated in *C. albicans* but not in *S. cerevisiae* (*STE4*, *CEK1*, *RCE1*, and *RAM1*), only one gene is up-regulated in *S. cerevisiae* but not *C. albicans* (*FAR1*).

Genes Associated with Filamentation

To form long mating tubes, *C. albicans* appears to have selectively placed filamentation-associated genes under the regulation of pheromone. Bennett et al. (11) found that several genes either involved in the regulation of filamentation or regulated by filamentation were up-regulated in **a** cells by α-pheromone. Zhao et al. (137a) also identified several genes up-regulated either in mating mixtures of **a/a** and α/α cells or induced in **a/a** cells by α-pheromone. The genes identified in the two studies included *CPH1*, *CEK1*, *FRG23*, *RBT1*, *HWP1*, *SAP4*, *RBT2*, *SAP6*, *SAP5*, *ECE1*, *CAF1*, and *RIM101*. Zhao et al. (137a) also demonstrated that the gene *YWP1*, previously identified as down-regulated during filamentation (108), was also down-regulated in mating mixtures. None of these genes are similarly regulated during mating in *S. cerevisiae*, primarily because there are no orthologs of them in the *S. cerevisiae* genome.

Genes Associated with Switching

To mate, a cell must switch from white to opaque (64, 80). The transition from white to opaque has been demonstrated to involve the up-regulation of opaque-specific genes and down-regulation of white-specific genes (7, 44, 82, 83, 111, 112, 114–116, 132, 138; D. Sanglard, F. Ischer, M. Monod, S. Dogra, R. Prasad, and J. Bille, *Abstr. Am. Soc. Microbiol. Conf. Candida Candidiasis, abstr.* C27, 1999). Lan et al. (57), using microarray analyses, identified 152 genes that were down-regulated and 221 genes that were up-regulated in the the white-opaque transition. As Tsong et al. (127) point out, this may exceed the number of genes that are regulated by the mating genes. One or more of the many genes up-regulated or down-regulated in the white-opaque transition must play a role in or be necessary for mating competence, a specific trait of

opaque-phase cells (80). In addition, one must wonder if any of the unique cytological traits of opaque cells play a role in mating competence, especially the unique pimples on the cell surface and the large cell vacuole containing vesicles (2, 3). The opaque cell wall pimples contain channels and are associated with unique cytoplasmic structures that have been visualized by transmission electron microscopy (2).

Lockhart et al. (65), using Northern analysis, demonstrated that the three opaque-specific genes *OP4*, *SAP1*, and *SAP3* were down-regulated when opaque cells were treated with α-pheromone. However, they found that *CDR3* was not similarly down-regulated. Zhao et al. (137a) subsequently demonstrated that *OP4*, *SAP1*, and *SAP3* were down-regulated in mating mixtures of natural **a/a** and α/α opaque cells. Because the Northern blot hybridization signals for the three down-regulated genes were low to negligible in mating mixtures, one can conclude that the three genes were also down-regulated in α/α cells by **a**-pheromone. Zhao et al. (137a) further demonstrated that the white-specific genes *WH11* and *EFG1* were not coordinately up-regulated by pheromone, indicating that pheromone did not induce a complete switch from opaque back to white.

Pheromone-induced down-regulation of opaque-specific genes suggests that although the expression of these genes is necessary for attaining mating competency (64, 80), expression of select opaque-specific genes may be incompatible with the actual mating process once it has begun. The absence of concomitant up-regulation of white-specific genes suggests that they are not necessary for the pheromone response and that the regulation of phase-specific gene expression is controlled differently during switching and mating. Interestingly, although many opaque-phase-specific genes are down-regulated by pheromone, a conjugation tube that fails to fuse forms an apical bud that is also in the opaque phase and capable of responding to later rounds of pheromone (65). Since fusion generates an **a**/α cell, presumably capable of expressing Mtla1p-Mtlα2p, one wonders when all opaque-specific genes are turned off and all white-specific genes are turned on in the switch from opaque to white. Perhaps it requires two generations, as is the case in temperature-induced opaque-to-white mass conversion (111, 112).

C. ALBICANS WHITE CELLS ARE PARTIALLY RESPONSIVE TO PHEROMONE

Since *C. albicans* must switch from white to opaque in order to shmoo and mate, one might expect white cells to be completely unresponsive to pheromone. Surprisingly, this is not the case. Lockhart et al. (65) found that while α-pheromone does not induce shmooing, it selectively up-regulates several mating-associated genes in white a/a (P37005) cells. Transcript levels of both STE2, which encodes the α-pheromone receptor, and STE4, which encodes the G-protein β subunit, were found to be up-regulated by α-pheromone in white cells as well as in opaque cells, and to a similarly induced level. On the other hand, FIG1 and KAR4, the S. cerevisiae orthologs involved in controlling mating efficiency (31) and karyogamy (50), respectively, were induced by α-pheromone in opaque cells but not in white cells. Because this result was so unexpected, pheromone induction of white cells was repeated in two independent a/a strains (T. Srikantha, K. Daniels, C. Pujol, S. Lockhart, and D. R. Soll, unpublished data). This analysis demonstrated once again that silent mating-associated genes are pheromone induced in white cells, just as they are in opaque cells. It is hard to believe that this selective pheromone response in white cells serves no purpose in mating.

RETURNING TO A DIPLOID STATE AFTER C. ALBICANS FUSION

Tzung et al. (128), in a bioinformatic analysis of the emerging C. albicans genome sequence, found that while orthologs for all of the genes necessary for the pheromone response in S. cerevisiae were present, orthologs for many of the genes necessary for meiosis, recombination, and synaptonemal complex formation could not be identified. They suggested that alternatives to meiosis may function in C. albicans for genetic exchange. However, orthologs to many of the genes involved in S. cerevisiae meiosis were found in C. albicans, including IME2, a kinase necessary for premeiotic DNA replication; genes in the RIM101 pathway that activate IME1; a transcription regulator involved in the activation of early meiotic genes, MEK1, which is meiosis specific; and AMA1, an activator of the anaphase-promoting complex (128). Orthologs were also found of S. cerevisiae genes involved in sister chromatid cohesion, strand invasion, Holliday junction formation, mismatch repair, and sporulation (128). The presence in C. albicans of orthologs to so many meiosis- and sporulation-related S. cerevisiae genes suggests that C. albicans may indeed be capable of undergoing meiosis and that either orthologs of missing genes may have undergone major changes, so that they were missed, or alternative genes perform similar functions. For these reasons, meiosis should continue to be searched for in C. albicans.

C. albicans appears to be an obligate diploid. Although aneuploidy has been documented (125), a haploid strain or a strain even close to haploid has never been described. In the original C. albicans mating experiments (46, 70), it was demonstrated that fusants, identified by complementation, contained one nucleus and elevated DNA levels (i.e., higher than diploid). Chen et al. (20) also demonstrated that fusants contained more DNA than parent diploid strains did. It was therefore incumbent to demonstrate meiosis, if it occurred. Bennett et al. (11) investigated whether tetraploid mating products underwent meiosis. They generated a tetraploid strain with markers on chromosomes 1, 3, and 5 and then analyzed segregation and chromosome loss under a variety of conditions that included eight different media and three temperatures (22, 30, and 37°C). They found that tetraploids lost chromosomes in S. cerevisiae presporulation medium (1% yeast extract, 0.8% peptone, 10% glucose) in an efficient manner at higher temperature. After 8 days in presporulation medium, cells had efficiently reduced chromosomes 1 and 3 to the diploid state. The time-dependent loss of chromosomes appeared to occur randomly. This was not true for growth in rich growth medium, suggesting that medium composition was important in this process. Growth on sorbose medium (47) also results in chromosome loss not limited to chromosome 5. Bennett and Johnson (10) concluded that presporulation medium and sorbose medium induce random loss of chromosomes. They further suggested that chromosome loss in tetraploid cells in response to stress may have evolved as an alternative pathway to meiosis. However, they did not rule out the possibility that meiosis could be induced by as yet unidentified conditions.

SKIN FACILITATES C. ALBICANS MATING

Given that C. albicans is both a commensal and a pathogen of humans, one would expect mating to occur at 37°C. Yet, in three studies it was reported that mating, assessed either by complementation or at the cellular level, did not occur efficiently at 37°C (10, 65, 70). One might consider the possibility that since mating is dependent on a switch from white to opaque and the opaque phenotype is unstable at 37°C, the increased temperature simply drives cells into the white phase. However, Lockhart et al. (65) found that if pheromone was added to opaque cells and the temperature was raised to 37°C, no shmoos formed. Instead, cells formed buds. Pheromone did not arrest cells at 37°C. These observations suggested that at least the initial mating steps of shmooing and fusion may not be able to occur at 37°C, which is the body temperature of the human host.

Lachke et al. (55) considered the possibility that rather than mating in the body, C. albicans might mate outside on the body surface. They based this prediction on two observations. First, Kvaal et al. (52) had demonstrated that when opaque cells were applied to newborn mouse skin under a gauze patch for 24 h, there was intense colonization, but when white cells were applied, there was little colonization. Hence, opaque cells were far more virulent than white cells in this model. Second, the temperature of mouse skin is approximately 32°C (55), roughly 3°C below the threshold temperature of 35°C above which mass conversion of opaque to white is induced (111). Lachke et al. (55) therefore tested mating on skin by applying mixtures of opaque **a/a** and α/α cells. This was accomplished by simply saturating a gauze pad with cells, taping the pad to the skin of a newborn mouse, and incubating for 24 h. They found that skin facilitated mating. Opaque cells were able to extend conjugation tubes over two cells in length and efficiently find a cell of the opposite mating type, suggesting that the skin environment facilitated the generation of pheromone gradients (Fig. 6). Lachke et al. (55) obtained fusion of close to 50% of the cells in **a/a**-plus-α/α mixtures on some skin patches, compared to a maximum of 25% fusion in shaking cultures. They obtained shmooing on average in 73% of the cells in an applied mixture and fusion of over half of shmooed cells. Lachke et al. (55) concluded that skin represented a unique facilitator of

mating and might even be the natural site. However, they noted that only on rare occasions did fusants on skin form the first daughter bud (55). Hence, one wonders if mating begins on the skin and continues at another body locale. What is titillating about the observations by Lachke et al. (55) is that they suggest that the mating process may be intimately dependent on host-pathogen interactions.

MATING TYPE AND VIRULENCE OF *C. ALBICANS*

Unlike *S. cerevisiae*, *C. albicans* loses alternative mating information when it spontaneously undergoes *MTL* homozygosis. Spontaneously generated **a/a** cells lose *MTL*α information, and spontaneously generated α/α cells lose *MTL***a** information. Several **a**/α strains have been observed to undergo spontaneous *MTL* homozygosis in Lee's medium (60), modified by the method of Bedell and Soll (9). In addition, media can be developed that appear to induce *MTL* homozygosis (47, 70), suggesting that medium composition and perhaps stress conditions may induce *MTL* homozygosis. Hence, it seems reasonable to conclude that **a**/α strains of *C. albicans* must continuously generate *MTL*-homozygous strains. There does not appear, however, to be a similar clear route from the *MTL*-homozygous to

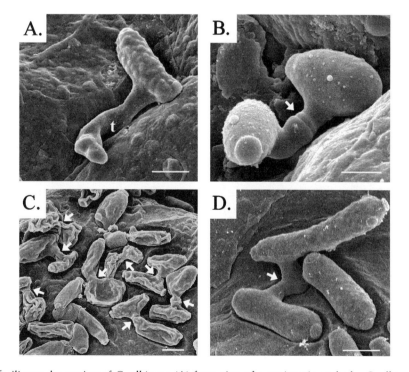

Figure 6. Skin facilitates the mating of *C. albicans*. (A) formation of a conjugation tube by *C. albicans* along skin of a newborn mouse. (B) Fusion between a *C. albicans* **a/a** and an α/α cell on skin. (C) Lower-magnification image reveals high frequency fusion of *C. albicans* **a/a** and α/α cells on skin. (D) Fusion between a *C. albicans* **a/a** and a *C. dubliniensis* α/α cell on skin. Arrows point to fusions. t, long conjugation tube. Scale bars, 2 μm.

MTL-heterozygous state. One might propose that this route is mating, but studies of population structure have demonstrated that *C. albicans* reproduces primarily by clonal propagation (35, 94, 137). Populations, however, are rarely purely "sexual" or "clonal" (74). DNA fingerprinting studies have separated *C. albicans* strains into five major clades (I, II, III, SA, and E) and a number of minor clades (109). These clades exhibit geographical enrichment, but for the most part they overlap in geographical regions ranging from South Africa to Europe to the United States (13, 96, 98). The fact that overlapping clades maintain genetic integrity supports the conclusion that propagation is primarily clonal. However, in almost every study of population structure, there are hints of recombination (93). Perhaps the most important study related to genetic exchange is that by Anderson et al. (1). They analyzed nuclear DNA and mitochondrial polymorphisms and found that while 45 of 48 strains presented congruent polymorphisms, in which mitochondrial haplotypes corresponded to distinct clusters of nuclear genotypes deduced from a dendrogram analysis, 3 presented incompatible DNA polymorphisms, demonstrating infrequent past genetic exchange (1). Although one immediately considers mating the source of recombination, other mechanisms can also effect recombination. Regardless of the mechanism, the combined results of a variety of studies indicate that mating is rare. Hence, there appears to be no clear route from the *MTL*-homozygous to *MTL*-heterozygous state, as there is in the opposite direction. How, then, does *C. albicans* remain primarily *MTL* heterozygous in nature? One possibility is that **a**/α strains are more competitive than their **a**/**a** or α/α offspring for the niches they inhabit, in this case the colonized host. **a**/α strains may therefore be more competitive than their *MTL*-homozygous offspring in both commensalism and virulence. To begin to test this possibility, Lockhart et al. (67) compared the virulence of **a**/α strains with that of their spontaneous **a**/**a** and α/α offspring by estimating average time of host death in the mouse model of systemic infection. They found that three unrelated **a**/α strains were far more virulent than their **a**/**a** or α/α offspring. While the cells of the three parent strains, P37037, P37039, and P75063, caused death in 50% of infected mice after 10, 12, and 5 days, respectively, their *MTL*-homozygous offspring caused death in no more than 20% of infected mice after 16 days (67). More revealing were experiments in which **a**/α cells were coinjected with *MTL*-homozygous offspring in a 1:1 ratio. Of eight combinations tested, which included four **a**/α parent strains and their *MTL*-homozygous offspring, the **a**/α parent strain predominated in the kidneys at the

time of host death or sacrifice. **a**/α parent cells made up between 80 and 99% of cells in the kidneys at the time of extreme morbidity or death.

The results obtained by Lockhart et al. (67) indicate that **a**/α strains are more virulent than their **a**/**a** or α/α offspring when injected individually into the mouse model of systemic infection and have a competitive advantage over their *MTL*-homozygous offspring when injected in combination. These results suggest that the mating system in *C. albicans* is conserved in nature by the superior colonizing ability of **a**/α over their **a**/**a** and α/α offspring. This conclusion will be greatly strengthened by proving that **a**/α strains also have a competitive advantage over their **a**/**a** or α/α offspring in the commensal state if a model can be identified for this purpose.

The **a**/α advantage, although apparently true for *MTL*-homozygous offspring of natural **a**/α strains, may not be true for established *MTL*-homozygous strains in nature. The strain WO-1, which is α/α, was isolated from the blood of an infected bone marrow transplant recipient (107). This strain eventually killed this patient. Subsequent studies of its virulence in the mouse model of systemic infection revealed that animals injected with the same concentration of cells used in the studies performed by Lockhart et al. (67) exhibited 50% survival after 6 days (53). Hence, the natural α/α strain WO-1 appears to be more virulent than spontaneously generated **a**/**a** or α/α offspring of the **a**/α strains studied by Lockhart et al. (67). There exists the possibility that *MTL*-homozygous strains established in nature may have undergone compensatory changes that increase their virulence.

The results obtained by Lockhart et al. (67) therefore suggest that the **a**/α genotype provides a higher level of virulence and a competitive advantage over *MTL*-homozygous offspring. Lockhart et al. (67) further demonstrated that generating an isogenic **a**/**a**/α2 strain from an **a**/**a** strain resulted in a competitive advantage over its parent strain when analyzed in 1:1 mixing experiments in the mouse model of systemic infection. These latter results suggest that the Mtl**a**1p-Mtlα2p complex may not only suppress mating and white-opaque switching but also may confer virulence and the competitive advantage to **a**/α cells. However, because of the demonstration by Wu et al. (135) that the homologs of chromosome 5 harbor *MTL***a**- and *MTL*α-linked alleles of genes outside the *MTL* locus, another hypothesis must be entertained, namely, that heterozygosity of other genes along chromosome 5 provide increased virulence and a competitive advantage. Indeed, seven of the eight *MTL*-homozygous offspring tested by Lockhart et al. (67) arose by loss of one chromosome 5 homolog followed by duplication of the retained homolog, resulting in homozygosis of all genes along chromosome 5. There is a large body of literature demonstrating that in many cases, both alleles of a gene are necessary for function. Interallelic complementation or synergism has been demonstrated in both *Drosophila* (81) and mice (120). Adding to the above reservations, Legrand et al. (61) observed karyotypic rearrangements assessed by contour-clamped homogeneous electric field electrophoresis in a majority of natural *MTL*-homozygous strains, suggesting that additional chromosomal and perhaps gene expression changes accompany *MTL* homozygosis. Therefore, mouse model experiments involving the selective deletion or addition of individual *MTL* genes must be performed to definitively prove the hypothesis that Mtla1p-Mtlα2p regulates virulence and competitive advantage.

MTL GENOTYPE OF *C. ALBICANS* AND DRUG SUSCEPTIBILITY

Rustad et al. (103) analyzed the *MTL* genotype of 96 clinical isolates, consisting of 50 that were azole resistant or less susceptible to azoles than average isolates of *C. albicans* and 46 azole-susceptible isolates. They found that of the total collection, 84 were **a**/α, 6 were **a**/**a**, and 6 were α/α. Remarkably, of the 12 *MTL*-homozygous isolates, 11 were either resistant or less susceptible to azole, suggesting that *MTL*-homozygous strains had a greater propensity to be azole resistant or less susceptible. The proportion of natural *MTL* homozygotes in the collection of resistant or less susceptible strains was 26%, far higher than that measured by Lockhart et al. (63) in a general collection of strains. However, when Rustad et al. (103) compared *MTL*a1 and *MTL*α2 null mutants with strains containing both *MTL*a1 and *MTL*α2 derived from the same parent strain, CAI4, the former strains, which could not form an Mtla1p-Mtlα2p complex, exhibited no difference in azole susceptibility (103). These latter results suggested that azole resistance and lowered susceptibility associated with natural *MTL*-homozygous strains could very well have been due to homozygosis of genes other than the *MTL* genes on chromosome 5.

Pujol et al. (97) used an opposite strategy to assess the relationship between the *MTL* genotype and drug resistance. They first analyzed drug susceptibility in 17 natural **a**/**a** and α/α isolates and found that only one strain exhibited increased resistance to fluconazole as well as other tested azoles. They next examined *MTL*-homozygous progeny of three natural **a**/α strains. They again found no correlation between *MTL* homozygosis and drug resistance.

Legrand et al. (61) also found no correlation between fluconazole resistance and *MTL* homozygosity. In light of these results, those of Rustad et al. (103) suggest that *MTL*-homozygous strains may be better at developing azole resistance, presumably through exposure to azoles, than are *MTL*-heterozygous strains, but that *MTL* homozygosis does not directly result in drug resistance.

C. DUBLINIENSIS CAN MATE WITH *C. ALBICANS*

In the early 1990s, a number of atypical strains of *C. albicans* collected from human immunodeficiency virus-positive individuals exhibited unusual DNA fingerprints when assessed by a number of different genetic fingerprinting methods (14, 75, 87, 98, 124). Coleman, Sullivan, and coworkers (123) subsequently grouped these strains into a new species, *C. dubliniensis*. This species has been demonstrated to be closely related to *C. albicans* based on a number of genetic criteria, including ribosomal DNA sequences (123). Retention of the repeat sequence RPS (48) further suggested that the two species diverged quite recently in evolutionary history. Because of the high level of relatedness, Pujol et al. (95) examined whether *C. dubliniensis* possessed the same mating-type gene configuration as *C. albicans*. They probed Southern blots of 82 natural strains of *C. dubliniensis* with the *MTL*a1, *MTL*α1, and *MTL*α2 genes of *C. albicans*. They found that 67% were *MTL* heterozygous (**a**/α), 21% were *MTL*a homozygous (**a**/**a**), and 12% were *MTL*α homozygous (α/α). The proportion of *MTL*-homozygous strains in this general *C. dubliniensis* collection was 10-fold higher than that in a general collection of *C. albicans* strains (63). Pujol et al. (95) also found that a majority of *MTL*-homozygous strains were members of one clade, group I.

The *MTL*α locus was found to be organized in a manner similar to that of *C. albicans*. The nucleotide sequences of the five genes harbored by the locus ranged between 88 and 90% identical to that of *C. albicans*, while the deduced amino acid sequences were between 97 and 99.5% similar. Data from the Pathogen Sequencing Unit at the Wellcome Trust Sanger Institute (www.sanger.ac.uk/Projects/C_dubliniensis/) revealed that the *MTL*a locus of *C. dubliniensis* was also similar to that of *C. albicans* in configuration and sequence.

Pujol et al. (95) also found that, as was the case in *C. albicans*, only opaque cells of *C. dubliniensis* could mate. They demonstrated that in mating mixtures and in **a**/**a** cell cultures treated with the chemically synthesized *C. albicans* 14-mer α-pheromone (65), *C. dubliniensis* formed extremely long conjugation tubes, like

C. albicans. They also demonstrated that *C. dubliniensis* underwent the same sequence of cellular events as *C. albicans* during the mating progress. They used vital dyes to stain a/a and α/α cells and nuclear stains to describe the nuclear events associated with nuclear fusion and first daughter cell formation. The events were similar to those that have been observed during *C. albicans* mating (K. Daniels, S. Lockhart, and D. R. Soll, unpublished data). The nuclei of parent a/a and α/α cells migrated into the conjugation bridge after tube fusion, were juxtaposed at the time of daughter cell evagination, fused, and then divided, with one nucleus remaining in the tube and the other entering the growing daughter cell. However, *C. dubliniensis* a/a and α/α cells fused in suspension cultures with far lower efficiency than *C. albicans* did. This was demonstrated to be due to the lack of clumping, since fusion between *C. dubliniensis* a/a and α/α cells increased fivefold in dense, undisturbed standing cultures (95). Given the similarity of the two mating systems, Pujol et al. (95) tested whether *C. albicans* and *C. dubliniensis* could mate. They found that *C. dubliniensis* a/a cells mated with *C. albicans* α/α cells and did so with high efficiency because the mating cultures of mixed species underwent clumping. However, given the integrity of the two species in nature, it seems very unlikely that mating between them occurs in nature. Alternatively, they may mate in nature but their offspring may not be competitive.

MATING SYSTEM OF *C. GLABRATA*

As was the case for *C. albicans*, it had generally been assumed that *C. glabrata* lacked a sexual cycle. Phylogenetic studies based on ribosomal DNA sequences, however, demonstrated that *C. glabrata* was far more closely related to *S. cerevisiae* than to *C. albicans* (8, 18, 51), suggesting that if it had a sexual system, it would more probably be like that of *S. cerevisiae*. In 2003, two studies, quite different in approach, demonstrated that *C. glabrata* possessed mating loci configured very much like that of *S. cerevisiae*. Using genome survey sequencing, Wong et al. (134) identified orthologs of 31 *S. cerevisiae* genes that functioned exclusively in the mating process or meiosis. Mating genes included *MFα2*, *STE13*, and *STE6*. Orthologs were identified for the genes encoding the MAP kinases involved in the pheromone response pathway, *STE12* and *FAR1* (134). Finally, they identified the *MATα1* and *MATα2* genes oriented divergently at the same locus, as in *S. cerevisiae*. Using a different approach, Srikantha et al. (116) developed degenerate primers that encompassed the highly conserved carboxy-terminal homeodomain of approximately 125 bp of the *S. cerevisiae*

MATα2 open reading frame (ORF), which is also highly conserved in the MTLα2 ORF of *C. albicans*. This sequence is also conserved in the *MATa2* gene of *S. cerevisiae* and *Kluyveromyces lactis*. These primers were used to clone three mating-type-like *MTL* loci (*MTL1*, *MTL2*, and *MTL3*) from *C. glabrata*. They found that each locus contained a pair of *MTL* genes with respective coding regions on the complementary Crick and Watson DNA strands. They demonstrated that of 38 unrelated *C. glabrata* strains, all but one were either a/a/α or a/α/α, suggesting that, just as in the case of *S. cerevisiae*, there very probably existed one *MTLa* and one *MTLα* silent locus and one expressed locus that is either *MTLa* or *MTLα*. Northern blot analysis revealed that strains that contained α genes at the *MTL1* locus expressed *MTLα1* and strains that contained a genes at the *MTL1* locus expressed *MTLα1*, indicating that *MTL1* was the expression locus and the correlate of the *MAT* locus of *S. cerevisiae*. The three mating loci of *C. glabrata*, therefore, appear to represent a cassette system like that in *S. cerevisiae*. However, only the expression locus *MTL1* and one of the silent loci, *MTL3*, are located in the same chromosome (116). The second silent locus, *MTL2*, is located on a separate chromosome, suggesting that a chromosomal translocation may have occurred in the *C. glabrata* lineage after divergence from an ancestral species. This is in contrast to *S. cerevisiae*, in which the expression locus, *MAT*, and both silent loci, *HML* and *HMR*, are located on the same chromosome.

Butler et al. (17) further demonstrated that the boundary between the Y and Z1 regions, where sequences of *MTLa1* connect to flanking sequences, contained a cleavage site for the endonuclease HO. For *MTLα1*, this site was altered but the alteration did not affect function. Based on this observation, Butler et al. (17) searched for and identified the *HO* endonuclease gene in *C. glabrata*, which has 57% deduced amino acid identity to the HO protein of *S. cerevisiae*, including the motifs essential for mating-type switching in *S. cerevisiae*. The presence of this apparently functional *HO* gene suggested that mating-type switching occurred. Two studies demonstrated that this was indeed the case. Butler et al. (17) demonstrated by PCR analysis of *C. glabrata* strain RND13 (129, 130), an *MTLa1* strain, that while a strong signal was obtained with *MTLa1*-specific primers, as expected, a weak but clear signal was obtained with *MTLa1*-specific primers. They also demonstrated that the *HO* site was cleaved and repaired. They obtained a similar result with strain CBS138, an *MTLα1* strain, in which they obtained a weak signal with *MTLa1*-specific primers. These results suggested that mating-type switching occurs at low frequency in vitro. Brockert et al. (15) presented evidence that mating-type switching also occurred in vivo. They analyzed individual isolates from different body locations of three vaginitis patients infected with *C. glabrata*. Each isolate was fingerprinted with the complex fingerprinting probe Cg6 (66). The collection of strains from each patient proved to represent a single strain. For two patients, P2 and P3, Southern analysis with the probe FuncP2, an oligonucleotide that discriminates the *MTL* genotype of *C. glabrata*, revealed that all collected isolates of each strain exhibited the same *MTL* genotype. However, for the third patient, P1, there was a switch from **a** to α in one isolate, whereas the remaining isolates were **a**.

Hence, the results obtained by Wong et al. (134) and Srikantha et al. (116) demonstrated that *C. glabrata* possesses three mating-type-like loci configured in a cassette system like that of *S. cerevisiae*, and the results of Butler et al. (17) and Brockert et al. (15) demonstrated that *C. glabrata* underwent mating-type switching. However, there has been no report of mating demonstrated at either the genetic or cellular level of analysis. Interestingly, Pujol et al. (95) found that all of the 120 natural *C. glabrata* strains analyzed in their study of *MTL* genotype were either **a** or α, suggesting that all strains in nature are haploid. This differs from *S. cerevisiae*, which in nature is primarily diploid (22). There has therefore been no demonstration of an *MTL*-heterozygous strain and no report of sporulation or meiosis in *C. glabrata*. It would be naive to think that *C. glabrata* maintains a complex and functional mating-type switching system, and undergoes mating-type switching, without selective pressure. It therefore seems likely that mating and meiosis will eventually be demonstrated in *C. glabrata*.

MATING IN *C. GLABRATA*, AS IN *C. ALBICANS*, MAY BE RARE

de Meeus et al. (28) first analyzed a collection of 52 *C. glabrata* isolates by multilocus enzyme electrophoresis for population structure. They found it to be primarily clonal but did not rule out recombination. Dodgson et al. (29), using multilocus sequence typing (MLST), subsequently demonstrated that 109 isolates in a geographically diverse collection separated into five robust clades, which exhibited some geographical specificity. Dodgson et al. (29) identified synaptomorphic alleles, shared by all members of a specific group, which supported a clonal form of propagation. In a follow-up population genetics analysis of 165 isolates, Dodgson et al. (30), again using MLST, demonstrated a predominantly clonal population structure. However, phylogenetic incompatibilities indicated

that recombination does occur, but infrequently. Indeed, such tests provide neither a timetable for when recombination occurred nor the mechanism for recombination (93). Hence, as in the case of *C. albicans*, if mating does occur, it must be a rare event. In contrast to *C. albicans*, however, *C. glabrata* maintains its mating system by virtue of the redundancies in the silent loci. However, one must wonder why *S. cerevisiae* is diploid in nature, while *C. glabrata* is haploid, and whether this impacts the mating process.

CONCLUSION

Recently, several analyses of the configuration, structure, and function of mating genes, as well as the mating process, have been performed in order to develop an evolutionary tree for the hemiascomycetes. It is clear from these analyses that while *C. albicans* diverges early along the tree from *S. cerevisiae*, *C. glabrata* diverges quite late. In this chapter, the mating systems of *C. albicans* and *C. glabrata* have been compared to that of *S. cerevisiae* since the last of these has been so intensely studied over the past three decades and thus provides a contextual framework for interpreting the first two. While discussing both similarities and dissimilarities, the latter have been emphasized, particularly in light of their possible integration into the host-dependent life histories resulting from commensalism and pathogenesis. In the case of *C. albicans*, fundamental differences exist in the configuration of the mating loci, in mating-type genes, and in the mating processes they regulate. In addition, the *C. albicans* mating process includes an essential switch from white to opaque that involves the regulation of hundreds of genes, possibly more genes than are regulated by mating-type pheromones. There is no correlate in the mating process of *S. cerevisiae* or, for that matter, any other hemiascomycete. Furthermore, mating pheromones in turn selectively down-regulate opaque-specific genes and up-regulate a number of hypha-specific genes. This chapter has argued that incorporation of both the white-opaque switching program and the filamentation program into the mating process may be due at least in part to the intimate relationship *C. albicans* has developed with its host (72). There should be little question that developmentally, and in terms of gene expression, the mating process of *C. albicans* is more complex than that of *S. cerevisiae*. Added to this is the discovery that hemoglobin, and most likely other environmental cues that are host dependent, regulate mating-type genes and in turn switching and mating competency, again possibly without correlates in *S. cerevisiae*. It should, however, be clear in this review that although discoveries related to *C. albicans* mating have been made at a fast pace over the past 5 years, many fundamental questions remain unanswered. Is there meiosis? Why must a cell switch from white to opaque in order to mate? Why do pheromone-induced conjugation tubes grow so long? What signals besides hemoglobin affect mating-type gene expression, switching, and the mating process? Do environmental niches other than skin facilitate different stages of the mating process? Does the Mtla1p-Mtlα2p complex regulate virulence? No doubt, given the fast pace of discovery in the past 5 years, these questions will be rapidly answered.

In contrast to *C. albicans*, the configuration and structure of mating-type genes and loci are highly similar to those of *S. cerevisiae*. It therefore seems likely that the regulatory circuitry and the genes regulated by pheromone will also be similar. Since mating-type switching has been demonstrated in vivo and in vitro and since intact mating-specific and meiosis-specific genes have been identified in the genome, it also seems likely that both mating-type-dependent fusion and meiosis will eventually be demonstrated. However, several additional questions must be addressed. First, why are natural strains haploid rather than diploid like *S. cerevisiae*? Second, why hasn't fusion been reported, given the recent attention paid to mating? Finally, given the similarities in both commensalism and pathogenesis between *C. albicans* and *C. glabrata*, can it be that *C. glabrata* mating, like that of *C. albicans*, has also become intimately intertwined with its host? Answers to these questions, unfortunately, may be obtained at a far slower pace than those to the questions posed for *C. albicans*, given the disproportionate research attention paid the latter.

Finally, one must wonder, for both *C. albicans* and *C. glabrata*, why complex mating systems are maintained, given the predominantly clonal population structure of both. Why devote so many genes, especially in *C. albicans*, to such a rare event? There may be more than one answer to this question. First, it may be that mating and associated recombination, although rare, is essential for the continued survival of the species. Second, it may be that the mating process has become integral to pathogenesis.

Acknowledgments. The work performed in my laboratory on infectious fungi was funded by NIH grants AI2392 and DE014219. I am indebted to T. Srikantha, S. Lockhart, and C. Pujol for reviewing the manuscript and to K. Daniels and J. Collins for generating figures.

REFERENCES

1. Anderson, J. B., C. Wickens, M. Khan, L. E. Cowen, N. Federspiel, T. Jones, and L. M. Kohn. 2001. Infrequent genetic exchange and recombination in the mitochondrial genome of *Candida albicans*. *J. Bacteriol.* **183:**865–872.

2. Anderson, J. M., R. Mihalik, and D. R. Soll. 1990. Ultrastructure and antigenicity of the unique cell wall pimple of the *Candida* opaque phenotype. *J. Bacteriol.* **172:**224–235.

3. Anderson, J. M., and D. R. Soll. 1987. Unique phenotype of opaque cells in the white-opaque transition of *Candida albicans*. *J. Bacteriol.* **169:**5579–5588.

4. Aramayo, R., and R. L. Metzenberg. 1996. Meiotic transvection in fungi. *Cell* **86:**103–113.

5. Astell, C. R., L. Ahlstrom-Jonasson, M. Smith, K. Tatchell, K. A. Nasmyth, and B. D. Hall. 1981. The sequence of the DNAs coding for the mating-type loci of *Saccharomyces cerevisiae*. *Cell* **27:**15–23.

6. Astrom, S. U., A. Kegel, J. O. Sjostrand, and J. Rine. 2000. *Kluyveromyces lactis* Sir2p regulates cation sensitivity and maintains a specialized chromatin structure at the cryptic alpha-locus. *Genetics* **156:**81–91.

7. Balan, I., A. M. Alarco, and M. Raymond. 1997. The *Candida albicans* CDR3 gene codes for an opaque-phase ABC transporter. *J. Bacteriol.* **179:**7210–7218.

8. Barns, S. M., D. J. Lane, M. L. Sogin, C. Bibeau, and W. G. Weisburg. 1991. Evolutionary relationships among pathogenic *Candida* species and relatives. *J. Bacteriol.* **173:**2250–2255.

9. Bedell, G., and D. R. Soll. 1979. The effects of low concentrations of zinc on the growth and dimorphism of *Candida albicans*: evidence for zinc-resistant and zinc-sensitive pathways for mycelium formation. *Infect. Immun.* **26:**348–354.

10. Bennett, R. J., and A. D. Johnson. 2003. Completion of a parasexual cycle in *Candida albicans* by induced chromosome loss in tetraploid strains. *EMBO J.* **22:**2505–2515.

11. Bennett, R. J., M. A. Uhl, M. G. Miller, and A. D. Johnson. 2003. Identification and characterization of a *Candida albicans* mating pheromone. *Mol. Cell. Biol.* **23:**8189–8201.

12. Bennett, R. J., M. G. Miller, P. R. Chua, M. E. Maxon, and A. D. Johnson. 2005. Nuclear fusion occurs during mating in *Candida albicans* and is dependent on the KAR3 gene. *Mol. Microbiol.* **55:**1046–1059.

13. Blignaut, E., C. Pujol, S. Lockhart, S. Joly, and D. R. Soll. 2002. Ca3 fingerprinting of *Candida albicans* isolates from human immunodeficiency virus-positive individuals reveals a new clade in South Africa. *J. Clin. Microbiol.* **40:**826–836.

14. Boerlin, P., F. Boerlin-Petzold, C. Durussel, M. Ado, J.-L. Pagani, J.-P. Chave, and J. Bille. 1995. Cluster of oral atypical *Candida albicans* isolates in a group of human immunodeficiency virus-positive drug users. *J. Clin. Microbiol.* **33:**1129–1135.

15. Brockert, P. J., S. A. Lachke, T. Srikantha, C. Pujol, R. Galask, and D. R. Soll. 2003. Phenotypic switching and mating-type switching of *Candida glabrata* at sites of colonization. *Infect. Immun.* **12:**7109–7118.

16. Bucking-Throm, E., W. Duntze, L. H. Hartwell, and T. R. Manney. 1973. Reversible arrest of haploid yeast cells in the initiation of DNA synthesis by a diffusible sex factor. *Exp. Cell Res.* **76:**99–110.

17. Butler, G., C. Kenny, A. Fagan, C. Kurischko, C. Gaillardin, and K. H. Wolfe. 2004. Evolution of the MAT locus and its Ho endonuclease in yeast species. *Proc. Natl. Acad. Sci. USA* **101:**1632–1637.

18. Cai, J., I. N. Roberts, and M. D. Collins. 1996. Phylogenetic relationships among members of the ascomycetous yeast genera *Brettanomyces*, *Debaryomyces*, *Dekkera*, and *Kluyveromyces* deduced by small-subunit rRNA gene sequences. *Int. J. Syst. Bacteriol.* **46:**542–549.

19. Campbell, D. A. 1973. Kinetics of the mating-specific aggregation in *Saccharomyces cerevisiae*. *J. Bacteriol.* **116:**323–330.

20. Chen, J., J. Chen, S. Lane, and H. Liu. 2002. A conserved mitogen-activated protein kinase pathway is required for mating in *Candida albicans*. *Mol. Microbiol.* **46:**1335–1344.

21. Chenevert, J. 1994. Cell polarization directed by extracellular cues in yeast. *Mol. Biol. Cell* **5:**1169–1175.

22. Clemons, K. V., P. Park, J. H. McCusker, M. J. McCullough, R. W. Davis, and D. A. Stevens. 1997. Application of DNA typing methods and genetic analysis to epidemiology and taxonomy of *Saccharomyces* isolates. *J. Clin. Microbiol.* **35:**1822–1828.

23. Coppin, E., R. Debuchy, S. Arnaise, and M. Picard. 1997. Mating-types and sexual development in filamentous ascomycetes. *Microbiol. Mol. Biol. Rev.* **61:**411–428.

24. Cross, F., L. H. Hartwell, C. Jackson, and J. B. Konopka. 1988. Conjugation in *Saccharomyces cerevisiae*. *Annu. Rev. Cell Biol.* **4:**429–457.

25. Csank, C., and K. Haynes. 2000. *Candida glabrata* displays pseudohyphal growth. *FEMS Microbiol. Lett.* **189:**115–120.

26. Daniels, K. J., S. R. Lockhart, P. Sundstrum, and D. R. Soll. 2003. During *Candida albicans* mating, the adhesin Hwp1 and the first daughter bud localize to the a/a portion of the conjugation bridge. *Mol. Biol. Cell* **14:**4920–4930.

27. Debuchy, R., S. Arnaise and G. Lecellier. 1993. The *mat*–allele of *Podospora anserina* contains three regulatory genes required for the development of fertilized female organs. *Mol. Gen. Genet.* **241:**667–673.

28. de Meeus, T., F. Renaud, E. Mouveroux, J. Reynes, G. Galeazzi, M. Mallie and J. M. Bastide. 2002. Genetic structure of *Candida glabrata* populations in AIDS and non-AIDS patients. *J. Clin. Microbiol.* **40:**2106–2109.

29. Dodgson, A. R., C. Pujol, D. W. Denning, D. R. Soll, and A. J. Fox. 2003. Multilocus sequence typing of *Candida glabrata* reveals geographically enriched clades. *J. Clin. Microbiol.* **41:**5709–5717.

30. Dodgson, A. R., C. Pujol, M. A. Pfaller, D. W. Denning, and D. R. Soll. 2005. Evidence for recombination in *Candida glabrata*. *Fungal Genet.* **42:**233–243.

31. Erdman, S., L. Lin, M. Malczynski, and M. Snyder. 1998. Pheromone-regulated genes required for yeast mating differentiation. *J. Cell Biol.* **140:**461–483.

32. Ferreira, A. V., S. Saupe, and N. L. Glass. 1996. Transcriptional analysis of the *mtA* idiomorph of *Neurospora crassa* identifies two genes in addition to *mtA-1*. *Mol. Gen. Genet.* **250:**767–774.

33. Fidel, P. L., J. A. Vazquez, and J. D. Sobel. 1999. *Candida glabrata*: review of epidemiology, pathogenesis and clinical disease with comparison to *C. albicans*. *Clin. Microbiol. Rev.* **12:**80–96.

34. Goutte, C., and A. D. Johnson. 1994. Recognition of a DNA operator by a dimer composed of two different homeodomain proteins. *EMBO J.* **13:**1434–1442.

35. Gräser, Y., M. Volovsek, J. Arrington, G. Schönian, W. Presber, T. G. Mitchell and R. Vilgalys. 1996. Molecular markers reveal that population structure of the human pathogen *Candida albicans* exhibits both clonality and recombination. *Proc. Natl. Acad. Sci. USA* **93:**12473–12477.

36. Haber, J. E. 1998. Mating-type gene switching in *Saccharomyces cerevisiae*. *Annu. Rev. Genet.* **32:**561–599.

37. Harashima, S., A. M. Miller, K. Tanaka, K.-I. Kusumoto, K.-I. Tanaka, Y. Mukai, K. Nasmyth, and Y. Oshima. 1989. Mating-type control in *Saccharomyces cerevisiae*: isolation and characterization of mutants defective in repression by aα2. *Mol. Cell. Biol.* **9:**4523–4530.

38. Hazen, K. C. 1995. New and emerging yeast pathogens. *Clin. Microbiol. Rev.* **8:**462–478.

39. Heid, P., E. Voss, and D. R. Soll. 2002. 3D-DIASemb: a computer-assisted system for reconstructing and motion analyzing in 4D every cell and nucleus in a developing embryo. *Dev. Biol.* **245:**329–347.

40. Herskowitz, I. 1988. Life cycle of the budding yeast *Saccharomyces cerevisiae*. *Microbiol. Rev.* **52:**536–553.

41. Herskowitz, I. 1995. MAP kinase pathways in yeast: for mating and more. *Cell* **80:**187–197.

42. Herskowitz, I., and R. E. Jensen. 1991. Putting the HO gene to work: practical uses for mating-type switching. *Methods Enzymol.* **194:**132–146.

43. Herskowitz, I., J. Rine, and J. N. Strathern. 1992. Mating-type determination and mating-type interconversion in *Saccharomyces cerevisiae*, p. 583–656. *In* E. W. Jones, J. R. Pringle, and J. R. Broach (ed.), *The Molecular and Cellular Biology of the Yeast Saccharomyces*. Cold Spring Harbor Laboratory Press, Cold Spring Harbor, N.Y.

44. Hube, B., M. Monod, D. Schofield, A. Brown, and N. Gow. 1994. Expression of seven members of the gene family encoding aspartyl proteinases in *Candida albicans*. *Mol. Microbiol.* **14:**87–99.

45. Hull, C. M., and A. D. Johnson. 1999. Identification of a mating-type-like locus in the asexual pathogenic yeast *Candida albicans*. *Science* **285:**1271–1275.

46. Hull, C. M., R. M. Raisner, and A. D. Johnson. 2000. Evidence for mating of the "asexual" yeast *Candida albicans* in a mammalian host. *Science* **289:**307–310.

47. Janbon, G., F. Sherman, and E. Rustcheko. 1999. Appearance and properties of L-sorbose-utilizing mutants of *Candida albicans* obtained on a selective plate. *Genetics* **153:**653–664.

48. Joly, S., C. Pujol, and D. R. Soll. 2002. Microevolutionary changes and chromosomal translocations are more frequent at RPS loci in *Candida dubliniensis* than in *Candida albicans*. *Infect. Genet. Evol.* **2:**19–37.

49. Keleher, C. A., S. Passmore, and A. D. Johnson. 1989. Yeast repressor alpha2 binds to its operator with yeast protein Mcm1. *Mol. Cell. Biol.* **9:**5228–5230.

50. Kurihara, L. J., B. G. Stewart, A. E. Gammie, and M. D. Rose. 1996. Kar4p, a karyogamy-specific component of the yeast pheromone response pathway. *Mol. Cell. Biol.* **16:**3990–4002.

51. Kurtzman, D. P., and C. J. Roberts. 1998. Identification and phylogeny of ascomycetous yeasts from analysis of nuclear large subunit (265) ribosomal DNA partial sequences. *Antoine Leeuwenhoek* **73:**331–371.

52. Kvaal, C., S. A. Lachke, T. Srikantha, K. Daniels, J. McCoy, and D. R. Soll. 1999. Misexpression of the opaque phase-specific gene *PEP1* (*SAP1*) in the white

phase of *Candida albicans* confers increased virulence in a mouse model of cutaneous infection. *Infect. Immun.* 67:6652–6662.

53. Kvaal, C. A., T. Srikantha, and D. R. Soll. 1997. Misexpression of the white phase-specific gene *WH11* in the opaque phase of *Candida albicans* affects switching and virulence. *Infect. Immun.* 65:4468–4475.

54. Lachke, S. A., S. Joly, K. Daniels, and D. R. Soll. 2002. Phenotypic switching and filamentation in *Candida glabrata*. *Microbiology* 148:2661–2674.

55. Lachke, S. A., S. R. Lockhart, K. J. Daniels, and D. R. Soll. 2003. Skin facilitates *Candida albicans* mating. *Infect. Immun.* 71:4970–4976.

56. Lachke, S. A., T. Srikantha, L. Tsai, K. Daniels, and D. R. Soll. 2000. Phenotypic switching in *Candida glabrata* involves phase-specific regulation of the metallotheionein gene *MT-II* and the newly discovered hemolysin gene *HLP*. *Infect. Immun.* 68:884–895.

57. Lan, C. Y., G. Newport, L. A. Murillo, T. Jones, S. Scherer, R. W. Davis, and N. Agabian. 2002. Metabolic specialization associated with phenotypic switching in *Candida albicans*. *Proc. Natl. Acad. Sci. USA* 99:14907–14912.

58. Leberer, E., D. Dignard, D. Harcus, L. Hougan, M. Whiteway, and D. Y. Thomas. 1993. Cloning of *Saccharomyces cerevisiae STE5* as a suppressor of a Ste20 protein kinase mutant: structural and functional similarity of Ste5 to Far1. *Mol. Gen. Genet.* 241:241–254.

59. Leberer, E., D. Harcus, I. D. Broadbent, K. L. Clark, D. Dignard, K. Ziegelbauer, A. Schmidt, N. A. R. Gow, A. J. P. Brown, and D. Y. Thomas. 1996. Signal transduction through homologs of the Ste20p and Ste7p protein kinases can trigger hyphal formation in the pathogenic fungus *Candida albicans*. *Proc. Natl. Acad. Sci. USA* 93:13217–13222.

60. Lee, K. L., H. R. Buckley, and C. C. Campbell. 1975. An amino acid liquid synthetic medium for development of mycelial and yeast forms of *Candida albicans*. *Sabouraudia* 13:148–153.

61. Legrand, M., P. Lephart, A. Forsche, F.-M. C. Mueller, T. Walsh, P. T. Magee, and B. B. Magee. 2004. Homozygosity at the *MTL* locus in clinical strains of *Candida albicans*: karyotypic rearrangements and tetraploid formation. *Mol. Microbiol.* 52:1451–1462.

62. Leibowitz, M. J., and R. B. Wickner. 1976. A chromosomal gene required for killer plasmid expression, mating, and spore maturation in *Saccharomyces cerevisiae*. *Proc. Natl. Acad. Sci. USA* 73:2061–2065.

63. Lockhart, S. R., C. Pujol, K. Daniels, M. Miller, A. Johnson, and D. R. Soll. 2002. In *Candida albicans*, white-opaque switchers are homozygous for mating-type. *Genetics* 162:737–745.

64. Lockhart, S. R., K. J. Daniels, R. Zhao, D. Wessels, and D. R. Soll. 2003. Cell biology of mating in *Candida albicans*. *Eukaryot. Cell* 2:49–61.

65. Lockhart, S. R., R. Zhao, K. J. Daniels, and D. R. Soll. 2003. α-Pheromone-induced shmooing and gene regulation require white-opaque switching during *Candida albicans* mating. *Eukaryot. Cell* 2:847–855.

66. Lockhart, S. R., S. Joly, C. Pujol, J. Sobel, M. Pfaller, and D. R. Soll. 1997. Development and verification of fingerprinting probes for *Candida glabrata*. *Microbiology* 143:3733–3746.

67. Lockhart, S. R., W. Wu, J. Radke, and D. R. Soll. 2005. Increased virulence and competitive advantage of a/α over a/a or α/α offspring conserves the mating system of *Candida albicans*. *Genetics* 169:1883–1890.

68. Lockhart, S. R., S. Joly, K. Vargas, J. Swails-Wenger, L. Enger, and D. R. Soll. 1999. Natural defenses against *Candida* colonization breakdown in the oral cavities of the elderly. *J. Dent. Res.* 78:857–868.

69. MacKay, V., and T. R. Manney. 1974. Mutations affixing sexual conjugation in *Saccharomyces cerevisiae*. I. Isolation and phenotypic characterization of non-mating mutants. *Genetics* 76:255–271.

70. Magee, B. B., and P. T. Magee. 2000. Induction of mating in *Candida albicans* by construction of MTLa and MTLalpha strains. *Science* 289:310–313.

71. Magee, B. B., M. Legrand, A. M. Alarco, M. Raymond, and P. T. Magee. 2002. Many of the genes required for mating in *Saccharomyces cerevisiae* are also required for mating in *Candida albicans*. *Mol. Microbiol.* 46:1345–1351.

72. Magee, P. T., and B. B. Magee. 2004. Through a glass opaquely: the biological significance of mating in *Candida albicans*. *Curr. Opin. Microbiol.* 7:661–665.

73. Marsh, L., A. M. Neiman, and I. Herskowitz. 1991. Signal transduction during pheromone response in yeast. *Annu. Rev. Cell Biol.* 7:699–728.

74. Maynard Smith, J., N. H. Smith, M. O'Rourke, and B. G. Spratt. 1993. How clonal are bacteria? *Proc. Natl. Acad. Sci. USA* 90:4384–4388.

75. McCullough, M., B. Ross, and P. Reade. 1995. Characterization of genetically distinct subgroup of *Candida albicans* strains isolated from oral cavities of patients infected with human immunodeficiency virus. *J. Clin. Microbiol.* 33:696–700.

76. McCusker, J. H., K. V. Clemons, D. A. Stevens, and R. W. Davis. 1994. Genetic characterization of pathogenic *Saccharomyces cerevisiae* isolates. *Genetics* 136:1261–1269.

77. McGrath, J. P., and A. Varshavsky. 1989. The yeast *STE6* gene encodes a homologue of the mammalian multidrug resistance P-glycoprotein. *Nature* 340:400–404.

78. McGuire, I. C., R. E. Marra, B. G. Turgeon, and M. G. Milgroom. 2001. Analysis of mating-type genes in the chestnut blight fungus, *Cryphonectria parasitica*. *Fungal Genet. Biol.* 34:131–144.

79. Meluh, P. B., and M. D. Rose. 1990. KAR3, a kinesin-related gene required for yeast nuclear fusion. *Cell* 60:1029–1041.

80. Miller, M. G., and A. D. Johnson. 2002. White-opaque switching in *Candida albicans* is controlled by mating-type locus homeodomain proteins and allows efficient mating. *Cell* 110:293–302.

81. Mongelard, F., M. Labrador, E. M. Baxter, T. I. Gerasimova, and V. G. Corces. 2002. Trans-splicing as a novel mechanism to explain interallelic complementation in *Drosophila*. *Genetics* 160:1481–1487.

82. Morrow, B., T. Srikantha, and D. R. Soll. 1992. Transcription of the gene for a pepsinogen, *PEP1*, is regulated by white-opaque switching in *Candida albicans*. *Mol. Cell. Biol.* 12:2997–3005.

83. Morrow, B., T. Srikantha, J. Anderson, and D. R. Soll. 1993. Coordinate regulation of two opaque-specific genes during white-opaque switching in *Candida albicans*. *Infect. Immun.* 61:1823–1828.

84. Mortimer, R. K., P. Romano, G. Suzzi, and M. Polsinelli. 1994. Genome renewal: a new phenomenon revealed from a genetic study of 43 strains of *Saccharomyces cerevisiae* derived from natural fermentation of grape musts. *Yeast* 10:1543–1552.

85. **Nasmyth, K.** 1983. Molecular analysis of a cell lineage. *Nature* **302:**670–676.

86. **Odds, F. C.** 1988. *Candida and Candidosis,* 2nd ed. Bailliere Tindall, London, United Kingdom.

87. **Odds, F., J. Schmidt, and D. R. Soll.** 1990. Epidemiology of *Candida* infections in AIDS. p. 67–74. *In* H. V. Bossche (ed.), *Mycoses in AIDS Patients.* Plenum Press, New York, N.Y.

88. **Oh, S.-H., G. Cheng, J. Nuessen, R. Jajko, K. Yeater, X. Zhao, C. Pujol, D. R. Soll, and L. L. Hoyer.** 2005. Functional specificity of *Candida albicans* Als3p proteins and clade specificity of *ALS3* alleles discriminated by the number of copies of the tandem repeat sequence in the central domain. *Microbiology* **151:**673–681.

89. **Panwar, S. L., M. Legrand, D. Dignard, M. Whiteway, and P. T. Magee.** 2003. MFα1, the gene encoding the α mating pheromone of *Candida albicans. Eukaryot. Cell* **2:**1350–1360.

90. **Pendrak, M. L., S. S. Yan, and D. D. Roberts.** 2004. Hemoglobin regulates expression of an activator of mating-type locus α genes in *Candida albicans. Eukaryot. Cell.* **3:**764–775.

91. **Pfaller, M. A., R. N. Jones, S. A. Messer, M. B. Edmond, R. P. Wenzel, and S. P. Group.** 1998. National surveillance of nosocomial blood stream infection due to species of *Candida* other than *Candida albicans*: frequency of occurrence and antifungal susceptibility in the SCOPE Program. *Diagn. Microbiol. Infect. Dis.* **30:**121–129.

92. **Praekelt, U. M., and P. A. Meacock.** 1990. *HSP12*, a new small heat shock gene of *Saccharomyces cerevisiae*: analysis of structure, regulation and function. *Mol. Gen. Genet.* **223:**97–106.

93. **Pujol, C., A. R. Dodgson, and D. R. Soll.** 2005. Population genetics of ascomycetes pathogenic to humans and animals. *In* J.-P. Xu (ed.), *Evolutionary Genetics of Fungi,* in press. Horizon Scientific Press, Norwich, United Kingdom.

94. **Pujol, C., J. Reynes, F. Renaud, M. Raymond, M. Tibayrenc, F. J. Ayala, F. Janbon, M. Mallie, and J. M. Bastide.** 1993. The yeast *Candida albicans* has a clonal mode of reproduction in a population of infected human immuno-deficiency virus-positive patients. *Proc. Natl. Acad. Sci. USA* **90:**9456–9459.

95. **Pujol, C., K. J. Daniels, T. Srikantha, S. R. Lockhart, J. Geiger, and D. R. Soll.** 2004. The two closely related species *Candida albicans* and *Candida dubliniensis* can mate. *Eukaryot. Cell* **3:**1015–1027.

96. **Pujol, C., M. Pfaller, and D. R. Soll.** 2002. Ca3 fingerprinting of *C. albicans* bloodstream isolates from the United States, Canada, South America and Europe reveals a European clade. *J. Clin. Microbiol.* **40:**2729–2740.

97. **Pujol, C., S. A. Messer, M. A. Pfaller, and D. R. Soll.** 2003. Drug resistance is not directly affected by mating-type locus zygosity in *Candida albicans. Antimicrob. Agents Chemother.* **47:**1207–1212.

98. **Pujol, C., S. Joly, S. R. Lockhart, S. Noel, M. Tibayrenc, and D. R. Soll.** 1997. Parity among the randomly amplified polymorphic DNA method, multilocus enzyme electrophoresis, and southern blot hybridization with the moderately repetitive DNA probe Ca3 for fingerprinting *Candida albicans. J. Clin. Microbiol.* **35:**2348–2358.

99. **Raymond, M., D. Dignard, A. M. Alarco, N. Mainville, B. B. Magee, and D. Y. Thomas.** 1998. A Ste6p/P-glycoprotein homologue from the asexual yeast *Candida albicans* transports the a-factor mating pheromone in *Saccharomyces cerevisiae. Mol. Microbiol.* **27:**587–598.

100. **Rikkerink, E. H., B. B. Magee, and P. T. Magee.** 1988. Opaque-white phenotype transition: a programmed morphological transition in *Candida albicans. J. Bacteriol.* **170:**895–899.

101. **Roberts, C. J., B. Nelson, M. J. Marton, R. Stoughton, M. R. Meyer, H. A. Bennett, Y. D. He, H. Dai, W. L. Walker, T. R. Hughes, M. Tyers, C. Boone, and S. H. Friend.** 2000. Signaling and circuitry of multiple MAPK pathways revealed by a matrix of global gene expression profiles. *Science* **287:**873–880.

102. **Rose, M. D.** 1991. Nuclear fusion in yeast. *Annu. Rev. Microbiol.* **45:**539–567.

103. **Rustad, T. R., D. A. Stevens, M. A. Pfaller, and T. C. White.** 2002. Homozygosity at the *Candida albicans MTL* locus associated with azole resistance. *Microbiology* **148:**1061–1072.

104. **Sadhu, D., D. Hoekstra, M. J. McEachern, S. I. Reed, and J. B. Hicks.** 1992. A G-protein alpha subunit from asexual *Candida albicans* functions in the mating signal transduction pathway of *Saccharomyces cerevisiae* and is regulated by the a1-α2 repressor. *Mol. Cell. Biol.* **12:**1977–1985.

105. **Segall, J. E.** 1993. Polarization of yeast cells in spatial gradients of alpha mating factor. *Proc. Natl. Acad. Sci. USA* **190:**8332–8336.

106. **Sena, E. P., D. N. Radin, and S. Fogel.** 1973. Synchronous mating in yeast. *Proc. Natl. Acad. Sci. USA* **70:**1373–1377.

107. **Slutsky, B., M. Staebell, J. Anderson, L. Risen, M. Pfaller, and D. R. Soll.** 1987. "White-opaque transition": a second high-frequency switching system in *Candida albicans. J. Bacteriol.* **169:**189–197.

108. **Sohn, K., C. Urban, H. Brunner, and S. Rupp.** 2003. EFG1 is a major regulator of cell wall dynamics in *Candida albicans* as revealed by DNA microarrays. *Mol. Microbiol.* **47:**89–102.

109. **Soll, D. R., and C. Pujol.** 2003. DNA fingerprinting *Candida albicans* clades. *FEMS Immunol. Med. Microbiol.* **39:**1–7.

110. **Soll, D. R.** 1992. High frequency switching in *Candida albicans. Clin. Microbiol. Rev.* **5:**183–203.

111. **Soll, D. R.** 2003. *Candida albicans,* p. 165–201. *In* A. Craig and A. Scherf (ed.), *Antigenic Variation.* Academic Press, Ltd., London, United Kingdom.

112. **Srikantha, T., and D. R. Soll.** 1993. A white-specific gene in the white-opaque switching system of *Candida albicans. Gene* **131:**53–60.

113. **Srikantha, T., L. K. Tsai, A. Klar, and D. R. Soll.** 2002. The histone deacetylase genes *HDA1* and *RPD3* play distinct roles in the regulation of high frequency phenotypic switching in *Candida albicans. J. Bacteriol.* **183:**4614–4625.

114. **Srikantha, T., L. Tsai, K. Daniels, and D. R. Soll.** 2000. *EFG1* null mutants of *Candida albicans* switch but cannot express the complete phenotype of white-phase budding cells. *J. Bacteriol.* **182:**1580–1591.

115. **Srikantha, T., L. Tsai, K. Daniels, L. Enger, K. Highley, and D. R. Soll.** 1998. The two-component hybrid kinase regulator *CaNIK1* of *Candida albicans. Microbiology* **144:**2715–2729.

116. **Srikantha, T., S. A. Lachke, and D. R. Soll.** 2003. Three mating-type-like loci in *Candida glabrata. Eukaryot. Cell* **2:**328–340.

117. Staab, J. F., S. D. Bradway, P. L. Fidel, and P. Sundstrum. 1999. Adhesive and mammalian transglutaminase substrate properties of *Candida albicans* Hwp1. *Science* 283:1535–1538.

118. Staab, J., C. A. Ferrer, and P. Sundstrum. 1996. Developmental expression of a tandemly repeated proline- and glutamine-rich amino acid motif on hyphal surfaces on *Candida albicans. J. Biol. Chem.* 27:6298–6305.

119. Staben, C., and C. Yanofsky. 1990. *Neurospora crassa* a mating-type region. *Proc. Natl. Acad. Sci. USA* 87:4917–4921.

120. Steingrimsson, E., H. Arnheiter, J. H. Hallsson, M. L. Lamoreux, N. G. Copeland, and N. A. Jenkins. 2003. Interallelic complementation at the mouse *Mitf* locus. *Genetics* 163:267–276.

121. Stone, R. L., V. Matarese, B. B. Magee, P. T. Magee, and D. A. Bernlohr. 1990. Cloning, sequencing and chromosomal assignment of a gene from *Saccharomyces cerevisiae* which is negatively regulated by glucose and positively by lipids. *Gene* 96:171–176.

122. Sudbery, P., N. Gow, and J. Berman. 2004. The distinct morphogenic states of *Candida albicans. Trends Microbiol.* 12:317–324.

123. Sullivan, D. J., T. Westerneng, K. Haynes, D. Bennett, and D. Coleman. 1995. *Candida dubliniensis* sp. nov.: phenotypic and molecular characterization of a novel species associated with oral candidosis in HIV-infected individuals. *Microbiology* 141:1507–1521.

124. Sullivan, D., D. Bennett, M. Henman, P. Harwood, S. Flint, F. Mulcahy, D. Shanley, and D. Coleman. 1993. Oligonucleotide fingerprinting of isolates of *Candida* species other than *C. albicans* and of atypical *Candida* species from immunodeficiency virus-positive and AIDS patients. *J. Clin. Microbiol.* 31:2124–2133.

125. Suzuki, T., S. Nishibayashi, T. Kuroiwa, T. Kanbe, and K. Tanaka. 1982. Variance of ploidy in *Candida albicans. J. Bacteriol.* 152:893–896.

126. Tatchell, K., K. A. Nasmyth, B. D. Hall, C. Astell, and M. Smith. 1981. *In vitro* mutation analysis of the mating-type locus in yeast. *Cell* 27:25–35.

127. Tsong, A. E., M. G. Miller, R. M. Raisner, and A. D. Johnson. 2003. Evolution of a combinatorial transcriptional circuit: a case study in yeasts. *Cell* 115:389–399.

128. Tzung, K. W., R. M. Williams, S. Scherer, N. Federspiel, T. Jones, N. Hansen, V. Bivolarevic, L. Huizar, C. Komp, R. Surzycki, R. Tamse, R. W. Davis, and N. Agabian. 2001. Genomic evidence for a complete sexual cycle in *Candida albicans. Proc. Natl. Acad. Sci. USA* 98:3249–3253.

129. Ueno, R., N. Urano, and S. Kimura. 2002. Effect of temperature and cell density on ethanol fermentation by a thermotolerant aquatic yeast strain isolated from a hot spring environment. *Fish. Sci.* 68:571–578.

130. Ueno, R., N. Urano, and S. Kimura. 2001. Characterization of thermotolerant, fermentative yeasts from hot spring drainage. *Fish. Sci.* 67:138–145.

131. Wessels, D., E. Voss, N. Von Bergen, R. Burns, J. Stites, and D. R. Soll. 1998. A computer-assisted system for reconstructing and interpreting the dynamic three-dimensional relationships of the outer surface, nucleus and pseudopods of crawling cells. *Cell Motil. Cytoskel.* 41:225–246.

132. White, T. C., H. Miyasaki, and N. Agabian. 1993. Three distinct secreted aspartyl proteinases in *Candida albicans. J. Bacteriol.* 175:6126–6135.

133. Wilkinson, L. E., and J. R. Pringle. 1974. Transient G1 arrest of *S. cerevisiae* cells of mating-type alpha by a factor produced by cells of mating-type a. *Exp. Cell Res.* 89:175–187.

134. Wong, S., M. A. Fares, W. Zimmermann, G. Butler, and K. H. Wolfe. 2003. Evidence from comparative genomics for a complete sexual cycle in the 'asexual' pathogenic yeast *Candida glabrata. Genome Biol.* 4:R10.

135. Wu, W., C. Pujol, S. R. Lockhart, and D. R. Soll. Mechanisms of mating-type homozygosis in *C. albicans. Genetics*, in press.

136. Wu, X., C. Wu, and J. E. Haber. 1997. Rules of donor preference in *Saccharomyces* mating-type gene switching revealed by a competition assay involving two types of recombination. *Genetics* 147:399–407.

137. Xu, J., T. G. Mitchell, and R. Vilgalys. 1999. PCR-restriction fragment length polymorphism (RFLP) analyses reveal both extensive clonality and local genetic differences in *Candida albicans. Mol. Ecol.* 8:59–73.

137a. Zhao, R., K. J. Daniels, S. R. Lockhart, K. M. Yeater, L. L. Hoyer, and D. R. Soll. 2005. Unique aspects of gene expression during *Candida albicans* mating and possible G1 dependency. *Euk. Cell* 4:1175–1190.

138. Zhao, R., S. Lockhart, and D. R. Soll. 2002. The role of Tup1 in switching, phase maintenance and phase-specific gene expression in *Candida albicans. Eukaryot. Cell* 1:353–365.

Molecular Principles of Fungal Pathogenesis
Edited by Joseph Heitman et al.
©2006 ASM Press, Washington, D.C.

Chapter 8

Evolution of Human-Pathogenic Fungi: Phylogenies and Species

JOHN W. TAYLOR

Not many of us open medical mycology texts to the chapter on systematics; even I am attracted first to the lurid photos of exotic diseases and the morbid fascination that they provoke. But, sooner or later, we must confront fungal evolution to answer a fundamental question, "Which fungus is distressing my patient?" Simply knowing that a fungus is responsible is not enough, because treatment differs dramatically for different fungi. Neither is knowledge of a few famous pathogenic fungi enough, because today so many fungi can be a threat to the immunodeficient patient. In modern medicine there is no substitute for a broad knowledge of fungi, and the knowledge must go beyond simply knowing the names of fungi. The evolutionary relationships also must be considered because only a few medically important fungi have been thoroughly investigated, and one must often make inferences about the behavior of poorly studied fungi from their better-studied relatives. Obviously, the inference is only as valuable as the evolutionary relationship is accurate.

Our knowledge of the pattern of evolution in fungi has increased dramatically over the past 15 years due to (i) the need to know more about fungal evolution, which was created by the increase in the number of immunodeficient humans owing to the tragedy of AIDS and the promise of organ and bone marrow transplantation, (ii) the spread of cladistic theory (81), and (iii) the technical advances of PCR amplification of DNA (158, 212) and automated DNA sequencing.

The expansion of the population of immunodeficient humans has allowed many fungi to become human pathogens. There are 399 species listed in the *Atlas of Clincal Fungi* (34), and more are reported every month; however, in the interest of brevity, only the 10% that are most commonly encountered as agents of mycoses are considered here (Table 1).

Evolutionary studies of fungi have proceeded at both ends of the phylogenetic tree, i.e., the extremely old speciation events that mark the deep divergences joining the major fungal clades and the recent speciation events that give rise to new species. The deeper divergences are becoming better known as more genes are used (117), but it is proving easier to define major clades than to understand the exact order in which they branch (156). Phylogenetics based on nucleic acid variation is blind to the sexual habits of fungi, and so both meiosporic and mitosporic fungi are treated together and the best-known name, whether it applies to the mitosporic state (anamorph) or the meiosporic state (teleomorph), is used here. Coming forward in time to the most recent divergences at the tips of phylogenetic trees, one finds many new fungal phylogenetic species, whose recognition by concordance of gene genealogies (Fig. 1) was pioneered for fungi with human pathogens (192–194).

A new epidemiological approach, multilocus sequence typing (MLST), has been associated mostly with pathogenic bacteria (120), but there are many fungal examples (191) and none is more fully developed than that for *Penicillium marneffei* (44). MLST has the potential to be the approach that brings together philosophically and geographically disparate research groups working on the same fungus. Some modification of MLST may be needed for fungi, where sequence of housekeeping genes may lack the needed variation; microsatellites, for example, may prove more useful and lead to multilocus microsatellite typing (MLMT) (44, 191).

The studies of species and populations tell us which taxa and genetically differentiated groups are worth identifying, but clinicians still must make the identifications. At the present, it seems as if every research group has devised its own method and only

---

John W. Taylor • Department of Plant and Microbial Biology, University of California, Berkeley, CA 94720-3102.

Table 1. Fungi and other agents of the most common human mycoses

Fungi

<u>Ascomycota</u>
- Pezizomycotina
  - Eurotiomycetes
    - Onygenales
      - Uncinocarpus clade
        - *Coccidioides immitis*
        - *Coccidioides posadasii*
      - Ajellomycetaceae
        - *Histoplasma capsulatum*
        - *Blastomyces dermatitidis*
        - *Paracoccidioides brasiliensis*
        - *Lacazia loboi*
      - Arthromdermataceae
        - *Epidermophyton floccosum*
        - *Trichophyton rubrum*
        - *Trichophyton interdigitale*
        - *Trichophyton schoenleinii*
        - *Trichophyton tonsurans*
        - *Trichophyton violaceum*
        - *Microsporum audouinii*
        - *Microsporum canis*
  - Chaetothyriomycetes
    - Herpotrichiellaceae
      - *Exophiala jeanselmei*
      - *Exophiala dermatitidis*
      - *Cladophialophora bantiana*
      - *Cladophialophora carrionii*
      - *Phialophora verrucosa*
      - *Phialophora americana*
      - *Fonsecaea pedrosii*
  - Dothidiomycetes
    - Dothidiales
      - *Hortea werneckii*
      - *Neotestudina rosatii*
    - Pleosporales
      - *Madurella grisea*
  - Sordariomycetes
    - Sordariales
      - *Madurella mycetomatis*
    - Ophiostomatales
      - *Sporothrix schenckii*
    - Microascales
      - *Pseudallescheria boydii*
      - *Scedosporium prolificans*
    - Hypocreales
      - *Fusarium solani*
      - *Fusarium oxysporum*
- Saccharomycotina
  - Candida albicans clade
    - *Candida albicans*
    - *Candida dubliniensis*
  - Saccharomyces clade
    - *Candada glabrata*
    - *Candida krusei*
    - *Candida tropicalis*
  - Clavispora-Metchnikowia clade
    - *Clavispora lusitaniae*
- Taphrinomycotina
  - *Pneumocystis jiroveci*

Continues

Table 1. *Continued*

<u>Basidiomycota</u>
- Hymenomycetes
  - *Cryptococcus neoformans*
  - *Trichosporon asahii*
  - *Trichosporon inkin*
  - *Trichosporon mucoides*
  - *Trichosporon cutaneum*
  - *Trichosporon ovoide*
- Ustilaginomycetes
  - *Malassezia furfur*

<u>Zygomycota</u>
- Mucorales
  - *Rhizopus oryzae*
- Entomophthorales
  - *Conidiobolus coronatus*

<u>Chytridiomycota</u>
- Chytridiales
  - *Batrachochytrium dendrobatidis*

Stramenopila
<u>Oomycota</u>
- *Pythium insidiosum*

Mezomycetozoea
<u>Dermocystidia</u>
- *Rhinosporidium seeberi*

its members can implement it (mea culpa). A global solution to the problem of clinical identification, at least using DNA variation, may be at hand thanks to microarray technology. Researchers working with bacteria have used ribosomal DNA (rDNA) to identify species collected in air samples (214), and a system for fungi involving a sufficiently variable region that is easily PCR amplified, perhaps the internal transcribed spacer (ITS) of the rDNA, could provide rapid identification using one or a few sets of PCR primers. Once the fungus was identified, and cultures were available, then MLST or MLMT could follow.

As noted at the outset, the field of fungal evolution has made tremendous strides in the past 15 years, but bigger challenges and opportunities await mycologists in the form of genomics, where the comparative approach so familiar to evolutionary biologists will

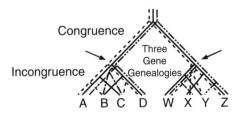

Figure 1. Phylogenetic species recognition by congruence of gene genealogies. Within recombining species, different genes have conflicting genealogies. Between genetically isolated species, drift and selection lead to fixation of ancestral polymorphism and congruence of genealogies. The point where congruence gives way to conflict is the point where species are recognized (arrows). Reprinted from reference 194 with permission.

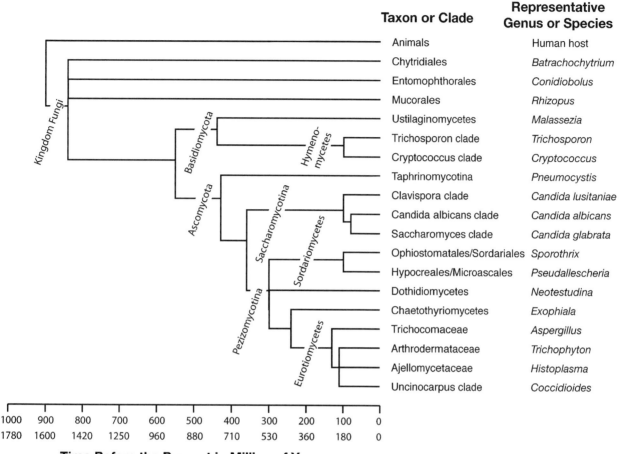

Figure 2. Phylogenetic tree of fungi pathogenic to humans or animals fit to geologic time. A maximum-likelihood phylogeny with an enforced molecular clock based on small-subunit rDNA sequence was used to estimate divergence times for diverse fungi (9), and to this phylogeny were added taxa of commonly encountered pathogenic fungi. Two geologic timescales are used, one reflecting a divergence between animals and fungi at 900 MYA (9) and the other reflecting a divergence at 1,600 MYA (79). Estimates of divergence times are subject to considerable uncertainty, the least of which is due to differences in the estimated time of divergence between animals and fungi (92).

provide a key tool to unraveling the evolution of fungi and their virulence.

Speaking of virulence, what better place to begin than with the Onygenales, the fungal group that has most of the bad actors, those fungi capable of causing systemic mycoses in otherwise healthy individuals. From there, we will work outward at increasing phylogenetic distances, moving through other groups of Ascomycota, then to Basidiomycota, and ending with the earliest diverging fungal groups, Zygomycota and Chytridiomycota (Fig. 2).

PHYLUM ASCOMYCOTA

Two of every three described fungi are members of the Ascomycota, which is divided into three principal groups, all of which harbor pathogens: Taphrinomycotina, Saccharomycotina, and Pezizomycotina.

In Pezizomycotina, pathogens are found in Eurotiomycetes, Sordariomycetes, Chaetothyriomycetes, and Dothidiomycetes. In Eurotiomycetes, pathogens are found in Eurotiales and Onygenales (181). The Onygenales contains the Trichocomaceae, Arthrodermataceae, and our starting place, the Ajellomycetaceae and the Uncinocarpus clade.

Subphylum Pezizomycotina
Class Eurotiomycetes
ORDER ONYGENALES

The fungi that cause systemic disease in otherwise healthy humans are found in the Onygenales and are classified in the genera *Coccidioides*, *Histoplasma*, *Blastomyces*, *Paracoccidioides*, and *Lacazia* (11, 15, 83, 145, 146). The Ajellomycetaceae (65, 69, 203) comprises the last four genera, along with *Emmonsia* species, one of which has been reported to cause disease in immunodeficient humans (39). Nearby is the Uncinocarpus clade, home to *Coccidioides* species (15, 145, 168).

The most infamous member of the group is *Coccidioides immitis*, which was the only human pathogenic fungus to be considered a "select agent of terrorism" by an act of the United States Congress. The first MLST study of fungi showed that this fungus harbored two phylogenetic species, *C. immitis* in California and northern Mexico and *C. posadasii* in the rest of the range of this New World endemic fungus (48, 99). Easily recognized phenotypic differences between the two species were not initially apparent; however, subsequent research has discovered that only *C. immitis* can make hyphae at 37°C (M. Orbach, personal communication). The *Coccidioides* species also were the first fungi subjected to MLMT (Color Plate 5), which was used to determine that there were at least two genetically differentiated populations in *C. immitis* and at least three in *C. posadasii* (46, 47). Although coccidioidomycosis is acquired only from the environment, the distribution of microsatellite genotypes in South America, Mexico, and Texas suggests that the South American populations represent a relatively recent introduction (8,000 to 130,000 years ago), perhaps aided by human migration (47). The database of nine microsatellites typed for 167 individuals permits the assignment of newly collected isolates (whether obtained from clinics or the environment [62]) to species and to population (50). Given that the two species diverged as many as 10 million years ago (MYA), many loci can be used for species identification (12). Although *Coccidioides* species are mitosporic, population genetic tests of association of alleles among loci in both species have supported a significant role for recombination (18, 49).

Family Ajellomycetaceae. Nearly as infamous as *Coccidioides* is the genus *Histoplasma*. MLST analysis of 130 individuals from the three traditional varieties, *H. capsulatum* var. *capsulatum*, *H. capsulatum* var. *duboisii,* and *H. capsulatum* var. *farciminosum,* showed that there are at least eight species in the genus and that the present concept of varieties is insupportable (90, 91). For example, the African clade of *Histoplasma* contains fungi that both do and do not show the disease phenotype of *H. capsulatum* var. *duboisii,* and in *H. capsulatum* var. *farciminosum* the isolates collected from horses and related mammals have arisen from three other genetically isolated clades (91). Most of the isolates received as *H. capsulatum* var. *farciminosum* were European or Asian and had identical genotypes, a finding that is consistent with their production of skin lesions and associated transmission from animal to animal. In fact, the entire Eurasian clade emerges from within the large, tropical South American clade and therefore has originated more recently than the older radiation of the other species, some 3 to 13 MYA (91). As with *C. posadasii*, perhaps humans were involved in the transport of *Histoplasma*, but in this case it was a migration out of Latin America to Europe. Humans are not the only agents of long-distance dispersal in *Histoplasma*; bats certainly are involved and may have helped spread members of the Latin American clades to Mexico and vice versa (196, 197). Microsatellites have been discovered for *Histoplasma* and used to show genetic differentiation within one of two North American clades as well as between North and South American clades (21). As would be expected for a fungus that makes both mitospores and meiospores (*Ajellomyces capsulatus* is the meiosporic state) (102, 103), population genetic analysis of MLST data has supported significant recombination (90).

The third North American member of this group is *Blastomyces dermatitidis*, the mitosporic state of the meiosporic fungus *Ajellomyces dermatitidis*. A population genetic study of 59 isolates from North America, India, and Africa, using restriction fragment length polymorphisms (RFLPs) of the rDNA region and PCR fingerprinting, found that there were genetically isolated groups within the species and that genetic differentiation could be associated with geographic distances as short as that between the northern and southern United States of America (125). Although the fungus is capable of sexual reproduction (126), it also was found that the same genotype could predominate at one geographic locale, presumably through clonal spread (125).

Although *Coccidioides* and *Histoplasma* are found in Latin America, the most important mycotic species in this continent is *Paracoccidioides brasiliensis* (160). Calcagno et al. (19) used randomly amplified polymorphic DNAs (RAPDs) to show that there was genetic variation in this species and that the variation seemed to correlate with the geographic origin of the isolates. One interesting environmental reservoir for the species is the armadillo, and molecular data have been used to confirm the identity of isolates from these animals (7). A very recent MLST study has found what appear to be three phylogenetic species in *Paracoccidioides*, and two of them are sympatric (123a). Additionally, microsatellite markers have been described and used to ask if clinical isolates are genetically distinct from environmental isolates; as it turned out, they are not (137). The microsatellite markers certainly will be useful in future population genetic studies of this fungus from every corner of South America.

Rounding out the group is *Lacazia loboi*, an enigmatic pathogen if ever there was one (189). It causes infections of humans and freshwater dolphins (78) and cannot be cultivated. DNA of *L. loboi* is unstable in clinical samples, and only rapid freezing in liquid nitrogen permitted the recovery of sufficient DNA for phylogenetic studies (83). The first case of mycosis caused by *L. loboi* in the United States has been reported, but it almost certainly was acquired during travel in Venezuela (17). The difficulty in obtaining good-quality DNA samples has hampered population studies of this fungus, but an intrepid medical mycologist tackling this fungus might be rewarded by an interesting evolutionary history.

From the Ajellomycetaceae and Uncinocarpus clade, we now move back to the node that joins them with the Arthrodermataceae, some 120 MYA, and from that node forward in time to the home of the ascomycete dermatophytic fungi.

Family Arthrodermataceae. Among the many species in the family Arthrodermataceae, eight are considered important associates of humans (34, 93): two species of *Microsporum*, *M. audouinii* and *M. ferrugineum*, and six species of *Trichophyton*, *T. tonsurans*, *T. interdigitale*, *T. rubrum*, *T. violaceum*, *T. schoenleinii*, and *T. concentricum*. In addition, it is impossible to treat this group without considering the zoophilic species, *M. canis*. All of these names apply to mitosporic states, and where there is a meiosporic state it is in the genus *Arthroderma*. Close relationships among mitosporic and mitosporic species are seen for *A. simii* with *T. schoenleinii* and for *A. vanbreuseghemii* with *T. tonsurans* and *T. interdigitale* (93). *A. benhamiae* has a close relationship with *T. concentricum*, and *A. otae* is a close relative of *M. audouinii*, *M. canis*, and *M. ferrugineum*. Mitosporic dermatophytes without close meiosporic relatives include *T. rubrum* and *T. violaceum*.

Summerbell has hypothesized that the mitosporic, human dermatophytes have evolved from meiosporic species associated with animals that live in the soil and that the human dermatophytes now are spread exclusively by mitospores (184). In framing his hypothesis, Summerbell knew that human dermatophyte individuals typically are exclusively, or nearly exclusively, of one mating type (MAT) and that there is very little variation among individuals (as judged by ITS sequence, fingerprints of repeated DNA, or amplification fragment length polymorphism [AFLP] analysis) (84). Subsequently, a very thorough study from Y. Gräser's laboratory of over 200 individuals of the *A. otae* complex (*M. canis*, *M. ferrugineum*, and *M. audouinii*), using 10 polymorphic microsatellites and 3 sequenced coding regions, provided even more

compelling evidence for the relationship of clonal lineages to sexual species (93). *A. otae* individuals are heavily biased to *MAT* (−), with only a few *MAT* (+) individuals known from Japan, and the genetic distance between the two mating types is considerable. *M. canis* individuals are genetically nearly indistinguishable from *A. otae MAT* (−), and both *M. canis* and *A. otae MAT* (−) genotypes are quite distinct from that of *A. otae MAT* (+). It seems likely that *M. canis* is a single clonal lineage derived from one *A. otae MAT* (−) individual. Classification of clonal lineages that lack nearly identical sexual relatives is not straightforward, because each tiny branch could merit its own name (139). If the mitosporic lineage has a close, but genotypically distinct, sexual relative, it is possible to define a mitosporic species in relation to the meiosporic relative (194) (Fig. 3). However, if the clonal lineage emerges from a sexual species, some might argue that the clonal lineage is one, long-lived individual (see the discussion of *H. capsulatum* var. *farciminosum* above). To complicate matters further, neither *M. ferrugineum* nor *M. audouinii* has as close a sexual relative as does *M. canis*, and their closest meiosporic relative is *A. otae* mating type (+). Gräser and colleagues favor the interpretation that meiosporic *A. otae* has spawned three clonal species, *M. canis*, *M. audouinii*, and *M. ferrugineum*, implying that *A. otae* harbors considerable variation (93). It is possible, however, that *A. otae* contains more than one species, a hypothesis that would become testable only if populations of *A. otae* are found with more balanced distributions of mating-type alleles.

A benefit of the research described above is that classification of *Microsporum* species has been simplified by synonymizing taxa that are phylogenetically close or indistinguishable so that fewer taxa remain; e.g., *M. canis* (60), *T. rubrum* and *T. violaceum* (59),

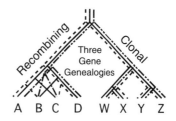

Figure 3. Phylogenetic species recognition applied to species that are exclusively clonal. Exclusively clonal species would not exhibit the change from congruent to conflicting gene genealogies seen in recombining species, which would make species recognition arbitrary. However, if clonal species are uncommon compared to recombining species, then clonal species can be recognized in comparison to their nearest recombining relatives, in this case a sister species. Reprinted from reference 193 with permission.

T. schoenleinii (149), and other species (121). The data used for classification also provide aids to the identification of dermatophytes, in particular: regions of the rDNA, e.g., ITS (121) or intergenic spacers (76), or of protein-coding genes, e.g., those encoding topoisomerase (88) or chitin sythase (89), or fingerprints of repetitive DNA (84). In search of ever more polymorphic markers, microsatellites now are being developed for *T. rubrum* and *T. violaceum* (144), in addition to those used for the *Microsporum* research discussed above.

We now travel back to 130 MYA to the node joining Onygenales to Eurotiales and from there forward on the branch leading to the Trichocomaceae, home to the genera *Aspergillus* and *Penicillium*.

ORDER EUROTIALES

Family Trichocomaceae. *Aspergillus* species form a monophyletic clade in or near the *Eupenicillium* group of penicillia, and *Aspergillus* species have become important opportunistic pathogens as the number of immunodeficient hosts has increased. There is a real and as yet unmet need for a broad, multigene phylogeny of *Aspergillus* species (208, 211). *A. fumigatus* is by far the most important human pathogen, followed by *A. flavus* and *A. terreus* (34). Protein-coding gene phylogenetics of the *A. fumigatus* group have shown that this mitosporic fungus has a close meiosporic relative, *Neosartorya fischerii* (56), and similar studies have shown that *A. flavus* has a close meiosporic relative, *Petromyces alliaceus* (207). Phylogenetic species recognition of *A. flavus* found four cryptic species (57), and later analysis of reproductive structures and aflatoxin production showed that these four phylogenetic species could be recognized by a combination of toxin production and sclerotium size (55). One of the cryptic species produces both aflatoxins B and G, and it appears to be more common in the Southern Hemisphere (147, 200).

Fingerprinting studies using repeated DNA failed to find any cryptic species in *A. fumigatus* or any correlation between fungal genotype and either pathogenicity or geographic origin (27). Researchers performing a subsequent study of Canadian isolates by using microsatellite markers came to the same conclusion (157), as did the workers performing a European study involving presumed single-nucleotide polymorphisms scored by PCR amplification (188). However, a study of *A. fumigatus* from French hospitals, using microsatellites as well as RAPDs and single-nucleotide polymorphisms, concluded that isolates did differ by geographic origin (10). A recent study using the DNA sequence flanking four microsatellites and a global sample of *A. fumigatus*

found one new cryptic species (it and *A. fumigatus* are sympatric) and no population structure in *A. fumigatus*. As a result, identical genotypes are found on different continents and there is no evidence that certain genotypes are associated with pathogenicity (148a). *A. fumigatus* appears to be a globally distributed fungus with no geographic population structure. Workers performing a similar and contemporary study using a different collection of global isolates and a different set of nucleic markers have arrived at the same conclusion (145a). It seems as if *A. fumigatus* is unlike any other fungal species that has been examined by a thorough MLST approach, although it is impossible to rule out the existence of a structured population somewhere on Earth from which this global expansion has emanated. Even if the latter scenario should prove to be true, the size of the clonal expansion would be unprecedented. The third *Aspergillus* species, *A. terreus*, has not been the subject of a thorough phylogenetic or population genetic study, but studies using an RAPD approach have begun and show considerable variation in the genotypes of isolates collected from different cystic fibrosis patients (24), but similar genotypes in serial isolates from the same patient (187).

Although both *A. fumigatus* and *A. flavus* are mitosporic, analyses of the sequenced genes used to search for phylogenetic species in both taxa have found evidence for recombination, in addition to evidence for clonality (57, 148a). In the case of *A. fumigatus*, one mating-type gene was found from a search of the sequenced genome (148) and the other was found in a second individual (40), indicating that the fungus may achieve recombination by sexual reproduction through outbreeding (heterothallism). The aforementioned phylogenetic study of *A. fumigatus* and its relatives (56) indicated that heterothallic *Aspergillus* species in the *A. fumigatus* group evolved from homothallic relatives. This result is surprising because heterothallism is ancestral in Ascomycota as a whole, but in this group of *Aspergillus* species, homothallism appears to be ancestral. The recent completion of genome sequences for *A. fumigatus* (138a), *A. oryzae* (117a), and *A. nidulans* (54a) should provide the information to address questions about the evolution of reproductive modes (5, 41) and about pathogenesis. Here, the ability of *Aspergillus* species to make secondary compounds that are toxic to humans (e.g., aflatoxin) (14, 20, 206) not only makes these fungi different from yeasts or Basidiomycota, which, for example, seem to lack genes for polyketide synthases (100), but also provides them with the potential for a very different mode of virulence.

From *Aspergillus*, we move to the closely related genus *Penicillium*. Several *Penicillium* species are capable of causing disease in immunodeficient humans, but only *P. marneffei* presents a significant danger to public health (34). This fungus is not a member of the clade of well-known *Penicillium* species, i.e., those that make penicillin and that may have meiosporic states in the genus *Eupenicillium*, but, rather, is in the clade with meiosporic states in the genus *Talaromyces* (114). No member of the Eupenicillium clade is known to grow as single cells, but *P. marneffei* apparently can convert to a single-celled form. Able to produce single cells at 37°C, *P. marneffei* has become the third most important opportunistic pathogen in Thailand, behind only tuberculosis and cryptococcosis (185). The fungus was described from a captive bamboo rat, and a recent study of *P. marneffei* infection in India showed that nearly 10% of some rat species are infected with the fungus (72). *P. marneffei* had been studied by a variety of molecular approaches (RFLP, M-13 fingerprinting, electrophoretic karyotyping, and ITS-PCR [72]), but the recent discovery and application of microsatellite markers shows the potential of these highly polymorphic markers to reveal the biogeography of a fungus (44, 45, 108).

Fisher et al. have found 20 useful microsatellite markers in *P. marneffei* and used them with an MLMT approach to study 24 isolates from Asia, including individuals from China, Vietnam, Indonesia, and India (44). They found two well-separated clades, one from China plus India and the other from Vietnam plus Indonesia. There is no evidence for recombination, which could be due to its absence or to fine-scale population structure. Arguing in favor of clonality are the lack of observed sexual reproduction in *P. marneffei* and the fact that identical MLMT genotypes have been found in rats caught at the same locations and in humans and rats living in the same area (72). Of course, it also is possible that *P. marneffei* is clonal in part of its geographic range and sexual in another, as has been seen for a number of plant-pathogenic fungi (193), as well as for *Cryptococcus neoformans* var. *grubii*, as described below (113). To address the question of reproductive mode, we need more isolates from throughout the area of endemic infection. There is every prospect that more isolates will be found and typed and that analyses of the growing data set will answer questions about *P. marneffei* reproduction, speciation, and population structure. The MLMT approach is ideal for this research because different research groups can pool their microsatellite data using the *P. marneffei* MLMT website (44). Not incidentally, the MLMT data are ideal for typing new isolates; the markers are so variable that finding two isolates with identical

genotypes makes it almost certain that they are from the same individual (45).

From the Trichocomaceae, we move back through the Eurotiales and Eurotiomycetes to the node that joins them all to the Chaethothyriomycetes, some 250 MYA, and then forward to the modern members of this group of darkly pigmented fungi that cause diseases known as phaeohyphomycoses and chromoblastomycoses.

Class Chaetothyriomycetes

The fungi that cause chromoblastomycosis or phaeohyphomycosis are found in the family Herpotrichiellaceae of the Chaetothyriomycetes and have melanized cell walls. They can cause disease in otherwise healthy humans, but in order to do so they require a wound to bypass the defense offered by skin (16, 22). They are becoming more common, and more dangerous in terms of dissemination, due to the rise in the number of immunocompromised hosts. Although these fungi are far more common in the tropics, they are not unknown in the Northern Hemisphere (16).

Chaetothyriomycetes is one of two great groups of fissitunicate ascomycetes that make flask-shaped ascoma, the other being Dothidiomycetes, home to many plant-pathogenic fungi. de Hoog et al. (34) consider 10 species of Chaetothyriomycetes to be important human pathogens in that they are common or particularly severe. They are found in five genera: *Exophiala* with *E. oligosperma* (the description of which reduced the medical importance of *E. jeanselmei* sensu stricto [33]), *E. dermatitidis*, and *E. spinifera*; *Phialophora* with *P. verrucosa*, *P. americana*, and *P. europaea*; *Cladophialophora* with *C. carrionii* and *C. bantiana*; *Ramichloridium* with *R. mackenziei*; and *Fonsecaea* with *F. pedrosoi*. Another fungus growing in skin and making darkly pigmented hyphae is *Hortaea werneckii*, the agent of tinea nigra; this fungus, however, is a member of the Dothidiomycetes (along with a *Madurella* species to be mentioned below).

Phylogenetic studies of Chaetothyriomycetes, performed using regions of the rDNA repeat, have shown that the features of conidiation used to define the different genera in this class are not reliable indicators of phylogenetic relationships (1, 202). In some cases, one individual fungus can make both annelloconidia (associated with *Exophiala*) and sympoduloconidia (associated with *Rhinocladiella*) (30, 33). Therefore, the genera *Phialophora*, *Exophiala*, *Chadophialophora*, and *Rhinocladiella* need revision, and all of these mitosporic taxa have close relatives or meiosporic states among meiosporic Chaetothyriomycetes in the genus *Capronia*.

Comparisons of rapidly evolving sequences of large collections of isolates of these fungi are now appearing and have resulted in (i) a better understanding of described *Exophiala* species (122), (ii) the discovery of cryptic *Exophiala* species (210), and (iii) the discovery of a cryptic *Fonsecaea* species (6, 32). Given that considerable variation is seen within these taxa, it seems likely that phylogenetic species recognition studies involving many individuals and several protein-coding genes would discover even more species.

From the Chaetothyriomycetes, we move back in time, past even their divergence from the Eurotiomycetes, to a point between 270 and 330 MYA when these two classes diverged from Dothidiomycetes, Pezizomycetes, Lecanoromycetes, and our next destination, the Sordariomycetes.

Class Sordariomycetes

In the class Sordariomycetes are the fungi responsible for mycetomas and sporotrichosis. These filamentous ascomycetes can infect otherwise healthy humans, but only if they are implanted past the skin barrier by trauma. Among the many fungi capable of causing this type of disease, five species are considered important (34). Four are Sordariomycetes, *Sporothrix schenckii*, *Pseudallescheria boydii*, *Scedosporium prolificans*, and *Madurella mycetomatis*, but one is a fissitunicate ascomycete, *Madurella grisea* (Pleosporales). Of course, these fungi also cause disease in immunocompromised hosts, where dissemination of the fungus from the initial wound is the greater threat.

There are two main clades of Sordariomycetes, one with Ophistomatales and Sordariales and the other with Hypocreales and Microascales. In the Ophiostomatales is *S. schenckii*, which causes sporotrichosis, an infection of lymph nodes or skin. This mitosporic fungus is the sister species to the meiosporic fungus *Ophiostoma stenoceras* plus species in the *O. nigrocarpum* complex (28). In the Sordariales is found *M. mycetomatis*, the principal agent of black-grained mycetoma, which is most common in Africa (31). In the Microascales is found *P. boydii* (whose anamorphic state is *Scedosporium apiospermum*), the principal agent of white-grained mycetoma, which is common in North America (86). Also in this order is *S. prolificans*, which affects the skin and joints (66). *S. prolificans* is a close relative of *P. boydii* and a closer relative of meiosporic *Petriella* species (86). Molecular identification schemes have been proposed for *M. mycetomatis* (116), *S. prolificans* (170), and *P. boydii* (153), and all show genetic variation among individuals of each species. In the case of *P. boydii*, studies of cystic fibrosis patients have shown that although this fungus is difficult to isolate from air, a single patient can host multiple isolates (23, 29, 221). Here, again, are fungi that would be candidates for phylogenetic species recognition. Another agent of black-grained mycetoma is *M. grisea*, which, as mentioned above, is a dothidiomycete along with the genus *Hortaea* (31).

The last order of Sordariomycetes hosting pathogens is the Hypocreales, home to the genus *Fusarium*. Species of *Fusarium* can cause infection in immunodeficient hosts, and the outcome is grave. The most commonly reported species is *F. solani*, which embraces at least 25 species (140), but there are also many cases involving the *F. oxysporum* complex (also very speciose) and *F. verticilliodes* (likewise very speciose) (37). In addition, fungi classified in other mitosporic genera, i.e., *Cylindrocarpon lichenicola* and *Acremonium falciforme*, have proved to be members of the *F. solani* clade (3). *Fusarium solani* has been recovered from the water supplies of hospitals where *Fusarium* infections have been reported (3), as have isolates of *F. oxysporum* (143) and *F. verticilloides* (63).

Fusarium species are far better known as plant pathogens, and a very thorough study of *F. oxysporum* complex isolates from humans, set in the context of plant and other environmental isolates, has demonstrated the value of a broad sampling of individuals combined with multiple, sequenced loci (143). The authors studied 88 patient isolates and 18 environmental isolates from four continents, using DNA sequences from protein-coding genes, nuclear rDNA, and mitochondrial rDNA, as well as 173 biallelic AFLP loci. Each of the four previously known clades of *F. oxysporum* (141) proved to harbor human-pathogenic isolates, although the vast majority came from just one of the clades. Among these individuals was a large clone, as judged by its DNA sequence, which proved to embrace seven smaller clones when examined with the more polymorphic AFLP loci. The largest of the AFLP clones are found both in North America and Europe, and all but one of the AFLP clones contained both human and environmental isolates. The low variation among individuals is consistent with the clonal spread of these fungi; alas, such low variation makes it difficult to determine the source of infection when environmental isolates from within and without hospitals have the same genotype as patient isolates. Arguments have been made for waterborne infection (3), airborne infection, or carriage of *F. solani* prior to elective immunosuppression (152); all seem possible.

Now, we will proceed toward the divergence of Pezizomycotina and Saccharomycotina, some 370 MYA, and then move forward in time to the Saccharomycotina, home to the most famous fungus, *Saccharomyces cerevisiae*, and the most important opportunistic pathogenic fungus, *Candida albicans*.

Subphylum Saccharomycotina

The Saccharomycotina contains a group of yeasts that, in what is now a recurrent theme, have become increasingly important human pathogens as the population of immunodeficient hosts has grown. Among the many yeasts that have been found in humans, six are considered to be most important (34): *Candida albicans, C. dubliniensis, C. glabrata, C. krusei, C. tropicalis,* and *C. lusitaniae* (= *Clavispora lusitaniae*). These yeasts belong to three clades, the *Saccharomyces cerevisiae* clade (*C. glabrata* and *C. krusei*), the *Candida albicans* clade (*C. albicans, C. dublinensis,* and *C. tropicalis*), and the *Clavispora-Metschnikowia* clade (*C. lusitaniae*) (101). These yeasts are part of the normal commensal mycota of humans, and, recently, PCR fingerprinting has been used to show that different body regions have different complements of species (the mouth has the lowest diversity), that the community composition of *Candida* species changes over time on a single host individual, and that isolates may be transmitted among family members (87).

C. albicans is the most important of the yeasts because it is the most common opportunistic pathogen of immunodeficient hosts. It was in *C. albicans* that the first phylogenetically recognized, cryptic fungal species was found, *C. dubliniensis* (183). Studies of nucleic acid variation in *C. albicans* have uncovered genetically isolated populations in Africa (51) that may prove to be another cryptic species, *C. africana* (199). Other studies have found variation that correlates with geography; for example, an RFLP study showed five clades (124), a DNA fingerprinting study found five groups (172), and a comparison of microsatellite and single nucleotide polymorphisms found a recently evolved North American clade. As Soll and Pujol (172) wrote, "these findings lay to rest the idea that one strain represents all strains of *C. albicans,* support the need for a worldwide analysis of populations structure and clade-specific phenotypic characteristics, and demonstrate that in the future, pathogenic characteristics must be analyzed in representatives from all five clades." Clearly, *C. albicans* would be an excellent candidate for a global MLST (190) or MLMT effort.

The possibility of sexual reproduction in *C. albicans* was raised as a result of multilocus electrophoresis studies (151) and supported by studies of nucleic acid variation (61). Studies of mitochondrial DNA variation supported a hypothesis of ancient recombination followed by exclusively clonal reproduction in present-day populations (4). Developmental biologists have shown that *C. albicans* can be induced to mate in mice or in vitro, but meiosis has not been detected (85, 118). Two recent reviews tell the story very well (86a, 119). *C. albicans* individuals are normally diploid, and after mating they become tetraploid, only to then lose chromosomes to restore diploidy. The ability to mate is correlated with a switch in colony morphology from white to opaque. Natural diploid individuals that have both mating-type alleles produce a heterodimer of the mating-type-like locus (*MTL*) a1 and MTLα gene products, which prevents the switch from white to opaque colony morphology. Preventing the transition from white to opaque severely inhibits mating because white cells mate at a frequency three orders of magnitude lower than that of opaque cells. When individuals lose *MTL*a or *MTL*α and become either hemi- or homozygous for one *MTL*, they switch to the opaque form and mate; it has just been shown that plasmogamy is followed by karyogamy (8a). The loss of alleles seems to be caused by mitotic recombination or gene conversion, which makes it more difficult to interpret population studies aimed at detecting recombination (190). Interestingly, for a human commensal, growth at 37°C promotes the switch from opaque to white and the inhibition of mating. Even more interestingly (and somewhat contradictory), mating proceeds better on the skin of mammals at 32°C than it does in the laboratory at 37°C (107). The relationship of mating competence, mating itself, and pathogenicity is complex and surely will be the focus of many future studies (171). As a taste of what is to come, a recent comparison of these processes in *C. albicans* and *S. cerevisiae* shows that developmental paths and circuits have evolved to employ different components to achieve similar developmental outcomes in these two yeasts (201).

C. dubliniensis, the aforementioned sister species of *C. albicans,* also has been found to mate and even can mate with *C. albicans*; in fact, interspecific mating can be more efficient than intraspecific mating (150). *C. albicans* is a more virulent pathogen than *C. dubliniensis* (182), and a recent comparative study of the genomes of these two species by competitive DNA hybridizaton to a *C. albicans* microarray identified suites of genes missing from *C. dubliniensis;* these missing genes might help explain the superior virulence of *C. albicans* (135).

C. glabrata, the closest pathogenic relative of *S. cerevisiae,* also has been the subject of population genetic studies. A study of French isolates by using multilocus enzyme electrophoresis and RAPD markers found some population structure in the form of reduced gene flow between isolates collected in Paris and those collected in Montpellier, as well as an association among markers consistent with clonal reproduction in this haploid yeast (36). However, analysis of the genome of *C. glabrata* has discovered representatives of most of the genes known to be involved in *S. cerevisiae* sexual reproduction, indicating that the last word on reproduction in *C. glabrata* has yet to be written (215). DNA sequence variation also showed population structure in *C. glabrata* between North and South America (161). An MLST scheme has been developed for *C. glabrata*, using six sequenced loci, and has found six clades that correlate with the geographic origin of the isolates; e.g., clade I is most common in Europe, clade IV is most common in Japan, and clade III is most common in North America (38).

C. krusei and *C. lusitaniae* have not received the attention paid to their more clinically important relatives, but population genetic studies have begun that use both sequenced markers (106) and microsatellites (167).

We now leave the Saccharomycotina to move back to the earliest divergence in the Ascomycota at 430 MYA, the one that leads to the Taphrinomycotina, an assemblage of species that have very long branches and whose monophyletic nature is not as well supported as that of the other two subphyla, Saccharomycotina and Pezizomycotina.

Subphylum Taphrinomycotina

In the Taphrinomycotina is the model fungus, *Schizosaccharomyces pombe*, and the species of *Pneumocystis*, which are found in the lungs of mammals and one of which causes human disease if the host becomes immunodeficient (169). Until the late 1980s, these fungi were considered to be protists, but comparison of rDNA sequences showed them to be fungi (42, 175); to be fair, earlier work by electron microscopists already had found the truth (209). *Pneumocystis* species recognition and phylogenetic relationships are based on nucleic acid variation, undoubtedly because none of the species can be cultivated axenically. There is a close correlation between species of *Pneumocystis* and their mammalian host; the most famous species, *P. carinii*, is found in rats, as is a more recently described species, *P. wakefieldii* (26). Mice have a species, *P. murina* (95), and there are many species associated with primates (35), including the human-associated species, *P. jiroveci*. Phylogenetic species recognition using a combination of rDNA regions and protein-coding genes is now the norm for this genus, and studies of multiple isolates from primates (macaques) have shown that two phylogenetic species of *Pneumocystis* can be associated with a single macaque species and vice versa (73). *Pneumocystis* species and their mammalian hosts are a study in coevolution, and dates for divergences among the fungal species correlate with those proposed for the mammalian hosts (96). MLST

would be very difficult with *Pneumocystis* species because individuals cannot be cultivated. Instead, typing has focused on the rDNA ITS sequence. A recent study of South African *P. jiroveci* ITS sequences (155) showed how useful even one sequenced allele can be: individual hosts can be infected by up to six *P. jiroveci* strains, the same *P. jiroveci* individual can infect several host individuals, and analysis of the ITS sequences shows that recombination is a likely event in its evolution, consistent with earlier electron microscopic reports of meiosis in *Pneumocystis* (123).

We now move back along the branch that unites all Ascomycota to its divergence from the Basidiomycota, about 550 MYA, and then come forward in time to the Basidiomycota.

PHYLUM BASIDIOMYCOTA

In the phylum Basidiomycota is found about one-third of the described fungi, but only a few basidiomycete species cause disease in otherwise healthy humans. Basidiomycota comprise three major lineages, one with the rusts and other simple-septate basidiomycetes (Urediniomycetes), a second with the monocot smuts (Ustilaginomycetes), and a third (Hymenomycetes) with the hymenium-forming basidiomycetes, including the "jelly" fungi, bracket fungi, and mushrooms (117, 186).

Class Hymenomycetes

The best-known basidiomycete human pathogens are found in the Hymenomycetes. Within this class there are two lineages, the subclass Hymenomycetidae, with the orders Dacrymycetales, Auriculariales, Aphyllophorales, and Agaricales, and the subclass Tremellomycetidae, with the order Tremellales. In the Tremellomycetidae are found two genera of pathogenic basidiomycetes, *Filobasidiella* and *Trichosporon*. *Filobasidiella neoformans*, better known by the name for its mitosporic state, *Cryptococcus neoformans*, is the agent of cryptococcosis and is the best-known basidiomycete human pathogen. There are five serotypes of *C. neoformans*, A, B, C, D, and AD, which are partitioned among three genetically isolated groups, *C. neoformans* var. *neoformans* (serotype D), *C. neoformans* var. *grubii* (serotoype A) (52), and *C. neoformans* var. *gattii* (serotypes B and C), and individuals that are hybrids between *C. neoformans* var. *neoformans* and *C. neoformans* var. *grubii* (serotype AD) (13, 110, 217). The divergence times between the varieties are estimated to be in the tens of millions of years, long enough for them to be considered phylogenetic species (219). In fact, *C. neoformans* var. *gattii*, long ago, was described as a species, *C. bacillisporus* (104), and the proposal to conserve *C. gattii* as the name for this taxon emphasizes its true rank (105).

C. neoformans var. *neoformans* is known to mate in cultivation, and individuals of mating types **a** and α are found in nature, although mating type α (*MAT*α) is far more commonly encountered (*MAT*α: *MAT*a, 49:1). In *C. neoformans* var. *grubii* the distribution of *MAT* is even more skewed; in fact, this species was known only from *MAT*α individuals until recently. However, the ability of *C. neoformans* var. *grubii MAT*α individuals to hybridize with *C. neoformans* var. *neoformans MAT*a individuals indicated that *C. neoformans* var. *grubii* maintained the capacity for sexual reproduction (80). The discovery of *C. neoformans* var. *grubii MAT*a individuals (112) led to laboratory demonstrations of mating in this variety (138), and population genetic analysis of multiple loci in serotype AD hybrids supported a history of recombination in nature for both parent varieties (218). The recent discovery of a *C. neoformans* var. *grubii* population in Botswana in which *MAT*a individuals account for ca. 10% of the population and for which population genetic tests are consistent with recombination (113) indicates that this fungus is a sexual basidiomycete. The discovery of a natural, sexual *Cryptococcus* population was a first for the genus. A previous study of Australian *C. gattii* had discovered a population with a balance of both mating types, but population genetic analysis detected only clonal reproduction (77). When *C. neoformans* var. *grubii* was known only from *MAT*α individuals, it was hypothesized that only fungi of that mating type were virulent. However, when a *MAT*a individual was discovered, experimental infections of mice showed that the *MAT*a individual also was virulent (97, 138). Adherents of phylogenetic species recognition would consider each variety to be a species, whereas those who favor biological species recognition might retain varieties *grubii* and *neoformans*, because they can mate (80). However, it seems likely that these two lineages were kept apart by geographic isolation and that when their allopatric distribution was changed to one of sympatry (at least for *MAT*α individuals of *C. neoformans* var. *grubii*), hybrid progeny followed. As more populations are discovered, the taxonomic situation is certain to become more complex; e.g., the aforementioned discovery of recombination in Botswanan isolates also uncovered several genetically isolated populations (113). The failure of some *C. neoformans* var. *grubii* pairs to mate is consistent with cryptic species (chapter 2, this volume), and recent investigations of *C. gattii* suggest that it, too, embraces several genetically isolated lineages (94, 98, 176).

The global effort made to genotype *Cryptococcus* individuals by using repetitive DNA sequences is unprecedented in pathogenic fungi, and phylogenetic analysis of the genotypes shows good agreement with the varieties (109, 131). Given the global distribution of this fungus (130) and its sexual nature, *Cryptococcus* would be another ideal candidate for a large, multinational, fungal MLST approach (176). *C. neoformans* also is becoming a model for basidiomycete evolution, judging from recent studies of mutation, mating loci (53, 111), mitochondrial inheritance (220), and genotype by environment studies (216). *Cryptococcus* also was instrumental in the development of a very interesting hypothesis about the origins of pathogenicity resting in interactions between pathogenic yeasts, such as *Cryptococcus*, and simple eukaryote hosts, such as soil amoebae (173, 174).

Leaving the genus *Cryptococcus*, we move back to its divergence from the genus *Trichosporon* and then move forward in time to the several species of this genus of keratinophilic fungi known for their ability to grow on hair or skin. *Trichosporon* species are known to make only mitospores, and so assignment of *Trichosporon* species to the Basidiomycotina could not rely on sexual characteristics. The chemical composition and laminated form of the cell wall suggested that this fungus was a basidiomycete, and this hypothesis was confirmed by comparison of partial 28S rRNA gene sequences (68). From phenotype, site of infection, DNA-DNA hybridization, and 28S rRNA gene sequence, at least five *Trichosporon* species capable of infecting humans have been proposed (71). Two cause white piedra, one on pubic hair, *T. inkin*, and the other on capital hair, *T. ovoides*; three cause systemic disease, *T. mucoides*, *T. asahii*, and *T. cutaneum*; a sixth was isolated from skin, *T. asteroides*. The *Atlas of Clinical Fungi* considers *T. inkin*, *T. asahii*, and *T. mucoides* to be the most commonly encountered human pathogenic species in the genus (34). The familiar *T. beigelli* is now considered a doubtful name because it was improperly described, and isolates formerly classified as *T. beigelli* should be referred to one of the other species. *Trichosporon* phylogeny (43, 133) and taxonomy (71) are well studied, with new species being added with regularity. Methods have been designed for the identification of medically important *Trichosporon* species by substrate utilization (132) or by sequencing of rDNA ITS (177). However, all species recognition in *Trichosporon* is by phenotype, and no studies have been conducted to examine genetic variation within species or to recognize species by phylogenetics.

Moving back in time along the branch leading to the divergence between Hymenomycetes and Ustilaginomycetes at 440 MYA, we then join the Ustilaginomycetes and move forward in time to the genus *Malassezia*.

Class Ustilaginomycetes

In the Ustilaginomycetes (8) is found another genus of animal pathogenic yeasts, *Malassezia*. Species of *Malassezia* can cause disease of the stratum corneum of the body and head, mostly in the tropics (25, 67, 74). Many isolates are lipid dependent and can also cause deeper mycoses in temperate climates if the host is taking lipid intravenously. Traditionally, isolates were classified into two species, *M. furfur* and *M. pachydermatis*, which were thought to be restricted to humans and animals, respectively. However, studies of nuclear large-subunit rRNA gene sequences, morphology, growth substrate, and catalase production have shown that there were at least seven species, all of which could be found in human clinical samples (although *M. furfur* was by far the most common) and several of which also were associated with animals (70). The seven species recognized by Guého and colleagues are *M. furfur*, *M. pachydermatis*, *M. sympodialis*, *M. globosa*, *M. obtusa*, *M. restricta*, and *M. slooffiae*. New species continue to be described, e.g., *M. dermatis*, *M. yanatoensis*, and *M. japonica* (178–180), and both phenotype and molecular methods of identification have been proposed (58, 75, 198). Species recognition is by phenotype and by phylogenetic relationships using variable regions of the rDNA repeat. Intraspecific variation of protein electrophoretic mobility (134) and of chitin synthase gene sequence in *M. pachydermatis* (2) suggest that species recognition by concordance of gene genealogies might discover more genetically isolated groups.

Other basidiomycetes that occasionally have been reported to cause disease include *Rhodotorula rubra*, a member of the Urediniomycetes clade, *Ustilago* spp., which are found on the Ustilaginomycetes clade, and *Coprinus cinereus* and *Schizophyllum commune*, hymenomycetes in the Agaricales and Aphyllophorales, respectively. That mushrooms such as *Coprinus* or wood decay fungi such as *Schizophyllum* can cause human disease shows that almost any fungus is a potential danger to a host whose immune system is sufficiently degraded.

Finished with the Basidiomycota, we move back past its divergence with the Ascomycota at 550 MYA to divergences involving clades of morphologically simpler Zygomycota and Chytridiomycota between 660 and 830 MYA.

PHYLUM ZYGOMYCOTA

Molecular phylogenetic studies have shown that neither Zygomycota nor Chytridiomycota is likely to be monophyletic. It is prudent, therefore, to consider theses fungi to be members of well-supported, order-level clades (165). In Zygomycota, there are two prominent orders, Mucorales and Entomophthorales, that harbor human pathogens. Many members of the Mucorales may be responsible for mycoses, but *Rhizopus oryzae* most commonly is the cause. The phylogenetic relationships of *Rhizopus* species to other Mucorales have been analyzed (142), and genetic variation within *Rhizopus* species has been detected (159, 204). The other order harboring pathogens, Entomophthorales, is well known for its pathogens of insects, and one species, *Conidiobolus coronatus*, is known to cause a subcutaneous infection in humans and animals (64, 154). Species of *Conidiobolus* have not been recognized phylogenetically, and there are suggestions that the species contain considerable variation (54, 166, 213). Another genus capable of causing human disease, *Basidiobolus*, is classified in the Entomophthorales, but phylogenetic studies of 18S rDNA have shown that it does not belong there and suggest that, instead, it may belong in the Chytridiales (136). However, more recent research with both 18S and 28S rDNAs and a number of protein coding genes indicates that *Basidiobolus* is a member of the Entomophthorales (T. Y. James, R. J. Vilgalys, J. E. Longcore, S. E. Mozley-Standridge, and the Assembling the Fungal Tree of Life Working Group, abstract, *Inoculum* 56:28, 2005). Population genetic and phylogenetic studies of *Basidiobolus* species indicate that human pathogenic isolates are distinct from those that are saprobic (138).

PHYLUM CHYTRIDIOMYCOTA

Going back along the entomophthoralean branch, we come to the polytomy of zygo- and chytridiomycete orders at about 830 MYA. Picking the lineage leading to the Chytridiales, we head toward the most important animal pathogenic chytrid, *Batrachochytrium dendrobatidis*, which appears to be responsible for the worldwide amphibian decline (115). With the return of *Basidiobolus* species to the Entomophthorales, no chytrid is known to cause human disease. It had been suggested that *Rhinosporidium seeberi*, the mycotic agent of rhinosporidiosis, was a chytridiomycete; however, it is not a fungus but a mesomycetozoean (82, 129). With the Chytridiales, we have completed our phylogenetic tour of fungal pathogens of humans. The only pathogen studied by mycologists that we have missed is *Pythium insidiosum*, a member of the Oomycota, a phylum now recognized to belong to the Stramenopila, home also to brown algae and diatoms. As with the fungi, studies of *P. insidiosum* have focused on phylogenetics (164), population structure (162), and molecular means of identification (163, 205). Conversely, there is a group of fungal or near-fungal pathogens that have not been studied by mycologists: Microsporidia (94a, 198a). These obligate parasites of many animal groups are capable of causing human disease (51a, 94b), and methods of their identification using nucleic acid variation are being developed (139a).

CONCLUSION

If this short treatment of the evolution of fungi pathogenic on humans has whetted your appetite for more information, you should examine recent works on the place of fungi in *Assembling the Tree of Life* (195), the *Mycota* volumes on fungal systematics (127, 128), and, of course, the *Atlas of Clinical Fungi* (34).

Acknowledgments. G. Sybren de Hoog, Yvonne Gräser, and Joe Heitman provided much needed scholarly advice and editing. This chapter was prepared with support from the National Science Foundation (DEB0516511, DEB0316710) and the National Institutes of Health, National Institute of Allergy and Infectious Disease (AI37232).

REFERENCES

1. **Abliz, P., K. Fukushima, K. Takizawa, and K. Nishimura.** 2004. Identification of pathogenic dematiaceous fungi and related taxa based on large subunit ribosomal DNA D1/D2 domain sequence analysis. *FEMS Immunol. Med. Microbiol.* **40:**41–49.

2. **Aizawa, T., R. Kano, Y. Nakamura, S. Watanabe, and A. Hasegawa.** 2001. The genetic diversity of clinical isolates of *Malassezia pachydermatis* from dogs and cats. *Med. Mycol.* **39:**329–334.

3. **Anaissie, E. J., R. T. Kuchar, J. H. Rex, A. Francesconi, M. Kasai, F.-M. C. Muller, M. Lozano-Chiu, R. C. Summerbell, M. C. Dignani, S. J. Chanock, and T. J. Walsh.** 2001. Fusariosis associated with pathogenic *Fusarium* species colonization of a hospital water system: a new paradigm for the epidemiology of opportunistic mold infections. *Clin. Infect. Dis.* **33:**1871–1878.

4. **Anderson, J. B., C. Wickens, M. Khan, L. E. Cowen, N. Federspiel, T. Jones, and L. M. Kohn.** 2001. Infrequent genetic exchange and recombination in the mitochondrial genome of *Candida albicans*. *J. Bacteriol.* **183:**865–872.

5. **Archer, D. B., and P. S. Dyer.** 2004. From genomics to post-genomics in *Aspergillus. Curr. Opin. Microbiol.* **7:**499–504.

6. **Attili, D. S., G. S. de Hoog, and A. A. Pizzirani-Kleiner.** 1998. rDNA-RFLP and ITSI sequencing of species of the genus *Fonsecaea*, agents of chromoblastomycosis. *Med. Mycol.* **36:**219–225.

7. **Bagagli, E., A. Sano, K. I. Coelho, S. Alquati, M. Miyaji, Z. Pires De Camargo, G. M. Gomes, M. Franco, and M. R. Montenegro.** 1998. Isolation of *Paracoccidioides brasiliensis* from armadillos (*Dasypus novemcinctus*) captured in an endemic area of paracoccidioidomycosis. *Am. J. Trop. Med. Hyg.* **58:**505–512.

8. **Begerow, D., R. Bauer, and T. Boekhout.** 2000. Phylogenetic placements of ustilaginomycetous anamorphs as deduced from nuclear LSU rDNA sequences. *Mycol. Res.* **104:**53–60.

8a. **Bennett, R. J., M. G. Miller, P. R. Chua, M. E. Maxon, and A. D. Johnson.** 2005. Nuclear fusion occurs during

mating in *Candida albicans* and is dependent on the KAR3 gene. *Mol. Microbiol.* **55:**1046–1059.

9. **Berbee, M. L., and J. W. Taylor.** 2001. Fungal molecular evolution: gene trees and geologic time, p. 229–245. *In* D. McLaughlin, E. McLaughlin, and P. Lemke (eds.), *The Mycota*: *Systematics and Evolution*, vol. VII, part B. Springer-Verlag KG, Berlin, Germany.

10. **Bertout, S., F. Renaud, R. Barton, F. Symoens, J. Burnod, M. A. Piens, B. Lebeau, M. A. Viviani, F. Chapuis, J. M. Bastide, R. Grillot, and M. Mallie.** 2001. Genetic polymorphism of *Aspergillus fumigatus* in clinical samples from patients with invasive aspergillosis: investigation using multiple typing methods. *J. Clin. Microbiol.* **39:**1731–1737.

11. **Bialek, R., A. Ibricevic, A. Fothergill, and D. Begerow.** 2000. Small subunit ribosomal DNA sequence shows *Paracoccidioides brasiliensis* closely related to *Blastomyces dermatitidis. J. Clin. Microbiol.* **38:**3190–3193.

12. **Bialek, R., J. Kern, T. Herrmann, R. Tijerina, L. Cecenas, U. Reischl, and G. M. Gonzalez.** 2004. PCR assays for identification of *Coccidioides posadasii* based on the nucleotide sequence of the antigen 2/proline-rich antigen. *J. Clin. Microbiol.* **42:**778–783.

13. **Boekhout, T., B. Theelen, M. Diaz, J. W. Fell, W. C. J. Hop, E. C. A. Abeln, F. Dromer, and W. Meyer.** 2001. Hybrid genotypes in the pathogenic yeast *Cryptococcus neoformans. Microbiology* **147:**891–907.

14. **Bok, J. W., and N. P. Keller.** 2004. LaeA, a regulator of secondary metabolism in *Aspergillus* spp. *Eukaryot. Cell* **3:**527–535.

15. **Bowman, B. H., T. J. White, and J. W. Taylor.** 1996. Evolutionary relationships of human pathogenic fungi: multiple origins of pathogenicity in the fungal order Onygenales. *Mol. Phylogenet. Evol.* **6:**89–96.

16. **Brandt, M. E., and D. W. Warnock.** 2003. Epidemiology, clinical manifestations, and therapy of infections caused by dematiaceous fungi. *J. Chemother.* **15:**36–47.

17. **Burns, R. A., J. S. Roy, C. Woods, A. A. Padhye, and D. W. Warnock.** 2000. Report of the first human case of lobomycosis in the United States. *J. Clin. Microbiol.* **38:**1283–1285.

18. **Burt, A., D. A. Carter, G. L. Koenig, T. J. White, and J. W. Taylor.** 1996. Molecular markers reveal cryptic sex in the human pathogen *Coccidioides immitis. Proc. Natl. Acad. Sci. USA* **93:**770–773.

19. **Calcagno, A. M., G. Nino-Vega, F. San-Blas, and G. San-Blas.** 1998. Geographic discrimination of *Paracoccidioides brasiliensis* strains by randomly amplified polymorphic DNA analysis. *J. Clin. Microbiol.* **36:**1733–1736.

20. **Calvo, A. M., R. A. Wilson, J. W. Bok, and N. P. Keller.** 2002. Relationship between secondary metabolism and fungal development. *Microbiol. Mol. Biol. Rev.* **66:**447–459.

21. **Carter, D. A., J. W. Taylor, B. Dechairo, A. Burt, G. L. Koenig, and T. J. White.** 2001. Amplified single-nucleotide polymorphisms and a (GA)$_n$ microsatellite marker reveal genetic differentiation between populations of *Histoplasma capsulatum* from the Americas. *Fungal Genet. Biol.* **34:**37–48.

22. **Chabasse, D.** 2002. Phaeohyphomycetes agents of phaeohyphomycosis: emerging fungi. *J. Mycol. Med.* **12:**65–85.

23. **Cimon, B., J. Carrere, J. F. Vinatier, J. P. Chazalette, D. Chabasse, and J. P. Bouchara.** 2000. Clinical significance of *Scedosporium apiospermum* in patients with cystic fibrosis. *Eur. J. Clin. Microbiol. Infect. Dis.* **19:**53–56.

24. Cimon, B., R. Zouhair, F. Symoens, J. Carrere, D. Chabasse, and J. P. Bouchara. 2003. *Aspergillus terreus* in a cystic fibrosis clinic: environmental distribution and patient colonization pattern. *J. Hosp. Infect.* **53**:81–82.

25. Crespo Erchiga, V., and V. Delgado Florencio. 2002. *Malassezia* species in skin diseases. *Curr. Opin. Infect. Dis.* **15**:133–142.

26. Cushion, M. T., S. P. Keely, and J. R. Stringer. 2004. Molecular and phenotypic description of *Pneumocystis wakefieldiae* sp nov., a new species in rats. *Mycologia* **96**:429–438.

27. Debeaupuis, J.-P., J. Sarfati, V. Chazalet, and J.-P. Latgé. 1997. Genetic diversity among clinical and environmental isolates of *Aspergillus fumigatus*. *Infect. Immun.* **65**:3080–3085.

28. de Beer, Z. W., T. C. Harrington, H. F. Vismer, B. D. Wingfield, and M. J. Wingfield. 2003. Phylogeny of the *Ophiostoma stenoceras-Sporothix schenckii* complex. *Mycologia* **95**:434–441.

29. Defontaine, A., R. Zouhair, B. Cimon, J. Carrere, E. Bailly, F. Symoens, M. Diouri, J. N. Hallet, and J. P. Bouchara. 2002. Genotyping study of *Scedosporium apiospermum* isolates from patients with cystic fibrosis. *J. Clin. Microbiol.* **40**:2108–2114.

30. de Hoog, G. S. 1977. *Rhinocladiella* and allied genera. *Stud. Mycol.* **15**:1–144.

31. de Hoog, G. S., D. Adelmann, A. O. A. Ahmed, and A. van Belkum. 2004. Phylogeny and typification of *Madurella mycetomatis*, with a comparison of other agents of eumycetoma. *Mycoses* **47**:121–130.

32. de Hoog, G. S., D. Attili-Angelis, V. A. Vicente, A. Van den Ende, and F. Queiroz-Telles. 2004. Molecular ecology and pathogenic potential of *Fonsecaea* species. *Med. Mycol.* **42**:405–416.

33. de Hoog, G. S., V. Vicente, R. B. Caligiorne, S. Kantarcioglu, K. Tintelnot, A. van den Ende, and G. Haase. 2003. Species diversity and polymorphism in the *Exophiala spinifera* clade containing opportunistic black yeast-like fungi. *J. Clin. Microbiol.* **41**:4767–4778.

34. de Hoog, S. H., J. Guarro, J. Gene, and M. J. Figueras. 2000. *Atlas of Clinical Fungi.* CBS and Unveresitat Roviera i Virgili, Utrecht, The Netherlands, and Reus, Spain.

35. Demanche, C., M. Berthelemy, T. Petit, B. Polack, A. E. Wakefield, E. Dei-Cas, and J. Guillot. 2001. Phylogeny of *Pneumocystis carinii* from 18 primate species confirms host specificity and suggests coevolution. *J. Clin. Microbiol.* **39**:2126–2133.

36. de Meeus, T., F. Renaud, E. Mouveroux, J. Reynes, G. Galeazzi, M. Mallie, and J. M. Bastide. 2002. Genetic structure of *Candida glabrata* populations in AIDS and non-AIDS patients. *J. Clin. Microbiol.* **40**:2199–2206.

37. Dignani, M. C., and E. Anaissie. 2004. Human fusariosis. *Clin. Microbiol. Infect.* **10**:67–75.

38. Dodgson, A. R., C. Pujol, D. W. Denning, D. R. Soll, and A. J. Fox. 2003. Multilocus sequence typing of *Candida glabrata* reveals geographically enriched clades. *J. Clin. Microbiol.* **41**:5709–5717.

39. Drouhet, E., E. Guého, S. Gori, M. Huerre, F. Provost, M. Borgers, and B. Dupont. 1998. Mycological, ultrastructural and experimental aspects of a new dimorphic fungus *Emmonsia pasteuriana* sp. nov. isolated from a cutaneous disseminated mycosis in AIDS. *J. Mycol. Med.* **8**:64–77.

40. Dyer, P. S., and M. Paoletti. 2005. Reproduction in *Aspergillus fumigatus*: sexuality in a supposedly asexual species? *Med. Mycol.* **43**(Suppl. 1):S7–S14.

41. Dyer, P. S., M. Paoletti, and D. B. Archer. 2003. Genomics reveals sexual secrets of *Aspergillus.* *Microbiology* **149**:2301–2303.

42. Edman, J. C., J. A. Kovacs, H. Masur, D. V. Santi, H. J. Elwood, and M. L. Sogin. 1988. Ribosomal RNA sequence shows *Pneumocystis carinii* to be a member of the Fungi. *Nature* **334**:519–522.

43. Fell, J. W., and G. Scorzetti. 2004. Reassignment of the basidiomycetous yeasts *Trichosporon pullulans* to *Guehomyces pullulans* gen. nov., comb. nov. and *Hyalodendron lignicola* to *Trichosporon lignicola* comb. nov. *Int. J. Syst. Evol. Microbiol.* **54**:995–998.

44. Fisher, M. C., D. Aanensen, S. de Hoog, and N. Vanittanakom. 2004. Multilocus microsatellite typing system for *Penicillium marneffei* reveals spatially structured populations. *J. Clin. Microbiol.* **42**:5065–5069.

45. Fisher, M. C., S. de Hoog, and N. Vanittanakom. 2004. A highly discriminatory multilocus microsatellite typing (MLMT) system for *Penicillium marneffei*. *Mol. Ecol. Notes* **4**:515–518.

46. Fisher, M. C., G. Koenig, T. J. White, and J. W. Taylor. 2000. A test for concordance between the multilocus genealogies of genes and microsatellites in the pathogenic fungus *Coccidioides immitis*. *Mol. Biol. Evol.* **17**:1164–1174.

47. Fisher, M. C., G. L. Koenig, T. J. White, G. San-Blas, R. Negroni, I. G. Alvarez, B. Wanke, and J. W. Taylor. 2001. Biogeographic range expansion into South America by *Coccidioides immitis* mirrors New World patterns of human migration. *Proc. Natl. Acad. Sci. USA* **98**:4558–4562.

48. Fisher, M. C., G. L. Koenig, T. J. White, and J. W. Taylor. 2002. Molecular and phenotypic description of *Coccidioides posadasii* sp. nov., previously recognized as the non-California population of *Coccidioides immitis*. *Mycologia* **94**:73–84.

49. Fisher, M. C., G. L. Koenig, T. J. White, and J. W. Taylor. 2000. Pathogenic clones versus environmentally driven population increase: analysis of an epidemic of the human fungal pathogen *Coccidioides immitis*. *J. Clin. Microbiol.* **38**:807–813.

50. Fisher, M. C., B. Rannala, V. Chaturvedi, and J. W. Taylor. 2002. Disease surveillance in recombining pathogens: multilocus genotypes identify sources of human *Coccidioides* infections. *Proc. Natl. Acad. Sci. USA* **99**:9067–9071.

51. Forche, A., G. Schonian, Y. Gräser, R. Vilgalys, and T. G. Mitchell. 1999. Genetic structure of typical and atypical populations of *Candida albicans* from Africa. *Fungal Genet. Biol.* **28**:107–125.

51a. Franzen, C., and A. Muller. 2001. Microsporidiosis: human diseases and diagnosis. *Microbes Infect.* **3**:389–400.

52. Franzot, S. P., I. F. Salkin, and A. Casadevall. 1999. *Cryptococcus neoformans* var. *grubii*: separate varietal status for *Cryptococcus neoformans* serotype A isolates. *J. Clin. Microbiol.* **37**:838–840.

53. Fraser, J. A., and J. Heitman. 2004. Evolution of fungal sex chromosomes. *Mol. Microbiol.* **51**:299–306.

54. Freimoser, F. M., S. Screen, G. Hu, and R. S. Leger. 2003. EST analysis of genes expressed by the zygomycete pathogen *Conidiobolus coronatus* during growth on insect cuticle. *Microbiology* **149**:1893–1900.

54a. Galagan, J. E., et al. 2005. Sequencing of *Aspergillus nidulans* and comparative analysis with *A. fumigatus* and *A. oryzae*. *Nature* **438**:1105–1115.

55. Geiser, D. M., J. W. Dorner, B. W. Horn, and J. W. Taylor. 2000. The phylogenetics of mycotoxin and

sclerotium production in *Aspergillus flavus* and *Aspergillus oryzae*. *Fungal Genet. Biol.* **31:**169–179.

56. **Geiser, D. M., J. C. Frisvad, and J. W. Taylor.** 1998. Evolutionary relationships in *Aspergillus* section Fumigata inferred from partial beta-tubulin and hydrophobin DNA sequences. *Mycologia* **90:**831–845.

57. **Geiser, D. M., J. I. Pitt, and J. W. Taylor.** 1998. Cryptic speciation and recombination in the aflatoxin-producing fungus *Aspergillus flavus*. *Proc. Natl. Acad. Sci. USA* **95:**388–393.

58. **Giusiano, G., S. Bustillo, M. Mangiaterra, and G. DeLuca.** 2003. Identification of *Malassezia* species by PCR-REA. *Rev. Arg. Microbiol.* **35:**162–166.

59. **Gräser, Y., J. Kuhnisch, and W. Presber.** 1999. Molecular markers reveal exclusively clonal reproduction in *Trichophyton rubrum*. *J. Clin. Microbiol.* **37:**3713–3717.

60. **Gräser, Y., A. F. A. Kuijpers, M. El Fari, W. Presber, and G. S. de Hoog.** 2000. Molecular and conventional taxonomy of the *Microsporum canis* complex. *Med. Mycol.* **38:**143–153.

61. **Gräser, Y., M. Volovsek, J. Arrington, G. Schonian, W. Presber, T. G. Mitchell, and R. Vilgalys.** 1996. Molecular markers reveal that population structure of the human pathogen *Candida albicans* exhibits both clonality and recombination. *Proc. Natl. Acad. Sci. USA* **93:**12473–12477.

62. **Greene, D. R., G. Koenig, M. C. Fisher, and J. W. Taylor.** 2000. Soil isolation and molecular identification of *Coccidioides immitis*. *Mycologia* **92:**406–410.

63. **Grigis, A., C. Farina, F. Symoens, N. Nolard, and A. Goglio.** 2000. Nosocomial pseudo-outbreak of *Fusarium verticillioides* associated with sterile plastic containers. *Infect. Control Hosp. Epidemiol.* **21:**50–52.

64. **Grooters, A. M.** 2003. Pythiosis, lagenidiosis, and zygomycosis in small animals. *Vet. Clin. North Am. Small Anim. Pract.* **33:**695–720.

65. **Guarro, J., and J. Cano.** 2002. Phylogeny of Onygenalean fungi of medical interest. *Stud. Mycol.* **47:**1–4.

66. **Guarro, J., A. S. Kantarcioglu, R. Horré, J. L. Rodriguez-Tudela, M. C. Estrella, J. Berenguer, and G. S. de Hoog.** *Scedosporium* infection, an emerging fungal disease entity. *Med. Mycol.*, in press.

67. **Guého, E., T. Boekhout, H. R. Ashbee, J. Guillot, A. Van Belkum, and J. Faergemann.** 1998. The role of *Malassezia* species in the ecology of human skin and as pathogens. *Med. Mycol.* **36:**220–229.

68. **Guého, E., L. Improvisi, R. Christen, and G. S. de Hoog.** 1993. Phylogenetic relationships of *Cryptococcus neoformans* and some related basidiomycetous yeasts determined from partial large subunit rRNA sequences. *Antonie Leeuwenhoek Int. J. Gen. Mol. Microbiol.* **63:**175–189.

69. **Guého, E., M. C. Leclerc, G. S. de Hoog, and B. Dupont.** 1997. Molecular taxonomy and epidemiology of *Blastomyces* and *Histoplasma* species. *Mycoses* **40:**69–81.

70. **Guého, E., G. Midgley, and J. Guillot.** 1996. The genus *Malassezia* with description of four new species. *Antonie Leeuwenhoek Int. J. Gen. Mol. Microbiol.* **69:**337–355.

71. **Guého, E., M. T. Smith, and G. S. de Hoog.** 1998. *Trichosporon* Behrend, p. 854–872. *In* C. P. Kurtzman and J. W. Fell (ed.), *The Yeasts: a Taxonomic Study*, 4th ed. Elsevier, Amsterdam, The Netherlands.

72. **Gugnani, H., M. C. Fisher, A. Paliwal-Johsi, N. Vanittanakom, I. Singh, and P. S. Yadav.** 2004. Role of *Cannomys badius* as a natural animal host of *Penicillium marneffei* in India. *J. Clin. Microbiol.* **42:**5070–5075.

73. **Guillot, J., C. Demanche, K. Norris, H. Wildschutte, F. Wanert, M. Berthelemy, S. Tataine, E. Dei-Cas, and R. Chermette.** 2004. Phylogenetic relationships among *Pneumocystis* from Asian macaques inferred from mitochondrial rRNA sequences. *Mol. Phylogenet. Evol.* **31:**988–996.

74. **Gupta, A. K., R. Batra, R. Bluhm, T. Boekhout, and T. L. Dawson.** 2004. Skin diseases associated with *Malassezia* species. *J. Am. Acad. Dermatol.* **51:**785–798.

75. **Gupta, A. K., T. Boekhout, B. Theelen, R. Summerbell, and R. Batra.** 2004. Identification and typing of *Malassezia* species by amplified fragment length polymorphism and sequence analyses of the internal transcribed spacer and large-subunit regions of ribosomal DNA. *J. Clin. Microbiol.* **42:**4253–4260.

76. **Gupta, A. K., Y. Kohli, and R. C. Summerbell.** 2002. Exploratory study of single-copy genes and ribosomal intergenic spacers for rapid distinction of dermatophytes. *Stud. Mycol.* **47:**87–96.

77. **Halliday, C. L., and D. A. Carter.** 2003. Clonal reproduction and limited dispersal in an environmental population of *Cryptococcus neoformans* var. *gattii* isolates from Australia. *J. Clin. Microbiol.* **41:**703–711.

78. **Haubold, E. M., C. R. Cooper, J. W. Wen, M. R. McGinnis, and D. F. Cowan.** 2000. Comparative morphology of *Lacazia loboi* (syn. *Loboa loboi*) in dolphins and humans. *Med. Mycol.* **38:**9–14.

79. **Heckman, D. S., D. M. Geiser, B. R. Eidell, R. L. Stauffer, N. L. Kardos, and S. B. Hedges.** 2001. Molecular evidence for the early colonization of land by fungi and plants. *Science* **293:**1129–1133.

80. **Heitman, J., B. Allen, J. A. Alspaugh, and K. J. Kwon-Chung.** 1999. On the origins of congenic MATα and MATa strains of the pathogenic yeast *Cryptococcus neoformans*. *Fungal Genet. Biol.* **28:**1–5.

81. **Hennig, W.** 1966. *Phylogenetic Systematics.* University of Illinois Press, Urbana.

82. **Herr, R. A., L. Ajello, J. W. Taylor, S. N. Arseculeratne, and L. Mendoza.** 1999. Phylogenetic analysis of *Rhinosporidium seeberi*'s 18S small-subunit ribosomal DNA groups this pathogen among members of the protoctistan Mesomycetozoa clade. *J. Clin. Microbiol.* **37:**2750–2754.

83. **Herr, R. A., E. J. Tarcha, P. R. Taborda, J. W. Taylor, L. Ajello, and L. Mendoza.** 2001. Phylogenetic analysis of *Lacazia loboi* places this previously uncharacterized pathogen within the dimorphic onygenales. *J. Clin. Microbiol.* **39:**309–314.

84. **Howell, S. A., R. J. Barnard, and F. Humphreys.** 1999. Application of molecular typing methods to dermatophyte species that cause skin and nail infections. *J. Med. Microbiol.* **48:**33–40.

85. **Hull, C. M., R. M. Raisner, and A. D. Johnson.** 2000. Evidence for mating of the "asexual" yeast *Candida albicans* in a mammalian host. *Science* **289:**307–310.

86. **Issakainen, J., J. Jalava, J. Saari, and C. K. Campbell.** 1999. Relationship of *Scedosporium prolificans* with *Petriella* confirmed by partial LSU rDNA sequences. *Mycol. Res.* **103:**1179–1184.

86a. **Johnson, A.** 2003. The biology of mating in *Candida albicans*. *Natl. Rev. Microbiol.* **1:**106–116.

87. **Kam, A. P., and J. P. Xu.** 2002. Diversity of commensal yeasts within and among healthy hosts. *Diagn. Microbiol. Infect. Dis.* **43:**19–28.

88. Kanbe, T., Y. Suzuki, A. Kamiya, T. Mochizuki, M. Kawasaki, M. Fujihiro, and A. Kikuchi. 2003. Species-identification of dermatophytes *Trichophyton*, *Microsporum* and *Epidermophyton* by PCR and PCR-RFLP targeting of the DNA topoisomerase II genes. *J. Dermatol. Sci.* **33**:41–54.

89. Kano, R., Y. Nakamura, S. Watanabe, and A. Hasegawa. 2002. Chitin synthase 1 and 2 genes of dermatophytes. *Stud. Mycol.* **47**:49–55.

90. Kasuga, T., J. W. Taylor, and T. J. White. 1999. Phylogenetic relationships of varieties and geographical groups of the human pathogenic fungus *Histoplasma capsulatum* darling. *J. Clin. Microbiol.* **37**:653–663.

91. Kasuga, T., T. J. White, G. Koenig, J. McEwen, A. Restrepo, E. Castaneda, C. da Silva Lacaz, E. M. Heins-Vaccari, R. S. De Freitas, R. M. Zancope-Oliveira, Z. Qin, R. Negroni, D. A. Carter, Y. Mikami, M. Tamura, M. L. Taylor, G. F. Miller, N. Poonwan, and J. W. Taylor. 2003. Phylogeography of the fungal pathogen *Histoplasma capsulatum*. *Mol. Ecol.* **12**:3383–3401.

92. Kasuga, T., T. J. White, and J. W. Taylor. 2002. Estimation of nucleotide substitution rates in Eurotiomycete fungi. *Mol. Biol. Evol.* **19**:2318–2324.

93. Kaszubiak, A., S. Klein, G. S. de Hoog, and Y. Gräser. 2004. Population structure and evolutionary origins of *Microsporum canis*, *M. ferrugineum* and *M. audouinii*. *Med. Mycol.* **4**:179–186.

94. Katsu, M., S. Kidd, A. Ando, M. L. Moretti-Branchini, Y. Mikami, K. Nishimura, and W. Meyer. 2004. The internal transcribed spacers and 5.8S rRNA gene show extensive diversity among isolates of the *Cryptococcus neoformans* species complex. *FEMS Yeast Res.* **4**:377–388.

94a. Keeling, P. J. 2003. Congruent evidence from alpha-tubulin and beta-tubulin gene phylogenies for a zygomycete origin of microsporidia. *Fung. Genet. Biol.* **38**:298–309.

94b. Keeling, P. J., and N. M. Fast. 2003. Microsporidia: biology and evolution of highly reduced intracellular parasites. *Annu. Rev. Microbiol.* **56**:93–116.

95. Keely, S. P., J. M. Fischer, M. T. Cushion, and J. R. Stringer. 2004. Phylogenetic identification of *Pneumocystis murina* sp. nov., a new species in laboratory mice. *Microbiology* **150**:1153–1165.

96. Keely, S. P., J. M. Fischer, and J. R. Stringer. 2003. Evolution and speciation of *Pneumocystis*. *J. Eukaryot. Microbiol.* **50**:624–626.

97. Keller, S. M., M. A. Viviani, M. C. Esposto, M. Cogliati, and B. L. Wickes. 2003. Molecular and genetic characterization of a serotype A MATa *Cryptococcus neoformans* isolate. *Microbiology* **149**:131–142.

98. Kidd, S. E., F. Hagen, R. L. Tscharke, M. Huynh, K. H. Bartlett, M. Fyfe, L. MacDougall, T. Boekhout, K. J. Kwon-Chung, and W. Meyer. 2004. A rare genotype of *Cryptococcus gattii* caused the cryptococcosis outbreak on Vancouver Island (British Columbia, Canada). *Proc. Natl. Acad. Sci. USA* **101**:17258–17263.

99. Koufopanou, V., A. Burt, and J. W. Taylor. 1997. Concordance of gene genealogies reveals reproductive isolation in the pathogenic fungus *Coccidioides immitis*. *Proc. Natl. Acad. Sci. USA* **94**:5478–5482.

100. Kroken, S., N. L. Glass, J. W. Taylor, O. C. Yoder, and B. G. Turgeon. 2003. Phylogenomic analysis of type I polyketide synthase genes in pathogenic and saprobic ascomycetes. *Proc. Natl. Acad. Sci. USA* **100**:15670–15675.

101. Kurtzman, C. P., and C. J. Robnett. 1997. Identification of clinically important ascomycetous yeasts based on nucleotide divergence in the 5′ end of the large-subunit (26S) ribosomal DNA gene. *J. Clin. Microbiol.* **35**:1216–1223.

102. Kwon-Chung, K. J. 1972. Sexual stage of *Histoplasma capsulatum*. *Science* **175**:326.

103. Kwon-Chung, K. J. 1973. Studies on *Emmonsiella capsulata* 1. Heterothallism and development of the ascocarp. *Mycologia* **65**:109–121.

104. Kwon-Chung, K. J., J. E. Bennett, and T. S. Theodore. 1978. *Cryptococcus bacillisporus* sp. nov.: serotype B-C of *Cryptococcus neoformans*. *Int. J. Syst. Bacteriol.* **28**:616–620.

105. Kwon-Chung, K. J., T. Boekhout, and J. W. Fell. 2002. Proposal to conserve the name *Cryptococcus gattii* against *C. hondurianus* and *C. bacillisporus* (Basidiomycota, Hymenomycetes, Tremellomycetidae). *Taxon* **51**:804–806.

106. Lachance, M. A., H. M. Daniel, W. Meyer, G. S. Prasad, S. P. Gautam, and K. Boundy-Mills. 2003. The D1/D2 domain of the large-subunit rDNA of the yeast species *Clavispora lusitaniae* is unusually polymorphic. *FEMS Yeast Res.* **4**:253–258.

107. Lachke, S. A., S. R. Lockhart, K. J. Daniels, and D. R. Soll. 2003. Skin facilitates *Candida albicans* mating. *Infect. Immun.* **71**:4970–4976.

108. Lasker, B. A., and Y. P. Ran. 2004. Analysis of polymorphic microsatellite markers for typing *Penicillium marneffei* isolates. *J. Clin. Microbiol.* **42**:1483–1490.

109. Lemmer, K., D. Naumann, B. Raddatz, and K. Tintelnot. 2004. Molecular typing of *Cryptococcus neoformans* by PCR fingerprinting, in comparison with serotyping and Fourier transform infrared-spectroscopy-based phenotyping. *Med. Mycol.* **42**:135–147.

110. Lengeler, K. B., G. M. Cox, and J. Heitman. 2001. Serotype AD strains of *Cryptococcus neoformans* are diploid or aneuploid and are heterozygous at the mating-type locus. *Infect. Immun.* **69**:115–122.

111. Lengeler, K. B., D. S. Fox, J. A. Fraser, A. Allen, K. Forrester, F. S. Dietrick, and J. Heitman. 2002. Mating-type locus of *Cryptococcus neoformans*: a step in the evolution of sex chromosomes. *Eukaryot. Cell* **1**:704–718.

112. Lengeler, K. B., P. Wang, G. M. Cox, J. R. Perfect, and J. Heitman. 2000. Identification of the MATa mating-type locus of *Cryptococcus neoformans* reveals a serotype A MATa strain thought to have been extinct. *Proc. Natl. Acad. Sci. USA* **97**:14455–14460.

113. Litvintseva, A. P., R. E. Marra, K. Nielsen, J. Heitman, R. Vilgalys, and T. G. Mitchell. 2003. Evidence of sexual recombination among *Cryptococcus neoformans* serotype A isolates in sub-Saharan Africa. *Eukaryot. Cell* **2**:1162–1168.

114. Lobuglio, K. F., and J. W. Taylor. 1995. Phylogeny and PCR identification of the human pathogenic fungus *Penicillium marneffei*. *J. Clin. Microbiol.* **33**:85–89.

115. Longcore, J. E., A. P. Pessier, and D. K. Nichols. 1999. *Batrachochytrium dendrobatidis* gen et sp nov, a chytrid pathogenic to amphibians. *Mycologia* **91**:219–227.

116. Lopes, M. M., G. Freitas, and P. Boiron. 2000. Potential utility of random amplified polymorphic DNA (RAPD) and restriction endonuclease assay (REA) as typing systems for *Madurella mycetomatis*. *Curr. Microbiol.* **40**:1–5.

117. Lutzoni, F., F. Kauff, C. J. Cox, D. McLaughlin, G. Celio, B. Dentinger, M. Padamsee, D. Hibbett, T. Y. James, E. Baloch, M. Grube, V. Reeb, V. Hofstetter, C. Schoch, A. E. Arnold, J. Miadlikowska, J. Spatafora, D. Johnson, S. Hambleton, M. Crockett, R. Shoemaker,

G. H. Sung, R. Lucking, T. Lumbsch, K. O'Donnell, M. Binder, P. Diederich, D. Ertz, C. Gueidan, K. Hansen, R. C. Harris, K. Hosaka, Y. W. Lim, B. Matheny, H. Nishida, D. Pfister, J. Rogers, A. Rossman, I. Schmitt, H. Sipman, J. Stone, J. Sugiyama, R. Yahr, and R. Vilgalys. 2004. Assembling the fungal tree of life: progress, classification and evolution of subcellular traits. *Am. J. Bot.* **91:**1446–1480.

117a. Machida, M., et al. 2005. Genome sequencing and analysis of *Aspergillus oryzae*. *Nature* **438:**1157-1161.

118. Magee, B. B., and P. T. Magee. 2000. Induction of mating in *Candida albicans* by construction of MTLa and MTL alpha strains. *Science* **289:**310–313.

119. Magee, P. T., and B. B. Magee. 2004. Through a glass opaquely: the biological significance of mating in *Candida albicans*. *Curr. Opin. Microbiol.* **7:**661–665.

120. Maiden, M. C. J., J. A. Bygraves, E. Feil, G. Morelli, J. E. Russell, R. Urwin, Q. Zhang, J. J. Zhou, K. Zurth, D. A. Caugant, I. M. Feavers, M. Achtman, and B. G. Spratt. 1998. Multilocus sequence typing: a portable approach to the identification of clones within populations of pathogenic microorganisms. *Proc. Natl. Acad. Sci. USA* **95:**3140–3145.

121. Makimura, K., Y. Tamura, T. Mochizuki, A. Hasegawa, Y. Tajiri, R. Hanazawa, K. Uchida, H. Saito, and H. Yamaguchi. 1999. Phylogenetic classification and species identification of dermatophyte strains based on DNA sequences of nuclear ribosomal internal transcribed spacer 1 regions. *J. Clin. Microbiol.* **37:**920–924.

122. Matos, T., G. Haase, A. H. G. G. van den Ende, and G. S. de Hoog. 2003. Molecular diversity of oligotrophic and neurotropic members of the black yeast genus *Exophiala*, with accent on *E. dermatitidis*. *Antonie Leeuwenhoek Int. J. Gen. Mol. Microbiol.* **83:**293–303.

123. Matsumoto, Y., and Y. Yoshida. 1984. Sporogeny in *Pneumocystis carinii*: synaptonemal complexes and meiotic nuclear divisions observed in precysts. *J. Protozool.* **31:**420–428.

123a. Matute, D. R., J. G. McEwen, R. Puccia, B. A. Montes, G. San-Blas, E. Bagagli, J. T. Rauscher, A. Restrepo, F. Morais, G. Niño-Vega, and J. W. Taylor. 2006. Cryptic speciation and recombination in the fungus *Paracoccidioides brasiliensis* as revealed by gene genealogies. *Mol. Biol. Evol.* **23:**65–73.

124. McCullough, M. J., K. V. Clemons, and D. A. Stevens. 1999. Molecular epidemiology of the global and temporal diversity of *Candida albicans*. *Clin. Infect. Dis.* **29:**1220–1225.

125. McCullough, M. J., A. F. DiSalvo, K. V. Clemons, P. Park, and D. A. Stevens. 2000. Molecular epidemiology of *Blastomyces dermatitidis*. *Clin. Infect. Dis.* **30:** 328–335.

126. McDonough, E. S. 1970. Blastomycosis: epidemiology and biology of its etiologic agent *Ajellomyces dermatitidis*. *Mycopathol. Mycol. Appl.* **41:**195–201.

127. McLaughlin, D. J., E. G. McLaughlin, and P. A. Lemke (ed.). 2001. *The Mycota: Systematics and Evolution*, vol. VII, part A. Springer-Verlag KG, Berlin, Germany.

128. McLaughlin, D. J., E. G. McLaughlin, and P. A. Lemke (ed.). 2001. *The Mycota: Systematics and Evolution*, vol. VII, part B. Springer-Verlag KG, Berlin, Germany.

129. Mendoza, L., J. W. Taylor, and L. Ajello. 2002. The class Mesomycetozoea: a group of microorganisms at the animal-fungal boundary. *Annu. Rev. Microbiol.* **56:**315–344.

130. Meyer, W., A. Castaneda, S. Jackson, M. Huynh, and E. Castaneda. 2003. Molecular typing of IberoAmerican *Cryptococcus neoformans* isolates. *Emerg. Infect. Dis.* **9:**189–195.

131. Meyer, W., K. Marszewska, M. Amirmostofian, R. P. Igreja, C. Hardtke, K. Methling, M. A. Viviani, A. Chindamporn, S. Sukroongreung, M. A. John, D. H. Ellis, and T. C. Sorrell. 1999. Molecular typing of global isolates of *Cryptococcus neoformans* var. *neoformans* by polymerase chain reaction fingerprinting and randomly amplified polymorphic DNA—a pilot study to standardize techniques on which to base a detailed epidemiological survey. *Electrophoresis* **20:**1790–1799.

132. Middelhoven, W. J. 2003. Identification of clinically relevant *Trichosporon* species. *Mycoses* **46:**7–11.

133. Middelhoven, W. J., G. Scorzetti, and J. W. Fell. 2004. Systematics of the anamorphic basidiomycetous yeast genus *Trichosporon* Behrend with the description of five novel species: *Trichosporon vadense, T. smithiae, T. dehoogii, T. scarabaeorum* and *T. gamsii*. *Int. J. Syst. Evol. Microbiol.* **54:**975–986.

134. Midreuil, F., J. Guillot, E. Guého, F. Renaud, M. Mallie, and J.-M. Bastide. 1999. Genetic diversity in the yeast species *Malassezia pachydermatis* analysed by multilocus enzyme electrophoresis. *Int. J. System. Bacteriol.* **49:**1287–1294.

135. Moran, G., C. Stokes, S. Thewes, B. Hube, D. C. Coleman, and D. Sullivan. 2004. Comparative genomics using *Candida albicans* DNA microarrays reveals absence and divergence of virulence-associated genes in *Candida dubliniensis*. *Microbiology* **150:**3363–3382.

136. Nagahama, T., H. Sato, M. Shimazu, and J. Sugiyama. 1995. Phylogenetic divergence of the entomophthoralean fungi: evidence from nuclear 18s ribosomal RNA gene sequence. *Mycologia* **87:**203–209.

137. Nascimento, T., R. Martinez, A. R. Lopes, L. A. D. Bernardes, C. P. Barco, M. H. S. Goldman, J. W. Taylor, J. G. McEwen, M. P. Nobrega, F. G. Nobrega, and G. H. Goldman. 2004. Detection and selection of microsatellites in the genome of *Paracoccidioides brasiliensis* as molecular markers for clinical and epidemiological studies. *J. Clin. Microbiol.* **42:**5007–5014.

138. Nielsen, M., G. M. Cox, P. Wang, D. L. Toffaletti, J. R. Perfect, and J. Heitman. 2003. Sexual cycle of *Cryptococcus neoformans* var. *grubii* and virulence of congenic a and alpha isolates. *Infect. Immun.* **71:** 4831–4841.

138a. Nierman, W. C., et al. 2005. Genomic sequence of the pathogenic and allergenic filamentous fungus *Aspergillus fumigatus*. *Nature* **438:**1151–1156.

139. Nixon, K. C., and Q. D. Wheeler. 1990. An amplification of the phylogenetic species concept. *Cladistics* **6:**211–223.

139a. Notermans, D. W., R. Peek, M. D. de Jong, E. M. Wentink-Bonnema, R. Boom, and T. van Gool. 2005. Detection and identification of *Enterocytozoon bieneusi* and *Encephalitozoon* species in stool and urine specimens by PCR and differential hybridization. *J. Clin. Microbiol.* **43:**610–614.

140. O'Donnell, K. 2000. Molecular phylogeny of the *Nectria haematococca-Fusarium solani* species complex. *Mycologia* **92:**919–938.

141. O'Donnell, K., H. C. Kistler, E. Cigelink, and R. C. Ploetz. 1998. Multiple evolutionary origins of the fungus causing Panama disease of banana: concordant evidence from nuclear and mitochondrial gene genealogies. *Proc. Natl. Acad. Sci. USA* **95:**2044–2049.

142. O'Donnell, K., F. M. Lutzoni, T. J. Ward, and G. L. Benny. 2001. Evolutionary relationships among mucoralean fungi (Zygomycota): evidence for family polyphyly on a large scale. *Mycologia* 93:286–297.

143. O'Donnell, K., D. A. Sutton, M. G. Rinaldi, K. C. Magnon, P. A. Cox, S. G. Revankar, S. Sanche, D. M. Geiser, J. H. Juba, J. A. H. van Burik, A. Padhye, E. J. Anaissie, A. Francesconi, T. J. Walsh, and J. S. Robinson. 2004. Genetic diversity of human pathogenic members of the *Fusarium oxysporum* complex inferred from multilocus DNA sequence data and amplified fragment length polymorphism analyses: evidence for the recent dispersion of a geographically widespread clonal lineage and nosocomial origin. *J. Clin. Microbiol.* 42:5109–5120.

144. Ohst, T., S. de Hoog, W. Presber, V. Stavrakieva, and Y. Gräser. 2004. Origins of microsatellite diversity in the *Trichophyton rubrum-T. violaceum* clade (dermatophytes). *J. Clin. Microbiol.* 42:4444–4448.

145. Pan, S., L. Sigler, and G. T. Cole. 1994. Evidence for a phylogenetic connection between *Coccidioides immitis* and *Uncinocarpus reesii* (Onygenaceae). *Microbiology* 140:1481–1494.

145a. Paoletti, M., C. Rydholm, E. U. Schwier, M. J. Anderson, G. Szakacs, F. Lutzoni, J.-P. Debeaupuis, J.-P. Latgé, D. W. Denning, and P. S. Dyer. 2005. Evidence for sexuality in the opportunistic fungal pathogen *Aspergillus fumigatus*. *Curr. Biol.* 15:1242–1248.

146. Peterson, S. W., and L. Sigler. 1998. Molecular genetic variation in *Emmonsia crescens* and *Emmonsia parva*, etiologic agents of adiaspiromycosis, and their phylogenetic relationship to *Blastomyces dermatitidis* (*Ajellomyces dermatitidis*) and other systemic fungal pathogens. *J. Clin. Microbiol.* 36:2918–2925.

147. Pildain, M. B., G. Vaamonde, and D. Cabral. 2004. Analysis of population structure of *Aspergillus flavus* from peanut based on vegetative compatibility, geographic origin, mycotoxin and sclerotia production. *Int. J. Food Microbiol.* 93:31–40.

148. Poggeler, S. 2002. Genomic evidence for mating abilities in the asexual pathogen *Aspergillus fumigatus*. *Curr. Genet.* 42:153–160.

148a. Pringle, A., D. M. Baker, J. L. Platt, J. P. Wares, J.-P. Latgé, and J. W. Taylor. 2005. Cryptic speciation in the cosmopolitan and clonal human pathogenic fungus *Aspergillus fumigatus*. *Evolution* 59:1886–1899.

149. Probst, S., G. S. de Hoog, and Y. Gräser. 2002. Development of DNA markers to explore host shifts in dermatophytes. *Stud. Mycol.* 47:57–74.

150. Pujol, C., K. J. Daniels, S. R. Lockhart, T. Srikantha, J. B. Radke, J. Geiger, and D. R. Soll. 2004. The closely related species *Candida albicans* and *Candida dubliniensis* can mate. *Eukaryot. Cell* 3:1015–1027.

151. Pujol, C., J. Reynes, F. Renaud, M. Raymond, M. Tibayrenc, F. J. Ayala, F. Janbon, M. Mallie, and J. M. Bastide. 1993. The yeast *Candida albicans* has a clonal mode of reproduction in a population of infected human immunodeficiency virus-positive patients. *Proc. Natl. Acad. Sci. USA* 90:9456–9459.

152. Raad, I., J. Tarrand, H. Hanna, M. Albitar, E. Janssen, M. Boktour, G. Bodey, M. Mardani, R. Hachem, D. Kontoyiannis, E. Whimbey, and K. Rolston. 2002. Epidemiology, molecular mycology, and environmental sources of *Fusarium* infection in patients with cancer. *Infect. Control Hos. Epidemiol.* 23:532–537.

153. Rainer, J., G. S. de Hoog, M. Wedde, Y. Gräser, and S. Gilges. 2000. Molecular variability of *Pseudallescheria boydii*, a neurotropic opportunist. *J. Clin. Microbiol.* 38:3267–3273.

154. Ribes, J. A., C. L. Vanover-Sams, and D. J. Baker. 2000. Zygomycetes in human disease. *Clin. Microbiol. Rev.* 13:236–301.

155. Robberts, F. J. L., L. D. Liebowitz, and L. J. Chalkley. 2004. Genotyping and coalescent phylogenetic analysis of *Pneumocystis jiroveci* from South Africa. *J. Clin. Microbiol.* 42:1505–1510.

156. Rokas, A., B. L. Williams, N. King, and S. B. Carroll. 2003. Genome-scale approaches to resolving incongruence in molecular phylogenies. *Nature* 425:798–804.

157. Rosehart, K., M. H. Richards, and M. J. Bidochka. 2002. Microsatellite analysis of environmental and clinical isolates of the opportunist fungal pathogen *Aspergillus fumigatus*. *J. Med. Microbiol.* 51:1128–1134.

158. Saiki, R. K., D. H. Gelfand, S. Stoffel, S. J. Scharf, R. Higuchi, G. T. Horn, K. B. Mullis, and H. Erlich. 1988. Primer-directed enzymatic amplification of DNA with a thermostable DNA polymerase. *Science* 239:487–491.

159. Saito, K., A. Saito, M. Ohnishi, and Y. Oda. 2004. Genetic diversity in *Rhizopus oryzae* strains as revealed by the sequence of lactate dehydrogenase genes. *Arch. Microbiol.* 182:30–36.

160. San-Blas, G., G. Nino-Vega, and T. Iturriaga. 2002. *Paracoccidioides brasiliensis* and paracoccidioidomycosis: molecular approaches to morphogenesis, diagnosis, epidemiology, taxonomy and genetics. *Med. Mycol.* 40:225–242.

161. Sanson, G. F. O., and M. R. S. Briones. 2000. Typing of *Candida glabrata* in clinical isolates by comparative sequence analysis of the cytochrome *c* oxidase subunit 2 gene distinguishes two clusters of strains associated with geographical sequence polymorphisms. *J. Clin. Microbiol.* 38:227–235.

162. Schurko, A., L. Mendoza, A. de Cock, and G. R. Klassen. 2003. Evidence for geographic clusters: molecular genetic differences among strains of *Pythium insidiosum* from Asia, Australia and the Americas are explored. *Mycologia* 95:200–208.

163. Schurko, A. M., L. Mendoza, A. de Cock, J. E. J. Bedard, and G. R. Klassen. 2004. Development of a species-specific probe for *Pythium insidiosum* and the diagnosis of pythiosis. *J. Clin. Microbiol.* 42:2411–2418.

164. Schurko, A. M., L. Mendoza, C. A. Levesque, N. L. Desaulniers, W. A. M. De Cock, and G. R. Klassen. 2003. A molecular phylogeny of *Pythium insidiosum*. *Mycol. Res.* 107:537–544.

165. Schussler, A., D. Schwarzott, and C. Walker. 2001. A new fungal phylum, the Glomeromycota: phylogeny and evolution. *Mycol. Res.* 105:1413–1421.

166. Sharma, N. L., V. K. Mahajan, and P. Singh. 2003. Orofacial conidiobolomycosis due to *Conidiobolus incongruus*. *Mycoses* 46:137–140.

167. Shemer, R., Z. Weissman, N. Hashman, and D. Kornitzer. 2001. A highly polymorphic degenerate microsatellite for molecular strain typing of *Candida krusei*. *Microbiology* 147:2021–2028.

168. Sigler, L., A. L. Flis, and J. W. Carmichael. 1998. The genus *Uncinocarpus* (Onygenaceae) and its synonym *Brunneospora*: new concepts, combinations and connections to anamorphs in *Chrysosporium*, and further

evidence of relationship with *Coccidioides immitis*. *Can. J. Bot.* **76**:1624–1636.

169. Smulian, A. G. 2001. *Pneumocystis carinii*: genetic diversity and cell biology. *Fungal Genet. Biol.* **34**:145–154.

170. Sole, M., J. Cano, J. L. Rodriguez-Tudela, J. Ponton, D. A. Sutton, R. Perrie, J. Gene, V. Rodriguez, and J. Guarro. 2003. Molecular typing of clinical and environmental isolates of *Scedosporium prolificans* by inter-simple-sequence-repeat polymerase chain reaction. *Med. Mycol.* **41**:293–300.

171. Soll, D. R. 2004. Mating-type locus homozygosis, phenotypic switching and mating: a unique sequence of dependencies in *Candida albicans*. *Bioessays* **26**:10–20.

172. Soll, D. R., and C. Pujol. 2003. *Candida albicans* clades. *FEMS Immunol. Med. Microbiol.* **39**:1–7.

173. Steenbergen, J. N., J. D. Nosanchuk, S. D. Malliaris, and A. Casadevall. 2004. Interaction of *Blastomyces dermatitidis*, *Sporothrix schenckii*, and *Histoplasma capsulatum* with *Acanthamoeba castellanii*. *Infect. Immun.* **72**:3478–3488.

174. Steenbergen, J. N., H. A. Shuman, and A. Casadevall. 2001. *Cryptococcus neoformans* interactions with amoebae suggest an explanation for its virulence and intracellular pathogenic strategy in macrophages. *Proc. Natl. Acad. Sci. USA* **98**:15245–15250.

175. Stringer, S. L., K. Hudson, M. A. Blase, P. D. Walzer, M. T. Cushion, and J. R. Stringer. 1989. Sequence from ribosomal RNA of *Pneumocystis carinii* compared to those of four fungi suggests an ascomycetous affinity. *J. Protozool.* **36**:14S–16S.

176. Sugita, T., R. Ikeda, and T. Shinoda. 2001. Diversity among strains of *Cryptococcus neoformans* var. *gattii* as revealed by a sequence analysis of multiple genes and a chemotype analysis of capsular polysaccharide. *Microbiol. Immunol.* **45**:757–768.

177. Sugita, T., A. Nishikawa, R. Ikeda, and T. Shinoda. 1999. Identification of medically relevant *Trichosporon* species based on sequences of internal transcribed spacer regions and construction of a database for *Trichosporon* identification. *J. Clin. Microbiol.* **37**:1985–1993.

178. Sugita, T., M. Tajima, M. Takashima, M. Amaya, M. Saito, R. Tsuboi, and A. Nishikawa. 2004. A new yeast, *Malassezia yamatoensis*, isolated from a patient with seborrheic dermatitis, and its distribution in patients and healthy subjects. *Microbiol. Immunol.* **48**:579–583.

179. Sugita, T., M. Takashima, M. Kodama, R. Tsuboi, and A. Nishikawa. 2003. Description of a new yeast species, *Malassezia japonica*, and its detection in patients with atopic dermatitis and healthy subjects. *J. Clin. Microbiol.* **41**:4695–4699.

180. Sugita, T., M. Takashima, T. Shinoda, H. Suto, T. Unno, R. Tsuboi, H. Ogawa, and A. Nishikawa. 2002. New yeast species, *Malassezia dermatis*, isolated from patients with atopic dermatitis. *J. Clin. Microbiol.* **40**:1363–1367.

181. Sugiyama, M., R. C. Summerbell, and T. Mikawa. 2002. Molecular phylogeny of onygenalean fungi based on small subunit (SSU) and large subunit (LSU) ribosomal DNA sequences. *Stud. Mycol.* **47**:5–23.

182. Sullivan, D. J., G. P. Moran, E. Pinjon, A. Al-Mosaid, C. Stokes, C. Vaughan, and D. C. Coleman. 2004. Comparison of the epidemiology, drug resistance mechanisms, and virulence of *Candida dubliniensis* and *Candida albicans*. *FEMS Yeast Res.* **4**:369–376.

183. Sullivan, D. J., T. J. Westerneng, K. A. Haynes, D. E. Bennett, and D. C. Coleman. 1995. *Candida dubliniensis* sp. nov.: phenotypic and molecular characterization of a novel species associated with oral candidosis in HIV-infected individuals. *Microbiology* **141**:1507–1521.

184. Summerbell, R. C. 2002. What is the evolutionary and taxonomic status of asexual lineages in the dermatophytes? *Stud. Mycol.* **47**:97–101.

185. Supparatpinyo, K., C. Khamwan, V. Baosoung, K. E. Nelson, and T. Sirisanthana. 1994. Disseminated *Penicillium marneffei* infection in southeast Asia. *Lancet* **344**:110–113.

186. Swann, E. C., and J. W. Taylor. 1993. Higher taxa of basidiomycetes: an 18S rRNA gene perspective. *Mycologia* **85**:923–936.

187. Symoens, F., J. P. Bouchara, S. Heinemann, and N. Nolard. 2000. Molecular typing of *Aspergillus terreus* isolates by random amplification of polymorphic DNA. *J. Hosp. Infect.* **44**:273–280.

188. Symoens, F., J. Burnod, B. Lebeau, M. A. Viviani, M. A. Piens, A. M. Tortorano, N. Nolard, F. Chapuis, and R. Grillot. 2002. Hospital-acquired *Aspergillus fumigatus* infection: can molecular typing methods identify an environmental source? *J. Hosp. Infect.* **52**:60–67.

189. Taborda, P. R., V. A. Taborda, and M. R. McGinnis. 1999. *Lacazia loboi* gen. nov., comb. nov., the etiologic agent of lobomycosis. *J. Clin. Microbiol.* **37**:2031–2033.

190. Tavanti, A., N. A. R. Gow, M. C. J. Maiden, F. C. Odds, and D. J. Shaw. 2004. Genetic evidence for recombination in *Candida albicans* based on haplotype analysis. *Fungal Genet. Biol.* **41**:553–562.

191. Taylor, J. W., and M. C. Fisher. 2003. Fungal multilocus sequence typing—it's not just for bacteria. *Curr. Opin. Microbiol.* **6**:351–356.

192. Taylor, J. W., D. M. Geiser, A. Burt, and V. Koufopanou. 1999. The evolutionary biology and population genetics underlying fungal strain typing. *Clin. Microbiol. Rev.* **12**:126–146.

193. Taylor, J. W., D. J. Jacobson, and M. C. Fisher. 1999. The evolution of asexual fungi: reproduction, speciation and classification. *Annu. Rev. Phytopathol.* **37**:197–246.

194. Taylor, J. W., D. J. Jacobson, S. Kroken, T. Kasuga, D. M. Geiser, D. S. Hibbett, and M. C. Fisher. 2000. Phylogenetic species recognition and species concepts in fungi. *Fungal Genet. Biol.* **31**:21–32.

195. Taylor, J. W., J. W. Spatafora, K. O'Donnell, F. Lutzoni, T. James, D. S. Hibbett, D. M. Geiser, T. D. Bruns, and M. Blackwell. 2004. The fungi, p. 171–194. *In* J. Cracraft and M. J. Donoghue (ed.), *Assembling the Tree of Life*. Oxford University Press, Oxford, United Kingdom.

196. Taylor, M. L., C. B. Chavez-Tapia, and M. R. Reyes-Montes. 2000. Molecular typing of *Histoplasma capsulatum* isolated from infected bats, captured in Mexico. *Fungal Genet. Biol.* **30**:207–212.

197. Taylor, M. L., C. B. Chavez-Tapia, R. Vargas-Yanez, G. Rodriguez-Arellanes, G. R. Pena-Sandoval, C. Toriello, A. Perez, and M. R. Reyes-Montes. 1999. Environmental conditions favoring bat infection with *Histoplasma capsulatum* in Mexican shelters. *Am. J. Trop. Med. Hyg.* **61**:914–919.

198. Theelen, B., M. Silvestri, E. Guého, A. van Belkum, and T. Boekhout. 2001. Identification and typing of *Malassezia* yeasts using amplified fragment length

polymorphism (AFLPTm), random amplified polymorphic DNA (RAPD) and denaturing gradient gel electrophoresis (DGGE). *FEMS Yeast Res.* **1**:79–86.

198a. Thomarat, F., C. P. Vivares, and M. Gouy. 2004. Phylogenetic analysis of the complete genome sequence of *Encephalitozoon cuniculi* supports the fungal origin of microsporidia and reveals a high frequency of fast-evolving genes. *J. Mol. Evol.* **59**:780–791.

199. Tietz, H. J., M. Hopp, A. Schmalreck, W. Sterry, and V. Czaika. 2001. *Candida africana* sp nov., a new human pathogen or a variant of *Candida albicans*? *Mycoses* **44**:437–445.

200. Tran-Dinh, N., and D. Carter. 2000. Characterization of microsatellite loci in the aflatoxigenic fungi *Aspergillus flavus* and *Aspergillus parasiticus*. *Mol. Ecol.* **9**: 2170–2172.

201. Tsong, A. E., M. G. Miller, R. M. Raisner, and A. D. Johnson. 2003. Evolution of a combinatorial transcriptional circuit: a case study in yeasts. *Cell* **115**:389–399.

202. Untereiner, W. A., and F. A. Naveau. 1999. Molecular systematics of the Herpotrichiellaceae with an assessment of the phylogenetic positions of *Exophiala dermatitidis* and *Phialophora americana*. *Mycologia* **91**:67–83.

203. Untereiner, W. A., J. A. Scott, F. A. Naveau, L. Sigler, J. Bachewich, and A. Angus. 2004. The Ajellomycetaceae, a new family of vertebrate-associated Onygenales. *Mycologia* **96**:812–821.

204. Vagvolgyi, C., H. Heinrich, K. Acs, and T. Papp. 2004. Genetic variability in the species *Rhizopus stolonifer*, assessed by random amplified polymorphic DNA analysis. *Antonie Leeuwenhoek Int. J. Gen. Mol. Microbiol.* **86**:181–188.

205. Vanittanakom, N., J. Supabandhu, C. Khamwan, J. Praparattanapan, S. Thirach, N. Prasertwitayakij, W. Louthrenoo, S. Chiewchanvit, and N. Tananuvat. 2004. Identification of emerging human-pathogenic *Pythium insidiosum* by serological and molecular assay-based methods. *J. Clin. Microbiol.* **42**:3970–3974.

206. Varga, J., K. Rigo, S. Kocsube, B. Farkas, and K. Pal. 2003. Diversity of polyketide synthase gene sequences in *Aspergillus* species. *Res. Microbiol.* **154**:593–600.

207. Varga, J., B. Toth, E. Kevei, A. Palagyi, and Z. Kozakiewicz. 2000. Analysis of genetic variability within the genus Petromyces. *Antonie Leeuwenhoek Int. J. Gen. Mol. Microbiol.* **77**:83–89.

208. Varga, J., Z. Vida, B. Toth, F. Debets, and Y. Horie. 2000. Phylogenetic analysis of newly described *Neosartorya* species. *Antonie Leeuwenhoek Int. J. Gen. Mol. Microbiol.* **77**:235–239.

209. Vavra, J., and K. Kucera. 1970. *Pneumocystis carinii* Delanoe, its ultrastructure and ultrastructural affinities. *J. Protozool.* **17**:463–483.

210. Vitale, R. G., and G. S. de Hoog. 2002. Molecular diversity, new species and antifungal susceptibilities in the *Exophiala spinifera* clade. *Med. Mycol.* **40**:545–556.

211. Wang, L., K. Yokoyama, M. Miyaji, and K. Nishimura. 1998. The identification and phylogenetic relationship of pathogenic species of *Aspergillus* based on the mitochondrial cytochrome *b* gene. *Med. Mycol.* **36**:153–164.

212. White, T. J., T. Bruns, S. Lee, and J. Taylor. 1990. Amplification and direct sequencing of fungal ribosomal RNA genes for phylogenetics. p. 315–322. *In* M. Innis, D. Gelfand, J. Sninsky, and T. White (ed.), *PCR Protocols: a Guide to Methods and Applications.* Academic Press, Inc., Orlando, Fla.

213. Wieloch, W., M. Sacharczuk, M. I. Bogus, and K. Jaszczak. 2004. A study for minisatellitic markers of *Conidiobolus coronatus*' pathogenicity to *Galleria mellonella* larvae. *J. Invertb. Pathol.* **85**:63–69.

214. Wilson, K. H., W. J. Wilson, J. L. Radosevich, T. Z. DeSantis, V. S. Viswanathan, T. A. Kuczmarski, and G. L. Andersen. 2002. High-density microarray of small-subunit ribosomal DNA probes. *Appl. Environ. Microbiol.* **68**:2535–2541.

215. Wong, S., M. A. Fares, W. Zimmermann, G. Butler, and K. H. Wolfe. 2003. Evidence from comparative genomics for a complete sexual cycle in the 'asexual' pathogenic yeast *Candida glabrata*. *Genome Biol.* **4**:R10.1–R10.9.

216. Xu, J. P. 2004. Genotype-environment interactions of spontaneous mutations for vegetative fitness in the human pathogenic fungus *Cryptococcus neoformans*. *Genetics* **168**:1177–1188.

217. Xu, J. P., G. Z. Luo, R. J. Vilgalys, M. E. Brandt, and T. G. Mitchell. 2002. Multiple origins of hybrid strains of *Cryptococcus neoformans* with serotype AD. *Microbiology* **148**:203–212.

218. Xu, J. P., and T. G. Mitchell. 2003. Comparative gene genealogical analyses of strains of serotype AD identify recombination in populations of serotypes A and D in the human pathogenic yeast *Cryptococcus neoformans*. *Microbiology* **149**:2147–2154.

219. Xu, J. P., R. Vilgalys, and T. G. Mitchell. 2000. Multiple gene genealogies reveal recent dispersion and hybridization in the human pathogenic fungus *Cryptococcus neoformans*. *Mol. Ecol.* **9**:1471–1481.

220. Yan, Z., and J. P. Xu. 2003. Mitochondria are inherited from the MATa parent in crosses of the basidiomycete fungus *Cryptococcus neoformans*. *Genetics* **163**: 1315–1325.

221. Zouhair, R., A. Defontaine, C. Ollivier, B. Cimon, F. Symoens, J. N. Hallet, J. Deunff, and J. P. Bouchara. 2001. Typing of *Scedosporium apiospermum* by multilocus enzyme electrophoresis and random amplification of polymorphic DNA. *J. Med. Microbiol.* **50**:925–932.

Molecular Principles of Fungal Pathogenesis
Edited by Joseph Heitman et al.
©2006 ASM Press, Washington, D.C.

Chapter 9

Establishment of Cell Identity in Pathogenic Fungi

Christina M. Hull

CELL IDENTITY: WHAT IS IT, AND WHY IS IT IMPORTANT?

The establishment of cell identity is one of the most important processes that virtually all cells undertake. All cells, from single-celled microbes to developing neurons, need to follow specific directions to carry out their designated functions (1). For example, in the development of a mammalian embryo, the expression of key genes at specific times is crucial for proper cell positioning and the correct formation of an organism (e.g., neural tube cells need to form a nervous system, and mesoderm develops into multiple organs and tissues). In these cases, every cell contains the same set of genes, and cell identity is established by the expression of different combinations of those genes in response to environmental and other developmental signals. Through differential gene expression, cells can adopt dramatically different fates and contribute to the complexity of an organism with the information contained in a single, homogeneous genome.

Another strategy for establishing cell identity has been adopted by many organisms for the determination of distinct sexes. In these organisms, different cells encode distinct genes. The expression of one or more of these cell-type-specific factors changes the properties and fate of the cell of interest. For example, in humans and mice, the expression of the sex-determining gene *SRY* from the Y chromosome specifies the male sex (49). Thus, XY individuals are male, and XX individuals are female. Organisms from algae to humans have adopted this strategy, underscoring the importance of regulating crucial cell-cell interactions during sexual development and maintaining a sexual cycle (30). The expression of sex-specific factors is also the mechanism commonly used by fungi to maintain distinct sexes, known as mating types (12). Unlike their multicellular eukaryotic relatives that contain entire chromosomes dedicated to housing sex-determining information, mating-type information in fungi is embedded within otherwise homologous chromosomes. This specialized region of the genome is called the mating-type or *MAT* locus, and it encodes the factors essential for distinguishing mating types from one another. Traditionally, the term "cell identity" in fungi has referred to the establishment and maintenance of cell types formed during sexual development. Here we examine how cell identity influences mating-type determination, particularly in fungal pathogens, but we also explore cases where cell identity plays roles outside of mating type, affects cell morphology, and influences pathogenesis.

Specifically, we explore the different strategies by which fungal pathogens have organized their *MAT* loci and put gene-specific regulators to use in maintaining the sexual state. From there, we discuss the other influences of cell identity on pathogen behavior. We begin with a description of cell type determination in the budding yeast *Saccharomyces cerevisiae* and use this as a platform for exploring mechanisms in the human fungal pathogens *Candida albicans* and *Cryptococcus neoformans* as well as the plant fungal pathogen *Ustilago maydis*. We conclude with a short description of the influence of cell identity on the behaviors of several other plant and human fungal pathogens and how cell identity in fungi is evolving to encompass a more diverse array of fungal behaviors.

S. CEREVISIAE: A MODEL FOR CELL IDENTITY DETERMINATION

Cell type determination has been studied most thoroughly in the model eukaryote *S. cerevisiae*. Nearly all of the studies of mating types in fungi have depended on the fundamental principles of

Christina M. Hull • Departments of Biomolecular Chemistry and Medical Microbiology & Immunology, University of Wisconsin—Madison, Madison, WI 53706.

gene regulation discovered by exploring the *S. cerevisiae* sexual cycle. In *S. cerevisiae* two mating types interact with one another to carry out a complete sexual cycle that includes both mating and sporulation. The process is initiated when two haploid cells of opposite mating types (**a** and α) encounter one another and fuse to form the **a**/α diploid cell type. This diploid cell then has properties distinct from either haploid cell that make it competent to undergo meiosis and spore formation. Distinct cell identities are essential for this process: haploid **a** and α cells must recognize each other as potential mating partners, and after mating, the diploid cell must adopt properties distinct from haploid cells to prevent further fusion and allow sexual development (meiosis and sporulation) to occur (15).

The maintenance of these three distinct cell types is controlled by the actions of transcription factors encoded at the *MAT* locus. The *S. cerevisiae MAT* locus is less than 1 kb in size; in **a** cells *MAT***a** encodes the homeodomain protein **a**1, and in α cells *MAT*α encodes the α domain protein α1 and the homeodomain protein α2 (Fig. 1A). In α cells, α1 and α2 act in concert with other proteins and are responsible for activating α-specific genes (αsg) and repressing **a**-specific genes (asg) respectively, thus specifying the α mating type. **a**1 has no known role in haploid **a** cells, establishing **a** as the "default" mating type; however, in the **a**/α diploid cell, **a**1 and α2 interact with one another to form a novel regulatory complex. This heterodimeric transcriptional regulatory complex represses genes important in haploid cells known as haploid-specific genes (hsg), including α1 (Fig. 1B). This regulation allows meiosis and sporulation to occur under nutrient-limiting conditions (15, 24).

The specific genes regulated by each of the *MAT*-encoded transcription factors have been identified through extensive molecular and genetic analyses (15). In particular, the interactions of **a**1 and α2 with one another and their targets have been elucidated. **a**1 and α2 are homeodomain transcription factors that belong to a large class of DNA binding proteins essential for development in all eukaryotes. **a**1-α2 binds to a specific sequence in the promoters of haploid-specific genes (consensus, TGATGTN$_8$ACATGA) (14) and recruits the global repressor complex Tup1-Ssn6, which mediates the repression of these genes (28). Through the direct repression of ~30 genes in the genome, **a**1-α2 specifies the diploid cell type with novel properties from either haploid cell type and thus acts as a crucial regulator of cell type and development (13). This well-studied regulatory circuit is the paradigm on which other transcriptional regulatory circuits have been modeled and has laid a strong

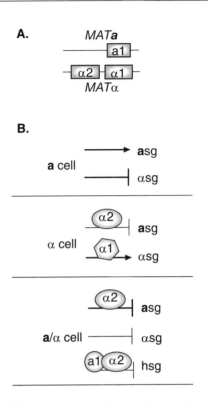

Figure 1. Cell type determination in *S. cerevisiae*. (A) The mating-type (*MAT*) locus of *S. cerevisiae* contains two alleles (<1 kb). *MAT***a** from **a** cells encodes the homeodomain regulator **a**1. *MAT*α from α cells encodes the α-domain protein α1 and the homeodomain protein α2. In diploid cells both *MAT***a** and *MAT*α are present. (B) The transcriptional regulatory circuit of *S. cerevisiae*. In **a** cells the protein **a**1 is produced but has no known effect, **a**-specific genes (asg) are constitutively active, and the α-specific genes (αsg) are not expressed. In α cells, α2 represses **a**-specific genes and α1 activates α-specific genes, thus establishing the α cell type. In **a**/α diploid cells, α2 continues to repress **a**-specific genes, but it also interacts with **a**1 to form a transcriptional regulatory complex that represses haploid-specific genes (hsg), including α1. This repression leads to the specification of the diploid cell type, making further sexual development (i.e., meiosis and sporulation) possible.

foundation for the study of mating-type determination in other fungi.

C. ALBICANS: LESSONS FROM THE MOST COMMON HUMAN FUNGAL PATHOGEN

Nowhere have the pivotal discoveries made about mating type in *S. cerevisiae* been more useful in the fungal world than in the study of the yeast *C. albicans*. *C. albicans* is the most common human fungal pathogen, causing both mucosal and systemic infections, particularly in immunocompromised individuals (43). *C. albicans* differs from many of its fungal counterparts because it is an obligate diploid. No stable haploid state has been identified for this organism, and no naturally occurring sexual cycle has been observed (7). Curiously,

A.

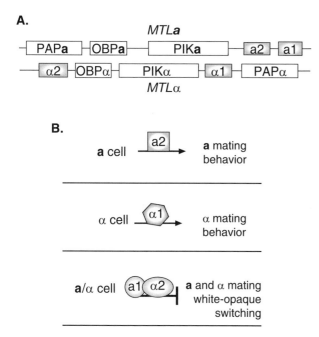

Figure 2. Cell type determination in *C. albicans.* (A) The mating-type-like *(MTL)* locus of *C. albicans* (~9 kb) contains transcriptional regulators and three other gene types. *MTLa* encodes the homeodomain protein **a**1 and the HMG box protein **a**2. *MTLα* encodes the α-domain protein α1 and the homeodomain protein α2. Both *MTLa* and *MTLα* contain diverged alleles of a poly(A) polymerase *(PAP)*, an oxysterol binding protein *(OBP)*, and a phosphotidylinositol kinase *(PIK)*. The roles of *PAP*, *OBP*, and *PIK* in cell type determination are not known. (B) The transcriptional circuit of *C. albicans.* Cells containing only **a** information (**a**/**a** or **a**/Δ at the *MTL* locus) mate as **a** cells because the predicted transcription factor **a**2 activates genes required for **a**-type mating. Cells containing only α information (α/α or Δ/α at the *MTL* locus) mate as α cells because the predicted transcription factor α1 activates genes required for α-type mating. Cells containing both **a** and α information (**a**/α at the *MTL* locus) do not mate with other cells because **a**1 and α2 work in concert to repress genes required for mating and white-opaque switching. Cells that cannot switch from white to opaque cannot mate, and this switch is repressed by **a**1-α2.

however, *C. albicans* contains a mating-type-like *(MTL)* locus that is nearly 9 kb in size and encodes homologs of the cell type regulators found in *S. cerevisiae* (along with several genes not seen before in *MAT* loci) (Fig. 2A) (20). *MTLa* contains the factors **a**1 and **a**2 (an HMG box-containing DNA binding protein); *MTLα* contains α1 and α2 (20, 50). Like the proteins in *S. cerevisiae,* these regulators confer specific cell identities in different contexts.

For example, by altering *C. albicans* cells to contain only **a** or α information (creating strains hemizygous or homozygous at *MTL*), this diploid can be induced to behave as though it were haploid. Under the proper conditions, cells containing only *MTLa* will fuse with cells containing only *MTLα* to create a

tetraploid. This mating reaction has been observed to occur both in vivo and in vitro and is dependent on the configurations of the transcriptional regulators (21, 39). Through careful analysis of the transcriptional profiles of regulator deletion strains, the roles of **a**1, **a**2, α1, and α2 have been delineated. Interestingly, the regulatory circuit differs significantly from that of *S. cerevisiae* in several respects (50). First, *C. albicans* encodes the HMG box protein **a**2. This protein is conserved in the *MAT* loci of other ascomycetes but appears to have been lost in *S. cerevisiae* (6). While the **a**1 protein has no apparent regulatory role on its own (as in *S. cerevisiae*), **a**2 is responsible for establishing the "**a**" cell type in cells that contain only *MTLa* information, allowing them to exhibit **a** mating behavior. In cells containing *MTLα*, α1 establishes the "α" cell type, allowing them to mate as α cells (similar to its role in *S. cerevisiae*), but the α2 protein has no regulatory role in α cells. In cells containing both *MTLa* and *MTLα* (most isolates of *C. albicans*), **a**1 and α2 act as they do in *S. cerevisiae* diploid cells. That is, **a**1 and α2 act in concert to regulate transcription and establish the **a**/α state (Fig. 2B) (50). In these ways, the transcriptional regulators encoded by the *MTL* locus establish cell type and facilitate mating; however, unlike in most fungi, distinct mating types in *C. albicans* are not the only requirements for efficient mating. The cells must also be of the proper morphological type.

Over 15 years ago, it was discovered that some isolates of *C. albicans* undergo a reversible switch from cells that form domed, white colonies to those that form flat, opaque colonies (46, 47). These changes in colony morphology are the result of changes in cell morphology; white cells are round, budding yeast, and opaque cells are elongated and stippled. These cell types also have different pathogenic properties. While white cells are relatively efficient in colonization in a murine tail vein injection model, opaque cells are not. Opaque cells do, however, cause infection in a murine model of cutaneous infection (31). A surprising finding was that the switch from white to opaque is required for efficient mating of *C. albicans* **a** and α cells. Mating was found to be approximately 10^6 times more efficient between opaque cells of opposite mating types than between white cells of opposite mating types (42). Even more fascinating was the finding that this change from white to opaque is controlled by the actions of transcriptional regulators at the *MTL* locus. In a tour de force of microarray analysis, Tsong et al. determined the transcriptional profiles for cells with targeted deletions of the *MTL* regulator components in every possible combination in both the white and opaque phases (where applicable) (50). This extensive analysis allowed the identification of genes involved in mating, those involved in white-opaque switching, and those involved in both. It was found that **a**1-α2 in *C. albicans* represses only seven genes directly, many of which are homologs of the haploid-specific genes in *S. cerevisiae*. What was striking is that in cells lacking either **a**1 or α2 (i.e., capable of switching from white to opaque), the expression patterns of several hundred genes were altered, and the expression of **a**-specific and α-specific genes occurs only in the opaque phase. The **a**1-α2 heterodimer regulates white-opaque switching, and this switching is essential for establishing mating type (Fig. 2B) (50). The system to establish mating types in other fungi has been altered in *C. albicans* to make the specification of mating type dependent on morphological type.

Why an organism would marry the regulation of these distinct pathways is not at all clear. However, it has been suggested that under some conditions (in the host organs or bloodstream) there may be a disadvantage to undergoing mating (23). Perhaps it is only advantageous to mate outside of the host (on the skin), so that linking switching and mating provides a safeguard against mating behaviors that could be detrimental in hostile tissue sites and safer in more neutral arenas. In fact, mating between opaque-phase cells has been observed to be particularly efficient on mouse skin, suggesting that mating might be a natural part of the *C. albicans* life cycle (34). Although the details of the *C. albicans* lifestyle remain to be determined, evidence is mounting that cell identity (in both mating and switching) is playing an important role in the life cycle and pathogenic properties of the most common of the human fungal pathogens.

U. MAYDIS: A MODEL PLANT FUNGAL PATHOGEN

A case in which there is a clear and strong relationship between cell identity and pathogenesis is in the corn smut and model plant pathogen *U. maydis*. *U. maydis* is a basidiomycete fungus with a tetrapolar mating system; there are four mating types that result from the presence of two distinct *MAT* loci. As is typical of many basidiomycete fungi, including the mushrooms *Coprinus cinereus* and *Schizophyllum commune*, *U. maydis* contains one *MAT* locus (*b*), which encodes homeodomain DNA binding proteins (bE and bW), and a second, unlinked locus (*a*), which encodes mating-type-specific pheromones and pheromone receptors (Fig. 3A) (9, 29). In the haploid state, cell type is distinguished by the expression of different pheromones and pheromone receptors

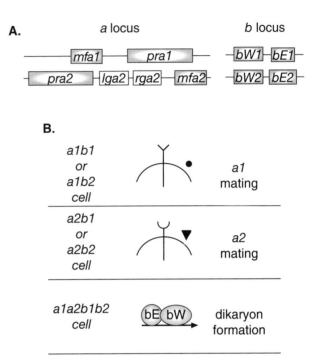

Figure 3. Cell type determination in *U. maydis*. (A) There are two independent, unlinked mating-type loci in *U. maydis*. One locus (*a*) contains the cell-type-specific pheromone gene *mfa1* and pheromone receptor gene *pra1* in one allele (*a1*, ~4.6 kb), and the other allele (*a2*, ~8.5 kb) contains the pheromone gene *mfa2* and pheromone receptor gene *pra2*. *a2* also contains *lga2* and *rga2*, whose products show no similarity to other proteins and play no apparent role in mating but do seem to affect pathogenesis by controlling mitochondrial function (4). The second locus (*b*, ~3.6 kb) encodes the homeodomain proteins bE and bW, of which there are 25 different forms, resulting in many different mating types in a tetrapolar mating-type system. (B) Regulation of cell identity in *U. maydis*. Half circles represent haploid cell surfaces. Haploid cell types are specified by the expression of *mfa1* (●) and *pra1* (forked receptor on cell surface) or *mfa2* (▼) and *pra2* (round receptor on cell surface). Soluble pheromones are sensed by surface receptors, activating cell fusion when two cells of the opposite mating type encounter one another. After cell fusion, the components of the *b* locus (bE and bW) interact with one another in specific combinations (e.g., bW1-bE2) to indicate the dikaryotic state and make the cells competent for further sexual development in corn plants.

(*mfa1/pra1* and *mfa2/pra2*). This mechanism readily specifies which cells are compatible for fusion (2). When haploid cells sense one another, they initiate mating and the upregulation of gene expression at the *b* locus in preparation for fusion (52). Once haploid cells fuse with one another, the components of the *b* locus (bE-bW) interact to direct the formation of a new cell type, the dikaryon. In this cell type, the haploid nuclei do not fuse with one another but are instead maintained as individual nuclei until later in development. This dikaryotic cell is specified when a heterodimeric complex of the homeodomain proteins bE and bW (akin to **a**1-α2) interacts to create a novel transcriptional regulatory complex (Fig. 3B). Because every cell contains both bE and bW, development into a dikaryon is limited to cells that contain a bE-bW complex composed of proteins from distinct alleles of bE and bW (e.g., bE1-bW2 or bE2-bW1) (3). There are 25 documented alleles of the genes encoding the b proteins, presenting the opportunity for hundreds of allele combinations (9). For the proteins to interact, they must have compatible interaction domains in the amino terminus of each protein (independent of the homeodomain DNA binding region) (26). Once a compatible regulatory pair is formed, the bE-bW complex binds specific promoter sequences and activates transcription. By using differential expression techniques and recent microarray analysis, over 200 genes have been identified as b regulated; however, thus far, only 3 have been identified as direct targets (25). The promoters of these directly regulated genes (dik6, polX, and lga2) contain a specific binding site dubbed the bbs (for "b binding sequence") to which bE-bW binds (from lga2: TTCAT**GATG**AGAAGTGTG**ACAG**ACT-GTGC) (5, 44). It appears that binding by bE-bW triggers a regulatory cascade in which a limited number of genes are regulated directly and the remaining genes are controlled indirectly by downstream factors. This regulation not only establishes the dikaryon but also leads to subsequent sexual development in plants. The dikaryon can develop only on the corn plant, and it is this part of the life cycle that begins the process of plant invasion and pathogenesis. Thus, the bE-bW complex can be viewed as a necessary control of the switch from saprophytic to biotrophic growth and a trigger of the pathogenic phase of the U. maydis life cycle (25). It remains to be determined how the signals from bE-bW and the corn plant are transmitted and integrated and which target genes represent factors directly involved in pathogenesis. However, recent advances in the molecular biology of U. maydis, such as a complete genome sequence and whole-genome microarrays (25), will rapidly facilitate revelations about this complex and interesting regulatory circuit and elucidate the relationship between cell type determination and pathogenesis.

C. NEOFORMANS: A MODEL HUMAN FUNGAL PATHOGEN

Another fungal pathogen in which cell identity determination has come to the fore is in the basidiomycete fungus C. neoformans. C. neoformans is a human fungal pathogen that causes meningoencephalitis, primarily in immunocompromised individuals (8). It is also a haploid, budding yeast with two stable mating types, a and α, that mate with one another and undergo sexual development. The factors that control this process are encoded at the MAT locus, but in this case the locus architecture is quite different from that of other fungi. Even though C. neoformans is a basidiomycete, it contains only a single MAT locus, which encodes components homologous to those found in the unlinked loci of other basidiomycetes (i.e., homeodomain proteins, pheromones, and pheromone receptors). The C. neoformans MAT locus is much larger than most (over 100 kb) and also contains many genes never seen before in MAT loci (11, 27, 35). Alleles of each of the genes in the locus are present in both MATa and MATα with two exceptions (Fig. 4A). The sex inducer genes SXI1α and SXI2a are distinct from one another and, unlike the other genes in the locus, do not represent divergent alleles of a common ancestor gene (SXI1α is encoded by MATα only, and SXI2a is encoded by MATa only). The SXI genes both encode predicted homeodomain transcription factors similar to a1-α2 and bE-bW (16).

Sexual development in C. neoformans takes place when haploid a and α cells encounter each other under nutrient-limiting conditions and fuse (19, 36). In this case, as in U. maydis, the transcriptional regulators appear to exert their effects after cell fusion, suggesting that haploid cell identity is controlled by another mechanism. The most likely possibility is that mating-type-specific pheromones and pheromone receptors specify the different haploid mating types; MFa pheromone is produced by a cells, and MFα is produced by α cells, and cell-type-specific receptors sense the pheromone of a mating partner (45). Unlike in U. maydis, however, where MAT encodes distinct pheromones and receptors, the predicted pheromones (MFa and MFα) and pheromone receptors (STE3a and STE3α) in C. neoformans are only moderately diverged from one another, similar to the other 15 genes in MAT that have alleles in both MATa and MATα. There are two obvious possibilities for specifying haploid cell identity: either the pheromone and pheromone receptor alleles in each mating type have diverged from one another sufficiently to confer pheromone-receptor binding specificity, or there are other factors or mechanisms at play in determining haploid cell types. The mechanism by which haploid cell identity is established in C. neoformans remains to be elucidated. Once haploid cells have fused with one another, the homeodomain proteins Sxi1α and Sxi2a are transcriptionally upregulated and initiate formation of the dikaryon (16).

It is not yet known exactly how Sxi1α and Sxi2a act at the molecular level, but it has been shown in a two-hybrid assay that they interact with one another and that the ectopic expression of either protein in a

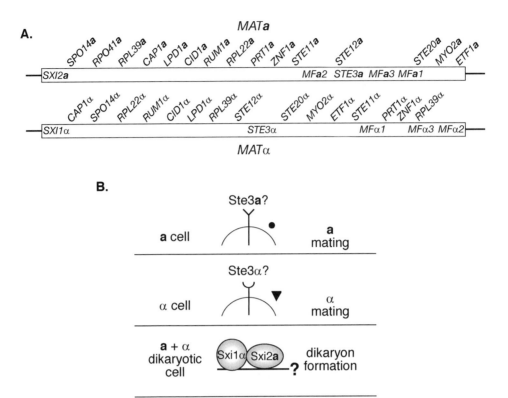

Figure 4. Cell type determination in *C. neoformans*. (A) The mating-type (*MAT*) locus of *C. neoformans* is unusually large (>100 kb) and contains many genes whose functions in cell type determination are unknown. A schematic of the *MAT* locus for *C. neoformans* var. *neoformans* (serotype D) is shown. *MATa* encodes three copies of the MFa pheromone, the predicted pheromone receptor Ste3a, and the homeodomain protein Sxi2a. *MATα* encodes three copies of the MFα pheromone, the predicted pheromone receptor Ste3α, and the homeodomain protein Sxi1α. Fifteen additional genes have related but diverged alleles in both *MATa* and *MATα*. (B) Regulation of cell identity in *C. neoformans*. The determinants of haploid cell identity are unknown; however, a likely mechanism is the expression of cell-type-specific pheromones and receptors. The expression of *MFa* and the pheromone receptor *STE3a* could specify the **a** mating type, and the expression of *MFα* and the pheromone receptor *STE3α* could specify the α mating type. Soluble pheromones would be sensed by surface receptors, activating cell fusion when two cells of the opposite mating type encounter one another. After cell fusion, the homeodomain proteins Sxi1α and Sxi2a are predicted to interact with one another and regulate transcription through an unknown mechanism to establish the dikaryotic state. The newly designated dikaryon is then competent to undergo extended filamentous growth, meiosis, and sporulation.

haploid of the opposite mating type initiates diploid sexual development (16, 18). That is, expression of *SXI1α* in **a** cells or expression of *SXI2a* in α cells results in the induction of complete sexual development including meiosis and sporulation in those haploid cells. In all cases, both Sxi1α and Sxi2a must be present for sexual development to occur. These findings lead to a model in which Sxi1α and Sxi2a interact with one another and bind DNA as a complex to regulate genes responsible for controlling the dikaryotic state and subsequent sexual development (as has been seen in other fungi) (Fig. 4B) (16). Although it remains to be determined how Sxi1α and Sxi2a exert their effects at the molecular level, the recent sequencing of several *C. neoformans* genomes and the development of whole-genome microarrays will rapidly facilitate the elucidation of the molecular pathways that are required for establishing both

haploid and dikaryon cell identities crucial for the maintenance of the sexual cycle (38).

Another way in which cell identity is important in *C. neoformans* is in its ecology. There has been a long-standing observation that α strains of *C. neoformans* vastly outnumber **a** strains in both clinical and environmental isolates (32). There is no clear explanation for why natural populations would be skewed toward a single mating type. In addition, in some backgrounds, the α mating type is more virulent than the **a** mating type. In a murine model of infection, a congenic *C. neoformans* strain pair gave different virulence outcomes depending on mating type (33, 35). These experiments spurred the cloning of the *MAT* locus in an effort to understand how cell identity determinants could be affecting virulence. Although there is an association between the α mating type and virulence, the connection is not yet clear.

As it stands, virulence and/or fitness is somehow linked to mating type, suggesting that the factors that control cell identity (or genes that are linked to those factors) play a role in pathogenesis. It is interesting, however, that Sxi1α does not appear to play a role in virulence (17), underscoring its role in sexual development (rather than haploid reproduction) and indicating that other α-specific factors affecting virulence remain to be identified. Understanding the determination of α cell identity in *C. neoformans* promises to reveal at least some of the important properties of *C. neoformans* associated with virulence.

An unusual finding regarding *C. neoformans* and cell identity is that α cells have the capacity to essentially ignore the constraints of cell identity and engage in sexual development in the absence of an **a** mating partner. This process, called monokaryotic fruiting, occurs under extreme nutrient limitation and desiccation and results in the formation of filaments, basidia, and spores that resemble sexual structures (53). It has been discovered that fruiting involves the fusion of two haploid α cells or nuclei in a manner that is promoted by mating factors (37). The fused α/α diploid cells then initiate a filamentation and sporulation pathway similar to that of sexual development. Furthermore, through restriction fragment length polymorphism analysis using a recently developed meiotic map (37, 41), it was shown that spores formed during fruiting are recombinant and appear to be the products of meiosis. Thus, α cells have the ability to carry out a sexual development process in the absence of a mating partner of opposite mating type. In addition, the cell-type-specific factor Sxi1α, required for classical sexual development, is not necessary for monokaryotic fruiting (18). It is unclear what roles **a**-α sexual development and α-α monokaryotic fruiting play in the natural life cycle of the organism, and it remains to be determined how haploid cell identity is controlled and when it is important. What is clear is that *C. neoformans* has the potential to develop in the absence of an opposite mating type and that this ability may explain the prevalence of α strains in the environment and suggest a mechanism by which α strains can have increased fitness and survival rates over their **a** counterparts in the environment.

CELL IDENTITY IN OTHER FUNGAL PATHOGENS

In addition to the budding yeasts and basidiomycetes, the *MAT* loci of many filamentous ascomycetes, including *Neurospora crassa* and numerous plant pathogens such as *Cochliobolus heterostrophus* (corn blight), *Cryphonectria parasitica* (chestnut blight), *Gibberella zeae* (wheat blight), and *Magnaporthe grisea* (rice blast), have been studied (10). The *MAT* locus of the corn blight fungus *C. heterostrophus* is representative of many of the pathogenic ascomycetes in which one mating type (*MAT1-1*) encodes an α domain protein similar to the *S. cerevisiae* α1 and the other mating type (*MAT1-2*) encodes an HMG box protein (51). Although *MAT* architecture has been identified in numerous filamentous ascomycetes, there are no known direct targets of the *MAT*-encoded factors, and the mechanisms by which they act remain largely unknown (48). How cell identity is established in the pathogenic filamentous ascomycetes remains an area open to investigation.

The mechanisms of mating-type determination in most of the human fungal pathogens are largely unknown. Mating types have been identified in some, but there is virtually no information about how mating-type identity is established. Where there is more information, however, is on other types of cell identity. As we saw with *C. albicans*, there are types of cell identity in fungi besides mating type that influence life cycle and pathogenesis. This is highlighted in the infectious cycles of dimorphic fungi such as *Histoplasma capsulatum*, *Blastomyces dermatitidis*, and *Coccidioides immitis*. In these fungi, there is a dimorphic switch in growth pattern and morphology that is dependent on environmental signals, and the disease process in this case is highly dependent on cell type (40). This type of regulation is not thought to be dependent on cell-type-specific factors and is more similar to the developmental processes of multicellular organisms. In response to environmental cues, cell type is altered, and this change is essential for pathogenesis.

This dimorphic switch has been characterized in the primary fungal pathogen *H. capsulatum* (54). There is a specific switch from a filamentous, infectious form in the environment to a budding-yeast form in the host. The filamentous fragments (or spores) are inhaled, and in response to the host environment, the cells undergo a switch to bud form growth. The ability to make this switch is essential for disease progression, but how these cell types are established and maintained is largely unknown. Hypha-specific and bud-specific genes have been identified (22), and understanding the regulation of these morphology-specific genes will reveal factors essential for the control of the cell identities and the developmental and/or pathogenic pathways in which they participate.

SUMMARY

Establishing and maintaining distinct cell types is an essential part of a cell's life, and there is no exception for fungi. Diverse fungal pathogens from *C. albicans* to *C. neoformans* are under pressure to devise strategies to maintain their sexual cycles, but they must also respond rapidly to changes in their surroundings and protect themselves against hostile host environments. As we continue to explore the basis of cell identity in the context of mating type and sexual development, we are finding links to other kinds of cell identities and the varied responses that accompany them. Inevitably, we will continue to discover new strategies undertaken by the pathogenic fungi to develop and integrate multiple cell identities in the effort to survive.

REFERENCES

1. Alberts, B., A. Johnson, J. Lewis, M. Raff, K. Roberts, and P. Walter. 2002. *Molecular Biology of the Cell*. Garland Science, Oxford, United Kingdom.

2. Banuett, F. 1995. Genetics of *Ustilago maydis*, a fungal pathogen that induces tumors in maize. *Annu. Rev. Genet.* 29:179–208.

3. Banuett, F., and I. Herskowitz. 1994. Morphological transitions in the life cycle of *Ustilago maydis* and their genetic control by the *a* and *b* loci. *Exper. Mycol.* 18:247–266.

4. Bortfeld, M., K. Auffarth, R. Kahmann, and C. W. Basse. 2004. The *Ustilago maydis* a2 mating-type locus genes *lga2* and *rga2* compromise pathogenicity in the absence of the mitochondrial p32 family protein Mrb1. *Plant Cell* 16:2233–2248.

5. Brachmann, A., G. Weinzierl, J. Kamper, and R. Kahmann. 2001. Identification of genes in the bW/bE regulatory cascade in *Ustilago maydis*. *Mol. Microbiol.* 42:1047–1063.

6. Butler, G., C. Kenny, A. Fagan, C. Kurischko, C. Gaillardin, and K. H. Wolfe. 2004. Evolution of the *MAT* locus and its HO endonuclease in yeast species. *Proc. Natl. Acad. Sci. USA* 101:1632–1637.

7. Calderone, R. (ed.). 2001. *Candida and Candidiasis*. ASM Press, Washington, D.C.

8. Casadevall, A., and J. R. Perfect. 1998. *Cryptococcus neoformans*. ASM Press, Washington, D.C.

9. Casselton, L. A., and N. S. Olesnicky. 1998. Molecular genetics of mating recognition in basidiomycete fungi. *Microbiol. Mol. Biol. Rev.* 62:55–70.

10. Coppin, E., R. Debuchy, S. Arnaise, and M. Picard. 1997. Mating types and sexual development in filamentous ascomycetes. *Microbiol. Mol. Biol. Rev.* 61:411–428.

11. Fraser, J. A., S. Diezmann, R. L. Subaran, A. Allen, K. B. Lengeler, F. S. Dietrich, and J. Heitman. 2004. Convergent evolution of chromosomal sex-determining regions in the animal and fungal kingdoms. *PLoS Biol.* 2:e384.

12. Fraser, J. A., and J. Heitman. 2003. Fungal mating-type loci. *Curr. Biol.* 13:R792–R795.

13. Galitski, T., A. J. Saldanha, C. A. Styles, E. S. Lander, and G. R. Fink. 1999. Ploidy regulation of gene expression. *Science* 285:251–254.

14. Goutte, C., and A. D. Johnson. 1994. Recognition of a DNA operator by a dimer composed of two different homeodomain proteins. *EMBO J.* 13:1434–1442.

15. Herskowitz, I., J. Rine, and J. Strathern. 1992. Mating-type determination and mating-type interconversion in *Saccharomyces cerevisiae*, p. 583–656. *In* E. W. Jones, J. R. Pringle, and J. R. Broach (ed.), *The Molecular and Cellular Biology of the Yeast* Saccharomyces, vol. 2. *Gene Expression*. Cold Spring Harbor Laboratory Press, Cold Spring Harbor, N.Y.

16. Hull, C. M., M.-J. Boily, and J. Heitman. 2004. The cell type-specific homeodomain proteins Sxi1α and Sxi2a coordinately regulate sexual development in *Cryptococcus neoformans*. *Eukaryot. Cell.* 4:526–535.

17. Hull, C. M., G. M. Cox, and J. Heitman. 2004. The α-specific cell identity factor Sxi1α is not required for virulence of *Cryptococcus neoformans*. *Infect. Immun.* 72:3643–3645.

18. Hull, C. M., R. C. Davidson, and J. Heitman. 2002. Cell identity and sexual development in *Cryptococcus neoformans* are controlled by the mating-type-specific homeodomain protein Sxi1α. *Genes Dev.* 16:3046–3060.

19. Hull, C. M., and J. Heitman. 2002. Genetics of *Cryptococcus neoformans*. *Annu. Rev. Genet.* 36:557–615.

20. Hull, C. M., and A. D. Johnson. 1999. Identification of a mating type-like locus in the asexual pathogenic yeast *Candida albicans*. *Science* 285:1271–1275.

21. Hull, C. M., R. M. Raisner, and A. D. Johnson. 2000. Evidence for mating of the "asexual" yeast *Candida albicans* in a mammalian host. *Science* 289:307–310.

22. Hwang, L., D. Hocking-Murray, A. K. Bahrami, M. Andersson, J. Rine, and A. Sil. 2003. Identifying phase-specific genes in the fungal pathogen *Histoplasma capsulatum* using a genomic shotgun microarray. *Mol. Biol. Cell.* 14:2314–2326.

23. Johnson, A. 2003. The biology of mating in *Candida albicans*. *Nat. Rev. Microbiol.* 1:106–116.

24. Johnson, A. D. 1995. Molecular mechanisms of cell-type determination in budding yeast. *Curr. Opin. Genet. Dev.* 5:552–558.

25. Kahmann, R., and J. Kamper. 2004. *Ustilago maydis*: how its biology relates to pathogenic development. *New Phytol.* 164:31–42.

26. Kamper, J., M. Reichmann, T. Romeis, M. Bolker, and R. Kahmann. 1995. Multiallelic recognition: nonself-dependent dimerization of the bE and bW homeodomain proteins in *Ustilago maydis*. *Cell* 81:73–83.

27. Karos, M., Y. C. Chang, C. M. McClelland, D. L. Clarke, J. Fu, B. L. Wickes, and K. J. Kwon-Chung. 2000. Mapping of the *Cryptococcus neoformans MATα* locus: presence of mating type-specific mitogen-activated protein kinase cascade homologs. *J. Bacteriol.* 182:6222–6227.

28. Keleher, C. A., M. J. Redd, J. Schultz, M. Carlson, and A. D. Johnson. 1992. Ssn6-Tup1 is a general repressor of transcription in yeast. *Cell* 68:709–719.

29. Kronstad, J. W., and C. Staben. 1997. Mating type in filamentous fungi. *Annu. Rev. Genet.* 31:245–276.

30. Kurvari, V., N. V. Grishin, and W. J. Snell. 1998. A gamete-specific, sex-limited homeodomain protein in *Chlamydomonas*. *J. Cell Biol.* 143:1971–1980.

31. Kvaal, C., S. A. Lachke, T. Srikantha, K. Daniels, J. McCoy, and D. R. Soll. 1999. Misexpression of the opaque-phase-specific gene *PEP1* (*SAP1*) in the white phase of *Candida albicans* confers increased virulence in a mouse model of cutaneous infection. *Infect. Immun.* 67:6652–6662.

32. Kwon-Chung, K. J., and J. E. Bennett. 1978. Distribution of α and a mating types of *Cryptococcus neoformans* among natural and clinical isolates. *Am. J. Epidemiol.* 108:337–340.

33. **Kwon-Chung, K. J., J. C. Edman, and B. L. Wickes.** 1992. Genetic association of mating types and virulence in *Cryptococcus neoformans. Infect. Immun.* **60:**602–605.

34. **Lachke, S. A., S. R. Lockhart, K. J. Daniels, and D. R. Soll.** 2003. Skin facilitates *Candida albicans* mating. *Infect. Immun.* **71:**4970–4976.

35. **Lengeler, K. B., D. S. Fox, J. A. Fraser, A. Allen, K. Forrester, F. S. Dietrich, and J. Heitman.** 2002. Mating-type locus of *Cryptococcus neoformans:* a step in the evolution of sex chromosomes. *Eukaryot. Cell* **1:**704–718.

36. **Lengeler, K. B., and J. Heitman.** 2002. *Cryptococcus neoformans* as a model fungal pathogen, p. 513–557. *In* H. D. Osiewacz (ed.), *Molecular Biology of Fungal Development.* Marcel Dekker, Inc., New York, N.Y.

37. **Lin, X., C. M. Hull, and J. Heitman.** 2005. Sexual reproduction between partners of the same mating type in *Cryptococcus neoformans. Nature* **434:**1017–1021.

38. **Loftus, B. J., E. Fung, P. Roncaglia, D. Rowley, P. Amedeo, et al.** 2005. The genome of the basidiomycetous yeast and human pathogen *Cryptococcus neoformans. Science* **307:**1321–1324.

39. **Magee, B. B., and P. T. Magee.** 2000. Induction of mating in *Candida albicans* by construction of *MTL*a and *MTL*α strains. *Science* **289:**310–313.

40. **Maresca, B., and G. S. Kobayashi.** 2000. Dimorphism in *Histoplasma capsulatum* and *Blastomyces dermatitidis,* p. 201–206. *In* J. F. Ernst and A. Schmidt (ed.), *Dimorphism in Human Pathogenic and Apathogenic Yeasts.* Karger, Basel, Switzerland.

41. **Marra, R. E., J. C. Huang, E. Fung, K. Nielsen, J. Heitman, R. Vilgalys, and T. G. Mitchell.** 2004. A genetic linkage map of *Cryptococcus neoformans* variety *neoformans* serotype D (*Filobasidiella neoformans*). *Genetics* **167:**619–631.

42. **Miller, M. G., and A. D. Johnson.** 2002. White-opaque switching in *Candida albicans* is controlled by mating-type locus homeodomain proteins and allows efficient mating. *Cell* **110:**293–302.

43. **Odds, F. C.** 1988. *Candida and Candidosis: a Review and Bibliography,* 2nd ed. Bailliere Tindall, London, United Kingdom.

44. **Romeis, T., A. Brachmann, R. Kahmann, and J. Kamper.** 2000. Identification of a target gene for the bE-bW homeodomain protein complex in *Ustilago maydis. Mol. Microbiol.* **37:**54–66.

45. **Shen, W. C., R. C. Davidson, G. M. Cox, and J. Heitman.** 2002. Pheromones stimulate mating and differentiation via paracrine and autocrine signaling in *Cryptococcus neoformans. Eukaryot. Cell* **1:**366–377.

46. **Slutsky, B., M. Staebell, J. Anderson, L. Risen, M. Pfaller, and D. R. Soll.** 1987. "White-opaque transition": a second high-frequency switching system in *Candida albicans. J. Bacteriol.* **169:**189–197.

47. **Soll, D. R., S. R. Lockhart, and R. Zhao.** 2003. Relationship between switching and mating in *Candida albicans. Eukaryot. Cell* **2:**390–397.

48. **Souza, C. A., C. C. Silva, and A. V. Ferreira.** 2003. Sex in fungi: lessons of gene regulation. *Genet. Mol. Res.* **2:**136–147.

49. **Tilmann, C., and B. Capel.** 2002. Cellular and molecular pathways regulating mammalian sex determination. *Recent Prog. Horm. Res.* **57:**1–18.

50. **Tsong, A. E., M. G. Miller, R. M. Raisner, and A. D. Johnson.** 2003. Evolution of a combinatorial transcriptional circuit: a case study in yeasts. *Cell* **115:**389–399.

51. **Turgeon, B. G., H. Bohlmann, L. M. Ciuffetti, S. K. Christiansen, G. Yang, W. Schafer, and O. C. Yoder.** 1993. Cloning and analysis of the mating type genes from *Cochliobolus heterostrophus. Mol. Gen. Genet.* **238:**270–284.

52. **Urban, M., R. Kahmann, and M. Bolker.** 1996. Identification of the pheromone response element in *Ustilago maydis. Mol. Gen. Genet.* **251:**31–37.

53. **Wickes, B. L., M. E. Mayorga, U. Edman, and J. C. Edman.** 1996. Dimorphism and haploid fruiting in *Cryptococcus neoformans*: association with the α-mating type. *Proc. Natl. Acad. Sci. USA.* **93:**7327–7331.

54. **Woods, J. P.** 2002. *Histoplasma capsulatum* molecular genetics, pathogenesis, and responsiveness to its environment. *Fungal. Genet. Biol.* **35:**81–97.

Molecular Principles of Fungal Pathogenesis
Edited by Joseph Heitman et al.
©2006 ASM Press, Washington, D.C.

Chapter 10

Signal Transduction in the Interactions of Fungal Pathogens and Mammalian Hosts

MALCOLM WHITEWAY AND CATHERINE BACHEWICH

The ability of fungal pathogens to interact with their hosts depends on the capacity of the pathogen to recognize the external environment and respond appropriately to the signals generated by this environment. The recognition of these environmental signals involves such components as receptor proteins, signal transduction networks, and intracellular changes in cellular behaviour. Because the ability to properly respond to the environment is a critical cellular function, the pathways and responses regulating host-pathogen interactions often show conservation with components of networks controlling other cellular behaviors. This chapter is an attempt to highlight both the generalities and the unique features of signaling pathways controlling interactions between fungal pathogens and mammalian hosts.

RESPONSE OF THE PATHOGEN TO THE HOST

The signaling pathways that contribute to virulence in fungal pathogens of humans have been identified in large part through in vitro studies, where the pathogens have been tested under various culture conditions that mimic the host environment. Environmental signals such as temperature, carbon source, pH, oxygen content, and mechanical and physical cues and elements, including soluble factors, activate a variety of signaling pathways. These in turn regulate several aspects of fungal virulence, including changes in morphology, cell surface composition, and protein secretion (75). Several of the regulatory networks identified also function in vivo, when the pathogen is in contact with host cells, thus supporting the validity of the in vitro investigations. Of the common fungal pathogens of humans, *Candida albicans*, *Cryptococcus neoformans*, and *Aspergillus fumigatus*

have been the most extensively investigated. Although these pathogens have a common mammalian host, they differ in specific sites of infection and modes of virulence. Despite these variations, common signaling pathways, including cyclic AMP (cAMP), mitogen-activated protein kinase (MAPK), Ca^{2+}/calcineurin, pH, and two-component signaling networks are utilized by the pathogens in different ways to respond to the host and regulate pathogenesis. Although many other signaling factors contribute to virulence in the fungi, this chapter focuses on the major common pathways described above, highlighting both shared features and variations on themes among the three major human pathogens.

Overview of Life Cycle, Host Environment, and Virulence Traits of *C. albicans*, *C. neoformans*, and *A. fumigatus*

C. albicans is one of the most prevalent fungal pathogens in humans and ranks as the fourth most common microbe involved in lethal septicemia in the United States (48). *C. albicans* is a commensal in humans with healthy immune systems and occupies mucosal surfaces of the mouth, gut, and vagina. However, under immunocompromised conditions, it can proliferate and cause mucosal and fatal systemic infections. *C. albicans* has many developmental options, and cell differentiation is an important virulence-contributing factor. *C. albicans* can differentiate between budding yeast, elongated yeast called pseudohyphae, or highly polarized, continuously elongating cells called hyphae (Fig. 1A). The ability to change between these cell types may enhance the virulence capacity of the organism (80, 112). Cell differentiation is stimulated by environmental conditions, where high temperature (37°C), the presence

Malcolm Whiteway • Health Sector, Biotechnology Research Institute, National Research Council of Canada, 6100 Royalmount Ave., Montreal, QC, H4P 2R2, Canada. **Catherine Bachewich** • Biology Department, Concordia University, 7141 Sherbrooke St. West, Montreal, QC, H4B 1R6, Canada.

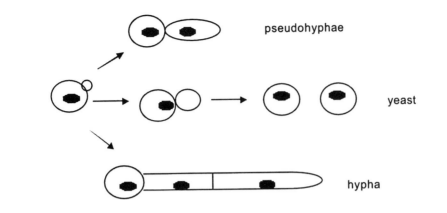

A

pseudohyphae

yeast

hypha

C. albicans

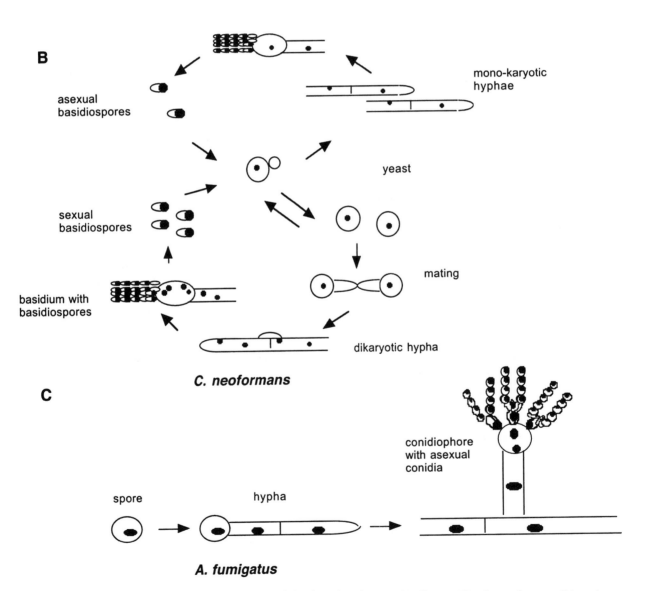

B

mono-karyotic hyphae

asexual basidiospores

yeast

sexual basidiospores

mating

basidium with basidiospores

dikaryotic hypha

C. neoformans

C

conidiophore with asexual conidia

spore

hypha

A. fumigatus

Figure 1. Cell types and developmental stages of the fungal pathogens *C. albicans* (A), *C. neoformans* (B), and *A. fumigatus* (C).

of serum, high pH, and different carbon sources stimulate hyphal growth whereas low temperature (30°C), low pH, and the absence of serum favor yeast development (18). Differences in cell shape and cell surface composition are proposed to be advantageous in different environments of the host. Yeast phase cells may be more adept at dissemination in the bloodstream, while hyphal filaments may be useful for tissue penetration. The various stages of a *C. albicans* infection have not been fully characterized but are proposed to involve adhesion to host cells, penetration of host tissue or engulfment by host cells, and cell proliferation or further dissemination within the host (112). Although *C. albicans* is diploid and does not undergo meiosis, it has high homology across the genome to the model yeast *Saccharomyces cerevisiae*. Thus, much of what we know about signaling networks in this organism originated with or was aided by studies of *S. cerevisiae* (18). Several virulence factors have been identified, such as the cell wall proteins Ece1p and Hwp1p, as well as secreted proteases, including the aspartyl proteases of the Sap family (18). Many of these factors are specifically associated with the hyphal phase, but both yeast and hyphal cells are required for full virulence, since mutants locked in a yeast or filamentous form are less virulent (22, 112).

C. neoformans is also an opportunistic pathogen and can cause cryptococcal meningitis in immunocompromised individuals. *C. neoformans* is a dimorphic basidiomycete that is commonly found as a saprotroph in bird dung. Haploid basidiospores can differentiate into yeast cells (75) (Fig. 1B), which divide through budding but can also mate to form dikaryotic hyphae in response to pheromone. These hyphae in turn develop basidia that give rise to sexual basidospores. Under nitrogen starvation conditions, the haploid yeast can also germinate to form haploid hyphae, which form asexual basidiospores. In vitro, the diploid filamentous form occurs at 24°C while the yeast form is favored at 37°C. While the filamentous growth stages occur in the environment, the yeast and basidiospores can be inhaled into the lungs and represent the infectious form of the organism. Once in the lungs, the organism can either proliferate, remain dormant, or disseminate to other organs, particularly the central nervous system. Aspects of *C. neoformans* virulence include the production of a thick polysaccharide capsule and deposition of the pigment melanin around the yeast cell (68); cells lacking capsule and melanin are nonvirulent. Virulence is also linked to mating in *C. neoformans*.

A. fumigatus is a filamentous, saprotrophic ascomycete that gives rise to airborne conidia (Fig. 1C).

Unlike *C. albicans* and *C. neoformans, A. fumigatus* is not dimorphic and does not form yeast. Inhalation of spores does not cause disseminated infection in healthy individuals, but if not cleared by the immune system, the spores can germinate and form mycelial colonies in the lungs (70). This can lead to respiratory or systemic disease, particularly in neutropenic individuals. *A. fumigatus* also produces the pigment melanin, which is thought to protect the fungus against reactive oxygen species of the host and thus to aid virulence. *pksP*, a polyketide synthase, is important in melanin biosynthesis, and absence of *pksP* results in reduced virulence (69). Polyketide synthases are known virulence factors in other fungal pathogens, including those infecting plants.

Signaling Pathways

cAMP

cAMP-based signaling contributes to a diversity of cellular processes and is used by many fungal pathogens to respond to the host and regulate many aspects of virulence. Much of our understanding of cAMP-based signaling pathways has come from investigations of *S. cerevisiae*. cAMP acts as a second messenger and is produced from ATP through the plasma membrane-bound enzyme adenylyl cyclase, called Cyr1p or Cdc35p, in *S. cerevisiae*. cAMP activates protein kinase A (PKA), which is composed of catalytic subunits Tpk1p, Tpk2p, and Tpk3p and the regulatory subunit Bcy1p (87, 123). A conformational change in PKA on binding of cAMP releases the regulatory subunits, allowing the catalytic subunits to phosphorylate downstream targets (20). Adenylyl cyclase activity in *S. cerevisiae* is activated by carbon and nitrogen sensing and regulated via the GTPases Ras2p and Ras1p, the Gα subunit Gpa2 (124), which is functionally linked to the glucose receptor Gpr1p, and Cap1p/Srv1p (115). cAMP is negatively regulated by the phosphodiesterases Pde1p and Pde2p (122). In addition to sensing nutrients and regulating growth, the pathway is involved in stress responses and in pseudohyphal development (20).

The cAMP pathway is one of the most extensively investigated signaling pathways in *C. albicans* and regulates many aspects of growth and virulence. Inhibition of adenylyl cyclase or phosphodiesterase activity with drugs, or the exogenous application of cAMP, can influence the ability of *C. albicans* yeast cells to form hyphae (29, 31, 65, 95, 108). In addition, cAMP levels have been reported to fluctuate during the yeast-to-hyphal transition, although the precise pattern of change is controversial (65, 108). Several components of the pathway have been cloned and characterized, including Cdc35p, Gpa2p,

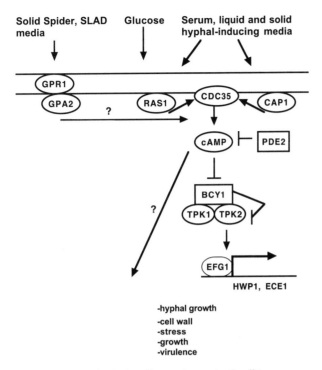

Solid Spider, SLAD media

Glucose

Serum, liquid and solid hyphal-inducing media

-hyphal growth
-cell wall
-stress
-growth
-virulence

Figure 2. cAMP signaling pathways in *C. albicans*.

Tpk2p, Pde2p, Gpr1p, Ras1p, and Cap1p (Fig. 2). The core factor, Cdc35p, is not essential (104). However, the homozygous deletion strain grows slowly, exhibits cell wall defects based on sensitivity to zymolyase, calcofluor, and osmotic shock (58), is unable to form hyphae under any hypha-inducing condition, and is avirulent in a mouse model of infection (104). The catalytic subunit of PKA in *C. albicans*, Tpk2p, contributes to cAMP signaling, and overexpression can drive filament formation (118). However, the mutant phenotype does not resemble that of *CDC35*, where Tpk2p is required for hyphal growth only under a limited number of conditions (36), and the mutant strain was partially avirulent. A Tpk1 homologue has been identified that may have a redundant function with Tpk2p, but the PKA activity in the *tpk2/tpk2* mutant, which presumably was provided by Tpk1p, was only 10% of that demonstrated by wild-type strains. Since Tpk2p is not required for hyphal growth under optimum inducing conditions, such as serum-containing liquid medium at 37°C, it was proposed that hyphal growth occurs in a PKA-independent pathway under this condition, yet Cdc35p is essential for serum-induced hyphal growth. The homologue to the regulatory subunit *BCY1* could not be deleted, suggesting that it is essential (28). However, a *bcy1/BCY1*, *tpk2/tpk2* mutant, which had constitutive PKA activity, was unable to germinate in some hypha-inducing media. Tpk1p was also delocalized in this mutant

background compared to *tpk2/tpk2* and wild-type strains, suggesting that Bcy1p contributes to localization and thus activity of PKA. The phosphodiesterase Pde2p contributes to cAMP signaling, since a *pde2/pde2* mutant had a higher level of intracellular cAMP (14, 65) and appeared to constitutively produce elongated cells (14), supporting the importance of cAMP signaling in morphogenic switching.

C. albicans contains a single *RAS* homologue, Ras1p, which was placed upstream of Cdc35p in the signaling pathway (Fig. 2) since the absence of *RAS* also prevented hyphal formation under most inducing conditions (51, 72), overexpression of *RAS1* could not rescue the *cdc35/cdc35* phenotype, and addition of cAMP suppressed the *ras/ras* mutant (104). Recent transcription profiling work (58) demonstrated that most of Ras1p function at the level of influencing transcription was mediated through Cdc35p. However, Cdc35p does not appear to be regulated exclusively by Ras1p, because the *cdc35/cdc35* strain demonstrated unique transcription signatures compared to the *ras1/ras1* strain. *C. albicans* contains homologues of Gpr1 and Gpa2 (88, 109), but deletions of these genes resulted in strains that were only partially defective in hyphal formation under solid-medium conditions and were not affected in virulence. Addition of cAMP rescued the *gpa2/gpa2* and *gpr1/gpr1* mutant (83, 88), and the latter was also suppressed by overexpression of *GPA2*. Cdc35p may also be regulated by the *CAP1* homologue in *C. albicans*, since the *cap1/cap1* mutant could not form hyphae under many hypha-inducing conditions, and this defect was rescued with the addion of cAMP. The strain was also avirulent (15). The Cap1p homologue Srv2p in *S. cerevisiae* is predicted to activate adenylate cyclase activity through interactions with both Ras2p and Cyr1p (115). The *cap1/cap1* mutant of *C. albicans* partially resembled the *ras1/ras1* mutant in that pseudohypha-like cells formed in serum-containing liquid medium (15), although the composition of the medium was not identical. If Ras1p and Cap1 exclusively regulate Cdc35p, the *ras1/ras1 cap1/cap1* double mutant should have as resistant a phenotype to all hypha-inducing conditions as the *cdc35/cdc35* mutant, but such a strain has not been reported. The way in which Cap1p, Ras1p, and other potential upstream factors contribute to activation of Cdc35p and, in turn, are activated themselves is not known. cAMP signaling is critical for serum-induced hyphal development, but the mechanism by which its effects are mediated under this condition is also not fully understood.

A major proposed target of the cAMP pathway is the transcription factor Efg1p. Efg1p contains a conserved PKA phosphorylation site, and over-expression of Efg1p can suppress the Tpk2p phenotype (118), suggesting that it is a downstream target of the cAMP pathway. The contribution of Efg1p to signaling and virulence is complex, since it appears to occupy many places in different signaling pathways (49). Transcription profiles of the *efg1/efg1* mutant challenged with hypha-inducing media had only some similarity to those of the *cdc35/cdc35* mutant under the same conditions, indicating that not all of Cdc35p function is exerted through Efg1p (58). In this respect, Cdc35p was shown to be specifically involved in growth, cell wall metabolism, and stress responses. However, the *cdc35/cdc35*, *efg1/efg1*, and *ras1/ras1* deletion strains challenged with serum demonstrated similar defects in up-regulation of hypha-specific transcripts such as *HWP1*, *ECE1*, and the *SAP* genes, for example, suggesting that these factors are downstream targets of a main branch of the cAMP signaling pathway. The cAMP pathway thus may contribute to virulence through regulating growth, morphogenesis, and expression of virulence-related factors. Consistent with this, the *cdc35/cdc35* mutant was hypersensitive to azole drugs normally used to combat *C. albicans* infections, possibly due to its inability to up-regulate the multidrug transporter gene *CDR1* (63).

The cAMP signaling pathway also contributes greatly to virulence in *C. neoformans*. The adenylyl cyclase gene *CAC1* is similar to *CDC35* in *C. albicans*, in that the deletion did not affect cell viability (6). There was no defect in vegetative growth, in contrast to the *cdc35/cdc35* mutant of *C. albicans*, but the deletion strain failed to produce melanin and capsule, could not mate, and was avirulent (6). Cac1 responds to glucose, similar to Cdc35p in *S. cerevisiae*. Other components of the cAMP pathway, including the Gα subunit Gpa1, the Cac1-intercating factor Aca1, and the catalytic subunits of PKA, including Pka1 and Pka2, have been identified (Fig. 3). Strains lacking Gpa1 did not produce melanin or the capsule and were avirulent, but production of virulence factors could be restored by the addition of cAMP (5). Aca1 physically associates with Cac1 and is required for filamentous morphogenesis and virulence (12). The two catalytic subunits of PKA demonstrate strain-dependent functions. In serotype A, absence of *PKA1* resulted in sterile strains that did not produce melanin or the capsule and were avirulent in mouse models (59), whereas Pka2 mediated melanin and capsule formation, as well as mating, in serotype D (59). Despite the decrease in virulence factors in serotype D, the strain was still virulent, and absence of *PKA1* or both subunits in this strain background also did not influence

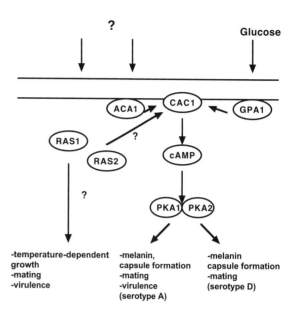

Figure 3. cAMP signaling pathways in *C. neoformans*.

virulence. Since Pka2 was not required for virulence in serotype A, the results demonstrate that the PKA subunits in *C. neoformans* have specialized functions and may act on different downstream factors or differentially regulate similar factors in the different strains (59). *C. neoformans* contains two Ras homologues, Ras1 and Ras2. Ras1 is required for mating and for growth at high temperature. Although the *ras1* mutant produced melanin and a capsule, the strain was avirulent and sterile, and exogenous cAMP could rescue defects only in adhesion, not in mating or in temperature-dependent growth (129). Ras2 is not essential for virulence or growth but may have some redundant function since its overexpression could suppress the *ras1* defects. Thus, Ras may contribute to cAMP as well as other signaling pathways in *C. neoformans*. Melanin and capsule are thus regulated by cAMP signaling, but direct targets of the signaling pathway, as well as the upstream receptors, have not been identified.

In *A. fumigatus*, the cAMP-PKA pathway is composed of the adenylyl cyclase ACYA, the Gα subunits GPAA and GPAB, and the PKA catalytic subunits PKAC1 and PKAC2 (77, 78) (Fig. 4). Conidiation and growth were reduced in strains lacking *acyA* and *pkaC1*, while conidiation alone was reduced in the cells lacking *gpaB*. PKA activity was detected in extracts of the *acyA* and *gpaB* strains but not in the *pkaC1* deletion strain, suggesting that PKAC1 was the predominant form of PKA in *A. fumigatus* (78). The pathway contributes to the regulation of the virulence factor *pksP*, a polyketide synthase that is necessary for production of pigment for conidia. Expression of a *pksPp-lacZ* construct was

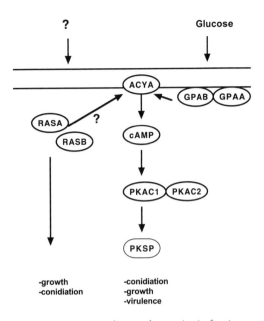

Figure 4. cAMP signaling pathways in *A. fumigatus.*

reduced in strains lacking *pkaC1* or *gpaB*, and these deletion strains were also reduced in virulence. Gene expression patterns in *A. fumigatus* cells that were incubated with endothelial cells demonstrated that a cAMP-regulatory subunit and a *RAS* homologue were among the up-regulated genes (102). *A. fumigatus* contains two *RAS* homologues, and deletions of the genes lead to differential defects in conidiation and growth; however, a link between RAS and cAMP signaling has not been established in this pathogen, nor has the role of RAS in virulence (52).

MAPK

The MAPK signaling pathway is used by most organisms to transduce extracellular signals to regulate growth and differentiation (131). *S. cerevisiae* contains several MAPK pathways that mediate mating, pseudohyphal growth, stress, and nutrient sensing. The core pathway consists of a cascade of kinase activities, where a MAPKKK activates a MAPKK that in turn regulates the MAPK. The upstream activating factors (input) and downstream targets (output) vary, although there is cross talk and shared components between the pathways. The MAPK mating pathway in *S. cerevisiae* has been extensively investigated and serves as a model. The pathway is activated by pheromone, which occupies a heterotrimeric G-protein-coupled receptor on the cell surface. The released Gβγ subunit then activates the MAPK pathway by interacting with a scaffold protein, Ste5p, and the p21-activated kinase (PAK) Ste20p. These elements, together with Ste50p, lie upstream of and activate the MAPKKK Ste11p, which in turn

stimulates the MAPKK Ste7, followed by the MAPK Fus3p and Kss1p. These in turn regulate the transcription of many genes required for mating through the transcription factor Ste12p. In the MAPK pathway required for filamentous growth, the PAK Ste20 is alternatively activated through Ras2p and Cdc42p activity, but Ste12p, along with other downstream transcription factors, contributes to the output response of filamentous growth. The high-osmotic-growth (HOG) MAPK pathway is required for responding to stress, including changes in osmolarity. Homologues of the MAPK core components constitute this pathway, including Ssk2p, Ste11p (MAPKKK), Pbs2p (MAPKK), and Hog1p (MAPK) (131).

Among the fungal pathogens of humans, *C. albicans* contains the most extensively characterized MAPK pathway, which contributes to the yeast-to-hypha transition and virulence under certain conditions (38, 39, 71). The PAK Cst20p, the MAPKK Hst7p, the MAPK Cek1p (130), and the Ste12-like transcription factor Cph1p (79) have been identified, and epistasis experiments suggest that these factors function in an order comparable to that of their homologs in *S. cerevisiae* (39, 71) (Fig. 5). A *STE11* homologue exists but has not been investigated. Deletions of *CST20*, *HST7*, and *CPH1* result in hyphal defects on a type of starvation medium called spider medium but have no effect on hyphal growth in the presence of serum or in most liquid or other solid media. However, the *cek1/cek1* deletion strain demonstrated additional phenotypes, such as growth rate defects in serum-containing medium and, similar to the *cst20/cst20* deletion strain, a significant defect in virulence in mouse models. In contrast, absence of *HST7* and *CPH1* did not affect virulence, suggesting that the upstream regulatory PAK and downstream MAPK may contribute to or be utilized by other pathways (39, 71). The Cpp1p phosphatase negatively regulates the MAPK pathway; deletion of *CPP1* results in hyperfilamentous growth, which is dependent on the presence of *CEK1* (39). The mutant also reduces virulence, supporting the idea that switching between cell forms could contribute to virulence potential. *CST20* and *CPP1* were up-regulated in thrush samples taken from human immunodeficiency virus-infected patients, supporting a potential in vivo role of these signaling components during infection (33). *C. albicans* also contains a HOG-like MAPK pathway (2, 3) (Fig. 5). The *hog1/hog1* deletion strain demonstrated alterations in the cell wall and was sensitive to changes in osmolarity but also to oxidative stress, in contrast to the *hog1* mutant of *S. cerevisiae*. The *hog1/hog1* mutant also demonstrated enhanced hyphal growth in response to serum, unlike other MAPK mutant

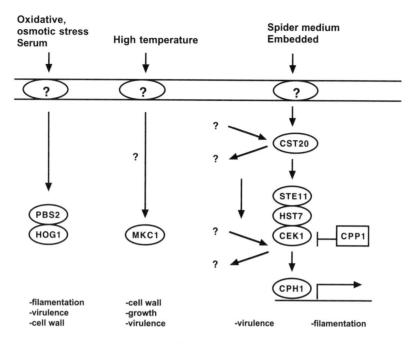

Figure 5. MAPK pathways in *C. albicans*.

strains, and strongly affected virulence. Since pathogens encounter many stresses within the host, factors regulating stress responses are expected to play important roles in virulence. The MAPKK Pbs2p was also required for adapting to osmotic and oxidative stress in *C. albicans,* suggesting that it lies upstream of Hog1p (8). The HOG MAPK pathway is regulated by a two-component signaling network (see Fig. 8), which is discussed in the next section. A third MAPK gene in *C. albicans, MKC1,* is related to the homologue *SLT2* of *S. cerevisiae,* and the deletion strain demonstrates cell wall defects, reduced growth rate at high temperature, and reduced virulence in a mouse model (46, 93). The MAPK signaling pathways in *C. albicans* thus are predicted to mediate different signals from the host environment, including a diversity of stresses and starvation.

 C. neoformans contains a MAPK pathway that responds to pheromone and regulates mating and virulence. The pathway consists of the G protein β subunit GPB1, the PAKs Ste20a, Ste20α, and Pak1, the MAPKKK Ste11a and Ste11α, the MAPKK Ste7, the MAPK Cpk1, and the MAPK target Ste12a and Ste12α (34, 40, 75, 127, 128) (Fig. 6). Pak1, Gpb1, Ste7, and Cpk1 are found in all cells, while Ste11, Ste20, and Ste12 are cell type specific (40). For example, Ste20a and Ste20α are encoded by the mating locus and play strain-specific roles in mating, differentiation, and virulence, whereas Pak1 is not mating type specific. Deletion of *STE20α* in serotype A reduced virulence and mating and impaired capsule

formation in vivo, while deletion of *STE20α* or *STE20*a had no effect on virulence in serotype D. In contrast, deletion of *PAK1* attenuated virulence in both serotypes and is not encoded by the mating-type locus (127). Double deletions between *PAK1* and the *STE20* PAKs were synthetically lethal. The *gpb1* deletion is sterile, and *CPK1* overexpression suppresses this defect. Ras1 is proposed to lie upstream of the Gpb1-MAPK pathway because the *ras1* deletion strain was avirulent, and its defects in mating were rescued by overexpression of Gpb1 (4). *C. neoformans* also contains a homologue of the MAPK Hog1, which is important for virulence and differentiation in serotype A (13), and a homologue of the MAPK Slt2, called Mpk1, which is required for growth at high temperature (66). The deletion strain is sensitive to cell wall synthesis inhibitors and demonstrates reduced virulence. Since Ras1 was also important for growth at high temperature, it may contribute to the regulation of this pathway.

 The MAPK homologue SakA has been identified in *A. fumigatus* (134) (Fig. 7) and is involved in responding to osmotic stress. The deletion strain is viable but fails to up-regulate genes such as *msnA, ptpA,* and *pbsA* on exposure to osmotic shock. Germlings lacking the gene arrested growth on increases in medium osmolarity, and conidia germinated more slowly on minimal medium. Other components of the MAPK pathway and the contribution of SAKA to virulence have not been investigated.

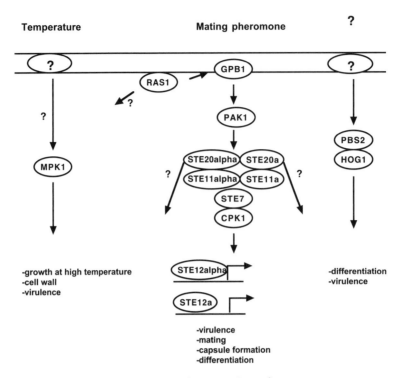

Figure 6. MAPK pathways in *C. neoformans*.

Two-component signaling

Two-component histidine kinase (HK) phospho-relay systems function to mediate extracellular signals with intracellular responses in bacteria, slime molds, fungi, and plants. The pathway regulates many processes, including differentiation, chemotaxis, and osmosensing, for example (60). Since similar signaling networks have not been identified to date in animal cells, the two-component signaling pathway factors are attractive targets for antipathogen drugs (30). The signaling system is composed of a HK, which autophosphorylates a histidine group in response to external signals. The phosphorylation is transferred to an aspartic residue in a response regulator domain, which in turn leads to activation of transcription cascades, often through regulation of a MAPK pathway. In *S. cerevisiae*, two-component signaling is involved in osmosensing and upstream regulation of the HOG MAPK pathway. A sensor transmits a signal to the plasma membrane-associated HK Sln1p, which autophosphorylates under low-osmolarity conditions. The phosphorylation is relayed through an intermediate, Ypd1p, then to a cytoplasmic response regulator, Ssk1p (99), which maintains the HOG pathway as inactive. Under high-osmolarity conditions, Sln1p is not phosphorylated, which leads to activation of the HOG MAPK pathway and growth.

C. albicans contains three HK, including Sln1p, Nik1p/Cos1p, and Chk1p (25, 92), as well as homologues of Ypd1p (26) and the response regulators Ssk1p and Skn7p (27, 117) (Fig. 8). The HK contribute to virulence in mouse models and differentially influence hyphal formation in the presence of serum, with Chk1p having the strongest influence (135). In contrast to Sln1p in *S. cerevisiae,* deletion strains of the HK in *C. albicans* demonstrated only minor sensitivity to osmotic conditions. Sln1p was located in the plasma membrane, but Nik1p and Chk1p were cytosolic, suggesting that they may

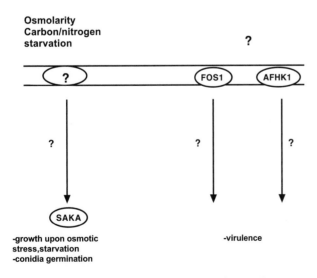

Figure 7. MAPK and two-component signaling pathways in *A. fumigatus*.

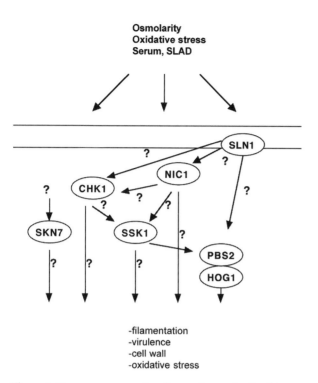

Osmolarity
Oxidative stress
Serum, SLAD

-filamentation
-virulence
-cell wall
-oxidative stress

Figure 8. Two-component signaling pathways in *C. albicans*.

function downstream of Sln1p. A *sln1/sln1 nic1/nic1* double mutant could not be constructed, but deleting *SLN1* or *NIK1* in the *chk1/chk1* deletion mutant partially restored filament formation in the presence of serum, leading to the conclusion that Sln1p and Nic1p may lie upstream of Chk1p, but the HK also probably function in separate pathways. The Chk1p pathway appears to function in oxidative stress and may involve the response regulator Ssk1p. Strains lacking *CHK1* or *SSK1* are sensitive to oxidants, are avirulent in a mouse model, and show flocculation in inducing medium (3, 24). Hog1p was modified in response to oxidative stress in an Ssk1p-dependent manner (32), and downstream transcriptional responses of an *ssk1* mutant under oxidative-stress conditions have been reported (32). The *ssk1* mutant is also similar to the *nic1* mutant in its inability to form hyphae on certain solid media, suggesting that Ssk1p may lie downstream of two independent HK (27). Another response regulator, Skn7p, is also required for growth on serum-containing agar and is sensitive to hydrogen peroxide but does not influence virulence and does not target Hog1p (117), suggesting that the response regulators mediate distinct pathways. Thus, two-component signaling in *C. albicans* mediates some environmental stresses but also participates in morphogenesis and virulence. The downstream effectors, upstream sensors, and relationships among the pathways are far from understood.

A. fumigatus contains two HK, FOS-1 and AFHK1 (Fig. 7). Deletion of *fos-1* reduced virulence in a mouse model of aspergilliosis (35), while *afhk1* expression was up-regulated in cells isolated from the host (47). The sensors, effectors, and extent of cross talk between two-component signaling networks remain to be determined.

pH pathways

Signaling pathways that govern responses to extracellular pH have been identified in *S. cerevisiae*, *A. nidulans*, and *C. albicans*. Since the pathways are fungus specific and are not found in animal cells, and since extracellular pH can vary greatly in different environments of the host, pH-responsive components may serve as useful targets for antifungal drugs, much like two-component signaling pathway factors. The conserved pH-responsive pathway in *S. cerevisiae* and *A. nidulans* is composed of the Rim101/PacC zinc finger transcription factor, which must be cleaved to an active form to allow growth at alkaline pH. Pathway components required for processing include Rim21p/PalH, Rim9p/PalI, Rim8p/PalF, Rim20p/PalA, and Rim13p/PalB. The order in which the factors function is not known, but Rim21p and Rim9p may be transmembrane proteins that sense extracellular pH and Rim13p is a calpain protease that may be responsible for cleaving Rim101p (41). The pathway regulates sporulation and conidiation in these organisms but has other functions in the pathogen *C. albicans*. *C. albicans* grows in acidic environments within the gastrointestinal tract and vagina and in more alkaline environments in the blood and in the oral region, demonstrating the critical importance of pH regulatory systems in the lifestyle of this organism. *C. albicans* contains homologues of Rim101p, Rim20p, Rim8p, and Rim13p (Fig. 9), and defects in the genes prevent hyphal formation at alkaline pH (43, 76) and growth of yeast at pH 10. In addition, several mutants are less virulent (42). Targets of the pH pathway include *PHR1* (111) and *PRA1* (113), which are expressed above pH 5.5, and *PHR2*, which is expressed below pH 5.0 (89). Phr1p and Phr2p are β-1-3 and β-1-6 glucan cross-linking factors, while Pra1p is a gene of unknown function. Consistent with their expression patterns, Phr1p is required for growth at alkaline pH whereas Phr2p is required for growth at acidic pH. Absence of *PHR1* or *PHR2* also compromised virulence in environments that supported the pH requirement of the factors, including blood versus vaginal/stomach models (45, 57). In the absence of *RIM101* or other components of the pH pathway, *PHR1* and *PRA1* are not expressed at alkaline pH, but *PHR2* is induced,

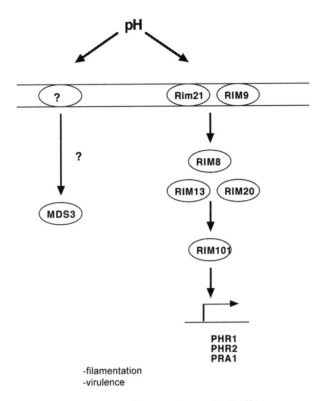

Figure 9. pH-regulatory pathways in *C. albicans*.

suggesting that the pH pathway also represses acid-induced genes while inducing alkali-associated genes at alkaline pH (43). Additional genes potentially regulated by Rim101p were identified using microarray analysis (17, 101) and included cell wall and secreted factors, as well as several genes involved in iron metabolism. Although the nature of the activating signal has not been identified, it was suggested that starvation for cations such as iron, which becomes restricted at alkaline pH, may be a contributing factor. It is thus intriguing that iron metabolism genes were highly modulated in the *rim101* mutant (17). The alkaline pH pathway may be regulated in part by the ESCRT protein complexes, which form multivesicular bodies for protein sorting to the vacuole or lysosome. ESCRT factors are required for alkali-induced filamentation in a Rim101p-dependent manner (132).

C. albicans contains another pH-responsive pathway that is independent of Rim101p signaling, which involves the factor Mds3p (44) (Fig. 9). Strains lacking *MDS3* cannot form filaments at alkaline pH, have reduced virulence in a mouse model, and fail to express hypha-specific factors Hwp1 and Ece1p. The *mds3/mds3 rim101/rim101* double mutant had even greater effects on virulence and growth, and *RIM101* expression could not suppress the *mds3* defects, suggesting that the factors acted in independent pathways. The *S. cerevisiae*

homologues of Mds3p are involved in meiosis but also influence responses to alkaline pH (44). Comparable to the Rim101p pathway, the direct downstream targets and upstream regulation of the Mds3p pathway remain to be elucidated in *C. albicans*.

Ca²⁺/calmodulin

Ca^{2+} is a second messenger that exerts its effects in part through binding calmodulin. Ca^{2+}-bound calmodulin in turn undergoes a conformational change and is then able to activate calmodulin-dependent kinases or the serine/threonine phosphatase calcineurin. Ca^{2+}/calmodulin signaling is involved in diverse cellular processes, including the cell cycle, polarized growth, and adaptations to stress (67).

Ca^{2+}/calmodulin signaling modules are conserved in fungi, and fungal pathogens, such as *C. albicans* and *C. neoformans*, utilize the pathway for growth and virulence under different conditions (Fig. 10). In *C. albicans*, absence of *CMP1*, encoding the catalytic subunit of calcineurin, resulted in increased sensitivity to fluconozole, decreased virulence, and death in the presence of serum (11, 97, 110). Alterations in colony morphology under certain conditions were reported in one examination of the disruption phenotype (110), but subsequent studies

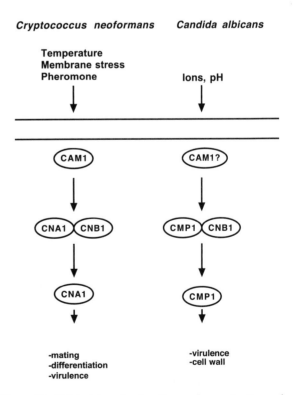

Figure 10. Ca^{2+}/calcineurin signaling pathways in *C. neoformans* and *C. albicans*.

identified no defects in the ability to form hyphae (11). However, the mutant was hypersensitive to alkaline pH and increasing concentrations of ions but not to temperature (11). Deletion of the regulatory subunit *CNB1* resulted in avirulence and prevented colonization in a mouse model but did not lead to defects in forming hyphae (19). Downstream targets of the signaling pathway in *C. albicans* are not known, but Crz1p, a target of calcineurin in *S. cerevisiae*, is not required for virulence, and the deletion phenotype is different from that of calcineurin in *C. albicans* (98).

In *C. neoformans*, lack of the catalytic subunit A (CnA1) or regulatory subunit B (CnB1) of calcineurin reduces virulence, growth at 37°C, mating and hyphal elongation (37, 53, 96). Calcineurin also contributes to maintaining cell wall integrity, along with a MAPK pathway (66). The calmodulin gene, *CAM1*, has been isolated and is also required for growth at 37°C (67). Thus, Ca^{2+}/calmodulin signaling appears to be mediated by stress conditions, but the upstream regulation, including receptors of the pathway, as well as downstream targets, remains to be determined.

Other Signal Transduction Networks Involved in Pathogen Responses to the Host

Several other signaling networks are also associated with pathogen responses to the host, including a variety of metabolic pathways. Global transcription profiling of *C. albicans* cells undergoing the yeast-to-hypha transition in response to serum and high temperature demonstrated modulation of many metabolic factors, while transcription profiling of *C. albicans* cells phagocytosed by macrophages demonstrated up-regulation of components of the glyoxylate cycle, DNA damage and repair, oxidative stress, and gluconeogenesis pathways (81). Cells exposed to human blood demonstrated similar modulation (54). Amino acid starvation was reported to occur in *C. albicans* cells ingested by neutrophils (107), which is consistent with the fact that an amino acid starvation response stimulates filamentous growth in *C. albicans* in a Gcn4p-, Ras-cAMP-dependent manner (125).

Summary

Although many different signal transduction cascades mediate responses of pathogens to their hosts, the most extensively investigated networks involve cAMP, MAPK, two-component HK, pH, and Ca^{2+}/calmodulin signaling. The pathways can mediate different environmental conditions, such as temperature, stress, and presence of serum, within the different pathogens and induce different responses, including changes in cell morphology and expression of particular virulence factors. Although the extent

of cross talk between the diverse pathways and the identity of upstream receptors and downstream direct targets is not well understood, the emergence of annotated genome sequences coupled with conserved aspects of the pathways should further our understanding of how individual pathogens utilize signal transduction networks for virulence.

RESPONSE OF THE HOST TO THE PATHOGEN

Compared to the situation with respect to the pathogen signaling pathways, we are at a very early point in our understanding of the host side of the signaling networks controlling the interaction between fungal pathogens and their mammalian targets. When a pathogen associates with a host organism, it encounters many different host cell types, and these different mammalian cells can respond in dramatically different ways to the pathogen. Thus, any interaction observed at the organismal level is multifactorial, and we are far from being able to define a signaling "pathway" for the response of a mammal to a potentially pathogenic fungus. Attempts to refine the question by examining the interaction between the pathogen and individual host cell types are also at an early stage. Such studies represent a relatively crude attempt at reductionism; it is clearly unrealistic to expect, for example, that the measured interactions between a fungal pathogen and a macrophage growing in a cell culture dish will fully reflect the process that occurs in the bloodstream of an infected host.

If we hope to define the host side of the host-fungus interaction in terms of classic signal transduction processes, we will have to identify the receptors used by the host to identify the presence of the pathogen, the components that are essential to transmit this recognition to the intracellular machinery involved in changing the behavior of the host in response to the pathogen, and the elements within the host that effect the behavioral changes. This understanding will have to be extended to the molecular level for each of the cells or subtypes of host cells that are influenced by the pathogen, but ultimately the consequences of all these effects would have to be visualized within the context of the whole organism. This investigative process is further complicated by the fact that the pathogen can have different interactions with the host. Depending on host genotype and phenotype, and involving many other considerations including the presence of other diseases, a fungal pathogen such as *C. albicans* can be a commensal, can cause superficial mucosal infections, or can create life-threatening systemic infections. Approaches to address these questions can range from experiments where host organisms are directly investigated to identify factors that influence the interactions with fungal pathogens to experiments where specific individual genes within a unique cultured cell population are monitored for their behavior in the presence of pathogens.

A primary interaction between the mammalian host and a fungal pathogen occurs at the level of the innate immune system. The relationship between the host immune system and fungal pathogens has been extensively reviewed (105, 106), including chapters in this volume, and thus we will not go into it in depth. The importance of this interaction to the host can be readily seen in the increased virulence of pathogens in hosts with compromised immune systems; for example, the virulence of the essentially ubiquitous *A. fumigatus* can be highly dependent on the status of the host immune system (56). The innate immune system plays a critical role in the initial response to fungal pathogens, and within this system, pattern recognition receptors (PRRs) have been implicated as the components that serve as the initial detectors of the pathogens. Among these PRRs are the Toll-like receptors (TLRs), named for their structural similarity to the *Drosophila melanogaster* Toll protein. An initial appreciation of the role of the Toll pathway in innate immunity came from pioneering studies of *D. melanogaster* (74); recently, direct tests with *C. albicans* (1) and *C. neoformans* (7) have shown the involvement of Toll in the response of flies to human fungal pathogens.

TLR Signaling

A variety of recent studies have addressed the question of function and specificity of the various TLRs in fungal pathogen response. There are currently at least 10 TLR family members in mammalian cells, and these receptors have been implicated in detecting a variety of molecular characteristics of a number of distinct pathogens. TLR2 and TLR4 appear to be the most important for the recognition of fungal pathogens. Both TLR2 and TLR4 play a role in the recognition of *A. fumigatus* (21) and *C. albicans* (94), while *C. neoformans*, through its polysaccharide capsule consisting of glucuronoxylomannan, appears to be recognized uniquely by TLR4 (116). However, many points still need to be clarified about the role of TLRs in the process of fungal pathogen recognition. The consequences of stimulation of the TLR molecules differ based on the cell type and the pathogen encountered. For example, although specific TLR4-defective mice show higher fungal burdens after *C. albicans* infections, suggesting a defective response, tumor necrosis factor alpha (TNF-α) and interleukin-1β production was not affected during the infection process. However, production of other chemokines in response

to *C. albicans* infection does appear to be TLR4 dependent (94). Therefore, the pathway/response pattern for specific cell types and specific responses still needs to be fully clarified; there does not appear to be a simple linear pathway leading from a specific TLR to a common set of responses. As well, the actual molecular species that are linked to the fungal pathogen and recognized by the PRRs are still under investigation. TLR2 has been implicated in the recognition of *C. albicans* phospholipomannan (64), while TLR4, in conjunction with CD14, appears important for the identification of unmodified *C. albicans* mannans (120).

A potential signaling network downstream of the TLRs has been identified, and its function in regulating the innate immune response to fungal pathogens is under study (Fig. 11) The picture of this network is based on work with model systems such as *Drosophila* and on analysis of signaling in mammalian cells (105, 121). While many details of the signaling system remain to be worked out, a general framework has been established. Activation of the receptor can regulate the association of adaptor proteins to the intracellular domain of the receptor. The intracellular domain of the TLR receptors contains a TIR (Toll/interleukin-1 receptor) motif; this motif is shared with a number of cytoplasmic adaptors such as MyD88 (myeloid differentiation marker 88), TIRAP (TIR domain containing adaptor protein) (also known as MAL [MyD88-adaptor-like]), and TICAM-1 (TIR-domain-containing adaptor molecule). The association between the receptors and the adaptors appears to take place through homotypic interactions involving these TIR domains (133). Adaptor proteins such as MyD88 serve to connect the activated receptors through a member of the IRAK (interleukin-1 receptor-associated kinase) serine/threonine kinase family. The MyD88-IRAK interaction takes place through homotypic module interactions as well, in this case involving death domain modules on the two molecules. IRAK kinases signal through TRAF6 (tumor necrosis factor receptor-associated factor 6) to transcription factors such as NF-κB and AP1 to stimulate cytokine expression. As can be seen by the names of the factors, the elements are not exclusive to the innate immune system signaling network, and defining the specific combinations of elements used as pathways when particular cell types interact with a particular pathogen will involve extensive work.

Recent efforts are serving to further define the roles of specific components of this network in the response to fungal pathogens. In one study, macrophages from MyD88-defective mice were tested for their response, measured by the production

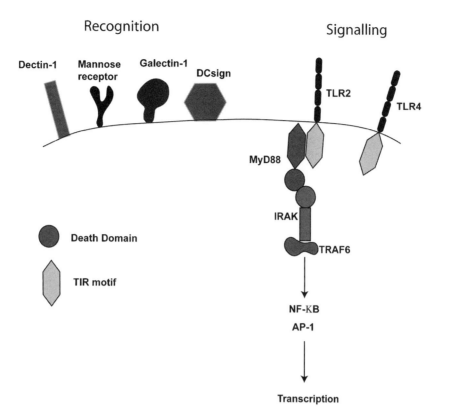

Figure 11. Recognition and TLR signaling involved in host detection and response to fungal pathogens. Many mammalian cell surface molecules are involved in the identification of fungal pathogens. These include proteins such as TLR2 and TLR4, which are connected to intracellular signaling pathways, and a variety of recognition proteins that interact primarily with components of fungal surfaces. These recognition receptors can be highly specific; recent evidence has been provided that dectin-1 on the surface of macrophages can distinguish hyphal from yeast forms of *C. albicans*. The signaling pathways involve protein-protein interactions; TIR domains on the TLRs interact with homologous domains on the MyD88 molecule, which in turn associates with IRAK serine/threonine kinase family members through shared motifs called death domains. Activated IRAKs signal through TRAF6 to transcription regulators such as AP1 and NF-kB. The relative importance of the surface receptors differs depending on the specific host cell type and the fungal pathogen detected.

of cytokines and by the killing of the pathogen, to both *A. fumigatus* and *C. albicans*. Intriguingly, although the MyD88-defective macrophage had a reduced ability to phagocytose and kill *Candida* cells, they were as proficient as wild-type macrophages in the killing of *A. fumigatus* conidia. In addition, induced cytokine production appeared to be pathogen specific; TNF-α production in response to either the yeast or hyphal form of *C. albicans* was reduced in the MyD88-defective macrophages, while the low level of TNF-α induced by *Aspergillus* hyphae was unaffected by the MyD88 state of the cells (86). This defines potential differences in the requirement for MyD88 in responses to specific fungal pathogens. However, analysis of the virulence of the pathogens *C. albicans* and *A. fumigatus* measured as the survival of infected mice showed that MyD88-defective mice were inherently more susceptible to infection with both pathogens (16). This difference in the role of MyD88 between the response of

a single cell type and the response at the level of the whole organism is not unexpected, but it underlines the need to analyze such interactions between the host and the pathogen at many levels to get a complete picture of the signaling process.

As yet, little is known about the involvement of further downstream components in the response to fungal pathogens. Components such as the IRAKs and TRAF6, which are predicted to be downstream in the pathway leading to the activation of inflammatory cytokines, have not yet been directly implicated in responses to fungal pathogens. However, the observation that cytokines are in fact activated in response to fungal infections suggests that these elements will be involved. Knockout mice and small interfering RNA treatments of cultured cells will provide the opportunity to test the specific details of the information transfer to transcription factors such as NF-κB and AP1 during the response of mammalian cells to fungal pathogens.

Surface Receptors

In addition to the TLR proteins, other surface molecules play a role in the recognition of fungal pathogens (100). The mannose receptor (MR), previously known as the macrophage mannose receptor, has been implicated in the recognition of *C. albicans*. For example, expression of the mannose receptor in Cos-1 cells permitted these cells to recognize and phagocytose *Candida* cells (50). In addition, various strategies to reduce MR activity, such as the use of antisense oligonucleotides or treatment of cells with MR ligands to compete with pathogen cells, suggested an important role of the MR in pathogen recognition (85, 136). However, direct analysis of MR-defective mice injected with *C. albicans* showed very little difference in the infection process (73). This result highlights the different effects seen in specific cells and the whole organism when it comes to the response to fungal pathogens. The ability of the macrophages from the MR-defective mice to deal with *C. albicans* was efficiently blocked by the addition of β-glucan, suggesting that a β-glucan receptor plays a critical role in the protection against *Candida* infection in these mice (73). Further experiments are necessary to identify the complete collection of receptors that act in the identification of *C. albicans* at the organismal level.

The MR is also implicated in the uptake and processing of *C. neoformans* by dendritic cells (119). As well, growing collections of other surface molecules are implicated in pathogen recognition. These include dendritic cell-specific ICAM-grabbing nonintegrin (DC-SIGN), which is important in the recognition of *A. fumigatus* (114), galectin-3, which plays a role in the recognition of *C. albicans* β-1,2-oligomannosides (55), and dectin-1, which is a key receptor for β-glucans (23). The molecular function of these receptor proteins still remains to be determined. In particular, it remains to be established if the receptors serve simply as binding surfaces for physical interactions between the pathogens and the host or whether the association between the ligand and the receptor transmits a signal to the intracellular machinery of the mammalian cells.

Sensitivity Determinants

The approach of building from known immunity determinants and their signaling networks has proven an important strategy to identifying elements involved in the ability of mammalian hosts to respond to fungal pathogens. An alternative method is to investigate directly the components needed by the host to properly respond to a fungal pathogen by identifying genetic variation in host response and establishing the host genes involved. This approach has also been successfully applied to the analysis of innate immunity components involved in host resistance to *C. albicans* (126). Genetic mapping has identified *Carg1* and *Carg2* as *C. albicans* susceptibility loci, but the cellular function defined by these loci remains to be established (9). However, the primary locus directly implicated in pathogen susceptibility has been complement C5. In a tail vein injection model of mouse systemic candidiasis, significant survival differences were observed between C5 complement-defective and C5 complement-proficient strains (10). A similar involvement of C5 in the pathology of *C. neoformans* had been observed previously (103). The *app1* mutant of *C. neoformans* is defective in an antiphagocytotic protein; mutants lacking this protein show reduced virulence in the C5-defective A/Jcr strains but hypervirulence in mouse strains defective in T and NK cells (82). Recently, an analysis of the *C. albicans*-sensitive, complement C5-defective A/J inbred mouse strain suggested that a dysregulated cytokine response was responsible for the heightened pathogen sensitivity (91). A systematic analysis of sensitivity loci should provide a direct way to identify components of signaling pathways critical for controlling the overall virulence of the pathogen.

Transcriptional Profiling

The postgenomic era promises to increase our level of understanding of the signaling networks involved in the response of mammals to invading pathogens. For example, the use of DNA microarrays will allow the consequences of the interaction between specific host cells and fungal pathogens to be investigated on a comprehensive scale. *C. albicans* was one subject of a pioneering transcriptional profiling study of a set of different pathogens interacting with dendritic cells (61). In this investigation, purified human monocytes were differentiated into dendritic cells during a week-long incubation with granulocyte-macrophage colony-stimulating factor and interleukin-4. These induced dendritic cells were then challenged with *Escherichia coli*, influenza virus, and *C. albicans*, and the similarities and differences in the pattern of expression of the human cells were monitored over a time course. This study detected a set of 166 genes whose expression was commonly modulated by the three pathogens; in general, *C. albicans* regulated a subset of the bacterially induced genes and did not modulate any specific fungal pathogen genes. Examination of the cellular processes that were transcriptionally affected by the pathogens showed that genes encoding functions involved in phagocytosis and pathogen recognition were downregulated early in the dendritic cell-pathogen interaction while cytokine expression and the expression of genes implicated in morphogenesis were up-regulated. Overall, this investigation suggested that *C. albicans* was a relatively unexceptional pathogen with respect to the dendritic cell response. The transcriptional response to the fungal pathogen was intermediate in speed between the fast response to the bacteria and a slower response to the virus, and overall the response resembled a weaker version of the bacterially triggered transcription profile (61).

A further investigation into the transcriptional response of host cells to the presence of a fungal pathogen involved the use of differentiated HL60 granulocytoids and *C. albicans* (90). In this study, undifferentiated HL60 cells that were unable to kill *C. albicans* cells were compared with the dimethyl formamide-differentiated cells that were capable of phagocytosis, inhibition of hyphal differentiation, and pathogen killing. The presence of *Candida* cells triggered the activation of a variety of genes, including regulators of the inflammatory response. In addition, high pathogen-to-host-cell ratios resulted in the down-regulation of genes implicated in pathogen killing, as though the *C. albicans* cells were blocking their own destruction. Many of these responses were also seen in freshly purified human neutrophils, providing evidence that the results were not an artifact of the in vitro differentiated cell model (90).

In a separate study, we have investigated the transcriptional response of cultured mouse macrophage (RAW264.7) to the presence of *C. albicans* (A. Marcil and M. Whiteway, unpublished data). These macrophages are capable of engulfing and killing the *C. albicans*, although overall the pathogen-macrophage interaction has only a moderate effect on the viability of both cell types (84). Analysis of the expression of macrophage genes in the presence of the differentiating pathogen reveals that a limited set of genes were modulated, with the majority being up-regulated. At a multiplicity of infection of 1, where *Candida* could survive macrophage phagocytosis, we observed an up-regulation of transcripts associated with hypoxia and stress and a down-regulation of transferrin receptor mRNA. We also noted the absence of a proinflammatory cytokine response, although transcripts associated with lipopolysaccharide stimulation were up-regulated. This demonstrates that the presence of *C. albicans* could manipulate and redirect the host response and also indicates an important role for iron in *Candida* pathogenesis.

Recently, studies have been undertaken to analyze the transcriptional readout of cells responding not to the pathogen but, rather, to components of the pathogen. The initial study of dendritic cells analyzed the response to mannan (61). More recently, studies have investigated the response of peripheral blood mononuclear cells stimulated with *Candida* cell wall glucan; the regulated genes encoded effectors with well-characterized proinflammatory properties (62). Such studies underline the potential of DNA microarrays in defining responses of host cells to fungal pathogens. Because a transcription profile can provide a multifaceted readout of the consequences of the interaction, it will provide a useful tool in the analysis of the host-pathogen interactions. The detail in the assay may also provide insight into the molecular pathways involved in the response. Such reductionism may eventually lead to a detailed picture of the cellular response to each pathogen, but the number of combinations of host cell types and pathogen components makes this an enormous undertaking. The hope is that broad patterns of response will be identifiable as these studies develop; too much specificity will overwhelm the researcher in a torrent of data.

SUMMARY

The developing tools of knockout mice, genome sequences, cultured cell lines, and DNA microarrays are having a profound impact on our ability to define the interaction between mammalian host cells and fungal pathogens. We are reaching the point where the global transcription pattern of a particular mammalian cell type can be measured in response to a fungal pathogen, and then the consequences of genetic manipulation of both the host cell and the pathogen can be determined for comparison. For example, we can ask what are the genes that come on in a human dendritic cell in response to *A. nidulans*, and what happens when those dendritic cells lack MyD88? We can determine if there is a difference in the response of a TLR2-defective macrophage to a capsule-defective *C. neoformans* strain compared to a wild-type pathogen. We can ask about the role of a specific IRAK kinase in signaling the presence of *C. albicans* and whether that role changes when the *C. albicans* cell is trapped in the yeast form. Ultimately, we will find if the answers to any of these questions allow us to better understand the progression of a fungal disease in a human patient. The tools to address such questions are here or soon to arrive, the questions themselves are exciting and answerable, and the potential payoffs for understanding the signaling pathways in fungal pathogens are profound.

Acknowledgments. We acknowledge support from members of the Whiteway and Bachewich labs, especially Anne Marcil for unpublished observations. This work was supported by the NRC Genomics and Health Initiative and by Canadian Institutes of Health Research grant MOP-42516 to M.W. This is National Research Council publication 4740.

REFERENCES

1. **Alarco, A. M., A. Marcil, J. Chen, B. Suter, D. Thomas, and M. Whiteway.** 2004. Immune-deficient *Drosophila melanogaster*: a model for the innate immune response to human fungal pathogens. *J. Immunol.* **172:** 5622–5628.

2. **Alonso-Monge, R., F. Navarro-Garcia, G. Molero, R. Diez-Orejas, M. Gustin, J. Pla, M. Sanchez, and C. Nombela.** 1999. Role of the mitogen-activated protein kinase Hog1p in morphogenesis and virulence of *Candida albicans. J. Bacteriol.* **181:**3058–3068.

3. **Alonso-Monge, R., F. Navarro-Garcia, E. Roman, A. I. Negredo, B. Eisman, C. Nombela, and J. Pla.** 2003. The Hog1 mitogen-activated protein kinase is essential in the oxidative stress response and chlamydospore formation in *Candida albicans. Eukaryot. Cell* **2:**351–361.

4. **Alspaugh, J. A., L. M. Cavallo, J. R. Perfect, and J. Heitman.** 2000. RAS1 regulates filamentation, mating and growth at high temperature of *Cryptococcus neoformans. Mol. Microbiol.* **36:**352–365.

5. **Alspaugh, J. A., J. R. Perfect, and J. Heitman.** 1997. *Cryptococcus neoformans* mating and virulence are regulated by the G-protein alpha subunit GPA1 and cAMP. *Genes Dev.* **11:**3206–3217.

6. **Alspaugh, J. A., R. Pukkila-Worley, T. Harashima, L. M. Cavallo, D. Funnell, G. M. Cox, J. R. Perfect, J. W. Kronstad, and J. Heitman.** 2002. Adenylyl cyclase functions downstream of the Gα protein Gpa1 and controls mating and pathogenicity of *Cryptococcus neoformans. Eukaryot. Cell* **1:**75–84.

7. **Apidianakis, Y., L. G. Rahme, J. Heitman, F. M. Ausubel, S. B. Calderwood, and E. Mylonakis.** 2004. Challenge of *Drosophila melanogaster* with *Cryptococcus neoformans* and role of the innate immune response. *Eukaryot. Cell* **3:**413–419.

8. **Arana, D. M., C. Nombela, R. Alonso-Monge, and J. Pla.** 2005. The Pbs2 MAP kinase kinase is essential for the oxidative-stress response in the fungal pathogen *Candida albicans. Microbiology* **151:**1033–1049.

9. **Ashman, R. B., A. Fulurija, and J. M. Papadimitriou.** 1998. A second *Candida albicans* resistance gene (*Carg2*) regulates tissue damage, but not fungal clearance, in sub-lethal murine systemic infection. *Microb. Pathog.* **25:**349–352.

10. **Ashman, R. B., J. M. Papadimitriou, A. Fulurija, K. E. Drysdale, C. S. Farah, O. Naidoo, and T. Gotjamanos.** 2003. Role of complement C5 and T lymphocytes in pathogenesis of disseminated and mucosal candidiasis in susceptible DBA/2 mice. *Microb. Pathog.* **34:**103–113.

11. **Bader, T., B. Bodendorfer, K. Schroppel, and J. Morschhauser.** 2003. Calcineurin is essential for virulence in *Candida albicans. Infect. Immun.* **71:**5344–5354.

12. **Bahn, Y. S., J. K. Hicks, S. S. Giles, G. M. Cox, and J. Heitman.** 2004. Adenylyl cyclase-associated protein Aca1 regulates virulence and differentiation of *Cryptococcus neoformans* via the cyclic AMP-protein kinase A cascade. *Eukaryot. Cell* **3:**1476–1491.

13. **Bahn, Y. S., K. Kojima, G. M. Cox, and J. Heitman.** 2005. Specialization of the HOG pathway and its impact on differentiation and virulence of *Cryptococcus neoformans. Mol. Biol. Cell* **16:**2285–2300.

14. **Bahn, Y. S., J. Staab, and P. Sundstrom.** 2003. Increased high-affinity phosphodiesterase *PDE2* gene expression in germ tubes counteracts CAP1-dependent synthesis of cyclic AMP, limits hypha production and promotes virulence of *Candida albicans. Mol. Microbiol.* **50:**391–409.

15. **Bahn, Y. S., and P. Sundstrom.** 2001. *CAP1*, an adenylate cyclase-associated protein gene, regulates bud-hypha transitions, filamentous growth, and cyclic AMP levels and is required for virulence of *Candida albicans. J. Bacteriol.* **183:**3211–3223.

16. **Bellocchio, S., C. Montagnoli, S. Bozza, R. Gaziano, G. Rossi, S. S. Mambula, A. Vecchi, A. Mantovani, S. M. Levitz, and L. Romani.** 2004. The contribution of the Toll-like/IL-1 receptor superfamily to innate and adaptive immunity to fungal pathogens in vivo. *J. Immunol.* **172:**3059–3069.

17. **Bensen, E. S., S. J. Martin, M. Li, J. Berman, and D. A. Davis.** 2004. Transcriptional profiling in *Candida albicans* reveals new adaptive responses to extracellular pH and functions for Rim101p. *Mol. Microbiol.* **54:**1335–1351.

18. **Berman, J., and P. E. Sudbery.** 2002. *Candida albicans*: a molecular revolution built on lessons from budding yeast. *Nat. Rev. Genet.* **3:**918–930.

19. **Blankenship, J. R., F. L. Wormley, M. K. Boyce, W. A. Schell, S. G. Filler, J. R. Perfect, and J. Heitman.** 2003. Calcineurin is essential for *Candida albicans* survival in serum and virulence. *Eukaryot. Cell* **2:**422–430.

20. **Borges-Walmsley, M. I., and A. R. Walmsley.** 2000. cAMP signalling in pathogenic fungi: control of dimorphic switching and pathogenicity. *Trends Microbiol.* **8:**133–141.

21. **Braedel, S., M. Radsak, H. Einsele, J. P. Latge, A. Michan, J. Loeffler, Z. Haddad, U. Grigoleit, H. Schild, and H. Hebart.** 2004. Aspergillus fumigatus antigens activate innate immune cells via toll-like receptors 2 and 4. *Br. J. Haematol.* **125:**392–399.

22. **Braun, B. R., and A. D. Johnson.** 1997. Control of filament formation in *Candida albicans* by the transcriptional repressor TUP1. *Science* **277:**105–109.

23. **Brown, G. D., J. Herre, D. L. Williams, J. A. Willment, A. S. Marshall, and S. Gordon.** 2003. Dectin-1 mediates the biological effects of beta-glucans. *J. Exp. Med.* **197:**1119–1124.

24. **Calera, J. A., and R. Calderone.** 1999. Flocculation of hyphae is associated with a deletion in the putative *CaHK1* two-component histidine kinase gene from *Candida albicans. Microbiology* **145:**1431–1442.

25. **Calera, J. A., G. H. Choi, and R. A. Calderone.** 1998. Identification of a putative histidine kinase two-component phosphorelay gene (*CaHK1*) in *Candida albicans. Yeast* **14:**665–674.

26. **Calera, J. A., D. Herman, and R. Calderone.** 2000. Identification of *YPD1*, a gene of *Candida albicans* which encodes a two-component phosphohistidine intermediate protein. *Yeast* **16:**1053–1059.

27. **Calera, J. A., X. J. Zhao, and R. Calderone.** 2000. Defective hyphal development and avirulence caused by a deletion of the *SSK1* response regulator gene in Candida albicans. *Infect. Immun.* **68:**518–525.

28. **Cassola, A., M. Parrot, S. Silberstein, B. B. Magee, S. Passeron, L. Giasson, and M. L. Cantore.** 2004. *Candida albicans* lacking the gene encoding the regulatory subunit of protein kinase A displays a defect in hyphal formation and an altered localization of the catalytic subunit. *Eukaryot. Cell* **3:**190–199.

29. **Castilla, R., S. Passeron, and M. L. Cantore.** 1998. *N*-Acetyl-D-glucosamine induces germination in *Candida*

albicans through a mechanism sensitive to inhibitors of cAMP-dependent protein kinase. *Cell Signal.* **10:**713–719.

30. **Catlett, N. L., O. C. Yoder, and B. G. Turgeon.** 2003. Whole-genome analysis of two-component signal transduction genes in fungal pathogens. *Eukaryot. Cell* **2:**1151–1161.

31. **Chattaway, F. W., P. R. Wheeler, and J. O'Reilly.** 1981. Involvement of adenosine 3':5'-cyclic monophosphate in the germination of blastospores of *Candida albicans*. *J. Gen. Microbiol.* **123:**233–240.

32. **Chauhan, N., D. Inglis, E. Roman, J. Pla, D. Li, J. A. Calera, and R. Calderone.** 2003. *Candida albicans* response regulator gene *SSK1* regulates a subset of genes whose functions are associated with cell wall biosynthesis and adaptation to oxidative stress. *Eukaryot. Cell* **2:**1018–1024.

33. **Cheng, S., C. J. Clancy, M. A. Checkley, M. Handfield, J. D. Hillman, A. Progulske-Fox, A. S. Lewin, P. L. Fidel, and M. H. Nguyen.** 2003. Identification of *Candida albicans* genes induced during thrush offers insight into pathogenesis. *Mol. Microbiol.* **48:**1275–1288.

34. **Clarke, D. L., G. L. Woodlee, C. M. McClelland, T. S. Seymour, and B. L. Wickes.** 2001. The *Cryptococcus neoformans STE11α* gene is similar to other fungal mitogen-activated protein kinase kinase kinase (MAPKKK) genes but is mating type specific. *Mol. Microbiol.* **40:**200–213.

35. **Clemons, K. V., T. K. Miller, C. P. Selitrennikoff, and D. A. Stevens.** 2002. fos-1, a putative histidine kinase as a virulence factor for systemic aspergillosis. *Med. Mycol.* **40:**259–262.

36. **Cloutier, M., R. Castilla, N. Bolduc, A. Zelada, P. Martineau, M. Bouillon, B. B. Magee, S. Passeron, L. Giasson, and M. L. Cantore.** 2003. The two isoforms of the cAMP-dependent protein kinase catalytic subunit are involved in the control of dimorphism in the human fungal pathogen *Candida albicans*. *Fungal Genet. Biol.* **38:**133–141.

37. **Cruz, M. C., D. S. Fox, and J. Heitman.** 2001. Calcineurin is required for hyphal elongation during mating and haploid fruiting in *Cryptococcus neoformans*. *EMBO J.* **20:**1020–1032.

38. **Csank, C., C. Makris, S. Meloche, K. Schroppel, M. Rollinghoff, D. Dignard, D. Y. Thomas, and M. Whiteway.** 1997. Derepressed hyphal growth and reduced virulence in a VH1 family-related protein phosphatase mutant of the human pathogen *Candida albicans*. *Mol. Biol. Cell* **8:**2539–2551.

39. **Csank, C., K. Schroppel, E. Leberer, D. Harcus, O. Mohamed, S. Meloche, D. Y. Thomas, and M. Whiteway.** 1998. Roles of the *Candida albicans* mitogen-activated protein kinase homolog, Cek1p, in hyphal development and systemic candidiasis. *Infect. Immun.* **66:**2713–2721.

40. **Davidson, R. C., C. B. Nichols, G. M. Cox, J. R. Perfect, and J. Heitman.** 2003. A MAP kinase cascade composed of cell type specific and non-specific elements controls mating and differentiation of the fungal pathogen *Cryptococcus neoformans*. *Mol. Microbiol.* **49:**469–485.

41. **Davis, D.** 2003. Adaptation to environmental pH in *Candida albicans* and its relation to pathogenesis. *Curr. Genet.* **44:**1–7.

42. **Davis, D., J. E. Edwards, Jr., A. P. Mitchell, and A. S. Ibrahim.** 2000. *Candida albicans* RIM101 pH response pathway is required for host-pathogen interactions. *Infect. Immun.* **68:**5953–5959.

43. **Davis, D., R. B. Wilson, and A. P. Mitchell.** 2000. RIM101-dependent and -independent pathways govern pH responses in *Candida albicans*. *Mol. Cell. Biol.* **20:**971–978.

44. **Davis, D. A., V. M. Bruno, L. Loza, S. G. Filler, and A. P. Mitchell.** 2002. *Candida albicans* Mds3p, a conserved regulator of pH responses and virulence identified through insertional mutagenesis. *Genetics* **162:**1573–1581.

45. **De Bernardis, F., F. A. Muhlschlegel, A. Cassone, and W. A. Fonzi.** 1998. The pH of the host niche controls gene expression in and virulence of *Candida albicans*. *Infect. Immun.* **66:**3317–3325.

46. **Diez-Orejas, R., G. Molero, F. Navarro-Garcia, J. Pla, C. Nombela, and M. Sanchez-Perez.** 1997. Reduced virulence of *Candida albicans MKC1* mutants: a role for mitogen-activated protein kinase in pathogenesis. *Infect. Immun.* **65:**833–837.

47. **Du, C., R. Li, S. Ma, and D. Wang.** 2002. Cloning of *Aspergillus fumigatus* histidine kinase gene fragment and its expression during invasive infection. *Mycopathologia* **153:**5–10.

48. **Edmond, M. B., S. E. Wallace, D. K. McClish, M. A. Pfaller, R. N. Jones, and R. P. Wenzel.** 1999. Nosocomial bloodstream infections in United States hospitals: a three-year analysis. *Clin. Infect. Dis.* **29:**239–244.

49. **Ernst, J. F.** 2000. Transcription factors in *Candida albicans*—environmental control of morphogenesis. *Microbiology.* **146:**1763–1774.

50. **Ezekowitz, R. A., K. Sastry, P. Bailly, and A. Warner.** 1990. Molecular characterization of the human macrophage mannose receptor: demonstration of multiple carbohydrate recognition-like domains and phagocytosis of yeasts in Cos-1 cells. *J. Exp. Med.* **172:**1785–1794.

51. **Feng, Q., E. Summers, B. Guo, and G. Fink.** 1999. Ras signaling is required for serum-induced hyphal differentiation in *Candida albicans*. *J. Bacteriol.* **181:**6339–6346.

52. **Fortwendel, J. R., J. C. Panepinto, A. E. Seitz, D. S. Askew, and J. C. Rhodes.** 2004. *Aspergillus fumigatus rasA* and *rasB* regulate the timing and morphology of asexual development. *Fungal Genet. Biol.* **41:**129–139.

53. **Fox, D. S., M. C. Cruz, R. A. Sia, H. Ke, G. M. Cox, M. E. Cardenas, and J. Heitman.** 2001. Calcineurin regulatory subunit is essential for virulence and mediates interactions with FKBP12-FK506 in *Cryptococcus neoformans*. *Mol. Microbiol.* **39:**835–849.

54. **Fradin, C., M. Kretschmar, T. Nichterlein, C. Gaillardin, C. d'Enfert, and B. Hube.** 2003. Stage-specific gene expression of *Candida albicans* in human blood. *Mol. Microbiol.* **47:**1523–1543.

55. **Fradin, C., D. Poulain, and T. Jouault.** 2000. β-1,2-Linked oligomannosides from *Candida albicans* bind to a 32-kilodalton macrophage membrane protein homologous to the mammalian lectin galectin-3. *Infect. Immun.* **68:**4391–4398.

56. **Gerson, S. L., G. H. Talbot, S. Hurwitz, B. L. Strom, E. J. Lusk, and P. A. Cassileth.** 1984. Prolonged granulocytopenia: the major risk factor for invasive pulmonary aspergillosis in patients with acute leukemia. *Ann. Intern. Med.* **100:**345–351.

57. **Ghannoum, M. A., B. Spellberg, S. M. Saporito-Irwin, and W. A. Fonzi.** 1995. Reduced virulence of *Candida albicans PHR1* mutants. *Infect. Immun.* **63:**4528–4530.

58. **Harcus, D., A. Nantel, A. Marcil, T. Rigby, and M. Whiteway.** 2004. Transcription profiling of cyclic

AMP signaling in *Candida albicans. Mol. Biol. Cell* **15:**4490–4499.

59. **Hicks, J. K., C. A. D'Souza, G. M. Cox, and J. Heitman.** 2004. Cyclic AMP-dependent protein kinase catalytic subunits have divergent roles in virulence factor production in two varieties of the fungal pathogen *Cryptococcus neoformans. Eukaryot. Cell* **3:**14–26.

60. **Hoch, J. A.** 2000. Two-component and phosphorelay signal transduction. *Curr. Opin. Microbiol.* **3:**165–170.

61. **Huang, Q., D. Liu, P. Majewski, L. C. Schulte, J. M. Korn, R. A. Young, E. S. Lander, and N. Hacohen.** 2001. The plasticity of dendritic cell responses to pathogens and their components. *Science* **294:**870–875.

62. **Ishibashi, K., N. N. Miura, Y. Adachi, N. Ogura, H. Tamura, S. Tanaka, and N. Ohno.** 2004. DNA array analysis of altered gene expression in human leukocytes stimulated with soluble and particulate forms of *Candida* cell wall beta-glucan. *Int. Immunopharmacol.* **4:**387–401.

63. **Jain, P., I. Akula, and T. Edlind.** 2003. Cyclic AMP signaling pathway modulates susceptibility of *Candida* species and *Saccharomyces cerevisiae* to antifungal azoles and other sterol biosynthesis inhibitors. *Antimicrob. Agents Chemother.* **47:**3195–3201.

64. **Jouault, T., S. Ibata-Ombetta, O. Takeuchi, P. A. Trinel, P. Sacchetti, P. Lefebvre, S. Akira, and D. Poulain.** 2003. *Candida albicans* phospholipomannan is sensed through toll-like receptors. *J. Infect. Dis.* **188:**165–172.

65. **Jung, W. H., and L. I. Stateva.** 2003. The cAMP phosphodiesterase encoded by *CaPDE2* is required for hyphal development in *Candida albicans. Microbiology* **149:**2961–2976.

66. **Kraus, P. R., D. S. Fox, G. M. Cox, and J. Heitman.** 2003. The *Cryptococcus neoformans* MAP kinase Mpk1 regulates cell integrity in response to antifungal drugs and loss of calcineurin function. *Mol. Microbiol.* **48:**1377–1387.

67. **Kraus, P. R., and J. Heitman.** 2003. Coping with stress: calmodulin and calcineurin in model and pathogenic fungi. *Biochem. Biophys. Res. Commun.* **311:**1151–1157.

68. **Kwon-Chung, K. J., and J. C. Rhodes.** 1986. Encapsulation and melanin formation as indicators of virulence in *Cryptococcus neoformans. Infect. Immun.* **51:**218–223.

69. **Langfelder, K., B. Jahn, H. Gehringer, A. Schmidt, G. Wanner, and A. A. Brakhage.** 1998. Identification of a polyketide synthase gene (*pksP*) of *Aspergillus fumigatus* involved in conidial pigment biosynthesis and virulence. *Med. Microbiol. Immunol.* (Berlin) **187:**79–89.

70. **Latge, J. P.** 2001. The pathobiology of *Aspergillus fumigatus. Trends Microbiol.* **9:**382–389.

71. **Leberer, E., D. Harcus, I. D. Broadbent, K. L. Clark, D. Dignard, K. Ziegelbauer, A. Schmidt, N. A. Gow, A. J. Brown, and D. Y. Thomas.** 1996. Signal transduction through homologs of the Ste20p and Ste7p protein kinases can trigger hyphal formation in the pathogenic fungus *Candida albicans. Proc. Natl. Acad. Sci. USA* **93:**13217–13222.

72. **Leberer, E., D. Harcus, D. Dignard, L. Johnson, S. Ushinsky, D. Y. Thomas, and K. Schroppel.** 2001. Ras links cellular morphogenesis to virulence by regulation of the MAP kinase and cAMP signalling pathways in the pathogenic fungus *Candida albicans. Mol. Microbiol.* **42:**673–687.

73. **Lee, S. J., N. Y. Zheng, M. Clavijo, and M. C. Nussenzweig.** 2003. Normal host defense during systemic candidiasis in mannose receptor-deficient mice. *Infect. Immun.* **71:**437–445.

74. **Lemaitre, B., E. Nicolas, L. Michaut, J. M. Reichhart, and J. A. Hoffmann.** 1996. The dorsoventral regulatory gene cassette spatzle/Toll/cactus controls the potent antifungal response in *Drosophila* adults. *Cell* **86:**973–983.

75. **Lengeler, K. B., R. C. Davidson, C. D'Souza, T. Harashima, W. C. Shen, P. Wang, X. Pan, M. Waugh, and J. Heitman.** 2000. Signal transduction cascades regulating fungal development and virulence. *Microbiol. Mol. Biol. Rev.* **64:**746–785.

76. **Li, D., V. Gurkovska, M. Sheridan, R. Calderone, and N. Chauhan.** 2004. Studies on the regulation of the two-component histidine kinase gene *CHK1* in *Candida albicans* using the heterologous *lacZ* reporter gene. *Microbiology* **150:**3305–3313.

77. **Liebmann, B., S. Gattung, B. Jahn, and A. A. Brakhage.** 2003. cAMP signaling in Aspergillus fumigatus is involved in the regulation of the virulence gene *pksP* and in defense against killing by macrophages. *Mol. Genet. Genomics* **269:**420–435.

78. **Liebmann, B., M. Muller, A. Braun, and A. A. Brakhage.** 2004. The cyclic AMP-dependent protein kinase a network regulates development and virulence in *Aspergillus fumigatus. Infect. Immun.* **72:**5193–5203.

79. **Liu, H., J. Kohler, and G. R. Fink.** 1994. Suppression of hyphal formation in *Candida albicans* by mutation of a *STE12* homolog. *Science* **266:**1723–1726.

80. **Lo, H. J., J. R. Kohler, B. DiDomenico, D. Loebenberg, A. Cacciapuoti, and G. R. Fink.** 1997. Nonfilamentous *C. albicans* mutants are avirulent. *Cell* **90:**939–949.

81. **Lorenz, M. C., J. A. Bender, and G. R. Fink.** 2004. Transcriptional response of *Candida albicans* upon internalization by macrophages. *Eukaryot. Cell* **3:**1076–1087.

82. **Luberto, C., B. Martinez-Marino, D. Taraskiewicz, B. Bolanos, P. Chitano, D. L. Toffaletti, G. M. Cox, J. R. Perfect, Y. A. Hannun, E. Balish, and M. Del Poeta.** 2003. Identification of App1 as a regulator of phagocytosis and virulence of *Cryptococcus neoformans. J. Clin. Investig.* **112:**1080–1094.

83. **Maidan, M. M., L. De Rop, J. Serneels, S. Exler, S. Rupp, H. Tournu, J. M. Thevelein, and P. Van Dijck.** 2005. The G protein-coupled receptor Gpr1 and the Gα protein Gpa2 act through the cAMP-protein kinase A pathway to induce morphogenesis in *Candida albicans. Mol. Biol. Cell* **16:**1971–1986.

84. **Marcil, A., D. Harcus, D. Y. Thomas, and M. Whiteway.** 2002. *Candida albicans* killing by RAW 264.7 mouse macrophage cells: effects of *Candida* genotype, infection ratios, and gamma interferon treatment. *Infect. Immun.* **70:**6319–6329.

85. **Marodi, L., S. Schreiber, D. C. Anderson, R. P. MacDermott, H. M. Korchak, and R. B. Johnston, Jr.** 1993. Enhancement of macrophage candidacidal activity by interferon-gamma. Increased phagocytosis, killing, and calcium signal mediated by a decreased number of mannose receptors. *J. Clin. Investig.* **91:**2596–2601.

86. **Marr, K. A., S. A. Balajee, T. R. Hawn, A. Ozinsky, U. Pham, S. Akira, A. Aderem, and W. C. Liles.** 2003. Differential role of MyD88 in macrophage-mediated

responses to opportunistic fungal pathogens. *Infect. Immun.* **71:**5280–5286.

87. **Matsumoto, K., I. Uno, Y. Oshima, and T. Ishikawa.** 1982. Isolation and characterization of yeast mutants deficient in adenylate cyclase and cAMP-dependent protein kinase. *Proc. Natl. Acad. Sci. USA* **79:** 2355–2359.

88. **Miwa, T., Y. Takagi, M. Shinozaki, C. W. Yun, W. A. Schell, J. R. Perfect, H. Kumagai, and H. Tamaki.** 2004. Gpr1, a putative G-protein-coupled receptor, regulates morphogenesis and hypha formation in the pathogenic fungus *Candida albicans. Eukaryot. Cell* **3:**919–931.

89. **Muhlschlegel, F. A., and W. A. Fonzi.** 1997. *PHR2* of *Candida albicans* encodes a functional homolog of the pH-regulated gene *PHR1* with an inverted pattern of pH-dependent expression. *Mol. Cell. Biol.* **17:** 5960–5967.

90. **Mullick, A., M. Elias, P. Harakidas, A. Marcil, M. Whiteway, B. Ge, T. J. Hudson, A. W. Caron, L. Bourget, S. Picard, O. Jovcevski, B. Massie, and D. Y. Thomas.** 2004. Gene expression in HL60 granulocytoids and human polymorphonuclear leukocytes exposed to *Candida albicans. Infect. Immun.* **72:**414–429.

91. **Mullick, A., M. Elias, S. Picard, L. Bourget, O. Jovcevski, S. Gauthier, A. Tuite, P. Harakidas, C. Bihun, B. Massie, and P. Gros.** 2004. Dysregulated inflammatory response to *Candida albicans* in a C5-deficient mouse strain. *Infect. Immun.* **72:**5868–5876.

92. **Nagahashi, S., T. Mio, N. Ono, T. Yamada-Okabe, M. Arisawa, H. Bussey, and H. Yamada-Okabe.** 1998. Isolation of CaSLN1 and CaNIK1, the genes for osmosensing histidine kinase homologues, from the pathogenic fungus *Candida albicans. Microbiology* **144:**425–432.

93. **Navarro-Garcia, F., R. Alonso-Monge, H. Rico, J. Pla, R. Sentandreu, and C. Nombela.** 1998. A role for the MAP kinase gene *MKC1* in cell wall construction and morphological transitions in *Candida albicans. Microbiology* **144:**411–424.

94. **Netea, M. G., C. A. Van Der Graaf, A. G. Vonk, I. Verschueren, J. W. Van Der Meer, and B. J. Kullberg.** 2002. The role of toll-like receptor (TLR) 2 and TLR4 in the host defense against disseminated candidiasis. *J. Infect. Dis.* **185:**1483–1489.

95. **Niimi, M.** 1996. Dibutyryl cyclic AMP-enhanced germ tube formation in exponentially growing *Candida albicans* cells. *Fungal Genet. Biol.* **20:**79–83.

96. **Odom, A., S. Muir, E. Lim, D. L. Toffaletti, J. Perfect, and J. Heitman.** 1997. Calcineurin is required for virulence of *Cryptococcus neoformans. EMBO J.* **16:**2576–2589.

97. **Onyewu, C., J. R. Blankenship, M. Del Poeta, and J. Heitman.** 2003. Ergosterol biosynthesis inhibitors become fungicidal when combined with calcineurin inhibitors against *Candida albicans, Candida glabrata,* and *Candida krusei. Antimicrob. Agents Chemother.* **47:**956–964.

98. **Onyewu, C., F. L. Wormley, Jr., J. R. Perfect, and J. Heitman.** 2004. The calcineurin target, Crz1, functions in azole tolerance but is not required for virulence of *Candida albicans. Infect. Immun.* **72:**7330–7333.

99. **Posas, F., S. M. Wurgler-Murphy, T. Maeda, E. A. Witten, T. C. Thai, and H. Saito.** 1996. Yeast HOG1 MAP kinase cascade is regulated by a multistep phosphorelay mechanism in the SLN1-YPD1-SSK1 "two-component" osmosensor. *Cell* **86:**865–875.

100. **Poulain, D., and T. Jouault.** 2004. Candida albicans cell wall glycans, host receptors and responses: elements for a decisive crosstalk. *Curr. Opin. Microbiol.* **7:**342–349.

101. **Ramon, A. M., and W. A. Fonzi.** 2003. Diverged binding specificity of Rim101p, the *Candida albicans* ortholog of PacC. *Eukaryot. Cell* **2:**718–728.

102. **Rhodes, J. C., B. G. Oliver, D. S. Askew, and T. W. Amlung.** 2001. Identification of genes of *Aspergillus fumigatus* up-regulated during growth on endothelial cells. *Med. Mycol.* **39:**253–260.

103. **Rhodes, J. C., L. S. Wicker, and W. J. Urba.** 1980. Genetic control of susceptibility to *Cryptococcus neoformans* in mice. *Infect. Immun.* **29:**494–499.

104. **Rocha, C. R., K. Schroppel, D. Harcus, A. Marcil, D. Dignard, B. N. Taylor, D. Y. Thomas, M. Whiteway, and E. Leberer.** 2001. Signaling through adenylyl cyclase is essential for hyphal growth and virulence in the pathogenic fungus *Candida albicans. Mol. Biol. Cell* **12:**3631–3643.

105. **Roeder, A., C. J. Kirschning, R. A. Rupec, M. Schaller, and H. C. Korting.** 2004. Toll-like receptors and innate antifungal responses. *Trends Microbiol.* **12:**44–49.

106. **Romani, L., F. Bistoni, and P. Puccetti.** 2002. Fungi, dendritic cells and receptors: a host perspective of fungal virulence. *Trends Microbiol.* **10:**508–514.

107. **Rubin-Bejerano, I., I. Fraser, P. Grisafi, and G. R. Fink.** 2003. Phagocytosis by neutrophils induces an amino acid deprivation response in *Saccharomyces cerevisiae* and *Candida albicans. Proc. Natl. Acad. Sci. USA* **100:**11007–11012.

108. **Sabie, F. T., and G. M. Gadd.** 1992. Effect of nucleosides and nucleotides and the relationship between cellular adenosine 3′:5′-cyclic monophosphate (cyclic AMP) and germ tube formation in *Candida albicans. Mycopathologia* **119:**147–156.

109. **Sanchez-Martinez, C., and J. Perez-Martin.** 2002. Gpa2, a G-protein alpha subunit required for hyphal development in *Candida albicans. Eukaryot. Cell* **1:**865–874.

110. **Sanglard, D., F. Ischer, O. Marchetti, J. Entenza, and J. Bille.** 2003. Calcineurin A of *Candida albicans*: involvement in antifungal tolerance, cell morphogenesis and virulence. *Mol. Microbiol.* **48:**959–976.

111. **Saporito-Irwin, S. M., C. E. Birse, P. S. Sypherd, and W. A. Fonzi.** 1995. *PHR1*, a pH-regulated gene of *Candida albicans*, is required for morphogenesis. *Mol. Cell. Biol.* **15:**601–613.

112. **Saville, S. P., A. L. Lazzell, C. Monteagudo, and J. L. Lopez-Ribot.** 2003. Engineered control of cell morphology in vivo reveals distinct roles for yeast and filamentous forms of *Candida albicans* during infection. *Eukaryot. Cell* **2:**1053–1060.

113. **Sentandreu, M., M. V. Elorza, R. Sentandreu, and W. A. Fonzi.** 1998. Cloning and characterization of *PRA1*, a gene encoding a novel pH-regulated antigen of *Candida albicans. J. Bacteriol.* **180:**282–289.

114. **Serrano-Gomez, D., A. Dominguez-Soto, J. Ancochea, J. A. Jimenez-Heffernan, J. A. Leal, and A. L. Corbi.** 2004. Dendritic cell-specific intercellular adhesion molecule 3-grabbing nonintegrin mediates binding and internalization of *Aspergillus fumigatus* conidia by dendritic cells and macrophages. *J. Immunol.* **173:**5635–5643.

115. **Shima, F., T. Okada, M. Kido, H. Sen, Y. Tanaka, M. Tamada, C. D. Hu, Y. Yamawaki-Kataoka, K. Kariya, and T. Kataoka.** 2000. Association of yeast adenylyl

cyclase with cyclase-associated protein CAP forms a second Ras-binding site which mediates its Ras-dependent activation. *Mol. Cell. Biol.* **20**:26–33.

116. Shoham, S., C. Huang, J. M. Chen, D. T. Golenbock, and S. M. Levitz. 2001. Toll-like receptor 4 mediates intracellular signaling without TNF-alpha release in response to *Cryptococcus neoformans* polysaccharide capsule. *J. Immunol.* **166**:4620–4626.

117. Singh, P., N. Chauhan, A. Ghosh, F. Dixon, and R. Calderone. 2004. SKN7 of *Candida albicans*: mutant construction and phenotype analysis. *Infect. Immun.* **72**:2390–2394.

118. Sonneborn, A., D. P. Bockmuhl, M. Gerads, K. Kurpanek, D. Sanglard, and J. F. Ernst. 2000. Protein kinase A encoded by *TPK2* regulates dimorphism of *Candida albicans*. *Mol. Microbiol.* **35**:386–396.

119. Syme, R. M., J. C. Spurrell, E. K. Amankwah, F. H. Green, and C. H. Mody. 2002. Primary dendritic cells phagocytose *Cryptococcus neoformans* via mannose receptors and Fcγ receptor II for presentation to T lymphocytes. *Infect. Immun.* **70**:5972–5981.

120. Tada, H., E. Nemoto, H. Shimauchi, T. Watanabe, T. Mikami, T. Matsumoto, N. Ohno, H. Tamura, K. Shibata, S. Akashi, K. Miyake, S. Sugawara, and H. Takada. 2002. *Saccharomyces cerevisiae*- and *Candida albicans*-derived mannan induced production of tumor necrosis factor alpha by human monocytes in a CD14- and Toll-like receptor 4-dependent manner. *Microbiol. Immunol.* **46**:503–512.

121. Takeda, K., and S. Akira. 2004. TLR signaling pathways. *Semin. Immunol.* **16**:3–9.

122. Thevelein, J. M., and J. H. de Winde. 1999. Novel sensing mechanisms and targets for the cAMP-protein kinase A pathway in the yeast *Saccharomyces cerevisiae*. *Mol. Microbiol.* **33**:904–918.

123. Toda, T., S. Cameron, P. Sass, M. Zoller, and M. Wigler. 1987. Three different genes in *S. cerevisiae* encode the catalytic subunits of the cAMP-dependent protein kinase. *Cell* **50**:277–287.

124. Toda, T., I. Uno, T. Ishikawa, S. Powers, T. Kataoka, D. Broek, S. Cameron, J. Broach, K. Matsumoto, and M. Wigler. 1985. In yeast, RAS proteins are controlling elements of adenylate cyclase. *Cell* **40**:27–36.

125. Tripathi, G., C. Wiltshire, S. Macaskill, H. Tournu, S. Budge, and A. J. Brown. 2002. Gcn4 co-ordinates morphogenetic and metabolic responses to amino acid starvation in *Candida albicans*. *EMBO J.* **21**:5448–5456.

126. Tuite, A., A. Mullick, and P. Gros. 2004. Genetic analysis of innate immunity in resistance to *Candida albicans*. *Genes. Immun.* **5**:576–587.

127. Wang, P., C. B. Nichols, K. B. Lengeler, M. E. Cardenas, G. M. Cox, J. R. Perfect, and J. Heitman. 2002. Mating-type-specific and nonspecific PAK kinases play shared and divergent roles in *Cryptococcus neoformans*. *Eukaryot. Cell* **1**:257–272.

128. Wang, P., J. R. Perfect, and J. Heitman. 2000. The G-protein beta subunit GPB1 is required for mating and haploid fruiting in *Cryptococcus neoformans*. *Mol. Cell. Biol.* **20**:352–362.

129. Waugh, M. S., C. B. Nichols, C. M. DeCesare, G. M. Cox, J. Heitman, and J. A. Alspaugh. 2002. Ras1 and Ras2 contribute shared and unique roles in physiology and virulence of *Cryptococcus neoformans*. *Microbiology* **148**:191–201.

130. Whiteway, M. 2000. Transcriptional control of cell type and morphogenesis in *Candida albicans*. *Curr. Opin. Microbiol.* **3**:582–588.

131. Xu, J. R. 2000. Map kinases in fungal pathogens. *Fungal Genet. Biol.* **31**:137–152.

132. Xu, W., F. J. Smith, Jr., R. Subaran, and A. P. Mitchell. 2004. Multivesicular body-ESCRT components function in pH response regulation in *Saccharomyces cerevisiae* and *Candida albicans*. *Mol. Biol. Cell* **15**:5528–5537.

133. Xu, Y., X. Tao, B. Shen, T. Horng, R. Medzhitov, J. L. Manley, and L. Tong. 2000. Structural basis for signal transduction by the Toll/interleukin-1 receptor domains. *Nature* **408**:111–115.

134. Xue, T., C. K. Nguyen, A. Romans, and G. S. May. 2004. A mitogen-activated protein kinase that senses nitrogen regulates conidial germination and growth in *Aspergillus fumigatus*. *Eukaryot. Cell* **3**:557–560.

135. Yamada-Okabe, T., T. Mio, N. Ono, Y. Kashima, M. Matsui, M. Arisawa, and H. Yamada-Okabe. 1999. Roles of three histidine kinase genes in hyphal development and virulence of the pathogenic fungus *Candida albicans*. *J. Bacteriol.* **181**:7243–7247.

136. Yamamoto, Y., T. W. Klein, and H. Friedman. 1997. Involvement of mannose receptor in cytokine interleukin-1β (IL-1β), IL-6, and granulocyte-macrophage colony-stimulating factor responses, but not in chemokine macrophage inflammatory protein 1β (MIP-1β), MIP-2, and KC responses, caused by attachment of *Candida albicans* to macrophages. *Infect. Immun.* **65**:1077–1082.

Molecular Principles of Fungal Pathogenesis
Edited by Joseph Heitman et al.
©2006 ASM Press, Washington, D.C.

Chapter 11

Function and Regulation of Adhesin Gene Families in *Saccharomyces cerevisiae*, *Candida albicans*, and *Candida glabrata*

Irene Castaño, Alejandro De Las Peñas, and Brendan P. Cormack

In fungi, the cell wall plays a primary role contributing to the structural integrity of the cell; in addition to this structural role, the cell wall is, by definition, the interface between the yeast and the environment. In adapting to different environments including, for pathogens, the environment of the mammalian host, fungi have engineered diversity at the cell surface in part by elaborating species-specific gene families of cell wall proteins. In this chapter, we examine the overlapping and divergent function and regulation of some surface glycoprotein gene families in *Saccharomyces cerevisiae* and its pathogenic cousins *Candida glabrata* and *Candida albicans*.

CELL WALL STRUCTURE

The fungal surface glycoproteins that form the subject of this review are so called glycosylphosphatidylinositol (GPI)-linked cell wall proteins (GPI-CWPs). To understand certain functional aspects of these proteins, we start with a brief review of the fungal cell wall.

The fungal cell wall is a network that includes three main polysaccharide classes—glucose polymers (β glucans; about 60% of total), mannose polymers (N linked or O linked to cell wall proteins; about 40% of total), and GlcNac polymers (chitin; about 1 to 2% of total). In *S. cerevisiae*, the β glucans include the major β1,3 glucan, which is present in chains up to 1,500 residues in length, and β1,6 glucan, present in chains of 150 to 200 residues (46). The main class of cell wall proteins, the GPI-CWPs are attached directly to β1,6 glucan, which is itself connected to β1,3 glucan (reviewed in reference 38). GPI-CWPs, β1,3 glucan chains, and β1,6 glucan chains are present in the cell wall in approximately equal numbers (reviewed in reference 42), consistent with a model in which a primary organizing block of the cell wall is a unit module consisting of GPI-CWPs attached to β1,6 glucan attached to β1,3 glucan (38, 42, 54a).

This overall model of cell wall structure and cell wall protein attachment to the cell wall, derived primarily from studies of *S. cerevisiae*, is likely to apply quite well to the cell wall organization of other ascomycetes including the pathogenic *Candida* species. For *C. glabrata* and *C. albicans*, in silico analysis suggests that between 75 and 100 GPI-CWPs are encoded in each genome (14, 90). In addition, proteomic analysis has made clear that *C. albicans* (13) and *C. glabrata* (P. de Groot, personal communication) cell walls both contain many covalently cross-linked proteins of the GPI-CWP class. Analysis of GPI-CWPs in *C. glabrata* and *C. albicans* shows that the overall linkage of GPI-CWPs to β1,6 glucan and β1,3 glucan is conserved in these species (see, for example, references 18, 32, 36, and 37).

S. cerevisiae, *C. albicans*, and *C. glabrata* encode a variety of GPI-CWPs that play accessory roles, functioning primarily as adhesins, facilitating yeast-yeast interactions or yeast adherence to a variety of surfaces (Table 1).

These accessory proteins in *S. cerevisiae* include the sexual agglutinins (54, 55, 69), the agglutination-modifying Fig2p protein (34, 95), and the flocculins encoded by the *FLO* genes (reviewed in references 87 and 88).

Adherence of *Candida* species to host cells is thought to be important in the pathogenesis of these organisms. In *C. albicans*, some genes implicated in adherence, including *INT1* (21) and the *CSH1* hydrophobin gene (74), are not predicted to encode GPI-CWPs and fall beyond the limited scope of this chapter. Most characterized adhesins, however, are

Irene Castaño and Alejandro De Las Peñas • Instituto Potosino de Investigacion Cientifica y Tecnologica, Division de Biologia Molecular, 78216 San Luis Potosi, San Luis Potosi, Mexico. **Brendan P. Cormack** • Department of Molecular Biology and Genetics, Johns Hopkins University School of Medicine, Baltimore, MD 21205.

Table 1. GPI-CWP with roles in adherence

GPI-CWP adhesin and organism	Function
S. cerevisiae	
Aga1, Sag1, Fig2	Sexual agglutination
Flo1, Flo5, Flo9, Flo10	Flocculation
Flo10	Filamentation, flocculation
Flo11	Pseudohyphal growth, filamentation
C. albicans	
Hwp1	Adherence
Eap1	Adherence
Als1	Adherence, filamentation
Als3, Als5, Als6	Adherence
C. glabrata	
Epa1, Epa7	Adherence
Epa6	Adherence, biofilm formation

Figure 1. Domain structure of GPI-CWP adhesins. Four domains are shown, all required for function of the adhesins. SS indicates the signal sequence; GPI indicates the C-terminal sequence required for covalent attachment of the GPI anchor to the mature C terminus. The effector domain confers specificity of adherence. C-terminal to the effector domain, a region rich in Ser/Thr residues, which are glycosylated, acts as a spacer between the GPI anchor and the effector domain.

predicted to be GPI-CWPs (reviewed in reference 82) including those encoded by *HWP1* (80), *EAP1* (52), and members of the *ALS* family, made up of eight genes (31). In *C. glabrata,* the *EPA* family encodes a family of lectins important in binding host cells (9, 11, 15) and implicated in biofilm formation (33).

FUNGAL ADHESIN STRUCTURE AND FUNCTION

This chapter focuses on three large gene families encoding GPI-CWPs: the *FLO* genes of *S. cerevisiae,* the *ALS* genes of *C. albicans,* and the *EPA* genes of *C. glabrata.* By virtue of functional differences and differences in regulation, individual members of each family contribute to differences in the potential of cells for adherence to other yeast, adherence to host cells, and adherence to inert surfaces. We first briefly review what is known about the functional differences of the adhesins encoded by different members of each gene family.

The *FLO* genes, the *EPA* genes, and the *ALS* genes all share key structural elements: an N-terminal signal sequence, an N-terminal effector domain, a Ser/Thr-rich repeated central domain, and a C-terminal hydrophobic signal that directs the addition of a GPI anchor in the endoplasmic reticulum (Fig. 1). The N-terminal effector domains have been demonstrated to provide specificity of adherence for the *FLO* genes (44), the *ALS* genes (73), and the *EPA* genes (18). The GPI anchor is required for cross-linking of GPI-CWPs to the cell wall in *S. cerevisiae* (91), *C. glabrata* (18), and *C. albicans* (60). The C-terminal Ser/Thr-rich domains are extensively glycosylated and function at least in part as a spacer between the

site of cell wall cross-linking and the environment, although additional roles in modulating adhesin function are not ruled out (18, 31, 59).

While the overall protein structure is conserved, the sequences of the N-terminal effector domains differ significantly among these three families. Data for all three gene families, detailed below, support the contention that different family members encode proteins with different adherence specificities (Table 1). The *S. cerevisiae FLO* family is made up of five genes: *FLO1, FLO5, FLO9, FLO10,* and *FLO11.* These fall into two groups. The first, including *FLO1, FLO5, FLO9,* and *FLO10,* are found near telomeres and encode lectins which bind mannose in the cell walls of other yeast; they play primary roles in mediating cell-cell adherence resulting in flocculation (27, 84). By contrast, *FLO11* plays a primary role in mediating pseudohyphal growth (49, 58), although it can also mediate some Ca^{2+}-dependent flocculation (27, 57). *FLO10* expression, like *FLO11* expression, promotes filamentation (27), adherence to plastic and agar, and biofilm formation (27, 67). Consistent with these differences in protein function, the N-terminal effector domains of Flo1, Flo5, Flo9 and Flo10 are highly related to one another at the sequence level, while that of Flo11 is divergent.

In *C. albicans,* there are eight *ALS* genes which have, in the N-terminal effector domain, between 45 and 90% amino acid identity (reviewed in reference 31). Als1 overexpression promotes flocculation and filamentation (19), and heterologous expression of Als1, Als3, Als5, and Als6 in *S. cerevisiae* confers adherence to mammalian cells and extracellular matrix proteins (20, 23, 73). Als1 and Als5 display differences in binding to different peptides (43), and Als5 and Als6 demonstrate notably different patterns of adherence to different cell types and extracellular matrix protein substrates (73).

In *C. glabrata,* the *EPA* genes form a large family. The exact number of *EPA* genes is not known, and there are likely to be strain-to-strain differences. The completed genome sequence of *C. glabrata* strain CBS138 includes 17 *EPA*-related sequences, many of which are localized to subtelomeric regions; this number probably does not represent the total repertoire: the repeated nature of the telomeric regions of the chromosome, as well as the high degree of homology between *EPA* family members, complicates the assembly of definitive sequence for the subtelomeric regions in general and the *EPA* family in particular. Strain BG2, our standard laboratory strain, has 23 confirmed sequences corresponding to complete *EPA*-related N-terminal effector domains; for 8 of these 23, we know that the sequence corresponds to a predicted full-length GPI-CWP (9). The *C. glabrata EPA* genes have more than 40% amino identity in the N-terminal effector domains. At least three, *EPA1*, *EPA6*, and *EPA7* encode lectins that mediate adherence to human cells by binding host cell carbohydrate; Epa2, Epa3, Epa4, and Epa5, when expressed heterologously at the *S. cerevisiae* cell surface, apparently do not mediate binding to epithelial cells (9). *EPA6*, unlike *EPA1*, also mediates adherence to plastic and facilitates biofilm formation in vitro (33).

TRANSCRIPTIONAL REGULATION OF ADHESIN GENE FAMILIES

As outlined above, different *FLO*, *ALS*, or *EPA* genes encode adhesins with different properties or specificities. Given this fact, expression of different family members at the cell surface will contribute to alterations in cell surface properties of the yeast. How, then, is transcription of individual members in these gene families regulated, and what is the functional consequence of that regulation? While most of the information available on the control of transcription of fungal adhesin genes comes from studies of *FLO* gene transcription in *S. cerevisiae*, some details of the regulation of the *EPA* genes and *ALS* genes are also known. Presaging the complexity of

the transcriptional regulation of the adhesin genes, the promoters of several of these genes are large; in the case of *FLO11*, *cis*-acting control sequences are found across 2.8 kb of the upstream intergenic region (70). In the case of *EPA1*, *cis*-acting positive regulatory sequences have been mapped 1.7 kb upstream of the open reading frame (ORF) (11).

Analysis of adhesin gene regulation in *S. cerevisiae* and in *Candida* species has revealed significant overlap in the signals that induce transcription of these gene families as well as the mechanistic basis for that regulation. In addition, at least in *S. cerevisiae* and *C. glabrata*, adhesin genes are subject to epigenetic regulation, in which distinct subsets of genetically identical cells show metastable and heritable differences in adhesin expression. As reviewed below, insight into the mechanistic basis of genetic and epigenetic regulation in *S. cerevisiae* and *Candida* can shed light on pathways regulating adhesin expression in the host.

S. cerevisiae

The transcriptional regulation of *FLO* gene expression is complex. *S. cerevisiae* strains flocculate, and transcription of the *FLO1-FLO5-FLO9* group of flocculins is induced in response to a variety of environmental signals, including nutrient deprivation (reviewed in reference 87). *FLO11* transcription is key to pseudohyphal growth in *S. cerevisiae* and is regulated by multiple transcription factors including Tec1, Ste12, Flo8, Sfl1, and Phd1 (Table 2). *FLO11* transcription is induced in response to nutritional signals, including nitrogen starvation, which are transduced by multiple signaling cascades, including the cyclic AMP/protein kinase A (cAMP/PKA) and mitogen-activated protein kinase pathways (reviewed in reference 22). The Flo8 DNA binding protein which activates *FLO11* (45) and *FLO10* (28) also activates *FLO1* and *FLO9* (and probably *FLO5* as well) (45). The deletion of *FLO8*

Table 2. Transcriptional regulators of *FLO*, *ALS*, and *EPA* genes

Organism and CWP gene	Signaling	Regulators
S. cerevisiae		
FLO1, FLO5, FLO9	Limiting carbon source	Flo8, Tup1/Ssn6, Hda1, Srb10
FLO10	Ras (Ira1p/Ira2p), cAMP/PKA (Tpk2)	Flo8, Sfl1, Tec1, Hst1 Hst2, Sir3
FLO11	Nitrogen starvation, cAMP/PKA (Tpk2)	Flo8, Sfl1, Ste12, Tec1, Phd1, Tup1/Ssn6, Hda1
C. albicans		
ALS1, ALS2, ALS3	cAMP/PKA	Efg1, Tup1, Nrg1, Rfg1
C. glabrata		
EPA1, EPA6, EPA7	Nicotinic acid limitation	Sir2, Sir3, Sir4, Rap1, Rif1, Yak1, Sfl1(?)

in many laboratory backgrounds renders these cells nonflocculating (56).

Strains, even those with *FLO8* deleted, that lack *TUP1* or *SSN6* become flocculent (40, 53) and derepress the transcription of *FLO1* (85), *FLO9* (25), and *FLO11* (10), implicating the Tup1-Ssn6 complex quite generally in *FLO* gene repression. *SSN6-TUP1*, a global transcriptional repressor, cannot bind DNA on its own but, rather, is recruited to different promoters by promoter-specific DNA binding proteins such as Mig1 (glucose-repressed genes), Rox1 (oxygen-regulated genes), or Crt1 (DNA damage-regulated genes) (reviewed in reference 77). Globally, Ssn6-Tup1 represses transcription, both by interacting with the Mediator complex (26, 29, 47, 65, 94) and by altering chromatin structure by recruiting histone deacetylases, of which the most important may be Hda1 (5, 12, 89, 93). Both mechanisms of repression are potentially important for *FLO* gene repression since *FLO1* is induced in cells lacking *HDA1* or *SRB10* (a subunit of mediator) (25, 96). The chromatin structure of the *FLO1* promoter is significantly altered in a *SSN6-TUP1*-dependent fashion (17). How is Ssn6-Tup1 recruited to *FLO* gene promoters? Ssn6-Tup1 is probably recruited to the *FLO11* promoter via interaction with the DNA binding protein Sfl1 (10). *SFL1* deletion results in transcriptional derepression of *FLO11* (68) and *FLO10* (28). *FLO10* and *FLO11* transcription, unlike *FLO1* transcription, is regulated by cAMP levels via the PKA subunit Tpk2 (62, 68). Tpk2 interacts physically with the Sfl1 repressor and the Flo8 activator and reciprocally regulates the association of these two DNA binding proteins with the *FLO11* promoter (63, 68). Activation of the PKA (Tpk2) pathway results in phosphorylation of both Sfl1 and Flo8, preventing binding of Sfl1 and promoting binding of Flo8 to DNA.

C. albicans

In *C. albicans*, orthologues of many of the regulators of *S. cerevisiae* flocculin gene expression are known to be implicated in transcriptional control of the *ALS* genes. The *EFG1* gene, an orthologue of *S. cerevisiae PHD1* (and, like *PHD1*, implicated in regulation of filamentation and hyphal specific genes) is a downstream target of cAMP/PKA signaling (4, 79) and is required for transcriptional induction of *ALS1* (7, 19) and *ALS3* (51). *tup1Δ* strains show derepression of *ALS1* (orf19.5741) (7), *ALS2* (orf19.2355), and *ALS3* (orf19.1816) (35). The transcription factors that recruit Tup1 to the *ALS* genes are not known, although the repressors Nrg1 and Rfg1 may be implicated since *ALS2* and *ALS3* are

derepressed in *nrg1* and *rfg1* backgrounds as well as in the *tup1* background (35).

C. glabrata

The regulators of *EPA* gene expression in *C. glabrata* are largely unknown. Sfl1 may be implicated in *EPA* gene regulation: transposon insertions in the *SFL1* gene render the cells hyperadherent, consistent with a role for Sfl1 (which represses *FLO* gene expression in *S. cerevisiae*) in normally repressing *C. glabrata EPA* gene transcription (B. P. Cormack, unpublished data).

The current data, therefore, are consistent with there being a broad overlap in the signaling pathways and transcriptional circuitry controlling *FLO* gene expression in *S. cerevisiae* and regulation of the *ALS* and *EPA* genes in *C. albicans* and *C. glabrata*.

GENOME ORGANIZATION AND EPIGENETIC CONTROL OF ADHESIN GENE EXPRESSION

The *FLO1*, *FLO5*, *FLO9*, and *FLO10* genes of *S. cerevisiae* are carried in subtelomeric loci, while *FLO11* is located 49 kb from the telomere (Fig. 2). In *C. glabrata*, *EPA9* and *EPA10* (ORFs CAGL0A01386 and CAGL0A1265) (http://cbi.labri.fr/Genolevures/elt/CAGL) are located internal to the chromosome in both strain BG2 and strain CBS138; one additional *EPA* homologue (CAGL0C00847) is chromosome-internal in CBS138; most of the other *EPA* genes map close to the telomeres in strain CBS138 and strain BG2. Subtelomeric regions in yeast are regions of rapid genome evolution (41), and the subtelomeric location of the *FLO* and *EPA* genes is likely to be due in part to the evolutionary expansion of the family by gene duplication (48). In *S. cerevisiae*, telomeres form clusters at the nuclear periphery (2, 24), providing a perfect scenario for nonallelic homologous recombination that could lead to the expansion of the gene family. In this regard, the structure of the *FLO* loci in *S. cerevisiae* and the *EPA* loci in *C. glabrata* exhibits substantial duplication in the intergenic regions flanking the ORFs. For example, *FLO1*, *FLO5*, and *FLO9* have long stretches of homology extending from their 3' end toward the telomere (Fig. 2). For *FLO1* and *FLO5*, the DNA homology extends from the TAA for 21.3 kb, interrupted only by an insertion at the *FLO5* locus of a full-length Ty1 element (at position +15). The *FLO9* homology to the *FLO1* and *FLO5* loci extends for 11.1 kb from the 3' end toward the telomere. There are no equivalent stretches of homology between the *FLO1-FLO5-FLO9* group and the *FLO10* or *FLO11* loci. The *FLO* genes do not have homology in their 5' intergenic regions.

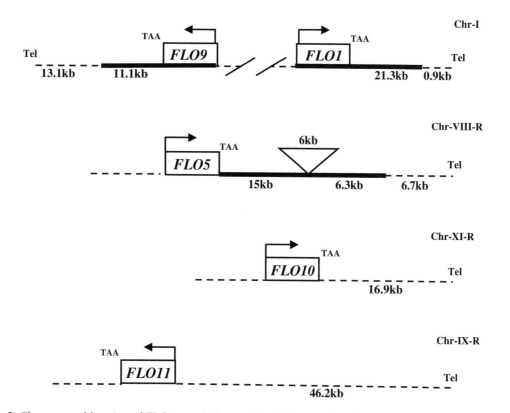

Figure 2. Chromosomal location of *FLO* genes in *S. cerevisiae*. Distances from the beginning of the telomeric repeats are indicated. Thick bars represent regions of homology between the *FLO1*, *FLO5*, and *FLO9* loci.

Similarly, in *C. glabrata*, some *EPA* genes exhibit long stretches of DNA homology 5′ and 3′ to the ORFs (Fig. 3). *EPA4* and *EPA5* form a near-perfect inverted repeat, with both genes being essentially identical, including 5′ and 3′ regions. The identical segments are separated by a 1.8-kb unique region. In addition, the 5′ untranslated regions of *EPA2* and *EPA4-EPA5* have homology extending 160 bp upstream of the ATG. The regions between *EPA6* and *EPA7* and the telomere are essentially identical, and the two genes have substantial homology in their promoters for at least 4 kb upstream from the ATG (in both strains BG2 and CBS138).

In addition to considerations of evolutionary expansion of the gene families, one functional consequence of the subtelomeric localization of the *FLO* and *EPA* genes is that both families are subject to epigenetic regulation that imposes a layer of regulation on these gene families in addition to the genetic control outlined above. Both *FLO* and *EPA* genes have been demonstrated to be subject to metastable chromatin-based silencing that limits their expression to a subset of cells in a genetically identical population of cells. Epigenetic control of expression appears to be a trait of only the *FLO* and *EPA*

gene families; the *C. albicans ALS* genes do not appear to be located in subtelomeric regions, and there is no indication that they are subject to epigenetic control.

SILENCING AND TELOMERE POSITION EFFECT IN *S. CEREVISIAE*

In *S. cerevisiae*, several regions of the chromosome are subject to epigenetic silencing in which the chromatin structure assumes a repressive conformation that limits transcription within the silenced region. Silencing in *S. cerevisiae* is the subject of an excellent and comprehensive review (71). Silencing of three regions in the *S. cerevisiae* genome has been analyzed in detail: the silent mating-type loci, the rDNA arrays, and the subtelomeric regions of the chromosomes. Many of the same proteins are required for repression of these three loci. Transcriptional repression involves the assembly of a heterochromatic state encompassing the silenced region. At the silent mating loci (HMR and HML), silencing depends on discrete sequence elements called silencers that act to recruit the silencing machinery to the locus. The silencers at HMR and HML include binding sites for the transcription factors Rap1 and Abf1 as well as for the origin recognition complex (ORC). These proteins in

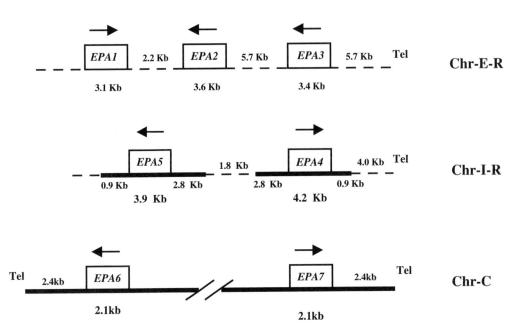

Figure 3. Chromosomal location of *EPA* genes in strain BG2. Distances from the beginning of the telomeric repeats are indicated. Thick bars represent homology between different *EPA* loci: the *EPA4* and *EPA5* loci are essentially identical to each other across 7.6 kb, including the ORFs and the indicated 5′ and 3′ intergenic regions. The *EPA6* and *EPA7* loci are highly homologous for the ORFs, for the 2.4 kb between the genes and the telomere, as well as for at least 4 kb upstream of the gene.

turn recruit the Sir (for "silent information regulator") proteins Sir1, Sir2, Sir3, and Sir4: Sir1 binds ORC (Orc1 has significant homology to Sir3), Sir4 binds Sir1 as well as Rap1, and Sir3 binds both Rap1 and Sir4. Sir2 binds Sir4 and is probably recruited as part of a Sir4-Sir2 complex. In the current model, Sir2 possesses the key catalytic activity of the Sir complex. Sir2 is an NAD$^+$-dependent histone deacetylase and deacetylates histones H3 and H4 of a targeted nucleosome. This in turn is thought to provide high-affinity binding sites for Sir3 and Sir4, which play a structural role in the repressed chromatin. The ability of Sir3-Sir4-Sir2 to bind to hypoacetylated histones allows for the spread of the Sir complex across a nucleosome region, as adjacent nucleosomes are sequentially deacetylated by Sir2 and bound by Sir3-Sir4-Sir2.

Marker genes inserted immediately adjacent to telomeric repeats are subject to a form of silencing called telomere position effect (TPE), which requires Sir2, Sir3, and Sir4 but not Sir1. The role of Sir1 at the silent mating loci seems limited to its capacity to bind ORC and aid in the recruitment of the other silencing proteins to the silencer. At the telomeres, the Sir2-Sir3-Sir4 complex is recruited by interaction with both Rap1, a DNA binding protein which binds the telomeric repeats, and Hdf1 (yKu70), which is also associated with the telomere (61, 86). Spreading from the telomere into the subtelomeric region requires Sir2,

Sir3, and Sir4, and marker genes present in this region are transcriptionally repressed. Other genes with primary roles in telomere metabolism are implicated in silencing as well. In particular, Rif1 and Rif2, which act to negatively regulate telomere length, play a role in silencing since deletions of Rif1 or Rif2 increase the silencing of subtelomeric marker genes and decrease the silencing of the silent mating loci (30, 92). Thus, in the current models, silencing of marker genes adjacent to truncated telomeres and silencer-mediated silencing at the HML and HMR loci are mechanistically highly related, with significant differences only in initial recruitment of the Sir complex to DNA.

EPIGENETIC REGULATION OF *FLO* GENE EXPRESSION

Four of the five *FLO* genes, *FLO1*, *FLO5*, *FLO9*, and *FLO10*, are located close to telomeres (Fig. 2). Analysis of *FLO* gene expression in strain E1287b uncovered a role for epigenetic regulation in *FLO* expression (28). In this strain, only *FLO11* is transcribed, while the telomeric *FLO* genes are transcriptionally silent. However, under conditions of nitrogen starvation, which induces *FLO11* transcription, the Flo11 protein is made only in a subset of cells. To analyze this heterogeneous expression, strains were made in which the *FLO11* ORF was replaced with the *URA3* ORF. Cells that express the *URA3* gene are URA$^+$, while those that do not are URA⁻ and are resistant to the uracil analogue 5-fluoroorotic acid (5-FOA). The strain carrying URA3 fused to the FLO11 promoter grew both on plates lacking uracil and on 5-FOA plates, showing that in a population of genetically identical cells, some express URA3 and others do not. Cells could switch between the Ura⁺ and 5-FOAʳ states since, for example, a 5-FOAʳ colony yielded Ura⁺ and 5-FOAʳ cells. This metastable expression of FLO11, measured by the URA3 reporter or directly by fluorescence, is similar to the Sir-mediated silencing of the silent mating loci or subtelomeric genes. However, deletion of the SIR genes, including SIR2, SIR3, and SIR4, or HDF1 (yKU70) or HDF2 (yKU80) had no effect on silencing of FLO11. By contrast, deletion of the transcriptional repressor SFL1 or the histone deacetylase HDA1 did eliminate silencing of FLO11. This suggests a model in which epigenetic silencing of FLO11 is initiated by recruitment of Hda1 to the FLO11 promoter through the transcriptional corepressor Sfl1 in complex with the Ssn6-Tup1 corepressor. Consistent with this model, silencing of FLO11 required sequences in the FLO11 promoter (presumably including Sfl1 binding sites). Silencing of the FLO11 transgene does depend on genomic location since moving the FLO11-URA3 construct from its normal subtelomeric location (ca. 46 kb form the telomere) to the URA3 locus (about 12 kb from the centromere) resulted in loss of silencing. In this, FLO11 silencing resembles TPE; however, the promoter specificity in FLO11 silencing distinguishes FLO11 epigenetic regulation from TPE, which silences telomeric transgenes irrespective of their precise promoter sequence (71).

Epigenetic regulation of FLO genes is not limited to FLO11. At a high frequency, strain E1287b generates stable colony variants that show expression of FLO10. These stable colony variants are the result of high-frequency inactivating mutations in IRA1 or IRA2, which encode Ras GTPase-activating proteins. In the ira backgrounds, FLO10 is transcribed and is subject to epigenetic silencing: in strains in which the FLO10 ORF was replaced with the URA3 ORF, cells grew into Ura⁺ or 5-FOAʳ colonies, indicating the existence of two states in which FLO10 is transcriptionally silent or transcriptionally active. As in the case of FLO11, silencing of FLO10 displayed promoter and regional specificity: other transgenes inserted at the FLO10 locus were not silenced, and silencing was lost when the FLO10-URA3 construct was moved to the centromere-proximal URA3 locus. For FLO10, perhaps reflecting the proximity of the FLO10 locus to the telomere (17 kb), silencing required SIR3, HDF1 (yKU70), and HDF2 (yKU80). Surprisingly, however, there was no requirement in FLO10 silencing for the histone deacetylase SIR2. Rather, the

histone deacetylase activity provided by Sir2 for classic TPE is provided at FLO10 by Hst1 and Hst2, so called "sirtuins," which are NAD⁺-dependent deacetylase proteins related to Sir2 (6). Chromatin immuno-precipitation experiments demonstrated a physical association of Hst1 and Hst2 with the FLO10 promoter. Hst1 and Hst2 are probably recruited to the FLO10 promoter via interaction with the repressor Sfl1: deletion of SFL1 results in loss of FLO10-URA3 silencing and a significant drop in Hst1 and Hst2 association with the FLO10 promoter (28).

Analysis of FLO gene transcription reveals unexpected complexity in chromatin regulation in yeast. While the epigenetic regulation of FLO10 and FLO11 expression appears similar and some key regulators (Sfl1) are shared, the requirement for chromatin-modifying proteins which underlie the silencing mechanism is quite different—Hda1 in the case of FLO11 and Hst1-Hst2 and Sir3 in the case of FLO10. This complexity probably foreshadows similar complexity in the regulation of other fungal adhesin families.

EPIGENETIC REGULATION OF EPA GENE EXPRESSION

As mentioned above, most genes of the EPA family in C. glabrata are localized to subtelomeric regions. Four loci containing seven EPA genes have been studied in detail: the EPA1 cluster, the EPA4/EPA5 cluster, EPA6, and EPA7. The EPA1 cluster is localized at the right end of chromosome E and is composed of EPA1 (20.7 kb from the telomere), EPA2 (14.8 kb from the telomere), and EPA3 (5.7 kb from the telomere) (Fig. 3). This region is under the control of chromatin-based silencing since the URA3 marker gene placed in the intergenic regions between EPA1 and EPA2 (20.6 kb from the telomere), between EPA2 and EPA3 (14.8 kb from the telomere), or between EPA3 and the telomere (1.3 kb from the telomere) showed silencing of the marker at all the positions tested; that is, each strain was phenotypically Ura⁺ and 5-FOAʳ, indicating that some cells of each strain express URA3, allowing growth on media lacking uracil, and some cells do not express URA3, permitting growth on media containing 5-FOA (15). In the case of the URA3 insertion between EPA1 and EPA2 (the farthest from the telomere), silencing was significantly reduced compared to the insertions closer to the telomere, possibly marking the border of the silenced region. A URA3 insertion upstream of EPA1 (30 kb from the telomere) was not silenced at all (Fig. 4). Silencing of this region is dependent on Rap1p, Rif1p, and Sir3p, since mutations in these genes allowed expression of the intergenic URA3 genes, thus abolishing the ability to grow in the presence of 5-FOA.

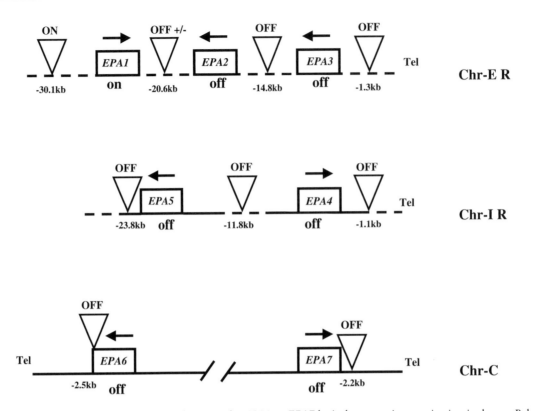

Figure 4. Silencing of the *EPA1* to *EPA7* loci. For the *EPA1* to *EPA7* loci, the genomic organization is shown. Below each gene is the transcriptional status of each gene grown under normal laboratory conditions. The triangles represent *URA3* gene insertions made at each locus; below each insertion is the distance of the *URA3* translational start from the telomeric repeats; above each triangle is the status of the correponding *URA3* gene insertion in a wild-type cell, as measured by the ability to grow on 5-FOA plates. "off" means that the corresponding strain was 5-FOA[r]; "on" means that the corresponding strain was 5-FOA[s]. For the insertion between *EPA1* and *EPA2*, "off +/−" indicates that the strain yielded a small number of 5-FOA[r] colonies.

At the right end of chromosome I, *EPA4* and *EPA5* form a near-perfect inverted repeat (15). *EPA5* and *EPA4* are 15.6 and 4 kb, respectively, from the telomere. As in the *EPA1* cluster, the *EPA4/EPA5* cluster is also subject to subtelomeric silencing as measured both by expression of *URA3* marker gene and by mRNA steady-state levels. As with the *EPA1* cluster, the *EPA4/EPA5* cluster silencing depends on *RAP1*, *RIF1*, and *SIR3*.

The *EPA2*, *EPA3*, and *EPA4/EPA5* promoters are presumably subject to additional layers of regulation in addition to the silencing described above, since it appears that additional regulators are needed for expression: while deletion of *SIR3* results in increased RNA transcript levels detectable by reverse transcription-PCR, the level of expression is very low relative to that of *EPA1* or *ACT1*, for example (15). Furthermore, fusion of the *EPA2*, *EPA3*, or *EPA4/EPA5* promoters to green fluorescent protein showed no expression, as measured by fluorescence, even though these fusion constructs were carried on plasmids, where they are not subject to TPE (unpublished data). This situation is similar to the case of *FLO10*

in *S. cerevisiae*. Without mutational inactivation of *IRA1* or *IRA2*, which probably affects signaling through the PKA-dependent pathway, *FLO10* is not expressed even in strains lacking *SIR3* or *HST1* or *HST2* (28). *FLO10* is thus subject to distinct levels of regulation—silencing on the one hand, and PKA-mediated activation on the other. Similarly, in our current model for transcriptional regulation of *EPA2*, *EPA3*, and *EPA4/EPA5*, additional activators are required, even in the absence of Sir-mediated silencing.

The *EPA6* and *EPA7* genes are 2.43 and 2.49 kb away from their telomeres, respectively, in strain BG2 (Fig. 3) (9). In strain CBS138, they are at opposite ends of chromosome C. *EPA6* and *EPA7* are subject to telomeric silencing, as measured by silencing of a *URA3* marker gene inserted at either locus (9, 33). Like *EPA2*, *EPA3*, and *EPA4/EPA5*, *EPA6* and *EPA7* are not normally expressed in a wild-type cell; however, unlike *EPA2*, *EPA3*, and *EPA4/EPA5*, the *EPA6* and *EPA7* promoters are constitutively expressed when placed on a plasmid or in strains with silencing mutations (9, 16, 33). Silencing of the

EPA6 and *EPA7* loci is dependent on at least *SIR3, SIR4, RIF1,* and *RAP1* (9, 33) because deletions of these genes derepress *EPA6* and *EPA7* transcription and abolish the silencing of *URA3* genes placed at the *EPA6* or *EPA7* loci. Strains carrying mutations in silencing genes (*sir3, sir4, rap1-21,* and *rif1*) show a hyperadherent phenotype (9), due in part to derepression of the *EPA6* and *EPA7* genes, both of which can mediate adherence when expressed in *S. cerevisiae*. In addition, the Yak1 kinase has been implicated in regulation of *EPA6* and *EPA7* transcription, although the mechanistic details of this regulation are not known (33).

How is silencing of the telomeric *EPA* genes brought about? The extent of silencing is dependent on the distance from the telomere; that is, repression decreases as the distance from the telomere increases (9). Constructs at the *EPA7* locus in which the *URA3* gene is placed 14 kb away from the telomere showed relatively less silencing than did constructs in which *URA3* was placed only 5 kb away from the telomere. The extent of silencing of the native *EPA6* and *EPA7* genes appears also to be dependent on the distance from the telomere. In wild-type cells, the transcriptional start sites of the *EPA6* and *EPA7* genes are approximately 4.6 kb from the telomere, where they are silenced. Large insertions at the *EPA6* or *EPA7* locus that place the transcriptional start site 18.5 kb away from the telomere result in a hyperadherent phenotype, presumably because silencing is relieved. However, small insertions that place the transcriptional start of *EPA7* approximately 7.7 kb from the telomere are not hyperadherent, suggesting that the genes remain largely silenced.

RIF1, PROTOSILENCERS, AND TPE

As stated above, deletion of *RIF1* in *S. cerevisiae* leads to increased silencing of a marker gene placed at a truncated telomere while decreasing silencing at the silent mating loci (30). The model to explain this derives from the observation that telomeres in the *rif1Δ* strain are much longer than in a wild-type strain. Since the telomeric repeats recruit Rap1 and Sir proteins to the telomere, the longer telomeres are proposed to titrate Sir proteins from the silent mating loci, thus reducing silencing; markers placed adjacent to the telomere, by contrast, are subject to increased silencing, presumably because of the greater Sir protein accumulation at the longer telomeres (8, 30).

In *C. glabrata*, by contrast, a *rif1Δ* strain shows decreased silencing at subtelomeric loci both for *URA3* marker genes and for the native *EPA* genes themselves (9, 33). In this, the behavior of *EPA* gene silencing mirrors the derepression of the silent mating

loci in a *S. cerevisiae rif1Δ* background. One possibility is that *C. glabrata* Rif1 might act directly in silencing. Even in *S. cerevisiae*, Rif1 probably plays a role in the structure or silencing of natural subtelomeric regions in *S. cerevisiae* because Rif1 has been localized by chromatin immunoprecipitation to regions up to 13 kb from the telomeric repeats in *S. cerevisiae* (75). An alternative potential explanation for the loss of *EPA* silencing in a *C. glabrata rif1Δ* background is that that there might be silencers associated with subtelomeric *EPA* genes, similar to those present at the *S. cerevisiae* silent mating loci. These would be subject to titration by elongated telomeres in a *rif1Δ* strain. In this regard, our preliminary data suggest that discrete silencer elements do in fact contribute to *EPA* gene silencing. Discrete sequences at native telomeres in *S. cerevisiae* have been mapped and shown to be required for telomeric silencing (66). We have mapped a silencer in *C. glabrata* between *EPA3* and the telomere which functions to silence transcription even when placed 27.5 kb away from the telomere (A. De Las Peñas and B. Cormack, unpublished data). It seems likely that, as in *S. cerevisiae* (66), the landscape of individual *C. glabrata* telomeres and an interplay between the telomeric repeats themselves and other subtelomeric silencer or protosilencer elements is likely to affect the extent of TPE and conceivably the regulation of silenced genes.

PHYSIOLOGICAL REGULATION OF EPIGENETIC SILENCING

Epigenetic regulation of the *FLO* genes and *EPA* genes has the result that in a population of wild-type yeast, the expression of particular adhesins is heterogeneous. In the case of *FLO11*, under inducing conditions of nitrogen starvation, adjacent cells in a colony, which presumably receive highly similar environmental cues, can show differential induction of *FLO11*, reflecting the two possible states (on and off) for *FLO11* expression. *FLO11*-expressing cells form pseudohyphae, whereas cells not expressing *FLO11* stay in the yeast form. Elimination of *HDA1*, which abrogates silencing of the *FLO11* locus, results in homogeneous constitutive *FLO11* expression and increased pseudohyphal formation in the colony (28). Thus, epigenetic control in essence limits the expression of *FLO11* to a subset of cells receiving an inducing environmental signal. While mutation of *HDA1* in the case of *FLO11*, or *HST1* or *HST2* in the case of *FLO10*, relieves this epigenetic silencing, it is not known if there are environmental signals that antagonize silencing, thereby increasing, for example, the percentage of cells that express *FLO11* in response to a constant environmental signal. Phosphorylation of Sir3 (which reduces Sir3 activity) has been implicated in regulation of silent chromatin in response to environmental stress, and, indeed, subtelomerically encoded cell wall proteins of the *PAU* gene family show increased expression in response to environmental stress (1, 81). It is not known if Sir3 phosphorylation plays any role in regulating *FLO10* gene expression.

PHYSIOLOGICAL REGULATION OF SILENT *EPA* GENE EXPRESSION DURING URINARY TRACT INFECTION

The host environment can directly regulate epigenetic silencing of *EPA* genes during infection, resulting in increased expression of certain *EPA* genes normally expressed in only a small number of yeast. Transcriptional induction of *EPA6* (normally strongly silenced) during infection was assessed using a recombinational IVET (in vitro expression technology) reporter that permanently marks cells with a Hyg^r phenotype should the *EPA6* promoter be transcribed (16). When this strain was cultured in laboratory media, approximately 1% of cells were Hyg^r, suggesting that in laboratory broth there is little induction of *EPA6*. After infection of the urinary tract, initiated by transurethral delivery of *C. glabrata* into the bladder, up to 50% of organisms recovered from the bladder or from the kidneys were Hyg^r, suggesting induction of *EPA6* in the urinary tract. Transcriptional induction of *EPA6* also occurred after growth in urine and was shown to be due to limitation for the vitamin nicotinic acid (niacin) (NA). How does NA limitation affect *EPA* gene transcription? NA is a precursor of NAD$^+$. In *C. glabrata*, NA is an essential nutrient, since *C. glabrata* has lost the de novo NAD$^+$ biosynthetic pathway (http://cbi.labri.fr/Genolevures/elt/CAGL), encoded in *S. cerevisiae* and *C. albicans* by the *BNA* genes (64). The current model for *C. glabrata*, therefore, is that limitation for NA in the environment is tied directly to a drop in NAD$^+$ cofactor levels; since Sir2 requires NAD$^+$ (50, 76, 83), Sir2 activity is compromised and there is a loss of subtelomeric silencing.

In the urinary tract at a minimum, environmental limitation for NA appears to contribute to derepression of silent *EPA* genes. Whether NA limitation also induces *EPA* gene transcription in other host environments has not been fully investigated; when the same reporter strain was used for bloodstream infections and cells were recovered from the liver, kidneys, or spleen at 4 or 7 days postinfection (via tail vein injection), there was no significant increase in the number of Hyg^r colonies, indicating that *EPA6* is not significantly induced during bloodstream infection. There are no current data for *C. glabrata* vaginal colonization or colonization of the gastrointestinal tract; however, it is very important to obtain such data.

The observation that NA limitation in the urinary tract influences *EPA* gene expression may have clinical implications. *C. glabrata* is a common cause of nosocomial funguria, accounting, in one large study, for 16% of cases (39). *EPA1*, *EPA6*, and *EPA7*, which all are transcriptionally derepressed by NA limitation, mediate adherence to cultured human bladder and kidney epithelial cells, and mutant strains lacking *EPA1*, *EPA6*, and *EPA7* are compromised for colonization of the bladder in experimental murine urinary tract infection (16). Moreover, the presence of indwelling urinary catheters is an important risk factor for funguria (78); the fact that *EPA6* expression is important for in vitro biofilm formation (33) suggests that *C. glabrata* colonization of urinary catheters may also be impacted by urinary NA levels.

Are Adhesin Genes in Other Fungal Species Regulated by NA Availability?

There is no indication that the *ALS* genes in *C. albicans* are subtelomerically located; and *C. albicans* is an NA prototroph, making it likely that this mechanism does not apply to *ALS* gene regulation. In wild-type *S. cerevisiae*, NA limitation first affects Hst1 activity, since the affinity for NAD$^+$ is lower for Hst1 than for other sirtuins (3). Since Hst1 normally represses the *BNA* genes, loss of Hst1 activity (by NA limitation) results in derepression of the *BNA* genes and induction of the de novo NAD$^+$ biosynthetic pathway, thus maintaining NAD$^+$ homeostasis. In classical marker-based silencing assays, NA limitation therefore does not affect silencing in wild-type *S. cerevisiae* (72). However, since NA limitation induces the *BNA* pathway via its effects on Hst1, it seems possible that silencing of *FLO10*, which depends on Hst1 and Hst2 (28), might be directly affected by environmental levels of NA.

CONCLUSION

The yeast cell surface forms a key interface between the yeast and its environment, and fungal species have adapted to different environments in part by altering their cell surface. This diversity in cell surface properties is reflected in the *FLO*, *ALS*, and *EPA* gene families, not only in the extreme sequence divergence between the three families but also in the expansion within each family of the number of genes. The regulation, by genetic and epigenetic mechanisms, of different family members encoding proteins with different physical properties or specificities, allows access to a reservoir of cell surface diversity in response to different environments including, for pathogens, the environment of the host. Understanding the importance and role of each family will require the development of a detailed picture of both the transcriptional regulation and functional specificity of each gene in the family.

REFERENCES

1. Ai, W., P. G. Bertram, C. K. Tsang, T. F. Chan, and X. F. Zheng. 2002. Regulation of subtelomeric silencing during stress response. *Mol. Cell* 10:1295–1305.

2. Andrulis, E. D., A. M. Neiman, D. C. Zappulla, and R. Sternglanz. 1998. Perinuclear localization of chromatin facilitates transcriptional silencing. *Nature* 394:592–595.

3. Bedalov, A., M. Hirao, J. Posakony, M. Nelson, and J. A. Simon. 2003. NAD$^+$-dependent deacetylase Hst1p controls biosynthesis and cellular NAD$^+$ levels in *Saccharomyces cerevisiae*. *Mol. Cell. Biol.* 23:7044–7054.

4. Bockmuhl, D. P., and J. F. Ernst. 2001. A potential phosphorylation site for an A-type kinase in the Efg1 regulator protein contributes to hyphal morphogenesis of *Candida albicans*. *Genetics* 157:1523–1530.

5. Bone, J. R., and S. Y. Roth. 2001. Recruitment of the yeast Tup1p-Ssn6p repressor is associated with localized decreases in histone acetylation. *J. Biol. Chem.* 276:1808–1813.

6. Brachmann, C. B., J. M. Sherman, S. E. Devine, E. E. Cameron, L. Pillus, and J. D. Boeke. 1995. The *SIR2* gene family, conserved from bacteria to humans, functions in silencing, cell cycle progression, and chromosome stability. *Genes Dev.* 9:2888–2902.

7. Braun, B. R., and A. D. Johnson. 2000. TUP1, CPH1 and EFG1 make independent contributions to filamentation in *Candida albicans*. *Genetics* 155:57–67.

8. Buck, S. W., and D. Shore. 1995. Action of a RAP1 carboxy-terminal silencing domain reveals an underlying competition between HMR and telomeres in yeast. *Genes Dev.* 9:370–384.

9. Castano, I., S. J. Pan, M. Zupancic, C. Hennequin, B. Dujon, and B. P. Cormack. 2005. Telomere length control and transcriptional regulation of subtelomeric adhesins in *Candida glabrata*. *Mol. Microbiol.* 55:1246–1258.

10. Conlan, R. S., and D. Tzamarias. 2001. Sfl1 functions via the co-repressor Ssn6-Tup1 and the cAMP-dependent protein kinase Tpk2. *J. Mol. Biol.* 309:1007–1015.

11. Cormack, B. P., N. Ghori, and S. Falkow. 1999. An adhesin of the yeast pathogen *Candida glabrata* mediating adherence to human epithelial cells. *Science* 285:578–582.

12. Davie, J. K., D. G. Edmondson, C. B. Coco, and S. Y. Dent. 2003. Tup1-Ssn6 interacts with multiple class I histone deacetylases in vivo. *J. Biol. Chem.* 278:50158–50162.

13. de Groot, P. W., A. D. de Boer, J. Cunningham, H. L. Dekker, L. de Jong, K. J. Hellingwerf, C. de Koster, and F. M. Klis. 2004. Proteomic analysis of *Candida albicans* cell walls reveals covalently bound carbohydrate-active enzymes and adhesins. *Eukaryot. Cell* 3:955–965.

14. de Groot, P. W., K. J. Hellingwerf, and F. M. Klis. 2003. Genome-wide identification of fungal GPI proteins. *Yeast* 20:781–796.

15. De Las Penas, A., S. J. Pan, I. Castano, J. Alder, R. Cregg, and B. P. Cormack. 2003. Virulence-related surface glycoproteins in the yeast pathogen *Candida glabrata* are encoded in subtelomeric clusters and subject to RAP1- and

SIR-dependent transcriptional silencing. *Genes Dev.* 17:2245–2258.

16. Domergue, R., I. Castano, A. De Las Peñas, M. Zupancic, V. Lockatell, J. R. Hebel, D. Johnson, and B. P. Cormack. 2005. Nicotinic acid limitation regulates silencing of *Candida* adhesins during UTI. *Science* 308:866–870.

17. Fleming, A. B., and S. Pennings. 2001. Antagonistic remodelling by Swi-Snf and Tup1-Ssn6 of an extensive chromatin region forms the background for *FLO1* gene regulation. *EMBO J.* 20:5219–5231.

18. Frieman, M. B., J. M. McCaffery, and B. P. Cormack. 2002. Modular domain structure in the *Candida glabrata* adhesin Epa1p, a β1,6 glucan-cross-linked cell wall protein. *Mol. Microbiol.* 46:479–492.

19. Fu, Y., A. S. Ibrahim, D. C. Sheppard, Y. C. Chen, S. W. French, J. E. Cutler, S. G. Filler, and J. E. Edwards, Jr. 2002. *Candida albicans* Als1p: an adhesin that is a downstream effector of the EFG1 filamentation pathway. *Mol. Microbiol.* 44:61–72.

20. Fu, Y., G. Rieg, W. A. Fonzi, P. H. Belanger, J. E. Edwards, Jr., and S. G. Filler. 1998. Expression of the *Candida albicans* gene ALS1 in *Saccharomyces cerevisiae* induces adherence to endothelial and epithelial cells. *Infect. Immun.* 66:1783–1786.

21. Gale, C. A., C. M. Bendel, M. McClellan, M. Hauser, J. M. Becker, J. Berman, and M. K. Hostetter. 1998. Linkage of adhesion, filamentous growth, and virulence in *Candida albicans* to a single gene, INT1. *Science* 279:1355–1358.

22. Gancedo, J. M. 2001. Control of pseudohyphae formation in *Saccharomyces cerevisiae*. *FEMS Microbiol. Rev.* 25:107–123.

23. Gaur, N. K., and S. A. Klotz. 1997. Expression, cloning, and characterization of a *Candida albicans* gene, ALA1, that confers adherence properties upon *Saccharomyces cerevisiae* for extracellular matrix proteins. *Infect. Immun.* 65:5289–5294.

24. Gotta, M., T. Laroche, A. Formenton, L. Maillet, H. Scherthan, and S. M. Gasser. 1996. The clustering of telomeres and colocalization with Rap1, Sir3, and Sir4 proteins in wild-type *Saccharomyces cerevisiae*. *J. Cell. Biol.* 134:1349–1363.

25. Green, S. R., and A. D. Johnson. 2004. Promoter-dependent roles for the Srb10 cyclin-dependent kinase and the Hda1 deacetylase in Tup1-mediated repression in Saccharomyces cerevisiae. *Mol. Biol. Cell* 15:4191–4202.

26. Gromoller, A., and N. Lehming. 2000. Srb7p is a physical and physiological target of Tup1p. *EMBO J.* 19:6845–6852.

27. Guo, B., C. A. Styles, Q. Feng, and G. R. Fink. 2000. A *Saccharomyces* gene family involved in invasive growth, cell-cell adhesion, and mating. *Proc. Natl. Acad. Sci. USA* 97:12158–12163.

28. Halme, A., S. Bumgarner, C. Styles, and G. R. Fink. 2004. Genetic and epigenetic regulation of the *FLO* gene family generates cell-surface variation in yeast. *Cell* 116:405–415.

29. Han, S. J., J. S. Lee, J. S. Kang, and Y. J. Kim. 2001. Med9/Cse2 and Gal11 modules are required for transcriptional repression of distinct group of genes. *J. Biol. Chem.* 276:37020–37026.

30. Hardy, C. F., L. Sussel, and D. Shore. 1992. A RAP1-interacting protein involved in transcriptional silencing and telomere length regulation. *Genes Dev.* 6:801–814.

31. Hoyer, L. L. 2001. The ALS gene family of *Candida albicans*. *Trends Microbiol.* 9:176–180.

32. Hoyer, L. L., R. Fundyga, J. E. Hecht, J. C. Kapteyn, F. M. Klis, and J. Arnold. 2001. Characterization of agglutinin-like sequence genes from non-*albicans Candida* and phylogenetic analysis of the ALS family. *Genetics* 157:1555–1567.

33. Iraqui, I., S. Garcia-Sanchez, S. Aubert, F. Dromer, J. M. Ghigo, C. d'Enfert, and G. Janbon. 2005. The Yak1p kinase controls expression of adhesins and biofilm formation in *Candida glabrata* in a Sir4p-dependent pathway. *Mol. Microbiol.* 55:1259–1271.

34. Jue, C. K., and P. N. Lipke. 2002. Role of Fig2p in agglutination in *Saccharomyces cerevisiae. Eukaryot. Cell* 1:843–845.

35. Kadosh, D., and A. D. Johnson. 2005. Induction of the *Candida albicans* filamentous growth program by relief of transcriptional repression: a genome-wide analysis. *Mol. Biol. Cell* 16:2903–2912.

36. Kapteyn, J. C., L. L. Hoyer, J. E. Hecht, W. H. Muller, A. Andel, A.J. Verkleij, M. Makarow, H. Van Den Ende, and F. M. Klis. 2000. The cell wall architecture of *Candida albicans* wild-type cells and cell wall-defective mutants. *Mol. Microbiol.* 35:601–611.

37. Kapteyn, J. C., R. C. Montijn, G. J. Dijkgraaf, H. Van den Ende, and F. M. Klis. 1995. Covalent association of β-1,3-glucan with β-1,6-glucosylated mannoproteins in cell walls of *Candida albicans. J. Bacteriol.* 177:3788–3792.

38. Kapteyn, J. C., H. Van Den Ende, and F. M. Klis. 1999. The contribution of cell wall proteins to the organization of the yeast cell wall. *Biochim. Biophys. Acta* 1426:373–383.

39. Kauffman, C. A., J. A. Vazquez, J. D. Sobel, H. A. Gallis, D. S. McKinsey, A. W. Karchmer, A. M. Sugar, P. K. Sharkey, G. J. Wise, R. Mangi, A. Mosher, J. Y. Lee, W. E. Dismukes, and The National Institute for Allergy and Infectious Diseases (NIAID) Mycoses Study Group. 2000. Prospective multicenter surveillance study of funguria in hospitalized patients. *Clin. Infect. Dis.* 30:14–18.

40. Keleher, C. A., M. J. Redd, J. Schultz, M. Carlson, and A. D. Johnson. 1992. Ssn6-Tup1 is a general repressor of transcription in yeast. *Cell* 68:709–719.

41. Kellis, M., N. Patterson, M. Endrizzi, B. Birren, and E. S. Lander. 2003. Sequencing and comparison of yeast species to identify genes and regulatory elements. *Nature* 423:241–254.

42. Klis, F. M., L. H. Caro, J. H. Vossen, J. C. Kapteyn, A. F. Ram, R. C. Montijn, M. A. Van Berkel, and H. Van den Ende. 1997. Identification and characterization of a major building block in the cell wall of *Saccharomyces cerevisiae. Biochem. Soc. Trans.* 25:856–860.

43. Klotz, S. A., N. K. Gaur, D. F. Lake, V. Chan, J. Rauceo, and P. N. Lipke. 2004. Degenerate peptide recognition by *Candida albicans* adhesins Als5p and Als1p. *Infect. Immun.* 72:2029–2034.

44. Kobayashi, O., N. Hayashi, R. Kuroki, and H. Sone. 1998. Region of FLO1 proteins responsible for sugar recognition. *J. Bacteriol.* 180:6503–6510.

45. Kobayashi, O., H. Yoshimoto, and H. Sone. 1999. Analysis of the genes activated by the *FLO8* gene in *Saccharomyces cerevisiae. Curr. Genet.* 36:256–261.

46. Kollar, R., B. B. Reinhold, E. Petrakova, H. J. Yeh, G. Ashwell, J. Drgonova, J. C. Kapteyn, F. M. Klis, and E. Cabib. 1997. Architecture of the yeast cell wall. β(1→6)-Glucan interconnects mannoprotein, β(1→)3-glucan, and chitin. *J. Biol. Chem.* 272:17762–17775.

47. Kuchin, S., and M. Carlson. 1998. Functional relationships of Srb10-Srb11 kinase, carboxy-terminal domain kinase CTDK-I, and transcriptional corepressor Ssn6-Tup1. *Mol. Cell. Biol.* 18:1163–1171.

48. Lafontaine, I., G. Fischer, E. Talla, and B. Dujon. 2004. Gene relics in the genome of the yeast *Saccharomyces cerevisiae. Gene* 335:1–17.

49. Lambrechts, M. G., F. F. Bauer, J. Marmur, and I. S. Pretorius. 1996. Muc1, a mucin-like protein that is regulated by Mss10, is critical for pseudohyphal differentiation in yeast. *Proc. Natl. Acad. Sci. USA* 93:8419–8424.

50. Landry, J., A. Sutton, S. T. Tafrov, R. C. Heller, J. Stebbins, L. Pillus, and R. Sternglanz. 2000. The silencing protein SIR2 and its homologs are NAD-dependent protein deacetylases. *Proc. Natl. Acad. Sci. USA* 97:5807–5811.

51. Leng, P., P. R. Lee, H. Wu, and A. J. Brown. 2001. Efg1, a morphogenetic regulator in *Candida albicans*, is a sequence-specific DNA binding protein. *J. Bacteriol.* 183:4090–4093.

52. Li, F., and S. P. Palecek. 2003. *EAP1*, a *Candida albicans* gene involved in binding human epithelial cells. *Eukaryot. Cell* 2:1266–1273.

53. Lipke, P. N., and C. Hull-Pillsbury. 1984. Flocculation of *Saccharomyces cerevisiae tup1* mutants. *J. Bacteriol.* 159:797–799.

54. Lipke, P. N., and J. Kurjan. 1992. Sexual agglutination in budding yeasts: structure, function, and regulation of adhesion glycoproteins. *Microbiol. Rev.* 56:180–194.

54a.Lipke, P. N., and R. Ovalle. 1998. Cell wall architecture in yeast: new structure and new challenges. *J. Bacteriol.* 180:3735–3740.

55. Lipke, P. N., D. Wojciechowicz, and J. Kurjan. 1989. AGα1 is the structural gene for the *Saccharomyces cerevisiae* α-agglutinin, a cell surface glycoprotein involved in cell-cell interactions during mating. *Mol. Cell. Biol.* 9:3155–3165.

56. Liu, H., C. A. Styles, and G. R. Fink. 1996. *Saccharomyces cerevisiae* S288C has a mutation in *FLO8*, a gene required for filamentous growth. *Genetics* 144:967–978.

57. Lo, W. S., and A. M. Dranginis. 1996. *FLO11*, a yeast gene related to the *STA* genes, encodes a novel cell surface flocculin. *J. Bacteriol.* 178:7144–7151.

58. Lo, W. S., and A. M. Dranginis. 1998. The cell surface flocculin Flo11 is required for pseudohyphae formation and invasion by *Saccharomyces cerevisiae. Mol. Biol. Cell* 9:161–171.

59. Loza, L., Y. Fu, A. S. Ibrahim, D. C. Sheppard, S. G. Filler, and J. E. Edwards, Jr. 2004. Functional analysis of the *Candida albicans ALS1* gene product. *Yeast* 21:473–482.

60. Mao, Y., Z. Zhang, and B. Wong. 2003. Use of green fluorescent protein fusions to analyse the N- and C-terminal signal peptides of GPI-anchored cell wall proteins in *Candida albicans. Mol. Microbiol.* 50:1617–1628.

61. Mishra, K., and D. Shore. 1999. Yeast Ku protein plays a direct role in telomeric silencing and counteracts inhibition by rif proteins. *Curr. Biol.* 9:1123–1126.

62. Pan, X., and J. Heitman. 1999. Cyclic AMP-dependent protein kinase regulates pseudohyphal differentiation in *Saccharomyces cerevisiae. Mol. Cell. Biol.* 19:4874–4887.

63. Pan, X., and J. Heitman. 2002. Protein kinase A operates a molecular switch that governs yeast pseudohyphal differentiation. *Mol. Cell. Biol.* 22:3981–3993.

64. Panozzo, C., M. Nawara, C. Suski, R. Kucharczyka, M. Skoneczny, A. M. Becam, J. Rytka, and C. J. Herbert. 2002. Aerobic and anaerobic NAD⁺ metabolism in *Saccharomyces cerevisiae. FEBS Lett.* 517:97–102.

65. Papamichos-Chronakis, M., R. S. Conlan, N. Gounalaki, T. Copf, and D. Tzamarias. 2000. Hrs1/Med3 is a Cyc8-Tup1

corepressor target in the RNA polymerase II holoenzyme. *J. Biol. Chem.* **275:**8397–8403.

66. Pryde, F. E., and E. J. Louis. 1999. Limitations of silencing at native yeast telomeres. *EMBO J.* **18:**2538–2550.

67. Reynolds, T. B., and G. R. Fink. 2001. Bakers' yeast, a model for fungal biofilm formation. *Science* **291:**878–881.

68. Robertson, L. S., and G. R. Fink. 1998. The three yeast A kinases have specific signaling functions in pseudohyphal growth. *Proc. Natl. Acad. Sci. USA* **95:**13783–13787.

69. Roy, A., C. F. Lu, D. L. Marykwas, P. N. Lipke, and J. Kurjan. 1991. The AGA1 product is involved in cell surface attachment of the *Saccharomyces cerevisiae* cell adhesion glycoprotein α-agglutinin. *Mol. Cell. Biol.* **11:**4196–4206.

70. Rupp, S., E. Summers, H. J. Lo, H. Madhani, and G. Fink. 1999. MAP kinase and cAMP filamentation signaling pathways converge on the unusually large promoter of the yeast *FLO11* gene. *EMBO J.* **18:**1257–1269.

71. Rusche, L. N., A. L. Kirchmaier, and J. Rine. 2003. The establishment, inheritance, and function of silenced chromatin in *Saccharomyces cerevisiae. Annu. Rev. Biochem.* **72:**481–516.

72. Sandmeier, J. J., I. Celic, J. D. Boeke, and J. S. Smith. 2002. Telomeric and rDNA silencing in *Saccharomyces cerevisiae* are dependent on a nuclear NAD⁺ salvage pathway. *Genetics* **160:**877–889.

73. Sheppard, D. C., M. R. Yeaman, W. H. Welch, Q. T. Phan, Y. Fu, A. S. Ibrahim, S. G. Filler, M. Zhang, A. J. Waring, and J. E. Edwards, Jr. 2004. Functional and structural diversity in the Als protein family of *Candida albicans. J. Biol. Chem.* **279:**30480–30489.

74. Singleton, D. R., J. Masuoka, and K. C. Hazen. 2001. Cloning and analysis of a *Candida albicans* gene that affects cell surface hydrophobicity. *J. Bacteriol.* **183:**3582–3588.

75. Smith, C. D., D. L. Smith, J. L. DeRisi, and E. H. Blackburn. 2003. Telomeric protein distributions and remodeling through the cell cycle in *Saccharomyces cerevisiae. Mol. Biol. Cell* **14:**556–570.

76. Smith, J. S., C. B. Brachmann, I. Celic, M. A. Kenna, S. Muhammad, V. J. Starai, J. L. Avalos, J. C. Escalante-Semerena, C. Grubmeyer, C. Wolberger, and J. D. Boeke. 2000. A phylogenetically conserved NAD⁺-dependent protein deacetylase activity in the Sir2 protein family. *Proc. Natl. Acad. Sci. USA* **97:**6658–6663.

77. Smith, R. L., and A. D. Johnson. 2000. Turning genes off by Ssn6-Tup1: a conserved system of transcriptional repression in eukaryotes. *Trends Biochem. Sci.* **25:**325–330.

78. Sobel, J. D., C. A. Kauffman, D. McKinsey, M. Zervos, J. A. Vazquez, A. W. Karchmer, J. Lee, C. Thomas, H. Panzer, W. E. Dismukes, and The National Institute of Allergy and Infectious Diseases (NIAID) Mycoses Study Group. 2000. Candiduria: a randomized, double-blind study of treatment with fluconazole and placebo. *Clin. Infect. Dis.* **30:**19–24.

79. Sonneborn, A., D. P. Bockmuhl, M. Gerads, K. Kurpanek, D. Sanglard, and J. F. Ernst. 2000. Protein kinase A encoded by *TPK2* regulates dimorphism of *Candida albicans. Mol. Microbiol.* **35:**386–396.

80. Staab, J. F., S. D. Bradway, P. L. Fidel, and P. Sundstrom. 1999. Adhesive and mammalian transglutaminase substrate properties of *Candida albicans* Hwp1. *Science* **283:**1535–1538.

81. Stone, E. M., and L. Pillus. 1996. Activation of an MAP kinase cascade leads to Sir3p hyperphosphorylation and strengthens transcriptional silencing. *J. Cell. Biol.* **135:**571–583.

82. Sundstrom, P. 2002. Adhesion in *Candida* spp. *Cell. Microbiol.* **4:**461–469.

83. Tanny, J. C., and D. Moazed. 2001. Coupling of histone deacetylation to NAD breakdown by the yeast silencing protein Sir2: evidence for acetyl transfer from substrate to an NAD breakdown product. *Proc. Natl. Acad. Sci. USA* **98:**415–420.

84. Teunissen, A. W., and H. Y. Steensma. 1995. Review: the dominant flocculation genes of *Saccharomyces cerevisiae* constitute a new subtelomeric gene family. *Yeast* **11:**1001–1013.

85. Teunissen, A. W., J. A. van den Berg, and H. Y. Steensma. 1995. Transcriptional regulation of flocculation genes in *Saccharomyces cerevisiae. Yeast* **11:**435–446.

86. Tsukamoto, Y., J. Kato, and H. Ikeda. 1997. Silencing factors participate in DNA repair and recombination in *Saccharomyces cerevisiae. Nature* **388:**900–903.

87. Verstrepen, K. J., G. Derdelinckx, H. Verachtert, and F. R. Delvaux. 2003. Yeast flocculation: what brewers should know. *Appl. Microbiol. Biotechnol.* **61:**197–205.

88. Verstrepen, K. J., T. B. Reynolds, and G. R. Fink. 2004. Origins of variation in the fungal cell surface. *Natl. Rev. Microbiol.* **2:**533–540.

89. Watson, A. D., D. G. Edmondson, J. R. Bone, Y. Mukai, Y. Yu, W. Du, D. J. Stillman, and S. Y. Roth. 2000. Ssn6-Tup1 interacts with class I histone deacetylases required for repression. *Genes Dev.* **14:**2737–2744.

90. Weig, M., L. Jansch, U. Gross, C. G. De Koster, F. M. Klis, and P. W. De Groot. 2004. Systematic identification in silico of covalently bound cell wall proteins and analysis of protein-polysaccharide linkages of the human pathogen *Candida glabrata. Microbiology* **150:**3129–3144.

91. Wojciechowicz, D., C. F. Lu, J. Kurjan, and P. N. Lipke. 1993. Cell surface anchorage and ligand-binding domains of the *Saccharomyces cerevisiae* cell adhesion protein α-agglutinin, a member of the immunoglobulin superfamily. *Mol. Cell. Biol.* **13:**2554–2563.

92. Wotton, D., and D. Shore. 1997. A novel Rap1p-interacting factor, Rif2p, cooperates with Rif1p to regulate telomere length in *Saccharomyces cerevisiae. Genes Dev.* **11:**748–760.

93. Wu, J., N. Suka, M. Carlson, and M. Grunstein. 2001. TUP1 utilizes histone H3/H2B-specific HDA1 deacetylase to repress gene activity in yeast. *Mol. Cell* **7:**117–126.

94. Zaman, Z., A. Z. Ansari, S. S. Koh, R. Young, and M. Ptashne. 2001. Interaction of a transcriptional repressor with the RNA polymerase II holoenzyme plays a crucial role in repression. *Proc. Natl. Acad. Sci. USA* **98:**2550–2554.

95. Zhang, M., D. Bennett, and S. E. Erdman. 2002. Maintenance of mating cell integrity requires the adhesin Fig2p. *Eukaryot. Cell* **1:**811–822.

96. Zhang, Z., and J. C. Reese. 2004. Redundant mechanisms are used by Ssn6-Tup1 in repressing chromosomal gene transcription in *Saccharomyces cerevisiae. J. Biol. Chem.* **279:**39240–39250.

Reference strain

Mixed *bcr1/bcr1-GFP* + Reference strain

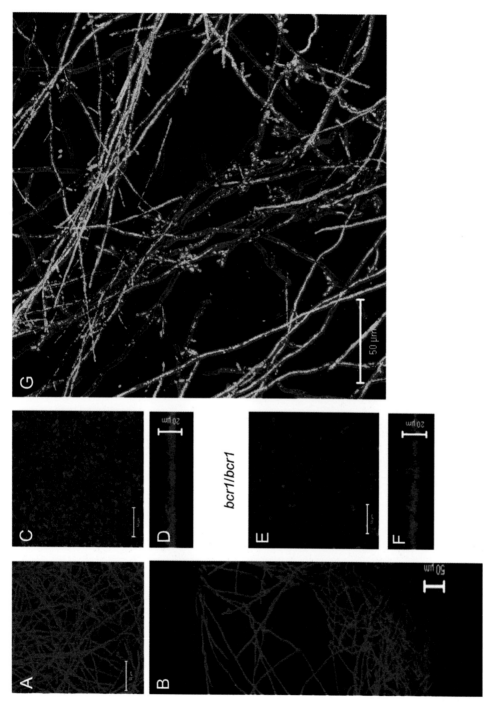

tec1/tec1

bcr1/bcr1

Color Plate 1 (chapter 3). Biofilm production by reference strain, *tec1/tec1* mutant, and *bcr1/bcr1* mutant strains. (A to F) Confocal scanning laser microscopy top views (A, C, and E) and side views (B, D, and F; all oriented with the substrate at the bottom of the figure) are shown. Biofilms were stained with concanavalin A conjugate for visualization. (G) Top view of a mixed biofilm containing the reference strain and a GFP-expressing *bcr1/bcr1* mutant. This panel is an overlay of concanavalin A conjugate (red) and GFP (green) images. Scale bars represent 50 μm (A, B, C, E, and G) or 20 μm (D and F).

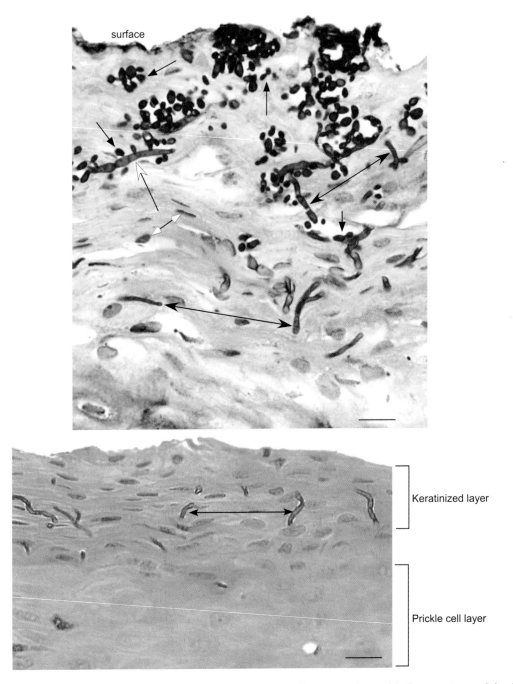

surface

Keratinized layer

Prickle cell layer

Color Plate 2 (chapter 4). Invasion of human lingual tissue by *C. albicans*. (A) Yeast (black arrows), pseudohyphae (open arrow), and true hyphae (double black arrow) are present. Yeast forms appear to bud off of pseudohyphae, hyphae, or other yeasts and are frequently associated with vacuolated regions of tissue. Hyphae appear embedded in deeper layers of tissue, whereas all morphologies are seen in regions proximal to the lingual surface. (B) *C. albicans* hyphae are distributed primarily in superficial keratinized layers, barely impinging on the spinous layer of stratified squamous epithelium. Paraffin-embedded sections were stained with periodic acid-Schiff reagent to visualize fungi and were counterstained with hematoxylin for characterization of host cells. Keratinocyte nuclei (double open arrows in A) appear violet, whereas keratinized regions are pink. Scale bars, 20 μm.

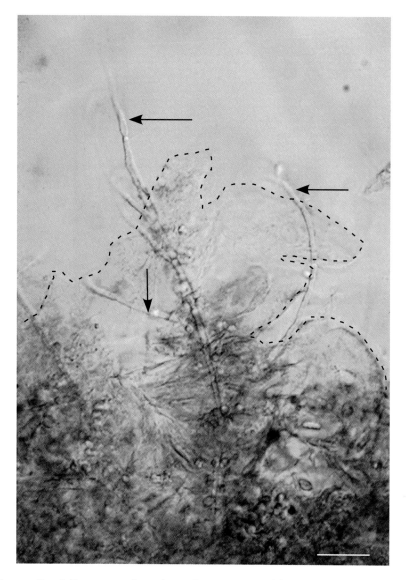

Color Plate 3 (chapter 4). Saline mount of pseudomembranous material from gingiva of a human immunodeficiency virus-positive patient. Epithelial cells are fastened together by *C. albicans* hyphae (arrows). The border of the mass of epithelial cells is indicated by a dotted line. Scale bar, 20 μm.

Mel⁻ *(MATa* parent)

Wild type (*MATα* parent)

F1 progeny showing the Mendelian inheritance of Mel phenotype

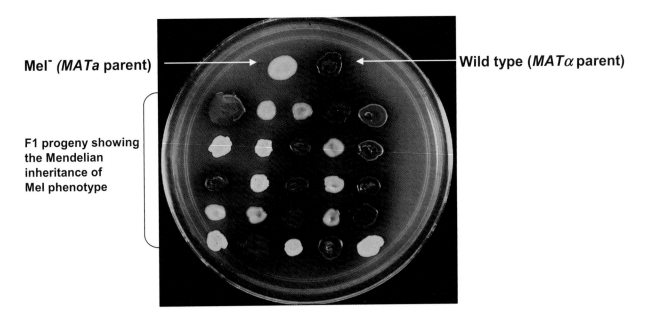

Color Plate 4 (chapter 5). Segregation of the Mel⁻ phenotype. The *MATa* parent is deficient in melanin formation (Mel⁻) and grows as white colonies on birdseed agar. The *MATα* parent is wild type and produces melanin on birdseed agar, which results in a brown colony. Mating between the two strains yields progeny in which the Mel⁻ and Mel⁺ phenotypes segregate independently from mating type and roughly in a 1:1 distribution.

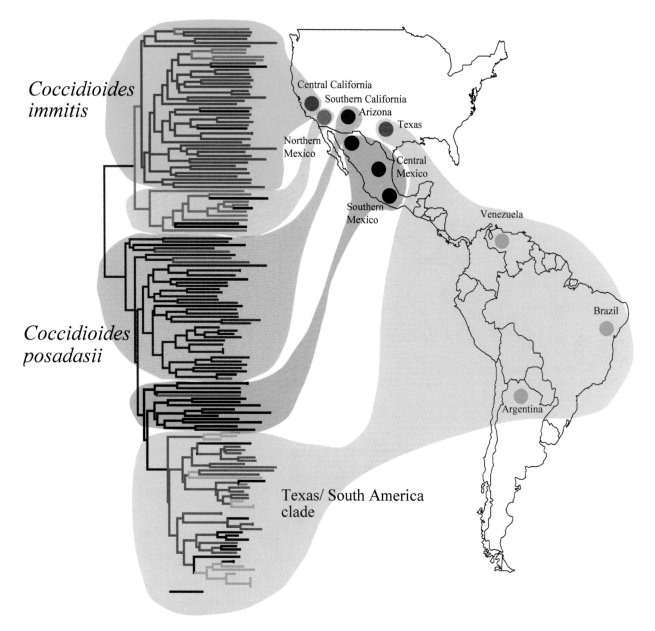

Color Plate 5 (chapter 8). Phylogenetic tree of 147 *Coccidioides* individuals based on MLMT using nine microsatellites. At least five genetically differentiated populations are identified in two species: *C. immitis* with Central California and Southern California populations and *C. posadasii* with Arizona, Mexico, and Texas-Latin American populations. Population genetic analysis of Latin American genotypes supports their relatively recent emigration from Texas (47). Reprinted from reference 191 with permission.

Color Plate 6 (chapter 22). *C. albicans* invasion of the chick CAM. Hyphae penetrate the membrane and invade the chick embryo. Yeast forms of the fungus proliferate on the membrane surface. Periodic acid-Schiff stain. Reprinted from reference 46 with permission.

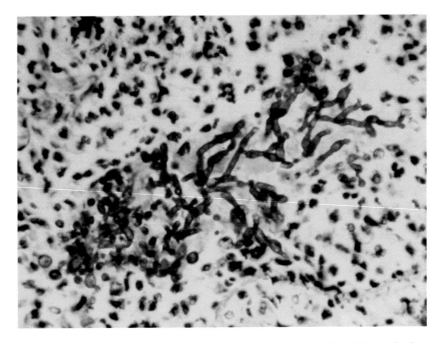

Color Plate 7 (chapter 22). *C. albicans* invading guinea pig kidneys. The fungal cells exhibit multiple morphologies as they appear to adapt, micrometer by micrometer, to the immediate local microenvironment. Periodic acid-Schiff stain.

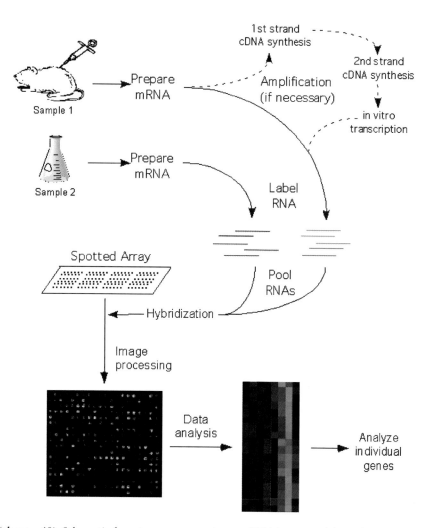

Color Plate 8 (chapter 40). Schematic for microarray experiments. RNA is prepared from distinct populations of cells, in this example an in vivo sample and an in vitro control condition. If necessary, the RNA is amplified (which should be done on both samples) and labeled fluorescently. It is then hybridized to a spotted microarray. Data analysis can include hierarchical clustering (shown) or a variety of other statistical analyses.

Molecular Principles of Fungal Pathogenesis
Edited by Joseph Heitman et al.
©2006 ASM Press, Washington, D.C.

Chapter 12

Fungal Biofilms: Agents of Disease and Drug Resistance

GORDON RAMAGE, MAHMOUD A. GHANNOUM, AND JOSÉ L. LÓPEZ-RIBOT

Our perception of microorganisms as unicellular life-forms is primarily based on the pure-culture mode of growth: since microorganisms can be diluted to a single cell and studied in liquid culture, this mode of growth has overwhelmingly predominated in the study of microbial physiology and pathogenesis in the research laboratory. However, the majority of microbes in their natural habitats persist attached to surfaces within a structured biofilm ecosystem and not as free-floating (planktonic) organisms (20, 21, 29). Biofilms can be defined as structured microbial communities attached to surfaces and encased within a matrix of exopolymeric materials (21). The biofilm lifestyle results in phenotypic characteristics that are markedly different from that of planktonic existence, such as increased resistance to antimicrobial agents and protection from host defenses (22, 27, 33, 40). Recently, there has been an increasing recognition of the role that microbial biofilms play in human medicine, and it has been estimated that about 65% of all human microbial infections involve biofilm formation (20, 22, 27, 28, 31, 46, 57). Bacterial biofilms and their role in disease have been investigated in detail over 15 to 20 years, but research on fungal biofilms has traditionally lagged. However, the last few years have witnessed an increased interest in the study of fungal biofilms. Although other fungal species have been associated with biofilm formation, this chapter concentrates on *Candida albicans* since it is the fungal system that has received the most attention to date regarding its biofilm-forming ability (17, 18, 30, 31, 65, 82, 85).

BIOFILMS AS AGENTS OF DISEASE: *CANDIDA* BIOFILMS IN HUMAN MEDICINE

Candida species are normal commensals of humans, and this fact facilitates their encounter with most implanted biomaterials and compromised biological surfaces. The increase in the incidence of *Candida* infections in the last decades has almost paralleled the increase and widespread use of a broad range of medical implant devices, mainly in populations with impaired host defenses. Indeed, a majority of infections caused by *Candida* spp., particularly *C. albicans*, are associated with biofilm formation (30, 31, 60, 65). Thus, the ability to form biofilms is intimately associated with the ability to cause infection, and, as such, biofilm formation should be considered an important virulence factor during candidiasis (65). Devices such as shunts, prostheses (voice, heart valve, knee, etc.), stents, implants (lens, breast, denture, etc.) endotracheal tubes, pacemakers, and various types of catheters all support colonization and biofilm formation by *Candida* (recently reviewed in reference 60). Of note, intravenous catheters are the leading cause of nosocomial bloodstream infections. Importantly, yeasts (mainly *Candida*) are the third leading cause of intravascular catheter-related infections, with the second highest colonization-to-infection rate and the overall highest crude mortality (25). Several aspects of *Candida* biofilms make their formation on host tissues and implanted biomaterials a clinically relevant issue. These are their increased resistance to antifungal agents, their ability to withstand host immunity, and their role as reservoirs for persistent sources of infectious microorganisms. Once *Candida* biofilms are formed on these devices, therapeutic interventions only rarely achieve clinical success and removal of the infected device is normally required. The net effect is that *Candida* biofilms adversely impact the health of these patients and result in significant increases in patient morbidity, duration of hospitalization, and hospital costs (13, 93, 94, 97).

Gordon Ramage • Department of Biological and Biomedical Sciences, Glasgow Caledonian University, Glasgow, G4 0BA, United Kingdom.   **Mahmoud A. Ghannoum** • Center for Medical Mycology, Department of Dermatology, University Hospitals of Cleveland and Case Western Reserve University, Cleveland, OH 44106-5028.   **José L. López-Ribot** • Department of Biology, The University of Texas at San Antonio, San Antonio, TX, 78249.

STRUCTURAL CHARACTERISTICS OF *C. ALBICANS* BIOFILMS

C. albicans biofilm formation proceeds through distinct developmental phases involving the transition from blastospores to well-defined cellular communities encased in a polysaccharide matrix. These phases are characterized by initial adherence of yeast cells, followed by development of a monolayer and microcolony formation, filamentation, proliferation, and maturation accompanied by production of extracellular material (17, 48, 85) (Fig. 1).

To colonize and form biofilms, fungal cells must first adhere to suitable biological or innate surfaces. Surface properties affecting initial adhesion include chemical properties, hydrophobicity, and surface roughness (16). More specific adherence events may also be mediated by cell surface proteins such as those encoded by members of the *ALS* family of adhesins and *EAP1* (17, 42, 69). After implantation, most biomedical devices acquire a glycoproteinaceous conditioning film that can also affect binding (35, 43), and certainly *C. albicans* is capable of adhering to a wide range of extracellular matrix proteins (16). In addition, *C. albicans* can interact with bacterial species that may represent early colonizers of biomedical devices and physiological surfaces, giving rise to mixed fungal-bacterial biofilms (52, 74).

Different microscopy techniques have been implemented for the observation of mature biofilms to provide a detailed analysis of their architectural features. As seen in Fig. 2, scanning electron microscopy (SEM) allowed a detailed examination of biofilms and indicated that mature *C. albicans* biofilms consisted of a dense network of yeasts, hyphae, and pseudohyphae (48, 85). Typically, yeast cells predominate in the basal layers of biofilms whereas filamentous forms are more prevalent in outer layers (17, 30). The nondestructive nature of confocal scanning laser microscopy (CSLM) allows the visualization of fully hydrated living biofilms. CSLM demonstrated that mature biofilms display a complex three-dimensional structure, with spatial heterogeneity and an architecture of microcolonies with ramifying water channels (17, 85). Overall, the structural features of mature biofilms seem to represent an optimal spatial arrangement for influx of nutrients and disposal of waste products for sessile cells within the biofilm. Biofilm architecture may differ depending on the substrate on which biofilms are formed (17, 91). Also, different strains of *C. albicans* and different *Candida* spp. vary in their ability to form biofilms (63, 70, 81).

Although the structural features of biofilm formation have been studied mostly using in vitro models, it is important to note that the general characteristics of model biofilms seem to be similar to those formed in vivo. Two different animal models of catheter-associated *Candida* infections have been described recently, and visualization of the resulting in vivo-formed biofilms indicate similar architectural features to those formed in vitro (4, 89). Also, in a limited number of cases, biofilms have been recovered directly from patients and visualized using different microscopy techniques, mainly SEM (17, 24, 80, 88). Again, these data suggest that in vitro model systems are analogous to in vivo events and may be clinically relevant.

ROLE OF MORPHOGENETIC CONVERSIONS IN *C. ALBICANS* BIOFILM FORMATION

C. albicans has the ability to reversibly switch between yeast and filamentous forms, and these morphogenetic conversions are important for pathogenicity (14, 15, 41, 87). The activity of signaling pathways, which includes key transcriptional regulators, is an important factor in filamentation. As discussed above, *C. albicans* biofilms are composed of a mixture of yeast cells and filamentous elements. Baillie and Douglas reported that biofilms formed by a strain defective in hypha formation consisted of densely packed yeast cells forming an atypical biofilm that was easily detached from the substratum by washing (11). This will argue for a role of hyphae

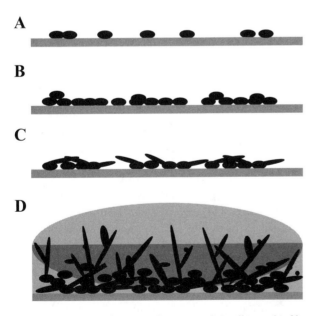

Figure 1. Diagram depicting the stages of *C. albicans* biofilm formation. (A) Initial attachment of yeast cells. (B) Formation of a basal monolayer of yeast cells. (C) Proliferation of microcolonies and filamentation. (D) Further filamentation during maturation and encasement within exopolymeric material (note that synthesis of EPS starts early during biofilm formation [not depicted]).

Figure 2. SEM image of a mature (48-h) *C. albicans* biofilm. Biofilms are composed of yeast cells, pseudohyphae, and hyphae. Also visible are the remains of the exopolymeric matrix after dehydration due to SEM procedures. Bar, 10 μm.

in providing the structural integrity necessary for biofilm architecture. Thus, dimorphism may not be an absolute prerequisite for biofilm formation but might be necessary for the development of a spatially organized structure (30). To begin to dissect the molecular pathways governing filamentation that are involved in biofilm formation, Ramage et al. used a series of genetically defined *C. albicans* mutant strains, characterized by their respective abilities to filament under different environmental conditions, and tested their ability to form biofilms (84). In this way, defects in filamentation and biofilm formation observed using this set of isogenic strains could be attributed to individual genes with key morphogenetic roles. Microscopic analysis of biofilms revealed that Δ*cph1*, Δ*hst7*, and Δ*cst20* mutants were able to filament and form structured biofilms displaying three-dimensional architecture similar to those formed by wild-type strains. These observations are not surprising considering that strains with these mutations were unable to filament on solid media but were able to filament in response to other environmental cues (59). However, the single Δ*efg1* and double Δ*cph1* Δ*efg1* mutants were unable to filament and did not form fully structured biofilms but, rather, formed poor biofilms composed mainly of sparse monolayers of elongated cells (84). The Δ*cph1* Δ*efg1* double mutant was also profoundly deficient for colonizing plasma-coated polyurethane catheters (68) but was still able to form a thin biofilm on glass

containing mostly blastospores (36). By using the Δ*cph1* Δ*efg1* double mutant, these authors were also able to identify a cluster of biofilm-related genes that were expressed independently of hypha formation, with a preponderance of genes implicated in the sulfur-amino acid biosynthesis pathway (36). Together, these results demonstrate that the Efg1p regulator protein is a key factor in the formation and subsequent development of *C. albicans* biofilms on most biological and artificial surfaces, although this role is probably intimately linked to its filamentation defect.

The biofilm-forming ability of several other *C. albicans* mutant strains has been evaluated. In *C. albicans*, Gcn4p controls many amino acid starvation responses, some of which lead to hyphal development (92), and a high proportion of genes involved in amino acid biosynthesis that are targets of Gcn4p are differentially expressed in biofilm cells (36). Thus, Garcia-Sanchez et al. tested the ability of a *C. albicans* Δ*gcn4* null strain to form biofilms and found that Gcn4p is required for normal biofilm formation (36). Disruption of *NOT4* leads to defects in hypha formation, and biofilm development does not readily occur for the Δ*not4* null mutant (61). Likewise, deletion of the *C. albicans ACE2* gene, encoding a key regulator of cell wall metabolism, leads to morphogenetic defects, and mutant cells are defective in adherence to plastic and generate biofilms with distinctly different morphology from wild-type cells (56).

ROLE OF QUORUM SENSING

Cell-cell signaling, in particular quorum sensing, is fundamental to microbial biofilm formation by preventing and controlling unnecessary overpopulation and competition for nutrients (29, 58, 73, 76). This carries important repercussions in the infectious process, dissemination, and the establishment of distal sites of infection. The spatial heterogeneity and complexity observed in intact *C. albicans* biofilms seems to indicate that cell-cell communication and highly regulated gene expression systems are important for biofilm formation in this fungus. It has been shown that farnesol (3,7,11-trimethyl-2,6,10-dodecatriene-1-ol) acts as a quorum-sensing molecule and inhibits filamentation in *C. albicans* (53). Subsequent experiments demonstrated that preincubation with high concentrations of farnesol almost completely inhibited *C. albicans* biofilm formation, with this effect being dependent on concentration and initial adherence time. Moreover, supernatants recovered from mature biofilms inhibited the filamentation of planktonic *C. albicans*, thus indicating that a morphogenetic autoregulatory compound, most probably farnesol, is produced in situ in biofilms (79). Another quorum-sensing/autoregulatory molecule with a role in growth and morphogenesis in *C. albicans* is tyrosol [2,4-(hydroxyphenyl)-ethanol], which is found in conditioned medium from high-density cultures (19, 23). In contrast to farnesol, tyrosol stimulates filamentation. It has been recently reported that both quorum sensing and biofilm formation in *C. albicans* are regulated by the two-component signal transduction protein Chk1p (62). Together, these results indicate that quorum-sensing and morphogenesis in *C. albicans* must be under complex positive and negative control by environmental conditions.

Another interesting set of experiments comes from a more ecological perspective that addresses the interactions between *C. albicans* and *Pseudomonas aeruginosa*. It was first described that *P. aeruginosa* forms a dense biofilm on *C. albicans* filaments and kills the fungus but that the bacteria neither bind to nor kill yeast forms of *C. albicans* (50). Subsequent experiments indicated that a *P. aeruginosa* quorum-sensing molecule (3-oxo-C_{12}-homoserine lactone) influences *C. albicans* morphology (51). This molecule and structurally related compounds with a 12-carbon chain length are able to suppress filamentation in *C. albicans* and also induce *C. albicans* filaments to revert to growth as yeast cells (51).

ANTIFUNGAL DRUG RESISTANCE IN *CANDIDA* BIOFILMS

Antifungal treatment strategies are limited to a small but growing armamentarium of compounds used to treat infections ranging from superficial to invasive mycoses (45, 90). Examples of these compounds include azoles, polyenes, and the recently introduced echinocandins (96). From the clinical perspective, the most important feature of *Candida* biofilms is their high levels of resistance to conventional antifungal therapy (30, 31, 55, 86). Several groups have demonstrated that the *Candida* biofilm lifestyle leads to dramatically increased levels of resistance to the most commonly used antifungal agents, fluconazole and amphotericin B (AMB) (7–9, 17, 18, 64, 81–83, 85, 86). It has been reported that *C. albicans* biofilms are up to 4,000 times more resistant to fluconazole than are planktonic, free-floating cells (85). It is important to note that standard NCCLS broth dilution techniques for antifungal susceptibility testing that use planktonic populations will not allow a prediction of antifungal efficacy against *Candida* biofilms. Thus, this may be one of the main reasons for the lack of correlation between results of antifungal susceptibility testing, as determined by NCCLS guidelines, and clinical outcome in patients with these types of infections (37–39). However, determining the effectiveness of different antifungal agents in this setting has important clinical implications in that it may guide therapeutic decisions that potentially may affect the outcome of patients with these difficult-to-treat infections. This may be facilitated by the recent development of a microtiter plate-based biofilm model that is compatible with the 96-well platform technology for the determination and standardization of antifungal susceptibility testing in *Candida* biofilms (82).

Newer antifungal agents, such as the echinocandins and liposomal formulations of AMB, show increased activity against *Candida* biofilms (5, 6, 64, 83). Kuhn et al. demonstrated that the newer triazoles (voriconazole and ravuconazole) were ineffective against preformed biofilms but that lipid formulations of AMB (liposomal AMB and AMB lipid complex) and echinocandins (caspofungin and micafungin) showed activity against *Candida* biofilms (64). Moreover, preincubation of *C. albicans* cells with sub-MIC levels of the antifungals (for example voriconazole) decreased the ability of cells to subsequently form biofilms (64). Bachmann et al. reported that caspofungin displayed potent in vitro activity against sessile *C. albicans* cells within biofilms, with sessile MICs for a number of *C. albicans* strains well within its therapeutic range (6). SEM and CSLM observations of caspofungin-treated biofilms indicated an effect on cellular morphology and metabolic status of cells within the biofilms (6). Also, coating of biomaterials with caspofungin had an inhibitory effect on subsequent biofilm development by *C. albicans* (6). Importantly, these were the first reports showing significant activity of any antifungal drug against *Candida* biofilms. Other echinocandins, such as anidulafungin and micafungin, were also shown to be effective against *Candida* growing as biofilm (M. Schinabeck and M. Ghannoum, *Program Abstr. 44th Intersci. Conf. Antimicrob. Agents Chemother.*, abstr. M-1141, 2004).

A more recent study examined the activity of several simulated antifungal treatment modalities for eradication of catheter-related bloodstream *Candida* infections by using a one-compartment pharmacodynamic model (67). Different antifungal regimens varied in their ability to suppress the growth of *Candida* species, with AMB regimens being superior to azole agents (voriconazole and fluconazole). However, no single agent demonstrated an ability to completely eradicate the *Candida* biofilms. Interestingly, a minocycline-EDTA flush solution was able to prevent *C. albicans* colonization of catheter materials and also to fully eradicate a preformed *C. albicans* biofilm at high minocycline concentrations (77). Likewise, aspirin and other prostaglandin inhibitors at therapeutic concentrations showed activity against preformed *C. albicans* biofilms (2). Thus, these innovative approaches may represent interesting alternatives for the prevention of catheter-related candidemia.

Due to the expanding arsenal of antifungal drugs, it is conceivable that combination therapy could become an important approach to fungal infections, including those of biofilm etiology, over the next coming years. A recent study examined the in vitro activity of combinations of antifungal agents against *C. albicans* biofilms (5). Use of a checkerboard microtiter method to determine the fractional inhibitory concentration (FIC) and the FIC indices of the paired combinations of antifungals revealed that all the combinations of paired antifungal agents had indifferent interactions. However, a trend for an additive effect was observed with the AMB-caspofungin combination, and a tendency toward an antagonistic effect was found with the fluconazole-caspofungin combination, which was more evident at high fluconazole concentrations. A subsequent series of experiments was performed with time-kill methods to study the pharmacodynamic properties of antifungal combinations against preformed biofilms. These experiments confirmed the antagonistic effect of high fluconazole doses with caspofungin. However, when AMB was coincubated with caspofungin during challenge of *C. albicans* biofilms, there was a more rapid and complete killing of the biofilms than could be attained by therapeutic concentrations of caspofungin alone (5).

MOLECULAR BASIS OF ANTIFUNGAL DRUG RESISTANCE IN *CANDIDA* BIOFILMS

Different mechanisms may be responsible for the intrinsic resistance of biofilms to antimicrobials. These include (i) effects of the biofilm matrix on penetration of drugs; (ii) decreased growth rate and nutrient limitation; (iii) expression of resistance genes, particularly those encoding efflux pumps; and (iv) presence of "persister" cells (30, 31, 55, 65, 66). However, the current consensus is that biofilm resistance is a complex phenomenon that cannot be explained by one mechanism alone; instead, it is multifactorial and phase specific and may involve different molecular mechanisms of resistance from those displayed by planktonic cells. Some of these are discussed below in reference to their impact on *C. albicans* biofilm resistance. The matrix or exopolymeric substance (EPS) is the substance that is produced by and envelops sessile communities of cells. The majority of studies of the EPS have been performed with bacterial biofilms, and these have indicated that the EPS can act as a barrier of diffusion to antimicrobial agents (54) and/or as an ion exchange resin to bind charged antibiotic molecules (43, 44). Thus, it has been implicated as a potential contributor to sessile recalcitrance. In *C. albicans* biofilms, matrix production increased dramatically when developing biofilms were subjected to a liquid flow than when they encountered static conditions. In any case, the two types of biofilms were equally resistant to antifungal drugs (10, 47). However, for biofilms formed under flow conditions, it was shown that intact biofilms were 20% more resistant to AMB than were resuspended cells that had lost their matrix during resuspension (9). It has also been demonstrated that cells adhering for as little as 1 to 2 h (when matrix material is not detectable) display high levels of fluconazole resistance. Subsequent experiments with resuspended sessile cells recovered from mature biofilms indicated that when tested as freefloating cells, they were also resistant to fluconazole challenge, but not to the same level as mature biofilms (78). In mixed fungal-bacterial biofilms, the extracellular polymer produced by bacterial cells retarded the diffusion of the antifungal agents; however, poor penetration did not account for the drug resistance of *Candida* biofilm cells (1, 3). Overall, these results seem to indicate that the EPS plays a partial role in sessile resistance but that other factors are likely involved too.

It has been reported that cells in bacterial biofilms are somewhat metabolically quiescent, which will impact antibiotic action since some antibiotics are dependent on cells actively undergoing cell division (34, 54). Several studies have examined the effects of growth rate and nutrient limitation in relation to drug resistance in *C. albicans*. Baillie and Douglas (8, 9) used a perfused biofilm fermentor model and demonstrated that mature biofilms were resistant to AMB at all growth rates tested and also at different levels of nutrient limitation. It has also been demonstrated that cells within *C. albicans* biofilms are not quiescent, since they are metabolically active as indicated by a colorimetric reduction assay (18, 82, 85). In addition, Chandra et al. reported that progression of drug resistance was associated with an increase in the metabolic activity of the developing biofilm and not with slower growth, which clearly indicates that drug resistance develops over time, coincident with biofilm maturation (17, 18). An intriguing observation is that anaerobic growth leads to high levels of antifungal drug resistance in *C. albicans*; this could contribute to the intrinsic resistance in cells within the biofilms, where oxygen limitation may occur (32).

The previous observations point to the existence of other mechanisms and suggest that physiological changes in biofilm cells with respect to their planktonic counterparts occur as a consequence of different patterns of gene expression that are specific to the biofilm lifestyle, as recently demonstrated using DNA macroarrays (36). Under planktonic conditions, one of the main mechanisms of azole resistance in *C. albicans* is through active efflux of these drugs mediated by ABC transporters and major facilitators (95). Importantly, multiple antifungal agents can be substrates for some of these transporters, and thus their overexpression leads to cross-resistance among different drugs. The susceptibility profiles of *C. albicans* biofilms, including their cross-resistance to multiple azole derivatives, suggest that efflux pumps may contribute to the resistance of biofilms. Indeed, the expression of genes encoding both types of efflux pump were up-regulated during the different phases of biofilm development, both in vitro and in vivo (4, 72, 75, 78). Interestingly, mutant strains deficient in efflux pumps and hypersusceptible to fluconazole when grown in a planktonic mode retained a resistant phenotype during biofilm growth, although they were more susceptible during the early adherence phase (72, 75, 78). Again, these findings seem to suggest that drug resistance in biofilms is complex and probably involves multiple mechanisms.

These analyses suggest that *C. albicans* efflux pumps are essential factors that have evolved independently of azole drugs. The main role for these transport systems is to allow the uptake of essential nutrients, excretion of metabolic and deleterious products, and communication between cells and the environment (12, 26, 49, 71). Because of the highly structured but heterogeneous biofilm environment, displaying efficient transport systems may allow individual cells within the biofilm to maintain an appropriate local physicochemical environment required for growth and survival. A most intriguing feature of multidrug transporters is their ability to bind to a range of structurally unrelated drugs; remarkably, these same transporter systems have also been found to play a role in resistance to antibiotics. Therefore, it is conceivable that, in a biofilm environment, the efflux pumps encoded by the *CDR1, CDR2,* and *MDR1* genes are employed primarily as a means of cellular detoxification. Thus, physiological overexpression of these genes originally associated with a sessile existence may coincidentally impact antimicrobial susceptibility. Additional sterol analyses showed that ergosterol levels were significantly decreased at intermediate and mature phases compared to those in early-phase biofilms (75).

CONCLUSIONS AND FUTURE PERSPECTIVES

The formation of adherent, highly organized, exopolymer-enshrouded candidal biofilms is one of the most important factors in the pathogenesis of candidiasis, and scientific interest in candidal biofilms has exploded in the last few years. The biofilm mode of growth results in antifungal drug resistance and protection from host defenses, which carry important clinical repercussions. Drug resistance in sessile cells occurs though multifaceted and phase-specific mechanisms that contribute to the overall antifungal resistance of fungal biofilms. In addition, it is likely that biofilm infections involve bacteria and various *Candida* species in a polymicrobial consortium. In these complex communities, cell-cell communication, across both kingdom and species, may modulate gene expression, including drug resistance mechanism. Molecular studies of biofilm formation have begun to shed light on the driving forces behind the transition to the biofilm mode of existence, including quorum sensing, which in the future may offer a potential therapeutic strategy. In the meantime, new antifungal agents and preventive strategies offer new hope for the treatment of these recalcitrant infections. Future studies should focus on the biofilms in vivo, probe other *Candida* species, particularly *C. arapsilosis*, and investigate materials to identify surfaces that could inhibit or retard biofilm formation. More knowledge is needed about how biofilms form in host tissues and how biofilm formation impacts the host-parasite interactions.

Acknowledgments. We gratefully acknowledge the support of the Texas Higher Education Coordinating Board and the

National Institutes of Health. J.L.L.-R. is the recipient of a New Investigator Award in Molecular Pathogenic Mycology from the Burroughs Wellcome Fund. M.A.G. is a recipient of the "Freedom to Discover Award" from Bristol-Myers Squibb Foundation, NIH grant AI035097-09, and Scientist Development Grant 0335313N from the American Heart Association.

REFERENCES

1. Adam, B., G. S. Baillie, and L. J. Douglas. 2002. Mixed species biofilms of *Candida albicans* and *Staphylococcus epidermidis. J. Med. Microbiol.* 51:344–349.
2. Alem, M. A., and L. J. Douglas. 2004. Effects of aspirin and other nonsteroidal anti-inflammatory drugs on biofilms and planktonic cells of *Candida albicans. Antimicrob. Agents Chemother.* 48:41–47.
3. Al-Fattani, M. A., and L. J. Douglas. 2004. Penetration of *Candida* biofilms by antifungal agents. *Antimicrob. Agents Chemother.* 48:3291–3297.
4. Andes, D., J. Nett, P. Oschel, R. Albrecht, K. Marchillo, and A. Pitula. 2004. Development and characterization of an in vivo central venous catheter *Candida albicans* biofilm model. *Infect. Immun.* 72:6023–6031.
5. Bachmann, S. P., G. Ramage, K. VandeWalle, T. F. Patterson, B. L. Wickes, and J. L. Lopez-Ribot. 2003. Antifungal combinations against *Candida albicans* biofilms in vitro. *Antimicrob. Agents Chemother.* 47:3657–3659.
6. Bachmann, S. P., K. VandeWalle, G. Ramage, T. F. Patterson, B. L. Wickes, J. R. Graybill, and J. L. Lopez-Ribot. 2002. In vitro activity of caspofungin against *Candida albicans* biofilms. *Antimicrob. Agents Chemother.* 46:3591–3596.
7. Baillie, G. S., and L. J. Douglas. 1999. *Candida* biofilms and their susceptibility to antifungal agents. *Methods Enzymol.* 310:644–656.
8. Baillie, G. S., and L. J. Douglas. 1998. Effect of growth rate on resistance of *Candida albicans* biofilms to antifungal agents. *Antimicrob. Agents Chemother.* 42:1900–1905.
9. Baillie, G. S., and L. J. Douglas. 1998. Iron-limited biofilms of *Candida albicans* and their susceptibility to amphotericin B. *Antimicrob. Agents Chemother.* 42:2146–2149.
10. Baillie, G. S., and L. J. Douglas. 2000. Matrix polymers of *Candida* biofilms and their possible role in biofilm resistance to antifungal agents. *J. Antimicrob. Chemother.* 46:397–403.
11. Baillie, G. S., and L. J. Douglas. 1999. Role of dimorphism in the development of *Candida albicans* biofilms. *J. Med. Microbiol.* 48:671–679.
12. Balzi, E., and A. Goffeau. 1995. Yeast multidrug resistance: the PDR network. *J. Bioenerg. Biomembr.* 27:71–76.
13. Beck-Sague, C., and W. R. Jarvis. 1993. Secular trends in the epidemiology of nosocomial fungal infections in the United States, 1980–1990. National Nosocomial Infections Surveillance System. *J. Infect. Dis.* 167:1247–1251.
14. Brown, A. J., and N. A. Gow. 1999. Regulatory networks controlling *Candida albicans* morphogenesis. *Trends Microbiol.* 7:333–338.
15. Brown, A. J. P. 2002. Morphogenetic signaling pathways in *Candida albicans*, p. 95–106. *In* R. A. Calderone (ed.), Candida *and* Candidiasis. ASM Press, Washington, D.C.
16. Chaffin, W. L., J. L. Lopez-Ribot, M. Casanova, D. Gozalbo, and J. P. Martinez. 1998. Cell wall and secreted proteins of *Candida albicans*: identification, function, and expression. *Microbiol. Mol. Biol. Rev.* 62:130–180.
17. Chandra, J., D. M. Kuhn, P. K. Mukherjee, L. L. Hoyer, T. McCormick, and M. A. Ghannoum. 2001. Biofilm formation by the fungal pathogen *Candida albicans*: development, architecture, and drug resistance. *J. Bacteriol.* 183:5385–5394.
18. Chandra, J., P. K. Mukherjee, S. D. Leidich, F. F. Faddoul, L. L. Hoyer, L. J. Douglas, and M. A. Ghannoum. 2001. Antifungal resistance of candidal biofilms formed on denture acrylic in vitro. *J. Dent. Res.* 80:903–908.
19. Chen, H., M. Fujita, Q. Feng, J. Clardy, and G. R. Fink. 2004. Tyrosol is a quorum-sensing molecule in *Candida albicans. Proc. Natl. Acad. Sci. USA* 101:5048–5052.
20. Costerton, J. W., K. J. Cheng, G. G. Geesey, T. I. Ladd, J. C. Nickel, M. Dasgupta, and T. J. Marrie. 1987. Bacterial biofilms in nature and disease. *Annu. Rev. Microbiol.* 41:435–464.
21. Costerton, J. W., Z. Lewandowski, D. E. Caldwell, D. R. Korber, and H. M. Lappin-Scott. 1995. Microbial biofilms. *Annu. Rev. Microbiol.* 49:711–745.
22. Costerton, J. W., P. S. Stewart, and E. P. Greenberg. 1999. Bacterial biofilms: a common cause of persistent infections. *Science* 284:1318–1322.
23. Cremer, J., V. Vatou, and I. Braveny. 1999. 2,4-(hydroxyphenyl)-ethanol, an antioxidative agent produced by *Candida* spp., impairs neutrophilic yeast killing in vitro. *FEMS Microbiol. Lett.* 170:319–325.
24. Cristobal, R., C. E. Edmiston, Jr., C. L. Runge-Samuelson, H. A. Owen, J. B. Firszt, and P. A. Wackym. 2004. Fungal biofilm formation on cochlear implant hardware after antibiotic-induced fungal overgrowth within the middle ear. *Pediatr. Infect. Dis. J.* 23:774–778.
25. Crump, J. A., and P. J. Collignon. 2000. Intravascular catheter-associated infections. *Eur. J. Clin. Microbiol. Infect. Dis.* 19:1–8.
26. Del Sorbo, G., H. Schoonbeek, and M. A. De Waard. 2000. Fungal transporters involved in efflux of natural toxic compounds and fungicides. *Fungal Genet. Biol.* 30:1–15.
27. Donlan, R. M. 2001. Biofilm formation: a clinically relevant microbiological process. *Clin. Infect. Dis.* 33:1387–1392.
28. Donlan, R. M. 2001. Biofilms and device-associated infections. *Emerg. Infect. Dis.* 7:277–281.
29. Donlan, R. M. 2002. Biofilms: microbial life on surfaces. *Emerg. Infect. Dis.* 8:881–890.
30. Douglas, L. J. 2003. *Candida* biofilms and their role in infection. *Trends Microbiol.* 11:30–36.
31. Douglas, L. J. 2002. Medical importance of biofilms in *Candida* infections. *Rev. Iberoam. Micol.* 19:139–143.
32. Dumitru, R., J. M. Hornby, and K. W. Nickerson. 2004. Defined anaerobic growth medium for studying *Candida albicans* basic biology and resistance to eight antifungal drugs. *Antimicrob. Agents Chemother.* 48:2350–2354.
33. Dunne, W. M., Jr. 2002. Bacterial adhesion: seen any good biofilms lately? *Clin. Microbiol. Rev.* 15:155–166.
34. Evans, D. J., M. R. Brown, D. G. Allison, and P. Gilbert. 1990. Susceptibility of bacterial biofilms to tobramycin: role of specific growth rate and phase in the division cycle. *J. Antimicrob. Chemother.* 25:585–591.
35. Francois, P., P. Vaudaux, and P. D. Lew. 1998. Role of plasma and extracellular matrix proteins in the physiopathology of foreign body infections. *Ann. Vasc. Surg.* 12:34–40.
36. Garcia-Sanchez, S., S. Aubert, I. Iraqui, G. Janbon, J. M. Ghigo, and C. d'Enfert. 2004. *Candida albicans* biofilms:

a developmental state associated with specific and stable gene expression patterns. *Eukaryot. Cell* 3:536–545.

37. **Ghannoum, M. A.** 1997. Susceptibility testing of fungi and correlation with clinical outcome. *J. Chemother.* 9(Suppl. 1):19–24.

38. **Ghannoum, M. A., J. H. Rex, and J. N. Galgiani.** 1996. Susceptibility testing of fungi: current status of correlation of in vitro data with clinical outcome. *J. Clin. Microbiol.* 34:489–495.

39. **Ghannoum, M. A., and L. B. Rice.** 1999. Antifungal agents: mode of action, mechanisms of resistance, and correlation of these mechanisms with bacterial resistance. *Clin. Microbiol. Rev.* 12:501–517.

40. **Gilbert, P., J. Das, and I. Foley.** 1997. Biofilm susceptibility to antimicrobials. *Adv. Dent. Res.* 11:160–167.

41. **Gow, N. A., A. J. Brown, and F. C. Odds.** 2002. Fungal morphogenesis and host invasion. *Curr. Opin. Microbiol.* 5:366–371.

42. **Green, C. B., G. Cheng, J. Chandra, P. Mukherjee, M. A. Ghannoum, and L. L. Hoyer.** 2004. RT-PCR detection of *Candida albicans* ALS gene expression in the reconstituted human epithelium (RHE) model of oral candidiasis and in model biofilms. *Microbiology* 150:267–275.

43. **Gristina, A. G.** 1987. Biomaterial-centered infection: microbial adhesion versus tissue integration. *Science* 237:1588–1595.

44. **Gristina, A. G., Y. Shibata, G. Giridhar, A. Kreger, and Q. N. Myrvik.** 1994. The glycocalyx, biofilm, microbes, and resistant infection. *Semin. Arthroplasty* 5:160–170.

45. **Groll, A. H., and T. J. Walsh.** 2002. Antifungal chemotherapy: advances and perspectives. *Swiss Med. Wkly.* 132:303–311.

46. **Hall-Stoodley, L., J. W. Costerton, and P. Stoodley.** 2004. Bacterial biofilms: from the natural environment to infectious diseases. *Nat. Rev. Microbiol.* 2:95–108.

47. **Hawser, S. P., G. S. Baillie, and L. J. Douglas.** 1998. Production of extracellular matrix by *Candida albicans* biofilms. *J. Med. Microbiol.* 47:253–256.

48. **Hawser, S. P., and L. J. Douglas.** 1994. Biofilm formation by *Candida* species on the surface of catheter materials in vitro. *Infect. Immun.* 62:915–921.

49. **Higgins, C. F.** 1992. ABC transporters: from microorganisms to man. *Annu. Rev. Cell Biol.* 8:67–113.

50. **Hogan, D. A., and R. Kolter.** 2002. *Pseudomonas-Candida* interactions: an ecological role for virulence factors. *Science* 296:2229–2232.

51. **Hogan, D. A., A. Vik, and R. Kolter.** 2004. A *Pseudomonas aeruginosa* quorum-sensing molecule influences *Candida albicans* morphology. *Mol. Microbiol.* 54:1212–1223.

52. **Holmes, A. R., R. D. Cannon, and H. F. Jenkinson.** 1995. Interactions of *Candida albicans* with bacteria and salivary molecules in oral biofilms. *J. Ind. Microbiol.* 15:208–213.

53. **Hornby, J. M., E. C. Jensen, A. D. Lisec, J. J. Tasto, B. Jahnke, R. Shoemaker, P. Dussault, and K. W. Nickerson.** 2001. Quorum sensing in the dimorphic fungus *Candida albicans* is mediated by farnesol. *Appl. Environ. Microbiol.* 67:2982–2992.

54. **Hoyle, B. D., J. Jass, and J. W. Costerton.** 1990. The biofilm glycocalyx as a resistance factor. *J. Antimicrob. Chemother.* 26:1–5.

55. **Jabra-Rizk, M. A., W. A. Falkler, and T. F. Meiller.** 2004. Fungal biofilms and drug resistance. *Emerg. Infect. Dis.* 10:14–19.

56. **Kelly, M. T., D. M. MacCallum, S. D. Clancy, F. C. Odds, A. J. Brown, and G. Butler.** 2004. The *Candida albicans*

CaACE2 gene affects morphogenesis, adherence and virulence. *Mol. Microbiol.* 53:969–983.

57. **Khardori, N., and M. Yassien.** 1995. Biofilms in device-related infections. *J. Ind. Microbiol.* 15:141–147.

58. **Kjelleberg, S., S. Molin, M. B. Miller, and B. L. Bassler.** 2002. Is there a role for quorum sensing signals in bacterial biofilms? Quorum sensing in bacteria. *Curr. Opin. Microbiol.* 5:254–258.

59. **Kohler, J. R., and G. R. Fink.** 1996. *Candida albicans* strains heterozygous and homozygous for mutations in mitogen-activated protein kinase signaling components have defects in hyphal development. *Proc. Natl. Acad. Sci. USA* 93:13223–13228.

60. **Kojic, E. M., and R. O. Darouiche.** 2004. *Candida* infections of medical devices. *Clin. Microbiol. Rev.* 17:255–267.

61. **Krueger, K. E., A. K. Ghosh, B. P. Krom, and R. L. Cihlar.** 2004. Deletion of the *NOT4* gene impairs hyphal development and pathogenicity in *Candida albicans*. *Microbiology* 150:229–240.

62. **Kruppa, M., B. P. Krom, N. Chauhan, A. V. Bambach, R. L. Cihlar, and R. A. Calderone.** 2004. The two-component signal transduction protein Chk1p regulates quorum sensing in *Candida albicans*. *Eukaryot. Cell* 3:1062–1065.

63. **Kuhn, D. M., J. Chandra, P. K. Mukherjee, and M. A. Ghannoum.** 2002. Comparison of biofilms formed by *Candida albicans* and *Candida parapsilosis* on bioprosthetic surfaces. *Infect. Immun.* 70:878–888.

64. **Kuhn, D. M., T. George, J. Chandra, P. K. Mukherjee, and M. A. Ghannoum.** 2002. Antifungal susceptibility of *Candida* biofilms: unique efficacy of amphotericin B lipid formulations and echinocandins. *Antimicrob. Agents Chemother.* 46:1773–1780.

65. **Kumamoto, C. A.** 2002. *Candida* biofilms. *Curr. Opin. Microbiol.* 5:608–611.

66. **Lewis, K.** 2001. Riddle of biofilm resistance. *Antimicrob. Agents Chemother.* 45:999–1007.

67. **Lewis, R. E., D. P. Kontoyiannis, R. O. Darouiche, I. I. Raad, and R. A. Prince.** 2002. Antifungal activity of amphotericin B, fluconazole, and voriconazole in an in vitro model of *Candida* catheter-related bloodstream infection. *Antimicrob. Agents Chemother.* 46:3499–3505.

68. **Lewis, R. E., H. J. Lo, I. I. Raad, and D. P. Kontoyiannis.** 2002. Lack of catheter infection by the *efg1/efg1 cph1/cph1* double-null mutant, a *Candida albicans* strain that is defective in filamentous growth. *Antimicrob. Agents Chemother.* 46:1153–1155.

69. **Li, F., and S. P. Palecek.** 2003. *EAP1*, a *Candida albicans* gene involved in binding human epithelial cells. *Eukaryot. Cell* 2:1266–1273.

70. **Li, X., Z. Yan, and J. Xu.** 2003. Quantitative variation of biofilms among strains in natural populations of *Candida albicans*. *Microbiology* 149:353–362.

71. **Marger, M. D., and M. H. Saier, Jr.** 1993. A major superfamily of transmembrane facilitators that catalyse uniport, symport and antiport. *Trends Biochem. Sci.* 18:13–20.

72. **Mateus, C., S. A. Crow, Jr., and D. G. Ahearn.** 2004. Adherence of *Candida albicans* to silicone induces immediate enhanced tolerance to fluconazole. *Antimicrob. Agents Chemother.* 48:3358–3366.

73. **Miller, M. B., and B. L. Bassler.** 2001. Quorum sensing in bacteria. *Annu. Rev. Microbiol.* 55:165–199.

74. **Millsap, K. W., R. Bos, H. C. van der Mei, and H. J. Busscher.** 2001. Adhesive interactions between voice

prosthetic yeast and bacteria on silicone rubber in the absence and presence of saliva. *Antonie Leeuwenhoek* **79:**337–343.

75. **Mukherjee, P. K., J. Chandra, D. M. Kuhn, and M. A. Ghannoum.** 2003. Mechanism of fluconazole resistance in *Candida albicans* biofilms: phase-specific role of efflux pumps and membrane sterols. *Infect. Immun.* **71:**4333–4340.

76. **O'Toole, G., H. B. Kaplan, and R. Kolter.** 2000. Biofilm formation as microbial development. *Annu. Rev. Microbiol.* **54:**49–79.

77. **Raad, I., I. Chatzinikolaou, G. Chaiban, H. Hanna, R. Hachem, T. Dvorak, G. Cook, and W. Costerton.** 2003. In vitro and ex vivo activities of minocycline and EDTA against microorganisms embedded in biofilm on catheter surfaces. *Antimicrob. Agents Chemother.* **47:**3580–3585.

78. **Ramage, G., S. Bachmann, T. F. Patterson, B. L. Wickes, and J. L. Lopez-Ribot.** 2002. Investigation of multidrug efflux pumps in relation to fluconazole resistance in *Candida albicans* biofilms. *J. Antimicrob. Chemother.* **49:**973–980.

79. **Ramage, G., S. P. Saville, B. L. Wickes, and J. L. Lopez-Ribot.** 2002. Inhibition of *Candida albicans* biofilm formation by farnesol, a quorum-sensing molecule. *Appl. Environ. Microbiol.* **68:**5459–5463.

80. **Ramage, G., K. Tomsett, B. L. Wickes, J. L. Lopez-Ribot, and S. W. Redding.** 2004. Denture stomatitis: a role for *Candida* biofilms. *Oral Surg. Oral Med. Oral Pathol. Oral Radiol. Endod.* **98:**53–59.

81. **Ramage, G., K. Vande Walle, B. L. Wickes, and J. L. Lopez-Ribot.** 2001. Biofilm formation by *Candida dubliniensis. J. Clin. Microbiol.* **39:**3234–3240.

82. **Ramage, G., K. Vande Walle, B. L. Wickes, and J. L. Lopez-Ribot.** 2001. Standardized method for in vitro antifungal susceptibility testing of *Candida albicans* biofilms. *Antimicrob. Agents Chemother.* **45:**2475–2479.

83. **Ramage, G., K. VandeWalle, S. P. Bachmann, B. L. Wickes, and J. L. Lopez-Ribot.** 2002. In vitro pharmacodynamic properties of three antifungal agents against preformed *Candida albicans* biofilms determined by time-kill studies. *Antimicrob. Agents Chemother.* **46:**3634–3636.

84. **Ramage, G., K. VandeWalle, J. L. Lopez-Ribot, and B. L. Wickes.** 2002. The filamentation pathway controlled by the Efg1 regulator protein is required for normal biofilm formation and development in *Candida albicans. FEMS Microbiol. Lett.* **214:**95–100.

85. **Ramage, G., K. VandeWalle, B. L. Wickes, and J. L. Lopez-Ribot.** 2001. Characteristics of biofilm formation by *Candida albicans. Rev. Iberoam. Micol.* **18:**163–170.

86. **Ramage, G., B. L. Wickes, and J. L. Lopez-Ribot.** 2001. Biofilms of *Candida albicans* and their associated resistance to antifungal agents. *Am. Clin. Lab.* **20:**42–44.

87. **Saville, S. P., A. L. Lazzell, C. Monteagudo, and J. L. Lopez-Ribot.** 2003. Engineered control of cell morphology in vivo reveals distinct roles for yeast and filamentous forms of *Candida albicans* during infection. *Eukaryot. Cell* **2:**1053–1060.

88. **Sbarbati, A., V. Fanos, P. Bernardi, and L. Tato.** 2001. Rapid diagnosis of fungal infection of intravascular catheters in newborns by scanning electron microscopy. *Scanning* **23:**376–378.

89. **Schinabeck, M. K., L. A. Long, M. A. Hossain, J. Chandra, P. K. Mukherjee, S. Mohamed, and M. A. Ghannoum.** 2004. Rabbit model of *Candida albicans* biofilm infection: liposomal amphotericin B antifungal lock therapy. *Antimicrob. Agents Chemother.* **48:**1727–1732.

90. **Sheehan, D. J., C. A. Hitchcock, and C. M. Sibley.** 1999. Current and emerging azole antifungal agents. *Clin. Microbiol. Rev.* **12:**40–79.

91. **Suci, P. A., G. G. Geesey, and B. J. Tyler.** 2001. Integration of Raman microscopy, differential interference contrast microscopy, and attenuated total reflection Fourier transform infrared spectroscopy to investigate chlorhexidine spatial and temporal distribution in *Candida albicans* biofilms. *J. Microbiol. Methods* **46:**193–208.

92. **Tripathi, G., C. Wiltshire, S. Macaskill, H. Tournu, S. Budge, and A. J. Brown.** 2002. Gcn4 co-ordinates morphogenetic and metabolic responses to amino acid starvation in *Candida albicans. EMBO J.* **21:**5448–5456.

93. **Viudes, A., J. Peman, E. Canton, P. Ubeda, J. L. Lopez-Ribot, and M. Gobernado.** 2002. Candidemia at a tertiary-care hospital: epidemiology, treatment, clinical outcome and risk factors for death. *Eur. J. Clin. Microbiol. Infect. Dis.* **21:**767–774.

94. **Wey, S. B., M. Mori, M. A. Pfaller, R. F. Woolson, and R. P. Wenzel.** 1988. Hospital-acquired candidemia. The attributable mortality and excess length of stay. *Arch. Intern. Med.* **148:**2642–2645.

95. **White, T. C., K. A. Marr, and R. A. Bowden.** 1998. Clinical, cellular, and molecular factors that contribute to antifungal drug resistance. *Clin. Microbiol. Rev.* **11:**382–402.

96. **Wiederhold, N. P., R. E. Lewis, M. D. Johnson, and J. R. Perfect.** 2003. The echinocandin antifungals: an overview of the pharmacology, spectrum and clinical efficacy. Caspofungin: first approved agent in a new class of antifungals. *Expert Opin. Investig. Drugs* **12:**1313–1333.

97. **Wilson, L. S., C. M. Reyes, M. Stolpman, J. Speckman, K. Allen, and J. Beney.** 2002. The direct cost and incidence of systemic fungal infections. *Value Health* **5:**26–34.

Chapter 13

Molecular Basis of Fungal Adherence to Endothelial and Epithelial Cells

SCOTT G. FILLER, DONALD C. SHEPPARD, AND JOHN E. EDWARDS, JR.

Fungi must adhere to host cells in order to colonize the host and cause invasive disease. Because adherence is a key step in the pathogenesis of many infectious diseases, the adherence of human fungal pathogens to a variety of host cells has been investigated for several decades. However, until recently, the fungal adhesins and their host cell ligands remained unknown. Early investigators used traditional biochemical methods to characterize fungal adhesins. However, this approach was not able to identify adhesins at the amino acid level, principally because most fungal adhesins are covalently linked to the cell wall and it was difficult to extract these adhesins without denaturing them. In the 1990s an alternate approach involving genomics and surrogate genetics was employed to identify fungal adhesins. This approach has been highly successful and has enabled many different fungal adhesins to be identified and characterized. This chapter focuses on fungal adhesins that have been characterized at the genetic level and that mediate adherence to host constituents.

C. ALBICANS ADHERENCE

Candida albicans can be isolated from the oropharynx of at least 40% of normal, asymptomatic individuals (38). This organism is the cause of the majority of cases of oropharyngeal candidiasis in patients with human immunodeficiency virus infection and AIDS, Sjögren's syndrome, diabetes mellitus, and head and neck cancers (51–53, 56, 76). Furthermore, *C. albicans* is the most common cause of hematogenously disseminated candidiasis and vulvovaginal candidiasis (26, 47, 54). Because *C. albicans* inhabits a variety of host niches, it must adhere to many different host substrates. Therefore, it is not surprising that this organism has multiple different adhesins. These adhesins include Als proteins, Hwp1p, Eap1p, Csh1p, and other less well characterized proteins.

Proteins Encoded by the *ALS* Gene Family

In 1995, Hoyer et al. (35) screened a *C. albicans* cDNA library by differential hybridization to search for hypha-specific genes. They discovered a gene whose product had homology to *Saccharomyces cerevisiae* AGα1p and named the gene *ALS1* (for "agglutinin-like sequence 1"). Because AGα1p is a glycoprotein that mediates cell-to-cell adherence of yeast cells during mating, it was speculated that Als1p might function as an adhesin in *C. albicans*. Subsequently, Fu et al. (13) transformed a *C. albicans* genomic library into the usually nonadherent *S. cerevisiae* and selected for clones that exhibited increased adherence to human vascular endothelial cells. One highly adherent clone was found to express *ALS1*. At approximately the same time, Guar and Klotz (18) used a similar approach to identify *C. albicans* genes that mediated adherence to extracellular matrix proteins. They identified a gene, *ALA1* (subsequently renamed *ALS5*), whose product mediated adherence to fibronectin, laminin, and collagen. Collectively, these early results substantiated the hypothesis that the proteins encoded by *ALS1* and other members of the *ALS* gene family are adhesins for host constituents.

General characteristics of the *ALS* gene family

The primary structure of Als1p is typical of many fungal adhesins (Fig. 1). The N terminus contains a putative 18-amino-acid signal peptide. It also contains the binding region of the molecule (see below). The central region consists of a variable

Scott G. Filler and John E. Edwards, Jr. • Los Angeles Biomedical Research Institute at Harbor-UCLA Medical Center, Torrance, and the David Geffen School of Medicine at UCLA, Los Angeles, California. Donald C. Sheppard • Departments of Microbiology and Immunology, Medicine, McGill University, Montreal, Quebec, Canada.

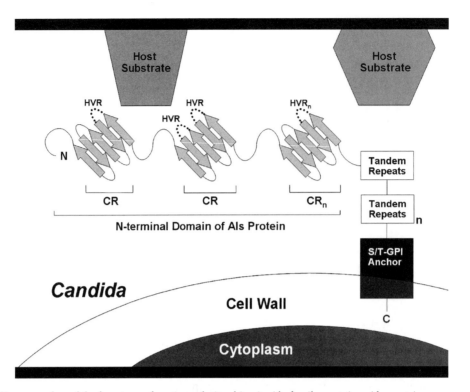

Figure 1. Conceptual model of structure-function relationships in Als family proteins. Als proteins are composed of three general components: an N-terminal domain, serine/threonine-rich tandem repeats, and a serine/threonine-rich C-terminal domain containing a GPI anchor that is bound to the *C. albicans* cell wall. As illustrated, Als proteins contain multiple conserved antiparallel β-sheet regions (CR_{1-n}) that are separated by extended spans, characteristic of the immunoglobulin superfamily. Projecting from the β-sheet domains are loop/coil structures containing the hypervariable regions (HVRs). The three-dimensional physicochemical properties of specific Als protein hypervariable regions probably govern interactions with host substrates that confer adhesive and invasive functions to *C. albicans*. For illustrative purposes, only three N-terminal β-sheet/coil domains and their respective conserved region and hypervariable region components are shown.

number of 36-amino-acid repeats that are arranged in tandem. Collectively, these repeats are rich in serine and threonine and contain many consensus sites for both O and N glycosylation. The C terminus also contains multiple serine and threonine residues, as well as a glycosylphosphatidylinositol (GPI) anchorage site. There is an isoleucine residue at the ω-5 site in the GPI anchorage sequence, suggesting that Als1p is processed like other GPI-linked proteins that are expressed on the cell surface (44). As the protein traffics from the interior of the cell to the cell surface, the GPI anchorage sequence is predicted to be cleaved from the molecule, after which the truncated protein is covalently cross-linked to β-1,6-glucans in the cell wall (27).

ALS1 is part of a gene family that is characterized by the presence of conserved tandem repeats (33). The *ALS* gene family contains at least eight members, *ALS1, ALS2, ALS3, ALS4, ALS5, ALS6, ALS7,* and *ALS9* (18, 31–34, 81). *ALS8* was originally thought to be a unique gene but was subsequently found to be an allele of *ALS3* (79). All

characterized *ALS* genes are predicted to encode surface proteins with the same basic structure as Als1p. Furthermore, most Als proteins have been localized to the cell surface by either indirect immunofluorescence or proteomic analysis of cell wall extracts (9, 12, 34).

A general characteristic of the members of the *ALS* gene family is their variability in size (46, 78, 81). This variability is due principally to the number of tandem repeats, which differs both among alleles of an individual *ALS* gene and among different *ALS* genes. It is very common for the same strain to contain two alleles of an individual *ALS* gene that significantly differ in size (12, 81). Alleles of *ALS1* and *ALS3* with more tandem repeats have greater adhesive function than those with fewer tandem repeats (44, 46). Why *C. albicans* strains frequently possess both a long and a short allele of these genes is currently unknown.

Different *ALS* genes are expressed under different conditions (12, 25, 35). Factors studied so far that influence the expression of one or more *ALS* genes include the growth medium, the morphological form of the organism, and the growth phase. Also, different strains of *C. albicans* express some *ALS* genes under different conditions from other strains, and some strains appear to lack certain *ALS* genes (31, 33). *ALS* genes are expressed in vitro in reconstituted human epithelium and in biofilms (23). The expression of these genes has also been detected in mouse models of hematogenously disseminated candidiasis and vaginal candidiasis and in vaginal isolates obtained from humans (8, 24, 29). Although each *ALS* gene is expressed at some level in some isolates, *ALS1*, *ALS3*, and *ALS9* appear to be the most highly and consistently expressed genes in vivo.

Role of the *ALS* gene family in *C. albicans* adherence

The binding specificities of the different Als proteins have been studied by two complementary approaches. One approach has been to construct strains of *C. albicans* in which both alleles of a specific *ALS* gene are disrupted. The adherence of these null mutant strains to various host substrates is then measured (12, 79, 80). The advantage of this approach is that the function of each *ALS* gene is studied in the organism in which it is normally expressed. Thus, posttranslational processing of the protein is normal. However, there are several disadvantages to this approach. First, disruption of both copies of a single *ALS* gene frequently results in only a partial reduction in adherence. This incomplete phenotype may be due to functional redundancy among the different Als proteins, as well as compensatory up-regulation of other *ALS* genes in response to disruption of the *ALS* gene of interest. Second, the adherence assay must be performed under conditions in which the *ALS* gene of interest is normally expressed. Ideally, the growth conditions should be chosen so that other *ALS* genes with similar substrate specificity are not expressed. However, such growth conditions are often difficult or impossible to identify. Third, at least one *ALS* gene, *ALS2*, may be essential, and so it has been necessary to test the phenotype of a heterozygous *ALS2/als2Δ* strain in which the remaining *ALS2* allele is under the control of a tightly regulated *MAL2* promoter (80).

Another approach to studying the function of different *ALS* genes is to express individual genes in *S. cerevisiae* (13, 17–20, 31, 39, 44, 50, 63). A major advantage of this approach is that nonflocculating strains of *S. cerevisiae* adhere very poorly to most host substrates. This low baseline adherence greatly facilitates the detection of adherence mediated by an Als protein. Another advantage is that different *ALS* genes can easily be placed under the control of the same promoter in *S. cerevisiae*. Therefore, one can compare the function of different Als proteins in *S. cerevisiae* by using strains grown under the same conditions. There are some limitations to analyzing the function of Als proteins that are heterologously expressed in *S. cerevisiae*. One potential problem is that the CUG codon is interpreted as specifying a serine in *C. albicans* and a leucine in *S. cerevisiae* (57, 58). Virtually all *ALS* genes contain at least one CUG codon. Another potential issue is that posttranslational processing of Als proteins may be different in *S. cerevisiae* than in *C. albicans*.

The best strategy to thoroughly analyze the function of Als proteins is to study them in both *C. albicans* and *S. cerevisiae*. To date, the adherence profiles of mutants of *C. albicans* that lack *ALS1*, *ALS2*, *ALS3*, and *ALS4* have been reported (12, 79, 80). Also, strains of *S. cerevisiae* that express *ALS1*, *ALS3*, *ALS5*, *ALS6*, *ALS7*, and *ALS9* have been constructed and analyzed (13, 17–20, 31, 39, 44, 50, 63). The two approaches have yielded generally congruent results. For example, *als1Δ/als1Δ* and *als3Δ/als3Δ* null mutants have reduced adherence to endothelial cells (12, 79), and *S. cerevisiae* strains that express either Als1p or Als3p have markedly enhanced adherence to these host cells (13, 44, 63). However, only an *als3Δ/als3Δ* mutant, but not an *als1Δ/als1Δ* null mutant, had decreased adherence to oral epithelial cells, and both *als1Δ/als1Δ* and *als3Δ/als3Δ* null mutants adhered normally to fibronectin (79). In contrast, *S. cerevisiae* expressing either Als1p or Als3p had increased adherence to both oral epithelial cells and fibronectin (13, 63). The probable explanation for these discordant results is that adhesins other than Als1p or Als3p also mediate binding to epithelial cells and fibronectin. These other adhesins obscure the effects of the lack of Als1p or Als3p in the *C. albicans* null mutants.

A comprehensive study of the binding specificity of Als proteins expressed in *S. cerevisiae* found that that Als1p, Als3p, and Als5p mediate adherence to a broad variety of host substrates, including endothelial cells, oral epithelial cells, collagen, fibronectin, and laminin (63). Als6p binds to collagen, and Als9p binds to laminin. Als7p did not bind to any of the substrates tested, and so the substrate specificity of this protein remains unknown. The binding specificities of Als2p and Als4p were not investigated in this study. However, an *als2Δ/PMAL2-ALS2* strain of *C. albicans* and an *als4Δ/als4Δ* null mutant both have decreased adherence to endothelial cells (80). Therefore, Als2p and Als4p probably mediate adherence to this host cell.

In addition to mediating adherence to host substrates, Als5p has an interesting property of binding to itself. This self-recognition is triggered by the adherence to an extracellular matrix protein (50). For example, when a single cell of *S. cerevisiae* expressing Als5p binds to a bead coated with fibronectin, there is a change in the conformation of Als5p such that it binds to Als5p on adjacent cells. This process results in the formation of large aggregates of cells that surround a single fibronectin-coated bead. These results suggest a model for Als5p-mediated aggregation in which an adhesion-triggered change in the conformation of Als5p propagates around the cell surface, forming ordered aggregation-competent regions. *S. cerevisiae* strains expressing Als1p also form aggregates. However, the formation of these aggregates does not require the presence of other ligands (39).

Structural analysis of Als proteins

Considerable effort has been made to identify the regions of the Als proteins that mediate binding to specific ligands. Als1p has structural and functional similarity to the *S. cerevisiae* AGα1p (35). The binding region of AGα1p is localized within its N terminus (77), and the N terminus of AGα1p has homology to the N terminus of Als1p. This similarity suggests that the ligand binding site of the Als proteins is also located in the N terminus. This hypothesis is supported by data from Loza et al. (44), who found that deletions or amino acid substitutions in the N terminus of Als1p abrogated the adherence function of this protein. They also found that a monoclonal antibody directed against the N terminus of Als1p blocked adherence. Furthermore, Sheppard et al. (63) discovered that Als5p and Als6p had distinctly different substrate binding profiles when expressed on the surface of *S. cerevisiae*. The binding specificity of one protein could be transferred to the other via the N-terminal portion of the protein. Collectively, these results indicate that the N terminus of the Als proteins contains the ligand binding domain.

Based on these results, the N terminus of Als proteins has been the focus of molecular modeling and structural analysis, which have been performed mainly by M. R. Yeaman and W. Welch (63). N-terminal sequence analysis and alignment of the N termini of multiple Als proteins reveal that there are regions of conserved sequences alternating with regions of hypervariable sequences. Structural models of the N-terminal domains of all Als proteins indicate that they contain multiple antiparallel β-sheet domains containing minor α-helical and turn components. The presence of these features indicates that the Als proteins are members of the immunoglobulin superfamily. Projecting from the β-sheet domains are loop/coil structures, which contain the hypervariable

regions of the predicted amino acid sequence (Fig. 1) (63). These modeling results have been substantiated by circular dichroism and Fourier transform infrared spectrometric analysis of the N-terminal domain of Als1p and by circular dichroism analysis of the N-terminal domain of Als5p (31, 63). Further refinement of the structural model of the Als proteins requires additional experimentation, including X-ray crystallography.

Of great interest, virtually all the Als proteins are predicted to have structural similarity to proteins of other microorganisms that mediate adherence or invasion. These proteins include collagen binding protein of *Staphylococcus aureus*, clumping factor of *S. aureus*, invasin/integrin binding protein of *Yersinia pseudotuberculosis*, surface layer protein of *Methanosarcina mazei*, and neuraminidase of influenza virus type B (63). The similarity of adhesin structure in these diverse organisms suggests that this common adhesin motif arose by convergent evolution.

Als3p mediates invasion of endothelial cells

The finding that Als proteins shared structural similarity to the invasin protein of *Y. pseudotuberculosis* suggested the possibility that one or more of these proteins could mediate the invasion of *C. albicans* into endothelial cells. It is known that *C. albicans* invades endothelial cells in vitro by inducing its own endocytosis (2, 11, 48, 49). Furthermore, this endocytosis is triggered in part by the binding of *C. albicans* to N-cadherin on the endothelial cell surface (49). Sheppard et al. (63) examined whether *S. cerevisiae* strains expressing different Als proteins could be endocytosed by endothelial cells in vitro. They found that a strain expressing Als3p was avidly endocytosed by these cells. Interestingly, strains expressing Als1p or Als5p adhered to endothelial cells as well as the Als3p-expressing strain did, but these strains were endocytosed only weakly. Therefore, adherence and endocytosis of *C. albicans* are distinct processes. The discovery that Als3 induces endothelial cell endocytosis has recently been substantiated by the finding that an *als3Δ/als3Δ* null mutant strain of *C. albicans* does not bind to N-cadherin and is endocytosed very poorly by endothelial cells in vitro (Q. T. Phan, C. L. Myers, Y. Fu, H. Park, D. C. Sheppard, J. E. Edwards, Jr., and S. G. Filler, *Abstr. 105th Gen. Meet. Am. Soc. Microbiol. 2005*, abstr. F-059, 2005). These results indicate that an additional function of at least one Als protein is to act as an invasin.

Diversity of ligands for the Als proteins

To identify common motifs that were bound by Als1p and Als5p, Klotz et al. studied the adherence of

S. cerevisiae expressing either of these proteins to polyethylene glycol beads coated with a library of random 7-mer peptides (39). They discovered that Als1p and Als5p recognize a broad range of target ligands. A common ligand sequence was "τ φ +" with τ representing an amino acid with high turn propensity (A, D, G, K, N, P, or S), φ representing a bulky hydrophobic or aromatic amino acid (F, H, I, L, M, T, V, W, or Y), and + representing R or K. The ability of Als1p and Als5p to bind to a multitude of peptide ligands indicates that these adhesins are degenerate in that their protein or peptide targets are neither highly specific nor highly organized. Furthermore, the large number of proteins in the Als family, along with their diversity in structure, function, and expression, enables *C. albicans* to adhere to a large repertoire of targets in a variety of host microenvironments. Furthermore, some *ALS* genes mutate easily, perhaps enabling the organisms to evade host defense mechanisms. Support for this hypothesis is demonstrated in part by studies of *ALS7*, which suggest that this gene is hypermutable and that the mutations can be caused by selective pressure (78).

Als proteins as vaccine targets

C. albicans adhesins and invasins are promising vaccine targets because they are expressed on the fungal surface and because blocking their function is likely to prevent adherence to and invasion of host cells. Ibrahim et al. (37) found that mice vaccinated intraperitoneally with a recombinant N-terminal domain of Als1p had improved survival after being challenged intravenously with a lethal inoculum of *C. albicans*. The protective effect of this vaccine is mediated by stimulation of cell-mediated rather than humoral immunity against *C. albicans*. Recently, these investigators have found that this vaccine is even more efficacious when administered subcutaneously (66). In fact, it protected both immunocompetent and neutropenic mice from hematogenously disseminated candidiasis. Also, it protected corticosteroid-treated mice from oropharyngeal candidiasis. These results provide strong support for the concept of fungal adhesins as therapeutic targets.

Hwp1p

In 1996, Staab et al. (69) used an antiserum against surface constituents of *C. albicans* hyphae to screen a *C. albicans* cDNA library. They identified a gene that they named *HWP1* (for "hyphal wall protein 1"). The protein encoded by *HWP1* is typical of an adhesin in that the N terminus contains a signal peptide followed by the binding region. In the middle of the protein is a series of 29-amino-acid tandem repeats that are rich in serine and threonine.

The C-terminal portion of the protein contains a GPI anchor site (67, 70). Hwp1p is unique in that its N-terminal region is similar to that of small proline-rich proteins of mammalian cells (68). These proteins serve as substrates for transglutaminases of keratinized epithelia, where they are covalently cross-linked to form stable cell-cell attachments between individual epithelial cells. Staab et al. (68) discovered that Hwp1p can also act as a substrate for epithelial cell transglutaminases, which results in the covalent binding of *C. albicans* to the surface of buccal epithelial cells. An *hwp1Δ/hwp1Δ* null mutant has reduced adherence to buccal epithelial cells in vitro (68) and has attenuated virulence in a mouse model of oropharyngeal and esophageal candidiasis (71). Thus, Hwp1p plays an important role in the pathogenesis of oroesophageal disease.

HWP1 is expressed by hyphae but not blastospores (62, 69). It was originally thought that Hwp1p was essential for normal hyphal formation because an *hwp1Δ/hwp1Δ* mutant strain had reduced ability to form hyphae in vitro and in the kidneys of mice (62, 74). Consistent with this defect in hyphal formation, *hwp1Δ/hwp1Δ* strains were found to have markedly attenuated virulence in the mouse model of hematogenously disseminated candidiasis (68, 74). However, it was subsequently discovered that these defects in hyphal formation and virulence were largely attributable to the chromosomal location of the *hisG-URA3-hisG* construct that was used to disrupt *HWP1*. Two *hwp1Δ/hwp1Δ* strains that contained *URA3* either at its native locus or at the *HWP1* locus but without flanking *hisG* sequences had normal virulence in the mouse model of hematogenously disseminated infection (55, 61). However, an *hwp1Δ/hpw1Δ* strain containing *URA3* integrated at the *ENO1* locus still had attenuated virulence (72). The reason for these differing results is uncertain but may involve differences in the inocula and strain of mice that were used. Nevertheless, it is clear that although Hwp1p is expressed only on hyphae, it is not required for normal hyphal formation. Furthermore, Hwp1p is probably more important for the pathogenesis of oroesophageal candidiasis than of hematogenously disseminated candidiasis.

Eap1p

The *C. albicans* gene *EAP1* (for "enhanced adherence to polystyrene 1") was identified as encoding an adhesin by heterologous expression in *S. cerevisiae* (42). *S. cerevisiae* strains expressing *EAP1* had significantly increased adherence to untreated polystyrene petri dishes under flow conditions. These cells also had increased adherence to a human embryonic kidney cell line under static conditions. The predicted structure of Eap1p is similar to that of other adhesins, with a signal peptide at the N terminus, serine- and threonine-rich tandem repeats in the central region, and a GPI anchorage sequence at the C terminus. Eap1p has homology to *C. albicans* Hwp1p, as well as *S. cerevisiae* Flo11p and Aga1p. In fact, in *S. cerevisiae*, *EAP1* can complement *flo11Δ* defects and restore agar invasion and filamentous growth. Therefore, *EAP1* appears to be a functional homolog of *FLO11*.

In *C. albicans*, *EAP1* expression is under the control of the transcription factor, Efg1p (42). An *efg1Δ/efg1Δ* strain did not express *EAP1* and adhered weakly to the human embryonic kidney cell line. The forced expression of *EAP1* in the *efg1Δ/efg1Δ* strain restored adherence to at least wild-type levels (42). Whether Eap1p mediates adherence to other host constituents and contributes to *C. albicans* virulence has not yet been determined.

Csh1p

When *C. albicans* blastospores are grown to stationary phase at 23°C, they become hydrophobic. In contrast, cells grown at 37°C become hydrophilic (22). Hydrophobic cells adhere more avidly than hydrophilic cells to activated human umbilical vein endothelial cells under flow conditions (22). Adherence of hydrophobic cells can be partially blocked by a monoclonal antibody directed against a hydrophobic protein. *C. albicans* germ tubes are also hydrophobic, and their adherence to the extracellular matrix proteins fibronectin and laminin is also partially inhibited by this monoclonal antibody (45).

Singelton et al. (65) determined that this monoclonal antibody bound to the protein encoded by *CSH1* (for "cell surface hydrophobicity 1"). They initially determined that a *csh1Δ/csh1Δ* null mutant had decreased cell surface hydrophobicity and reduced adherence to fibronectin. The predicted structure of Csh1p differs from other fungal adhesins in that it does not have a signal peptide, tandem repeats, or GPI anchorage sequence. Csh1p most closely resembles a family of aryl alcohol dehydrogenases (65). Although the majority of Csh1p is located intracellularly, a small percentage of this protein is expressed on the cell surface (64).

Subsequently, this group reported that the *csh1Δ/csh1Δ* null mutant had apparently changed after being stored as a frozen stock. The cell surface hydrophobicity and fibronectin adherence of this mutant returned to wild-type levels, even though the mutant still contained no detectable Csh1p. This result suggests that other hydrophobic proteins can compensate for the lack of Csh1p.

Adh1p

The alcohol dehydrogenase Adh1p may also have adhesive properties. *C. albicans* adheres avidly to fibronectin, and antibodies to the human fibronectin receptor ($\alpha_5\beta_1$) block *C. albicans* adherence to fibronectin (41). To identify the *C. albicans* homolog of the human fibronectin receptor, Klotz et al. screened a *C. albicans* cDNA library with a polyclonal antiserum to the human fibronectin receptor (40). They identified multiple positive clones, which were also recognized by polyclonal antiserum against the human vitronectin receptor ($\alpha_v\beta_3$). These clones were found to contain the *C. albicans ADH1* gene. Both *C. albicans* Adh1p produced by in vitro translation and purified Adh1p from *S. cerevisiae* were recognized by the polyclonal antisera. *C. albicans* Adh1p was also shed into the culture medium, suggesting that it may be present on the cell surface. Although these data are intriguing, the exact role of Adh1p in the adherence of *C. albicans* to extracellular matrix proteins remains to be determined by gene disruption experiments.

Pra1p

Casanova et al. (5) identified a 58-kDa mannoprotein expressed on the cell surface of *C. albicans* blastospores and hyphae that bound to fibrinogen in a ligand overlay assay. A partial cDNA sequence of this protein was originally named *FBP1* (43). However, it was later determined that this protein was actually Pra1p (60). Pra1p is expressed on the surface of *C. albicans* cells grown at neutral pH. It has homology to the *Aspergillus fumigatus* Asp F2 protein, which binds to laminin (see below). Pra1p plays a role in pH-dependent hyphal formation because a *pra1Δ/pra1Δ* mutant has defective hyphal formation at neutral pH. Unfortunately, the adherence of this mutant has not yet been tested, and so the role of Pra1p in the adherence of intact *C. albicans* is currently unknown.

Int1p

Monoclonal antibodies directed against the human complement receptor 3 ($\alpha_M\beta_2$) bind to the surface of *C. albicans* cells and block adherence to epithelial cells. Gale et al. (14) screened a *C. albicans* cDNA library with a cDNA probe from the transmembrane domain of human α_M. They identified a *C. albicans* gene that they named *INT1*. This gene was initially thought to encode an integrin-like adhesin (16). Subsequently, it was determined that the C terminus of Int1p has significant homology to Bud4p of *S. cerevisiae* (15). Int1p colocalizes with septins intracellularly and is important for determining axial budding patterns. Its major function appears to be the regulation of polarized growth and morphogenesis. Whether Int1p also functions as an adhesin is uncertain.

OTHER *CANDIDA* SPECIES

With the exception of *Candida glabrata*, little is known about the mechanisms by which non-*albicans* species of *Candida* adhere to host constituents. *C. glabrata* has a family of GPI-anchored adhesins encoded by the *EPA* family of genes. The overall structure of proteins encoded by this gene family is similar to that of other GPI-linked fungal adhesins. The *EAP* gene family is discussed in detail in chapter 11.

Fragments of genes with homology to the *C. albicans ALS* gene family have been detected in *Candida dubliniensis* and *Candida tropicalis* (30). However, it is not yet known if the proteins encoded by these genes function as adhesins. *C. dubliniensis* also contains a homolog of *CSH1*; however, the contribution of this protein to *C. dubliniensis* hydrophobicity and adherence remains to be determined (28).

A. FUMIGATUS ADHERENCE

Several studies have demonstrated that *Aspergillus fumigatus* conidia and hyphae adhere specifically to a variety of host substrates including laminin, fibrinogen, fibronectin, and pulmonary epithelial cells (3, 4, 10, 21, 75). Surprisingly, however, the fungal adhesins that mediate these interactions remain largely undefined. Indeed, only two proteins have been identified that probably mediate adherence to extracellular matrix proteins.

RodA

The first *A. fumigatus* adhesin gene to be identified, *rodA*, encodes the major hydrophobin found in conidia. Disruption of *rodA* results in a strain with conidia that are much less hydrophobic than wild-type conidia. The *rodAΔ* conidia have significantly reduced adherence to collagen and bovine serum albumin but not to laminin or primary rat pulmonary epithelial cells. Also, a *rodAΔ* strain has normal virulence in the immunosuppressed-mouse model of invasive pulmonary aspergillosis (73). These results suggest that adhesins other than RodA also mediate the adherence of conidia to host constituents.

Asp F2

Asp F2 was originally identified in a search for allergens, using sera from patients with allergic bronchopulmonary aspergillosis (1). This protein is produced in large amounts by hyphae of *A. fumigatus* and represents up to 20 to 40% of total protein in these cells. Expression of Asp F2 in vitro is suppressed by zinc (59). Asp F2 has significant homology to Pra1 of *C. albicans* (60). Recombinant Asp F2 binds to immobilized laminin (1). However, because an *A. fumigatus* strain deficient in Asp F2 has not been constructed, the role of Asp F2 in adherence and virulence remains incompletely determined.

B. DERMATITIDIS ADHERENCE

One identified adhesin in *B. dermatitidis* is encoded by *BAD1*. This very interesting adhesin binds to host cell phagocytes and inhibits the production of tumor necrosis factor alpha. Bad1p is discussed in detail in chapter 27.

C. NEOFORMANS ADHERENCE

Although adherence to and invasion of endothelial cells and other host constituents is probably central to the pathogenesis of cryptococcal meningitis, very little is known about the mechanisms by which these processes occur. Ibrahim et al. (36) discovered that unencapsulated strains of *C. neoformans* adhere to and invade human umbilical vein endothelial cells in vitro. However, encapsulated strains, even those with very small capsules, adhere minimally to these cells and do not invade them. This latter finding is significant because virtually all pathogenic strains of *C. neoformans* possess a capsule. Chen et al. (7) found that both encapsulated and unencapsulated organisms adhere to and invade brain microvascular endothelial cells in vitro. Moreover, *C. neoformans* invasion of brain endothelial cells has been observed in mice with cryptococcal meningitis (6). Nevertheless, the surface structures of *C. neoformans* that mediate adherence to and invasion of brain endothelial cells are currently unknown.

SUMMARY AND CONCLUSIONS

The identification and characterization of fungal adhesins have been greatly facilitated by the development of powerful molecular biology tools and the recently completed fungal genome-sequencing projects. The adhesins of *C. albicans* are the best-studied adherence proteins of all human fungal pathogens. Extrapolating from the existing data, it can be predicted that most fungi will be found to have multiple adhesins with distinct yet overlapping substrate specificities. These adhesins may be encoded by one or more gene families. Furthermore, many important fungal adhesins will probably be GPI-linked proteins with serine- and threonine-rich tandem repeats in their central domains. Identification of these adhesins is important because it provides insight into the mechanisms of fungal pathogenicity, as well as potential therapeutic targets.

REFERENCES

1. Banerjee, B., P. A. Greenberger, J. N. Fink, and V. P. Kurup. 1998. Immunological characterization of Asp f 2, a major allergen from *Aspergillus fumigatus* associated with allergic bronchopulmonary aspergillosis. *Infect. Immun.* 66:5175–5182.

2. Belanger, P. H., D. Johnston, R. A. Fratti, M. Zhang, and S. G. Filler. 2002. Endocytosis of *Candida albicans* by vascular endothelial cells is associated with tyrosine phosphorylation of specific host cell proteins. *Cell. Microbiol.* 4:805–812.

3. Bouchara, J. P., A. Bouali, G. Tronchin, R. Robert, D. Chabasse, and J. M. Senet. 1988. Binding of fibrinogen to the pathogenic *Aspergillus* species. *J. Med. Vet. Mycol.* 26:327–334.

4. Bouchara, J. P., M. Sanchez, A. Chevailler, A. Marot-Leblond, J. C. Lissitzky, G. Tronchin, and D. Chabasse. 1997. Sialic acid-dependent recognition of laminin and fibrinogen by *Aspergillus fumigatus* conidia. *Infect. Immun.* 65:2717–2724.

5. Casanova, M., J. L. Lopez-Ribot, C. Monteagudo, A. Llombart-Bosch, R. Sentandreu, and J. P. Martinez. 1992. Identification of a 58-kilodalton cell surface fibrinogen-binding mannoprotein from *Candida albicans*. *Infect. Immun.* 60:4221–4229.

6. Chang, Y. C., M. F. Stins, M. J. McCaffery, G. F. Miller, D. R. Pare, T. Dam, M. Paul-Satyasee, K. S. Kim, and K. J. Kwon-Chung. 2004. Cryptococcal yeast cells invade the central nervous system via transcellular penetration of the blood-brain barrier. *Infect. Immun.* 72:4985–4995.

7. Chen, S. H., M. F. Stins, S. H. Huang, Y. H. Chen, K. J. Kwon-Chung, Y. Chang, K. S. Kim, K. Suzuki, and A. Y. Jong. 2003. *Cryptococcus neoformans* induces alterations in the cytoskeleton of human brain microvascular endothelial cells. *J. Med. Microbiol.* 52:961–970.

8. Cheng, G., K. Wozniak, M. A. Wallig, P. L. Fidel, Jr., S. R. Trupin, and L. L. Hoyer. 2005. Comparison between *Candida albicans* agglutinin-like sequence gene expression patterns in human clinical specimens and models of vaginal candidiasis. *Infect. Immun.* 73:1656–1663.

9. de Groot, P. W., A. D. de Boer, J. Cunningham, H. L. Dekker, L. de Jong, K. J. Hellingwerf, C. de Koster, and F. M. Klis. 2004. Proteomic analysis of *Candida albicans* cell walls reveals covalently bound carbohydrate-active enzymes and adhesins. *Eukaryot. Cell* 3:955–965.

10. DeHart, D. J., D. E. Agwu, N. C. Julian, and R. G. Washburn. 1997. Binding and germination of *Aspergillus fumigatus* conidia on cultured A549 pneumocytes. *J. Infect. Dis.* 175:146–150.

11. Filler, S. G., J. N. Swerdloff, C. Hobbs, and P. M. Luckett. 1995. Penetration and damage of endothelial cells by *Candida albicans*. *Infect. Immun.* 63:976–983.

12. Fu, Y., A. S. Ibrahim, D. C. Sheppard, Y. C. Chen, S. W. French, J. E. Cutler, S. G. Filler, and J. E. Edwards. 2002. *Candida albicans* Als1p: an adhesin that is a downstream effector of the *EFG1* filamentation pathway. *Mol. Microbiol.* 44:61–72.

13. Fu, Y., G. Rieg, W. A. Fonzi, P. H. Belanger, J. E. Edwards, Jr., and S. G. Filler. 1998. Expression of the *Candida albicans* gene ALS1 in *Saccharomyces cerevisiae* induces adherence to endothelial and epithelial cells. *Infect. Immun.* 66:1783–1786.

14. Gale, C., D. Finkel, N. Tao, M. Meinke, M. McClellan, J. Olson, K. Kendrick, and M. Hostetter. 1996. Cloning and expression of a gene encoding an integrin-like protein in *Candida albicans*. *Proc. Natl. Acad. Sci. USA* 93:357–361.

15. Gale, C., M. Gerami-Nejad, M. McClellan, S. Vandoninck, M. S. Longtine, and J. Berman. 2001. *Candida albicans* Int1p interacts with the septin ring in yeast and hyphal cells. *Mol. Biol. Cell* 12:3538–3549.

16. Gale, C. A., C. M. Bendel, M. McClellan, M. Hauser, J. M. Becker, J. Berman, and M. K. Hostetter. 1998. Linkage of adhesion, filamentous growth, and virulence in *Candida albicans* to a single gene, *INT1*. *Science* 279:1355–1358.

17. Gaur, N. K., and S. A. Klotz. 2004. Accessibility of the peptide backbone of protein ligands is a key specificity determinant in *Candida albicans* SRS adherence. *Microbiology* 150:277–284.

18. Gaur, N. K., and S. A. Klotz. 1997. Expression, cloning, and characterization of a *Candida albicans* gene, *ALA1*, that confers adherence properties upon *Saccharomyces cerevisiae* for extracellular matrix proteins. *Infect. Immun.* 65:5289–5294.

19. Gaur, N. K., S. A. Klotz, and R. L. Henderson. 1999. Overexpression of the *Candida albicans* ALA1 gene in *Saccharomyces cerevisiae* results in aggregation following attachment of yeast cells to extracellular matrix proteins, adherence properties similar to those of *Candida albicans*. *Infect. Immun.* 67:6040–6047.

20. Gaur, N. K., R. L. Smith, and S. A. Klotz. 2002. *Candida albicans* and *Saccharomyces cerevisiae* expressing *ALA1/ALS5* adhere to accessible threonine, serine, or alanine patches. *Cell Commun. Adhes.* 9:45–57.

21. Gil, M. L., M. C. Penalver, J. L. Lopez-Ribot, J. E. O'Connor, and J. P. Martinez. 1996. Binding of extracellular matrix proteins to *Aspergillus fumigatus* conidia. *Infect. Immun.* 64:5239–5247.

22. Glee, P. M., J. E. Cutler, E. E. Benson, R. F. Bargatze, and K. C. Hazen. 2001. Inhibition of hydrophobic protein-mediated *Candida albicans* attachment to endothelial cells during physiologic shear flow. *Infect. Immun.* 69:2815–2820.

23. Green, C. B., G. Cheng, J. Chandra, P. Mukherjee, M. A. Ghannoum, and L. L. Hoyer. 2004. RT-PCR detection of *Candida albicans* ALS gene expression in the reconstituted human epithelium (RHE) model of oral candidiasis and in model biofilms. *Microbiology* 150:267–275.

24. Green, C. B., X. Zhao, and L. L. Hoyer. 2005. Use of green fluorescent protein and reverse transcription-PCR to monitor *Candida albicans* agglutinin-like sequence gene expression in a murine model of disseminated candidiasis. *Infect. Immun.* 73:1852–1855.

25. Green, C. B., X. Zhao, K. M. Yeater, and L. L. Hoyer. 2005. Construction and real-time RT-PCR validation of *Candida albicans* PALS-GFP reporter strains and their use in flow cytometry analysis of *ALS* gene expression in budding and filamenting cells. *Microbiology* 151:1051–1060.

26. Hajjeh, R. A., A. N. Sofair, L. H. Harrison, G. M. Lyon, B. A. Arthington-Skaggs, S. A. Mirza, M. Phelan, J. Morgan, W. Lee-Yang, M. A. Ciblak, L. E. Benjamin, L. T. Sanza, S. Huie, S. F. Yeo, M. E. Brandt, and D. W. Warnock. 2004. Incidence of bloodstream infections due to *Candida* species and in vitro susceptibilities of isolates collected from 1998 to 2000 in a population-based active surveillance program. *J. Clin. Microbiol.* 42:1519–1527.

27. Hamada, K., H. Terashima, M. Arisawa, N. Yabuki, and K. Kitada. 1999. Amino acid residues in the omega-minus region participate in cellular localization of yeast glycosylphosphatidylinositol-attached proteins. *J. Bacteriol.* 181:3886–3889.

28. Hazen, K. C. 2004. Relationship between expression of cell surface hydrophobicity protein 1 (CSH1p) and surface hydrophobicity properties of *Candida dubliniensis*. *Curr. Microbiol.* **48:**447–451.

29. Hoyer, L. L., J. Clevenger, J. E. Hecht, E. J. Ehrhart, and F. M. Poulet. 1999. Detection of Als proteins on the cell wall of *Candida albicans* in murine tissues. *Infect. Immun.* **67:**4251–4255.

30. Hoyer, L. L., R. Fundyga, J. E. Hecht, J. C. Kapteyn, F. M. Klis, and J. Arnold. 2001. Characterization of agglutinin-like sequence genes from non-*albicans Candida* and phylogenetic analysis of the *ALS* family. *Genetics* **157:**1555–1567.

31. Hoyer, L. L., and J. E. Hecht. 2001. The *ALS5* gene of *Candida albicans* and analysis of the Als5p N-terminal domain. *Yeast* **18:**49–60.

32. Hoyer, L. L., and J. E. Hecht. 2000. The *ALS6* and *ALS7* genes of *Candida albicans*. *Yeast* **16:**847–855.

33. Hoyer, L. L., T. L. Payne, M. Bell, A. M. Myers, and S. Scherer. 1998. *Candida albicans ALS3* and insights into the nature of the *ALS* gene family. *Clin. Microbiol. Rev.* **33:**451–459.

34. Hoyer, L. L., T. L. Payne, and J. E. Hecht. 1998. Identification of *Candida albicans ALS2* and *ALS4* and localization of als proteins to the fungal cell surface. *J. Bacteriol.* **180:**5334–5343.

35. Hoyer, L. L., S. Scherer, A. R. Shatzman, and G. P. Livi. 1995. *Candida albicans ALS1*: domains related to a *Saccharomyces cerevisiae* sexual agglutinin separated by a repeating motif. *Mol. Microbiol.* **15:**39–54.

36. Ibrahim, A. S., S. G. Filler, M. S. Alcouloumre, T. R. Kozel, J. E. Edwards, Jr., and M. A. Ghannoum. 1995. Adherence to and damage of endothelial cells by *Cryptococcus neoformans* in vitro: role of the capsule. *Infect. Immun.* **63:**4368–4374.

37. Ibrahim, A. S., B. J. Spellberg, V. Avenissian, Y. Fu, S. G. Filler, and J. E. Edwards, Jr. 2005. Vaccination with recombinant N-terminal domain of Als1p improves survival during murine disseminated candidiasis by enhancing cell-mediated, not humoral, immunity. *Infect. Immun.* **73:**999–1005.

38. Kleinegger, C. L., S. R. Lockhart, K. Vargas, and D. R. Soll. 1996. Frequency, intensity, species, and strains of oral *Candida* vary as a function of host age. *J. Clin. Microbiol.* **34:**2246–2254.

39. Klotz, S. A., N. K. Gaur, D. F. Lake, V. Chan, J. Rauceo, and P. N. Lipke. 2004. Degenerate peptide recognition by *Candida albicans* adhesins Als5p and Als1p. *Infect. Immun.* **72:**2029–2034.

40. Klotz, S. A., M. L. Pendrak, and R. C. Hein. 2001. Antibodies to alpha5beta1 and alpha(v)beta3 integrins react with *Candida albicans* alcohol dehydrogenase. *Microbiology* **147:**3159–3164.

41. Klotz, S. A., and R. L. Smith. 1991. A fibronectin receptor on *Candida albicans* mediates adherence of the fungus to extracellular matrix. *J. Infect. Dis.* **163:**604–610.

42. Li, F., and S. P. Palecek. 2003. *EAP1*, a *Candida albicans* gene involved in binding human epithelial cells. *Eukaryot. Cell* **2:**1266–1273.

43. Lopez-Ribot, J. L., P. Sepulveda, A. M. Cervera, P. Roig, D. Gozalbo, and J. P. Martinez. 1997. Cloning of a cDNA fragment encoding part of the protein moiety of the 58-kDa fibrinogen-binding mannoprotein of *Candida albicans*. *FEMS Microbiol. Lett.* **157:**273–278.

44. Loza, L., Y. Fu, A. S. Ibrahim, D. C. Sheppard, S. G. Filler, and J. E. Edwards, Jr. 2004. Functional analysis of the *Candida albicans ALS1* gene product. *Yeast* **21:**473–482.

45. Masuoka, J., G. Wu, P. M. Glee, and K. C. Hazen. 1999. Inhibition of *Candida albicans* attachment to extracellular matrix by antibodies which recognize hydrophobic cell wall proteins. *FEMS Immunol. Med. Microbiol.* **24:**421–429.

46. Oh, S. H., G. Cheng, J. A. Nuessen, R. Jajko, K. M. Yeater, X. Zhao, C. Pujol, D. R. Soll, and L. L. Hoyer. 2005. Functional specificity of *Candida albicans* Als3p proteins and clade specificity of *ALS3* alleles discriminated by the number of copies of the tandem repeat sequence in the central domain. *Microbiology* **151:**673–681.

47. Pfaller, M. A., R. N. Jones, G. V. Doern, H. S. Sader, S. A. Messer, A. Houston, S. Coffman, and R. J. Hollis. 2000. Bloodstream infections due to *Candida* species: SENTRY antimicrobial surveillance program in North America and Latin America, 1997–1998. *Antimicrob. Agents Chemother.* **44:**747–751.

48. Phan, Q. T., P. H. Belanger, and S. G. Filler. 2000. Role of hyphal formation in interactions of *Candida albicans* with endothelial cells. *Infect. Immun.* **68:**3485–3490.

49. Phan, Q. T., R. A. Fratti, N. V. Prasadarao, J. E. Edwards, Jr., and S. G. Filler. 2005. N-cadherin mediates endocytosis of *Candida albicans* by endothelial cells. *J. Biol. Chem.* **280:**10455–10461.

50. Rauceo, J. M., N. K. Gaur, K. G. Lee, J. E. Edwards, S. A. Klotz, and P. N. Lipke. 2004. Global cell surface conformational shift mediated by a *Candida albicans* adhesin. *Infect. Immun.* **72:**4948–4955.

51. Redding, S. W., R. C. Zellars, W. R. Kirkpatrick, R. K. McAtee, M. A. Caceres, A. W. Fothergill, J. L. Lopez-Ribot, C. W. Bailey, M. G. Rinaldi, and T. F. Patterson. 1999. Epidemiology of oropharyngeal *Candida* colonization and infection in patients receiving radiation for head and neck cancer. *J. Clin. Microbiol.* **37:**3896–3900.

52. Revankar, S. G., O. P. Dib, W. R. Kirkpatrick, R. K. McAtee, A. W. Fothergill, M. G. Rinaldi, S. W. Redding, and T. F. Patterson. 1998. Clinical evaluation and microbiology of oropharyngeal infection due to fluconazole-resistant *Candida* in human immunodeficiency virus-infected patients. *Clin. Infect. Dis.* **26:**960–963.

53. Rhodus, N. L., C. Bloomquist, W. Liljemark, and J. Bereuter. 1997. Prevalence, density, and manifestations of oral *Candida albicans* in patients with Sjogren's syndrome. *J. Otolaryngol.* **26:**300–305.

54. Richter, S. S., R. P. Galask, S. A. Messer, R. J. Hollis, D. J. Diekema, and M. A. Pfaller. 2005. Antifungal susceptibilities of *Candida* species causing vulvovaginitis and epidemiology of recurrent cases. *J. Clin. Microbiol.* **43:**2155–2162.

55. Sanchez, A. A., D. A. Johnston, C. Myers, J. E. Edwards, Jr., A. P. Mitchell, and S. G. Filler. 2004. Relationship between *Candida albicans* virulence during experimental hematogenously disseminated infection and endothelial cell damage in vitro. *Infect. Immun.* **72:**598–601.

56. Sangeorzan, J. A., S. F. Bradley, X. He, L. T. Zarins, G. L. Ridenour, R. N. Tiballi, and C. A. Kauffman. 1994. Epidemiology of oral candidiasis in HIV-infected patients:colonization, infection, treatment, and emergence of fluconazole resistance. *Am. J. Med.* **97:**339–346.

57. Santos, M. A., G. Keith, and M. F. Tuite. 1993. Nonstandard translational events in *Candida albicans* mediated by an unusual seryl-tRNA with a 5′-CAG-3′ (leucine) anticodon. *EMBO J.* **12:**607–616.

58. Santos, M. A., and M. F. Tuite. 1995. The CUG codon is decoded in vivo as serine and not leucine in *Candida albicans*. *Nucleic Acids Res.* **23:**1481–1486.

59. Segurado, M., R. Lopez-Aragon, J. A. Calera, J. M. Fernandez-Abalos, and F. Leal. 1999. Zinc-regulated biosynthesis of immunodominant antigens from *Aspergillus* spp. *Infect. Immun.* **67:**2377–2382.

60. Sentandreu, M., M. V. Elorza, R. Sentandreu, and W. A. Fonzi. 1998. Cloning and characterization of *PRA1*, a gene encoding a novel pH-regulated antigen of *Candida albicans*. *J. Bacteriol.* **180:**282–289.

61. Sharkey, L. L., W. L. Liao, A. K. Ghosh, and W. A. Fonzi. 2005. Flanking direct repeats of hisG alter *URA3* marker expression at the *HWP1* locus of *Candida albicans*. *Microbiology* **151:**1061–1071.

62. Sharkey, L. L., M. D. McNemar, S. M. Saporito-Irwin, P. S. Sypherd, and W. A. Fonzi. 1999. *HWP1* functions in the morphological development of *Candida albicans* downstream of *EFG1*, *TUP1*, and *RBF1*. *J. Bacteriol.* **181:**5273–5279.

63. Sheppard, D. C., M. R. Yeaman, W. H. Welch, Q. T. Phan, Y. Fu, A. S. Ibrahim, S. G. Filler, M. Zhang, A. J. Waring, and J. E. Edwards, Jr. 2004. Functional and structural diversity in the Als protein family of *Candida albicans*. *J. Biol. Chem.* **279:**30840–30849.

64. Singleton, D. R., and K. C. Hazen. 2004. Differential surface localization and temperature-dependent expression of the *Candida albicans* CSH1 protein. *Microbiology* **150:**285–292.

65. Singleton, D. R., J. Masuoka, and K. C. Hazen. 2001. Cloning and analysis of a *Candida albicans* gene that affects cell surface hydrophobicity. *J. Bacteriol.* **183:**3582–3588.

66. Spellberg, B. J., A. S. Ibrahim, V. Avenissian, S. G. Filler, C. L. Myers, Y. Fu, and J. E. Edwards, Jr. 2005. The anti-*Candida albicans* vaccine composed of the recombinant N terminus of Als1p reduces fungal burden and improves survival in both immunocompetent and immunocompromised mice. *Infect. Immun.* **73:**6191–6193.

67. Staab, J. F., Y. S. Bahn, C. H. Tai, P. F. Cook, and P. Sundstrom. 2004. Expression of transglutaminase substrate activity on *Candida albicans* germ tubes through a coiled, disulfide-bonded N-terminal domain of Hwp1 requires C-terminal glycosylphosphatidylinositol modification. *J. Biol. Chem.* **279:**40737–40747.

68. Staab, J. F., S. D. Bradway, P. L. Fidel, and P. Sundstrom. 1999. Adhesive and mammalian transglutaminase substrate properties of *Candida albicans* Hwp1. *Science* **283:**1535–1538.

69. Staab, J. F., C. A. Ferrer, and P. Sundstrom. 1996. Developmental expression of a tandemly repeated, proline-and glutamine- rich amino acid motif on hyphal surfaces on *Candida albicans*. *J. Biol. Chem.* **271:**6298–6305.

70. Staab, J. F., and P. Sundstrom. 1998. Genetic organization and sequence analysis of the hypha-specific cell wall protein gene *HWP1* of *Candida albicans*. *Yeast* **14:**681–686.

71. Sundstrom, P., E. Balish, and C. M. Allen. 2002. Essential role of the *Candida albicans* transglutaminase substrate, hyphal wall protein 1, in lethal oroesophageal candidiasis in immunodeficient mice. *J. Infect. Dis.* **185:**521–530.

72. Sundstrom, P., J. E. Cutler, and J. F. Staab. 2002. Reevaluation of the role of *HWP1* in systemic candidiasis by use of *Candida albicans* strains with selectable marker *URA3* targeted to the *ENO1* locus. *Infect. Immun.* **70:**3281–3283.

73. Thau, N., M. Monod, B. Crestani, C. Rolland, G. Tronchin, J. P. Latge, and S. Paris. 1994. Rodletless mutants of *Aspergillus fumigatus*. *Infect. Immun.* **62:**4380–4388.

74. Tsuchimori, N., L. L. Sharkey, W. A. Fonzi, S. W. French, J. E. Edwards, Jr., and S. G. Filler. 2000. Reduced virulence of *HWP1*-deficient mutants of *Candida albicans* and their interactions with host cells. *Infect. Immun.* **68:**1997–2002.

75. Wasylnka, J. A., and M. M. Moore. 2000. Adhesion of *Aspergillus* species to extracellular matrix proteins: evidence for involvement of negatively charged carbohydrates on the conidial surface. *Infect. Immun.* **68:**3377–3384.

76. Willis, A. M., W. A. Coulter, C. R. Fulton, J. R. Hayes, P. M. Bell, and P. J. Lamey. 1999. Oral candidal carriage and infection in insulin-treated diabetic patients. *Diabet. Med.* **16:**675–679.

77. Wojciechowicz, D., C. F. Lu, J. Kurjan, and P. N. Lipke. 1993. Cell surface anchorage and ligand-binding domains of the *Saccharomyces cerevisiae* cell adhesion protein α-agglutinin, a member of the immunoglobulin superfamily. *Mol. Cell. Biol.* **13:**2554–2563.

78. Zhang, N., A. L. Harrex, B. R. Holland, L. E. Fenton, R. D. Cannon, and J. Schmid. 2003. Sixty alleles of the *ALS7* open reading frame in *Candida albicans*: *ALS7* is a hypermutable contingency locus. *Genome Res.* **13:**2005–2017.

79. Zhao, X., S. H. Oh, G. Cheng, C. B. Green, J. A. Nuessen, K. Yeater, R. P. Leng, A. J. Brown, and L. L. Hoyer. 2004. *ALS3* and *ALS8* represent a single locus that encodes a *Candida albicans* adhesin; functional comparisons between Als3p and Als1p. *Microbiology* **150:**2415–2428.

80. Zhao, X., S. H. Oh, K. M. Yeater, and L. L. Hoyer. 2005. Analysis of the *Candida albicans* Als2p and Als4p adhesins suggests the potential for compensatory function within the Als family. *Microbiology* **151:**1619–1630.

81. Zhao, X., C. Pujol, D. R. Soll, and L. L. Hoyer. 2003. Allelic variation in the contiguous loci encoding *Candida albicans* ALS5, ALS1 and ALS9. *Microbiology* **149:** 2947–2960.

Chapter 14

Molecular Principles of Antifungal Drug Resistance

Dominique Sanglard and Theodore C. White

Fungal infections have increased in incidence and severity in the last decade, in part as the result of increased patient populations with compromised immune systems, such as AIDS patients, bone marrow transplant recipients, and solid-organ transplant recipients. The most common agents of fungal infections include *Candida*, *Cryptococcus*, and *Aspergillus*, since these are ubiquitous and commensal fungal colonizers. However, drugs developed for the treatment of fungal infections must be effective not only against these three genera but also against a wide range of fungal diseases. This review of fungal therapy and resistance focuses on the treatment of these three important fungal diseases, with an emphasis on *Candida*. However, the effectiveness of these therapies against other fungal infections also needs to be considered.

ANTIFUNGAL DRUGS AND THEIR MODE OF ACTION

Currently available antifungal drugs can be grouped into four main categories—polyenes, nucleic acid synthesis inhibitors, ergosterol biosynthesis inhibitors (EBI), and echinocandins. Each category of drug has a specific mode of action and a role in the treatment of fungal infections (reviewed in references 86 and 96).

Polyenes

The polyenes are a class of compounds with an amphipathic nature (one hydrophilic charged side of the molecule, and one hydrophobic uncharged side of the molecule). The polyenes target ergosterol, a sterol related to cholesterol, in fungal membranes. Ergosterol is the primary sterol in these fungal membranes, while cholesterol is the primary sterol in mammalian membranes. Polyenes work by binding to ergosterol and creating a pore that allows small molecules to diffuse across the membrane, resulting

in cell death. Hence, the polyenes are usually fungicidal at physiologic concentrations. There are two main polyenes: amphotericin B (AmB) and nystatin. AmB is the gold standard in the treatment of most fungal infections (*Candida*, *Cryptococcus*, and *Aspergillus*), especially in severe invasive infections where rapid response is needed. However, AmB has significant problems. Large doses are associated with nephrotoxicity, which limits its use. In addition, AmB is insoluble and so suspensions must be delivered intravenously. Use of nystatin is also limited by solubility issues. Both drugs can be used in a suspension for the treatment of oral fungal infections. Recently, lipid formulations of AmB have been developed which combine AmB with lipid structures to increase solubility and decrease toxicity. These formulations can be administered at higher AmB concentrations without toxicity, but they have a considerable expense.

5-Flucytosine

5-Flucytosine (5FC) is an antifungal drug that targets nucleic acid synthesis. The drug is fungus specific, since fungi and plants have a cytosine deaminase that converts 5FC into 5-fluorouracil, which is incorporated into DNA and RNA and inhibits cellular function and division. Mammalian cells do not have cytosine deaminase, and thus 5FC is not effective in these cells. 5FC is usually used in combination with polyenes or other antifungal agents in the treatment of fungal infections since it has a high frequency of developing resistance (see below).

Ergosterol Biosynthesis Inhibitors

EBI are a diverse group of antifungal agents that include azoles, morpholines, thiocarbamates, and allylamines. All of these drugs work by inhibiting the biosynthesis of ergosterol. Lack of ergosterol in the plasma membrane results in loss of membrane function and loss of fluidity that inhibits cell growth

Dominique Sanglard • Institute of Microbiology, University Hospital Lausanne, Rue de Bugnon 44, CH-1011 Lausanne, Switzerland.
Theodore C. White • Department of Pathobiology, School of Public Health and Community Medicine, University of Washington and Seattle Biomedical Research Institute, 307 Westlake Ave., N., Suite 500, Seattle, WA 98109-5219.

and division. Related methylated sterols resulting from inhibition of the pathway can replace ergosterol in the plasma membrane. However, these derivatives cannot substitute for ergosterol in its "sparking" function, where small amounts of ergosterol are necessary for the cell to proceed through the cell cycle.

The ergosterol pathway is divided into an early pathway that produces squalene from acetate and a late pathway that produces ergosterol from squalene. All of the EBI drugs inhibit steps in the late pathway. The allylamines (including terbinafine [Lamisil]) and thiocarbamates (including tolnaftate) inhibit the first step in the late pathway, the product of the *ERG1* gene. Terbinafine is used in the treatment of nail infections but is not currently used in systemic infections. The morpholines (including fenpropimorph, which is used primarily as a pesticide) inhibit the product of two genes, *ERG2* and *ERG24*, near the end of the late pathway.

The azoles are the most important class of EBI. They work by inhibiting the product of the *ERG11* gene, which demethylates lanosterol, an intermediate in the pathway. Inhibition of Erg11p results in the replacement of ergosterol with methylated sterols in the plasma membrane. There are two classes of azole drugs. The imidazoles, including ketoconazole, miconazole, and clotrimazole, have limited use for treatment of systemic infections but are used commonly for localized, surface infections and are commonly available over the counter. The triazoles, including fluconazole, voriconazole, and itraconazole, are now used primarily for systemic infections. Voriconazole is becoming the most common therapy for *Aspergillus* infections, replacing AmB. Because of its ability to cross the blood-brain barrier, fluconazole remains the best treatment for *Cryptococcus* infections of the brain and is important for long-term prophylaxis against reactivation of those infections.

Echinocandins

The echinocandins are the newest category of antifungal drug. The drugs work by inhibiting β-1,3-glucan synthase in the plasma membrane of the fungal cells. Glucan synthase is important for the production of glucan, an important component of the fungal cell wall. One echinocandin, caspofungin (Cancidas), is clinically available for treatment of *Candida* and *Aspergillus* infections, although it is not effective for *Cryptococcus* infections. Two other echinocandins, anidulafungin and micafungin, are in clinical trials.

CELLULAR MECHANISMS OF RESISTANCE

Standardized Testing

Resistance to standard antifungal drugs is regularly detected in a wide variety of fungal strains and species. It is important that clinical microbiology laboratories and basic researchers have a standardized method by which to test susceptibility to antifungal drugs. Such a standardized method was developed for yeasts (*Candida* and *Cryptococcus*) in the mid-1990s (61). A similar standardized method was subsequently developed for molds (*Aspergillus*) (60). Both methods determine the MIC of a specific drug for a particular fungal strain. An MIC is a specific value that reflects growth inhibition to 20% of the control at 48 h, not the 50% inhibition commonly used in other systems. Alternative methods have been developed for MIC determination, including the use of colorimetric dyes (XTT) (36) and the use of E-test strips, which are plastic strips containing a gradient of different antifungal drugs. E-test strips were recently approved for use in clinical microbiology laboratories for the determination of MIC.

Clinical Breakpoints

Once a standardized method for MIC determination was available, it became important to determine the clinical significance of MIC values, which is determined by finding clinical breakpoints. Clinical breakpoints have been identified for fluconazole, itraconazole, and 5FC in *Candida* spp. (61). For fluconazole, an MIC of ≥64 indicates that the strain is resistant, an MIC of ≤8 indicates that the strain is susceptible, and an MIC of 16 or 32 indicates that the strain is susceptible dose dependent (SDD) and may be treatable with high drug concentrations. For itraconazole, an MIC of ≥1.0 indicates that the strain is resistant, an MIC of ≤0.125 indicates that the strain is susceptible, and an MIC of 0.25 or 0.50 indicates that the strain is SDD. For 5FC, an MIC of ≥32 indicates that the strain is resistant, an MIC of ≤4 indicates that the strain is susceptible, and an MIC of 8 or 16 indicates that the strain is SDD. Interpretative breakpoints have not been determined for other drugs (AmB and voriconazole). The breakpoints for voriconazole are likely to be similar to those for itraconazole, based on distributions of MICs for strains and achievable drug concentrations in the blood.

The 90-60 Rule

Standardized MIC determinations and clinical breakpoints are not sufficient to determine how a strain will behave clinically. It is also important to consider the "90-60 rule" (72). This rule holds that susceptible isolates can be successfully treated 90% of the time while resistant isolates can be successfully treated 60% of the time. This has often been a criticism of antifungal drug susceptibility testing, but the rule holds for most of susceptibility testing, including that for most bacteria.

Intrinsic and Acquired Resistance

Clinical breakpoints and the 90-60 rule are used to predict how a "resistant" strain will behave clinically. It is also important to consider how the resistant strain arose. There are two types of drug resistance. A strain with primary or intrinsic resistance is resistant to a drug before it encounters the drug. Intrinsically resistant strains (and intrinsically hypersusceptible strains) are just the two extremes of a distribution of MIC in a random collection of isolates. The molecular basis of resistance in intrinsically resistant strains is difficult to identify or characterize, since strain differences unrelated to resistance will confuse the analysis. A strain with secondary or acquired resistance is a strain that has developed resistance in the presence of drug. The molecular basis of resistance of such a strain can be determined if a pretreatment susceptible parent of the strain is available for analysis. Matched sets of susceptible and resistant isolates of a single strain have been very useful in the identification of molecular mechanisms of resistance.

Stable and Transient Resistance

In addition to intrinsic and acquired resistance, a strain can be stably resistant, maintaining its resistance phenotype when grown in the absence of drug pressure, implying that the strain is genetically stable. Such strains are commonly found in clinical isolates from AIDS patients who have been receiving low doses of azole drugs for long periods (97). The alternative is that the strain is transiently resistant, losing its resistance phenotype when grown in the absence of the drug, thus indicating that the strain is phenotypically resistant but genotypically susceptible. Transient resistance has been found in clinical isolates from bone marrow transplant recipients who are receiving high doses of drug for short periods (54). Transient resistance might be caused by alterations in gene expression profiles or chromatin organization. A final possibility is that the resistant strain is less fit and that mutations that render the cell more susceptible would also increase its fitness. Susceptible cells will outgrow and outcompete the less fit, resistant strains. Such issues of fitness have been observed and are discussed in detail below.

Inducible and Heterogeneous Resistance

Most strains of *Candida* cannot be induced to develop drug resistance in vitro. However, there are some strains in which resistance can be induced (53). Such strains are likely to be predisposed to a resistance phenotype by alterations in resistance genes described below. Other strains of *Candida* and *Cryptococcus* display heterogeneous resistance, a phenotype in which the strain forms colonies at a low frequency (0.1 to 1%) on an agar plate containing drug at a concentration above the MIC for the strain (53, 57). However, the cells from these colonies are susceptible. When cells from these colonies are replated on agar plates containing the drug, again they form colonies at a low frequency. This phenomenon has been described for several bacterial systems.

Residual Growth

Determinations of MICs of azole drugs are complicated by the fungistatic nature of the azole drugs—they do not kill fungal cells, they just inhibit their growth. Therefore, at drug concentrations above the MIC for the strain, there is significant residual growth. This residual growth can interfere with the correct determination of an MIC, a phenomenon called "trailing growth." As originally defined, a strain shows trailing growth when the MIC of a drug at 24 h shows that it is susceptible but the MIC at 48 h shows that it is resistant. This is a significant clinical complication since trailing strains behave as susceptible strains in animal models (71).

Tolerance

The fungistatic nature of antifungal agents can be attributed to a phenomenon known as drug tolerance. Tolerance can be defined has the ability of yeast cells to survive at drug concentrations exceeding the MIC. Loss of tolerance can occur when a drug is fungicidal or when a fungistatic drug is converted into a fungicidal compound. Loss of tolerance to antifungal drugs has been demonstrated when cyclosporin A, an inhibitor of calcineurin activity, is used in combination with antifungal drugs against *Candida albicans*. For example, when cyclosporin A is given to *C. albicans* cultures containing fluconazole, loss of viability exceeds 99.9% (compared to the inoculum) over a 24-h incubation period, a feature not observed when cells are incubated with single compounds (51, 81). Interestingly, loss of tolerance to fluconazole is not associated with changes in drug susceptibility, since the presence of cyclosporin A in fluconazole susceptibility assays does not alter fluconazole MIC (51, 81). Thus, antifungal tolerance is governed by mechanisms different from those affecting drug susceptibility or resistance. Antifungal tolerance has another important consequence since it can determine the efficacy of drug treatment. In animal experiments (rat endocarditis with *C. albicans*), it was demonstrated that the fungicidal effect of the fluconazole-cyclosporin A combination could decrease the fungal load in some infected tissues (heart vegetations) to almost undetectable levels whereas treatment with fluconazole alone was only moderately decreasing the tissue fungal load (50). The molecular basis of antifungal tolerance is described below.

MOLECULAR MECHANISM OF RESISTANCE

Azole Drug Resistance

Yeast and fungal pathogens have developed several mechanisms enabling resistance to the azole class of antifungals. These mechanisms, summarized in Fig. 1, involve at least four different types of alterations.

(i) Alteration of antifungal transport by enhanced ATP-dependent efflux. Failure to accumulate azole antifungals has been identified as a cause of azole resistance in several posttreatment clinical fungal isolates. These isolates include yeast species such as *C. albicans, C. glabrata, C. krusei, C. dubliniensis,* and *Cryptococcus neoformans* (77). In azole-resistant isolates, genes encoding ATP binding cassette (ABC) transporters were upregulated compared to the corresponding azole-susceptible isolates. To date, ABC transporter genes that have been identified as upregulated in azole-resistant isolates include *CDR1* (for "*Candida* drug resistance 1") and *CDR2* in *C. albicans* (82, 85), *CdCDR1* in *C. dubliniensis* (58), a *CDR1* homologue in *C. tropicalis* (8), *CgCDR1* and *CgCDR2* in *C. glabrata* (15, 78, 79), and *CnAFR* in *C. neoformans* (65). In *Aspergillus fumigatus*, itraconazole is able to induce an ABC transporter gene, *atrF*. However, the role of this gene in the resistance of clinical isolates to itraconazole is still not established in detail (90). Heterologous expression of ABC transporter genes such as *CDR1* and *CDR2*, *CdCDR1*, *CgCDR1*, and *CgCDR2* in *Saccharomyces cerevisiae* conferred resistance to several azole derivatives (fluconazole, itraconazole, and ketoconazole) as well as a wide range of compounds including antifungals and metabolic inhibitors (59, 79, 82). With the sequence

data available from the genomes of yeast pathogens (*C. albicans* and *C. glabrata*), it is possible that other multidrug transporter genes involved in antifungal resistance will be characterized in the future.

(ii) Alteration of antifungal transport by enhanced efflux dependent on the membrane proton gradient. Several laboratories have also observed that, besides upregulation of ABC transporter genes, a multidrug transporter gene *CaMDR1* (for "multidrug resistance 1" and also previously known as *BEN*[r] for "benomyl resistance"), belonging to the family of major facilitators, was upregulated in some *C. albicans* azole-resistant yeast clinical isolates. In *C. albicans* and *C. dubliniensis* isolates with acquired azole resistance as a result of *MDR1* upregulation, deletion of *CaMDR1* or *CdMDR1* resulted in a sharp increase in azole susceptibility, thus supporting, by a genetic approach, the involvement of this specific gene in azole resistance (98, 99). Deletion of *CaMDR1* in an azole-susceptible laboratory strain did not result in a significant increase in azole susceptibility, consistent with the fact that *CaMDR1* is almost never expressed in this type of strain and more generally in azole-susceptible clinical isolates (83). Upregulation of a *CaMDR1*-like gene has been also observed in a fluconazole-exposed *C. tropicalis* isolate, which acquired cross-resistance to fluconazole and itraconazole (8). In most cases, azole resistance acquired in clinical situations by multidrug transporters in yeast pathogens is maintained over a large number of generations in vitro without drug selection. Azole resistance can, however, be a reversible phenomenon. Marr et al. (54) obtained *C. albicans* isolates that developed azole resistance from bone marrow transplant patients receiving fluconazole treatment. An

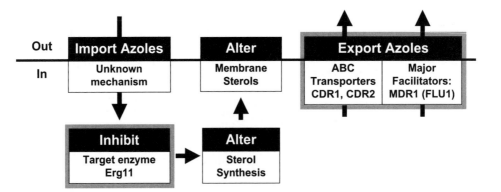

Figure 1. Azole interactions with fungal cells. Azole antifungals interact with a fungal cell in several ways. First, azole drugs enter the cell. There are conflicting data concerning the mechanisms of import. The azoles then inhibit the azole target enzyme Erg11. Inhibition of that enzyme alters sterol synthesis, which alters membrane sterols, replacing ergosterol with 14-α methyl sterols. Azoles are removed from the cell by two types of efflux pump, the ABC transporters, including Cdr1 and Cdr2, and the major facilitators, including Mdr1 and potentially Flu1. Mechanisms of resistance include alterations in *ERG11, CDR1, CDR2,* and *MDR1,* including mutations and changes in gene expression (boxes outlined in gray).

increase in the fluconazole MIC was coupled with upregulation of *CDR1*; the MIC was decreased in parallel with a decrease in *CDR1* expression in drug-free subculture. When azole-susceptible isolates from these patients were exposed in vitro to fluconazole, the isolates developed reversible azole resistance by the same *CDR1* upregulation mechanism.

(iii) Alteration of the target enzyme. Azole resistance can also be the result of alterations in the target enzyme, Erg11p, which demethylates lanosterol in the biosynthesis of ergosterol. Three types of alterations can occur with Erg11p. First, the gene can be overexpressed. In one study, 35% of the resistant isolates had increased levels of *ERG11* mRNA (64). Second, the gene can be mutated so that azoles are less active against the enzyme. In a survey of known alterations in *ERG11*, point mutations were identified in three broad "hot spots" in the enzyme, i.e., regions of the enzyme surrounding the active site (52). A small number of those mutations have been proven to cause resistance by alteration in *Candida* or expression in other systems. Other mutations have been identified only in resistant isolates and not in susceptible isolates and thus are likely to contribute to resistance. Additional point mutations have been found in both susceptible and resistant isolates; these may contribute to resistance in some genetic backgrounds and not others. Third, at least two groups have documented a gene conversion in which allelic differences in *ERG11* are eliminated as resistance develops (31, 94). It is postulated that this gene conversion occurs after one allele has been mutated to a resistant phenotype. Then drug pressure would select for a gene conversion event that resulted in the point mutation in both alleles of *ERG11*. *ERG11* mutations have been found in other fungal species: itraconazole resistance in *A. fumigatus* clinical isolates is associated with the occurrence of amino acid substitution in the *Cyp51A* product, which is the functional homolog of Erg11p in this fungal species. Interestingly, these mutations contribute only to resistance to itraconazole but not to resistance to voriconazole, which belongs to the new generation of triazole antifungals (28, 49, 54a). Analysis of *ERG11* from a clinical azole-resistant isolate of *C. neoformans* showed a point mutation linked to an amino acid substitution (G484S) that was not observed in the parent azole-susceptible isolates (73).

(iv) Alteration of the ergosterol biosynthetic pathway. Analysis of the sterol composition of some azole-resistant yeasts has revealed some alterations of enzymes involved in the ergosterol biosynthetic pathway. Accumulation of ergosta-7,22-dienol-3β-ol was observed in two separate azole-resistant *C. albicans* clinical isolates, which is a feature consistent with an absence of sterol $\Delta^{5,6}$-desaturase activity encoded by *ERG3* (63). Interestingly, azole resistance was coupled with resistance to AmB, which was expected because of the absence of ergosterol. The role of *ERG3* in azole resistance is implied by the observation that exposure of yeast to azoles inhibits Erg11p and thus results in accumulation of 14α-methylated sterols and 14α-methylergosta-8,24(28)-dien-3β,6α-diol. Formation of this latter sterol metabolite is thought to be catalyzed by the *ERG3* gene product, and thus inactivation of *ERG3* suppresses toxicity and causes azole resistance. Loss-of-function mutations in *ERG3* alleles from the known *C. albicans* azole-resistant Darlington strain were characterized recently (56). Unfortunately, the effect of these mutations on azole resistance was masked by other azole resistance mechanisms in this strain (56). However, the deletion of *ERG3* in a *C. albicans* laboratory strain results in azole resistance (84), and thus one can assume that azole resistance in the Darlington strain is due to inactive *ERG3* alleles. In vitro deletion analysis of other genes in the ergosterol biosynthetic pathway, along with biochemical sterol analysis of resistant isolates, has suggested that mutations in other enzymes in the pathway can result in changes in azole susceptibility. However, to date, molecular changes in other *ERG* genes have not been demonstrated in resistant clinical isolates.

Other Antifungal Drug Resistance

Resistance to AmB has been identified in some clinical isolates, although the mechanism by which these isolates became resistant is not known. In addition, some laboratory strains with AmB resistance have been constructed (84). Most of these isolates (clinical or laboratory) have a significant reduction of ergosterol in their plasma membrane, thus contributing to AmB resistance since the AmB target is absent.

Resistance to 5FC is common in patients receiving 5FC therapy. Common resistance mutations include mutation of the cytosine deaminase, which converts 5FC to 5-fluorouracil, and mutation in the uracil phosphoribosyltransferase (UPRT), an enzyme important for nucleic acid synthesis. In a recent study, primary resistance to 5FC in 25 *C. albicans* strains was investigated by identifying and sequencing the genes *FCA1*, *FUR1*, *FCY21*, and *FCY22*, which code for cytosine deaminase, UPRT, and two purine-cytosine permeases, respectively. These proteins are involved in pyrimidine salvage and 5FC metabolism. An association between a polymorphic nucleotide and resistance to 5FC was found within *FUR1*, where the substitution of cytosine for thymine at nucleotide position 301 results in the replacement of arginine with cysteine at amino acid position 101 in UPRT. Isolates that are homozygous for this mutation display increased levels of resistance to 5FC, whereas heterozygous isolates have reduced susceptibility (see also below). A single resistant isolate, lacking the above polymorphism in *FUR1*, has a homozygous polymorphism in *FCA1* that results in a glycine-to-aspartate substitution at position 28 in cytosine deaminase (39). Another resistance mutation can occur in the cytosine permease, which imports cytosine and 5FC into the cell. Recently, mutation in the permease demonstrated a correlation between permease mutation and fluconazole uptake, perhaps the result of competition between 5FC and fluconazole (62).

The echinocandin caspofungin has only recently become clinically available, and so the number of strains with caspofungin resistance, with or without cross-resistance to other echinocandins, is limited. There is some indication from *Candida* and from *Aspergillus* that efflux pumps related to the *CDR* gene family might be involved in caspofungin resistance, although there are data that contradict this result (88). There are clear data that point mutations in the glucan synthase in *S. cerevisiae* can cause resistance (29), although such mutations have yet to be detected in clinical isolates.

COMPLICATING FACTORS

Calcineurin

As mentioned above, inhibition of calcineurin activity by cyclosporin A affects tolerance to antifungal agents in *C. albicans*. The construction of mutants lacking the genes encoding calcineurin subunits A and B (*CNA* [*CMP1*] for calcineurin A or *CNB* [*CMP2*] for calcineurin B) verified that tolerance was mediated by calcineurin activity and its calcium-dependent activation pathway (23, 81). *CNA* mutants were killed by exposure to fluconazole or other agents such as terbinafine in a similar fashion to cyclosporin A-treated wild-type cells (81). This feature was also verified in *S. cerevisiae*: calcium could modulate the activity of azoles, and deletion of genes involved in the calcium-signaling pathway (*CNA1* encoding calcineurin A subunit or *CRZ1* encoding a calcineurin-dependent transcription factor) affected azole activity. Furthermore, exposure of *S. cerevisiae* to azole or terbinafine resulted in transcriptional upregulation of genes *FKS2* and *PMR1* known to be Ca^{2+} regulated (30). How the fungicidal effect of the fluconazole-cyclosporin A combination is obtained in *C. albicans* is not known in any detail. Taking *S. cerevisiae* as a model, it has

been reported that stress resulting from drug exposure engages the unfolded protein response in the endoplasmic reticulum. This in turn activates the uptake of extracellular calcium via the Cch1p-Mid1p Ca^{2+} channel, thus enabling the activation of the calcineurin pathway. This process is known as the calcium cell survival pathway (10, 11). It seems that this pathway is conserved in yeast human pathogens including *C. albicans* and *C. krusei*. Consistent with this observation is that the presence of an intact *CCH1* in *C. glabrata* was shown to be critical for survival in the presence of azoles and for increase of the intracellular calcium concentration in response to antifungal exposure (43).

In addition to its involvement in antifungal tolerance, calcineurin activity is needed for other cellular processes in *C. albicans*. Besides its putative role in regulating cell morphology in different growth media, one remarkable feature is that calcineurin is essential for virulence in animal models, as shown in separate studies (4, 9, 81). Calcineurin mutants lacking calcineurin activity were unable to colonize target organs (kidneys) after intravenous tail injection. Complementing this phenotype, calcineurin mutants were very sensitive to in vitro incubation in plasma or serum, thus perhaps partially explaining their lack of virulence in animals (9, 81).

High-Frequency Azole Resistance and Mitochondrial Function

Acquisition of azole resistance in *C. albicans* is a known clinical phenomenon. Repeated subculturing of azole-resistant clinical isolates in drug-free media does not usually result in a loss of resistance, thus indicating that acquisition of resistance is due to genetic alteration rather to transient adaptation. Attempts to acquire azole resistance in vitro in this yeast species by repeated subcultures in azole-containing media have not been successful, although there are a few exceptions (3). In vitro azole exposure has resulted mainly in rapidly reversible phenotypes (17). However, when similar in vitro azole exposure experiments are carried out with other species such as *C. glabrata*, the outcome is that stable azole-resistant isolates are obtained with much higher frequency. Azole resistance under these conditions was reported to occur at high frequency (2×10^{-4} to 4×10^{-4}) (15, 78). Phenotypic analysis of the obtained clones (HFAR mutants, for "high-frequency azole resistance") showed that they could not metabolize glycerol, a nonfermentable substrate whose assimilation is dependent on the existence of fully functional mitochondria. Therefore, it is not surprising that most HFAR mutants have lost their mitochondrial DNA. Loss of mitochondrial functions was also detected by others in *C. glabrata*, either by azole exposure or by treatment with respiratory inhibitors and DNA-intercalating agents (ethidium bromide) (14). Loss of mitochondrial function is known in *S. cerevisiae* and is readily visible in growth media by the resurgence of "petite" mutants. One of the characteristics of *C. glabrata* HFAR mutants and *S. cerevisiae* petite mutants is that they upregulate ABC transporters. For example, *C. glabrata* HFAR mutants upregulate *CgCDR1* and *CgCDR2* to high levels and thus can explain their azole resistance. This hypothesis is supported by the fact that mutants lacking *CgCDR1* and *CgCDR2* and subsequently exposed to ethidium bromide to eliminate mitochondrial DNA are not resistant to azoles (78, 79). Genome-wide transcriptional profiling of petite mutants of *S. cerevisiae* has shown that, in addition to the ABC transporter *PDR5*, several other ABC transporter genes were upregulated compared to those in the wild type. *PDR3*, a transcription factor involved in multidrug resistance in *S. cerevisiae*, was largely responsible for the upregulation of genes belonging to the *PDR* network (27, 44). The association between the loss of mitochondrial functions and azole resistance has not yet been established in *C. albicans*. One of the problems with *C. albicans* is that it is a member of the group of petite-negative yeasts (as opposed to petite-positive yeasts) in which the loss of mitochondrial function is considered lethal. However, some *C. albicans* respiratory mutants with mitochondrial lesions were reported, and therefore their azole susceptibility can be tested (33, 35).

The clinical relevance of loss of mitochondrial function has not yet been established in great detail. There is a single report on the isolation of a mitochondrial mutant of *C. glabrata* that was isolated from a patient receiving azole therapy (12). Therefore, one can assume that mitochondrial mutants are very rare in clinical settings. One likely explanation is that such mutants have reduced fitness compared to the wild type and therefore do not survive in the host environment.

Biofilms

Many fungal species including *C. albicans* form biofilms on solid surfaces. *Candida* species are known to form biofilms on teeth, dentures, catheters, artificial heart valves, and a variety of other surfaces. In many biological systems, biofilms are more resistant than cells growing in suspension (planktonic cells). Similar observations have been made for *Candida*—that biofilms are more resistant to antifungal drugs than are planktonic cells (5, 70). However, there are technical issues regarding the testing of drug susceptibility in a biofilm, since the

testing procedure is not the standard NCCLS procedure. It is clear that efflux pumps associated with drug resistance are overexpressed in biofilm cells (69), although cells in which several of these efflux pumps have been deleted are able to form resistant biofilms.

Mating

In recent years, the mating-type locus (*MTL*) of *C. albicans* has been defined and characterized. Most *C. albicans* strains are diploid and heterozygous at the *MTL* locus, having both **a** and α loci. However, homozygotes (**a**/**a** or α/α) do exist in clinical isolates and laboratory strains. These homozygous strains have a phenotype called phenotypic switching, in which they alternate between white cells and opaque cells. The opaque cell is the cell type that is able to undergo mating in which two diploid cells fuse into a tetraploid, which loses chromosomes until the cell returns to a diploid state (for details, see chapters 6 and 7).

A correlation has been described between homozygosity at the mating locus and drug resistance (76). Not all *MTL* homozygotes are resistant (66), but *MTL* homozygotes appear with increased frequency in collections of resistant isolates (25% or more of the resistant strains [76]). This correlation with *MTL* homozygosity may be the result of genetic linkage of the *TAC1* transcription factor (see below) and the *MTL* locus, as well as a gene conversion/mitotic recombination event similar to that described for *ERG11* (see above).

Synergy, Antagonism, and Cross-Resistance

The ability of antifungal drugs to treat a clinical infection can be dramatically affected by other drugs being used in combination, including other antifungals, other antimicrobials, and immune suppressive drugs. Most important is the use of combinations of antifungal drugs. The use of 5FC with AmB is a standard therapy for many infections since 5FC contributes antifungal activity but cannot be used alone because resistance to 5FC is easily generated. 5FC can also be used in combination with azole drugs. The effectiveness of using both azoles and AmB is a topic of much debate. The drugs can be synergistic or antagonistic in vitro, depending on the drug concentrations used, and can be at least additive in animal models (91). The clinical effectiveness of the combination requires further study.

Cross-resistance within the azoles is a phenomenon that is not well appreciated, in part because of the differences in clinical breakpoints. Cross-resistance cannot be evaluated by using MICs, since different drugs have different clinical breakpoints. When such breakpoints are considered in cross-resistance, there is significant azole cross-resistance (95), suggesting that no azole should be useful in the treatment of an azole-resistant strain. Reports of the successful use of one azole to treat a strain resistant to another azole are most probably a consequence of the 90-60 rule (72).

TRANSCRIPTIONAL REGULATION OF RESISTANCE

Antifungal resistance is often associated with changes in the transcription of genes, some of them being involved directly in resistance. For example, azole resistance is linked to the upregulation of multidrug transporters belonging to the class of ABC transporters and major facilitators and also to the upregulation of *ERG11* encoding the azole target. The molecular analysis of several azole-resistant isolates has revealed that these genes are in most cases not coordinately regulated but that each is regulated differentially (31, 47, 80, 85, 93). This suggests that these genes are regulated by distinct regulatory pathways. Here we summarize the current knowledge about the regulation of resistance in *C. albicans* (summarized in Table 1) and in other yeast species.

Regulation of ABC Transporters

As mentioned above, *CDR1* and *CDR2* are major ABC transporters involved in azole resistance in *C. albicans*. To identify factors involved in their regulation, genetic screens have been carried out in *S. cerevisiae*, assuming that the regulatory circuits for regulation of ABC transporters share some degree of equivalence between the two yeast species. A search was undertaken for *C. albicans* genes that were able to complement the absence of *PDR1* and *PDR3* in *S. cerevisiae*. These genes encode Zn_2Cys_6 zinc finger proteins involved in the regulation of *PDR5*, a major ABC transporter with functional similarity to *CDR1* and *CDR2* (6, 7). This screen led to the identification of *FCR1* (for "fluconazole resistance") and *FCR3* (92). *FCR1* encodes a protein belonging to the family of Zn_2Cys_6 zinc finger proteins. Its presence in an *S. cerevisiae PDR1 PDR3* mutant restores *PDR5* expression. The disruption of *FCR1* in *C. albicans* resulted in mutants with decreased susceptibility to fluconazole, thus suggesting that *FCR1* acted as a negative regulator of fluconazole susceptibility. The target genes of *FCR1* in *C. albicans* and the DNA segment targeted by this transcription factor are not yet known (92). On the other hand, *FCR3* encodes a transcription factor belonging to the family of bZip transcription factors. The involvement of this gene in regulating drug resistance genes in *C. albicans* is not yet known (100). *S. cerevisiae* has been used recently to isolate an additional regulator of ABC transporters by using a one-hybrid assay in which the *CDR1* promoter region was placed in fusion as a prey with the *lacZ* reporter system. A gene encoding a transcription factor (*NDT80*), which appears to regulate *CDR* expression, has been identified in *C. albicans* (19). The gene encodes a transcription factor (Ndt80p) that was able to activate the expression of the reporter system (19). This gene was named *NDT80* since it has similarity to the *S. cerevisiae* gene *NDT80*, which encodes a meiosis-specific protein. Disruption of *NDT80* affects basal expression levels of *CDR1* in *C. albicans* and reduces the ability of this gene to be upregulated when cells are exposed to miconazole. The DNA binding site for this protein in the *CDR1* promoter region has not yet been determined (19).

Due to the release of the entire *C. albicans* genome, other approaches were undertaken to isolate regulators of ABC transporter genes in *C. albicans*. Given the large number of putative transcription factors in this yeast species, the signatures of *cis*-acting regulatory elements of the *CDR1* and *CDR2* promoters were first analyzed to recognize typical DNA binding patterns of yeast transcription factors. It is known that the promoters of these genes contain a common drug-responsive element (DRE) that is essential for increasing *CDR1* and *CDR2* expression when *C. albicans* is exposed to antifungals or steroid hormones (26). This element is also required for the high-level expression of both genes in an azole-resistant isolate. This element has the nucleotide sequence 5'-<u>CGGA</u>(A/T)<u>ATCGGATA</u>TTTTTTTT-3'. It contains two repeated elements (underlined) with a CGG triplet. Such a triplet in *cis*-acting elements is recognized in yeast by the family of Zn_2Cys_6 transcription factors (37, 45, 55, 87). For *C. albicans*, genomic data revealed 72 open reading frames (ORF) with the Zn_2Cys_6 signature situated at their N termini, and thus it is possible that one of these ORF is binding to the DRE of *CDR1* and *CDR2* (20). A further element helped to identify such a protein: a cluster of three genes with a Zn_2Cys_6 signature is arranged in

Table 1. Transcription factors regulating resistance

Target genes[a]	Transcription factors	Regulation[b]	Reference
CDR1, CDR2	TAC1	Pos	20
CDR1	NDT80	Pos	19
MDR1	CAP1	Pos	2
ERG11	UPC2	Pos	89
ND[c]	FCR1	Neg	92
ND	FCR3	ND	100

[a]Genes listed together are known to be coregulated by the same transcription factor.

[b]The transcription factor has a negative (Neg) or positive (Pos) effect on the gene's transcription. That regulation can be direct or indirect.

[c]ND, not determined.

tandem approximately 14 kb upstream of the mating locus (*MTL*) of *C. albicans*. As mentioned above, mating-locus homozygosity is linked to the development of azole resistance in *C. albicans* (76), and therefore one hypothesis is that one of these genes is involved in this phenomenon by directly regulating the expression of *CDR1* and *CDR2*. Deletion of one of the genes in this cluster happened to abolish *CDR1* and *CDR2* upregulation in *C. albicans*. This gene was called *TAC1* (for "transcriptional activator of *CDR* genes") (20). Consistent with this finding, Tac1p was also shown to bind the previously identified DRE. Furthermore, a *TAC1* allele isolated from an azole-resistant *C. albicans* clinical isolate was found to confer resistance to azoles after introduction in an azole-susceptible isolate, thus suggesting that azole resistance mediated by *CDR1* and *CDR2* upregulation can be due to mutations in *TAC1* (20).

Separate studies reported additional regulatory elements in the promoter of *CDR1*. In one study, two elements, one called SRE1 (for "steroid-responsive element") (−677 to −648) and the other called SRE2 (−628 to −598) were identified and shown to be responsible for progesterone and estradiol induction of *CDR1*. Both SRE1 and SRE2 were specific for steroids, since they did not respond to other drugs, such as cycloheximide, miconazole, or terbinafine (42). The other study reported a negative regulatory element (NRE) situated at bp −1147 (with respect to the transcription start site). A protein of 55 kDa that binds to this element was isolated (32). However, these elements (SRE1, SRE2, and NRE) were different from the above-mentioned DRE, and the identification of the transcription factor(s) binding to these elements awaits further characterization.

Regulation of Major Facilitators

To date, *CaMDR1* is the only member of the major facilitator family that is known to be involved in azole resistance. Another member, *FLU1* (for "fluconazole resistance"), has been reported, but the expression of this gene does not correlate with azole resistance in several clinical isolates (16). Azole resistance mediated by *CaMDR1* is thought to involve mutation in *trans*-acting factors (98); however, their identification has not yet been reported. The *cis*-acting regulatory elements on the *CaMDR1* promoter also await a detailed analysis. However, putative regulators have been suggested from different studies. A *C. albicans* transcription factor of the bZip family (*CAP1*) was isolated by functional complementation in *S. cerevisiae*. This gene confers fluconazole resistance in *S. cerevisiae* by overexpression of *FLR1*, which has similarity to *CaMDR1* (1). Deletion of *CAP1* in *C. albicans* did not alter azole

susceptibility but resulted in increased susceptibility to hydrogen peroxide, suggesting that *CAP1* is involved in the response to oxidative stress (2). Disruption of *CAP1* in strain FR2, a strain selected in vitro for fluconazole resistance and constitutively overexpressing *CaMDR1*, did not suppress *CaMDR1* expression. This suggests either that *CAP1* is not involved in *CaMDR1* regulation or that deletion of *CAP1* is not sufficient to alter *CaMDR1* expression in this strain (2).

ERG11 Regulation

In *S. cerevisiae*, several genes in the late part of the ergosterol pathway are regulated by two genes, *ScUPC2* and *ScECM22*. These genes are members of the transcription factor family with the Zn_2Cys_6 DNA binding motif. The gene structure includes a transcription factor domain at the N terminus and a transmembrane domain at the C terminus. One hypothesis is that the transmembrane domain anchors the protein in the endoplasmic reticulum and that the transcription factor domain is released by cleavage from the transmembrane domain after an appropriate signal, such as the lack of ergosterol in the membrane. Once released, the transcription factor domain can enter the nucleus and activate *ERG* genes. This model is based on the SREBP model for cholesterol regulation in mammalian cells, described by Brown and Goldstein (13). A *C. albicans* gene with homology to both *S. cerevisiae* genes, *CaUPC2*, has been identified and characterized (89). The gene deletion is hypersusceptible to all ergosterol biosynthesis inhibitors, including the azoles. The deletion mutant has reduced ergosterol in the plasma membrane and has reduced uptake of sterols from the media. In addition, the deletion does not increase the expression of *ERG11* or other *ERG* genes in response to azole drugs, which does occur in the parental strain, suggesting that *UPC2* is necessary for activation of these genes. A *UPC2* core binding sequence is conserved between *Candida* and *Saccharomyces* and is found in most of the *ERG* genes in both the early and late parts of the ergosterol biosynthetic pathway. This core is found in the *ERG11* promoter, in the region identified as important for azole induction of *ERG11* expression.

Microarrays and Omics

The availability of the *C. albicans* genome has led to new opportunities in the study of azole resistance in this yeast species. Transcript profiling experiments have been carried out with azole-susceptible and azole-resistant matched isolates from different origins (41, 74, 75) and with strains having acquired stable azole resistance after serial passages on fluconazole

(22). These studies can contribute to the understanding of the regulatory circuits behind the development of azole resistance and also to the identification of potential new resistance mechanisms. Analysis of genes differentially expressed between azole-susceptible and azole-resistant isolates in three separate studies and previously known to upregulate *CDR* genes (22, 41, 75) revealed a group of genes with more than 1.5-fold upregulation. These genes were *RTA3*, *IFU5*, and *GPX1*. These genes are also among those upregulated when *C. albicans* is exposed to fluphenazine (41), a condition previously known to include both *CDR1* and *CDR2* (26). Taking into account that *TAC1* is a known regulator of *CDR* genes, these microarray data were compared with those obtained with *C. albicans* *TAC1* mutants. It is remarkable that *TAC1* mutants exposed to fluphenazine could not regulate a similar group of genes (i.e., *CDR1*, *CDR2*, *IFU5*, and *RTA3*) (20). The promoter regions of these genes were inspected and showed the presence of *DRE*-like sequences. The presence of this element in the promoters of these genes could explain their coordinate regulation. It is possible that the *DRE*-like sequences interact with Tac1p to achieve activation of expression. Moreover, it is possible that currently available microarrays are missing some genes that can be coregulated with the group of *TAC1*-dependent genes. One likely candidate is *PDR16*, a gene encoding a putative phosphatidylinositol transfer protein. This gene was reported to be coregulated in an azole-resistant isolate with *CDR1* and *CDR2* (25). Because the *PDR16* promoter contains DRE-like sequences (M. Raymond, personal communication), this gene could be considered an additional *TAC1* target, a hypothesis that remains to be confirmed. Transcript profiling of genes differentially expressed in clinical strains known to upregulate *CaMDR1* also revealed a cluster of coordinately upregulated genes (more than 1.5-fold upregulation). Besides *CaMDR1*, these genes were *GRP2*, *IFD5*, *IPF5987*, and *SNZ1*. In addition to the antifungal drug resistance function of *CaMDR1*, the other genes have oxidoreductive functions (*GRP2* and *IFD5*) or are potentially involved in pyridoxine (vitamin B$_6$) synthesis (*IPF5987* and *SNZ1*). It is, however, interesting that the *S. cerevisiae* homologues of all these five genes respond to cell damage conditions. Therefore, the common regulation of these genes constitutes a necessary process in the development of drug resistance mediated by the "*CaMDR1*" pathway. Since no *cis*-acting regulatory elements in *CaMDR1* have been described until now, it is difficult to predict a common regulatory element between these coregulated genes. As in the case of fluphenazine and *CDR* genes, benomyl exposure can result in the upregulation of *CaMDR1*. Transcript-profiling experiments

with cells exposed to benomyl compared to cells of clinical strains upregulating *CaMDR1* revealed a set of coregulated genes, most of them involved in stress response. A group of these genes (*IFD1*, *IFD4*, *IFD5*, and *IFD7*) belongs to a family encoding proteins with similarity to putative aryl-alcohol dehydrogenases. Interestingly, the proteins encoded by these genes were identified in separate studies (38, 46) involving proteome analysis of *C. albicans* isolates upregulating *CaMDR1*, thus enabling a correlation between the transcriptional gene activity and translational activity.

Systematic Deletion Screens

Genome-wide screens of mutants have been undertaken for yeast pathogens including *C. albicans* and *C. glabrata* for the identification of genes involved in azole susceptibility or resistance. For *C. albicans*, mutants were obtained by first generating a random collection of DNA fragments interrupted by transposon mutagenesis in *Escherichia coli*. In this strategy, the DNA fragments containing interrupted ORF were transformed in *C. albicans* and homozygous mutants were selected by a two-step selection procedure (24). Only a limited number of homozygote mutants (217 mutants) were obtained in this strategy, and they were used in a secondary screen for changes in azole susceptibility. One of the mutants showed decreased susceptibility to fluconazole in a simple diffusion assay with E-test strips. The interrupted ORF was identified as *CKA2* since it showed similarity to the *S. cerevisiae* *CKA2* gene encoding casein kinase II. Decreased susceptibility of the *CKA2* mutant to fluconazole seemed to be due to a slight upregulation of *CDR1*. Interestingly, *RTA3*, a gene previously identified as upregulated in azole-resistant strains and also dependent on the transcription factor *TAC1*, was also dependent on the inactivation of *CKA2*. *CKA2* is therefore a negative regulator of *CDR1* and *RTA3* and could operate at still unknown levels with *TAC1* in a common pathway of activation of these genes (V. Bruno and A. Mitchell, personal communication).

For *C. glabrata*, strategies to establish random collection of mutants were facilitated since this yeast is haploid. Transposon mutagenesis has been directly followed by transformation of transposon-interrupted DNA fragments into *C. glabrata*. Mutant collections containing a wide range of mutants were thus generated (18). A mutant collection (9,216 random insertion mutants) was screened for fluconazole susceptibility, and several mutants (27 mutants) were identified. Homologues of three of these genes have been implicated in azole and/or drug resistance in *S. cerevisiae*: two of these belong to the family of ABC transporters and phosphatidylinositol transporters (*PDR5* and *PDR16*), and one is involved in retrograde signaling from the mitochondria to the nucleus (*RTG2*). The remaining 24 genes were involved in diverse cellular functions, including ribosomal biogenesis and mitochondrial function, activation of RNA polymerase II transcription, nuclear ubiquitin ligase function, cell wall biosynthesis, and calcium homeostasis. Two sets of mutants were of particular interest, since they involved azole tolerance and resistance mechanisms previously identified in *C. albicans* and *C. glabrata*. One set contained strains defective in a putative plasma membrane calcium channel (Cch1-Mid1) that were modestly more susceptible to fluconazole but showed a significant loss of viability (equivalent to a loss of tolerance) on prolonged fluconazole exposure, suggesting that calcium signaling is required for survival of azole stress in *C. glabrata*. The second set of mutants characterized in detail were defective in mitochondrial assembly and organization, and these exhibited very high levels of fluconazole resistance that were similar to those previously described for mutants with mutations of mitochondrial functions (78, 79). Further analysis of these mutants indicated that a mechanism exists in *C. glabrata* for reversible loss of mitochondrial function that does not involve the loss of mitochondrial genome and that *C. glabrata* can switch between states of mitochondrial competence and incompetence in response to fluconazole exposure.

CLINICAL ASPECTS OF RESISTANCE

Genus and Species

The interaction of antifungal drugs with the fungal cell is clearly specified by the genera and species that are causing the infection. Whole genera can be resistant to a specific fungal drug. For example, fluconazole is not active against *Aspergillus* and caspofungin is not active against *Cryptococcus*. These specificities are usually not an issue clinically because physicians are aware of these specificities and the fungal genera that they are treating.

Less obvious in clinical treatments are species that are resistant to antifungals. *Aspergillus terreus* and *Candida lusitaniae* are two species that are intrinsically resistant to AmB, while most species in their genera are susceptible. Similarly, *Candida krusei* and many strains of *C. glabrata* are intrinsically resistant to azoles, and strains of *C. glabrata* can develop azole resistance quickly. Again, most other species of *Candida* are susceptible to azole drugs. These strains are a clinical problem because standard diagnoses do not usually determine the species, and these resistant strains can be treated with the drug until a species determination is completed. *C. glabrata* and *C. krusei* infections are increasing in frequency in many patient populations, in part because of the regular use of azole drugs, especially fluconazole, in these patients.

Strains

Within a species, strains can differ significantly in their intrinsic resistance. A collection of strains will have a range of susceptibilities, resembling a bell-shaped curve, in which a few strains are highly susceptible and a few strains are highly resistant. The development of acquired resistance can be detected when the frequency of highly resistant isolates increases and deviates from a standard curve.

C. albicans reproduces primarily by clonal expansion, and there is little random mating (40, 48, 68). Therefore, each strain evolves its own susceptibility pattern, which can include resistance or hypersusceptibility to specific drugs. The transfer of a strain from one patient to another is relatively rare, but it has been documented (96). Usually, resistance is intrinsic or acquired in the patient's commensal strain.

Clades

C. albicans clinical isolates can be separated into five major clades, groups I, II, III, SA, and E. Several clades can exist side by side, suggesting that little interclade recombination can occur. It is possible that antifungal resistance could develop preferentially in one of these clades. Interestingly, resistance to flucytosine (defined as an MIC of ≥32 µg/ml), an antifungal used for the treatment of systemic *C. albicans* infections, is observed mainly in strains from clade I (67). The genetic basis for this clade-specific resistance was identified in the UPRT gene (*FUR1*), where a single nucleotide change from cytosine to thymine at position 301 was observed in *FUR1* alleles obtained from 5FC-resistant isolates. The 5FC MICs for strains without copies of the mutant allele were almost exclusively ≤0.25 µg/ml, those for strains with one copy of the mutant allele were ≥0.5 µg/ml, and those for strains with two copies of the mutant allele were ≥16 µg/ml. The presence of this allele is responsible for clade I-specific resistance to 5FC within the *C. albicans* population and thus by inference is likely to be the major underlying 5FC resistance mechanism in *C. albicans*.

Fitness

Resistance to specific antifungal agents is mediated by several mechanisms, and some of them have been resolved by the alteration of specific genes. For example, mutations in *TAC1* are linked to the upregulation of *CDR1* and *CDR2*, mutations in *ERG11* are associated to azole resistance, and mutations in *FUR1* are responsible for 5FC resistance in *C. albicans* (see above). The occurrence of these mechanisms in clinical isolates and their selection by drug exposure can have a cost for overall cellular metabolism. Antifungal resistance may have a negative impact on fitness and therefore reduce the competitiveness of the strain with wild-type isolates, especially when the drug selection is removed. The decrease in fitness of resistant strains will compromise their virulence. Some studies have addressed this question by testing the virulence of azole-resistant isolates compared to their azole-susceptible parental isolate. It was concluded that no direct relationship exists between the development of azole resistance and virulence. The trajectory of azole resistance and the associated genome changes has rather an unpredictable effect on fitness (34). In vitro studies have also addressed the relationship between the development of azole resistance and changes in fitness. Individual *C. albicans* colonies were subcultured in fluconazole-containing medium, and each developed individual trajectories in their development of azole resistance (21). Overall, these in vitro studies showed no direct relationship between the cost of developing azole resistance and changes in fitness as measured by competition assays. Unfortunately, the in vitro-isolated azole-resistant colonies were not tested in an animal model, and thus the clinical relevance of these in vitro experiments is still unanswered.

Factors Affecting Clinical Resistance

As described above, the genus, species, strain, clade, and fitness can contribute to the resistance of a strain both in vitro and in vivo. Additional fungal factors that affect the way in which a strain behaves clinically in response to drug can include the cell type (yeast, hyphae, and switch phenotype), biofilm formation, the size of the population, and potential population bottlenecks within the patient, such as replacing an infected catheter or sterilizing dentures.

Characteristics of the antifungal drug can also affect the clinical response. Azole drugs are fungistatic and inhibit only the growth of the cells, while polyenes such as AmB are fungicidal at physiological conditions and can sterilize a site. The dose administered to the patient, including quantity, frequency, schedule, and cumulative dose, as well as absorption, distribution, and metabolism of the drug, can affect clinical outcome. Drug-drug interactions can also be important. The effect of calcineurin has been described above. While fluconazole is relatively free of drug-drug interactions, many other antifungals have significant alterations when used in the presence of other drugs.

Finally, host factors can affect the clinical response to a drug. The best way to treat a fungal infection (susceptible or resistant) is to restore the patient's immune system. Antifungal drugs only work in combination with the host immune system to resolve a fungal infection. Additional host factors, including the site and severity of infection, the presence of foreign materials such as dentures and catheters, and abscess formation, where the drug may not have access to the fungal cells, can affect treatment success. Finally, one of the most important host factors in treating a fungal infection is the patient's compliance with the drug regimen. Noncompliance or partial compliance in taking the antifungal drug can jeopardize the effectiveness of the drug in treating the infection and can create conditions in which a resistant strain can evolve.

CONCLUSIONS AND FUTURE DIRECTIONS

Drug resistance is a common problem in many microbial systems, including the fungi. Resistance to antifungal drugs has been observed with the azoles and with 5FC. Resistance to 5FC is a common problem that excludes its use as a monotherapy. Resistance to azole drugs was most common in the 1990s in AIDS patients receiving considerable amounts of azoles. Resistance continues to be a problem in AIDS patients, bone marrow transplant recipients, and other patients with immune suppression. The azoles continue to be used in many clinical settings where the frequency of azole resistance has not been determined, especially in surgical wards. Currently, the most common manifestation of azole resistance is the increased frequency of *C. glabrata* and *C. krusei* infections in patient populations who receive azole therapy.

The molecular mechanisms of resistance of *C. albicans* to azole drugs, including alterations in the target enzyme and in drug efflux pumps, have been well defined (Fig. 1). To date, studies of other fungi including *Cryptococcus* and *Aspergillus* indicate that similar mechanisms of resistance are used in these fungi. These mechanisms are important not only for understanding resistance but also for understanding how a cell (susceptible or resistant) responds to antifungal exposure. A thorough understanding of these interactions will assist in the design of new or improved drugs that avoid the development of resistance by these mechanisms or that suppress the drug resistance phenotype in resistant cells.

The most recent research efforts have focused on the transcriptional regulation of the target enzyme and efflux pumps. By understanding the regulation of these genes, it may be possible in the future to alter the phenotype of a cell by altering its gene expression profile, making all cells more susceptible to the available antifungals.

Resistance to azole drugs is likely to increase in the future as imidazoles continue to be available over the counter, fluconazole becomes available as a generic drug, and patient populations continue to be treated with these drugs. Newer azole drugs, with broader specificity, are being developed and are in clinical trials. However, taking clinical breakpoints and the 90-60 rule into account, these new azoles are unlikely to be more useful in treating resistant infections than previous azoles such as fluconazole. The most obvious way to avoid resistance is to use a different drug, such as caspofungin, which is not affected by azole resistance mechanisms. Indeed, caspofungin use is increasing in frequency, and caspofungin should be effective in treating azole-resistant strains and species, although extended use will eventually lead to resistance, probably by as yet unidentified mechanisms. Resistance to drugs will always be a problem in fungal cells, as it will always be a problem in any pathogen. As researchers develop newer and more effective drugs, the fungal cells will always develop newer and more complicated mechanisms of resistance. We as researchers will be analyzing mechanisms of resistance in the fungi for a long time to come.

REFERENCES

1. Alarco, A. M., I. Balan, D. Talibi, N. Mainville, and M. Raymond. 1997. AP1-mediated multidrug resistance in *Saccharomyces cerevisiae* requires *FLR1* encoding a transporter of the major facilitator superfamily. *J. Biol. Chem.* 272:19304–19313.

2. Alarco, A. M., and M. Raymond. 1999. The bZip transcription factor Cap1p is involved in multidrug resistance and oxidative stress response in *Candida albicans*. *J. Bacteriol.* 181:700–708.

3. Albertson, G. D., M. Niimi, R. D. Cannon, and H. F. Jenkinson. 1996. Multiple efflux mechanisms are involved in *Candida albicans* fluconazole resistance. *Antimicrob. Agents Chemother.* 40:2835–2841.

4. Bader, T., B. Bodendorfer, K. Schroppel, and J. Morschhauser. 2003. Calcineurin is essential for virulence in *Candida albicans*. *Infect. Immun.* 71:5344–5354.

5. Baillie, G. S., and L. J. Douglas. 2000. Matrix polymers of *Candida* biofilms and their possible role in biofilm resistance to antifungal agents. *J. Antimicrob. Chemother.* 46:397–403.

6. Balzi, E., and A. Goffeau. 1994. Genetics and biochemistry of yeast multidrug resistance. *Biochim. Biophys. Acta* 1187:152–162.

7. Balzi, E., M. Wang, S. Leterme, L. Van Dyck, and A. Goffeau. 1994. PDR5, a novel yeast multidrug resistance-conferring transporter controlled by the transcription regulator PDR1. *J. Biol. Chem.* 269:2206–2214.

8. Barchiesi, F., D. Calabrese, D. Sanglard, L. Falconi Di Francesco, F. Caselli, D. Giannini, A. Giacometti, S. Gavaudan, and G. Scalise. 2000. Experimental induction of fluconazole resistance in *Candida tropicalis* ATCC 750. *Antimicrob. Agents Chemother.* 44:1578–1584.

9. Blankenship, J. R., F. L. Wormley, M. K. Boyce, W. A. Schell, S. G. Filler, J. R. Perfect, and J. Heitman. 2003. Calcineurin is essential for *Candida albicans* survival in serum and virulence. *Eukaryot. Cell* 2:422–430.

10. Bonilla, M., and K. W. Cunningham. 2003. Mitogen-activated protein kinase stimulation of Ca^{2+} signaling is required for survival of endoplasmic reticulum stress in yeast. *Mol. Biol. Cell* 14:4296–4305.

11. Bonilla, M., K. K. Nastase, and K. W. Cunningham. 2002. Essential role of calcineurin in response to endoplasmic reticulum stress. *EMBO J.* 21:2343–2353.

12. Bouchara, J. P., R. Zouhair, S. Le Boudouil, G. Renier, R. Filmon, D. Chabasse, J. N. Hallet, and A. Defontaine. 2000. In-vivo selection of an azole-resistant petite mutant of *Candida glabrata*. *J. Med. Microbiol.* 49:977–984.

13. Brown, M. S., and J. L. Goldstein. 1997. The SREBP pathway: regulation of cholesterol metabolism by proteolysis of a membrane-bound transcription factor. *Cell* 89:331–340.

14. Brun, S., C. Aubry, O. Lima, R. Filmon, T. Berges, D. Chabasse, and J. P. Bouchara. 2003. Relationships between respiration and susceptibility to azole antifungals in *Candida glabrata*. *Antimicrob. Agents. Chemother.* 47:847–853.

15. Brun, S., T. Berges, P. Poupard, C. Vauzelle-Moreau, G. Renier, D. Chabasse, and J. P. Bouchara. 2004. Mechanisms of azole resistance in petite mutants of *Candida glabrata*. *Antimicrob. Agents Chemother.* 48:1788–1796.

16. Calabrese, D., J. Bille, and D. Sanglard. 2000. A novel multidrug efflux transporter gene of the major facilitator superfamily from *Candida albicans* (*FLU1*) conferring resistance to fluconazole. *Microbiology* 146:2743–2754.

17. Calvet, H. M., M. R. Yeaman, and S. G. Filler. 1997. Reversible fluconazole resistance in *Candida albicans*: a potential in vitro model. *Antimicrob. Agents Chemother.* 41:535–539.

18. Castano, I., R. Kaur, S. Pan, R. Cregg, L. Penas Ade, N. Guo, M. C. Biery, N. L. Craig, and B. P. Cormack. 2003. Tn7-based genome-wide random insertional mutagenesis of *Candida glabrata*. *Genome Res.* 13:905–915.

19. Chen, C. G., Y. L. Yang, H. I. Shih, C. L. Su, and H. J. Lo. 2004. CaNdt80 is involved in drug resistance in *Candida albicans* by regulating *CDR1*. *Antimicrob. Agents Chemother.* 48:4505–4512.

20. Coste, A. T., M. Karababa, F. Ischer, J. Bille, and D. Sanglard. 2004. TAC1, transcriptional activator of *CDR* genes, is a new transcription factor involved in the regulation of *Candida albicans* ABC transporters *CDR1* and *CDR2*. *Eukaryot. Cell* 3:1639–1652.

21. Cowen, L. E., L. M. Kohn, and J. B. Anderson. 2001. Divergence in fitness and evolution of drug resistance in experimental populations of *Candida albicans*. *J. Bacteriol.* 183:2971–2978.

22. Cowen, L. E., A. Nantel, M. S. Whiteway, D. Y. Thomas, D. C. Tessier, L. M. Kohn, and J. B. Anderson. 2002. Population genomics of drug resistance in *Candida albicans*. *Proc. Natl. Acad. Sci. USA* 99:9284–9289.

23. Cruz, M. C., A. L. Goldstein, J. R. Blankenship, M. Del Poeta, D. Davis, M. E. Cardenas, J. R. Perfect, J. H. McCusker, and J. Heitman. 2002. Calcineurin is essential for survival during membrane stress in *Candida albicans*. *EMBO J.* 21:546–559.

24. Davis, D. A., V. M. Bruno, L. Loza, S. G. Filler, and A. P. Mitchell. 2002. *Candida albicans* Mds3p, a conserved regulator of pH responses and virulence identified through insertional mutagenesis. *Genetics* 162:1573–1581.

25. De Deken, X., and M. Raymond. 2004. Constitutive activation of the *PDR16* promoter in a *Candida albicans* azole-resistant clinical isolate overexpressing *CDR1* and *CDR2*. *Antimicrob. Agents Chemother.* 48:2700–2703.

26. De Micheli, M., J. Bille, C. Schueller, and D. Sanglard. 2002. A common drug-responsive element mediates the upregulation of the *Candida albicans* ABC transporters *CDR1* and *CDR2*, two genes involved in antifungal drug resistance. *Mol. Microbiol.* **43**:1197–1214.

27. Devaux, F., E. Carvajal, S. Moye-Rowley, and C. Jacq. 2002. Genome-wide studies on the nuclear PDR3-controlled response to mitochondrial dysfunction in yeast. *FEBS Lett.* **515**:25–28.

28. Diaz-Guerra, T. M., E. Mellado, M. Cuenca-Estrella, and J. L. Rodriguez-Tudela. 2003. A point mutation in the 14α-sterol demethylase gene *cyp51A* contributes to itraconazole resistance in *Aspergillus fumigatus*. *Antimicrob. Agents Chemother.* **47**:1120–1124.

29. Douglas, C. M., J. A. Marrinan, W. Li, and M. B. Kurtz. 1994. A *Saccharomyces cerevisiae* mutant with echinocandin-resistant 1,3-β-D-glucan synthase. *J. Bacteriol.* **176**:5686–5696.

30. Edlind, T., L. Smith, K. Henry, S. Katiyar, and J. Nickels. 2002. Antifungal activity in *Saccharomyces cerevisiae* is modulated by calcium signalling. *Mol. Microbiol.* **46**:257–268.

31. Franz, R., S. L. Kelly, D. C. Lamb, D. E. Kelly, M. Ruhnke, and J. Morschhauser. 1998. Multiple molecular mechanisms contribute to a stepwise development of fluconazole resistance in clinical *Candida albicans* strains. *Antimicrob. Agents Chemother.* **42**:3065–3072.

32. Gaur, N. A., N. Puri, N. Karnani, G. Mukhopadhyay, S. K. Goswami, and R. Prasad. 2004. Identification of a negative regulatory element which regulates basal transcription of a multidrug resistance gene *CDR1* of *Candida albicans*. *FEMS Yeast Res.* **4**:389–399.

33. Geraghty, P., and K. Kavanagh. 2003. Disruption of mitochondrial function in *Candida albicans* leads to reduced cellular ergosterol levels and elevated growth in the presence of amphotericin B. *Arch. Microbiol.* **179**:295–300.

34. Graybill, J. R., E. Montalbo, W. R. Kirkpatrick, M. F. Luther, S. G. Revankar, and T. F. Patterson. 1998. Fluconazole versus *Candida albicans*: a complex relationship. *Antimicrob. Agents Chemother.* **42**:2938–2942.

35. Gyurko, C., U. Lendenmann, R. F. Troxler, and F. G. Oppenheim. 2000. *Candida albicans* mutants deficient in respiration are resistant to the small cationic salivary antimicrobial peptide histatin 5. *Antimicrob. Agents Chemother.* **44**:348–354.

36. Hawser, S. P., H. Norris, C. J. Jessup, and M. A. Ghannoum. 1998. Comparison of a 2,3-bis(2-methoxy-4-nitro-5-sulfophenyl)-5-[(phenyl-amino)carbonyl]-2*H*-tetrazolium hydroxide (XTT) colorimetric method with the Standardized National Committee for Clinical Laboratory Standards method of testing clinical yeast isolates for susceptibility to antifungal agents. *J. Clin. Microbiol.* **36**:1450–1452.

37. Hikkel, I., A. Lucau-Danila, T. Delaveau, P. Marc, F. Devaux, and C. Jacq. 2003. A general strategy to uncover transcription factor properties identifies a new regulator of drug resistance in yeast. *J. Biol. Chem.* **278**:11427–11432.

38. Hooshdaran, M. Z., K. S. Barker, G. M. Hilliard, H. Kusch, J. Morschhauser, and P. D. Rogers. 2004. Proteomic analysis of azole resistance in *Candida albicans* clinical isolates. *Antimicrob. Agents Chemother.* **48**:2733–2735.

39. Hope, W. W., L. Tabernero, D. W. Denning, and M. J. Anderson. 2004. Molecular mechanisms of primary resistance to flucytosine in *Candida albicans*. *Antimicrob. Agents Chemother.* **48**:4377–4386.

40. Hull, C. M., R. M. Raisner, and A. D. Johnson. 2000. Evidence for mating of the "asexual" yeast *Candida albicans* in a mammalian host. *Science* **289**:307–310.

41. Karababa, M., A. T. Coste, B. Rognon, J. Bille, and D. Sanglard. 2004. Comparison of gene expression profiles of *Candida albicans* azole-resistant clinical isolates and laboratory strains exposed to drugs inducing multidrug transporters. *Antimicrob. Agents Chemother.* **48**:3064–3079.

42. Karnani, N., N. A. Gaur, S. Jha, N. Puri, S. Krishnamurthy, S. K. Goswami, G. Mukhopadhyay, and R. Prasad. 2004. *SRE1* and *SRE2* are two specific steroid-responsive modules of *Candida* drug resistance gene 1 (*CDR1*) promoter. *Yeast* **21**:219–239.

43. Kaur, R., I. Castano, and B. P. Cormack. 2004. Functional genomic analysis of fluconazole susceptibility in the pathogenic yeast *Candida glabrata*: roles of calcium signaling and mitochondria. *Antimicrob. Agents Chemother.* **48**:1600–1613.

44. Koh, J. Y., P. Hajek, and D. M. Bedwell. 2001. Overproduction of PDR3 suppresses mitochondrial import defects associated with a *TOM70* null mutation by increasing the expression of *TOM72* in *Saccharomyces cerevisiae*. *Mol. Cell. Biol.* **21**:7576–7586.

45. Kren, A., Y. M. Mamnun, B. E. Bauer, C. Schuller, H. Wolfger, K. Hatzixanthis, M. Mollapour, C. Gregori, P. Piper, and K. Kuchler. 2003. War1p, a novel transcription factor controlling weak acid stress response in yeast. *Mol. Cell. Biol.* **23**:1775–1785.

46. Kusch, H., K. Biswas, S. Schwanfelder, S. Engelmann, P. D. Rogers, M. Hecker, and J. Morschhauser. 2004. A proteomic approach to understanding the development of multidrug-resistant *Candida albicans* strains. *Mol. Genet. Genomics* **271**:554–565.

47. Lopez–Ribot, J., R. K. McAtee, L. Lee, W. R. Kirkpatrick, T. C. White, D. Sanglard, and T. F. Patterson. 1998. Distinct patterns of gene expression associated with the development of fluconazole resistance in serial *Candida albicans* isolates from HIV-infected patients with oropharyngeal candidiasis. *Antimicrob. Agents Chemother.* **42**:2932–2937.

48. Magee, B. B., and P. T. Magee. 2000. Induction of mating in *Candida albicans* by construction of *MTL*a and *MTL*α strains. *Science* **289**:310–313.

49. Mann, P. A., R. M. Parmegiani, S. Q. Wei, C. A. Mendrick, X. Li, D. Loebenberg, B. DiDomenico, R. S. Hare, S. S. Walker, and P. M. McNicholas. 2003. Mutations in *Aspergillus fumigatus* resulting in reduced susceptibility to posaconazole appear to be restricted to a single amino acid in the cytochrome P450 14α-demethylase. *Antimicrob. Agents Chemother.* **47**:577–581.

50. Marchetti, O., J. M. Entenza, D. Sanglard, J. Bille, M. P. Glauser, and P. Moreillon. 2000. Fluconazole plus cyclosporine: a fungicidal combination effective against experimental endocarditis due to *Candida albicans*. *Antimicrob. Agents Chemother.* **44**:2932–2938.

51. Marchetti, O., P. Moreillon, M. P. Glauser, J. Bille, and D. Sanglard. 2000. Potent synergism of the combination of fluconazole and cyclosporine in *Candida albicans*. *Antimicrob. Agents Chemother.* **44**:2373–2381.

52. Marichal, P., L. Koymans, S. Willemsens, D. Bellens, P. Verhasselt, W. Luyten, M. Borgers, F. C. S. Ramaekers, F. C. Odds, and H. Vanden Bossche. 1999. Contribution

of mutations in the cytochrome P450 14α-demethylase (Erg11p, Cyp51p) to azole resistance in *Candida albicans*. *Microbiology* 45:2701–2713.

53. **Marr, K. A., C. N. Lyons, K. Ha, T. R. Rustad, and T. C. White.** 2001. Inducible azole resistance associated with a heterogeneous phenotype in *Candida albicans*. *Antimicrob. Agents Chemother.* 45:52–59.

54. **Marr, K. A., C. N. Lyons, T. R. Rustad, R. A. Bowden, and T. C. White.** 1998. Rapid, transient fluconazole resistance in *Candida albicans* is associated with increased mRNA levels of *CDR*. *Antimicrob. Agents Chemother.* 42:2584–2589.

54a. **Mellads, E., G. Garcia-Effron, L. Alcazar-Fuoli, M. Cuenca-Estrella, and J. L. Rodriguez-Tudela.** 2004. Substitutions at methionine 220 in the 14 alpha-sterol demethylase (Cyp51A) of *Aspergillus fumigatus* are responsible for resistance in vitro to azole antifungal drugs. *Antimicrob. Agents Chemother.* 48:2747–2750.

55. **Mendizabal, I., G. Rios, J. M. Mulet, R. Serrano, and I. F. de Larrinoa.** 1998. Yeast putative transcription factors involved in salt tolerance. *FEBS Lett.* 425:323–328.

56. **Miyazaki, Y., A. Geber, H. Miyazaki, D. Falconer, T. Parkinson, C. Hitchcock, B. Grimberg, K. Nyswaner, and J. E. Bennett.** 1999. Cloning, sequencing, expression and allelic sequence diversity of *ERG3* (C-5 sterol desaturase gene) in *Candida albicans*. *Gene* 236:43–51.

57. **Mondon, P., R. Petter, G. Amalfitano, R. Luzzati, E. Concia, I. Polacheck, and K. J. Kwon-Chung.** 1999. Heteroresistance to fluconazole and voriconazole in *Cryptococcus neoformans*. *Antimicrob. Agents Chemother.* 43:1856–1861.

58. **Moran, G. P., D. Sanglard, S. Donnelly, D. B. Shanley, D. J. Sullivan, and D. C. Coleman.** 1998. Identification and expression of multidrug transporters responsible for fluconazole resistance in *Candida dubliniensis*. *Antimicrob. Agents Chemother.* 42:1819–1830.

59. **Nakamura, K., M. Niimi, K. Niimi, A. R. Holmes, J. E. Yates, A. Decottignies, B. C. Monk, A. Goffeau, and R. D. Cannon.** 2001. Functional expression of *Candida albicans* drug efflux pump Cdr1p in a *Saccharomyces cerevisiae* strain deficient in membrane transporters. *Antimicrob. Agents Chemother.* 45:3366–3374.

60. **National Committee for Clinical Laboratory Standards.** 2002. *Reference Method for Broth Dilution Antifungal Susceptibility Testing of Molds*. Approved Standard. NCCLS document M38-A. National Committee for Clinical Laboratory Standards, Wayne, Pa.

61. **National Committee for Clinical Laboratory Standards.** 1997. *Reference Method for Broth Dilution Antifungal Susceptibility Testing of Yeasts*. Approved Standard. NCCLS document M27-A. National Committee for Clinical Laboratory Standards, Wayne, Pa.

62. **Noel, T., F. Francois, P. Paumard, C. Chastin, D. Brethes, and J. Villard.** 2003. Flucytosine-fluconazole cross-resistance in purine-cytosine permease-deficient *Candida lusitaniae* clinical isolates: indirect evidence of a fluconazole uptake transporter. *Antimicrob. Agents Chemother.* 47:1275–1284.

63. **Nolte, F. S., T. Parkinson, D. J. Falconer, S. Dix, J. Williams, C. Gilmore, R. Geller, and J. R. Wingard.** 1997. Isolation and characterization of fluconazole- and amphotericin B-resistant *Candida albicans* from blood of two patients with leukemia. *Antimicrob. Agents Chemother.* 44:196–199.

64. **Perea, S., J. L. Lopez Ribot, W. R. Kirkpatrick, R. K. McAtee, R. A. Santillan, M. Martinez, D. Calabrese,** D. Sanglard, and T. F. Patterson. 2001. Prevalence of molecular mechanisms of resistance to azole antifungal agents in *Candida albicans* strains displaying high-level fluconazole resistance isolated from human immunodeficiency virus-infected patients. *Antimicrob. Agents Chemother.* 45:2676–2684.

65. **Posteraro, B., M. Sanguinetti, D. Sanglard, M. La Sorda, S. Boccia, L. Romano, G. Morace, and G. Fadda.** 2003. Identification and characterization of a *Cryptococcus neoformans* ATP binding cassette (ABC) transporter-encoding gene, *CnAFR1*, involved in the resistance to fluconazole. *Mol. Microbiol.* 47:357–371.

66. **Pujol, C., S. A. Messer, M. Pfaller, and D. R. Soll.** 2003. Drug resistance is not directly affected by mating type locus zygosity in *Candida albicans*. *Antimicrob. Agents Chemother.* 47:1207–1212.

67. **Pujol, C., M. A. Pfaller, and D. R. Soll.** 2004. Flucytosine resistance is restricted to a single genetic clade of *Candida albicans*. *Antimicrob. Agents Chemother.* 48:262–266.

68. **Pujol, C., J. Reynes, F. Renaud, M. Raymond, M. Tibayrenc, F. J. Ayala, F. Janbon, M. Mallie, and J. M. Bastide.** 1993. The yeast *Candida albicans* has a clonal mode of reproduction in a population of infected human immunodeficiency virus-positive patients. *Proc. Natl. Acad. Sci. USA* 90:9456–9459.

69. **Ramage, G., S. Bachmann, T. F. Patterson, B. L. Wickes, and R. J. L. Lopez.** 2002. Investigation of multidrug efflux pumps in relation to fluconazole resistance in *Candida albicans* biofilms. *J. Antimicrob. Chemother.* 49:973–980.

70. **Ramage, G., K. vande Walle, B. L. Wickes, and J. L. Lopez-Ribot.** 2001. Standardized method for in vitro antifungal susceptibility testing of *Candida albicans* biofilms. *Antimicrob. Agents Chemother.* 45:2475–2479.

71. **Rex, J. H., P. W. Nelson, V. L. Paetznick, M. Lozano-Chiu, A. Espinel-Ingroff, and E. J. Anaissie.** 1998. Optimizing the correlation between results of testing in vitro and therapeutic outcome in vivo for fluconazole by testing critical isolates in a murine model of invasive candidiasis. *Antimicrob. Agents Chemother.* 42:129–134.

72. **Rex, J. H., and M. A. Pfaller.** 2002. Has antifungal susceptibility testing come of age? *Clin. Infect. Dis.* 35:982–989.

73. **Rodero, L., E. Mellado, A. C. Rodriguez, A. Salve, L. Guelfand, P. Cahn, M. Cuenca-Estrella, G. Davel, and J. L. Rodriguez-Tudela.** 2003. G484S amino acid substitution in lanosterol 14-α demethylase (*ERG11*) is related to fluconazole resistance in a recurrent *Cryptococcus neoformans* clinical isolate. *Antimicrob. Agents Chemother.* 47:3653–3656.

74. **Rogers, P. D., and K. S. Barker.** 2002. Evaluation of differential gene expression in fluconazole-susceptible and -resistant isolates of *Candida albicans* by cDNA microarray analysis. *Antimicrob. Agents Chemother.* 46:3412–3417.

75. **Rogers, P. D., and K. S. Barker.** 2003. Genome-wide expression profile analysis reveals coordinately regulated genes associated with stepwise acquisition of azole resistance in *Candida albicans* clinical isolates. *Antimicrob. Agents Chemother.* 47:1220–1227.

76. **Rustad, T. R., D. A. Stevens, M. A. Pfaller, and T. C. White.** 2002. Homozygosity at the *Candida albicans* MTL locus associated with azole resistance. *Microbiology* 148:1061–1072.

77. **Sanglard, D., and J. Bille.** 2002. Current understanding of the mode of action and of resistance mechanisms to

conventional and emerging antifungal agents for treatment of *Candida* infections, p. 349–383. *In* R. Calderone (ed.), *Candida and Candidiasis.* ASM Press, Washington, D.C.

78. **Sanglard, D., F. Ischer, and J. Bille.** 2001. Role of ATP-binding-cassette transporter genes in high-frequency acquisition of resistance to azole antifungals in *Candida glabrata. Antimicrob. Agents Chemother.* 45:1174–1183.

79. **Sanglard, D., F. Ischer, D. Calabrese, P. A. Majcherczyk, and J. Bille.** 1999. The ATP binding cassette transporter gene *CgCDR1* from *Candida glabrata* is involved in the resistance of clinical isolates to azole antifungal agents. *Antimicrob. Agents Chemother.* 43:2753–2765.

80. **Sanglard, D., F. Ischer, L. Koymans, and J. Bille.** 1998. Amino acid substitutions in the cytochrome P450 lanosterol 14α-demethylase (CYP51A1) from azole-resistant *Candida albicans* clinical isolates contributing to the resistance to azole antifungal agents. *Antimicrob. Agents Chemother.* 42:241–253.

81. **Sanglard, D., F. Ischer, O. Marchetti, J. Entenza, and J. Bille.** 2003. Calcineurin A of *Candida albicans*: involvement in antifungal tolerance, cell morphogenesis and virulence. *Mol. Microbiol.* 48:959–976.

82. **Sanglard, D., F. Ischer, M. Monod, and J. Bille.** 1997. Cloning of *Candida albicans* genes conferring resistance to azole antifungal agents: characterization of *CDR2*, a new multidrug ABC-transporter gene. *Microbiology* 143:405–416.

83. **Sanglard, D., F. Ischer, M. Monod, and J. Bille.** 1996. Susceptibilities of *Candida albicans* multidrug transporter mutants to various antifungal agents and other metabolic inhibitors. *Antimicrob. Agents Chemother.* 40:2300–2305.

84. **Sanglard, D., F. Ischer, T. Parkinson, D. Falconer, and J. Bille.** 2003. *Candida albicans* mutations in the ergosterol biosynthetic pathway and resistance to several antifungal agents. *Antimicrob. Agents Chemother.* 47:2404–2412.

85. **Sanglard, D., K. Kuchler, F. Ischer, J. L. Pagani, M. Monod, and J. Bille.** 1995. Mechanisms of resistance to azole antifungal agents in *Candida albicans* isolates from AIDS patients involve specific multidrug transporters. *Antimicrob. Agents Chemother.* 39:2378–2386.

86. **Sanglard, D., and F. C. Odds.** 2002. Resistance of *Candida* species to antifungal agents molecular mechanisms and clinical consequences. *Lancet Infect. Dis.* 2:73–85.

87. **Schjerling, P., and S. Holmberg.** 1996. Comparative amino acid sequence analysis of the C6 zinc cluster family of transcriptional regulators. *Nucleic Acids Res.* 24:4599–4607.

88. **Schuetzer-Muehlbauer, M., B. Willinger, G. Krapf, S. Enzinger, E. Presterl, and K. Kuchler.** 2003. The *Candida albicans* Cdr2p ATP-binding cassette (ABC) transporter confers resistance to caspofungin. *Mol. Microbiol.* 48:225–235.

89. **Silver, P. M., B. G. Oliver, and T. C. White.** 2004. Role of *Candida albicans* transcription factor Upc2p in drug resistance and sterol metabolism. *Eukaryot. Cell* 3:1391–1397.

90. **Slaven, J. W., M. J. Anderson, D. Sanglard, G. K. Dixon, J. Bille, I. S. Roberts, and D. W. Denning.** 2002. Induced expression of a novel *Aspergillus fumigatus* ABC transporter gene, atrF, in response to itraconazole. *Fungal Genet. Biol.* 36:199–206.

91. **Sugar, A. M., C. A. Hitchcock, P. F. Troke, and M. Picard.** 1995. Combination therapy of murine invasive candidiasis with fluconazole and amphotericin B. *Antimicrob. Agents Chemother.* 39:598–601.

92. **Talibi, D., and M. Raymond.** 1999. Isolation of a putative *Candida albicans* transcriptional regulator involved in pleiotropic drug resistance by functional complementation of a *pdr1 pdr3* mutation in *Saccharomyces cerevisiae. J. Bacteriol.* 181:231–240.

93. **White, T. C.** 1997. Increased mRNA levels of *ERG16*, *CDR*, and *MDR1* correlate with increases in azole resistance in *Candida albicans* isolates from a patient infected with human immunodeficiency virus. *Antimicrob. Agents Chemother.* 41:1482–1487.

94. **White, T. C.** 1997. The presence of an R467K amino acid substitution and loss of allelic variation correlate with an azole-resistant lanosterol 14α demethylase in *Candida albicans. Antimicrob. Agents Chemother.* 41:1488–1494.

95. **White, T. C., S. Holleman, F. Dy, L. F. Mirels, and D. A. Stevens.** 2002. Resistance mechanisms in clinical isolates of *Candida albicans. Antimicrob. Agents Chemother.* 46:1704–1713.

96. **White, T. C., K. A. Marr, and R. A. Bowden.** 1998. Clinical, cellular, and molecular factors that contribute to antifungal drug resistance. *Clin. Microbiol. Rev.* 11:382–402.

97. **White, T. C., M. A. Pfaller, R. G. Rinaldi, J. Smith, and S. W. Redding.** 1997. Stable azole drug resistance associated with a substrain of *Candida albicans* from an HIV-infected patient. *Oral Dis.* 3:S102–S109.

98. **Wirsching, S., S. Michel, G. Kohler, and J. Morschhauser.** 2000. Activation of the multiple drug resistance gene *MDR1* in fluconazole-resistant, clinical *Candida albicans* strains is caused by mutations in a *trans*-regulatory factor. *J. Bacteriol.* 182:400–404.

99. **Wirsching, S., S. Michel, and J. Morschhauser.** 2000. Targeted gene disruption in *Candida albicans* wild-type strains: the role of the *MDR1* gene in fluconazole resistance of clinical *Candida albicans* isolates. *Mol. Microbiol.* 36:856–865.

100. **Yang, X., D. Talibi, S. Weber, G. Poisson, and M. Raymond.** 2001. Functional isolation of the *Candida albicans FCR3* gene encoding a bZip transcription factor homologous to *Saccharomyces cerevisiae* Yap3p. *Yeast* 18:1217–1225.

II. MODEL SYSTEMS: HETEROLOGOUS HOSTS, MODEL PATHOGENS, AND FUNGAL-BACTERIAL INTERACTIONS

Molecular Principles of Fungal Pathogenesis
Edited by Joseph Heitman et al.
©2006 ASM Press, Washington, D.C.

Chapter 15

Heterologous Hosts and the Evolution and Study of Fungal Pathogenesis

Eleftherios Mylonakis, Stephen B. Calderwood, and Frederick M. Ausubel

A variety of human bacterial and fungal pathogens are highly virulent in diverse nonvertebrate hosts. In addition to providing an informative alternative approach for the in vivo study of microbial pathogenesis, these systems may provide paradigms that can help us understand the evolution and maintenance of microbial virulence (Table 1).

This chapter focuses on three invertebrate systems that have been used to study human fungal pathogens. First we discuss two genetically tractable hosts: the fly *Drosophila melanogaster* and the microscopic nematode *Caenorhabditis elegans* (Table 2). Then we discuss the use of the greater wax moth caterpillar *Galleria mellonella*, which provides certain interesting technical advantages, even though it does not have a sequenced genome. Finally, we briefly mention other nonmammalian model hosts that have been used for the study of pathogenesis due to bacteria or parasites, which workers may wish to consider for the study of fungal pathogens. Two other model hosts, *Acanthamoeba castellanii* and *Dictyostelium discoideum*, are discussed in chapter 16.

DROSOPHILA MELANOGASTER SYSTEM

Methods

A number of approaches have been used for the inoculation of *Drosophila* with pathogenic microorganisms. The first method is injection and can be used for human pathogens that cannot infect by a natural process. Injection involves one of the following:

- pricking the dorsal part of the fly thorax (or abdomen) with a sharp needle that has been dipped in a suspension of the pathogen; in our studies, larger numbers of microorganisms can be injected in the abdomen than into the thorax, but injection of the abdomen is associated with higher trauma-related mortality
- microinjection of a precise dose of microbes directly into the body cavity

Although the interaction between human fungal pathogens and *Drosophila* has not been studied extensively in nature, in the techniques implemented in the laboratory certain fungi can penetrate the cuticle of *Drosophila* and establish a lethal infection (18) while others kill the fly after ingestion or other "natural infection" (2). These "natural-infection" methods involve

- feeding *Drosophila* larvae or adults with a concentrated solution of microorganisms mixed with their food (for example, crushed banana containing a large concentration of the pathogen of interest)
- "spraying" the pathogen (for example, fungal spores) directly onto the fly exoskeleton
- "rolling" flies on agar plates that have been spread with the pathogen (for example, fungal spores) for 1 to 2 min ("rolling assay")

The Fly Fights Back: *D. melanogaster* Response to Pathogenic Fungi

In 1996, Lemaitre et al. demonstrated that the Toll receptor, previously known to play an essential role in *Drosophila* embryonic development, is required for antifungal defense of the fly (36) (Fig. 1). Subsequent work with *Drosophila* has shown that *Drosophila* innate immunity is mediated by the Toll and Imd signaling pathways (6, 21, 34, 43, 60). Flies with a mutation(s) involving components of the Toll pathway (but not the Imd pathway) are highly sensitive to natural infection by the entomopathogenic

Eleftherios Mylonakis and Stephen B. Calderwood • Division of Infectious Diseases, Massachusetts General Hospital, 55 Fruit Street, Gray 5, GRJ-504, Boston, MA 02114.   **Frederick M. Ausubel** • Wellman 10, Department of Molecular Biology, Massachusetts General Hospital, Boston, MA 02114.

Table 1. Shortcomings associated with mammalian model systems for the study of microbial pathogenesis

Difficulty of handling
High cost
Long reproductive cycles
Small brood sizes
Ethical considerations
Increasing practical difficulties associated with experiments on mammals
Mammalian responses to infection can be too complex to study, and adaptive responses sometimes obscure studies of the basic principles of innate immunity.
The technology for gene knockouts in mammals is still relatively time-consuming and expensive.

Table 2. Advantages of using *D. melanogaster* and *C. elegans* for the study of pathogenesis

Genomes have been sequenced.
They have short reproductive cycles and many progeny, accelerating experiments.
These hosts are amenable to forward and reverse genetics.
Genetic tools, such as RNAi and microarrays, are available (RNAi refers to the introduction of homologous double-stranded RNA to specifically inhibit the expression of the corresponding gene, resulting in null or hypomorphic phenotypes).
Virulence-related genes involved in mammalian infection also play a role in the interaction of fungal pathogens with these invertebrate organisms.
The host response in these invertebrate systems has demonstrated fundamental similarities to mammals.

fungus *Beauveria bassiana* (37) or injection of *Aspergillus fumigatus* (37), *Cryptococcus neoformans* (2), or *Candida albicans* (1), confirming that responses to fungi are largely activated by the Toll pathway (37).

- Systemic fungal infection leads to activation of the Toll receptor on the surface of fat body cells by a cleaved form of a cytokine-like protein, Spaetzle, which is present in the *Drosophila* hemolymph.
- The physical interaction between Spaetzle and Toll initiates an intracellular cascade that triggers signal transduction through the threonine-serine kinase Pelle. This leads to the phosphorylation and degradation of Cactus and subsequent nuclear translocation of the Rel family transcription factors Dorsal and Dif.

- The final step in this pathway is the synthesis of antimicrobial peptides.

In contrast to *Toll* mutants, which are impaired in survival following fungal infection, *imd* mutants are impaired in antibacterial responses, suggesting that distinct pathways are used to detect and respond to bacterial and fungal infection. However, recent evidence indicates that there is cross talk between the two pathways (6).

The discovery that the Toll pathway is required for defense against fungal infection in *Drosophila* was pivotal in studies of both mammalian and *Drosophila* immunity. The *Drosophila* Toll and Imd signaling pathways exhibit striking similarities to the vertebrate Toll-like receptor and tumor necrosis factor alpha pathways, respectively, which regulate NF-κB activity in vertebrates, suggesting common

Figure 1. Toll system response following fungal infection. A scanning electron micrograph shows a dead *Drosophila* adult with a mutation involving the Toll pathway that succumbed to infection by *A. fumigatus* and is covered with germinating hyphae. Wild-type flies are resistant to *A. fumigatus*. Reprinted from reference 36 with permission.

Table 3. Specific advantages and limitations of
D. melanogaster for the study of fungal pathogenesis

Advantages
   Drosophila has become a preferred model organism for
    deciphering animal innate immunity, and certain aspects
    of the host response in *D. melanogaster* are strikingly
    similar to mammalian host-pathogen interactions.
   Using robotic technologies, a library of double-stranded
    RNAs can be screened rapidly in specific cell-based
    assays to identify all the proteins encoded by the
    genome that interfere with a specific assay.
   A hemolymph sample can be obtained for the study of
    host responses.

Limitations
   The fly is very resistant to fungal pathogens, often necessitating
    the use of mutant lines for the study of fungal pathogenesis.
   Not all microbes have been shown to induce an immune
    response following natural infection.
   Flies are unable to survive at 37°C.

evolutionary roots (19). Because of the evolutionary conservation of innate mechanisms of host defense, *Drosophila* has emerged as a powerful model with which to study the genetic control of immune recognition and responses (Table 3) and a variety of resources are now available to assist workers in this field (Table 4).

Interestingly, the site of inoculation seems to play a role in the interaction of *Drosophila* with fungi, and this should be taken into consideration when planning an experiment. For example, mutation of the Toll immune response pathway did not play a role in the *Drosophila* defense response following ingestion of *C. neoformans* but did play a major role following injection (2). Also, flies that underwent "natural inoculation" with *alb1*, an *Aspergillus* mutant with attenuated virulence in a murine model of invasive aspergillosis (59), had improved survival compared with flies inoculated with wild-type *Aspergillus,* but there was no difference in survival following systemic infection with these two strains (39a).

Use of *D. melanogaster* To Study Fungal Virulence

Aspergillus fumigatus

In addition to the insights that the *Drosophila*-*Aspergillus* interactions have provided in understanding innate immunity, this system provides a useful model for the study of certain aspects of *Aspergillus* virulence. As noted above, the use of this system is limited by the strong innate immune response of *Drosophila* to *Aspergillus*. Thus, workers have had to use flies with mutations in the Toll pathway in order to study *Aspergillus* virulence. Bhabhra et al. used this system to study the virulence phenotype associated with a mutation in *A. fumigatus* CgrA (an ortholog of a yeast nucleolar protein that functions in ribosome synthesis). The *cgrA* mutant was significantly less virulent than the wild-type fungus or reconstituted strains in immunosuppressed mice that were challenged with an intranasal inoculum of conidia, and it was associated with smaller fungal colonies in lung tissue. Based on the in vitro characteristics of the *cgrA* mutant, these workers hypothesized that loss of *cgrA* would have little or no effect on virulence at 25°C. Indeed, the degree of attenuation was much less pronounced in a model of systemic *Drosophila* infection at 25°C (5).

Cryptococcus spp.

In addition to the work on the *Aspergillus*-*Drosophila* interaction, a number of other medically important fungi have been studied with respect to their effect on *Drosophila*. Apidianakis et al. found that ingestion of *C. neoformans* by *D. melanogaster* resulted in the death of the fly but ingestion of *Saccharomyces cerevisiae* or the nonpathogenic yeasts *Cryptococcus kuetzingii* or *Cryptococcus laurentii* did not (2). Moreover, the *C. neoformans* protein kinase A and RAS signal transduction pathways, previously shown to be involved in virulence in mammals, also play a role in killing *Drosophila* after ingestion, suggesting that there is an infection-like process that leads to killing of the fly after ingestion of *C. neoformans*. However, the Toll pathway appeared to play no role in this process, while, as noted above, after systemic infection the Toll pathway is necessary for clearance of *C. neoformans* cells when they are introduced directly into the

Table 4. Internet resources for the study of *D. melanogaster*

Category	URL
Introduction to *Drosophila*	http://www.ceolas.org/fly/intro.html
Discussion group	http://www.bio.net/hypermail/dros/
Virtual library	http://www.ceolas.org/fly/
Drosophila genome mirror sites	http://fbserver.gen.cam.ac.uk:7081
	http://flybase.bio.indiana.edu
	http://shigen.lab.nig.ac.jp:7081
Drosophila RNAi screening center at Harvard Medical School	http://www.flyrnai.org

hemolymph of *D. melanogaster*. These results suggest that for the interaction of *Drosophila* with *C. neoformans*, the Toll-mediated host responses, which are highly expressed following systemic infection, are not as important within the fly intestine (2).

Candida spp.

Alarco et al. have demonstrated that *Drosophila* devoid of functional Toll receptor is highly susceptible to *C. albicans* infection (1). After injection of *C. albicans*, infection in *Drosophila* spreads from the initial infection site to neighboring tissues, leading eventually to multiplication of the fungus in the *Drosophila* body. Proliferation of *C. albicans* within the fly is linked to morphological switching (i.e., during the morphological transition from a unicellular budding yeast to a filamentous form, a major virulence determinant of *C. albicans* [16, 35, 50] that is essential for mammalian pathogenesis) (1). Moreover, virulence traits involved in mammalian infection are also associated with killing of *D. melanogaster* by *C. albicans*; *Candida* strains with mutations involving genes that are essential for phenotypic switching demonstrate reduced virulence in the fly (1).

CAENORHABDITIS ELEGANS SYSTEM

Wild-type flies can clear impressive numbers of fungi (40), in some cases more than enough to cause fatal infection to a mouse (E. Mylonakis, unpublished data). This is not surprising if we consider that *Drosophila* lives in an environment rich in fungi. This is one of the reasons why workers turned to the soil-dwelling nematode *C. elegans* as an additional invertebrate host for the study of fungal pathogenesis (Table 5). There are numerous resources that assist scientists in this field (Table 6), and this system nicely complements the *Drosophila* model (Table 7).

Table 5. Specific advantages and limitations of *C. elegans* for the study of fungal pathogenesis

Advantages
RNA interference (RNAi) feeding library is available.
Nematodes are transparent and can be easily viewed with a dissecting microscope or by using Nomarski or confocal microsocopy.
The entire cell lineage of *C. elegans* has been described in exacting detail.
It is possible to create genetically identical populations by (simply) allowing a single *C. elegans* animal to produce progeny.
Limitations
The killing assay is based on feeding of the pathogen to the nematode, and systemic expansion of the pathogen within the nematode may not be part of the killing process.
The host reponse is not well studied.
The worm is unable to survive at 37°C.

Methods

The most commonly used wild-type strain of *C. elegans* for these studies is Bristol N2. The killing assay involves the simple transfer of *C. elegans* animals from a lawn of *Escherichia coli* (their most commonly used laboratory food source) to a lawn of the fungus to be tested. For example, in the *C. elegans*-*C. neoformans* killing assay, *C. neoformans* strains are inoculated into yeast peptone dextrose and grown with aeration. A 10-μl volume of the culture is spread on a 35-mm tissue culture plate containing brain heart infusion agar. The plates are then incubated overnight. An appropriate antibiotic (such as ampicillin or gentamicin) is added to the agar to selectively prevent growth of *E. coli* carried over when the worms are transferred to the yeast-containing plates. Nematodes are transferred to the lawn of *C. neoformans* via a flat-edged platinum wire pick. The plates are examined over several days for nematode viability by using a dissecting microscope. Worms are considered dead when

Table 6. Internet resources for study of *C. elegans*

Category	URL
Wormbase	http://www.wormbase.org (the Central Database Archive for *C. elegans*)
General	http://elegans.swmed.edu
Distribution of stocks of *C. elegans*	http://biosci.umn.edu/CGC/CGChomepage.htm
Worm genome sequence centers	http://www.sanger.ac.uk/Projects/C_elegans http://genome.wustl.edu/projects/celegans
Behavioral and structural anatomy of the worm	http://www.wormatlas.org
C. elegans II (CSHL Press, 1997)	http://www.ncbi.nlm.nih.gov/books/bv.fcgi?call=bv.View..ShowTOC &rid=ce2.TOC (courtesy of NCBI)

Table 7. Comparison of *D. melanogaster* and *C. elegans* systems for the study of pathogenesis

System	*D. melanogaster*	*C. elegans*
Sequenced and annotated genome	+ + +	+ + +
Short reproductive cycle	+ +	+ + +
Amenable to forward and reverse genetics	+	+ + +
Available genetic tools (RNAi library, microarrays, etc.)	+ + +	+ + +
Understanding of host response and similarity with mammal innate immunity	+ + +	+
Overall cost	+ +	+ + +
Microscopy	+	+ + +
Availability of mutants and protocols on the internet	+ + +	+ + +

they do not respond to touch with a platinum wire pick. Screening of pathogen mutant libraries to identify novel virulence traits is usually done using a three-stage screen with an increasing number of nematodes used for the killing assay at each stage (45–47) (Fig. 2).

C. elegans worms are self-fertilizing hermaphrodites. Some pathogens kill the nematodes before they can lay eggs. However, for studies of pathogens that do not prevent progeny production, nematodes must be transferred every 24 to 48 h to new plates that contain the microorganism being studied. This is a necessary procedure to prevent the original worms from being lost among the rapidly maturing progeny and allows the determination of survival of the original worms before they produce progeny.

The Worm Fights Back: *C. elegans* Response to Pathogenic Fungi

Despite possibly lacking cellular immune mechanisms, *C. elegans* does appear to have an innate immune system of defense (19). Virtual searches of the *C. elegans* genome sequence have revealed the presence of genes encoding presumptive homologues of a Toll receptor (*tol-1*), a Traf linker protein (*trf-1*), a Pelle (IRAK) kinase (*pik-1*), and a Cactus-(I-κB)-like inhibitory protein (*ikb-1*), but Rel-domain transcription factors such as *Drosophila* Dif, Dorsal, or Relish or mammalian NF-κB do not appear to be encoded in the *C. elegans* genome. Experiments using *C. elegans* pathogenesis models have failed to demonstrate a role for Toll signaling in the activation

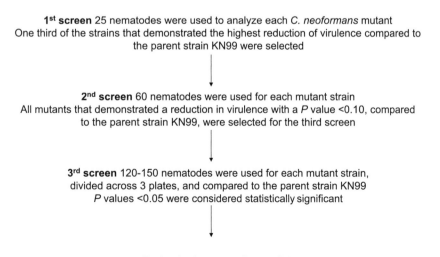

1st screen 25 nematodes were used to analyze each *C. neoformans* mutant
One third of the strains that demonstrated the highest reduction of virulence compared to the parent strain KN99 were selected

2nd screen 60 nematodes were used for each mutant strain
All mutants that demonstrated a reduction in virulence with a *P* value <0.10, compared to the parent strain KN99, were selected for the third screen

3rd screen 120-150 nematodes were used for each mutant strain, divided across 3 plates, and compared to the parent strain KN99
P values <0.05 were considered statistically significant

Evaluation in mammalian models

Figure 2. Algorithm outlining the approach for a screen of a mutant library of *C. neoformans* in *C. elegans*. Of 350 mutants tested, 7 with attenuated virulence were identified through three screens, and the attenuation persisted after the mutation was crossed back into a wild-type strain. Genetic analysis of one strain revealed an insertion in a gene homologous to *S. cerevisiae KIN1*, which encodes a serine/threonine protein kinase. *C. neoformans kin1* mutants exhibited significant defects in virulence in murine inhalation and hematogenous infection models and resulted in increased binding to alveolar and peritoneal macrophages. The *kin1* mutant phenotypes were complemented by the wild-type *KIN1* gene (47).

of a *C. elegans* innate immune response (51). On the other hand, *tol-1*, which is required for nematode development, similar to its role in *Drosophila*, plays a role in pathogen recognition and avoidance mechanisms against the pathogenic bacterium *Serratia marcescens* (51) and *C. neoformans* (Mylonakis, unpublished).

Although a *C. elegans* Toll signaling cascade does not appear to play a role in an immune response to bacterial pathogens, recently published work has shown that exposure of *C. elegans* to pathogenic bacteria does activate the transcription of genes known to be involved in defense responses in other species. Kim et al. performed a genetic screen to identify components of a presumptive *C. elegans* innate immune response pathway upstream of induced defense responses (30). This led to the finding that a *C. elegans* homolog of the mammalian p38 mitogen-activated protein kinase (53) is an important component of an apparent *C. elegans* defense response to pathogens.

Recently, two groups independently reported on the role of *tir-1*, a gene encoding an ortholog of SARM (sterile α and armadillo motifs), a highly conserved Toll-interleukin-1 resistance (TIR) domain protein, and demonstrated that TIR is required for *C. elegans* resistance to microbial pathogens (12, 39). Expression of two antimicrobial peptides, NLP-29 and NLP-31, was induced by infection with the fungus *Drechmeria coniospora* as well as by bacterial infection, and this induction was controlled in part by *tir-1* (12). Interestingly, NLP-31 (the product of gene *nlp-31*), when synthesized in vitro, was found to have strong antifungal activity, comparable to that of drosomycin, a much-studied antifungal peptide found in *Drosophila* (12).

Finally, Kato et al. demonstrated that two genes encoding the ASABF (*Ascaris suum* antibacterial factor)-type antimicrobial peptides, *abf-1* and *abf-2*, were present in *C. elegans*. Recombinant ABF-2 exhibited potent microbicidal activity against gram-positive and gram-negative bacteria and yeasts, such as *C. kruzei* (25–28). The tissue-specific distribution, estimated by immunofluorescence staining and transgenic analysis of a fusion to green fluorescent protein, suggested that ABF-2 contributes to surface defense in the pharynx (62).

Use of *C. elegans* To Study Fungal Virulence

Cryptococcus spp.

C. elegans animals grown on nonpathogenic fungal species such as *C. laurentii* and *C. kuetzingii* have a life span similar to (*C. laurentii*) or longer than (*C. kuetzingii*) those of worms fed on *E. coli* (45). However, all serotypes of *C. neoformans* kill *C. elegans* and render the nematode unable to produce progeny. *C. neoformans* cells accumulate in the intestine

of *C. elegans*, while the nonpathogenic fungus *C. laurentii* does not (Fig. 3). Of note is that acapsular mutants of *C. neoformans* kill *C. elegans* without accumulation in the intestine, suggesting that neither accumulation nor capsule is strictly necessary for killing. The *C. neoformans* polysaccharide capsule, as well as several *C. neoformans* genes previously shown to be involved in mammalian virulence, also plays a role in *C. elegans* killing (45). These include genes associated with signal transduction pathways (*GPA1*, *PKA1*, *PKR1*, and *RAS1*), laccase production (*LAC1*), and (in *C. neoformans* serotype D) the α mating type. Interestingly, a green fluorescent protein reporter fused to the mating pheromone MFα1 was expressed only in cryptococcal cells within

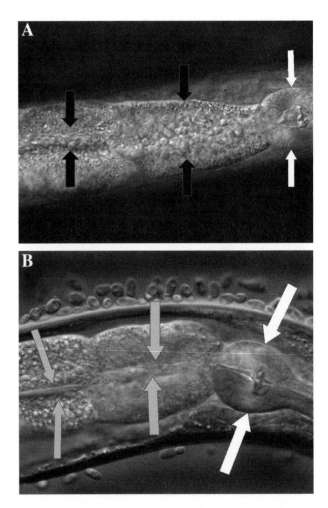

Figure 3. *C. neoformans* but not *C. laurentii* accumulates in the gastrointestinal tract of *C. elegans*. (A) Intact yeast cells present in the distended gastrointestinal tract after feeding for 24 h on *C. neoformans* strain H99. (B) No *C. laurentii* cells can be detected in the gastrointestinal tract after 24 h of feeding. The round structure (white arrows) is the pharyngeal grinder organ, which functions to disrupt ingested organisms. Black and gray arrows point to the intestinal lumen. Reprinted from reference 45 with permission.

the nematode intestine and only during the proliferative stage, analogous to expression of this gene promoter in the central nervous system of a rabbit model of cryptococcal meningitis (45).

One of the exciting aspects of the *C. elegans* model is that it can be used to screen libraries of random mutants in order to identify hypovirulent mutants that can then be evaluated in mammalian models. Recently, the first screen of fungal pathogens was completed. As was seen previously in screens of bacterial mutants in the *C. elegans* system (17, 32, 41, 58), approximately 2% of the fungal mutants tested were attenuated in *C. elegans* killing and hypovirulence persisted after the mutation was crossed back into a wild-type strain (Fig. 2). Genetic analysis of one strain revealed an insertion in a gene homologous to *S. cerevisiae KIN1*, which encodes a serine/threonine protein kinase. *C. neoformans kin1* mutants showed increased binding to alveolar and peritoneal macrophages, and this mutation resulted in significant defects in cryptococcal virulence in both the murine inhalation and hematogenous infection models (47).

Recently, Tang and coworkers screened a library of random insertional *C. neoformans* mutants for strains that permitted the production of *C. elegans* progeny and identified mutants that had a progeny-permissive phenotype. One of these strains had a mutation that was avirulent in an inhalation infection model in mice (58a). This progeny-based approach may provide an alternative approach to evaluate for fungal virulence.

Candida spp.

Preliminary results suggest that nematodes die within 48 h of feeding on lawns of several different *Candida* spp., including *C. albicans*. Interestingly, although there is no evidence for filamentation of *C. albicans* within the nematode, the *C. albicans efg1/efg1* and *efg1/efg1 cph1/cph1* mutants, known to be attenuated in systemic infections in mice, were also less virulent in *C. elegans* killing (C. Sifri, B. Enloe, E. Mylonakis, F. M. Ausubel, and S. B. Calderwood, *Abstr. Gen. Meet. 103rd Am. Soc. Microbiol. 2003*, abstr. 612, 2003).

GALLERIA MELLONELLA SYSTEM

C. elegans and *D. melanogaster* provide facile systems for the study of important elements of fungal biology and virulence. However, these systems also have practical limitations for the study of fungal pathogenesis:

- They do not allow the facile administration of exact inocula
- They can not survive at 37°C

- There are considerable difficulties in the study of antifungal agents in these models

To address these limitations, workers have turned to the greater wax moth *G. mellonella* for the study of pathogens (8, 18, 23, 42, 44, 47a, 54, 55). Highly inbred *G. mellonella* caterpillars are readily available from a variety of suppliers at a very modest cost and are large enough (about 275 to 350 mg and ca. 2.5 cm in length) to be easily inoculated by hand using a small syringe.

Methods

For virulence assays, a Hamilton syringe is typically used to inject inocula into individual *G. mellonella* larvae. Larvae can be temporarily immobilized by being placed at 4°C for 30 min before injection. An antibacterial agent is usually coadministered to prevent infection by bacteria naturally present on the surface of the larvae. After injection, larvae are incubated in plastic containers and the number of dead larvae is scored daily. A larva is considered dead when it displays no movement in response to touch. A mock inoculation with phosphate-buffered saline is performed in each experiment to monitor killing due to physical injury or infection by pathogenic contaminants.

The Moth Fights Back

The host response of *G. mellonella* to infection consists of structural and passive barriers as well as cellular and humoral responses that are performed by hemocytes within the hemolymph. Six types of hemocytes have been identified in *G. mellonella*, and the insect response includes phagocytosis by plasmatocytes and granulocytes (7) and nodulation by layers of hemocytes to encapsulate large invading pathogens (reviewed in reference 29). Interestingly, the kinetics of phagocytosis and microbial killing by *G. mellonella* hemocytes of *Candida* cells is similar to that by human neutrophils, and moth hemocytes probably have proteins homologous to gp91phox, p67phox, p47phox, and the GTP-binding protein rac 2 (4a).

Use of *G. mellonella* To Study Fungal Virulence

Candida spp.

G. mellonella can be killed by *C. albicans* and by a range of other *Candida* spp. but not by the yeast *S. cerevisiae* (11). In *C. albicans* there is significant correlation between the virulence in *G. mellonella* and mice (4, 7). More specifically, hyphal transition is associated with *Candida*-mediated killing of *G. mellonella* (15). Avirulent myosin-I-defective yeast cells are rapidly removed from the hemolymph in vivo, whereas virulent *C. albicans* strains grow rapidly in larval serum, limit nodulation, and produce toxins that damage hemocytes and the fat body. Of note is that insects infected with nonpathogenic *C. albicans* are more resistant to subsequent infection by wild-type yeast cells than were noninduced insects (15).

Aspergillus spp.

St. Leger et al. (57) used the interaction of *A. flavus* with *G. mellonella* to demonstrate the versatility of this human pathogen. Conidia of *A. flavus* were not virulent when applied to the surface of healthy caterpillars, but they killed the caterpillars (100% mortality within 48 h) when injected. Within 72 h of host death, dead caterpillars were covered in a thick coat of conidia. In contrast to *A. flavus*, *G. mellonella* was able to clear large inocula of *A. fumigatus* (57; Mylonakis, unpublished). Similar to *A. flavus*, spores of the four isolates of *A. fumigatus* studied by St. Leger et al. were rapidly phagocytosed, but unlike *A. flavus*, they failed to germinate within hemocytes and produced few (small lesions) or no symptoms, and no hyphal bodies were observed free in the hemolymph (57). Recently, Reeves et al. reported that *G. mellonella* is susceptible to *A. fumigatus* strain ATCC 26933 and suggested a role for gliotoxin, an immunosuppressive agent previously implicated in assisting tissue penetration, in the virulence of *A. fumigatus* in this model (52).

Cryptococcus spp.

C. neoformans also proliferates in *G. mellonella* and kills the caterpillars. The killing of *G. mellonella* larvae by *C. neoformans* is related to the number of CFU injected; the *C. neoformans* polysaccharide capsule, as well as several *C. neoformans* genes previously shown to be involved in mammalian virulence (*GPA1*, *PKA1*, and *RAS1*), also play a role in *G. mellonella* killing. Interestingly, there is a correlation between the efficacy of antifungals (alone or in combination) with findings from studies of humans. Administration of amphotericin B (1.5 mg/kg), flucytosine (20 mg/kg), or fluconazole (14 mg/kg) 4 h before or after inoculation prolongs survival, and the combination of amphotericin B with flucytosine is more effective than amphotericin B monotherapy (47a).

POSSIBLE NONMAMMALIAN MODELS FOR THE STUDY OF FUNGAL PATHOGENESIS

- *Anopheles gambiae*: The mosquito *A. gambiae* is the host of the malaria parasite *Plasmodium*, one of the most devastating human pathogens (9, 10, 14, 20, 31, 49).
- *Arabidopsis thaliana* (mustard weed): Significant similarities have been uncovered in the molecular mode of perception of pathogen-associated molecular patterns in

animals and plants, including the discovery of a leucine-rich repeat transmembrane plant receptor that recognizes eubacterial flagella (3, 48), making plants an interesting candidate for the study of pathogens.

- *Ciona intestinalis* (sea squirt) (13)
- *Danio rerio* (zebrafish), a vertebrate model that has adaptive immunity
- *Hyalophora cecropia* moth
- *Limulus polyphemus* (horseshoe crab) (22)
- *Manduca sexta* (tobacco hornworm) (24)
- *Periplaneta americana* (American cockroach) (61)

Multihost Pathogenesis System

An extension of the use of the individual alternative model systems described above is the use of multiple models in parallel for the study of host-pathogen interactions. This approach to cross-species study of virulence (referred to as the multihost pathogenesis system) has been used mostly in the study of bacterial pathogens. The "multihost" approach usually includes one invertebrate model for the screen and then a relevant mammalian host to confirm the phenotype (an example is detailed in Fig. 3). However, multiple nonmammalian models have been used for screens of bacterial systems, and the same approach could be used for fungi. For example, Kurz et al. used *C. elegans* to screen a bank of ca. 2,300 transposon-induced *S. marcescens* mutants and isolated 23 clones with attenuated virulence. Three of these clones exhibited decreased virulence in an insect model, and one of them was also markedly attenuated in virulence in a murine model (33).

CONCLUSIONS

The similarities between microbial pathogenesis in mammalian models and host-pathogen interactions in alternative model hosts suggest that certain virulence aspects of these pathogens may have developed during the interaction of these microbes with environmental predators (38, 56). In addition to providing a paradigm for understanding the origin and maintenance of fungal virulence, these models provide invaluable insights into host-pathogen interactions in general. Moreover, these models promote collaboration between scientists working on pathogenesis with scientists working on the biology of the invertebrate hosts. This interdisciplinary approach has the potential to advance scientific knowledge more rapidly. However, each of these alternative models provides differences, strengths, challenges, and limitations. As with every scientific endeavor, the challenge lies with the investigators who have to select the model and experimental approach to best evaluate their hypothesis.

Acknowledgments. Financial support was provided by a New Scholar Award in the Global Infectious Diseases from the Ellison Medical Foundation and a Pfizer fellowship in Medical Mycology from the Infectious Diseases Society of America, both to E.M.

REFERENCES

1. Alarco, A. M., A. Marcil, J. Chen, B. Suter, D. Thomas, and M. Whiteway. 2004. Immune-deficient *Drosophila melanogaster*: a model for the innate immune response to human fungal pathogens. *J. Immunol.* **172:**5622–5628.

2. Apidianakis, Y., L. G. Rahme, J. Heitman, F. M. Ausubel, S. B. Calderwood, and E. Mylonakis. 2004. Challenge of *Drosophila melanogaster* with *Cryptococcus neoformans* and role of the innate immune response. *Eukaryot. Cell* **3:**413–419.

3. Asai, T., G. Tena, J. Plotnikova, M. R. Willmann, W. L. Chiu, L. Gomez-Gomez, T. Boller, F. M. Ausubel, and J. Sheen. 2002. MAP kinase signalling cascade in Arabidopsis innate immunity. *Nature* **415:**977–983.

4. Bergin, D., M. Brennan, and K. Kavanagh. 2003. Fluctuations in haemocyte density and microbial load may be used as indicators of fungal pathogenicity in larvae of *Galleria mellonella*. *Microbes. Infect.* **5:**1389–1395.

4a. Bergin, D., E. P. Reeves, J. Renwick, F. B. Wientjes, and K. Kavanagh. 2005. Superoxide production in *Galleria mellonella* hemocytes: identification of proteins homologous to the NADPH oxidase complex of human neutrophils. *Infect. Immun.* **73:**4161–4170.

5. Bhabhra, R., M. D. Miley, E. Mylonakis, D. Boettner, J. Fortwendel, J. C. Panepinto, M. Postow, J. C. Rhodes, and D. S. Askew. 2004. Disruption of the *Aspergillus fumigatus* gene encoding nucleolar protein CgrA impairs thermotolerant growth and reduces virulence. *Infect. Immun.* **72:**4731–4740.

6. Brennan, C. A., and K. V. Anderson. 2004. Drosophila: the genetics of innate immune recognition and response. *Annu. Rev. Immunol.* **22:**457–483.

7. Brennan, M., D. Y. Thomas, M. Whiteway, and K. Kavanagh. 2002. Correlation between virulence of *Candida albicans* mutants in mice and *Galleria mellonella* larvae. *FEMS Immunol. Med. Microbiol.* **34:**153–157.

8. Choi, J. Y., C. D. Sifri, B. C. Goumnerov, L. G. Rahme, F. M. Ausubel, and S. B. Calderwood. 2002. Identification of virulence genes in a pathogenic strain of *Pseudomonas aeruginosa* by representational difference analysis. *J. Bacteriol.* **184:**952–961.

9. Christophides, G. K., D. Vlachou, and F. C. Kafatos. 2004. Comparative and functional genomics of the innate immune system in the malaria vector *Anopheles gambiae*. *Immunol. Rev.* **198:**127–148.

10. Christophides, G. K., E. Zdobnov, C. Barillas-Mury, E. Birney, S. Blandin, C. Blass, P. T. Brey, F. H. Collins, A. Danielli, G. Dimopoulos, C. Hetru, N. T. Hoa, J. A. Hoffmann, S. M. Kanzok, I. Letunic, E. A. Levashina, T. G. Loukeris, G. Lycett, S. Meister, K. Michel, L. F. Moita, H. M. Muller, M. A. Osta, S. M. Paskewitz, J. M. Reichhart, A. Rzhetsky, L. Troxler, K. D. Vernick, D. Vlachou, J. Volz, C. von Mering, J. Xu, L. Zheng, P. Bork, and F. C. Kafatos. 2002. Immunity-related genes and gene families in *Anopheles gambiae*. *Science* **298:**159–165.

11. Cotter, G., S. Doyle, and K. Kavanagh. 2000. Development of an insect model for the in vivo pathogenicity testing of yeasts. *FEMS Immunol. Med. Microbiol.* **27:**163–169.

12. Couillault, C., N. Pujol, J. Reboul, L. Sabatier, J. F. Guichou, Y. Kohara, and J. J. Ewbank. 2004. TLR-independent control of innate immunity in *Caenorhabditis elegans* by the TIR domain adaptor protein TIR-1, an ortholog of human SARM. *Nat. Immunol.* **5:**488–494.

13. Dehal, P., Y. Satou, R. K. Campbell, J. Chapman, B. Degnan, A. De Tomaso, B. Davidson, A. Di Gregorio, M. Gelpke, D. M. Goodstein, N. Harafuji, K. E. Hastings, I. Ho, K. Hotta, W. Huang, T. Kawashima, P. Lemaire, D. Martinez, I. A. Meinertzhagen, S. Necula, M. Nonaka, N. Putnam, S. Rash, H. Saiga, M. Satake, A. Terry, L. Yamada, H. G. Wang, S. Awazu, K. Azumi, J. Boore, M. Branno, S. Chin-Bow, R. DeSantis, S. Doyle, P. Francino, D. N. Keys, S. Haga, H. Hayashi, K. Hino, K. S. Imai, K. Inaba, S. Kano, K. Kobayashi, M. Kobayashi, B. I. Lee, K. W. Makabe, C. Manohar, G. Matassi, M. Medina, Y. Mochizuki, S. Mount, T. Morishita, S. Miura, A. Nakayama, S. Nishizaka, H. Nomoto, F. Ohta, K. Oishi, I. Rigoutsos, M. Sano, A. Sasaki, Y. Sasakura, E. Shoguchi, T. Shin-i, A. Spagnuolo, D. Stainier, M. M. Suzuki, O. Tassy, N. Takatori, M. Tokuoka, K. Yagi, F. Yoshizaki, S. Wada, C. Zhang, P. D. Hyatt, F. Larimer, C. Detter, N. Doggett, T. Glavina, T. Hawkins, P. Richardson, S. Lucas, Y. Kohara, M. Levine, N. Satoh, and D. S. Rokhsar. 2002. The draft genome of *Ciona intestinalis*: insights into chordate and vertebrate origins. *Science* **298:**2157–2167.

14. Dimopoulos, G., G. K. Christophides, S. Meister, J. Schultz, K. P. White, C. Barillas-Mury, and F. C. Kafatos. 2002. Genome expression analysis of *Anopheles gambiae*: responses to injury, bacterial challenge, and malaria infection. *Proc. Natl. Acad. Sci. USA* **99:**8814–8819.

15. Dunphy, G. B., U. Oberholzer, M. Whiteway, R. J. Zakarian, and I. Boomer. 2003. Virulence of *Candida albicans* mutants toward larval *Galleria mellonella* (Insecta, Lepidoptera, Galleridae). *Can. J. Microbiol.* **49:**514–524.

16. Felk, A., M. Kretschmar, A. Albrecht, M. Schaller, S. Beinhauer, T. Nichterlein, D. Sanglard, H. C. Korting, W. Schafer, and B. Hube. 2002. *Candida albicans* hyphal formation and the expression of the Efg1-regulated proteinases Sap4 to Sap6 are required for the invasion of parenchymal organs. *Infect. Immun.* **70:**3689–3700.

17. Garsin, D. A., C. D. Sifri, E. Mylonakis, X. Qin, K. V. Singh, B. E. Murray, S. B. Calderwood, and F. M. Ausubel. 2001. A simple model host for identifying Gram-positive virulence factors. *Proc. Natl. Acad. Sci. USA* **98:**10892–10897.

18. Hendrickson, E. L., J. Plotnikova, S. Mahajan-Miklos, L. G. Rahme, and F. M. Ausubel. 2001. Differential roles of the Pseudomonas aeruginosa PA14 *rpoN* gene in pathogenicity in plants, nematodes, insects, and mice. *J. Bacteriol.* **183:**7126–7134.

19. Hoffmann, J. A., and J. M. Reichhart. 2002. *Drosophila* innate immunity: an evolutionary perspective. *Nat. Immunol.* **3:**121–126.

20. Holt, R. A., G. M. Subramanian, A. Halpern, G. G. Sutton, R. Charlab, D. R. Nusskern, P. Wincker, A. G. Clark, J. M. Ribeiro, R. Wides, S. L. Salzberg, B. Loftus, M. Yandell, W. H. Majoros, D. B. Rusch, Z. Lai, C. L. Kraft, J. F. Abril, V. Anthouard, P. Arensburger, P. W. Atkinson, H. Baden, V. de Berardinis, D. Baldwin, V. Benes, J. Biedler, C. Blass, R. Bolanos, D. Boscus, M. Barnstead, S. Cai, A. Center, K. Chaturverdi, G. K. Christophides, M. A. Chrystal, M. Clamp, A. Cravchik,

V. Curwen, A. Dana, A. Delcher, I. Dew, C. A. Evans, M. Flanigan, A. Grundschober-Freimoser, L. Friedli, Z. Gu, P. Guan, R. Guigo, M. E. Hillenmeyer, S. L. Hladun, J. R. Hogan, Y. S. Hong, J. Hoover, O. Jaillon, Z. Ke, C. Kodira, E. Kokoza, A. Koutsos, I. Letunic, A. Levitsky, Y. Liang, J. J. Lin, N. F. Lobo, J. R. Lopez, J. A. Malek, T. C. McIntosh, S. Meister, J. Miller, C. Mobarry, E. Mongin, S. D. Murphy, D. A. O'Brochta, C. Pfannkoch, R. Qi, M. A. Regier, K. Remington, H. Shao, M. V. Sharakhova, C. D. Sitter, J. Shetty, T. J. Smith, R. Strong, J. Sun, D. Thomasova, L. Q. Ton, P. Topalis, Z. Tu, M. F. Unger, B. Walenz, A. Wang, J. Wang, M. Wang, X. Wang, K. J. Woodford, J. R. Wortman, M. Wu, A. Yao, E. M. Zdobnov, H. Zhang, Q. Zhao, et al. 2002. The genome sequence of the malaria mosquito *Anopheles gambiae*. *Science* **298**:129–149.

21. Hultmark, D. 2003. Drosophila immunity: paths and patterns. *Curr. Opin. Immunol.* **15**:12–19.

22. Inamori, K., S. Ariki, and S. Kawabata. 2004. A Toll-like receptor in horseshoe crabs. *Immunol. Rev.* **198**:106–115.

23. Jander, G., L. G. Rahme, and F. M. Ausubel. 2000. Positive correlation between virulence of *Pseudomonas aeruginosa* mutants in mice and insects. *J. Bacteriol.* **182**:3843–3845.

24. Kanost, M. R., H. Jiang, and X. Q. Yu. 2004. Innate immune responses of a lepidopteran insect, *Manduca sexta*. *Immunol. Rev.* **198**:97–105.

25. Kato, Y., T. Aizawa, H. Hoshino, K. Kawano, K. Nitta, and H. Zhang. 2002. abf-1 and abf-2, ASABF-type antimicrobial peptide genes in *Caenorhabditis elegans*. *Biochem. J.* **361**:221–230.

26. Kato, Y., M. Ito, K. Kawai, K. Nagata, and M. Tanokura. 2002. Determinants of ligand specificity in groups I and IV WW domains as studied by surface plasmon resonance and model building. *J. Biol. Chem.* **277**:10173–10177.

27. Kato, Y., and S. Komatsu. 1996. ASABF, a novel cysteine-rich antibacterial peptide isolated from the nematode *Ascaris suum*. Purification, primary structure, and molecular cloning of cDNA. *J. Biol. Chem.* **271**:30493–30498.

28. Kato, Y., K. Nagata, M. Takahashi, L. Lian, J. J. Herrero, M. Sudol, and M. Tanokura. 2004. Common mechanism of ligand recognition by group II/III WW domains: redefining their functional classification. *J. Biol. Chem.* **279**:31833–31841.

29. Kavanagh, K., and E. P. Reeves. 2004. Exploiting the potential of insects for in vivo pathogenicity testing of microbial pathogens. *FEMS Microbiol. Rev.* **28**:101–112.

30. Kim, D. H., R. Feinbaum, G. Alloing, F. E. Emerson, D. A. Garsin, H. Inoue, M. Tanaka-Hino, N. Hisamoto, K. Matsumoto, M. W. Tan, and F. M. Ausubel. 2002. A conserved p38 MAP kinase pathway in *Caenorhabditis elegans* innate immunity. *Science* **297**:623–626.

31. Kumar, S., G. K. Christophides, R. Cantera, B. Charles, Y. S. Han, S. Meister, G. Dimopoulos, F. C. Kafatos, and C. Barillas-Mury. 2003. The role of reactive oxygen species on *Plasmodium* melanotic encapsulation in *Anopheles gambiae*. *Proc. Natl. Acad. Sci. USA* **100**:14139–14144.

32. Kurz, C. L., S. Chauvet, E. Andres, M. Aurouze, I. Vallet, G. P. Michel, M. Uh, J. Celli, A. Filloux, S. De Bentzmann, I. Steinmetz, J. A. Hoffmann, B. B. Finlay, J. P. Gorvel, D. Ferrandon, and J. J. Ewbank. 2003. Virulence factors of the human opportunistic pathogen *Serratia marcescens* identified by in vivo screening. *EMBO J.* **22**:1451–1460.

33. Kurz, C. L., and J. J. Ewbank. 2003. *Caenorhabditis elegans*: an emerging genetic model for the study of innate immunity. *Nat. Rev. Genet.* **4**:380–390.

34. Lau, G. W., B. C. Goumnerov, C. L. Walendziewicz, J. Hewitson, W. Xiao, S. Mahajan-Miklos, R. G. Tompkins, L. A. Perkins, and L. G. Rahme. 2003. The *Drosophila melanogaster* toll pathway participates in resistance to infection by the gram-negative human pathogen *Pseudomonas aeruginosa*. *Infect. Immun.* **71**:4059–4066.

35. Leberer, E., K. Ziegelbauer, A. Schmidt, D. Harcus, D. Dignard, J. Ash, L. Johnson, and D. Y. Thomas. 1997. Virulence and hyphal formation of *Candida albicans* require the Ste20p-like protein kinase CaCla4p. *Curr. Biol.* **7**:539–546.

36. Lemaitre, B., E. Nicolas, L. Michaut, J. M. Reichhart, and J. A. Hoffmann. 1996. The dorsoventral regulatory gene cassette spatzle/Toll/cactus controls the potent antifungal response in *Drosophila* adults. *Cell* **86**:973–983.

37. Leulier, F., A. Rodriguez, R. S. Khush, J. M. Abrams, and B. Lemaitre. 2000. The *Drosophila* caspase Dredd is required to resist gram-negative bacterial infection. *EMBO Rep.* **1**:353–358.

38. Levitz, S. M. 2001. Does amoeboid reasoning explain the evolution and maintenance of virulence factors in *Cryptococcus neoformans*? *Proc. Natl. Acad. Sci. USA* **98**:14760–14762.

39. Liberati, N. T., K. A. Fitzgerald, D. H. Kim, R. Feinbaum, D. T. Golenbock, and F. M. Ausubel. 2004. Requirement for a conserved Toll/interleukin-1 resistance domain protein in the *Caenorhabditis elegans* immune response. *Proc. Natl. Acad. Sci. USA* **101**:6593–6598.

39a. Lionakis, M. S., R. E. Lewis, G. S. May, N. P. Wiederhold, N. D. Albert, G. Halder, and D. P. Kontoyiannis. 2005. Toll-deficient *Drosophila* flies as a fast, high-throughput model for the study of antifungal drug efficacy against invasive aspergillosis and *Aspergillus* virulence. *J. Infect. Dis.* **191**:1188–1195.

40. Loker, E. S., C. M. Adema, S. M. Zhang, and T. B. Kepler. 2004. Invertebrate immune systems—not homogeneous, not simple, not well understood. *Immunol. Rev.* **198**:10–24.

41. Mahajan-Miklos, S., M. W. Tan, L. G. Rahme, and F. M. Ausubel. 1999. Molecular mechanisms of bacterial virulence elucidated using a *Pseudomonas aeruginosa-Caenorhabditis elegans* pathogenesis model. *Cell* **96**:47–56.

42. Mak, P., D. Chmiel, and G. J. Gacek. 2001. Antibacterial peptides of the moth *Galleria mellonella*. *Acta. Biochim. Pol.* **48**:1191–1195.

43. Medzhitov, R. 2001. Toll-like receptors and innate immunity. *Nat. Rev. Immunol.* **1**:135–145.

44. Miyata, S., M. Casey, D. W. Frank, F. M. Ausubel, and E. Drenkard. 2003. Use of the *Galleria mellonella* caterpillar as a model host to study the role of the type III secretion system in *Pseudomonas aeruginosa* pathogenesis. *Infect. Immun.* **71**:2404–2413.

45. Mylonakis, E., F. M. Ausubel, J. R. Perfect, J. Heitman, and S. B. Calderwood. 2002. Killing of *Caenorhabditis elegans* by *Cryptococcus neoformans* as a model of yeast pathogenesis. *Proc. Natl. Acad. Sci. USA* **99**:15675–15680.

46. Mylonakis, E., F. M. Ausubel, R. J. Tang, and S. B. Calderwood. 2003. The art of serendipity: killing of *Caenorhabditis elegans* by human pathogens as a model of bacterial and fungal pathogenesis. *Expert Rev. Anti-Infect. Ther.* **1**:167–173.

47. Mylonakis, E., A. Idnurm, R. Moreno, J. El Khoury, J. B. Rottman, F. M. Ausubel, J. Heitman, and S. B. Calderwood. 2004. *Cryptococcus neoformans* Kin1 protein

kinase homologue, identified through a *Caenorhabditis elegans* screen, promotes virulence in mammals. *Mol. Microbiol.* **54:**407–419.

47a.Mylonakis, E., R. Moreno, J. El Khoury, A. Idnurm, J. Heitman, S. B. Calderwood, F. M. Ausubel, and A. Diener. 2005. *Galleria mellonella* as a model system to study *Cryptococcus neoformans* pathogenesis. *Infect. Immun.* **73:**3842–3850.

48. Nurnberger, T., F. Brunner, B. Kemmerling, and L. Piater. 2004. Innate immunity in plants and animals: striking similarities and obvious differences. *Immunol. Rev.* **198:**249–266.

49. Osta, M. A., G. K. Christophides, and F. C. Kafatos. 2004. Effects of mosquito genes on *Plasmodium* development. *Science* **303:**2030–2032.

50. Phan, Q. T., P. H. Belanger, and S. G. Filler. 2000. Role of hyphal formation in interactions of *Candida albicans* with endothelial cells. *Infect. Immun.* **68:**3485–3490.

51. Pujol, N., E. M. Link, L. X. Liu, C. L. Kurz, G. Alloing, M. W. Tan, K. P. Ray, R. Solari, C. D. Johnson, and J. J. Ewbank. 2001. A reverse genetic analysis of components of the Toll signaling pathway in *Caenorhabditis elegans*. *Curr. Biol.* **11:**809–821.

52. Reeves, E. P., C. G. Messina, S. Doyle, and K. Kavanagh. 2004. Correlation between gliotoxin production and virulence of *Aspergillus fumigatus* in *Galleria mellonella*. *Mycopathologia* **158:**73–79.

53. Rincon, M. 2001. MAP-kinase signaling pathways in T cells. *Curr. Opin. Immunol.* **13:**339–345.

54. Schuhmann, B., V. Seitz, A. Vilcinskas, and L. Podsiadlowski. 2003. Cloning and expression of gallerimycin, an antifungal peptide expressed in immune response of greater wax moth larvae, *Galleria mellonella*. *Arch. Insect Biochem. Physiol.* **53:**125–133.

55. Slepneva, I. A., D. A. Komarov, V. V. Glupov, V. V. Serebrov, and V. V. Khramtsov. 2003. Influence of fungal infection on the DOPA-semiquinone and DOPA-quinone production in haemolymph of *Galleria mellonella* larvae. *Biochem. Biophys. Res. Commun.* **300:**188–191.

56. Steenbergen, J. N., and A. Casadevall. 2003. The origin and maintenance of virulence for the human pathogenic fungus *Cryptococcus neoformans*. *Microbes Infect.* **5:**667–675.

57. St. Leger, R. J., S. E. Screen, and B. Shams-Pirzadeh. 2000. Lack of host specialization in *Aspergillus flavus*. *Appl. Environ. Microbiol.* **66:**320–324.

58. Tan, M. W., L. G. Rahme, J. A. Sternberg, R. G. Tompkins, and F. M. Ausubel. 1999. *Pseudomonas aeruginosa* killing of *Caenorhabditis elegans* used to identify *P. aeruginosa* virulence factors. *Proc. Natl. Acad. Sci. USA* **96:**2408–2413.

58a.Tang, R. J., J. Breger, A. Idnurm, K. Gerik, J. K. Lodge, J. Heitman, S. B. Calderwood, and E. Mylonakis. 2005. *Cryptococcus neoformans* gene involved in mammalian pathogenesis identified by a *C. elegans* progeny-based approach. *Infect. Immun.* **73:**8219–8225.

59. Tsai, H. F., Y. C. Chang, R. G. Washburn, M. H. Wheeler, and K. J. Kwon-Chung. 1998. The developmentally regulated alb1 gene of *Aspergillus fumigatus*: its role in modulation of conidial morphology and virulence. *J. Bacteriol.* **180:**3031–3038.

60. Tzou, P., E. De Gregorio, and B. Lemaitre. 2002. How *Drosophila* combats microbial infection: a model to study innate immunity and host-pathogen interactions. *Cur. Opin. Microbiol.* **5:**102–110.

61. Verrett, J. M., K. B. Green, L. M. Gamble, and F. C. Crochen. 1987. A hemocoelic *Candida* parasite of the American cockroach (Dictyoptera: Blattidae). *J. Econ. Entomol.* **80:**1205–1212.

62. Zhang, H., and Y. Kato. 2003. Common structural properties specifically found in the CSalphabeta-type antimicrobial peptides in nematodes and mollusks: evidence for the same evolutionary origin? *Dev. Comp. Immunol.* **27:**499–503.

Molecular Principles of Fungal Pathogenesis
Edited by Joseph Heitman et al.
©2006 ASM Press, Washington, D.C.

Chapter 16

Amoeba and Slime Mold: Hosts of Virulence Evolution

ARTURO CASADEVALL

Of the more than 1,500,000 estimated fungal species (21), only a few are associated with virulence to humans and animals (23). The human-pathogenic fungi comprise a phylogenetically diverse group of organisms that can also be divided depending on their major ecologic niche. The first group is found primarily in soils, vegetation, and decaying matter, includes the dimorphic fungi, and is represented by *Aspergillus* spp., *Blastomyces dermatitidis*, *Cryptococcus neoformans*, *Coccidioidis immitis*, *Histoplasma capsulatum*, and *Sporothrix schenckii*. These organisms are not considered commensal and have no apparent requirement for an animal host in their replication and survival, yet they are capable of causing life-threatening diseases in numerous animal hosts in specific circumstances that often involve impaired host immunity. The second group of pathogenic fungi is part of the human-associated flora and includes *Candida* spp., dermatophytes, and *Malassezia* spp. These fungi have an intimate association with their respective hosts, are transmitted from one host to another, and are associated with disease under conditions where their ecology is disturbed and/or impaired immunity. Both groups of fungal pathogens share their ecologic niches with other microbes and consequently compete for space and nutrients. However, given their very different ecologic sites, they are under different types of selection pressure by their environment and differ in their capacity for virulence in immunologically intact hosts.

The soil-inhabiting fungi can be recovered from their environmental niche with the capacity for virulence, a phenomenon that has been called "ready-made virulence" (7). These organisms can establish persistent and sometimes lethal infection with relatively small inocula in a variety of hosts. However, the most likely outcome of the interaction of most environmental fungi with animal hosts is asymptomatic infection without manifest disease. For example, skin reactivity studies have shown very high prevalence of infection with *H. capsulatum* in certain geographic areas, yet life-threatening histoplasmosis is a relatively rare disease in immunologically intact hosts (24). Similarly, serological studies in New York City have demonstrated a high prevalence of *C. neoformans* infection, yet cryptococcosis is rare (19). In contrast, the host-associated pathogenic fungi are usually associated with disease under conditions where host defenses are impaired. For example, *Candida albicans* is a common human commensal yet candidiasis is associated primarily with immunosuppressive states, broken integument, and disruptions of the endogenous flora caused by antimicrobial drugs. Both soil- and host-associated (commensal) fungi are in intimate contact with other microbes, including ameboid predators. In recent years numerous studies have implicated amoebae in the emergence and maintenance of virulence for various pathogenic microbes. This chapter summarizes the available information on amoeba-fungus interactions and speculates on the origins of fungal virulence for mammalian hosts.

THE AMOEBAE

The amoebae are an enormously diverse group of eukaryotic microorganisms that have in common the ability to move and feed using cellular pseudopodia (for reviews, see references 20, 37, and 49). Although this historical phenotypic description emphasizes their morphological appearance, the ameboid phenotype is found in a large polyphyletic group of organisms. Amoebae move by using pseudopods (false feet), are unicellular, and reproduce asexually by binary fission. Other organisms such as slime molds have ameboid cells that behave like amoebae in movement and feeding. There are two major types of amoebae, naked and testate, with

Arturo Casadevall • Department of Medicine, Albert Einstein College of Medicine, 1300 Morris Park Ave., Bronx, NY 10461.

the later having a mineral shell. Amoebae and slime molds are grouped under the kingdom Protozoa. Amoebae are capable of cyst formation under adverse conditions.

FUNGI AND AMOEBAE SHARE ECOLOGIC SITES

Amoebae and fungi are everywhere in the biosphere and share many ecologic sites where they have the opportunity to interact. Amoebae can be found in soils, freshwater, and oceans. River water contains large numbers of amoebae, and the prevalence of amoebae is proportional to that of bacteria, which presumably serve as food sources (15). Even arid soils contain large numbers of amoeba species, which are often found concentrated in soils containing shrubs (36). Analysis of the atmosphere in a desert city in Mexico revealed the presence of 57 strains of amoebae following 1 week of rehydration; 39% of these were members of the *Acanthamoeba* genus (38). Hence, desert-dwelling fungal species such as *C. immitis* should not be considered immune to interactions with amoebae. In fact, amoebae have been recovered from hot springs, attesting to their hardiness and thermal tolerance (41). Studies of edible mushrooms have shown that up to 96% of surfaces and internal tissues contain amoebae (27). Pathogenic and free-living amoebae are also found in the human nasal mucosa and mouth, sites that may also contain commensal fungi (35).

FUNGUS-AMOEBA INTERACTIONS

Amoebae are the major soil predators of microorganisms, especially bacteria (37). Amoebae and amoeboid cells of slime molds can ingest yeast cells, and the coexistence of these two microorganisms in many ecologic niches makes a compelling case for frequent interactions. Much of what we know about fungus-amoeba interactions comes from a handful of laboratory studies using a few species of amoebae and fungi. Some amoebae ingest yeast, whereas others do not. Comparison of algal, bacterial, and yeast food sources for the soil bacterium *Hartmannella glebae* revealed a preference for gram-negative bacteria over gram-positive bacteria and an inability to use yeasts, molds, or algae for growth (48). That study suggested that the preference for bacterial food sources reflected a lytic enzymatic activity that was not effective against yeasts and molds. In contrast, *Acanthamoeba castellanii* was initially recovered by Castellani from cultures of a nonpathogenic *Cryptococcus* sp. (9). A subsequent study with a different yeast (*Torulopsis famata*) established that *A. castellanii* ingested yeast cells and used them for food (30). Studies with pathogenic and

nonpathogenic *Naegleria* species demonstrated differences in their ability to phagocytose baker's yeast (46). When considering fungus-amoeba interactions, it is noteworthy that the number of amoeba species is estimated to be >100,000, that there is tremendous biodiversity in amoeboid species even within relatively small areas of soil, and that both the fungal and protozoal kingdoms are ancient in geological time. In fact, the interactions of pathogenic fungi and amoebae have been studied with only a handful of amoeboid species: *castellanii* (25, 44), *Dictyostelium discoideum* (42), and *Acanthamoeba polyphaga* (29). Hence, the available information on amoeba-fungus interactions is very limited relative to the potential for these types of organisms to interact in the environment.

Various studies have reported specific types of interactions and responses when amoebae and fungi are brought into contact in the laboratory. *A. castellanii* phagocytoses yeast cells through a mannose-type receptor that delivers the ingested particles to a lysosomal pathway for degradation and digestion (1) and stimulates phosphoinositide metabolism (2). *A. castellanii* can discriminate between digestible particles such as live yeast and indigestible particles such as chemically cross-linked yeast particles and latex beads (3). Yeast cells may produce soluble molecules that stimulate or interfere with amoeba phagocytic activity. Incubating *Entamoeba histolytica* with a suspension of *Saccharomyces boulardii* produced a significant inhibition of amoeba phagocytosis of erythrocytes (34). Amoeba phagocytosis can be induced by arachidonic acid and prostaglandins (33), a phenomenon that may be highly relevant to fungus-amoeba interactions since many species of fungi produce arachidonic acid derivatives and prostaglandins (31, 32). Although each of these observations was made under artificial laboratory conditions, they suggest certain specificity to amoeba-fungus interactions that buttresses the argument that these two types of organisms frequently interact in the environment.

INTERACTIONS OF AMOEBAE WITH PATHOGENIC FUNGI

In the mid-1950s, Castellani carried out an early study of the interaction of amoeba with a pathogenic yeast when he reported that *A. castellanii* phagocytosed and destroyed a strain of *C. neoformans* (10). In the late 1970s and early 1980s, Bulmer and colleagues carried out a remarkable series of studies of the interaction of amoebae and *C. neoformans* that provided much insight into this subject (5, 29). These experiments were inspired by the chance observation that attempts to recover *C. neoformans* from mouse feces often led to amoebal contamination of the isolation plates (29). After identifying the amoebae as *Acanthamoeba polyphaga*, they proceeded to study the interaction of the protozoa and fungal cells in detail (29). Incubation of *A. polyphaga* cysts with yeast cells was shown to induce rapid excystement, with the emerging trophozoites becoming rapidly engorged with ingested yeast cells as they migrated over the agar surface (29). These investigators also noted that surviving *C. neoformans* colonies were often composed of pseudohyphal cells and speculated that the yeast-hypha transition may provide an escape hatch for survival of this organism against amoeboid predators (29), similar to the role that *C. albicans* germ tube formation and filamentous growth play in escape from macrophages. The colonies of pseudohyphal cells were heterogeneous, and all were significantly less virulent for mice than was the parental strain (29). On the basis of these studies, Bulmer and his collaborators proposed that amoebae were important for the control of *C. neoformans* in the environment (39, 40).

Another impetus for the study of amoeba-fungus interactions came from the observation that *C. neoformans* had a unique and remarkable intracellular survival strategy in macrophages (44). The observations that *C. neoformans* (i) could replicate inside macrophages from various mammalian species (4, 13, 16); (ii) accumulated vesicles filled with polysaccharide during intracellular survival, indicating a unique pathogenic strategy (16, 47); and (iii) is a free-living organism which was pathogenic without an obvious need for animal host in either replication or survival suggested that environmental pressures were responsible for the emergence and maintenance of virulence. Furthermore, the idea that amoebae could be involved in this process was suggested by the fact that amoebae had been implicated in the virulence of such diverse bacteria as *Legionella pneumophila* and *Mycobacterium avium* (11, 12, 45).

In the late 1990s, our laboratory revisited the question of amoeba-fungus relationships and their potential impact in fungal virulence by analyzing the interaction of *A. castellanii* and *C. neoformans* (44). *A. castellanii* was chosen for study because this amoeba could be grown in axenic cultures and was already being extensively used to study interactions with *L. pneumophila*. For these experiments, phosphate-buffered saline was chosen as the medium since this buffered solution could not support either amoebal or fungal growth and since nutrients obtained by either contender would have to come from the other species. Incubation of amoebae with *C. neoformans* led to phagocytosis and ingestion of the yeast cells by the amoebae, but the interaction

usually resulted in fungal growth and death of the amoebae (44). Analysis of amoebal cells with ingested *C. neoformans* revealed the formation of intracellular vesicles that were highly reminiscent of those found in macrophages. This suggested that *C. neoformans* undermined amoebae by using a similar strategy to that used for intracellular survival in mammalian cells (44).

The connection between amoeba-*C. neoformans* interactions and mammalian virulence was further enhanced by the demonstration that three fungal phenotypes associated with virulence were also important for survival in amoebae: the polysaccharide capsule, phospholipase activity, and melanin (44). Encapsulated *C. neoformans* cells were able to grow in the presence of *A. castellanii*, whereas nonencapsulated cells were killed. *C. neoformans* phospholipase-deficient mutants were unable to grow in the presence of amoebae, whereas mutants complemented with the wild-type *PLB1* gene manifested growth comparable to that of the parental strain. A role for melanin was more difficult to establish since there was no apparent difference between the growth of melanized and nonmelanized encapsulated yeast cells in the presence of amoebae. However, melanin did provide a survival advantage to nonencapsulated cells when these were incubated with amoebae.

The results obtained by Steenbergen et al. (44) differed from those reported by other investigators (10, 29), in that the amoeba were killed by their interaction with *C. neoformans*. There are several potential explanations for this result. First, both the *A. castellanii* and *C. neoformans* strains used by Steenbergen et al. (44) had been maintained under laboratory conditions for years, and the amoeba was adapted to growth in axenic conditions. Second, Bulmer and colleagues used *A. polyphaga*, which is a different species from *A. castellanii* (5, 29). Third, the interaction between amoebae and fungi took place in saline, where the amoebae may have been at a relative disadvantage. Whatever the explanation, the different results serve to illustrate the fact that the outcome of fungus-amoeba interactions may vary depending on the species involved and possibly the individual strain used. In this regard, we note that a follow-up study analyzing the interaction of *C. neoformans* var. *gattii* with *A. castellanii* gave significantly different results from those obtained with the *C. neoformans* var. *neoformans* and *grubbii* strains (25). Specifically, phagocytosis of *C. neoformans* var. *gattii* cells was rare, a finding that was attributed to structural differences in the polysaccharide capsule (25). Hence, caution should be exercised when generalizing the findings from any one system consisting of amoeboid cells and fungi. Nevertheless, these studies illustrate the wide variety of outcomes that are possible in amoeba-fungus interactions and suggest the likelihood of great diversity in the types of interactions that can be expected between protozoa and fungi, given the large numbers of species of each kingdom that share ecologic niches.

The outcome of the interaction of *A. castellanii* with three dimorphic pathogenic fungi, *Blastomyces dermatitidis*, *Sporothrix schenckii*, and *Histoplasma capsulatum*, has also been reported (43). Like *C. neoformans*, each of these organisms killed and utilized *A. castellanii* as a source of food, but they differed in the degree to which the individual fungal species were ingested. *H. capsulatum* (Fig. 1) and *S. schenckii* were readily phagocytosed by *A. castellanii*, while only a

small number of *B. dermatitidis* yeast cells were ingested. In contrast, *A. castellanii* killed cells of *C. albicans* and a laboratory-adapted strain of *Saccharomyces cerevisiae*. The susceptibility of *C. albicans* to amoeba predation may reflect the fact that, as a human commensal adapted for survival among the endogenous microflora, it has not been selected to possess characteristics for resistance to amoebae. The susceptibility of the *S. cerevisiae* strain could have reflected many years of adaptation to the rich media used under laboratory conditions or an inherent vulnerability of this fungal species to amoeba predation. The potential association between these types of interactions and the emergence of virulence for mammals was strengthened by the observation that passage of an avirulent *H. capsulatum* isolate in *A. castellanii*

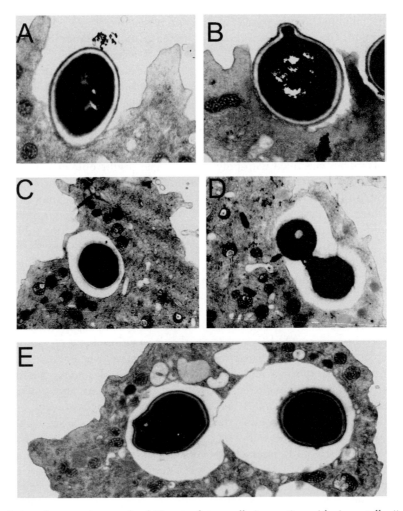

Figure 1. Transmission electron micrograph of *H. capsulatum* cells interacting with *A. castellanii*. (A and B) Two separate phagocytic events occurring 2 h after incubation of fungal cells with amoebae. (C and D) Yeast cells in a membrane-bound vacuole surrounding the fungal cell 2 h after infection of the amoeba suspension with fungal cells. (E) Two individual *H. capsulatum* fungal cells in separate phagocytic compartments, indicating two independent phagocytic events. Magnification, ×15,000 (panels A, B, and E) and ×12,000 (panels C and D). Reprinted from reference 43 with permission.

cells restored virulence as measured by the ability of the amoeba-passaged fungal cells to establish themselves in the lungs of mice and elicit significant inflammation.

A central problem in fungal pathogenesis is the phenomenon of dimorphism, whereby the organism can change between yeast and hyphal growth depending on external stimuli such as temperature. For many dimorphic fungi, the temperature trigger for the yeast-hypha transition coincides with ambient and mammalian temperatures such that these organisms produce hyphae and conidia at 25°C and yeast cells at 37°C. Hence, at mammalian temperatures, infection by dimorphic fungi usually results in yeast cells in tissue whereby in soils hyphal forms presumably predominate. An intriguing observation made during the studies of the interaction of dimorphic fungi was that coincubation of *H. capsulatum*, *S. schenckii*, or *B. dermatitidis* yeast with *A. castellanii* induced an increase in hyphal cells even when the protozoa and fungi were incubated at 37°C (43). For example, the proportion of *H. capsulatum* exhibiting a filamentous (hyphal, pseudohyphal) phenotype at 37°C increased to 12.5% at 24 h and 22.5% at 48 h whereas no comparable change was observed under conditions without amoebae (43). This observation, combined with the earlier finding that pseudohyphal *C. neoformans* was relatively resistant to predation by *A. polyphaga* (28), suggests that the phenomenon of fungal dimorphism may have been selected in evolution as a mechanism of escape from certain forms of predation by other microorganisms (6). For *C. neoformans*, the finding that the yeast-hypha transition can be a manifestation of phenotypic switching (18) suggests the possibility that pressures from amoeboid predators may have also contributive

powerful selective pressures for the emergence of the phenomenon of phenotypic switching.

SLIME MOLD-*C. NEOFORMANS* INTERACTIONS

D. discoideum is a social amoeba that lies in the interface between unicellular and multicellular organisms since the ameboid cells can aggregate to form multicellular stalk structures. This organism has proven to be an extremely useful system for the study of various cellular processes such as chemotaxis and cell-to-cell signaling because it is haploid and genetically tractable and has an advanced molecular toolbox. Analysis of the interaction of *D. discoideum* amoeboid cells with *C. neoformans* led to phagocytosis (Fig. 2) and produced results that closely replicated those observed with *A. castellanii* (42). Like the experience with *A. castellanii*, acapsular mutants of *C. neoformans* were not virulent for *D. discoideum*, highlighting the potential importance of the capsule in defense against amoeboid predators. However, when *D. discoideum* mutants defective in myosin VII expression or endosomal vesicle fusion were used as the host cells, the acapsular mutants were able to grow at the expense of the slime mold cells. This phenomenon provides a correlate for the phenomenon of microbial opportunism that has been described for animals with impaired immune system at the unicellular level. Presumably, these mutations impaired host defenses of *D. discoideum* cells such that they became vulnerable to acapsular cryptococci, which are not virulent for wild-type *D. discoideum* cells. These experiments illustrate the promise of this system for dissecting yeast cell interactions with phagocytic amoeboid cells and suggest that the identification of genes involved in the susceptibility and

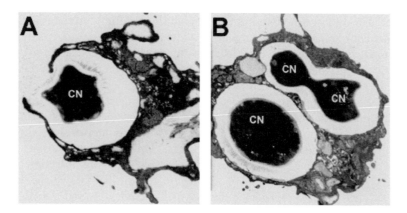

Figure 2. Transmission electron micrographs of *C. neoformans* cells interacting with *D. discoideum*. (A) *C. neoformans* is being engulfed by the pseudopods of a *D. discoideum* cell after 2 h of incubation of fungal and amoeboid cells. (B) Two individual phagocytic events by one *D. discoideum* cell 2 h postinfection. In one event, the *C. neoformans* cells are in membrane-bound vacuoles. The second event shows a budding *C. neoformans* cell in a vacuole. Magnification, ×24,000 (panel A) and ×18,000 (panel B). Reprinted from reference 42 with permission.

resistance of *D. discoideum* mutants could lead to the identification of mammalian homologs. One limitation of the *D. discoideum* system is that experiments must be conducted at ambient temperatures and that this precludes the study of genes expressed at the elevated temperatures found in mammals.

Passage of *C. neoformans* under laboratory conditions is sometimes associated with a loss of virulence for mice (17). Passage of an attenuated *C. neoformans* strain in culture with *D. discoideum* resulted in a marked increase in virulence for mice (42). Although the mechanism for increased virulence was not elucidated, passaged cells were noted to have significantly enlarged capsules and to manifest more rapid melanization than did the nonpassaged, less virulent strain. Hence, similar to the observation that passage of avirulent *H. capsulatum* in *A. castellanii* enhanced virulence (43), the observation that such a complex phenotype as virulence can be affected for *C. neoformans* by simple passage in amoeboid slime mold cells provides strong evidence implicating these interactions in the emergence of fungal virulence.

"DUAL-USE" VIRULENCE FACTORS: THE *C. NEOFORMANS* PARADIGM

C. neoformans is a soil- and possibly tree-associated fungus which has several well-defined virulence factors despite having no obvious requirement for animal virulence in replication or survival. For a microbe to be virulent in a mammalian host, it must be capable of damaging the host (8). *C. neoformans* is a nonspecific pathogen because it can cause disease in numerous hosts including many mammalian species and birds. Furthermore, *C. neoformans* can be recovered from its natural habitat with the capacity for mammalian virulence, and early investigators would simply inject environmental samples into mice to recover cryptococci. The mouse immune system would clear all the other contaminating microorganisms but not *C. neoformans*, which would establish itself in mouse tissues and could then be recovered in pure culture. These observations present a paradox since virulence is a complex trait and one would not expect that it would be maintained in the environment without strong selective pressures (6, 7). However, the observations with amoebae (44), slime molds (42), and nematodes (26) suggest that traits associated with virulence for mammals also play a role in promoting the survival of *C. neoformans* following interactions with environmental predators. According to this view, virulence factors for the soil-dwelling pathogenic fungi have "dual use" by enhancing the probability of day-to-day survival in the fungal environmental ecologic niche and also

allowing survival in the relatively unusual circumstances whereby the fungus would find itself in an animal host (7). As such, fungal virulence factors would originate under very different types of selection pressures from those that would be expected for host-associated and host-dependent pathogens, which require virulence factors for survival in infected hosts. Hence, the immunological studies reveal that the capsule of *C. neoformans* is antiphagocytic and may have similar functions in virulence to the capsule of pathogenic bacteria, but it is likely that this represents an example of convergent evolution whereby their functions are coincident since amoeboid predators utilize phagocytosis for nutrition and host immune cells rely on phagocytosis for host defense.

THOUGHTS ON THE ORIGIN AND MAINTENANCE OF VIRULENCE FOR PATHOGENIC FUNGI

Amoeba-fungus interactions cannot be the sole explanation of the origin of mammalian virulence among the soil-dwelling pathogenic fungi, given that only a small number of fungal species are pathogenic, that amoebae are ubiquitous in the environment, and that amoeba-fungus interactions must be exceedingly frequent. Clearly, one necessary characteristic for mammalian virulence is the ability of the microbe to survive and grow at 37°C. The importance of temperature as an animal antifungal defense strategy is illustrated by the observation that frogs can be cured of lethal infection by the chrytrid pathogen *Batrachochytrium dendrobatidis* by raising their temperatures to 37°C (22). Similarly, locusts can control infection by *Metarhizium anisopliae* var. *acridum* by raising body temperatures through behavioral changes (14). Since most soil fungi grow best at ambient temperatures, the vertebrate endothermy found in mammals almost certainly provides a barrier to pathogenicity for the overwhelming majority of fungal species (6). Similarly, the preference of most fungal species to grow best under acidic conditions suggests that the slightly alkaline pH of mammalian tissues is a less than favorable environment (6). Furthermore, attachment to host tissues is an essential component of successful infection, and adhesions have been described in many of the pathogenic fungi. Finally, infecting fungi must be able to resist the onslaught of the host immune system, which includes layered defenses consisting of innate and adaptive immunity, each with humoral and cellular components. Fungi that are under selection by amoeba predation, that grow at mammalian temperatures, that tolerate slightly alkaline pH, and that have suitable adhesions may be able to establish themselves in a mammalian host (6). In this regard, the role of amoeba predation in the maintenance of virulence may be to provide selection pressures which lead to the emergence of fungal traits that are also useful for surviving the onslaught of phagocytic cells that accompanies an inflammatory response.

REFERENCES

1. Allen, P. G., and E. A. Dawidowicz. 1990. Phagocytosis in *Acanthamoeba*. I. A mannose receptor is responsible for the binding and phagocytosis of yeast. *J. Cell. Physiol.* 145:508–513.

2. Allen, P. G., and E. A. Dawidowicz. 1990. Phagocytosis in *Acanthamoeba*. II. Soluble and insoluble mannose-rich ligands stimulate phosphoinositide metabolism. *J. Cell. Physiol.* 145:514–521.

3. Bowers, B., and T. E. Olszewski. 1983. *Acanthamoeba* discriminates internally between digestible and indigestible particles. *J. Cell. Biol.* 97:317–322.

4. Bulmer, G. S., and J. R. Tacker. 1975. Phagocytosis of *Cryptococcus neoformans* by alveolar macrophages. *Infect. Immun.* 11:73–79.

5. Bunting, L. A., J. B. Neilson, and G. S. Bulmer. 1979. *Cryptococcus neoformans*: gastronomic delight of a soil ameba. *Sabouraudia* 17:225–232.

6. Casadevall, A. 2005. Fungal virulence, vertebrate endothermy, and dinosaur extinction: is there a connection? *Fungal Genet. Biol.* 42:2132–2142.

7. Casadevall, A., J. D. Nosanchuk, and J. N. Steenbergen. 2003. 'Ready-made' virulence and 'dual-use' virulence factors in pathogenic enviromental fungi—the *Cryptococcus neoformans* paradigm. *Curr. Opin. Microbiol.* 112:1164–1175.

8. Casadevall, A., and L. Pirofski. 2003. The damage-response framework of microbial pathogenesis. *Nat. Microbiol. Rev.* 1:17–24.

9. Castellani, A. 1931. An amoeba growing in cultures of a yeast. *J. Trop. Med. Hyg.* 33:188–191.

10. Castellani, A. 1955. Phagocytic and destructive action of *Hartmanella castellanii (Amoeba castellanii)* on pathogenic encapsulated yeast-like fungus *Torulopsis neoformans (Cryptococcus neoformans)*. *Ann. Inst. Pasteur* 89:1–7.

11. Cirillo, J. D., S. L. Cirillo, L. Yan, L. E. Bermudez, S. Falkow, and L. S. Tompkins. 1999. Intracellular growth in *Acanthamoeba castellanii* affects monocyte entry mechanisms and enhances virulence of *Legionella pneumophila*. *Infect. Immun.* 67:4427–4434.

12. Cirillo, J. D., S. Falkow, L. S. Tompkins, and L. E. Bermudez. 1997. Interaction of *Mycobacterium avium* with environmental amoebae enhances virulence. *Infect. Immun.* 65:3759–3767.

13. Diamond, R. D., and J. E. Bennett. 1973. Growth of *Cryptococcus neoformans* within human macrophages in vitro. *Infect. Immun.* 7:231–236.

14. Elliot, S. L., S. Blanford, and M. B. Thomas. 2002. Host-pathogen interactions in a varying environment: temperature, behavioural fever and fitness. *Proc. R. Soc. Lond. Ser. B.* 269:1599–1607.

15. Ettinger, M. R., S. R. Webb, S. A. Harris, S. P. McIninch, C. Garman, and B. L. Brown. 2003. Distribution of free-living amoebae in James River, Virginia, USA. *Parasitol. Res.* 89:6–15.

16. Feldmesser, M., Y. Kress, P. Novikoff, and A. Casadevall. 2000. *Cryptococcus neoformans* is a facultative intracellular pathogen in murine pulmonary infection. *Infect. Immun.* 68:4225–4237.

17. Franzot, S. P., J. Mukherjee, R. Cherniak, L. Chen, J. S. Hamdan, and A. Casadevall. 1998. Microevolution of a standard strain of *Cryptococcus neoformans* resulting in differences in virulence and other phenotypes. *Infect. Immun.* 66:89–97.

18. Fries, B. C., D. L. Goldman, R. Cherniak, R. Ju, and A. Casadevall. 1999. Phenotypic switching in *Cryptococcus neoformans* strain 24067A associated with changes in virulence, polysaccharide structure, and cellular morphology. *Infect. Immun.* 67:6076–6083.

19. Goldman, D. L., H. Khine, J. Abadi, D. J. Lindenberg, L. Pirofski, R. Niang, and A. Casadevall. 2001. Serologic evidence for *Cryptococcus* infection in early childhood. *Pediatrics* 107:E66.

20. Greub, G., and D. Raoult. 2004. Microorganisms resistant to free-living amoebae. *Clin. Microbiol. Rev.* 17:413–433.

21. Hawksworth, D. L. 2001. The magnitude of fungal diversity: the 1.5 million species estimate revisited. *Mycol. Res.* 105:1422–1432.

22. Kluger, M. J., W. Kozak, C. A. Conn, L. R. Leon, and D. Soszynski. 1998. Role of fever in disease. *Ann. N. Y. Acad. Sci.* 856:224–233.

23. Kwon-Chung, K. J., and J. E. Bennett. 1992. *Medical Mycology*. Lea & Febiger, Philadelphia, Pa.

24. Leggiadro, R. J., G. S. Luedtke, A. Convey, L. Gibson, and F. F. Barrett. 1991. Prevalence of histoplasmosis in a midsouthern population. *South. Med. J.* 84: 1360–1361.

25. Malliaris, S. D., J. N. Steenbergen, and A. Casadevall. 2004. *Cryptococcus neoformans* var. *gattii* can exploit *Acanthamoeba castellanii* for growth. *Med. Mycol.* 42:149–158.

26. Mylonakis, E., F. M. Ausubel, J. R. Perfect, J. Heitman, and S. B. Calderwood. 2002. Killing of *Caenorhabditis elegans* by *Cryptococcus neoformans* as a model of yeast pathogenesis. *Proc. Natl. Acad. Sci. USA* 99:15675–15680.

27. Napolitano, J. J. 1982. Isolation of amoebae from edible mushrooms. *Appl. Environ. Microbiol.* 44:255–257.

28. Neilson, J. B., R. A. Fromtling, and G. S. Bulmer. 1981. Pseudohyphal forms of *Cryptococcus neoformans*: decreased survival in vivo. *Mycopathologia* 73:57–59.

29. Neilson, J. B., M. H. Ivey, and G. S. Bulmer. 1978. *Cryptococcus neoformans*: pseudohyphal forms surviving culture with *Acanthamoeba polyphaga*. *Infect. Immun.* 20:262–266.

30. Nero, L. C., M. G. Tarver, and L. R. Hedrick. 1964. Growth of *Acanthamoeba castellani* with the yeast *Torulopsis famata*. *J. Bacteriol.* 87:220–225.

31. Noverr, M. C., J. R. Erb-Downward, and G. B. Huffnagle. 2003. Production of eicosanoids and other oxylipins by pathogenic eukaryotic microbes. *Clin. Microbiol. Rev.* 16:517–533.

32. Noverr, M. C., G. B. Toews, and G. B. Huffnagle. 2002. Production of prostaglandins and leukotrienes by pathogenic fungi. *Infect. Immun.* 70:400–402.

33. Prusch, R. D., S. M. Goette, and P. Haberman. 1989. Prostaglandins may play a signal-coupling role during phagocytosis in *Amoeba proteus*. *Cell Tissue Res.* 255:553–557.

34. Rigothier, M. C., J. Maccario, and P. Gayral. 1994. Inhibitory activity of *Saccharomyces* yeasts on the adhesion of *Entamoeba histolytica* trophozoites to human erythrocytes in vitro. *Parasitol. Res.* 80:10–15.

35. Rivera, F., F. Medina, P. Ramirez, J. Alcocer, G. Vilaclara, and E. Robles. 1984. Pathogenic and free-living protozoa cultured from the nasopharyngeal and oral regions of dental patients. *Environ. Res.* 33:428–440.

36. Robinson, B. S., S. S. Bamforth, and P. J. Dobson. 2002. Density and diversity of protozoa in some arid Australian soils. *J. Eukaryot. Microbiol.* 49:449–453.

37. Rodriguez-Zaragoza, S. 1994. Ecology of free-living amoebae. *Crit. Rev. Microbiol.* 20:225–241.

38. Rodriguez-Zaragoza, S., F. Rivera, P. Bonilla, E. Ramirez, E. Gallegos, A. Calderon, R. Ortiz, and D. Hernandez. 1993. Amoebological study of the atmosphere of San Luis Potosi, SLP, Mexico. *J. Expo. Anal. Environ. Epidemiol.* 3(Suppl. 1):229–241.

39. Ruiz, A., R. A. Fromtling, and G. S. Bulmer. 1981. Distribution of *Cryptococcus neoformans* in a natural site. *Infect. Immun.* 31:560–563.

40. Ruiz, A., J. B. Neilson, and G. S. Bulmer. 1982. Control of *Cryptococcus neoformans* in nature by biotic factors. *Sabouraudia* 20:21–29.

41. Sheehan, K. B., J. A. Fagg, M. J. Ferris, and J. M. Henson. 2003. PCR detection and analysis of the free-living amoeba *Naegleria* in hot springs in Yellowstone and Grand Teton National Parks. *Appl. Environ. Microbiol.* 69:5914–5918.

42. Steenbergen, J. N., J. D. Nosanchuk, S. D. Malliaris, and A. Casadevall. 2003. *Cryptococcus neoformans* virulence is enhanced after intracellular growth in the genetically malleable host *Dictyostelium discoideum*. *Infect. Immun.* 71:4862–4872.

43. Steenbergen, J. N., J. D. Nosanchuk, S. D. Malliaris, and A. Casadevall. 2004. Interaction of *Blastomyces dermatitidis*, *Sporothrix schenckii*, and *Histoplasma capsulatum* with *Acanthamoeba castellanii*. *Infect. Immun.* 72:3478–3488.

44. Steenbergen, J. N., H. A. Shuman, and A. Casadevall. 2001. *Cryptococcus neoformans* interactions with amoebae suggest an explanation for its virulence and intracellular pathogenic strategy in macrophages. *Proc. Natl. Acad. Sci. USA* 98:15245–15250.

45. Swanson, M. S., and B. K. Hammer. 2000. *Legionella pneumophila* pathogenesis: a fateful journey from amoebae to macrophages. *Annu. Rev. Microbiol.* 54:567–613.

46. Thong, Y. H., A. Ferrante, and C. Shepherd. 1978. Phagocytic behaviour towards baker's yeast distinguishes pathogenic from non-pathogenic *Naegleria*. *Trans. R. Soc. Trop. Med. Hyg.* 72:207–209.

47. Tucker, S. C., and A. Casadevall. 2002. Replication of *Cryptococcus neoformans* in macrophages is accompanied by phagosomal permeabilization and accumulation of vesicles containing polysaccharide in the cytoplasm. *Proc. Natl. Acad. Sci. USA* 99:3165–3170.

48. Upadhyay, J. M. 1968. Growth and bacteriolytic activity of a soil amoeba, *Hartmannella glebae*. *J. Bacteriol.* 95:771–774.

49. Winiecka-Krusnell, J., and E. Linder. 2001. Bacterial infections of free-living amoebae. *Res. Microbiol.* 152:613–619.

Molecular Principles of Fungal Pathogenesis
Edited by Joseph Heitman et al.
©2006 ASM Press, Washington, D.C.

Chapter 17

Aspergillus nidulans: a Model for Elucidation of *Aspergillus fumigatus* Secondary Metabolism

Nancy Keller

Aspergillus is a genus of intense biological, industrial, agricultural, and medicinal importance. This genus represents a large family of fungi with over 185 recognized species (55). Members are distributed worldwide and occupy diverse ecological niches (38, 55). Most species are saprophytes that grow on a large number of substrates from plant and animal waste to pesticides and plasticizers, and thus they are very important in nutrient cycling and detoxification (38). However, a select few of these organisms are potent pathogens of animals and plants. Invasive pulmonary infection and dissemination to other end organs represent the most common and highly lethal *Aspergillus* disease state in immunocompromised humans. *Aspergillus* species cause additional morbidity in immunocompetent people, being implicated as a cause of fungal sinusitis, asthma, and allergic bronchopulmonary aspergillosis. The species responsible for more than 90% of human disease is *Aspergillus fumigatus* (14, 42, 49, 60).

A. *fumigatus* pathogenicity attributes include a combination of factors secreted from growing mycelia and terminal hyphal cells, cell wall structural components (including hydrophobins), and pigments that confer resistance to phagocytic killing (33–35, 40, 65–67). One set of putative *A. fumigatus* virulence attributes includes mycotoxins, secreted fungal secondary metabolites that are harmful to humans and animals (1, 46). Various metabolites produced by *A. fumigatus* which may act as toxins in the host cell include fumagillin, helvolic acid, fumitremorgins, phthioic acid, and gliotoxin (see http://www.aspergillus.man.ac.uk/indexhome.htm and references therein). Gliotoxin in particular has been fingered as a likely virulence factor due to its cytotoxic (22), genotoxic (47), and apoptosis-stimulating properties (39, 62).

Whereas *A. fumigatus* secondary metabolism has received little attention, great strides in understanding the molecular genetics of *Aspergillus* secondary metabolism have been achieved with the model *Aspergillus nidulans*. This fungus is one of the best described eukaryotic genetic systems and has been used to decipher the biology of the cell cycle, pathogenicity, drug resistance, human disease, and primary and secondary metabolism, among other topics. The available genomic sequence (http://www-genome.wi.mit.edu/annotation/fungi/aspergillus/index.html), useful vectors and DNA libraries (www.fgsc.net), and the existence of a sexual cycle (rare in *Aspergillus*) have greatly contributed to the ease of genetic manipulation of this species. This amenability of *A. nidulans* to genetic analysis makes it a powerful tool for examining important questions about the development and metabolism of *Aspergillus* species. This chapter reviews the contributions made by *A. nidulans* to the understanding of fungal secondary metabolism and the ways in which advances in our understanding of this species have spurred parallel studies of *A. fumigatus*.

FUNGAL SECONDARY METABOLISM

The most thorough insight into fungal secondary-metabolite regulation has arisen from studies of the mycotoxin sterigmatocystin (ST) (6) and the antibiotic penicillin (4) in *A. nidulans*. As covered below, critical advances in our understanding of fungal secondary metabolism include the discovery of penicillin (4) and ST biosynthetic gene clusters (6) and the discovery of a G-protein-mediated growth pathway regulating secondary-metabolite production in *A. nidulans* (32, 63). It is now apparent that structural genes required for secondary-metabolite production are usually clustered (37), that the regulation of the clustered genes is largely dependent on pathway-specific transcription factors (17, 51, 52) and global regulators (3), and that G-protein regulation of fungal secondary metabolism is

Nancy Keller • Department of Plant Pathology, University of Wisconsin–Madison, Madison, WI 53706.

likely to be a conserved phenomenon coupled with fungal sporulation (reviewed in references 9 and 63).

Cluster Motif

With possibly the exception of the penicillin metabolic cluster, the most thoroughly examined fungal secondary-metabolite gene clusters are those involved in mycotoxin biosynthesis, particularly the aflatoxin (AF) and ST biosynthetic clusters found in several *Aspergillus* spp. (Fig. 1A). Both carcinogenic metabolites are products of the same lengthy pathway, in which ST is the penultimate precursor of AF (6, 73). The AF cluster in *A. parasiticus* and *A. flavus* contains nearly 30 genes that constitute a cluster spanning more than 70 kb. Among these genes, 21 have been verified or predicted to encode biosynthetic enzymes, including fatty acid synthases, a polyketide synthase, monooxygenases, reductases, dehydrogenases, methyltransferases, an esterase, a desaturase, and an oxidase (45, 73). One of the genes in the cluster, *aflR*, encodes a binuclear zinc cluster [$Zn(II)_2Cys_6$] transcription factor regulating the transcription of the AF biosynthetic genes (12, 72). In *A. nidulans*, the 60-kb ST gene cluster consists of ca. 25 genes also regulated by *aflR* (6, 17, 74). The functions of most of the ST cluster genes have been determined and have shown that these genes are orthologs of AF cluster genes (31).

A.

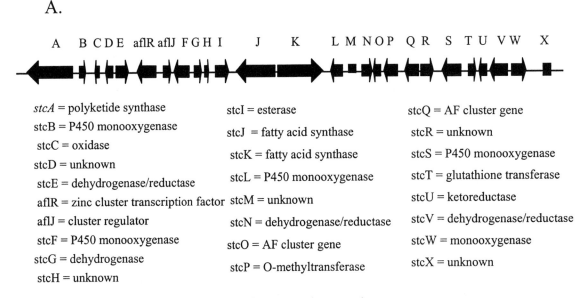

stcA = polyketide synthase
stcB = P450 monooxygenase
stcC = oxidase
stcD = unknown
stcE = dehydrogenase/reductase
aflR = zinc cluster transcription factor
aflJ = cluster regulator
stcF = P450 monooxygenase
stcG = dehydrogenase
stcH = unknown

stcI = esterase
stcJ = fatty acid synthase
stcK = fatty acid synthase
stcL = P450 monooxygenase
stcM = unknown
stcN = dehydrogenase/reductase
stcO = AF cluster gene
stcP = O-methyltransferase

stcQ = AF cluster gene
stcR = unknown
stcS = P450 monooxygenase
stcT = glutathione transferase
stcU = ketoreductase
stcV = dehydrogenase/reductase
stcW = monooxygenase
stcX = unknown

A. nidulans sterigmatocystin gene cluster

B.

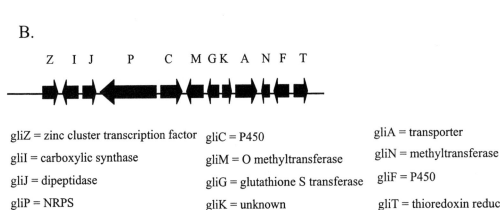

gliZ = zinc cluster transcription factor
gliI = carboxylic synthase
gliJ = dipeptidase
gliP = NRPS

gliC = P450
gliM = O methyltransferase
gliG = glutathione S transferase
gliK = unknown

gliA = transporter
gliN = methyltransferase
gliF = P450
gliT = thioredoxin reductase

A. fumigatus gliotoxin gene cluster

Figure 1. *A. nidulans* ST (A) and *A. fumigatus* gliotoxin (B) gene clusters. The function of each protein is listed. AF, aflatoxin.

The cluster motif has been useful in predicting secondary-metabolite gene clusters in the *A. nidulans* and *A. fumigatus* genomes, including the gliotoxin cluster as described below. Identification of signature secondary-metabolite enzymes (e.g., polyketide synthetases and nonribosomal peptide synthetases) in a contiguous array with other predictable cluster genes [e.g., oxidoreductases, methylases, and $Zn(II)_2Cys_6$ transcription factors] is a potent genomic motif of a natural-product pathway.

Transcriptional Regulation

Genes in secondary-metabolite clusters must be coregulated for obvious reasons of assembly line production. Coordinate regulation is largely explained by transcriptional control by pathway-specific regulatory factors (e.g., *aflR*) and global regulatory proteins including transcription factors mediating environmental signals (pH, carbon, and nitrogen) and the cluster-specific methylase, LaeA. This multilevel regulation by both specific- and broad-domain transcription factors ensures that secondary-metabolite pathways can respond to the demands of general cellular metabolism and the presence of specific pathway inducers. Here we focus on regulation by pathway-specific regulators and LaeA.

Pathway-specific transcription factors

Many but not all of the clusters contain genes encoding transcription factors that positively regulate gene expression. Perhaps the archetypal protein in this group is AflR, the $Zn(II)_2Cys_6$ domain protein required for AF and ST biosynthetic gene activation (12, 17, 72). Typical for this group of DNA binding proteins, AflR recognizes and binds to a palindromic sequence, 5'-$TCGN_5GCA$, found in the promoters of the AF and ST biosynthetic genes (16, 17, 50). A second binding site, 5'-TTAGGCCTAA, has also been reported for *A. flavus* and *A. parasiticus* and is considered important in autoregulation of *aflR* transcript in these species (13, 15, 50). Disruption of *aflR* eliminates the expression of structural genes (74), and modifications of its promoter region alter not only its own but subsequent cluster gene expression (17).

The $Zn(II)_2Cys_6$ domain is highly conserved in these fungus-specific transcription factors and is useful in identifying other proteins of this class (64). Identification of a $Zn(II)_2Cys_6$ protein in a secondary-metabolite cluster is predictive of a cluster regulator, such as that described below for the gliotoxin cluster. Once a $Zn(II)_2Cys_6$ protein is confirmed as a cluster-specific regulator, a search for a conserved palindromic sequence in the promoters of structural genes generally yields the likely binding motif.

LaeA

A novel mechanism of gene cluster regulation was uncovered by complementation of an *A. nidulans* ST mutant that was unable to express *aflR*. The complementing gene, termed *laeA* (for "loss of *aflR* expression"), encodes a nuclear protein whose closest identity is arginine and histone methyltransferases (3). Loss of LaeA function silences not only ST cluster expression but also a multitude of other metabolites including penicillin and numerous mycelial pigments in *A. nidulans* and gliotoxin in *A. fumigatus* (Fig. 2), whereas overexpression of *laeA* upregulates

A.

B.

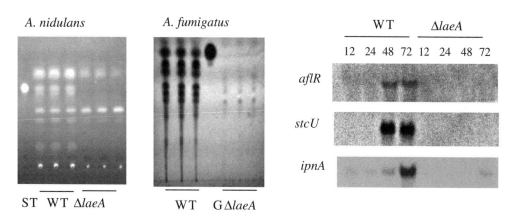

Figure 2. (A) Thin-layer chromatography of organic extracts of the *A. nidulans* and *A. fumigatus* Δ*laeA* mutants compared to the wild type (WT). ST, sterigmatocystin standard: G, gliotoxin standard. Note that both Δ*laeA* mutants are deficient in the production of many metabolites, not just ST (*A. nidulans*) or G (*A. fumigatus*). (B) Northern analysis of sterigmatocystin genes, *aflR* and *stcU*, in *A. nidulans* wild-type and Δ*laeA* strains at the 12-, 24-, 48-, and 72-h time points. Expression of both genes is repressed in the Δ*laeA* mutant. Reprinted from reference 3 with permission.

cluster gene expression. Furthermore, microarray examination of the *A. nidulans laeA* deletion and *laeA* overexpression strains clearly shows that LaeA transcriptionally regulates multiple novel secondary metabolite clusters (2a). The finding that LaeA regulates multiple clusters may support a coregulation model for clustering (77), possibly via chromatin remodeling of cluster loci. Putative LaeA orthologs are found in all filamentous and dimorphic fungi examined to date, but function has been determined only for *Aspergillus* spp.

G-Protein/Protein Kinase A Signaling

Many researchers have long noted a connection between fungal sporulation processes and secondary metabolite production (reviewed in reference 9). Formal genetic linkage of these disparate developmental processes was achieved when it was found that a conserved G-protein signaling cascade negatively regulated both spore and ST production in *A. nidulans* (Fig. 3) (32).

The initial study by Hicks et al. (32) revealed that an activated allele of an α subunit of a heterotrimeric G protein, FadA (75), repressed the expression of conidiation and ST genes. Activated FadA, FadA-GTP, is dissociated from its cognate Gβγ dimer, and both FadA-GTP and the dimer function to activate proliferative growth (53, 75). The β (SfaD) subunit (and probably the γ subunit) is also involved in conidiation (53), although its role in ST biosynthesis is still not clear.

FadA-GTP and Gβγ signaling has to be at least partially inactivated for conidiation and ST

biosynthesis to occur. This is achieved in part by FlbA, an RGS protein presumably enabling the intrinsic GTPase activity of FadA (75). Loss of *flbA* generates a similar conidiation- and ST-defective phenotype to activated FadA-GTP. Mutations in *sfaD* restore ST production in an *flbA* loss-of-function mutant (53, 75) and may suggest that the Gβγ dimer is also a negative regulator of ST production.

ST production and sporulation is also partially remediated in the *flbA* loss-of-function mutant by deletion of *pkaA* encoding the catalytic subunit of protein kinase A (58). These data provide evidence that the the FlbA-FadA signaling pathway regulating ST production and morphological development is partially mediated through PkaA (Fig. 3). Further genetic and biochemical studies of PkaA showed that it negatively regulated ST production by inhibiting *aflR* transcription (58); this transcriptional control was later found to be mediated through LaeA (3). PkaA also plays an important role in posttranscriptional regulation of AflR through phosphorylation (57). Mutation of the three putative PkaA phosphorylation sites in AflR remediated function and ST production in a heterologously expressed *aflR* allele normally inactivated by PkaA.

Conservation of this complex signaling pathway in regulating AF biosynthesis has been described to varying degrees for both *A. flavus* (44) and *A. parasiticus* (32, 54). Furthermore, the importance of G-protein signaling has been extended to penicillin biosynthesis in *A. nidulans* (63), cyclopiazonic acid production in *A. flavus* (44), trichothecene production in *Fusarium sporotrichioides* (63), pigment production in *Penicillium marneffei* (78), and pigment production in *Cryphonectria parasitica* (19, 36). These and other studies, including those with *A. fumigatus* (41), have also implicated G-protein signaling cascades as important in fungal virulence.

Ligands and Receptors

The importance of G-protein signaling in secondary metabolism and sporulation suggests a critical role for ligand-receptor pairs in these processes. One of the first extracellular signals described to regulate both asexual and sexual spore development is psi factor, the collective term for a series of oleic, linoleic, and linolenic acid-derived oxylipins, produced by *A. nidulans* (8, 10, 11, 43) and other fungal genera (5, 23, 61). Oxylipins, secondary metabolites derived through oxygenation of fatty acids and other lipids, mediate cell-cell and interspecies signaling for such diverse functions as quorum sensing in bacteria, defense responses in plants, innate immunity in mammals, and morphological differentiation and secondary metabolism in fungi (references 7, 29, and 48 and

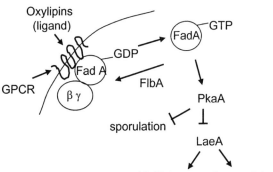

Figure 3. Proposed model of G-protein signaling in regulating sporulation and mycotoxin biosynthesis in *Aspergillus* spp. Oxylipin acts as a ligand to initiate the G-protein signaling cascade, which can lead to either suppression of toxin formation and sporulation (shown here) or activation of both processes (via different GPCR and different α subunits of the heterotrimeric G proteins). FadA, alpha subunit of a heterotrimeric G protein; FlbA, RGS protein; PkaA, protein kinase A; LaeA, global regulator of secondary metabolites.

references therein). Specific oxylipins, e.g., various prostaglandins, are ligands to G-protein-coupled receptors (GPCR), which are important in inflammatory and immune responses in mammals (27).

The fungal oxylipin biosynthetic pathway has been partially delineated in *A. nidulans*. Two psi-producing oxygenases (*ppoA* and *ppoC*) with antagonistic roles in *Aspergillus* development have been characterized (69, 71). Biochemical data implicate PpoA as the major producer of linoleic acid-derived oxylipins (71) and implicate PpoC in oleic acid oxylipin production (69). Deletion of *ppoA* increased asexual spore production, whereas deletion of *ppoC* increased sexual spore production. These results complemented previous physiological and biochemical studies that pointed out an important role for oxylipins in integrating mitotic and meiotic spore development (11). Most recently, a third oxygenase, *ppoB*, has been identified. Deletion of this gene greatly increases asexual spore production (70). ST biosynthesis is also affected in *ppo* mutants. The double Δ*ppoA* Δ*ppoC* mutant eliminated ST production, whereas deletion of *ppoB* greatly stimulated its synthesis (D. I. Tsitsigiannis and N. P. Keller, unpublished data). These opposite effects on ST are reminiscent of the differential effects of plant oxylipins on ST and AF biosynthesis (7).

Putative Ppo orthologs have been found in all filamentous fungi examined by genome database searches and have been characterized to some degree in *A. fumigatus* (see below) and *Fusarium*. Deletion of a *ppo* gene in *F. sporotrichoides* generated a strain impaired in both conidiation and T-toxin production (44), a phenotype reminiscent of certain *A. nidulans* *ppo* mutants. Current studies in our laboratory suggest a model where the different oxylipin products generated by Ppo oxygenases are secreted and function as ligands activating specific GPCR signaling cascades in *Aspergillus* and other fungi (Fig. 3). Interestingly, recent studies have identified three *A. nidulans* GPCR that impact asexual and sexual spore production (25, 26, 56), and efforts are under way to determine if Ppo products may be potential ligands for these receptors.

PROGRESS IN DELINEATION OF *A. FUMIGATUS* SECONDARY METABOLISM

Analysis of the *A. fumigatus* genome indicates the potential for upwards of 30 to 40 novel secondary metabolites, more than the current number of mycotoxins postulated as virulence factors. This section highlights our insight into regulation of these metabolites as deduced from parallel studies with *A. fumigatus* concurrent with investigations of secondary metabolism in *A. nidulans*.

Identification of a Gliotoxin Gene Cluster

Chemical analysis of gliotoxin shows it to belong to the epipolythiodioxopiperazine (ETP) class of fungal secondary metabolites (28). These compounds are characterized by a disulfide bridge across a diketopiperazine ring and are probably generated from nonribosomal peptide synthetases (NRPS), enzymes easily identified by sequence homology to known NRPSs. As implied above, the hallmark features of secondary-metabolite genes, including the conserved nature of the enzyme sequence and clustering of pathway genes, allows for the prediction of specific gene clusters with some certainty, provided that the DNA sequence is available for any given organism. Examination of the *A. fumigatus* sequence revealed the presence of many putative NRPSs, at least two of which were bimodular, as predicted for gliotoxin production. Sequence of contiguous genes on either side of these NRPSs suggested that both of these genes could be involved in gliotoxin biosynthesis.

The identity of "the" gliotoxin cluster has now been confidently predicted by means of comparison to another ETP cluster recently identified in the canola fungal pathogen *Leptosphaeria maculans* (20, 21). Disruption of a gene encoding a putative NRPS blocked production of the *L. maculans* ETP, sirodesmin, which is chemically similar in structure to gliotoxin. Gardiner et al. (21) searched the fungal databases for signs of other ETP clusters by blasting with *L. maculans* sirodesmin proteins and identified a putative gliotoxin gene cluster in *A. fumigatus* (20) (Fig. 1B). The putative cluster contains both enzymatic genes and a Zn(II)$_2$Cys$_6$-encoding gene, *gliZ*, predicted to be the pathway-specific regulator. Our laboratory has recently been successful in disrupting *gliZ* by replacing this gene with the *A. parasiticus pyrG* allele (N. P. Keller et al., unpublished data). The mutant does not produce gliotoxin, and efforts are under way to assess virulence attributes of this strain.

LaeA: a Novel Virulence Factor

As described above, an exciting advance in support of a global requirement for fungal secondary metabolite gene clustering has arisen from the identification of an *Aspergillus* methyltransferase, LaeA (3). Disruption of the gene in both *A. nidulans* and *A. fumigatus* resulted in strains that appeared nearly identical to the wild type in growth media, with the exception of loss of pigment production in the mycelium. Thin-layer chromatography examination of organic extracts of these strains showed them to be deficient in the production of many secondary metabolites including ST in *A. nidulans* and gliotoxin in *A. fumigatus* (Fig. 2). Furthermore, Northern analysis of Δ*laeA* and *OE::laeA A. nidulans* strains showed LaeA to be a transcriptional regulator of ST, penicillin, and lovastatin regulatory and biosynthetic genes. Further studies showed *laeA* expression to be regulated by both protein kinase A and Ras protein. These findings led to the model of LaeA as a global regulator of secondary-metabolite gene clusters (3).

Considering the numerous studies citing the potential role of gliotoxin in *A. fumigatus* pathogenicity, the lack of gliotoxin production in the *A. fumigatus* Δ*laeA* strain raised the possibility that this mutant could be affected in virulence. To address this possibility, the mutant strain was examined for virulence in a pulmonary murine model and further analyzed for interactions with macrophages and neutrophils. The Δ*laeA* mutant was impaired in the ability to cause fatal infection in the murine model, with very few fungal CFU recovered from the lungs of mice sacrificed 3 days after infection (2). Two striking alterations in Δ*laeA*-host cell interactions were that macrophage phagocytosis of Δ*laeA* conidia was much higher than that of wild-type conidia and that extracts from wild-type hyphae but not Δ*laeA* hyphae readily killed neutrophils in polymorphonuclear neutrophil viability assays (2). Gliotoxin production is found in hyphal tissues in *A. fumigatus* and may contribute to neutrophil killing, but its role in the altered macrophage-conidium interactions is less clear. Considering the increased phagocytosis of Δ*laeA* conidia, this might suggest that secondary metabolites other than gliotoxin are important in the conidial-macrophage interactions.

Prostaglandin Production and *ppo* Genes

Database comparisons of the three *A. nidulans* Ppo proteins to known proteins showed a significant similarity to prostaglandin (PG) synthases (COX-1 and COX-2), where homology ranged from 25 to 29% identity and 40 to 45% similarity for COX-2 paralogs (E values; 10^{-24} to 10^{-18}) and 25 to 26% identity and 38 to 40% similarity for COX-1 paralogs (E values; 2×10^{-18} to 5×10^{-18}). This was intriguing because recent studies suggest that PG production by eukaryotic microbes could be contributing to the infection process (see references 30 and 48 and references therein). Although these studies have shown that fungi, including *A. fumigatus*, are capable of producing PGs, no PG synthase has been identified in fungi.

PGs, along with leukotrienes, comprise a class of oxylipins called eicosanoids formed from C_{20} fatty acids (dihomo-γ-linolenic acid, arachidonic acid, and eicosanopentaenoic acid). Eicosanoids are critically involved in mammalian immune responses such as regulation of inflammation and allergic responses, as well as regulation of the cardiovascular system, reproduction, and renal function (18, 59). Mammalian prostaglandin synthases, and subsequent PG production, are activated by mechanical trauma or by specific growth factors, cytokines, and other abiotic or biotic stimuli, including pathogen invasion (18).

Examination of the three *A. nidulans ppo* mutants showed that loss of *ppoC* but not *ppoA* or *ppoB* resulted in a ca. 50% reduction in PG production (68). Database analysis of *A. fumigatus* showed it to contain three *ppo* genes (70); however, although each gene was phylogenetically grouped according to comparison to *A. nidulans* sequence, it was not possible to determine which, if any, produced PGs. Therefore, to evaluate the potential of any of the *A. fumigatus* Ppo proteins to produce PG, a vector was designed to simultaneously silence *Af-ppoA*, *Af-ppoB*, and *Af-ppoC* gene expression in the fungus. This was achieved by utilizing RNA interference technology (24), where inverted repeats of all three genes were placed in one vector (68). Incorporation of the RNA intereference vector into the *A. fumigatus* genome resulted in downregulation of all three *ppo* genes and a decrease in PG production (68). Interestingly, this mutant showed an increase in virulence in the murine pulmonary system as measured by a higher kill rate. These findings led to a proposal that host cell recognition of *A. fumigatus* PGs might accelerate host defense response (68).

CONCLUSION

The worldwide frequency and the profile of patients at risk for invasive aspergillosis by *A. fumigatus* continue to expand, owing to increased numbers of immunocompromised patients, particularly those afflicted with malignancy and organ dysfunction. However, our current understanding of the pathogenesis of this historically difficult-to-treat pathogen is meager. Many of the proposed *A. fumigatus* virulence attributes are secondary metabolites with extreme biological effects on host cells. Only recently have secondary-metabolite genes been cloned and their regulation articulated in this fungus. This chapter demonstrates the power of using the model ascomycete and related species *A. nidulans* in elucidating secondary metabolism in *A. fumigatus*. Initial studies of *laeA* and *ppo* function in *A. fumigatus* indicate a potent role for these secondary-metabolite genes in pathogenesis. A thorough understanding of the function of these and other secondary metabolism genes may assist in the development of future therapeutics.

REFERENCES

1. Bennett, J. W., and M. Klich. 2003. Mycotoxins. *Clin. Microbiol. Rev.* **16:**497–516.

2. Bok, J. W, S. A. Balajee, K. A. Marr, D. Andes, K. Fog Nielsen, J. C. Frisvad, and N. P. Keller. 2005. LaeA, a regulator of virulence determinants in *Aspergillus fumigatus.* *Eukaryot. Cell* **4:**1574–1582.

2a. Bok, J.-W., D. Hoffmeister, L. Maggio-Hall, R. Murillo, J. D. Glasner, and N. P. Keller. 2006. Genomic mining for *Aspergillus* natural products. *Chem. Biol.* **13:**31–37.

3. Bok, J. W., and N. P. Keller. 2004. LaeA, a regulator of secondary metabolism in *Aspergillus* spp. *Eukaryot. Cell* **3:**527–535.

4. Brakhage, A. A. 1998. Molecular regulation of β-lactam biosynthesis in filamentous fungi. *Microbiol. Mol. Biol. Rev.* **62:**547–585.

5. Brodowsky, I. D., and E. H. Oliw. 1993. Biosynthesis of 8R-hydroperoxylinoleic acid by the fungus *Laetisaria arvalis.* *Biochim. Biophys. Acta* **20:**68–72.

6. Brown, D. W., J. H. Yu, H. S. Kelkar, M. Fernandes, T. C. Nesbitt, N. P. Keller, T. H. Adams, and T. J. Leonard. 1996. Twenty-five coregulated transcripts define a sterigmatocystin gene cluster in *Aspergillus nidulans.* *Proc. Natl. Acad. Sci. USA* **93:**1418–1422.

7. Burow, G. B., T. C. Nesbitt, J. D. Dunlap, and N. P. Keller. 1997. Seed lipoxygenase products modulate *Aspergillus* mycotoxin biosynthesis. *Mol. Plant-Microbe Interact.* **10:**380–387.

8. Calvo, A., H. W. Gardner, and N. P. Keller. 2001. Genetic connection between fatty acid metabolism and sporulation in *Aspergillus nidulans.* *J. Biol. Chem.* **276:**25766–25774.

9. Calvo, A. M., R. A. Wilson, J. W. Bok, and N. P. Keller. 2002. Relationship between secondary metabolism and fungal development. *Microbiol. Mol. Biol. Rev.* **66:**447–459.

10. Champe, S. P., and A. A. E. El-Zayat. 1989. Isolation of a sexual sporulation hormone from *Aspergillus nidulans.* *J. Bacteriol.* **171:**3982–3988.

11. Champe, S. P., P. Rao, and A. Chang. 1987. An endogenous inducer of sexual development in *Aspergillus nidulans.* *J. Gen. Microbiol.* **133:**1383–1388.

12. Chang, P. K., J. W. Cary, D. Bhatnagar, T. E. Cleveland, J. W. Bennett, J. E. Linz, C. P. Woloshuk, and G. A. Payne. 1993. Cloning of the *Aspergillus parasiticus apa-2* gene associated with the regulation of aflatoxin biosynthesis. *Appl. Environ. Microbiol.* **59:**3273–3279.

13. Chang, P. K., J. Yu, D. Bhatnagar, and T. E. Cleveland. 1999. Repressor-AFLR interaction modulates aflatoxin biosynthesis in *Aspergillus parasiticus.* *Mycopathologia* **147:**105–112.

14. Denning, D. W., and D. A. Stevens. 1990. Antifungal and surgical treatment of invasive aspergillosis: review of 2,121 published cases. *Rev. Infect. Dis.* **12:**1147–1201.

15. Ehrlich, K. C., B. G. Montalbano, D. Bhatnagar, and T. E. Cleveland, 1998. Alteration of different domains in *aflR* affects aflatoxin pathway metabolism in *Aspergillus parasiticus* transformants. *Fungal Genet. Biol.* **23:**279–287.

16. Ehrlich, K. C., B. G. Montalbano, and J. W. Cary. 1999. Binding of the C_6-zinc cluster protein, AflR, to the promoters of aflatoxin pathway biosynthesis genes in *Aspergillus parasiticus.* *Gene* **230:**249–257.

17. Fernandes, M., N. P. Keller, and T. H. Adams. 1998. Sequence-specific binding by *Aspergillus nidulans* AflR, a C6 zinc cluster protein regulating mycotoxin biosynthesis. *Mol. Microbiol.* **28:**1355–1365.

18. Funk, C. D. 2001. Prostaglandins and leukotrienes: advances in eicosanoid biology. *Science* **294:**1871–1875.

19. Gao, S., and D. L. Nuss. 1996. Distinct roles for two G protein α subunits in fungal virulence, morphology, and reproduction revealed by targeted gene disruption. *Proc. Natl. Acad. Sci. USA* **93:**14122–14127.

20. Gardiner, D. M, and B. J. Howlett. 2005. Bioinformatic and expression analysis of the putative gliotoxin biosynthetic gene cluster of *Aspergillus fumigatus.* *FEMS Microbiol. Lett.* **248:**241–248.

21. Gardiner, D. M., A. J. Cozijnsen, L. M. Wilson, M. S. C. Pedras, and B. J. Howlett. 2004. The sirodesmin biosynthetic gene cluster of the plant pathogenic fungus *Leptosphaeria maculans.* *Mol. Microbiol.* **53:**1307–1318.

22. Grovel, O., Y. F. Pouchus, T. Robiou du Pont, M. Montagu, Z. Amzil, and J. Verbist. 2002. Ion trap MS(n) for identification of gliotoxin as the cytotoxic factor of a marine strain of *Aspergillus fumigatus* Fresenius. *J. Microbiol. Methods* **48:**171–179.

23. Hamberg, M., C. Su, and E. Oliw. 1998. Manganese lipoxygenase. Discovery of a bisallylic hydroperoxide as product and intermediate in a lipoxygenase reaction. *J. Biol. Chem.* **273:**13080–13088.

24. Hammond, T. M., and N. P. Keller. 2005. RNA silencing in *Aspergillus nidulans* is independent of RNA-dependent RNA polymerases. *Genetics* **169:**607–617.

25. Han, K.-H., J.-A. Seo, and J.-H. Yu, 2004. A putative G protein-coupled receptor negatively controls sexual development in *Aspergillus nidulans.* *Mol. Microbiol.* **51:**1333–1345.

26. Han, K.-H., J.-A. Seo, and J.-H. Yu. 2004. Regulators of G-protein signaling in *Aspergillus nidulans:* RgsA down regulates stress response and stimulates asexual sporulation through attenuation of GanB (Gα) signaling. *Mol. Microbiol.* **53:**529–540.

27. Hata, A. N., and R. M. Breyer. 2004. Pharmacology and signaling of prostaglandin receptors: multiple roles in inflammation and immune modulation. *Pharmacol. Ther.* **103:**147–166.

28. Herbert, R. B. 1989. *The Biosynthesis of Secondary-metabolites.* Chapman & Hall, Ltd., London, United Kingdom.

29. Herman, R. P. 1998. Oxylipin production and action in fungi and related organisms, p. 115–130. *In* A. F. Rowley, H. Kuhn, and T. Schewe (ed.), *Eicosanoids and Related Compounds in Plants and Animals.* Princeton University Press, Princeton, N.J.

30. Herman, R. P., and C. A. Herman. 1985. Prostaglandins or prostaglandin like substances are implicated in normal growth and development in oomycetes. *Prostaglandins* **29:**819–830.

31. Hicks, J. K., K. Shimizu, and N. P. Keller. 2002. Genetics and biosynthesis of aflatoxins and sterigmatocystin, p. 55–69. *In* F. Kempken and J. Bennett (ed.), *The Mycota XI.* Springer-Verlag, KG, Berlin, Germany.

32. Hicks, J. K., J.-H. Yu, N. P. Keller, and T. H. Adams. 1997. *Aspergillus* sporulation and mycotoxin production both require inactivation of the FadA Gα protein-dependent signaling pathway. *EMBO J.* **16:**4916–4923.

33. Hogan, L. H., B. S. Klein, and S. M. Levitz. 1996. Virulence factors of medically important fungi. *Clin. Microbiol. Rev.* **9:**469–488.

34. Jahn, B., F. Boukhallouk, J. Lotz, K. Langfelder, G. Wanner, and A. A. Brakhage. 2000. Interaction of human

phagocytes with pigmentless *Aspergillus* conidia. *Infect. Immun.* **68**:3736–3739.

35. **Jahn, B., A. Koch, A. Schmidt, G. Wanner, H. Gehringer, S. Bhakdi, and A. A. Brakhage.** 1997. Isolation and characterization of a pigmentless-conidium mutant of *Aspergillus fumigatus* with altered conidial surface and reduced virulence. *Infect. Immun.* **65**:5110–5117.

36. **Kasahara, S., and D. L. Nuss.** 1997. Targeted disruption of a fungal G-protein beta subunit gene results in increased vegetative growth but reduced virulence. *Mol. Plant-Microbe Interact.* **10**:984–993.

37. **Keller, N. P., and T. M. Hohn.** 1997. Metabolic pathway gene clusters in filamentous fungi. *Fungal Genet. Biol.* **21**:17–29.

38. **Klich, M. A., L. H. Tiffany, and G. Knaphus.** 1992. Ecology of the aspergilli in soils and litter, p. 329–353. *In* J. W. Bennett and M. A. Klich (ed.), *Aspergillus: Biology and Industrial Applications.* Butterworth-Heinemann, Boston, Mass.

39. **Kweon, Y. O., Y. H. Paik, B. Schnabl, T. Qian, J. J. Lemasters, and D. A. Brenner.** 2003. Gliotoxin-mediated apoptosis of activated human hepatic stellate cells. *J. Hepatol.* **39**:38–46.

40. **Latge, J. P.** 1999. *Aspergillus fumigatus* and aspergillosis. *Clin. Microbiol. Rev.* **12**:310–350.

41. **Liebmann, B., M. Mäller, A. Braun, and A. A. Brakhage.** 2004. The cyclic AMP-dependent protein kinase a network regulates development and virulence in *Aspergillus fumigatus*. *Infect. Immun.* **72**:5193–5203.

42. **Lin, S. J., J. Schranz, and S. M. Teutsch.** 2001. Aspergillosis case-fatality rate: systematic review of the literature. *Clin. Infect. Dis.* **32**:358–366.

43. **Mazur, P., K. Nakanishi, A. A. E. El-Zayat, and S. P. Champe.** 1991. Structure and synthesis of sporogenic psi factors from *Aspergillus nidulans*. *J. Chem. Soc. Chem. Commun.* **20**:1486–1487.

44. **McDonald, T., T. Devi, K. Shimizu, S.-C. Sim, and N. P. Keller.** 2004. Signaling events connecting mycotoxin biosynthesis and sporulation in *Aspergillus* and *Fusarium* spp., p. 139–147. *In* T. Yoshizawa (ed.), *New Horizon of Mycotoxicology for Assuring Food Safety.* Proceedings of the International Symposium of Mycotoxicology. Takamatsu, Kagawa, Japan.

45. **Minto, R. E., and C. A. Townsend.** 1997. Enzymology and molecular biology of aflatoxin biosynthesis. *Chem. Rev.* **97**:2537–2556.

46. **Nielsen, K. F., and J. Smedsgaard.** 2003. Fungal metabolite screening: database of 474 mycotoxins and fungal metabolites for dereplication by standardised liquid chromatography-UV-mass spectrometry methodology. *J. Chromatogr. Ser. A.* **1002**:111–136.

47. **Nieminen, S. M., J. Maki-Paakkanen, M. R. Hirvonen, M. Roponen, and A. von Wright.** 2002. Genotoxicity of gliotoxin, a secondary-metabolite of *Aspergillus fumigatus*, in a battery of short-term test systems. *Mutat. Res.* **520**:161–170.

48. **Noverr, M. C., J. R. Erb-Downward, and G. B. Huffnagle.** 2003. Production of eicosanoids and other oxylipins by pathogenic eukaryotic microbes. *Clin. Microbiol. Rev.* **16**:517–533.

49. **Patterson, T. F., W. R. Kirkpatrick, M. White, J. W. Hiemenz, J. R. Wingard, B. Dupont, M. G. Rinaldi, D. A. Stevens, J. R. Graybill, and the I3 *Aspergillus* Study Group.** 2000. Invasive aspergillosis. Disease spectrum, treatment practices, and outcomes. *Medicine* **79**:250–260.

50. **Payne, G. A., and M. P. Brown.** 1998. Genetics and physiology of aflatoxin biosynthesis. *Annu. Rev. Phytopathol.* **36**:329–362.

51. **Pedley, K. F., and J. D. Walton.** 2001. Regulation of cyclic peptide biosynthesis in a plant pathogenic fungus by a novel transcription factor. *Proc. Natl. Acad. Sci. USA* **98**:14174–14179.

52. **Proctor, R. H., T. M. Hohn, S. P. McCormick, and A. E. Desjardins.** 1995. Tri6 encodes an unusual zinc finger protein involved in regulation of trichothecene biosynthesis in *Fusarium sporotrichioides*. *Appl. Environ. Microbiol.* **61**:1923–1930.

53. **Rosén, S., J.-H. Yu, and T. H. Adams.** 1999. The *Aspergillus nidulans sfaD* gene encodes a G protein beta subunit that is required for normal growth and repression of sporulation. *EMBO J.* **18**:5592–5600.

54. **Roze, L. V., R. M. Beaudry, N. P. Keller, and J. E. Linz.** 2004. Regulation of aflatoxin synthesis by FadA/cAMP/protein kinase A signaling in *Aspergillus parasiticus*. *Mycopathologia* **158**:219–232.

55. **Samson, R. A.** 1992. Current taxonomic schemes of the genus *Aspergillus* and its teleomorphs. *Bio/Technology* **23**:355–390.

56. **Seo, J.-A., K.-H. Han, and J.-H. Yu.** 2004. The *gprA* and *gprB* genes encode putative G protein-coupled receptors required for self-fertilization in *Aspergillus nidulans*. *Mol. Microbiol.* **53**:1611–1623.

57. **Shimizu, K., J. K. Hicks, T. P. Huang, and N. P. Keller.** 2003. Pka, Ras and RGS protein interactions regulate activity of AflR, a $Zn(II)_2Cys_6$ transcription factor in *Aspergillus nidulans*. *Genetics* **165**:1095–1104.

58. **Shimizu, K., and N. P. Keller.** 2001. Genetic involvement of a cAMP-dependent protein kinase in a G protein signaling pathway regulating morphological and chemical transitions in *Aspergillus nidulans*. *Genetics* **157**:591–600.

59. **Smith, W. L., D. L. DeWitt, and R. M. Garavito.** 2000. Cyclooxygenases: structural, cellular, and molecular biology. *Annu. Rev. Biochem.* **69**:145–182.

60. **Stevens, D. A., V. L. Kan, M. A. Judson, V. A. Morrison, S. Dummer, D. W. Denning, J. E. Bennett, T. J. Walsh, T. F. Patterson, and G. A. Pankey.** 2000. Practice guidelines for diseases caused by *Aspergillus*. *Clin. Infect. Dis.* **30**:696–709.

61. **Su, C., M. Sahlin, and E. H. Oliw.** 1998. Kinetics of manganese lipoxygenase with a catalytic mononuclear redox center. *J. Biol. Chem.* **273**:20744–20751.

62. **Suen, Y. K., K. P. Fung, C. Y. Lee, and S. K. Kong.** 2001. Gliotoxin induces apoptosis in cultured macrophages via production of reactive oxygen species and cytochrome *c* release without mitochondrial depolarization. *Free Radic. Res.* **35**:1–10.

63. **Tag, A., J. Hicks, G. Garifullina, C. Ake, Jr., T. D. Phillips, M. Beremand, and N. Keller.** 2000. G-protein signalling mediates differential production of toxic secondary-metabolites. *Mol. Microbiol.* **38**:658–665.

64. **Todd, R. B., and A. Andrianopoulos.** 1997. Evolution of a fungal regulatory gene family: the $Zn(II)_2Cys_6$ binuclear cluster DNA binding motif. *Fungal Genet. Biol.* **21**:388–405.

65. **Tsai, H. F., Y. C. Chang, R. G. Washburn, M. H. Wheeler, and K. J. Kwon-Chung.** 1998. The developmentally regulated *alb1* gene of *Aspergillus fumigatus*: its role in modulation of conidial morphology and virulence. *J. Bacteriol.* **180**:3031–3038.

66. **Tsai, H. F., R. G. Washburn, Y. C. Chang, and K. J. Kwon-Chung.** 1997. *Aspergillus fumigatus arp1* modulates

conidial pigmentation and complement deposition. *Mol. Microbiol.* **26:**175–183.

67. Tsai, H. F., M. H. Wheeler, Y. C. Chang, and K. J. Kwon-Chung. 1999. A developmentally regulated gene cluster involved in conidial pigment biosynthesis in *Aspergillus fumigatus*. *J. Bacteriol.* **181:**6469–6477.

68. Tsitsigiannis, D. I., J. W. Bok, D. Andes, K. F. Nielsen, J. C. Frisvad, and N. P. Keller. 2005. *Aspergillus* cyclooxygenase-like enzymes are associated with prostaglandin production and virulence. *Infect. Immun.* **73:**4548–4559.

69. Tsitsigiannis, D. I., T. Kowieski, R. Zarnowski, and N. P. Keller. 2004. Endogenous lipogenic regulators of spore balance in *Aspergillus nidulans*. *Eukaryot. Cell* **3:**1398–1411.

70. Tsitsigiannis, D. I., T. Kowieski, R. Zarnowski, and N. P. Keller. 2005. Three putative oxylipin biosynthetic genes integrate sexual and asexual development in *Aspergillus nidulans*. *Microbiology* **151:**1809–1821.

71. Tsitsigiannis, D. I., R. Zarnowski, and N. P. Keller. 2004. The lipid body protein, PpoA, coordinates sexual and asexual sporulation in *Aspergillus nidulans*. *J. Biol. Chem.* **279:**11344–11353.

72. Woloshuk, C. P., K. R. Fount, J. F. Brewer, D. Bhatnagar, T. E. Cleveland, and G. A. Payne. 1994. Molecular characterization of *aflR*, a regulatory locus for aflatoxin biosynthesis. *Appl. Environ. Microbiol.* **60:**2408–2414.

73. Yu, J., P. K. Chang, K. C. Ehrlich, J. W. Cary, D. Bhatnagar, T. E. Cleveland, G. A. Payne, J. E. Linz, C. P. Woloshuk, and J. W. Bennett. 2004. Clustered pathway genes in aflatoxin biosynthesis. *Appl. Environ. Microbiol.* **70:**1253–1262.

74. Yu, J.-H., R. A. E. Butchko, M. Fernandes, N. P. Keller, T. J. Leonard, and T. H. Adams. 1996. Conservation of structure and function of the aflatoxin regulatory gene *aflR* from *Aspergillus nidulans* and *A. flavus*. *Curr. Genet.* **29:**549–555.

75. Yu, J.-H., J. Wieser, and T. H. Adams. 1996b. The *Aspergillus* FlbA RGS domain protein antagonizes G-protein signaling to block proliferation and allow development. *EMBO J.* **15:**5184–5190.

76. Yu, J.-H., S. Rosèn, and T. H. Adams. 1999. Extragenic suppressors of loss-of-function mutations in the *Aspergillus* FlbA regulator of G-protein signaling domain protein. *Genetics* **151:**97–105.

77. Zhang, Y.-Q., H. Wilkinson, N. P. Keller, and D. Tsitsigiannis. 2004. Secondary-metabolite gene clusters, p. 355–386. *In* Z. An (ed.), *Handbook of Industrial Microbiology*. Marcel Dekker, Inc., New York, N.Y.

78. Zuber, S., M. J. Hynes, and A. Andrianopoulos. 2003. The G-protein α-subunit GasC plays a major role in germination in the dimorphic fungus *Penicillium marneffei*. *Genetics* **164:**487–499.

Molecular Principles of Fungal Pathogenesis
Edited by Joseph Heitman et al.
©2006 ASM Press, Washington, D.C.

Chapter 18

Saccharomyces cerevisiae: an Emerging and Model Pathogenic Fungus

JOHN H. MCCUSKER

There are a variety of approaches—some of which are described in other chapters in this volume—to using *Saccharomyces cerevisiae* to tell us about the pathogenic fungi and fungal pathogenesis. This chapter also describes the use of *S. cerevisiae* to tell us about the pathogenic fungi and fungal pathogenesis but does so in a rather different way, focusing on the fact that *S. cerevisiae* is itself an emerging opportunistic pathogen. The fact that isolates of an extraordinarily well-studied model eukaryote are found clinically—and differ genotypically and phenotypically from nonclinical isolates and laboratory *S. cerevisiae* strains—offers a unique opportunity to address two questions. First, which genes or phenotypes distinguish nonpathogenic and nonclinical isolates of a species from pathogenic and clinical isolates of a species, in this case, *S. cerevisiae*? And second, how does a saprophyte such as *S. cerevisiae* become an emerging opportunistic pathogen?

This chapter first reviews the evidence (i.e., clinical reports and incidence) that *S. cerevisiae* is an emerging opportunistic pathogen, how and why *S. cerevisiae* is found clinically (e.g., acquisition and host factors), and species identification and typing. The chapter describes the analysis (e.g., genetic, phenotypic, and experimental infections) of clinical *S. cerevisiae* isolates and clinically derived *S. cerevisiae* strains. Because genetic diversity and the sexual cycle are likely to be important for pathogenic fungi in general and for emerging pathogens in particular, clinical and nonclinical isolates of *S. cerevisiae* are compared and contrasted. In addition to making some specific comparisons of *S. cerevisiae* with other pathogenic fungi, this chapter describes classical genetic analysis of the ability of *S. cerevisiae* to survive in and kill mice. Finally, the genetic complexity

of one likely virulence trait, high-temperature growth, is discussed. The analysis demonstrates the value of *S. cerevisiae* as a model pathogen, sheds light on the emergence of *S. cerevisiae* as an opportunistic pathogen, and illustrates issues likely to be relevant to other species of established and emerging pathogenic fungi.

S. CEREVISIAE: AN EMERGING OPPORTUNISTIC PATHOGEN

The model eukaryote *S. cerevisiae* is closely related to such important opportunistic pathogens as *Candida albicans* and *C. glabrata* (see, e.g., references 9, 50, and 53). Despite this close relationship, many find it surprising that *S. cerevisiae* is itself an emerging opportunistic pathogen (reviewed in references 42, 66, and 83). However, *S. cerevisiae* carriage is observed and *S. cerevisiae* has been found to cause disease that in some cases has been fatal.

The following is a list of *S. cerevisiae* infections (single patients, unless noted otherwise) described in the literature with, in some cases, other information (e.g., body site, AIDS status, nosocomial infection): systemic (ingestion of brewer's yeast) (45), septicemia (nosocomial) (32), peritonitis (postoperative) (29), oral plaques (31), scrotum (31), pleural effusion (oral consumption of brewer's yeast) (31), kidney abcess (31), urine (31), septicemia (AIDS patient) (95), vaginitis (with transmission to partner) (113), pneumonia (also oral mucosa and spleen, AIDS patient) (107), pneumonia (6), liver abcess and sepsis (6), disseminated infection (6), vaginitis (9 patients) (100), fungemia (20), fungemia (56), fungemia (4 patients) (72), vaginitis (10 patients) (1), empyema (17), septicemia (77), oral (10 AIDS patients) (92), fungemia (13), vaginitis (8 patients) (74), peritonitis

John H. McCusker • Department of Molecular Genetics and Microbiology, 3020, Duke University Medical Center, Durham, NC 27710.

(108), fungemia (108), vaginitis (3 patients) (3), esophagitis (AIDS patient) (48), fungemia and aortic graft infection (98), diarrhea (immunocompetent patient with ulcerative colitis) (14), and lung nodule (baker) (86).

As can be seen from the cases listed above, *S. cerevisiae* causes a variety of infections, and, while host factors are important (see below), *S. cerevisiae* infects (or can be recovered from) multiple patient types, e.g., human immunodeficiency virus (HIV) positive and negative, neutropenic, and nonneutropenic. *S. cerevisiae* also infects (or can be found in carriers in) both nonsterile and sterile body sites. Therefore, *S. cerevisiae*, or at least some isolates of it, can invade, persist, and proliferate, all of which are important characteristics for a pathogen.

WHAT IS THE INCIDENCE OF *S. CEREVISIAE* VERSUS OTHER PATHOGENIC FUNGI?

The clinical incidence of *S. cerevisiae* is unknown. As stated by Smith (99): "How often this yeast produces disease and how frequently such disease is associated with exogenous ingestion or other exposure is unknown. As with most infections, the interaction between inoculum, virulence, and host defense is the important determinant in the development of disease. Once individual *Saccharomyces* infections are well described in the literature, journals are unlikely to accept additional, essentially repetitive case reports. Thus the true incidence and prevalence of *Saccharomyces* infections are unknown."

Although definitive studies have not been performed, it has been suggested that the incidence of *S. cerevisiae* infections (or colonizations) is increasing (see, e.g., reference 84). If *S. cerevisiae* infections and colonizations are increasing, or are being increasingly noticed, the reasons may include (i) improved testing, (ii) an increasing number of immunodeficient and/or debilitated patients, (iii) an increased willingness to consider *S. cerevisiae* as an opportunistic pathogen, (iv) increases in the virulence or pathogenicity (or both) of *S. cerevisiae*, and/or (v) increased use of azole antimycotics (discussed below).

As shown in Table 1, as an emerging opportunistic pathogen, *S. cerevisiae* is isolated clinically less often than *C. glabrata* or *C. albicans*; variation in the frequency of isolation of *S. cerevisiae* in different studies may be due to differences in the geographic locations of the studies, patient/subject populations, body site(s) assayed, etc. However, while *C. albicans* and *C. glabrata* are isolated more frequently, *S. cerevisiae* can be isolated at significant frequencies.

HOW IS *S. CEREVISIAE* ACQUIRED?

There is evidence that *S. cerevisiae* infections (and/or carriage) can be acquired nosocomially (see, e.g., references 32, 76, 91, and 117), through use in

Table 1. Isolation of *C. albicans*, *C. glabrata*, and *S. cerevisiae* from different patient and subject populations and from different body sites

Subject or population and body site	No. of isolates, patients, or subjects	Frequency of isolation (%) of:			Reference
		C. albicans	*C. glabrata*	*S. cerevisiae*	
Cancer patients, multiple sites	3,340[a]	68.5%	10.5%	0.8%	47
Vaginitis	>2,000[c]	ND[b]	ND	<0.5%	100
Vaginitis	74[d]	67.5%	15.6%	65%	73
Vaginitis	18[d]	33%	38.8%	16.6%	3
Vaginal carriage	24[e]	54.1%	20.8%	16.6%	3
Vaginitis	186[d]	49.9%	22.3%	5.4%	1
Vaginal carriage	122[e]	19.7%	21.3%	19.7%	1
Vaginitis and vaginal carriage	513[c]	ND	ND	5.8%	84
HIV+ oral	77[d]	81%	9%	5%	92
HIV+/HIV−, pregnant/not pregnant, oral/vaginal/rectal	49[e]	73.5%	3.7%	2.5%	116
Mostly HIV−, oral	407[e]	34.6%	12%	23.3%	21

[a]Yeast isolates.
[b]ND, not determined or data not known.
[c]Number of patients.
[d]Number of patients from whom yeasts of any species were recovered.
[e]Number of subjects from whom yeasts of any species were recovered.

baking (see, e.g., references 57, 74, 84, and 113), sexually (and/or from person to person) (see, e.g., references 57 and 113), and orally by direct ingestion (see, e.g., references 31, 45, and 57). (With respect to oral ingestion of *S. cerevisiae* resulting in infection, see also the discussion of *S. boulardii*, below.) However, in most cases the origin of clinically observed *S. cerevisiae*, whether colonizing or infecting patients, is not known.

WHICH HOST FACTORS ARE ASSOCIATED WITH *S. CEREVISIAE* CARRIAGE AND/OR INFECTION?

Aside from consumption of *S. cerevisiae* (e.g., as a "health" food) or its use in baking, several host factors may be associated with *S. cerevisiae* infections or carriage (42), including neutropenia (2, 32, 72), venous catheterization (2, 20, 72), antibacterial chemotherapy (72), renal failure (20), and AIDS (48, 92, 95, 107). While infections and carriage are by no means limited to these classes of patients, immunodeficient, neutropenic, or debilitated hosts appear to be at increased risk for *S. cerevisiae* infections. With respect to *S. cerevisiae* vaginitis, host factors may include recurrent vaginal candidiasis as well as antimycotic and antibiotic treatment (100).

One final factor that is likely to be associated with the clinical presence of *S. cerevisiae* is the use of azole antimycotics. As is the case with many other emerging opportunistic pathogenic fungi, *S. cerevisiae* is relatively resistant to many azoles (see, e.g., references 3, 7, 81, 91, 100, 108, and 117). Indeed, while *S. cerevisiae* did not appear to be the sole cause of oral thrush in AIDS patients, Sangeorzan et al. (92) found that the recovery of *S. cerevisiae* increased after treatment with clotrimazole or fluconazole. In this sense, one can view the azole-treated host as an ecological niche that is open to exploitation by relatively azole-resistant species of fungi such as *S. cerevisiae*.

CAN THESE CLINICALLY OBSERVED *SACCHAROMYCES* ISOLATES BE IDENTIFIED AS *S. CEREVISIAE*?

The tests used in clinical microbiology laboratories to distinguish different species of fungi, which include assimilation tests and microscopic examination, are not sufficient to distinguish the *Saccharomyces* sensu stricto species, that is, the taxonomically accepted biological sibling species of *S. cerevisiae* which include *S. paradoxus*, *S. bayanus*, *S. cariocanus*, *S. kudriavzevii*, and *S. mikatae* (see, e.g., reference 50). Instead, more stringent criteria are required to differentiate the sibling species (see, e.g., reference 69). Clinical isolates identified in clinical

microbiology laboratories as *S. cerevisiae* have been subjected to a variety of DNA-typing analyses and, by this criterion, are members of the species *S. cerevisiae* (see, e.g., references 23, 43, 57, 84, and 117). In addition to identifying the isolates as *S. cerevisiae*, in every study DNA typing distinguished different isolates. With the exceptions of *S. boulardii* (discussed below), some vaginal isolates in one study (57), and some nosocomial isolates (76, 117), the *S. cerevisiae* clinical isolates were genetically highly diverse, suggesting both that there is a lack of clonality and that many members of the *S. cerevisiae* population are capable of colonizing or infecting humans.

As a further proof of species identity, we used segregants from *S. cerevisiae* clinical isolates (23, 57, 59) to perform a genetic species test similar to that used to identify and define species in the genus *Saccharomyces* (see, e.g., reference 69). Specifically, for *S. cerevisiae* clinical isolates that underwent meiosis and produced viable spores, we crossed segregants (haploid spores in the case of *HO* segregants) with a haploid laboratory strain of *S. cerevisiae*. In every case, the resulting diploids sporulated and the resulting spores showed high viability. (Note that the ability to perform such crosses allows the quantitative genetic analysis of traits such as the ability to grow at high temperatures [see below].) By this criterion, the clinical isolates are indeed members of the species *S. cerevisiae*.

The clinical recovery of *S. cerevisiae*, but not the other five taxonomically accepted biological *Saccharomyces* sensu stricto species, may be due to the frequent (in comparison to other species of *Saccharomyces*) encounters between *S. cerevisiae* and humans. Alternatively, interspecific phenotypic differences, such as the ability of *S. cerevisiae*—but not some of the sibling species of *Saccharomyces*—to grow at 37°C (see, e.g., references 69 and 106), may be important. *S. cerevisiae* is the only species of *Saccharomyces* to be observed clinically, with the possible exception of *S. boulardii* (see below).

S. BOULARDII

One special class of *Saccharomyces* infections involves *S. boulardii*. *S. boulardii*, which is either a species very closely related to *S. cerevisiae* or a strain of *S. cerevisiae*, is licensed for clinical use in Europe as a probiotic agent for the treatment of "traveler's disease" and antibiotic-associated diarrhea caused by *Clostridium difficile*; *S. boulardii* may also prove to be useful in the treatment of other enteric infections (reviewed in reference 28). While the mechanism of action with respect to treating enteric infections is not clear, orally administered *S. boulardii* colonizes the gut; *S. boulardii* administration also increases secretory immunoglobulin A production, the number of Kupffer cells, and tumor necrosis factor alpha (TNF-α), gamma interferon, and interleukin-12 (IL-12) levels in serum (11, 85, 90).

In some studies, brewer's/baker's yeast (*S. cerevisiae*) and *S. boulardii* have been reported to have similar effects with respect to (i) attenuation of *C. difficile*-induced colonic secretion (44) and (ii) probiotic efficacy (18, 94). In contrast, the results from other studies (25, 79) have been interpreted (61) as demonstrating a probiotic effect for *S. boulardii* but not for *S. cerevisiae*. The conflicting results may be due to differences in study design, in particular the use of different *S. cerevisiae* strains with distinct phenotypic properties.

Although it appears to be rare, patients given *S. boulardii* orally have developed *Saccharomyces* fungemias with the recovered yeast being indistinguishable from *S. boulardii* (8, 15, 16, 37, 51, 54, 71, 80, 82, 88, 89). The oral administration route of and gut colonization by *S. boulardii* suggests that *S. boulardii* can transit from the intestinal tract to the bloodstream to cause fungemias and also suggests a likely route of infection for *S. cerevisiae*. In addition to those patients given *S. boulardii* who became fungemic, there is also evidence of nosocomial transmission of *S. boulardii*; that is, patients not receiving therapeutic *S. boulardii* appear to have been infected with *S. boulardii* administered to other patients (see, e.g., references 15, 51, and 80). Consistent with the observations of *S. boulardii* (and *S. cerevisiae*) fungemias, the ability of *S. boulardii* to survive in both wild-type CD-1 and complement-deficient DBA/2N mice is substantial and comparable to that of some clinical isolates of *S. cerevisiae* (58) (see below).

There is some disagreement in the literature about whether *S. boulardii* is (i) a species distinct from *S. cerevisiae* (61) or (ii) a strain/isolate of *S. cerevisiae*. For *S. boulardii* to be accepted as a distinct species, a fungal taxonomist would have to publish a formal description of *S. boulardii* as has been done for other biological species of *Saccharomyces* (see, e.g., reference 69). In the absence of a published formal description of *S. boulardii* as a distinct species, the overall weight of evidence leans in favor of the hypothesis that *S. boulardii* is a strain/isolate of *S. cerevisiae* (see, e.g., references 30, 35, 43, 55, 58, 62, and 109). Indeed, in *The Yeasts, a Taxonomic Study*, Kurtzman and Fell (49) list *S. boulardii* as being a synonym of *S. cerevisiae*. Unfortunately, a genetic species test, such as that carried out for clinical *S. cerevisiae* isolates (23, 57, 59) and to differentiate sibling species in the genus *Saccharomyces* (69), cannot be carried out because *S. boulardii* (like

some wine [46] and clinical [23, 57] isolates of *S. cerevisiae*) does not mate or sporulate (undergo meiosis) (58).

SPORULATION AND SPORE VIABILITY OF *S. CEREVISIAE* CLINICAL ISOLATES

Because the ability to sporulate (undergo meiosis) (Spo+) is very important for detailed genetic analysis, the first step I take in analyzing clinical *S. cerevisiae* isolates is to determine if they sporulate and, if so, if they form viable spores. In one of our studies, 1 of 49 clinical *S. cerevisiae* isolates was Spo− (23). Similarly, Spo− *S. cerevisiae* wine strains are known (46). Of the 48 *S. cerevisiae* clinical isolates that were Spo+, 18 produced no viable spores. Of the 27 Spo+ *S. cerevisiae* clinical isolates that produced viable spores that were tested, 5 appeared to be tetraploid interspecific hybrids (i.e., allotetraploids) (23).

What might be the source of these *S. cerevisiae* clinical isolates that are allotetraploid or produce no viable spores? With respect to the allotetraploid clinical isolates, some industrial *Saccharomyces* strains that produce viable spores are allotetraploid interspecific hybrids between *S. cerevisiae* and the sibling species *S. bayanus* (see, e.g., reference 70). With respect to the clinical isolates that produce no viable spores, diploid interspecific hybrids (allodiploids), which include some industrial *Saccharomyces* strains (*S. pastorianus* is an *S. cerevisiae*-*S. bayanus* hybrid, discussed in reference 46), produce no viable spores (see, e.g., reference 69). In contrast, in the more natural wine or vineyard environment, a smaller proportion (relative to our results with clinical isolates) of natural wine isolates of *S. cerevisiae* produce no (or very few) viable spores (46). The presumably frequent human exposure to alloploid industrial *Saccharomyces* strains (e.g., the occupational, baking, and oral consumption of *S. cerevisiae* and its association with carriage and infection, as described above) may account for the recovery of allotetraploids and the larger (relative to natural wine isolates) proportion of *S. cerevisiae* clinical isolates that produce no viable spores.

While the reasons may differ, similar to many pathogenic fungi, some *S. cerevisiae* clinical (and nonclinical) isolates lack a full sexual cycle. Although they are of interest from epidemiological and population genetic perspectives, clinical *S. cerevisiae* isolates that lack a (full) sexual cycle are not readily amenable to further detailed genetic analysis. In contrast, highly detailed genetic analysis, including both Mendelian and quantitative genetic analyses, is possible with *S. cerevisiae* isolates and strains that have a full sexual cycle (see below).

THE SEXUAL CYCLE, HOMOTHALLISM/ HETEROTHALLISM, AND THE GENETIC STRUCTURE OF CLINICAL AND NONCLINICAL *S. CEREVISIAE* ISOLATES

One of the virtues of *S. cerevisiae* (relative to many pathogenic fungi) is that genetic analysis is greatly facilitated by its sexual cycle: *MATa/MATα* diploids can (with some exceptions) sporulate, tetrads can be dissected, and viable meiotic segregants can be genotyped and phenotyped. Therefore, for the clinical (and nonclinical) isolates of *S. cerevisiae* that produce viable spores and have a complete sexual cycle, it is possible to perform tetrad analysis and determine their genetic structures.

Because it is relevant to the sexual cycle and their genetic structure, it was important to determine if the clinical *S. cerevisiae* isolates that produced viable spores were homothallic or heterothallic: although heterothallic (*ho*) strains exist, *S. cerevisiae* is a naturally homothallic (*HO*) species. Haploid *ho* cells do not switch mating type and presumably are free to mate with unrelated strains. In contrast, haploid *HO* spores are able to switch mating type (in postgermination growth) and can self-diploidize to form *HO/HO MATa/MATα* diploids that (with the exception of the *MAT* locus) are homozygous throughout their entire genome (reviewed in references 68 and 102). Of 22 Spo⁺ viable spore-producing *S. cerevisiae* clinical isolates in the study discussed above (23), 17 were *HO/HO MATa/MATα* diploids, 3 were *ho/ho MATa/MATα* diploids, and 2 were *HO/ho MATa/MATα* diploids. Interestingly, similar proportions of *HO/HO*, *ho/ho*, and *HO/ho* are observed in wine or vineyard *S. cerevisiae* isolates (46, 65).

One might expect *HO/HO* isolates to be self-diploidized and therefore to lack any heterozygous mutations. However, although most clinical and nonclinical *S. cerevisiae* isolates are *HO/HO*, many wine or vineyard (46, 65) and clinical *S. cerevisiae* isolates (23, 57, 59) are heterozygous at one or more loci affecting obvious phenotypes, such as the ability to ferment different sugars. In the case of clinical isolates, there is also heterozygosity at multiple loci affecting a putative virulence trait, the ability to grow at high temperatures (59).

Given the ability of haploid *MATa* and *MATα* *HO* strains to self-diploidize and form *HO/HO MATa/MATα* diploids that are homozygous (except for *MAT*) over their entire genomes, what is the source of the heterozygosities in *S. cerevisiae* *HO/HO MATa/MATα* isolates? In wine or vineyard *S. cerevisiae* isolates, these heterozygosities are hypothesized to have arisen from mutations occurring during mitotic growth of diploid cells (see, e.g., references 63 and 65). However, while they are capable

of self-diploidizing, haploid *HO* spores can be crossed with other strains in the laboratory and, at some unknown frequency, are likely to cross with unrelated strains in nature. Therefore, while mutations undoubtedly occur in mitotically growing diploid cells, an alternative hypothesis for at least some of the heterozygosities in both wine/vineyard and clinical *S. cerevisiae* isolates is that these isolates are the products of crosses (outcrosses) between unrelated genetic backgrounds.

The frequency of outcrossing would presumably be reduced by homothallism, but the importance and frequency of the sexual cycle and the frequency of outcrossing in natural populations of *S. cerevisiae* is unknown. However, the outcrossing of unrelated *S. cerevisiae* genetic backgrounds in nature has the potential to generate a high degree of genotypic and phenotypic heterogeneity, which would be relevant to survival in different natural environments as well as to pathogenicity. Specifically, one appeal of the outcrossing hypothesis is that the diploid progeny of such crosses may exhibit heterosis, i.e., hybrid vigor or improved fitness. Below, I describe experimental crosses between unrelated *S. cerevisiae* genetic backgrounds that result in heterosis in terms of increased ability to grow at high temperatures and increased ability to survive in mice.

PHENOTYPIC DIFFERENCES BETWEEN CLINICAL AND NONCLINICAL ISOLATES OF *S. CEREVISIAE*: IN VITRO ADHERENCE, PSEUDOHYPHA FORMATION, AND HIGH-TEMPERATURE GROWTH

Certain phenotypes, such as the ability to adhere to host cells, undergo dimorphism (in the case of *S. cerevisiae*, to form pseudohyphae), and grow at high temperatures (>37°C), are considered to be important for fungal pathogenesis. Consistent with the importance of these factors and the importance of *S. cerevisiae* isolate-specific factors in infection, there were in vitro phenotypic differences between clinical and nonclinical isolates. First, although all of the *S. cerevisiae* isolates tested ($n = 7$) adhered to exfoliated buccal epithelial cells less than *C. albicans* did, clinical isolates of *S. cerevisiae* have been reported to adhere to buccal epithelial cells more strongly than nonclinical isolates (67). Second, while the proportion of clinical (13 of 15 isolates) versus nonclinical (6 of 10) *S. cerevisiae* isolates that formed pseudohyphae at 37°C was not statistically different ($P = 0.15$), the clinical isolates formed more extensive pseudohyphae (60). Finally, there was a highly significant correlation between clinical origin and the high-temperature-growth (Htg) phenotype, with the ability to grow at the highest temperature tested

(42°C) being found almost exclusively in clinical isolates (10 of 15 clinical isolates versus 1 of 10 non-clinical isolates [$P = 0.007$]) (60). These results demonstrate intraspecific heterogeneity in the ability of *S. cerevisiae* to adhere to host cells, undergo pseudo-hypha formation, and grow at high temperatures, and the clinical association suggests that these traits may play a role in *S. cerevisiae* pathogenesis. Genetic analysis of pseudohypha formation and high-temperature growth are described below, as well as the correlation of these phenotypes with *S. cerevisiae* survival in a CD-1 mouse model of infection.

EXPERIMENTAL INFECTIONS OF CLINICAL AND NONCLINICAL *S. CEREVISIAE* STRAINS AND ISOLATES: THE CD-1 MOUSE MODEL

Using outbred wild-type CD-1 mice, we established an *S. cerevisiae* infection model. After injection of up to 2×10^7 CFU of an *S. cerevisiae* strain or isolate into their tail veins, 4 to 10 mice were sacrificed at different postinfection time points; organs (brain, liver, kidneys, spleen, and lungs) were then removed and homogenized, and organ homogenates were plated to determine the CFU (22). There was little difference in any organ in the CFU of the clinical isolate YJM128 and the laboratory strain Y55 at 4 h postinfection. However, later postinfection time points (7, 14, and 28 days) revealed considerable differences (as much as ca. 1,000-fold in the brain on days 7 and 14 postinfection) in the number of CFU for YJM128 versus Y55, with YJM128 always being present in larger numbers. Of the five organs tested, yeast survival was highest in the brain. Proliferation of YJM128 on day 7 postinfection (a ca. fivefold increase over the 4-h postinfection CFU), but not Y55, was observed only in the brain.

Consistent with the importance of *S. cerevisiae* strain- and isolate-specific factors in infection, we found a very strong association between clinical origin and the ability to survive (persist) and proliferate in the brains of CD-1 mice (22), a phenotype that I refer to as in vivo survival-proliferation (Isp). In the 28 strains or isolates examined on day 14 postinfection, there was also a wide range (ca. 10^5-fold differences in CFU) in the Isp phenotypes, which (i) demonstrates intraspecific heterogeneity for Isp, (ii) is consistent with the heterogeneity in DNA typing, and (iii) suggests a high degree of genetic complexity, such as that found in quantitative (i.e., polygenic) traits.

Consistent with Isp being genetically complex and a quantitative trait, in experimental crosses between unrelated *S. cerevisiae* genetic backgrounds we found hybrid vigor (heterosis) for the Isp phenotype. First, two unrelated, prototrophic diploid laboratory genetic background strains, Y55 and YJM237, showed very limited survival and the lowest ability to survive in CD-1 mice, respectively. We crossed the S288c and Y55 genetic backgrounds to generate the Y55-S288c hybrid diploid YJM431. Testing in CD-1 mice showed that YJM431 had an Isp phenotype considerably greater than that of either parent (22), as well as an increased ability to grow at high temperatures (59, 60). Second, we crossed spores of the clinically derived YJM145 background (high Isp) with the S288c background to generate the YJM145-S288c hybrid diploid YJM274. YJM274 (and other YJM145-S288c hybrids) grow better at high temperatures (59, 103) and survive better than the YJM145 parent in the brains of CD-1 mice (H. Sinha and J. H. McCusker, unpublished data). Figure 1 (Fig. 1 of reference 103) depicts the

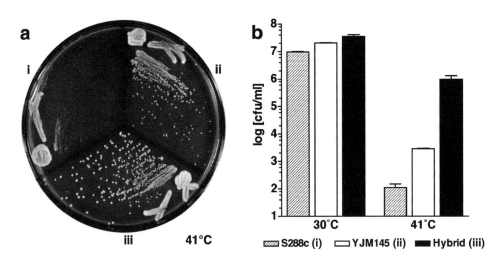

Figure 1. High-temperature growth phenotypes of S288c, YJM145, and S288c-YJM145 diploid strains as determined by a qualitative colony size assay (a) and a quantitative growth assay (b). Reprinted from reference 103 with permission.

Htg phenotypes of the S288c, YJM145, and S288c-YJM145 backgrounds as determined by both colony size and a more quantitative competition assay; note that the S288c-YJM145 hybrid strain displays heterosis. These results are consistent with the sexual cycle playing an important role in generating phenotypic diversity in natural populations. Conceptually, the increased Isp phenotype of the Y55-S288c and YJM145-S288c hybrids may be similar to observations made with *Toxoplasma gondii*, where some progeny from crosses between diverged lineages have increased virulence or oral transmission (see, e.g., references 40, 41, 104, and 110).

WHICH PHENOTYPES CORRELATE OR ASSOCIATE WITH Isp?

As discussed above, there was no statistically significant difference in the proportion of clinical versus nonclinical isolates that formed pseudohyphae. However, there was a significant correlation between the ability to form pseudohyphae in vitro at 37°C and the Isp phenotype; specifically, eight of eight strains or isolates with the highest Isp values formed pseudohyphae, in contrast to three of eight strains or isolates with the lowest Isp values ($P = 0.013$) (60). Figure 2 gives an example of robust in vitro pseudohypha formation by one high-Isp clinical *S. cerevisiae* isolate (YJM309). Further analysis of pseudohypha formation is discussed below.

Consistent with the correlation of Htg with clinical origin and the importance of high-temperature growth for fungal virulence, there was a significant correlation between the Isp and Htg phenotypes, with seven of eight strains or isolates with the highest Isp values being able to grow at 42°C, compared with none of eight strains or isolates with the lowest Isp values ($P < 0.001$) (60). However, the Isp-Htg correlation was not absolute. Specifically, the S288c background diploid YJM237 and the clinically derived diploid YJM454 are informative outliers: the two strains both grew poorly at 42°C (59, 60) but differed dramatically (by ca. 5×10^4-fold in CFU at 14 days postinfection) with respect to Isp (22). These results suggest that there are additional (i.e., non-Htg) contributors to the Isp phenotype.

Another phenotype important for Isp is mitochondrial genome maintenance (Mgm). Relative to isogenic ρ^+ (containing an intact mitochondrial genome) strains, we showed that ρ^0 (lacking the mitochondrial genome) *S. cerevisiae* strains were eliminated very rapidly in vivo; that is, ρ^0 strains had very poor Isp phenotypes (38). Weger et al. (111) showed a correlation between the ability of different *S. cerevisiae* strains or isolates to survive in vivo and their ability to maintain the mitochondrial genome when (i) grown at 37°C in vitro and (ii) recovered in vivo postinfection. The relationship between Mgm and Htg is not clear, although there is literature describing increased loss of the mitochondrial genome when some *S. cerevisiae* strains are grown at high temperatures (see, e.g., references 75, 96, and 97). Regardless of the relationship between Mgm and

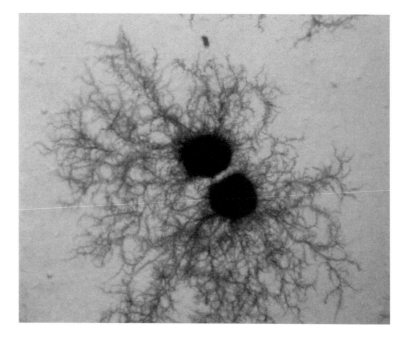

Figure 2. Pseudohypha formation of a clinically derived *S. cerevisiae* strain, YJM309, determined using a casein-containing medium. Reprinted from reference 60 with permission.

Htg, there is intraspecific heterogeneity for Mgm, an extensively studied process in *S. cerevisiae* (reviewed in reference 24) that is critical for Isp.

THE COMPLEMENT-DEFICIENT (C5⁻) MODEL OF *S. CEREVISIAE* INFECTION

Complement, a part of the innate immune system, plays an important role in defending the host against *C. albicans* (4, 5) and *Cryptococcus neoformans* (87) infections. Given the importance of complement, we wanted to determine if *S. cerevisiae*, which has a limited ability to survive in (and does not cause lethal infections in) complement-sufficient CD-1 and BALB/c mouse strains (12, 22), would (i) survive better in complement-deficient mice and (ii) kill complement-deficient mice. Using a tail vein infection model, we showed that 7 of 12 *S. cerevisiae* strains or isolates killed complement-deficient (C5⁻) DBA/2N mice; for the strains or isolates that killed C5⁻ mice, there was strain-to-strain (isolate-to-isolate) variation in the percentage of mice killed and in the time to kill mice. Therefore, as with Isp, Htg, Mgm, adherence, and pseudohypha formation, there is intraspecific heterogeneity in the ability of *S. cerevisiae* to kill C5⁻ mice.

In addition, (i) both the clinical isolate YJM128 and the laboratory strain Y55 survived better in C5⁻ B10.D2/oSnJ mice than in congenic C5⁺ B10.D2/nSnJ mice (12) and (ii) YJM128 (but not the laboratory strain Y55) killed B10.D2/oSnJ but not congenic C5⁺ B10.D2/nSnJ mice (12). We concluded that, similar to many pathogenic fungi, *S. cerevisiae* displays increased survival in and can cause lethal infections in C5⁻ hosts. The ability to cause lethal infections is a further demonstration of the pathogenic potential of *S. cerevisiae*.

Are *S. cerevisiae* strains or isolates that do not kill C5⁻ mice present in smaller numbers than strains or isolates that do kill C5⁻ mice? That is, is there any association between the lethal infection and Isp phenotypes? To address these questions, we infected DBA/2N mice (separately) with YJM128 and Y55 and determined the CFU in the first 3 days postinfection. Y55 (a laboratory strain, no lethal infections) and YJM128 (a clinical isolate, 75% lethal infections by 7 days, 60% lethal infections by 3 days) showed remarkably similar numbers of CFU and degrees of proliferation in the brains of C5⁻ DBA/2N mice (12), suggesting that the presence of viable yeast cells was not sufficient to cause lethal infections. Consistent with these results, in an examination of 12 strains (12, 22), there was no obvious association or significant (by Spearman's test) correlation of the Isp phenotype with the lethal-infection phenotype.

Given that YJM128 (lethal infections) and Y55 (no lethal infections) are present in approximately equal numbers in C5⁻ mice, how do (some) *S. cerevisiae* strains or isolates cause lethal infections? In some strains and isolates, *SSD1*, a gene that is polymorphic in the *S. cerevisiae* population (see, e.g., references 33 and 105), plays a role in causing lethal infections. Wheeler et al. showed that (i) a clinical *S. cerevisiae* isolate (CISC44) was *SSD1/ssd1* and (ii) while CISC44 did not kill DBA/2N mice, *ssd1/ssd1* derivatives of CISC44 did kill them (112).

What is the mechanism by which *ssd1*-deficient *S. cerevisiae* strains and isolates kill C5⁻ mice? As shown by Wheeler et al., compared to *SSD1* strains, *ssd1* mutants (i) have altered cell walls and (ii) induce the production of significantly more of the proinflammatory cytokines TNF-α, IL-6, and IL-1β; Wheeler et al. suggested that "the lethality was due to septic shock induced through overstimulation of the proinflammatory arm of the innate immune system" (112). The hypervirulent phenotype of *ssd1* mutants is both interesting and somewhat paradoxical since *ssd1* mutants show reduced growth at high temperature (112). In this sense, *ssd1* may be viewed as a pathoadaptive mutation (reviewed in references 36 and 101).

Are all of the *S. cerevisiae* strains and isolates that kill C5⁻ mice *ssd1* mutants? If so, one would expect an inverse association between the lethal-infection and Htg phenotypes. However, of the seven *S. cerevisiae* strains or isolates that killed C5⁻ DBA/2N mice (12), only one failed to grow at 37°C while all of the others grew at 39 and even 42°C. Although the temperature at which *ssd1* mutant strains fail to grow may be background dependent, these results suggest that *SSD1 S. cerevisiae* strains can cause lethal infections in C5⁻ mice. Indeed, although *ssd1/ssd1* derivatives showed increased virulence, EM93, which is *SSD1/SSD1*, killed DBA/2N mice (112). Note also that EM93, which was originally isolated from a rotting fig, constitutes ca. 88% of the S288c background (64) and, while EM93 killed DBA/2N mice (112), the S288c background diploid YJM237 did not (22). Therefore, while *SSD1* plays a role in some *S. cerevisiae* strains and isolates that kill C5⁻ mice, other genes that are polymorphic in the *S. cerevisiae* population are likely to play significant roles as well, which would suggest that the lethal-infection phenotype is a quantitative trait.

GENETIC ANALYSIS

Examination of Defined *S. cerevisiae* Mutants In Vivo

In a classical genetic approach to analyzing pathogenesis, we constructed (and reconstituted) multiple mutants and analyzed the ability of these mutants to survive in CD-1 mice. For this analysis, it was essential to use an *S. cerevisiae* genetic background that survived well in mice. As described above, the S288c laboratory genetic background has the poorest Isp phenotype of all strains and isolates tested while the clinically derived YJM145 genetic background has a high-Isp phenotype (22). Therefore, we chose to use the YJM145 genetic background.

We infected CD-1 mice with pools containing multiple isogenic (YJM145 genetic background) mutants and showed that (i) strains did not mate or undergo meiosis in vivo, (ii) the dominant drug resistance markers (39) that we used to make and distinguish different mutants were phenotypically neutral in vivo, and (iii) we could examine pools of up to 13 mutants. These are critical factors for examining quantitative traits, pathogenesis, and the relative ability of specific strains in pools to survive and proliferate in vivo.

We used competitive indices, a sensitive and quantifiable measure of in vivo survival and proliferation used extensively in bacterial (see, e.g., reference 19) and fungal (10) pathogenesis studies, to reliably determine the relative fitness of multiple mutants in pools. To date, we have determined the in vivo survival and proliferation phenotypes of 43 single mutants, 9 double mutants, 4 triple mutants, and 1 quadruple mutant (38; A. Goldstein and J. H. McCusker, unpublished data; J. M. Kingsbury and J. H. McCusker, unpublished data), including multiple auxotrophic mutations, drug resistance mutations, isocitrate lyase mutants, mutants deficient in respiration (*cox15Δ*, *mip1Δ*, and ρ⁰), and mutants deficient in pseudohypha formation/dimorphism. Although there are exceptions (species-specific mutant phenotypes are to be expected), the phenotypes of *S. cerevisiae* mutants are usually similar to those of the equivalent *C. albicans* and/or *C. neoformans* mutants, which speaks to the generality of fungal gene products and processes that are important for fungi in the host environment. The analysis of pseudohypha formation/dimorphism and its role in the survival of *S. cerevisiae* in the host environment is described in more detail below.

S. cerevisiae Pseudohypha Formation and Isp

The role of dimorphism in fungal pathogenesis appears to be species specific. For fungi such as *C. albicans*, the ability to undergo dimorphism in vivo is an important factor in pathogenesis. While there may be other important differences between yeast and (pseudo)hyphal cell types, yeast cells may be better able to disseminate in vivo while

(pseudo)hyphae may be better able to invade and attach to tissues and avoid being engulfed and killed by host cells (e.g., macrophages). Experimental evidence strongly supports the importance of in vivo dimorphism of *C. albicans* in pathogenesis (see, e.g., reference 93). In contrast, while *C. glabrata* has been reported to form pseudohyphae in vitro (27), this commonly observed pathogenic fungus does not form pseudohyphae in vivo and, like *S. cerevisiae*, does not kill wild-type, immunocompetent mice (34).

Two pieces of evidence suggested that pseudohypha formation would be important for *S. cerevisiae* survival in mice. First, the S288c genetic background, which is the least able of all strains or isolates tested to survive in vivo (22), has a naturally occurring *flo8* mutation that abolishes pseudohypha formation (52). Second, as described above, we found a correlation between pseudohypha formation (in vitro) and survival in the brains of CD-1 mice (60). However, the failure (as determined by histology) of YJM128 and Y55 to form pseudohyphae in mice (12, 22) raised a note of caution.

To determine the contribution of dimorphism to the survival of *S. cerevisiae* in vivo, we constructed *flo8Δ*, *flo8Δ tec1Δ*, and *flo11Δ* mutant strains in the clinically derived *S. cerevisiae* YJM145 genetic background, which makes profuse pseudohyphae in vitro (60), and used them to infect CD-1 mice. While all of the mutations abolished pseudohypha formation in vitro, we found that the *flo8Δ* and *flo8Δ tec1Δ* mutants had only slight in vivo survival defects while *flo11Δ* mutants were phenotypically neutral in vivo (38).

There are two ways to reconcile (i) the Isp-in vitro pseudohypha correlation, (ii) the failure of YJM128 and Y55 to form pseudohyphae in vivo, and (iii) the YJM145 near-normal *flo8Δ* and normal *flo11Δ* Isp phenotypes. First, there may be strain-to-strain (or isolate-to-isolate) variation in the ability of *S. cerevisiae* to form pseudohyphae in vivo; that is, some *S. cerevisiae* strains or isolates—but not YJM128, YJM145 (a YJM128 segregant), or Y55— may form pseudohyphae in vivo. Second, like its close relative *C. glabrata*, *S. cerevisiae* may not form pseudohyphae in vivo; in this case, a Flo8p- and Flo11p-independent process or pathway that is important for pseudohypha formation in vitro may also be important for survival in vivo.

Why is there such a difference between the closely related species *S. cerevisiae*, *C. glabrata*, and *C. albicans* in the ability to undergo dimorphism in vivo, the importance of dimorphism for virulence, and the degree of virulence? One hypothesis is that the in vivo phenotypes of the emerging pathogen *S. cerevisiae* (poor or brain-only survival-proliferation and [in C5$^+$ mice] nonlethal infections), which is similar to *C. glabrata*, are at least in part a consequence of the inability of *S. cerevisiae* to undergo dimorphism in vivo.

Why might *S. cerevisiae*—and *C. glabrata*—fail to undergo a dimorphic switch in vivo? In vitro, *S. cerevisiae* dimorphism occurs under low-nitrogen-source conditions (similar to *C. glabrata* [27]) and is further enhanced by poorly utilized carbon sources (reviewed in reference 78). In serum (presumably a close analog of in vivo conditions), glucose, which is readily utilizable, appears to be the sole carbon source and there is a relatively high concentration of nitrogen (26). Therefore, the environmental conditions in vivo are a likely explanation for the failure of *S. cerevisiae* (and *C. glabrata*) to undergo dimorphism in vivo, and this may in turn contribute to the relatively low pathogenicity of *S. cerevisiae* (and *C. glabrata*). In this sense, acquisition of the ability to undergo dimorphism in vivo may be one of several pathways that contribute to the emergence of some opportunistic pathogenic fungi.

Analysis of a Putative Virulence Trait—the Quantitative Trait of High-Temperature Growth

As discussed above, the Htg phenotype is found almost exclusively in *S. cerevisiae* clinical isolates and is highly correlated with the Isp phenotype. For these reasons, and because Htg is readily assayed in vitro, we chose to genetically analyze the Htg phenotype. When tested for Htg, the segregants of *S. cerevisiae* clinical isolates displayed a wide range of Htg phenotypes, consistent with these isolates being heterozygous at multiple loci affecting Htg and with Htg being a quantitative trait (59). However, because they are largely undefined, the genetic analysis of clinical isolates can be quite complicated. Therefore, we performed crosses between clinically derived Htg$^+$ strains (self-diploidized segregants of clinical isolates) and the well-defined Htg$^-$ S288c laboratory genetic background. The segregants from these crosses all displayed a wide range of Htg phenotypes, clearly demonstrating that the Htg phenotype is a genetically complex, polygenic, or quantitative trait (59).

How does one analyze a quantitative trait? Unfortunately, while they are ideal for the analysis of Mendelian traits where a single gene makes a major contribution to phenotype, the standard techniques of yeast genetics (e.g., complementation testing and cloning by complementation) are ill suited for the analysis of quantitative traits, where a single gene makes a small contribution to the phenotype. Instead, to analyze a quantitative trait, one must first map the quantitative trait loci (QTLs). To map Htg QTLs, we crossed the Htg$^+$ YJM145 and Htg$^-$ S288c

backgrounds, determined the Htg phenotypes of the segregants, and genotyped the segregants on Affymetrix arrays (103). (Affymetrix arrays readily detect single-nucleotide polymorphisms, thereby providing a very large number of markers for mapping studies [see, e.g., reference 115].)

In this first cross, we mapped one of the QTLs to a 32-kb region of chromosome 14; since a QTL contains multiple genes (in this QTL, 15 genes), we were then faced with the problem of identifying the quantitative-trait gene(s) (QTGs); each QTL will contain one or more QTGs. As a cautionary tale to those who hope to use similar techniques, expression analysis, presumed gene product relevance to Htg, and marker trait association all failed to identify candidate QTGs (103). Therefore, we developed reciprocal hemizygosity analysis (RHA) to identify QTGs. Using RHA, we compared the Htg phenotypes of pairs of isogenic, hemizygous S288c-YJM145 hybrids. For each gene in the QTL, one hybrid was deleted for the S288c-derived allele of that gene but contained the YJM145-derived allele, and the other hybrid was deleted for the YJM145-derived allele of that gene but contained the S288c-derived allele. Figure 3 (Fig. 4 of reference 103) depicts RHA and also shows the phenotypes of *MKT1*, *END3*, and *RHO2* reciprocal hemizygotes.

For 12 of the 15 genes in the QTL, we observed no difference in the Htg phenotypes of the reciprocal hemizygotes; that is, there were no phenotypically relevant differences between the alleles of these 12 genes so they were not QTGs. However, for *MKT1*, *END3*, and *RHO2*, we found allele-specific contributions to phenotype; specifically, the YJM145-derived alleles of *MKT1* and *RHO2* and the S288c-derived allele of *END3* made major contributions to the Htg phenotype and were QTGs. The complexity of the QTL was unexpected, as was the finding that the allele of *END3* from the Htg$^-$ S288c background contributed to the Htg phenotype. However, the *END3* result explained much of the observed heterosis (S288c-YJM145 hybrids are more Htg$^+$ than is YJM145) and the previous observation that segregants of S288c-YJM145 hybrids never grew as well as the S288c-YJM145 hybrid itself (59): *MKT1* and *END3* are contiguous genes and the Htg$^+$ alleles are in *trans*. In subsequent (unpublished) work, we have mapped and dissected three additional Htg QTLs, all of which contain more than one QTG, and identified another nine Htg QTGs. Our analysis of high-temperature growth illustrates the complexity of a quantitative trait that is relevant to pathogenesis. We were well served in this case by the molecular tools and facile genetics of *S. cerevisiae*. Together, these points underscore the significance of

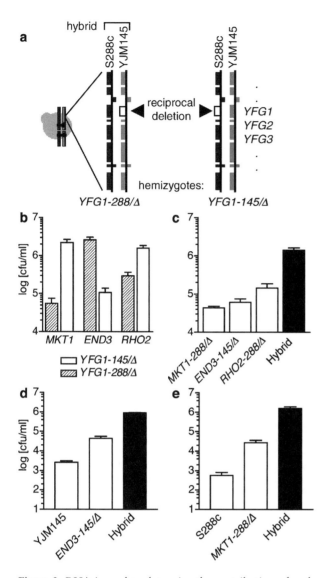

Figure 3. RHA is used to determine the contribution of each allele to a specific phenotype (in this case, high-temperature growth) and thereby identify quantitative-trait genes. Reprinted from reference 103 with permission.

clinical *S. cerevisiae* isolates, not as a major pathogen but as an emerging opportunistic pathogen and a readily analyzed model.

CONCLUSION

The phenotypic diversity in natural populations is the result of genetic diversity: i.e., alleles of multiple, segregating loci interact to produce a spectrum of phenotypes. The extreme ends of the phenotypic spectrum differ at multiple loci and represent genetically complex, polygenic, or quantitative traits. The genotypic and phenotypic diversity in natural populations plays a critical role in many species, including adaptation to novel environments. For fungi, when genotypic and phenotypic diversity

results in adaptation to one novel environment—the human body—it is a matter of particular concern since the result is the emergence of a pathogenic fungus.

Very little is known about the emergence of pathogenic fungi. Understanding the emergence of pathogenic fungi will aid our basic understanding of how fungi (i) survive and proliferate in the host environment and (ii) infect and kill the host. Our understanding of the emergence of pathogenic fungi will be facilitated by the study of a genetically tractable emerging pathogen, with a highly annotated genome sequence and genomic tools. As described in this chapter, *S. cerevisiae* is just such an emerging opportunistic pathogen.

The analysis of clinical (and nonclinical) isolates of *S. cerevisiae* benefits immensely from the use of sensitive, high-throughput genotyping technologies, such as the Affymetrix yeast genome arrays, that readily detect deletions, changes in gene copy numbers, and single-nucleotide polymorphisms across the entire genome (see, e.g., reference 114). Similarly, the analysis of other pathogenic fungi will be greatly facilitated by the development of high-throughput genotyping technologies. Recurring themes in this chapter are that (i) multiple traits contribute to *S. cerevisiae* pathogenicity and virulence, (ii) there is a high degree of genotypic and phenotypic diversity in *S. cerevisiae*, and (iii) the phenotypic diversity for each trait is largely quantitative. This is true for *S. cerevisiae* and will be true for other emerging opportunistic pathogenic fungi as well as for well-established pathogenic fungi.

REFERENCE

1. **Agatensi, L., F. Franchi, F. Mondello, R. L. Bevilacqua, T. Ceddia, F. De Bernardis, and A. Cassone.** 1991. Vaginopathic and proteolytic *Candida* species in outpatients attending a gynaecology clinic. *J. Clin. Pathol.* **44:**826–830.

2. **Anaissie, E., G. P. Bodey, H. Kantarjian, J. Ro, S. E. Vartivarian, R. Hopfer, J. Hoy, and K. Rolston.** 1989. New spectrum of fungal infections in patients with cancer. *Rev. Infect. Dis.***11:**369–378.

3. **Arzeni, D., M. Del Poeta, O. Simonetti, A. M. Offidani, L. Lamura, M. Balducci, N. Cester, A. Giacometti, and G. Scalise.** 1997. Prevalence and antifungal susceptibility of vaginal yeasts in outpatients attending a gynecological center in Ancona, Italy. *Eur. J. Epidemiol.***13:**447–450.

4. **Ashman, R. B.** 1998. *Candida albicans*: pathogenesis, immunity and host defence. *Res. Immunol.* **149:**281–288; discussion, 494–496.

5. **Ashman, R. B., E. M. Bolitho, and J. M. Papadimitriou.** 1993. Patterns of resistance to *Candida albicans* in inbred mouse strains. *Immunol. Cell Biol.* **71:**221–225.

6. **Aucott, J. N., J. Fayen, H. Grossnicklas, A. Morrissey, M. M. Lederman, and R. A. Salata.** 1990. Invasive infection with *Saccharomyces cerevisiae*: report of three cases and review. *Rev. Infect. Dis.* **12:**406–411.

7. Barchiesi, F., D. Arzeni, P. Compagnucci, L. F. Di Francesco, A. Giacometti, and G. Scalise. 1998. *In vitro* activity of five antifungal agents against clinical isolates of *Saccharomyces cerevisiae*. *Med. Mycol.* **36**:437–440.

8. Bassetti, S., R. Frei, and W. Zimmerli. 1998. Fungemia with *Saccharomyces cerevisiae* after treatment with *Saccharomyces boulardii*. *Am. J. Med.* **105**:71–72.

9. Bowman, B. H., J. W. Taylor, and T. J. White. 1992. Molecular evolution of the fungi: human pathogens. *Mol. Biol. Evol.* **9**:893–904.

10. Brown, J. S., A. Aufauvre–Brown, J. Brown, J. M. Jennings, H. Arst, Jr., and D. W. Holden. 2000. Signature-tagged and directed mutagenesis identify PABA synthetase as essential for *Aspergillus fumigatus* pathogenicity. *Mol. Microbiol.* **36**:1371–1380.

11. Buts, J. P., P. Bernasconi, J. P. Vaerman, and C. Dive. 1990. Stimulation of secretory IgA and secretory component of immunoglobulins in small intestine of rats treated with *Saccharomyces boulardii*. *Dig. Dis. Sci.* **35**:251–256.

12. Byron, J. K., K. V. Clemons, J. H. McCusker, R. W. Davis, and D. A. Stevens. 1995. Pathogenicity of *Saccharomyces cerevisiae* in complement factor five (C5)-deficient mice. *Infect. Immun.* **63**:478–485.

13. Cairoli, R., P. Marenco, R. Perego, and F. de Cataldo. 1995. *Saccharomyces cerevisiae* fungemia with granulomas in the bone marrow in a patient undergoing BMT. *Bone Marrow Transplant.* **15**:785–786.

14. Candelli, M., E. C. Nista, M. Nestola, A. Armuzzi, N. G. Silveri, G. Gasbarrini, and A. Gasbarrini. 2003. *Saccharomyces cerevisiae*-associated diarrhea in an immunocompetent patient with ulcerative colitis. *J. Clin. Gastroenterol.* **36**:39–40.

15. Cassone, M., P. Serra, F. Mondello, A. Girolamo, S. Scafetti, E. Pistella, and M. Venditti. 2003. Outbreak of *Saccharomyces cerevisiae* subtype boulardii fungemia in patients neighboring those treated with a probiotic preparation of the organism. *J. Clin. Microbiol.* **41**:5340–5343.

16. Cesaro, S., P. Chinello, L. Rossi, and L. Zanesco. 2000. *Saccharomyces cerevisiae* fungemia in a neutropenic patient treated with *Saccharomyces boulardii*. *Support Care Cancer* **8**:504–505.

17. Chertow, G. M., E. R. Marcantonio, and R. G. Wells. 1991. *Saccharomyces cerevisiae* empyema in a patient with esophago-pleural fistula complicating variceal scleropathy. *Chest* **99**:1518–1519.

18. Chia, J. K., S. M. Chan, and H. Goldstein. 1995. Baker's yeast as adjunctive therapy for relapses of *Clostridium difficile* diarrhea. *Clin. Infect. Dis.* **20**:1581.

19. Chiang, S. L., and J. J. Mekalanos. 1998. Use of signature-tagged transposon mutagenesis to identify *Vibrio cholerae* genes critical for colonization. *Mol. Microbiol.* **27**:797–805.

20. Cimolai, N., M. J. Gill, and D. Church. 1987. *Saccharomyces cerevisiae* fungemia: case report and review of the literature. *Diagn. Microbiol. Infect. Dis.* **8**:113–117.

21. Clayton, Y. M., J. Shao, and G. Brubaker. 1991. Oral yeast flora of a population in an area of Tanzania bordering an AIDS endemic zone. *East Afr. Med. J.* **68**:975–979.

22. Clemons, K. V., J. H. McCusker, R. W. Davis, and D. A. Stevens. 1994. Comparative pathogenesis of clinical and nonclinical isolates of *Saccharomyces cerevisiae*. *J. Infect. Dis.* **169**:859–867.

23. Clemons, K. V., P.-S. Park, J. H. McCusker, M. J. McCullough, R. W. Davis, and D. A. Stevens. 1997. Application of DNA typing methods and genetic analysis to epidemiology and taxonomy of *Saccharomyces* isolates. *J. Clin. Microbiol.* **35**:1822–1828.

24. Contamine, V., and M. Picard. 2000. Maintenance and integrity of the mitochondrial genome: a plethora of nuclear genes in the budding yeast. *Microbiol. Mol. Biol. Rev.* **64**:281–315.

25. Corthier, G., F. Dubos, and R. Ducluzeau. 1986. Prevention of *Clostridium difficile* induced mortality in gnotobiotic mice by *Saccharomyces boulardii*. *Can. J. Microbiol.* **32**:894–896.

26. Crispens, C. G. 1975. *Handbook of the Laboratory Mouse, Section IV: Blood*. p. 93–123. Charles C Thomas, Springfield, Ill.

27. Csank, C., and K. Haynes. 2000. *Candida glabrata* displays pseudohyphal growth. *FEMS Microbiol. Lett.* **189**:115–120.

28. Czerucka, D., and P. Rampal. 2002. Experimental effects of *Saccharomyces boulardii* on diarrheal pathogens. *Microbes Infect.* **4**:733–739.

29. Dougherty, S. H., and R. L. Simmons. 1982. Postoperative peritonitis caused by *Saccharomyces cerevisiae*. *Arch. Surg.* **117**:248–249.

30. Edwards-Ingram, L. C., M. E. Gent, D. C. Hoyle, A. Hayes, L. I. Stateva, and S. G. Oliver. 2004. Comparative genomic hybridization provides new insights into the molecular taxonomy of the *Saccharomyces sensu stricto* complex. *Genome Res.* **14**:1043–1051.

31. Eng, R. H. K., R. Drehmel, S. M. Smith, and E. J. C. Goldstein. 1984. *Saccharomyces cerevisiae* infections in man. *J. Med. Vet. Mycol.* **22**:403–407.

32. Eschete, M. L., and B. C. West. 1980. *Saccharomyces cerevisiae* septicemia. *Arch. Intern. Med.* **140**:1539.

33. Evans, D. R., and M. J. Stark. 1997. Mutations in the *Saccharomyces cerevisiae* type 2A protein phosphatase catalytic subunit reveal roles in cell wall integrity, actin cytoskeleton organization and mitosis. *Genetics* **145**:227–241.

34. Fidel, P. L., Jr., J. A. Vazquez, and J. D. Sobel. 1999. *Candida glabrata*: review of epidemiology, pathogenesis, and clinical disease with comparison to *C. albicans*. *Clin. Microbiol. Rev.* **12**:80–96.

35. Fietto, J. L., R. S. Araujo, F. N. Valadao, L. G. Fietto, R. L. Brandao, M. J. Neves, F. C. Gomes, J. R. Nicoli, and I. M. Castro. 2004. Molecular and physiological comparisons between *Saccharomyces cerevisiae* and *Saccharomyces boulardii*. *Can. J. Microbiol.* **50**:615–621.

36. Foreman-Wykert, A. K., and J. F. Miller. 2003. Hypervirulence and pathogen fitness. *Trends Microbiol.* **11**:105–108.

37. Fredenucci, I., M. Chomarat, C. Boucaud, and J. P. Flandrois. 1998. *Saccharomyces boulardii* fungemia in a patient receiving Ultra-levure therapy. *Clin. Infect. Dis.* **27**:222–223.

38. Goldstein, A. L., and J. H. McCusker. 2001. Development of *Saccharomyces cerevisiae* as a model pathogen. A system for the genetic identification of gene products required for survival in the mammalian host environment. *Genetics* **159**:499–513.

39. Goldstein, A. L., and J. H. McCusker. 1999. Three new dominant drug resistance cassettes for gene disruption in *Saccharomyces cerevisiae*. *Yeast* **15**:1541–1553.

40. Grigg, M. E., S. Bonnefoy, A. B. Hehl, Y. Suzuki, and J. C. Boothroyd. 2001. Success and virulence in

Toxoplasma as the result of sexual recombination between two distinct ancestries. *Science* 294:161–165.

41 Grigg, M. E., and Y. Suzuki. 2003. Sexual recombination and clonal evolution of virulence in *Toxoplasma*. *Microbes Infect.* 5:685–690.

42. Hazen, K. C. 1995. New and emerging yeast pathogens. *Clin. Microbiol. Rev.* 8:462–478.

43. Hennequin, C., A. Thierry, G. F. Richard, G. Lecointre, H. V. Nguyen, C. Gaillardin, and B. Dujon. 2001. Microsatellite typing as a new tool for identification of *Saccharomyces cerevisiae* strains. *J. Clin. Microbiol.* 39:551–559.

44. Izadnia, F., C. T. Wong, and S. A. Kocoshis. 1998. Brewer's yeast and *Saccharomyces boulardii* both attenuate *Clostridium difficile*-induced colonic secretion in the rat. *Dig. Dis. Sci.* 43:2055–2060.

45. Jensen, D. P., and D. L. Smith. 1976. Fever of unknown origin secondary to brewer's yeast ingestion. *Arch. Intern. Med.* 136:332–333.

46. Johnston, J. R., C. Baccari, and R. K. Mortimer. 2000. Genotypic characterization of strains of commercial wine yeasts by tetrad analysis. *Res. Microbiol.* 151:583–590.

47. Kiehn, T. E., F. F. Edwards, and D. Armstrong. 1980. The prevalence of yeasts in clinical specimens from cancer patients. *Am. J. Clin. Pathol.* 73:518–521.

48. Konecny, P., F. M. Drummond, K. N. Tish, and J. W. Tapsall. 1999. *Saccharomyces cerevisiae* oesophagitis in an HIV-infected patient. *Int. J. STD AIDS* 10:821–822.

49. Kurtzman, C. P., and J. W. Fell. 1998. *The Yeasts: a Taxonomic Study*, 4th ed. Elsevier, Amsterdam, The Netherlands.

50. Kurtzman, C. P., and C. J. Robnett. 2003. Phylogenetic relationships among yeasts of the 'Saccharomyces complex' determined from multigene sequence analyses. *FEMS Yeast Res.* 3:417–432.

51. Lherm, T., C. Monet, B. Nougiere, M. Soulier, D. Larbi, C. Le Gall, D. Caen, and C. Malbrunot. 2002. Seven cases of fungemia with *Saccharomyces boulardii* in critically ill patients. *Intensive Care Med.* 28:797–801.

52. Liu, H., C. A. Styles, and G. R. Fink. 1996. *Saccharomyces cerevisiae* S288C has a mutation in *FLO8*, a gene required for filamentous growth. *Genetics* 144:967–978.

53. Lott, T. J., R. J. Kuykendall, and E. Reiss. 1993. Nucleotide sequence analysis of the 5.8S rDNA and adjacent ITS2 region of *Candida albicans* and related species. *Yeast* 9:1199–1206.

54. Lungarotti, M. S., D. Mezzetti, and M. Radicioni. 2003. Methaemoglobinaemia with concurrent blood isolation of *Saccharomyces* and *Candida*. *Arch. Dis. Child. Fetal Neonatal Ed.* 88:F446.

55. Mallie, M., P.-N. Van, S. Bertout, C. Vaillant, and J.-M. Bastide. 2001. Genotypic study of *Saccharomyces boulardii* compared to the *Saccharomyces sensu stricto* complex species. *J. Mycol. Med.* 11:19–25.

56. Manzella, J. P., S. Shaffer, N. Agarwal, and J. A. Kellogg. 1989. *Saccharomyces cerevisiae* fungemia in a multiply traumatized patient. *J. Trauma* 29:129–130.

57. McCullough, M. J., K. V. Clemons, C. Farina, J. H. McCusker, and D. A. Stevens. 1998. Epidemiological investigation of vaginal *Saccharomyces cerevisiae* isolates by a genotypic method. *J. Clin. Microbiol.* 36:557–562.

58. McCullough, M. J., K. V. Clemons, J. H. McCusker, and D. A. Stevens. 1998. Species identification and virulence attributes of *Saccharomyces boulardii* (nom. inval.). *J. Clin. Microbiol.* 36:2613–2617.

59. McCusker, J. H., K. V. Clemons, D. A. Stevens, and R. W. Davis. 1994. Genetic characterization of pathogenic *Saccharomyces cerevisiae* isolates. *Genetics* 136:1261–1269.

60. McCusker, J. H., K. V. Clemons, D. A. Stevens, and R. W. Davis. 1994. *Saccharomyces cerevisiae* virulence phenotype in CD-1 mice is associated with the ability to grow at 42°C and form pseudohyphae. *Infect. Immun.* 62:5447–5455.

61. McFarland, L. V. 1996. *Saccharomyces boulardii* is not *Saccharomyces cerevisiae*. *Clin. Infect. Dis.* 22:200–201.

62. Mitterdorfer, G., H. K. Mayer, W. Kneifel, and H. Viernstein. 2002. Clustering of *Saccharomyces boulardii* strains within the species *S. cerevisiae* using molecular typing techniques. *J. Appl. Microbiol.* 93:521–530.

63. Mortimer, R. K. 2000. Evolution and variation of the yeast (*Saccharomyces*) genome. *Genome Res.* 10:403–409.

64. Mortimer, R. K., and J. R. Johnston. 1986. Genealogy of principal strains of the Yeast Genetic Stock Center. *Genetics* 113:35–43.

65. Mortimer, R. K., P. Romano, G. Suzzi, and M. Polsinelli. 1994. Genome renewal: a new phenomenon revealed from a genetic study of 43 strains of *Saccharomyces cerevisiae* derived from natural fermentation of grape musts. *Yeast* 10:1543–1552.

66. Murphy, A., and K. Kavanagh. 1999. Emergence of *Saccharomyces cerevisiae* as a human pathogen: implications for biotechnology. *Enzyme Microb. Technol.* 25:551–557.

67. Murphy, A. R., and K. A. Kavanagh. 2001. Adherence of clinical isolates of *Saccharomyces cerevisiae* to buccal epithelial cells. *Med. Mycol.* 39:123–127.

68. Nasmyth, K., and D. Shore. 1987. Transcriptional regulation in the yeast life cycle. *Science* 237:1162–1170.

69. Naumov, G. I., S. A. James, E. S. Naumova, E. J. Louis, and I. N. Roberts. 2000. Three new species in the *Saccharomyces sensu stricto* complex: *Saccharomyces cariocanus*, *Saccharomyces kudriavzevii* and *Saccharomyces mikatae*. *Int. J. Syst. Evol. Microbiol.* 50:1931–1942.

70. Naumov, G. I., E. S. Naumova, I. Masneuf, M. Aigle, V. I. Kondratieva, and D. Dubourdieu. 2000. Natural polyploidization of some cultured yeast *Saccharomyces sensu stricto*: auto- and allotetraploidy. *Syst. Appl. Microbiol.* 23:442–449.

71. Niault, M., F. Thomas, J. Prost, F. H. Ansari, and P. Kalfon. 1999. Fungemia due to *Saccharomyces* species in a patient treated with enteral *Saccharomyces boulardii*. *Clin. Infect. Dis.* 28:930.

72. Nielsen, H., J. Stenderup, and B. Bruun. 1990. Fungemia with Saccharomycetaceae. Report of four cases and review of the literature. *Scand. J. Infect. Dis.* 22:581–584.

73. Nyirjesy, P., S. M. Seeney, M. H. Grody, C. A. Jordan, and H. R. Buckley. 1995. Chronic fungal vaginitis: the value of cultures. *Am. J. Obstet. Gynecol.* 173:820–823.

74. Nyirjesy, P., J. A. Vazquez, D. D. Ufberg, J. D. Sobel, D. A. Boikov, and H. R. Buckley. 1995. *Saccharomyces cerevisiae* vaginitis: transmission from yeast used in baking. *Obstet. Gynecol.* 86:326–329.

75. Ogur, M., S. Ogur, and R. S. John. 1960. Temperature dependence of the spontaneous mutation rate to respiration deficiency in *Saccharomyces*. *Genetics* 45:189–194.

76. Olver, W. J., S. A. James, A. Lennard, A. Galloway, I. N. Roberts, T. C. Boswell, and N. H. Russell. 2002. Nosocomial transmission of *Saccharomyces cerevisiae* in bone marrow transplant patients. *J. Hosp. Infect.* 52:268–272.

77. Oriol, A., J. M. Ribera, J. Arnal, F. Milla, M. Batlle, and E. Feliu. 1993. *Saccharomyces cerevisiae* septicemia in a patient with myelodysplastic syndrome. *Am. J. Hematol.* 43:325–326.

78. Palecek, S. P., A. S. Parikh, and S. J. Kron. 2002. Sensing, signalling and integrating physical processes during *Saccharomyces cerevisiae* invasive and filamentous growth. *Microbiology* 148:893–907.

79. Pecquet, S., D. Guillaumin, C. Tancrede, and A. Andremont. 1991. Kinetics of *Saccharomyces cerevisiae* elimination from the intestines of human volunteers and effect of this yeast on resistance to microbial colonization in gnotobiotic mice. *Appl. Environ. Microbiol.* 57:3049–3051.

80. Perapoch, J., A. M. Planes, A. Querol, V. Lopez, I. Martinez-Bendayan, R. Tormo, F. Fernandez, G. Peguero, and S. Salcedo. 2000. Fungemia with *Saccharomyces cerevisiae* in two newborns, only one of whom had been treated with ultra-levura. *Eur J. Clin. Microbiol. Infect. Dis.* 19:468–470.

81. Pfaller, M. A., S. Messer, and R. N. Jones. 1997. Activity of a new triazole, Sch 56592, compared with those of four other antifungal agents tested against clinical isolates of *Candida* spp. and *Saccharomyces cerevisiae*. *Antimicrob. Agents. Chemother.* 41:233–235.

82. Piarroux, R., L. Millon, K. Bardonnet, O. Vagner, and H. Koenig. 1999. Are live *Saccharomyces* yeasts harmful to patients? *Lancet* 353:1851–1852.

83. Ponton, J., R. Ruchel, K. V. Clemons, D. C. Coleman, R. Grillot, J. Guarro, D. Aldebert, P. Ambroise-Thomas, J. Cano, A. J. Carrillo-Munoz, J. Gene, C. Pinel, D. A. Stevens, and D. J. Sullivan. 2000. Emerging pathogens. *Med. Mycol.* 38 (Suppl. 1):225–236.

84. Posteraro, B., M. Sanguinetti, G. D'Amore, L. Masucci, G. Morace, and G. Fadda. 1999. Molecular and epidemiological characterization of vaginal *Saccharomyces cerevisiae* isolates. *J. Clin. Microbiol.* 37:2230–2235.

85. Qamar, A., S. Aboudola, M. Warny, P. Michetti, C. Pothoulakis, J. T. LaMont, and C. P. Kelly. 2001. *Saccharomyces boulardii* stimulates intestinal immunoglobulin A immune response to *Clostridium difficile* toxin A in mice. *Infect. Immun.* 69:2762–2765.

86. Ren, P., S. Sridhar, and V. Chaturvedi. 2004. Use of paraffin-embedded tissue for identification of *Saccharomyces cerevisiae* in a baker's lung nodule by fungal PCR and nucleotide sequencing. *J. Clin. Microbiol.* 42:2840–2842.

87. Rhodes, J. C., L. S. Wicker, and W. J. Urba. 1980. Genetic control of susceptibility to *Cryptococcus neoformans* in mice. *Infect. Immun.* 29:494–499.

88. Rijnders, B. J., E. Van Wijngaerden, C. Verwaest, and W. E. Peetermans. 2000. Saccharomyces fungemia complicating *Saccharomyces boulardii* treatment in a nonimmunocompromised host. *Intensive Care Med.* 26:825.

89. Riquelme, A. J., M. A. Calvo, A. M. Guzman, M. S. Depix, P. Garcia, C. Perez, M. Arrese, and J. A. Labarca. 2003. *Saccharomyces cerevisiae* fungemia after *Saccharomyces boulardii* treatment in immunocompromised patients. *J. Clin. Gastroenterol.* 36:41–43.

90. Rodrigues, A. C., D. C. Cara, S. H. Fretez, F. Q. Cunha, E. C. Vieira, J. R. Nicoli, and L. Q. Vieira. 2000. *Saccharomyces boulardii* stimulates sIgA production and the phagocytic system of gnotobiotic mice. *J. Appl. Microbiol.* 89:404–414.

91. Salonen, J. H., M. D. Richardson, K. Gallacher, J. Issakainen, H. Helenius, O. P. Lehtonen, and J. Nikoskelainen. 2000. Fungal colonization of haematological patients receiving cytotoxic chemotherapy: emergence of azole-resistant *Saccharomyces cerevisiae*. *J. Hosp. Infect.* 45:293–301.

92. Sangeorzan, J. A., S. F. Bradley, X. He, L. T. Zarins, G. L. Ridenour, R. N. Tiballi, and C. A. Kauffman. 1994. Epidemiology of oral candidiasis in HIV-infected patients: colonization, infection, treatment, and emergence of fluconazole resistance. *Am. J. Med.* 97:339–346.

93. Saville, S. P., A. L. Lazzell, C. Monteagudo, and J. L. Lopez-Ribot. 2003. Engineered control of cell morphology in vivo reveals distinct roles for yeast and filamentous forms of *Candida albicans* during infection. *Eukaryot. Cell* 2:1053–1060.

94. Schellenberg, D., A. Bonington, C. M. Champion, R. Lancaster, S. Webb, and J. Main. 1994. Treatment of *Clostridium difficile* diarrhoea with brewer's yeast. *Lancet* 343:171–172.

95. Sethi, N., and W. Mandell. 1988. *Saccharomyces* fungemia in a patient with AIDS. *N. Y. State J. Med.* 88:278–279.

96. Sherman, F. 1959. The effects of elevated temperature on yeast. II. Induction of respiratory deficient mutants. *J. Cell. Comp. Physiol.* 54:37–52.

97. Simoes-Mendes, B., A. Madeira-Lopes, and N. van Uden. 1978. Kinetics of petite mutation and thermal death in *Saccharomyces cerevisiae* growing at superoptimal temperatures. *Z. Allg. Mikrobiol.* 18:275–279.

98. Smith, D., D. Metzgar, C. Wills, and J. Fierer. 2002. Fatal *Saccharomyces cerevisiae* aortic graft infection. *J. Clin. Microbiol.* 40:2691–2692.

99. Smith, D. L. 1996. Brewer's yeast as a cause of infection. *Clin. Infect. Dis.* 22:201.

100. Sobel, J. D., J. Vazquez, M. Lynch, C. Meriwether, and M. J. Zervos. 1993. Vaginitis due to *Saccharomyces cerevisiae*: epidemiology, clinical aspects and therapy. *Clin. Infect. Dis.* 16:93–99.

101. Sokurenko, E. V., D. L. Hasty, and D. E. Dykhuizen. 1999. Pathoadaptive mutations: gene loss and variation in bacterial pathogens. *Trends Microbiol.* 7:191–195.

102. Sprague, G. F., Jr., L. C. Blair, and J. Thorner. 1983. Cell interactions and regulation of cell type in the yeast *Saccharomyces cerevisiae*. *Annu. Rev. Microbiol.* 37:623–660.

103. Steinmetz, L. M., H. Sinha, D. R. Richards, J. I. Spiegelman, P. J. Oefner, J. H. McCusker, and R. W. Davis. 2002. Dissecting the architecture of a quantitative trait locus in yeast. *Nature* 416:326–330.

104. Su, C., D. Evans, R. H. Cole, J. C. Kissinger, J. W. Ajioka, and L. D. Sibley. 2003. Recent expansion of *Toxoplasma* through enhanced oral transmission. *Science* 299:414–416.

105. Sutton, A., D. Immanuel, and K. T. Arndt. 1991. The SIT4 protein phosphatase functions in late G_1 for progression into S phase. *Mol. Cell. Biol.* 11:2133–2148.

106. Sweeney, J. Y., H. A. Kuehne, and P. D. Sniegowski. 2004. Sympatric natural *Saccharomyces cerevisiae* and *S. paradoxus* populations have different thermal growth profiles. *FEMS Yeast Res.* 4:521–525.

107. Tawfik, O. W., C. J. Papasian, A. Y. Dixon, and L. M. Potter. 1989. *Saccharomyces cerevisiae* pneumonia in a

patient with acquired immune deficiency syndrome. *J. Clin. Microbiol.* **27:**1689–1691.

108. **Tiballi, R. N., J. E. Spiegel, L. T. Zarins, and C. A. Kauffman.** 1995. *Saccharomyces cerevisiae* infections and antifungal susceptibility studies by colorimetric and broth macrodilution methods. *Diagn. Microbiol. Infect. Dis.* **23:**135–140.

109. **van der Aa Kuhle, A., and L. Jespersen.** 2003. The taxonomic position of *Saccharomyces boulardii* as evaluated by sequence analysis of the D1/D2 domain of 26S rDNA, the ITS1-5.8S rDNA-ITS2 region and the mitochondrial cytochrome-c oxidase II gene. *Syst. Appl. Microbiol.* **26:**564–571.

110. **Volkman, S. K., and D. L. Hartl.** 2003. Parasitology. A game of cat and mouth. *Science* **299:**353–354.

111. **Weger, S. D., A. Ganji, K. V. Clemons, J. K. Byron, Y. Minn, and D. A. Stevens.** 2002. Correlation of the frequency of petite formation by isolates of *Saccharomyces cerevisiae* with virulence. *Med. Mycol.* **40:**161–168.

112. **Wheeler, R. T., M. Kupiec, P. Magnelli, C. Abeijon, and G. R. Fink.** 2003. A *Saccharomyces cerevisiae* mutant with increased virulence. *Proc. Natl. Acad. Sci. USA* **100:**2766–2770.

113. **Wilson, J. D., B. M. Jones, and G. R. Kinghorn.** 1988. Bread-making as a source of vaginal infection with *Saccharomyces cerevisiae*. Report of a case in a woman and apparent transmission to her partner. *Sex. Transm. Dis.* **15:**35–36.

114. **Winzeler, E. A., C. I. Castillo–Davis, G. Oshiro, D. Liang, D. R. Richards, Y. Zhou, and D. L. Hartl.** 2003. Genetic diversity in yeast assessed with whole-genome oligonucleotide arrays. *Genetics* **163:**79–89.

115. **Winzeler, E. A., D. R. Richards, A. R. Conway, A. L. Goldstein, S. Kalman, M. J. McCullough, J. H. McCusker, D. A. Stevens, L. Wodicka, D. J. Lockhart, and R. W. Davis.** 1998. Direct allelic variation scanning of the yeast genome. *Science* **281:**1194–1197.

116. **Xu, J., C. M. Boyd, E. Livingston, W. Meyer, J. F. Madden, and T. G. Mitchell.** 1999. Species and genotypic diversities and similarities of pathogenic yeasts colonizing women. *J. Clin. Microbiol.* **37:**3835–3843.

117. **Zerva, L., R. J. Hollis, and M. A. Pfaller.** 1996. In vitro susceptibility testing and DNA typing of *Saccharomyces cerevisiae* clinical isolates. *J. Clin. Microbiol.* **34:**3031–3034.

Molecular Principles of Fungal Pathogenesis
Edited by Joseph Heitman et al.
©2006 ASM Press, Washington, D.C.

Chapter 19

Fungal-Bacterial Interactions

DEBORAH A. HOGAN AND ROBERTO KOLTER

Opportunistic fungal pathogens reside in a wide range of environmental and host-associated niches where they live amidst numerous, often hundreds, of bacterial, archeal, and eukaryotic microbial species. The interactions within these communities probably influence almost every aspect of a fungal cell's life including its metabolism, morphology, virulence, growth, and survival. Since opportunistic fungal infections generally occur after a breach of some aspect of the host defense systems, it is not surprising that mixed infections by multiple pathogens are a common phenomenon. The effects of mixed-species versus single-species infections on the patient have not been well studied. This chapter describes current approaches used to investigate how interactions between bacteria and opportunistic bacterial pathogens affect fungal biology and virulence. By learning about the relationships between fungi and the bacteria that they encounter, we will be better able to predict the behavior of fungal pathogens and perhaps identify new strategies for combating or controlling fungal diseases.

A variety of approaches have been used by researchers to elucidate the mechanisms and biological importance of various fungal-bacterial interactions. In this chapter, we describe a number of in vitro systems that have been used to model different aspects of fungal biology in naturally occurring complex, host-associated communities. We have divided fungal-bacterial model systems into categories based on the in vivo community that they seek to model. First, we address fungal-bacterial interactions associated with opportunistic infections. Subsequently, we discuss models designed to elucidate interactions in the oral microflora, intestinal environment, and the female reproductive tract. In conclusion, we discuss the challenges and potential benefits of translating what has been learned with in vitro model systems to the analyses of in vivo fungal-bacterial communities.

FUNGAL-BACTERIAL INTERACTIONS IN OPPORTUNISTIC INFECTIONS

The frequent co-occurrence of bacteria and fungi in opportunistic infections is illustrated in a report by Hermann et al. which summarizes the microbial analyses of a large number of diverse clinical samples over the course of 3 years (33). Furthermore, this study found a positive correlation between the presence of *Candida* spp. and the occurrence of a specific subset of bacterial pathogens, including *Enterobacter* spp., *Pseudomonas aeruginosa*, and *Klebsiella pneumoniae*. These data emphasize the point that the persistence of a fungal species at a site of infection is determined by both its ability to interact with the host and its success in competing or acting synergistically with other microbes. While the effects of mixed fungal-bacterial infections on the host have not been well characterized, one can envision many ways by which these microbial interactions could impact virulence factor production, host immune responses, and/or susceptibility to antibiotic therapy. Below we describe several approaches that have been used to begin to characterize the molecular mechanisms that govern fungal-bacterial interactions in opportunistic infections.

Implant- and Catheter-Related Infections

Implanted medical devices, including urinary bladder catheters, endotracheal tubes, central venous catheters, mechanical heart valves, and peritoneal dialysis catheters, are particularly susceptible to colonization by *Candida* spp. and opportunistic bacterial pathogens such as *Staphylococcus epidermidis*, *Staphylococcus aureus*, *P. aeruginosa*, and *Enterococcus* spp. (reviewed in reference 81). The formation of microbial biofilms on medical devices is problematic because colonization can seed later infections as well as have a negative impact on the function of the implanted material.

Deborah A. Hogan • Department of Microbiology and Immunology, Dartmouth Medical School, Hanover, NH 03755. **Roberto Kolter** • Department of Microbiology and Molecular Genetics, Harvard Medical School, Boston, MA 02115.

Candida spp. are particularly frequent colonizers of implants that come in contact with the skin, such as central venous catheters. Using a sonication method to quantitate microbial colonization in 1,681 catheters over 3 years, it was found that 9.2% had *Candida* spp. present (predominantly *C. albicans*) (75). Strikingly, 68.4% of the *C. albicans*-positive catheters were taken from patients with a blood culture positive for *C. albicans* within the 48 h prior to catheter removal. This illustrates the potential for disseminated candidiasis resulting from catheter-related infections. The problem of microbial-biofilm colonization of implanted devices is further complicated by the fact that in biofilms, microbes can exhibit high levels of resistance to antibiotics (5, 51, 65). Thus, biofilm-related infections of implanted medical devices most often necessitate removal and replacement of the implanted material. The medical costs and additional patient trauma that are associated with replacement of contaminated medical devices, along with the increased risk for severe infection, make research on microbial biofilm formation and persistence of significant medical importance (81).

Microbial biofilms on catheters and other implanted devices often contain multiple microbial species (24), and it has been proposed that mixed-species biofilms are more stable and perhaps more resistant to conventional antibiotic therapy. To test this idea directly, Adam et al. (1) devised a system in which they compared the drug susceptibility of mixed fungal-bacterial biofilms to those composed of single species. Biofilms formed by *S. epidermidis*, *C. albicans*, or both organisms together were assessed using a catheter disk model in which an inoculated piece of catheter material is submerged in medium to allow an adherent biofilm to develop (1). The resulting single-species and mixed-species biofilms were tested for the susceptibility of component organisms to vancomycin, an antibacterial compound, and fluconazole, an antifungal agent. While *C. albicans* was only modestly "protected" from fluconazole inhibition within a mixed-species biofilm with an exopolymer-producing *S. epidermidis* strain, the presence of *C. albicans* conferred on *S. epidermidis* an almost twofold increase in its resistance to killing by vancomycin. The authors suggest that the combination of extracellular matrix material from multiple species may have an additional ability to impede the diffusion of antibiotics. Similar approaches could be used to assess the susceptibilities of different organisms in mixed fungal-bacterial biofilms to different antimicrobial agents and immune system components. Several other assays for the study of *C. albicans* biofilm formation (9, 74) and susceptibility of biofilms to antifungal agents (65, 80) have also been published and could be adapted for the study of mixed-species biofilms.

P. aeruginosa-*C. albicans* Interactions in Association with Cystic Fibrosis

Mixed bacterial-fungal infections are also observed in the respiratory sputum of patients who are predisposed to lung infections. Multispecies respiratory infections may be particularly relevant for patients with cystic fibrosis (CF), a disease that results from mutations in the CF transmembrane regulator chloride channel (19). Individuals with CF are highly susceptible to chronic, progressive pulmonary infections that severely damage lung tissues and most often lead to respiratory failure in early adulthood (64). The predominant colonizer of the lungs of CF patients is the bacterium *P. aeruginosa*, but fungi such as *C. albicans* and *Aspergillus fumigatus* are also observed and are often the most commonly identified opportunistic pathogen after *P. aeruginosa* in adult patients (6, 10, 32). While the potentially damaging effects of secondary fungal colonizers have not been specifically investigated, it has been shown that the presence of fungi in CF respiratory sputum of CF patients correlates with decreased lung function (60). Bronchopulmonary aspergillosis, a disease with asthma-like symptoms caused by *A. fumigatus*, is also a common problem associated with CF (56). The role, if any, of fungal-bacterial interactions in developing or exacerbating bronchopulmonary aspergillosis is not known. In general, when *P. aeruginosa* and *C. albicans* or *A. fumigatus* are found together in the lungs of CF patients, the numbers of fungal cells remain relatively small until patients are treated with antibacterial compounds that attenuate the *P. aeruginosa* infection. After antibacterial therapy, fungal levels in the sputum increase (17), consistent with previous reports that *P. aeruginosa* can suppress the proliferation of *C. albicans* (44, 45).

To better understand the relationship between *P. aeruginosa* and *C. albicans* in respiratory infections, an in vitro system was developed to characterize their interactions at the molecular level. In this assay, *P. aeruginosa* and *C. albicans* are grown separately in liquid culture and then mixed in nutrient-limited medium. The bacteria then readily attach to the fungal filaments and form a dense biofilm over the course of 24 to 72 h (Fig. 1A) (35). The outcome of this interaction is that the fungal filaments are killed as shown by both vital staining and viable counts (Fig. 1B) (35). Interestingly, *P. aeruginosa* neither attaches to nor kills yeast-form fungal cells. Thus, to characterize the killing activity of *P. aeruginosa* on *C. albicans* filamentous cells, the constitutively filamentous

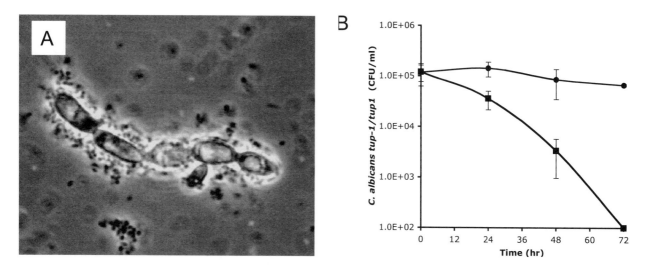

Figure 1. Interactions between *P. aeruginosa* and *C. albicans*. (A) *P. aeruginosa* biofilm surrounding a *C. albicans* *tup1/tup1* filament after 24 h of coincubation in conditioned M63 medium (35). (B) *C. albicans tup1/tup1* survival in (●) *P. aeruginosa* cell-free culture supernatant and (■) in conditioned medium containing 10^5 *P. aeruginosa* cells/ml.

tup1/tup1 mutant was used (15). Experiments using defined *P. aeruginosa* mutants indicate that *C. albicans* is killed in part by the action of some of the same virulence factors that have been implicated in human disease, such as phospholipase C, which degrades eukaryotic membrane lipids, and phenazines, which generate toxic reactive oxygen species. These data may indicate the potential for increased *P. aeruginosa* virulence factor production when *C. albicans* is also present in the sputum of CF patients.

When *P. aeruginosa* and *C. albicans* were cocultured, the presence of the bacteria promoted the growth of yeast-form cells despite conditions that would normally stimulate filamentation. To identify bacterial factors that repressed *C. albicans* morphology, a genetic screen was developed using a reporter strain created by Uhl and Johnson with a *lacZ* fusion to the promoter of *HWP1*, a gene specifically expressed during filamentous growth (82). Bacterial mutants that lacked the ability to alter fungal morphology were identified as those that did not generate a white zone of yeast-form cells after being plated on agar containing the *C. albicans HWP1-lacZ* fusion strain and the chromogenic substrate 5-bromo-4-chloro-3-indolyl-β-D-galactopyranoside (X-Gal) (36). Using this approach, it was found that mutants defective in the production of 3-oxo-C12 homoserine lactone, a cell-cell signaling molecule produced by *P. aeruginosa*, are unable to inhibit *C. albicans* filamentation. Microscopic analysis and real-time reverse transcriptase PCR analysis of morphology-specific markers of *C. albicans* in filament-inducing medium confirmed that exogenous

3-oxo-C12 homoserine lactone was sufficient for inhibiting fungal filamentation. Structurally related compounds with a 12-carbon chain length, e.g., C12-acyl homoserine lactone and dodecanol, also affected *C. albicans* filamentation at similar concentrations and with similar kinetics (36). In contrast, acylated homoserine lactones of chain lengths other than C12 did not affect fungal morphology. *P. aeruginosa* repression of *C. albicans* filamentation may explain how *P. aeruginosa* suppresses *C. albicans* growth in the host, although the in vivo morphology of the fungus in respiratory sputum containing *P. aeruginosa* has not been determined experimentally. Bacterial modulation of fungal morphology, and thus potentially virulence, is also observed in other fungal-bacterial interactions as described below. Together, these data might suggest a common type of interaction between microbial species that could have profound effects in the outcome of opportunistic infections involving bacteria and fungi. We may in fact find that strategies similar to those employed by bacteria have applications in the clinic. Perhaps the administration of compounds that block invasive filamentous growth will be able to protect susceptible patients from some types of fungal infections.

Aspergillus-Mycobacterium Interactions

Observations recorded in the clinical literature may provide important clues to additional fungal-bacterial interactions that can occur in the respiratory environment. For example, a report provides evidence for growth stimulation of *Mycobacterium* spp. during coculture on solid medium with opportunistic fungi (*Blastomyces* and *Aspergilllus*) (21). The mechanism for this growth stimulation remains unexplored. A number of clinical reports describe mixed *Aspergillus-Mycobacterium* infections or *Mycobacterium* superinfection following aspergillosis (3, 11, 40, 47). While these mixed infections may be explained by the coincidental infection of a susceptible host by multiple microbial species, it is also possible that fungal cells provide growth factors or environmental conditioning that enhances subsequent mycobacterial growth. The study of *Mycobacterium-Aspergillus* interactions in an in vitro model system might provide tools to better understand how these mycobacteria cause disease in the human lungs and how mixed bacterial-fungal infections could negatively impact the patient.

FUNGAL-BACTERIAL INTERACTIONS IN THE ORAL MICROFLORA

In most cases, the oral microbial community, which is composed of a complex network of microbe-microbe interactions (48), provides protection from invading pathogens. Upon a decline in function of the immune system or damage to the host's physical barriers, resident organisms have the potential to become pathogens. *C. albicans* is an example of one such organism. *C. albicans* is carried by a significant percentage of the healthy human population, although the precise numbers vary according to the study population and the sampling technique (70, 76). In susceptible individuals, such as the elderly, newborns, or immunocompromised individuals, *C. albicans* can also cause superficial infections that range in their severity. To understand and manipulate the colonization of *C. albicans* in different patient populations, researchers have devised a number of ways to study the physical interactions between fungi and the bacteria normally found in the oral cavity.

Electron microscopy analysis of *C. albicans* from the tongue and buccal mucosa of patients with oral candidiasis showed that yeast and bacterial cells were often found adhered to one another. In vitro studies showed that a ruthenium red-staining material covered the cellular surfaces involved in these interactions, suggestive of an extracellular matrix material that may contribute to cell-cell adhesion (55). Several methods have been described for the analysis of fungal-bacterial coadherence (46, 57). Coaggregation assays, in which different species are coincubated and then assessed for the formation of aggregate "clumps" that do not form in single-species incubations, can provide a useful tool for studying physical intermicrobial interactions (4). Chemical and biochemical treatments of the bacterial and fungal cells

prior to coaggregation experiments allow a determination of the nature of the components that govern adhesion. These types of analyses have shown that *C. albicans* binds to multiple strains of *Actinomyces* through a mechanism that involves protein components on the fungal surface and carbohydrate moieties on bacterial cells (30). Also, *C. albicans* interactions with *Fusobacterium* spp. (including *Fusobacterium nucleatum*, *F. periodontium*, and *F. sulci*) occur via carbohydrate components on the fungi and proteinacious bacterial structures (29).

C. albicans also binds to *Streptococcus gordonii*, a resident oral microbe, by using multiple factors including high-molecular-weight cell surface polypeptides and salivary adhesins (38). *C. albicans* adherence to other oral streptococci (*S. oralis* and *S. sanguis*) is promoted by the addition of salivary proteins. In studies with *S. mutans*, adhesion to *C. albicans* occurred only after bacterial growth on sucrose, but not glucose, indicating the importance of nutrient sources and environmental conditions in determining whether cells interact with one another (14). *C. albicans* also shows altered adhesive properties dependent on the level of aeration of the culture medium (58).

One of the next major steps in the research on interactions between oral bacteria and *C. albicans* is to better understand the nature of these relationships in the host and to determine if fungal-bacterial interactions can be manipulated to decrease fungal colonization or virulence. Because environmental conditions and nutrient availability can alter the adhesion of *C. albicans* to bacterial cells, perhaps changes in diet or topical treatments may be useful in decreasing fungal loads in association with the oral mucosa. If *C. albicans* cells are generally found in direct contact with certain populations of oral microbes, we may also learn more about fungal physiology when in association with humans by studying changes that occur in the fungus on adherence to oral microbes and exposure to their secreted products. Lastly, identification of physical interactions between bacteria and fungi may have biotechnological applications. For example, recombinant strains of *S. gordonii* that either secrete or display microbiocidal antibodies have been used to successfully treat *C. albicans* vaginitis in a rat model system (8). Because *S. gordonii* is a resident microbe in the oral environment, similar therapies may be effective in the treatment of oral thrush.

FUNGAL-BACTERIAL INTERACTIONS IN THE INTESTINAL MICROFLORA

In the intestinal tract of healthy humans, *C. albicans* exists as a benign commensal organism in the midst of hundreds of different microbial species (62). However, treatment with broad-spectrum antibiotics leads to decreased populations of bacterial components of the normal flora and increased levels of *C. albicans*; these alterations in the intestinal microflora often result in what is termed as antibiotic-associated diarrhea (20, 28, 31, 37, 41, 50, 63, 71, 72). Several studies report correlations between the decline of specific microbial populations and increases in the levels of *C. albicans* after treatment with certain antibiotics. For example, one study showed that antibiotics that caused a decrease in the numbers of enteric bacilli and gram-negative anaerobes resulted in concurrent overgrowth and dissemination of *C. albicans* in mice (77). In the same study, antibiotics that did not induce major changes in composition of the intestinal microflora did not lead to candidemia. Mouse model studies have also shown that treatment with certain antibiotics leads to increased *C. albicans* association with the intestinal wall (42). Furthermore, the introduction of *Lactobacillus acidophilus*, a member of the normal flora, into the intestinal tracts of immunocompromised gnotobiotic mice has been shown to be protective against systemic candidiasis (85). Studies such as these have led to an increased interest in the use of orally administered "probiotic" organisms, usually microbes of the gut floras found in healthy individuals, for the purpose of restoring the physiological and protective effects of the intestinal microflora community (73, 83).

The reasons why certain microbes are protective against fungal proliferation are unclear, but several hypotheses have been proposed to explain these clinical observations:

1. bacteria specifically inhibit the growth of *C. albicans* or limit its ability to invade the epithelium

2. bacteria physically compete with *C. albicans* for attachment sites on the host cells

3. bacteria compete with *C. albicans* for nutrients

4. bacteria induce changes in epithelial cells or the local immune system that protect against *C. albicans*

These hypotheses are not mutually exclusive, and the interactions between bacteria of the normal flora and *C. albicans* probably involve a combination of these factors.

A number of in vitro studies have tested the above hypotheses directly to better understand how bacteria control *C. albicans* in the intestinal environment. Experiments with cultured intestinal epithelial cells demonstrate the role of physical interactions in *Lactobacillus-Candida* interactions. Different *Lactobacillus plantarum* and *L. acidophilus* strains attach to human colonic epithelial cells by using a mannose-specific, proteinaceous adherence mechanism (2). Interestingly, the ability to adhere to the epithelium correlates with a strain's capacity to bind to the yeast *Saccharomyces cerevisiae*. Thus, *Lactobacillus* spp. with mannose-specific adherence mechanisms may both colonize the epithelium, thereby protecting the host from potentially pathogenic microbes, and aid in the clearance of yeasts by the formation of fungal-bacterial aggregates.

Organisms associated with intestinal health have also been shown to produce compounds that inhibit fungal growth. *L. plantarum* produces fungicidal compounds (phenyllactic acid and 4-hydroxyphenyllactic acid) (52, 53), and other *Lactobacillus* spp. produce hydrogen peroxide and organic acids that inhibit *C. albicans* growth (16, 25, 26, 34). Cultured dairy products appear to be effective in controlling *C. albicans* growth on voice prostheses (18). In vitro studies attributed the antifungal activities to the production of the surfactant surlactin by *Lactobacillus* species (84).

In a screen of intestinal isolates with anti-*Candida* activity isolated from mouse fecal pellets, we found that the only bacterial isolates that grew under our defined conditions (brain heart infusion agar at 37°C with atmospheric oxygen concentrations) and possessed antifungal properties were *Bacillus laterosporus* species (D. Hogan, A. Vik, R. Vazquez, and R. Kolter, unpublished data). Barsby et al. have identified an environmental strain of *B. laterosporus* that produced a spectrum of compounds with anti-*Candida* properties including cyclic decapeptides, basiliskamides, and acyldipeptides (7). Interestingly, commercially available preparations of probiotic organisms advertised as having anti-*Candida* activity contain *B. laterosporus*, indicating that other researchers most probably have also observed the antifungal properties of this organism. To our knowledge, studies demonstrating the effectiveness of such products for controlling *Candida* overgrowth in the human intestine, if they exist, have not been published.

As discussed above, bacteria are capable of altering fungal morphology. In the intestine, resident microbes may suppress fungal filamentation and thus prevent fungal infiltration of tissues leading to damage and inflammation. Using a microtiter-dish method that allows for easy monitoring of *C. albicans* morphology, Noverr and Huffnagel have shown that, at biologically relevant concentrations, butyric acid produced by lactic acid bacteria inhibits *C. albicans* germination (61). Several publications have reported similar effects (16, 34). Acetic acid and propionic acid were not able to block *C. albicans* germination, and the effect was not due to the creation of an acidic environment, since the organic acids were buffered to a neutral pH. Noverr and Huffnagel also showed that longer-chain fatty acids with structures similar to the eicosinoids produced by human cells can enhance filamentation (61). These data indicate that a wide spectrum of compounds from both eukaryotic and prokaryotic organisms may be acting in concert to modulate *C. albicans* morphology and growth.

BACTERIAL-FUNGAL INTERACTIONS IN THE FEMALE REPRODUCTIVE TRACT

One of the commonly noted risk factors for fungal vulvovaginitis in otherwise healthy women is treatment with bacterial antibiotics (43, 54, 66). Numerous studies have examined the relationship between the normal vaginal microflora and susceptibility to vaginal candidiasis, but a correlative relationship remains unclear (22, 79). Research on fungal-bacterial interactions that occur in the female reproductive tract has focused predominantly on *Lactobacillus* spp., the numerically dominant genus in the reproductive tracts of most premenopausal women (86). Lactobacilli have been proposed to influence vaginal *C. albicans* pathogenesis in a number of ways (12, 49). First, as described above, lactobacilli can bind to urogenital epithelial cells in vitro and in tight association with the vaginal walls in studies with human subjects (13, 23). Thus, endogenous *Lactobacillus* strains may block the attachment of potentially pathogenic fungi. The metabolic products of organisms from the vaginal microflora are thought to contribute to generating a low-pH environment in women of reproductive age (pH values of 4 to 4.5). Importantly, this pH range favors *C. albicans* growth in its noninvasive yeast morphology (27). Butyric acid also specifically inhibits *C. albicans* filamentation (16, 34, 61). A recently published synthetic medium that simulates vaginal conditions may allow for better analysis of the biologically relevant interactions that occur between bacteria and fungi in vitro (59).

Studies of the use of exogenous *Lactobacillus* spp. as therapy for candidiasis are not conclusive, but they show promise for the use of bacteria to combat fungal diseases (for reviews, see references 39 and 67). Because of the relative simplicity of the female reproductive tract in terms of environmental fluctuations and decreased microbial diversity (in comparison to the oral mucosa and the intestinal tract), the vaginal mucosa represents an important system for the dissection of the relationship between indigenous bacteria, fungal opportunistic pathogens, and the host immune response (Fig. 2).

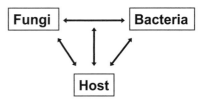

Figure 2. Complex interactions affecting host-associated microbial communities.

CONCLUSIONS

The majority of research on bacterial-fungal interactions has focused on the fungus *C. albicans*. Much less is known about the interactions between bacteria and other important fungal pathogens such as *Cryptococcus* spp. and *Aspergillus* spp. In the future, we may identify other bacterial-fungal pairs that have medical relevance. Furthermore, by studying the environmental lifestyle of potentially pathogenic fungi that are known to exist outside of the host, we may learn about different strategies that bacteria use to compete with fungal organisms. From the perspective of the bacterium, a fungus could represent a synergistic partner for the degradation of complex substrates, a competitor for scarce nutrients, a source of food, or the producer of lethal antibiotics. Survival of the bacterium can depend on its being able to control these interactions. Thus, it is likely that bacteria have evolved numerous ways to manipulate fungal behavior.

With respect to *C. albicans*, in vitro model systems have taught us much about the potential for the many diverse types of physical and chemical interactions that can occur between bacteria and fungi. The next challenge is to determine how these types of relationships modulate *C. albicans* interactions within human hosts. Correlative studies that examine the coincidence of fungi and a given bacterial species in the same biological sample are the first step in determining the importance of a given interaction. Subsequently, genetic and biochemical techniques can be applied to the comparison of biological samples that contain both interacting organisms to samples that contain one or the other. Thus, specific hypotheses regarding how one organism affects the other can be tested.

An alternative approach to the study of fungal-bacterial interactions within the microflora involves the use of germ-free or antibiotic-treated animals to investigate changes in fungal population size, morphology, and virulence in the presence and absence of commensal microbes. Studies of this nature have already been used to illustrate the importance of bacteria in controlling *C. albicans* in the mouse intestine (68). Similar approaches can now be used to investigate the specific molecular differences between *C. albicans* in the conventional mouse intestine and *C. albicans* in mice treated with different classes of antibiotics. Antibiotic-based studies may also be useful for creating model systems for the in vivo analysis of fungal-bacterial interactions on skin or on the mucosal surfaces of the mouth or female reproductive tract.

The studies described in this chapter have focused largely on the effects of bacteria on fungi. However, there is also a significant body of literature on the ways in which fungi alter bacterial physiology. For example, two recent studies have shown that yeast metabolic products, such as ethanol and acetic acid, can affect the physiology and stress responses of bacteria in mixed-species cultures (69, 78). There are many additional studies on the effects of nonlethal concentrations of fungally produced antimicrobial compounds on bacterial behavior. Much remains to be learned about the many ways that microbes can impact the physiology of adjacent species and how these interactions change under different environmental conditions. Ultimately, our goal is not only to predict how microbes interact with one another but also to determine the part that the host plays in governing and responding to different microbial communities.

REFERENCES

1. Adam, B., G. S. Baillie, and L. J. Douglas. 2002. Mixed species biofilms of *Candida albicans* and *Staphylococcus epidermidis. J. Med. Microbiol.* **51**:344–349.
2. Adlerberth, I., S. Ahrne, M. L. Johansson, G. Molin, L. A. Hanson, and A. E. Wold. 1996. A mannose-specific adherence mechanism in *Lactobacillus plantarum* conferring binding to the human colonic cell line HT-29. *Appl. Environ. Microbiol.* **62**:2244–2251.
3. Aksoy, D. Y., A. Turker, M. K. Altundag, H. Abali, M. Durusu, M. Erman, A. Uner, A. A. Sungur, S. Unal, and O. Uzun. 2003. Concomitant *Mycobacterium tuberculosis* and *Aspergillus niger* infection in a patient with acute myeloid leukemia. *Chemotherapy* **49**:264–266.
4. Bagg, J., and R. W. Silverwood. 1986. Coagglutination reactions between *Candida albicans* and oral bacteria. *J. Med. Microbiol.* **22**:165–169.
5. Baillie, G. S., and L. J. Douglas. 1998. Effect of growth rate on resistance of *Candida albicans* biofilms to antifungal agents. *Antimicrob. Agents Chemother.* **42**:1900–1905.
6. Bakare, N., V. Rickerts, J. Bargon, and G. Just-Nubling. 2003. Prevalence of *Aspergillus fumigatus* and other fungal species in the sputum of adult patients with cystic fibrosis. *Mycoses* **46**:19–23.
7. Barsby, T., M. T. Kelly, and R. J. Andersen. 2002. Tupuseleiamides and basiliskamides, new acyldipeptides and antifungal polyketides produced in culture by a *Bacillus laterosporus* isolate obtained from a tropical marine habitat. *J. Nat. Prod.* **65**:1447–1451.
8. Beninati, C., M. R. Oggioni, M. Boccanera, M. R. Spinosa, T. Maggi, S. Conti, W. Magliani, F. De Bernardis, G. Teti, A. Cassone, G. Pozzi, and L. Polonelli. 2000. Therapy of mucosal candidiasis by expression of an anti-idiotype in human commensal bacteria. *Nat. Biotechnol.* **18**:1060–1064.
9. Bernhardt, H., K. Zimmermann, and M. Knoke. 1999. The continuous flow culture as an *in vitro* model in experimental mycology. *Mycoses* **42**:29–32.
10. Bhargava, V., J. F. Tomashefski, Jr., R. C. Stern, and C. R. Abramowsky. 1989. The pathology of fungal infection and colonization in patients with cystic fibrosis. *Hum. Pathol.* **20**:977–986.
11. Bollert, F. G., P. J. Sime, W. MacNee, and G. K. Crompton. 1994. Pulmonary *Mycobacterium malmoense* and aspergillus infection: a fatal combination? *Thorax* **49**:521–522.
12. Boris, S., and C. Barbés. 2000. Role played by lactobacilli in controlling the population of vaginal pathogens. *Microbes. Infect.* **2**:543–546.
13. Boris, S., J. Suárez, F. Vázquez, and C. Barbés. 1998. Adherence of human vaginal lactobacilli to vaginal epithelial cells and interaction with uropathogens. *Infect. Immun.* **66**:1985–1989.
14. Branting, C., M. L. Sund, and L. E. Linder. 1989. The influence of *Streptococcus mutans* on adhesion of *Candida albicans* to acrylic surfaces in vitro. *Arch. Oral Biol.* **34**:347–353.
15. Braun, B. R., and A. D. Johnson. 1997. Control of filament formation in *Candida albicans* by the transcriptional repressor TUP1. *Science* **277**:105–109.
16. Braun, P. C., R. F. Hector, M. E. Kamark, J. T. Hart, and R. L. Cihlar. 1987. Effect of cerulenin and sodium butyrate on chitin synthesis in *Candida albicans. Can. J. Microbiol.* **33**:546–550.
17. Burns, J., J. Van Dalfsen, R. Shawar, K. Otto, R. Garber, J. Quan, A. Montgomery, G. Albers, B. Ramsey, and A. Smith. 1999. Effect of intermittent administration of inhaled tobramycin on respiratory microbial flora in patients with cystic fibrosis. *J. Infect. Dis.* **179**:1190–1196.
18. Busscher, H. J., R. H. Free, R. Van Weissenbruch, F. W. Albers, and H. C. Van Der Mei. 2000. Preliminary observations on influence of dairy products on biofilm removal from silicone rubber voice prostheses in vitro. *J. Dairy Sci.* **83**:641–647.
19. Cheng, S. H., R. J. Gregory, J. Marshall, S. Paul, D. W. Souza, G. A. White, C. R. O'Riordan, and A. E. Smith. 1990. Defective intracellular transport and processing of CFTR is the molecular basis of most cystic fibrosis. *Cell* **63**:827–834.
20. Danna, P. L., C. Urban, E. Bellin, and J. J. Rahal. 1991. Role of *Candida* in pathogenesis of antibiotic-associated diarrhoea in elderly inpatients. *Lancet* **337**:511–514.
21. Darling, W. M. 1976. Co-cultivation of mycobacteria and fungus. *Lancet* **ii**:740.
22. Demirezen, S. 2002. The Lactobacilli-Candida relationship in cervico-vaginal smears. *Cent. Eur. J. Public Health* **10**:97–99.
23. Domingue, P. A., K. Sadhu, J. W. Costerton, K. Bartlett, and A. W. Chow. 1991. The human vagina: normal flora considered as an *in situ* tissue-associated, adherent biofilm. *Genitourin. Med.* **67**:226–231.
24. Donlan, R. M., and J. W. Costerton. 2002. Biofilms: survival mechanisms of clinically relevant microorganisms. *Clin. Microbiol. Rev.* **15**:167–193.
25. Elmer, G. W., C. M. Surawicz, and L. V. McFarland. 1996. Biotherapeutic agents. A neglected modality for the treatment and prevention of selected intestinal and vaginal infections. *JAMA* **275**:870–876.

26. **Fitzsimmons, N., and D. R. Berry.** 1994. Inhibition of *Candida albicans* by *Lactobacillus acidophilus*: evidence for the involvement of a peroxidase system. *Microbios* **80:**125–133.

27. **Fonzi, W. A.** 2002. Role of pH response in *Candida albicans* virulence. *Mycoses* **45**(Suppl. 1):16–21.

28. **Giuliano, M., M. Barza, N. V. Jacobus, and S. L. Gorbach.** 1987. Effect of broad-spectrum parenteral antibiotics on composition of intestinal microflora of humans. *Antimicrob. Agents Chemother.* **31:**202–206.

29. **Grimaudo, N. J., and W. E. Nesbitt.** 1997. Coaggregation of *Candida albicans* with oral *Fusobacterium* species. *Oral Microbiol. Immunol.* **12:**168–173.

30. **Grimaudo, N. J., W. E. Nesbitt, and W. B. Clark.** 1996. Coaggregation of *Candida albicans* with oral *Actinomyces* species. *Oral Microbiol. Immunol.* **11:**59–61.

31. **Gupta, T. P., and M. N. Ehrinpreis.** 1990. *Candida*-associated diarrhea in hospitalized patients. *Gastroenterology* **98:**780–785.

32. **Haase, G., H. Skopnik, T. Groten, G. Kusenbach, and H. G. Posselt.** 1991. Long-term fungal cultures from sputum of patients with cystic fibrosis. *Mycoses* **34:**373–376.

33. **Hermann, C., J. Hermann, U. Munzel, and R. Ruchel.** 1999. Bacterial flora accompanying *Candida* yeasts in clinical specimens. *Mycoses* **42:**619–627.

34. **Hoberg, K. A., R. L. Cihlar, and R. A. Calderone.** 1983. Inhibitory effect of cerulenin and sodium butyrate on germination of *Candida albicans*. *Antimicrob. Agents Chemother.* **24:**401–408.

35. **Hogan, D. A., and R. Kolter.** 2002. *Pseudomonas-Candida* interactions: an ecological role for virulence factors. *Science* **296:**2229–2232.

36. **Hogan, D. A., A. Vik, and R. Kolter.** 2004. A *Pseudomonas aeruginosa* quorum sensing molecule inhibits *Candida albicans* filamentation. *Mol. Microbiol.* **54:**1212.

37. **Hogenauer, C., H. F. Hammer, G. J. Krejs, and E. C. Reisinger.** 1998. Mechanisms and management of antibiotic-associated diarrhea. *Clin. Infect. Dis.* **27:**702–710.

38. **Holmes, A. R., R. McNab, and H. F. Jenkinson.** 1996. *Candida albicans* binding to the oral bacterium *Streptococcus gordonii* involves multiple adhesin-receptor interactions. *Infect. Immun.* **64:**4680–4685.

39. **Jeavons, H. S.** 2003. Prevention and treatment of vulvovaginal candidiasis using exogenous *Lactobacillus*. *J. Obstet. Gynecol. Neonatal Nurs.* **32:**287–296.

40. **Johnston, I. D.** 1988. *Mycobacterium xenopi* infection and aspergilloma. *Tubercle* **69:**139–143.

41. **Kennedy, M. J., and P. A. Volz.** 1983. Dissemination of yeasts after gastrointestinal inoculation in antibiotic-treated mice. *Sabouraudia* **21:**27–33.

42. **Kennedy, M. J., and P. A. Volz.** 1985. Effect of various antibiotics on gastrointestinal colonization and dissemination by *Candida albicans*. *Sabouraudia* **23:**265–273.

43. **Kent, H. L.** 1991. Epidemiology of vaginitis. *Am. J. Obstet. Gynecol.* **165:**1168–1176.

44. **Kerr, J.** 1994. Inhibition of fungal growth by *Pseudomonas aeruginosa* and *Pseudomonas cepacia* isolated from patients with cystic fibrosis. *J. Infect.* **28:**305–310.

45. **Kerr, J. R., G. W. Taylor, A. Rutman, N. Hoiby, P. J. Cole, and R. Wilson.** 1999. *Pseudomonas aeruginosa* pyocyanin and 1-hydroxyphenazine inhibit fungal growth. *J. Clin. Pathol.* **52:**385–387.

46. **Kinder, S., and S. Holt.** 1994. Coaggregation between bacterial species. *Methods Enzymol.* **236:**254–269.

47. **Kohno, S., T. Hatano, H. Yamada, S. Maesaki, A. Yasuoka, M. Kaku, H. Koga, and K. Hara.** 1992. Superinfection of chronic necrotizing pulmonary aspergillosis by *Mycobacterium tuberculosis*. *Intern. Med.* **31:**540–543.

48. **Kolenbrander, P. E., R. N. Andersen, K. Kazmerzak, R. Wu, and R. J. Palmer, Jr.** 1999. Spatial organization of oral bacteria in biofilms. *Methods Enzymol.* **310:**322–332.

49. **Krasner, R. I., G. Young, and P. L. Yudkofsky.** 1956. Interactions of oral strains of *Candida albicans* and lactobacilli. *J. Bacteriol.* **72:**525–529.

50. **Krause, R., E. Schwab, D. Bachhiesl, F. Daxbock, C. Wenisch, G. J. Krejs, and E. C. Reisinger.** 2001. Role of *Candida* in antibiotic-associated diarrhea. *J. Infect. Dis.* **184:**1065–1069.

51. **Kuhn, D. M., and M. A. Ghannoum.** 2004. *Candida* biofilms: antifungal resistance and emerging therapeutic options. *Curr. Opin. Investig. Drugs* **5:**186–197.

52. **Lavermicocca, P., F. Valerio, A. Evidente, S. Lazzaroni, A. Corsetti, and M. Gobbetti.** 2000. Purification and characterization of novel antifungal compounds from the sourdough *Lactobacillus plantarum* strain 21B. *Appl. Environ. Microbiol.* **66:**4084–4090.

53. **Lavermicocca, P., F. Valerio, and A. Visconti.** 2003. Antifungal activity of phenyllactic acid against molds isolated from bakery products. *Appl. Environ. Microbiol.* **69:**634–640.

54. **Maccato, M. L., and R. H. Kaufman.** 1991. Fungal vulvovaginitis. *Curr. Opin. Obstet. Gynecol.* **3:**849–852.

55. **Marrie, T. J., and J. W. Costerton.** 1981. The ultrastructure of *Candida albicans* infections. *Can. J. Microbiol.* **27:**1156–1164.

56. **Mastella, G., M. Rainisio, H. K. Harms, M. E. Hodson, C. Koch, J. Navarro, B. Strandvik, and S. G. McKenzie.** 2000. Allergic bronchopulmonary aspergillosis in cystic fibrosis. A European epidemiological study. Epidemiologic Registry of Cystic Fibrosis. *Eur. Respir. J.* **16:**464–471.

57. **Millsap, K. W., R. Bos, H. C. Van Der Mei, and H. J. Busscher.** 2000. Dot assay for determining adhesive interactions between yeasts and bacteria under controlled hydrodynamic conditions. *J. Microbiol. Methods* **40:**225–232.

58. **Millsap, K. W., R. Bos, H. C. van der Mei, and H. J. Busscher.** 1999. Influence of aeration of *Candida albicans* during culturing on their surface aggregation in the presence of adhering *Streptococcus gordonii*. *FEMS Immunol. Med. Microbiol.* **26:**69–74.

59. **Moosa, M. Y., J. D. Sobel, H. Elhalis, W. Du, and R. A. Akins.** 2004. Fungicidal activity of fluconazole against *Candida albicans* in a synthetic vagina-simulative medium. *Antimicrob. Agents Chemother.* **48:**161–167.

60. **Navarro, J., M. Rainisio, H. K. Harms, M. E. Hodson, C. Koch, G. Mastella, B. Strandvik, and S. G. McKenzie.** 2001. Factors associated with poor pulmonary function: cross-sectional analysis of data from the ERCF. European Epidemiologic Registry of Cystic Fibrosis. *Eur. Respir. J.* **18:**298–305.

61. **Noverr, M. C., and G. B. Huffnagle.** 2004. Regulation of *Candida albicans* morphogenesis by fatty acid metabolites. *Infect. Immun.* **72:**6206–6210.

62. **Odds, F. C.** 1988. Candida and Candidosis, 2nd ed. Baillière Tindall, London, United Kingdom.

63. **Ponnuvel, K. M., R. Rajkumar, T. Menon, and V. S. Sankaranarayanan.** 1996. Role of *Candida* in indirect pathogenesis of antibiotic associated diarrhea in infants. *Mycopathologia* **135:**145–147.

64. **Rajan, S., and L. Saiman.** 2002. Pulmonary infections in patients with cystic fibrosis. *Semin. Respir. Infect.* **17:**47–56.

65. **Ramage, G., K. Vande Walle, B. L. Wickes, and J. L. Lopez-Ribot.** 2001. Standardized method for in vitro antifungal susceptibility testing of *Candida albicans* biofilms. *Antimicrob. Agents Chemother.* **45:**2475–2479.

66. **Reed, B. D.** 1992. Risk factors for Candida vulvovaginitis. *Obstet. Gynecol. Surv.* **47:**551–560.

67. **Reid, G.** 2001. Probiotic agents to protect the urogenital tract against infection. *Am. J. Clin. Nutr.* **73:**437S–443S.

68. **Rogers, T., and E. Balish.** 1976. Experimental *Candida albicans* infection in conventional mice and germfree rats. *Infect. Immun.* **14:**33–38.

69. **Romano, J., and R. Kolter.** 2005 *Pseudomonas-Saccharomyces* interactions: influence of fungal metabolism on bacterial physiology and survival. *J. Bacteriol.* **187:**940–948.

70. **Ruhnke, M.** 2002. Skin and mucous membrane infections, p. 307–326. *In* R. A. Calderone (ed.), *Candida and Candidiasis.* ASM Press, Washington, D.C.

71. **Samonis, G., A. Gikas, E. Anaissie, G. Vrenzos, S. Maraki, Y. Tselentis, and G. Bodey.** 2003. Prospective evaluation of effects of broad-spectrum antibiotics on gastrointestinal yeast colonization of humans. *Antimicrob. Agents Chemother.* **37:**51–53.

72. **Sanderson, P. J., and S. S. Bukhari.** 1991. *Candida* spp. and *Clostridium difficile* toxin-negative antibiotic-associated diarrhoea. *J. Hosp. Infect.* **19:**142–143.

73. **Sandine, W. E.** 1976. Roles of *Lactobacillus* in the intestinal tract. *J. Food Prot.* **42:**259–262.

74. **Schinabeck, M. K., L. A. Long, M. A. Hossain, J. Chandra, P. K. Mukherjee, S. Mohamed, and M. A. Ghannoum.** 2004. Rabbit model of *Candida albicans* biofilm infection: liposomal amphotericin B antifungal lock therapy. *Antimicrob. Agents Chemother.* **48:**1727–1732.

75. **Sherertz, R. J., I. I. Raad, A. Belani, L. C. Koo, K. H. Rand, D. L. Pickett, S. A. Straub, and L. L. Fauerbach.** 1990. Three-year experience with sonicated vascular catheter cultures in a clinical microbiology laboratory. *J. Clin. Microbiol.* **28:**76–82.

76. **Shin, E. S., S. C. Chung, Y. K. Kim, S. W. Lee, and H. S. Kho.** 2003. The relationship between oral *Candida* carriage and the secretor status of blood group antigens in saliva. *Oral Surg. Oral Med. Oral Pathol. Oral Radiol. Endod.* **96:**48–53.

77. **Sjovall, J., B. Huitfeldt, L. Magni, and C. E. Nord.** 1986. Effect of beta-lactam prodrugs on human intestinal microflora. *Scand. J. Infect. Dis. Suppl.* **49:**73–84.

78. **Smith, M. G., S. G. Des Etages, and M. Snyder.** 2004. Microbial synergy via an ethanol-triggered pathway. *Mol. Cell. Biol.* **24:**3874–3884.

79. **Sobel, J. D., and W. Chaim.** 1996. Vaginal microbiology of women with acute recurrent vulvovaginal candidiasis. *J. Clin. Microbiol.* **34:**2497–2499.

80. **Suci, P. A., and B. J. Tyler.** 2003. A method for discrimination of subpopulations of *Candida albicans* biofilm cells that exhibit relative levels of phenotypic resistance to chlorhexidine. *J. Microbiol. Methods* **53:**313–325.

81. **Thomas, J. G., G. Ramage, and J. L. Lopez-Ribot.** 2004. Biofilms and implant infections, p. 269–293. *In* M. Ghannoum and G. A. O'Toole (ed.), *Microbial Biofilms.* ASM Press, Washington, D.C.

82. **Uhl, M. A., and A. D. Johnson.** 2001. Development of *Streptococcus thermophilus lacZ* as a reporter gene for *Candida albicans. Microbiology* **147:**1189–1195.

83. **Vaughan, E. E., B. Mollet, and W. M. deVos.** 1999. Functionality of probiotics and intestinal lactobacilli: light in the intestinal tract tunnel. *Curr. Opin. Biotechnol.* **10:**505–510.

84. **Velraeds, M. M., B. van de Belt-Gritter, H. C. van der Mei, G. Reid, and H. J. Busscher.** 1998. Interference in initial adhesion of uropathogenic bacteria and yeasts to silicone rubber by a *Lactobacillus acidophilus* biosurfactant. *J. Med. Microbiol.* **47:**1081–1085.

85. **Wagner, R. D., C. Pierson, T. Warner, M. Dohnalek, M. Hilty, and E. Balish.** 2000. Probiotic effects of feeding heat-killed *Lactobacillus acidophilus* and *Lactobacillus casei* to *Candida albicans*-colonized immunodeficient mice. *J. Food Prot.* **63:**638–644.

86. **Zhou, X., S. J. Bent, M. G. Schneider, C. C. Davis, M. R. Islam, and L. J. Forney.** 2004. Characterization of vaginal microbial communities in adult healthy women using cultivation-independent methods. *Microbiology* **150:**2565–2573.

Molecular Principles of Fungal Pathogenesis
Edited by Joseph Heitman et al.
©2006 ASM Press, Washington, D.C.

Chapter 20

Amphibian Chytridiomycosis as an Emerging Infectious Disease of Wildlife: What Can We Learn from the Earliest Diverging Fungi?

TIMOTHY Y. JAMES AND RYTAS VILGALYS

THE EMERGENCE OF CHYTRIDIOMYCOSIS

Emerging infectious diseases are microbial diseases that are increasing in incidence, host, or geographical range, are newly evolved, or have only recently been discovered. Examples of emerging diseases of humans are many: AIDS, methicillin-resistant *Staphylococcus aureus* infection, West Nile virus infection, and others. Increased incidence rates of human diseases are often attributed to changes in modern lifestyles and demography, such as global travel, pollution, and overpopulation (38). Zoonoses are diseases that infect both wildlife and humans, and roughly half of human diseases are zoonotic (19). Until recently, little attention was paid to the possibility that wildlife populations could also be suffering from emerging diseases and that the emergence of wildlife diseases is also likely to be initiated by anthropogenic disturbance of animal ecosystem and habitat (18). The rapid expansion of such zoonotic diseases in wildlife populations may ultimately cause "spill-over" of pathogens into human populations (e.g., avian flu and Lyme disease [18]).

Amphibian populations have been precipitously declining in many regions worldwide since the 1970s (9, 25, 54). Furthermore, amphibian species are going extinct at a rate higher than that for other fauna, such as birds and mammals (51). Investigation by wildlife biologists into the cause of the declines has revealed that many factors (e.g., pollution, habitat loss, and invasive species) are likely to be involved (15).

Some regions with the greatest amphibian mortality in recent years include protected areas where the impacts of anthropomorphic disturbance are likely to be relatively minor, such as tropical montane forests in Central America and Queensland, Australia (31, 47). Exotic disease was suspected first in these regions because declines showed the temporal and geographic signatures of an epidemic (30) and mass mortalities were observed (32). A combination of microbiological and histological examination of dead and dying frogs from these regions revealed the presence of numerous sporangia of a Chytridiomycete fungus (chytrid) on the epidermis (7) (Fig. 1). Berger et al. (7) hypothesized that declines in both Central American and Australian frog populations were caused by the same fungal disease, chytridiomycosis. In total, 19 frog species from areas of dieoffs in Australia and Panama were positive by histological examination for the chytrid; in contrast, museum samples collected before the dieoffs were negative for signs of chytrid infection. Longcore et al. (34) have recently described the agent of chytridiomycosis as a new genus and species of chytrid, *Batrachochytrium dendrobatidis*. Since the discovery of chytridiomycosis in 1998, the pathogen has been found associated with population declines in western North America (13) and Europe (11). It has also been detected in Africa (45), New Zealand (4), and South America (35). This was the first case ever reported of vertebrate mortality caused by an infectious chytrid.

THE CHYTRIDIOMYCETES AS UNIQUE BASAL FUNGI

The Chytridiomycetes are a relatively poorly known and understudied phylum, in part due to their microscopic size and minimal effects on humans. There are currently five orders (Blastocladiales, Chytridiales, Monoblepharidales, Neocallimastigales, and Spizellomycetales) containing approximately 100 genera and 1,000 species. It is uncertain if the chytrids are a monophyletic group or rather an "evolutionary

Timothy Y. James and Rytas Vilgalys • Department of Biology, Duke University, Durham, NC 27708.

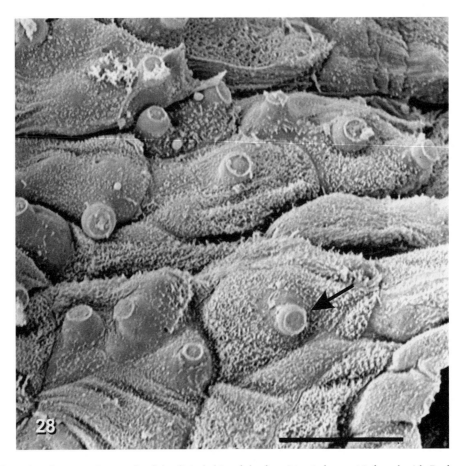

Figure 1. Scanning electron micrograph of the digital skin of the frog *Litoria lesueuri* infected with *B. dendrobatidis*. Shown is the superficial layer of skin, displaying marked roughening and numerous cells containing sporangia. The arrow indicates discharge tubes of the sporangia from which zoospores are released. Bar, 10 μm. Reprinted from reference 6 with permission.

grade" of early-diverging fungal lineages (27). Chytrids reproduce asexually through sac-like sporangia, which are typically produced on either a brief or extensive system of thread-like rhizoids that aid in anchoring to the substrate and absorption of nutrients. At maturity, the sporangia undergo cleavage into zoospores. When fully developed, the unicellular zoospores swim away to find a new host or substrate. The zoospore is the feature that makes chytrids unique among fungi. Zoospores are bound only by a plasmalemma and possess a whiplash, posterior flagellum that drives the cell forward. Zoospores often contain large lipid globules which are presumably used in supplying the zoospore with energy during the motile phase (21). Systematics of the group is based primarily on transmission electron microscopy of the zoospore. Most members of the Chytridiales have a single large lipid globule. *B. dendrobatidis* (Chytridiales) has several lipid globules and other traits that are atypical for this order (Fig. 2).

Most chytrids have a zygotic life cycle in which karyogamy is followed directly by meiosis. However, in *Allomyces* and some other Blastocladiales, the life cycle alternates between diploid and haploid thalli. Sexuality is unknown in the majority of chytrid species. Molecular population genetics suggests that many of these apparently asexual species may actually be vegetatively diploid (33, 37).

The earliest occurrence of chytrids in the fossil record is approximately 400 million years ago (52). It is unclear whether these earliest ancestors of fungi were saprophytic or parasitic (14). Modern chytrids can be saprophytic, parasitic, or both. Chytrids are generally thought to be ubiquitous but not abundant. They are easily isolated from aquatic ecosystems and soil throughout the world, using proper baiting techniques (3), and they are often found attached to moribund and living green algae. The diversity of possible hosts for chytrids is very broad: plants, algae, fungi, and small metazoans. The ecological

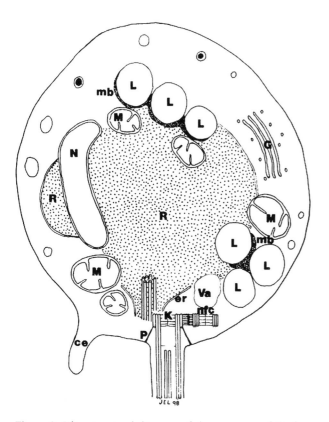

Figure 2. Ultrastructural diagram of the zoospore of *B. dendrobatidis*. ce, cytoplasmic extension; er, endoplasmic reticulum; G, Golgi apparatus; K, kinetosome; L, lipid globule; M, mitochondrion; mb, microbody; nfc, nonflagellated centriole; N, nucleus; P, prop; R, ribosomes; Va, vacuole. Reprinted from reference 34 with permission.

lability of the chytrids is also observed in other early-diverging fungal lineages (e.g., Entomophthorales, Zoopagales, and Microsporidia). The collective group of basal fungal lineages parasitizes the breadth of eukaryotic diversity, whereas more evolutionarily derived fungi (the Ascomycetes and Basidiomycetes) appear to favor associations with green plants over associations with animals or fungi, which occur more rarely. As parasites, chytrids generally "pick on things their own size," parasitizing organisms such as desmids, mosquitoes, and nematodes (44). Therefore, it came as a surprise in 1998 to find that a chytrid might be at the heart of global amphibian dieoffs, doing something that it was not supposed to do—parasitizing vertebrates, with a dramatic and unpredictable outcome.

As basal fungi, Chytridiomycetes have presumably retained many of the characteristics of the last common ancestor with other Fungi millions of years ago. The central characteristic that the chytrids have retained from their protistan ancestors is the flagellated zoospore (Fig. 2). The relationship of the chytrids to the remaining fungi might be considered analogous to the relationship of the choanoflagellates to the animal kingdom (14). Both chytrids and choanoflagellates are highly simplified protist-like organisms that are presumably very similar to the most recent common ancestors of their respective kingdoms. Gene phylogenies have confirmed the basal position of the chytrids in the fungal kingdom (27, 40) and place their time of divergence from the remaining fungi around 660 million years ago (5). Because of their basal position within the fungal kingdom, a greater understanding of characteristics of Chytridiomycetes, e.g., biochemistry and drug resistance, should elucidate fundamental principles that may apply to the whole of the kingdom Fungi.

PATHOLOGY AND ECOLOGY

Frogs with chytridiomycosis display lethargy, loss of righting response, skin sloughing, abnormal posture, and, in severe cases, skin ulceration (17). Sporangia can be seen on histological examination of the skin between digits and from the ventrum. The presence of sporangia is diagnostic for the chytrid. The fungus appears to be restricted to the skin and causes cell loss, erosion, and hyperkeratosis of the stratum corneum. In severe cases there is also hyperplasia of the stratum intermedium (7). The fungus spreads on the skin of the infected individual through the formation and release of zoospores (Fig. 1). The chytrid is thought to be acquiring nutrition from the host in the form of keratin because it infects only keratinized portions of the body (7, 17). The chytrid can be easily cultured on solid and liquid media containing tryptone, but a chemically defined medium is lacking, as is definitive proof of ability to grow solely on keratin (42). The chytrid can also be observed on the keratinized mouthparts of tadpoles but does not appear to cause mortality at this life stage (8).

For susceptible frog species, chytridiomycosis results in death in typically less than 1 month (7, 34, 39). Using inoculation of poison dart frogs (*Dendrobates* spp.) with zoospores, Koch's postulates have been satisfied (34, 39). How the fungus actually kills the host is unknown, but two primary hypotheses have been put forward (17): hyperplasia of the epidermis may impair the frog's ability to properly respire or osmoregulate, or a fungal toxin may be produced by the fungus and be absorbed by the host. It is clear that *B. dendrobatidis* can act as the sole infectious agent in dying frogs and is not merely creating an environment where an opportunistic infection by another pathogenic microbe can take hold (7).

Chytridiomycosis has a very broad host range, infecting many species of frogs and even salamanders.

Some frog species are highly susceptible and have suffered large population declines, while other, sympatric, species appear to be resistant (32, 46). Some of the species not harmed by the fungus can, however, carry the disease sublethally in small lesions (24). It has been hypothesized that these resistant frogs that can harbor chytridiomycosis could act as disease reservoirs and could have been the mechanism by which the fungus has been globally dispersed (35).

Variation in the susceptibility of different amphibian species suggests some variation in the host's ability to fight the fungus through its immune system. However, it is also possible that environmental or other factors that modulate the immune response may vary among host species and geographic regions (13). Amphibians, like other vertebrates, have innate and adaptive immunity, utilizing macrophages, neutrophils, and lymphocytes to respond to pathogen invasion. They are also capable of synthesizing a broad range of antimicrobial peptides that are secreted from granular glands in the skin. The mode of action of the small 12- to 46-amino-acid peptides is thought to be through disruption of the pathogen's cell membrane (57). There is some evidence that the peptides show some specificity for certain microbes (36), and some of the peptides are active against B. dendrobatidis (49, 50). Somehow, B. dendrobatidis is able to evade this first line of immune defense; environmental factors may contribute to the weakening of this response (49). On the other hand, variance in susceptibility could actually be related to other intrinsic differences between species, such as the ability of the chytrid to attach to and penetrate the host's epithelial cells or the ability of the chytrid to compete with the other microfauna associated with amphibian skin.

The areas and species affected by chytridiomycosis have a few factors in common. Geographic regions suffering catastrophic declines are typically upland rain forests with cool climates rather than humid lowland forests (17, 30). Also, the disease primarily impacts riparian rather than terrestrial frog species (32). These factors are intrinsically tied to the biology of B. dendrobatidis and of chytrids in general. Chytrids are very common in cool freshwater ecosystems. B. dendrobatidis has a similar temperature optimum, with maximal growth between 17 and 25°C, and is intolerant of elevated temperatures (42). In fact, exposing sick amphibians at 37°C for 16 h has been shown to be an effective way of treating the disease (56). Another fundamental aspect of chytrid biology is a reliance on water for dispersal to new substrates and hosts. B. dendrobatidis favors

stream-dwelling or stream-reproducing hosts, suggesting that these aquatic locations are the site of transmission from frog to frog.

For most human mycoses, the ability to grow at 37°C is clearly a virulence factor (26). The inability of B. dendrobatidis to grow at 37°C suggests that it would be unlikely for the fungus to make a host switch to homeothermic (warm-blooded) animals. In addition, the aquatic dispersal phase should prevent B. dendrobatidis from parasitizing wholly terrestrial organisms. Precedent exists, however, for parasitism of terrestrial animals by zoosporic fungi, as in the case of the oomycete Pythium insidiosum on animals in tropical and subtropical climates.

The localization of B. dendrobatidis to the skin of amphibians is analogous to the better-known fungal infections caused by mammalian dermatophytes (e.g., Trichophyton rubrum). Both mammalian dermatophytes and B. dendrobatidis can be transmitted through direct host-to-host contact, are considered to be keratinophilic, and may have low host specificity (55). Unlike B. dendrobatidis, most mammalian dermatophytes are not lethal; it is probably because amphibians respire through their skin that they are more sensitive to a cutaneous infection.

MOLECULAR EPIDEMIOLOGY

Chytridiomycosis is now considered to be an emerging infectious disease afflicting a wide range of frog species (19). Several hypotheses have been put forward to explain the rapid emergence of chytridiomycosis as a widespread and prevalent disease. One hypothesis suggests that the recent outbreak of chytridiomycosis was spawned by a recent anthropogenic introduction of the pathogen into a naive population, i.e., "pathogen pollution" (17). An alternate hypothesis is that the pathogen may have been endemic to regions of current amphibian declines but was either unnoticed or nonpathogenic until recently. Retrospective studies using museum specimens suggest that the disease was absent until recently and is still on the rise. These studies have determined that chytridiomycosis had emerged by 1974 in the United States and 1978 in Australia (13, 28).

Molecular tools are essential for pathogen surveillance and epidemiology, and amphibian-disease researchers are beginning to develop tools for sensitive detection and population genetics of B. dendrobatidis (2, 12, 37). Two goals of population genetic studies are to determine whether the disease has emerged due to recent introduction and to determine where the fungus came from and how has it been dispersed. Molecular phylogenetic studies of the Chytridiomycetes have not provided any clue to the origins of the disease; it is not clear how

B. dendrobatidis originated, because no sister taxon or close relative has been sampled (27). A strategy used in other pathosystems for determining the origin of a disease is to discover which geographic region has the highest genetic diversity. The geographic origins of some notorious diseases such as malaria and late blight of potato have been revealed, using molecular markers, by the observation of higher genetic diversity in these areas (1, 23).

Using multilocus sequence typing (MLST) (the genotyping of single-nucleotide polymorphisms by using DNA sequencing of housekeeping or anonymous loci), we recently addressed the molecular epidemiology of chytridiomycosis (37). Two critical points about *B. dendrobatidis* population genetics have emerged from this study. First, a global sample of 35 isolates revealed only three variable nucleotides among nearly 6,000 surveyed positions. This low level of genetic variation suggested that the coalescence time, and thus the expansion of the global sample, was very recent. These data supported the "recent-introduction" hypothesis over the "recently-turned-pathogenic" hypothesis. Even isolates from such geographically disparate regions as Panama and Australia had identical MLST genotypes. These data implicated the introduction of a single clone to these areas of mass dieoffs. Second, direct sequencing of PCR-amplified gene regions demonstrated multiple sequence types per strain at 4 of the 10 loci examined. These two sequence types were interpreted as two alleles of a diploid genotype. The observation that some loci showed individuals that were all or nearly all heterozygous (fixed heterozygosity) suggested that independent assortment was not occurring and satisfied one of the criteria of clonality (53). This result was in agreement with the absence of sexuality observed in culture or on the host. Taking these results together, we concluded that *B. dendrobatidis* is a recently dispersed clonal organism. However, the MLST dataset included some genotypic variation among the various isolates in the form of combinations of heterozygous and homozygous genotypes among loci. In addition, some heterozygous loci displayed more deviation from Hardy-Weinberg proportions than did others. One explanation for this pattern is that the variation among strains was generated by mitotic recombination, in much the same way as is suggested for *Candida albicans* (16). Location along the chromosome (i.e., distance from centromere) has been correlated with the amount of heterozygosity lost through somatic recombination in *C. albicans* (29); a similar process could be generating differences in heterozygosity among loci in *B. dendrobatidis*.

The striking lack of genetic variation in *B. dendrobatidis* is a major hindrance to determining the geographic origin of the current epidemic. Neither the MLST genotype data nor other molecular markers have provided any evidence for population subdivision based on geography or host (18, 20), nor is there any evidence of areas or hosts harboring higher levels of genetic diversity. The lack of genetic variation in *B. dendrobatidis* could limit its ability to counteract evolved resistance of its host. Recently, it has been reported that some amphibian populations that have suffered from population declines, presumably caused by chytridiomycosis, may be recovering and persisting with a stable demography and rate of infection (46). It is plausible that the hosts have evolved resistance to the pathogen. Perhaps the frog species which can recover from the initial epidemic can win the genetic arms race over their genetically depauperate opponents.

LESSONS LEARNED FROM A MODEL CHYTRID

The investigation of emerging infectious fungal diseases of wildlife can help elucidate principles shared among all pathogenic fungi. For example, it is still unknown if *B. dendrobatidis* possesses homologous mechanisms of pathogenesis with other fungi. Further, monitoring of wildlife diseases will be critical to predicting and mitigating future zoonoses. *B. dendrobatidis* represents a model pathogen of vertebrates from a novel clade of Fungi (Chytridiomycota) on an interesting host (amphibians) acting in a role that is quite familiar to medical mycologists (as a dermatophyte). As a model system, *B. dendrobatidis* informs us about general principles of mycoses, such as invasive growth and evasion of host immunity. *B. dendrobatidis* could also be an important model system in disease ecology, since much of the work on amphibian disease is strongly rooted in community ecology and disease surveillance. Finally, chytridiomycosis has the added advantage that both host and parasite are amenable to experimentation.

In just a few years, research on *B. dendrobatidis* has taught us a few salient points regarding disease ecology, disease management and policy, and fungal biology.

1. Sometimes cures can be very simple if they can take advantage of the basic biology of the parasite. The use of brief exposure to temperatures above those tolerated by *B. dendrobatidis* (37°C) to treat chytridiomycosis exploits the cool-temperature requirement of the organism (56). The reaction of *B. dendrobatidis* to elevated temperatures is probably similar to that of other Chytridiomycetes and

probably reflects their niche in the wild—cool aquatic ecosystems. These results emphasize that it is very important for medical mycologists to understand what fungi that cause opportunistic infections are doing in nature, as is being done with *Cryptococcus neoformans* (26).

2. There is a wealth of information in the specimens archived in museums and herbaria. Retrospective, histological analyses of frog specimens have been able to pinpoint the time at which the disease first appeared (7, 13, 28). In some cases even genetic analyses can be performed on DNA recovered from specimens. For example, analysis of DNA sequences obtained from potato plant samples deposited during the Irish potato blight (1845 to 1847) forced a reappraisal of earlier hypotheses concerning the intercontinental spread of the disease (48).

3. Globalization has caused the rapid spread and increased gene flow of infectious human diseases and may also be occurring for wildlife diseases (18, 19). Intercontinental transport and traffic of animals or plant and animal materials threaten to make pathogenic microbial species one single genetic population through incidental global translocation. Quarantines are needed to protect wildlife resources and to prevent the spread of future zoonoses.

4. Changes in the environment may trigger emerging infectious disease epidemics. Global climate change is thought to play a role in the disappearances of amphibians from Central America (43). Here, global warming is suspected of changing the precipitation patterns and increasing the population density of amphibians, potentially fostering the rapid spread of chytridiomycosis. How a frog species reacts to a disease agent is contingent on a number of things: temperature, pollutants, UV radiation, and the presence of competitors or predators (10, 15, 41). Exposure to a novel pathogen is likely to produce unpredictable results; in the worst case, such as with *B. dendrobatidis*, it may drive species to extinction.

5. Chytrid fungi may be vegetatively diploid and primarily asexual. Identification of asexual lineages can be facilitated by detecting fixed heterozygosity as in other parasitic protozoa and oomycetes (22, 53). However, as in *Candida albicans*, somatic recombination can work to generate novel genotypes by shuffling heterozygous genotypes.

Acknowledgments. We thank Jeri Parent for a critical reading of this manuscript. Work on *Batrachochytrium* in the laboratory of R.V. is funded in part by NSF-DEB-0213851. We thank Joyce Longcore and Lee Berger for permission to use their figures.

REFERENCES

1. Anderson, T. J. C., B. Haubold, J. T. Williams, J. G. Estrada-Franco, L. Richardson, R. Mollinedo, M. Bockarie, J. Mokili, S. Mharakurva, N. French, J. Whitworth, I. D. Velez, A. H. Brockman, F. Nosten, M. U. Ferreira, and K. P. Day. 2000. Microsatellite markers reveal a spectrum of population structures in the malaria parasite *Plasmodium falciparum*. *Mol. Biol. Evol.* 17:1467–1482.

2. Annis, S. L., F. P. Dastoor, H. Ziel, P. Daszak, and J. E. Longcore. 2004. A DNA-based assay identifies *Batrachochytrium dendrobatidis* in amphibians. *J. Wildl. Dis.* 40:420–428.

3. Barr, D. J. S. 1987. Isolation, culture and identification of Chytridiales, Spizellomycetales, and Hyphochytriales, p. 118–120. *In* M. S. Fuller and A. Jaworski (ed.), *Zoosporic Fungi in Teaching and Research*. Southeastern Publishing Corp., Athens, Ga.

4. Bell, B. D., S. Carver, N. J. Mitchell, and S. Pledger. 2004. The recent decline of a New Zealand endemic: how and why did populations of Archey's frog *Leiopelma archeyi* crash over 1996–2001? *Biol. Conserv.* 120:189–199.

5. Berbee, M. L., and J. W. Taylor. 2001. Fungal molecular evolution: gene trees and geologic time, p. 229–245. *In* D. J. McLaughlin, E. G. McLaughlin, and P. A. Lemke (ed.), *The Mycota*, vol. 7B. *Systematics and Evolution*. Springer-Verlag KG, Berlin, Germany.

6. Berger, L. 2001. *Diseases in Australian Frogs*. Ph.D. dissertation. James Cook University, Townsville, Australia.

7. Berger, L., R. Speare, P. Daszak, D. E. Green, A. A. Cunningham, C. L. Goggin, R. Slocombe, M. A. Ragan, A. D. Hyatt, K. R. McDonald, H. B. Hines, K. R. Lips, G. Marantelli, and H. Parkes. 1998. Chytridiomycosis causes amphibian mortality associated with population declines in the rain forests of Australia and Central America. *Proc. Natl. Acad. Sci. USA* 95:9031–9036.

8. Berger, L., R. Speare, and A. D. Hyatt. 1999. Chytrid fungi and amphibian declines: overview, implications and future directions, p. 23–33. *In* A. Campbell (ed.), *Declines and Disappearances of Australian Frogs*. Declines Environment Australia, Canberra.

9. Blaustein, A. R., and D. B. Wake. 1990. Declining amphibian populations: a global phenomenon. *Trends Ecol. Evol.* 5:203–204.

10. Blaustein, A. R., J. M. Romansic, J. M. Kiesecker, and A. C. Hatch. 2003. Ultraviolet radiation, toxic chemicals and amphibian population declines. *Divers. Distrib.* 9:123–140.

11. Bosch, J., I. Martínez-Solano, and M. García-París. 2001. Evidence of a chytrid fungus infection involved in the decline of the common midwife toad (*Alytes obstetricans*) in protected areas of central Spain. *Biol. Conserv.* 97:331–337.

12. Boyle, D. G., D. B. Boyle, V. Olsen, J. A. Morgan, and A. D. Hyatt. 2004. Rapid quantitative detection of chytridiomycosis (*Batrachochytrium dendrobatidis*) in amphibian samples using real-time Taqman PCR assay. *Dis. Aquat. Org.* 60:141–148.

13. Carey, C., N. Cohen, and L. A. Rollins-Smith. 1999. Amphibian declines: an immunological perspective. *Dev. Comp. Immunol.* 23:459–472.

14. Cavalier-Smith, T. 2001. What are fungi? p. 3–37. *In* D. J. McLaughlin, E. G. McLaughlin, and P. A. Lemke (ed.), *The Mycota*, vol. 7A. *Systematics and Evolution*. Springer-Verlag KG, Berlin, Germany.

15. Collins, J. P., and A. Storfer. 2003. Global amphibian declines: sorting the hypotheses. *Divers. Distrib.* 9:89–98.

16. Cowen, L. E., J. B. Anderson, and L. M. Kohn. 2002. Evolution of drug resistance in *Candida albicans*. *Annu. Rev. Microbiol.* **56:**139–165.

17. Daszak, P., L. Berger, A. A. Cunningham, A. D. Hyatt, D. E. Green, and R. Speare. 1999. Emerging infectious diseases and amphibian population declines. *Emerg. Infect. Dis.* **5:**735–748.

18. Daszak, P., A. A. Cunningham, and A. D. Hyatt. 2000. Emerging infectious diseases of wildlife—threats to biodiversity and human health. *Science* **287:**443–449.

19. Daszak, P., A. A. Cunningham, and A. D. Hyatt. 2001. Anthropogenic environmental change and the emergence of infectious diseases in wildlife. *Acta Trop.* **78:**103–116.

20. Daszak, P., A. A. Cunningham, and A. D. Hyatt. 2003. Infectious disease and amphibian population declines. *Divers. Distrib.* **9:**141–150.

21. Fuller, M. S. 1977. The zoospore, hallmark of the aquatic fungi. *Mycologia* **69:**1–20.

22. Goodwin, S. B., B. A. Cohen, and W. E. Fry. 1994. Panglobal distribution of a single clonal lineage of the Irish potato famine fungus. *Proc. Natl. Acad. Sci. USA* **91:**11591–11595.

23. Goodwin, S. B. 1997. The population genetics of *Phytophthora*. *Phytopathology* **87:**462–473.

24. Hanselmann, R., A. Rodríguez, M. Lampo, L. Fajardo-Ramos, A. A. Aguirre, A. M. Kilpatrick, J. P. Rodríguez, and P. Daszak. 2004. Presence of an emerging pathogen of amphibians in introduced bullfrogs *Rana catesbeiana* in Venezuela. *Biol. Conserv.* **120:**115–119.

25. Houlahan, J. E., C. S. Findlay, B. R. Schmidt, A. H. Meyer, and S. L. Kuzmin. 2000. Quantitative evidence for global amphibian population declines. *Nature* **404:**752–755.

26. Hull, C. M., and J. Heitman. 2002. Genetics of *Cryptococcus neoformans*. *Annu. Rev. Genet.* **36:**557–615.

27. James, T. Y., D. Porter, C. A. Leander, R. Vilgalys, and J. E. Longcore. 2000. Molecular phylogenetics of the Chytridiomycota supports the utility of ultrastructural data in chytrid systematics. *Can. J. Bot.* **78:**336–350.

28. Johnson, M. L., and R. Speare. 2003. Survival of *Batrachochytrium dendrobatidis* in water: quarantine and disease control implications. *Emerg. Infect. Dis.* **9:**922–925.

29. Jones, T., N. A. Federspiel, H. Chibana, J. Dungan, S. Kalman, B. B. Magee, G. Newport, Y. R. Thorstenson, N. Agabian, P. T. Magee, R. W. Davis, and S. Scherer. 2004. The diploid genome sequence of *Candida albicans*. *Proc. Natl. Acad. Sci. USA* **101:**7329–7334.

30. Laurance, W. F., K. R. McDonald, and R. Speare. 1996. Epidemic disease and the catastrophic decline of Australian rain forest frogs. *Conserv. Biol.* **10:**406–413.

31. Lips, K. R. 1998. Decline of a tropical montane amphibian fauna. *Conserv. Biol.* **12:**106–117.

32. Lips, K. R. 1999. Mass mortality and population declines of anurans at an upland site in western Panama. *Conserv. Biol.* **13:**117–125.

33. Liu, Y., M. Hodson, and B. Hall. 2003. Heterozygosity in chytrids. *Fungal Genet. Newsl.* **50(Suppl):**459.

34. Longcore, J. E., A. P. Pessier, and D. K. Nichols. 1999. *Batrachochytrium dendrobatidis* gen. et sp. nov., a chytrid pathogenic to amphibians. *Mycologia* **91:**219–227.

35. Mazzoni, R., A. A. Cunningham, P. Daszak, A. Apolo, E. Perdomo, and G. Speranza. 2003. Emerging pathogen of wild amphibians in frogs (*Rana catesbeiana*) farmed for international trade. *Emerg. Infect. Dis.* **9:**995–998.

36. Mor, A., K. Hani, and P. Nicolas. 1994. The vertebrate peptide antibiotics dermaseptins have overlapping structural features but target specific microorganisms. *J. Biol. Chem.* **269:**31635–31641.

37. Morehouse, E. A., T. Y. James, A. R. D. Ganley, R. Vilgalys, L. Berger, P. J. Murphy, and J. E. Longcore. 2003. Multilocus sequence typing suggests the chytrid pathogen of amphibians is a recently emerged clone. *Mol. Ecol.* **12:**395–403.

38. Morse, S. S. 1995. Factors in the emergence of infectious diseases. *Emerg. Infect. Dis.* **1:**7–15.

39. Nichols, D. K., E. W. Lamirande, A. P. Pessier, and J. E. Longcore. 2001. Experimental transmission of cutaneous chytridiomycosis in dendrobatid frogs. *J. Wildl. Dis.* **37:**1–11.

40. Paquin, B., M. J. Laforest, L. Forget, I. Roewer, Z. Wang, J. Longcore, and B. F. Lang. 1997. The fungal mitochondrial genome project: evolution of fungal mitochondrial genomes and their gene expression. *Curr. Genet.* **31:**380–395.

41. Parris, M. J., and J. G. Beaudoin. 2004. Chytridiomycosis impacts predator-prey interactions in larval amphibian communities. *Oecologia* **140:**626–632.

42. Piotrowski, J. S., S. L. Annis, and J. E. Longcore. 2004. Physiology of *Batrachochytrium dendrobatidis*, a chytrid pathogen of amphibians. *Mycologia* **96:**9–15.

43. Pounds, J. A., M. P. L. Fogden, and J. H. Campbell. 1999. Biological response to climate change on a tropical mountain. *Nature* **398:**611–615.

44. Powell, M. J. 1993. Looking at mycology with a Janus face: a glimpse at Chytridiomycetes active in the environment. *Mycologia* **85:**1–20.

45. Reed, K. D., G. R. Ruth, J. A. Meyer, and S. K. Shukla. 2000. *Chlamydia pneumoniae* infection in a breeding colony of African clawed frogs (*Xenopus tropicalis*). *Emerg. Infect. Dis.* **6:**196–199.

46. Retallick, R. W. R., H. McCallum, and R. Speare. 2004. Endemic infection of the amphibian chytrid fungus in a frog community post-decline. *PLoS Biol.* **2:**e351.

47. Richards, S. J., K. R. McDonald, and R. A. Alford. 1993. Declines in populations of Australia's endemic tropical forest frogs. *Pac. Conserv. Biol.* **1:**66–77.

48. Ristaino, J. B., C. T. Groves, and G. R. Parra. 2001. PCR amplification of the Irish potato famine pathogen from historic specimens. *Nature* **411:**695–697.

49. Rollins-Smith, L. A., L. K. Reinert, V. Miera, and J. M. Conlon. 2002. Antimicrobial peptide defenses of the Tarahumara frog, *Rana tarahumarae*. *Biochem. Biophys. Res. Commun.* **297:**361–367.

50. Rollins-Smith, L. A., C. Carey, J. M. Conlon, L. K. Reinert, J. K. Doersam, T. Bergman, J. Silberring, H. Lankinen, and D. Wade. 2003. Activities of temporin family peptides against the chytrid fungus (*Batrachochytrium dendrobatidis*) associated with global amphibian declines. *Antimicrob. Agents Chemother.* **47:**1157–1160.

51. Stuart, S. N., J. S. Chanson, N. A. Cox, B. E. Young, A. S. L. Rodrigues, D. L. Fischman, and R. W. Waller. 2004. Status and trends of amphibian declines and extinctions worldwide. *Science* **306:**1783–1786.

52. Taylor, T. N., and E. L. Taylor. 1997. The distribution and interactions of some Paleozoic fungi. *Rev. Palaeobot. Palynol.* **95:**83–94.

53. Tibayrenc, M., F. Kjellberg, and F. J. Ayala. 1990. A clonal theory of parasitic protozoa: the population structures of *Entamoeba*, *Giardia*, *Leishmania*, *Naegleria*, *Plasmodium*, *Trichomonas*, and *Trypanosoma* and their medical and taxonomical consequences. *Proc. Natl. Acad. Sci. USA* **87:**2414–2418.

54. Wake, D. B. 1991. Declining amphibian populations. *Science* **253:**860.

55. Weitzman, I., and R. C. Summerbell. 1995. The dermatophytes. *Clin. Microbiol. Rev.* **8:**240–259.

56. Woodhams, D. C., R. A. Alford, and G. Marantelli. 2003. Emerging disease of amphibians cured by elevated body temperature. *Dis. Aquat. Org.* **55:**65–67.

57. Zasloff, M. 2002. Antimicrobial peptides of multicellular organisms. *Nature* **415:**389–395.

III. SPECIFIC PATHOGENS

Molecular Principles of Fungal Pathogenesis
Edited by Joseph Heitman et al.
©2006 ASM Press, Washington, D.C.

Chapter 21

Cryptococcus neoformans: a Sugar-Coated Killer

JOHN R. PERFECT

A little over 100 years ago, *Cryptococcus neoformans* was first identified from peach juice, and very shortly after that report, the first clinical case of infection with this fungal pathogen was described when the yeast was isolated from a tibial lesion in a young German woman. Over the next century, this encapsulated heterobasidiomycetous fungus was extensively characterized as both a primary and secondary human and animal pathogen (143).

There has been an evolution in the nomenclature for this yeast, and the taxonomy continues to evolve with further genetic understanding. For instance, at least 38 other cryptococcal yeast species have been isolated from a variety of environmental niches, but except for an occasional clinical case of disease associated with *Cryptococcus laurentii* or *Cryptococcus albidus,* none of these other species except for *C. neoformans* appear to have the pathobiological tools to produce disease in mammals. Within the designation of *C. neoformans* there has been some divergence of the species over the last 30 million to 40 million years, and with the use of genetic comparisons, biochemical variations, epidemiological factors, and some physical differences in spore structures, taxonomic classifications have evolved. *C. neoformans* strains have classically been grouped into five serotypes, A, D, B, C, and A/D. The serotype differences reflect antigenic differences in the structure of the capsular polysaccharide and can be identified by either specific monoclonal or polyclonal antibodies, and kits are available for these serotyping tests. These five serotypes have been divided into three varieties (55): (i) *C. neoformans* var. *grubii* (serotype A), (ii) *C. neoformans* var. *neoformans* (serotype D), and (iii) *C. neoformans* var. *gattii* (serotypes B and C). The serotype A/D strains are primarily diploid yeasts resulting from unions between serotype A and D strains (101). Through recent genetic and molecular studies, it has been estimated that serotype A and D strains became genetically separated from each other over 18 million years ago and that serotype B and C strains are even more diverged from these other two serotypes. In fact, there is a proposal to make this variety into a separate species, *Cryptococcus bacillospora*, and most recently this name has been changed to *Cryptococcus gattii* (96). There are selective media which can distinguish this variety or species from the other two varieties because of biochemical differences such as the ability of var. *gattii* to utilize glycine and be resistant to concanavalin. With the sequencing of the genomes for all three varieties or species, the genetic differences can be fully appreciated, and taxonomic classifications may continue to evolve in the future. However, it is important to emphasize that there is some genetic separation between these varieties. Therefore, some minor consequences or variations between varieties will be uncovered when the genetic profile of the virulence composite is understood within each variety. Some of these virulence composite differences have already been discovered and are further discussed in this chapter. Furthermore, there have been some epidemiological studies which suggest that var. *gattii* infections more commonly involve immunocompetent individuals, with a tendency toward more impressive toruloma formation in the lungs or brain (161). On the other hand, all three cryptococcal varieties produce a clinically similar disease process which has been identified as cryptococcosis and the differences in the varieties will be responsible only for certain specific genetic nuances among their virulence or epidemiological compositions. The main principles for the genetic pathophysiology of cryptococcosis will probably be consistent among all three varieties, and in this chapter I will consider the yeasts to be different varieties or serotypes.

John R. Perfect • Division of Infectious Diseases, Department of Medicine, Duke University Medical Center, Durham, NC 27710.

C. neoformans has dramatically become a major worldwide pathogen which is no longer a rare medical curiosity. Prior to the mid-1950s, fewer than 300 cases of cryptococcosis had been reported in the medical literature, but today it ranks as one of the most common infectious disease agents to cause human meningoencephalitis. The yeast displays a wide host range for infection from small invertebrates to the water-dwelling porpoises, and in veterinary medicine cryptococcosis has become a well-described clinical entity which affects domestic cats and dogs and requires clinical management (136). This yeast is also particularly fascinating for the study of fungal pathogenesis because of its breadth of "at risk" hosts in which it produces disease. It has been considered both a primary and secondary pathogen. For instance, it can produce primary disease in apparently normal, immunocompetent individuals. On the other hand, it more commonly causes secondary disease in severely immunosuppressed hosts such as those with human immunodeficiency virus (HIV) infection, organ transplant recipients, patients with lymphoreticular malignancies, and those receiving corticosteroids for an underlying disease (143). It is apparent that this encapsulated yeast, which in its environmental niche may have evolved adaptive tools to protect itself from predatory scavengers such as soil amoebas, worms, and other environmental insults, has co-opted these protective mechanisms to quite successfully survive and produce disease in the harsh environment of the mammalian host (15). With infection, this yeast can produce a variety of clinical conditions or diseases. For instance, *C. neoformans* can infect the host without producing apparent disease in which it is either eliminated by a successful immune response or simply resides in a colonized state within the upper airway or respiratory tract of a mammalian host. On the other hand, *C. neoformans* can produce severe pulmonary disease (pneumonia to acute respiratory distress syndrome). From a localized infection, it can disseminate to reside in any organ within the human body. However, its profound propensity to disseminate and produce disease within the central nervous system of a host remains incompletely understood, but it is this feature that dominates the seriousness and difficulties in the management of this infection (Fig. 1).

Several recent research and clinical developments have elevated this yeast to the center stage for the study of fungal pathogenesis (142). It can be argued that *C. neoformans* is not the most common fungal pathogen since it is not an ascomycete but belongs to the less common human fungal pathogen group, the basidiomycetes. Futhermore, it possesses a

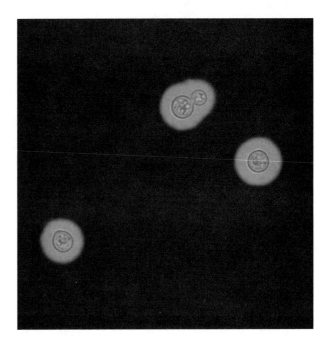

Figure 1. India ink preparation from cerebrospinal fluid, showing encapsulated yeasts (*C. neoformans*).

unique structural feature, a protective capsule, which is not common with other fungi. Finally, it only has a core group of investigators at present who study this pathogen. Despite these potential negatives for its general utility in aiding our understanding of fungal pathogenesis, this yeast does represent an ideal fungus to attempt to understand what it means to be a fungal pathogen at a molecular level. In the following discussion, seven major areas are examined to support the potential molecular insights into this pathogenic yeast which will allow us to identify drug targets, define drug resistance mechanisms, and/or prepare mutants or fungal products for protective fungal vaccines.

IMPORTANCE OF *C. NEOFORMANS* AS A HUMAN PATHOGEN

Since there is no routine clinical skin test to detect prior infection with this yeast, the precise magnitude and distribution of cryptococcosis remain uncertain. However, through serological studies which detect antibodies to *C. neoformans*, it has been found that the majority of adults possess anticryptococcal antibodies and most children acquire these antibodies to cryptococcal antigens before the age of 10 years in New York City (65). A laboratory-based skin-testing study suggested that most laboratory scientists who work with the fungus are positive for antibodies without evidence of clinical disease (158). These observations suggest that asymptomatic cryptococcal infections are frequent and occur at a relatively young age in persons in certain locales but that most cryptococcal infections do not produce disease because the host's immune system efficiently controls infection. However, recently a large outbreak of cryptococcal infections with disease did occur on Victoria Island in British Columbia (87). The strain was *C. neoformans* var. *gattii,* and it produced infection in apparently immunocompetent individuals. It is still unclear whether this outbreak was caused by a hypervirulent strain or simply resulted from a large-inoculum environmental exposure.

If documented clinical cases of cryptococcal disease are examined, the numbers are quite different. The best estimates for rates of cryptococcosis prior to the AIDS pandemic are from insurance records in Northern California from 1971 to 1980. In this report, an overall incidence of 0.8 case per million persons per year was found (116). It is important to note that this represents only one area of the United States and that during a similar period the Centers for Disease Control and Prevention observed a substantial rise in positive serological tests for cryptococcosis, which suggested that the incidence of cryptococcosis was rising and that, for the entire United States, rates may have been higher than the incidence in Northern California. As the HIV pandemic spread in the 1980s, a dramatic increase in reported cases of cryptococcosis followed. In 1992, during the peak of the AIDS epidemic in the United States, several urban areas had cryptococcosis rates that soared to over 5 cases per 100,000 persons (71). Then, during the widespread use of fluconazole and the emergence of highly active antiretroviral therapy (HAART) in the mid- to late 1990s, rates of symptomatic infections were substantially reduced in the United States to approximately 1 case per 100,000 persons per year, which is comparable to the incidence rate of another meningeal pathogen, *Neisseria meningitides* (179). In fact, today within the United Sates, cryptococcal meningitis commonly presents in a disadvantaged patient who is either undiagnosed with or untreated for HIV infection (118) or a patient receiving corticosteroids for an underlying medical condition. However, in less-developed countries such as those in sub-Saharan Africa, which continue to be ravaged by the frequent occurrence of HIV infection and lack of consistent HAART, 15 to 45% of those with advanced HIV infection will succumb to cryptococcosis (30, 176), and without treatment the mortality rate is 100% within the first 2 weeks of hospitalization (72). It is also important to recognize that HIV infection is not the only driving force behind the increased numbers of cryptococcosis cases.

Lymphoproliferative diseases have an increased risk of cryptococcosis, and in a large cancer hospital between 1989 and 1999 there was a reported incidence of cryptococcosis at 18 cases per 100,000 admissions (90). In one series of solid-organ transplant recipients, cryptococcosis occurred in approximately 2.8% of all transplant recipients with a mortality of 42% in those with cryptococcosis (83). It is predicted that in both cancer and transplant patients, the incidence of cryptococcosis may actually continue to rise since there is more frequent use of powerful immunosuppressive monoclonal antibodies such as alemtuzumab (Cam Path), which produces severe, prolonged CD4 lymphocytopenia, and CD4 cells are so critical in keeping dormant cryptococcal infections controlled.

In the past 20 to 30 years, there has been substantial progress in understanding the management of life-threatening cryptococcal meningitis. Prior to the advent of antifungal drugs, this central nervous system infection caused 100% mortality. There are now several antifungal drugs (amphotericin B, flucytosine, and fluconazole) available in combination and with sequential strategies to allow this otherwise uniformly fatal disease to be successfully managed in the majority of cases. In fact, with a series of clinical studies, there is now a routine strategy used to manage cryptococcal meningitis. Induction drug therapy starts with amphotericin B (0.7mg/kg/day) or the lipid product of amphotericin B (4 to 5 mg/kg/day) with or without the use of flucytosine. Then, after 2 to 3 weeks, the patient is switched to treatment with fluconazole in high doses (400 to 800 mg/day) for 8 weeks during the clearance phase of treatment (177). Finally, the patient is subjected to a suppressive regimen with fluconazole (200 to 400 mg/day) for 1 to 2 years. However, despite these drugs, clinical resistance still frequently occurs during the management of this infection, and in fact the acute mortality of cryptococcal meningitis in medically advanced countries still runs between 10 and 25% (143). It is apparent that this is not a rare infection, and it still produces substantial morbidity and mortality with the development of increased intracranial pressure (69) and host immune reconstitution syndromes (159). Its management needs further therapeutic improvement in efficient killing of the yeasts while controlling an appropriate immune response and abrogating complications of increased intracranial pressures. Therefore, further progress in the studies of molecular pathogenesis with this fungus may aid in drug discovery and new insights into pathogenesis for the development of optimal prevention and diagnostic strategies for this major human pathogen.

GENETIC SYSTEM

The life cycle of *C. neoformans* incorporates two distinct fungal forms: asexual and sexual (82). The asexual form is characterized by yeasts with or without buds and with variably sized capsules depending on the particular strain and its immediate environmental exposures. In human infections, the yeast is the form found in biological tissues and fluids; similarly, *C. neoformans* grows as a yeast in routine clinical cultures. It is generally classified clinically as a yeast, and the majority of the molecular pathogenesis studies have studied this specific form of the fungus.

The sexual form was first described by Kwon-Chung in 1975 (94). In the presence of *MATα* and *MATa* strains physically mixed on a minimally nutritious media such as V-8 juice agar and with some environmental desiccation, mating will occur with the formation of hyphae, true clamp connections, and basidia where meiosis and then the production of 1- to 2-μm basidiospores occurs. These spores are oval for serotype A and D strains and more elliptical and bacilliform for serotype B and C strains. It has also been shown that some *C. neoformans* strains, particularly *MATα* strains, have the potential to haploid fruit, in which they can produce hyphae and basidiospores on certain media without the process of mating with another strain (187).

The sexual cycle has become an interesting area for scientific exploration for the following reasons. First, the basidiospore is an ideal structure for aerosolization and possible inhalation and deposition into the host's airway for establishment of a pulmonary infection (167). However, despite its ideal size, the sexual structures (basidiospores or hyphae) have yet to be conclusively found in nature, although genetic population studies in certain areas of the world have shown recent recombination events suggesting that the sexual cycle occurs in nature despite most strains being clonal (106). In fact, it is not yet certain that the basidiospore would be a more infectious propagule compared to small, poorly encapsulated yeasts. Second, a significant epidemiological factor for the genetics of strains was the major bias toward *MATα* strains compared to *MATa* strains in both clinical and environmental strains (>90% of strains) (95). What is the reason behind this significant bias? In 1992, Kwon-Chung et al. made an interesting discovery that contributed to possibly answering this question, since they observed that during the creation of congenic strains the *MATα* strain was more virulent than the congenic *MATa* strain (93). These studies suggested that the mating-type loci might contain certain fitness genes for disease and thus might explain the dramatic bias among

clinical isolates in their mating locus, but it was less clear about the reason for bias in environmental isolates. Recent studies for the first time have found that when *tup1* (a global repressor gene) mutant is made there are phenotype differences between *MATα* and *MATa* strains, with the growth of *MATa* strains being more susceptible to environmental temperatures (100). The original work on the mating-type loci was performed in a *C. neoformans* var. *neoformans* strain (serotype D). On the other hand, Nielsen et al. were able to make congenic strains in *C. neoformans* var. *grubii* (serotype A) (133). In contrast to serotype D strains, the congenic pair of serotype A strains had no mating-type bias in regards to virulence (133). Therefore, the impact of the mating-type locus on the virulence composite is probably minimal at best and may occur only in certain strains or a specific serotype. Third, despite their uncertain importance in the control of virulence, the *MAT* loci have been sequenced and annotated; it is clear that these loci have had major implications in the control of morphology for this fungus. Focused investigations of sexual reproduction from pheromones to networks and pathways which control this cellular differentiation are now being performed on this heterothallic fungus (82). Fourth, with the use of congenic pairs of both *C. neoformans* var. *neoformans* and *C. neoformans* var. *grubii*, a major technical advance in molecular pathogenesis studies has resulted. Genetic crosses can now be used (i) to confirm specific mutants, (ii) to reduce any background mutations when null mutants are created by specific gene disruption strategies, and (iii) to allow confirmation of gene linkages to specific markers. Fifth, the sexual (perfect) state has been divided into two taxonomic species: *Filobasidiella neoformans* (serotypes A and D) and *Filobasidiella bacillospora* (serotypes B and C). It has been shown that within the serotype A and D strains, mating and viable progeny can be observed. It is less certain that all these serotypes (varieties) can efficiently mate and produce viable offspring through several generations. It is, however, apparent that most *C. neoformans* yeast strains in nature are haploid, but occasionally diploid strains have been found. Many of these strains have an A/D serotype and thus are diploid strains imitating parts of the genomes of both varieties (101). Finally, a meiotic map has been created for a serotype D strain, and this map has been extremely helpful in the alignment of the cryptococcal genome (112).

The 100-kb mating-type loci of *C. neoformans* have been sequenced and annotated. *C. neoformans* is heterothallic, with no silenced mating loci (56). Several genes in these loci have been studied for their impact on the virulence composite and morphogenesis of *C. neoformans* (39, 102, 186). However, there remain many unanswered questions regarding the linkage of sex and pathogenesis for many microorganisms, and *C. neoformans* has become an ideal model to help dissect its importance.

MOLECULAR BIOLOGY FOUNDATION

To fully investigate the details of molecular pathogenesis for cryptococcosis, it has been necessary to develop a molecular biology platform for the study of its virulence composite. In this respect, *C. neoformans* has had a decade of major technical growth in the creation of molecular tools which recently culminated in the completion of the entire genome sequence of several strains.

Initial molecular biology studies focused on distinguishing molecular strain differences with the use of karyotypes, repetitive elements, and eventually randomly amplified polymorphic DNAs, amplification fragment length polymorphisms, and PCR fingerprinting for strain genotyping (37, 144, 181). These molecular biology-based typing studies identified individual strains and thus led to clinical studies which distinguished relapsed infections from reinfections with a novel strain or a linkage between environmental and clinical isolates. Most recurrences of infection were found to be relapsed infection rather reinfection with a new strain (162). However, it has been observed that within a single infection, more than one unique strain may be present in some cases and microevolution of a strain might occur during a single infection (73, 168).

With these genome-wide molecular tools, it was possible to perform molecular evolutionary studies. It was determined that *C. neoformans* can be divided into several standard genotypes since many of the clinical and environmental strains appear to be separated into clonal lineages (117). However, in a group of isolates from Botswana, some recent genetic recombination between the strains appears to have occurred (106). In further specific studies it has been predicted that *C. neoformans* var. *neoformans* (serotype D) and *C. neoformans* var. *grubii* (serotype A) have been separated from each other genetically for over 18 million years (192), and the genetic separation between these varieties and *C. neoformans* var. *gattii* occurred even earlier. This is relevant to pathogenesis studies since gene networks and the importance of a specific gene to the virulence composite may differ between these varieties, and this prediction has been confirmed. For instance, several examples of virulence gene variance between varieties have already been discovered in *C. neoformans*. The *STE12* gene is important for the virulence composite of a serotype D strain, but the homologous gene in a serotype A strain appears to play no role in the virulence composite of this variety (23, 195). Conversely, the *STE20* gene is important for the serotype A strain virulence composite but has no impact on a serotype D strain (185). Even an entire signaling pathway may be used differently for the virulence composite between these closely related serotypes. For instance, the cyclic AMP (cAMP) pathway has a varietal difference for virulence such that the kinase gene, *PKA1*, is important for the virulence composite of a serotype A strain through its control of capsule formation and melanin production but this is not used for the virulence composite similarly in a serotype D strain (75). Another example of varietal difference in the molecular basis of virulence is in the above-described mating loci, in which the α mating locus impacts the virulence of a serotype D strain but not a serotype A strain (93, 133). It is clear from these examples that there is some recent evolutionary drift in the molecular virulence composite of *C. neoformans*.

A critical development in *C. neoformans* molecular biology has been the development of several DNA transformation systems. DNA transformation in *C. neoformans* has used electroporation, biolistics, or *Agrobacterium*-mediated DNA delivery with the use of either auxotrophic complementation (adenine or uracil) or dominant selection markers (hygromycin, nourseothricin, neomycin) (34, 38, 48, 77, 84, 115, 173). Electroporation of DNA has high-efficiency transformation in *C. neoformans* but with many episomal tranformants and very low homologous recombination events. Biolistic transformation has been used primarily for gene replacement events. Homologous DNA recombination with biolistics can average between 2 and 10% of transformants, but the targeting frequency can vary with the allele. Also, biolistic DNA transformation commonly produces stable transformants, but in some of these transformants there are multiple ectopic insertions of the construct. The *Agrobacterium*-mediated DNA transformation generally gives one single integrative event but much less homologous integration than the biolistics delivery. Since determination of homologous recombination is still a screening event, multiple strategies have been used to identify, confirm, and utilize essential genes in *C. neoformans*. Several of these strategies, from insertion of a second homologous gene into a strain to the use of diploid strains, have identified essential genes in *C. neoformans* such as *FKS1* (169), *TOP1* (41), *PKC1* (74), and *RAM1* (175). Furthermore, to analyze functions of essential genes, both antisense (67) and RNA interference (107) techniques have been successfully adapted to the molecular studies of this yeast.

A series of other molecular biological tools are available for use with *C. neoformans*. Episomal vectors are used in *C. neoformans,* but there is probably room for more development in vector designs for stable, usuable episomal vectors in all cryptococcal varieties and strains. For the detection of gene expression, both β-galactosidase and green fluorescent protein have been used in the study of *C. neoformans* as recorders of gene expression (42, 172) and regulated promoters are available (139, 188). In the era of genome-wide studies, *C. neoformans* has made substantial progress. Random signature-tagged mutant libraries have been created and used in pathobiological screens (84, 132). Finally, the completion of the sequencing, alignment, and annotation of several cryptococcal genomes by Stanford University, The Institute for Genomic Research, British Columbia Genome Center, and the Duke/Whitehead Institute Consortium (http://www.neo.genetics.duke.edu; http://www.tigr.org.tdby/e2kl/cnal; http://www/genome.stanford.edu; http://www.genome.ou.edu.cneo.html; http://www.bcgsc.bc.ca) will make rapid analysis and localization of genes possible.

A complete aligned and annotated *C. neoformans* var. *neoformans* genome of a serotype D strain was recently published (108). It was found that this strain had a genome size of approximately 20 Mb and contained approximately 6,500 intron-rich gene structures. These genes were arranged on 14 separate chromosomes. The genome contained many transposons which cluster at the centromeric region, and their presence may drive the karyotypic instability and the rapid genotypic and phenotypic variations which are characteristic of *C. neoformans* strains.

It was found in the genome studies that translocations and rearrangements can occur rapidly within this yeast. Consistent with the divergence of the phylum Basidiomycota from the ascomycetes ca. 900 million years ago, only 65% of *C. neoformans* genes have conserved sequence homologs to other completed fungal genomes. Ten percent of genes appear to be unique to *C. neoformans* based on the absence of identifiable homologs in current public databases, and the remaining 25% of genes match nonfungal sequences. In direct contrast to *Saccharomyces cerevisiae,* the *C. neoformans* genome shows an intron-rich gene structure with a transcriptional pattern which displays alternative splicing and antisense transcripts. This complexity in gene structure and transcription of *C. neoformans* might be a general feature of basidiomycetes, and further comparative studies of other fungal genomes may further elucidate how these genome structures relate to functions.

The power of this informational genome resource has substantially increased the potential of this fungal species as a molecular model for pathogenesis research. A recent benefit of the genome-sequencing projects has been the development of microarray transcriptional profiling for *C. neoformans* (91). The use of transcriptional profiling has now been successfully used in cryptococcal pathogenesis studies to make and direct hypotheses for further focused investigations. Earlier studies with the use of cDNA library subtraction protocols (2), differential-display reverse transcription-PCR (45), and serial analysis of gene expression (SAGE) (163, 164) analysis of *C. neoformans* have all been successfully used to collect and identify global transcriptional responses to both specific in vitro and in vivo environments. These studies have been able to direct further specific investigations of genes and certain pathways and determine their impact on pathobiology. The microarray technology allows for even more rapid generation of identifiable transcriptional data to make a new hypothesis or confirm or refute old ones. Since the development of a first-generation microarray, a more complete oligonucleotide microarray representing 5,000 to 6,000 open reading frames from both serotype A and D has been created for the scientific community use.

Several important features of global transcriptional profiling for pathogenesis research have already been demonstrated with *C. neoformans.* First, in examining high-temperature transcriptional profiles, a transcriptional regulatory gene (*MGA2*) was found to be up-regulated at 37°C by microarray analysis, and with the use of the microarray platform it was further observed that this gene, by creation of a null mutant, was linked to fatty acid synthesis and possibly membrane functions (91). There is now a link between the high-temperature growth response of *C. neoformans* and fatty acid synthesis, which might not have been intuitively linked without these profiles and creation of null mutants. The power of these profiling techniques is to identify genes or networks of genes which will help explain how *C. neoformans* will respond to specific external signals. It allows insight into how this yeast responds to its environment. In the study identifying the up-regulation of *MGA2* by high-temperature (37°C) growth, it was found that without *MGA2* the strain could not grow at this elevated temperature. There was a concordance between expression and function. However, studies of *C. neoformans* have also demonstrated that gene regulation does not always correlate with gene function. For instance, the *C. neoformans* isocitrate lyase gene (*ICL1*) is highly up-regulated in its expression at the site of a central nervous system

Table 1. Site-specific gene disruptants[a] in *C. neoformans* and their impact on virulence in animal models

Attenuated virulence	*ACA1* (8); *ADE2* (145); *AOX1* (2); *APP1*[b] (109); *CAC1* (5); *CAP10, CAP59, CAP60, CAP64* (17–22); *CAS2* (126); *CCN1* (29); *CLC1* (197); *CNA1* (137); *CNB1* (53); *CPA1* (182); *CPRa* (21); *CTS1* (52); *FHB1* (40); *GPA1* (4); *ILV2* (88); *LAC1* (155); *MAN1* (189); *MET3* (193); *MET6* (141); *MFα1, MFα2, MFα3* (160); *MPK1* (92); *PAK1* (185); *PKA1* (45); *PLB1* (32); *RAS1* (3); *SCH9* (183); *SKN7* (Wormley et al.; abstract); *SOD1* (31, 131); *SOD2* (62); *SPE3/LYS9* (89); *STE12D*[c] (23); *STE20A*[c] (185); *TPS1* (Wills et al.; abstract); *TSA1* (121); *UGD1* (70, 125); *URA5* (48); *URE1* (33); *USX1* (126); *VPH1* (49)
Hypervirulent	*APP1* (109), *CAS1* (85), *CRG1* (184), *PKR1A*[c] (45)
No impact	*CBP1* (66), *CCP1* (63), *CPA2* (182), *CPK1* (39), *FKB1* (36), *GPB1* (186), *ICL1* (154), *PAK1D*[c] (75), *PKA2* (36), *PKR1D*[c] (36), *RAS2* (182), *SET11* (39), *STE12A*[c] (195), *STE20D*[c] (185), *STE7* (39), *SXI1* (81)

[a]Genes are given names and numbers consistent with *Saccharomyces* homologs in many cases or designated letters taken from the references provided (which give specific descriptions of the genes).
[b]Impact on virulence is dependent on the host.
[c]A, serotype A; D, serotype D.

infection in rabbits but an *icl1* null mutant does not have an attenuated phenotype in the same model (154). The interpretation is that under the metabolic conditions of the cerebrospinal fluid in the host, the glyoxylate pathway in *C. neoformans* for utilizing two-carbon molecules is active and stimulated but, without this pathway, it appears that the yeast can still acquire enough energy from other sources to allow it to survive and cause disease in the host. It is essential that as global profiles are used in molecular pathogenesis studies, specific null mutants can be created to evaluate the specific phenotype (virulence) when a gene's importance is implicated by its transcriptional or translational expression profile. In *C. neoformans*, this combined strategy for evaluation can be readily carried out.

With our present molecular tools, the fulfillment of molecular postulates for "virulence" is now expected for all of the molecular pathogenesis studies

performed with *C. neoformans* (50). In this standard molecular strategy, the use of site-directed null mutants and the reconstituted mutant strains and wild-type strains are compared during infections in a standard, relevant animal model. Over 50 specific genes have been examined for their virulence potential with this site-directed gene disruption strategy. Table 1 contains a list of genes which have been studied and whose impact on the virulence composite has been identified. It is expected that this list will rapidly grow over the next decade. Two important features have been noted with creation of null mutants and animal studies. First, a focus on known (classical) virulence phenotypes such as capsule production, melanin synthesis, high-temperature growth, and intracellular growth has already yielded a series of identifiable genes which participate in these phenotypes, and generally these genes have an impact on the virulence composite (Table 2). Studies of

Table 2. Genes linked to the major virulence phenotypes of *C. neoformans*[a]

Capsule formulation	*ACA1, CAC1, CAP10, CAP59, CAP60, CAP64, CAS1, CAS2, CAS35, GPA1, MAN1, MET6, PKA1, PKR1, SCH9, SPE3/LYS9, STE12, STE20, USX1, VPH1*
Melanin production	*ACA1, CAC1, CLC1, GPA1, IPC1, LAC1, LAC1, MET3, PKA1, PKR1, STE12, VPH1*
Temp (high-temperature growth, 37–39°C)	*CCN1, CNA1, CNB1, CPA1, ILV2, MGA2, MPK1, RAS1, SOD2, SPE3/LYS9, STE20, TPS1, TPS2, TSA1, UGD1, VPH1*
Intracellular growth defect	*AOX1, FHB1, IPC1, PLB1, SKN7, SOD1*

[a]All genes are described in references listed in Table 1.

Table 3. Examples of the relative impact of gene disruptants on virulence in mice[a]

Severe[b]	CNA1, CNB1, ILV2, MAN1, MET3, RAS1, SOD2, TPS1
Moderate[c]	CAC1, GPA1, MET6, PAK1, PKA1, SPE3/LYS9
Mild[d]	AOX1, FHB1, PLB1, SOD1, STE 12D/20A, URE1

[a]All genes are described in references listed in Table 1.
[b]Complete elimination of yeast from host and full survival of animals.
[c]Reduction in yeast counts compared to wild type, with the majority of animals surviving.
[d]Statistically reduced yeast counts in tissue and prolonged survival compared to wild type, but with all animals eventually succumbing to infection.

C. neoformans are now beginning to indicate how these genes are networked with the use of signal transduction pathways and the specific gene mechanism(s) or features behind their virulence phenotype. Second, null mutants do have a quantitative effect on the virulence composite. While some *C. neoformans* genes are absolutely essential for survival in the host, others may have minor pathogenic effects in that the null mutants delay the killing of the host compared to the wild-type strain but nonetheless are still pathogenic. In Table 3, specific genes are listed in groups that reflect their relative impact on virulence composite in a standard murine model. This concept of a quantitative measurement of gene impact on the virulence phenotype has important implications in choosing an antifungal target (fungicidal versus fungistatic) or their use in vaccine strategies.

Finally, molecular biology studies have begun to help in defining a mechanism(s) for drug resistance in *C. neoformans*. It has been known that for azoles, there is a heteroresistance pattern for resistance development and many strains can adapt to azole exposure with selection of these azole resistant clones (124). The specific mechanism(s) behind this resistance pattern is not certain. However, several studies have identified specific mechanisms by which a cryptococcal strain has become azole resistant. For instance, some strains have point mutations in the azole target, the 14α-demethylase gene, *ERG11* (98, 152). Also, an ATP binding cassette (ABC) transporter gene (*AFR1*) has been identified, and its increased expression has been associated with fluconazole resistance (148). These studies demonstrate that both target alterations and drug efflux mechanisms are involved in *C. neoformans* drug resistance. In response to drug resistance and the need for new target development, cryptococcal mutants have now been studied to determine if another antifungal target which potentiates the present antifungal drugs could be identified. Two examples are the *mga2* and *met6* mutants, which appear to be hypersensitive to fluconazole (91, 141). However, many of the failures in treatment of cryptococcosis are actually related to "clinical" resistance, which is not necessarily identified by direct drug resistance of the isolate (i.e., high MICs of the treatment drug) but simply represents the inability to manage all aspects of the infection in a crippled host.

VIRULENCE PHENOTYPE

The definition of virulence is always a contentious term, but from a simplistic viewpoint, Webster's dictionary defines virulence as "the relative infectiousness of a microorganism causing disease or its ability to overcome the natural defenses of the host." This definition describes the quantitative nature of virulence and allows for the inclusion of specific properties in possession of the parasite to injure the host. Finally, it acknowledges that the host is a major factor in the expression of the virulence phenotype. In the study of *C. neoformans* and its virulence composite, this simple definition has been applied and a series of genes have met the three principles described above.

C. neoformans has several well-characterized virulence phenotypes which have been approached in their understanding by both genetic and molecular tools. Three classical phenotypes have been shown to be under genetic control: (i) capsule production, (ii) melanin formation, and (iii) the ability of the yeast to grow well at 37°C (13, 146).

The most distinctive feature of *C. neoformans* is its capsule, which is a polysaccharide structure (GXM) made of unbranched chains of α1,3-linked mannose units substituted with various xylosyl and β-glucuronyl groups depending on the specific serotype of the strain. The pathological mechanisms displayed by the capsule and the shedding of the polysaccharide into tissue and fluids are extensive. The polysaccharide amounts have formed the basis of one of the best serological diagnostic tests in human mycoses, but this complex sugar also has major implications for pathophysiology of infection; these include antiphagocytosis, complement depletion, antibody unresponsiveness, dysregulation of cytokine secretion, inhibition of leukocyte migration, enhancement of HIV infection, L-selectin and tumor necrosis factor receptor loss, induction of apoptosis, a possible factor in cerebral edema, and its negative charge for cellular interactions (13).

It has been shown that the capsule is a very dynamic structure, which can enlarge in response to several relevant environmental stimuli such as elevated ambient pCO_2 levels, low iron concentrations, and the presence of serum (68, 180, 196). Even in host tissue, the site of infection can exert some control over the capsule size of yeast strains (16, 151). It is important that the yeast strain has the ability to alter its capsule size in response to the environment since a mutant with a fixed capsule size can show attenuated virulence (68). The molecular biology and biochemistry of the capsule still has many gaps, but work has begun to understand its structure through careful nuclear magnetic resonance spectroscopy studies (27) and molecular biology. For instance, recently this polysaccharide structure was shown to be attached to the cell wall of the yeast by glucan bridges (14).

The initial molecular approaches to cryptococcal capsule studies focused on identifying genes that were important to the appearance of a capsule which encased the yeast. It has been known that there is not a precise correlation to the size of the capsule in vitro and the virulence of a strain (47). For instance, small-capsule strains may be more virulent than large-capsule strains and vice versa, and recent clinical studies showed that patients with poorly encapsulated strains in the lungs had the same outcome as those with prominent capsules in the tissue (174). However, it was apparent that if an investigator created an acapsular or severely hypocapsular mutant strain or a strain that could not regulate its capsule in response to external stimuli, the strain would be significantly less virulent than the parental strain. A series of capsule-deficient mutants were created with "unknown-function" genes such as *CAP10*, *CAP59*, *CAP60*, and *CAP64*; these mutant strains had no significant capsule, but when the gene was replaced in the strain, the capsule reappeared around the yeast. These poorly encapsulated mutants were routinely avirulent in animal models (4, 17–20, 22). Recent work has suggested that some of these mutants still produce polysaccharide but that there is a transport defect such that the polysaccharide does not attach to the cell wall to form its barrier protection (61). There have also been studies which have identified regulatory pathways that control capsule production. For instance, the cAMP pathway appears to control the capsular growth of *C. neoformans* and both the up- and down-regulation of capsule synthesis can be directly correlated with the virulence phenotype. For example, in a serotype A strain a *pka1* mutant which is missing a catalytic kinase in the cAMP pathway is hypocapsular and hypovirulent but the regulatory *pkr1* mutant, which allows an uncontrolled cAMP pathway, produces large capsules and is hypervirulent (45). These genes within the cAMP pathway do

control other virulence phenotypes, but capsule production is a major component of its virulence control. It has also been shown that an impact on the specific structure of the capsule can have profound effects on the pathobiology of the yeast. For instance, the *cas1* mutant has a capsular structure which lacks GXM O acetylation, and this change in capsule structure created a hypervirulent strain compared to the wild-type strain. Mechanistically, O acetylation and/or xylosylation is important for binding of anti-GXM monoclonal antibodies, for complement activation, and for tissue accumulation of GXM and inhibition of neutrophil recruitment, which, if altered, might create a host-pathogen imbalance. In further studies of the main capsular constituents (GXMs) of *C. neoformans*, a gene, *USX1* (encoding UDP-xylose synthase), was found to be necessary for capsule xylosylation; when it was not present, the virulence of *C. neoformans* was attenuated (126). Finally, the capsular state has been linked to cell wall formation. Mutants with mutations in the UDP-glucose dehydrogenase gene (*UGD1*) have shown that it is probable that glucuronic acid residues are important to the proteins involved in both capsule and cell wall biosynthesis and in the ability of the yeast to grow at 37°C (70).

The formation of melanin is observed in many fungi pathogenic for animals and plants (99). *C. neoformans* possesses prominent melanin production through a simple pathway involving a laccase enzyme that converts diphenolic compounds to melanin through a series of autoxidation steps (198). Early studies of "albino" mutants created by nonspecific mutagenesis showed that these strains were specifically less virulent than wild-type strains (97). In the early-gene disruption experiments, a laccase gene (*LAC1*) was targeted for disruption. The *lac1* mutant had the predicted hypovirulent phenotype (155). Subsequently, a second laccase gene (*LAC2*) in *C. neoformans* was identified with a less strong phenotype compared to that of the *LAC1* gene (120, 149), but it can complement *lac1* or *gpa1* mutants when overexpressed in them (149). Animal studies have shown that *LAC1* is transcribed in vivo (155) and melanin is formed in yeast cells within the host (153). Extensive studies have identified other genes which appear to have a direct impact on melanin production in *C. neoformans* (5, 45, 110, 149, 155, 193, 197). This importance of melaninization to the ability of the yeast to produce disease has led to a series of studies of its potential mechanism(s) for pathological significance. These include (i) its capacity to act as antioxidant, (ii) its use for cell wall integrity and charge, (iii) its interference with antifungal susceptibility, (iv) abrogation of antibody-mediated phagocytosis, and (v) protection from extreme temperatures (13). With the use of *lac1* mutants in mice, it appears that the importance of melanin might even be infection stage specific (135). Melanin may not be essential to the yeast infection during the lung phase of infection but may be extremely helpful for the yeast in its ability to escape the lungs and infect the central nervous system.

The ability to grow at 37°C is a simple and intuitively obvious virulence phenotype for an invasive mammalian pathogen. Of over 1.5 million different fungal species in the world biota, only approximately two dozen consistently cause human disease and thus have developed the prerequisite high-temperature (mammalian) growth phenotype. This important virulence phenotype of high-temperature growth is vividly illustrated among the many cryptococcal species, some which possess capsules and/or produce melanin. On the other hand, except for *C. neoformans* and an occasional *C. laurentii* and *C. albidus* strain, none of these species can grow consistently well at 37°C, and all of the non-*neoformans* cryptococcal species are generally considered to be nonpathogenic for humans. There also appears to be some evolutionary drift in high-temperature growth among the three cryptococcal varieties. For instance, although there may be overlap with some strains, in general, serotype B and C (variety *gattii*) and serotype D (variety *neoformans*) isolates appear to be more sensitive to growth inhibition and killing at high environmental temperatures (37 to 39°C) compared to serotype A isolates (variety *grubii*) (113). It is possible that the ability to grow at high environmental temperatures has been a driving force in the yeast's vegetative and pathological fitness. In some simple but insightful experiments, Xu examined the patterns of genotype-environment interactions of spontaneous mutations with *C. neoformans* for multiple clinical and environmental isolates (191). It was found that as spontaneous mutations accumulated during vegetative growth in several environments, the yeast clones maintained at 37°C showed less decline in vegetative fitness than did those maintained at 25°C. These results suggest that a high-temperature environment might be very important for the general fitness of the yeast to grow and that the *C. neoformans* virulence composite benefits from periodic exposure to high temperatures from either a certain environmental niche or its presence in avian or mammalian hosts.

The ability to grow well at 37°C in *C. neoformans* was first shown to be under genetic control and specifically linked to virulence by Kwon-Chung and Rhodes (97). With the use of genetic crosses and mutants, these investigators were able to link the high-temperature growth phenotype with genetic inheritance. From these initial genetic experiments, substantial progress in identifying the specific gene control over high-temperature growth has recently been made. One of the first molecular biology studies of pathogenesis in *C. neoformans* was the identification of a gene necessary for high-temperature growth. Odom et al. found that calcineurin A-null mutants grew normally at 30°C but were not viable at 37°C in vitro and were not pathogenic in several mammalian models of infection (137). Besides the potential discovery of a network required for signaling high-temperature growth of *C. neoformans*, an important feature of this initial molecular biology work was that it vividly illustrated the importance of directly studying the pathogenic yeast strains since there are significant evolutionary distances between other fungi. For instance, this phenotype for the signaling molecule, calcineurin, in *C. neoformans* would not have been predicted from studies with the model yeast, *S. cerevisiae*, since the calcineurin A mutant of *S. cerevisiae* does not have a temperature-sensitive (ts) growth phenotype. This work is an outstanding example of the importance of the need for direct study with the pathogen rather than making extrapolations in gene functions from surrogate model organisms.

Further study of genes associated with high-temperature growth in *C. neoformans* has taken two different directions. First, directed-gene studies with known pathways or with specific functions were identified and characterized. In this process of understanding the gene's importance, it was found that the null mutant possessed a ts phenotype. Several *C. neoformans* genes have been identified by using this strategy. For instance, *VPH1* (which encodes vacuolar ATPase) (49) and *UGD1* (which encodes UDP-glucose dehydrogenase involved in capsule synthesis) (70, 125) are involved in the ability of the yeast to grow at high temperatures. These results emphasize how high-temperature growth links to other virulence factors such as capsule formation, and these findings also begin to identify specific intracellular functions which allow adaptation to high-temperature growth. It has been shown for *C. neoformans* that even functional genes encoding components of basic amino acid metabolism such as *ILV2* and *SPE3/LYS9*, a chimeric gene, are required for *C. neoformans* to grow optimally at high temperatures (88, 89). Recent studies have also begun to identify genes in signaling pathways and networks, such as *RAS1*, *CNA1*, *CNB1*, *MPK1*, and *CTS1*, which need to be intact for high-temperature growth. Several *C. neoformans* genes have shown their importance at very high temperature (>37°C). For instance, genes such as *STE20α* and *CPA1* (which encodes cyclophilin) are required for very-high-temperature growth (39 to 40°C) but are dispensable for efficient growth at 37°C (182, 185).

The second method for the discovery of temperature-related genes is to directly attempt to discover them, and this strategy has been used successfully multiple times in *C. neoformans*. One strategy was to identify a ts strain or mutant and complement it with a gene library. An example of this strategy was the elegant work of Chung et al. when a ts strain isolated from a canine nasal cavity was complemented for its ts phenotype with a genome library and eventually the complementing gene, *CCN1*, was identified and characterized (29). This screening strategy is likely to yield many more ts genes, particularly with the signature-tagged library procedures adapted to *C. neoformans*.

A second strategy for *C. neoformans* is to predict temperature sensitivity from other model systems and to test for conservation of certain high-temperature growth genes and/or pathways. An example of this focus was the study of the stress protectant sugar, trehalose, in *C. neoformans*. It has been shown that a disruption in the pathway of trehalose in *S. cerevisiae* would lead to a yeast which was temperature sensitive for growth since this sugar can be used to stabilize proteins when cells are under stress. It was hypothesized that *C. neoformans* would also use this sugar, and, if so, it would identify a specific mechanism of how this yeast responds to high temperatures. *C. neoformans* mutants with mutations in *TPS1* (which encodes trehalose-6-phosphate synthase) and *TPS2* (which encodes trehalose-6-phosphate phosphatase), which are the two trehalose synthesis genes, were created and found to grow well at 30°C but not at 37°C (E. A. Wills, U. Himmelreich, E. Mylonakis, and J. R. Perfect, *Abstr. Inst. Soc. Hum. Anim. Mysol. Meet.*, abstr. 281, 2003). As predicted from other fungi, trehalose is important to high-temperature growth, but in these studies another principle was illustrated. The *tps1* mutant was inhibited in growth only at high temperature, but the *tps2* mutant was actually killed, suggesting that a block in intermediary metabolites such as an accumulation of trehalose-6-phosphate in the *tps2* mutant might add to the severity of the growth phenotype at high temperature. This observation within known pathways may have relevance to choosing specific antifungal targets, where the goal for a target might be a fungicidal response.

The third strategy for identifying high-temperature growth genes is to use transcriptional profiling to identify genes which are up- or down-regulated in response to environmental temperatures. For instance, genes are identified that are induced or repressed at 37°C and then genes are examined to determine if they participate in the growth of *C. neoformans* at this temperature. Although the temperature-regulated gene transcriptional profile can be complex and changes rapidly, a variety of methods including cDNA library subtraction, SAGE, differential-display reverse transcription-PCR, and microarrays have begun to identify genes and pathways important to high-temperature growth. For example, these screens have led to the identification of temperature-induced genes like *COX1* (which encodes cytochrome oxidase 1 subunit 1), *AOX1* (alternative oxidase), and *MGA2* (transcriptional regulatory gene) when *C. neoformans* is shifted from low to high temperature (2, 91, 170). With the mitochondrial gene *COX1*, it was found that there are dramatic differences during temperature changes in *COX1* transcription between var. *grubii* and var. *neoformans* strains, but this difference in *COX1* transcription did not affect the virulence difference observed between the two strains (2, 170). In fact, in these studies, the cryptococcal mitochondrial DNA between a virulent var. *grubii* strain and an avirulent var. *neoformans* strain did not appear to contain specific virulence genes (171). An *aox1* mutant did possess a mild defect in virulence. However, it did not possess a ts growth defect but had an antioxidant function (2). Conversely, the calcineurin A gene, which is necessary for growth at 37°C, is not regulated by temperature shifts. Thus, cryptococcal genes regulated by temperature are not always involved in high-temperature growth and genes involved in high-temperature growth are not always regulated by temperature. Despite this lack of a consistent direct correlation in regulation and growth of genes regulating temperature, the identification of how *C. neoformans* adapts to high temperatures through capturing transcript regulation has provided great insights into its general ability to become a pathogen.

Transcriptional profiling has provided two other examples of how it has been used to understand this pathogenic yeast. First, in the temperature-regulated profiles, it was clear that mitochondrial genes were up-regulated, and it was surmised that the function of this organelle was to provide energy for both synthesis and degradation of products during the stress of high-temperature growth (91). This focus on mitochondrial function and high-temperature growth was merged with an understanding of another stress function, oxidative stress, by examination of the *SOD2* gene. *SOD2* encodes a manganese superoxide dismutase which functions in the mitochondria. With the use of *sod2* mutants, it was shown that the ability to grow at high temperature was critically linked to the function of this protein and its handling of the mitochondrially generated oxidative stresses at high temperatures (62). The focus on the mitochondrial function at high temperatures was initiated by general, exploratory transcriptional profiling methods.

A second example of the use of transcriptional profiling to begin understanding the mechanisms behind high-temperature growth of *C. neoformans* is with the use of microarrays to carefully examine transcripts at 37°C compared to 25°C (91). In this study, the transcription factor gene *MGA2* was found to be highly up-regulated at 37°C and the null mutant had reduced growth at 37°C. *MGA2* is a probable homolog of an *S. cerevisiae* gene which is involved in regulating the transcriptional response to cold and hypoxia and specifically activates the expression of a fatty acid desaturase gene. With the use of comparative microarray transcriptional profiling between wild-type and *mga2* strains, it is apparent that *MGA2* in *C. neoformans* is also involved in fatty acid synthesis and thus demonstrates a focus for further studies of the relationship between high-temperature growth and fatty acids. During studies with the *mga2* mutant, it also displayed another phenotype, which was hypersensitivity to cell membrane stresses such as fluconazole sensitivity. These findings vividly illustrate that these networks for high-temperature growth are linked to other stress response pathways.

The ts genes in *C. neoformans* represent several important features in the molecular pathogenesis of cryptococci. First, all null ts mutants studied so far appear to be avirulent in mammalian models. However, despite the ts phenotype, the mutants might not be rapidly eliminated from the host. For instance, when inoculated intranasally, the *cna1* mutant persists in the lungs of mice for several weeks before eventually dying and produces inflammation but not a protective immune response (190). However, ts mutants might still be molecularly manipulated to produce a better immune response to act as live-vaccine candidates, and even within ts mutants, some may be eliminated faster than others. Therefore, each ts mutant must be studied in the mammalian host to determine if it would be a potent fungicidal or fungistatic target. Second, in temperature-regulatory genes, the disconnect between regulation and function can frequently occur but the use of transcriptional profiling remains an excellent tool for virulence gene discovery. Third, there may be some conservation between yeast species in their ts genes but *C. neoformans* will use genes differently. Fourth, ts genes in *C. neoformans* participate in a variety of functions including signaling networks, stress pathways, and basic metabolism. Finally, since studies of *S. cerevisiae* have identified over 70 genes with a ts

phenotype, the approximately 12 identified ts genes in *C. neoformans* are probably just a small fraction of the genes required for efficient growth at mammalian temperatures, and thus these genes represent many potential unique genetic weak spots for inhibition of *C. neoformans* growth in the host.

In addition to the classical virulence phenotypes, substantial progress has been made in identifying and studying other genes important to the virulence composite in urease production (33), phospholipase secretion (32), mannitol production (25), oxidative stress reduction (2, 31, 62, 121), and vacuolar acidification (49). The *ure1* mutant was observed to be attenuated in a pulmonary murine model but not in an intracisternal rabbit model (33). This finding suggested either a site-specific or host range-specific function for urease. Mechanistic support for urease site-specific importance has been provided by experiments which showed that cryptococcal urease contributes to central nervous system invasion by enhancing yeast sequestration within microcapillary beds within the brain during hematogenous spread (138). This yeast sequestration appears to facilitate the passage from the blood to the brain by *C. neoformans*. This theme of infection site (i.e., central nervous system invasion) importance of a gene has now been found for several other *C. neoformans* genes. For instance, the laccase gene (*LAC1*) is required for invasion of the brain but not for pulmonary persistence (135) and the antiphagocytic gene (*APP1*) can modulate the internalization of *C. neoformans* by professional phagocytes, which has an impact on the ability of an *app1* mutant to disseminate into the brain (109). Phospholipase (*PBL1*) can be used by *C. neoformans* to grow intracellularly, and this enzyme appears to be important as *C. neoformans* makes its way from a pulmonary-site infection through the lymph system into the blood (156). It is known that *C. neoformans* can grow in the host both as an intracellular and an extracellular yeast (51). Several genes have now been identified which aid the intracellular growth of this yeast within macrophages (2, 31, 32, 74; F. L. Wormley, G. M. Cox, and J. R. Perfect, *Abstr. 104th Gen. Meet. Am. Soc. Microbiol.* 2004, abstr. F.074, 2004). Acidification of the vacuolar compartment (*VPH1*) demonstrates how a single cellular function can have profound impact on several virulence factors such as capsule production, laccase and urease expression, and high-temperature growth (49).

C. neoformans is a respiratory yeast which requires oxygen for survival and must effectively protect itself from the harsh mammalian environment with its oxidative and nitrosative products (119). Substantial studies have begun to dissect mechanisms used by *C. neoformans* to combat these noxious agents. There appears to be a variety of genes producing protective enzymes in *C. neoformans,* and these systems appear to be redundant, so that the gene knockout strains for certain catalases and peroxidases have little impact on the virulence phenotype (63). However, some of these enzymes are important by themselves to pathobiology. For instance, the cytoplasmic, magnesium superoxide dismutase (encoded by *SOD1*) has an intracellular growth defect and a mildly attenuated virulence phenotype when the gene is disrupted (31). A thiol peroxide (encoded by *TSA1*) has been identified which protects *C. neoformans* against both oxidative and nitrosative damage, and a *tsa1* mutant is less virulent in animal models (121). An interesting feature is that this protection appears to be utilized critically at high temperatures since the *tsa1* mutant is temperature sensitive for growth at 37°C. Also, *C. neoformans* needs some protection from nitrosative damage in that a flavohemoglobin denitrosylase (encoded by *FHB1*) is required for optimal intracellular growth and full virulence of a strain and combines with superoxide dismutase (encoded by *SOD1*) to protect the yeast against host defenses (40). Finally, the principle of need for protection against not only exogenous but also endogenous oxidative products has been best acknowledged by studies of the mitochondrial superoxide dismutase (*SOD2*) in which its absence within the enclosed energy factory (mitochondria) dramatically and proportionally reduces the yeast's viability as the environmental temperature rises from 24 to 37°C until survival is not possible without *SOD2* at temperatures above 35°C (62).

There are several aspects of the virulence phenotype studies of *C. neoformans* which should be emphasized. First, most studies are presently performed on only a few well-characterized strains, and information may or may not relate to the general features of all cryptococcal isolates. Second, it is clear that *C. neoformans* has methods for rapid microevolution in that phenotypic switching of colony morphology can rapidly occur in a strain and can even be directed by antifungal therapy to certain morphological switching variants (57); it involves changes in both the virulence of the variant strains and their impact on the host responses (64, 147). The *Cryptococcus* genome is filled with many transposons (108), and it is clear that within a single infection there can be morphologically different yeast cells and detectable genetic variations (54, 73, 168). Third, within a single strain, virulence parameters can change by simple in vitro passage, which will attenuate a strain, or in vivo passage, which will increase virulence (146). Fourth, the impact of the

host genetic background may affect the virulence phenotype. Although the major histocompatibility complex appears to have only a minor impact on the virulence phenotype of cryptococcal strains (114), there are a series of murine strains and gene knockout strains which have a major influence on expression of the virulence phenotype (80, 194). For example, A/Jcr mice with a C5 complement deficiency show a much more rapid progression of infection compared to BALB/c mice (150). Fifth, there are genes in which there is a dosage effect or which act in synergy for their impact on the virulence composite (182). The *cya1* (cyclophilin A) mutant is attenuated in virulence, but the double-knockout *cya1 cya2* mutant is even less virulent than the *cya1* mutant despite the fact that the *cya2* mutant has no impact by itself on virulence. Finally, the virulence composite of var. *gattii* is just beginning to be investigated; many genes will be similarly important for the virulence composite in this variety like the other two varieties (31, 131). An example is the *SOD1* gene, which participates in the virulence composite in all three varieties. However, it is also likely that there will be some variations in how the networks of virulence are arranged for each variety. With the recent progress in the signaling pathways of virulence for var. *grubii* and *neoformans*, it will be interesting to compare differences and similarities that have evolved. Finally, drugs used for other medical conditions may have an impact on *C. neoformans* virulence factors. Two examples of this interaction are cyclosporin-tacrolimus, which inhibit high-temperature growth (137), and indinavir, which affect capsule and urease production by *C. neoformans* (123). Conversely, the virulence factors of *C. neoformans* like melanin can directly affect the antifungal drugs used to treat it (178).

IMMUNOLOGY

The host responses to the invasion of *C. neoformans* are relatively well understood due to several decades of outstanding, insightful studies into the immunological response to this yeast (103, 128). An attempt will be made to summarize this wealth of knowledge in relation to pathogenesis studies. In the 1950s, Baker and Haugen carried out autopsy studies which documented the existence of a primary lung-lymph node complex in individuals with and without a history of cryptococcosis (9). This finding allowed clinicians to make a hypothesis about the general pathophysiology of this infection. Infectious propagules (either small-capsule yeasts or basidiospores) are inhaled into the lungs. The fungal exposure can produce asymptomatic colonization of airways, symptomatic pneumonitis, or asymptomatic infection of the lungs and draining lymph nodes. It is likely that the majority of infections result in a strong cellular immune response, producing local granulomatous inflammation in tissue, which is essential for containment of infection. The fungus will either die or become dormant within the lung-lymph node complex. If the host has or develops specific or severe cell-mediated immune defects, this yeast can grow within the lungs or lymph nodes and disseminate outside the lungs with a propensity to invade the central nervous system. This proposed pathological scenario continues to remain intact and is supported by skin test data, serological studies, and epidemiological evidence (60). Much of the recent immunological work has been done in an attempt to understand the details of how *C. neoformans* growth is controlled in the host. *C. neoformans* infection presents the interesting paradox that normal immune mechanisms must be highly effective at containing this infection but in many cases may still be unable to eradicate the yeasts once established in tissues, even within immunocompetent hosts.

The mammalian host response to *C. neoformans* is multilayered and begins with a series of innate host factors (serum, saliva, and complement) that discourage this infection. Furthermore, creative studies over the last decade have shown that a humoral response with production of a specific set of antibodies can play an important role in the elimination or control of cryptococci in the body (12, 127). The most critical immune response to control of infection primarily focuses around cell-mediated immunity. For instance, the effector cells against the yeast include CD4$^+$ and CD8$^+$ lymphocytes, natural killer cells, and activated professional phagocytes that produce granulomatous inflammation (76, 104). Because granuloma formation is a result of a *Th1*-polarized response, there is a general requirement for tumor necrosis factor, gamma interferon, interleukin-2 (IL-2), IL-12, and inflammatory chemokines such as monocyte chemotatic protein 1 (MCP-1) and macrophage inflammatory protein 1α (MIP-1α) for recruitment of host cells to the site of infection (1, 78, 79, 86). A shift toward a TH2 response (IL-4, IL-10, certain antibodies, and eosinophilia) probably antagonizes the clearance of this yeast from host tissue. It is also clear from clinical observations that a continuing immune response is also important to keep the infection dormant. For example, the risk of disseminated cryptococcosis rises precipitously for patients with HIV infection when CD4 counts fall below 100 cells/μl of blood without an apparent exogenous exposure to yeasts (35).

This general understanding of the host response has led to more active investigations of how it can be manipulated in favor of the host. An initial polysaccharide-tetanus toxoid conjugate vaccine was used successfully to protect mice against *C. neoformans* challenge (43), but its further development has been delayed. However, with the advancement of molecular biology, several *C. neoformans* genes and encoded proteins have now been identified which may help elicit a protective immune response (10, 105, 111). With the use of attenuated strains of cryptococci, the possibility of live vaccines can also be considered. Furthermore, the understanding of the importance of the TH1 response to control of infection has led to studies with the use of recombinant gamma interferon for treatment of human cryptococcal meningitis. Early studies have been either case reports or a small series of patients and have given encouraging results regarding the use of gamma interferon as adjunctive therapy to help sterilize the cerebrospinal fluid (140). However, the precise use of gamma interferon remains uncertain as a definitive treatment strategy. The importance of these cytokines is further emphasized in recent molecular biology studies in which it has been possible to create *C. neoformans* strains which secrete the host gamma interferon (Wormley et al., *Abstr. 104th Gen. Meet. Am. Soc. Microbiol. 2004*). These strains induce an impressive TH1 response at the site of infection, and this appears to aid in the yeast's destruction. It opens up the possibility that these attenuated strains may be used to create protective immune responses in vaccine development or may even be used to deliver cytokines to the site of infection for treatment.

An effective, coordinated immune response is essential to prevent disease, and when it occurs, to effectively treat it. On the other hand, this is a complex infection, and during the age of HAART it has been reported that a rapid immune reconstitution syndrome (IRS) has complicated the management of patients (159). This syndrome is thought to result from enhanced partially reconstituted pathogen-specific cell-mediated immunity and rapid induction of proinflammatory cytokines, leading to an exaggerated inflammatory reaction as the retroviral infection is controlled. A too rapid and vigorous host response within the central nervous system can paradoxically have dire consequences for the host; this makes the control of immunomodulation critical. It is likely that the IRS occurs in other patient populations such as organ transplant recipients and makes the management of a patient receiving an antifungal agent confusing if the patient experiences new symptoms while receiving therapy. Is it progressive infection, or is it IRS?

Host risk factors identify the complex but important nature of the host in this infection. HIV infection interacts indirectly with *C. neoformans* through its impact on reducing cell-mediated immunity. Exogenous corticosteroids, with their myriad of immune dysfunctional effects, are major contributors to the appearance of cryptococcosis. Idiopathic CD4$^+$ lymphocytopenia and hyper-immunoglobulin M syndrome are associated with specific immune defects which allow *C. neoformans* to proliferate in the host. Sarcoidosis, diabetes, and chronic obstructive pulmonary disease bring together a variety of host factors which may be important to the control of infection (143). Other immunosuppressive factors, such as cyclosporine and tacrolimus, have complex interactions in that they are both immunosuppressive but also possess direct anticryptococcal activity. Generally, in the doses used for antirejection therapy, these immunosuppressants tend to put a patient at risk for infection rather than preventing it. The more frequent use of the immunomodulating monoclonal antibodies (such as infliximab and alemtuzumab) in clinical practice will probably become a risk factor for cryptococcosis since these antibodies block host cytokines (tumor necrosis factor) or host cells (CD52 cells) that are important for the control of this infection (7).

It is apparent from several studies that different *C. neoformans* strains such as those with switching colony morphology or strains with a specific null mutation can produce a variable immune response from the host (58, 134, 135, 147). For example, *C. neoformans* strains possess a series of products which have major influences on host immunity. Two factors are its secretion of GXM and phospholipases. GXM is the major polysaccharide component in the capsule and is found in a soluble form in tissue and fluids of the host. GXM has profound effects on a series of biological functions including an effect on macrophages. The sum of these effects is that it provides an environment to limit host inflammatory responses and induce a lipopolysaccharide-tolerant state (122) and could contribute to an immune reconstitution syndrome as its levels are lowered. GXM possibly helps with mannitol production by the yeast, which increases intracranial pressure. Despite its large amount in tissues and known impact on immune function, its total impact on infection is probably substantial, but it remains to be specifically quantitated. The second product, phospholipase, is more subtle in its effect on host immunity, but this effect may still contribute to the importance of this enzyme as a virulence factor. It was found with the *pb11* mutant that it exhibited a defect in both eicosanoid production (135) and intracellular growth (32). Host eicosanoids down-modulate macrophage functions, and thus a mutant which

cannot produce and release arachidonic acid from phospholipids by phospholipase action might be more susceptible to the antifungal defenses of macrophages; conversely, the wild-type strain might protect itself by using the enzyme to reduce macrophage functions.

ANIMAL MODELS

C. neoformans is widely reported to cause natural infections in a variety of animal species. Experimental cryptococcosis in animal models has tended to be associated with large inocula or some type of induced immunosuppression. However, one of the major advantages of using *C. neoformans* in fungal molecular pathogenesis studies is that the mammalian and nonmammalian models are so robust and clinically relevant to the study of both primary and secondary infections. There is outstanding flexibility in the types of strains, hosts, sites of infection, outcome end points, and treatment strategies that can be used, and the excellent reproducibility of infections makes experiments efficient. For example, both host survival and burden of infection through quantitation of yeast in tissue in animal models have been used in over 50 site-directed null mutants to determine virulence phenotypes. These models are consistent and have the ability to distinguish quantitative degrees of virulence impact. For example, a *sod1* mutant is hypovirulent with prolonged survival in mice, but it still kills all mice over a 90-day period (31). On the other hand, a *sod2* mutant is avirulent, with no yeast found in tissue after 1 week of infection and with 100% 90-day survival (62). A review of many null-mutant *C. neoformans* outcomes shows that animal models reveal genes which produce avirulence, hypovirulence, or hypervirulence or have no impact on the virulence composite (Table 1).

Routine mammalian models include rabbits, rats, guinea pigs, and mice. Since host range determinants appear to be wide, most specific site-directed gene (null) mutants have concordant impact on virulence in several models. However, there are occasional genes like *URE1* (encoding urease) which are important in one model but not another (138). It is clear that the animal models can identify the organ or site-specific importance of a gene and its encoding protein. In fact, the use of animal models will be essential in attempting to understand the mechanisms of the propensity for *C. neoformans* to establish infection in the brain. At least two mechanisms have been suggested: one in which yeasts are carried within macrophages across the blood-brain barrier (i.e., a Trojan horse scenario) (28, 109). Another proposed mechanism is that *C. neoformans* can penetrate or transmigrate through brain endothelial cells and then reach the brain or subarachnoid space (24). Further studies with yeast mutants and standard animal models should provide insights into why this yeast is the major central nervous system fungal pathogen of humans.

Many of the pathobiological studies have been performed with murine models given intranasal, intrapulmonary, intracerebral, or intravenous inoculation. Mice are attractive because of their immunology background, host genetic studies, and reduced costs and because they are generally more susceptible to *C. neoformans* infection than the more resistant rabbits, which require immunosuppression for the creation of disease. With the murine model, it is important to recognize how the genetic background can affect the outcome of infection. For instance, various immunodeficient mice, such as beige and nude mice, are very susceptible to wild-type *C. neoformans*. There are also more subtle differences, such that mice such as A/Jcr and DBA2 quickly die of acute cryptococcal pneumonia with small inocula of wild-type yeast cells while, in contrast, BALB/c mice survive longer and succumb to late progressive meningoencephalitis (150). The host background can even have a major influence on the importance of the virulence potential of a specific strain or mutant. For example, the *app1* mutant cannot produce an antiphagocytic protein and therefore is easily phagocytized, and in a standard murine strain (A/Jcr) it is hypovirulent. However, the same mutant, in the T and NK cell-deficient Tg epsilon mice, was found actually to be hypervirulent (109). With the continued use of transgenic mice and whole-gene mapping of mice, it is very likely that *C. neoformans* can be used as an ideal pathogen to probe the host genome for host-fungus susceptibility genes. Furthermore, these animal models have been used very effectively to determine the antifungal activity of many new and old antifungal compounds. The results of treatment of these animal models for antifungal drug activity have correlated very well with the clinical outcomes of these drugs in human trials.

For the expansion of host-fungus studies, several models of cryptococcosis have been developed in less complex organisms in which acquired immunity and high-temperature growth are not factors for the outcome. Since it has been hypothesized that some of the virulence factors which *C. neoformans* possesses were formed in response to predatory scavengers in its ecological niche, these alternative models were pursued (15). Amoebae (165), *Dictyostelium* (166), nematodes (*C. elegans*) (129), grubs (*Galleria*), and *Drosophila* (6) with *C. neoformans* infection have been studied as potential surrogates for mammalian infection models. In fact, the nematode model has shown a concordant virulence assessment for some mutants compared to the mammalian models (129). The model has been used to screen libraries of mutants for avirulent strains and, with the use of signature-tagged mutants, has identified a virulence gene. A *kin1* mutant (protein kinase) was identified as attenuated in the nematode model and then confirmed to be hypovirulent in the mammalian model (130). The alternate models do have some limitations in what they test in the virulence profile, but they are economical for large mutant screens and definitely can identify virulence genes through assessment of infectivity in mutants.

It is important to emphasize that even with the power of molecular biology combined with the relevant animal models, some issues still need to be addressed. First, the majority of molecular pathogenesis studies are performed with several prominent cryptococcal strains. As work branches out into using more wild-type strains (46), the virulence composite may be redefined. For instance, an adenine auxotroph (*ADE2*) caused disease in a severely immunosuppressed AIDS patient (157), but in molecular biology studies with the standard serotype A strain (H99), the *ade2* mutant auxotroph was severely attenuated (145) in both mice and rabbits. The host plays a major role in pathogenesis, and some paradigm-setting studies will have to be confirmed by using other wild-type strains. Second, the virulence composite of *C. neoformans* is dynamic, with measurable microevolution occurring in the laboratory (11, 27, 58, 59). This rapid loss or acquisition of virulence needs further study, but on a practical basis, strains need to be carefully stored and used under controlled conditions of in vitro growth and standard inocula. Failure to be aware of these issues will add to an already complex virulence composite of this pathogenic yeast.

CONCLUSION

In summary, the encapsulated yeast *C. neoformans* has developed into a major human pathogen over the last two decades. It causes a serious infection which we can treat, but it is still associated with substantial morbidity and mortality and is correctly called the sugar-coated killer. With its rise in importance, there has been a concurrent increase in the ability to molecularly study its pathobiology. In *C. neoformans*, all the keys are in place to unlock the mysteries of what it means to be a fungal pathogen. This basidiomycete has become a model pathogen for the study of molecular fungal pathogenesis, and although it has unique features, many of the paradigms discovered by its study will be used to understand other invasive mycoses.

REFERENCES

1. **Aguirre, K., E. A. Havell, G. W. Gibson, and L. L. Johnson.** 1995. Role of tumor necrosis factor and gamma interferon in acquired resistance to *Cryptococcus neoformans* in the central nervous system of mice. *Infect. Immun.* **63:**1725–1731.

2. **Akhter, S., H. C. McDade, J. M. Gorlach, G. Heinrich, G. M. Cox, and J. R. Perfect.** 2003. Role of alternative oxidase gene in pathogenesis of *Cryptococcus neoformans* variety *grubii*. *Infect. Immun.* **71:**4831–4841.

3. **Alspaugh, J. A., L. M. Cavallo, J. R. Perfect, and J. Heitman.** 2000. *RAS1* regulates filamentation, mating, and growth at high temperature of *Cryptococcus neoformans*. *Mol. Microbiol.* **36:**352–365.

4. **Alspaugh, J. A., J. R. Perfect, and J. Heitman.** 1997. *Cryptococcus neoformans* mating and virulence are regulated by the G-protein alpha subunit GPA1 and cAMP. *Genes Dev.* **11:**3206–3217.

5. **Alspaugh, J. A., R. Pukkila-Worley, T. Harashima, L. M. Cavallo, D. Funnell, G. M. Cox, J. R. Perfect, J. W. Kronstad, and J. Heitman.** 2002. Adenylyl cyclase functions downstream of the G-alpha protein Gpa1 and controls mating and pathogenicity of *Cryptococcus neoformans*. *Eukaryot. Cell* **1:**75–84.

6. **Apidianakis, Y., L. G. Rhame, J. Heitman, F. M. Ausubel, S. B. Calderwood, and E. Mylonakis.** 2004. Challenge of *Drosophila melanogaster* with *Cryptococcus neoformans* and role of the innate immune response. *Eukaryot. Cell* **3:**413–419.

7. **Arend, S. M., E. J. Kuijper, C. F. Allaart, W. H. Muller, and J. T. Van Dissel.** 2004. Cavitating pneumonia after treatment with infliximab and prednisone. *Eur. J. Clin. Microbiol. Infect. Dis.* **23:**638–641.

8. **Bahn, Y. S., J. K. Hicks, S. S. Giles, G. M. Cox, and J. Heitman.** 2004. Adenylyl cyclase-associated protein Aca1 regulates virulence and differentiation of *Cryptococcus neoformans* via the cyclic AMP-protein kinase A cascade. *Eukaryot. Cell* **3:**1476–1491.

9. **Baker, R. D., and R. K. Haugen.** 1955. Tissue changes and tissue diagnosis of cryptococcosis: a study of 26 cases. *Am. J. Pathol.* **25:**14.

10. **Biondo, C., C. Beninati, D. Delfino, M. Oggioni, G. Mancuso, A. Midiri, M. Bombaci, G. Tomaselli, and G. Teti.** 2002. Identification and cloning of a cryptococcal deacetylase that produces protective immune responses. *Infect. Immun.* **70:**2383–2391.

11. **Blasi, E., A. Brozzetti, D. Francisci, R. Neglia, G. Cardinal, F. Ristoni, V. Vidotto, and F. Baldelli.** 2001. Evidence of microevolution in a clinical case of recurrent *Cryptococcus neoformans* meningoencephalitis. *Eur. J. Clin. Microbiol.* **20:**535–543.

12. **Casadevall, A.** 1995. Antibody immunity and invasive fungal infections. *Infect. Immun.* **63:**4211–4218.

13. **Casadevall, A., and J. Perfect.** (ed.). 1998. *Cryptococcus neoformans.* ASM Press, Washington, D.C.

14. **Casadevall, A., and J. R. Perfect.** 1998. Virulence factors, p. 145–176. *In* A. J. Casadevall and J. R. Perfect (ed.), *Cryptococcus neoformans.* ASM Press, Washington, D.C.

15. **Casadevall, A., J. N. Steenbergen, and J. D. Nosanchuk.** 2003. "Ready made" virulence and "dual use" virulence factors in pathogenic environmental fungi—the *Cryptococcus neoformans* paradigm. *Curr. Opin. Microbiol.* **6:**332–337.

16. **Chang, Y., L. Penoyer, and K. J. Kwon-Chung.** 1996. The second capsule gene of *Cryptococcus neoformans* CAP64 is essential for virulence. *Infect. Immun.* **64:**1977–1983.

17. **Chang, Y. C., R. Cherniak, T. R. Kozel, D. L. Granger, L. C. Morris, L. C. Weinhold, and K. J. Kwon-Chung.** 1997. Structure and biological activities of acapsular *Cryptococcus neoformans* 602 complemented with CAP64 gene. *Infect. Immun.* **65:**1584–1592.

18. **Chang, Y. C., and K. J. Kwon-Chung.** 1994. Complementation of a capsule-deficiency mutation of *Cryptococcus neoformans* restores its virulence. *Mol. Cell. Biol.* **14:**4912–4919.

19. **Chang, Y. C., and K. J. Kwon-Chung.** 1998. Isolation of the third capsule-associated gene, *CAP60*, required for virulence in *Cryptococcus neoformans*. *Infect. Immun.* **66:**2230–2236.

20. **Chang, Y. C., and K. J. Kwon-Chung.** 1999. Isolation, characterization, and localization of a capsule-associated gene, *CAP10*, of *Cryptococcus neoformans*. *J. Bacteriol.* **181:**5636–5643.

21. **Chang, Y. C., G. F. Miller, and K. J. Kwon-Chung.** 2003. Importance of a developmentally regulated pheromone receptor of *Cryptococcus neoformans* for virulence. *Infect. Immun.* **71:**4953–4960.

22. **Chang, Y. C., L. A. Penoyer, and K. J. Kwon-Chung.** 1996. The second capsule gene of *Cryptococcus neoformans*, *CAP 64*, is essential for virulence. *Infect. Immun.* **64:**1977–1983.

23. **Chang, Y. C., L. A. Penoyer, and K. J. Kwon-Chung.** 2001. The second *STE12* homologue of *Cryptococcus neoformans* is *MAT*a-specific and plays an important role in virulence. *Proc. Natl. Acad. Sci. USA* **98:**3258–3263.

24. **Chang, Y. C., M. F. Stins, M. J. McCaffery, G. F. Miller, D. R. Pare, T. Dam, M. Paul-Satyaseela, K. S. Kim, K. J. Kwon-Chung, and M. Paul-Satyasee.** 2004. Cryptococcal yeast cells invade the central nervous system via transcellular penetration of the blood-brain barrier. *Infect. Immun.* **72:**4985–4995.

25. **Chaturvedi, V. P., T. Flynn, W. G. Niehaus, and B. Wong.** 1996. Stress tolerance and pathogenic potential of a mannitol-mutant of *Cryptococcus neoformans*. *Microbiology* **142:**937–943.

26. (Reference deleted.)

27. **Cherniak, R. L., L. C. Morris, T. Belay, E. D. Spitzer, and A. Casadevall.** 1995. Variation in the structure of glucuronoxylomannan in isolates from patients with recurrent cryptococcal meningitis. *Infect. Immun.* **63:**1899–1905.

28. **Chretien, F., O. Lortholary, I. Kansau, S. Neuville, F. Gray, and F. Dromer.** 2002. Pathogenesis of cerebral *Cryptococcus neoformans* infection after fungemia. *J. Infect. Dis.* **186:**522–530.

29. **Chung, S., P. Mondon, Y. C. Chang, and K. J. Kwon-Chung.** 2003. *Cryptococcus neoformans* with a mutation in the tetra tricopeptide repeat-containing gene, *CCN1*, causes subcutaneous lesions but fails to cause systemic infection. *Infect. Immun.* **71:**1988–1994.

30. **Clumeck, N., J. Sonnet, H. Taelman, F. Mascart-Lemone, M. De Bruyere, P. Vandeperre, J. Dasnoy, L. Marcelis, M. Lamy, C. Jonas, L. Eyckmans, H. Noel, M. Vanhaeverbeek, and J.-P. Butzler.** 1984. Acquired immunodeficiency syndrome in African patients. *N. Engl. J. Med.* **310:**492–497.

31. **Cox, G. M., T. S. Harrison, C. P. Taborda, H. C. McDade, G. Heinrich, A. Casadevall, and J. R. Perfect.** 2003. Superoxide dismutase influences the virulence of

Cryptococcus neoformans by affecting growth within macrophages. *Infect. Immun.* **71:**173–180.

32. Cox, G. M., H. C. McDade, S. C. Chen, S. C. Tucker, M. Gottfredsson, L. C. Wright, T. C. Sorrell, S. D. Leidich, A. Casadevall, M. A. Ghannoum, and J. R. Perfect. 2001. Extracellular phospholipase activity is a virulence factor for *Cryptococcus neoformans. Mol. Microbiol.* **39:**166–175.

33. Cox, G. M., J. Mukherjee, G. T. Cole, A. Casadevall, and J. R. Perfect. 2000. Urease as a virulence factor in experimental cryptococcosis. *Infect. Immun.* **68:**443–448.

34. Cox, G. M., D. L. Toffaletti, and J. R. Perfect. 1996. Dominant selection system for use in *Cryptococcus neoformans. J. Med. Vet. Mycol.* **34:**385–391.

35. Crowe, S. M., J. B. Carlin, K. I. Stewart, C. R. Lucas, and J. F. Hoy. 1991. Predictive value of CD4 lymphocyte numbers for the development of opportunistic infections and malignancies in HIV-infected persons. *J. Acquir. Immune Defic. Syndr.* **4:**770–776.

36. Cruz, M. C., L. Cavallo, J. Gorlach, J. R. Perfect, M. E. Cardenas, and J. Heitman. 1999. Rapamycin antifungal action is mediated via conserved complexes with FKBP12 and Tor kinase homologs in *C. neoformans. Mol. Cell. Biol.* **19:**4101–4112.

37. Currie, B. P., L. F. Freundlich, and A. Casadevall. 1994. Restriction fragment length polymorphism analysis of *Cryptococcus neoformans* isolates from environmental (pigeon excreta) and clinical sources in New York City. *J. Clin. Microbiol.* **32:**1188–1192.

38. Davidson, R. C., M. C. Cruz, R. A. Sia, B. Allen, J. A. Alspaugh, and J. Heitman. 2000. Gene disruption by biolistic transformation in serotype D strains of *Cryptococcus neoformans. Fungal Genet. Biol.* **29:**38–48.

39. Davidson, R. C., C. B. Nichols, G. M. Cox, J. R. Perfect, and J. Heitman. 2003. A MAP kinase cascade composed of cell type specific and non-specific elements controls mating and differentiation of the fungal pathogen *Cryptococcus neoformans. Mol. Microbiol.* **49:**469–485.

40. de Jesus-Berrios, M., L. Liv, J. C. Nussbaum, G. M. Cox, J. S. Stamler, and J. Heitman. 2003. Enzymes that counteract nitrosative stress promote fungal virulence. *Curr. Biol.* **13:**1963–1968.

41. Del Poeta, M., D. L. Toffaletti, T. H. Rude, C. C. Dykstra, J. Heitman, and J. R. Perfect. 1999. Topoisomerase 1 is essential in *Cryptococcus neoformans*: role in pathobiology and as an antifungal target. *Genetics* **152:**167–178.

42. Del Poeta, M., D. L. Toffaletti, T. H. Rude, S. D. Sparks, J. Heitman, and J. R. Perfect. 1999. *Cryptococcus neoformans* differential gene expression detected in vitro and in vivo with green fluorescent protein. *Infect. Immun.* **67:**1812–1820.

43. Devi, S. J. N. 1996. Preclinical efficacy of a glucuronoxylomannan-tetanus toxoid conjugate vaccine of *Cryptococcus neoformans* in a murine model. *Vaccine* **14:**841–842.

44. (Reference deleted.)

45. D'Souza, C. A., J. A. Alspaugh, C. Yue, T. Harashima, G. M. Cox, J. R. Perfect, and J. Heitman. 2001. Cyclic-AMP-dependent protein kinase controls virulence of the fungal pathogen, *Cryptococcus neoformans. Mol. Cell. Biol.* **21:**3179–3191.

46. D'Souza, C. A., F. Hagen, T. Boekhout, G. M. Cox, and J. Heitman. 2004. Investigation of the basis of virulence in serotype A strains of *Cryptococcus neoformans* from apparently immunocompetent individuals. *Curr. Genet.* **46:**92–102.

47. Dykstra, M. A., L. Friedman, and J. W. Murphy. 1977. Capsule size of *Cryptococcus neoformans*: control and relationship to virulence. *Infect. Immun.* **16:**129.

48. Edman, J. C., and K. J. Kwon-Chung. 1990. Isolation of the *URA5* gene from *Cryptococcus neoformans* var. *neoformans* and its use as a selective marker for transformation. *Mol. Cell. Biol.* **10:**4538–4544.

49. Erickson, T., L. Liu, A. Gueylkian, X. Zhu, J. Gibbons, and P. R. Williamson. 2001. Multiple virulence factors of *Cryptococcus neoformans* are dependent on VPH1. *Mol. Microbiol.* **42:**1121–1131.

50. Falkow, S. 1988. Molecular Koch's postulates applied to microbial pathogenicity. *Rev. Infect. Dis.* **10:**S274–S276.

51. Feldmesser, M., Y. Kress, P. Novikoff, and A. Casadevall. 2000. *Cryptococcus neoformans* is a facultative intracellular pathogen in murine pulmonary infection. *Infect. Immun.* **68:**4225–4237.

52. Fox D. S., G. M. Cox, and J. Heitman. 2003. Phospholipid-binding protein Cts1 controls septation and functions coordinately with calcineurin in *Cryptococcus neoformans. Eukaryot. Cell* **2:**1025–1035.

53. Fox, D. S., M. C. Cruz, R. A. Sia, H. Ke, G. M. Cox, M. E. Cardenas, and J. Heitman. 2001. Calcineurin regulatory subunit is essential for virulence and mediates interactions with FKBP12-FK506 in *Cryptococcus neoformans. Mol. Microbiol.* **39:**835–849.

54. Franzot, S. P., J. Mukherjee, R. Cherniak, L. Chen, J. S. Hamdan, and A. Casadevall. 1998. Microevolution of a standard strain of *Cryptococcus neoformans* resulting in differences in virulence and other phenotypes. *Infect. Immun.* **66:**89–97.

55. Franzot, S. P., I. F. Salkin, and A. Casadevall. 1999. *Cryptococcus neoformans* var. *grubii*: separate variety status for *Cryptococcus neoformans* serotype A isolates. *J. Clin. Microbiol.* **37:**838–840.

56. Fraser, J. A., and J. Heitman. 2003. Fungal mating-type loci. *Curr. Biol.* **13:**R792–R795.

57. Fries, B. C., E. Cook, X. Wang, and A. Casadevall. 2005. Effects of antifungal interventions on the outcome of experimental infections with phenotypic switch variants of *Cryptococcus neoformans. Antimicrob. Agents Chemother.* **49:**350–357.

58. Fries, B. C., D. L. Goldman, and A. Casadevall. 2002. Phenotypic switching in *Cryptococcus neoformans. Microbiol. Infect. Dis.* **4:**1345–1352.

59. Garcia-Hermoso, D., F. Dromer, and G. Janbon. 2004. *Cryptococcus neoformans* capsule structure evolution in vitro and during murine infection. *Infect. Immun.* **72:**3359–3365.

60. Garcia-Hermoso, D., G. Janbon, and F. Dromer. 1999. Epidemiological evidence for dormant *Cryptococcus neoformans* infection. *J. Clin. Microbiol.* **37:**3204–3209.

61. Garcia-Rivera, J., Y. C. Chang, C. J. Kwon-Chung, and A. Casadevall. 2004. CAP59 is involved in the extra cellular trafficking of capsular glucoronoxylomannan. *Eukaryot. Cell* **3:**385–392.

62. Giles, S. S., I. Batinic-Haberle, J. Perfect, and G.M. Cox. 2005. *Cryptococcus neoformans* mitochondrial superoxide dismutase: an essential link between antioxidant function and high-temperature growth. *Eukaryot. Cell* **4:**46–54.

63. Giles, S. S., J. R. Perfect, and G. M. Cox. 2005. Cytochrome *c* peroxidase contributes to the antioxidant

defense of *Cryptococcus neoformans*. *Fungal Genet. Biol.* **42**:20–29.

64. **Goldman, D. L., B. C. Fries, S. P. Franzot, L. Montella, and A. Casadevall.** 1998. Phenotypic switching in the human pathogenic fungus, *Cryptococcus neoformans*, is associated with changes in virulence and pulmonary inflammatory response in rodents. *Proc. Natl. Acad. Sci. USA* **95**:14967–14972.

65. **Goldman, D. L., H. Khine, and J. Abadi.** 2001. Serologic evidence for cryptococcus infection in early childhood. *Pediatrics* **107**:66.

66. **Gorlach, J., D. S. Fox, N. S. Cutler, G. M. Cox, J. R. Perfect, and J. Heitman.** 2000. Identification and characterization of a highly conserved calcineurin binding protein CBP1/calcipressin in *Cryptococcus neoformans*. *EMBO J.* **19**:3618–3629.

67. **Gorlach, J. M., H. C. McDade, J. R. Perfect, and G. M. Cox.** 2002. Antisense repression in *Cryptococcus neoformans* as a laboratory tool and potential antifungal strategy. *Microbiology* **148**:213–219.

68. **Granger, D. L., J. R. Perfect, and D. T. Durack.** 1985. Virulence of *Cryptococcus neoformans*: regulation of capsule synthesis by carbon dioxide. *J. Clin. Investig.* **76**:508–516.

69. **Graybill, J. R., J. Sobel, M. Saag, C. van der Horst, W. Powderly, G. Cloud, L. Riser, R. J. Hamil, and W. Dismukes.** 2000. Diagnosis and management of increased intracranial pressure in patients with AIDS and cryptococcal meningitis. *Clin. Infect. Dis.* **30**:47–54.

70. **Griffith, C. L., J. S. Klutts, L. Zhang, S. B. Levery, and T. L. Doering.** 2004. UDP-glucose dehydrogenase plays multiple roles in the biology of the pathogenic fungus *Cryptococcus neoformans*. *J. Biol. Chem.* **279**:51669–51676.

71. **Hajjman, A., L. A. Conn, D. S. Stephens, and the Cryptococcal Active Surveillance Group.** 1999. Cryptococcosis: population-based multistate active surveillance and risk factors in human immunodeficiency virus-infected persons. *J. Infect. Dis.* **179**:449–454.

72. **Hakim, J. G., I. T. Gangaidzo, and R. S. Heyderman.** 2000. Impact of HIV infection on meningitis in Harare, Zimbabwe: a prospective study of 406 predominantly adult patients. *AIDS* **14**:1401–1407.

73. **Haynes, K. A., D. J. Sullivan, D. C. Coleman, J. C. K. Clarke, R. Emilianus, J. Atkinson, and K. J. Cann.** 1995. Involvement of multiple *Cryptococcus neoformans* strains in a single episode of cryptococcosis and reinfection with novel strains in recurrent infection demonstrated by random amplification of polymorphic DNA and DNA fingerprinting. *J. Clin. Microbiol.* **33**:99–102.

74. **Heung, L. J., C. Luberto, A. Plowden, Y. A. Hannun, and M. Del Poeta.** 2004. The sphingolipid pathway regulates Pkc1 through the formation of diacylglycerol in *Cryptococcus neoformans*. *J. Biol Chem.* **279**:21144–21153.

75. **Hicks, J. K., C. D'Souza, G. M. Cox, and J. Heitman.** 2004. Cyclic AMP-dependent protein kinase catalytic subunits have divergent roles in virulence factor production in two varieties of the fungal pathogen *Cryptococcus neoformans*. *Eukaryot. Cell* **3**:14–28.

76. **Hill, J. O.** 1992. CD4[+] T cells cause multinucleated giant cells to form around *Cryptococcus neoformans* and confine the yeast within the primary site of infection in the respiratory tract. *J. Exp. Med.* **175**:1685–1695.

77. **Hua, J., J. D. Meyer, and J. K. Lodge.** 2000. Development of positive selectable markers for the

fungal pathogen *Cryptococcus neoformans*. *Clin. Diagn. Lab. Immunol.* **7**:125–128.

78. **Huffnagle, G. B., R. M. Strieter, and L. K. McNeil.** 1997. Macrophage inflammatory protein-1 alpha (MIP-alpha) is required for the efferent phase of pulmonary cell-mediated immunity to a *Cryptococcus neoformans* infection. *J. Immunol.* **159**:318–327.

79. **Huffnagle, G. B., T. R. Traynor, and R. A. McDonald.** 2000. Leukocyte recruitment during pulmonary *Cryptococcus neoformans* infection. *Immunopharmacology* **48**:231–236.

80. **Huffnagle, G. B., J. L. Yates, and M. F. Lipscomb.** 1991. T cell-mediated immunity in the lung: a *Cryptococcus neoformans* pulmonary infection model using SCID and athymic nude mice. *Infect. Immun.* **59**:1423–1433.

81. **Hull, C. M., G. M. Cox, and J. Heitman.** 2004. The alpha-specific cell identity factor Sxi1α is not required for virulence of *Cryptococcus neoformans*. *Infect. Immun.* **72**:3643–3645.

82. **Hull, C. M., and J. Heitman.** 2002. Genetics of *Cryptococcus neoformans*. *Annu. Rev. Genet.* **36**:557–615.

83. **Husain, S., M. M. Wagner, and N. Singh.** 2001. *Cryptococcus neoformans* infection in organ transplant recipients: variables influencing clinical characteristics and outcome. *Emerg. Infect. Dis.* **7**:375–381.

84. **Idnurm, A., J. L. Reedy, J. C. Nussbaum, and J. Heitman.** 2004. *Cryptococcus neoformans* virulence gene discovery through insertional mutagenesis. *Eukaryot. Cell* **3**:420–429.

85. **Janbon, G., U. Himmelreich, F. Moyrand, L. Improvisi, and F. Dromer.** 2001. Cas1p is a membrane protein necessary for the O-acetylation of the *Cryptococcus neoformans* capsular polysaccharide. *Mol. Microbiol.* **42**:453–467.

86. **Kawakami, K., M. Tohyama, and K. Teruya.** 1996. Contribution of interferon-gamma in protecting mice during pulmonary and disseminated infection with *Cryptococcus neoformans*. *FEMS Immunol. Med. Microbiol.* **13**:133–140.

87. **Kidd, S. E., F. Hagen, R. L. Tscharke, M. Huynh, K. H. Bartlett, M. Fyfe, L. Macdougall, T. Boekhout, K. J. Kwon-Chung, and W. Meyer.** 2004. A rare genotype of *Cryptococcus gattii* caused the cryptococcosis outbreak on Vancouver Island (British Columbia, Canada). *Proc. Natl. Acad. Sci. USA* **101**:17258–17263.

88. **Kingsbury, J. M., G. M. Cox, and J. H. McCusker.** 2004. *Cryptococcus neoformans* Ilv2p confers resistance to sulfometuron methyl and is required for survival at 37C and for virulence and survival in vivo. *Microbiology* **150**:1547–1558.

89. **Kingsbury, J. M., Z. Yang, T. M. Ganous, G. M. Cox, and J. H. McCusker.** 2004. A novel chimeric SPE3-LYS9 gene in the human pathogen *Cryptococcus neoformans*. *Eukaryot. Cell* **3**:752–763.

90. **Kontoyiannis, D. P., W. K. Peitsch, and B. T. Reddy.** 2001. Cryptococcosis in patients with cancer. *Clin. Infect. Dis.* **32**:145–150.

91. **Kraus, P. R., M. J. Boily, S. S. Giles, J. E. Stajich, A. Allen, G. M. Cox, F. S. Dietrich, J. R. Perfect, and J. Heitman.** 2004. Identification of *Cryptococcus neoformans* temperature-regulated genes with a genomic-DNA microarray. *Eukaryot. Cell* **3**:1249–1260.

92. **Kraus, P. R., D. S. Fox, G. M. Cox, and J. Heitman.** 2003. The *Cryptococcus neoformans* MAP kinase homolog Mpk1 regulates cell integrity in response to

antifungal drugs and loss of calcineurin function. *Mol. Microbiol.* 48:1377–1387.

93. Kwon-Chung, C. J., J. C. Edman, and B. L. Wickes. 1992. Genetic association of mating types and virulence in *Cryptococcus neoformans. Infect. Immun.* 60:602–605.

94. Kwon-Chung, K. J. 1975. Filobasidiella the perfect state of *Cryptococcus neoformans. Mycologia* 67:1197–1200.

95. Kwon-Chung, K. J., and J. E. Bennett. 1978. Distribution of "alpha" and "a" mating types of *Cryptococcus neoformans* among natural and clinical isolates. *Am. J. Med.* 108:337–340.

96. Kwon-Chung, K. J., T. Boekhout, J. W. Fell, and M. Diaz. 2002. Proposal to conserve the name *Cryptococcus gattii* against *C. hondurianus* and *C. bacillisporus* (*Basidiomycota, Hymenomycetes, Tremellomycetidae*). *Taxon* 51:804–806.

97. Kwon-Chung, K. J., and J. C. Rhodes. 1986. Encapsulation and melanin formation as indicators of virulence in *Cryptococcus neoformans. Infect. Immun.* 51:218–223.

98. Lamb, D. C., A. Corran, B. C. Baldwin, J. Kwon-Chung, and S. L. Kelly. 1995. Resistant P45051 A1 activity in azole antifungal tolerant *Cryptococcus neoformans* from AIDS patients. *FEBS Lett.* 368:326–330.

99. Langfelder, K., M. Streibel, B. John, G. Haase, and A. Brakhage. 2003. Biosynthesis of fungal melanins and their importance for human pathogenic fungi. *Fungal Genet. Biol.* 38:143–158.

100. Lee, H., Y. C. Chang, and K. J. Kwon-Chung. 2005. TUP1 disruption reveals biological differences between *MATa* and *MAT* alpha strains of *Cryptococcus neoformans. Mol. Microbiol.* 55:1222–1232.

101. Lengeler, K. B., G. M. Cox, and J. Heitman. 2001. Serotype AD strains of *Cryptococcus neoformans* are diploid or aneuploid and are heterozygous at the mating-type locus. *Infect. Immun* 69:115–122.

102. Lengeler, K. B., P. Wang, G. M. Cox, J. R. Perfect, and J. Heitman. 2000. Identification of the *MATa* mating-type locus of *Cryptococcus neoformans* reveals a serotype A *MATa* strain thought to have been extinct. *Proc. Natl. Acad. Sci. USA* 97:14455–14460.

103. Levitz, S. M. 1992. Overview of host defenses in fungal infections. *Clin. Infect. Dis.* 14:S37–S42.

104. Levitz, S. M. 1994. Macrophage-*Cryptococcus* interactions, p. 533–543. *In* B. S. Zwilling and T. K. Eisenstein (ed.), *Macrophage-Pathogen Interactions.* Marcel Dekker, Inc., New York, N.Y.

105. Levitz, S. M., S. Nong, M. K. Mansour, C. Huang, and C. Specht. 2001. Molecular characterization of a mannoprotein with homology to chitin deacetylases that stimulates T-cell responses to *Cryptococcus neoformans. Proc. Natl. Acad. Sci. USA* 98:10422–10427.

106. Litvintseva, A. P., R. E. Marra, K. Nielsen, J. Heitman, R. Vilgalys, and T. G. Mitchell. 2003. Evidence of sexual recombination among *Cryptococcus neoformans* serotype A isolates in sub-Saharan Africa. *Eukaryot. Cell* 2:1162–1168.

107. Liu, H., T. R. Cottrell, L. M. Pierini, W. E. Goldman, and T. L. Doering. 2002. RNA interference in the pathogenic fungus *Cryptococcus neoformans. Genetics* 160:463–470.

108. Loftus, B. J., E. Fung, P. Roncaglia, D. Rowley, P. Amedeo, D. Bruno, J. Vamathevan, M. Miranda, I. J. Anderson, J. A. Fraser, J. E. Allen, I. E. Bosdet, M. R. Brent, R. Chiu, T. L. Doering, M. J. Donlin, C. A. D'Souza, D. S. Fox, V. Grinberg, J. Fu, M. Fukushima,

B. J. Haas, J. C. Huang, G. Janbon, S. J. Jones, H. L. Koo, M. I. Krzywinski, J. K. Kwon-Chung, K. B. Lengeler, R. Maiti, M. A. Marra, R. E. Marra, C. A. Mathewson, T. G. Mitchell, M. Pertea, F. R. Riggs, S. L. Salzberg, J. E. Schein, A. Shvartsbeyn, H. Shin, M. Shumway, C. A. Specht, B. B. Suh, A. Tenney, T. R. Utterback, B. L. Wickes, J. R. Wortman, N. H. Wye, J. W. Kronstad, J. K. Lodge, J. Heitman, R. W. Davis, C. M. Fraser, and R. W. Hyman. 2005. The genome of the basidiomycetous yeast and human pathogen *Cryptococcus neoformans. Science* 307:1321–1324.

109. Luberto, C., D. Taraskiewicz, B. Martinez-Marino, B. Bolanos, P. Chitano, D. L. Toffaletti, G. M. Cox, J. R. Perfect, Y. A. Hannun, E. Balish, and M. Del Poeta. 2003. IPC1 regulates the phagocytosis of *Cryptococcus neoformans* by macrophages through modulation of the antiphagocytic protein (APP1) phagocytosis and pathogenicity: when host takes control. *J. Clin. Investig.* 112:1080–1094.

110. Luberto, C., D. L. Toffaletti, E. A. Wills, S. C. Tucker, A. Casadevall, J. R. Perfect, Y. A. Hannum, and M. Del Poeta. 2001. Roles for inositol-phosphorylceramide synthase 1 (IPC1) in pathogenesis of *Cryptococcus neoformans. Genes Dev.* 15:201–212.

111. Mandel, M. A., G. G. Grace, K. I. Osborn, F. Schafer, J. W. Murphy, M. J. Orbach, and J. N. Galgiani. 2002. The *Cryptococcus neoformans* gene *DHA1* encodes an antigen that elicits a delayed-type hypersensitivity reaction in immune mice. *Infect. Immun.* 68:6196–6201.

112. Marra, R. E., J. C. Huang, E. Fung, K. Nielsen, J. Heitman, R. Vilgalys, and T. G. Mitchell. 2004. A genetic linkage map of *Cryptococcus neoformans* variety *neoformans* serotype D (*Filobasidiella neoformans*). *Genetics* 167:619–631.

113. Martinez, L. R., J. Garcia-Rivera, and A. Casadevall. 2001. *Cryptococcus neoformans* var. *neoformans* (serotype D) strains are more susceptible to heat than *C. neoformans* var. *grubii* (serotype A strains). *J. Clin. Microbiol.* 39:3365–3367.

114. McClelland, E. E., F. R. Adler, D. L. Granger and W. K. Potts. 2004. Major histocompatibility complex controls the trajectory but not host-specific adaptation during virulence evolution of the pathogenic fungus *Cryptococcus neoformans. Proc. R. Soc. Lond. Ser. B.* 271:1557–1564.

115. McDade, H. C., and G. M. Cox. 2002. A new dominant selectable marker for use in *Cryptococcus neoformans. Med. Mycol.* 39:151–154.

116. McNeill, J. I., and V. L. Kan. 1995. Decline in the incidence of cryptococcosis among HIV-related patients. *J. Acquir. Immune Defic. Syndr. Hum. Retrovirol.* 206:207–208.

117. Meyer, W., A. Castaneda, S. Jackson, M. Huynh, and E. Castaneda. 2003. Molecular typing of Ibero American *Cryptococcus neoformans* isolates. *Emerg. Infect. Dis.* 9:189–195.

118. Mirza, S., M. Phelan, D. Rimland, E. Graviss, R. Hamill, M. E. Brandt, T. Gardner, M. Sattah, G. Ponce de Leon, and H. R. A. Baughman. 2002. The changing epidemiology of cryptococcosis: an update from population-based active surveillance in 2 large metropolitan areas, 1992–2000. *Clin. Infect. Dis.* 36:789–794.

119. Missall, T. A., J. K. Lodge, and J. E. McEwen. 2004. Mechanisms of resistance to oxidative and nitrosative stress: implications for fungal survival in mammalian hosts. *Eukaryot. Cell* 3:835–846.

120. **Missall, T. A., J. M. Moran, J. A. Corbett, and J. K. Lodge.** 2005. Distinct stress responses of two functional laccases in *Cryptococcus neoformans* are revealed in the absence of the thiol-specific antioxidant Tsa1. *Eukaryot. Cell* **4**:202–208.

121. **Missall, T. A., M. E. Pusateri, and J. K. Lodge.** 2004. Thiol peroxidase is critical for virulence and resistance to nitric oxide and peroxide in the fungal pathogen, *Cryptococcus neoformans*. *Mol. Microbiol.* **51**: 1447–1448.

122. **Monari, C., F. Bistoni, A. Casadevall, E. Pericolini, D. Pietrella, T. R. Kozel, and A. Vecchiarelli.** 2005. Glucuronoxylomannan, a microbial compound, regulates expression of costimulatory molecules and production of cytokines in macrophages. *J. Infect. Dis.* **191**:127–137.

123. **Monari, C., E. Pericolini, G. Bistoni, E. Cenci, F. Bistoni, and A. Vecchiarelli.** 2005. Influence of indinavir on virulence and growth of *Cryptococcus neoformans*. *J. Infect. Dis.* **191**:307–311.

124. **Mondon, P., R. Petter, G. Amalfitano, R. Luzzati, E. Concia, I. Polacheck, and K. J. Kwon-Chung.** 1999. Heteroresistance to fluconazole and voriconazole in *Cryptococcus neoformans*. *Antimicrob. Agents Chemother.* **43**:1856–1861.

125. **Moyrand, F., and G. Janbon.** 2004. UGD1, encoding the *Cryptococcus neoformans* UDP-glucose dehydrogenase, is essential for growth at 37°C and for capsule biosynthesis. *Eukaryot. Cell* **3**:1601–1608.

126. **Moyrand, F., B. Klaproth, U. Himmelreich, F. Dromer, and G. Janbon.** 2002. Isolation and characterization of capsule structure mutant strains of *Cryptococcus neoformans*. *Mol. Microbiol.* **45**:837–849.

127. **Mukherjee, J., M. D. Sharff, and A. Casadevall.** 1992. Protective murine monoclonal antibodies to *Cryptococcus neoformans*. *Infect. Immun.* **60**: 4534–4541.

128. **Murphy, J. W.** 1992. Cryptococcal immunity and immunostimulation. *Adv. Exp. Med. Biol.* **319**:225–230.

129. **Mylonakis, E., F. M. Ausubel, J. R. Perfect, J. Heitman, and S. B. Calderwood.** 2002. Killing of *Caenorhabditis elegans* by *Cryptococcus neoformans* as a model of yeast pathogenesis. *Proc. Natl. Acad. Sci. USA* **99**: 15675–15680.

130. **Mylonakis, E., A. Idnurm, R. Moreno, J. El Khoury, J. B. Rottman, F. M. Ausubel, J. Heitman, and S. B. Calderwood.** 2004. *Cryptococcus neoformans* Kin1 protein kinase homologue, identified through a *Caenorhabditis elegans* screen, promotes virulence in mammals. *Mol. Microbiol.* **54**:407–419.

131. **Narasipura, S. D., J. G. Ault, M. J. Behr, V. Chaturvedi, and S. Chaturvedi.** 2003. Characterization of Cu, Zn superoxide dismutase (*SOD1*) gene knock-out mutant of *Cryptococcus neoformans* var. *gattii*: role in biology and virulence. *Mol. Microbiol.* **47**:1681–1694.

132. **Nelson, R. T., J. Hua, B. Pryor, and J. K. Lodge.** 2001. Identification of virulence mutants of the fungal pathogen *Cryptococcus neoformans* using signature-tagged mutagenesis. *Genetics* **157**:935–947.

133. **Nielsen, K., G. M. Cox, P. Wang, D. L. Toffaletti, J. R. Perfect, and J. Heitman.** 2003. Sexual cycle of *Cryptococcus neoformans* var. *grubii* and virulence of congenic a and alpha isolates. *Infect. Immun.* **71**:4831–4841.

134. **Noverr, M. C., G. M. Cox, J. R. Perfect, and G. B. Huffnagle.** 2003. Role of *PLB1* in pulmonary inflammation and cryptococcal eicosanoid production. *Infect. Immun.* **71**:1538–1547.

135. **Noverr, M. C., P. R. Williamson, R. S. Fajardo, and G. B. Huffnagle.** 2004. *CNLAC1* is required for extrapulmonary dissemination of *Cryptococcus neoformans* but not pulmonary persistence. *Infect. Immun.* **72**:1693–1699.

136. **O'Brien, C. R., M. B. Krockenberger, D. I. Wigney, P. Martin, and R. Malik.** 2004. Retrospective study of feline and canine cryptococcosis in Australia from 1981 to 2001: 195 cases. *Med. Mycol.* **42**:449–460.

137. **Odom, A., S. Muir, E. Lim, D. L. Toffaletti, J. R. Perfect, and J. Heitman.** 1997. Calcineurin is required for virulence of *Cryptococcus neoformans*. *EMBO J.* **16**: 2576–2589.

138. **Olszewski, M. A., M. C. Noverr, G. H. Chen, G. B. Toews, G. M. Cox, J. R. Perfect, and G. B. Huffnagle.** 2004. Urease expression by *Cryptococcus neoformans* promotes microvascular sequestration, thereby enhancing central nervous system invasion. *Am. J. Pathol.* **164**:1761–1771.

139. **Ory, J. J., C. L. Griffith, and T. L. Doering.** 2004. An efficiently regulated promoter system for *Cryptococcus neoformans* utilizing the CTR4 promoter. *Yeast* **21**:919–926.

140. **Pappas, P. G., B. Bustamante, E. Ticona, R. J. Hamill, P. C. Johnson, A. Reboli, J. Aberg, R. Hasbun, and H. H. Hsu.** 2004. Recombinant interferon-gamma 1b as adjunctive therapy for AIDS-related acute cryptococcal meningitis. *J. Infect. Dis.* **189**:2185–2191.

141. **Pascon, R. C., T. M. Ganous, J. M. Kingsbury, G. M. Cox, and J. H. McCusker.** 2004. *Cryptococcus neoformans* methionine synthase: expression analysis and requirement for virulence. *Microbiology* **150**: 3013–3023.

142. **Perfect, J. R.** 2005. *Cryptococcus neoformans*: a sugar-coated killer with designer genes. *FEMs Immunol. Med. Microbiol.* **45**:395–404.

143. **Perfect, J. R., and A. Casadevall.** 2002. Cryptococcosis. *Infect. Dis. Clin. North Am.* **16**:837–874.

144. **Perfect, J. R., N. Ketabchi, G. M. Cox, C. I. Ingram, and C. Beiser.** 1993. Karyotyping of *Cryptococcus neoformans* as an epidemiological tool. *J. Clin. Microbiol.* **31**:3305–3309.

145. **Perfect, J. R., D. L. Toffaletti, and T. H. Rude.** 1993. The gene encoding for phosphoaminoimidazole carboxylase is essential for growth of *Cryptococcus neoformans* in cerebrospinal fluid. *Infect. Immun.* **61**: 4446–4451.

146. **Perfect, J. R.** 2003. Genetic requirements for virulence in *Cryptococcus neoformans*. *Mycota* **12**:89–112.

147. **Pietrella, D., B. Fries, P. Cupo, F. Bistoni, A. Casadevall, and A. Vecchiarelli.** 2003. A phenotypic switching of *Cryptococcus neoformans* can influence the outcome of the human immune response. *Microbiology* **5**:513–522.

148. **Posteraro, B., M. Sanguinetti, D. Sanglard, M. La Sorda, S. Boccia, L. Romano, G. Morace, and G. Fadda.** 2003. Identification and characterization of a *Cryptococcus neoformans* ATP binding cassette (ABC) transporter-encoding gene, *CnAFR1*, involved in the resistance to fluconazole. *Mol. Microbiol.* **47**:357–371.

149. **Pukkila-Worley, R., Q. D. Gerrald, P. R. Kraus, M. J. Boily, M. J. Davis, S. S. Giles, G. M. Cox, J. Heitman, and J. A. Alspaugh.** 2005. Transcriptional network of multiple capsule and melanin genes governed by the

Cryptococcus neoformans cyclic AMP cascade. *Eukaryot. Cell* **4:**190–201.

150. Rhodes, J. C., L. S. Wicker, and W. J. Urba. 1980. Genetic control of susceptibility to *Cryptococcus neoformans* in mice. *Infect. Immun.* **29:**494–499.

151. Rivera, J., M. Feldmesser, M. Cammer, and A. Casadevall. 1998. Organ-dependent variation of capsule thickness in *Cryptococcus neoformans* during experimental murine infection. *Infect. Immun.* **66:**5027–5030.

152. Rodero, L., E. Mellado, A. C. Rodriguez, A. Salve, L. Guelfand, P. Cahn, M. Cuenca-Estrella, G. Davel, and J. L. Rodriguez-Tudela. 2003. G484S amino acid substitution in lanosterol 14-alpha demethylase (ERG11) is related to fluconazole resistance in a recurrent *Cryptococcus neoformans* clinical isolate. *Antimicrob. Agents Chemother.* **47:**3653–3656.

153. Rosas, A. L., J. D. Nosanchuk, and M. Feldmesser. 2000. Synthesis of polymerized melanin by *Cryptococcus neoformans* in infected rodents. *Infect. Immun.* **68:**2845–2853.

154. Rude, T. H., D. L. Toffaletti, G. M. Cox, and J. R. Perfect. 2002. Relationship of the glyoxylate pathway to the pathogenesis of *Cryptococcus neoformans*. *Infect. Immun.* **70:**5684–5694.

155. Salas, S. D., J. E. Bennett, K. J. Kwon-Chung, J. R. Perfect, and P. R. Williamson. 1996. Effect of the laccase gene, *CNLAC1*, on virulence of *Cryptococcus neoformans*. *J. Exp. Med.* **184:**377–386.

156. Santangelo, R., H. Zoellner, T. Sorrell, C. Wilson, C. Donald, J. Djordjevic, Y. Shounan, and L. Wright. 2004. Role of extracellular phospholipases and mononuclear phagocytes in dissemination of cryptococcosis in a murine model. *Infect. Immun.* **72:**2229–2239.

157. Schiappa, D., A. Gueylkian, J. A. Alspaugh, J. R. Perfect, and P. R. Williamson. 2002. Myelotropism and auxotrophy of a pigmented *Cryptococcus neoformans*. *Med. Mycol.* **40:**1–6.

158. Schimpff, S. C., and J. E. Bennett. 1975. Abnormalities in cell-mediated immunity in patients with *Cryptococcus neoformans* infection. *J. Allergy Clin. Immunol.* **55:**430–441.

159. Shelburne, S. A., III, and R. J. Hamill. 2003. The immune reconstitution inflammatory syndrome. *AIDS Rev.* **5:**67–79.

160. Shen, W. C., R. C. Davidson, G. M. Cox, and J. Heitman. 2002. Pheromones stimulate mating and differentiation via paracrine and autocrine signaling in *Cryptococcus neoformans*. *Eukaryot. Cell* **1:**366–377.

161. Speed, B., and D. Dunt. 1995. Clinical and host differences between infections with the two varieties of *Cryptococcus neoformans*. *Clin. Infect. Dis.* **21:**28–34.

162. Spitzer, E. D., S. G. Spitzer, L. F. Freundlich, and A. Casadevall. 1993. Persistence of initial infection in recurrent *Cryptococcus neoformans* meningitis. *Lancet* **341:**595–596.

163. Steen, B. R., T. Lian, S. Zuyderduyn, W. K. MacDonald, M. Marra, S. J. M. Jones, and J. W. Kronstad. 2002. Temperature-related transcription in the pathogenic fungus *Cryptococcus neoformans*. *Genome Res.* **12:**1386–1400.

164. Steen, B. R., S. Zuyderduyn, D. L. Toffaletti, M. Marra, S. J. Jones, J. R. Perfect, and J. Kronstad. 2003. *Cryptococcus neoformans* gene expression during experimental cryptococcal meningitis. *Eukaryot. Cell* **2:**1336–1349.

165. Steenbergen, J. N., H. A. Shuman, and A. Casadevall. 2001. *Cryptococcus neoformans* interactions with amoebae suggest an explanation for its virulence and intracellular pathogenic strategy in macrophages. *Proc. Natl. Acad. Sci. USA* **98:**15245–15250.

166. Steenbergen, J. N., H. A. Shuman, and A. Casadevall. 2003. *Cryptococcus neoformans* is enhanced after growth in the genetically malleable host, *Dictyostelium discoideum*. *Infect. Immun.* **71:**4862–4872.

167. Sukroongreung, S., K. Kitiniyom, C. Nilakul, and S. Tantimavanich. 1998. Pathogenicity of basidiospores of *Filobasidiella neoformans* var. *neoformans*. *Med. Mycol.* **36:**419–424.

168. Sullivan, D., K. Haynes, G. Moran, D. Shanley, and D. Coleman. 1996. Persistence, replacement, and microevolution of *Cryptococcus neoformans* strains in recurrent meningitis in AIDS patients. *J. Clin. Microbiol.* **34:**1739–1744.

169. Thompson, J. R., C. M. Douglas, N. Li, K. J. Chang, B. Pramanik, X. Yuan, T. H. Rude, D. L. Toffaletti, J. R. Perfect, and M. Kurtz. 1999. A glucan synthase FKS1 homolog in *Cryptococcus* is a single copy and encodes an essential function. *J. Bacteriol.* **181:**444–453.

170. Toffaletti, D. L., M. Del Poeta, T. H. Rude, F. S. Dietrich, and J. R. Perfect. 2003. Regulation of cytochrome C oxidase subunit 1 (COX1) expression in *Cryptococcus neoformans* by temperature and host environment. *Microbiology* **149:**1041–1049.

171. Toffaletti, D. L., K. Nielsen, F. Dietrich, J. Heitman, and J. R. Perfect. 2004. *Cryptococcus neoformans* mitochondrial genomes from serotype A and D strains do not influence virulence. *Curr. Genet.* **46:**193–204.

172. Toffaletti, D. L., and J. R. Perfect. 1997. Study of *Cryptococcus neoformans* actin gene regulation with a beta-galactosidase-actin fusion. *J. Med. Vet. Mycol.* **35:**313–320.

173. Toffaletti, D. L., T. H. Rude, S. A. Johnston, D. T. Durack, and J. R. Perfect. 1993. Gene transfer in *Cryptococcus neoformans* using biolistic delivery of DNA. *J. Bacteriol.* **175:**1405–1411.

174. Torres, H. A., V. G. Prieto, I. I. Raad, and D. P. Kontoyiannis. 2005. Proven pulmonary cryptococcosis due to capsule-deficient *Cryptococcus neoformans* does not differ clinically from proven pulmonary cryptococcosis due to capsule-intact *Cryptococcus neoformans*. *Mycoses* **48:**21–24.

175. Vallim, M. A., L. Fernandes, and J. A. Alspaugh. 2004. The *RAM1* gene encoding a protein-farnesyltransferase beta-subunit homologue is essential in *Cryptococcus neoformans*. *Microbiology* **150:**1925–1935.

176. Van de Perre, P., P. Lepage, and P. Kestelyn. 1984. Acquired immunodeficiency syndrome in Rwanda. *Lancet* **ii:**62–65.

177. van der Horst, C. M., M. S. Saag, G. A. Cloud, and National Institute of Allergy and Infectious Diseases Mycoses Study Group and AIDS Clinical Trials Group. 1997. Treatment of cryptococcal meningitis associated with the acquired immunodeficiency syndrome. *N. Engl. J. Med.* **337:**15–21.

178. van Duin, D., A. Casadevall, and J. D. Nosanchuk. 2002. Melanization of *Cryptococcus neoformans* and *Histoplasma capsulatum* reduces their susceptibilities to amphotericin B and caspofungin. *Antimicrob. Agents Chemother.* **46:**3394–3400.

179. van Elden, L. J., A. M. Walenkamp, and M. M. Lipovsky. 2000. Declining number of patients with cryptococcosis in the Netherlands in the era of highly active antiretroviral therapy. *AIDS* **14:**2787–2788.

180. Vartivarian, S. E., E. J. Anaissie, R. E. Cowart, H. A. Sprigg, M. J. Tingler, and E. J. Jacobson. 1993. Regulation of cryptococcal capsular polysaccharide by iron. *J. Infect. Dis.* **167:**186–190.

181. Velegraki, A., V. G. Klosses, A. Kansovzidov, S. Smilakov, A. Mitroussia-Ziouua, and N. J. Legakis. 2004. Prospective use of RFLP analysis on amplified *Cryptococcus neoformans URA5* gene sequences for rapid identification of varieties and serotypes in clinical samples. *Med. Mycol.* **39:**409–417.

182. Wang, P., M. E. Cardenas, G. M. Cox, J. R. Perfect, and J. Heitman. 2001. Two cyclophilin A homologs with shared and distinct functions are expressed in *Cryptococcus neoformans. EMBO Rep.* **21:**511–518.

183. Wang, P., G. M. Cox, and J. Heitman. 2004. A Sch9 protein kinase homologue controlling virulence independently of the cAMP pathway in *Cryptococcus neoformans. Curr. Genet.* **46:**247–255.

184. Wang, P., J. Cutler, J. King, and D. Palmer. 2004. Mutation of the regulator of G protein signaling Crg1 increases virulence in *Cryptococcus neoformans. Eukaryot. Cell* **3:**1028–1035.

185. Wang, P., C. S. Nichols, K. B. Lengeler, M. E. Cardenas, G. M. Cox, J. R. Perfect, and J. Heitman. 2002. Mating type specific and nonspecific PAK kinases play shared and divergent role in cytokinesis, differentiation and virulence of *Cryptococcus neoformans. Eukaryot. Cell* **1:**257–272.

186. Wang, P., J. R. Perfect, and J. Heitman. 2000. The G-protein beta subunit GPB1 is required for mating and haploid fruiting in *Cryptococcus neoformans. Mol. Cell. Biol.* **20:**352–362.

187. Wickes, B. L., M. E. Mayorga, U. Edman, and J. C. Edman. 1996. Dimorphism and haploid fruiting in *Cryptococcus neoformans*: association with the alpha-mating type. *Proc. Natl. Acad. Sci. USA* **93:**7327–7331.

188. Wickes, B., and J. Edman. 1995. The *Cryptococcus neoformans GAL7* gene and its use as an inducible promoter. *Mol. Microbiol.* **16:**1099–1109.

189. Wills, E. A., I. S. Roberts, M. Del Poeta, J. Rivera, A. Casadevall, G. M. Cox, and J. R. Perfect. 2001.

Identification and characterization of the *Cryptococcus neoformans* phosphomannose isomerase-encoding gene, *MAN1*, and its impact on pathogenicity. *Mol. Microbiol.* **40:**610–620.

190. Wormley, F. L., G. M. Cox, and J. R. Perfect. 2005. Evaluation of host immune responses to pulmonary cryptococcosis using calcineurin mutant strain. *Microb. Pathog.* **38:**113–123.

191. Xu, J. 2004. Genotype-environment interactions of spontaneous mutations for vegetative fitness in the human pathogenic fungus *Cryptococcus neoformans. Genetics* **168:**1177–1188.

192. Xu, J., R. Vilgalys, and T. G. Mitchell. 2002. Multiple gene genealogies reveal recent dispersion and hybridization in the human fungus, *Cryptococcus neoformans. Mol. Ecol.* **9:**1471–1481.

193. Yang, Z., R. C. Pascon, J. A. Alspaugh, G. M. Cox, and J. H. McCusker. 2002. Molecular and genetic analysis of the *Cryptococcus neoformans MET3* gene and a *met3* mutant. *Microbiology* **148:**2617–2625.

194. Yauch, L. E., M. K. Mansour, S. Shoham, J. B. Rottman, and S. M. Levitz. 2004. Involvement of CD14, Toll-like receptors 2 and 4, and MyD88 in the host response to the fungal pathogen *Cryptococcus neoformans* in vivo. *Infect. Immun.* **72:**5373–5382.

195. Yue, C., L. Cavallo, J. A. Alspaugh, G. M. Cox, J. R. Perfect, and J. Heitman. 1999. The *STE12α* homolog is required for haploid filamentation but dispensable for mating and virulence in *Cryptococcus neoformans. Genetics* **153:**1601–1615.

196. Zaragoza, O., and A. Casadevall. 2004. Experimental modulation of capsule size in *Cryptococcus neoformans. Biol. Proc. Online* **6:**10–15.

197. Zhu, X., and P. R. Williamson. 2003. A CLC-type chloride channel gene is required for laccase activity and virulence in *Cryptococcus neoformans. Mol. Microbiol.* **50:**1271–1281.

198. Zhu, X., and P. R. Williamson. 2004. Role of laccase in the biology and virulence of *Cryptococcus neoformans. FEMS Yeast Res.* **5:**1–10.

Molecular Principles of Fungal Pathogenesis
Edited by Joseph Heitman et al.
©2006 ASM Press, Washington, D.C.

Chapter 22

Toward a Molecular Understanding of *Candida albicans* Virulence

Frank C. Odds, Neil A. R. Gow, and Alistair J. P. Brown

Several chapters in this volume address specific aspects of the pathobiology of *Candida* species. In this chapter we focus on *Candida albicans*, taking a more holistic view of the strengths and weaknesses of the various approaches that have been taken to examine *C. albicans* virulence factors. It is well known that putative *C. albicans* virulence factors include adhesins, the secretion of hydrolytic enzymes, cellular morphogenesis, and phenotypic switching. Over 10 years ago it was suggested that these virulence factors might contribute differentially at the various stages during the establishment of superficial and deep-seated *Candida* infections (27, 93). Since then it has become fashionable to suggest that the transition from commensalism to pathogenesis depends on the balance between *C. albicans* virulence factors and host immune defenses. We argue that *C. albicans* virulence is more complex than this. Additional parameters should be considered, including the multiplicity of infection pathologies, the diverse range of predisposed hosts, and the potential for subtle variations in the molecular virulence factors among different *C. albicans* isolates.

In addition, we review briefly the multiplicity of infection models and the molecular tools that have been developed for the dissection of *C. albicans* virulence. We argue that these models are often simplistic, and must be better defined at a molecular level and that *C. albicans* mutants that are tested in these models should be more tightly controlled. The aims here must be to establish which models reflect a specific aspect of *C. albicans* pathobiology with a reasonable degree of accuracy and to create mutants that minimize secondary phenotypic effects.

Finally, while applauding the rapid advances that have been made in our understanding of *C. albicans* pathobiology, we highlight weaknesses in current

molecular approaches and suggest possible solutions. Part of the problem lies in the perceived view of what defines a virulence factor. In this chapter we argue in favor of the recent suggestion that virulence factors are expressed at the cell surface and contribute directly to fungus-host interactions (95). However, we also highlight the importance of *C. albicans* fitness attributes for virulence (16). The overall aim here must be to increase the precision with which we dissect the roles of specific *C. albicans* virulence and fitness factors.

BACKGROUND: *CANDIDA* SPECIES, *CANDIDA* INFECTIONS, AND *CANDIDA* PATHOGENICITY

Current taxonomies list almost 200 species of ascomycetous, asexual yeasts in the genus *Candida*. Among these, only *C. albicans*, *C. dubliniensis*, *C. glabrata*, *C. krusei*, *C. lusitaniae*, *C. parapsilosis*, and *C. tropicalis* are encountered with regularity as gut commensals and opportunistic pathogens of humans. Other *Candida* species that are occasionally reported as causes of infection are either infrequently contacted by humans or, in some cases, have been misidentified. *C. albicans* remains the species most commonly associated with infections, but successful prophylactic and therapeutic use of antifungal agents through the 1990s has substantially reduced the incidence of serious *C. albicans* infections, resulting in a higher prevalence of other species in recent surveys (129, 137). There are considerable regional and even institutional differences in the mix of *Candida* species causing serious infections (110).

The range of pathological types of *Candida* infection is considerable. Oral infections, of several different forms, arise in infants, the elderly, and patients infected with human immunodeficiency

Frank C. Odds, Neil A. R. Gow, and Alistair J. P. Brown • Aberdeen Fungal Group, School of Medical Sciences, Institute of Medical Sciences, University of Aberdeen, Foresterhill, Aberdeen, Scotland AB25 2ZD.

virus. *Candida* vulvovaginitis is seen mainly in females of child-bearing age; infections of the male genitalia are much less common although not unknown. Cutaneous infections with *Candida* species arise on areas of skin that are moistened by occlusion or maceration. All these superficial forms of *Candida* infection are typically easy to diagnose and, with rare exceptions, easy to treat. The same is not true of *Candida* infections of deep tissues, where the fungus gains access to the bloodstream, most often from the gut but also via catheters and other devices, and is disseminated via the blood to the viscera. This situation occurs in hosts with considerable immune compromise, for example patients with neutropenia, carcinoma, major abdominal trauma or surgery, and several other underlying conditions. Major risk factors statistically associated with development of disseminated infection include central venous catheters, age older than 65 years, surgery within the past 3 months, and immunosuppressive therapy of any type (48). Therefore, the etiology of *Candida* disease is multifaceted, and it would be astonishing if the same range of virulence attributes played the same roles under these diverse conditions.

The clinical picture of multiple infection pathologies in a wide range of predisposed hosts is an important but frequently overlooked consideration in the scientific investigation of pathogenicity in *Candida* species. The simple view of infection as the outcome of a contest between microbial virulence factors and host defenses has to be expanded to take account of the different types of *Candida* infection. While it is true to generalize that a depletion of one or more host defenses is the principal factor encouraging commensal *Candida* cells to become invasive, the nature of the defense mechanism differs with the site of infection. Clinical observation of patients with HIV or chronic mucocutaneous *Candida* infection indicates that impairment of T-lymphocyte functions tends to favor the invasion of superficial sites rather than deep tissues while a fall in neutrophil count facilitates the access of *Candida* cells to the bloodstream. Host immune status is therefore the dominant trigger for tissue invasion by a *Candida* species; nevertheless, there are sound reasons to investigate virulence attributes in the study of the pathogenesis of *Candida* infections.

Intravenous challenge of small rodents, typically mice, reproducibly reveals a species-level rank order of *Candida* pathogenicity in which *C. albicans* causes more severe pathological consequences (measured in terms of survival and/or fungal tissue burden) than *C. tropicalis*, with other *Candida* species unable to infect experimental animals unless they are heavily

immunosuppressed (92). (While dictionaries define "virulence" and "pathogenicity" as synonyms, we use the term "pathogenicity" to apply to species and "virulence" to apply to strains within a species.) Within the species *C. albicans*, strains can be found that are naturally attenuated in mouse virulence (49, 57, 72, 96, 102). Mutants of virulent *C. albicans* strains can be created that are attenuated in lethality for immunologically intact mice (see below). It is therefore reasonable to study "virulence" at the molecular level in *C. albicans* without necessarily and inevitably considering host immunity.

What is less evident is the meaning of the term "virulence" in the context of microbial pathogenesis. A strong case has been made that virulence should be defined as the capacity to cause damage to a host (23). The semantic advantage of this definition is that it accommodates equally the consequences of interaction of a highly pathogenic microbe with an immunologically intact host and an inherently avirulent pathogen with an immunocompromised host. Thus, in *C. albicans*, virulence factors can be determined as attributes whose absence in the fungus eliminates or reduces a measurable parameter of damage in a defined host system in vivo. In an opportunistic pathogen such as *C. albicans*, virulence factors are unlikely to be either highly potent or required at all stages of the complete infectious process (colonization, epithelial penetration, vascular dissemination, and endothelial and parenchymal penetration); what is being sought is the set of molecular virulence factors—expressed gene products—that contribute significantly to each stage of infectious damage.

MEASUREMENT OF *CANDIDA* VIRULENCE

If the consequence of virulence is host damage and the aim is to assess the roles of specific molecular virulence factors in inflicting this damage, then experimental models of *Candida* infection should permit quantification of such damage. The most widely used animal model of hematogenously disseminated *Candida* infection (candidemia) involves intravenous challenge of mice with an inoculum that is lethal in virulent, wild-type strains. Damage is measured in terms of survival rates and/or fungal burdens in target organs such as the kidneys. This model bypasses the most common infection route by which *Candida* species pass from the gut into the bloodstream of immunocompromised humans, but it may be argued that it reasonably mimics the situation of a nosocomial infection whose origin is a venous catheter. Its scientific advantage is that it offers high reproducibility and is currently by far the most widely used assay of *C. albicans* virulence. Table 1 summarizes issues of

Table 1. Notes on virulence assays in the murine intravenous-challenge
model of disseminated *Candida* infection

Issue	Potential solutions
Consistency/reproducibility of data	Inbred mouse strains should be used.
Preparation of fungal cells	Physiological status of the fungal cells affects their virulence in vivo. Therefore, all strains must be in equivalent growth states when harvested for analysis.
Inoculum size	The range of inoculum sizes that lead to consistent infection outcomes is narrow, and mouse susceptibility can vary slightly between batches. It is best to pretest a small sample from each animal batch and adjust the challenge dose appropriately.
Survival end points	Humane considerations usually prohibit allowing animals to die of infection. Carefully designed criteria for euthanasia can balance the need for a scientific "survival" end point with ethical considerations.
Fungal burden measurements	Viable counting, though simple, is vulnerable to artifacts when multicelled hyphae generate single colonies in culture. Real-time PCR measurements may come to replace viable counts in time. Detection limits for burden measurements should be stated routinely: tissue sterilization can be demonstrated only if the entire tissue is somehow cultured. Expression of results as logarithms requires careful consideration of how to handle negative culture results.
Histopathology	The extra effort involved in examining sections of infected tissue can be rewarded if differential morphological development and host responses are noted between strains.
Growth defects of mutant in vitro	A mutant that displays a growth defect in vitro will almost certainly show attenuated virulence in vivo whether or not the disrupted gene has a role in virulence. It is highly questionable whether it is any longer supportable to use animals on testing such mutants in vivo.

concern about the mouse intravenous-challenge model. Alternatives have been described in which severely immunocompromised or neonatal mice are challenged by gastrointestinal (5, 105) or intranasal (69) instillation of the fungus. Although such models are better mimics of human disseminated disease, both the frequency at which the fungus disseminates to the viscera and the burdens established in the viscera show high variability; hence more animals need to be infected to allow statistically powered analyses of damage outcomes. Superficial *Candida* infections can be modeled, with vaginal infections typically established in mice (36) or rats (28) and oral infections established in rats (141).

Models of *Candida* infection in vivo have been described for wax moth larvae (15, 31) and immunodeficient fruit flies (1). These may be of use for investigation of specific types of fungus-host interaction, but the immune responses of such hosts differ substantially from that of mammals. Hence, in terms of assessing virulence per se, these models are inherently less valuable than those in rodents, even with the limitations of the latter. An alternative model in vivo that may come closer to mimicking a mammalian system involves infection of the chick chorioallantoic membrane (CAM) (46). In this model, hyphal forms are seen to penetrate the membrane while yeast forms proliferate on the outer surface of the membrane (Color Plate 6).

Many approaches exist for investigation of damage caused to systems ex vivo. Examples include models based on reconstituted oral (7, 47, 115, 117), esophageal (11, 67), and vaginal (114) human

epithelium or reconstituted epidermis (116, 118), the use of tissues such as isolated endothelial cells (109), and the analysis of phagocytosis by isolated neutrophils and macrophages or by cultured macrophage cell lines (68, 70, 76, 107). All of these allow investigation of specific aspects of host-*Candida* interaction, but they cannot ultimately provide the level of information available in whole-animal models.

If, however, the aim is to assess the specific molecular role of a particular virulence factor, at least some of these ex vivo experimental models of *Candida* infection may have considerable potential. For example, if one wants to test whether a particular factor is essential for adherence or tissue penetration, this is possible using ex vivo or in vitro models. The key issue here is the extent to which the model reflects the actual in vivo situation during disease establishment and progression. For example, does the ability of a *Candida* strain to penetrate the chick CAM reflect its ability to invade oral or vaginal epithelia? In this model, non-hypha-forming mutants were noninvasive but still killed the CAM tissue (46). To what extent does phagocytosis by cultured macrophage cell lines in vitro accurately reflect the ability of macrophages to kill the fungus in vivo? Ideally, models in vitro and ex vivo should be subjected to detailed validation in an attempt to define exactly which aspects of the infection process they are able to mimic accurately. Determination of the molecular processes of disease in model systems will be facilitated by the development of robust transcript profiling technologies that allow the comparison of global patterns of fungal gene expression in various infected tissues. These can then inform detailed comparisons of the spatial and temporal regulation of global fungal expression patterns in different infection models ex vivo. The aim must be to establish a set of models that define specific aspects of *Candida* infections. This process has started with the transcript profiling of fungal cells undergoing phagocytosis ex vivo (70, 107). Such studies will be usefully augmented with cytokine profiling and the use of proteomic analyses, perhaps facilitated by microdissection techniques, to compare the physiological and immunological responses of host cells at the point of invasion. Efforts to validate models ex vivo will reduce but not entirely remove the need for animal experimentation: ultimately, key experiments must be performed in vivo if we are to fully understand fungal virulence, since no other test system is likely to confront the fungus with the same host cells and molecules that interact with the fungus in the intact animal.

EXPERIMENTAL APPROACHES TO DISSECTING VIRULENCE IN *C. ALBICANS*

In addition to the development of well-defined infection models, the molecular dissection of *C. albicans* virulence depends on the generation of well-defined mutants. However, molecular genetic experimentation with *C. albicans* is made relatively difficult, in comparison with budding *Saccharomyces cerevisiae*, by the biology of the fungus. It is a diploid organism with no exploitable sexual cycle (10). *C. albicans* has been observed to mate to generate tetraploid products (55, 58, 75, 79, 121). Furthermore, diploid recombinants can be isolated from these tetraploids, thereby completing a parasexual cycle (9). Although the *C. albicans* genome contains many of the genes needed to complete a meiotic cycle (132), meiosis has not been demonstrated in *C. albicans*. Hence, forward genetics has proven difficult: early attempts to analyze morphogenetic mutations were hampered by the inability to clean up these mutations using genetic crosses. Even reverse genetics was problematic for many years because no dominant selectable marker was available to facilitate gene disruption, and the potential for unwanted changes to be introduced into the genome during gene disruption procedures is considerable.

For many years, the sole dependable method for disruption of both alleles of a *C. albicans* gene was the so-called "URA-blaster" approach. This utilizes the uridine auxotroph host strain, CAI-4, with both copies of the *URA3* gene disrupted. CAI-4 (*ura3/ura3*) was generated from the wild-type clinical isolate SC5314 via a *URA3/ura3* heterozygous intermediate,

CAF2-1 (37). The *ura3* background of CAI-4 provides a selectable marker for gene disruption. A cassette containing *C. albicans URA3* flanked by tandem copies of a bacterial *hisG* gene (*hisG-URA3-hisG*) is used to generate a disruption cassette containing 5′ and 3′ sequences homologous to the gene to be disrupted. This disruption cassette is then used to transform CAI-4. Heterozygous transformants, with the *hisG-URA3-hisG* cassette integrated into one allele of the target gene, are chosen from the uridine prototrophic transformants. These transformants are plated on a medium containing 5′-fluoroorotic acid, to select for *ura3* segregants that have arisen at random via homologous recombination of the tandem *hisG* repeats and hence loop-out of the *URA3* gene. The procedure is then repeated to disrupt the second allele of the target gene, thereby generating a homozygous null mutant. The entire procedure can be recapitulated to disrupt other genes, in each case recycling the single genetic marker.

More rapid gene disruption technologies have been developed, many of which retain basic principles of "URA blasting." Mitchell's group constructed an alternative recyclable *URA3* gene disruption cassette (135). The advantage of this cassette is that it is relatively short and hence can be PCR amplified with ease. However, it still relies on homologous recombination between tandem repeats in the cassette to achieve *URA3* recycling. Meanwhile, Morschhauser et al. (80) developed a *URA3* flipper cassette. This cassette exploits the *S. cerevisiae* FLP-mediated site-specific recombination system, which facilitates efficient recycling of the *URA3* marker during gene disruption.

Mitchell's group (136) also developed a PCR-based gene disruption strategy for *C. albicans* that is based on the approach used by Wach et al. (134) for gene disruption in *S. cerevisiae*. This convenient approach involves the PCR amplification of a selectable marker by using relatively long oligonucleotide primers containing homology to the 5′ and 3′ ends of the target gene. The resultant PCR product, which contains the selectable marker flanked by about 100 bp of homology to the target locus, is used to transform *C. albicans*. Sequential rounds of transformation with different markers allow both alleles of a target locus to be disrupted without the need for an intervening marker-recycling step. However, this method depends on having a *C. albicans* host strain with more than one selectable marker. Commonly used strains for this "rapid hypothesis testing" approach are RM1000 (*his1/his1 ura3/ura3*) and BWP17 (*arg4/arg4 his1/his1 ura3/ura3*) (33, 136). Both are still derived from the clinical isolate SC5314, which limits virulence experimentation to a

Table 2. Notes on the construction of *C. albicans* knockout mutants for virulence studies

Issue	Potential solutions
Attenuated virulence of the host strain	Even following reintroduction of the markers, this is the case for most of the standard strains for molecular analysis of *C. albicans*, with the possible exception of CAI-4 (72). Ensure that proper control strains, which differ only with respect to the target gene, are generated in the same background. Use dominant selectable markers to create the mutation in a clinical isolate.
Residual activity from the gene knockout	Delete the open reading frame from the start to stop codons.
Secondary ectopic integration	Use diagnosis by Southern blotting as well as PCR.
Secondary mutations	Reintegrate the gene under the control of its own promoter; do this at the target locus (ensuring the presence of a new restriction site to distinguish the new reintegrant from the old heterozygote); alternatively, reintroduce the gene using a plasmid that is known to avoid position effects (such as CIp10). Use a conditional mutant (e.g., *tet-ABC1*). Analyze several mutants that have been constructed independently.
Marker position effects	Reintroduce the marker(s) (e.g., *URA3*) on a plasmid that is known to avoid position effects (such as CIp10). Use a dominant selectable marker (e.g., *SAT1*).

single strain background. This problem is likely to be circumvented by the development of cassettes containing dominant selectable markers that can be used for gene disruption in clinical isolates (101) (Table 2). A second problem is that heterogeneity in the target sequence can inhibit transformation (139), particularly when short flanking regions are being used to target the cassette. A third problem is that mutants made using these approaches carry the selectable markers at the target locus (e.g., *abc1::HIS1/abc1::URA3*). Hence these mutants are potentially subject to the marker position effects discussed below. A fourth problem is that it is not possible to generate viable null mutants for essential genes.

This last problem has been addressed in several ways. First, an artificial *UAU1* marker was developed to test whether a gene is essential (33). This marker carries a functional *ARG4* gene flanked by 5′ and 3′ regions of *URA3*. Homologous recombination between these regions results in the formation of a functional *URA3* gene and the loss of the *ARG4* locus. Therefore, the test involves the disruption of one allele of the target locus with the *UAU1* cassette. Uri⁺ Arg⁺ segregants are then selected. These are presumed to have arisen via gene conversion to generate an *abc1::UAU1/abc1::UAU1* segregant, followed by homologous recombination to form an *abc1::UAU1/abc1::URA3* cell. If the *ABC1* gene is essential for viability, no Uri⁺ Arg⁺ segregants can survive without retaining a wild-type copy of this target gene (33). An FLP-based system has also been described for the generation of conditional lethal *C. albicans* mutants (78).

These approaches provide useful tests of essentiality. However, they do not provide viable null mutants for analysis. Therefore, regulatable promoters have been developed to facilitate the generation of conditional null mutants. These include glucose-repressible *PCK1* and *MAL* promoters and the methionine/cysteine-repressible *MET3* promoter (4, 22, 66, 82). However, the heterologous tetracycline-repressible promoter system is particularly useful because this allows *tet-ABC1* fusions to be repressed in vivo by the administration of doxycycline to animals (89). This system is being exploited effectively in screens of doxycycline-conditional *C. albicans* mutants for potential antifungal drug targets (104). This is discussed in more detail in chapter 42.

A second issue that has hindered the molecular dissection of *C. albicans* is the nonstandard decoding of the CUG codon as serine instead of leucine (112). Hence, most standard reporter genes such as *Escherichia coli lacZ* and the firefly luciferase and jellyfish green fluorescent protein (GFP) gene cannot be expressed functionally in *C. albicans*. This problem has been overcome by the development of specialized reporter genes for *C. albicans*, including *Kluyveromyces lactis LAC4*, *Renilla reniformis* luciferase, codon-optimized GFP, and *Streptococcus thermophilus lacZ* (25, 65, 123, 133). Numerous convenient cassettes have been generated with these reporters (see, for example, references 43 and 44). The *Renilla* luciferase gene is arguably the most sensitive reporter and hence is suited to promoter dissection in *C. albicans*. The codon-optimized GFP has the advantage that it facilitates the analysis of expression levels in individual *C. albicans* cells in vivo as well as in vitro (6, 25).

C. albicans molecular genetics has been revolutionized by genome sequencing. The public *C. albicans* genome-sequencing program was undertaken at the Stanford DNA Sequencing and Technology Center (59). This led to the construction of websites that have greatly facilitated the development of

C. albicans genomics (http://www-sequence.stanford. edu/group/candida/; http://genome-www.stanford. edu/fungi/Candida/; http://genolist.pasteur.fr/ CandidaDB (30). Together, the sequence data and genome annotation provided a strong platform for groups to initiate *C. albicans* transcript profiling, proteomics, and large-scale gene deletion studies (18, 24, 26, 38, 39, 42, 50, 63, 64, 84, 85, 90, 97, 104, 107, 130, 140). Arguably one of the greatest impacts of the new *C. albicans* genomics era has been the ability to examine the global roles of specific genes in a relatively unbiased fashion. Clearly this is vital when assessing the roles of specific genes in *C. albicans* virulence. The challenge here is to distinguish the primary effects of mutations or environmental signals on the fungal cell from their secondary and tertiary effects in the complex datasets generated in genomics experiments. Incisive experimental design and the application of stringent controls are essential to achieve this. The impact of molecular genetics and genomics on our understanding of *C. albicans* pathobiology is discussed further in chapters 6 and 40.

VIRULENCE FACTORS OF *C. ALBICANS*: OVERVIEW

A set of putative virulence attributes in *C. albicans* have been defined using the combination of molecular approaches and infection models described above. Chapters 4 and 11 through 14 of this volume focus on the roles of specific virulence attributes. Hence these are only briefly summarized here in the context of dissecting *C. albicans* virulence.

Flexibility to adapt to a changing microenvironment would seem to be an obvious requirement for a fungus able to infect as many different tissues as *C. albicans*. The property known as phenotypic switching, in which subpopulations of *C. albicans* cells can alter their metabolism, morphologies, and cell surface properties at relatively high frequencies, probably enhances flexible responses to change and serves as a "master regulator" of some, if not many, of the properties commonly regarded as factors of virulence in the fungus (94, 120). Phenotypic switching mechanisms also play a role in adaptation of *C. albicans* cells for a mating process that might be part of a full meiotic cycle that has not yet been defined (58, 121).

In order to initiate processes of damage to a mammalian host, *C. albicans* must first be able to adhere to epithelial surfaces. As with any microbe, *C. albicans* cells possess innate abilities for cell-to-cell binding resulting from their gross surface hydrophobicity (32). However, studies over many years indicate that the surface of *C. albicans* bears adhesin molecules that confer a high level of fungal stickiness

for epithelial and other surfaces (21). Many of these adhesins are encoded by members of gene families, such as the adhesin-like sequence (*ALS*) family (52). The members of these families are likely to display some functional redundancy as well as differential spatial and temporal regulation. Hence, determining their individual contributions to *C. albicans* virulence requires the careful phenotypic dissection of strains that carry mutations in multiple members of the gene family of interest. In addition, the glycans of many cell wall proteins may participate directly in adhesion reactions, and these cannot be eliminated by disruption even of whole families of cell wall proteins (21).

Penetrative damage to epithelial surfaces may result from mechanically applied pressures or from the activities of hydrolytic enzymes: both have been studied in *C. albicans* for many years (45, 87). If circulating neutrophils and tissue macrophages are major lines of defense against dissemination of *C. albicans*, the fungus may possess molecular camouflage and/or antiphagocytic mechanisms that enhance its ability to survive in the presence of neutrophils (93). Colonization and penetration from endothelial locations to tissue parenchyma probably involve mechanisms similar to those of epithelial penetration that lead to disease. Common to all these "classical" virulence attributes is a superficial or exocellular location. Adhesins and molecules that mimic host components must be situated at the fungal cell surface, and hydrolytic enzymes must be surface bound or secreted to exert their effects externally. Even the morphological change from yeast forms to hyphae, thought by many to explain mechanical penetrative properties in *C. albicans*, is the result of structural changes in the cell wall.

At the molecular level, the search for *C. albicans* virulence factors has been based principally on the approach often known as "molecular Koch's postulates" (35). If specific disruption of a single gene (both alleles in a diploid organism such as *C. albicans*) generates a mutant attenuated in its ability to cause damage in an animal model, and if reintegration of a copy of the gene restores virulence in the same model, the gene can be claimed to encode a virulence factor. The conceptual difficulty with this approach is that any gene essential for normal physiological functioning of *C. albicans* is likely to emerge as a putative virulence factor; clearly, the definition of a virulence factor based on molecular Koch's postulates is too generous because it does not differentiate fitness in vitro and fitness in vivo. A definition is needed that distinguishes genes essential for the ability of *C. albicans* to damage host tissue in vivo from genes that contribute to fungal growth. It has been suggested that the term "virulence factor" be reserved for molecules that interact directly with host components and that molecules regulating the expression of such factors be termed "virulence regulators" (95). The term "fitness attribute" could then be used to define molecules that are required for fungal growth (16). This goes some way to removing "essential" genes from lists of putative virulence factors but does not constitute a solution to a problem that, in truth, extends beyond semantic considerations. If the aim is to define the interactions between the fungus and its host, it is appropriate to focus much attention on cell surface virulence attributes. However, if the aim is to understand disease establishment and progression, scientific horizons must include fitness as well as virulence attributes.

GENE DISRUPTION AND VIRULENCE TESTING IN *C. ALBICANS*

Since the early 1980s, considerable research based on strain attenuation by specific gene disruption has generated a long list of putative virulence factors (more than 70 are listed in reference 14, and many are reviewed in reference 91). These include molecules involved in almost every conceivable aspect of cell structure, central metabolism, physiology, and gene regulation. In the majority of cases, the experimental procedure to determine the virulence role of a gene product has involved "*URA*-blaster" gene disruption, followed by assessment of virulence in terms of survival and/or tissue burdens of fungi by intravenous challenge in mice. This experimental approach is subject to criticism. In some cases no reintegrant strain was constructed to demonstrate regain of virulence. (In other words, a wild-type copy of the gene was not reintroduced into the genome of the null mutant to control for the possible effects of secondary mutations that might have arisen during the construction of the null mutant.) In other cases the homozygous null mutant expressed *URA3* ectopically at the locus of the second disrupted allele. This is problematic because the level of *URA3* expression at a disrupted locus is often considerably lower than in wild-type *C. albicans* isolates, so that virulence attenuation in the mutant may have resulted from suboptimal *URA3* expression rather than from the effects of the gene disruption (14, 125, 127). It is now recognized that these position effects can be circumvented by the reintroduction of the *URA3* marker on particular plasmids derived from CIp10 (14, 83). Alternatively, the *URA3* gene and/or the disrupted gene can be restored at their natural loci. In such experiments the choice of the promoter of the reconstituted gene is also relevant. In most cases the natural promoter is used, but assumptions

have to be made about the amount of 3′ sequence to include upstream that will achieve normal transcriptional regulation. In all cases the worry remains that the reconstituted genotype is never perfectly isogenic at all loci other than the disrupted site.

Other concerns involve the choice of animal strain tested (Table 1). Outbred mouse strains, although cheap to purchase, are increasingly not favored for serious scientific investigation; virtually all studies now published in journals such as *Immunology*, *Journal of Immunology*, and *Cancer Research* involve inbred rodent strains. However, inbred mouse strains differ in their susceptibility to intravenous *C. albicans* challenge; each typically has a specific genetic alteration in one or more components of its immune function. This means that a *C. albicans* mutant found to be attenuated in virulence in a particular inbred mouse strain may appear normal in a different inbred mouse strain. Such differences could convey useful information on the pathogenic role played by a particular *C. albicans* gene. However, the emphasis of most publications devoted to the discovery of putative virulence factors is on the molecular science of the gene disruption and the in vitro phenotypes of the mutants. The mouse challenge is often presented minimally, even to the extent that detailed conditions of this part of the experimentation are incompletely described. Yet, for example, the physiological status of the *C. albicans* inoculum is known to be important in determining mouse survival (96). It is most exceptional for an investigator to test mutants in more than one strain of mouse or more than one challenge model. However, De Bernardis et al. showed convincingly that a mutant disrupted in *PHR1* was attenuated in virulence in an intravenous-challenge model but normally virulent in a vaginal-infection model while the opposite applied to a *PHR2* mutant (29). This evidence alone should be sufficient to encourage investigators to test mutants in more than a single animal model as a means of learning more about the virulence role of gene products. Ethical considerations and the high costs of work with animals probably restrict the extent to which the pathological consequences of gene disruptions can be routinely followed up in detail.

One possible conclusion from the very large numbers of putative virulence factors discovered by gene disruption approaches is that *C. albicans* is highly susceptible to the consequences of many different gene disruptions. What is being affected by these disruptions may be the fitness of mutants to thrive in the most general sense. Mouse lethality relates directly to growth rate in *C. albicans* mutants (103) and varies with the medium used to prepare a challenge inoculum (96). For a given mouse strain, the "window" of dosage in which all animals challenged with *C. albicans* develop an infection and succumb with measurably damaged target organs is narrow (73). Too low a challenge and the infection is readily cleared; too high a challenge and the animals are overwhelmed within hours. It is therefore possible that a mutant with a relatively minor loss of fitness might display large differences in survival times or tissue burdens by intravenous challenge. Few *C. albicans* mutants are entirely avirulent. The term "attenuated" implies an effect in which deaths and/or tissue infections occur but at a lower rate than is caused by the wild type. One approach that might facilitate the differentiation of fungus virulence attributes from effects dependent on host susceptibility is to challenge animals with elevated doses that lead to the same overall survival times as the wild type. It would then be possible to compare the pathological appearance of mutant and wild-type strains in infected tissues (96). An alternative approach would be to examine the competition of wild-type and mutant strains in vivo.

"WELL-PROVEN" VIRULENCE FACTORS IN *C. ALBICANS*

Implicit in the foregoing discussion is the notion that many of the gene products published as putative virulence factors of *C. albicans* may not merit this description. However, a number of molecules have been studied well beyond molecular Koch's postulates experimentation and are well established by multiple lines of evidence as contributors to the ability of the fungus to cause damage to its host. Some prominent examples follow.

Since its first discovery 40 years ago (126), the ability of *C. albicans* to express secreted aspartyl proteinase (Sap) enzymes has been studied as a probable virulence attribute. The evidence for Saps as virulence factors comes from many sources, including nonspecific (62, 74) and specific (54, 111) mutagenesis to create proteinase-deficient strains. It must, however, be acknowledged that many of these studies were completed before the era in which reconstituted mutants were routinely used as controls, and the attenuated nature of these mutants remains to be confirmed unequivocally. The demonstration that Sap production is a property of the two most pathogenic species, *C. albicans* and *C. tropicalis*, but is seldom found in other *Candida* species (108) strengthens the case for Saps as virulence factors. This evidence was reinforced by the characterization of the expression patterns of individual members of the *SAP* gene family (from *SAP1* to *SAP8*) in superficial and visceral forms of *C. albicans* infection in

vivo and ex vivo (see references 53, 86, and 87 for reviews).

Study of adhesins in *C. albicans* dates back many decades, and *C. albicans* is known to bind to many host cell surfaces as well as to several types of plastic (92). Gene disruption experiments and complementation of poorly adherent *S. cerevisiae* strains with genes from a *C. albicans* library have identified the products of the genes *ALS1*, *ALS5*, *HWP1*, and *INT1* as surface proteins involved in the adhesion of the fungus to various host cell surfaces (20). Expression of several of these genes has been monitored in experimental infections ex vivo (47), and the properties of Int1p as a putative adhesin and virulence factor have been extensively investigated by Hostetter and colleagues (51). With the exception of *ALS5*, each of the adhesin-encoding genes has satisfied molecular Koch's postulates requirements with respect to the mouse intravenous-challenge model (8, 40, 41, 124, 131). This observation is of particular interest since it clearly indicates that adhesion mechanisms play a role beyond that of epithelial colonization. The genes *MNT1* and *PMT1* have both been found important for adhesion to host surfaces in vitro and wild-type lethality for mice (19, 81, 128), but these encode mannosyl transferases. Hence, their virulence roles may be indirect; the adhesins proper are more likely to be the mannan products of action of Mnt1p and Pmt1p. Poulain's group has made a convincing case for the role of phosphorylated α- and β-1,2-linked oligomannan polysaccharides as adhesins and virulence factors in *C. albicans* (98). It is clear that, while the ability to adhere to host surfaces can be described as a virulence attribute per se, the exact role of each adhesin at different temporal and anatomical stages in the natural history of host damage caused by *C. albicans* requires more detailed elucidation.

The most intensely studied *C. albicans* virulence factor deserves mention here, although its real significance remains uncertain. Morphological interchange between yeast, pseudohyphal, and hyphal forms is one of the earliest recognized properties of *C. albicans*. The dogma that hyphal forms are invasive while yeast forms are noninvasive has been perpetrated in the literature for very many years. However, unambiguous evidence supporting this view remains unavailable. Several studies purport to indicate that virulence is dependent on morphogenesis (for example, see references 68 and 113). However, these studies employed regulatory mutations with pleiotropic effects on numerous cellular processes in addition to morphogenesis. The problem lies not in these data but in the interpretation of these data. How can we distinguish properties dependent on the shape of the

C. albicans cells from properties resulting from differential collateral expression of particular molecules (notably surface adhesins) in yeast and hyphal forms? At least three signaling pathways regulate genes involved in morphological change (17, 34), and disruption experiments with several components of these pathways have generated mutants attenuated in mouse virulence. It is therefore inappropriate to make simplistic assumptions about one morphological form being intrinsically more or less virulent than another (45, 106).

An attribute of virulence that has recently gained attention involves oxidative stress responses and the ability of *C. albicans* to survive phagocytosis. Since an ability to degrade hydrogen peroxide and reactive oxygen species is an obvious mechanism for the fungus to combat phagocytic killing, it is perhaps unsurprising that many oxidative stress response components have now been described as virulence attributes that satisfy molecular Koch's postulates, including *SOD1* (56), *SOD5* (77), and *CAT1/CCT1* (88, 138), and evidence has been provided for the role of *SSK1* as a major regulator of oxidative stress responses in *C. albicans* (24). The inactivation of the stress-activated mitogen-activated protein kinase *HOG1*, which plays a role in oxidative stress responses (3, 119), also attenuates the virulence of *C. albicans* (2).

Survival following phagocytosis also appears to be dependent on the metabolic flexibility of *C. albicans*. Lorenz and Fink (71) showed that genes encoding glyoxylate cycle enzymes were upregulated in *S. cerevisiae* and *C. albicans* cells ingested by macrophages and that disruption of *C. albicans* *ICL1*, encoding isocitrate lyase, created a mutant attenuated in mouse lethality. They speculated that a glucose-poor environment within the phagolysosome induced glyoxylate cycle enzymes as a means of generating glucose from the fatty acids present within the macrophages. β-Oxidation is the method of degradation of fatty acids to acetyl coenzyme A, which feeds into the glyoxylate cycle. This has now been borne out by transcript profiling studies of *C. albicans* cells following phagocytosis (70) and by several other studies which have provided evidence for a virulence role for various β-oxidation enzymes (77, 99, 100). Changes in central metabolism have also been associated with phenotypic switching (63), another putative virulence factor (60, 61).

MULTIPLE VIRULENCE FACTORS: WHY NOT DIFFERENTIAL STRAIN VIRULENCE?

One point should be self-evident from the preceding discussion: virulence in *C. albicans* is largely a function of processes that depend on cooperative gene functions, rather being than the result of the action of any single gene. Individual processes may play a critical role at individual stages of infection, but no one process can account for the multiple facets of *C. albicans* infection: commensal adhesion and colonization, epithelial-cell penetration (superficial infection), survival of phagocytic attack in tissues and fluids, endothelial-cell penetration, and tissue invasion (Fig. 1). Cutler, in 1991, expressed this notion in terms of a "virulence set hypothesis" (27), but with our current state of knowledge it is important to appreciate that each individual virulence process can vary in terms of the genes expressed to support it. Even the simple concept of secretion of proteolytic activity as a virulence factor must now be viewed as secretion of the products of one or more of at least eight genes encoding Saps, each of which is expressed differently in different microenvironments and growth forms. Adhesion, morphogenesis, and oxidative stress, as expressed by a given population of *C. albicans* cells within an infected tissue, are likely to differ in fine detail according to the particular subset of genes currently expressed that can influence these properties. Indeed, it is not too far-fetched to suggest that closely adjacent individual cells could differ in the details of their transcriptome. This would explain the not uncommon histopathological picture of *C. albicans*-infected tissue in which the fungus appears to be altering its morphology and, presumably, other properties as well, with each new micrometer of invasive extension (Color Plate 7). It would also help to explain apparently contradictory observations about the expression profiles of fungal cells in ex vivo infection models. For example, both glycolytic and gluconeogenic metabolism were apparently activated following the exposure of *C. albicans* cells to human blood ex vivo (39). The simplest explanation for this was that the fungal cell population was heterogeneous: phagocytosed fungal cells had activated gluconeogenesis, whereas cells remaining in the plasma had activated glycolysis. Hence, to understand fungal responses to complex host environments, we must examine the behavior of individual cells, not just cell populations. Recent technological advances have made this possible through the application of the GFP reporter in single-cell profiling (6, 25). The powerful combination of transcript and single-cell profiling will indubitably advance our understanding of fungal responses within the host.

If multiple factors interact to promote *C. albicans* virulence, why should different strains not display differential virulence? Molecular technology for differentiation of *C. albicans* isolates has now reached the point at which multilocus sequence

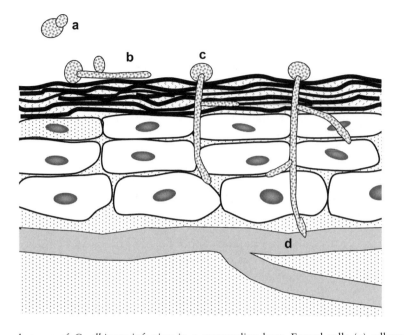

Figure 1. Temporal stages of *C. albicans* infection in a mammalian host. Fungal cells (a) adhere to, and become commensal colonizers of, an epithelial surface (b). Given local deterioration in host defense, the fungi can penetrate epithelial layers (c); they may be able to invade as far as the bloodstream (d), leading to dissemination of fungal propagules, which may adhere to and penetrate vascular endothelia and thus gain access to deep tissues.

typing (MLST) can be routinely applied to epidemiological investigations of the species (13). By both MLST (12, 127a) and DNA fingerprinting (122), *C. albicans* population structure is revealed as a set of at least four major clades of closely related strains that contain around two-thirds of all isolates. Our own work suggests that the clades differ significantly in the proportions of isolates from superficial and deep tissue sources (127a), suggesting that the clades may also differ in their virulence properties. The clades may differ with respect to the expression of multiple genes required to invade and damage a mammalian host. To date, little virulence research has been done with isolates other than the clinical isolate SC5314 and its derivatives. Now that dominant selective markers have been devised for gene disruption experimentation with *C. albicans* (80, 101), the future search for molecular virulence attributes in the fungus needs to involve a greater diversity of clinical isolates, representing the whole population structure of the species, to determine the rank order of importance of the many factors whose individual contribution affects the fitness of the fungus to zdamage the host.

Acknowledgment. We are grateful to our many friends and colleagues in the *Candida* field for stimulating discussions. Also, we thank the British Society for Antimicrobial Chemotherapy, the European Commission, Merck & Co., Inc., the U.K. Biotechnology and Biological Sciences and Medical Research Councils, and the Wellcome Trust for supporting our work in this area in recent years.

REFERENCES

1. Alarco, A. M., A. Marcil, J. Chen, B. Suter, D. Thomas, and M. Whiteway. 2004. Immune-deficient *Drosophila melanogaster*: a model for the innate immune response to human fungal pathogens. *J. Immunol.* **172:** 5622–5628.
2. Alonso-Monge, R., F. Navarro-Garcia, G. Molero, R. Diez-Orejas, M. Gustin, J. Pla, M. Sanchez, and C. Nombela. 1999. Role of the mitogen-activated protein kinase hog1p in morphogenesis and virulence of *Candida albicans. J. Bacteriol.* **181:**3058–3068.
3. Alonso-Monge, R., F. Navarro-Garcia, E. Roman, A. I. Negredo, B. Eisman, U. Nombela, and J. Pla. 2003. The Hog1 mitogen-activated protein kinase is essential in the oxidative stress response and chlamydospore formation in *Candida albicans. Eukaryot. Cell* **2:**351–361.
4. Backen, A. C., I. D. Broadbent, R. W. Fetherston, J. D. C. Rosamond, N. F. Schnell, and M. J. R. Stark. 2000. Evaluation of the CaMAL2 promoter for regulated expression of genes in *Candida albicans. Yeast* **16:**1121–1129.
5. Balish, E., H. I. Filutowicz, and T. D. Oberly. 1990. Correlates of cell-mediated immunity in *Candida albicans*-colonized gnotobiotic mice. *Infect. Immun.* **58:**107–113.
6. Barelle, C. J., C. L. Manson, D. M. MacCallum, F. C. Odds, N. A. R. Gow, and A. J. P. Brown. 2004. GFP as a quantitative reporter of gene regulation in *Candida albicans. Yeast* **21:**333–340.

7. Bartie, K. L., D. W. Williams, M. J. Wilson, A. J. C. Potts, and M. A. O. Lewis. 2004. Differential invasion of *Candida albicans* isolates in an in vitro model of oral candidosis. *Oral Microbiol. Immunol.* 19:293–296.

8. Bendel, C. M., K. M. Kinneberg, R. P. Jechorek, C. A. Gale, S. L. Erlandsen, M. K. Hostetter, and C. L. Wells. 1999. Systemic infection following intravenous inoculation of mice with *Candida albicans int1* mutant strains. *Mol. Genet. Metab.* 67:343–351.

9. Bennett, R. J., and A. D. Johnson. 2003. Completion of a parasexual cycle in *Candida albicans* by induced chromosome loss in tetraploid strains. *EMBO J.* 22:2505–2515.

10. Berman, J., and P. E. Sudbery. 2002. *Candida albicans*: a molecular revolution built on lessons from budding yeast. *Nat. Rev. Genet.* 3:918–930.

11. Bernhardt, J., D. Herman, M. Sheridan, and R. Calderone. 2001. Adherence and invasion studies of *Candida albicans* strains, using in vitro models of esophageal candidiasis. *J. Infect. Dis.* 184:1170–1175.

12. Bougnoux, M.-E., D. M. Aarensen, S. Morand, M. Théraud, B. G. Spratt, and C. d'Enfert. 2004. Multilocus sequence typing of *Candida albicans*: strategies, data exchange and applications. *Infect. Genet. Evol.* 4:243–252.

13. Bougnoux, M. E., A. Tavanti, C. Bouchier, N. A. R. Gow, A. Magnier, A. D. Davidson, M. C. J. Maiden, C. d'Enfert, and F. C. Odds. 2003. Collaborative consensus for optimized multilocus sequence typing of *Candida albicans*. *J. Clin. Microbiol.* 41:5265–5266.

14. Brand, A., D. M. MacCallum, A. J. P. Brown, N. A. R. Gow, and F. C. Odds. 2004. Ectopic expression of *URA3* can influence the virulence phenotypes and proteome of *Candida albicans* but can be overcome by targeted reintegration of *URA3* at the *RPS10* locus. *Mol. Microbiol.* 3:900–909.

15. Brennan, M., D. Y. Thomas, M. Whiteway, and K. Kavanagh. 2002. Correlation between virulence of *Candida albicans* mutants in mice and *Galleria mellonella* larvae. *FEMS Immunol. Med. Microbiol.* 34:153–157.

16. Brown, A. J. P. 2005. Integration of metabolism with virulence in *Candida albicans*. *In* A. J. P. Brown (ed.), *Fungal Genomics* (Mycota XIII), in press. Springer-Verlag KG, Heidelberg, Germany.

17. Brown, A. J. P. 2002. Morphogenetic signaling pathways in *Candida albicans*, p. 95–106. *In* R. A. Calderone (ed.), *Candida and Candidiasis*. ASM Press, Washington, D.C.

18. Bruneau, J. M., I. Maillet, E. Tagat, R. Legrand, F. Supatto, C. Fudali, J. P. Le Caer, V. Labas, D. Lecaque, and J. Hodgson. 2003. Drug induced proteome changes in *Candida albicans*: comparison of the effect of beta(1,3) glucan synthase inhibitors and two triazoles, fluconazole and itraconazole. *Proteomics* 3:325–336.

19. Buurman, E. T., C. Westwater, B. Hube, A. J. P. Brown, F. C. Odds, and N. A. R. Gow. 1998. Molecular analysis of Camnt1p, a mannosyl transferase important for adhesion and virulence of *Candida albicans*. *Proc. Nat. Acad. Sci. USA* 95:7670–7675.

20. Calderone, R. A., and W. A. Fonzi. 2001. Virulence factors of *Candida albicans*. *Trends Microbiol.* 9:327–335.

21. Calderone, R. A., and N. A. R. Gow. 2002. Host recognition by *Candida* species, p. 67–86. *In* R. A. Calderone (ed.), *Candida and Candidiasis*. ASM Press, Washington, D.C.

22. Care, R. S., J. Trevethick, K. M. Binley, and P. E. Sudbery. 1999. The *MET3* promoter: a new tool for *Candida albicans* molecular genetics. *Mol. Microbiol.* 34:792–798.

23. Casadevall, A., and L. A. Pirofski. 2001. Host-pathogen interactions: the attributes of virulence. *J. Infect. Dis.* 184:337–344.

24. Chauhan, N., D. Inglis, E. Roman, J. Pla, D. M. Li, J. A. Calera, and R. Calderone. 2003. *Candida albicans* response regulator gene *SSK1* regulates a subset of genes whose functions are associated with cell wall biosynthesis and adaptation to oxidative stress. *Eukaryot. Cell* 2:1018–1024.

25. Cormack, B. P., G. Bertram, M. Egerton, N. A. R. Gow, S. Falkow, and A. J. P. Brown. 1997. Yeast-enhanced green fluorescent protein (yEGFP)—a reporter of gene expression in *Candida albicans*. *Microbiology* 143:303–311.

26. Cowen, L. E., A. Nantel, M. S. Whiteway, D. Y. Thomas, D. C. Tessier, L. M. Kohn, and J. B. Anderson. 2002. Population genomics of drug resistance in *Candida albicans*. *Proc. Nat. Acad. Sci. USA* 99: 9284–9289.

27. Cutler, J. E. 1991. Putative virulence factors of *Candida albicans*. *Annu. Rev. Microbiol.* 45:187–218.

28. De Bernardis, F., R. Lorenzini, and A. Cassone. 1999. Rat model of *Candida* vaginal infection, p. 735–740. *In* O. Zak and M. A. Sande (ed.), *Handbook of Animal Models of Infection*. Academic Press, Inc., San Diego, Calif.

29. De Bernardis, F., F. A. Muhlschlegel, A. Cassone, and W. A. Fonzi. 1998. The pH of the host niche controls gene expression in and virulence of *Candida albicans*. *Infect. Immun.* 66:3317–3325.

30. d'Enfert, C., S. Goyard, S. Rodriguez-Arnaveilhe, L. Frangeul, L. Jones, F. Tekaia, O. Bader, L. Castillo, A. Dominguez, J. Ernst, C. Fradin, C. Gaillardin, S. Garcia-Sanchez, P. de Groot, B. Hube, F. Klis, S. Krishnamurthy, D. Kunze, M.-C. Lopez, A. Mavor, N. Martin, I. Moszer, D. Onésime, J. Perez Martin, R. Sentandreu, and A. J. P. Brown. 2004. CandidaDB: a genome database for *Candida albicans* pathogenomics. *Nucleic Acids Res.* 33:D353–357.

31. Dunphy, G. B., U. Oberholzer, M. Whiteway, R. J. Zakarian, and I. Boomer. 2003. Virulence of *Candida albicans* mutants toward larval *Galleria mellonella* (Insecta, Lepidoptera, Galleridae). *Can. J. Microbiol.* 49:514–524.

32. Ener, B., and L. J. Douglas. 1992. Correlation between cell-surface hydrophobicity of *Candida albicans* and adhesion to buccal epithelial cells. *FEMS Microbiol. Lett.* 99:37–42.

33. Enloe, B., A. Diamond, and A. P. Mitchell. 2000. A single-transformation gene function test in diploid *Candida albicans*. *J. Bacteriol.* 182:5730–5736.

34. Ernst, J. F. 2000. Transcription factors in *Candida albicans*—environmental control of morphogenesis. *Microbiology* 146:1763–1774.

35. Falkow, S. 1988. Molecular Koch's postulates applied to microbial pathogenicity. *Rev. Infect. Dis.* 10 (Suppl. 2): S274–S276.

36. Fidel, P. L., and J. D. Sobel. 1999. Murine models of *Candida* vaginal infections, p. 742–748. *In* O. Zak and M. A. Sande (ed.), *Handbook of Animal Models of Infections*. Academic Press, Inc., San Diego, Calif.

37. Fonzi, W., and M. Irwin. 1993. Isogenic strain construction and gene mapping in *Candida albicans*. *Genetics* 134:717–728.

38. Fradin, C., and B. Hube. 2003. Tissue infection and site-specific gene expression in *Candida albicans*. *Adv. Appl. Microbiol.* 53:271–290.

39. Fradin, C., M. Kretschmar, T. Nichterlein, C. Gaillardin, C. d'Enfert, and B. Hube. 2003. Stage-specific gene expression of *Candida albicans* in human blood. *Mol. Microbiol.* 47:1523–1543.

40. Fu, Y., A. S. Ibrahim, D. C. Sheppard, Y. C. Chen, S. W. French, J. E. Cutler, S. G. Filler, and J. E. Edwards. 2002. *Candida albicans* Als1p: an adhesin that is a downstream effector of the *EFG1* filamentation pathway. *Mol. Microbiol.* 44:61–72.

41. Gale, C. A., C. M. Bendel, M. Mcclellan, M. Hauser, J. M. Becker, J. Berman, and M. K. Hostetter. 1998. Linkage of adhesion, filamentous growth, and virulence in *Candida albicans* to a single gene, *Int1*. *Science* 279:1355–1358.

42. Garcia-Sanchez, S., S. Aubert, I. Iraqui, G. Janbon, J. M. Ghigo, and C. d'Enfert. 2004. *Candida albicans* biofilms: a developmental state associated with specific and stable gene expression patterns. *Eukaryot. Cell* 3:536–545.

43. Gerami-Nejad, M., J. Berman, and C. A. Gale. 2001. Cassettes for PCR-mediated construction of green, yellow, and cyan fluorescent protein fusions in *Candida albicans*. *Yeast* 18:859–864.

44. Gola, S., R. Martin, A. Walther, A. Dunkler, and J. Wendland. 2003. New modules for PCR-based gene targeting in *Candida albicans*: rapid and efficient gene targeting using 100 bp of flanking homology region. *Yeast* 20:1339–1347.

45. Gow, N. A. R., A. J. P. Brown, and F. C. Odds. 2002. Fungal morphogenesis and host invasion. *Curr. Opin. Microbiol.* 5:366–371.

46. Gow, N. A. R., Y. Knox, C. A. Munro, and W. D. Thompson. 2003. Infection of chick chorioallantoic membrane (CAM) as a model for invasive hyphal growth and pathogenesis of *Candida albicans*. *Med. Mycol.* 41:331–338.

47. Green, C. B., G. Cheng, J. Chandra, P. Mukherjee, M. A. Ghannoum, and L. L. Hoyer. 2004. RT-PCR detection of *Candida albicans ALS* gene expression in the reconstituted human epithelium (RHE) model of oral candidiasis and in model biofilms. *Microbiology* 150:267–275.

48. Hajjeh, R. A., A. N. Sofair, L. H. Harrison, G. M. Lyon, B. A. Arthington-Skaggs, S. A. Mirza, M. Phelan, J. Morgan, W. Lee-Yang, M. A. Ciblak, L. E. Benjamin, L. T. Sanza, S. Huie, S. F. Yeo, M. E. Brandt, and D. W. Warnock. 2004. Incidence of bloodstream infections due to *Candida* species and in vitro susceptibilities of isolates collected from 1998 to 2000 in a population-based active surveillance program. *J. Clin. Microbiol.* 42:1519–1527.

49. Hasenclever, H. F., and W. O. Mitchell. 1961. Antigenic studies of *Candida*. III. Comparative pathogenicity of *Candida albicans* group A, group B and *Candida stellatoidea*. *J. Bacteriol.* 82:578–581.

50. Hernandez, R., C. Nombela, R. Diez-Orejas, and C. Gil. 2004. Two-dimensional reference map of *Candida albicans* hyphal forms. *Proteomics* 4:374–382.

51. Hostetter, M. K. 2000. RGD-mediated adhesion in fungal pathogens of humans, plants and insects. *Curr. Opin. Microbiol.* 3:344–348.

52. Hoyer, L. L., R. Fundyga, J. E. Hecht, J. C. Kapteyn, F. M. Klis, and J. Arnold. 2001. Characterization of agglutinin-like sequence genes from non-*albicans Candida* and phylogenetic analysis of the *ALS* family. *Genetics* 157:1555–1567.

53. Hube, B., and J. Naglik. 2001. *Candida albicans* proteinases: resolving the mystery of a gene family. *Microbiology* 147:1997–2005.

54. Hube, B., D. Sanglard, F. C. Odds, D. Hess, M. Monod, W. Schafer, A. J. P. Brown, and N. A. R. Gow. 1997. Disruption of each of the secreted aspartyl proteinase genes *SAP1*, *SAP2*, and *SAP3* of *Candida albicans* attenuates virulence. *Infect. Immun.* 65:3529–3538.

55. Hull, C. M., R. M. Raisner, and A. D. Johnson. 2000. Evidence for mating of the "asexual" yeast *Candida albicans* in a mammalian host. *Science* 289:307–310.

56. Hwang, C. S., Y. U. Baek, H. S. Yim, and S. O. Kang. 2003. Protective roles of mitochondrial manganese-containing superoxide dismutase against various stresses in *Candida albicans*. *Yeast* 20:929–941.

57. Isenberg, H. D., J. Allerhand, J. I. Berkman, and D. Goldberg. 1963. Immunological and toxic differences between mouse-virulent and mouse-avirulent *Candida albicans*. *J. Bacteriol.* 86:1010–1018.

58. Johnson, A. 2003. The biology of mating in *Candida albicans*. *Nat. Rev. Microbiol.* 1:106–116.

59. Jones, T., N. A. Federspiel, H. Chibana, J. Dungan, S. Kalman, B. B. Magee, G. Newport, Y. R. Thorstenson, N. Agabian, P. T. Magee, R. W. Davis, and S. Scherer. 2004. The diploid genome sequence of *Candida albicans*. *Proc. Nat. Acad. Sci. USA* 101:7329–7334.

60. Kvaal, C., S. A. Lachke, T. Srikantha, K. Daniels, J. McCoy, and D. R. Soll. 1999. Misexpression of the opaque-phase-specific gene *PEP1* (*SAP1*) in the white phase of *Candida albicans* confers increased virulence in a mouse model of cutaneous infection. *Infect. Immun.* 67:6652–6662.

61. Kvaal, C. A., T. Srikantha, and D. R. Soll. 1997. Misexpression of the white-phase-specific gene *wh11* in the opaque phase of *Candida albicans* affects switching and virulence. *Infect. Immun.* 65:4468–4475.

62. Kwon-Chung, K. J., D. Lehman, C. Good, and P. T. Magee. 1985. Genetic evidence for role of extracellular proteinase in virulence of *Candida albicans*. *Infect. Immun.* 49:571–575.

63. Lan, C. Y., G. Newport, L. A. Murillo, T. Jones, S. Scherer, R. W. Davis, and N. Agabian. 2002. Metabolic specialization associated with phenotypic switching in *Candida albicans*. *Proc. Nat. Acad. Sci. USA* 99:14907–14912.

64. Lane, S., C. Birse, S. Zhou, R. Matson, and H. P. Liu. 2001. DNA array studies demonstrate convergent regulation of virulence factors by Cph1, Cph2, and Efg1 in *Candida albicans*. *J. Biol. Chem.* 276:48988–48996.

65. Leuker, C. E., A.-M. Hahn, and J. F. Ernst. 1992. β-Galactosidase of *Kluyveromyces lactis* (Lac4p) as reporter of gene expression in *Candida albicans* and C. *tropicalis*. *Mol. Gen. Genet.* 235:235–241.

66. Leuker, C. E., A. Sonneborn, S. Delbruck, and J. F. Ernst. 1997. Sequence and promoter regulation of the *pck1* gene encoding phosphoenolpyruvate carboxykinase of the fungal pathogen *Candida albicans*. *Gene* 192:235–240.

67. Li, D. M., J. Bernhardt, and R. Calderone. 2002. Temporal expression of the *Candida albicans* genes *CHK1* and *CSSK1*, adherence, and morphogenesis in a model of reconstituted human esophageal epithelial candidiasis. *Infect. Immun.* 70:1558–1565.

68. Lo, H. J., J. R. Köhler, D. Di, D. Loebenberg, A. Cacciapuoti, and G. R. Fink. 1997. Nonfilamentous C. *albicans* mutants are avirulent. *Cell* 90:939–949.

69. Londono, P., X. M. Gao, F. Bowe, W. L. Mcpheat, G. Booth, and G. Dougan. 1998. Evaluation of the intranasal challenge route in mice as a mucosal model

for *Candida albicans* infection. *Microbiology* 144: 2291–2298.

70. Lorenz, M. C., J. A. Bender, and G. R. Fink. 2004. Transcriptional response of *Candida albicans* upon internalization by macrophages. *Eukaryot. Cell* 3: 1076–1087.

71. Lorenz, M. C., and G. R. Fink. 2001. The glyoxylate cycle is required for fungal virulence. *Nature* 412:83–86.

72. Louria, D. B., R. G. Brayton, and G. Finkel. 1963. Studies on the pathogenesis of experimental *Candida albicans* infections in mice. *Sabouraudia* 2:271–283.

73 MacCallum, D. M., and F. C. Odds. 2005. Temporal events in the intravenous challenge model for experimental *Candida albicans* infections in female mice. *Mycoses* 48:151–161.

74. Macdonald, F., and F. C. Odds. 1983. Virulence for mice of a proteinase-secreting strain of *Candida albicans* and a proteinase-deficient mutant. *J. Gen. Microbiol.* 129:421–438.

75. Magee, B. B., and P. T. Magee. 2000. Induction of mating in *Candida albicans* by construction of MTLa and MTLα strains. *Science* 289:310–313.

76. Marcil, A., D. Harcus, D. Y. Thomas, and M. Whiteway. 2002. *Candida albicans* killing by RAW 264.7 mouse macrophage cells: effects of *Candida* genotype, infection ratios, and gamma interferon treatment. *Infect. Immun.* 70:6319–6329.

77. Martchenko, M., A. M. Alarco, D. Harcus, and M. Whiteway. 2004. Superoxide dismutases in *Candida albicans*: transcriptional regulation and functional characterization of the hyphal-induced *SOD5* gene. *Mol. Biol. Cell* 15:456–467.

78. Michel, S., S. Ushinsky, B. Klebl, E. Leberer, D. Thomas, M. Whiteway, and J. Morschhauser. 2002. Generation of conditional lethal *Candida albicans* mutants by inducible deletion of essential genes. *Mol. Microbiol.* 46:269–280.

79. Miller, M. G., and A. D. Johnson. 2002. White-opaque switching in *Candida albicans* is controlled by mating-type locus homeodomain proteins and allows efficient mating. *Cell* 110:293–302.

80. Morschhauser, J., S. Michel, and P. Staib. 1999. Sequential gene disruption in *Candida albicans* by FLP-mediated site-specific recombination. *Mol. Microbiol.* 32:547–556.

81. Munro, C. A., S. Bates, E. T. Buurman, H. B. Hughes, D. M. MacCallum, G. Bertram, A. Atrih, M. A. J. Ferguson, J. M. Bain, A. Brand, S. Hamilton, C. Westwater, L. M. Thomson, A. J. P. Brown, F. C. Odds, and N. A. R. Gow. 2005. Mnt1p and Mnt2p of *Candida albicans* are partially redundant α-1,2-mannosyltranferases that participate in O-linked mannosylation and are required for adhesion and virulence. *J. Biol. Chem.* 280:1051–1060.

82. Munro, C. A., K. Winter, A. Buchan, K. Henry, J. M. Becker, A. J. P. Brown, C. E. Bulawa, and N. A. R. Gow. 2001. Chs1 of *Candida albicans* is an essential chitin synthase required for synthesis of the septum and for cell integrity. *Mol. Microbiol.* 39:1414–1426.

83. Murad, A. M. A., P. R. Lee, I. D. Broadbent, C. J. Barelle, and A. J. P. Brown. 2000. CIp10, an efficient and convenient integrating vector for *Candida albicans*. *Yeast* 16:325–327.

84. Murad, A. M. A., C. d'Enfert, C. Gaillardin, H. Tournu, F. Tekaia, D. Talibi, D. Marechal, V. Marchais, J. Cottin, and A. J. P. Brown. 2001. Transcript profiling in *Candida albicans* reveals new cellular functions for the transcriptional repressors CaTup1, CaMig1 and CaNrg1. *Mol. Microbiol.* 42:981–993.

85. Murad, A. M. A., P. Leng, M. Straffon, J. Wishart, S. Macaskill, D. MacCallum, N. Schnell, D. Talibi, D. Marechal, F. Tekaia, C. d'Enfert, C. Gaillardin, F. C. Odds, and A. J. P. Brown. 2001. NRG1 represses yeast-hypha morphogenesis and hypha-specific gene expression in *Candida albicans*. *EMBO J.* 20:4742–4752.

86. Naglik, J., A. Albrecht, O. Bader, and B. Hube. 2004. *Candida albicans* proteinases and host/pathogen interactions. *Cell. Microbiol.* 6:915–926.

87. Naglik, J. R., S. J. Challacombe, and B. Hube. 2003. *Candida albicans* secreted aspartyl proteinases in virulence and pathogenesis. *Microbiol. Mol. Biol. Rev.* 67: 400–428.

88. Nakagawa, Y., T. Kanbe, and I. Mizuguchi. 2003. Disruption of the human pathogenic yeast *Candida albicans* catalase gene decreases survival in mouse-model infection and elevates susceptibility to higher temperature and to detergents. *Microbiol. Immunol.* 47: 395–403.

89. Nakayama, H., T. Mio, S. Nagahashi, M. Kokado, M. Arisawa, and Y. Aoki. 2000. Tetracycline-regulatable system to tightly control gene expression in the pathogenic fungus *Candida albicans*. *Infect. Immun.* 68: 6712–6719.

90. Nantel, A., D. Dignard, C. Bachewich, D. Harcus, A. Marcil, A. P. Bouin, C. W. Sensen, H. Hogues, M. van het Hoog, P. Gordon, T. Rigby, F. Benoit, D. C. Tessier, D. Y. Thomas, and M. Whiteway. 2002. Transcription profiling of *Candida albicans* cells undergoing the yeast-to-hyphal transition. *Mol. Biol. Cell* 13: 3452–3465.

91. Navarro-Garcia, F., M. Sanchez, C. Nombela, and J. Pla. 2001. Virulence genes in the pathogenic yeast *Candida albicans*. *FEMS Microbiol. Rev.* 25:245–268.

92. Odds, F. C. 1988. *Candida and Candidosis*, 2nd ed. Bailliere Tindall, London, United Kingdom.

93. Odds, F. C. 1994. Pathogenesis of *Candida* infections. *J. Am. Acad. Dermatol.* 31:S2–S5.

94. Odds, F. C. 1997. Switch of phenotype as an escape mechanism of the intruder. *Mycoses* 40:9–12.

95. Odds, F. C., R. A. Calderone, B. Hube, and C. Nombela. 2003. Virulence in *Candida albicans*: views and suggestions from a peer-group workshop. *ASM News* 69: 54–55.

96. Odds, F. C., L. Van Nuffel, and N. A. R. Gow. 2000. Survival in experimental *Candida albicans* infections depends on inoculum growth conditions as well as animal host. *Microbiology* 146:1881–1889.

97. Pitarch, A., M. Sanchez, C. Nombela, and C. Gil. 2002. Sequential fractionation and two-dimensional gel analysis unravels the complexity of the dimorphic fungus *Candida albicans* cell wall proteome. *Mol. Cell. Proteomics* 1:967–982.

98. Poulain, D., and T. Jouault. 2004. *Candida albicans* cell wall glycans, host receptors and responses: elements for a decisive crosstalk. *Curr. Opin. Microbiol.* 7:342–349.

99. Prigneau, O., A. Porta, and B. Maresca. 2004. *Candida albicans CTN* gene family is induced during macrophage infection: homology, disruption and phenotypic analysis of *CTN3* gene. *Fungal Genet. Biol.* 41:783–793.

100. Prigneau, O., A. Porta, J. A. Poudrier, S. Colonna-Romano, T. Noel, and B. Maresca. 2003. Genes

involved in beta-oxidation, energy metabolism and gly-oxylate cycle are induced by *Candida albicans* during macrophage infection. *Yeast* 20:723–730.

101. Reuss, O., A. Vik, R. Kolter, and J. Morschhauser. 2004. The *SAT1* flipper, an optimized tool for gene disruption in *Candida albicans*. *Gene* 341:119–127.

102. Richardson, M. D., and H. Smith. 1981. Resistance of virulent and attenuated strains of *Candida albicans* to intracellular killing by human and mouse phagocytes. *J. Infect. Dis.* 144:557–564.

102. Rieg, G., Y. Fu, A. S. Ibrahim, X. Zhou, S. G. Filler, and J. E. Edwards. 1999. Unanticipated heterogeneity in growth rate and virulence among *Candida albicans* *AAF1* null mutants. *Infect. Immun.* 67:3193–3198.

103. Roemer, T., B. Jiang, J. Davison, T. Ketela, K. Veillette, A. Breton, F. Tandia, A. Linteau, S. Sillaots, C. Marta, N. Martel, S. Veronneau, S. Lemieux, S. Kauffman, J. Becker, R. Storms, C. Boone, and H. Bussey. 2003. Large-scale essential gene identification in *Candida albicans* and applications to antifungal drug discovery. *Mol. Microbiol.* 50:167–181.

105. Rogers, T. J., and E. Balish. 1980. Immunity to *Candida albicans*. *Microbiol. Rev.* 44:660–682.

106. Romani, L., F. Bistoni, and P. Puccetti. 2003. Adaptation of *Candida albicans* to the host environment: the role of morphogenesis in virulence and survival in mammalian hosts. *Curr. Opin. Microbiol.* 6:338–343.

107. Rubin-Bejerano, I., I. Fraser, P. Grisafi, and G. R. Fink. 2003. Phagocytosis by neutrophils induces an amino acid deprivation response in *Saccharomyces cerevisiae* and *Candida albicans*. *Proc. Natl. Acad. Sci. USA* 100:11007–11012.

108. Rüchel, R., K. Uhlemann, and B. Böning. 1983. Secretion of acid proteinases by different species of the genus *Candida*. *Zentbl. Bakteriol. Mikrobiol. Hyg. 1 Abt Origi. A* 255:537–578.

109. Sanchez, A. A., D. A. Johnston, C. Myers, J. E. Edwards, A. P. Mitchell, and S. G. Filler. 2004. Relationship between *Candida albicans* virulence during experimental hematogenously disseminated infection and endothelial cell damage in vitro. *Infect. Immun.* 72:598–601.

110. Sandven, P. 2000. Epidemiology of candidemia. *Rev. Iberoam. Micol.* 17:73–81.

111. Sanglard, D., B. Hube, M. Monod, F. C. Odds, and N. A. R. Gow. 1997. A triple deletion of the secreted aspartyl proteinase genes *sap4*, *sap5*, and *sap6* of *Candida albicans* causes attenuated virulence. *Infect. Immun.* 65:3539–3546.

112. Santos, M., G. Keith, and M. Tuite. 1993. Non-standard translational events in *Candida albicans* mediated by an unusual seryl-transfer RNA with a 5′-CAG-3′ anti-codon. *EMBO J.* 12:607–616.

113. Saville, S. P., A. L. Lazzell, C. Monteagudo, and J. L. Lopez-Ribot. 2003. Engineered control of cell morphology in vivo reveals distinct roles for yeast and filamentous forms of *Candida albicans* during infection. *Eukaryot. Cell* 2:1053–1060.

114. Schaller, M., M. Bein, H. C. Korting, S. Baur, G. Hamm, M. Monod, S. Beinhauer, and B. Hube. 2003. The secreted aspartyl proteinases Sap1 and Sap2 cause tissue damage in an in vitro model of vaginal candidiasis based on reconstituted human vaginal epithelium. *Infect. Immun.* 71:3227–3234.

115. Schaller, M., U. Boeld, S. Oberbauer, G. Hamm, B. Hube, and H. C. Korting. 2004. Polymorphonuclear leukocytes (PMNs) induce protective Th1-type cytokine epithelial responses in an in vitro model of oral candido-sis. *Microbiology* 150:2807–2813.

116. Schaller, M., E. Januschke, C. Schackert, B. Woerle, and H. C. Korting. 2001. Different isoforms of secreted aspartyl proteinases (Sap) are expressed by *Candida albicans* during oral and cutaneous candidosis in vivo. *J. Med. Microbiol.* 50:743–747.

117. Schaller, M., H. C. Korting, W. Schafer, J. Bastert, W. C. Chen, and B. Hube. 1999. Secreted aspartic proteinase (Sap) activity contributes to tissue damage in a model of human oral candidosis. *Mol. Microbiol.* 34:169–180.

118. Schaller, M., R. Mailhammer, and H. C. Korting. 2002. Cytokine expression induced by *Candida albicans* in a model of cutaneous candidosis based on reconstituted human epidermis. *J. Med. Microbiol.* 51:672–676.

119. Smith, D. A., S. Nicholls, B. A. Morgan, A. J. P. Brown, and J. Quinn. 2004. A conserved stress-activated protein kinase regulates a core stress response in the human pathogen *Candida albicans*. *Mol. Biol. Cell* 15:4179–4190.

120. Soll, D. R. 2002. *Candida* commensalism and virulence: the evolution of phenotypic plasticity. *Acta Trop.* 81:101–110.

121. Soll, D. R. 2004. Mating-type locus homozygosis, phe-notypic switching and mating: a unique sequence of dependencies in *Candida albicans*. *Bioessays* 26:10–20.

122. Soll, D. R., and C. Pujol. 2003. *Candida albicans* clades. *FEMS Immunol. Med. Microbiol.* 39:1–7.

123. Srikantha, T., A. Klapach, W. W. Lorenz, L. K. Tsai, L. A. Laughlin, J. A. Gorman, and D. R. Soll. 1996. The sea pansy *Renilla reniformis* luciferase serves as a sensi-tive bioluminescent reporter for differential gene expres-sion in *Candida albicans*. *J. Bacteriol.* 178:121–129.

124. Staab, J. F., S. D. Bradway, P. L. Fidel, and P. Sundstrom. 1999. Adhesive and mammalian transglutaminase sub-strate properties of *Candida albicans* Hwp1. *Science* 283:1535–1538.

125. Staab, J. F., and P. Sundstrom. 2003. *URA3* as a selec-table marker for disruption and virulence assessment of *Candida albicans* genes. *Trends Microbiol.* 11:69–73.

126. Staib, F. 1965. Serum-proteins as nitrogen source for yeast-like fungi. *Sabouraudia* 4:187–193.

127. Sundstrom, P., J. E. Cutler, and J. F. Staab. 2002. Reevaluation of the role of *HWP1* in systemic candidia-sis by use of *Candida albicans* strains with selectable marker *URA3* targeted to the *ENO1* locus. *Infect. Immun.* 70:3281–3283.

127a. Tavanti, A., A. D. Davidson, M. J. Fordyce, N. A. R. Gow, M. C. J. Maiden, and F. C. Odds. 2005. *Candida albicans*: populations and properties determined by multi-locus sequence typing. *J. Clin. Microbiol.* 44:5601–5613.

128. Timpel, C., S. Strahlbolsinger, K. Ziegelbauer, and J. F. Ernst. 1998. Multiple functions of Pmt1p-mediated pro-tein O-mannosylation in the fungal pathogen *Candida albicans*. *J. Biol. Chem.* 273:20837–20846.

129. Trick, W. E., S. K. Fridkin, J. R. Edwards, R. A. Hajjeh, and R. P. Gaynes. 2002. Secular trend of hospital-acquired candidemia among intensive care unit patients in the United States during 1989–1999. *Clin. Infect. Dis.* 35:627–630.

130. Tsong, A. E., M. G. Miller, R. M. Raisner, and A. D. Johnson. 2003. Evolution of a combinatorial transcrip-tional circuit: a case study in yeasts. *Cell* 115:389–399.

131. Tsuchimori, N., L. L. Sharkey, W. A. Fonzi, S. W. French, J. E. Edwards, and S. G. Filler. 2000. Reduced virulence of *HWP1*-deficient mutants of *Candida*

albicans and their interactions with host cells. *Infect. Immun.* **68**:1997–2002.

132. **Tzung, K. W., R. M. Williams, S. Scherer, N. Federspiel, T. Jones, N. Hansen, V. Bivolarevic, L. Huizar, C. Komp, R. Surzycki, R. Tamse, R. W. Davis, and N. Agabian.** 2001. Genomic evidence for a complete sexual cycle in *Candida albicans*. *Proc. Nat. Acad. Sci. USA* **98**:3249–3253.

133. **Uhl, M. A., and A. D. Johnson.** 2001. Development of *Streptococcus thermophilus lacZ* as a reporter gene for *Candida albicans*. *Microbiology* **147**:1189–1195.

134. **Wach, A., A. Brachat, R. Pohlmann, and P. Philippsen.** 1994. New heterologous modules for classical or PCR-based gene disruptions in *Saccharomyces cerevisiae*. *Yeast* **10**:1793–1808.

135. **Wilson, R. B., D. Davis, B. M. Enloe, and A. P. Mitchell.** 2000. A recyclable *Candida albicans* URA3 cassette for PCR product-directed gene disruptions. *Yeast* **16**:65–70.

136. **Wilson, R. B., D. Davis, and A. P. Mitchell.** 1999. Rapid hypothesis testing with *Candida albicans* through gene disruption with short homology regions. *J. Bacteriol.* **181**:1868–1874.

137. **Wisplinghoff, H., H. Seifert, R. P. Wenzel, and M. B. Edmond.** 2003. Current trends in the epidemiology of nosocomial bloodstream infections in patients with hematological malignancies and solid neoplasms in hospitals in the United States. *Clin. Infect. Dis.* **36**:1103–1110.

138. **Wysong, D. R., L. Christin, A. M. Sugar, P. W. Robbins, and R. D. Diamond.** 1998. Cloning and sequencing of a *Candida albicans* catalase gene and effects of disruption of this gene. *Infect. Immun.* **66**:1953–1961.

139. **Yesland, K., and W. A. Fonzi.** 2000. Allele-specific gene targeting in *Candida albicans* results from heterology between alleles. *Microbiology* **146**:2097–2104.

140. **Yin, Z. K., D. Stead, L. Selway, J. Walker, I. Riba-Garcia, T. McInerney, S. Gaskell, S. G. Oliver, P. Cash, and A. J. P. Brown.** 2004. Proteomic response to amino acid starvation in *Candida albicans* and *Saccharomyces cerevisiae*. *Proteomics* **4**:2425–2436.

141. **Zhao, X. J., G. E. McElhaney-Feser, W. H. Bowen, M. F. Cole, S. E. Broedel, and R. L. Cihlar.** 1996. Requirement for the *Candida albicans* fas2 gene for infection in a rat model of oropharyngeal candidiasis. *Microbiology* **142**:2509–2514.

Molecular Principles of Fungal Pathogenesis
Edited by Joseph Heitman et al.
©2006 ASM Press, Washington, D.C.

Chapter 23

Molecular Determinants of *Histoplasma capsulatum* Pathogenesis

JON P. WOODS

Interaction of the fungal pathogen *Histoplasma capsulatum* with the mammalian host can result in a wide range of clinical presentations, from asymptomatic through fatal infection, from active and acute through persistent and subclinical syndromes, and from localized pulmonary through extensively disseminated disease. The organism was originally found by Samuel Darling in histopathologic specimens about a century ago, hence the term "Darling's disease" for histoplasmosis. It was named for its localization in histiocytes (an archaic term for macrophages), its resemblance to protozoal malaria parasites (plasmodia), and the apparent presence of a capsule. In fact, only one of these three descriptors was accurate: *H. capsulatum* does occupy intracellular compartments in macrophages, but it is a fungus that is not encapsulated. Based on several aspects of biology, infection, and pathogenesis, *H. capsulatum* and histoplasmosis are fungal equivalents of the bacterium *Mycobacterium tuberculosis* and tuberculosis (14, 16, 18, 66, 84, 85).

Both organisms are facultative intracellular pathogens of macrophages that must modulate or survive the harsh environment in this professionally antimicrobial host cell. Infection by both organisms occurs by the respiratory route via inhalation of particles from the environment, can be very common, and establishes an initial pulmonary infection but can readily disseminate throughout the mononuclear phagocytic system (e.g., liver, spleen, and bone marrow). Active disease or persistent subclinical infection can occur, with containment of the infection resulting from the development of cell-mediated immune responses, particularly antigen-specific CD4[+] T lymphocyte-mediated activation of macrophages. Such cell-mediated immunity can be detected by skin testing with microbial antigens reflecting delayed-type

hypersensitivity as a marker of past exposure. Even when active infection is contained by the host, sterile clearance typically does not occur. Instead, incompletely understood persistent subclinical infection is established in which the microbe is relatively quiescent and/or microbial replication is balanced by host clearance. Reactivation disease can occur if the host becomes immunocompromised, such as by human immunodeficiency virus infection, cancer, or immunosuppressive therapy in transplant recipients or for inflammatory syndromes (14, 16, 18, 66, 84, 85).

An important distinction is the source of infection. *M. tuberculosis* is an obligate human pathogen that is highly infectious in respiratory secretions and is typically contagious, or acquired from other infected humans. In contrast, histoplasmosis is not contagious except in very rare cases of transmission via transplantation of infected organs, a feature shared with the majority of fungal diseases. Instead, *H. capsulatum* is acquired from the soil environment, in which it displays a normally saprobic rather than parasitic lifestyle. This process involves a dimorphic switch from the infectious but nonpathogenic mold morphotype to a host-adapted, pathogenic yeast morphotype, a phenomenon that is discussed in more detail below. In the soil, *H. capsulatum* is a successful denizen of a complex, competitive, polymicrobial ecosystem containing a plethora of viral, bacterial, protozoal, and other fungal microbes (14, 16, 18, 66, 84, 85). Although it infects other eukaryotic microbes such as amoebas in the laboratory (75), this fungus does not have an obligatory interaction with mammalian hosts as a requisite part of a life cycle, and it is likely that the majority of the planetary biomass of this fungus never interacts with a mammalian host. Therefore, the ability of *H. capsulatum*

Jon P. Woods • Department of Medical Microbiology and Immunology, University of Wisconsin—Madison, Madison, WI 53706.

Table 1. Specific components implicated in *H. capsulatum* pathogenesis

Component	Cellular localization	Function
Calcium-binding protein encoded by *CBP1*	Extracellular	Binds calcium; calcium acquisition or signaling role?
Yps3 encoded by *YPS3*	Cell surface, extracellular	EGF homolog; binds chitin; adhesion, signaling, or immunomodulatory role?
HSP60 encoded by *HIS62*	Cell surface, intracellular	Mitochondrial chaperone; adhesion to host cells; T-lymphocyte and macrophage stimulation
Siderophores	Extracellular	Iron acquisition
Ferric reductive activities	Cell surface, extracellular	Iron acquisition
Hemin-binding activity	Cell surface	Iron acquisition
α-1,3-Glucan resulting from *AGS1*	Cell surface	Affects colony morphology, broth growth, and virulence

to be a successful human pathogen and its considerable array of abilities that are well suited for pathogenesis provide a fascinating evolutionary issue. Clearly, *H. capsulatum* is active in promoting its survival in the host: while live yeasts are virulent in animal and cell culture models of infection, killing, weakening by antibiotic treatment, or inhibition of protein synthesis renders the fungus susceptible to degradation and clearance by host cells or animals (61). This chapter focuses on the pathogenic biology of *H. capsulatum*, including a selection of both mechanisms and specific components; the latter all involve products secreted extracellularly and/or found on the cell surface (Table 1). Other chapters in this volume cover host responses to the organism and recent developments in genomic knowledge and approaches to its study.

CLASSIFICATION

H. capsulatum is an ascomycete that is phylogenetically very closely related to other systemic dimorphic fungal pathogens, particularly *Blastomyces dermatitidis*, to which it has near identity in several genes including ribosomal DNA (rDNA) and chitin synthase genes as well as virtually identical total G+C content (5, 7). This fungus mates; *H. capsulatum* is the anamorphic classification, while the teleomorph or perfect form is known as *Ajellomyces capsulatus*. Sex occurs in the mold form of the fungus and is heterothallic, involving two mating types (47–49). Many phenotypic and genotypic typing approaches have been used to differentiate *H. capsulatum* strains. Three variants have been described based on host predilection, clinical presentation, geographic distribution, or serology: variant *duboisii* in Africa, variant *farciminosum* in horses, and variant *capsulatum* for the majority of isolates (14, 38, 66, 84, 85). Restriction fragment length polymorphism (RFLP) analyses of mitochondrial DNA and several nuclear genes, particularly rDNA, allowed division of the

species into several classes which correspond roughly to geographic distribution or some clinical or biological features. RFLP class 1 strains have been isolated from AIDS patients or other severely immunocompromised individuals, are relatively thermosensitive, and show relatively low virulence in infection models. Most North American clinical isolates are in RFLP class 2, which show high relative thermotolerance and virulence. Most Central and South American isolates are in RFLP class 3, which show intermediate relative thermotolerance and virulence (40, 73, 74, 82). Recent molecular phylogenetic examination of population structure in a large number of clinical and environmental isolates has led to a proposal that *H. capsulatum* be reclassified as at least seven separate species (38).

ENVIRONMENTAL NICHE

Based mainly on the occurrence of clinical disease or evidence of exposure in people rather than isolation from the environment, *H. capsulatum* is thought to have a worldwide distribution but is particularly associated with the Mississippi and Ohio River valleys, i.e., the U.S. Midwest and South, where the exposure prevalence is very high. The mold form of the fungus thrives in moist rich soil in temperate climates and is particularly closely associated with soil containing bird or bat guano. The association with bird feces is shared with *Cryptococcus neoformans*, perhaps representing a common theme in medical mycology, although cryptococci are more frequently associated with pigeon droppings and *H. capsulatum* is associated with starling guano. Due to the frequent residency of bats in caves, spelunkers are at particular risk for acquiring histoplasmosis (14, 66, 84, 85).

DIMORPHIC SWITCH

In the soil, *H. capsulatum* exists as multinucleate branched hyphae or mycelia giving rise to two types of spores: tuberculate macroconidia 8 to 14 μm in diameter and microconidia 2 to 5 μm in diameter. Biological differences between the large and small spores have not been extensively characterized, and any adaptive advantage for making two kinds of asexual spores has not been determined. Although macroconidia and even mycelial fragments are potentially infectious, it is likely that microconidia are the major infectious propagule due to their small size, making them readily aerosolizable and able to penetrate to the small airways of mammalian lungs by inhalation. This saprobic, infectious, but nonpathogenic morphotype can also be grown in the laboratory at room temperature. In the host or in laboratory culture at 37°C, *H. capsulatum* exists as a uninucleate oval budding yeast, 2 to 4 μm in diameter, with a narrow bud neck (66, 84, 85). Once inhaled, conidia could give rise to yeasts prior to entry of the yeasts into host cells or conidia could be phagocytosed and then undergo the dimorphic transition inside host cells (57). A temperature shift is sufficient for the morphogenic transition between mycelia and yeasts, but other environmental stimuli are involved and may override the thermal signal. Sulfhydryl compounds and cyclic AMP (cAMP) levels are two such identified contributing factors (35, 54, 69). cAMP levels are about fivefold higher in mycelia than in yeast, and cAMP or cAMP agonists induce the morphogenic switch from yeast to mycelia even at 37°C. Additionally, whereas the mycelium-yeast conversion requires prolonged exposure to new environmental conditions in laboratory culture, the transition occurs more rapidly in the host, possibly due to the contribution of multiple environmental signals that have been only partially characterized.

Regulatory pathways governing the dimorphic switch have not been identified, although the recent determination of *H. capsulatum* genome sequences and use of microarray approaches hold promise for identifying such transcriptional circuits (37). Genes governing or essential for the dimorphic transition have not been definitively identified. Despite this lack of specific mechanistic knowledge, a requirement for the dimorphic transition in *H. capsulatum* virulence has been demonstrated. Mycelia treated with the sulfhydryl blocking agent *p*-chloromercuriphenylsulfonic acid (PCMS) are heritably locked in the mycelial morphotype and are unable to convert to yeasts, and these cells are unable to cause productive animal infection (55, 56). The mechanism of the effect of this chemical treatment is unknown, but these results provide supportive evidence for the importance of morphogenesis in *H. capsulatum* pathogenesis. These results have been the principal basis for considering that the switch from an environmental mold form to a host-adapted form,

usually yeast, is critical for pathogenesis of all dimorphic fungi causing systemic disease, also including *B. dermatitidis*, *Coccidioides immitis* and *C. posadasii*, *Paracoccidioides brasiliensis*, and *Sporothrix schenckii*. However, without further understanding of the PCMS effect on a molecular level, it cannot be excluded that another feature involved in pathogenesis is responsible and the effect on morphogenesis is incidental.

MORPHOTYPE-SPECIFIC GENE EXPRESSION

Given that the dimorphic switch is such a central part of *H. capsulatum* biology and pathogenesis, much attention has been focused on the identification of genes with differential expression related to this process. Transcriptional profiling is another approach to define these regulatory circuits. One rationale for this quest is that cataloging such genes may contribute to identifying underlying mechanisms for the switch. Another rationale is that yeast-specific genes may contribute to virulence since this morphotype is the pathogenic one adapted to the host, and the encoded proteins may provide new vaccine or antibiotic targets. A few morphotype-specific genes have been identified that show undetectable or very low expression in the other morphotype. Two yeast-specific genes, *CBP1* and *YPS3*, are discussed individually below. One mold-specific gene, *MS8*, has been subjected to gene disruption and to forced misexpression in yeast, both of which resulted in growth alterations. The *ms8* null mutant was able to grow as mycelia but showed aberrant morphology in this form (79). A larger number of *H. capsulatum* genes have shown more modest differential expression between the two morphotypes or altered expression during the dimorphic switch. A calmodulin gene (24) and a homolog of *CDC2* which encodes a regulatory protein kinase (19) were expressed more highly in yeast, whereas both α- and β-tubulin genes (31) were expressed more highly in mycelia, and an actin gene showed a transient decrease in expression during conversion of mycelia to yeast (25). Recent use of a microarray approach resulted in identification of a considerably larger number of genes with relatively greater yeast or mycelial expression as well as quantitation of the differences in expression (37). Because the dimorphic switch involves many metabolic and biochemical as well as structural changes, it is not surprising to find alteration in expression for genes involved in "housekeeping," cell cycle, and signaling functions.

The *CBP1* Gene

H. capsulatum was shown to release a small calcium-binding protein into culture medium (4).

This protein was purified, and the encoding gene, *CBP1*, was cloned and subsequently shown to display yeast-specific expression (3). Green fluorescent protein reporter fusions were used to confirm yeast-specific expression and demonstrate expression during infection (46). Presumptive promoter region truncations fused to a *lacZ* reporter were used to delineate upstream regulatory regions important in the transcriptional control of expression (64). The *CBP1* gene was disrupted by allelic replacement, and the *cbp1* null mutant was complemented by resupply of the wild-type gene. These strains were used to fulfill Koch's molecular postulates, demonstrating an essential role for *CBP1* in cell culture and animal models of infection (72). Interestingly, yeasts, which express the calcium-binding protein, were shown to be more resistant to EGTA-mediated calcium limitation than were mycelia, which do not express the protein (4). Testing of null mutant and complemented yeast strains confirmed the linkage of the EGTA resistance phenotype with functional *CBP1* gene expression rather than with the yeast morphotype (72). This result raises the possibility that the calcium-binding protein plays a role in calcium acquisition akin to the function of siderophores in iron acquisition, which is discussed more below. However, an alternate or additional role of the calcium-binding protein in influencing calcium-mediated signaling processes of fungal or host cells has not been excluded.

The *YPS3* Gene

The *YPS3* gene was originally identified by a differential hybridization approach to the isolation of genes expressed by *H. capsulatum* yeasts and not mycelia. This yeast phase-specific expression is regulated at the transcriptional level (41). Southwestern blotting has been used to identify morphotype-specific nuclear proteins that bind the upstream presumptive promoter region of the *YPS3* locus. In particular, the protein p30M expressed specifically by *H. capsulatum* mycelia has been proposed as a candidate repressor of *YPS3* expression (1). The *YPS3* gene shows interesting interstrain expression differences. RFLP class 2 strains initiate *YPS3* transcription relatively soon after a temperature change stimulating the dimorphic switch from mycelial to yeast growth, and continue to express the gene while growing as yeasts at 37°C. RFLP class 3 strains are slower to initiate *YPS3* transcription during the dimorphic transition, and this expression is transient, such that these strains do not transcribe the gene while growing stably as yeasts. RFLP class 1 strains do not express the *YPS3* gene at all, although they possess the genetic locus (6, 41). Thus, *YPS3* gene expression is roughly correlated

with relatively greater thermotolerance and virulence associated with RFLP class differences. Obviously this observation is not definitive or mechanistic, and the function and potential role of *YPS3* in either of these phenotypes have not been established.

The encoded Yps3 protein is approximately 20 kDa in size and is secreted extracellularly as well as localized to the cell wall and exposed on the cell surface (6, 39, 83). At least one mechanism for reaching this location is specific binding of released Yps3 back to the cell surface via interaction with the cell wall carbohydrate polymer chitin (6). The protein shows homology to epidermal growth factor (EGF) motifs from a variety of organisms across a broad phylogenetic spectrum (6). EGF-like proteins frequently function in adhesion between cells or to extracellular matrix components, or in intercellular signaling (2, 15).

The relationship of the *H. capsulatum YPS3* gene and encoded Yps3 protein to the *B. dermatitidis BAD1* gene and encoded Bad1 protein previously known as the WI-1 antigen (32, 42) is particularly intriguing. As mentioned above, the dimorphic systemic pathogenic fungi *H. capsulatum* and *B. dermatitidis* are phylogenetically very closely related, although they show important differences in a variety of biological, pathogenic, and clinical features. Both *YPS3* and *BAD1* show yeast-specific expression controlled at a transcriptional level in their respective fungi (3, 46, 64, 67, 68). Their upstream presumptive promoter regions share several specific sequence motifs, and heterologous promoter-reporter fusion experiments were used to demonstrate conservation of a *cis*-acting mechanism for yeast-specific expression in these two fungi, although *trans*-acting regulatory components have not been identified (1, 67, 68).

In addition to this genetic regulatory similarity, the Yps3 and Bad1 proteins have some interesting shared and distinct features. The predicted 20-kDa Yps3 protein has a short N-terminal hydrophobic domain consistent with a signal sequence, and the rest of the protein has EGF homology (6). The predicted Bad1 protein is 120 kDa in size and consists of three domains: a short N-terminal hydrophobic domain consistent with a signal sequence, a C-terminal domain also with EGF homology, and, between them, the majority of this large protein, consisting of tandem repeats of a short amino acid sequence homologous to the invasin protein of the bacterial pathogen *Yersinia pestis* (9, 32, 42). The N- and C-terminal domains of Bad1 align closely with the entirety of Yps3, which is lacking any region similar to the internal invasin-like repeat region constituting most of Bad1. The conservation of upstream promoter sequences and N- and C-terminal domain-coding regions is consistent with allelism between the *YPS3* and *BAD1* loci (Fig. 1). Perhaps the Bad1 internal invasin-like region was lost or, alternatively, newly introduced into an ancestral Yps3 progenitor. In addition to these genetic observations, there is functional similarity between Yps3 and Bad1 in that both are released extracellularly and bind back to yeast cell surface chitin (6–9). For Bad1, this binding is mediated by the C-terminal EGF-like domain (9), consistent with conservation of this effect with Yps3.

B. dermatitidis BAD1 has been characterized much more extensively than *H. capsulatum YPS3* by using purified protein and construction of null mutant and complemented strains to fulfill molecular Koch's postulates for linkage to phenotypic effects. Bad1 is an adhesin that mediates yeast binding to host cell CR3 and CD14 (58); it is an immunosuppressive molecule that represses host tumor necrosis factor alpha production by both transforming growth factor β-dependent and -independent mechanisms (26), and it is an essential virulence determinant for animal infection (10). These observations may provide clues to functions and effects of Yps3, but it should be noted that the adhesin and immunosuppressive properties of Bad1 are related to the major central invasin-like domain, which is absent from Yps3. The only function demonstrated for the C-terminal EGF-like domain of Bad1 that is shared with and corresponds to most of Yps3 is the ability to bind to cell surface chitin and thus achieve surface localization (6, 9).

ENTRY INTO HOST CELLS AND ESTABLISHMENT OF INTRACELLULAR PARASITISM

Before (as conidia) or after (as yeast) the dimorphic conversion in the host, *H. capsulatum* cells may enter host cells by different routes, which may influence their subsequent intracellular fate and the outcome of the interaction. These phenomena are particularly important for an organism such as *H. capsulatum*, an intracellular pathogen of professionally antimicrobial phagocytic cells which may either serve as a safe harbor or a hostile environment. *H. capsulatum* cells can be coated with complement components, antibody, or complement bound to antibody and internalized via traditional host-mediated opsonophagocytosis by using complement or Fc receptors. Alternately, specific fungal surface components (adhesins) can bind to specific host surface molecules (receptors). The latter mechanism may be a microbial approach to determining the intracellular compartment in which the fungus resides and avoiding host antimicrobial defenses such as an oxidative burst and acidification.

The best known host cell for *H. capsulatum* intracellular pathogenesis is the macrophage, in which it first resides in a membrane-bound phagosome regardless of its entry mechanism. Depending on experimental conditions such as host cell type and activation status, this phagosome may undergo fusion with host lysosomes or persist as a modified phagosome without phagolysosomal fusion (20, 60, 76, 78). *H. capsulatum* yeasts and microconidia can bind to macrophages via the CD18 or β_2-integrin family of mammalian surface glycoprotein receptors (11, 57). These heterodimeric molecules pair different α chains with the same β_2 molecule: αL or CD11a to yield LFA-1, αM or CD11b to yield CR3 or Mac-1, and αX or CD11c to yield CR4 or P150, 95. The fungal surface adhesin responsible for binding host β_2 integrin receptors is the heat shock protein HSP60 (53), which is discussed in more detail below. Macrophages frequently provide a hospitable environment for *H. capsulatum*. In contrast, *H. capsulatum* yeasts bind a different receptor,

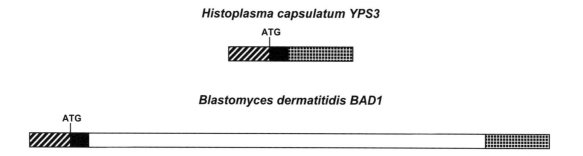

Figure 1. Comparison of *H. capsulatum YPS3* and *B. dermatitidis BAD1* genetic loci. Symbols: hatched bar, promoter region with conserved regulatory sequence motifs; solid bar, region encoding N-terminal hydrophobic signal sequence domain; white bar, region encoding tandem invasin-like repeat domains; cross-hatched bar, region encoding the C-terminal EGF-like domain.

VLA-5 or the fibronectin receptor, on the surface of host dendritic cells, which provide a more hostile, destructive environment for intracellular *H. capsulatum* (28).

The HSP60 Protein

The identification of HSP60, encoded by the *HIS62* gene, as the adhesin for binding host cell surface β_2-integrins (53) is interesting on several levels. One is its cellular localization. Members of the eukaryotic HSP60 family are typically mitochondrial chaperones, and presumably *H. capsulatum* HSP60 plays this role and has this intracellular localization, although these features have not been experimentally reported. Its additional presence and function on the cell surface are not unique among heat shock proteins but are distinctive, and its secretion mechanism has not been characterized. *H. capsulatum* HSP60 is also interesting for its biological effects and functions. This protein was originally identified as a highly immunogenic antigen and in fact was the first *H. capsulatum* protein shown to serve as a protective antigen. Vaccination with HSP60 protected mice from lethal and sublethal histoplasmosis, and this effect correlated with its ability to stimulate host CD4$^+$ T lymphocytes (17, 29, 30). Additionally, *H. capsulatum* HSP60 was shown to directly stimulate macrophages, which could also contribute to its vaccine efficacy by virtue of an adjuvant effect (17). The addition of an adhesin role in binding host cell surface β_2-integrins makes for more complex implications. This protein has functions and effects that could facilitate *H. capsulatum* pathogenesis as well as providing benefits for the host in combating infection.

MICROENVIRONMENTAL pH MODULATION

The host endosomal maturation or phagolysosome fusion process normally involves acidification via a compartment membrane proton pump/ATPase. *H. capsulatum* modulates the pH of its intracellular microenvironment, maintaining a more neutral condition. Intracellular compartments containing control zymosan particles or dead *H. capsulatum* yeasts were at a considerably lower pH than those containing viable *H. capsulatum* yeasts (21). This outcome is predicted to be adaptive for *H. capsulatum* survival by avoiding the direct deleterious effects of acid as well as the acidic pH optima of host hydrolytic enzymes necessary for degradation of ingested particles. The mechanism for this pH modulation by *H. capsulatum* is not known but could involve reversal of the acidification by secretion of neutralizing compounds such as polyamines or inhibition of the acidification via modification of the compartment by relative exclusion of the host membrane proton pump. The latter mechanism is utilized by *M. tuberculosis* (77), and there is evidence for its occurring with *H. capsulatum* under some experimental conditions (60, 76). However, an acidic pH could be important for *H. capsulatum* iron acquisition, which is discussed further below, leading to a proposal that *H. capsulatum* actively acidifies its intracellular microenvironment (76).

IRON ACQUISITION

A ubiquitous innate host defense against microbial invaders in general and *H. capsulatum* in particular is limitation of the essential nutrient iron. Iron acquisition is a process lying at the competitive interface between the host and the pathogen. Virtually all living organisms including microbes require iron for growth, generally due to its participation in essential redox reactions related to its existence in reduced Fe^{2+} and oxidized Fe^{3+} states. There is very little free soluble iron in an animal; its concentration is below the minimum requirement for viability of most microbes. Mammalian host iron metabolism includes storage or transport bound to constitutively expressed proteins such as transferrin that may make it inaccessible to microbes. Infection or other inflammatory stimuli further induce iron sequestration as part of an acute-phase response. For a pathogen to be successful, it must have countermechanisms for acquiring iron in the host microenvironment in which it exists. Typically, these microbial iron acquisition mechanisms are specifically induced or upregulated when the cell senses iron limitation. This precise regulation is necessary because iron, as well as being essential, can be toxic in excess due to the generation of free radicals that damage membranes, nucleic acids, and enzymes (34, 84, 85).

Iron limitation by any of several approaches restricts *H. capsulatum* growth in laboratory medium (50, 59, 62, 80, 81). Although *H. capsulatum* iron acquisition mutants have not been isolated for testing in infection models, there are several lines of evidence supporting the use of iron limitation as a host defense against *H. capsulatum* infection and the essential nature of iron acquisition for *H. capsulatum* pathogenesis. Several laboratories have shown that *H. capsulatum* yeasts replicate within nonactivated mouse peritoneal or human monocyte-derived macrophages in cell culture (22, 23, 50, 51, 59–62, 76). The iron chelator deferrioxamine suppresses *H. capsulatum* growth in these host cells. This deferrioxamine-mediated intracellular growth inhibition is reversed by iron-laden holotransferrin but not by iron-free apotransferrin (50, 59, 62). This experimental approach is somewhat artificial since deferrioxamine is not normally present during infection and would be unlikely to be used as an antimicrobial agent. However, similar results were found in examining natural cytokine-mediated host responses to *H. capsulatum*. Gamma interferon-activated mouse macrophages inhibit intracellular *H. capsulatum* growth. Similar to the situation with deferrioxamine, this gamma interferon-mediated intracellular growth inhibition is reversed by iron-laden holotransferrin but not by iron-free apotransferrin (50, 51, 59, 62). Thus, iron limitation appears to be one of the host defense mechanisms used against *H. capsulatum* during natural infection that must be overcome by the fungus in order for it to be virulent.

No *H. capsulatum* iron acquisition genes or regulatory genes involved in sensing and responding to environmental iron availability have been characterized. However, a multitude of iron acquisition approaches, i.e., components or activities potentially contributing to iron acquisition, have been identified in this fungus. These strategies fall into four general areas: siderophores, acidic pH, reductive activities, and receptors for host iron-binding compounds (34, 84, 85) (Fig. 2). None of these approaches is unique to *H. capsulatum* since there are other microbial examples for each; however, the coexistence of such a plethora in a single microbe is distinctive. These mechanisms could operate redundantly to satisfy the essential iron requirement, could be used separately in different environments, or could be coexpressed and interact cooperatively in particular situations.

Secretion of siderophores was first reported for *H. capsulatum* over 20 years ago (12, 13). This mechanism is used by many microbes and entails the release of small chelating compounds that bind iron, most frequently the oxidized Fe^{3+} form, and make the scavenged iron available to the microbe. It is useful in polymicrobial aerobic environments such as soil, in which free bioavailable iron is scarce although total iron is abundant since oxidized iron forms complexes that are insoluble in aqueous solution at neutral pH. However, siderophores can also compete with iron-binding host molecules during infection and have been shown to liberate transferrin-bound iron (81). *H. capsulatum* secretes several siderophore types of the hydroxamate class under iron-limited conditions in vitro, including dimerum acid, coprogen B, and fusarinine (12, 13, 36). Of course, transfer of iron from transferrin to siderophores simply substitutes one extracellular iron-binding molecule for another. Some additional mechanism(s) must mediate the transport of iron into the *H. capsulatum* cell. Some microbes express siderophore receptors on the cell surface which bind the iron-laden siderophore, but such

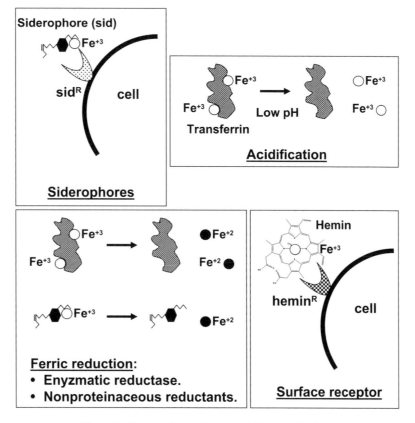

Figure 2. *H. capsulatum* iron acquisition mechanisms.

molecules have not yet been identified in *H. capsulatum*. Ferrisiderophore binding may be followed by internalization of the entire complex and removal of iron within the cell or by reductive iron removal (discussed more below).

The residence of *H. capsulatum* in an intracellular compartment at acidic pH provides a second opportunity for iron acquisition from transferrin, either passively exploiting a host phagolysosome acidification process or perhaps actively participating to lower the pH of a modified phagosome. Each transferrin molecule has the capacity to bind two molecules of Fe^{3+} iron at neutral pH, resulting in holotransferrin at 100% iron saturation. One Fe^{3+} is released at slightly acidic pH (ca. 6 to 6.5), and the second is released at lower pH (ca. 5 to 5.5). Thus, acidification of a phagolysosome or phagosome containing transferrin as well as *H. capsulatum* could cause iron to be released and made available to the fungus. This mechanism is the normal route used by the host for cellular iron metabolism. Mammalian cells express surface transferrin receptors that bind the iron-laden protein to form a complex which is then internalized by endocytosis. Endosomal acidification results in release of the bound ferric iron for intracellular utilization and recycling of the transferrin receptors back to the surface (84, 85). Experimental evidence supporting the exploitation of this mechanism by *H. capsulatum* to acquire iron is provided by the activity of the weakly alkalinizing lysosomotropic agent chloroquine against the fungus during cell culture or animal infection (59).

A third general mechanism for iron acquisition is reduction of Fe^{3+} iron to Fe^{2+} iron. Since transferrin and most microbial siderophores, including those made by *H. capsulatum*, specifically bind Fe^{3+} iron, reduction results in the release of iron from these molecules as well as operating on inorganic Fe^{3+} iron salts. *H. capsulatum* expresses at least three distinct ferric reducing activities: an extracellular ferric reductase enzyme that utilizes reduced glutathione as a cofactor, extracellular low-molecular-weight nonproteinaceous ferric reductants, and cell surface ferric reducing activity (80). Such components have been found in other fungi; e.g., cell membrane ferric reductases in *Saccharomyces cerevisiae* (52) and extracellular ferric reductants in *C. neoformans* (63), but the variety expressed by *H. capsulatum* is distinctive. All of these activities were expressed under iron-replete conditions but upregulated by iron limitation in vitro (80). The *H. capsulatum* ferric reductase could use several host and soil iron compounds as substrates for enzymatic activity, including holotransferrin, hemin, and siderophores of the type produced by *H. capsulatum* or xenosiderophores produced by other microbes (81). In the host during

infection, this reductive approach could thus exploit transferrin-bound iron directly or interact with other *H. capsulatum* iron acquisition mechanisms, including ferric iron bound to *H. capsulatum* siderophores and ferric iron released from transferrin at acidic pH.

Finally, *H. capsulatum* cell surface binding of the host iron-containing compound hemin has been reported (27), although a specific receptor has not been identified and subsequent steps such as removal of iron at the surface versus internalization of the entire complex have not been characterized.

ROUGH-SMOOTH PHENOTYPIC VARIATION, α-1,3-GLUCAN, AND *AGS1*

Some but not all *H. capsulatum* strains undergo spontaneous variation first identified in colony morphology but also affecting several other important phenotypes. In general, RFLP class 1 and 3 strains show this variation while RFLP class 2 strains do not. In all cases of this variation that have been studied, clinical isolates show a rough colony morphology on solid medium and grow as large clumps of cells in broth culture. Spontaneously arising variants exhibit a smooth colony morphology and dispersed broth growth (22, 23, 43, 45). The one demonstrated biochemical difference between the rough and smooth variants is the presence of the cell wall carbohydrate polymer α-1,3-glucan in rough-variant cells and its absence in smooth-variant cells (44). Importantly, rough variants are relatively more virulent in animal and macrophage infection models than are smooth variants of the same strain (22, 23, 43). This phenomenon is shared with two other dimorphic systemic fungal pathogens, *B. dermatitidis* (33) and *P. brasiliensis* (70, 71), which also have been reported to show spontaneous or environmentally induced variation in cell wall α-1,3-glucan correlated with a change in relative virulence in infection models.

The mechanisms of changes in expression and the phenotypic effects have not been determined for these variants. The observations that *H. capsulatum* smooth variants are relatively more virulent than their rough parents in an epithelial cell line and that there was apparent induction or selection of smooth variants after infection of this cell line with rough-variant cells led to speculation that this variation may be involved with exploitation of different host cell reservoirs by the fungus during active or persistent infection (22, 23, 45). However, although cell wall α-1,3-glucan is associated with virulence in strains which express it, RFLP class 2 strains do not express this polymer and are phenotypically "smooth," yet fully virulent in animals, macrophages, and epithelial cells (22, 23). Therefore, the "rough" phenotype and α-1,3-glucan are not universally necessary for

H. capsulatum pathogenesis. These *H. capsulatum* studies, as well as those with *B. dermatitidis* and *P. brasiliensis*, used nonisogenic strains, including spontaneous or environmentally induced variants that were not constructed by molecular genetic manipulation of specific genes. More recent work has identified the biosynthetic gene *AGS1* encoding α-1,3-glucan synthase. In a rough, RFLP class 3 strain, disruption of this gene by allelic replacement to construct an *ags1* null mutant and suppression of *AGS1* expression by RNA interference both resulted in loss of α-1,3-glucan expression and diminution of virulence (65). These results provided the first direct linkage, on a molecular level, among a gene, expression of this cell wall carbohydrate polymer, and virulence. However, the mechanism for the effect of α-1,3-glucan on pathogenesis has not been demonstrated; it is also not known whether the natural variation in α-1,3-glucan expression shown in systemic dimorphic fungal pathogens involves the *AGS1* gene or other as yet unidentified mechanisms.

CONCLUSION

H. capsulatum is a biologically interesting inhabitant of soil and mammalian hosts, a clinically significant cause of respiratory and systemic infection, and an excellent fungal model of dimorphic cell development and facultative intracellular pathogenesis. It is well suited to parasitize mammalian hosts, and a plethora of properties and activities consistent with this ability have been identified, although generally the genes responsible for expressing or regulating these features have not been identified and mutants have not been isolated. Conversely, several genes with important and interesting characteristics have been studied, but generally the function of these genes has not been entirely elucidated. There are thus both compelling motivations and ripe opportunities for continuing studies with new tools such as molecular genetic and genomic approaches.

REFERENCES

1. Abidi, F. E., H. Roh, and E. J. Keath. 1998. Identification and characterization of a phase-specific, nuclear DNA binding protein from the dimorphic pathogenic fungus *Histoplasma capsulatum*. *Infect. Immun.* **66**:3867–3873.
2. Appella, E., I. T. Weber, and F. Blasi. 1988. Structure and function of epidermal growth factor-like regions in proteins. *FEBS Lett.* **231**:1–4.
3. Batanghari, J. W., G. S. Deepe, Jr., E. D. Cera, and W. E. Goldman. 1998. *Histoplasma* acquisition of calcium and expression of *CBP1* during intracellular parasitism. *Mol. Microbiol.* **27**:531–539.
4. Batanghari, J. W., and W. E. Goldman. 1997. Calcium dependence and binding in cultures of *Histoplasma capsulatum*. *Infect. Immun.* **65**:5257–5261.

5. Bawdon, R. E., R. G. Garrison, and L. R. Fina. 1972. Deoxyribonucleic acid base composition of the yeastlike and mycelial phases of *Histoplasma capsulatum* and *Blastomyces dermatitidis*. *J. Bacteriol.* **111**:593–596.
6. Bohse, M. L, and J. P. Woods. 2005. Surface localization of the Yps3p protein of *Histoplasma capsulatum*. *Eukaryot. Cell* **4**:685–693.
7. Bowman, B. H., J. W. Taylor, and T. J. White. 1992. Molecular evolution of the fungi: human pathogens. *Mol. Biol. Evol.* **9**:893–904.
8. Brandhorst, T. T., and B. S. Klein. 2000. Cell wall biogenesis of *Blastomyces dermatitidis*. *J. Biol. Chem.* **275**: 7925–7934.
9. Brandhorst, T. T., M. Wuthrich, B. Finkel-Jimenez, and B. Klein. 2003. A C-terminal EGF-like domain governs BAD1 localization to the yeast surface and fungal adherence to phagocytes, but is dispensable in immune modulation and pathogenicity of *Blastomyces dermatitidis*. *Mol. Microbiol.* **48**:53–65.
10. Brandhorst, T. T., M. Wüthrich, T. Warner, and B. Klein. 1999. Targeted gene disruption reveals an adhesin indispensable for pathogenicity of *Blastomyces dermatitidis*. *J. Exp. Med.* **189**:1207–1216.
11. Bullock, W. E., and S. D. Wright. 1987. Role of the adherence-promoting receptors, CR3, LFA-1, and p150,95, in binding of *Histoplasma capsulatum* by human macrophages. *J. Exp. Med.* **165**:195–210.
12. Burt, W. R. 1982. Identification of coprogen B and its breakdown products from *Histoplasma capsulatum*. *Infect. Immun.* **35**:990–996.
13. Burt, W. R., A. L. Underwood, and G. L. Appleton. 1981. Hydroxamic acid from *Histoplasma capsulatum* that displays growth factor activity. *Appl. Environ. Microbiol.* **42**:560–563.
14. Cano, M. V. C., and R. A. Hajjeh. 2001. The epidemiology of histoplasmosis: a review. *Semin. Respin. Infect.* **16**:109–118.
15. Davis, C. G. 1990. The many faces of epidermal growth factor repeats. *New Biol.* **2**:410–419.
16. Deepe, G. S., Jr. 2000. Immune response to early and late *Histoplasma capsulatum* infections. *Curr. Opin. Microbiol.* **3**:359–362.
17. Deepe, G. S., Jr., R. Gibbons, G. D. Brunner, and F. J. Gomez. 1996. A protective domain of heat-shock protein 60 from *Histoplasma capsulatum*. *J. Infect. Dis.* **174**: 828–834.
18. Deepe, G. S., Jr., and R. A. Seder. 1998. Molecular and cellular determinants of immunity to *Histoplasma capsulatum*. *Res. Immunol.* **149**:397–406.
19. Di Lallo, G., S. Gargano, and B. Maresca. 1994. The *Histoplasma capsulatum cdc2* gene is transcriptionally regulated during the morphologic transition. *Gene* **140**:51–57.
20. Eissenberg, L. G., and W. E. Goldman. 1988. Fusion of lysosomes with phagosomes containing *Histoplasma capsulatum*: use of fluresceinated dextran. *Adv. Exp. Med. Biol.* **239**:53–61.
21. Eissenberg, L. G., W. E. Goldman, and P. H. Schlesinger. 1993. *Histoplasma capsulatum* modulates the acidification of phagolysosomes. *J. Exp. Med.* **177**:1605–1611.
22. Eissenberg, L. G., S. Poirier, and W. E. Goldman. 1996. Phenotypic variation and persistence of *Histoplasma capsulatum* yeasts in host cells. *Infect. Immun.* **64**: 5310–5314.
23. Eissenberg, L. G., J. L. West, J. P. Woods, and W. E. Goldman. 1991. Infection of P388D1 macrophages and

respiratory epithelial cells by *Histoplasma capsulatum*: selection of avirulent variants and their potential role in persistent histoplasmosis. *Infect. Immun.* 59:1639–1646.

24. el-Rady, J., and G. Shearer, Jr. 1996. Isolation and characterization of a calmodulin-encoding cDNA from the pathogenic fungus *Histoplasma capsulatum*. *J. Med. Vet. Mycol.* 34:163–169.

25. el-Rady, J., and G. Shearer, Jr. 1997. Cloning and analysis of an actin-encoding cDNA from the dimorphic pathogenic fungus *Histoplasma capsulatum*. *J. Med. Vet. Mycol.* 35:159–166.

26. Finkel-Jimenez, B., M. Wuthrich, and B. S. Klein. 2002. BAD1, an essential virulence factor of *Blastomyces dermatitidis*, suppresses host TNF-α production through TGF-β-dependent and -independent mechanisms. *J. Immunol.* 168:5746–5755.

27. Foster, L. A. 2002. Utilization and cell-surface binding of hemin by *Histoplasma capsulatum*. *Can. J. Microbiol.* 48:437–442.

28. Gildea, L. A., R. E. Morris, and S. L. Newman. 2001. *Histoplasma capsulatum* yeasts are phagocytosed via very late antigen-5, killed, and processed for antigen presentation by human dendritic cells. *J. Immunol.* 166:1049–1056.

29. Gomez, F. J., R. Allendoerfer, and G. S. Deepe, Jr. 1995. Vaccination with recombinant heat shock protein 60 from *Histoplasma capsulatum* protects mice against pulmonary histoplasmosis. *Infect. Immun.* 63:2587–2595.

30. Gomez, F. J., A. M. Gomez, and G. S. Deepe, Jr. 1991. Protective efficacy of a 62-kilodalton antigen, HIS-62, from the cell wall and cell membrane of *Histoplasma capsulatum* yeast cells. *Infect. Immun.* 59:4459–4464.

31. Harris, G. S., E. J. Keath, and J. Medoff. 1989. Expression of α- and β-tubulin genes during dimorphic-phase transitions of *Histoplasma capsulatum*. *Mol. Cell. Biol.* 9:2042–2049.

32. Hogan, L. H., S. Josvai, and B. S. Klein. 1995. Genomic cloning, characterization, and functional analysis of the major surface adhesin WI-1 on *Blastomyces dermatitidis* yeasts. *J. Biol. Chem.* 270:30725–30732.

33. Hogan, L. H., and B. S. Klein. 1994. Altered expression of surface α-13-glucan in genetically related strains of *Blastomyces dermatitidis* that differ in virulence. *Infect. Immun.* 62:3543–3546.

34. Howard, D. H. 1999. Acquisition, transport, and storage of iron by pathogenic fungi. *Clin. Microbiol. Rev.* 12:394–404.

35. Howard, D. H., N. Dabrowa, V. Otto, and J. Rhodes. 1980. Cysteine transport and sulfite reductase activity in a germination-defective mutant of *Histoplasma capsulatum*. *J. Bacteriol.* 141:417–421.

36. Howard, D. H., R. Rafie, A. Tiwari, and K. F. Faull. 2000. Hydroxamate siderophores of *Histoplasma capsulatum*. *Infect. Immun.* 68:2338–2343.

37. Hwang, L., D. Hocking-Murray, A. K Bahrami, M. Andersson, J. Rine, and A. Sil. 2003. Identifying phase-specific genes in the fungal pathogen *Histoplasma capsulatum* using a genomic shotgun microarray. *Mol. Biol. Cell* 14:2314–2326.

38. Kasuga, T., T. J. White, G. Koenig, J. McEwen, A. Restrepo, E. Castaneda, C. da Silva Lacaz, E. M. Heins-Vaccari, R. S. de Freitas, R. M. Zancope-Oliveira, Z. Qin, R. Negroni, D. A. Carter, Y. Mikami, M. Tamura, M. L. Taylor, G. F. Miller, N. Poonwan, and J. W. Taylor. 2003. Phylogeography of the fungal pathogen *Histoplasma capsulatum*. *Mol. Ecol.* 12:3383–3401.

39. Keath, E. J., and F. E. Abidi. 1994. Molecular cloning and sequence analysis of *yps-3*, a yeast phase-specific gene in the dimorphic fungal pathogen *Histoplasma capsulatum*. *Microbiology* 140:759–767.

40. Keath, E. J., G. S. Kobayashi, and G. Medoff. 1992. Typing of *Histoplasma capsulatum* by restriction fragment length polymorphisms in a nuclear gene. *J. Clin. Microbiol.* 30:2104–2107.

41. Keath, E. J., A. A. Painter, G. S. Kobayashi, and G. Medoff. 1989. Variable expression of a yeast-phase-specific gene in *Histoplasma capsulatum* strains differing in thermotolerance and virulence. *Infect. Immun.* 57:1384–1390.

42. Klein, B. S. 2000. Molecular basis of pathogenicity in *Blastomyces dermatitidis*: the importance of adhesion. *Curr. Opin. Microbiol.* 3:339–343.

43. Klimpel, K. R., and W. E. Goldman. 1987. Isolation and characterization of spontaneous avirulent variants of *Histoplasma capsulatum*. *Infect. Immun.* 55:528–533.

44. Klimpel, K. R., and W. E. Goldman. 1988. Cell walls from avirulent variants of *Histoplasma capsulatum* lack α-(1, 3)-glucan. *Infect. Immun.* 56:2997–3000.

45. Kügler, S., T. S. Sebghati, L. G. Eissenberg, and W. E. Goldman. 2000. Phenotypic variation and intracellular parasitism by *Histoplasma capsulatum*. *Proc. Natl. Acad. Sci. USA* 97:8794–8798.

46. Kügler, S., B. Young, V. L. Miller, and W. E. Goldman. 2000. Monitoring phase-specific gene expression in *Histoplasma capsulatum* with telomeric GFP fusion plasmids. *Cell. Microbiol.* 2:537–547.

47. Kwon-Chung, K. J. 1972. *Emmonsiella capsulata*: perfect state of *Histoplasma capsulatum*. *Science* 177:368–369.

48. Kwon-Chung, K. J. 1973. Studies on *Emmonsiella capsulata*. I. Heterothallism and development of the ascocarp. *Mycologia* 65:109–121.

49. Kwon-Chung, K., R. J. Weeks, and H. W. Larsh. 1974. Studies on *Emmonsiella capsulata* (*Histoplasma capsulatum*). II. Distribution of the two mating types in 13 endemic states of the United States. *Am. J. Epidemiol.* 99:44–49.

50. Lane, T. E., B. A. Wu-Hsieh, and D. H. Howard. 1991. Iron limitation and the gamma interferon-mediated antihistoplasma state of murine macrophages. *Infect. Immun.* 59:2274–2278.

51. Lane, T. E., B. A. Wu-Hsieh, and D. H. Howard. 1993. Gamma interferon cooperates with lipopolysaccharide to activate mouse splenic macrophages to an antihistoplasma state. *Infect. Immun.* 61:1468–1473.

52. Lesuisse, E., and P. Labbe. 1989. Reductive and nonreductive mechanisms of iron assimilation by the yeast *Saccharomyces cerevisiae*. *J. Gen. Microbiol.* 135:257–263.

53. Long, K. H., F. J. Gomez, R. E. Morris, and S. L. Newman. 2003. Identification of heat shock protein 60 as the ligand on *Histoplasma capsulatum* that mediates binding to CD18 receptors on human macrophages. *J. Immunol.* 170:487–494.

54. Medoff, J., E. Jacobson, and G. Medoff. 1981. Regulation of dimorphism in *Histoplasma capsulatum* by cyclic AMP. *J. Bacteriol.* 145:1452–1455.

55. Medoff, G., G. S. Kobayashi, A. Painter, and S. Travis. 1987. Morphogenesis and pathogenicity of *Histoplasma capsulatum*. *Infect. Immun.* 55:1355–1358.

56. Medoff, G., M. Sacco, B. Maresca, D. Schlessinger, A. Painter, G. S. Kobayashi, and L. Carratu. 1986. Irreversible block of the mycelial-to-yeast phase transition of *Histoplasma capsulatum*. *Science* 231:476–479.

57. Newman, S. L., C. Bucher, J. Rhodes, and W. E. Bullock. 1990. Phagocytosis of *Histoplasma capsulatum* yeasts and microconidia by human cultured macrophages and alveolar macrophages. Cellular cytoskeleton requirement for attachment and ingestion. *J. Clin. Investig.* 85:223–230.

58. Newman, S. L., S. Chaturvedi, and B. S. Klein. 1995. The WI-1 antigen of *Blastomyces dermatitidis* yeasts mediates binding to human macrophage CD11b/CD18 (CR3) and CD14. *J. Immunol.* 154:753–761.

59. Newman, S. L., L. Gootee, G. Brunner, and G. S. Deepe, Jr. 1994. Chloroquine induces human macrophage killing of *Histoplasma capsulatum* by limiting the availability of intracellular iron and is therapeutic in a murine model of histoplasmosis. *J. Clin. Investig.* 93:1422–1429.

60. Newman, S. L., L. Gootee, C. Kidd, G. M. Ciraolo, and R. Morris. 1997. Activation of human macrophage fungistatic activity against *Histoplasma capsulatum* upon adherence to type 1 collagen matrices. *J. Immunol.* 158:1779–1786.

61. Newman, S. L., L. Gootee, R. Morris, and W. E. Bullock. 1992. Digestion of *Histoplasma capsulatum* yeasts by human macrophages. *J. Immunol.* 149:574–580.

62. Newman, S. L., L. Gootee, V. Stroobant, H. van der Goot, and J. R. Boelaert. 1995. Inhibition of growth of *Histoplasma capsulatum* yeast cells in human macrophages by the iron chelator VUF 8514 and comparison of VUF 8514 with deferoxamine. *Antimicrob. Agents Chemother.* 39:1824–1829.

63. Nyhus, K. J., A. T. Wilborn, and E. S. Jacobson. 1997. Ferric iron reduction by *Cryptococcus neoformans*. *Infect. Immun.* 65:434–438.

64. Patel, J. B., J. W. Batanghari, and W. E. Goldman. 1998. Probing the yeast phase-specific expression of the *CBP1* gene of *Histoplasma capsulatum*. *J. Bacteriol.* 180:1786–1792.

65. Rappleye, C. A., J. T. Engle, and W. E. Goldman. 2004. RNA interference in *Histoplasma capsulatum* demonstrates a role for α-(1,3)-glucan in virulence. *Mol. Microbiol.* 53:153–165.

66. Retallack, D. M., and J. P. Woods. 1999. Molecular epidemiology, pathogenesis, and genetics of the dimorphic fungus *Histoplasma capsulatum*. *Microbes Infect.* 1:817–825.

67. Rooney, P. J., and B. S. Klein. 2004. Elements necessary for transcriptional activation of BAD1 in the yeast phase of *Blastomyces dermatitidis*. *Eukaryot. Cell* 3:785–794.

68. Rooney, P. J., T. D. Sullivan, and B. S. Klein. 2001. Selective expression of the virulence factor BAD1 upon morphogenesis to the pathogenic yeast form of *Blastomyces dermatitidis*: evidence for transcriptional regulation by a conserved mechanism. *Mol. Microbiol.* 39:875–889.

69. Sacco, M., B. Maresca, B. V. Kumar, G. S. Kobayashi, and G. Medoff. 1981. Temperature- and cyclic nucleotide-induced phase transitions of *Histoplasma capsulatum*. *J. Bacteriol.* 146:117–120.

70. San-Blas, G., E. San-Blas, E. Ormaechea, and L. E. Serrano. 1977. Cell wall analysis of an adenine requiring mutant of the yeast-like form of *Paracoccidiodes brasiliensis* strain IVIC Pb9. *Sabouraudia* 15:297–303.

71. San-Blas, G., F. San-Blas, and L. E. Serrano. 1977. Host-parasite relationships in the yeastlike form of *Paracoccidiodes brasiliensis* strain IVIC Pb9. *Infect. Immun.* 15:343–346.

72. Sebghati, T. S., J. T. Engle, and W. E. Goldman. 2000. Intracellular parasitism by *Histoplasma capsulatum*: fungal virulence and calcium dependence. *Science* 290:1368–1372.

73. Spitzer, E., B. A. Lasker, S. J. Travis, G. S. Kobayashi, and G. Medoff. 1989. Use of mitochondrial and ribosomal DNA polymorphisms to classify clinical and soil isolates of *Histoplasma capsulatum*. *Infect. Immun.* 57:1409–1412.

74. Steele, P. E., G. F. Carle, G. S. Kobayashi, and G. Medoff. 1989. Electrophoretic analysis of *Histoplasma capsulatum* chromosomal DNA. *Mol. Cell. Biol.* 9:983–987.

75. Steenbergen, J. N., J. D. Nosanchuk, S. D. Malliaris, and A. Casadevall. 2004. Interaction of *Blastomyces dermatitidis*, *Sporothrix schenckii*, and *Histoplasma capsulatum* with *Acanthamoeba castellanii*. *Infect. Immun.* 72:3478–3488.

76. Strasser, J. E., S. L. Newman, G. M. Ciraolo, R. E. Morris, M. L. Howell, and G. E. Dean. 1999. Regulation of the macrophage vacuolar ATPase and phagosome-lysosome fusion by *Histoplasma capsulatum*. *J. Immunol.* 162:6148–6154.

77. Sturgill-Koszycki, S., P. H. Schlesinger, P. Chakraborty, P. L. Haddix, H. L. Collins, A. K. Fok, R. D. Allen, S. L. Gluck, J. Heuser, and D. G. Russell. 1994. Lack of acidification in *Mycobacterium* phagosomes produced by exclusion of the vesicular proton-ATPase. *Science* 263:678–681.

78. Taylor, M. L., M. E. Espinosa-Schoelly, R. Iturbe, B. Rico, J. Casasola, and F. Goodsaid. 1989. Evaluation of phagolysosome fusion in acridine orange stained macrophages infected with *Histoplasma capsulatum*. *Clin. Exp. Immunol.* 75:466–470.

79. Tian, X., and G. Shearer, Jr. 2002. The mold-specific *MS8* gene is required for normal hypha formation in the dimorphic pathogenic fungus *Histoplasma capsulatum*. *Eukaryo Cell* 1:249–256.

80. Timmerman, M. M., and J. P. Woods. 1999. Ferric reduction is a potential iron acquisition mechanism for *Histoplasma capsulatum*. *Infect. Immun.* 67:6403–6408.

81. Timmerman, M. M., and J. P. Woods. 2001. Potential role for extracellular glutathione-dependent ferric reductase in utilization of environmental and host ferric compounds by *Histoplasma capsulatum*. *Infect. Immun.* 69:7671–7678.

82. Vincent, R., R. Goewert, W. E. Goldman, G. S. Kobayashi, A. M. Lambowitz, and G. Medoff. 1986. Classification of *Histoplasma capsulatum* isolates by restriction fragment polymorphisms. *J. Bacteriol.* 165:813–818.

83. Weaver, C. H., K. C. F. Sheehan, and E. J. Keath. 1996. Localization of a yeast phase-specific gene product to the cell wall in *Histoplasma capsulatum*. *Infect. Immun.* 64:3048–3054.

84. Woods, J. P. 2002. *Histoplasma capsulatum* molecular genetics, pathogenesis, and responsiveness to its environment. *Fungal Genet. Biol.* 35:81–97.

85. Woods, J. P., E. L. Heinecke, J. W. Luecke, E. Maldonado, J. Z. Ng, D. M. Retallack, and M. M. Timmerman. 2001. Pathogenesis of *Histoplasma capsulatum*. *Semin. Respir. Infect.* 16:91–101.

Molecular Principles of Fungal Pathogenesis
Edited by Joseph Heitman et al.
©2006 ASM Press, Washington, D.C.

Chapter 24

Molecular Determinants of Virulence in *Aspergillus fumigatus*

JUDITH C. RHODES AND AXEL A. BRAKHAGE

The opportunistic pathogen *Aspergillus fumigatus* is a ubiquitous fungus that can cause diseases that range in severity from mild allergic rhinitis to life-threatening invasive pulmonary and disseminated infection. In immunocompetent hosts, the hundreds of conidia that are breathed in every day are quickly and efficiently killed (53). Unfortunately, in patients with compromised innate immunity, invasive disease may result from exposure to the mold. Although not all the mechanisms used by this versatile fungus to cause disease have been elucidated, investigators have begun to unravel some of the key processes and pathways that appear to be necessary for the full expression of virulence by *A. fumigatus*. Genes whose disruption or deletion causes a decrease in virulence are listed in Table 1.

INTERACTION OF *A. FUMIGATUS* CONIDIA WITH THE HOST

Because of their abundance in air, it is estimated that several hundred *A. fumigatus* conidia are continuously inhaled during routine daily activities (17, 31, 37). In immunosuppressed patients, the lung is the site of infection by *A. fumigatus*. In immunocompetent individuals, mucociliary clearance and phagocytic defense normally prevent the disease. Alveolar macrophages are the major resident cells of the lung alveoli; they, along with neutrophils (which are actively recruited during inflammation), are the major cells in the phagocytosis of *A. fumigatus*. Macrophages kill conidia by producing reactive oxygen species (81). Conidia escaping alveolar macrophages grow out and are attacked by neutrophils, the second line of phagocytic cells. Hyphae are considered to be the primary target for neutrophils. However, polymorphonuclear leukocytes are also able to kill resting or swollen conidia (53). Contact between neutrophils

and hyphae triggers a respiratory burst, secretion of reactive oxygen intermediates, and degranulation (22, 43, 56, 57, 68, 92). A severely depressed immune system, however, provides an opportunity for conidia to germinate and invade lung tissue (reviewed in references 26 and 88).

ANIMAL MODELS

Animal models of invasive aspergillosis were reviewed by Latgé in 1999 (53). Since the model used for virulence testing may significantly influence the outcome of an experiment, a brief description of the different aspergillosis models most commonly used in the literature is presented. Models for invasive aspergillosis can be classified by species of host, usually mice, rabbits, or rats; by route of inoculation, usually intravenous, intranasal, or inhalation; and by immune status, usually immunocompetent or immunosuppressed with a glucorticoid alone or in combination with a cytotoxic drug or immunodepleting antibody.

Mice are the most frequently used hosts for pathogenesis studies. Intravenous injection of conidia into immunocompetent or immunocompromised mice is a model system that has been used widely for testing of pharmacological agents and less commonly for virulence testing (16, 20). Because intravenous inoculation does not mimic the natural route of infection, investigators have turned to intranasal and inhalational routes to introduce conidia into immunosuppressed mice (4, 6, 47, 61, 77, 95, 99, 101, 104). Immunosuppression has been induced by administration of glucorticoids such as cortisone acetate and triamcinolone acetonide, alone or in combination with cyclophosphamide, vinblastine, or monoclonal antibody RB6-8C, to induce neutropenia (5, 6, 64, 77, 99, 102). The use of the

Judith C. Rhodes • Department of Pathology and Laboratory Medicine, University of Cincinnati College of Medicine, Cincinnati, OH 45267-0529.   **Axel A. Brakhage** • Department of Molecular and Applied Microbiology, Leibniz-Institute for Natural Products Research and Infection Biology, Hans-Knoell-Institute, D-07745 Jena, Germany.

Table 1. *A. fumigatus* gene deletion or disruption mutants reported to show a decrease in virulence

Gene	Model	Virulence	Reference(s)
cat1/cat2	Rat i.t.[a]	Reduced	79
cgrA	Mouse i.n.[b], fly[c]	Reduced	6
chsC/chsG	Mouse i.n.	Reduced	65
cpcA	Mouse i.n.	Reduced	47
fos1	Mouse i.v.[d]	Reduced	18
gpaB	Mouse i.n.	Reduced	61
laeA	Mouse i.n.	Reduced	7a
lysF	Mouse i.n.	Avirulent	60
pabaA	Mouse i.v., mouse i.n.	Avirulent	13, 90
pkaC1	Mouse i.n.	Avirulent	61
pksP (alb1)	Mouse i.v.	Reduced	50, 106
pyrG	Mouse i.n.	Avirulent	19
rasB	Mouse i.n.	Reduced	27a
rhbA	Mouse i.n.	Reduced	77

[a] Intratracheal inoculation of conidia in agarose beads into immunosuppressed rats.

[b] Intranasal inoculation of conidia into immunosuppressed mice.

[c] Intra-abdominal inoculation of *spz⁻ Drosophila*.

[d] Intravenous inoculation of immunosuppressed or immunocompetent mice.

neutrophil-depleting antibody, either alone or in combination with cyclophosphamide or vinblastine, or cortisone acetate alone has been shown to markedly influence the required inoculum, as well as the inflammatory response, histopathology, and likely cause of death, in two different mouse strains (5, 102). Use of an inhalational chamber appears to allow conidia to be deposited into the alveolar spaces, which better mimics the situation in human disease, than does intranasal instillation, in which most of the conidia are found in bronchioles (95, 101, 102). One benefit of using the intranasal inoculation route for comparison studies between mutant and wild-type organisms is that the intranasal route allows comparison of virulence in mixed infections, which has been reported to be a more sensitive technique for detecting virulence differences in *A. fumigatus* (2, 91).

A rabbit model in which cytosine arabinoside and methylprednisilone are used for immunosuppression, along with intratracheal instillation of conidia, has been developed (28). This model is very useful when repeated sampling of infected animals is required to monitor diagnostic and prognostic markers, especially in evaluating potential antifungal agents. A rat model has been validated for histopathological studies, and it also has the ability to differentiate between isolates with disparate virulence in a mouse model (79, 96).

There is considerable diversity among the models reported in the literature. Each offers its own set of strengths and weaknesses. Therefore, investigators must carefully evaluate their own requirements for an experimental model, as well as literature reports on pathogenicity, since experimental outcome may be influenced by the host, the immunosuppressive regimen, and the route of inoculation.

GENES INVOLVED IN THE BIOSYNTHESES OF NUTRIENTS AND REQUIRED FOR GROWTH IN THE HOST

The importance of biosyntheses of primary metabolites for the growth of the fungus in host tissue has been addressed by a number of investigators. For example, *p*-aminobenzoic acid-requiring mutants are nonpathogenic in a murine infection model of invasive pulmonary aspergillosis (90). The *A. fumigatus pabaA* gene encodes *p*-aminobenzoic acid synthase. The complete inability of *pabaA* deletion strains to cause lethal infections in mice confirmed the importance of the folate synthesis pathway for in vivo survival of this pathogen (13). The same applies for the *A. fumigatus* UMP biosynthesis pathway. Most of the conidia of a *pyrG* mutant defective in orotidine-5'-monophosphate decarboxylase, and therefore deficient in UMP biosynthesis, did not germinate in the lungs of infected mice. Consequently, the mutant was nonpathogenic (19). The Δ*pyrG* strain has proven to be a useful transformation recipient, however, allowing for selection for uracil auxotrophy and counter selection with 5-fluoroorotic acid (110). A similar observation was made for the *lysF* gene of *A. fumigatus*, encoding an enzyme of lysine biosynthesis. The amino acid lysine is an essential amino acid for humans. By contrast, fungi can synthesize this amino acid via the α-aminoadipate pathway. Since humans lack most of the biosynthesis enzymes, fungal lysine biosynthesis might be a potential target for the development of antimycotic drugs. Activities of five of the (at least) seven enzymes, i.e., homocitrate synthase, homoisocitrate dehydrogenase, α-aminoadipate reductase, saccharopine reductase, and saccharopine dehydrogenase of the α-aminoadipate pathway, were found in wild-type cells of *A. fumigatus* (30). However, until now only one of the corresponding genes has been characterized in *A. fumigatus*, i.e., the *lysF* gene encoding homoaconitase, which is responsible for conversion of homoaconitate to homoisocitrate (60, 111). Deletion of the *A. fumigatus lysF* gene led to a lysine-auxotrophic mutant. In a low-dose intranasal-infection mouse model of invasive aspergillosis, the Δ*lysF* mutant was avirulent. These data suggest that lysine was limited in the lungs. Therefore, lysine biosynthesis or at least a functional homoaconitase is essential for invasive growth of *A. fumigatus* in the lung tissue of infected mice. In agreement with these results for *A. fumigatus*, mixed-inoculum infections have revealed that the growth of lysine-auxotrophic *A. nidulans lysA* strains, which were deficient in saccharopine dehydrogenase, was significantly less prolific than that of prototrophic strains in the lungs of neutropenic mice. However, this reduction was not reflected in a significant difference in the survival of mice inoculated with the auxotrophic mutant strain alone (105). Possibly, the latter result reflects a difference between *A. nidulans* and *A. fumigatus*. For example, the number of conidia required to cause infection in neutropenic mice by *A. nidulans* is about 1,000-fold greater than for *A. fumigatus*. In support of the importance of lysine prototrophy in fungal pathogenesis, auxotrophic strains of *Candida albicans* with a requirement for lysine have reduced virulence following intravenous inoculation (62, 94). Taken together, the results indicated that *p*-aminobenzoic acid, uridine/uracil, and lysine are limiting in the lungs of mice, thereby preventing virulence of the respective mutant conidia.

GENES INVOLVED IN GROWTH CONTROL

Several genes involved in sensing of the environment and control of growth in the environment have been investigated for their roles in mediating virulence in *A. fumigatus*. The rheb GTPase, RhbA, plays a role in sensing the nitrogen available in the environment (76). Steady-state levels of *rhbA* mRNA are markedly increased by nitrogen, but not carbon, starvation. When *rhbA* is deleted, the growth of the mutant is equal to the growth of the wild type on complex media and on minimal medium that contains ammonium as a nitrogen source, whereas the mutant grows more slowly than does the wild type on minimal medium containing nitrate as the nitrogen source (77). A further phenotype of the Δ*rhbA* mutant is hypersensitivity to the TOR (target of rapamycin) inhibitor rapamycin, suggesting that RhbA functions in the TOR pathway, which is known to play an important role in controlling ribosomal function and the translational machinery. Further work with *Drosophila* has confirmed that rheb is upstream of TOR kinase and functions via the S6 ribosomal kinase in the TOR-mediated control of nutrient sensing and growth (80). Although mutants that lack RhbA are not avirulent in an immunosuppressed-mouse model, both overall mortality and fungal lesion size are decreased in mice that received the Δ*rhbA* mutant compared with the wild type as an intranasal challenge (77). The mitogen-activated protein kinase, SakA, is also involved in nitrogen sensing, and mRNA levels of *sakA* are elevated by both nitrogen and carbon starvation (113). Mutants lacking SakA fail to respond to osmotic stress but have

wild-type growth regardless of the nitrogen source. However, compared with the wild type, conidial germination is accelerated in the mutants in the presence of nitrate or nitrite. In contrast to the results obtained with the Δ*rhbA* mutant, the Δ*sakA* mutant showed no change in virulence when tested in mice (G. S. May, T. Xue, and G. Reyes, *Abstr. Advances Against Aspergillosis*, p. 29, 2004).

The cross-pathway control system of amino acid biosynthesis is another pathway involved with growth control that has been studied for its role in virulence (47). The bZIP transcription factor CpcA, an ortholog of Gcn4p in *Saccharomyces cerevisiae*, enhances the transcription of CpcA targets in response to amino acid starvation and other stresses. Mutants of *A. fumigatus* that lack a functional copy of *cpcA* are hypersensitive to the amino acid analog 5-methyltryptophan, which induces a starvation response in the wild type. Although the mutants grow equally well on ammonium or nitrate, they fail to derepress pathways involved in amino acid biosynthesis when grown under conditions of amino acid starvation. In two different murine models, the Δ*cpcA* mutants display decreased virulence as measured by cumulative mortality and competitive growth (47). Although perhaps not a virulence factor in the classical bacterial sense, pathways involved in nutrient sensing and utilization clearly contribute to the nutritional versatility of *A. fumigatus*, which in turn contributes to its ability to function as an effective opportunistic pathogen.

THERMOTOLERANCE GENES

Thermotolerance is thought to be a property that sets *A. fumigatus* apart from other species in the genus and contributes to its being the most common opportunistic pathogen in the genus *Aspergillus* (53). With the availability of microarrays, there is significant interest in determining the transcriptional profile of the organism when grown at 37°C (74a). Sequencing of cDNAs prepared from hyphae grown at 37°C has generated a list of numerous genes, many homologous to genes involved in growth control in other organisms (93). However, initial analysis of these transcripts does not suggest a gene or pathway that might be responsible for the thermotolerant phenotype. Two reports in the literature have described genes that do contribute to thermotolerance in *A. fumigatus*. *THTA* is a gene with unknown function that was isolated by complementation of a mutant unable to grow at 42°C (16). No difference was seen in the protein levels of ThtA in cultures grown at 37 or 48°C. Deletion of the gene recapitulated the phenotype of the original mutant. The deletion mutant and the wild type were tested for virulence by intranasal instillation of conidia in neutropenic mice and by intravenous injection in immunocompetent mice. In both models, the virulence was equivalent between the wild type and the mutant, suggesting that although the gene was required for growth at 42°C, virulence at physiological temperature was not affected. The second gene, *cgrA*, is orthologous to a gene in *S. cerevisiae*, *CGR1*, that is involved in ribosome biogenesis (7, 72). Deletion of *cgrA* yields a mutant with a significantly reduced growth rate and delayed germination at 37°C. At 25°C, the Δ*cgrA* mutant is slightly delayed in germination but displays wild-type radial growth. This mutant was tested in two models for its virulence. In a murine intranasal model, the mutant displayed a significant decrease in virulence compared with the wild type. However, in a *Drosophila* model of infection, in which the flies are kept at 25°C, the virulence phenotype was still slightly reduced, but the kinetics of progressive mortality was much closer to that obtained with wild-type *A. fumigatus* (6). This suggests that the ability of conidia to germinate efficiently, which requires a burst of protein synthesis, is a key step in initiation of infection and that an initial delay in germination is sufficient to influence the outcome of virulence experiments.

CHITIN SYNTHASE GENES

Chitin is a linear polymer of N-acetylglucosamine with high tensile strength that makes up a portion of the cell wall of *A. fumigatus*, as well as other fungi. The cell wall provides support for the fungal cells and allows interaction with the environment (87). Therefore, mutations in genes that contribute to the synthesis of the cell wall might be expected to be deleterious to the virulence of the fungus. However, perhaps because the wall is so important to the organism, there is marked redundancy in chitin synthase (CS) genes in *A. fumigatus*. At least seven CSs have been identified in *A. fumigatus*, *chsA* through *chsG*, and they represent all six classes of CS genes (87). *chsA*, *chsB*, *chsC*, and *chsD* deletion mutants do not have any obvious phenotypic defects in morphology or development. For this group of CS genes, virulence studies have been performed only with *chsD* deletion mutants, which showed wild-type virulence when tested in a neutropenic-mouse model (65–67, 99). There has been no report of attempts to disrupt the *chsF* gene. However, deletion of either *chsE* or *chsG* results in striking phenotypic defects, both macroscopically and microscopically. The *chsE* gene is closely related to the *CHS3* gene of yeast; deletion of the *chsE* gene in *A. fumigatus* results in a decrease in the amount of

chitin in the cell walls, in swellings along the hyphae, and in a block in conidiation (4). Even though the conidiation can be partially restored by growth on osmotically stabilized medium, the morphology of the resulting conidiophores and phialides is aberrant. Conidia formed under these conditions germinate normally, and colonies initiated from them have normal rates of radial growth. Despite the morphological defects in the hyphae of the Δ*chsE* mutant, no loss of virulence was detected, and the same bulbous morphology was seen in the hyphae in histological sections from the infected lungs as was seen in vitro (4). Deletion of *chsG* yielded mutants with dense colonies showing reduced radial growth and increased hyphal branching. A Δ*chsE* Δ*chsG* double mutant was made, which had a phenotype that showed features of both single mutants, but that strain was not tested for virulence (66). A second double mutant, Δ*chsC* Δ*chsG*, has also been reported (65). In this case the mutant phenotype of the double mutant was not distinguishable from that of the Δ*chsG* mutant. Both Δ*chsG* and Δ*chsC* Δ*chsG* mutants had reduced radial growth and reduced growth in liquid medium. Branching was increased and probably contributed to the dense colony morphology of the mutants. Conidiation was not affected in these mutants, and the conidia that were produced germinated normally. When these mutants were tested in a neutropenic-mouse model, development of respiratory disease was delayed in mice that received the double mutant and the cumulative mortality was significantly reduced in those animals (65). However, the double mutant was still able to cause invasive disease in the mice, and histology of the lesions reflected the more highly branched morphology seen in vitro. Therefore, because synthesis of chitin is of key importance for *A. fumigatus*, it has developed a highly redundant set of CSs (3). Although this complicates the analysis of these genes as potential virulence factors in this organism, it undoubtedly serves the fungus well.

PIGMENTATION GENES AND PATHWAY

Conidia of *A. fumigatus* contain a 1,8-dihydroxynaphthalene-like melanin. It forms the pigment and plays a protective role against environmental conditions and in pathogenicity. In general, melanins are macromolecules formed by oxidative polymerization of phenolic or indolic compounds. Often the resulting pigments are brown or black, but many other colors have also been observed. Melanins are also hydrophobic and negatively charged. Several different types of melanin have been identified in fungi to date. The two most important types are DHN-melanin (named for one of the pathway intermediates, 1,8-dihydroxynaphthalene) and DOPA-melanin (named for one of the precursors, L-3,4-dihydroxyphenylalanine). Both types of melanin have been implicated in pathogenesis, perhaps by providing protection against enzymatic lysis or by scavenging toxic oxygen radicals (reviewed in references 32, 33, 39, and 52). Furthermore, in several fungi, including *A. fumigatus*, the cyclic AMP (cAMP) signal transduction pathway was shown to be involved in pathogenicity (59, 61, 75). A link to the expression of at least one of the melanin biosynthesis genes in *A. fumigatus* also appears to exist (see below).

BIOSYNTHESIS PATHWAY OF THE DHN-MELANIN IN *A. FUMIGATUS*

The DHN-melanin biosynthesis pathway (Fig. 1) is based on genetic and biochemical evidence obtained from *Verticillium dahliae* and *Wangiella dermatitidis* and recent genetic and biochemical data from *A. fumigatus* (11, 50, 106–109, 112). A cluster of six genes was identified in *A. fumigatus*, and all were shown to be involved in DHN-melanin biosynthesis (Fig. 2) (11, 50, 106, 108). The *pksP* (*alb1*) gene encodes a type I polyketide synthase. *pksP* mutants of *A. fumigatus* have white conidia, while the wild-type conidia are grayish-green. Also, *pksP*

mutants have a smooth surface; the ornaments characteristic of wild-type conidia are lacking (41). Homologs of *pksP* have been identified in several other fungi, including the *wA* gene of *A. nidulans*, *ALM* of *Alternaria alternata*, *pks1* of *Colletotrichum lagenarium*, *WdPKS1* of *W. dermatitidis*, and *alb1* of *Magnaporthe grisea*, all of which are involved in pigment biosynthesis (reviewed in reference 52). PksP (*A. fumigatus*), WA (*A. nidulans*), and PKS1 (*C. lagenarium*) were overproduced in *Aspergillus oryzae*. These experiments showed that whereas PKS1 of *C. lagenarium* used malonyl coenzyme A as starter and extender units for the synthesis of 1,3,6,8-tetrahydroxynaphthalene (1,3,6,8-THN), for the *Aspergillus* species the polyketide formed was a heptaketide naphthopyrone. This was unexpected since only 1,3,6,8-THN but not naphthopyrone could be detected in *A. fumigatus* by thin-layer chromatography. This observation could be explained when Tsai et al. (107) found that the product of the *A. fumigatus ayg1* gene, AYG1, is able to convert naphthopyrone to 1,3,6,8-THN, by chain-length shortening. An enzyme with activity similar to AYG1 in melanin biosynthesis has not been found in any other organism (29). The *arp1* gene encodes a scytalone dehydratase which catalyzes the dehydration of scytalone to 1,3,8-trihydroxynaphthalene (108).

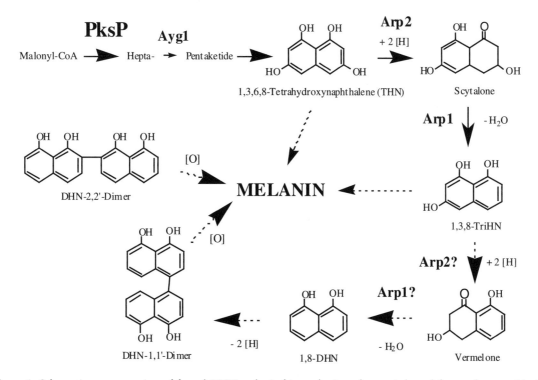

Figure 1. Schematic representation of fungal DHN-melanin biosynthesis pathways (adapted from references 11, 13, 51, 104, and 107). Probable reaction types are indicated ([O], oxidation; [H], reduction; $-H_2O$, dehydration). Arrows with dashed lines indicate probable reactions. CoA, coenzyme A.

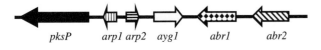

Figure 2. Pentaketide-like melanin biosynthesis gene cluster of *A. fumigatus*. Adapted from reference 11.

Deletion of this gene resulted in *A. fumigatus* colonies with redish-pink conidia. A further dehydration step, possibly also catalyzed by scytalone dehydratase, leads to the intermediate 1,8-dihydroxynaphthalene (DHN), for which this pathway was named. Subsequent steps are thought to involve a dimerization of the 1,8-DHN molecules, followed by polymerization, possibly catalyzed by a laccase. A laccase-encoding gene (*abr2*) and a multicopper oxidase-encoding gene (*abr1*) have been identified in *A. fumigatus*, but it is not yet clear at which point of the pathway they are involved (109). The involvement of the *abr1* and *abr2* genes in the formation of the grayish-green spore color of *A. fumigatus* was shown by disruption of each gene, which led to altered conidial color phenotypes (brown conidia) (109).

This is a general model for DHN-melanin biosynthesis but the pathway may vary in individual fungi, e.g., in *A. fumigatus*. Interestingly, several by-products of the fungal DHN-melanin pathway have antibacterial or immunosuppressive properties (89).

MELANIN IN PATHOGENESIS

The proposed functions of fungal melanins include protection against UV irradiation, enzymatic lysis, oxidants, and in some instances extremes of temperature. Also, melanins have been shown to bind metals, function as a physiological redox buffer (thereby possibly acting as a sink for harmful unpaired electrons), provide structural rigidity to cell walls, and store water and ions (thus helping to prevent desiccation) (reviewed in references 14 and 39). *Cryptococcus neoformans* is probably the most prominent fungus among opportunistic pathogens known to make DOPA-melanin. In most of the pentaketide-melanin-forming fungal pathogens such as *A. fumigatus*, *Sporothrix schenckii*, and the dematiaceous fungi in general, the melanin pathway plays a significant role in their infectivity (reviewed in reference 32).

A number of results indicate that both DHN-melanin and DOPA-melanin are able to quench reactive oxygen species (ROS), one of the defense mechanisms of the human immune system, and also able to interfere with phagocytosis to some degree (reviewed in references 12 and 32). In *A. fumigatus*,

a mutant producing pigmentless conidia showed a reduced virulence (41). The mutant could be complemented by the *pksP* gene (50, 106). The complete absence of pentaketide-melanin, as in the case of *pksP* mutants, resulted in a severe reduction in virulence. The *pksP* mutant of *A. fumigatus* is significantly more sensitive to hydrogen peroxide and sodium hypochlorite than is the wild-type strain. Also, the mutant strain is more susceptible to damage by murine macrophages in vitro. As in other cases, it was shown that melanin-containing conidia are able to quench ROS derived from human granulocytes (40, 41). These results indicated that conidial DHN-melanin of *A. fumigatus* is involved in protecting conidia from the host immune response in which ROS are important for eliminating fungal conidia (reviewed in references 52 and 53).

Although the green pigment of the nonpathogenic *A. nidulans* conidia does not seem to be synthesized via a DHN-melanin pathway (e.g., its synthesis cannot be inhibited by tricyclazole), it acts as a protective agent against oxidant-based host defense mechanisms and thus contributes to the relative resistance of conidia to neutrophil attack, as described for *A. fumigatus* (40). Therefore, resistance to ROS does not explain why *A. fumigatus* conidia can be pathogenic while this is rarely the case with *A. nidulans* conidia. One attractive hypothesis is that besides the pigment, the *pksP* gene product of *A. fumigatus* is involved in the production of another compound that is immunosuppressive. This could be the DHN-melanin pigment, an intermediate or byproduct of the DHN-melanin, or even another compound in whose biosynthesis PKSP is involved (reviewed in references 10, 12, and 52). This hypothesis is further supported by the notion that the presence of a functional *pksP* gene in *A. fumigatus* conidia is associated with an inhibition of the fusion of phagosomes and lysosomes in human monocyte-derived macrophages (42). Other pathways involving polyketide synthases have been shown to synthesize two different active products (reviewed in reference 52).

The presence of melanins per se does not define a human-pathogenic fungus because several nonpathogenic fungi also are able to synthesize melanins. Therefore, additional virulence determinants are required, such as the ability of the fungus to grow at 37°C or possibly the production of melanins at specific stages of the infectious process or the presence of melanins in certain cell types or structures like conidia and appressoria, respectively.

ANTIOXIDANTS

Genes encoding catalases, catalase-peroxidases, and Cu,Zn superoxide dismutases have been identified in *A. fumigatus* as potential virulence determinants (15, 34, 35, 79). The Cu,Zn superoxide dismutase is produced in culture, is recognized by human sera from infected patients, and is stable under physiological conditions (34–36). To date, gene deletion studies have not been reported for this superoxide dismutase. The sequences of three different catalase genes, two mycelial and one conidial, have been entered into GenBank for *A. fumigatus*, and as many as four others have been identified in the genome (15, 79, 103). Deletion of the gene encoding Cat1, a mycelial catalase that was identified because it is an immunodominant antigen, resulted in no change in sensitivity to hydrogen peroxide and no change in virulence in a murine model (15). Although mutants lacking CatA, the conidial catalase, were more sensitive to hydrogen peroxide, their sensitivity to killing by alveolar macrophages in vitro and their virulence in animals were similar to that of the wild type (79). A double mutant, in which both mycelial catalases, *cat1* and *cat2*, were deleted, had no detectable constitutive catalase activity and slightly increased sensitivity to hydrogen peroxide, but in tests of in vitro neutrophil killing, there was no difference from the wild type. However, this Δ*cat1* Δ*cat2* mutant did have reduced virulence in a rat model of invasive pulmonary aspergillosis, as judged by histological progression of the lesions (79, 96). Although it is intuitive to think that enzymes involved in resisting oxidative damage play an important part in the pathogenesis of aspergillosis, the redundancy and layers of control of the expression of the genes in this class may make their contributions difficult to demonstrate in deletion mutants.

PROTEINASES

Many different proteinases from *A. fumigatus* have been described, and many mutants with deletions of these enzymes have been reported in the literature. Because this topic has been previously reviewed (69), only a limited number of studies are discussed here to make specific points. The production of elastinolytic proteinases was correlated with virulence in environmental and clinical isolates; the production of these enzymes during infection was demonstrated by detecting antibody responses to the enzymes in patients (46, 71, 83, 85). Although proteinase production was demonstrated in the tissue of experimentally infected animals, no evidence of elastinolysis was found in infected human lung tissue, despite the ability of several described enzymes to degrade elastin and collagen in vitro (21, 54, 63). A mutant that failed to produce detectable elastase in vitro was produced by chemical mutagenesis; that mutant had decreased virulence in a mouse model

(45). However, when single-gene deletion mutants that lacked an alkaline proteinase, a metalloproteinase, an aspartic proteinase, or an intracellular metalloproteinase were tested in mouse (intranasal) or guinea pig (intravenous) models of invasive aspergillosis, no difference in virulence was found (38, 70, 84, 99, 104). A double deletion mutant that lacked genes for producing both the alkaline proteinase and the metalloproteinase was also constructed and also demonstrated wild-type virulence in a murine model of invasive aspergillosis (44). Although an argument could still be made that one or more of these proteinases play a role in aspergilloma or in allergic bronchopulmonary aspergillosis, the experimental data for invasive aspergillosis clearly do not support a role for invasive disease.

TOXINS

Toxins of two classes have been studied for their contribution to the virulence of *A. fumigatus*. The first are the toxins that fall in the 18-kDa or mitogillin family: Aspf1 antigen, mitogillin, and restrictocin (1, 49, 78, 98). These proteins appear to have identical molecular weights, but they may vary in a single amino acid. They are cytotoxic in tissue culture assays, inhibit protein synthesis, and have ribonucleolytic activity (1, 78, 98). They are produced during infection and can be identified in the urine of infected patients (49). In addition, both immunoglobulin G and E antibodies to Aspf1 have been detected in patient sera (1). However, the *aspf1* and *res* genes have been deleted by two different groups of investigators; in both cases, the deletion mutants had wild-type virulence in murine models of invasive aspergillosis (78, 98). Therefore, it does not appear that toxins of this class play a significant role in the pathogenesis of invasive disease.

Gliotoxin is a toxin produced by *A. fumigatus* that mediates the in vitro inhibition of phagocytosis by macrophages and induces T-cell cytotoxicity (73, 74). Experimental infections in turkey poults and in mice have both been used to illustrate that the metabolite can be produced at the site of infection and detected in the serum (58, 86). Indeed, gliotoxin could be detected in the sera from patients with confirmed invasive aspergillosis caused by both *A. fumigatus* and *A. terreus* (58). Using the *Galleria* model to assay virulence, four clinical strains of *A. fumigatus* and one of *A. niger* were compared for production of gliotoxin, growth rate, elastase and catalase activities, and virulence. Virulence in the larvae was correlated with gliotoxin production in vitro (82). Because these are not isogenic strains, it is difficult to prove that the correlation is causative.

Finally, an *A. fumigatus* strain in which the *laeA* (for "loss of aflatoxin expression") gene was deleted was constructed and evaluated in a murine model. LaeA is a transcriptional regulator of toxin synthesis in *Aspergillus* (8), and the mutant had deficient synthesis of gliotoxin, fumagillin, and helvoic acid. In a murine model of invasive pulmonary aspergillosis, the mutant had significantly reduced virulence compared with the wild type (7a). However, because of the regulatory nature of this gene, the effects of its ablation are likely to be highly pleiotropic, making it difficult to attribute the loss of virulence to the loss of gliotoxin alone. Therefore, gliotoxin continues to be studied both as a potential diagnostic marker

and as a potential virulence factor in invasive aspergillosis.

cAMP SIGNAL TRANSDUCTION AND VIRULENCE IN *A. FUMIGATUS*

Recently, elements of the cAMP signaling pathway of *A. fumigatus* were identified (59, 61, 75) (Fig. 3). These include the adenylate cyclase gene *acyA*; the G-protein α subunit gene *gpaB*; *pkaC1*, encoding the predominant form of the catalytic subunit of protein kinase A; and *pkaR*, encoding the regulatory subunit of protein kinase A (PKA). *pkaC1* deletion mutants were severely impaired in sporulation and growth. The Δ*pkaC1* phenotypes resembled

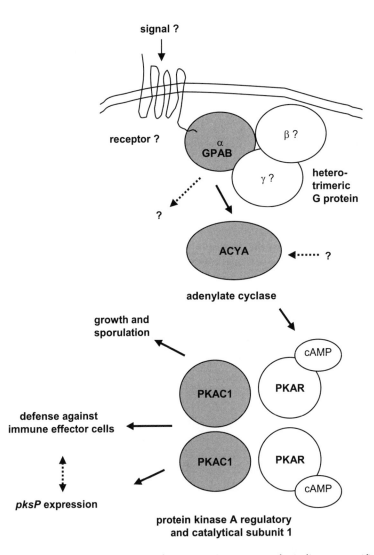

Figure 3. Model of cAMP signal transduction in *A. fumigatus*. Question marks indicate unverified components and reactions. Dashed arrows mark unverified relationships. Proteins deduced from cloned genes are labeled with circles shaded in gray. It is expected that cross talk between the cAMP regulatory network and other regulatory networks exists. Some of these potential interactions, which were deduced from current data, are indicated by questions marks. Adapted form reference 11.

those of the *A. fumigatus* adenylate cyclase mutant (Δ*acyA*), which was also reduced in growth but, in contrast, showed almost no conidiation on agar plates (59, 61). With respect to sporulation, this phenotype was also observed for the *A. fumigatus* Δ*gpaB* mutant, which is defective in a stimulatory Gα subunit. Most probably, deletion of the *gpaB* gene led to a decreased activation of adenylate cyclase and therefore to reduced cAMP levels. The Δ*gpaB* mutant produced only half as many conidia as the wild-type strains did. Wild-type phenotypes of both Δ*acyA* and Δ*gpaB* mutants were restored by addition of dibutyryl-cAMP to the medium (59). As expected, this was not the case for the Δ*pkaC1* mutants, because PKA acts downstream of cAMP synthesis. This finding was confirmed by the result that there was no measurable PKA activity in crude extracts of the Δ*pkaC1* deletion strains. Moreover, no PKA activity was detectable in either the Δ*gpaB* or Δ*acyA* mutants, most probably due to decreased levels or lack of cAMP in these strains. PKA activity in the Δ*acyA* and Δ*gpaB* deletion mutants was partially restored by addition of cAMP to the assay mixtures. These results indicated that PkaC1 activity depends on the cAMP level and, in addition, that PkaC1 acts downstream of GpaB and AcyA in the same pathway (61).

Interestingly, the observed Δ*pkaC1* phenotypes only partially resembled those of *A. nidulans* Δ*pkaA* and *A. niger* Δ*pkaC* mutants that are mutated in PKA catalytic subunit genes with highest similarity to *A. fumigatus pkaC1*. The last two mutants displayed reduced growth rates on agar plates and delayed germination (27, 97, 100). Deletion of *pkaA* in *A. nidulans* led to hyperconidiation, whereas *pkaA* overexpression resulted in decreased sporulation (97). This result is in contrast to the findings reported for *A. fumigatus*. The Δ*pkaC1* strains showed drastically reduced sporulation, indicating that asexual development of conidia is regulated differently at the level of PKA in both fungi. Taken together, PKA mutants of fungi closely related to *A. fumigatus* exhibited similar growth phenotypes but different phenotypes with respect to sporulation.

The key question concerns the importance of cAMP-PKA signaling for virulence. Therefore, different aspects were considered. (i) A possible influence of elements of the cAMP-PKA signaling cascade on the expression of the pathogenicity determinant-encoding gene *pksP* was studied. (ii) The importance of these elements for the pathogenicity of the fungus in human monocyte-derived macrophages and a low-dose animal infection model was investigated. As shown previously, the *pksP* gene is expressed mainly during sporulation but also in outgrowing hyphae of *A. fumigatus* when conidia germinated in the lungs

of immunocompromised mice (50, 51, 106, 109). In Δ*pkaC1* mutant strains, reduced *pksP*p-*lacZ* expression was measured in hyphae under standard growth conditions. This result agreed well with previous findings that *pksP* expression is decreased in an *A. fumigatus* Δ*gpaB* mutant (59). The Δ*gpaB* effect is most likely to be due to the lowered cAMP level and therefore to the lack of stimulation of adenylate cyclase activity by the stimulating G$_{(s)}$α subunit GpaB. Lower cAMP levels then led to decreased activation of PKA. However, the significantly higher *pksP*p-*lacZ* expression in the Δ*pkaC1* mutants than in the Δ*gpaB* strain suggested that in addition to the GpaB-cAMP/PKA network, a GpaB-dependent, PKA-independent signaling cascade exists, influencing *pksP* expression (Fig. 3).

The deletion of *acyA*, *gpaB*, or *pkaC1* led to drastically attenuated virulence of *A. fumigatus* in an optimized inhalative, low-dose mouse infection model for invasive aspergillosis (60, 61; B. Liebmann and A. A. Brakhage, unpublished data). In fact, the mutants were virtually avirulent. This result is in agreement with the finding that conidia of *A. fumigatus* cAMP signaling mutants, Δ*gpaB* and Δ*acyA*, were killed more efficiently by human monocyte-derived macrophages than were conidia of the wild-type strains (59). However, the decreased virulence of both *acyA* and *pkaC1* mutants needs to be seen in the context of their poor growth. Survival of mice infected with the mutant may have resulted from their delayed germination or reduced growth rate. Similar observations were made for PKA mutants of *C. neoformans* and *Ustilago maydis* (23, 25). However, in contrast to the *pkaC1* and *acyA* mutants, the Δ*gpaB* mutant did not show any growth retardation (59). Nevertheless, the virulence of the Δ*gpaB* strain was strongly attenuated, confirming the importance of cAMP signaling for virulence in *A. fumigatus* (61). Moreover, the *A. fumigatus* virulence determinant-encoding gene *pksP* was also regulated by cAMP-PKA signaling, and a decreased expression of *pksP*, as shown in both the Δ*gpaB* and the Δ*pkaC1* mutant, could contribute to better killing of the respective mutants in the animal model. Signaling through cAMP was found to control virulence and development in several human-pathogenic fungi such as *Candida albicans* and *C. neoformans* and plant-pathogenic fungi such as *U. maydis* or *M. grisea* (reviewed in references 9, 24, 48, 52, and 55).

In summary, the involvement of cAMP signaling in virulence appears to be a common theme in plant- and human-pathogenic fungi. For *A. fumigatus* it will be interesting to discover which other genes are regulated by the cAMP-PKA network and whether other pathogenicity determinants will be found among them.

CONCLUSION

A. fumigatus is a complex eukaryotic organism that functions as a successful opportunistic pathogen. The ability of the organism to behave in this manner is controlled by the careful regulation of pathways involved in nutrient sensing, growth control, cAMP-PKA signal transduction, and pigmentation. The number of factors defined so far is likely to be expanded exponentially over the next several years as genomic and proteomic methods are applied to *A. fumigatus* virulence. Based on our current knowledge, the only certainty is that this versatile organism will surprise us with its novel pathways, as well as its novel usage of conserved pathways.

REFERENCES

1. Arruda, L. K., T. A. E. Platts-Mills, J. W. Fox, and M. D. Chapman. 1990. *Aspergillus fumigatus* allergen I, a major IgE-binding protein, is a member of the mitogillin family of cytotoxins. *J. Exp. Med.* **172:**1529–1532.

2. Aufauvre-Brown, A., J. S. Brown, and D. W. Holden. 1998. Comparison of virulence between clinical and environmental isolates of *Aspergillus fumigatus*. *Eur. J. Clin. Microbiol. Infect. Dis.* **17:**778–780.

3. Aufauvre-Brown, A., E. Mellado, N. A. Gow, and D. W. Holden. 1995. A multigene family related to chitin synthase genes in yeast in the opportunistic pathogen *Aspergillus fumigatus*. *Mol. Gen. Genet.* **246:**353–359.

4. Aufauvre-Brown, A., E. Mellado, N. A. R. Gow, and D. W. Holden. 1997. *Aspergillus fumigatus chsE*: a gene related to *CHS3* of *Saccharomyces cerevisiae* and important for hyphal growth and conidiophore development but not pathogenicity. *Fungal Genet. Biol.* **21:**141–152.

5. Balloy, V., M. Huerre, J.-P. Latgé, and M. Chignard. 2005. Differences in patterns of infection and inflammation for corticosteroid treatment and chemotherapy in experimental invasive pulmonary aspergillosis. *Infect. Immun.* **73:**494–503.

6. Bhabhra, R., M. D. Miley, E. Mylonakis, D. Boettner, J. Fortwendel, J. C. Panepinto, M. Postow, J. C. Rhodes, and D. S. Askew. 2004. Disruption of the *Aspergillus fumigatus* gene encoding nucleolar protein CgrA impairs thermotolerant growth and reduces virulence. *Infect. Immun.* **72:**4731–4740.

7. Boettner, D., N. Huebner, J. C. Rhodes, and D. S. Askew. 2001. Molecular cloning of *Aspergillus fumigatus cgrA*, the ortholog of a conserved fungal nucleolar protein. *Med. Mycol.* **39:**517–521.

7a. Bok, J. W., S. A. Balajee, K. A. Marr, D. Andes, K. F. Nielsen, J. C. Frisvad, and N. P. Keller. 2005. LaeA, a regulator of morphogenetic fungal virulence factors. *Eukaryot. Cell* **4:**1574–1582.

8. Bok, J. W., and N. P. Keller. 2004. LaeA, a regulator of secondary metabolism in *Aspergillus* spp. *Eukaryot. Cell* **3:**527–535.

9. Bölker, M. 1998. Sex and crime: heterotrimeric G proteins in fungal mating and pathogenesis. *Fungal Genet. Biol.* **25:**143–156.

10. Brakhage, A. A., and K. Langfelder. 2002. Menacing mold: the molecular biology of *Aspergillus fumigatus*. *Annu. Rev Microbiol.* **56:**433–455.

11. Brakhage, A. A., K. Langfelder, G. Wanner, A. Schmidt, and B. Jahn. 1999. Pigment biosynthesis and virulence. *Contrib. Microbiol.* **2:**205–215.

12. Brakhage, A. A., and B. Liebmann. 2005. *Aspergillus fumigatus* conidial pigment and cAMP signal transduction: significance for virulence. *Med. Mycol.* **43:** 575–582.

13. Brown, J. S., A. Aufauvre-Brown, J. Brown, J. M. Jennings, H. Arst, and D. W. Holden. 2000. Signature-tagged and directed mutagenesis identify PABA synthetase as essential for *Aspergillus fumigatus* pathogenicity. *Mol. Microbiol.* **36:**1371–1380.

14. Butler, M. J., and A. W. Day. 1998. Fungal melanins: a review. *Can. J. Microbiol.* **44:**1115–1136.

15. Calera, J. A., S. Paris, M. Monod, A. J. Hamilton, J.-P. Debeaupuis, M. Diaquin, R. Lopez-Medrano, F. Leal, and J.-P. Latgé. 1997. Cloning and disruption of the antigenic catalase gene of *Aspergillus fumigatus*. *Infect. Immun.* **65:**4718–4724.

16. Chang, Y. C., H.-F. Tsai, M. Karos, and K. J. Kwon-Chung. 2004. *THTA*, a thermotolerance gene of *Aspergillus fumigatus*. *Fungal Genet. Biol.* **41:**888–896.

17. Chazalet, V., J.-P. Debeaupuis, J. Sarfati, J. Lortholary, P. Ribaud, P. Shah, M. Cornet, H. V. Thien, E. Gluckman, G. Brücker, and J-P. Latgé. 1998. Molecular typing of environmental and patient isolates of *Aspergillus fumigatus* from various hospital settings. *J. Clin. Microbiol.* **36:**1494–1500.

18. Clemons, K. V., T. K. Miller, C. P. Selitrennikoff, and D. A. Stevens. 2002. *fos-1*, a putative histidine kinase as a virulence factor for systemic aspergillosis. *Med. Mycol.* **40:**259–262.

19. d'Enfert, C., M. Diaquin, A. Delit, N. Wuscher, J. Debeaupuis, M. Huerre, and J. Latgé. 1996. Attenuated virulence of uridine-uracil auxotrophs of *Aspergillus fumigatus*. *Infect. Immun.* **64:**4401–4405.

20. Denning, D. W., L. Hall, M. Jackson, and S. Hollis. 1995. Efficacy of D0870 compared with those of intraconazole and amphotericin B in two murine models of invasive aspergillosis. *Antimicrob. Agents Chemother.* **39:**1809–1814.

21. Denning, D. W., P. N. Ward, L. E. Fenelon, and E. W. Benbow. 1992. Lack of vessel wall elastinolysis in human invasive pulmonary aspergillosis. *Infect. Immun.* **60:**5153–5156.

22. Diamond, R. D., and R. A. Clark. 1982. Damage of *Aspergillus fumigatus* and *Rhizopus oryzae* hyphae by oxidative and nonoxidative microbicidal products of human neutrophils in vitro. *Infect. Immun.* **38:**487–495.

23. D'Souza, C., J. Alspaugh, C. Yue, T. Harashima, G. Cox, J. Perfect, and J. Heitman. 2001. Cyclic AMP-dependent protein kinase controls virulence of the fungal pathogen *Cryptococcus neoformans*. *Mol. Cell. Biol.* **21:**3179–3191.

24. D'Souza, C. A., and J. Heitman. 2001. Conserved cAMP signaling cascades regulate fungal development and virulence. *FEMS Microbiol. Rev.* **25:**349–364.

25. Dürrenberger, F., K. Wong, and J. W. Kronstad. 1998. Identification of a cAMP-dependent protein kinase catalytic subunit required for virulence and morphogenesis in *Ustilago maydis*. *Proc. Natl. Acad. Sci. USA* **95:** 5684–5689.

26. Ellis, M. 1999. Therapy of *Aspergillus fumigatus*-related diseases, p. 105–129. *In* A. A. Brakhage, B. Jahn, and A. Schmidt (ed.), *Aspergillus fumigatus*: *Biology, Clinical Aspects and Molecular Approaches to Pathogenicity*. Karger, Basel, Switzerland.

27. Fillinger, S., M.-K. Chaveroche, K. Shimizu, N. Keller, and C. d'Enfert. 2002. cAMP and ras signalling independently control spore germination in the filamentous fungus *Aspergillus nidulans*. *Mol. Microbiol.* **44:** 1001–1016.

27a. Fortwendel, J. R., W. Zhao, R. Bhabra, S. Parks, D. S. Perlin, D. S. Askew, and J. C. Rhodes. 2005. A fungus-specific ras homolog contributes to the hyphal growth and virulence of *Aspergillus fumigatus*. *Eukaryot. Cell* **4:**1982–1989.

28. Francis, P., J. W. Lee, A. Hoffman, J. Peter, A. Francesconi, J. Bacher, J. Shelhamer, P. A. Pizzo, and T. J. Walsh. 1994. Efficacy of unilamellar liposomal amphotericin B in treatment of pulmonary aspergillosis in persistently granulocytopenic rabbits: the potential role of bronchoalveolar D-mannitol and serum galactomannan as markers of infection. *J. Infect. Dis.* **169:**356–368.

29. Fujii, I., Y. Yasuoka, H.-F. Tsai, Y. C. Chang, K. J. Kwon-Chung, and Y. Ebizuka. 2004. Hydrolytic polyketide shortening by Ayg1p, a novel enzyme involved in fungal melanin biosynthesis. *J. Biol. Chem.* **279:** 44613–44620.

30. Garrad, R. C., and J. K. Bhattacharjee. 1992. Lysine biosynthesis in selected pathogenic fungi: characterization of lysine auxotrophs and the cloned *LYS1* gene of *Candida albicans*. *J. Bacteriol.* **174:**7379–7384.

31. Goodley, J. M., Y. M. Clayton, and R. J. Hay. 1994. Environmental sampling for aspergilli during building construction on a hospital site. *J. Hosp. Infect.* **26:** 27–35.

32. Haase, G., and A. A. Brakhage. 2004. Melanized fungi infecting humans, p. 67–88. *In* J. E. Domer and G. S. Kobayashi (ed.), *Human Fungal Pathogens*, vol. XII. Springer-Verlag KG, Berlin, Germany.

33. Hamilton, A. J., and B. L. Gomez. 2002. Melanins in fungal pathogens. *J. Med. Microbiol.* **51:**189–191.

34. Hamilton, A. J., M. D. Holdom, and L. Jeavons. 1996. Expression of the Cu,Zn superoxide dismutase of *Aspergillus fumigatus* as determined by immunochemistry and immunoelectron microscopy. *FEMS Immunol. Med. Microbiol.* **14:**95–102.

35. Holdom, M. D., R. J. Hay, and A. J. Hamilton. 1996. The Cu,Zn superoxide dismutases of *Aspergillus flavus*, *Aspergillus niger*, *Aspergillus nidulans*, and *Aspergillus terreus*: purification and biochemical comparison with the *Aspergillus fumigatus* Cu,Zn superoxide dismutase. *Infect. Immun.* **64:**3326–3332.

36. Holdom, M. D., B. Lechenne, R. J. Hay, A. J. Hamilton, and M. Monod. 2000. Production and characterization of recombinant *Aspergillus fumigatus* Cu,Zn superoxide dismutase and its recognition by immune human sera. *J. Clin. Microbiol.* **38:**558–562.

37. Hospenthal, D. R., K. J. Kwon-Chung, and J. E. Bennett. 1998. Concentrations of airborne *Aspergillus* compared to the incidence of invasive aspergillosis: lack of correlation. *Med. Mycol.* **36:**165–168.

38. Ibrahim-Granet, O., and C. D'Enfert. 1997. The *Aspergillus fumigatus mepB* gene encodes an 82 kDa intracellular metalloproteinase structurally related to mammalian thimet oligopeptidases. *Microbiology* **143:** 2247–2253.

39. Jacobson, E. S. 2000. Pathogenic roles for fungal melanins. *Clin. Microbiol. Rev.* **13:**708–717.

40. Jahn, B., F. Boukhallouk, J. Lotz, K. Langfelder, G. Wanner, and A. A. Brakhage. 2000. Interaction of human phagocytes with pigmentless *Aspergillus* conidia. *Infect. Immun.* **68:**3736–3739.

41. Jahn, B., A. Koch, A. Schmidt, G. Wanner, H. Gehringer, S. Bhakdi, and A. A. Brakhage. 1997. Isolation and characterization of a pigmentless-conidium mutant of *Aspergillus fumigatus* with altered conidial surface and reduced virulence. *Infect. Immun.* **65:**5110–5117.

42. Jahn, B., K. Langfelder, U. Schneider, C. Schindel, and A. A. Brakhage. 2002. PKSP-dependent reduction of phagolysosome fusion and intracellular kill of *Aspergillus fumigatus* conidia by human monocyte-derived macrophages. *Cell. Microbiol.* **4:**793–803.

43. Jahn, B., A. Stueben, and S. Bhakdi. 1996. Colorimetric susceptibility testing of *Aspergillus fumigatus*: comparison of menandione-augmented 3-(4,5-dimethyl-2-thiazolyl)-2,5-diphenyl-2H-tetrazolium bromide and alamar blue tests. *J. Clin. Microbiol.* **34:**2039–2041.

44. Jaton-Ogay, K., S. Paris, M. Huerre, M. Quadroni, R. Falchetto, G. Togni, J.-P. Latgé, and M. Monod. 1994. Cloning and disruption of the gene encoding an extracellular metalloprotease of *Aspergillus fumigatus*. *Mol. Microbiol.* **14:**917–928.

45. Kolattukudy, P. E., J. D. Lee, L. M. Rogers, P. Zimmerman, S. Ceselski, B. Fox, B. Stein, and E. A. Copelan. 1993. Evidence for possible involvement of an elastinolytic serine protease in aspergillosis. *Infect. Immun.* **61:**2357–2368.

46. Kothary, M. H., J. T. Chase, and J. D. Macmillan. 1984. Correlation of elastase production by some strains of *Aspergillus fumigatus* with ability to cause pulmonary invasive aspergillosis in mice. *Infect. Immun.* **43:** 320–325.

47. Krappmann, S., E. M. Bignell, U. Reichard, T. Rogers, K. Haynes, and G. H. Braus. 2004. The *Aspergillus fumigatus* transcriptional activator CpcA contributes significantly to the virulence of this fungal pathogen. *Mol. Microbiol.* **52:**785–799.

48. Kronstad, J., D. DeMaria, D. Funnell, R. D. Laidlaw, N. Lee, M. deSa, and M. Ramesh. 1998. Signaling via cAMP in fungi: interconnections with mitogen-activated protein kinase pathways. *Arch. Microbiol.* **170:**395–404.

49. Lamy, B., M. Moutaouakil, J.-P. Latgé, and J. Davies. 1991. Secretion of a potential virulence factor, a fungal ribonucleotoxin, during human aspergillosis infections. *Mol. Microbiol.* **5:**1811–1815.

50. Langfelder, K., B. Jahn, H. Gehringer, A. Schmidt, G. Wanner, and A. A. Brakhage. 1998. Identification of a polyketide synthase gene (*pksP*) of *Aspergillus fumigatus* involved in conidial pigment biosynthesis and virulence. *Med. Microbiol. Immunol.* **187:**79–89.

51. Langfelder, K., B. Philippe, B. Jahn, J.-P. Latgé, and A. A. Brakhage. 2001. Differential expression of the *Aspergillus fumigatus pksP* gene detected in vitro and in vivo with green fluorescent protein. *Infect. Immun.* **69:**6411–6418.

52. Langfelder, K., M. Streibel, B. Jahn, G. Haase, and A. A. Brakhage. 2003. Biosynthesis of fungal melanins and their importance for human pathogenic fungi. *Fungal. Genet. Biol.* **38:**143–158.

53. Latgé, J.-P. 1999. *Aspergillus fumigatus* and aspergillosis. *Clin. Microbiol. Rev.* **12:**310–350.

54. Lee, J. D., and P. E. Kolattukudy. 1995. Molecular cloning of the cDNA and gene for an elastinolytic aspartic proteinse from *Aspergillus fumigatus* and evidence of its secretion by the fungus during invasion of the host lung. *Infect. Immun.* **63:**3796–3803.

55. Lengeler, K. B., R. C. Davidson, C. D'Souza, T. Harashima, W.-C. Shen, P. Wang, X. Pan, M. Waugh, and J. Heitman. 2000. Signal transduction cascades regulating fungal development and virulence. *Microbiol. Mol. Biol. Rev.* **64:**746–785.

56. Levitz, S. M., and R. D. Diamond. 1985. A rapid colorimetric assay of fungal viability with tetrazolium salt MTT. *J. Infect. Dis.* **152:**938–945.

57. Levitz, S. M., and T. P. Farrell. 1990. Human neutrophil degranulation stimulated by *Aspergillus fumigatus*. *J. Leukoc. Biol.* **47:**170–175.

58. Lewis, R. E., N. P. Wiederhold, J. Chi, X. Y. Han, K. V. Komanduri, D. P. Kontoyiannis, and R. A. Prince. 2005. Detection of gliotoxin in experimental and human aspergillosis. *Infect. Immun.* **73:**635–637.

59. Liebmann, B., S. Gattung, B. Jahn, and A. A. Brakhage. 2003. cAMP signaling in *Aspergillus fumigatus* is involved in the regulation of the virulence gene *pksP* and in defense against macrophage killing by macrophages. *Mol. Genet. Genomics* **269:**420–435.

60. Liebmann, B., T. W. Mühleisen, M. Müller, M. Hecht, G. Weidner, A. Braun, M. Brock, and A. A. Brakhage. 2004. Deletion of the *Aspergillus fumigatus* lysine biosynthesis gene *lysF* encoding homoaconitase leads to attenuated virulence in a low-dose model mouse infection model of invasive aspergillosis. *Arch. Microbiol.* **181:**378–383.

61. Liebmann, B., M. Müller, A. Braun, and A. A. Brakhage. 2004. The cyclic AMP-dependent protein kinase A network regulates development and virulence in *Aspergillus fumigatus*. *Infect. Immun.* **72:**5193–5203.

62. Manning, M., C. B. Snoddy, and R. A. Fromtling. 1984. Comparative pathogenicity of auxotrophic mutants of *Candida albicans*. *Can. J. Microbiol.* **30:**31–35.

63. Markaryan, A., I. Morozova, H. Yu, and P. E. Kolattukudy. 1994. Purification and characterization of an elastinolytic metalloproteinse from *Aspergillus fumigatus* and immunoelectron microscopic evidence of secretion of this enzyme by the fungus invading the murine lung. *Infect. Immun.* **62:**2149–2157.

64. Mehrad, B., R. M. Strieter, T. A. Moore, W. C. Tsai, S. A. Lira, and T. J. Standiford. 1999. CXC chemokine receptor-2 ligands are necessary components of neutrophil-mediated host defense in invasive pulmonary aspergillosis. *J. Immunol.* **163:**6086–6094.

65. Mellado, E., A. Aufauvre-Brown, N. A. Gow, and D. W. Holden. 1996. The *Aspergillus fumigatus chsC* and *chsG* genes encode class III chitin synthases with different functions. *Mol. Microbiol.* **20:**667–679.

66. Mellado, E., G. Dubreucq, P. Mol, J. Sarfati, S. Paris, M. Diaquin, D. W. Holden, J. L. Rodriguez-Tudela, and J. P. Latgé. 2003. Cell wall biogenesis in a double chitin synthase mutant (*chsG⁻/chsE⁻*) of *Aspergillus fumigatus*. *Fungal Genet. Biol.* **38:**98–109.

67. Mellado, E., C. A. Specht, P. W. Robbins, and D. W. Holden. 1996. Cloning and characterization of *chsD*, a chitin synthase-like gene of *Aspergillus fumigatus*. *FEMS Microbiol. Lett.* **143:**69–76.

68. Meshulam, T., S. M. Levitz, L. Christin, and R. D. Diamond. 1995. A simplified new assay for assessment of fungal cell damage with the tetrazolium dye, (2,3)-bis-(2-methoxy-4-nitro-5-sulphenyl)-(2H)-tetrazolium-5-carboxyanilide (XTT). *J. Infect. Dis.* **172:**1153–1156.

69. Monod, M., K. Jaton-Ogay, and U. Reichard. 1999. *Aspergillus fumigatus*-secreted proteases as antigenic factors and virulence factors., p. 182–192. *In* A. A. Brakhage, B. Jahn, and A. Schmidt (ed.), *Aspergillus fumigatus. Biology, Clinical Aspects and Molecular Approaches to Pathogenicity*, vol. 2. Karger, Basel, Switzerland.

70. Monod, M., S. Paris, J. Sarfati, K. Jaton-Ogay, P. Ave, and J.-P. Latgé. 1993. Virulence of alkaline protease-deficient mutants of *Aspergillus fumigatus*. *FEMS Microbiol. Lett.* **106:**39–46.

71. Moser, M., G. Menz, K. Blaser, and R. Crameri. 1994. Recombinant expression and antigenic properties of a 32-kilodalton extracellular alkaline protease, representing a possible vilulence factor from *Aspergillus fumigatus*. *Infect. Immun.* **62:**936–942.

72. Moy, T. I., D. Boettner, J. C. Rhodes, P. A. Silver, and D. S. Askew. 2002. Identification of a role for *Saccharomyces cerevisiae* Cgr1p in pre-rRNA processing and 60S ribosome subunit synthesis. *Microbiology* **148:**1081–1090.

73. Mullbacher, A., and R. D. Eichner. 1984. Immunosuppression in vitro by a metabolite of a human pathogenic fungus. *Proc. Natl. Acad. Sci. USA* **81:**3835–3837.

74. Mullbacher, A., P. Waring, and R. D. Eichner. 1985. Identification of an agent in culture of *Aspergillus fumigatus* displaying anti-phagocytic and immunomodulating activity *in vitro*. *J. Gen. Microbiol.* **131:**1251–1258.

74a. Nierman, W. C., G. S. May, H. S. Kim, M. J. Anderson, D. Chen, and D. W. Henning. 2005. What the *Aspergillus* genomes have told us. *Med. Mycol.* **43:**S3–S5.

75. Oliver, B. G., J. C. Panepinto, J. R. Fortwendel, D. L. Smith, D. S. Askew, and J. C. Rhodes. 2001. Cloning and expression of *pkaC* and *pkaR*, the genes encoding the cAMP-dependent protein kinase of *Aspergillus fumigatus*. *Mycopathologia* **154:**85–91.

76. Panepinto, J. C., B. G. Oliver, T. W. Amlung, D. S. Askew, and J. C. Rhodes. 2002. Expression of the *Aspergillus fumigatus rheb* homologue, *rhbA*, is induced by nitrogen starvation. *Fungal Genet. Biol.* **36:**207–214.

77. Panepinto, J. C., B. G. Oliver, J. R. Fortwendel, D. L. H. Smith, D. S. Askew, and J. C. Rhodes. 2003. Deletion of the *Aspergillus fumigatus* gene encoding the ras-related protein RhbA reduces virulence in a model of invasive pulmonary aspergillosis. *Infect. Immun.* **71:**2819–2826.

78. Paris, S., M. Monod, M. Diaquin, B. Lamy, L. K. Arruda, P. J. Punt, and J.-P. Latgé. 1993. A transformant of *Aspergillus fumigatus* deficient in the antigenic cytotoxin ASPF1. *FEMS Microbiol. Lett.* **111:**31–36.

79. Paris, S., D. Wysong, J.-P. Debeaupuis, K. Shibuya, B. Philippe, R. D. Diamond, and J.-P. Latgé. 2003. Catalases of *Aspergillus fumigatus*. *Infect. Immun.* **71:**3551–3562.

80. Patel, P. H., N. Thapar, L. Guo, M. Martinez, J. Maris, C.-L. Gau, J. A. Lengyei, and F. Tamanoi. 2003. *Drosophila* Rheb GTPase is required for cell cycle progression and cell growth. *J. Cell Sci.* **116:**3601–3610.

81. Philippe, B., O. Ibrahim-Granet, M. C. Prévost, M. A. Gougerot-Pocodalo, M. S. Perez, A. VanderMeeren, and J. P. Latgé. 2003. Killing of *Aspergillus fumigatus* by alveolar macrophages is mediated by reactive oxidant intermediates. *Infect. Immun.* **71:**3034–3042.

82. Reeves, E. P., C. G. M. Messina, S. Doyle, and K. Kavanagh. 2004. Correlation between gliotoxin production and virulence of *Aspergillus fumigatus* in *Galleria mellonella*. *Mycopathologia* **158:**73–79.

83. Reichard, U., S. Büttner, H. Eiffert, F. Staib, and R. Rüchel. 1990. Purification and characterisation of an extracellular serine proteinase from *Aspergillus fumigatus* and its detection in tissue. *J. Med. Microbiol.* **33:**243–251.

84. Reichard, U., M. Monod, F. Odds, and R. Rüchel. 1997. Virulence of an aspergillopepsin-deficient mutant of *Aspergillus fumigatus* and evidence for another aspartic proteinase linked to the fungal cell wall. *J. Med. Vet. Mycol.* **35:**189–196.

85. Rhodes, J. C., R. B. Bode, and C. M. McCuan-Kirsch. 1988. Elastase production in clinical isolates of *Aspergillus. Diagn. Microbiol. Infect. Dis.* **10:**165–170.

86. Richard, J. L., and M. C. DeBey. 1995. Production of gliotoxin during the pathogenic state in turkey poults by *Aspergillus fumigatus* Fresenius. *Mycopathologia* **129:**111–115.

87. Roncero, C. 2002. The genetic complexity of chitin synthesis in fungi. *Curr. Genet.* **41:**367–378.

88. Rüchel, R., and U. Reichard. 1999. Pathogenesis and clinical presentation of aspergillosis, p. 1–43. *In* A. A. Brakhage, B. Jahn, and A. Schmidt (ed.), *Aspergillus fumigatus: Biology, Clinical Aspects and Molecular Approaches to Pathogenicity.* Karger, Basel, Switzerland.

89. Sakemi, S., T. Inagaki, and K. Kaneda. 2000. CJ-12,371 and CJ-12,372, two novel DNA gyrase inhibitors. Fermentation, isolation, structural elucidation and biological activities. *J. Antibiot.* **48:**134–142.

90. Sandhu, D. K., R. S. Sandhu, Z. U. Khan, and V. N. Damodaran. 1976. Conditional virulence of a *p*-aminobenzoic acid requiring mutant of *Aspergillus fumigatus. Infect. Immun.* **13:**527–532.

91. Sarfati, J., M. Diaquin, J. P. Debeaupuis, A. Schmidt, D. Lecaque, A. Beauvais, and J. P. Latgé. 2002. A new experimental murine aspergillosis model to identify strains of *Aspergillus fumigatus* with reduced virulence. *Jpn. J. Med. Mycol.* **43:**203–213.

92. Schaffner, A., C. E. Davis, T. Schaffner, M. Markert, H. Douglas, and A. I. Braude. 1986. *In vitro* susceptibility of fungi to killing by neutrophil granulocytes discriminates between primary pathogenicity and opportunism. *J. Clin. Investig.* **78:**511–524.

93. Shankar, J., S. Nigam, S. Saxena, T. Madan, and P. U. Sarma. 2004. Identification and assignment of function to the genes of *Aspergillus fumigatus* expressed at 37°C. *J. Eukaryot. Microbiol.* **51:**428–432.

94. Shepherd, M. G. 1985. Pathogenicity of morphological and auxotrophic mutants of *Candida albicans* in experimental infections. *Infect. Immun.* **50:**541–544.

95. Sheppard, D. C., G. Rieg, L. Y. Chiang, S. G. Filler, J. E. Edwards, Jr., and A. S. Ibrahim. 2004. Novel inhalational murine model of invasive pulmonary aspergillosis. *Antimicrob. Agents Chemother.* **48:**1908–1911.

96. Shibuya, K., M. Takaoka, K. Uchida, M. Wakayama, H. Yamaguchi, K. Takahashi, S. Paris, J.-P. Latgé, and S. Naoe. 1999. Histopathology of experimental invasive pulmonary aspergillosis in rats: pathological comparison of pulmonary lesions induced by specific virulent factor deficient mutants. *Microb. Pathog.* **27:**123–131.

97. Shimizu, K., and N. P. Keller. 2001. Genetic involvement of a cAMP-dependent protein kinase in a G protein signaling pathway regulating morphological and chemical transitions in *Aspergillus nidulans. Genetics* **157:**591–600.

98. Smith, J. M., J. E. Davies, and D. W. Holden. 1993. Construction and pathogenicity of *Aspergillus fumigatus* mutants that do not produce the ribotoxin restrictocin. *Mol. Microbiol.* **9:**1071–1077.

99. Smith, J. M., C. M. Tang, S. VanNoorden, and D. W. Holden. 1994. Virulence of *Aspergillus fumigatus* double mutants lacking restrictocin and an alkaline protease

in a low-dose model of invasive pulmonary aspergillosis. *Infect. Immun.* **62:**5247–5254.

100. Štaudohar, M., M. Benčina, P. van deVondervoort, H. Panneman, M. Legiša, J. Visser, and G. Ruijter. 2002. Cyclic AMP-dependent protein kinase is involved in morphogenesis in *Aspergillus niger. Microbiology* **148:**2635–2645.

101. Steinbach, W. J., J. D. K. Benjamin, S. A. Trasi, J. L. Miller, W. A. Schell, A. K. Zaas, W. M. Foster, and J. R. Perfect. 2004. Value of an inhalational model of invasive aspergillosis. *Med. Mycol.* **42:**417–425.

102. Stephens-Romero, S. D., A. J. Mednick, and M. Feldmesser. 2005. The pathogenesis of fatal outcome in murine pulmonary aspergillosis depends on the neutrophil depletion strategy. *Infect. Immun.* **73:**114–125.

103. Takasuka, T., N. M. Sayers, M. J. Anderson, E. W. Benbow, and D. W. Denning. 1999. *Aspergillus fumigatus* catalases: cloning of an *Aspergillus nidulans* homologue and evidence for at least three catalases. *FEMS Immunol. Med. Microbiol.* **23:**125–133.

104. Tang, C. M., J. Cohen, T. Krausz, S. VanNoorden, and D. W. Holden. 1993. The alkaline protease of *Aspergillus fumigatus* is not a virulence factor in two murine models of invasive pulmonary aspergillosis. *Infect. Immun.* **61:**1650–1656.

105. Tang, C. M., J. M. Smith, J. H. N. Arst, and D. W. Holden. 1994. Virulence studies of *Aspergillus nidulans* mutants requiring lysine or *p*-benzoic acid in invasive pulmonary aspergillosis. *Infect. Immun.* **62:**5255–5260.

106. Tsai, H.-F., Y. C. Chang, R. G. Washburn, M. H. Wheeler, and K. J. Kwon-Chung. 1998. The developmentally regulated *alb1* gene of *Aspergillus fumigatus:* its role in modulation of conidial morphology and virulence. *J. Bacteriol.* **180:**3031–3038.

107. Tsai, H.-F., I. Fujii, A. Watanabe, M. H. Wheeler, Y. C. Chang, Y. Yasuoka, Y. Ebizuka, and K. J. Kwon-Chung. 2001. Pentaketide melanin biosynthesis in *Aspergillus fumigatus* requires chain-length shortening of a heptaketide precursor. *J. Biol. Chem.* **276:**29292–29298.

108. Tsai, H.-F., R. G. Washburn, Y. C. Chang, and K. J. Kwon-Chung. 1997. *Aspergillus fumigatus arp1* modulates conidial pigment and complement deposition. *Mol. Microbiol.* **26:**175–183.

109. Tsai, H.-F., M. H. Wheeler, Y. C. Chang, and K. J. Kwon-Chung. 1999. A developmentally regulated gene cluster involved in conidial pigment biosynthesis in *Aspergillus fumigatus. J. Bacteriol.* **181:**6469–6477.

110. Weidner, G., C. d'Enfert, A. Koch, P. C. Mol, and A. A. Brakhage. 1998. Development of a homologous transformation system for the human pathogenic fungus *Aspergillus fumigatus* based on the *pyrG* gene encoding orotidine 5′-monophosphate decarboxylase. *Curr. Genet.* **33:**378–385.

111. Weidner, G., B. Steffan, and A. A. Brakhage. 1997. The *Aspergillus nidulans lysF* gene encodes homoaconitase, an enzyme involved in the fungus-specific lysine biosynthesis pathway. *Mol. Gen. Genet.* **255:**237–247.

112. Wheeler, M. H., and A. A. Bell. 1988. Melanins and their importance in pathogenic fungi, p. 338–387. *In* M. M. McGinnis (ed.), *Current Topics in Medical Mycology,* vol. 2. Springer-Verlag, Inc., New York, N.Y.

113. Xue, T., C. K. Nguyen, A. Romans, and G. S. May. 2004. A mitogen-activated protein kinase that senses nitrogen regulates conidial germination and growth in *Aspergillus fumigatus. Eukaryot. Cell* **3:**557–560.

Molecular Principles of Fungal Pathogenesis
Edited by Joseph Heitman et al.
©2006 ASM Press, Washington, D.C.

Chapter 25

Pathogenesis of *Pneumocystis*

A. George Smulian and Melanie T. Cushion

THE ORGANISM(S)

The Genus *Pneumocystis*

Although first described as a protozoan, species in the genus *Pneumocystis* are now considered members of the fungal kingdom. These organisms have been detected by histological or genetic methods in almost every mammalian species in which they have been sought. *Pneumocystis* organisms are found primarily in mammalian lungs, where they grow as microscopic yeast-like organisms with a life cycle that appears to be composed of asexual binary fission and sexual reproduction, resulting in asci with eight ascospores. A hyphal phase has not been identified. *Pneumocystis* can be detected in small numbers in hosts with intact immune systems, although the precise kinetics of this host-pathogen relationship are only now being systematically studied. In hosts with compromised immune status, the organisms can fill the lung alveoli and cause a lethal pneumonia.

In 1909, Carlos Chagas described the life cycle of a trypanosome he called *Schizotrypanum cruzi*, which incorporated developmental stages of both the *Trypanosoma* and the *Pneumocystis* genera (16). Through the work of the Delanoës at the Pasteur Institute, it soon became known and accepted that *Pneumocystis* represents a distinct parasite (22). Subsequently, in 1913 Chagas revised the life cycle of what he now called *Trypanosoma cruzi* to exclude the *Pneumocystis* stages. In the late 1980s, comparative sequence analyses of the nuclear small subunit of rRNA provided definitive evidence for the fungal nature of *Pneumocystis* (24, 99), which was subsequently borne out by additional genetic analyses. Sequence comparisons of *Pneumocystis* isolated from the lungs of different mammalian species and animal infectivity studies showed that the *Pneumocystis* populations infecting these different hosts were as

distinct from one another as bona fide fungal species. A provisional nomenclature system was proposed in 1994 which recognized these distinct *Pneumocystis* populations as "special forms" (93), in accordance with the *International Code of Botanical Nomenclature*, the set of guidelines that direct the fungal naming process (49). Since then, four species of *Pneumocystis* have been described in the literature from 1999 to the present: *Pneumocystis jirovecii* (from humans) (28), *Pneumocystis carinii* (from rats) (28), *Pneumocystis wakefieldiae* (from rats) (18), and *Pneumocystis murina* (from mice) (53). Questions have arisen as to the validity of the names of some species, especially the use of *P. jirovecii* as the name for the species found in humans, rather than *P. carinii*. These concerns are currently being addressed by investigators and the Committee on Fungi.

Members of the genus *Pneumocystis* have been placed in the fungal phylum Ascomycota, subphylum Taphrinomycotina (O. E. Eriksson and Winka 1997), Order Pneumocystidales (O. E. Eriks. 1994), Class Pneumocystidomycetes (sensu O. E. Erikss. & Winka 1997), Family Pneumocystidaceae (O. E. Erikss. 1994), Genus *Pneumocystis* (Delanoë & Delanoë 1912) (24a, 25) (Table 1).

Basic Biology

Life cycle

The morphologies of organisms found in the lungs of various mammalian hosts are quite similar. Three developmental forms of *Pneumocystis* are generally recognized: the trophic form (1 to 5 μm), the "precyst" or sporocyte (4 to 8 μm), and the "cysts" or asci (5 to 10 μm) (Fig. 1). The life cycle of *Pneumocystis* is thought to involve an asexual phase of binary fission by the trophic forms and a sexual phase involving mating types and the progression to

A. George Smulian • University of Cincinnati College of Medicine, VAMC, 3200 Vine Street, Cincinnati, OH 45220.
Melanie T. Cushion • Department of Internal Medicine, Division of Infectious Diseases, University of Cincinnati College of Medicine, 231 Albert Sabin Way, Cincinnati, OH 45267-0560.

Table 1. Taxonomic hierarchy of *Pneumocystis*[a]

Level	Tax on
Kingdom	Fungi
Phylum	Ascomycota
Subphylum	Taphrinomycotina
Order	Pneumocystidales
Class	Pneumocystidomycetes
Family	*Pneumocystidaceae*
Genus	*Pneumocystis*
Species	*jirovecii, carinii, wakefieldiae, murina*

[a]Hierarchy above the genus level is taken from reference 25.

ascus formation. Unfortunately, no species of *Pneumocystis* can be grown outside the mammalian lung, and verification of the life cycle, like many other aspects of its research, has been impeded by the lack of an in vitro growth system. Microscopic observations of *Pneumocystis* organisms in mammalian lungs, kinetic studies in limited in vitro culture systems, and, more recently, the presence of gene homologs associated with mating, meiosis, and sporulation in the genome of *P. carinii* have contributed to a better, but not complete, understanding of its reproductive processes (19). A proposed life cycle is shown in Fig. 2.

Transmission and epidemiology

Noticeably absent from Fig. 2 are the mechanisms of entry into and exit from the mammalian host, presence of potential reservoirs, or whether an environmental amplification cycle exists. Some

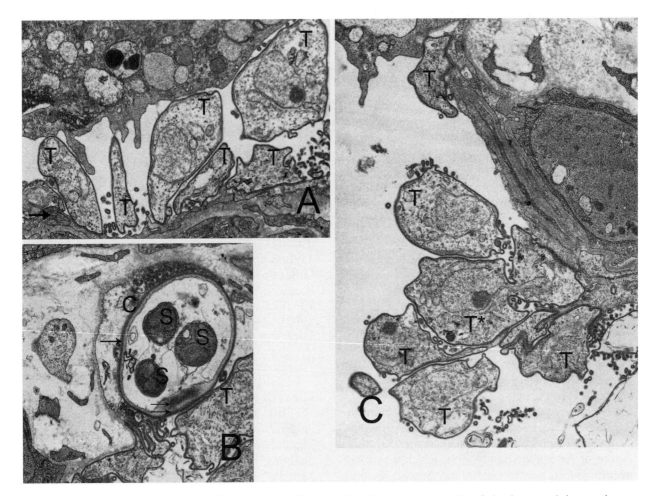

Figure 1. Morphological forms of *Pneumocystis*. Transmission electron micrographs of developmental forms of *P. carinii* are shown. (A) Trophic forms (T) in the alveolar lumen; most of these are attached to type I pneumocytes (arrow). The large cell at the top of the photograph is a macrophage (not fully shown). (B) A cyst/ascus (C) containing three ascospores (S). The single arrow indicates the double membrane and expanded periplasmic space; the double arrows indicate an area of increased periplasmic space often seen in this stage. A large trophic form is visible at the bottom right. (C) A cluster of trophic forms (T) adhered to one another and anchored by attachment to the type I pneumocytes extend into the alveolar lumen. The trophic form (T*) may be undergoing binary fission. Magnification, ×10,000. Courtesy of M. T. C and Susan G. Langreth, Uniformed Services University of the Health Sciences, Bethesda, Md.

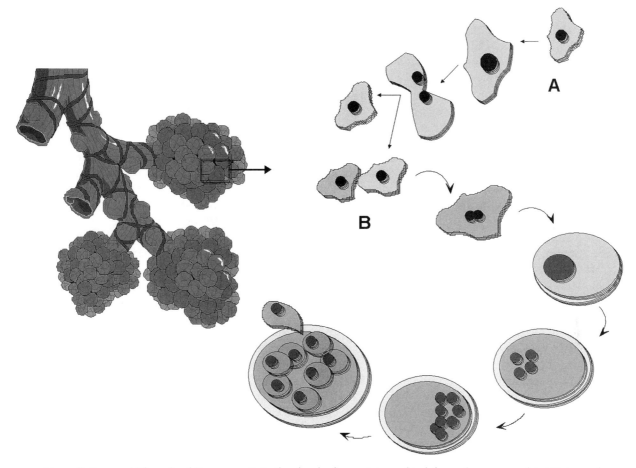

Figure 2. Proposed life cycle of *Pneumocystis* in the alveolar lumen (pictured in left panel as a rectangle). (A) Asexual binary fission. A trophic form replicates its nuclear content, separates by binary fission, and produces two genetically identical progeny. (B) Sexual replication. The two proposed mating types of *P. carinii* are represented by the light gray and dark gray cytoplasms. The two mating types fuse and undergo karyogamy to produce a diploid zygote or "precyst." Meiosis proceeds with a reduction division followed by a mitotic replication, resulting in four nuclei. An additional mitotic replication of the haploid nuclei results in eight nuclei, which are then sequestered into eight individual spores or daughter forms. It is assumed that the spores are released from the mature cyst (ascus) and become new vegetative trophic forms. Figure created with SmartDraw v.7.0.

aspects of the infection's natural history are better known than others. There is good evidence that the infection is acquired by an airborne route (43), but the agent of transmission has not been identified. The manner in which the infectious agent is released from the host is not known; it is also unknown whether the particle is immediately infectious or requires an additional period outside the lungs. Although amplification of *Pneumocystis* DNA from the environment has been reported (45), there is no evidence that an amplification cycle exists outside the mammal.

Recent studies have detected *Pneumocystis* DNA in neonatal human and rat populations, but an assessment of potential reservoirs in the general population is only in its infancy. Significantly, there is no information about replication within the lungs of nonimmunocompromised mammals or its transmission or acquisition among intact hosts. The

evidence on hand supports transmission by immunocompetent individuals or those with subclinical infections. Over 75% of the human populations surveyed in the United States and Europe are seropositive for *Pneumocystis* antigens by 4 years of age, implying that the organisms are ubiquitous, or at least plentiful, and that exposure is common (113). The sheer number of *Pneumocystis*-seropositive immunocompetent individuals argues against the fulminant *Pneumocystis* pneumonia (PCP)-infected host as the primary source of infection, since these patients are rare (in relation to the general public) and are often hospitalized or have limited access to the general population. More recent data strongly support the concept of host-to-host transmission. A geographic study of *P. jirovecii* genotypes showed that the genotype with a double mutation in the dihydropteroate synthase (DHPS) gene was the most prevalent in the populations surveyed, replacing the wild type (6). The fact that these mutations were not observed before 1993 but had spread over the globe throughout the preceding 10 years argues for a selective pressure and host-dependent spread (52, 76). Direct evidence of transmission among immunologically intact hosts is provided by the rat animal model of PCP. Infection with *Pneumocystis* is widespread in healthy nonimmunosuppressed rats available from commercial vendors (46), arguing for highly efficient transmission and maintenance of the infection within these healthy populations. The transmission of the organism from mother to offspring occurs within 1 to 2 h after birth in rats (47), and animals within an infected colony remain positive for *P. carinii* amplicons throughout their residence.

Studies now being conducted with human populations, both immunosuppressed and nonimmunsuppressed, point to certain reservoirs of infection in the general population, including patients with underlying disease states besides human immunodeficiency virus infection (HIV) or overt immunosuppression, neonates, and possibly pregnant women. This is an active and concerted area of research, and basic epidemiological principles for *Pneumocystis* infection should emerge within the next few years.

Evolution of Drug Resistance

Failure of both treatment of and prophylaxis for *Pneumocystis* has been reported (97). Organism-mediated drug resistance and host factors, such as medication compliance, inadequate or excessive immune response, and erratic drug absorption, probably play a role in clinical failure. The exact role of organism resistance is difficult to assess in the absence of a continuous-culture system.

Failure to respond to drugs such as trimethoprim-sulfamethoxazole (TMP-SMX) (97), other sulfonamide/dihydrofolate reductase (DHFR) inhibitor combinations (78), and atovaquone (112) has been noted. TMP-SMX remains the first-line regimen for both treatment of and prophylaxis for PCP (14). Widespread use of this agent as both short-term treatment of infections such as urinary tract infections and long-term use such as for PCP prophylaxis raise concerns regarding the development of drug resistance. The prevalence of TMP-SMX resistance in other organisms such as *Escherichia coli*, *Staphylococcus aureus*, and members of the family *Enterobacteraciae* has increased during the 1990s in both HIV- and non-HIV-infected populations. In these and other organisms, resistance to the sulfonamide component is most commonly associated with amino acid substitutions in regions of DHPS, the target of sulfonamides, that are critical for *p*-aminobenzoic acid binding (3).

Mutations in the gene encoding DHPS in *P. jirovecii* have been extensively studied. Sulfa medications appear to exert a selective pressure in *Pneumocystis* since nonsynonymous DHPS mutations are more common in patients who have previously been exposed to this class of drugs (42). The most common mutations described result in amino acid substitutions at position 55 or 57 or both. Based on the crystal structure of the *E. coli* DHPS enzyme, mutations at these positions are predicted to affect the substrate binding pocket of the enzyme (1). Similar point mutations conferring drug resistance have been described in *Plasmodium falciparum* and *Mycobacterium leprae* (97).

The inability to reliably culture *P. jirovecii* prevents direct examination of the causal nature of these mutations in prophylaxis or treatment failure, and their role can only be inferred from a correlation with clinical outcome in patients with PCP. Studies have consistently demonstrated an association between the use of TMP-SMX, dapsone, or sulfadoxine-pyrimethamine for PCP prophylaxis and the presence of DHPS gene mutations (42). DHPS mutations were also seen at a lower frequency in patients without reported exposure to sulfonamides. The frequency of DHPS mutations varied geographically, suggesting a role for prescription of antibiotics for other infections and person-to-person transmission (7). Recent reports of the isolation of *P. jirovecii* harboring DHPS mutations from individuals with chronic bronchitis raise concerns about a larger drug-resistant reservoir population (13, 77). Several studies have examined the effects of DHPS gene mutations on clinical outcomes such as death and PCP treatment failure with TMP-SMX or related regimens. One study found the presence of a DHPS mutation to be the strongest predictor of death, with a threefold-increased risk of death at 3 months compared to patients infected with strains with wild-type DHPS (7). In another study, DHPS mutations were associated with a twofold-increased risk of failure of treatment with TMP-SMP or dapsone and trimethoprim (7). In both of these studies, the majority of patients infected with *P. jirovecii* strains with DHPS mutations responded to TMP-SMX therapy. Several additional studies failed to demonstrate any adverse clinical outcome associated with the presence of DHPS mutations. These findings may be explained by the fact that similar mutations in *P. falciparum* DHPS result in a two- to threefold increase in the SMX MIC, which may be sufficient to result in prophylaxis failure; however, treatment failure may be avoided because of the high doses of TMP-SMX used for treatment.

More recently the DHPS gene of *P. jirovecii* has been used to functionally complement DHPS-deficient strains of *E. coli* and *Saccharomyces cerevisiae* (48, 75). The *S. cerevisiae* model was also used to examine the role of the T55A and P57S mutations in sulfa resistance. Findings consistent with altered substrate binding were noted. Interestingly, resistance varied with different sulfonamides, suggesting that these models may be used to predict alternate treatment options for DHPS mutants. Increased *p*-aminobenzoic acid synthesis was also suggested as a mechanism of sulfonamide resistance.

Three studies have correlated mutations in DHFR, the target of pyrimethamine and the trimethoprim component of TMP-SMX, with unsuccessful PCP prophylaxis (68, 78, 101). Two studies failed to demonstrate any nonsynonymous mutations in conserved regions of the enzyme. A recent study found nonsynonymous mutations in putative active sites of DHFR in isolates from 9 of 15 patients who failed to respond to a PCP prophylaxis regimen containing a DHFR inhibitor compared to 2 of 18 patients who did not receive such prophylaxis (78). These results suggest that DHFR mutations may contribute to *P. jirovecii* resistance.

Atovaquone, a second-line agent in the prophylaxis for and treatment of PCP, has also been associated with prophylaxis failures. Mutations in cytochrome *b* gene sequences have been identified in *P. jirovecii* isolated from individuals failing to respond to atovaquone prophylaxis (51). When similar mutations were introduced into the gene encoding cytochrome *b* of *S. cerevisiae*, resistance to atovaquone was noted, with increases in the 50% inhibitory concentration from 25 to >500 nM with some described mutations (55). These studies provide a structural and molecular basis for atovaquone drug resistance based on altered binding of atovaquone to the cytochrome bc_1 complex.

THE DISEASE

PCP is a disease associated with immunocompromised individuals. Initially described as the cause of interstitial plasma cell pneumonia in malnourished infants housed in overcrowded conditions, it gained prominence as the most common opportunistic infection in HIV-infected individuals (106). The major presenting symptoms in the immunocompromised host are shortness of breath, fever, and a nonproductive cough. Non-HIV-infected patients generally report symptoms for 1 to 2 weeks prior to seeking medical attention, while in HIV-infected individuals the onset is more gradual and symptoms may be present for weeks to months (56). Despite the more gradual onset of symptoms, the organismal burden is higher in HIV-infected patients and the lung damage is less severe. Despite these generalizations, the clinical picture is quite varied and may be further influenced by prior use of prophylaxis.

The chest radiograph classically exhibits bilateral diffuse infiltrates extending from the perihilar regions, but atypical findings such as unilateral infiltrates, nodules, lymphadenopathy, and effusions may be seen. In patients receiving medical care, a heightened index of suspicion leads to more frequent early diagnosis while the radiograph may still be equivocal or normal. High-resolution computed tomography may be useful under these circumstances (83). Extrapulmonary spread was considered a rare event in the pre-HIV era but has been more frequently detected in severely immunocompromised hosts. The main sites for involvement are lymph nodes, spleen, bone marrow, gastrointestinal tract, eyes, thyroid, and kidneys.

The cardinal pathological finding is the formation of a foamy eosinophilic exudate filling the pulmonary alveoli. The foamy exudate is composed of *Pneumocystis* organisms and host proteins. The organisms remain extracellular throughout all stages of infection and are killed following phagocytosis by activated macrophages in the presence of T cells. The host inflammatory response is usually inconspicuous, with scanty mononuclear cell infiltrating. By electron microscopy, *Pneumocystis* organisms can be seen tightly adherent to type I pneumocytes but not to type II cells.

HOST-PARASITE INTERACTIONS

Pneumocystis Antigens

Immunoblotting studies identified the major immunoglobulin G-reactive antigens of *Pneumocystis* from humans and animal models to be present at two regions with distinct molecular masses (113). The major surface glycoproteins (MSGs) migrate between 95 and 120 kDa, and a second group migrate at 35 to 55 kDa and probably contains many discrete antigens.

Within the genome of a given *Pneumocystis* species, about 100 different MSGs and related genes reside at the telomeric ends of each of the linear chromosomes (98). The proteins encoded by the *MSG* family of genes are localized to the surface of the organism and evoke strong humoral and cell-mediated responses (113). It is thought that the MSGs function in a variety of ways by promoting adherence. These molecules bind to host lung cells by interacting with proteins such as fibronectin, vitronectin, and surfactant protein A (65, 73, 79), some of which may aid in circumventing the immune response. *Pneumocystis* life cycle stages form tightly bound aggregates within the lung alveoli, presumably mediated by MSG-MSG interactions. A role for MSGs as adhesins during the mating response has been suggested recently (19). Investigation of the molecular mechanisms involved in the expression of this family of genes has identified a sequence common to nearly all MSG mRNAs, the upstream conserved sequence (UCS), which is present in a single copy per organism genome and resides on a single chromosome (98). It is thought that the *MSG* genes are transcribed only when adjacent to the UCS and are translocated from subtelomeric regions on each chromosome via reciprocal exchange, gene conversion, or a recombination event (Fig. 3). The UCS encodes a signal peptide thought to aid in chaperoning the molecule through the endoplasmic reticulum and Golgi for processing and ultimately delivering the molecules to the cell surface, where they are the most abundant surface protein.

The MSGs are heavily glycosylated and contain protective B- and T-cell epitopes (113). Immunization trials in rodents using native or recombinant MSG have produced conflicting results; in some cases a protective effect was observed, while in other studies no such protection occurred (31, 102, 104). Besides the multigene family of MSGs located at the telomeric ends of the chromosomes, other repetitive gene families are present. These include the kexin-like subtilisin protease (PRT) family of genes, which encode proteins that are probably involved in processing of signal peptide on the nascent MSG (3), and the MSR (MSG-related proteins) genes, which encode proteins similar to but distinct from the MSG (54). It should be noted that the PRT genes exist as a multiple gene family in *Pneumocystis* species infecting rats but as only a single copy in *P. murina* (63) and *P. jirovecii* (60), illustrating that there are significant differences among the species of *Pneumocystis*. Very little is known about the protective or nonprotective humoral responses to the PRT and MSR gene products since they have been the subject of few studies. A similar subtelomeric arrangement

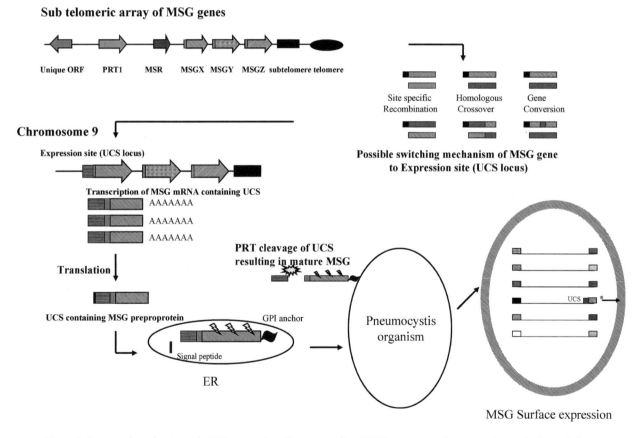

Figure 3. Proposed mechanism of MSG expression. Genes encoding MSG are arranged in arrays in a subtelomeric location on the ends of each chromosome. The arrays also contain genes encoding the MSG-related protein (MSR) and protease (PRT). One MSG gene present at an expression site (UCS locus) on chromosome 9 is actively transcribed. MSG gene translocation to the expression site may involve sequence-specific recombination, homologous crossover, or gene conversion. All transcribed MSG mRNAs contain the UCS sequence and result in the production of a preprotein, which is postulated to be processed through the endoplasmic reticulum (ER). Cleavage of a signal peptide would result in a proprotein which undergoes posttranslational modification including attachment of a glycosyl phosphatidylinositol (GPI) anchor. PRT1 proteases may be involved in the cleavage of the UCS sequence from the proMSG protein to result in the surface expression of the mature MSG protein. (Adapted from reference 98.)

of genes encoding proteins involved in organism adhesion has now been described for other fungal species; these include the *FLO1, FLO10,* and *FLO11* floculin genes of *S. cerevisiae* and the *ALS* genes of *Candida albicans.*

The other major antigen group migrates as a broad band between 35 and 45 kDa in *P. jirovecii* and between 45 and 55 kDa in *P. carinii* and *P. wakefieldiae.* One gene encoding an antigen in the 45- to 55-kDa region of *P. carinii* (p55) was cloned and shown to be a single-copy gene which hybridized to a chromosome band of about 670 kb on Southern-blotted electrophoretic karyotypes (89). The function of this antigen is unclear, but it has been shown to elicit a cell-mediated immune response in rats previously exposed to *Pneumocystis* (105). A 7-amino-acid repeat motif that maps to the C-terminal portion of the molecule was shown to be an immunodominant epitope of this antigen. The cells responding to p55 were shown to be CD4$^+$ cells that secrete a Th1 cytokine pattern. Additional closely related genes, some of which are also transcribed, have recently been identified, but their role in the immune response has not been studied (69). The corresponding p55 gene from *P. murina* was recently cloned and characterized and shown to be present in a single copy as well (69).

Innate Immunity

Alveolar macrophages are the first line of defense against *Pneumocystis* and the principal effector cell in clearing the organism from the lungs (111). Alveolar macrophage function is impaired in HIV-infected patients as well as in cancer and transplant patients receiving immunosuppressive drugs. However, activated macrophages, in the absence of CD4 cells, are unable to control *Pneumocystis* infection (37). Also needed for *Pneumocystis* clearance is the urokinase-type plasminogen activator, which aids the recruitment of inflammatory cells (10). An enolase produced by *Pneumocystis* can activate plasminogen and thus may impair the host plasminogen activator (27). *Pneumocystis* recognition and adherence to macrophages occur by multiple pathways involving MSG and β-glucan in the organism, extracellular matrix and surfactant proteins, mannose, and Fc receptors on the cells (34). HIV downregulates mannose receptor expression, which results in decreased binding and uptake of *Pneumocystis* (57). Recently the macrophage Dectin-1 β-glucan receptor was shown to be important in the phagocytosis and subsequent killing of *Pneumocystis* (94). Blockage of Dectin-1 binding with an anti-Dectin-1 antibody inhibited *Pneumocystis* binding to macrophages and reduced macrophage MIP2 production. Macrophages ingest, degrade, and kill *Pneumocystis*, releasing cytokines such as tumor necrosis factor alpha (TNF-α),

eicosanoids, and reactive oxidants (62, 110). Although nitric acid is released by macrophages, its role in host defenses against *Pneumocystis* is unclear (23, 88). HIV changes the macrophage cytokine response, which may play a role in impaired *Pneumocystis* clearance (57). *Pneumocystis* itself impairs phagocytosis by promoting shedding of the mannose receptor, whereas the administration of a soluble mannose immunoadhesin counteracts this process (90).

Alterations in the surfactant system also appear to be important in innate immune responses to *Pneumocystis*. There is a fall in the amounts of surfactant phospholipids (mainly phosphatidycholine) due to inhibition of phospholipid secretion mediated by MSG (66, 82, 100). Changes in the surfactant proteins include a decline in the levels of SP-B and SP-C and a rise in the levels of SP-A and SP-D (11). SP-A- and SP-D-deficient mice demonstrated enhanced susceptibility to *Pneumocystis* infection. SP-A-deficient mice demonstrated attenuated production of proinflammatory cytokines and reactive oxygen-nitrogen species (4). In contrast, SP-D-deficient mice developed higher organism burdens despite having increased expression of inducible nitric oxide synthase and higher levels of reactive nitrogen species (5). The pathogenic significance of surfactant abnormalities is supported by reports that have shown that the administration of surfactant is of benefit in the treatment of pneumocystosis in humans (40).

Humoral Responses

The importance of the humoral immune response in the control of *Pneumocystis* infection was predicted by its occurrence in hosts with specific immunodeficiencies and established by experimental-animal studies. *Pneumocystis* infection has been found in B-cell-deficient mice and other animals (67, 70) and in patients with hypogammaglobulinemia (114). Administration of specific monoclonal antibodies or polyclonal hyperimmune sera raised against *Pneumocystis* decreased the organism burden in immunosuppressed-animal models (30, 85). Intratracheal administration of organisms as a means of immunizing CD4-depleted mice was shown to afford protection against subsequent *Pneumocystis* challenge (37). Addition of specific antisera and complement to macrophages in vitro led to a rapid binding, internalization, and degradation of the organisms (91).

It is clear from early serological studies and later surveys using more sophisticated detection techniques that *Pneumocystis* elicits a strong humoral response. The indirect fluorescent-antibody assay, the enzyme-linked immunosorbent assay, and the complement fixation assay with whole or fractionated *Pneumocystis* organisms or crude antigen preparations showed that most of the human population had antibodies to

Pneumocystis (113). Immunoblotting techniques permitted an assessment of responses to specific antigens (92). The consensus of such survey studies provided an appreciation of the early exposure by humans to *P. jirovecii*. By the age of 3 years, most children have developed antibodies to the organism (92). Geographic surveys of adult exposures showed that 76% (of 680 samples) reacted with *P. jirovecii* antigens by immunoblotting techniques (90). Most (60%) of the reactivity was observed in the area of 35 to 45 kDa, while only about 30% of the sera reacted at ca. 95 kDa, where the MSG family migrates. The frequency of reactivity in the MSG region varied significantly by geographic region, suggesting that different strains of *P. jirovecii* may be predominant in a given area or that distinct MSG isotypes predominate in a region. Development of serodiagnostic assays has suffered because of the apparent ubiquitous presence of *Pneumocystis* and concomitant seroreactivity of most of the human population. Surprisingly, most serological surveys have shown the frequency and/or level of antibodies to *Pneumocystis* among HIV-infected patients and other immunocompromised hosts were similar to those in healthy controls (92). This frequent presence of antibodies in most human populations has contributed to the view that humoral immunity is not a significant factor in the host defense against *Pneumocystis* infection.

More recent studies have used recombinant technology to further probe the complexity of the humoral response in humans. Workers in one laboratory expressed a highly conserved region of the carboxy terminus of an MSG gene that was recognized by all serum samples and thus could not discriminate among patients with active disease, those with previous episodes, and healthy controls (74). In a separate study, three overlapping fragments, MSG A, MSG B, and MSG C, spanning an entire human MSG gene were generated (21). Recognition of MSG A and MSG B was significantly lower in HIV-positive patients than normal healthy controls. MSG C was recognized by patients with previous episodes of PCP, and the responses did not vary by geographic location. These results hold promise for the development of future diagnostic tools because different regions of the MSG molecule were shown to identify previously unrecognized differences among populations.

Cell-Mediated Immunity

Impaired cellular immunity has long been considered to be the most important predisposing factor in the development of PCP (106). The clearest evidence for the role of defective cell-mediated immunity in the development of pneumocystosis in humans is based on persons infected with HIV. The risk that adult HIV-infected patients will develop PCP increases greatly when circulating CD4 cell numbers fall below $200/mm^3$ (106). Because CD4 counts are much higher in young children than in adults, different criteria must be used. The presence of other clinical complications of HIV (e.g., fever and oral candidiasis) increases the risk of pneumocystosis independent of the CD4 count. Cases of pneumocystosis associated with low CD4 counts have been encountered in cancer patients receiving cytotoxic drugs, in adults with idiopathic CD4 T lymphopenia, and in otherwise healthy individuals with subtle T-cell defects (59). The correlation between the number of CD4 cells in peripheral blood or bronchoalveolar lavage (BAL) fluid with the severity or outcome of pneumonia remains controversial, with some studies suggesting a poor outcome with lower CD4-cell levels, but other studies have failed to confirm this. In a recent series, peripheral CD4-cell numbers did not correlate with outcome but low CD4-cell levels in BAL fluid did indicate a worse prognosis (2).

Animal models allow modulation of individual components of the immune response and allow direct testing of hypotheses regarding the development of protective immunity against pneumocystosis (8). Most experimental models rely on the provocation of a defect in cell-mediated immunity. Pneumocystosis may also be induced in normal rodents by the administration of corticosteroids, and corticosteroid-induced immunosuppression models have been used for over three decades (44). Protein malnutrition, ethanol administration, and an immature immune system also impair host defenses against *Pneumocystis* (20, 29). Naturally occurring outbreaks of PCP have occurred in mice with severe combined immunodeficiency disease (SCID mice) and athymic (nude) mice and rat colonies (115). $CD4^+$ T cells are central to the immune response, as evidenced in both cell depletion and reconstitution experiments, and by knockout mice (35, 38, 87). Monoclonal antibody-mediated CD4-cell depletion renders mice susceptible to *Pneumocystis* infection either by environmental exposure or by direct intratracheal inoculation of *Pneumocystis* organisms. SCID mice with pneumocystosis clear the infection following reconstitution with $CD4^+$ T cells (84). Depletion of $CD4^+$ T cells from the reconstituting T cells abrogates this effect (38). The specific *Pneumocystis* epitopes recognized by $CD4^+$ T cells have not been well characterized. T cells of humans, rats, and mice proliferate in response to MSG in vitro (103). Splenic T cells from rats previously exposed to *Pneumocystis* also proliferate in response to the p55 antigen (105). CD4 cells mediate their effects via control of a wide range of cytokine and chemokine response in the lungs. When infected SCID mice are immunologically reconstituted with splenocytes, for example, lung expression of the cytokines interleukin-1α (IL-1α), IL-1β, IL-3, IL-6, gamma interferon (INF-γ), TNF-α, and TNF-β and the chemokines lymphotactin, macrophage inflammatory protein 1α (MIP-1α), MIP-1β, MIP-2, and monocyte chemotactic protein 1 (MCP-1) is increased (118). Clearance may also be effected by administration of *Pneumocytis*-primed dendritic cells and hyperimmune serum in the absence of CD4$^+$ T cells (85, 121).

The role of CD8$^+$ T cells in clearance and control of *Pneumocystis* infection is less well defined. CD8 accumulate with CD4 cells in the lungs of immunologically intact mice following *Pneumocystis* inoculation. CD8 cells do not appear to play a role in elimination of infection in the SCID mouse model; however, CD8$^+$ T cells appear to play some role in the control of infection in other models (9, 84). Depletion of CD8 cells alone does not result in susceptibility to *Pneumocystis*, whereas depletion of CD4$^+$ and CD8$^+$ T cells results in a heavier organism burden than does CD4$^+$ T cell depletion alone. Some of the CD8$^+$ T cells recruited to the lungs in response to *Pneumocystis* infection demonstrate a T cytotoxic-1 phenotype as evidenced by the production of IFN-γ (72). The CD8 cells effect clearance of the organism in adoptive-transfer studies. In contrast, CD8$^+$ T cells with non-T-cytotoxic-1 phenotypes lacked effector activity and contributed to lung injury on adoptive transfer. γδ T cells also contribute to host defenses against *Pneumocystis*; however, their specific role is less well defined. Recent studies with mice deficient in γδ T cells demonstrated more rapid and complete clearance of infection. The augmented resolution was associated with elevated IFN-γ levels produced in BAL fluid by CD8$^+$ T cells (95). Accessory molecules involved in the interaction of T cells with antigen-presenting cells, such as CD40 and CD154 (CD40 ligand), are also important (33, 117).

There is now a considerable body of evidence showing that the immune/inflammatory response to *Pneumocystis* can be harmful as well as helpful in the host lung. These effects are complex and depend to some degree on the experimental model being used. Effective clearance of *Pneumocystis* is dependent on CD4$^+$ T cells and cytotoxic CD8$^+$ T cells, but these same T-cell subsets may result in inflammation-mediated lung injury. CD4 cells, particularly those expressing the CD25$^-$ phenotype, are a major contributor responsible for these inflammatory changes (41). Immunologic reconstitution of infected SCID mice with immune splenocytes results in clearance of *Pneumocystis* that is associated with a hyper-inflammatory response composed of increased levels

of proinflammatory cytokines and chemokines and reduced oxygenation and compliance (119). The administration of large numbers of MSG-sensitized splenocytes or CD4 cells to rats with corticosteroid-induced pneumocystosis results in clinical illness and cytokine cascade along with a reduction in *Pneumocystis* burden (102). The presence of corticosteroids appears not to have an influence on the deleterious effects of the inflammatory response in this model. Factors that can ameliorate these effects include hyperimmune serum and CD8 cells. On the other hand, work with another mouse model has shown that mice depleted of CD4 cells and exposed to *Pneumocystis* develop pneumonia, impaired respiration, and a hyperinflammatory response characterized by the influx of CD8 cells and neutrophils (119). Depletion of CD8 cells abrogates this response. CD8$^+$ T cells have also been implicated in the derangements of surfactant function that occur during PCP. *Pneumocystis* infection reduces the levels of the hydrophobic surfactant proteins B and C, leading to increased alveolar surface tension, atelectasis, and hypoxia. Inflammation induced by CD8$^+$ T cells is associated with decreased lung compliance and decreased oxygenation. In SCID mice reconstituted with CD8-depleted splenocytes, the surfactant dysfunction and the physiological derangements are abrogated.

Exposure to *Pneumocystis* or its antigens stimulates the production of a multitude of cytokines. Two proinflammatory cytokines, TNF-α and IL-1, are important in host defenses against the organism, particularly in the early stages of the infections (34). IL-1, TNF, and nitrous oxide are insufficient to prevent the development of natural *Pneumocystis* infection in T-cell-deficient mice such as class II major histocompatibility complex-deficient or RAG-1$^{-/-}$ mice (36). IL-6, another proinflammatory cytokine, has been produced in response to *Pneumocystis*, but its contributions to host resistance to the organism are unclear (17). Infected mice treated with an anti-IL-6 antibody have significantly more inflammatory cells in the lungs, suggesting that IL-6 might downregulate inflammation associated with *Pneumocystis* infection in mice. IFN-γ and granulocyte-macrophage colony-stimulating factor are important contributors to host defense by macrophage activation or in cooperation with TNF-α (34). The role of IFN-γ is particularly complex. One study has shown that the deletion of IFN-γ or TNF-α or IFN-γ receptor genes did not hinder the clearance of *Pneumocystis* infection from the lungs; however, deletion of both genes led to severe infection (86). IFN-γ also influences the host inflammatory response (34). IL-10 also modulates the host inflammatory response, but no role in host defense against *Pneumocystis* has been found for IL-4 or granulocyte colony-stimulating factor (80).

Although much information has been generated regarding immune responses to *Pneumocystis* in human subjects, the complexity of human disease often precludes unambiguous interpretation. Studies of cellular immune function have shown that HIV-infected patients have a decline in peripheral blood lymphocyte proliferative responses to whole *Pneumocystis* or MSG with progression of the disease and a fall in the number of CD4 cells (103). A similar decline occurs in the Th1-like cytokine response (IFN-α) but not in the Th2-like cytokine response (IL-4). Patients who have recovered from pneumocystosis exhibit higher proliferative and IL-4 responses to MSG than do HIV-infected patients at a similar stage of the infection who have never had pneumocystosis. Thus, *Pneumocystis* patients retain enough CD4 memory cells to recognize the organism but exhibit a shift from a Th1- to a Th2-like response with progression of HIV.

The role of other cells in the host defenses against *Pneumocystis* is poorly understood. One study has shown that neutrophils from *Pneumocystis* patients stimulated with the organism have an impaired respiratory burst compared with the cells of healthy controls (111). *Pneumocystis* can also activate NK cells (116). A role for NK cells has been suggested by reports of the occurrence of pneumocystosis in HIV-infected or other immunodeficient patients who have NK cells decreased in either numbers or function (12).

The contribution of the host inflammatory response to lung damage in HIV-infected patients with pneumocystosis has been suggested by studies which have correlated increased numbers of neutrophils and levels of IL-8 in BAL fluid with more severe pneumonia and a worse prognosis (113). IL-8 functions as a potent chemoattractant, and its release is mediated by host inflammatory-cell interaction with MSG (113). Alterations in eicosanoids, TNF-α, IL-1, other cytokines, and inflammatory mediators have also been noted in these studies and other reports; however, the pathogenic significance of these changes is unclear. HIV-infected patients with PCP also frequently experience a worsening of respiratory function soon after receiving antimicrobial drugs, and the prompt administration of corticosteroids can ameliorate or prevent this outcome and improve survival (61). It is thought that the beneficial effects of corticosteroids are due to their anti-inflammatory properties or their effects on surfactant components, yet studies examining these issues have produced inconsistent results.

The interaction of HIV and *Pneumocystis* with CD4 and other T cells in humans is of considerable potential interest but has received only limited attention. One report has suggested that HIV depletes *Pneumocystis*-specific T-cell clones horizontally (i.e., by reducing the number of memory cells in the progeny without affecting the number of clones) (64). On the other hand, *Pneumocystis* infection enhances HIV replication in the lungs and may accentuate the depletion of CD4 cells (58). The number of CD8 cells, which can be infected with HIV, increase in the BAL fluid of HIV patients with pneumocystosis, but the pathogenic significance of this accumulation is unknown.

Potential Virulence Factors

A lack of clonal populations and of the ability to manipulate the genetics of *Pneumocystis* has prevented the demonstration of specific virulence factors. However, it may be postulated that some of the unusual characteristics of the organism may have developed or been maintained as virulence factors. Two such factors include the complex system of variable surface antigens and the single ribosomal gene locus.

The *Pneumocystis* genome contains a single copy of the nuclear ribosomal genes in the ribosomal gene locus (32). Experimentally induced reductions in the number of copies of ribosomal genes have been associated with prolonged doubling time and reduced metabolic activity in organisms such as *S. cerevisiae*. It may thus be postulated that the single copy of the ribosomal gene cluster in *Pneumocystis*, permitting low-level but sustainable growth, may be maintained as an adaptive factor supporting long-term commensal existence with its mammalian host.

The complex surface antigens of *Pneumocystis* may also be considered potential virulence factors. The best understood of these antigens is the MSG family described above. Variation in surface antigen expression is achieved by translocation of a different MSG gene into the UCS expression by recombination or gene conversion (98). This results in the expression of a new MSG proprotein which undergoes modification allowing surface expression. Like other microbes that "switch" their antigen coat, such as *Trypanosoma* and *Borrelia* spp., this process may provide *Pneumocystis* with a means of evading immune surveillance. The well-known adherent properties of these surface glycoproteins may also promote the creation of impenetrable organism clusters that deflect host attempts at eradication, while in turn permitting close communication among the clustered organisms. Other variable antigens, such as the subtilisin-like protease family expressed on the surface and the p55 family of antigens, may also act as virulence factors by promoting immune evasion.

New Concepts in the Host-Parasite Relationship

Since the outbreaks of the 1980s, the demographics of HIV infection have changed worldwide. The medically underserved, minority, and female populations in the United States are now considered at greatest risk (50, 81). In these populations, PCP remains a leading opportunistic infection (26). In countries considered to be "developing" (i.e., those in Africa, Asia, the Philippines, and Central and South America), the number of individuals infected with HIV has increased dramatically, with one-third of the population being infected in some sub-Saharan countries. Although rare in Africa during the first decade of the AIDS pandemic, the incidence of PCP has risen dramatically as it has in other developing countries, reaching a 25 to 80% coinfection rate with tuberculosis and having a mortality rate ranging from 20 to 80% (26).

Besides the HIV-infected host, PCP is occurring more frequently in other patient groups that in some cases were rarely affected previously (Table 2). Chronic immunosuppression induced by immunosuppressive agents or underlying disease states marks the most common susceptible patient populations. Paradoxically, PCP has been found in both HIV-negative and -positive patients who have undergone immune reconstitution after tapering of steroids or after initiation of highly active antiretroviral therapy (120).

Perhaps associated with the occurrence of *P. jirovecii* in previously rarely affected subpopulations may be its ability to colonize individuals with intact or slightly debilitated immune systems. Apparent asymptomatic carriage has been shown for both healthy and HIV-infected individuals. In a recent study, *P. jirovecii* was detected by PCR in 17 (18%) of 93 consecutive BAL fluid samples from HIV-negative patients undergoing diagnostic bronchoscopy (71). Analysis of potential risk factors showed that only glucocorticoid use (and not lung cancer or chronic obstructive pulmonary disease [COPD]) was associated with the presence of *P. jirovecii*. Other factors appear to favor colonization with *P. jirovecii*. In a large study of 367 patients with suspected bacterial pneumonia, carriage of *P. jirovecii* (detected by PCR) was statistically associated with old age, concurrent disease, and, again, steroid treatment (39). *P. jirovecii* was detected in 36.7% of patients with severe COPD (GOLD stage IV) compared to 5.3% of smokers with normal lung functions or less severe COPD (stages 0 to III) and 9.1% of control subjects (77). Moreover, a link was established between the severity of airflow obstruction and the presence of *P. jirovecii*, suggesting a role for these organisms in the pathogenesis of severe COPD.

The state of pregnancy may also facilitate asymptomatic carriage. A study of 33 pregnant healthy women in their third trimester and 28 healthy women within 15 days of their menstrual cycle showed the presence of *P. jirovecii* DNA in 5 (15.5%) of deep nasal swabs from the pregnant women compared to none in the nonpregnant women (109). It was postulated that the slightly compromised immune system of pregnant women may foster these low-level infections.

The primary infection was once thought to be asymptomatic. However, recent studies have shown a high prevalence of *P. jirovecii* in neonates with respiratory symptoms (107), calling into question this widespread assumption. Moreover, reports of the presence of *P. jirovecii* in cases of sudden infant death syndrome (15, 108) also suggest that the organism may be more virulent than once thought.

Table 2. Underlying conditions leading to development of PCP

Underlying state	Reference(s)
Transplantation recipients	44a, 59
Patients receiving high-dose corticosteroids for brain neoplasms and inflammatory and collagen vascular disorders	44a, 59, 106
Asthma	44a
COPD	13, 77
Cystic fibrosis	44a
Epstein-Barr virus infection	44a
Lupus erythematosus	44a
Rheumatoid arthritis	44a
Thyroiditis	44a
Ulcerative colitis	44a

REFERENCES

1. Achari, A., D. O. Somers, J. N. Champness, P. K. Bryant, J. Rosemond, and D. K. Stammers. 1997. Crystal structure of the anti-bacterial sulfonamide drug target dihydropteroate synthase. *Nat. Struct. Biol.* 4:490–497.
2. Agostini, C., F. Adami, L. W. Poulter, D. Israel-Biet, M. Freitas e Costa, A. Cipriani, R. Sancetta, M. C. Lipman, K. Juvin, A. D. Teles-Araujo, P. Cadrobbi, G. Masarotto, and G. Semenzato. 1997. Role of bronchoalveolar lavage in predicting survival of patients with human immunodeficiency virus infection. *Am. J. Respir. Crit. Care Med.* 156:1501–1507.
3. Ambrose, H. E., S. P. Keely, E. M. Aliouat, E. Dei-Cas, A. E. Wakefield, R. F. Miller, and J. R. Stringer. 2004. Expression and complexity of the PRT1 multigene family of *Pneumocystis carinii*. *Microbiology* 150:293–300.
4. Atochina, E. N., J. M. Beck, A. M. Preston, A. Haczku, Y. Tomer, S. T. Scanlon, T. Fusaro, J. Casey, S. Hawgood, A. J. Gow, and M. F. Beers. 2004. Enhanced lung injury and delayed clearance of *Pneumocystis carinii* in surfactant protein A-deficient mice: attenuation of cytokine responses and reactive oxygen-nitrogen species. *Infect. Immun.* 72:6002–6011.
5. Atochina, E. N., A. J. Gow, J. M. Beck, A. Haczku, A. Inch, H. Kadire, Y. Tomer, C. Davis, A. M. Preston,

F. Poulain, S. Hawgood, and M. F. Beers. 2004. Delayed clearance of *Pneumocystis carinii* infection, increased inflammation, and altered nitric oxide metabolism in lungs of surfactant protein-D knockout mice. *J. Infect. Dis.* **189:**1528–1539.

6. Beard, C. B., J. L. Carter, S. P. Keely, L. Huang, N. J. Pieniazek, I. N. Moura, J. M. Roberts, A. W. Hightower, M. S. Bens, A. R. Freeman, S. Lee, J. R. Stringer, J. S. Duchin, C. del Rio, D. Rimland, R. P. Baughman, D. A. Levy, V. J. Dietz, P. Simon, and T. R. Navin. 2000. Genetic variation in *Pneumocystis carinii* isolates from different geographic regions: implications for transmission. *Emerg. Infect. Dis.* **6:**265–272.

7. Beard, C. B., P. Roux, G. Nevez, P. M. Hauser, J. A. Kovacs, T. R. Unnasch, and B. Lundgren. 2004. Strain typing methods and molecular epidemiology of *Pneumocystis* pneumonia. *Emerg. Infect. Dis.* **10:**1729–1735.

8. Beck, J. M., and A. G. Harmsen. 1998. Lymphocytes in host defense against *Pneumocystis carinii*. *Semin. Respir. Infect.* **13:**330–338.

9. Beck, J. M., R. L. Newbury, B. E. Palmer, M. L. Warnock, P. K. Byrd, and H. B. Kaltreider. 1996. Role of CD8⁺ lymphocytes in host defense against *Pneumocystis carinii* in mice. *J. Lab. Clin. Med.* **128:**477–487.

10. Beck, J. M., A. M. Preston, and M. R. Gyetko. 1999. Urokinase-type plasminogen activator in inflammatory cell recruitment and host defense against *Pneumocystis carinii* in mice. *Infect. Immun.* **67:**879–884.

11. Beers, M. F., E. N. Atochina, A. M. Preston, and J. M. Beck. 1999. Inhibition of lung surfactant protein B expression during *Pneumocystis carinii* pneumonia in mice. *J. Lab. Clin. Med.* **133:**423–433.

12. Bonagura, V. R., S. Cunningham-Rundles, B. L. Edwards, N. T. Ilowite, J. F. Wedgwood, and D. J. Valacer. 1989. Common variable hypogammaglobulinemia, recurrent *Pneumocystis carinii* pneumonia on intravenous gamma-globulin therapy, and natural killer deficiency. *Clin. Immunol. Immunopathol.* **51:**216–231.

13. Calderon, E., C. de la Horra, F. J. Medrano, A. Lopez-Suarez, M. A. Montes-Cano, N. Respaldiza, J. Elvira-Gonzalez, J. Martin-Juan, A. Bascunana, and J. M. Varela. 2004. *Pneumocystis jirovecii* isolates with dihydropteroate synthase mutations in patients with chronic bronchitis. *Eur. J. Clin. Microbiol. Infect. Dis.* **23:**545–549.

14. Centers for Disease Control and Prevention. 1999. USPHS/IDSA guidelines for the prevention of opportunistic infections in persons infected with human immunodeficiency virus. U.S. Public Health Service (USPHS) and Infectious Diseases Society of America (IDSA). *Infect. Dis. Obstet. Gynecol.* **8:**5–74.

15. Chabe, M., S. L. Vargas, I. Eyzaguirre, E. M. Aliouat, A. Follet-Dumoulin, C. Creusy, L. Fleurisse, C. Recourt, D. Camus, E. Dei-Cas, and I. Durand-Joly. 2004. Molecular typing of *Pneumocystis jirovecii* found in formalin-fixed paraffin-embedded lung tissue sections from sudden infant death victims. *Microbiology* **150:**1167–1172.

16. Chagas, C. 1909. Nova tripanozomiaze humana. *Mem. Inst. Oswaldo Cruz* **1:**159–218.

17. Chen, W., E. A. Havell, F. Gigliotti, and A. G. Harmsen. 1993. Interleukin-6 production in a murine model of *Pneumocystis carinii* pneumonia: relation to resistance and inflammatory response. *Infect. Immun.* **61:**97–102.

18. Cushion, M. T., S. Keely, and J. R. Stringer. 2004. Molecular and phenotypic description of *Pneumocystis wakefieldiae* sp. nov., a new species in rats. *Mycologia* **96:**429–438.

19. Cushion, M. T. 2004. Pneumocystis: unraveling the cloak of obscurity. *Trends Microbiol.* **12:**243–249.

20. D'Souza, N. B., J. F. Mandujano, S. Nelson, W. R. Summer, and J. E. Shellito. 1995. Alcohol ingestion impairs host defenses predisposing otherwise healthy mice to *Pneumocystis carinii* infection. *Alcohol Clin. Exp. Res.* **19:**1219–1225.

21. Daly, K. R., C. J. Fichtenbaum, R. Tanaka, M. J. Linke, R. O'Bert, T. D. Thullen, M. S. Hui, A. G. Smulian, and P. D. Walzer. 2002. Serologic responses to epitopes of the major surface glycoprotein of *Pneumocystis jirovecii* differ in human immunodeficiency virus-infected and uninfected persons. *J. Infect. Dis.* **186:**644–651.

22. Delanoë, P., and M. Delanoë. 1912. Sur les rapports des kystes des Carinii des poumon des rats avec le *Trypanosoma lewisii*. *C. R. Acad. Sci.* (Paris) **155:**658–660.

23. Downing, J. F., D. L. Kachel, R. Pasula, and W. J. Martin. 1999. Gamma interferon stimulates rat alveolar macrophages to kill *Pneumocystis carinii* by L-arginine- and tumor necrosis factor-dependent mechanisms. *Infect. Immun.* **67:**1347–1352.

24. Edman, J. C., J. A. Kovacs, H. Masur, D. V. Santi, H. J. Elwood, and M. L. Sogin. 1988. Ribosomal RNA sequence shows *Pneumocystis carinii* to be a member of the fungi. *Nature* **334:**519–522.

24a. Eriksson, O. E. 1994. *Pneumocystis carinii*, a parasite in lungs of mammals, referred to as a new family and order (*Pneumocystidaceae, Pneumocystidales, Ascomycota*). *Syst. Ascomycetum* **13:**165–180.

25. Eriksson, O. E., and K. Winka. 1997. Families and higher taxa of Ascomycota. *Myconet* **1:**17–24.

26. Fisk, D. T., S. Meshnick, and P. H. Kazanjian. 2003. *Pneumocystis carinii* pneumonia in patients in the developing world who have acquired immunodeficiency syndrome. *Clin. Infect. Dis.* **36:**70–78.

27. Fox, D., and A. G. Smulian. 2001. Plasminogen-binding activity of enolase in the opportunistic pathogen *Pneumocystis carinii*. *Med. Mycol.* **39:**495–507.

28. Frenkel, J. K. 1999. *Pneumocystis* pneumonia, an immunodeficiency-dependent disease (IDD): a critical historical overview. *J. Eukaryot. Microbiol.* **46:**89S–92S.

29. Garvy, B. A., and A. G. Harmsen. 1996. Susceptibility to *Pneumocystis carinii* infection: host responses of neonatal mice from immune or naive mothers and of immune or naive adults. *Infect. Immun.* **64:**3987–3992.

30. Gigliotti, F., and W. T. Hughes. 1988. Passive immunoprophylaxis with specific monoclonal antibody confers partial protection against *Pneumocystis carinii* pneumonitis in animal models. *J. Clin. Investig.* **81:**1666–1668.

31. Gigliotti, F., J. A. Wiley, and A. G. Harmsen. 1998. Immunization with *Pneumocystis carinii* gpA is immunogenic but not protective in a mouse model of *P. carinii* pneumonia. *Infect. Immun.* **66:**3179–3182.

32. Giuntoli, D., S. L. Stringer, and J. R. Stringer. 1994. Extraordinarily low number of ribosomal RNA genes in *P. carinii*. *J. Eukaryot. Microbiol.* **41:**88S.

33. Grewal, I. S., J. Xu, and R. A. Flavell. 1995. Impairment of antigen-specific T-cell priming in mice lacking CD40 ligand. *Nature* **378:**617–620.

34. Hahn, P. Y., and A. H. Limper. 2003. The role of inflammation in respiratory impairment during *Pneumocystis carinii* pneumonia. *Semin. Respir. Infect.* **18:**40–47.

35. Hanano, R., and S. H. Kaufmann. 1999. *Pneumocystis carinii* pneumonia in mutant mice deficient in both TCRalphabeta and TCRgammadelta cells: cytokine and antibody responses. *J. Infect. Dis.* **179:**455–459.

36. Hanano, R., K. Reifenberg, and S. H. Kaufmann. 1998. Activated pulmonary macrophages are insufficient for resistance against *Pneumocystis carinii. Infect. Immun.* **66:**305–314.

37. Harmsen, A. G., W. Chen, and F. Gigliotti. 1995. Active immunity to *Pneumocystis carinii* reinfection in T-cell-depleted mice. *Infect. Immun.* **63:**2391–2395.

38. Harmsen, A. G., and M. Stankiewicz. 1990. Requirement for CD4$^+$ cells in resistance to *Pneumocystis carinii* pneumonia in mice. *J. Exp. Med.* **172:**937–945.

39. Helweg-Larsen, J., J. S. Jensen, B. Dohn, T. L. Benfield, and B. Lundgren. 2002. Detection of *Pneumocystis* DNA in samples from patients suspected of bacterial pneumonia—a case-control study. *BMC. Infect. Dis.* **2:**28.

40. Herting, E., O. Moller, J. H. Schiffmann, and B. Robertson. 2002. Surfactant improves oxygenation in infants and children with pneumonia and acute respiratory distress syndrome. *Acta Paediatr.* **91:**1174–1178.

41. Hori, S., T. L. Carvalho, and J. Demengeot. 2002. CD25$^+$ CD4$^+$ regulatory T cells suppress CD4$^+$ T cell-mediated pulmonary hyperinflammation driven by *Pneumocystis carinii* in immunodeficient mice. *Eur. J. Immunol.* **32:**1282–1291.

42. Huang, L., K. Crothers, C. Atzori, T. Benfield, R. Miller, M. Rabodonirina, and J. Helweg-Larsen. 2004. Dihydropteroate synthase gene mutations in *Pneumocystis* and sulfa resistance. *Emerg. Infect. Dis.* **10:**1721–1728.

43. Hughes, W. T. 1982. Natural mode of acquisition for de novo infection with *Pneumocystis carinii. J. Infect. Dis.* **145:**842–848.

44. Hughes, W. T. 1989. Animal models for *Pneumocystis carinii* pneumonia. *J. Protozool.* **36:**41–45.

44a. Hughes, W. T. 2004. *Pneumocystis* Pneumonitis in non-HIV-infected Patients: Update, p. 407–434. *In* M. T. Cushion and P. D. Walzer (ed.), *Pneumocystis Pneumonia.* Marcel Dekker, Inc., New York, N.Y.

45. Icenhour, C. R., and M. T. Cushion. 2001. Putative transmissive form of *Pneumocystis carinii* f. sp. *carinii. J. Eukaryot. Microbiol.* **Suppl.:**139S–140S.

46. Icenhour, C. R., S. L. Rebholz, M. S. Collins, and M. T. Cushion. 2001. Widespread occurrence of *Pneumocystis carinii* in commercial rat colonies detected using targeted PCR and oral swabs. *J. Clin. Microbiol.* **39:**3437–3441.

47. Icenhour, C. R., S. L. Rebholz, M. S. Collins, and M. T. Cushion. 2002. Early acquisition of *Pneumocystis carinii* in neonatal rats as evidenced by PCR and oral swabs. *Eukaryot. Cell* **1:**414–419.

48. Iliades, P., S. R. Meshnick, and I. G. Macreadie. 2004. Dihydropteroate synthase mutations in *Pneumocystis jirovecii* can affect sulfamethoxazole resistance in a *Saccharomyces cerevisiae* model. *Antimicrob. Agents Chemother.* **48:**2617–2623.

49. International Association for Plant Taxonomy. 2000. *International Code of Botanical Nomenclature.* Lubrecht & Cramer Ltd, Konigstein, Germany.

50. Justice, A. C., C. S. Landefeld, S. M. Asch, A. L. Gifford, C. C. Whalen, and K. E. Covinsky. 2001. Justification for a new cohort study of people aging with and without HIV infection. *J. Clin. Epidemiol.* **54(suppl. 1):**S3–S8.

51. Kazanjian, P., W. Armstrong, P. A. Hossler, C. H. Lee, L. Huang, C. B. Beard, J. Carter, L. Crane, J. Duchin, W. Burman, J. Richardson, and S. R. Meshnick. 2001.

Pneumocystis carinii cytochrome *b* mutations are associated with atovaquone exposure in patients with AIDS. *J. Infect. Dis.* **183:**819–822.

52. Kazanjian, P. H., D. Fisk, W. Armstrong, Q. Shulin, H. Liwei, Z. Ke, and S. Meshnick. 2004. Increase in prevalence of *Pneumocystis carinii* mutations in patients with AIDS and *P. carinii* pneumonia, in the United States and China. *J. Infect. Dis.* **189:**1684–1687.

53. Keely, S. P., J. M. Fischer, M. T. Cushion, and J. R. Stringer. 2004. Phylogenetic identification of *Pneumocystis murina* sp. nov., a new species in laboratory mice. *Microbiology* **150:**1153–1165.

54. Keely, S. P., H. Renauld, A. E. Wakefield, M. T. Cushion, A. G. Smulian, N. Fosker, A. Fraser, D. Harris, L. Murphy, C. Price, M. A. Quail, K. Seeger, S. Sharp, C. J. Tindal, T. Warren, E. Zuiderwijk, B. G. Barrell, J. R. Stringer, and N. Hall. 2005. Gene arrays at *Pneumocystis carinii* telomeres. *Genetics* **170:** 1589–1600.

55. Kessl, J. J., P. Hill, B. B. Lange, S. R. Meshnick, B. Meunier, and B. L. Trumpower. 2004. Molecular basis for atovaquone resistance in *Pneumocystis jirovecii* modeled in the cytochrome bc_1 complex of *Saccharomyces cerevisiae. J. Biol. Chem.* **279:**2817–2824.

56. Kovacs, J. A., J. W. Hiemenz, A. M. Macher, D. Stover, H. W. Murray, J. Shelhamer, H. C. Lane, C. Urmacher, C. Honig, and D. L. Longo. 1984. *Pneumocystis carinii* pneumonia: a comparison between patients with the acquired immunodeficiency syndrome and patients with other immunodeficiencies. *Ann. Intern. Med.* **100:**663–671.

57. Koziel, H., Q. Eichbaum, B. A. Kruskal, P. Pinkston, R. A. Rogers, M. Y. Armstrong, F. F. Richards, R. M. Rose, and R. A. Ezekowitz. 1998. Reduced binding and phagocytosis of *Pneumocystis carinii* by alveolar macrophages from persons infected with HIV-1 correlates with mannose receptor downregulation. *J. Clin. Investig.* **102:**1332–1344.

58. Koziel, H., S. Kim, C. Reardon, X. Li, R. Garland, P. Pinkston, and H. Kornfeld. 1999. Enhanced in vivo human immunodeficiency virus-1 replication in the lungs of human immunodeficiency virus-infected persons with *Pneumocystis carinii* pneumonia. *Am. J. Respir. Crit. Care Med.* **160:**2048–2055.

59. Kulke, M. H., and E. A. Vance. 1997. *Pneumocystis carinii* pneumonia in patients receiving chemotherapy for breast cancer. *Clin. Infect. Dis.* **25:**215–218.

60. Kutty, G., and J. A. Kovacs. 2003. A single-copy gene encodes Kex1, a serine endoprotease of *Pneumocystis jirovecii. Infect. Immun.* **71:**571–574.

61. Lane, H. C., B. E. Laughon, J. Falloon, J. A. Kovacs, R. T. J. Davey, M. A. Polis, and H. Masur. 1994. NIH conference. Recent advances in the management of AIDS-related opportunistic infections. *Ann. Intern. Med.* **120:**945–955.

62. Laursen, A. L., J. Rungby, and P. L. Andersen. 1995. Decreased activation of the respiratory burst in neutrophils from AIDS patients with previous *Pneumocystis carinii* pneumonia. *J. Infect. Dis.* **172:**497–505.

63. Lee, L. H., F. Gigliotti, T. W. Wright, P. J. Simpson-Haidaris, G. A. Weinberg, and C. G. Haidaris. 2000. Molecular characterization of KEX1, a kexin-like protease in mouse *Pneumocystis carinii. Gene* **242:** 141–150.

64. Li, P. G., D. Fenoglio, L. Bottone, P. Terranova, E. Pontali, F. Caroli, M. Seri, J. C. Cailliez, G. Koopman,

R. Accolla, F. del Galdo, G. Abbate, R. de Palma, and F. Manca. 2002. Preservation of clonal heterogeneity of the *Pneumocystis carinii*-specific CD4 T cell repertoire in HIV infected, asymptomatic individuals. *Clin. Exp. Immunol.* **128**:155–162.

65. Limper, A. H., J. E. Standing, O. A. Hoffman, M. Castro, and L. W. Neese. 1993. Vitronectin binds to *Pneumocystis carinii* and mediates organism attachment to cultured lung epithelial cells. *Infect. Immun.* **61**:4302–4309.

66. Lipschik, G. Y., J. F. Treml, S. D. Moore, and M. F. Beers. 1998. *Pneumocystis carinii* glycoprotein A inhibits surfactant phospholipid secretion by rat alveolar type II cells. *J. Infect. Dis.* **177**:182–187.

67. Lobetti, R. 2000. Common variable immunodeficiency in miniature dachshunds affected with *Pneumonocystis carinii* pneumonia. *J. Vet. Diagn. Investig.* **12**:39–45.

68. Ma, L., L. Borio, H. Masur, and J. A. Kovacs. 1999. *Pneumocystis carinii* dihydropteroate synthase but not dihydrofolate reductase gene mutations correlate with prior trimethoprim-sulfamethoxazole or dapsone use. *J. Infect. Dis.* **180**:1969–1978.

69. Ma, L., G. Kutty, Q. Jia, and J. A. Kovacs. 2003. Characterization of variants of the gene encoding the p55 antigen in *Pneumocystis* from rats and mice. *J. Med. Microbiol.* **52**:955–960.

70. Marcotte, H., D. Levesque, K. Delanay, A. Bourgeault, R. de la Durantaye, S. Brochu, and M. C. Lavoie. 1996. *Pneumocystis carinii* infection in transgenic B cell-deficient mice. *J. Infect. Dis.* **173**:1034–1037.

71. Maskell, N. A., D. J. Waine, A. Lindley, J. C. Pepperell, A. E. Wakefield, R. F. Miller, and R. J. Davies. 2003. Asymptomatic carriage of *Pneumocystis jirovecii* in subjects undergoing bronchoscopy: a prospective study. *Thorax* **58**:594–597.

72. McAllister, F., C. Steele, M. Zheng, E. Young, J. E. Shellito, L. Marrero, and J. K. Kolls. 2004. T cytotoxic-1 CD8$^+$ T cells are effector cells against pneumocystis in mice. *J. Immunol.* **172**:1132–1138.

73. McCormack, F. X., A. L. Festa, R. P. Andrews, M. Linke, and P. D. Walzer. 1997. The carbohydrate recognition domain of surfactant protein A mediates binding to the major surface glycoprotein of *Pneumocystis carinii*. *Biochemistry* **36**:8092–8099.

74. Mei, Q., R. E. Turner, V. Sorial, D. Klivington, C. W. Angus, and J. A. Kovacs. 1998. Characterization of major surface glycoprotein genes of human *Pneumocystis carinii* and high-level expression of a conserved region. *Infect. Immun.* **66**:4268–4273.

75. Meneau, I., D. Sanglard, J. Bille, and P. M. Hauser. 2004. *Pneumocystis jirovecii* dihydropteroate synthase polymorphisms confer resistance to sulfadoxine and sulfanilamide in *Saccharomyces cerevisiae*. *Antimicrob. Agents Chemother.* **48**:2610–2616.

76. Montes-Cano, M. A., C. de la Horra, J. Martin-Juan, J. M. Varela, R. Torronteras, N. Respaldiza, F. J. Medrano, and E. J. Calderon. 2004. *Pneumocystis jirovecii* genotypes in the Spanish population. *Clin. Infect. Dis.* **39**:123–128.

77. Morris, A., F. C. Sciurba, I. P. Lebedeva, A. Githaiga, W. M. Elliott, J. C. Hogg, L. Huang, and K. A. Norris. 2004. Association of chronic obstructive pulmonary disease severity and *Pneumocystis* colonization. *Am. J. Respir. Crit. Care Med.* **170**:408–413.

78. Nahimana, A., M. Rabodonirina, J. Bille, P. Francioli, and P. M. Hauser. 2004. Mutations of *Pneumocystis jirovecii*

dihydrofolate reductase associated with failure of prophylaxis. *Antimicrob. Agents Chemother.* **48**: 4301–4305.

79. Pottratz, S. T., J. Paulsrud, J. S. Smith, and W. J. Martin. 1991. *Pneumocystis carinii* attachment to cultured lung cells by pneumocystis gp120, a fibronectin binding protein. *J. Clin. Investig.* **88**:403–407.

80. Qureshi, M. H., A. G. Harmsen, and B. A. Garvy. 2003. IL-10 modulates host responses and lung damage induced by *Pneumocystis carinii* infection. *J. Immunol.* **170**:1002–1009.

81. Rabeneck, L., T. Menke, M. S. Simberkoff, et al. 2001. Using the national registry of HIV-infected veterans in research: lessons for the development of disease registries. *J. Clin. Epidemiol.* **54**:1195–1203.

82. Rice, W. R., F. M. Singleton, M. J. Linke, and P. D. Walzer. 1993. Regulation of surfactant phosphatidylcholine secretion from alveolar type II cells during *Pneumocystis carinii* pneumonia in the rat. *J. Clin. Investig.* **92**:2778–2782.

83. Richards, P. J., L. Riddell, R. H. Reznek, P. Armstrong, A. J. Pinching, and J. M. Parkin. 1996. High resolution computed tomography in HIV patients with suspected *Pneumocystis carinii* pneumonia and a normal chest radiograph. *Clin. Radiol.* **51**:689–693.

84. Roths, J. B., and C. L. Sidman. 1992. Both immunity and hyperresponsiveness to *Pneumocystis carinii* result from transfer of CD4$^+$ but not CD8$^+$ T cells into severe combined immunodeficiency mice. *J. Clin. Investig.* **90**:673–678.

85. Roths, J. B., and C. L. Sidman. 1993. Single and combined humoral and cell-mediated immunotherapy of *Pneumocystis carinii* pneumonia in immunodeficient scid mice. *Infect. Immun.* **61**:1641–1649.

86. Rudmann, D. G., A. M. Preston, M. W. Moore, and J. M. Beck. 1998. Susceptibility to *Pneumocystis carinii* in mice is dependent on simultaneous deletion of IFN-gamma and type 1 and 2 TNF receptor genes. *J. Immunol.* **161**:360–366.

87. Shellito, J., V. V. Suzara, W. Blumenfeld, J. M. Beck, H. J. Steger, and T. H. Ermak. 1990. A new model of *Pneumocystis carinii* infection in mice selectively depleted of helper T lymphocytes. *J. Clin. Invest.* **85**: 1686–1693.

88. Shellito, J. E., J. K. Kolls, R. Olariu, and J. M. Beck. 1996. Nitric oxide and host defense against *Pneumocystis carinii* infection in a mouse model. *J. Infect. Dis.* **173**:432–439.

89. Smulian, A. G., J. R. Stringer, M. J. Linke, and P. D. Walzer. 1992. Isolation and characterization of a recombinant antigen of *Pneumocystis carinii*. *Infect. Immun.* **60**:907–915.

90. Smulian, A. G., D. W. Sullivan, M. J. Linke, N. A. Halsey, T. C. Quinn, A. P. MacPhail, M. A. Hernandez-Avila, S. T. Hong, and P. D. Walzer. 1993. Geographic variation in the humoral response to *Pneumocystis carinii*. *J. Infect. Dis.* **167**:1243–1247.

91. Smulian, A. G., and S. A. Theus. 1994. Cellular immune response in *Pneumocystis carinii* infection. *Parasitol. Today* **10**:229–231.

92. Smulian, A. G., and P. D. Walzer. 1994. Serological studies of *Pneumocystis carinii* infection, p. 141–154. *In* P. D. Walzer (ed.), *Pneumocystis carinii Pneumonia*. Marcel Dekker, Inc., New York, N.Y.

93. Society of Protozoologists. 1994. Revised nomenclature for *Pneumocystis carinii*. The *Pneumocystis* workshop. *J. Eukaryot. Microbiol.* **41**:121s–122s.

94. Steele, C., L. Marrero, S. Swain, A. G. Harmsen, M. Zheng, G. D. Brown, S. Gordon, J. E. Shellito, and J. K. Kolls. 2003. Alveolar macrophage-mediated killing of *Pneumocystis carinii* f. sp. *muris* involves molecular recognition by the Dectin-1 beta-glucan receptor. *J. Exp. Med.* **198**:1677–1688.

95. Steele, C., M. Zheng, E. Young, L. Marrero, J. E. Shellito, and J. K. Kolls. 2002. Increased host resistance against *Pneumocystis carinii* pneumonia in γδ T-cell-deficient mice: protective role of gamma interferon and CD8⁺ T cells. *Infect. Immun.* **70**:5208–5215.

96. Stehle, S. E., R. A. Rogers, A. G. Harmsen, and R. A. Ezekowitz. 2000. A soluble mannose receptor immunoadhesin enhances phagocytosis of *Pneumocystis carinii* by human polymorphonuclear leukocytes in vitro. *Scand. J. Immunol.* **52**:131–137.

97. Stein, C. R., C. Poole, P. Kazanjian, and S. R. Meshnick. 2004. Sulfa use, dihydropteroate synthase mutations, and *Pneumocystis jirovecii* pneumonia. *Emerg. Infect. Dis.* **10**:1760–1765.

98. Stringer, J. R., and S. P. Keely. 2001. Genetics of surface antigen expression in *Pneumocystis carinii*. *Infect. Immun.* **69**:627–639.

99. Stringer, S. L., J. R. Stringer, M. A. Blase, P. D. Walzer, and M. T. Cushion. 1989. *Pneumocystis carinii*: sequence from ribosomal RNA implies a close relationship with fungi. *Exp. Parasitol.* **68**:450–461.

100. Su, T. H., V. Natarajan, D. L. Kachel, M. A. Moxley, W. J. Longmore, and W. J. Martin. 1996. Functional impairment of bronchoalveolar lavage phospholipids in early *Pneumocystis carinii* pneumonia in rats. *J. Lab. Clin. Med.* **127**:263–271.

101. Takahashi, T., T. Endo, T. Nakamura, H. Sakashitat, K. Kimurat, K. Ohnishit, Y. Kitamura, and A. Iwamoto. 2002. Dihydrofolate reductase gene polymorphisms in *Pneumocystis carinii* f. sp. *hominis* in Japan. *J. Med. Microbiol.* **51**:510–515.

102. Theus, S. A., R. P. Andrews, P. Steele, and P. D. Walzer. 1995. Adoptive transfer of lymphocytes sensitized to the major surface glycoprotein of *Pneumocystis carinii* confers protection in the rat. *J. Clin. Investig.* **95**:2587–2593.

103. Theus, S. A., N. Sawhney, A. G. Smulian, and P. D. Walzer. 1998. Proliferative and cytokine responses of human T lymphocytes isolated from human immunodeficiency virus-infected patients to the major surface glycoprotein of *Pneumocystis carinii*. *J. Infect. Dis.* **177**:238–241.

104. Theus, S. A., A. G. Smulian, P. Steele, M. J. Linke, and P. D. Walzer. 1998. Immunization with the major surface glycoprotein of *Pneumocystis carinii* elicits a protective response. *Vaccine* **16**:1149–1157.

105. Theus, S. A., D. W. Sullivan, P. D. Walzer, and A. G. Smulian. 1994. Cellular responses to a 55-kilodalton recombinant *Pneumocystis carinii* antigen. *Infect. Immun.* **62**:3479–3484.

106. Thomas, C. F., Jr., and A. H. Limper. 2004. Pneumocystis pneumonia. *N. Engl. J. Med.* **350**:2487–2498.

107. Vargas, S. L., W. T. Hughes, M. E. Santolaya, A. V. Ulloa, C. A. Ponce, C. E. Cabrera, F. Cumsille, and F. Gigliotti. 2001. Search for primary infection by *Pneumocystis carinii* in a cohort of normal, healthy infants. *Clin. Infect. Dis.* **32**:855–861.

108. Vargas, S. L., C. A. Ponce, W. T. Hughes, A. E. Wakefield, J. C. Weitz, S. Donoso, A. V. Ulloa, P. Madrid, S. Gould, J. J. Latorre, R. Avila, S. Benveniste, M. Gallo, J. Belletti, and R. Lopez. 1999. Association of primary *Pneumocystis carinii* infection and sudden infant death syndrome. *Clin. Infect. Dis.* **29**:1489–1493.

109. Vargas, S. L., C. A. Ponce, C. A. Sanchez, A. V. Ulloa, R. Bustamante, and G. Juarez. 2003. Pregnancy and asymptomatic carriage of *Pneumocystis jirovecii*. *Emerg. Infect. Dis.* **9**:605–606.

110. Vassallo, R., T. J. Kottom, J. E. Standing, and A. H. Limper. 2001. Vitronectin and fibronectin function as glucan binding proteins augmenting macrophage responses to *Pneumocystis carinii*. *Am. J. Respir. Cell Mol. Biol.* **25**:203–211.

111. Vassallo, R., C. F. Thomas, Jr., Z. Vuk-Pavlovic, and A. H. Limper. 2000. Mechanisms of defence in the lung: lessons from *Pneumocystis carinii* pneumonia. *Sarcoidosis Vasc. Diffuse Lung Dis.* **17**:130–139.

112. Walker, D. J., A. E. Wakefield, M. N. Dohn, R. F. Miller, R. P. Baughman, P. A. Hossler, M. S. Bartlett, J. W. Smith, P. Kazanjian, and S. R. Meshnick. 1998. Sequence polymorphisms in the *Pneumocystis carinii* cytochrome *b* gene and their association with atovaquone prophylaxis failure. *J. Infect. Dis.* **178**:1767–1775.

113. Walzer, P. D. 2004. Immunological features of *Pneumocystis* infection in humans, p. 451–477. *In* M. T. Cushion and P. D. Walzer (ed.), *Pneumocystis Pneumonia*. Marcel Dekker, Inc., New York, N.Y.

114. Walzer, P. D., C. K. Kim, and M. T. Cushion. 1989. *Pneumocystis carinii*, p. 83–178. *In* P. D. Walzer and R. M. Genta (ed.), *Parasitic Infections in the Immunocompromised Host*. Marcel Dekker, Inc., New York, N.Y.

115. Walzer, P. D., C. K. Kim, M. J. Linke, C. L. Pogue, M. J. Huerkamp, C. E. Chrisp, A. V. Lerro, S. K. Wixson, E. Hall, and L. D. Shultz. 1989. Outbreaks of *Pneumocystis carinii* pneumonia in colonies of immunodeficient mice. *Infect. Immun.* **57**:62–70.

116. Warschkau, H., H. Yu, and A. F. Kiderlen. 1998. Activation and suppression of natural cellular immune functions by *Pneumocystis carinii*. *Immunobiology* **198**:343–360.

117. Wiley, J. A., and A. G. Harmsen. 1995. CD40 ligand is required for resolution of *Pneumocystis carinii* pneumonia in mice. *J. Immunol.* **155**:3525–3529.

118. Wright, T. W., C. J. Johnston, A. G. Harmsen, and J. N. Finkelstein. 1999. Chemokine gene expression during *Pneumocystis carinii*-driven pulmonary inflammation. *Infect. Immun.* **67**:3452–3460.

119. Wright, T. W., R. H. Notter, Z. Wang, A. G. Harmsen, and F. Gigliotti. 2001. Pulmonary inflammation disrupts surfactant function during *Pneumocystis carinii* pneumonia. *Infect. Immun.* **69**:758–764.

120. Wu, A. K., V. C. Cheng, B. S. Tang, I. F. Hung, R. A. Lee, D. S. Hui, and K. Y. Yuen. 2004. The unmasking of *Pneumocystis jirovecii* pneumonia during reversal of immunosuppression: case reports and literature review. *BMC Infect. Dis.* **4**:57.

121. Zheng, M., J. E. Shellito, L. Marrero, Q. Zhong, S. Julian, P. Ye, V. Wallace, P. Schwarzenberger, and J. K. Kolls. 2001. CD4⁺ T cell-independent vaccination against *Pneumocystis carinii* in mice. *J. Clin. Investig.* **108**:1469–1474.

Molecular Principles of Fungal Pathogenesis
Edited by Joseph Heitman et al.
©2006 ASM Press, Washington, D.C.

Chapter 26

Virulence Mechanisms of *Coccidioides*

GARRY T. COLE, JIANMIN XUE, KALPATHI SESHAN, PRISCILA BORRA, RICARDO BORRA,
ERIC TARCHA, RUTH SCHALLER, JIEH-JUEN YU, AND CHIUNG-YU HUNG

Coccidioidomycosis has been referred to as an occupational disease, since risk of infection with the soil-inhabiting pathogen for people who work in the regions of endemicity is significantly increased by frequent exposure to outdoor dust (150). Farmers and harvesters, construction laborers, field archeologists, military personnel, as well as the large number of individuals who engage in recreational activities in areas of the southwestern United States, are most likely to be exposed to the airborne spores of *Coccidioides* (22, 50, 219). Coccidioidomycosis has also been reported to be a frequent laboratory-acquired respiratory disease (112, 203). Plate cultures of the saprobic phase of *Coccidioides* grown in laboratory incubators produce large numbers of spores (arthroconidia), which can easily contaminate the environment if not handled properly. Live cultures of *Coccidioides* spp. must be maintained within a biosafety level 3 (BSL3) containment facility because of the highly infectious nature of the arthroconidia (158). The Centers for Disease Control and Prevention has included *Coccidioides* spp. on its list of restricted agents, along with viruses, bacteria, and toxins which are considered to be potential bioweapons (www.nih.gov/od/ors/ds/pubs/appendxa .html) (57, 60, 67). Research involving these agents is currently regulated under the U.S. Public Health Security and Bioterrorism Preparedness and Response Act of 2002.

Two species of this ascomycetous fungus are distinguished on the basis of molecular and biogeographical differences: *C. immitis* is found primarily in the San Joaquin Valley of California, while *C. posadasii* is widespread throughout regions of endemicity in the Americas (76). Within the United States, these regions extend from western Texas to southern California and as far north as Dinosaur National Monument in northeastern Utah (22). The fungus also propagates in soil of semiarid areas of northern Mexico and Central and South America, including Guatemala, Honduras, Colombia, Venezuela, Paraguay, and Argentina (161). Dense human populations exist throughout these regions, including some of the most rapidly growing communities in the United States. The southwestern "sunbelt" attracts large numbers of elderly citizens to retirement villages, as well as many visitors from other countries. There are now more people at risk of coccidioidal infection than 50 years ago, mainly because of a surge in the number of "immunonaive" immigrants to the southwestern United States (179). This is well illustrated by the dramatic increase in the number of reported cases of coccidioidomycosis in Arizona over the last 10 years (7, 23). Crum et al. (51) documented 223 cases of coccidioidomycosis at the Naval Medical Center in San Diego between 1994 and 2002. This was the largest cohort of *Coccidioides* infections reported at that facility since the 1950s. Of these patients, 58% presented with isolated pulmonary disease, 14% were seroreactive without clear evidence of dissemination, 22% had confirmed disseminated coccidioidal infection, and 5% had unclassified disease. The risk of systemic infection was indicated to be highest in non-Caucasian races; African American patients more often develop coccidioidomycosis that disseminates to bone, while Filipinos are more prone to cutaneous and central nervous system disease. Relapse of disseminated coccidioidal infection among members of the cohort was 24%, and the risk was highest (71%) for those with central nervous system disease.

Although coccidioidomycosis can be a life-threatening disease, about half of the people who are infected present with only mild symptoms and no

Garry T. Cole, Jianmin Xue, Kalpathi Seshan, Priscila Borra, Ricardo Borra, Eric Tarcha, Ruth Schaller, Jieh-Juen Yu, and Chiung-Yu Hung • South Texas Center for Emerging Infectious Diseases, Dept. of Biology, University of Texas at San Antonio, San Antonio, TX 78249.

medical intervention is sought. On the other hand, infection of many of these individuals is subsequently demonstrable by positive skin test and/or X-ray detection of a pulmonary cavity or nodule (160). A study conducted with 58 children who lived in a region of endemicity and demonstrated a skin test reaction to coccidioidin also revealed that 50% had X-ray evidence of an inflammatory process or calcific densities in their lungs (48). Coccidioidin is a soluble fraction derived from mycelial cultures of *Coccidioides* and has been used for immunoassays of human populations presumably exposed to the soil pathogen (183, 184). For individuals who present with clear symptoms of coccidioidal infection, an atypical pneumonia is most common and is characterized by such features as pleuritic chest pain, cough, fever, malaise, rash, sore throat, headache, arthralgia, myalgia, and anorexia. These early manifestations of infection, none of which are in themselves diagnostic of coccidioidomycosis, typically develop within 1 to 4 weeks after exposure to the fungus (160). Most infected persons recover during the next few weeks to several months, while 5 to 10% show evidence of active disease based on positive results of serologic testing and chest radiographs. Fewer than 5% of infected individuals develop severe disseminated or fatal disease (118). Since there is no person-to-person transmission of *Coccidioides*, with the rare exception of maternal-fetal transmission (24, 130, 168), it would appear that coccidioidal infection of the mammalian host is an evolutionary dead end for the pathogen. On the contrary, a substantial and growing body of clinical evidence suggests that *Coccidioides* can persist in host tissue and reactivation of coccidioidomycosis may occur years after the original insult, especially if infected individuals contract an immunocompromising disease, undergo elective immunosuppression for organ transplantation, or are otherwise subjected to long-term drug therapy that compromises their cellular immunity (15, 25, 116, 134, 140, 168, 185, 198, 206, 215, 217, 218, 222). Certain microbial pathogens, like *Coccidioides*, have evolved pathogenicity determinants that allow them to coexist with their host rather than exacerbate the infection process and intensify the host response (139). The virulence of *Coccidioides* appears to be calibrated, such that the pathogen typically does not kill the host but successfully utilizes available resources with cost to the host in terms of prolonged morbidity. Although coccidioidal infections exhibit a range of clinical forms, a large number of human cases may also result in cryptic coccidioidomycosis, as has been described for other respiratory mycoses reviewed in this volume and elsewhere (29, 54, 89, 213).

Coccidioides infections are characterized by both an intracellular and extracellular relationship with the host. The period during which fungal cells reside within the phagocyte is relatively brief and is restricted primarily because of size constraints of the pathogen. Arthroconidia are barrel-shaped cells that measure 2.5 to 3 μm in width by 3 to 6 μm in length and are small enough to reach the alveoli if inhaled. In vivo-germinated conidia undergo isotropic growth which, if uninhibited by host defenses, give rise to large, multinucleate cells (spherules) that typically range from 40 to 80 μm in diameter but can be as large as 120 μm (191). The contents of mature spherules undergo a complex process of differentiation to yield an average of 200 to 300 endospores, each with a diameter of approximately 4 to 6 μm. Each endospore released from the maternal spherule that survives within the host undergoes isotropic growth and gives rise to a second generation of endosporulating spherules (Fig. 1) (30). In vitro studies have suggested that only a small percentage of the parasitic cells that are small enough to be engulfed by phagocytes (i.e., arthroconidia, germinated arthroconidia ["spherule initials"], and endospores) are killed (62). It is evident on the basis of our current understanding of coccidioidal infections that the fungus has evolved under the selective pressure of a complex array of mammalian defenses brought to bear against it to establish a niche as a successful respiratory pathogen. The purpose of this chapter is to examine the various mechanisms utilized by *Coccidioides* to survive within the hostile environment of the host.

Probably all mammals that reside in the regions of endemicity are at risk of contracting coccidioidal infection (80, 95). Burrowing rodents (kangaroo rats and pocket mice), armadillos, wild dogs, desert foxes, and coyotes have been suggested to play roles in harboring and spreading the pathogen in regions of endemic infection in North and South America (69, 113, 162, 177). Death and subsequent decay of *Coccidioides*-infected animals under natural conditions culminates in conversion of the parasitic phase of the fungus into its saprobic form, thus recharging the soil with mycelia that yield profuse "arthroconidial blooms." The conidia are air dispersed and may be inhaled by new mammalian hosts, or they may persist in the soil and remain viable for long periods. An intriguing proposal has been presented that soil-borne dimorphic fungi which are causative agents of human disease may have evolved pathogenic strategies from environmental interactions with phagocytic microorganisms (187). Phagocytic predators in the soil can engulf fungal spores and mycelial elements.

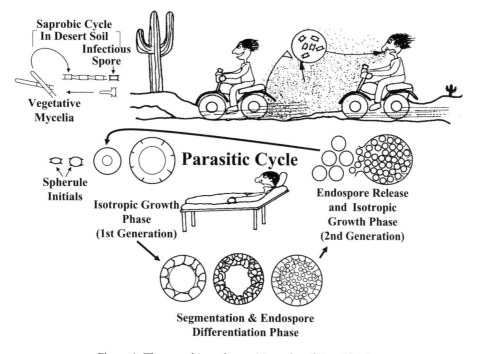

Figure 1. The saprobic and parasitic cycles of *Coccidioides*.

Laboratory studies have demonstrated that these interactions could occur with such environmental pathogens as *Blastomyces dermatitidis*, *Sporothrix schenckii*, and *Histoplasma capsulatum*. It has been suggested that predatory amebae in the soil could exert selective pressures which favor fungal attributes that confer survival advantages in animal hosts. Although the ability to grow at 37°C would seem to be an additional requirement for fungal virulence in mammals, it is interesting that endosporulating spherules of *Coccidioides* have been detected in lesions isolated from the lungs of a Sonoran gopher snake (194), which argues that the parasitic phase is not restricted to homoiotherms (39). Vegetative mycelia of *Coccidioides* most probably do not survive in the uppermost layers of desert soil during the hot summer months but instead reside in cooler subsurface zones, where nutrients are available to the microorganism. In fact, studies have shown that *Coccidioides* is more frequently found several inches below the surface of the soil, where temperatures are significantly lower (178) and nitrogen levels are higher (209). Long periods of drought in the desert Southwest, as have occurred over the past few years (40), combined with wind and seismic activity in regions of endemicity, generate dust storms which transport *Coccidioides* arthroconidia over long distances, potentially to human populations not previously exposed to the pathogen (164). Thus, *Coccidioides*

appears to have successfully adapted to a wide range of environmental conditions during its evolution as a saprophyte in alkaline desert soil and as a pathogen in the mammalian host.

PARASITIC CYCLE OF *COCCIDIOIDES*

Much of the discussion which follows in this chapter is dependent on an understanding of the events that occur during conidiogenesis and stages of parasitic cell development of *Coccidioides* within the mammalian host. We have therefore provided a detailed description of the unique parasitic cycle of this respiratory pathogen. Fortunately, the parasitic phase can be grown in vitro using a defined glucose-salts medium (126). Inhalation of arthroconidia is the most common route by which the host insult occurs, although percutaneous inoculation with *Coccidioides* spores or mycelia can also result in systemic infection. Conidiogenesis takes place by disarticulation of septate hyphae. Alternate cells of the chain undergo a degenerative process, which appears to be a highly regulated event (38). The arthroconidial chain produced in nature at the soil surface, and in vitro on agar plates, is easily disrupted by air currents. Spore dispersal is enhanced by the existence of an outer, hydrophobic cell wall (Fig. 2A). Electron microscopy studies of arthroconidia have revealed that the outer conidial wall layer, which is derived from the original hyphal wall, contains hydrophobins (118, 157). These are cysteine-rich proteins that

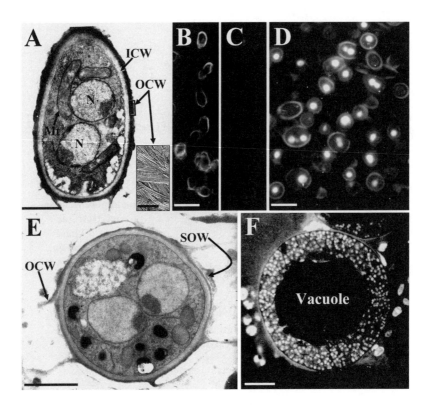

Figure 2. Morphological features of arthroconidia and spherule initials. (A) Thin section of arthroconidium. The outer conidial wall (OCW) contains hydrophobins which appear as fascicles of rodlets (inset) at the cell surface. ICW, inner conidial wall; Mt, mitochondrion; N, nucleus. Bar, 1μm (inset, 20 nm). (B and C) Arthroconidia grown on plate culture containing glucose-salts medium (126) supplemented with L-DOPA (1 mM) were reacted with either a monoclonal antibody raised against *C. neoformans* melanin (6D2) (B) or a control monoclonal antibody raised against *C. neoformans* glucuronoxylomannan (2D10) (C), both provided by J. Nosanchuk. The cells were subsequently exposed to fluorescein isothiocyanate-conjugated goat anti-mouse Ig. Bar in panel B represents 2 μm. (D) Mithramycin-stained spherule initials (germinated arthroconidia) grown at 37°C in the presence of murine tracheal explants. Bar, 4 μm. (E) Thin section of spherule initial, grown as described in panel D, revealing an early stage in differentiation of the lipid-rich spherule outer wall (SOW) layer. Remnants of the outer conidial wall (OCW) are visible. Bar, 2 μm. (F) Mithramycin-stained, multinucleate spherule which had nearly completed its isotropic growth phase. Note the presence of a large, central vacuole. Bar, 10 μm.

typically occur as linear crystalline structures (rodlets; inset in Fig. 2A) at the surface of air-dispersed conidia produced by most filamentous fungi (211). Although the amino acid sequence of hydrophobins is diverse, they have three common characteristics: all are small proteins (96 to 157 residues), all have eight cysteines arranged in a conserved pattern, and all have a similar hydrophobicity profile (190). The rodlet layer probably serves as a hydration barrier at the surface of arthroconidia, which would contribute to air dispersal and permit the cells to retain water and remain viable in the desert soil for extended periods. For uptake of water and stimulation of germination to occur, either the outer hydrophobic wall layer is physically disrupted or the cells are hydrated by exposure to compounds in the respiratory tract with detergent-like properties (e.g., surfactant complexes) (36, 108, 147). If conidia

are stripped of their outer wall, they remain viable but lose hydrophobicity (35). Further discussion of this pivotal phase of conidial germination follows in our examination of mechanisms of resistance of arthroconidia to host innate defenses. Arthroconidia obtained from different isolates of *Coccidioides* also show a range of brown to dark brown pigmentation in their wall (104, 125). Melanin-binding monoclonal antibodies (152), kindly provided by J. Nosanchuk (Albert Einstein College of Medicine), were used to determine whether this pigment could be detected in the conidial wall. Preliminary results of immunofluorescence studies suggest that wall-associated melanin is present as discontinuous patches (Fig. 2B and C). It is not known whether the pigment is localized in the outer hydrophobic sleeve, the inner conidial wall layer, or both.

Initiation of parasitic cell differentiation is signaled by isotropic growth of the arthroconidia (Fig. 1). This occurs in vitro when the propagules are incubated at 39°C in sealed flasks containing a defined glucose-salts medium to which 20% CO_2 is added. Glucose and ammonium acetate are the sole sources of organic carbon and nitrogen in the culture medium (126). CO_2 is essential for in vitro development of the parasitic phase and prevention of reversion to the mycelial phase. This same critical role of CO_2 in the morphogenesis of *Coccidioides* was demonstrated by using subcutaneously implanted dialysis chambers in mice (121). Arthroconidia converted to mycelia within the chambers unless CO_2 supplementation was provided (CO_2 partial pressure of 20 to 80 mm Hg), in which case endosporulating spherules were produced. The effect of CO_2 could not be reproduced with either $NaHCO_3$ or other buffers. It seems that CO_2 at a partial pressure found in normal host tissues is essential for parasitic-phase development and that, in contradiction to an earlier report, host phagocytes are not needed for such conversion in tissue fluids (83). The physiological role of CO_2 in the differentiation of parasitic cells is unknown. Arthroconidia were incubated ex vivo in our laboratory by using murine tracheal explants, as described for other host-pathogen studies (212). After 4 to 6 h in a CO_2 incubator (5%) at 37°C, the majority of fungal cells had rounded up and contained one or two nuclei, as revealed by mithramycin A staining (13) (Fig. 2D). Thin sections of these first-generation "spherule initials" (Fig. 2E) showed remnants of the fractured outer conidial wall and the presence of an underlying amorphous spherule outer wall (SOW) layer, which is lipid rich (37, 100). During the subsequent 36 to 96 h of incubation, the spherules undergo rapid isotropic growth and nuclear division (Fig. 2F). A large central vacuole is consistently present during this period of differentiation. Its role as a storage organelle and source of internal turgor pressure during the isotropic growth phase seems likely but has not yet been explored.

Once the spherule has nearly completed its diametric expansion, the multinucleate cell begins the process of compartmentalization of its cytoplasm. This is initiated by invagination of the plasma membrane and coordinated synthesis of new wall material (Fig. 3A to C). Both the triggering of isotropic growth of arthroconidia and synchronous ingrowth of the spherule wall (118) correlate with peaks of ornithine decarboxylase (ODC) activity in vitro (93). We have shown that conversion of *Coccidioides* arthroconidia to spherule initials as well as onset of the segmentation phase can be blocked by addition of 1,4-diamino-2-butanone, an ODC inhibitor, to

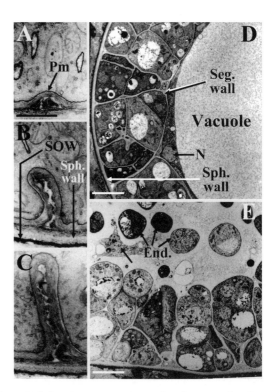

Figure 3. Spherule segmentation and early stage of endospore differentiation. (A to C) Thin sections of spherules which show early stages of segmentation wall formation. Pm, plasma membrane; Sph. wall, spherule wall. Bar, 1 μm. (D) Thin section of segmented spherule showing compartments formed by growth and fusion of invaginated segmentation wall (Seg. wall). Note the residual cytoplasm trapped between vacuolar membrane and compartments. Bar, 5 μm. (E) Thin section of spherule at early stage of endosporulation. Note that the vacuolar membrane has ruptured, spheroidal endospores (End.) have been released from compartments, and the segmentation wall has partially disappeared. Arrow indicates cytoplasmic debris. Bar, 4 μm.

cultures just prior to the respective stages of parasitic cell differentiation (93). ODC catalyzes the highly regulated pathway of polyamine biosynthesis from arginine and ornithine. Polyamines (putrescine, spermidine, and spermine) can be transported in and out of the cytoplasm and are abundant in rapidly dividing cells (123, 146, 207). They are aliphatic nitrogenous compounds, considered to be important growth regulators because of their influence on cell division and differentiation even at millimolar concentrations (105, 122, 123). Endogenous pools of polyamines regulate the growth and differentiation of fungal pathogens (173). In addition, polyamine production and accumulation have been suggested to play important roles in the response of the mammalian host to microbial infection (208). It is possible that host-derived polyamines released during the infection process could be taken up by the pathogen and utilized for its growth in situ. For example, the dimorphic phenotype of *Ustilago maydis*, which undergoes differentiation within infected tissue of corn plants, requires polyamine concentrations significantly higher than those necessary to sustain mycelial growth (94); the polyamines may be derived from host polyamine reservoirs during pathogenesis. Similar conditions could exist during *Coccidioides* parasitic cell development at sites of infection in the lungs. More discussion of this follows later in this chapter.

Segmentation wall growth and differentiation proceed within the spherule while the vacuolar membrane remains intact (Fig. 3D). Each compartment is still multinucleate at this stage. Ingrowth and fusion of the segmentation wall result in exclusion of some of the spherule cytoplasm from the compartments that later differentiate into endospores (Fig. 3E). This sacrifice of biomass is suggestive of an inefficient process of differentiation. On the other hand, it is possible that the residual cytoplasmic debris, which is eventually released from ruptured spherules, includes components which elicit an intense host inflammatory response that could restrict growth of the pathogen but could also partly account for lung tissue damage at sites of infection. An explanation of this point is presented later in our discussion of *Coccidioides* urease activity and related increase in

pH at infection sites. As compartments begin to differentiate into endospores, several structural features are evident: the central vacuole disappears, presumably as a result of the rupture of its membrane, and the released endospores begin to assume a spherical shape. The space previously occupied by the central vacuole within the still intact, maternal spherule accommodates the thin-walled, fragile endospores during their early stage of development (Fig. 3E). Particulate material and degenerate cytoplasmic organelles are clearly visible within the lumen of the spherule at this stage. Also evident is that the segmentation wall which previously separated compartments has largely disappeared (Fig. 4A). The intact segmentation wall shows high affinity for lectins which are known to bind to *N*-acetylglucosamine polymers, the structural components of chitin (34, 118). A possible explanation for the loss of the segmentation wall is that it is digested by chitinases secreted by the endospore initials. If this is the case, the thin endospore wall must be protected against self-digestion. In fact, we have found that the wall of the endospore initials shows low affinity for chitin-binding lectins, suggesting that it is composed of another structural polymer such as β-1,3-glucan. The wall of the maternal spherule becomes thinner during this developmental stage but remains intact. On the basis of the speculation that chitinases play a role in

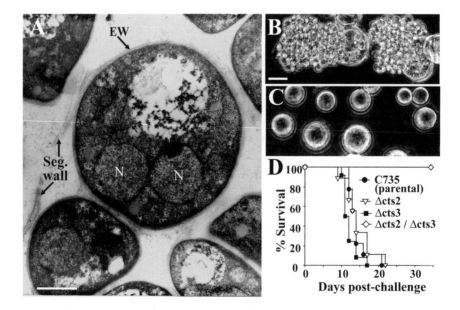

Figure 4. Digestion of the segmentation wall is essential for endospore maturation. (A) Thin section of endospores contained within the maternal spherule. Note the fragments of digested segmentation wall (Seg. wall). EW, endospore wall. Bar, 1 μm. (B and C) Light micrographs of endosporulating spherules (B) and sterile spherules (C) derived from cultures of the wild-type (parental) strain of *C. posadasii* and the Δcts2/Δcts3 mutant strain, respectively. Bar, 20 μm. (D) Survival plot of BALB/c mice challenged intranasally with arthroconidia derived from the parental strain (C735) or arthroconidia isolated from the single or double chitinase gene knockout strains. Mice were challenged separately with 50 arthroconidia of the parental, Δcts2, and Δcts3 strains and with 200 conidia of the Δcts2/Δcts3 strain.

endosporulation, we cloned all seven members of the *Coccidioides* family of chitinase genes (*CTS1* through *CTS7*) and examined their expression during the parasitic cycle (X. Chen, C. Y. Hung, and G. T. Cole, *Abstr. 103rd Gen. Meet. Am. Soc. Microbiol.* 2003, abstr. F-42, 2003). The results revealed that both *CTS2* and *CTS3* transcript levels were elevated during spherule segmentation. The predicted structure of the Cts2 protein is homologous to that of the *Saccharomyces cerevisiae* chitinase, which, when deleted, yielded a mutant that was unable to sporulate (124, 169). Separate deletions of the *CTS2* and *CTS3* genes resulted in mutants which showed delay in endosporulation in vitro compared to the parental strain but endosporulated in vivo and remained highly virulent in mice (Chen et al., *Abstr. 103rd Gen. Meet. Am. Soc. Microbiol.*, 2003). On the other hand, the double mutant (Δ*cts2*/Δ*cts3*) revealed normal saprobic-phase growth with formation of viable arthroconidia but produced sterile spherules in vitro and in vivo (compare Fig. 4B and C). Infection with the double mutant was asymptomatic in mice even when the animals were challenged with four times the normal lethal inoculum of arthroconidia by the intranasal route. The mice survived beyond 30 days postinfection (Fig. 4D). On the basis of these results, it appears that endosporulation and dissemination of the pathogen from sites of coccidioidal infection are essential for establishment of disease. The potential use of the Δ*cts2*/Δ*cts3* mutant strain as a live vaccine against coccidioidomycosis is currently under investigation (39).

Endospores grow isotropically while still contained within the maternal spherule and eventually cause the spherule wall to rupture. The endospores remain thin walled at this stage and are presumably vulnerable to host cellular defenses. As stated above, the host mounts an intense inflammatory response to infection sites upon exposure to the contents of the ruptured spherule (39). Neutrophils are the predominant host cells that target the pathogen at this stage; some of these cells migrate into the lumen of the spherule (Fig. 5A). The fungal products and host factors responsible for this chemotaxis of inflammatory cells are undefined. In spite of the intense host response to ruptured spherules, endospores are rarely observed within phagocytes, suggesting that the fungal cells somehow evade host detection. As endospores grow isotropically within the maternal spherule, an amorphous, lipid-rich outer wall layer (SOW) is produced (81), which is similar in composition to the SOW produced at the surface of germinated arthroconidia (compare Fig. 2E). More discussion of the relevance of this lipid layer to the course of disease follows later in this chapter. Spherules release a multitude of progeny (Fig. 5B), which permit the fungus to disseminate and establish new sites of infection. In vitro-grown, second-generation spherules typically reach their full size (approximately 40 to 80 μm in diameter) by 6 to 7 days (Fig. 1) but mature more quickly (3 to 4 days) in vivo. The fully grown parasitic cells are too large to be engulfed by host phagocytes, and, as suggested above, this is an important passive defense mechanism of *Coccidioides* that contributes to its survival within the host.

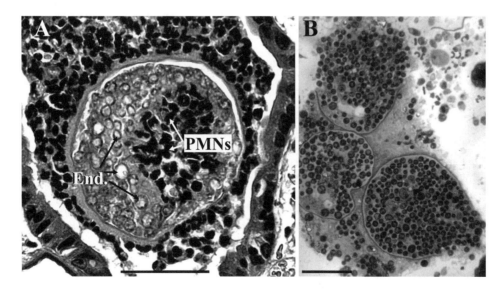

Figure 5. Endospore release and host response. (A) Hemotoxylin-eosin-stained paraffin section through a terminal bronchiole of a mouse lung infected with *C. posadasii* (C735 strain), showing an abundance of neutrophils (PMNs) which have surrounded and entered a ruptured spherule. Bar, 20 μm. (B) Periodic acid-Schiff-stained paraffin section of mature, ruptured spherules in an infected murine lung. Bar, 20 μm.

RESISTANCE OF ARTHROCONIDIA TO HOST INNATE DEFENSES

Neutrophil and macrophage defenses are critical for clearance of fungal infections (137). Dendritic cells (DCs) also play a role in phagocytosis and killing of fungi but are most influential as antigen-presenting cells that initiate and modulate the host adaptive immune response. Alveolar macrophages are likely to be among the first innate cells to encounter *Coccidioides*. However, in vitro studies have indicated that the majority of arthroconidia survive in the presence of nonactivated macrophages (31, 62). Cox (42) has pointed out that dissemination of *Coccidioides* in susceptible people of African American and Filipino ancestry occurs relatively early after pulmonary infection, which the author suggested is an indication of a defect in their alveolar macrophage function. An impressive feature of coccidioidal infection in BALB/c mice is that intranasal challenge with as few as 10 viable arthroconidia leads to death of the animals within about 3 weeks (181). In the mouse, as in humans, susceptibility and resistance to fungal infection depends on genetic predisposition, age, and relative state of immunocompetence (8, 114, 119). Among different inbred strains of mice, the mean lethal dose of *Coccidioides* arthroconidia varies by as much as 100-fold. DCs produced by BALB/c mice are less effective in enhancing cellular immune response to microbial infection than are DCs of C57BL/6 mice (133). This may be a contributing factor to the higher susceptibility of the BALB/c strain than of the C57BL/6 strain to coccidioidomycosis (46, 74, 136). It is reasonable to assume that the initial interaction between arthroconidia and phagocytes is pivotal for the establishment of coccidioidal infection. It also seems likely from the discussion of the structural features of arthroconidia that these cells are equipped with inherent factors that permit their survival in the presence of host phagocytes. One such factor that was previously mentioned is the presence of an impermeable outer conidial wall layer which is rich in hydrophobins (118) and has been suggested to have antiphagocytic properties (62). Conidia produced by a mutant strain of *Aspergillus fumigatus* which lacked this hydrophobin-containing cell surface layer are more sensitive to killing by alveolar macrophages than are conidia produced by the parental strain (166). The hydrophobin-rich outer wall layer of *Coccidioides* arthroconidia may serve as a passive barrier to destructive enzymes and oxidative products released by host defense cells. Removal of this outer layer by passing conidia through a Ribi cell fractionator (33) renders the cells more susceptible to phagocytosis, but the efficiency of killing by nonactivated phagocytes was not significantly improved compared to that of intact conidia (62). This latter observation suggested that soluble conidial wall components released from the intermural space and/or the inner conidial wall may affect the host phagocyte response (32, 33). We have shown that intact conidia exposed in vitro to 1% dipalmitoyl phosphatidylcholine (DPPC; commercial surfactant) undergo hydration and concomitant release of protein from their cell wall (37). Comparable hydration of the conidial surface and release of cell wall proteins may also occur in vivo on exposure of arthroconidia to surfactant complexes, which could influence host phagocyte response to *Coccidioides* infection. DPPC produced in alveoli is known to interact with pulmonary surfactant, a mixture of lipids and proteins which can affect the function of alveolar macrophages (49, 147). Arthroconidia of *Coccidioides* that reach the distal airspaces of the lung are most probably phagocytosed by alveolar macrophages in the presence of pulmonary surfactant A (SP-A), a dominant component of the surfactant complex and an important contributor to host defense mechanisms. SP-A has been reported to facilitate phagocytosis (85) by opsonizing bacteria, fungi, and viruses to stimulate the oxidative burst by phagocytes, and to modulate proinflammatory cytokine production by phagocytic cells (135). On the other hand, contradictory data on the role of pulmonary SP-A have been presented. It has been suggested that SP-A can assume an anti-inflammatory role by inhibiting macrophage production of reactive oxygen intermediates through reduction in NADPH oxidase activity (49). Under these latter conditions, SP-A binding to the surface of arthroconidia, perhaps as an aggregate with DPPC and other surfactant components, could contribute to survival of the fungal cells.

A mechanism by which phagocytic cells kill ingested microorganisms is by production of superoxide radicals in phagolysosomes. On the other hand, certain microbial pathogens can effectively defend themselves against oxidative products of host cells. Superoxide dismutase (SOD) is a metalloenzyme produced by bacterial and fungal pathogens that detoxifies oxygen radicals by conversion of superoxide to hydrogen peroxide and oxygen (44, 82). Further breakdown of the H_2O_2 in yeast occurs in the presence of catalase, glutathione peroxidase, and thioredoxin peroxidase (21, 92). The natural function of SODs is to protect cells from endogenously generated superoxide anion, which is a by-product of normal aerobic respiration. A mutant of *Cryptococcus neoformans* which revealed a marked decrease in SOD activity was shown to be more susceptible than the parental strain to reactive oxygen species in a cell-free system and to growth inhibition within the phagolysosome of nonactivated macrophages (44). We have reported that extracts of *Coccidioides* arthroconidia can inhibit in vitro production of superoxide anion by rat alveolar macrophages (31). Not surprisingly, BLASTx searches of the translated genomic database of *Coccidioides* (56) (www.tigr.org) have identified homologs of reported fungal Cu/Zn SODs (84 to 89% sequence similarity), catalases (76 to 91% similarity), glutathione peroxidases (78 to 85% similarity), and thioredoxin peroxidases (68 to 93% similarity). Release of these antioxidants by the pathogen may be facilitated by exposure of arthroconidia to surfactant complexes as described above.

Our preliminary studies of pigment production by *Coccidioides* have suggested that the conidial wall layer contains melanin. Since melanins are known to absorb UV light and prevent photoinduced cell damage (97), production of this pigment by spores exposed to sunlight at the surface of desert soil may provide an important survival function. Cell wall-associated melanin also contributes to the virulence of bacterial and fungal pathogens (151, 155, 223, 224). In *C. neoformans* and *Exophiala dermatitidis*, melanins have a strong antioxidant property. Melanized cells exhibit more resistance to oxygen- and nitrogen-derived radicals than nonmelanized cells do, and the wall-associated pigment of fungal pathogens contributes significantly to microbial protection against the oxidative burst of neutrophils. Melanization of *C. neoformans* cells also interferes with phagocytosis. The wall-associated pigment of *Coccidioides* (Fig. 2B) may be a contributing factor to the reported antiphagocytic property of the outer arthroconidial wall. Fungi can synthesize melanin via phenoloxidases (e.g., laccase activity) and/or the polyketide synthase pathway (214). Laccase catalyzes the synthesis of melanin from diverse phenolic substrates, including dopamine (224). Production of pigment associated with the conidial wall of *Coccidioides* was enhanced by growth of the fungus on media containing L-3,4-dihydroxyphenylalanine (L-DOPA). On the basis of these results, we conducted a BLASTx search (3) of the *C. posadasii* translated genomic database for the existence of laccase homologs. A predicted homolog was identified with 53 to 60% amino acid sequence similarity to reported fungal laccases (131, 180, 216). We suggest that the pathway of melanin biosynthesis in *Coccidioides* is comparable to that in *C. neoformans* and requires phenoloxidase activity (224).

RESISTANCE OF FIRST-GENERATION SPHERULE INITIALS TO HOST DEFENSES

Nitric oxide (NO) produced by macrophages and DCs is another important component of the host oxidative attack directed against microbial pathogens (68). The fungicidal and fungistatic effects of NO against a wide range of fungal pathogens have been reported (1, 2, 16, 18, 58, 72, 90, 175). However, the importance of host production of NO as a mechanism of defense against *Coccidioides* has not been explored. We have employed a macrophage-free system (58) to test the sensitivity of arthroconidia, spherule initials harvested from 48-h parasitic-phase cultures, and segmented/endosporulating spherules from 144-h first-generation cultures to an NO generator, 3-morpholinosydnonimine (SIN-1). The latter was selected because of its ability to release both NO and superoxide (71), thus mimicking the generation of nitrogen and oxygen radicals by activated macrophages. These two radicals combine to form the powerful oxidant peroxynitrite (ONOO$^-$) (11), which is capable of oxidizing known biomolecules of microbial cells (129). The inhibitory effects of SIN-1 on fungal growth are critically dependent on the biological environment in which the oxidant is present (70), and the compound can be efficiently detoxified provided that antioxidant defenses of the fungal cells are intact (58). Sustained exposure to this molecule at high concentrations eventually leads to the depletion of antioxidants, allowing the attack of ONOO$^-$ on cell targets. We exposed *C. posadasii* to SIN-1 in vitro at concentrations of 0.15 to 5 mM for either 3 or 6 h (Fig. 6). The results were quite surprising: arthroconidia were significantly more sensitive to SIN-1 exposure than either spherule initials or segmented/endosporulating spherules at concentrations between 0.15 and 1.25 mM. In the course of a natural infection, arthroconidia would be expected to encounter nonactivated macrophages, which do not produce large amounts of nitric oxide (44). Perhaps in the case of the conidia, therefore, resistance to macrophage clearance is more dependent on the physical barrier of the cell wall as well as the protective effects of superoxide dismutase as discussed above. On the other hand, both spherule initials and segmented/endosporulating spherules apparently have mechanisms which can more efficiently detoxify NO, as demonstrated in the macrophage-free system. It has been suggested that *C. neoformans* reduces NO activity in macrophages via NO consumption involving its polysaccharide capsule and/or flavohemoglobin (44, 195). Two recently identified enzymes of *C. neoformans*, flavohemoglobin denitrosylase and *S*-nitrosoglutathione reductase,

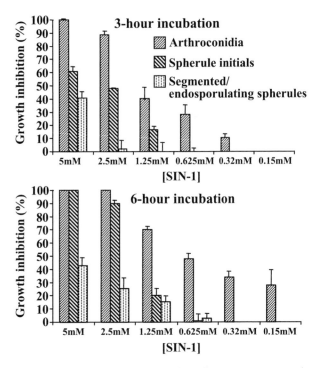

Figure 6. Sensitivity of *Coccidioides* cell types to nitric oxide exposure. Arthroconidia, spherule initials (germinated arthroconidia after 48 h of incubation in glucose-salts medium), and segmented/endosporulating spherules (from 144-h parasitic-phase cultures) were isolated, washed in PBS, and transferred to 96-well plates (10^5 cells in 100 μl of PBS per well), to which different amounts of SIN-1 were added as indicated. The 96-well plates were sealed and incubated in the absence of light at 37°C for 3 or 6 h. Control cells were incubated in PBS alone. Serial dilutions of the cell suspensions were plated onto GYE medium to determine number of viable fungal cells, which were recorded as CFU. The data are presented as percentage of growth inhibition, which was calculated on the basis of the formula [1 − (CFU of SIN-1-exposed cells ÷ CFU of control cell suspension)] ×100%.

apparently consume and metabolize NO and protect yeast from nitrosative challenge in vitro (55). When mutations in flavohemoglobin and superoxide dismutase were combined in *C. neoformans*, pathogenicity was markedly attenuated. *Candida albicans* produces a flavohemoglobin under control of the *CaYHB1* gene, and this substance is responsible for NO consumption and detoxification. Expression of *CaYHB1* mRNA is induced by cell exposure to NO and NO-generating agents, and disruption of this gene increased the sensitivity of the opportunistic pathogen to NO-mediated growth inhibition (201). Flavohemoglobins have been suggested to be involved in a conserved mechanism of protection in all microorganisms (44, 132). A single gene which encodes a homolog of a bacterial flavohemoglobin (96) has been identified in the *C. posadasii* genome database.

Pretreatment of macrophages with host cytokines (gamma interferon [IFN-γ] and/or tumor necrosis factor alpha [TNF-α]) has been reported to enhance the ability of the phagocytes to kill *Coccidioides* arthroconidia and endospores in vitro (10). However, other investigators have demonstrated a very modest killing of arthroconidia (20 to 30%) by peripheral blood mononuclear cells and have shown that this process could not be modulated by either IFN-γ or TNF-α (6). In the latter study, it was also observed that as first-generation spherules develop in vitro, they become increasingly more resistant to macrophage growth inhibition and killing. These inconsistencies in the literature prompted us to reexamine the nature of cytokine response of nonactivated macrophages to *Coccidioides* in vitro. We focused our studies on TNF-α production, since this inflammatory cytokine has been shown to play roles both in the inhibition of microbial growth and granuloma formation in mice (12, 78, 117). TNF-α knockout mice challenged with *Mycobacterium* exhibited higher microbial loads in infected organs, exacerbated IFN-γ responses as evaluated in the sera of infected animals, and reduced host survival (77). In a study of a hypervirulent strain of *M. tuberculosis* it was shown that downregulation of the murine innate immune response to infection was at least partly due to production of a bacterial wall-associated phenolic glycolipid (PGL) (172). In vitro studies using isolated murine bone marrow-derived macrophages revealed that PGL exposure was responsible for reduced levels of production of the proinflammatory cytokines TNF-α, interleukin-6 (IL-6), and IL-12. Disruption of PGL biosynthesis resulted in loss of the hypervirulent phenotype. Resistance of young spherules to macrophage growth inhibition and killing may be related to synthesis of the lipid-rich SOW layer at the parasitic cell surface. Although we have not detected PGL in the lipid fractions isolated from *C. posadasii* spherules (37, 100, 102), phospholipids and glycosphingolipids were detected by thin-layer chromatographic analysis and determined to be major components of the SOW layer (Fig. 7A). We have tested whether the lipid fraction of SOW associated with intact spherules influences the macrophage inflammatory response in vitro. We performed a series of assays in which a nonactivated murine macrophage cell line (RAW 264.7) was exposed to different *Coccidioides* cell types and cell fractions (Fig. 7B). Each fungal morphotype was incubated with an equal number of macrophages for 4 h as reported (109), and the total amount of TNF-α protein in the supernatant was measured by an enzyme-linked immunosorbent assay (ELISA). Arthroconidia

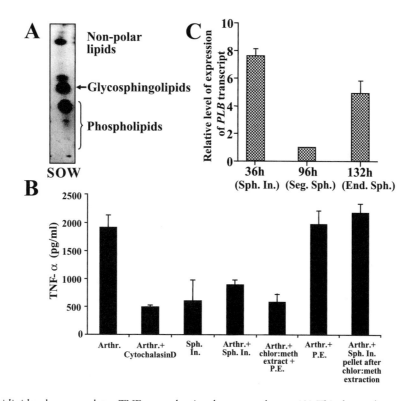

Figure 7. *Coccidioides* down-regulates TNF-α production by macrophages. (A) Thin-layer chromatographic separation of SOW polar lipids using chloroform-acetic acid-methanol-water (150:50:10:4.4) as the solvent. (B) Arthroconidia (Arth.) and/or spherule initials (Sph.In) (as described in the legend to Fig. 6) were harvested, washed in PBS, and coincubated with a macrophage cell line (RAW 2647, American Type Culture Collection, Manassas, Va.) (1:1 ratio) for 4 h in Dulbecco modified Eagle medium as reported (109). The supernatants were isolated, and the TNF-α was quantified by ELISA using an OptEIA mouse TNF-α detection kit (PharMingen, San Diego, Calif.). In a separate experiment, macrophages were exposed to cytochalasin D (2.5 μg/ml of 1% dimethyl sulfoxide [final concentration]; Biosource, Camarillo, Calif.) for 2 h prior to incubation with fungal cells in order to inhibit phagocytosis. Also, in a separate experiment, spherule initials were extracted with chloroform-methanol (2:1; high-performance liquid chomatography grade) (chlor:meth extract) for 30 min at 4°C as reported previously (79), the extract supernatant was collected by centrifugation, and the lipid layer was aspirated, dried in an N₂ stream, and resuspended in petroleum ether (PE). No protein could be detected in this extract by SDS-PAGE. Aliquots of the resolubilized extract were tested for their effect on TNF-α production by macrophages as described in the text. The extracted, intact spherule initials were also tested for their influence on TNF-α production. (C) Quantitative real-time PCR analysis of *C. posadasii PLB* expression in vitro during stages of parasitic cell development. Relative amounts of *PLB* transcript produced by spherule initials (Sph. In.) and endosporulating spherules (End. Sph.) were compared to that of segmented spherules (Seg. Sph.), which was assigned an arbitrary value of 1.

freshly isolated from agar plates stimulated macrophages to secrete significant amounts of TNF-α, in spite of the apparent inability of the host cells to disrupt the cell wall (118) or inhibit growth of the pathogen (62). A possible explanation for this is that the indigestible, outer conidial wall and associated melanin serve as "foreign-body"-like material and stimulated an inflammatory response (44). We determined that the up-regulation of TNF-α production required the engulfment of arthroconidia by macrophages; an increase in the amount of cytokine above the baseline level was not observed when the fungal and host cells were separated by a porous membrane (pore size, 0.4 μm) or when phagocytosis was inhibited by exposure of the macrophages to

cytochalasin D (2.5 μg/ml). In contrast to the phagocyte response after arthroconidial engulfment, macrophages incubated with spherule initials (Fig. 2E) failed to induce elevated levels of TNF-α secretion and significantly fewer fungal cells were phagocytosed. When a mixture of equal numbers of arthroconidia and spherule initials was coincubated with macrophages, a marked reduction in TNF-α production was observed compared to the results of incubating macrophages with arthroconidia alone. This inhibition of cytokine production was reproduced with a chloroform-methanol (2:1) extract of the spherule initials. The lipid fraction was obtained from spherules grown in vitro for 48 h and then subjected to chloroform-methanol extraction as previously described (79). Sodium dodecyl sulfate-polyacrylamide gel electrophoresis (SDS-PAGE) separation of the extract revealed the absence of protein components. The dried, total lipid extract from 2×10^6 viable cells was resuspended in petroleum ether, transferred as aliquots of equal volume to the wells of a 24-well tissue plate, and allowed to evaporate before adding the tissue culture media. Arthroconidia and RAW 264.7 cells (1:1) were then transferred to the wells, and the supernatants were assayed 4 h later for levels of TNF-α production. Control assays were conducted by incubation of macrophages with arthroconidia plus evaporated petroleum ether alone or with arthroconidia plus an equal number of intact, chloroform-methanol-extracted spherule initials. The results of these preliminary studies suggested that lipid components of the first-generation spherules, possibly associated with the SOW layer, confer the ability of *Coccidioides* to inhibit the release of a key inflammatory effector molecule by cells of the host innate immune defense system. Comparable results were obtained using chloroform-methanol extracts of mature, pre-endosporulating, and endosporulating spherules which were also harvested from parasitic-phase cultures and tested as above. Earlier investigators have suggested that *Coccidioides* can engender host anergy (5, 173). This is the first report of a defined mechanism of suppression of cellular immunity by *C. posadasii*.

Cationic lipids (including phospholipids) inhibit the production of TNF-α by macrophages and thereby demonstrate anti-inflammatory activity (75). Phosphatidylethanolamine suppresses the host immune response to Gross virus surface antigen by activation of macrophage prostaglandin E_2 synthesis (9), which in turn has an inhibitory effect on TNF-α production (202). Eicosanoids (e.g., prostaglandins and leukotrienes) are potent regulators of the host immune response (153). Prostaglandins inhibit the T-helper 1 (Th1) pathway of the immune response and instead promote Th2-type responses (14, 186, 189). The Th2 pathway is characterized by elevated levels of secretion of IL-4, IL-5, IL-6, and IL-10 by the Th2 subset of lymphocytes, stimulation of B cells to produce antibody, activation of mast cells and eosinophils, and partial down-regulation of cellular immune responses (149). Activated Th1 lymphocytes, on the other hand, secrete elevated levels of IL-2, IL-12, TNF-α, and IFN-γ and stimulate cell-mediated immunity against fungal infections. Host cell-derived eicosanoids play an important role in balancing Th1 and Th2 pathways of the host immune response and thereby minimize tissue damage which can result from intense, chronic inflammation. Another potential source of eicosanoids during

an infection is the pathogen itself. Fungus-derived eicosanoids may be exported and could function as host immune regulators, shifting the balance toward a dominant Th2 response and promotion of chronic disease (153). Th2 immunodominance compromises host protection against coccidioidal infection and exacerbates the course of disease (39). Arachidonic acid within intracellular lipid pools of inflammatory cells (mast cells, eosinophils, and monocytes) can be utilized for host eicosanoid biosynthesis (197). Host cell phospholipases participate in the regulation of eicosanoid production by the liberation of fatty acid precursors (e.g., arachidonic acid, dihomo-γ-linolenic acid, eicosanopentaenoic acid) from these phospholipid pools (167). Fungal pathogens have been reported to secrete phospholipases, which have been implicated in virulence via their destruction of host cell membranes (61, 84, 86, 148). Noverr et al. (153) have also presented evidence that a cryptococcal phospholipase (Plb1) may enhance the survival of yeast in the presence of macrophages by production of fungal eicosanoids, which can down-regulate host TNF-α production and modulate the immune response to the advantage of the pathogen. Eicosanoids are membrane diffusible and can bind to intracellular macrophage receptors, resulting in deactivation of the phagocytes. Numerous species of pathogenic fungi produce eicosanoids (154), and they probably all produce phospholipases. A single phospholipase B gene (*PLB*) has been identified in the *C. posadasii* genome database, and its peaks of mRNA expression were detected during arthroconidial germination (spherule initial formation) and endospore differentiation (Fig. 7C). Based on the reported role of Plb1 in cryptococcal eicosanoid production, it is possible that the phospholipase of *C. posadasii* may target the SOW layer of parasitic cells to release fatty acid precursors for synthesis of fungal eicosanoids, which, in the presence of macrophages, could lead to suppression of the inflammatory response. Prostaglandins and leukotrienes may be produced by *Coccidioides* either within the phagosome and/or during the pathogen's extracellular association with the host, as demonstrated by *C. neoformans* (153). The spherule product(s) responsible for the down-regulation of TNF-α production shown in Fig. 7B, as well as the host signal transduction pathway which is modulated by exposure to the fungal ligand(s), is currently under investigation in our laboratory.

ENDOSPORE AVOIDANCE OF HOST DETECTION

Antibody from patients with confirmed coccidioidal infection shows high affinity for the crude, lipid-rich SOW membranous fraction of parasitic cells which is released into the culture medium (32, 37, 100). The component of SOW responsible for this immunoreactivity is a single glycoprotein, designated SOWgp (100, 102). The purified antigen is also a potent stimulant of murine immune T-cell and human peripheral blood monocytic cell proliferation (32, 102). The peripheral blood monocytic cells were obtained from individuals living in regions of endemic coccidioidomycosis who showed skin test-positive reaction to spherulin, a commercially available parasitic-phase antigen of *Coccidioides* used to measure the delayed-type hypersensitivity response (42, 103). SOWgp has been identified as an immunodominant, parasitic cell surface antigen that elicits both antibody-mediated and cellular immune responses in patients with coccidioidal infection (100). The glycoprotein is associated in an unknown manner with the phospholipid and glycosphingolipid components of SOW at the surface of parasitic cells. SOWgp consists of a signal peptide and propeptide, a proline- and aspartic acid-rich tandem repeat motif, and a glycosylphosphatidylinositol (GPI) anchor signal consensus sequence (102) (Fig. 8A). The repeat domain contains four to six repeat sequences

(fungal strain dependent), each approximately 47 residues in length, and is responsible for patient antibody reactivity with SOWgp (Fig. 8B). High titers of antibody reactive with the bacterially expressed, recombinant repeat fragment of the glycoprotein have been demonstrated by ELISA to be present in sera of patients with confirmed coccidioidal infection (Fig. 8C). Exposure of SOWgp at the parasitic cell surface, therefore, would be expected to bind antibody and enhance opsonization, phagocytosis, and clearance of endospores from sites of infection. In defiance of these immune defenses, however, the pathogen has evolved a mechanism which permits the fragile endospores to evade host detection. The peaks of *SOWgp* gene expression and glycoprotein production during first-generation parasitic cell development occur during the isotropic growth phase of spherules (Fig. 1) but then sharply decrease during endospore production (102). Crude SOW fractions collected from the media of endosporulation-phase cultures lacked detectable SOWgp but instead revealed a catalytically active, 34-kDa metalloproteinase (Mep1) that belongs to the metzincin superfamily (101, 171). In vitro studies have demonstrated that the recombinant Mep1 enzyme can efficiently

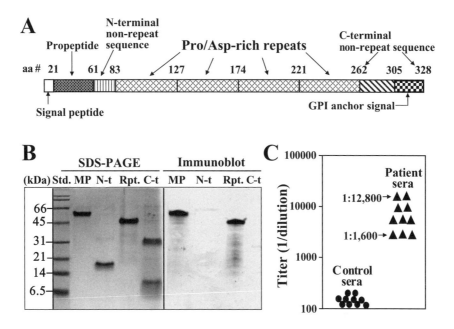

Figure 8. SOWgp is an immunodominant, parasitic cell surface antigen. (A) Structure of the SOWgp antigen (Silveira isolate). (B) Bacterialy expressed, full-length recombinant SOWgp protein (mature protein [MP]) and fragments of SOWgp (83-residue N-terminal fragment [N-t], 179-residue repeat fragment [Rpt.], and 67-residue C-terminal fragment [C-t]) were separated by SDS-PAGE and tested in an immunoblot assay for reactivity with antiserum from a patient with confirmed coccidioidomycosis. The lower-molecular-mass band in the lane containing the C-t fragment is an *Escherichia coli* peptide contaminant. (C) Results of ELISA showing reactivity of the recombinant repeat fragment (Rpt.) of SOWgp with control human sera and sera from patients with confirmed coccidioidal infection. The concentration of the recombinant repeat protein bound to wells of the microtiter plate was 10 ng/well. Goat anti-human IgG (H+L) conjugated to peroxiclase was used for detection of adsorbed antibody. Antibody titers were determined as reported previously (128).

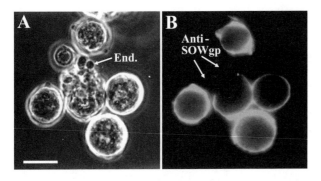

Figure 9. Mep1 digestion of SOWgp permits endospores to evade host detection. A phase contrast micrograph (A) and a matching immunofluorescence light micrograph (B) of a spherule which has released its endospores are shown. The cells in panel B were incubated with antibody raised against purified, native SOWgp. Note the absence of fluorescent antibody reactivity with the surface of endospores, contrasting with the high avidity of antibody for the walls of the ruptured and nonruptured spherules. Bar, 40 μm.

digest purified SOWgp. Temporal expression of the *C. posadasii MEP1* gene is restricted, at least in vitro, to the early period of endospore formation. Very small amounts of the gene-specific transcript were detected before or after this stage of development (101). It appears, therefore, that SOWgp production is tightly regulated during the parasitic cycle as a result of both transcriptional control of *SOWgp* gene expression and phase-specific digestion of the glycoprotein. The impact of these regulatory processes is that endospores which emerge from ruptured spherules lack the immunodominant antigen at their cell surface (Fig. 9) and so the parasitic cells evade host antibody recognition, opsonization, and engulfment by phagocytes. In addition to this evasive mechanism, we have suggested in the above discussion that a phospholipase (Plb), expressed during the endosporulation phase, may be responsible for release of fatty acids from the SOW layer which could contribute to fungus-host eicosanoid synthesis and partial suppression of the inflammatory response. Together, these mechanisms of evasion of host detection and down-regulation of the inflammatory response may contribute significantly to the persistence of coccidioidal infections in the lungs of the infected host.

IMMUNOMODULATION OF HOST RESPONSE IN FAVOR OF *COCCIDIOIDES* SURVIVAL

T-cell immunity is considered essential for defense against coccidioidomycosis; both clinical and experimental data support this assumption (39, 47). No evidence has been provided to date that antibody-mediated immunity protects against a natural coccidioidal infection. In fact, a rising titer of anti-*Coccidioides* antibody in patient sera typically correlates with a poor clinical outcome (163, 165, 183). We have shown that high titers of patient anti-SOWgp antibody described above correlate with dissemination and severity of disease (100–102). The fact that SOWgp also stimulates a proliferative response of $CD4^+$ T cells in vitro (32, 100) suggested to us that this immunodominant antigen may direct Th2 cells to secrete cytokines that enhance B cells to produce antibody. The results of three separate experiments support this concept (Fig. 10). C57BL/6 mice were challenged intranasally with a sublethal inoculum of *C. posadasii* (35 viable arthroconidia), and total T lymphocytes were isolated from the spleens at 30 days post-challenge. The T cells were then tested in vitro for the proliferative response to the chromatographically purified native SOWgp, and supernatants of the cultured T cells were examined by ELISA for cytokine production. The native antigen stimulated T-cell proliferation and production of IL-6, IFN-γ, and IL-10 (Fig. 10A). It appears, based on the nature of cytokine responses to the full-length glycoprotein, that host exposure to SOWgp stimulates a mixed Th1-Th2 immune response. In the second experiment, B-cell knockout mice ($Igh-6^{-/-}$ [B6.129S2-$Igh-6^{tm1Cgn/J}$]; Jackson Laboratory, Bar Harbor, Maine) and normal black mice ($Igh-6^{+/+}$ [C57BL/6]) were each challenged intranasally with a sublethal inoculum of *C. posadasii* arthroconidia as above. At 30 days postinfection, total T cells were isolated from the spleens and stimulated in vitro with the 45-kDa recombinant repeat protein fragment of SOWgp described above (Fig. 8B). The T cells from both groups of mice proliferated in response to exposure to the peptide, but the supernatants of the peptide-stimulated T lymphocytes isolated from normal mice revealed significantly higher levels of production of inflammatory and Th2-type cytokines (IL-6 and IL-10) than did the T-cell culture supernatants derived from the B-cell knockout mice (Fig. 10B). These data suggest that B cells contribute to the stimulation of a Th2 immune response to coccidioidal infection and may also participate as SOWgp antigen-presenting cells during the course of disease. In the third experiment, we determined the anti-SOWgp isotype produced by C57BL/6 mice which were first immunized with the recombinant repeat protein fragment of SOWgp and then challenged intranasally with a lethal inoculum (80 arthroconidia) of *C. posadasii*. Control mice were immunized with phosphate-buffered saline (PBS) prior to challenge. Sera were collected 1 day before inoculation with arthroconidia and at 10 and 20 days post-challenge. Comparison of immunoglobulin titers

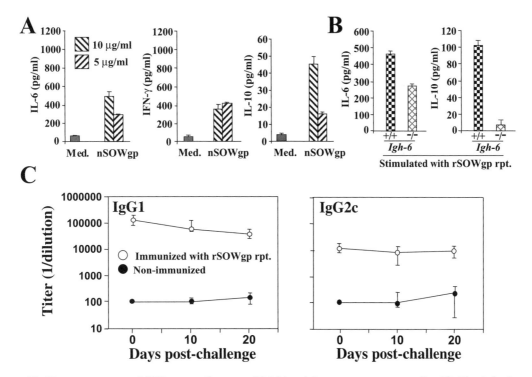

Figure 10. Host exposure to SOWgp contributes to Th2-biased immune response to *Coccidioides* infection. (A) C57BL/6 mice were challenged by the intranasal route with 35 viable arthroconidia of *C. posadasii* (C735 strain) and euthanized 30 days later. Total splenocytes were pooled from three infected animals, and T cells were isolated as reported previously (128). The T cells were transferred to RPMI plus 10% FBS, stimulated with two different concentrations of purified native SOWgp, and incubated for 48 h. Control cells were incubated in medium alone. The supernatants were tested for the concentration of selected cytokines as reported previously (220). (B) Normal C57BL/6 mice (*Igh-6*$^{+/+}$) or B-cell knockout mice (*Igh-6*$^{-/-}$) were challenged as in panel A, their T cells were isolated and stimulated with the recombinant repeat fragment of SOWgp (rSOWgp rpt.) (10 μg/ml), and the concentrations of selected cytokines were determined as above. (C) Results of ELISA with mouse sera obtained from *Coccidioides*-infected animals immunized with the rSOWgp repeat fragment (1 μg/dose; twice) versus nonimmunized controls. Mice were challenged intranasally 4 weeks after immunization, and their antibody titers against the recombinant repeat protein were determined as reported previously (101).

(IgG1 versus IgG2c) reactive with the purified, recombinant repeat protein fragment of SOWgp was conducted by ELISA (Fig. 10C). C57BL/6 mice have the *Igh1-b* allele and do not express the gene for IgG2a; instead, they express the IgG2c isotype (138). The approximately 10-fold higher titers of IgG1 than IgG2c antibody detected in immunized and challenged mice prior to and after infection is consistent with stimulation of a Th2 pathway of immune response (128). Mice that were not immunized with SOWgp showed low titers of antibody to the *Coccidioides* antigen because the animals typically became moribund within 2 weeks and died by 3 weeks post-challenge. In this respect, the murine model of coccidioidomycosis is not an ideal simulation of the human disease and immune response to *Coccidioides* infection. Nevertheless, the combined results of these experiments suggest that host exposure to the immunodominant, parasitic cell surface antigen (SOWgp) helps to direct a Th2 pathway of immunity to infection. As

stated above, the latter is not effective in mounting a protective response to *Coccidioides*.

A functionally comparable, immunodominant 43-kDa glycoprotein (gp43) is produced at the surface of *Paracoccidioides brasiliensis* yeast (20, 196). BALB/c mice infected with *P. brasiliensis* or immunized with gp43 produce high titers of IgG1-, IgG2a-, IgG3-, and IgE-specific antibody (192). The gp43 antigen has also been shown to induce delayed-type hypersensitivity in animals and humans (176). Murine studies of paracoccidioidomycosis in susceptible B10.A mice have demonstrated that gp43 is presented primarily by B cells, resulting in activation of a Th2 subset of T lymphocytes (52). On the other hand, gp43 was reported to be preferentially presented by macrophages and stimulated Th1 lymphocytes in the resistant A/Sn strain of mice. In both murine strains, macrophages and B cells were able to present gp43, and the purified antigen stimulated T-cell proliferation in vitro equally in the two strains in spite of the different tendencies of the antigen-presenting cells (APCs) to activate Th1 or Th2 subsets of T lymphocytes. It has been proposed that host susceptibility versus resistance to paracoccidioidomycosis is significantly influenced by the haplotype of the mouse strain, as well as the nature and function of the APCs (macrophages, DCs, B cells) involved in gp43 presentation to lymphocytes (52, 73, 88, 192, 193). We have reported that BALB/c mice immunized subcutaneously with recombinant SOWgp (rSOWgp) and then challenged by the intraperitoneal route with a lethal inoculum of arthroconidia of *C. posadasii* appeared to mount a Th1 response to infection and were partially protected against coccidioidomycosis (118). On the other hand, the same strain of mice immunized subcutaneously with comparable doses of rSOWgp did not protect the animals against infection via the natural, intranasal route. This same contradictory outcome between the two routes of challenge has also been reported for mice immunized with another *Coccidioides* antigen (Ag2/Pra), which has been promoted as a vaccine candidate (39, 110, 111, 120). Although intraperitoneal challenge has technical advantages over intranasal inoculation of arthroconidia in the murine model of coccidioidomycosis (e.g., reproducibility of infectious dose, less variation in fungal burden between infected mice), significant differences apparently exist in the nature of the response of the immunized host to infection and in the disease outcome following inoculation by these two routes. Differences in the dominant APCs that respond to intraperitoneal versus intranasal infections, as well as differences in T lymphocyte stimulation during the systemic versus mucosal immune response to *Coccidioides*, could influence the balance between Th1 and Th2 pathways of host immunity (204).

To further test the importance of host exposure to SOWgp during coccidioidomycosis, we first deleted the gene which encodes this polypeptide and then determined whether the mutant (Δsowgp) strain

had diminished virulence compared to the parental strain (102). Loss of *SOWgp* expression resulted in reduced virulence of the mutant strain when used to intranasally challenge C57BL/10J mice, while the revertant strain containing the restored, wild-type *SOWgp* gene was as virulent as the parental isolate. We used a microarray approach to initially explore the differences between the host response to infection with the parental versus the Δsowgp strain. The murine genome microarray (U74A chip; Affymetrix, Santa Clara, Calif.) was hybridized with total labeled cRNA derived from the lungs of *C. posadasii*-infected or noninfected C57BL/6 mice. Infected mice were challenged intranasally with equal numbers of arthroconidia isolated from plate cultures of either the parental or Δsowgp strain. Noninfected, control mice were inoculated with PBS by the intranasal route. In each case, total RNA was isolated from excised lungs at 6 days after the intranasal inoculation. We chose this early stage of disease onset because we have found that by this time mice had generated a well-defined profile of Th1- and Th2-type cytokine responses to infection (220). The RNA was reverse transcribed using SuperScript II reverse transcriptase (Invitrogen, Carlsbad, Calif.) and a poly(T) nucleotide primer containing a T_7 RNA polymerase promoter as described previously (93). The single-stranded cDNA was converted to the double-stranded form using a SuperScript Double-stranded cDNA Synthesis kit (Invitrogen). Biotinylated antisense target cRNA was then synthesized by in vitro transcription and used to hybridize with the gene chip as reported previously (26). Gene transcript levels were determined from data image files, using algorithms in the Microarray Suite version 5.0 software (Affymetrix). The expression levels of all genes on the three separate arrays were compared pairwise, and selected data are summarized in Table 1. Differences of twofold or greater were interpreted as a reflection of significant change in the level of gene expression. The data indicate that significantly

Table 1. Affymetrix mouse array 6 days post-challenge (U74A chip)

Mouse gene	Signal intensity			Fold increase in intensity
	Noninfected	Parental strain	Δsowgp strain	Parental/Δsowgp (log ratio)
Small inducible cytokine A2	2.6	131.8	32.6	3.7
Macrophage inflammatory protein 2	2.4	112.3	21.7	3.2
IFN-β, fibroblast	449.7	944.9	302.4	2.8
Monokine (induced by IFN-γ)	2.6	283.5	109.6	2.6
Monocyte chemoattractant (MCP2)	9.5	440.1	170.5	2.6
Small inducible cytokine A9	170.8	650.6	192.5	2.5
IL-6	1.4	31.1	5.1	7.0
Arginase	7.1	1350.3	78.6	17.1

higher degrees of up-regulation of expression of IL-6 and other inflammation-related cytokines and chemokines in mice infected with the parental strain occurred compared to that in mice infected with the mutant strain. The marked elevation of IL-6 production in the former is consistent with our observations of an in vitro proliferative response and IL-6 production by immune T cells upon exposure to the purified SOWgp antigen. IL-6 plays a key role in acute inflammation (115). However, under chronic inflammatory conditions, elevated IL-6 levels may have a negative impact on the host due to its role in favoring mononuclear cell accumulation at sites of infection and enhancement of cellular immune responses, which can lead to tissue damage and exacerbation of disease (115, 199, 200). Advanced stages of coccidioidomycosis in mice are characterized by the appearance of large numbers of macrophages at sites of infection. Thin sections of spherules in murine lung tissue typically reveal macrophages apparently in the process of engulfment of the lipid-rich SOW layer, which is shed from the surface of the parasitic cells (Fig. 11A). At this stage in the course of disease, macrophages most probably play a pivotal role in presentation of SOWgp peptides to T lymphocytes and, based on our earlier discussion of this

immunodominant antigen, stimulate a Th2-biased immune response. Persistently high levels of IL-6 have been observed in bronchoalveolar lavage (BAL) fluids of BALB/c and C57BL/6 mice infected with the parental strain of *C. posadasii* between day 4 post-challenge and death of the animals (Fig. 11B). In contrast, mice infected with the Δsowgp strain showed a significantly lower level and apparently more regulated production of IL-6 during the course of infection. Production of IFN-γ by mice infected with the parental or Δsowgp strain was comparable, except that mice challenged with the mutant showed an extended period of elevated production of the Th1-type cytokine. High IL-5 production is another predictor of a Th2 response and is a known eosinophil differentiation factor which recruits large numbers of eosinophils to sites of coccidioidal infection (39, 182). Eosinophilia is an indicator of active disease and has been found in approximately 25% of patients diagnosed with coccidioidomycosis (118). Coincident with eosinophilia in *Coccidioides*-infected mice is the hyperproduction of IgE, which in patients correlates with depressed cellular immune status and disease severity (43). Huffnagle et al. (99) have shown that development of chronic eosinophilia in the lungs of *C. neoformans*-infected C57BL/6 mice is a

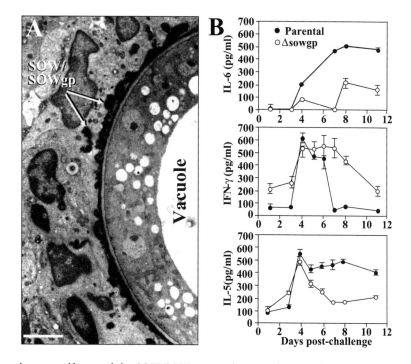

Figure 11. Macrophage engulfment of the SOW/SOWgp complex stimulates production of Th2-type cytokines. (A) Thin section of presegmented spherule showing the surface layer of SOW plus SOWgp (stained with lipophilic osmium tetroxide). Fragments of this stained layer have been shed and engulfed by phagocytes. Bar, 4 μm. (B) Groups of BALB/c mice were challenged separately by the intranasal route with an equal number of viable arthroconidia (approximately 80) derived from either the parental strain (C735) or the Δsowgp mutant strain. BAL fluid samples were collected at various times post-challenge, and the concentration of selected cytokines at each time point was determined by ELISA as reported previously (220). BAL fluid samples from three mice per time point were examined.

nonprotective immune response that causes significant lung pathology. A marked difference in BAL fluid IL-5 concentrations between the two groups of *Coccidioides*-infected mice in Fig. 11B is evident between 5 and 11 days post-challenge, providing additional evidence that exposure to SOWgp modulates host immunity in a Th2-biased manner. Persistence of an immune response to coccidioidal infection in which the balance is shifted in the direction of the Th2 pathway offers a distinct advantage to the pathogen by compromising the host cellular immune defenses.

INDUCED HOST ARGINASE I PRODUCTION COMPROMISES DEFENSES AGAINST INFECTION

Results of our microarray studies, presented in Table 1, also revealed a significant increase in the level of expression of arginase in mice infected with the parental strain compared to that in animals infected with the Δsowgp mutant. Macrophages express two arginase isoforms, arginase I and arginase II (205). Arginase I is located in the cytosol, while arginase II is a mitochondrial enzyme. The hydrolysis of L-arginine to L-ornithine by arginase I provides the substrate for ODC, which, as discussed earlier in this chapter, is a key enzyme in the polyamine biosynthetic pathway and putative regulator of parasitic cell differentiation (107). Also discussed was the notion that the availability of host-derived L-ornithine at sites of fungal infection may promote pathogen growth and proliferation by providing a pool of the monoamine, which could be utilized for synthesis of L-glutamine, L-proline, and polyamines via metabolic pathways of the parasitic cell (205). Arginase I expression is induced by macrophages on exposure to Th2-type cytokines, such as IL-4 and IL-10. DCs also up-regulate arginase I expression and arginase activity on Th2 stimulation (149). Numerous reports of microbially induced production of arginase by macrophages have appeared in the parasitology literature (63, 65, 107). For example, the humoral response to cruzipain, a major antigen produced by *Trypanosoma cruzi*, induces a Th2 cytokine profile which enhances arginase production by macrophages and increases the susceptibility of the host to infection (87). Treatment of macrophages derived from *Leishmania major*-infected BALB/c mice with IL-4, IL-10, or transforming growth factor β induced arginase I production, which resulted in an increase in the number of intracellular amastigotes (107). Although this regulatory pathway of the host response to parasitic infection has been extensively investigated, no comparable studies of fungus-host interactions have been conducted. Our studies have provided evidence that a similar sequence of events involving arginase I production by host inflammatory cells occurs in murine lungs infected with *C. posadasii*.

Arginase I competes with inducible nitric oxide synthase (iNOS) in macrophages for the common substrate, L-arginine (143, 149). The balance between the two enzyme activities is competitively regulated by Th1 and Th2 lymphocytes via their secreted cytokines. Th1 cells secrete IFN-γ, which induces macrophages to produce iNOS, whereas Th2 cells induce arginase I expression and arginase activity (17). High levels of expression of arginase I by host inflammatory cells result in a major reduction in the basal levels of NO, which would be expected to compromise host defense against *Coccidioides* infection (59, 127, 175, 201). In addition, this shift in the balance of enzyme activity within macrophages in favor of arginase results in up-regulation of urea production, which, in the presence of microbial urease, could generate a localized alkaline microenvironment at sites of infection and potentially jeopardize the ability of host defense cells to function properly (53, 145).

A summary of the steps in the host metabolic pathway influenced by elevated levels of macrophage and DC arginase activity is presented in Fig. 12. We have evidence that *Coccidioides* can modulate the levels of arginase I, iNOS, and urea production in infected lungs of mice, presumably as a result of the Th2-biased immune response. Lung homogenates from C57BL/6 mice infected with the parental strain of *C. posadasii* at 7 days post-challenge showed a marked increase in *ARGI* gene expression compared to normal, noninfected mice and a significantly higher level of gene expression than mice infected

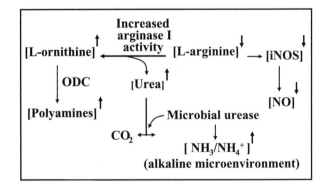

Figure 12. Summary of metabolic events which occur in murine macrophages in response to Th2-type cytokine stimulation. A shift in balance between arginase I and iNOS activity in the direction of the former results in depletion of arginine and reduction in NO production but increased synthesis of ornithine and urea. Fungal uptake of urea and the concomitant increase in intracellular and extracellular urease activity results in high concentrations of NH_3/NH_4^+ secreted by the pathogen at sites of infection.

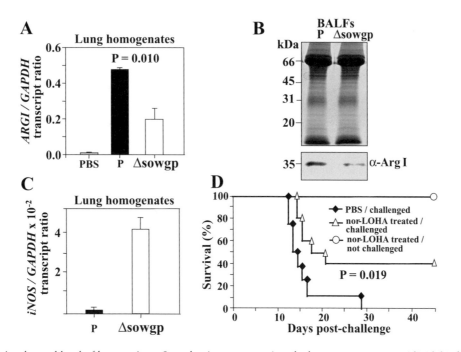

Figure 13. An elevated level of host arginase I production compromises the host response to coccidioidal infection. (A) Quantitative real-time PCR of murine arginase I (*ARGI*) expression in the lungs of C57BL/6 mice at 7 days after intranasal infection with 80 arthroconidia isolated from either the parental (P) or mutant (Δsowgp) strain of *C. posadasii*, compared to expression of the constitutive murine glyceraldehyde-3-phosphate dehydrogenase (*GAPDH*) gene. Control mice were challenged with PBS alone. The data are presented as the mean values from analyses of three mice for each group. The *P* value indicates a significant difference between the two groups of infected C57BL/6 mice. (B) BAL fluid samples collected from C57BL/6 mice at 7 days after intranasal infection with 80 arthroconidia from either the parental (P) or mutant (Δsowgp) strain were separated by SDS-PAGE and immunoblotted using anti-arginase I (α-Arg I) monoclonal antibody (1:500; BD Biosciences PharMingen, Franklin Lakes, N.J.). (C) Quantitative real-time PCR of murine iNOS gene expression in the lungs of mice infected with either the parental (P) or mutant (Δsowgp) strain of *C. posadasii*. Infection conditions were the same as in the experiment in panel A. (D) BALB/c mice treated with an inhibitor of arginase I activity (nor-LOHA; Alexis, San Diego, Calif.) prior to and after intranasal challenge with a lethal inoculum of *C. posadasii* (80 arthroconidia) showed a significant increase in percent survival compared to untreated mice. Noninfected control mice were treated with nor-LOHA alone. Administration of nor-LOHA was via the intraperitoneal route, beginning 1 h prior to infection (6 mg/kg of body weight in 100 μl of PBS; lipopolysaccharide free), followed by intraperitoneal administration of the same amount of nor-LOHA once per day for 45 days post-challenge.

with the Δsowgp strain (Fig. 13A). This difference between the two groups of infected mice was confirmed by immunoblot analysis of respective BAL fluid samples (Fig. 13B). Although we focused our assays on the levels of expression of murine *ARGI*, it is also possible that release of fungal arginase within the macrophage phagosomes could have an inhibitory effect on iNOS activity by consumption of the pool of host arginine (66). We have cloned the *ARG* gene of *Coccidioides* (159) and determined that its level of expression during the infection cycle within murine lung tissue is constitutive. In vitro studies of the interaction between endospores and macrophages are needed to determine whether fungal arginase production is increased after engulfment of the pathogen and whether it influences host cell production of iNOS and NO. We have shown that

expression of the host iNOS-encoding gene in lung homogenates of mice infected with the parental strain of *C. posadasii* at 7 days post-challenge is very low (Fig. 13C). In sharp contrast, the level of iNOS expression in mice infected with the Δsowgp strain is approximately ninefold higher, which is consistent with our argument that host exposure to SOWgp contributes to a Th2-biased response to infection and skews the balance between arginase and iNOS activities in favor of the former. Given that host arginase activation may result in decreased levels of nitric oxide production in macrophages and permit intracellular survival of the fungal pathogen, it follows that a physiological inhibitor of arginase (e.g., N^{ω}-hydroxy-nor-L-arginine [nor-LOHA]) should contribute to pathogen clearance and host survival (106, 143, 188). Intraperitoneal treatment of

Coccidioides-infected BALB/c mice with nor-LOHA significantly increased host survival (Fig. 13D), suggesting that arginase activity plays an important role in disease development.

As suggested in Fig. 12, arginase activity at sites of infection can result in accumulation of urea, which could serve as substrate for the urease of *Coccidioides* that has been detected (221). Cryosections of *Coccidioides*-infected and noninfected lung tissue were extracted with PBS, and the urea concentrations in the extracts were estimated on the basis of a phenol-hypochlorite reaction (Bertholet reaction) (141, 210). Equal numbers of cryosections were prepared, and 400-μl portions of pooled extracts from lung sections of each group of mice were assayed. The results revealed that the concentrations of urea in extracts from noninfected and Δsowgp-infected mice were approximately 150 to 200 μM while the urea concentration from mice infected with the parental strain was approximately 630 μM. We have shown that *URE* gene expression is up-regulated during the endosporulation phase of the parasitic cycle of *C. posadasii* and that the enzyme is secreted in vitro at the time of spherule rupture and endospore release (141). The enzyme is present in the cytoplasmic debris that remains within the lumen of the spherule after endospore differentiation. Urease catalyzes the hydrolysis of urea to produce ammonia and carbamate. The latter is further hydrolyzed to yield carbonic acid and ammonia. At physiological pH, the carbonic acid proton dissociates and ammonia molecules become protonated, with a resultant increase in pH (144). Urease is an important virulence factor in several bacterial pathogens (19, 53). In *Helicobacter pylori*, the urease protein associates with the surface of bacterial cells, apparently as a result of its release from bacterial cells that had undergone autolysis (64). The active enzyme is a major virulence factor of *H. pylori*; it generates ammonia from gastric urea, which in turn injures the gastric mucosa either directly by forming ammonium hydroxide or indirectly by stimulating a persistent inflammatory response. Although urease activity has been detected in several genera of medically important fungi (e.g., *Aspergillus*, *Candida*, *Cryptococcus*, *Rhodotorula*, and *Trichosporon*), only in *Coccidioides* and *Cryptococcus* has it been suggested to play a role in pathogenesis (27, 28, 41). Mice infected with a urease knockout strain of *C. neoformans* showed significant reduction in inflammatory response at sites of blood-to-brain invasion (156). The authors suggested that the enzyme contributes to central nervous system invasion by enhancing yeast sequestration within microcapillary beds in the brain during hematogenous

dissemination. Deletion of the *URE* gene from *C. posadasii* also resulted in a significant reduction in pathogenicity of the mutant strain in mice. It has been suggested that ammonia and enzymatically active urease released from spherules during the parasitic cycle of *C. posadasii* contribute to host tissue damage, which exacerbates the severity of coccidioidal infection and enhances the virulence of this human respiratory pathogen (142).

SUMMARY AND DISCUSSION OF RESEARCH OPPORTUNITIES

We have presented evidence that *Coccidioides* is well equipped with mechanisms to withstand an attack by the sophisticated innate and acquired immune defense systems of the mammalian host. It also appears that *Coccidioides* can manipulate signaling of the inflammatory response to its advantage. A pivotal first step for in vivo survival of the microbe is to complete its first generation of parasitic cell development, which results in a 200- to 300-fold increase in the number of infectious cells (endospores) per inhaled arthroconidium. *Coccidioides* alternates between an intracellular and extracellular relationship with phagocytes (Fig. 14). In vitro studies of macrophage interaction with *Coccidioides* conidia indicate that the majority of fungal cells survive and differentiate into spherule initials (Fig. 14A). This may be partly due to the presence of melanin in the conidial wall layer, which could serve as a barrier to oxidative products of the host. It is also likely that the fungal cells are able to effectively detoxify oxygen radicals by secretion of neutralizing enzymes, such as SOD and catalase, as reported for other fungi. This is a fertile area for research on the nature of the encounter between fungus and host during the onset of coccidioidomycosis, as well as histoplasmosis, blastomycosis, and cryptococcosis in which studies of animal models of the respective mycoses have essentially ignored the events of conidium-phagocyte interaction.

Arthroconidial germination results in the activation of virulence factors which are also focused on the neutralization of oxidative responses of host phagocytes. In fact, spherule initials have been shown in vitro to be more resistant to the NO donor, 3-morpholinosydnonimine (SIN-1) than are arthroconidia. Up-regulation of SOD and flavohemoglobin production has been suggested to contribute to NO consumption and detoxification during *Candida* and *Cryptococcus* infections. Similar mechanisms of resistance to host defenses may be at work in *Coccidioides*. Spherule initials also partially inhibit the production of the proinflammatory cytokine, TNF-α, and are apparently more efficient in avoiding

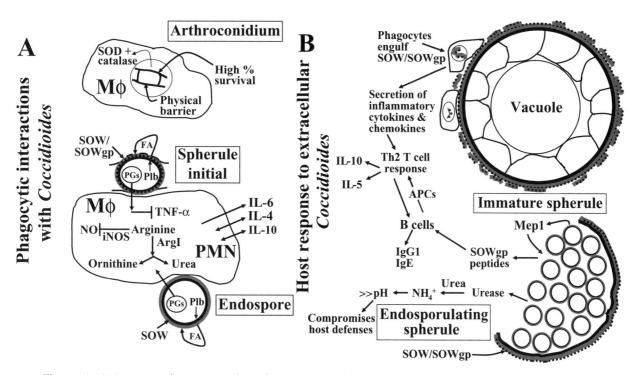

Figure 14. (A) Summary of putative virulence factors expressed by arthroconidia, spherule initials, and endospores of *Coccidioides* during their interaction with host phagocytes. (B) Summary of putative virulence factors expressed by mature spherules and endosporulating spherules of *Coccidioides* during their extracellular association with the host. ArgI, host arginase I; FA, fatty acids; Mϕ, macrophage; PGs, prostaglandins; Plb, phospholipase B; Mep1, *Coccidioides* secreted metalloproteinase; PMN, polymorphonuclear neutrophil.

phagocytosis than arthroconidia when exposed to immune macrophages in vitro. The ability of spherule initials to down-regulate the innate immune response to infection may contribute significantly to the pathogenicity of *Coccidioides*. Possibly related to the suppression of TNF-α production during the differentiation of spherule initials is the appearance of a lipid-rich layer at the surface of these fungal cells. Chloroform-methanol extracts of in vitro-grown spherule initials exhibited suppression of the macrophage-derived cytokine production at a level equal to that of the intact fungal cells. The SOW layer is composed largely of phospholipids and glycosphingolipids and an immunodominant glycoprotein (SOWgp). The latter was not present in the bioreactive lipid fraction which down-regulated macrophage TNF-α production. The component(s) of the chloroform-methanol extract responsible for suppression of this key inflammatory response, and the nature of the macrophage signal transduction pathway involved in cytokine suppression, are unknown.

The lipid-rich SOW layer of parasitic cells of *Coccidioides* has been suggested to be a major virulence factor. Its high phospholipid content may be a source of fatty acid precursors utilized by the pathogen and/or host phagocyte for synthesis of

eicosanoids (e.g., prostaglandins, leukotrienes), which have been shown to be potent modulators of the immune response to fungal and other microbial infections. Release of fatty acid precursors utilized for eicosanoid biosynthesis in *C. neoformans* appears to be regulated by phospholipase activity. *Coccidioides* has a single phospholipase, and peaks of expression of the *PLB* gene occur during arthroconidium germination and endospore formation when the fungal cells are most prone to phagocytosis. Prostaglandins deactivate macrophages and promote a Th2 pathway of immune response to infection. The latter is characterized by elevated levels of antibody production, activation of mast cells and eosinophils, and partial suppression of cellular immunity. A biased Th2 response to *Coccidioides* insult compromises host protection and exacerbates the course of disease. Synthesis of the phospholipid-rich SOW layer by spherule initials and developing endospores, combined with the parasitic phase-regulated expression of phospholipase B, provides the conditions necessary for production of eicosanoids during periods of the parasitic cycle when the fungus is in need of protection against host phagocytes (Fig. 14A).

Spherule initials and endospores which survive the attack of host innate defenses undergo isotropic growth and become too large to be engulfed by phagocytes. As spherules mature, they synthesize and shed SOW components at their cell surface. The dominant antigenic component of SOW is a glycoprotein (SOWgp), which binds antibody and elicits an immune T-cell response in patients with confirmed coccidioidomycosis. A 47-amino-acid sequence, which occurs as a tandem repeat within the SOWgp polypeptide, contains B-cell-dominant epitopes. Exposure of this antigen at the surface of endospores in patients with active disease would be expected to enhance opsonization and phagocytosis. Instead, the endospores which emerge from ruptured spherules are devoid of SOWgp. We have provided evidence that expression of an endosporulation stage-specific metalloproteinase (Mep1) is responsible for digestion of SOWgp and for the ability of the fragile, newly formed endospores to evade host detection.

Repeated exposure of host innate and acquired immune cells to SOWgp, and possibly the products of Mep1 digestion of SOWgp presented during endosporulation, has been suggested to elicit a Th2-biased response. B cells may play an important role in this process by functioning as APCs in the differentiation of Th2 lymphocytes (Fig. 14B). Several lines of evidence derived from in vitro studies of T-cell activation support this proposal. We have also argued that repeated exposure to SOWgp results in a persistently elevated level of IL-6 secretion, which results in the maintenance of an intense inflammatory response at sites of infection. The latter may cause host tissue damage, which further exacerbates the disease. Thus, production of SOWgp at the parasitic cell surface apparently contributes to the modulation of host immune response which results in an advantage to the pathogen. A comparable account of host response to the immunodominant cell surface antigen of *P. brasiliensis* (gp43) has been reported, which suggests that stimulation of a dominant Th2 response which fails to protect the host against infection may be a conserved virulence strategy among fungal agents of respiratory disease. In the final section of this chapter, we have examined the impact of a biased Th2 pathway of the immune response to *Coccidioides* infection on macrophage function. We focused our attention on arginase I production since there is extensive evidence in the parasitology literature suggesting that overproduction of this macrophage cytosolic enzyme during the course of infection can compromise the host's ability to clear the pathogen. High levels of arginase I activity can deplete the pool of arginine in macrophages, which is the common substrate for ArgI and iNOS. We suggest that reduced iNOS activity and NO production

jeopardizes the microbicidal effectiveness of macrophages and DCs. Arginine hydrolysis by arginase also results in ornithine and urea production, both of which can be metabolized by microbial pathogens. We have shown that ornithine is utilized in vitro by *Coccidioides* during polyamine biosynthesis, it is under the stringent regulation of ODC activity, and its uptake appears to play a key role in parasitic cell differentiation. *Coccidioides* produces a urease which is released to infection sites during spherule rupture and endospore release. We have suggested that the *Coccidioides* urease hydrolyzes both fungus- and host-derived urea to generate an alkaline microenvironment within abscesses, which can further compromise the host phagocytic response to the coccidioidal insult (Fig. 14B).

This discussion of virulence mechanisms of *Coccidioides* is intended to stimulate research, since the majority of factors described here are in need of further investigation. Laboratory workers and clinicians who have experience in studying and treating coccidioidomycosis know that the causative agent of this disease is an exceptionally formidable pathogen. Nevertheless, development of a human vaccine against the respiratory mycosis is under way (39, 45) and identification of novel targets for the production of more effective antifungal drugs is in progress (4, 91, 98, 170), both offering promise to the hundreds of thousands of people who are exposed to *Coccidioides* in the United States, Mexico, and Central and South America each year.

Acknowledgments. We are grateful for grant support from the National Institute of Allergy and Infectious Diseases, National Institutes of Health (AI19149 and AI37232), and the California HealthCare Foundation.

REFERENCES

1. **Aguirre, K. M., and G. W. Gibson.** 2000. Differing requirement for inducible nitric oxide synthase activity in clearance of primary and secondary *Cryptococcus neoformans* infection. *Med. Mycol.* 38:343–353.
2. **Allendoerfer, R., and G. S. Deepe, Jr.** 2000. Regulation of infection with *Histoplasma capsulatum* by TNFR1 and -2. *J. Immunol.* 165:2657–2664.
3. **Altschul, S. F., T. L. Madden, A. A. Schaffer, J. Zhang, Z. Zhang, W. Miller, and D. J. Lipman.** 1997. Gapped BLAST and PSI-BLAST: a new generation of protein database search programs. *Nucleic Acids Res.* 25:3389–3402.
4. **Ampel, N. M.** 2004. Combating opportunistic infections: coccidioidomycosis. *Expert Opin. Pharmacother.* 5:255–261.
5. **Ampel, N. M.** 1999. Delayed-type hypersensitivity, in vitro T-cell responsiveness and risk of active coccidioidomycosis among HIV-infected patients living in the coccidioidal endemic area. *Med. Mycol.* 37:245–250.
6. **Ampel, N. M., G. C. Bejarano, and J. N. Galgiani.** 1992. Killing of *Coccidioides immitis* by human peripheral blood mononuclear cells. *Infect. Immun.* 60:4200–4204.

7. **Ampel, N. M., D. G. Mosley, B. England, P. D. Vertz, K. Komatsu, and R. A. Hajjeh.** 1998. Coccidioidomycosis in Arizona: increase in incidence from 1990 to 1995. *Clin. Infect. Dis.* 27:1528–1530.

8. **Arruda, C., R. C. Valente-Ferreira, A. Pina, S. S. Kashino, R. A. Fazioli, C. A. C. Vaz, M. Franco, A. C. Keller, and V. L. G. Calich.** 2004. Dual role of interleukin-4 (IL-4) in pulmonary paracoccidioidomycosis: endogenous IL-4 can induce protection or exacerbation of disease depending on the host genetic pattern. *Infect. Immun.* 72:3932–3940.

9. **Bakouche, Q., F. David, and D. Gerlier.** 1987. Impairment of immunogenicity by antigen presentation in liposomes made from dimyristoylphosphatidylethanolamine linked to the secretion of prostaglandins by macrophages. *Eur. J. Immunol.* 17:1839–1842.

10. **Beaman, L.** 1987. Fungicidal activation of murine macrophages by recombinant gamma interferon. *Infect. Immun.* 55:2951–2955.

11. **Beckman, J. S., T. W. Beckman, J. Chen, P. A. Marshall, and B. A. Freeman.** 1990. Apparent hydroxyl radical production by peroxynitrite: implications for endothelial injury from nitric oxide and superoxide. *Proc. Natl. Acad. Sci. USA* 87:1620–1624.

12. **Bermudez, L. E., and L. S. Young.** 1988. Tumor necrosis factor, alone or in combination with IL-2, but not IFN-gamma, is associated with macrophage killing of *Mycobacterium avium* complex. *J. Immunol.* 140:3006–3013.

13. **Bernander, R., T. Stokke, and E. Boye.** 1998. Flow cytometry of bacterial cells: comparison between different flow cytometers and different DNA stains. *Cytometry* 31:29–36.

14. **Betz, M., and B. S. Fox.** 1991. Prostaglandin E2 inhibits production of Th1 lymphokines but not of Th2 lymphokines. *J. Immunol.* 146:108–113.

15. **Blair, J. E., D. D. Douglas, and D. C. Mulligan.** 2003. Early results of targeted prophylaxis for coccidioidomycosis in patients undergoing orthotopic liver transplantation within an endemic area. *Trans. Infect. Dis.* 5:3–8.

16. **Bocca, L., S. Valenti, C. M. Cuttica, R. Spaziante, G. Giordano, and M. Giusti.** 2000. Nitric oxide biphasically modulates GH secretion in cultured cells of GH-secreting human pituitary adenomas. *Min. Endocrin.* 25:55–59.

17. **Bronte, V., P. Serafini, C. De Santo, I. Marigo, V. Tosello, A. Mazzoni, D. M. Segal, C. Staib, M. Lowel, G. Sutter, M. Colombo, and P. Zanovello.** 2003. IL-4-induced arginase 1 suppresses alloreactive T cells in tumor-bearing mice. *J. Immunol.* 170:270–278.

18. **Brummer, E., and D. A. Stevens.** 1995. Antifungal mechanisms of activated murine bronchoalveolar or peritoneal macrophages for *Histoplasma capsulatum*. *Clin. Exp. Immunol.* 102:65–67.

19. **Burall, L. S., J. M. Harro, X. Li, C. V. Lockatell, S. D. Himpsl, J. R. Hebel, D. E. Johnson, and H. L. T. Mobley.** 2004. *Proteus mirabilis* genes that contribute to pathogenesis of urinary tract infection: identification of 25 signature-tagged mutants attenuated at least 100-fold. *Infect. Immun.* 72:2922–2938.

20. **Camargo, Z. P., C. P. Taborda, E. G. Rodrigues, and L. R. Travassos.** 1991. The use of cell-free antigens of *Paracoccidioides brasiliensis* in serological tests. *J. Med. Vet. Mycol.* 29:31–38.

21. **Carmel-Harel, O., and G. Storz.** 2000. Roles of the glutathione- and thioredoxin-dependent reduction systems in the *Escherichia coli* and *Saccharomyces cerevisiae* responses to oxidative stress. *Annu. Rev. Microbiol.* 54:439–461.

22. **Centers for Disease Control and Prevention.** 2001. Coccidioidomycosis in workers at an archeologic site—Dinosaur National Monument, Utah, June–July 2001. *Morb. Mortal. Wkly. Rep.* 50:1005–1008.

23. **Centers for Disease Control and Prevention.** 1996. Coccidioidomycosis—Arizona, 1990–1995. *Morb. Mortal. Wkly. Rep.* 45:1069–1073.

24. **Charlton, V., K. Ramsdell, and S. Sehring.** 1999. Intrauterine transmission of coccidioidomycosis. *Pediatr. Infect. Dis. J.* 18:561–563.

25. **Chaturvedi, V., R. Ramani, S. Gromadzki, B. Rodeghier, H. G. Chang, and D. L. Morse.** 2000. Coccidioidomycosis in New York State. *Emerg. Infect. Dis.* 6:25–29.

26. **Chtanova, T., R. A. Kemp, A. P. Sutherland, F. Ronchese, and C. R. Mackay.** 2001. Gene microarrays reveal extensive differential gene expression in both CD4$^+$ and CD8$^+$ type 1 and type 2 T cells. *J. Immunol.* 167:3057–3063.

27. **Cole, G. T.** 1997. Ammonia production by *Coccidioides immitis* and its possible significance to the host-fungus interplay, p. 247–263. *In* H. Vanden Bossche, D. A. Stevens, and F. C. Odds (ed.), *Host-Fungus Interplay: Proceedings of the 5th Symposium on Topics in Mycology*. Plenum Press, New York, N.Y.

28. **Cole, G. T.** 1993. *Coccidioides immitis*: resistance to host defense mechanisms. *Clin. Adv. Treat. Fungal Infect.* 4:1–5, 11, 12.

29. **Cole, G. T.** 2003. Fungal pathogenesis, p. 20–45. *In* E. Anaissie, M. R. McGinnis, and M. A. Pfaller (ed.), *Clinical Mycology*. Churchill Livingstone, Inc., New York, N.Y.

30. **Cole, G. T., C.-Y. Hung, and N. Delgado.** 2002. Parasitic phase-specific gene expression in *Coccidioides*. *ASM News* 68:603–611.

31. **Cole, G. T., and T. N. Kirkland.** 1991. Conidia of *Coccidioides immitis*: their significance in disease initiation, p. 403–443. *In* G. T. Cole and H. C. Hoch (ed.), *The Fungal Spore and Disease Initiation in Plants and Animals*. Plenum Press, New York, N.Y.

32. **Cole, G. T., T. N. Kirkland, M. Franco, S. Zhu, L. Yuan, S. H. Sun, and V. M. Hearn.** 1988. Immunoreactivity of a surface wall fraction produced by spherules of *Coccidioides immitis*. *Infect. Immun.* 56:2695–2701.

33. **Cole, G. T., T. N. Kirkland, and S. H. Sun.** 1987. An immunoreactive, water-soluble conidial wall fraction of *Coccidioides immitis*. *Infect. Immun.* 55:657–667.

34. **Cole, G. T., and E. J. Pishko.** 1995. Possible roles of wall hydrolases in the morphogenesis of *Coccidioides immitis*. *Can. J. Bot.* 73 (Suppl. 1):S1132–S1141.

35. **Cole, G. T., L. M. Pope, M. Huppert, S. H. Sun, and P. Starr.** 1983. Ultrastructure and composition of conidial wall fractions of *Coccidioides immitis*. *Exp. Mycol.* 7:297–318.

36. **Cole, G. T., and R. A. Samson.** 1984. The conidia, p. 66–103. *In* Y. Al-Doory and J. Domson (ed.), *Mould Allergy*. Lea & Febiger, Philadelphia. Pa.

37. **Cole, G. T., K. R. Seshan, M. Franco, E. Bukownik, S. H. Sun, and V. M. Hearn.** 1988. Isolation and morphology of an immunoreactive outer wall fraction produced by spherules of *Coccidioides immitis*. *Infect. Immun.* 56:2686–2694.

38. **Cole, G. T., and S. H. Sun.** 1985. Arthroconidium-spherule-endospore transformation in *Coccidioides*

immitis, p. 281–333. *In* P. Szaniszlo (ed.), *Dimorphism.* Plenum Press, New York, N.Y.

39. Cole, G. T., J.-M. Xue, C. N. Okeke, E. J. Tarcha, V. Basrur, R. A. Schaller, R. A. Herr, J.-J. Yu, and C.-Y. Hung. 2004. A vaccine against coccidioidomycosis is justified and attainable. *Med. Mycol.* **42:**189–216.

40. Cook, E. R., C. A. Woodhouse, C. M. Eakin, D. M. Meko, and D. W. Stahle. 2004. Long-term aridity changes in the Western United States. *Science* **306:**1015–1018.

41. Cox, G. M., J. Mukherjee, G. T. Cole, A. Casadevall, and J. R. Perfect. 2000. Urease as a virulence factor in experimental cryptococcosis. *Infect. Immun.* **68:**443–448.

42. Cox, R. A. 1983. Cell-mediated immunity, p. 61–98. *In* D. H. Howard (ed.), *Fungi Pathogenic for Humans and Animals*, part B. *Pathogenicity and Detection*, vol. 3. Marcel Dekker, Inc., New York, N.Y.

43. Cox, R. A., B. S. Baker, and D. A. Stevens. 1982. Specificity of immunoglobulin E in coccidioidomycosis and correlation with disease involvement. *Infect. Immun.* **37:**609–616.

44. Cox, R. A., T. S. Harrison, H. C. McDade, C. Taborda, G. Heinrich, A. Casadevall, and J. R. Perfect. 2003. Superoxide dismutase influences the virulence of *Cryptococcus neoformans* by affecting growth within macrophages. *Infect. Immun.* **71:**173–180.

45. Cox, R. A., and D. M. Magee. 2004. Coccidioidomycosis: host response and vaccine development. *Clin. Microbiol. Rev.* **17:**804–839.

46. Cox, R. A., and D. M. Magee. 1995. Production of tumor necrosis factor alpha, interleukin-1 alpha, and interleukin-6 during murine coccidioidomycosis. *Infect. Immun.* **63:**4178–4180.

47. Cox, R. A., and D. M. Magee. 1998. Protective immunity in coccidioidomycosis. *Res. Immunol.* **149:** 417–428; discussion, 506–517.

48. Cox, R. A., J. R. Vivas, A. Gross, G. Lecara, E. Miller, and E. Brummer. 1976. In vivo and in vitro cell-mediated responses in coccidioidomycosis. I. Immunologic responses of persons with primary, asymptomatic infections. *Am. Rev. Respir. Dis.* **114:**937–943.

49. Crowther, J. E., V. K. Kutala, P. Kuppusamy, J. S. Ferguson, A. A. Beharka, J. L. Zweier, F. X. McCormack, and L. S. Schlesinger. 2004. Pulmonary surfactant protein A inhibits macrophage reactive oxygen intermediate production in response to stimuli by reducing NADPH oxidase activity. *J. Immunol.* **172:**6866–6874.

50. Crum, N. F., E. R. Lederman, B. R. Hale, M. L. Lim, and M. R. Wallace. 2003. A cluster of disseminated coccidioidomycosis cases at a US military hospital. *Mil. Med.* **168:**460–464.

51. Crum, N. F., E. R. Lederman, C. M. Stafford, J. S. Parrish, and M. R. Wallace. 2004. Coccidioidomycosis: a descriptive survey of a reemerging disease. Clinical characteristics and current controversies. *Medicine* **83:**149–175.

52. de Almeida, S. R., J. Z. de Moraes, Z. P. de Camargo, J. L. Gesztesi, M. Mariano, and J. D. Lopes. 1998. Pattern of immune response to GP43 from *Paracoccidioides brasiliensis* in susceptible and resistant mice is influenced by antigen-presenting cells. *Cell. Immunol.* **190:**68–76.

53. de Bernard, M., A. Cappon, G. Del Giudice, R. Rappuoli, and C. Montecucco. 2004. The multiple cellular activities of the VacA cytotoxin of *Helicobacter pylori*. *Int. J. Med. Microbiol.* **293:**589–597.

54. Deepe, G. S. 1996. *Histoplasma capsulatum*: darling of the river valleys. *ASM News* **63:**599.

55. de Jesus-Berrios, M., L. Liu, J. C. Nussbaum, G. M. Cox, J. S. Stamler, and J. Heitman. 2003. Enzymes that counteract nitrosative stress promote fungal virulence. *Curr. Biol.* **13:**1963–1968.

56. Delgado, N., J. Xue, J.-J. Yu, C.-Y. Hung, and G. T. Cole. 2003. A recombinant beta-1,3-glucanosyltransferase homolog of *Coccidioides posadasii* protects mice against coccidioidomycosis. *Infect. Immun.* **71:**3010–3019.

57. Deresinski, S. 2003. *Coccidioides immitis* as a potential bioweapon. *Semin. Respir. Infect.* **18:**216–219.

58. Diez-Orejas, R., G. Molero, M. A. Moro, C. Gil, C. Nombela, and M. Sanchez-Perez. 2001. Two different NO-dependent mechanisms account for the low virulence of a non-mycelial morphological mutant of *Candida albicans*. *Med. Mycol. Immunol.* **189:**153–160.

59. Diniz, S. N., P. S. Cisalpino, A. T. Freire, D. N. Silva-Teixeira, C. Contigli, V. Rodrigues Junior, and A. M. Goes. 2001. In vitro granuloma formation, NO production and cytokines profile from human mononuclear cells induced by fractionated antigens of *Paracoccidioides brasiliensis*. *Hum. Immunol.* **62:** 799–808.

60. Dixon, D. M. 2001. *Coccidioides immitis* as a select agent of bioterrorism. *J. Appl. Microbiol.* **91:**602–605.

61. Dolan, J. W., A. C. Bell, B. Hube, M. Schaller, and T. F. Warner. 2004. *Candida albicans* PLD1 activity is required for full virulence. *Med. Mycol.* **42:**439–447.

62. Drutz, D. J., and M. Huppert. 1983. Coccidioidomycosis: factors affecting the host-parasite interaction. *J. Infect. Dis.* **147:**372–390.

63. Duleu, S., P. Vincendeau, P. Courtois, S. Semballa, I. Lagroye, S. Daulouede, J. L. Boucher, K. T. Wilson, B. Veyret, and A. P. Gobert. 2004. Mouse strain susceptibility to trypanosome infection: an arginase-dependent effect. *J. Immunol.* **172:**6298–6303.

64. Dunn, B. E., and S. H. Phadnis. 1998. Structure, function and localization of *Helicobacter pylori* urease. *Yale J. Biol. Med.* **71:**63–73.

65. Dzik, J. M., B. Golos, E. Jagielska, Z. Zielinski, and E. Walajtys-Rode. 2004. A non-classical type of alveolar macrophage response to *Trichinella spiralis* infection. *Parasite Immunol.* **26:**197–205.

66. Elnekave, K., R. Siman-Tov, and S. Ankri. 2003. Consumption of L-arginine mediated by *Entamoeba histolytica* L-arginase (EhArg) inhibits amoebicidal activity and nitric oxide production by activated macrophages. *Parasite Immunol.* **25:**597–608.

67. Enserink, M., and D. Malakoff. 2001. Bioterrorism. Congress weighs select agent update. *Science* **294:**1438.

68. Eriksson, S., B. J. Chambers, and M. Rhen. 2003. Nitric oxide produced by murine dendritic cells is cytotoxic for intracellular *Salmonella enterica* sv. Typhimurium. *Scand. J. Immunol.* **58:**493–502.

69. Eulalio, K. D., R. L. de Macedo, M. A. Cavalcanti, L. M. Martins, M. S. Lazera, and B. Wanke. 2001. *Coccidioides immitis* isolated from armadillos (*Dasypus novemcinctus*) in the state of Piaui, northeast Brazil. *Mycopathologia* **149:**57–61.

70. Fass, U., K. Panickar, K. Williams, K. Nevels, D. Personett, and M. McKinney. 2004. The role of glutathione in nitric oxide donor toxicity to SN56 cholinergic neuron-like cells. *Brain Res.* **1005:**90–100.

71. Feelisch, M., J. Ostrowski, and E. Noack. 1989. On the mechanism of NO release from sydnonimines. *J. Cardiovasc. Pharmacol.* **14**(Suppl. 11):13–22.

72. Fernandes, K. S., A. L. Coelho, L. M. Lopes Bezerra, and C. Barja-Fidalgo. 2000. Virulence of *Sporothrix schenckii* conidia and yeast cells, and their susceptibility to nitric oxide. *Immunology* **101:**563–569.

73. Ferreira, K. S., J. D. Lopes, and S. R. Almeida. 2003. Regulation of T helper cell differentiation in vivo by GP43 from *Paracoccidioides brasiliensis* provided by different antigen-presenting cells. *Scand. J. Immunol.* **58:**290–297.

74. Fierer, J., L. Walls, F. Wright, and T. N. Kirkland. 1999. Genes influencing resistance to *Coccidioides immitis* and the interleukin-10 response map to chromosomes 4 and 6 in mice. *Infect. Immun.* **67:**2916–2919.

75. Filion, M. C., and N. C. Phillips. 1997. Anti-inflammatory activity of cationic lipids. *Br. J. Pharmacol.* **122:**551–557.

76. Fisher, M. C., G. L. Koenig, T. J. White, and J. W. Taylor. 2002. Molecular and phenotypic description of *Coccidioides posadasii* sp. nov., previously recognized as the non-California population of *Coccidioides immitis.* *Mycologia* **94:**73–84.

77. Florido, M., and R. Appelberg. 2004. Granuloma necrosis during *Mycobacterium avium* infection does not require tumor necrosis factor. *Infect. Immun.* **72:**6139–6141.

78. Flynn, J. L., M. M. Goldstein, J. Chan, K. J. Triebold, K. Pfeffer, C. J. Lowenstein, R. Schreiber, T. W. Mak, and B. R. Bloom. 1995. Tumor necrosis factor-alpha is required in the protective immune response against *Mycobacterium tuberculosis* in mice. *Immunity* **2:**561–572.

79. Folch, J., M. Lees, and G. H. S. Stanley. 1957. A simple method for the isolation and purification of total lipides from animal tissues. *J. Biol. Chem.* **226:**497–509.

80. Fowler, M. E., D. Pappagianis, and I. Ingram. 1992. Coccidioidomycosis in llamas in the United States: 19 cases (1981–1989). *J. Am. Vet. Med. Assoc.* **201:**1609–1614.

81. Frey, C. L., and D. J. Drutz. 1986. Influence of fungal surface components on the interaction of *Coccidioides immitis* with polymorphonuclear neutrophils. *J. Infect. Dis.* **153:**933–943.

82. Fridovich, I. 1995. Superoxide radical and superoxide dismutases. *Annu. Rev. Biochem.* **64:**97–112.

83. Galgiani, J. N., R. Hayden, and C. M. Payne. 1982. Leukocyte effects on the dimorphism of *Coccidioides immitis.* *J. Infect. Dis.* **146:**56–63.

84. Ganendren, R., F. Widmer, V. Singhal, C. Wilson, T. Sorrell, and L. Wright. 2004. In vitro antifungal activities of inhibitors of phospholipases from the fungal pathogen *Cryptococcus neoformans.* *Antimicrob. Agents Chemother.* **48:**1561–1569.

85. Gaynor, C. D., F. X. McCormack, D. R. Voelker, S. E. McGowan, and L. S. Schlesinger. 1995. Pulmonary surfactant protein A mediates enhanced phagocytosis of *Mycobacterium tuberculosis* by a direct interaction with human macrophages. *J. Immunol.* **155:**5343–5351.

86. Ghannoum, M. A. 1998. Extracellular phospholipases as universal virulence factor in pathogenic fungi. *Jpn. J. Med. Mycol.* **39:**55–59.

87. Giordanengo, L., N. Guinazu, C. Stempin, R. Fretes, F. Cerban, and S. Gea. 2002. Cruzipain, a major *Trypanosoma cruzi* antigen, conditions the host immune response in favor of parasite. *Eur. J. Immunol.* **32:**1003–1011.

88. Godoy, L. C., M. Mariano, and J. D. Lopes. 2003. Immunity and hypersensitivity to gp43 antigen in susceptible and resistant mice infected with *Paracoccidioides brasiliensis.* *Med. Mycol.* **41:**427–436.

89. Goldman, D. L., S. C. Lee, A. J. Mednick, L. Montella, and A. Casadevall. 2000. Persistent *Cryptococcus neoformans* pulmonary infection in the rat is associated with intracellular parasitism, decreased inducible nitric oxide synthase expression, and altered antibody responsiveness to cryptococcal polysaccharide. *Infect. Immun.* **68:**832–838.

90. Gonzalez, A., W. De Gregori, D. Velez, A. Restrepo, and L. E. Cano. 2000. Nitric oxide participation in the fungicidal mechanism of gamma interferon-activated murine macrophages against *Paracoccidioides brasiliensis* conidia. *Infect. Immun.* **68:**2546–2552.

91. Gonzalez, G. M., R. Tijerina, L. K. Najvar, R. Bocanegra, M. G. Rinaldi, and J. R. Graybill. 2004. Efficacies of amphotericin B (AMB) lipid complex, AMB colloidal dispersion, liposomal AMB, and conventional AMB in treatment of murine coccidioidomycosis. *Antimicrob. Agents Chemother.* **48:**2140–2143.

92. Grant, C. M., G. Perrone, and I. W. Dawes. 1998. Glutathione and catalase provide overlapping defenses for protection against hydrogen peroxide in the yeast *Saccharomyces cerevisiae.* *Biochem. Biophys. Res. Commun.* **253:**893–898.

93. Guevara-Olvera, L., C.-Y. Hung, J.-J. Yu, and G. T. Cole. 2000. Sequence, expression and functional analysis of the *Coccidioides immitis* ODC (ornithine decarboxylase) gene. *Gene* **242:**437–448.

94. Guevara-Olvera, L., B. Xoconostle-Cazares, and J. Ruiz-Herrera. 1997. Cloning and disruption of the ornithine decarboxylase gene of *Ustilago maydis:* evidence for a role of polyamines in its dimorphic transition. *Microbiology* **143:**2237–2245.

95. Harrison, W. R., C. F. Merbs, and C. R. Leathers. 1991. Evidence of coccidioidomycosis in the skeleton of an ancient Arizona Indian. *J. Infect. Dis.* **164:**436–437.

96. Haselkorn, R. 2003. The complete genome sequence of *Chromobacterium violaceum* reveals remarkable and exploitable bacterial adaptability. *Proc. Natl. Acad. Sci. USA* **100:**11660–11665.

97. Hill, Z. H. 1992. The function of melanin or 6 people examine an elephant. *Bioessays* **14:**49–56.

98. Hsu, G., J. T. Napier, R. A. Prince, J. Chi, and D. R. Hospenthal. 2004. Treatment of meningeal coccidioidomycosis with caspofungin. *J. Antimicrob. Chemother.* **54:**292–294.

99. Huffnagle, G. B., M. B. Boyd, N. E. Street, and M. F. Lipscomb. 1998. IL-5 is required for eosinophil recruitment, crystal deposition, and mononuclear cell recruitment during a pulmonary *Cryptococcus neoformans* infection in genetically susceptible mice (C57BL/6). *J. Immunol.* **160:**2393–2400.

100. Hung, C.-Y., N. M. Ampel, L. Christian, K. R. Seshan, and G. T. Cole. 2000. A major cell surface antigen of *Coccidioides immitis* which elicits both humoral and cellular immune responses. *Infect. Immun.* **68:**584–593.

101. Hung, C. Y., K. R. Seshan, J.-J. Yu, R. Schaller, J. Xue, V. Basrur, M. Gardner, and G. T. Cole. 2005. A metalloproteinase of *Coccidioides posadasii* contributes to evasion of host detection. *Infect. Immun.* **73:**6689–6703.

102. Hung, C.-Y., J.-J. Yu, K. R. Seshan, U. Reichard, and G. T. Cole. 2002. A parasitic phase-specific adhesin of *Coccidioides immitis* contributes to the virulence of this respiratory fungal pathogen. *Infect. Immun.* **70:**3443–3456.

103. Huppert, M. 1983. Antigens used for measuring immunological reactivity, p. 219–302. *In* D. Howard

(ed.), *Fungi Pathogenic for Humans and Animals. Part B. Pathogenicity and Detection*. Marcel Dekker, Inc., New York, N.Y.

104. **Huppert, M., H. B. Levine, S. H. Sun, and E. T. Peterson.** 1967. Resistance of vaccinated mice to typical and atypical strains of *Coccidioides immitis*. *J. Bacteriol.* **94:**924–927.

105. **Igarashi, K., and K. Kashiwagi.** 2000. Polyamines: mysterious modulators of cellular functions. *Biochem. Biophys. Res. Commun.* **271:**559–564.

106. **Iniesta, V., L. C. Gomez-Nieto, and I. Corraliza.** 2001. The inhibition of arginase by N(omega)-hydroxy-L-arginine controls the growth of *Leishmania* inside macrophages. *J. Exp. Med.* **193:**777–784.

107. **Iniesta, V., L. C. Gomez-Nieto, I. Molano, A. Mohedano, J. Carcelen, C. Miron, C. Alonso, and I. Corraliza.** 2002. Arginase I induction in macrophages, triggered by Th2-type cytokines, supports the growth of intracellular *Leishmania* parasites. *Parasite Immunol.* **24:**113–118.

108. **Inselman, L. S., A. Chander, and A. R. Spitzer.** 2004. Diminished lung compliance and elevated surfactant lipids and proteins in nutritionally obese young rats. *Lung* **182:**101–117.

109. **Isowa, N., A. M. Xavier, E. Dziak, M. Opas, D. I. McRitchie, A. S. Slutsky, S. H. Keshavjee, and M. Liu.** 1999. LPS-induced depolymerization of cytoskeleton and its role in TNF-alpha production by rat pneumocytes. *Am. J. Physiol.* **277:**606–615.

110. **Jiang, C., D. M. Magee, F. D. Ivey, and R. A. Cox.** 2002. Role of signal sequence in vaccine-induced protection against experimental coccidioidomycosis. *Infect. Immun.* **70:**3539–3545.

111. **Jiang, C., D. M. Magee, T. N. Quitugua, and R. A. Cox.** 1999. Genetic vaccination against *Coccidioides immitis*: comparison of vaccine efficacy of recombinant antigen 2 and antigen 2 cDNA. *Infect. Immun.* **67:**630–635.

112. **Johnson, J. E., J. E. Perry, and F. R. Fekety.** 1964. Laboratory-acquired coccidioidomycosis: report of 210 cases. *Ann. Intern. Med.* **60:**941–956.

113. **Johnson, L. R., E. J. Herrgesell, A. P. Davidson, and D. Pappagianis.** 2003. Clinical, clinicopathologic, and radiographic findings in dogs with coccidioidomycosis: 24 cases (1995–2000). *J. Am. Vet. Med. Assoc.* **222:**461–466.

114. **Kajiwara, H., M. Saito, S. Ohga, T. Uenotsuchi, and S. Yoshida.** 2004. Impaired host defense against *Sporothrix schenckii* in mice with chronic granulomatous disease. *Infect. Immun.* **72:**5073–5079.

115. **Kaplanski, G., V. Marin, F. Montero-Julian, A. Mantovani, and C. Farnarier.** 2003. IL-6: a regulator of the transition from neutrophil to monocyte recruitment during inflammation. *Trends Immunol.* **24:**25–29.

116. **Kauffman, C. A.** 2002. Endemic mycoses in patients with hematologic malignancies. *Semin. Respir. Infect.* **17:**106–112.

117. **Kindler, V., A. P. Sappino, G. E. Grau, P. F. Piguet, and P. Vassalli.** 1989. The inducing role of tumor necrosis factor in the development of bactericidal granulomas during BCG infection. *Cell* **56:**731–740.

118. **Kirkland, T. N., and G. T. Cole.** 2002. Coccidioidomycosis: pathogenesis, immune response and vaccine development, p. 365–399. *In* R. A. Calderone and L. C. Cihlar (ed.), *Fungal Pathogenesis: Principles and Applications*. Marcel Dekker, Inc., New York, N.Y.

119. **Kirkland, T. N., and J. Fierer.** 1983. Inbred mouse strains differ in resistance to lethal *Coccidioides immitis* infection. *Infect. Immun.* **40:**912–916.

120. **Kirkland, T. N., F. Finley, K. I. Orsborn, and J. N. Galgiani.** 1998. Evaluation of the proline-rich antigen of *Coccidioides immitis* as a vaccine candidate in mice. *Infect. Immun.* **66:**3519–3522.

121. **Klotz, S. A., D. J. Drutz, M. Huppert, S. H. Sun, and P. L. DeMarsh.** 1984. The critical role of CO_2 in the morphogenesis of *Coccidioides immitis* in cell-free subcutaneous chambers. *J. Infect. Dis.* **150:**127–134.

122. **Kubo, S., A. Tamori, S. Nishiguchi, T. Omura, H. Kinoshita, K. Hirohashi, T. Kuroki, and S. Otani.** 1998. Relationship of polyamine metabolism to degree of malignancy of human hepatocellular carcinoma. *Oncol. Rep.* **5:**1385–1388.

123. **Kubo, S., A. Tamori, H. Tanaka, S. Takemura, T. Shuto, K. Hirohashi, H. Kinoshita, and S. Nishiguchi.** 2004. Polyamine metabolism and recurrence after resection for hepatocellular carcinoma. *Hepato-Gastroenterology* **5:**208–210.

124. **Kuranda, M., and P. Robbins.** 1991. Chitinase is required for cell separation during growth of *Saccharomyces cerevisiae*. *J. Biol. Chem.* **266:**19758–19767.

125. **Kwon-Chung, K. J., and J. E. Bennett.** 1992. *Medical Mycology*. Lea & Febiger, Philadelphia, Pa.

126. **Levine, H. B.** 1961. Purification of the spherule-endospore phase of *Coccidioides immitis*. *Sabouraudia* **1:**112–115.

127. **Li, H., C. J. Meininger, J. R. J. Hawker, T. E. Haynes, D. Kepka-Lenhart, S. K. Mistry, S. M. J. Morris, and G. Wu.** 2001. Regulatory role of arginase I and II in nitric oxide, polyamine, and proline syntheses in endothelial cells. *Am. J. Physiol. Ser. E* **280:**E75–E82.

128. **Li, K., J.-J. Yu, C.-Y. Hung, P. F. Lehmann, and G. T. Cole.** 2001. Recombinant urease and urease DNA of *Coccidioides immitis* elicit an immunoprotective response against coccidioidomycosis in mice. *Infect. Immun.* **69:**2878–2887.

129. **Linares, E., S. Giorgio, R. A. Mortara, C. X. Santos, A. T. Yamada, and O. Augusto.** 2001. Role of peroxynitrite in macrophage microbicidal mechanisms in vivo revealed by protein nitration and hydroxylation. *Free Radic. Biol. Med.* **30:**1234–1242.

130. **Linsangan, L. C., and L. A. Ross.** 1999. *Coccidioides immitis* infection of the neonate: two routes of infection. *Pediatr. Infect. Dis. J.* **18:**171–173.

131. **Litvintseva, A. P., and J. M. Henson.** 2002. Cloning, characterization, and transcription of three laccase genes from *Gaeumannomyces graminis* var. *tritici*, the take-all fungus. *Appl. Environ. Microbiol.* **68:**1305–1311.

132. **Liu, L., M. Zeng, A. Hausladen, J. Heitman, and J. S. Stamler.** 2000. Protection from nitrosative stress by yeast flavohemoglobin. *Proc. Natl. Acad. Sci. USA* **97:**4672–4676.

133. **Liu, T., T. Matsuguchi, N. Tsuboi, T. Yajima, and Y. Yoshikai.** 2002. Differences in expression of toll-like receptors and their reactivities in dendritic cells in BALB/c and C57BL/6 mice. *Infect. Immun.* **70:**6638–6645.

134. **Logan, J. L., J. E. Blair, and J. N. Galgiani.** 2001. Coccidioidomycosis complicating solid organ transplantation. *Semin. Resp. Infect.* **16:**251–256.

135. **MacNeill, C., T. M. Umstead, D. S. Phelps, Z. Lin, J. Floros, D. A. Shearer, and J. Weisz.** 2004. Surfactant protein A, an innate immune factor, is expressed in the vaginal mucosa and is present in vaginal lavage fluid. *Immunology* **111:**91–99.

136. **Magee, D. M., and R. A. Cox.** 1995. Roles of gamma interferon and interleukin-4 in genetically determined

resistance to *Coccidioides immitis*. *Infect. Immun.* 63:3514–3519.

137. Mansour, M. K., and S. M. Levitz. 2002. Interactions of fungi with phagocytes. *Curr. Opin. Microbiol.* 5:359–365.

138. Martin, R. M., J. L. Brady, and A. M. Lew. 1998. The need for IgG2c specific antiserum when isotyping antibodies from C57BL/6 and NOD mice. *J. Immunol. Methods* 212:187–192.

139. Merrell, D. S., and S. Falkow. 2004. Frontal and stealth attack strategies in microbial pathogenesis. *Nature* 430:250–256.

140. Minamoto, G., and D. Armstrong. 1988. Fungal infections in AIDS. Histoplasmosis and coccidioidomycosis. *Infect. Dis. Clin. North Am.* 2:447–456.

141. Mirbod, F., R. A. Schaller, and G. T. Cole. 2002. Purification and characterization of urease isolated from the pathogenic fungus *Coccidioides immitis*. *Med. Mycol.* 40:35–44.

142. Mirbod-Donnovan, F., R. Schaller, C.-Y. Hung, J. Xue, U. Reichard, and G. T. Cole. 2006. Urease produced by *Coccidioides posadasii* contributes to the virulence of this respiratory pathogen. *Infect. Immun.* 74:504–515.

143. Moali, C., J. L. Boucher, M. A. Sari, D. J. Stuehr, and D. Mansuy. 1998. Substrate specificity of NO synthases: detailed comparison of L-arginine, homo-L-arginine, their N-omega-hydroxy derivatives, and N-omega-hydroxynor-L-arginine. *Biochemistry* 37:10453–10460.

144. Mobley, H. L., M. D. Island, and R. P. Hausinger. 1995. Molecular biology of microbial ureases. *Microbiol. Rev.* 59:451–480.

145. Montecucco, C., E. Papini, M. de Bernard, and M. Zoratti. 1999. Molecular and cellular activities of *Helicobacter pylori* pathogenic factors. *FEBS Lett.* 452:16–21.

146. Morgan, D. M. 1999. Polyamines. An overview. *Mol. Biotechnol.* 11:229–250.

147. Morrow, M. R., N. Abu-Libdeh, J. Stewart, and K. M. W. Keough. 2003. Interaction of pulmonary surfactant protein SP-A with DPPC/egg-PG bilayers. *Biophys. J.* 85:2397–2405.

148. Mukherjee, P. K., K. R. Seshan, S. D. Leidich, J. Chandra, G. T. Cole, and M. A. Ghannoum. 2001. Reintroduction of the PLB1 gene into *Candida albicans* restores virulence in vivo. *Microbiology* 147:2585–2597.

149. Munder, M., K. Eichmann, J. M. Moran, F. Centeno, G. Soler, and M. Modolell. 1999. Th1/Th2-regulated expression of arginase isoforms in murine macrophages and dendritic cells. *J. Immunol.* 163:3771–3777.

150. Nicas, M., and A. Hubbard. 2002. A risk analysis for airborne pathogens with low infectious doses: application to respirator selection against *Coccidioides immitis* spores. *Risk Anal.* 22:1153–1163.

151. Nosanchuk, J. D., and A. Casadevall. 2003. The contribution of melanin to microbial pathogenesis. *Cell. Microbiol.* 5:203–223.

152. Nosanchuk, J. D., B. L. Gomez, S. Youngchim, S. Diez, P. Aisen, R. M. Zancope-Oliveira, A. Restrepo, A. Casadevall, and A. J. Hamilton. 2002. *Histoplasma capsulatum* synthesizes melanin-like pigments in vitro and during mammalian infection. *Infect. Immun.* 70:5124–5131.

153. Noverr, M. C., G. M. Cox, J. R. Perfect, and G. B. Huffnagle. 2003. Role of PLB1 in pulmonary inflammation and cryptococcal eicosanoid production. *Infect. Immun.* 71:1538–1547.

154. Noverr, M. C., G. B. Toews, and G. B. Huffnagle. 2002. Production of prostaglandins and leukotrienes by pathogenic fungi. *Infect. Immun.* 70:400–402.

155. Noverr, M. C., P. R. Williamson, R. S. Fajardo, and G. B. Huffnagle. 2004. CNLAC1 is required for extrapulmonary dissemination of *Cryptococcus neoformans* but not pulmonary persistence. *Infect. Immun.* 72:1693–1699.

156. Olszewski, M. A., M. C. Noverr, G.-H. Chen, G. B. Toews, G. M. Cox, J. R. Perfect, and G. B. Huffnagle. 2004. Urease expression by *Cryptococcus neoformans* promotes microvascular sequestration, thereby enhancing central nervous system invasion. *Am. J. Pathol.* 164:1761–1771.

157. Paananen, A., E. Vuorimaa, M. Torkkeli, M. Penttila, M. Kauranen, O. Ikkala, H. Lemmetyinen, R. Serimaa, and M. B. Linder. 2003. Structural hierarchy in molecular films of two class II hydrophobins. *Biochemistry* 42:5253–5258.

158. Padhye, A. A., J. E. Bennett, M. R. McGinnis, L. Sigler, A. Fliss, and I. F. Salkin. 1998. Biosafety considerations in handling medically important fungi. *Med. Mycol.* 36:258–265.

159. Pan, S., M. Zhang, and G. T. Cole. 1995. Isolation and characterization of the arginase-encoding gene (*arg*) from *Coccidioides immitis*. *Gene* 154:115–118.

160. Pappagianis, D. 1996. Clinical presentation of infectious entities, p. 9–11. *In* H. Einstein (ed.), *Coccidioidomycosis*. National Foundation for Infectious Disease, Washington, D.C.

161. Pappagianis, D. 1980. Epidemiology of coccidioidomycosis, p. 63–85. *In* D. A. Stevens (ed.), *Coccidioidomycosis: a Text*. Plenum Press, New York, N.Y.

162. Pappagianis, D. 1988. Epidemiology of coccidioidomycosis. *Curr. Top. Med. Mycol.* 2:199–238.

163. Pappagianis, D. 2001. Serologic studies in coccidioidomycosis. *Semin. Respir. Infect.* 16:242–250.

164. Pappagianis, D., and H. Einstein. 1978. Tempest from Tehachapi takes toll on *Coccidioides* conveyed aloft and afar. *West. J. Med.* 129:527–530.

165. Pappagianis, D., and B. L. Zimmer. 1990. Serology of coccidioidomycosis. *Clin. Microbiol. Rev.* 3:247–268.

166. Paris, S., J.-P. Debeaupuis, R. Crameri, M. Carey, F. Charles, M. C. Prevost, C. Schmitt, B. Philippe, and J. P. Latge. 2003. Conidial hydrophobins of *Aspergillus fumigatus*. *Appl. Environ. Microbiol.* 69:1581–1588.

167. Peters-Golden, M. 1997. Lipid mediator synthesis by lung macrophages, p. 151–182. *In* M. F. Lipscomb and S. W. Russell (ed.), *Lung Macrophages and Dendritic Cells in Health and Disease*, vol. 102. Marcel Dekker, Inc., New York, N.Y.

168. Peterson, C. M., K. Schuppert, P. C. Kelly, and D. Pappagianis. 1993. Coccidioidomycosis and pregnancy. *Obstet. Gynecol. Surv.* 48:149–156.

169. Pishko, E. J., T. N. Kirkland, and G. T. Cole. 1995. Isolation and characterization of two chitinase-encoding genes (*cts1, cts2*) from the fungus *Coccidioides immitis*. *Gene* 167:173–177.

170. Proia, L. A., and A. R. Tenorio. 2004. Successful use of voriconazole for treatment of *Coccidioides meningitis*. *Antimicrob. Agents Chemother.* 48:2341.

171. Rawlings, N. D., D. P. Tolle, and A. J. Barrett. 2004. MEROPS: the peptidase database. *Nucleic Acids Res.* 32(Database issue):D160–D164.

172. Reed, M. B., P. Domenech, C. Manca, H. Su, A. K. Barczak, B. N. Kreiswirth, G. Kaplan, and C. E. Barry III. 2004. A glycolipid of hypervirulent tuberculosis strains that inhibits the innate immune response. *Nature* 431:84–87.

173. Reyna-Lopez, G. E., and J. Ruiz-Herrera. 2004. Specificity of DNA methylation changes during fungal

dimorphism and its relationship to polyamines. *Curr. Microbiol.* **48:**118–123.

174. Richards, J. O., N. M. Ampel, and D. F. Lake. 2002. Reversal of coccidioidal anergy in vitro by dendritic cells from patients with disseminated coccidioidomycosis. *J. Immunol.* **169:**2020–2025.

175. Rivera, J., J. Mukherjee, L. M. Weiss, and A. Casadevall. 2002. Antibody efficacy in murine pulmonary *Cryptococcus neoformans* infection: a role for nitric oxide. *J. Immunol.* **168:**3419–3427.

176. Rodrigues, E., and L. Travassos. 1994. Nature of the reactive epitopes in *Paracoccidioides brasiliensis* polysaccharide antigen. *J. Med. Vet. Mycol.* **32:**77–81.

177. Rubensohn, M., and S. Stack. 2003. Coccidiomycosis in a dog. *Can. Vet. J.* **44:**159–160.

178. Saubolle, M. A. 1996. Life cycle and epidemiology of *Coccidioides immitis*, p. 1–9. *In* H. E. Einstein and A. Catenzaro (ed.), *Coccidioidomycosis*. National Foundation for Infectious Diseases, Washington, D.C.

179. Saubolle, M. A., and J. Sutton. 1994. Coccidioidomycosis: centennial year on the North American Continent. *Clin. Microbiol. Newsl.* **16:**137–144.

180. Schouten, A., L. Wagemakers, F. L. Stefanato, R. M. van der Kaaij, and J. A. L. van Kan. 2002. Resveratrol acts as a natural profungicide and induces self-intoxication by a specific laccase. *Mol. Microbiol.* **43:**883–894.

181. Shubitz, L., T. Peng, R. Perrill, J. Simons, K. Orsborn, and J. N. Galgiani. 2002. Protection of mice against *Coccidioides immitis* intranasal infection by vaccination with recombinant antigen 2/PRA. *Infect. Immun.* **70:**3287–3289.

182. Simon, H.-U. 2002. The neutralization of interleukin-5 as a therapeutic concept in allergic inflammation. *Sarcoid. Vasc. Diffuse Lung Dis.* **19:**25–28.

183. Smith, C., M. Saito, and S. Simmons. 1956. Patterns of 39,500 serologic tests in coccidioidomycosis. *JAMA* **160:**546–552.

184. Smith, C., E. Whiting, E. Baker, H. Rosenberger, R. Beard, and M. Saito. 1948. The use of coccidioidin. *Am. Rev. Tuberc.* **57:**330–360.

185. Smith, M. A., A. E. Anderson, and K. Kostroff. 1994. An unusual case of coccidioidomycosis. *J. Clin. Microbiol.* **32:**1063–1064.

186. Snijdewint, F., P. Kalinski, E. Wierenga, J. Bos, and M. Kapsenberg. 1993. Prostaglandin E2 differentially modulates cytokine secretion profiles of human T helper lymphocytes. *J. Immunol.* **150:**5321–5329.

187. Steenbergen, J. N., J. D. Nosanchuk, S. D. Malliaris, and A. Casadevall. 2004. Interaction of *Blastomyces dermatitidis*, *Sporothrix schenckii*, and *Histoplasma capsulatum* with *Acanthamoeba castellanii*. *Infect. Immun.* **72:**3478–3488.

188. Stempin, C., T. Tanos, O. A. Coso, and F. M. Cerban. 2004. Arginase induction promotes *Trypanosoma cruzi* intracellular replication in Cruzipain-treated J774 cells through the activation of multiple signaling pathways. *Eur. J. Immunol.* **34:**200–209.

189. Strassmann, G., V. Patil-Koota, F. Finkelman, M. Fong, and T. Kambayashi. 1994. Evidence for the involvement of interleukin 10 in the differential deactivation of murine peritoneal macrophages by prostaglandin E2. *J. Exp. Med.* **180:**2365–2370.

190. Stringer, M. A., and W. E. Timberlake. 1993. Ceratoulmin, a toxin involved in Dutch elm disease, is a fungal hydrophobin. *Plant Cell* **5:**145–146.

191. Sun, S. H., G. T. Cole, D. J. Drutz, and J. L. Harrison. 1986. Electron-microscopic observations of the *Coccidioides immitis* parasitic cycle in vivo. *J. Med. Vet. Mycol.* **24:**183–192.

192. Taborda, C. P., M. A. Juliano, R. Puccia, M. Franco, and L. R. Travassos. 1998. Mapping the T-cell epitope in the major 43-kilodalton glycoprotein of *Paracoccidioides brasiliensis* which induces a Th-1 response protective against fungal infection in BALB/c mice. *Infect. Immun.* **66:**786–793.

193. Taborda, C. P., C. R. Nakaie, E. M. Cilli, E. G. Rodrigues, L. S. Silva, M. F. Franco, and L. R. Travassos. 2004. Synthesis and immunological activity of a branched peptide carrying the T-cell epitope of gp43, the major exocellular antigen of *Paracoccidioides brasiliensis*. *Scand. J. Immunol.* **59:**58–65.

194. Timm, K. I., R. J. Sonn, and B. D. Hultgren. 1988. Coccidioidomycosis in a Sonoran Gopher snake *Pituophismelanoleusus affinis*. *J. Med. Vet. Mycol.* **26:**101–104.

195. Trajkovic, V., S. Stepanovic, I. Samardzic, V. Jankovic, V. Badovinac, and M. Stojkovic. 2000. *Cryptococcus neoformans* neutralizes macrophage and astrocyte derived nitric oxide without interfering with inducible nitric oxide synthase induction or catalytic activity—possible involvement of nitric oxide consumption. *Scand. J. Immunol.* **51:**384–391.

196. Travassos, L. R., R. Puccia, P. Cisalpino, C. Taborda, E. G. Rodrigues, M. Rodrigues, J. F. Silveira, and I. C. Almeida. 1995. Biochemistry and molecular biology of the main diagnostic antigen of *Paracoccidioides brasiliensis*. *Arch. Med. Res.* **26:**297–304.

197. Triggiani, M., A. Oriente, G. de Crescenzo, G. Rossi, and G. Marone. 1995. Biochemical functions of a pool of arachidonic acid associated with triglycerides in human inflammatory cells. *Int. Arch. Allergy Immunol.* **107:**261–263.

198. Tripathy, U., G. L. Yung, J. M. Kriett, P. A. Thistlethwaite, D. P. Kapelanski, and S. W. Jamieson. 2002. Donor transfer of pulmonary coccidioidomycosis in lung transplantation. *Ann. Thorac. Surg.* **73:**306–308.

199. Tsujimoto, M., M. Sawaki, M. Sakamoto, K. Mikasa, K. Hamada, K. Maeda, S. Teramoto, K. Mori, K. Ueda, N. Narita, and E. Kita. 1997. The evaluation of interleukin-6 (IL-6) and tumor necrosis factor alpha (TNF-alpha) level in peripheral blood of patients with chronic lower respiratory tract infection. *J. Jpn. Assoc. Infect. Dis.* **71:**430–436.

200. Tsunoda, I., N. D. Tolley, D. J. Theil, J. L. Whitton, H. Kobayashi, and R. S. Fujinami. 1999. Exacerbation of viral autoimmune animal models for multiple sclerosis by bacterial DNA. *Brain Pathol.* **9:**481–493.

201. Ullmann, B. D., H. Myers, W. Chiranand, A. L. Lazzell, Q. Zhao, L. A. Vega, J. L. Lopez-Ribot, P. R. Gardner, and M. C. Gustin. 2004. Inducible defense mechanism against nitric oxide in *Candida albicans*. *Eukaryot. Cell* **3:**715–723.

202. Vassiliou, E., H. Jing, and D. Ganea. 2003. Prostaglandin E2 inhibits TNF production in murine bone marrow-derived dendritic cells. *Cell. Immunol.* **223:**120–132.

203. Verghese, S., D. Arjundas, K. C. Krishnakumar, P. Padmaja, D. Elizabeth, A. A. Padhye, and D. W. Warnock. 2002. Coccidioidomycosis in India: report of a second imported case. *Med. Mycol.* **40:**307–309.

204. **Vigna, A. F. G., L. C. Godoy, S. Rogerio de Almeida, M. Mariano, and J. D. Lopes.** 2002. Characterization of B-1b cells as antigen presenting cells in the immune response to gp43 from *Paracoccidioides brasiliensis* in vitro. *Immunol. Lett.* **83:**61–66.

205. **Vincendeau, P., A. P. Gobert, S. Daulouede, D. Moynet, and M. D. Mossalayi.** 2003. Arginases in parasitic diseases. *Trends Parasitol.* **19:**9–12.

206. **Walker, M. P., C. Z. Brody, and R. Resnik.** 1992. Reactivation of coccidioidomycosis in pregnancy. *Obstet. Gynecol.* **79:**815–817.

207. **Wallace, H. M., A. V. Fraser, and A. Hughes.** 2003. A perspective of polyamine metabolism. *Biochem. J.* **376:**1–14.

208. **Walters, D. R.** 2003. Polyamines and plant disease. *Phytochemistry* **64:**97–107.

209. **Walvoord, M. A., F. M. Phillips, D. A. Stonestrom, R. D. Evans, P. C. Hartsough, B. D. Newman, and R. G. Striegle.** 2003. A reservoir of nitrate beneath desert soils. *Science* **302:**1021–1024.

210. **Weatherburn, M.** 1967. Phenol-hydrochlorite reaction for determination of ammonia. *Anal. Chem.* **39:** 971–974.

211. **Wessels, J. G. H.** 1993. Wall growth, protein excretion and morphogenesis in fungi. *New Phytol.* **123:**397–413.

212. **Westerberg, S. C., C. B. Smith, B. B. Wiley, and C. Jensen.** 1972. Mycoplasma-virus interrelationships in mouse tracheal organ cultures. *Infect. Immun.* **5:** 840–846.

213. **Wheat, L. J.** 1988. Systemic fungal infections: diagnosis and treatment. I. Histoplasmosis. *Infect. Dis. Clin. North Am.* **2:**841–859.

214. **Wheeler, M., and A. Bell.** 1988. Melanins and their importance in pathogenic fungi. *Curr. Top. Med. Mycol.* **2:**338–387.

215. **Williams, F. M., V. Markides, J. Edgeworth, and A. J. Williams.** 1998. Reactivation of coccidioidomycosis in a fit American visitor. *Thorax* **53:**811–812.

216. **Williamson, P. R.** 1994. Biochemical and molecular characterization of the diphenol oxidase of *Cryptococcus neoformans. J. Bacteriol.* **176:**656–664.

217. **Woo, J. H., J. S. Lee, D. W. Lee, S. Y. Jin, D. W. Kim, and W. G. Lee.** 1996. A case of disseminated coccidioidomycosis—autopsy report. *J. Korean Med. Sci.* **11:**258–264.

218. **Wright, P. W., D. Pappagianis, M. Wilson, A. Louro, S. A. Moser, K. Komatsu, and P. G. Pappas.** 2003. Donor-related coccidioidomycosis in organ transplant recipients. *Clin. Infect. Dis.* **37:**1265–1269.

219. **Wu, J., A. J. Linscott, A. Oberle, and M. Fowler.** 2003. Pathology case of the month. Occupational hazard? Coccidioidomycosis (*Coccidioides immitis*). *J. La. State Med. Soc.* **155:**187–188.

220. **Xue, J.-M., C.-Y. Hung, J.-J. Yu, and G. T. Cole.** 2005. Immune response of vaccinated and non-vaccinated mice to *Coccidioides posadasii* infection. *Vaccine* **20:**3535–3544.

221. **Yu, J.-J., S. L. Smithson, P. W. Thomas, T. N. Kirkland, and G. T. Cole.** 1997. Isolation and characterization of the urease gene (*URE*) from the pathogenic fungus *Coccidioides immitis. Gene* **198:**387–391.

222. **Zalatnai, A., J. Zala, and G. Sandor.** 1998. Coccidioidomycosis in Hungary. The first import case. *Pathol. Oncol. Res.* **4:**147–151.

223. **Zhu, X., J. Gibbons, J. Garcia-Rivera, A. Casadevall, and P. R. Williamson.** 2001. Laccase of *Cryptococcus neoformans* is a cell wall-associated virulence factor. *Infect. Immun.* **69:**5589–5596.

224. **Zhu, X., and P. R. Williamson.** 2004. Role of laccase in the biology and virulence of *Cryptococcus neoformans. FEMS Yeast Res.* **5:**1–10.

Molecular Principles of Fungal Pathogenesis
Edited by Joseph Heitman et al.
©2006 ASM Press, Washington, D.C.

Chapter 27

Blastomyces dermatitidis Cell Surface Determinants and Their Application in Vaccine Development

BRUCE KLEIN

Blastomyces dermatitidis is a pathogenic ascomycete (family Onygenaceae) which exists in nature as a sporulating mold (19) and converts to a large (15- to 20-μm) thick-walled yeast form at elevated temperatures. This dimorphic fungus is closely related to *Histoplasma capsulatum, Chrysosporium parvum, Coccidioides immitis* (6, 40), and *Paracoccidioides brasiliensis* (6). Although the ecology of *B. dermatitidis* is still the subject of debate (19), these phylogenetic relatives of *B. dermatitidis* grow naturally in soils, often involving or requiring the presence of animal excreta. The favored substrate of *B. dermatitidis* appears to be material with a high organic content, with the assumption frequently being made that it, too, must grow within the soil. Some studies suggest that decaying wood might be its reservoir (3, 19), however, whereas others have pointed out the strong association between bodies of water and the exogenous saprophytic source of *B. dermatitidis* (34).

Conidia constitute the infectious form of this organism. After inhalation into the lungs of animals or humans, these swell and germinate into yeast, the form required for pathogenicity and proliferation. Unchecked proliferation in turn results in blastomycosis, a systemic mycosis that can become life-threatening when undiagnosed or untreated. Acute blastomycosis is occasionally self-limiting but more frequently progresses to cause severe pulmonary disease with the potential for dissemination to other organs, skin, and bone (15). These tissues become inflamed as neutrophils and mononuclear phagocytes are recruited to control the infection. Despite this robust response, phagocytosis by macrophages does not inhibit the replication of yeast (35), and in some cases it is stimulatory (25). Several weeks after the onset of an infection, the host develops a delayed-type hypersensitivity (DTH) response with activated T-cell proliferation. In a murine model of blastomycosis, these T cells confer protection when transferred into naïve animals whereas serum does not, suggesting that antigen-specific T cells are the key to an effective host response (12, 35, 45).

B. dermatitidis is endemic in the central United States in and around the Ohio and Mississippi River Valleys, although blastomycosis has been reported in parts of Canada and in the southeastern United States as well. The 100th meridian is commonly cited as the westernmost boundary of this area of endemic infection, but the actual extent of its range is uncertain since mycoses are frequently unreported in the United States and Canada. Sporadic cases have been verified in other regions throughout the world, but these are usually deemed to have stemmed from exposure during time spent in areas of endemic infection or exposure to contaminated materials originating in these areas (18, 19). African varieties of *B. dermatitidis* have been isolated and described, but these strains lack antigens characteristic of North American *B. dermatitidis* strains (32, 38) and produce a variant pattern of blastomycosis consisting largely of chronic cutaneous lesions (15; Editorial, *Lancet* i:25–26, 1989). It has been suggested that African *B. dermatitidis* may in fact be a separate species, bearing a closer taxonomic resemblance to *P. brasiliensis* (4, 22; Editorial, *Lancet* i:25–26, 1989).

PATHOGENESIS OF INFECTION

Thermally regulated dimorphism represents the single most strongly defining trait of *B. dermatitidis* and related systemic dimorphic fungi. These agents reversibly differentiate between mold and yeast phases via a morphologic transition, which can be controlled and induced in vitro by growing the fungi at 25°C (mold) or 37°C (yeast). Medoff and colleagues (39, 42–44, 53) demonstrated that the

Bruce Klein • Department of Pediatrics, Internal Medicine, and Medical Microbiology and Immunology, and the University of Wisconsin Comprehensive Cancer Center, University of Wisconsin—Madison, Madison, WI 53792.

temperature-induced transition is accompanied by a shared and well-characterized sequence of biochemical events, as follows. After temperature increase, there is a rapid decline in intracellular ATP concentration that follows uncoupling of oxidative phosphorylation. This is followed by a progressive decrease in respiration rate (stage 1). After 24 to 40 h, cells enter a dormant period (stage 2) that can last as long as 4 to 6 days. In stage 3, cytochrome components are restored, normal respiration is resumed, yeast phase-specific cysteine oxidase is induced, and the transition to yeast morphology is completed.

It has been accepted for some time that the transition to yeast is a requirement for pathogenesis. Medoff et al. demonstrated this phenomenon by treating *H. capsulatum* mycelia with the sulfhydryl inhibitor *p*-chloromercuriphenylsulfonic acid, which prevents conversion to the yeast form, and then showing that inoculation with this material can no longer cause disease in a murine model of infection (44). In addition, it has been shown that conidia are vulnerable to killing and clearance by macrophages and neutrophils, and conversion to the yeast form affords some measure of protection against this host immune response (20, 59). These findings imply that yeast phase-specific factors are essential for pathogenicity of dimorphic fungi. Developing an understanding of the functions and regulation of such factors is integral to understanding the pathogenesis of disease and could serve as the foundation for innovative approaches to the treatment of fungal infections.

Surprisingly, relatively few phase-specific genes have been identified in *B. dermatitidis* and other dimorphic fungi. Perhaps the most well studied phase-specific gene in *B. dermatitidis* is *BAD1* (for "*Blastomyces* adhesion"; formerly termed WI-1). As described in the next section, several of the molecular and genetic tools applicable to *B. dermatitidis* have been developed to aid in the study of this yeast-phase-specific virulence factor. Another yeast-phase-specific gene in *B. dermatitidis*, termed *bys1* (for "*Blastomyces* yeast specific"), was identified by differential hybridization screening of a cDNA library. *bys1* encodes an 18-kDa protein with a hydrophobic N terminus and two 34-amino-acid domains with similar 9-amino-acid repeating motifs (14). The function of *bys1* is unknown, and its regulation has not been studied. A yeast-phase-specific protein of *H. capsulatum*, termed CBP1, is a calcium-binding protein and an essential virulence determinant (56). Another yeast-phase-specific gene in *H. capsulatum*, *yps-3*, has been studied extensively (33), but its function and role in pathogenesis have yet to be elucidated. A few additional phase-specific genes have

been detected in other dimorphic fungi including the opportunistic fungal pathogen *Candida albicans*.

α-Glucan is an additional differentially expressed product of *B. dermatitidis*. Although the genes encoding it are not fully known, AGS1 [α-(1,3)-glucan synthase] is required for its production in *H. capsulatum* (49). The cell wall glucan content of *B. dermatitidis* yeast is 95% α-glucan (31), which is situated in the outer wall. In contrast, the proportion of α- and β-glucan is more evenly distributed in the cell walls of the mold. Hence, these polymers are differentially expressed in the yeast and mold forms of the fungus. In *B. dermatitidis*, as well as in *P. brasiliensis* and chemotype II strains of *H. capsulatum* (21), the loss of yeast cell wall α-(1,3)-glucan in genetically related variants correlates with the loss of virulence in animal and cell culture models of infection (26, 28). RNA interference has firmly established a role for the polymer in *H. capsulatum* virulence (49). It is possible that α-(1,3)-glucan functions in a protective manner or that its effect on strain virulence may have more to do with structural alterations of the cell wall (e.g., allowing proper surface presentation or secretion of proteins). The relationship of α-(1,3)-glucan to virulence is intriguing and warrants additional study, especially since it has been observed independently in three dimorphic fungal pathogens (28).

TOOLS FOR GENETIC MANIPULATION OF *B. DERMATITIDIS*

Since there are a number of yeast-phase-specific components that may be responsible for promoting virulence in *B. dermatitidis*, the challenge for researchers will lie in exploring their functions individually. The comparison of wild-type strains with strains lacking specific yeast-phase characteristics requires reliable, precise methods of genetic manipulation. Investigation of pathogenesis in *B. dermatitidis* has thus required the adaptation of a wide variety of molecular biology tools to this task (Table 1).

High-voltage electric pulse has remained the standard method for introducing exogenous DNA into *B. dermatitidis* since the protocol was first applied successfully in the mid-1990s (29). The *Escherichia coli* hygromycin phosphotransferase gene (*hph*), coding for hygromycin B resistance, serves as the staple dominant marker for the transformation of *B. dermatitidis* (29). Chlorimuron ethyl resistance, encoded by the *sur* (sulfonyl urea resistance) gene from *Magnaporthe grisea* (7, 61), is an equally effective dominant marker for the transformation of *B. dermatitidis* and has been particularly useful in the retransformation of previously manipulated, hygromycin-resistant strains. In

Table 1. Approaches available for genetic manipulation of *B. dermatitidis*[a]

Advance	Tool/example	Reference
Dominant selection	Hygromycin B	29
	Chlorimuron ethyl	7
	G418 (neomycin)	Unpublished
	Nourseothrecin	Unpublished
Marked strains	Uracil auxotrophs (*ura5*)	60
Transformation	Electrotransformation	29
	Agrobacterium-mediated transfer	60
Reporter genes	β-Galactosidase (*lacZ*)	51, 52
	β-Glucuronidase (*gus*)	Unpublished
	Green fluorescent protein	Unpublished
Homologous recombination	Allelic replacement at *BAD1* locus	7
Heterologous gene regulation	*Blastomyces* BAD1 in *Histoplasma*	51, 52
Genetically proven virulence factors	BAD1	7, 9

[a]Adapted from reference 8a with permission.

addition, G418 and nourseothrecin have been successfully employed (P. J. Rooney and T. D. Sullivan, unpublished data), yielding three viable options for antibiotic resistance-based transformation schemes.

Auxotrophic strains of *B. dermatitidis* have not been available until recently, when strains with spontaneous and UV-induced mutations in the *ura5* gene for uracil production were isolated by using 5-fluoroorotic acid selection (60). Complementation of this auxotrophy has proven a more efficient selection than antibiotic resistance, presumably because the latter subjects yeast to highly toxic conditions immediately following the stress of electroporation. Even so, the transformation efficiency obtained by electroporation remains low (range, 1 to 42 CFU/μg of transforming DNA) (29). In instances where numerous transformants are to be screened, a more efficient gene transfer method would be required for reasons of practicality.

Agrobacterium tumefaciens-mediated gene transfer, developed originally for the transformation of plant species, is effective in a variety of different fungi: transformation of *Saccharomyces* (13), *Fusarium, Neurospora,* and *Aspergillus* species (17), and *C. immitis* (2) has been described. This system was adapted for use with *B. dermatitidis* (51, 52, 60), routinely generating transformants with a frequency of up to 10^3 per 10^7 target yeast. The preferred fate of the transforming DNA appears to be integration into a single chromosomal locus. Development of this technique thus offers the dual advantages of single-site integration, which is desirable for transformation-mediated mutagenesis strategies, and a consistently high yield of transformants. Whether this system will allow the facile generation of homologous genetic integrants in *B. dermatitidis*, as was found to be the

case with *Saccharomyces cerevisae* transformation (13), remains to be seen.

Homologous recombination takes place in *B. dermatitidis*, but its rate relative to the usual mode of illegitimate recombination is uncertain. In efforts to knock out the *BAD1* gene by homologous recombination, the *hph* selection marker was placed downstream of a truncated span of *BAD1* promoter. This strategy may have functioned in essence as a promoter trap, selecting specifically for transformants in which the gene integrated adjacent to a promoter sequence of sufficient vigor to induce robust antibiotic resistance. Homologous replacements of *BAD1* represented approximately 1% of hygromycin B-resistant transformants in this system (7). Targeted knockout of individual genes, while attainable by more conventional techniques, may be rendered more tractable through employment of such a strategy.

As techniques for genetic manipulation of *B. dermatitidis* are developed and refined, so too are reporters for studies of gene regulation. β-Galactosidase activity can be measured in extracts of *B. dermatitidis* transformed with this reporter gene, and BAD1–β-galactosidase reporter fusions have been used to confirm that BAD1 is transcriptionally regulated (51, 52). β-Glucuronidase and green fluorescence reporters are also functional in *B. dermatitidis* (P. J. Rooney, J. D. Sullivan, and B. S. Klein, unpublished data).

NEW AVENUES OF INVESTIGATION ESTABLISH THE IMPORTANCE OF BAD1

The investigation and manipulation of the BAD1 genetic locus provides an example of the utility of our new armamentarium of genetic tools. The gene for BAD1 was successfully deleted from

B. dermatitidis strain ATCC 26199 by allelic replacement, using a vector coding for hygromycin B resistance. Using the previously described promoter trap system, 1 strain out of the 100 surveyed was found to be devoid of the BAD1 adhesin. PCR analysis and Southern hybridization verified that transformed strain no. 55 (*bad1::hph*) lacked the coding region for the gene. It had a diminished capacity to bind to macrophages and sections of murine lung tissue, compared to the parental strain and, most importantly, was highly attenuated in a murine model of blastomycosis (7).

Following its characterization, strain 55 was again subjected to genetic transformation, this time with a cloned copy of the *BAD1* gene on a vector coding for chlorimuron ethyl resistance. A transformant was selected in which the deleted *BAD1* gene had been restored. In this strain, 4-55, the production of BAD1 approximated that seen in the parent ATCC 26199 strain. Whereas strain 55 had been rendered avirulent following knockout of the *BAD1* gene, strain 4-55 demonstrated restored virulence in the murine model and adhesive properties on a par with the parental strain (ATCC 26199). The indispensable role of BAD1 in virulence was thus highlighted through the agency of several molecular genetic tools newly available for use in studies of *B. dermatitidis*.

Sequence analysis of the *BAD1* gene predicts a product consisting of a short amino-terminal region, a 24-amino-acid repeat, present in 30 tandem copies within the core of the protein, and a carboxyl-terminal region with homology to epidermal growth factor (27). The tandem repeat core of BAD1, which constitutes over 75% of the protein's primary structure, has an affinity for host CR3 on human macrophages (48). The C-terminal epidermal growth factor domain mediates binding to chitin fibrils on the surface of the BAD1 knockout strain 55 (9). This affinity may explain one mechanism by which the organism situates the adhesin: secretion to the extracellular milieu followed by a binding back to exposed chitin fibrils on the cell surface (8). BAD1 has also been found to self-associate in the presence of calcium ions (8). This suggests that under the conditions present in the lungs, chitin-bound BAD1 should begin to capture unassociated BAD1, assembling layers of it at the cell surface. Since α-glucan is also found on the surface of *B. dermatitidis* yeast, the structural relationship between it and exposed chitin fibrils and BAD-1 are a matter of conjecture. Surface α-glucan might serve a protective function by partially masking chitin fibrils, preventing the hyperaccumulation of BAD1 that has been observed in avirulent, glucan-deficient strains of *B. dermatitidis* (36).

MECHANISMS OF BAD1 ACTIVITY IN PATHOGENICITY OF *B. DERMATITIDIS*

Figure 1 represents a model illustrating various points at which BAD1 influences the pathogenesis of infection. BAD1 promotes the binding of *B. dermatitidis* yeast to host cells and tissue (7, 37). Complement type 3 (CD11b/CD18) receptors are the known primary anchor points for this BAD1-mediated binding (48). BAD1 binding to phagocytes also modulates the host immune responses. For example, BAD1 downregulates the production of tumor necrosis factor alpha (TNF-α), which would normally be stimulated in response to an infection (23). A BAD1 knockout strain evokes the robust production of TNF-α in vivo in an experimental model of infection and in vitro during coculture with macrophages and neutrophils, whereas the isogenic, wild-type strain, as well as soluble, purified BAD1, actually curtails TNF-α production in parallel experiments (Fig. 2). This perturbation of proinflammatory cytokine production is associated with a skewed inflammatory response dominated by neutrophils rather than by T cells and other mononuclear cells found in mature granulomas; it leads to unchecked progression of *B. dermatitidis* infection in experimental blastomycosis (23). In the murine model of pulmonary blastomycosis, gene therapy with a TNF-α-producing recombinant adenovirus provides a salutary effect, ameliorating the course of the infection (23). Hence, in addition to promoting the attachment of the fungus to cells and tissue, BAD1 enhances pathogenicity by allowing *B. dermatitidis* to avoid at least one critical host immune response by modulating it in a manner that favors fungal survival over host survival.

BAD1 modulates TNF-α production in part through induction of transforming growth factor β (TGF-β) (24), which antagonizes phagocyte TNF-α production at the posttranscriptional level (8). Surface BAD1 on *B. dermatitidis* induces the release of TGF-β from macrophages and neutrophils in vitro (Fig. 3), and this cytokine is responsible for suppression of TNF-α in response to the fungal cell. Antibody neutralization of TGF-β in these assays is thus able to fully restore the production of TNF-α. Nevertheless, soluble BAD1 also suppresses the production of TNF-α in vitro, and, in contrast to the above, this suppression cannot be relieved by antibody that neutralizes TGF-β (Fig. 3). In murine pulmonary blastomycosis, TGF-β levels in the alveoli are increased severalfold in response to virulent wild-type *B. dermatitidis* compared with the response to an isogenic bad1 knockout strain. Nevertheless, administration of anti-TGF-β neutralizing antibody to these animals fails to restore concomitantly suppressed TNF-α levels in the lungs, even though

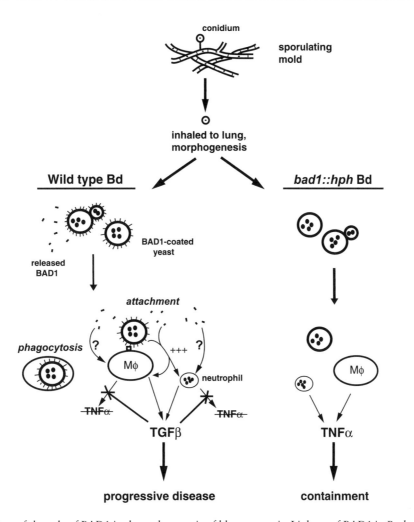

Figure 1. Overview of the role of BAD1 in the pathogenesis of blastomycosis. Linkage of BAD1 in *B. dermatitidis* with the phase transition of mold to yeast, adhesion, and modulation and evasion of host immunity is shown. After morphogenesis from mold to yeast (top), effects of BAD1 expression on wild-type yeast are shown on the left side of the figure, whereas effects of the null mutation in the isogenic knockout strain are shown on the right. BAD1 on wild-type yeast suppresses TNF-α production, which leads to progressive infection (23). Surface BAD1 suppresses TNF-α production via TGF-β (24), whereas soluble BAD1 suppresses TNF-α independently of TGF-β. BAD1 thus acts in multiple key ways to foster infection and disease progression. Adapted from reference 8a with permission.

levels of TGF-β are sharply curtailed and are comparable to those in mice infected with the bad-1 knockout strain. Since soluble BAD1 accumulates in the lung alveoli of mice infected with virulent *B. dermatitidis*, these findings argue that soluble BAD-1 contributes to TNF-α suppression and virulence enhancement and that TGF-β-independent mechanisms of BAD-1 suppression of TNF-α may shape the outcome of infection.

During cell wall biogenesis, BAD1 is released extracellularly before it binds back to the yeast cell surface (8). Post-release reassociation with the yeast surface is a surprisingly simple method by which the BAD1 adhesin might evenly coat cell surfaces. The fact that cell-associated and soluble forms of the BAD1 protein exist side by side raises the question

whether one form is predominantly responsible for the observed contribution to *B. dermatitidis* virulence.

Unexpectedly, recent findings suggest that the binding of yeast to host cells via surface BAD1 (the adhesive function of BAD1) is dispensable in the requisite role of BAD1 in pathogenesis (9). Deletion of the C-terminal epidermal growth factor-like domain (ΔC term protein) profoundly affects BAD1 function, leading to nonassociation with yeast, extracellular accumulation, and impaired yeast adherence to macrophages. Despite the absence of ΔC term protein on cell surfaces and impaired binding of these strains to macrophages, the defects have no discernible impact on the pathogenicity of *B. dermatitidis* in vivo. The ΔC term strains (and protein) retain the ability to

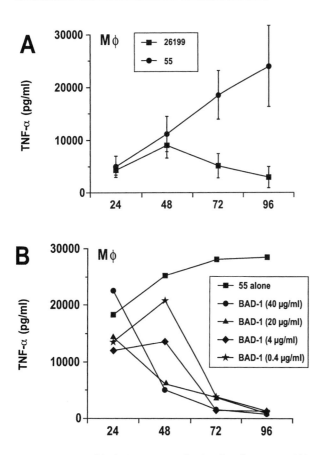

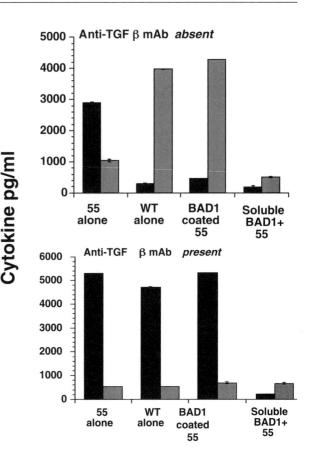

Figure 2. BAD1 blocks TNF-α production by phagocytes. (A) Isogenic BAD1 plus and minus strains (ATCC 26199 and knockout strain 55) were cocultured in vitro with peritoneal macrophages, and TNF-α levels were measured at serial time points postincubation (denoted on the *x* axis from 24 to 96 h). Wild-type strain ATCC 26199 blocks TNF-α production, and knockout strain 55 stimulates TNF-α. (B) Soluble purified BAD1 was added in vitro to wells where knockout strain 55 induced TNF-α production by macrophages. BAD1 blocked TNF-α production in a concentration-dependent manner. Note that BAD1 had previously been named WI-1. Adapted from reference 23 with permission.

Figure 3. BAD1 suppresses TNF-α via TGF-β-dependent and -independent mechanisms. Isogenic BAD1 plus and minus strains (wild-type strain ATCC 26199 and BAD1 knockout 55) or BAD1-coated knockout yeasts were cocultured in vitro with peritoneal macrophages, and levels of TNF-α (black bars) and TGF-β (gray bars) were measured after 24 to 48 h. (Top) Wild-type yeast or BAD1-coated knockout yeast induced TGF-β and suppressed TNF-α. Soluble BAD1 failed to induce TGF-β but nevertheless suppressed TNF-α. (Bottom) Anti-TGF-β neutralized the product and restored TNF-α suppressed by yeast cell surface BAD1. Anti-TGF-β failed to restore TNF-α suppressed in response to soluble BAD1. Thus, surface BAD1 suppresses TNF-α in a TGF-β-dependent manner and soluble BAD1 suppresses TNF-α in a TGF-β-independent manner. Adapted from reference 24 with permission.

suppress TNF-α production by phagocytes in vitro, which might explain how they maintain pathogenicity in vivo. To explore this, TNF-α levels were measured in the pulmonary compartment after infection with ΔC term strains. During the initial 72 h of infection, when lung CFU were comparable for the strains tested, TNF-α levels were severalfold lower with wild-type yeast and ΔC term yeast compared with BAD1-null strain 55. Given the evidence of cytokine profiles, survival curves, and binding studies, it appears that BAD1 may be able to exert its effect on virulence independently of adherence, and possibly largely by deviating the host immune response.

BAD1 modulates TNF-α production through interaction with one or more of its established binding receptors (CR3, p150, 95, and CD14). For example, soluble BAD1 enters macrophages rapidly, within several minutes, in a receptor-dependent manner (10). Blockade of receptor recycling and expression inhibits the entry of soluble BAD-1 into the cells and suppression of TNF-α. Soluble BAD1 internalization via CR3 appears to be crucial for its pathogenic mode of action. Cytochalasin D treatment of cells, so that BAD1 binds them but does not enter, or delivery of soluble BAD1 into cells via FcR reverses BAD1 suppression of TNF-α. Similarly, CR3$^{-/-}$ macrophages from knockout mice are resistant to BAD1 suppression of TNF-α in vitro; remarkably, these mice retain concomitantly higher levels of

TNF-α in the lungs following infection with virulent *B. dermatitidis,* and they are relatively resistant to the infection compared to their wild-type CR3$^{+/+}$ littermates. These findings suggest that BAD1 of *B. dermatitidis* exploits CR3 receptors on macrophages to suppress TNF-α production in a manner that tilts the balance in favor of the pathogen over the host. Other pathogens such as *Leishmania major* and *Bordetella pertusis* have been reported to exploit CR3 receptors by alternate mechanisms (47, 54).

PHASE-REGULATED EXPRESSION OF BAD1

Investigators have long postulated that virulence factors are up-regulated during the phase transition of systemic dimorphic fungi, but evidence in support of this concept has been lacking. This idea was explored with regard to BAD1. BAD1 is found only on the surface (or as a secreted product) of the yeast form of *B. dermatitidis.* It is undetectable on the surfaces of the hyphal filaments that make up the mold form, and, despite its association with the process of infection, neither is it present on the surface of conidia (51). BAD1 is not a prerequisite for the phase transition to yeast, since strains from which BAD1 has been deleted can make the transition in both directions with no apparent hindrance (51). Given the established pathogenic importance of BAD1, these observations help to explain how the thermally triggered phase transition leads to establishment of a virulent state.

Once the conversion to yeast has begun, BAD1 rapidly accumulates on the surface of the organism. An analysis of yeast and mycelial morphotypes employing both Northern hybridization and reporter genes fused to the *BAD1* promoter has shown that the phase-specific regulation of *BAD1* is tightly controlled at the level of transcription (51). Hence, this virulence gene has proven to be under a transcriptional control that is tied, either directly or indirectly, to the temperature-induced transition from mold to yeast.

Many African strains of *B. dermatitidis* are naturally devoid of BAD1, since they lack the gene for this protein (38). When the BAD1 coding region, under the control of its native promoter, was transferred into an African strain of *B. dermatitidis,* the heterologously expressed BAD1 retained phase-specific expression and control at the level of transcription (51). The same phenomenon was observed when the *BAD1* gene was transferred into *H. capsulatum,* which also lacks a native *BAD1* locus (51), indicating that there is conservation of the phase-specific mechanism of transcriptional control. Sequence analysis of the upstream region of *BAD1* has identified regions of homology to the promoter of the *H. capsulatum* yeast-phase-specific gene *yps-3*. A shared 63-bp region is essential for up-regulation of BAD1 in the yeast phase; it harbors two essential, putative transcription factor-binding sites: CREB and myb (52). It would seem likely that other, as yet unsequenced yeast-phase-specific genes could be controlled by the same transcriptional regulatory mechanism.

DELAYED-TYPE HYPERSENSITIVITY: T CELLS AND THEIR PRODUCTS CORRELATE WITH RESISTANCE

Cellular immunity is critical in acquired resistance to *B. dermatitidis*. Mice immunized with either viable or merthiolate-killed yeast demonstrate peak DTH responses that coincide temporally with maximal resistance to a lethal challenge with live *B. dermatitidis* given intranasally (45), intraperitoneally (58), and intravenously (16). This functional capability can be transferred passively to naïve mice via spleen cells of immunized mice but not via serum (55). These results illustrate that the maturation of cell-mediated immunity is a vital aspect of host resistance to this fungal infection.

DEVELOPMENT OF ATTENUATED *B. DERMATITIDIS* FOR VACCINATION

The finding that targeted disruption of BAD1 abolishes the pathogenicity of *B. dermatitidis* (7) has been exploited to create a fungal vaccine and study its mechanism(s) of action. Administration of viable yeast of this attenuated strain successfully vaccinates against lethal pulmonary experimental infection due to isogenic and nonisogenic strains from diverse geographic regions (62). This recombinant attenuated vaccine induces DTH and polarized type 1 cytokine responses, which are linked with resistance. Immunization subcutaneously or by the multiple routes, intranasally, subcutaneously, and intravenously, prompts strong DTH, whereas intranasal immunization yields marginal DTH (Fig. 4). Immunization subcutaneously or by the combined route reduces lung CFU by between two and three log units compared to controls. Immunization intranasally reduces lung CFU marginally, but the reduction is significant compared to controls. In survival analyses, 80% of mice immunized subcutaneously and 50% immunized by the combined route survive and most of the survivors demonstrate evidence of sterilizing immunity in the lungs. Thus, vaccine-induced protection in the lungs is most effective when the BAD1 null vaccine strain is administered subcutaneously, but protective efficacy in the lungs is minimal when the vaccine is delivered at the site of natural infection.

The parental, wild-type strain ATCC 26199 is not suitable for vaccination. That strain disseminates from the skin injection site to the lungs, where

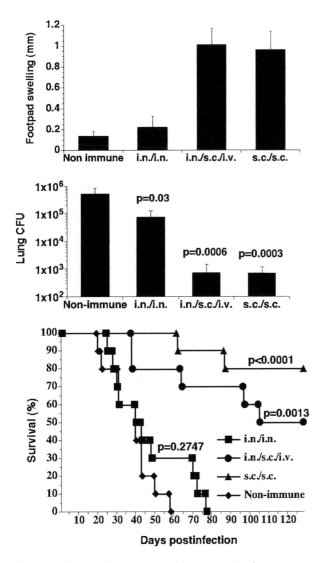

Figure 4. Acquired resistance to blastomycosis after vaccination with a recombinant, live, attenuated strain of *B. dermatitidis*. (Top) DTH responses. C57BL/6 mice were vaccinated intranasally (i.n.), subcutaneously (s.c.), and intravenously (i.v.). After immunization, mice were injected with 10^5 dead yeast cells of strain 55 into one footpad or phosphate-buffered saline (PBS) into the other. Footpad swelling was measured 24 h later. Error bars show the standard error of the mean. (Center) Lung infection. At 2 weeks after immunization, mice were infected intranasally with 10^4 CFU of wild-type ATCC 26199 yeast. Lung CFU were measured 3 weeks later. Error bars show the standard error of the mean. *P* values are for comparison of each immunized group versus nonimmune controls. (Bottom) Survival analysis. Groups of C57BL/6 mice were infected as above and monitored for survival. *P* values are for comparison of each immunized group versus nonimmune mice. Reprinted from reference 62 with permission.

infection progresses. Morozumi et al. (45) reported that subcutaneous administration of live ATCC 26199 yeast protected mice against pulmonary blastomycosis. In contrast to our study, those

authors reported that ATCC 26199 yeast injected subcutaneously into each of two sites was cleared by 4 weeks after injection. The discrepancy between their findings and ours might be explained by recent findings about spontaneous variation of ATCC strain 26199. Mutants of strain ATCC 26199 have arisen spontaneously by serial passage in vitro: ATCC 60915 is an attenuated mutant (11), and ATCC 60916 is avirulent (46). Phenotypic alterations recently found in these mutants include partial or complete loss of surface α-(1,3)-glucan (26) and changes in the amount of surface and secreted BAD1 (36). To guard against loss of pathogenicity in *B. dermatitidis*, including strain ATCC 26199, we monitor virulence in mice and measure surface α-(1,3)-glucan and BAD1 routinely during passage. The wild-type strain 26199 used for vaccination by Morozumi et al. (45) could have mutated spontaneously and lost virulence, which would have escaped attention since phenotypes associated with loss of virulence had not been delineated at that time. It should be emphasized that spontaneous attenuated mutants of dimorphic fungi are often unstable and frequently revert to the virulence phenotype after passage in serum in vitro or in animals (28). This fact underscores the importance of using genetically defined and stable mutations for attenuated vaccine strains of dimorphic fungi rather than using spontaneous, uncharacterized mutants. Parenthetically, spontaneous variation does not account for loss of virulence of recombinant strain 55. Resupply of BAD1 in *trans* restores pathogenicity to strain 55, which establishes that loss of BAD1, not another defect, is responsible for loss of virulence (7).

Additional attributes make the recombinant, attenuated strain 55 a promising vaccine and source of protective antigens. Vaccination induces substantial resistance: the lung CFU count 2 to 3 weeks postinfection is reduced by 3 log units or more; nearly all immunized mice survive a lethal challenge; and most surviving mice show sterilizing immunity. Importantly, vaccinated mice resist challenge with both the isogenic wild-type strain and nonisogenic strains from various geographical regions. The vaccine strain 55 thus shares an immunodominant, protective antigen(s) with nonisogenic strains of *B. dermatitidis*. Consequently, mutation of a pathogenetic locus in a dimorphic fungus has resulted in an attenuated vaccine strain, which, in work outlined below, was used to elucidate fungal and host elements requisite for vaccine immunity.

ELUCIDATION OF THE MECHANISMS OF VACCINE IMMUNITY

To elucidate the elements requisite for vaccine resistance, a series of approaches have been used:

knockout mice, depletion of cellular subsets, and neutralization of products from vaccinated mice during the expression phase of immunity. For example, to define the cell population(s) responsible for vaccine resistance, athymic, nude (nu/nu) mice, TCR-$\alpha^{-/-}$ mice, μ-chain-deficient mice, and wild-type mice were investigated (63). Vaccinated nude mice and TCR-$\alpha^{-/-}$ mice fail to resist lethal pulmonary infection with virulent wild-type yeast. In contrast, vaccinated μ-chain-deficient mice and their wild-type littermates resist reinfection. Thus, $\alpha\beta$ TCR$^+$ T cells are required for vaccine immunity.

To explore the role of $\alpha\beta$ T-cell subsets in vaccine immunity, depletion studies were done (63). CD4 and CD8 T cells were antibody depleted alone or together during the expression phase of the vaccine immunity, defined as the period after infection. CD4-depleted mice had significantly more lung CFU than did immunoglobulin G (IgG)-treated rat controls and died rapidly after infection, whereas controls appeared healthy for an extended period after infection and up to the time the study was terminated. In contrast, mice depleted of CD8 cells did not differ significantly from IgG-treated rat controls, as measured by lung CFU and survival. CD4 and CD8 double-depleted animals had significantly more lung CFU and a shorter mean survival than did CD4-depleted mice, and they were as vulnerable as unvaccinated mice. Thus, CD4 cells are chiefly responsible for vaccine resistance in immunocompetent hosts; CD8 cells contribute if CD4 cells are absent.

During the above studies, unexpected plasticity was detected in how the immune system defends against fungi. For example, neutralization of type 1 cytokines gamma interferon (IFN-γ) and TNF-α during the expression phase significantly impaired vaccine immunity in immunocompetent mice. In contrast, knockout mice lacking either of these cytokines showed levels of vaccine resistance as high as that in wild-type mice (Fig. 5) (63). In these immunodeficient mice, other cytokines compensated—TNF-α in IFN-$\gamma^{-/-}$ mice and granulocyte-macrophage colony-stimulating factor (GM-CSF) in TNF-$\alpha^{-/-}$ mice—during the induction phase of vaccine immunity. These findings pointed to the possibility that residual elements of immunity in immunodeficient hosts could be educated during the induction phase of immunity to compensate for elements that were instrumental in vaccine resistance in immunocompetent hosts. This concept was tested in studies described below by investigating whether CD4 T cells were essential or dispensable in vaccine immunity to pathogenic fungi.

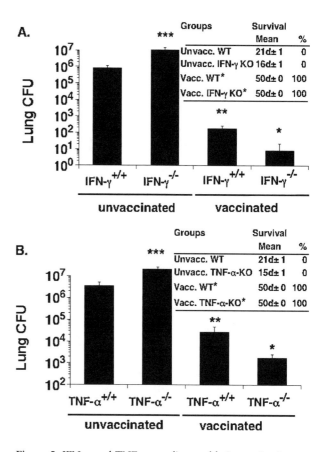

Figure 5. IFN-γ and TNF-α are dispensable in vaccine immunity. Knockout and wild-type mice were analyzed for lung CFU 2 weeks after infection or monitored for survival. Mean lung CFU ± standard error of the mean and mean survival time ± standard error of the mean are shown. %, percentage alive 50 days postinfection. *, $P < 0.007$ versus unvaccinated controls; **, $P < 0.012$ versus vaccinated transgenic mice; ***, $P < 0.005$ versus unvaccinated wild-type mice. Reprinted from reference 63 with permission.

DISPENSABILITY OF CD4 T CELLS IN VACCINE IMMUNITY AGAINST *B. DERMATITIDIS* AND *H. CAPSULATUM* IN IMMUNODEFICIENT HOSTS

Depletion of T-cell subsets during the period of vaccination (defined as the induction phase of vaccine immunity) and throughout the period postinfection gave striking results (64). Mice that were depleted of either CD4 or CD8 T cells acquired levels of vaccine immunity similar to that of rat IgG-treated controls, as assessed by lung CFU analysis (Fig. 6A). Additionally, in the absence of CD4 T cells, vaccination greatly prolonged survival and a significant proportion of those animals survived and acquired sterilizing immunity. Animals depleted of both CD4 and CD8 T cells were as susceptible as unvaccinated controls. Thus, in this model, T cells are required for acquisition of vaccine immunity but

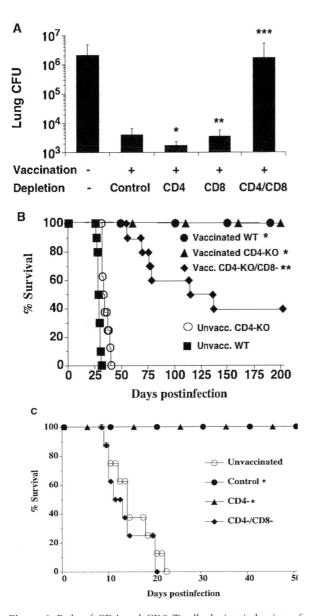

Figure 6. Role of CD4 and CD8 T cells during induction of vaccine immunity. (A) Lung CFU. Vaccinated mice were depleted of T cells during vaccination and *B. dermatitidis* infection. *, *P* = 0.1983 versus IgG control; **, *P* = 0.9825 versus control; ***, *P* = 0.7 versus unvaccinated mice. (B) Survival of vaccinated, CD4$^{-/-}$ mice and wild-type mice. Vaccinated CD4$^{-/-}$/CD8$^-$ mice were depleted of CD8 cells after vaccination and through *B. dermatitidis* infection. *, *P* < 0.0001 versus unvaccinated mice; **, *P* = 0.008 versus vaccinated CD4$^{-/-}$ mice. (C) CD8 T cells mediate vaccine immunity to *H. capsulatum*. Anti-CD4$^-$/anti-CD8 denotes vaccinated CD4-depleted mice, depleted of CD8 T cells after vaccination and through infection. *, *P* < 0.0002 versus unvaccinated mice or anti-CD4$^-$/anti-CD8-treated mice. Reprinted from reference 64 with permission.

CD4 T cells appear to be dispensable if they are absent during the induction of vaccine immunity. CD8 T cells are essential in a CD4-T-cell-deficient host.

In the study described above, vaccine immunity evoked in the absence of CD4 cells was durable and persisted for at least 8 weeks postvaccination. CD4-depleted mice that were rested for 8 weeks after vaccination (but maintained CD4 deficient) remained highly resistant to lethal infection. The majority of CD4-depleted mice acquired sterilizing immunity, and the few that did not do so had only 20 to 200 CFU in the lungs 25 days postinfection. In contrast, unvaccinated mice appeared moribund, with several million lung CFU at this time point. All vaccinated wild-type mice had cleared the infection. Thus, resting of vaccinated CD4-depleted mice indicated durable immunity and even greater resistance to infection.

Because antibody-mediated depletion of CD4 cells might not be complete in studies such as the one above and because any residual CD4 cells might provide sufficient help for other effector cells, vaccine immunity in CD4$^{-/-}$ mice was also investigated. Remarkably, all vaccinated CD4 T-cell-knockout mice survived an extended period postinfection (Fig. 6B); 9 of 11 had in fact acquired sterilizing immunity, and the other 2 had small numbers of CFU in their lungs (200 and 280 CFU). Survival data generated with class II knockout mice were similar to those observed for CD4-depleted wild-type mice. Thus, the results obtained using mice congenitally deficient in CD4 cells supported the data using CD4 cell depletion and the notion that CD4 T cells are dispensable.

The findings just described would be most significant if they could be generalized and extended to other pathogenic fungi or microbes. An experimental model of pulmonary histoplasmosis, a frequent cause of life-threatening, opportunistic infection in AIDS patients and others with compromised immunity, was used to explore this question. All mice vaccinated in either the presence or absence of CD4 T cells survived lethal pulmonary challenge with a wild-type virulent strain, G217B, and acquired sterilizing immunity by 50 days postinfection (Fig. 6C) (65). Unvaccinated mice and CD4-depleted mice depleted of CD8 cells after infection died rapidly. Thus, vaccine resistance can be raised against pulmonary histoplasmosis in the absence of CD4 cells. When CD4 T cells are absent, CD8 T cells are required during the induction phase and participate as effectors during the efferent phase of vaccine immunity to pulmonary *B. dermatitidis* and *H. capsulatum* infection.

MECHANISMS OF CD8 T-CELL ACTION IN VACCINE IMMUNITY

CD8 T cells could mediate vaccine effector functions in three distinct or overlapping ways: (i) by production of type 1 cytokines; (ii) by release of cytotoxic granules containing perforin, granulysin, or related molecules; and (iii) by an unconventional mechanism that involves class I major histocompatibility complex (MHC)-independent, direct binding of T cells to yeast, followed by one or more of the effector functions noted above.

Type 1 cytokines IFN-γ, TNF-α, and GM-CSF are instrumental in CD8 vaccine immunity to *B. dermatitidis* (64). First, lung cytokine levels (transcript and protein) correlate directly with the influx of CD8 T cells into this organ following infection in vaccinated CD4-depleted mice. Second, on intracellular cytokine staining, these lung CD8 T cells selectively express these products. Lastly, in functional studies, neutralization of these cytokines significantly impairs CD8 cell vaccine immunity. In contrast, perforin$^{-/-}$ mice depleted of CD4 cells acquired vaccine immunity as well as wild-type mice did (M. Wuethrich and B. S. Klein, unpublished data). Thus, perforin is dispensable in CD8 vaccine immunity. Nevertheless, perforin or other cytotoxic molecules could still contribute to CD8 effector mechanisms, and we have found that in vitro stimulation of in vivo-primed CD8 cells enhances the expression of cytotoxic granule products (granzyme B) in the cells (unpublished data). Lastly, vaccine-induced CD8 cells require class I MHC for vaccine immunity (see below) (64), in contrast to the results of in vitro studies of *Cryptococcus neoformans* and *C. albicans* (41), where T cells exert direct antifungal effector functions in the absence of MHC.

These results establish the significant role and function of CD8 cells in vaccine-generated immunity to fungi in immunodeficient hosts. The findings also raise questions that must be answered to fully understand the mechanisms that are operative. Among the questions are: what cell(s) and processing pathway(s) cross-prime and -present exogenous fungal antigens on class I MHC during CD8 vaccine immunity, and, in particular, how do they do so in the absence of CD4 T-cell help?

IMPORTANCE OF CROSS-PRESENTATION AND -PRIMING OF CD8 T CELLS

An enigma is how CD8 cells become activated during infections with *B. dermatitidis*, a predominantly extracellular pathogen, and with *H. capsulatum*, an intracellular pathogen that resides in the phagolysosome with no demonstrated access to the cytosol. Because of the reported precedent of direct antifungal activity of T cells (41), the requirement for MHC class I was explored. Adoptively transferred CD8 T cells protected wild-type mice, but not β$_2$m$^{-/-}$ mice lacking MHC class I. Thus, CD8 immunity against *B. dermatitidis* requires MHC class I for

antigen presentation and protection (64). These results suggest that exogenous fungal antigens are being processed for cross-presentation and -priming of class I-restricted CD8 cells as part of vaccine immunity to these fungi. The cellular and molecular mechanisms that govern this physiologically relevant aspect of immunity to fungi are worthy of investigation.

Bone marrow-derived antigen-presenting cells harbor a unique pathway for MHC class I presentation of exogenous antigens. This pathway permits cross-presentation of pathogen or pathogen-infected cells and priming of cytotoxic T-lymphocyte responses to microbial infections. All three bone marrow-derived antigen-presenting cells (i.e., macrophages, dendritic cells [DC], and B cells) have been reported to present class I-restricted, exogenously derived antigens in vitro. In vivo, however, CD8 cross-priming persists in the absence of B cells (57). Moreover, only DC are able to induce naïve cytotoxic T lymphocytes to respond by cross-priming (50), presumably because DC are best able to supply the necessary costimulatory signals. Recently, it was shown formally that DC were responsible for in vivo priming of CD8 T cells by exogenous microbial antigens. Using a novel diphtheria toxin-based system that allows the inducible, short-term ablation of DC in vivo Jung et al. (30) showed that DC are required to cross-prime CD8 precursors in vivo in experimental models of infection with *Listeria monocytogenes* and *Plasmodium yoelii*. For these reasons, DC are a worthy starting point for investigations of the cellular and molecular mechanisms behind cross-presentation of exogenous fungal antigen for priming antifungal CD8 cells that confer vaccine immunity.

CONCLUSIONS

B. dermatitidis produces a primary progressive pneumonia in immunocompetent hosts but can also behave as an opportunistic pathogen. The primary pathogenic potential of this fungus is unusual among the systemic dimorphic fungi, where infections are often asymptomatic or mild and self-limited. Although conidia of the fungus are sensitive to phagocytosis and killing by mononuclear and polymorphonuclear leukocytes, conversion to the yeast enhances the resistance of *B. dermatitidis* to these cellular defenses. A reason for the enhanced pathogenicity of the yeast is the expression of BAD1, which is a major virulence factor expressed during the transition. BAD1 impairs the activity of phagocytes against the yeast, blunts proinflammatory cytokine responses, and shapes the quality of adaptive immunity. Targeted deletion of BAD1 abolishes the pathogenicity of the fungus and has led to the development of a live attenuated vaccine strain, which is effective in a murine model and is under study in field trials with dogs. Creation of this strain has made possible an investigation into the mechanisms of protective immunity. Resistance in healthy hosts depends on T cells rather than B cells and antibody, and CD4[+] T cells that produce type 1 cytokine (TNF-α, IFN-γ, and GM-CSF) are crucial in the expression of vaccine immunity. In immunodeficient hosts, there is sufficient redundancy of the immune system that loss of these cytokines can be compensated. Even the loss of type 1 cytokine-producing CD4[+] T cells can be compensated by CD8[+] T cells during the induction of vaccine immunity. Hence, vaccines may induce resistance even in patients with impaired immunity, suggesting that vaccines can be developed for immunodeficient hosts. Understanding antigen processing and presentation mechanisms and the development of memory immunity under these unusual circumstances will allow the rational design of safe, effective vaccines that harness natural or unnatural lung immunity against B. dermatitidis and other fungal pathogens.

Acknowledgments. Work in my laboratory is supported by grants from the U.S. Public Health Service. Portions of this chapter have been excerpted from reference 8a with permission.

REFERENCES

1. Abidi, F. E., H. Roh, and E. J. Keath. 1998. Identification and characterization of a phase-specific nuclear DNA binding protein from the dimorphic pathogenic fungus *Histoplasma capsulatum*. *Infect. Immun.* 66:3867–3873.
2. Abuodeh, R. O., M. J. Orbach, M. A. Mandel, A. Das, and J. N. Galgiani. 2000. Genetic transformation of *Coccidioides immitis* facilitated by *Agrobacterium tumefaciens*. *J. Infect. Dis.* 181:2106–2110.
3. Baumgardner, D. J., and D. P. Paretsky. 1999. The *in vitro* isolation of *Blastomyces dermatitidis* from a woodpile in north central Wisconsin, USA. *Med. Mycol.* 37:163–168.
4. Bialek, R., A. Ibricevic, A. Fothergill, and D. Begerow. 2000. Small-subunit ribosomal DNA sequence shows *Paracoccidioides brasiliensis* closely related to *Blastomyces dermatitidis*. *J. Clin. Microbiol.* 38:3190–3193.
5. Bogdan, C., J. Paik, Y. Vodovotz, and C. Nathan. 1992. Contrasting mechanisms for suppression of macrophage cytokine release by transforming growth factor-beta and interleukin-10. *J. Biol. Chem.* 15: 23301–23308.
6. Bowman, B. H., J. W. Taylor, and T. J. White. 1992. Molecular evolution of the fungi: human pathogens. *Mol. Biol. Evol.* 9:893–904.
7. Brandhorst, T. T., M. Wuethrich, T. Warner, and B. S. Klein. 1999. Targeted gene disruption reveals an adhesin indispensable for pathogenicity of *Blastomyces dermatitidis*. *J. Exp. Med.* 189:1207–1216.
8. Brandhorst, T. T., and B. S. Klein. 2000. Cell wall biogenesis of *Blastomyces dermatitidis*. *J. Biol. Chem.* 275:7925–7934.
8a. Brandhorst, T. T., P. J. Rooney, T. Sullivan, and B. S. Klein. 2002. Using new genetic tools to study the pathogenesis of *Blastomyces dermatitidis*. *Trends Microbiol.* 10:25–30.
9. Brandhorst, T. T., M. Wuethrich, B. Finkel-Jimenez, and B. S. Klein. 2003. A C-terminal EGF-like domain governs BAD1 localization to the yeast surface and fungal adherence to phagocytes, but is dispensable in immune modulation and pathogenicity of *Blastomyces dermatitidis*. *Mol. Microbiol.* 48:53–65.
10. Brandhorst, T. T., M. Wuethrich, B. Finkel-Jimenez, T. Warner, and B. S. Klein. 2004. Exploiting type 3 complement receptor for TNF-α suppression, immune evasion, and progressive pulmonary fungal infection. *J. Immunol.* 173:7444–7453.
11. Brass, C., C. M. Volkmann, D. E. Philpott, H. P. Klein, C. J. Halde, and D. A. Stevens. 1982. Spontaneous mutant of *Blastomyces dermatitidis* attenuated in virulence for mice. *Sabouraudia* 20:145–158.
12. Brummer, E., P. A. Morozumi, P. T. Vo, and D. A. Stevens. 1982. Protection against pulmonary blastomycosis: adoptive transfer with T-lymphocytes, but not serum, from resistant mice. *Cell. Immunol.* 73:349–359.
13. Bundock, P., A. den Dulk-Ras, A. Beijersbergen, and P. J. Hooykaas. 1995. Trans-kingdom T-DNA transfer from *Agrobacterium tumefaciens* to *Saccharomyces cerevisiae*. *EMBO J.* 14:3206–3214.
14. Burg, E. F., III, and L. H. Smith. 1994. Cloning and characterization of *bys1*, a temperature-dependent cDNA specific to the yeast phase of the pathogenic dimorphic fungus *Blastomyces dermatitidis*. *Infect. Immun.* 62:2521–2528.
15. Causey, W. A., and G. D. Campbell. 1992. Clinical aspects of blastomycosis, p.165–188. *In* Y. Al-Doory and A. F. Disalvo (ed.), *Blastomycosis*. Plenum Publishing Corp., New York, N.Y.
16. Cozad, G. C., and C. T. Chang. 1980. Cell-mediated immunoprotection in blastomycosis. *Infect. Immun.* 28:398–403.
17. de Groot, M. J. A., P. Bundock, P. J. Hooykaas, and A. G. Beijersbergen. 1998. *Agrobacterium tumefaciens*-mediated transformation of filamentous fungi. *Nat. Biotechnol.* 16:839–842.
18. De Groote, M. A., R. Bjerke, H. Smith, and L. V. Rhodes III. 2000. Expanding epidemiology of blastomycosis: clinical features and investigation of 2 cases in Colorado. *Clin. Infect. Dis.* 30:582–584.
19. Disalvo, A. F. 1992. Ecology of *Blastomyces dermatitidis*, p. 43–73. *In* Y. Al-Doory and A. F. Disalvo (ed.), *Blastomycosis*. Plenum Publishing Corp., New York, N.Y.
20. Drutz, D. J., and C. L. Frey. 1985. Intracellular and extracellular defenses of human phagocytes against *Blastomyces dermatitidis* conidia and yeasts. *J. Lab. Clin. Med.* 105:737–750.
21. Eissenberg, L. G., and W. E. Goldman. 1991. *Histoplasma* variation and adaptive strategies for parasitism: new perspectives on histoplasmosis. *Clin. Microbiol. Rev.* 4:411–421.
22. Gueho, E., M. C. Leclerc, G. S. de Hoog, and B. Dupont. 1997. Molecular taxonomy and epidemiology of *Blastomyces* and *Histoplasma* species. *Mycoses* 40:69–81.
23. Finkel-Jimenez, B. E., M. Wuethrich, T. T. Brandhorst, and B. S. Klein. 2001. Down-regulation of phagocyte TNF-α production by the WI-1 adhesin imparts pathogenicity on *Blastomyces dermatitidis*. *J. Immunol.* 166:2665–2673.
24. Finkel-Jimenez, B., M. Wuethrich, and B. S. Klein. 2002. BAD1, an essential virulence factor of *Blastomyces dermatitidis*, suppresses host TNF-α production through TGF-β-dependent and -independent mechanisms. *J. Immunol.* 168:5746–5755.

25. Giles, S., B. S. Klein, and C. Czuprynski. 1999. The effect of canine macrophages on the adherence and growth of *Blastomyces dermatitidis* yeast: evidence of a soluble factor that enhances the growth of *B. dermatitidis* yeast. *Microb. Pathog.* 27:395–405.

26. Hogan, L. H., and B. S. Klein. 1994. Altered expression of surface α-1,3-glucan in genetically related strains of *Blastomyces dermatitidis* that differ in virulence. *Infect. Immun.* 62:3543–3546.

27. Hogan, L. H., S. Josvai, and B. S. Klein. 1995. Genomic cloning, characterization, and functional analysis of the major surface adhesin WI-1 on *Blastomyces dermatitidis* yeasts. *J. Biol. Chem.* 270:30725–30732.

28. Hogan, L. H., B. S. Klein, and S. M. Levitz. 1996. Virulence factors of medically important fungi. *Clin. Microbiol. Rev.* 9:469–488.

29. Hogan, L. H., and B. S. Klein. 1997. Transforming DNA integrates at multiple sites in the dimorphic fungal pathogen *Blastomyces dermatitidis*. *Gene* 186:219–226.

30. Jung, S., D. Unutmaz, P. Wong, G. Sano, K. De los Santos, T. Sparwasser, S. Wu, S. Vuthoori, K. Ko, F. Zavala, E. G. Pamer, D. R. Littman, and R. A. Lang. 2002. In vivo depletion of CD11c⁺ dendritic cells abrogates priming of CD8⁺ T cells by exogenous cell-associated antigens. *Immunity* 17:211–220.

31. Kanetsuna, F., and L. M. Carbonell. 1971. Cell wall composition of the yeast-like and mycelial forms of *Blastomyces dermatitidis*. *J. Bacteriol.* 106:946–948.

32. Kaufman, L., P. G. Standard, R. J. Weeks, and A. A. Padhye. 1983. Detection of two *Blastomyces dermatitidis* serotypes by exoantigen analysis. *J. Clin. Microbiol.* 18:110–114.

33. Keath, E. J., and F. E. Abidi. 1994. Molecular cloning and sequence analysis of *yps-3*, a yeast-phase-specific gene in the dimorphic fungal pathogen *Histoplasma capsulatum*. *Microbiology* 140:759–767.

34. Klein, B. S., J. M. Vergeront, R. J. Weeks, U. N. Kumar, G. Mathai, B. Varkey, L. Kaufman, R. W. Bradsher, J. F. Stoebig, and J. P. Davis. 1986. Isolation *of Blastomyces dermatitidis* in soil associated with a large outbreak of blastomycosis in Wisconsin. *N. Engl. J. Med.* 314:529–534.

35. Klein, B. S. 1992. Immunology of blastomycosis, p. 133–163. *In* Y. Al-Doory and A. F. Disalvo (ed.), *Blastomycosis*. Plenum Publishing Corp., New York, N.Y.

36. Klein, B. S., S. Chaturvedi, L. H. Hogan, J. M. Jones, and S. L. Newman. 1994. Altered expression of surface protein WI-1 in genetically related strains of *Blastomyces dermatitidis* that differ in virulence regulates recognition of yeasts by human macrophages. *Infect. Immun.* 62:3536–3542.

37. Klein, B. S., and S. L. Newman. 1996. Role of cell surface molecules on *Blastomyces dermatitidis* in host:pathogen interactions. *Trends Microbiol.* 4:246–251.

38. Klein, B. S., B. D. Aizenstein, and L. H. Hogan. 1997. African strains of *Blastomyces dermatitidis* that do not express surface adhesin WI-1. *Infect. Immun.* 65:1505–1509.

39. Lambowitz, A. M., G. S. Kobayashi, A. Painter, and G. Medoff. 1983. Possible relationship of morphogenesis in pathogenic fungus, *Histoplasma capsulatum*, to heat shock response. *Nature* 303:806–808.

40. Leclerc, M. C., H. Philippe, and E. Gueho. 1994. Phylogeny of dermatophytes and dimorphic fungi based on large subunit ribosomal RNA sequence comparison. *J. Med. Vet. Mycol.* 32:331–341.

41. Levitz, S. M., H. L. Mathews, and J. W. Murphy. 1995. Direct antimicrobial activity of T cells. *Immunol. Today* 16:387–391.

42. Maresca, B., and G. S. Kobayashi. 1989. Dimorphism in *Histoplasma capsulatum*: a model for the study of cell differentiation in pathogenic fungi. *Microbiol. Rev.* 53:186–209.

43. Medoff, G., A. Painter, and G. S. Kobayashi. 1987. Mycelial- to yeast-phase transitions of the dimorphic fungi *Blastomyces dermatitidis* and *Paracoccidioides brasiliensis*. *J. Bacteriol.* 169:4055–4060.

44. Medoff, G., M. Sacco, B. Maresca, D. Schlessinger, A. Painter, G. S. Kobayashi, and L. Carratu. 1986. Irreversible block of the mycelial-to-yeast phase transition of *Histoplasma capsulatum*. *Science* 231:476–479.

45. Morozumi, P. A., E. Brummer, and D. A. Stevens. 1982. Protection against pulmonary blastomycosis: correlation with cellular and humoral immunity in mice after subcutaneous nonlethal infection. *Infect. Immun.* 37:670–678.

46. Morrison, C. J., and D. A. Stevens. 1991. Mechanisms of fungal pathogenicity: correlation of virulence in vivo, susceptibility to killing by polymorphonuclear neutrophils in vitro, and neutrophil superoxide anion induction among *Blastomyces dermatitidis* isolates. *Infect. Immun.* 59:2744–2749.

47. Mosser, D. M., and P. J. Edelson. 1987. The third component of complement (C3) is responsible for the intracellular survival of *Leishmania major*. *Nature* 327:329–331.

48. Newman, S. L., S. Chaturvedi, and B. S. Klein. 1995. The WI-1 antigen of *Blastomyces dermatitidis* yeasts mediates binding to human macrophage CD11b/CD18 (CR3) and CD14. *J. Immunol.* 154:753–761.

49. Rappleye, C. A., J. T. Engle, and W. E. Goldman. 2004. RNA interference in *Histoplasma capsulatum* demonstrates a role for α-(1,3)-glucan in virulence. *Mol. Microbiol.* 53:153–165.

50. Ronchetti, A., P. Rovere, G. Iezzi, G. Galati, S. Heltai, M. P. Protti, M. P. Garancini, A. A. Manfredi, C. Rugarli, and M. Bellone. 1999. Immunogenicity of apoptotic cells in vivo: role of antigen load, antigen-presenting cells, and cytokines. *J. Immunol.* 163:130–136.

51. Rooney, P. J., T. Sullivan, and B. S. Klein. 2001. Selective expression of the virulence factor BAD1 upon morphogenesis to the pathogenic yeast form of *Blastomyces dermatitidis*: evidence for transcriptional regulation by a conserved mechanism. *Mol. Microbiol.* 39:875–889.

52. Rooney, P. J., and B. S. Klein. 2004. Sequence elements necessary for transcriptional activation of BAD1 in the yeast phase of *Blastomyces dermatitidis*. *Eukaryot. Cell* 3:785–794.

53. Sacco, M., G. Medoff, A. M. Lambowitz, B. V. Kumar, G. S. Kobayashi, and A. Painter. 1983. Sulfhydryl induced respiratory "shunt" pathways and their role in morphogenesis in the fungus, *Histoplasma capsulatum*. *J. Biol. Chem.* 258:8223–8230.

54. Saukkonen, K., C. Cabellos, M. Burroughs, S. Prasad, and E. Tuomanen. 1991. Integrin mediated localization of *Bordetella pertussis* within macrophages: role in pulmonary colonization. *J. Exp. Med.* 173:1143–1149.

55. Scillian, J. J., G. C. Cozad, and H. D. Spencer. 1974. Passive transfer of delayed hypersensitivity to *Blastomyces dermatitidis* between mice. *Infect. Immun.* 10:705–711.

56. Sebghati, T. S., J. T. Engle, and W. E. Goldman. 2000. Intracellular parasitism by *Histoplasma capsulatum*: fungal virulence and calcium dependence. *Science* 290:1368–1372.

57. Shen, H., J. K. Whitmire, X. Fan, D. J. Shedlock, S. M. Kaech, and R. Ahmed. 2003. A specific role for B cells in the generation of CD8 T cell memory by recombinant *Listeria monocytogenes*. *J. Immunol.* 170:1443–1451.

58. **Spencer, H. D., and G. C. Cozad.** 1973. Role of delayed hypersensitivity in blastomycosis of mice. *Infect. Immun.* **7:**329–334.

59. **Sugar, A. M., and M. Picard.** 1991. Macrophage- and oxidant-mediated inhibition of the ability of live *Blastomyces dermatitidis* conidia to transform to the pathogenic yeast phase: implications for the pathogenesis of dimorphic fungal infections. *J. Infect. Dis.* **163:**371–375.

60. **Sullivan, T. D., P. J. Rooney, and B. S. Klein.** 2002. *Agrobacterium tumefaciens* integrates transfer DNA into single chromosomal sites of dimorphic fungi and yields homokaryotic progeny from multinucleate yeast. *Eukaryot. Cell* **1:**895–905.

61. **Sweigard, J.** 1997. A series of vectors for fungal transformation. *Fungal Genet. Newsl.* **44:**52–53.

62. **Wüthrich, M., H. I. Filutowicz, and B. S. Klein.** 2000. Mutation of the WI-1 gene yields an attenuated *Blastomyces dermatitidis* strain that induces host resistance. *J. Clin. Investig.* **106:**1381–1389.

63. **Wüthrich, M., H. I. Filutowicz, T. Warner, and B. S. Klein.** 2002. Requisite elements in vaccine immunity to *Blastomyces dermatitidis*: plasticity uncovers vaccine potential in immune-deficient hosts. *J. Immunol.* **169:**6969–6976.

64. **Wüthrich, M., H. I. Filutowicz, T. Warner, G. S. Deepe, Jr., and B. S. Klein.** 2003. Vaccine immunity to pathogenic fungi overcomes the requirement for CD4 help in exogenous antigen presentation to CD8$^+$ T cells: implications for vaccine development in immune-deficient hosts. *J. Exp. Med.* **197:**1405–1416.

Molecular Principles of Fungal Pathogenesis
Edited by Joseph Heitman et al.
©2006 ASM Press, Washington, D.C.

Chapter 28

Virulence Factors in Black Molds with Emphasis on Melanin, Chitin, and *Wangiella* as a Molecularly Tractable Model

PAUL J. SZANISZLO

Wangiella dermatitidis is an asexual polymorphic mold that is arguably the most studied black fungal pathogen of humans. This emphasis results from its designation as the paradigm for phaeohyphomycosis, the emerging mycosis it causes, and its growing status as a model for the numerous other black fungi known to cause human disease (28, 74, 81, 120, 123, 124). The latter is particularly useful for medical mycology because increasing numbers of infections by each species are being diagnosed, the list of agents continues to lengthen, and the mycoses caused by most black fungi are notoriously troublesome, often require surgery, and frequently are associated with failed antifungal therapy and death (2, 4, 13, 15, 16, 18, 24, 29, 36, 47, 49, 52, 55, 56, 57, 64, 68, 69, 71, 77–81, 86, 88, 94, 96, 97, 99, 102–104, 108, 119, 130–133). A model is also useful because, while no one species produces a large number of infections each year, collectively the black fungi give rise to many worldwide cases of primary and opportunistic disease. Most common among these are the well-characterized forms of cutaneous and subcutaneous primary phaeohyphomycosis, chromoblastomycosis, and black grain mycetoma, in addition to the many secondary types of phaeohyphomycosis that continue to emerge and manifest as serious systemic, neurotropic, and cerebral disease. The clinical disease diversity and relatively low clinical incidence of mycosis attributable to any one black fungus predict that detailed study of even one representative of each disease type is unlikely. They further suggest that the model approach is not only useful but also essential for increasing our understanding of the black fungal pathogens of humans. The purpose of this chapter is to provide a brief overview of these fungi, summarize certain salient historical facts about *W. dermatitidis*, including the main attributes that contributed to its rise to model status, and then review results of recent molecular genetic studies that are primarily, but not exclusively, aimed at the cell wall virulence factors melanin and chitin. The advances described herein are intended to motivate others to determine whether the molecular findings with the model have relevance to the other, less well studied black mold pathogens.

THE BLACK PATHOGENIC MOLDS: WHAT ARE THEY AND WHY THE CONFUSION?

Dark brown to black fungi that cause human infections are known by many names. To emphasize general morphology, they are variously called pathogenic black molds, black pathogenic Hyphomycetes, and pathogenic black yeasts. Use of the black mold and black Hyphomycetes designations is to indicate that the vegetative structures of the fungi under discussion are those of monomophic, conidiogenous black molds or alternative morphotypes or natural mutant forms of known or suspected dimorphic or polymorphic conidiogenous black molds (27, 39). When the black yeast designation is used, it is intended to suggest that the yeast cell is the predominant morphotype of the dimorphic or polymorphic conidiogenous black molds in question, or their only known morphotype (14, 26, 27, 79). Not infrequently, the same fungi are described as dematiaceous or phaeoid molds, Hyphomycetes, yeasts, or fungi (25, 27, 31, 39, 40, 80, 85). Substitution of "dematiaceous" or "phaeoid" for "black" is for mycological reasons, with "dematiaceous" reflecting the historical inclusion of such darkly pigmented fungi in the Hyphomycetes form-family Dematiaceae of the Fungi Imperfecti and the newer term "phaeoid" reflecting the opinion that "dematiaceous" is a traditional misnomer (69, 81, 83, 105). However, placement of fungi

Paul J. Szaniszlo • Section of Molecular Genetics and Microbiology and Institute of Cell and Molecular Biology, The University of Texas at Austin, Austin, TX 78712-0162.

in the Dematiaceae taxon was never uniform. This often led to the grouping of species that are basically hyaline molds and that become darkly pigmented only late in culture or only produce certain darkly pigmented cell types together with species that are darkly pigmented in all life cycle phases (51, 69). It also led to the exclusion of some species, because their conidiogenous mold phases are not readily apparent or are not known to exist. Because of these ambiguities, which continue to this day, contemporary investigators have begun to define the black fungal pathogens of humans as those that are darkly pigmented in all stages of their life cycles by the constitutive polymerization of 1,8-dihydroxynaphthalene (1,8-DHN) into melanin in their cell walls (23, 31, 85, 142). Phylogenetic analyses of ribosomal DNA (rDNA) sequences largely support this evolving convention and tend to confirm that most constitutively black pathogenic fungi are best circumscribed by the teleomorphic order Chaetotheriales of the Ascomycota whereas those with 1,8-DHN melanin more closely associated with environmental induction or certain developmental stages are likely to be members of the Ascomycota orders Dothidiales or Pleosporales (51, 118). The molecular analyses also show that the most common black fungal pathogens of humans, such as those defined by the Fungi Imperfecti genera *Cladophialophora* (previously *Cladosporium*), *Exophiala*, *Fonsecea*, *Phialophora*, *Ramichloridium*, *Rhinocladiella*, and *Wangiella*, no matter what they are collectively called, are closely related members of a distinct clade of conidiogenous black molds, or derived forms of conidiogenous black molds, which have their closest relationships to the ascomycetous teleomorphic genus *Capronia* of the Herpotrichiellaceae (51). It is this particular fungal clade for which *W. dermatitidis* is the best model.

W. DERMATITIDIS: BRIEF HISTORY, AND WHY TWO CURRENT ACCEPTABLE NAMES?

The 1,8-DHN melanin biosynthetic pathway was first identified among the black fungal pathogens of humans in *W. dermatitidis* almost 50 years after the initial isolation and description of the species in 1937 by Kwaiichero Kano (45, 65). Kano reported that his isolate, which he named *Hormiscium dermatitidis*, came from a skin lesion of a Japanese woman who had chromoblastomycosis. Subsequent descriptions by others of morphologically similar clinical isolates from patients with cutaneous and disseminated disease, however, suggested that none, including the Kano strain, was an agent of chromoblastomycosis and that all were better accommodated by another genus (6, 14, 19, 40, 82). Of these two problems, the first was resolved in 1974 by

Ajello et al., who coined the term "phaeohyphomycosis" to describe cutaneous, subcutaneous, and systemic mycoses caused by black fungi that form septate hyphal elements in lesions (2). The definition of phaeohyphomycosis was then periodically broadened to convey the notion that subcutaneous phaeohyphomycotic lesions are cystic and not granulomatous in the manner of those produced in chromoblastomycosis and that when the fungus is observed in a phaeohyphomycotic lesion, it rarely has the characteristics of the thick-walled, muriform (multicellular) morphotypes called sclerotic cells and sclerotic bodies that are so closely associated with the subcutaneous lesions produced in humans by a chromoblastomycosis agent (20, 69, 81, 84). In contrast, the question of what is the best genus name for isolates determined to be the same species as the Kano strain persists (51). The result is that some investigators currently use the binomial *W. dermatitidis* and others use the binomial *Exophiala dermatitidis* for the same strains. This practice is further complicated by the previous assignments of the same strains to the Fungi Imperfecti genera *Aureobasidium*, *Fonsecaea*, *Hormiscium*, *Hormodendrum*, *Phialophora*, *Rhinocladiella*, and *Torula* (63, 69, 115).

The nomenclatural uncertainty about what is the best Fungi Imperfecti species name for isolates similar to the Kano strain results from their highly variable natures (28, 63, 78). From a taxonomic perspective, the main variable is the tendency of the strains to produce conidia by both phialidic and annellidic mechanisms and in different proportions (28). This unique characteristic led McGinnis to create the new genus *Wangiella* for the strains (82). However, de Hoog instead assigned them to the existing genus *Exophiala*, which accommodates other black molds that produce annellidically borne conidia (27). Because the resulting new combinations of *W. dermatitidis* and *E. dermatitidis* were both properly entered into the literature in the same year, each has equal taxonomic priority. For this reason, and although a strong case can be made for the use of either species name, the binomial *W. dermatitidis* is used in this chapter. The author chooses to do this because he has consistently applied that name to the parental strain (NIH strain 8656, ATCC 34100 [CBS525.76]) used in his work, after originally using the binomial *Phialophora dermatitidis*, and with the caveat he recognizes that strains of this species subjected to rDNA phylogenetic analysis cluster closely with many *Exophiala* species (48, 51, 100, 101, 109, 125). Possibly the indecision about which binomial is most appropriate for the strains having the characteristics of the original Kano strain will not be resolved to everyone's satisfaction until ruled upon by appropriate international committees or until its teleomorph, if it exists, is discovered.

ATTRIBUTES THAT CONTRIBUTED TO THE MODEL CONCEPT

Phenotypic variability in *W. dermatitidis* is even more pronounced during vegetative growth than during conidiogenesis. In the manner of a number of other human-pathogenic fungi, *W. dermatitidis* expresses two basic types of polarized growth that lead to bud formation by cells of its yeast phase or apical extension by its hyphal phase, respectively. Additionally, a third type of nonpolarized, isotropic growth is sometimes expressed, which produces a variety of spherically enlarged cell types that may become multicellular by internal septation (28, 63, 67, 69, 95, 98, 122, 123, 125). Variation in the degree of simultaneous expression of the three basic growth modes produces a number of more specific morphotypes (Fig. 1).

Among the many morphotypes of *W. dermatitidis*, the most common is a rather ordinary-appearing yeast cell that is produced in most nutrient-rich culture media and in nutrient-rich tissue environments, such as those of the blood and brain. The predominance of the yeast morphotype is most responsible for the frequent use of *W. dermatitidis* as a model for the black mold pathogens (48, 58, 67, 92, 98, 101, 109–112, 124, 145–147). This is because the yeast morphotype can be produced in homogeneous populations and its growth can be exactly quantified by standard single-cell techniques instead of the less accurate and more time-consuming methods needed to evaluate mold growth. In addition, the yeast morphotype can be synchronized for cell cycle studies or otherwise manipulated experimentally to carry out coordinated phase transitions to hyphae or to isotropic forms (Fig. 2) (58, 59, 101, 109–112). Each morphotype can then be used for studies of its biology and virulence, as well as to determine the regulatory factors and mechanisms responsible for the transitions of one morphotype to another. Furthermore, because the yeast morphotype of *W. dermatitidis* is uninucleate and haploid, it is very amenable to mutagenesis by chemicals, UV light, and various molecular genetic methods (22, 23, 44, 45, 106, 109, 120). By these mutagenic means, numerous auxotrophic, temperature-sensitive (ts), morphological, and knockout (Δ) mutants became available, many of which have helped to establish the *W. dermatitidis* model and to identify its virulence determinants. Last, but certainly not least, the yeast cell wall is easily digested enzymatically, which among other things facilitates parasexual genetic analyses by protoplast fusions of this asexual species

and molecular transformations (22, 106, 149). However, in contrast to the methods used with many other filamentous fungi, protoplasting is not required for the molecular transformation of *W. dermatitidis*. Instead, the intact yeast cell is easily transformed by electroporation (70, 152).

By subjecting yeast populations to conditions that are less nutrient rich or do not support maximum yeast growth rates, a number of factors were identified that can induce *W. dermatitidis* to produce one or the other of its many alternative morphotypes (Fig. 2) (28, 67, 122, 123, 125, 126). The identification of these factors led to the realization that the polymorphism of *W. dermatitidis* is not a liability but instead is a major asset that can be exploited in a model black mold pathogen (123). The hypothesis that resulted is as follows: because this single fungus can be induced to form in vitro any morphotype of any other black pathogenic mold, a discovery of a factor responsible for that induction in the model, or about the morphotype itself, would have relevance to one or more of the other species. This ability to study each morphotype in the absence of another is particularly important for a model black mold pathogen, because some are hyphal in nature but exist almost exclusively as sclerotic cells and sclerotic bodies in human tissue, others are hyphal both in nature and in tissue, and still others, such as *W. dermatitidis* itself, exist predominantly as yeast in nature but are seen as mixtures of yeast, hyphae, pseudohyphae, moniliform hyphae, and isotropically enlarged cells and multicellular bodies in tissue (Table 1).

A seminal early discovery that suggested the potential of a model concept for *W. dermatitidis* was the serendipitous discovery that when its yeast cells are cultured in extremely acidic (pH 2.5), rich media, they stop reproductive growth by budding but nonetheless remain able to continue growth by cell enlargement, carry out mitoses, and often execute cytokinesis (125). The isotropically enlarged morphotypes that result are usually thick-walled, often multinucleate, and frequently internally partitioned with one septum or multiple intersecting septa. The similarity of these morphotypes to the sclerotic cells and sclerotic bodies of chromoblastomycosis agents observed in human tissue lesions prompted additional studies, which showed that extreme acidity also induces those predominately hyphal fungi to produce relatively homogeneous populations of sclerotic cells and sclerotic bodies in vitro (20, 122, 125). Although the mechanistic basis for the inductions was not obvious, these results suggested that *W. dermatitidis* was a model not only for agents of phaeohyphomycosis but also for agents of chromoblastomycosis (122, 125). Supporting this concept were results from light and

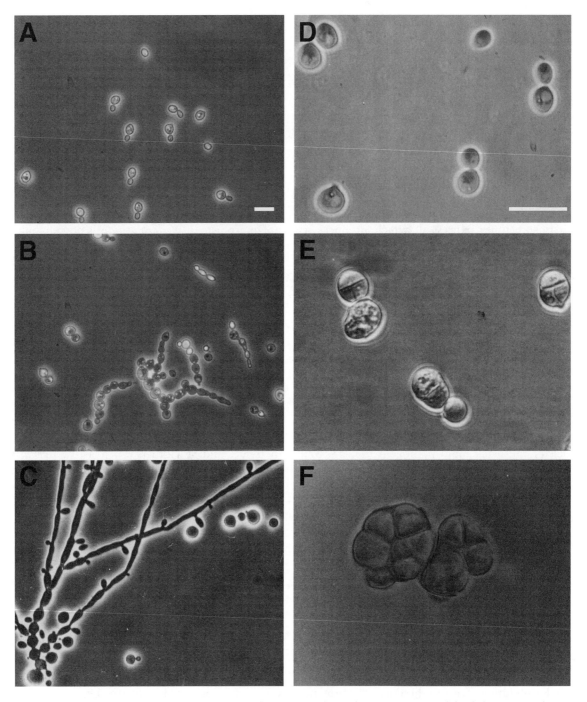

Figure 1. Representative morphotypes of *W. dermatitidis*. (A) Logarithmic-phase yeast; (B) pseudohyphal yeast; (C) sclerotic cell that first produced a branching moniliform hypha which in turn produced true hyphae, both of which are producing yeast-like cells (blastospores) at septal regions that are in various stages of disarticulation; (D) cells in panel A at about ×3 higher magnification; (E) isotropically enlarging sclerotic cells and sclerotic bodies in various stages of septation; (F) a multicellular sclerotic body produced in acid-rich medium undergoing fission. The bar in panel A represents about 5 μm and also applies to panels B and C, whereas the bar in panel D represents about 15 μm and also applies to panels E and F.

electron microscopy studies which showed that the isotropic morphotypes of both types of fungi are virtually identical (20, 122, 125, 126). Additional results from physiological studies, which as yet have failed to identify another black pathogen that produces a sclerotic morphotype in response to extreme acidity, are also supportive. For example, common mechanisms of developmental control for both types of pathogens were indicated by finding that addition of proper Ca^{2+} concentrations to rich acidic media

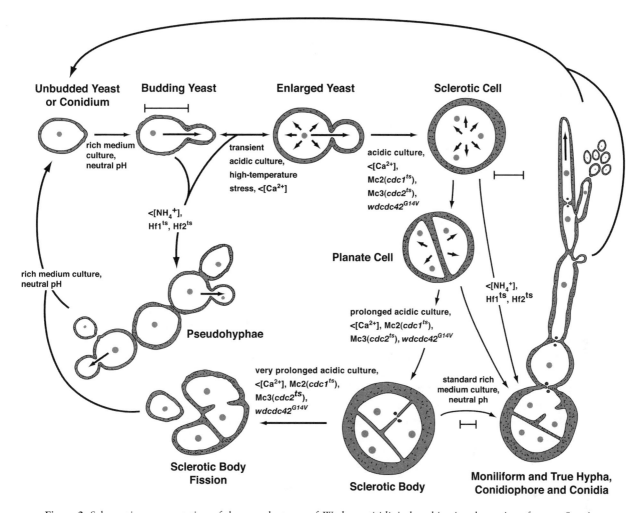

Figure 2. Schematic representation of the morphotypes of *W. dermatitidis* induced in vitro by various factors. Starting with a polarized budding yeast cell or conidium, some conditions induce pseudohypha development because of the lack of yeast cell separation while others induce the loss of polarity that leads to transient or prolonged isotropic development. The latter produces the morphotypes known as sclerotic cells, planate cells, and sclerotic bodies. Return of the isotropic forms to conditions conducive to normal yeast growth usually leads to polarity reestablishments that result in moniliform and true hyphal outgrowths. Blastoconidia are then produced directly from hyphae and phialoconidia and annelloconidia are produced from conidiophores, both of which can reestablish budding yeast populations. The morphotypes are not drawn to scale (note the differences in scale bars): the sclerotic cells and sclerotic bodies are often three to many times larger than mother yeast cells or conidia and usually have thicker walls enriched with both melanin and chitin. The ~ in the cell walls indicates chitin. Adapted from reference 120, Fig. 1.

reverses the inhibition of yeast budding in *W. dermatitidis* and of hyphal apical extension by the chromoblastomycosis fungi and that withholding Ca^{2+} at pH 6.5 can induce both pathogens to produce sclerotic morphotypes (67, 90). The coupling of the morphological, cytological, and physiological results with the knowledge that the pathogenic black molds are constitutively melanized by 1,8-DHN melanin, are members of a unique clade, and other results too numerous to review here suggests that *W. dermatitidis* will continue as a model for these fungi for some time. This last contention is supported by the knowledge that no other black mold besides *W. dermatitidis* is

yet molecularly tractable and amenable to molecular transformation, targeted gene disruption, molecular complementation, and site-specific gene expression experiments, all of which are being used, as reviewed below, to study the biology and virulence of this model.

1,8-DHN MELANIN IS A VIRULENCE FACTOR

Hypopigmented Mel Mutants

Elucidation of the 1,8-DHN melanin biosynthetic pathway in *W. dermatitidis* initially resulted from the use of the specific pathway inhibitor

Table 1. Selected etiological agents of mycoses caused by black molds[a]

Mycosis	Etiological agent(s)	Vegetative phenotypes (in vivo and/or in vitro)
Black grain mycetoma	*Exophiala jeanselmei*	Yeasts, hyphae
Phaeohyphomycosis	*Exophiala bergeri*	Yeasts, hyphae
	Exophiala jeanselmei	Yeasts, hyphae
	Exophiala spinifera	Yeasts, hyphae, pseudohyphae
	Phialophora parasitica	Yeasts (rare), hyphae
	Phialophora richardsiae	Hyphae
	Ramichoridiam mackenziei	Hyphae
	Wangiella dermatitidi	Yeasts, pseudohyphae, moniliform and true hyphae, sclerotic cells, planate cells, sclerotic bodies
	Cladophialophora bantiana	Hyphae
Chromoblastomycosis	*Sarcinomyces phaeomuriformis*	Sclerotic cells, sclerotic bodies
	Cladophialophora carrioni	Hyphae, sclerotic cells, sclerotic bodies
	Fonsecaea pedrsosi	Hyphae, sclerotic cells, sclerotic bodies
	Phialophora verrucosa	Hyphae, sclerotic cells, sclerotic bodies
	Rhinocladiella aquaspersa	Hyphae, sclerotic cells, sclerotic bodies

[a]Data from references 51, 69, 83, and 84.

tricyclazole and the identification of key pathway intermediates, secretion products, and enzymes of the wild-type strain and a few mutant strains (Mel) with obvious pigmentation defects (45). This approach, whose principles were established during the initial description of the pathway in plant pathogenic fungi (5, 142), subsequently allowed its identification in numerous other species, including numerous other black mold pathogens (3, 10, 127). Unfortunately, identification of the pathway in other black molds is based mostly on results with tricyclazole, and not from studies of new hypopigmented Mel mutants. This near-exclusive use of tricyclazole is surprising, because with *W. dermatitidis*, Mel mutants often arise spontaneously or are easily derived by chemical and UV light mutagenesis (23, 44, 45). Hypopigmented Mel mutants of *W. dermatitidis* are also easily identified visually because they form colonies that are albino or shades of red-brown, tan, light brown, or dark brown but not as black as the wild type. The lack of color or the production of the different shades depends on where in the 1,8-DHN melanin biosynthetic pathway the mutation causes the enzymatic block and the types of pathway intermediates that a mutant accumulates and secretes (35). The dearth of Mel mutants of other species is even more surprising, because all hypopigmented mutants of *W. dermatitidis* are less virulent when tested in mouse models of acute infection and are less resistant to killing by human neutrophils (31–35). Thus, at this point, it can only be assumed that the 1,8-DHN melanin deposited in the cell walls of other black pathogens affects virulence in the same ways as it does in the model.

The early studies of melanin biosynthesis in *W. dermatitidis* suggested that the first cyclized product in the 1,8-DHN melanin pathway is 1,3,6,8-tetrahydroxynapthhalene (1,3,6,8-THN). The penta-ketide 1,3,6,8-THN is then converted by enzymatic steps consisting of reduction to scytalone, dehydration of the scytalone to 1,3,8-trihydroxynapthalene (1,3,8-THN), reduction of the 1,3,8-THN to vermelone, and dehydration of the vermelone to 1,8-DHN, which is then polymerized by an uncharacterized oxidation reaction in the cell wall to 1,8-DHN melanin (23, 45). Evidence for the cell wall location of the 1,8-DHN melanin in *W. dermatitidis* is largely from transmission electron microscopy and chemical studies of isolated cell walls. The transmission electron micrographs show that an outer wall layer present in the wild type, which is presumed to be made electron opaque by the 1,8-DHN melanin, is absent from the walls of an albino (*mel3*) mutant (143). The opaque layer can be restored by providing the albino mutant with a pathway intermediate synthesized downstream from the mutational block. Providing additional evidence is the observation that all the dark pigment in the yeast and sclerotic morphotypes, which ranges from 5 to 25% by weight, respectively, remains with the cell walls during isolation (124). Thus, the 1,8-DHN melanin in *W. dermatitidis* not only is a virulence factor but, more precisely, is a cell wall-related virulence factor. The 1,8-DHN melanin probably protects by neutralizing oxidants in tissue environments, such as might be encountered by *W. dermatitidis* exposed to oxidative bursts in neutrophils (60, 61, 114).

Decreased virulence due to defective 1,8-DHN melanin biosynthesis in *W. dermatitidis* was initially detected in the hypopigmented Mel strain *mel3*, which is albino because of an inability to synthesize 1,3,6,8-THN (33, 35). Similar virulence decreases were subsequently detected in other strains, including *mel1*, which is red-brown because it cannot convert 1,3,8-THN to scytalone and thus accumulates and secretes scytalone and a variety of branch pathway metabolites, and *mel2*, which is more brown than black because it is unable to polymerize 1,8-DHN into 1,8-DHN melanin (32, 35). While strongly suggesting that an inability to synthesize 1,8-DHN melanin is responsible for the decreased virulence of hypopigmented Mel strains, these results were not considered conclusive evidence that 1,8-DHN melanin is a virulence factor, because the Mel mutants could not be genetically complemented. To circumvent this problem initially, two elegant physiological experiments were carried out (31). In one, the *mel3* albino mutant was turned black by feeding it the downstream 1,8-DHN melanin pathway intermediate scytalone and then demonstrating that the mutant had regained virulence, whereas in the other, the wild-type strain was shown to lose virulence when its ability to synthesize 1,8-DHN melanin was inhibited with tricyclazole (31). More recently, even stronger evidence for the importance of 1,8-DHN melanin to virulence in *W. dermatitidis* was obtained with a different set of hypopigmented Mel strains (43). In this case, the new Mel strains were derived by molecular genetic disruption of *WdPKS1*, the gene that encodes the polyketide synthase (WdPks1p) proposed to cyclize 1,3,6,8-THN in the 1,8-DHN biosynthetic pathway (43). The resulting *wdpks1Δ* mutants are albino, less virulent in mouse survival tests, and less resistant to killing by neutrophils and have all the other characteristics of the older *mel3* strain, including the ability to grow at wild-type rates at both 25 and 37°C (35, 43, 114). More importantly, molecular complementation of the disruption strain with the wild-type *WdPKS1* gene, which was cloned both by a marker rescue strategy and also by cosmid complementation of *mel3*, produces strains with wild-type characteristics, including comparable abilities to kill mice and to resist killing by neutrophils (43). These data represent the best formal proof that the 1,8-DHN melanin of at least one black mold pathogen of humans is a virulence factor.

WdPKS1 is the only 1,8-DHN melanin biosynthetic gene of *W. dermatitidis* characterized in detail (43). Like its homologs in other fungi, *WdPKS1* is very large, has introns, and encodes a type I polyketide synthase with conserved domains for a β-ketoacyl synthase, an acetyl-malonyl transferase, two acyl carrier proteins, and a thioesterase. It is assumed that other 1,8-DHN melanin biosynthesis structural genes and their encoded proteins would be similar to those found in other fungi and that mutants with those genes disrupted would have virulence decreases

similar to those of other hypopigmented Mel strains. Nonetheless, attempts to clone those genes have been largely without success (42). One exception is *WdYG1*, a gene with homology to *AYG1* of *Aspergillus fumigatus* (Q. Cheng, M. H. Wheeler, and P. J. Szaniszlo, unpublished data). As with *WdPKS1*, *WdYG1* was cloned by marker rescue, but in this case from a hypopigmented Mel strain produced by the ectopic integration of a vector designed to disrupt *WdURA5*, which encodes the orotidine monophosphate pyrophosphorylase (WdUra5p) of *W. dermatitidis* (152). Further study of this mutant was prompted by its secretion of pigments ranging in color from yellow to brown depending on the culture medium. Among these, one has the spectral characteristics of the heptaketide pigment YWA1, which in *A. fumigatus* is the actual product of its polyketide synthase (128). The shortening of YWA1 is then carried out by Ayg1p to generate 1,3,6,8-THN, the pentaketide that in *A. fumigatus* is an upstream precursor of its green conidial pigment. The discovery of a gene in *W. dermatitidis* that encodes an ortholog of Ayg1p indicates that a ring-shortening step is not unique to the conidial pigments of some aspergilli but may be a more universal step in the biosynthesis of 1,8-DHN melanin.

It is important to note that some albino Mel strains become pink to orange when exposed to light of relatively high intensity for extended periods (23, 44). The pigments responsible include the carotenoids torulene and torularhodin. Although no photo-induced carotenoid biosynthetic system has been studied extensively in any dematiaceous fungus, tests of one *W. dermatitidis* albino strain (Mel4/*mel3-2*, ATCC 58058) and the wild type showed that light-grown yeast cells of both are more resistant to the killing effects of UV radiation than are dark-grown cells. However, these tests also showed that of the two types of pigments, the 1,8-DHN melanin of *W. dermatitidis* contributes more to photoprotection than do the carotenoids. Comparisons of the efficacy of the two pigment types in protecting *W. dermatitidis* against the killing effects of oxidative bursts in human neutrophils similarly suggest that the carotenoids contribute little to protection whereas significant protection is afforded by 1,8-DHN melanin (114).

Hyperpigmented Mel Mutants

1,8-DHN melanin biosynthesis in *W. dermatitidis* is considered constitutive because the fungus is darkly pigmented in all life cycle phases. Nonetheless, some data suggest that the product of *WdPKS1* participates in a cell wall integrity pathway, because mutants with defects in cell wall biosynthesis, or the wild-type strain induced to carry out yeast-to-hyphal-form or yeast-to-sclerotic-form transitions, are often hyperpigmented (35, 42, 73, 120, 140). It is suspected that a feedback loop is sometimes freed from its basal expression level when the cell wall is perturbed, which allows WdPKS1 to be transcribed at higher levels (Z. Wang and P. J. Szaniszlo, unpublished data). Although the hyperpigmentation could potentially increase virulence or resistance to phagocytic killing, few experiments have addressed these possibilities directly and all of those have yielded negative results. For example, no virulence difference is apparent between the wild-type strain and the ts morphological mutant Mc3 (wdcdc2, ATCC 38716), even though this mutant converts in mice from the yeast morphotype of the inoculum to the sclerotic morphotype, because of the higher mouse body temperature, and sclerotic cells and sclerotic bodies have considerably more melanin (34, 124). In contrast, an albino variant of Mc3 (Mc3W) (23) is less virulent than either the wild type or the Mc3 strains in identical tests. Similarly, a ts mutant, Hf1, which converts from a yeast to a hyphal morphology at temperatures of infection, also seems no more virulent in mice than does the wild type, even though it too produces darker colonies in culture (87; N. D. P. McIntosh, J. R. Graybill, and P. J. Szaniszlo, unpublished data). Finally, no loss of virulence is associated with the disruption of certain chitin synthase genes (WdCHS), such as WdCHS1 and WdCHS4, which encode two of the five chitin synthases (WdChsp) known to exist in W. dermatitidis, even though the resulting mutants are also hyperpigmented (140; S. Kauffman, J. M. Becker, and P. J. Szaniszlo, unpublished data). In spite of these negative results, it remains tempting to suggest that hyperpigmentation increases the virulence of W. dermatitidis. Possibly the detection of increased virulence requires better assays.

While increased virulence is not correlated with the disruption of either WdCHS1 or WdCHS4, yeast cells of wdchs1Δ and wdchs4Δ mutants are not only hyperpigmented but also more pseudohyphal or multiply budded, respectively, than those of the wild type. These data suggest that 1,8-DHN melanin biosynthesis participates in a cell wall integrity pathway activated by abnormal chitin biosynthesis (120, 140). Supporting this possibility is the finding that mutants with both WdCHS1 and WdCHS2 disrupted in the same background or with WdCHS5 disrupted alone, which are even more aberrant morphologically than wdchs1Δ and wdchs4 mutants, are also hyperpigmented (73, 120). However, to date only one gene of W. dermatitidis has been cloned from a strain selected for study because it is hyperpigmented as a result of insertional mutagenesis (42). The derived protein of this particular gene, which was cloned by marker rescue, has the highest amino acid identity (44%) to Mot1p, which in Saccharomyces cerevisiae is an essential ATP-dependent negative regulator of basal transcription and at least in vitro removes TATA box-binding protein from TATA sites (1, 120). Surprisingly, unlike its counterpart in S. cerevisiae, WdMOT1 is not essential for viability. This has allowed the wdmot1Δ mutant to be characterized beyond hyperpigmentation. In this respect, such mutants produce considerably more invasive hyphae and express WdPKS1 at considerably higher levels than the wild type does (42). Apparently, the product of WdMOT1 acts to repress filamentous growth in W. dermatitidis and, when defective, allows abnormal cell wall synthesis patterns that lead to filamentation and to hyperpigmentation.

VIRULENCE FACTOR POTENTIAL OF CHITIN

When experiments are extended beyond 25 or 30 days, deaths among mice infected with hypopigmented Mel strains of W. dermatitidis often occur in a slow, sometimes linear fashion (31–35). Even surviving mice frequently show signs of central nervous system damage, such as chronically running in circles or exhibiting torticollis and ataxia. However, no significant differences are apparent in the number of CFU cultured from the brains of mice infected with either the wild type or the albino mel3 strain or, for that matter, the number and sizes of the lesions produced. This observation implies that factors other than 1,8-DHN melanin contribute to the virulence of W. dermatitidis. Among the many possibilities, chitin is receiving the most attention (120, 121). The original rationale for directing attention to chitin originated from three early observations. First, the cell wall content of this homopolymer of $\beta(1\rightarrow4)$-linked N-acetylglucosamine, in the manner of melanin, is significantly enriched in most of the alternative morphotypes of W. dermatitidis compared to that of its budding yeast (21, 121, 124). Second, the additional chitin is delocalized from predominantly yeast septal regions to all cell wall locations in hyphae and sclerotic forms (21, 53, 59, 87, 121). Third, even the walls of budding yeast are frequently enriched with dispersed chitin. This happens when budding yeast are temporarily arrested in normal cell cycle progression, which usually induces a period of isotropic growth with its concomitant burst of delocalized chitin synthesis (59, 121, 126). These outcomes all indicate that chitin could act as a virulence factor by adding additional strength to the cell walls of W. dermatitidis whenever conditions are encountered in hosts that temporarily retard rapid yeast growth or induce transitions of yeast cells to other morphotypes.

Host encounters that slow rapid yeast growth and bring about chitin enrichments in *W. dermatitidis* are probably common. Tissue sections from patients and animals with chronic cutaneous and subcutaneous infections frequently show yeast and hyphae with abnormally thickened walls (49, 56, 78, 79, 81, 95). Even brain and other tissues from systemically infected humans and animals with rapidly progressing disease often show many types of thick-walled forms, which are variously described as yeast, chains and clusters of yeast, sclerotic cells, planate cells, and cells with internal septa (69, 78, 79, 95). Whether these different in vivo tissue forms are actually enriched with chitin is not known, but chemical and cytochemical analyses of identical morphotypes produced in vitro indicate that they are (21, 121, 124). For example, chemical analyses of isolated cell walls of sclerotic forms and of intact sclerotic forms show that, compared to log-phase yeast, fivefold or more enrichment with chitin is typical. Various microscopic analyses also indicate that the sclerotic morphotypes are enriched significantly with chitin. These include analyses by transmission electron microscopy with a chitin-specific chitinase–colloidal-gold affinity label and by light microscopy with the fluorescent chitin probe calcofluor, both of which clearly show that the added chitin is well dispersed in all cell wall areas of the isotropic forms (21, 53, 121). Similar studies have not been done with hyphae beyond showing with calcofluor that they too are enriched with chitin in all cell wall areas (87).

Support for the hypothesis that chitin enrichments actually strengthen the cell walls of some *W. dermatitidis* morphotypes first came from studies with polyoxin (21). With this specific inhibitor of chitin synthases, chitin synthesis inhibition does not significantly retard total growth even though normal septum formation and cell separations of yeast are frequently altered, leading to the production of pseudohypha-like forms. Apparently the small amounts of localized chitin present in rapidly dividing yeast can be absent without drastically affecting the basic cell wall integrity to the extent that budding cells lyse or burst. In contrast, inhibition of chitin synthesis results in the death of enlarging sclerotic forms by causing focal-point cell wall ruptures. In this case, when the walls are not strengthened by chitin during isotropic expansion, the sclerotic forms become so fragile that they rapidly burst under the weight of microscope coverslips and in the absence of osmotic stabilization. This finding suggests that the β-glucans and other components of the yeast wall compensate, albeit imperfectly, for the absence of chitin but that the walls of the expanding isotropic forms require chitin strengthening to maintain integrity. While no similar studies have been carried out with hyphae, it is suspected that some chitin strengthening is required to at least maintain hyphal tip integrity during periods of rapid apical extension. Furthermore, because hyphae develop from isotropic forms, it is likely that chitin strengthening is also important to any cell capable of producing a hyphal outgrowth.

CHITIN SYNTHASES AS VIRULENCE FACTORS

Each of the five known WdChsp isozymes of *W. dermatitidis* is a member of a different class, according to the system first established by Bowen et al. and later extended by others (8, 30, 93, 117, 120): WdChs1p (class II), WdChs2p (class I), WdChs3p (class III), WdChs4p (class IV), and WdChs5p (class V). They also are members of two families, with WdChs1p, WdChs2p, and WdChs3p being members of one and WdChs4p and WdChs5p being members of the other (72, 93, 113). In this second respect, then, *W. dermatitidis* is similar to *S. cerevisiae*, which also has multiple chitin synthases distributed between the two families of isozymes (93, 113). However, *W. dermatitidis* is very different from *S. cerevisiae* because the latter has only three chitin synthases and is devoid of enzymes of the class III and class V types (11, 93).

The *WdCHS* structural genes are all very large, have open reading frames ranging in size from 2,658 to 5,658 bp, and, with one exception, have one or two small introns (120). Numerous gene disruption and one-step gene replacement experiments have produced what may be the largest collection of isogenic chitin synthase structural gene mutants (*wdchs*Δ) of a single species (Table 2). Included in the collection are numerous independently derived strains with each *WdCHS* gene disrupted singly, two *WdCHS* genes disrupted in all 10 possible double combinations, and three *WdCHS* genes disrupted in 2 combinations in the wild-type background. The collection also includes a small number of strains with one *WdCHS* gene disrupted in the ts Mc3 (*wdcdc2*) and Hf1 strains, which at the restrictive temperature produce the sclerotic and hyphal morphotypes, respectively (137). Tests of the *wdchs*Δ disruption mutants in mouse models suggest that the products of four of the five *WdCHS* genes directly or indirectly influence virulence. In two cases, a temperature sensitivity is imparted to disruption strains, which then cannot grow well or at all at temperatures of infection, whereas in a third case, the virulence of a disruption strain is reduced without affecting its ability to grow at higher temperatures (73, 74, 139). However, in no case is the mechanism

Table 2. Summary of chitin synthase gene disruptions in wild-type *Wangiella*[a]

Disruption strain	Isozyme class(es) affected	Gene marker(s) used[b]	Phenotypic characteristics produced
wdchs1Δ	II	hph/ble	Short hyperpigmented yeast chains at 25 and 37°C
wdchs2Δ	I	hph/WdURA5	No morphological change, but drastically reduced Chsp activity at 25 and 37°C
wdchs3Δ	III	hph	No morphological change, but reduced Chsp activity at 37°C
wdchs4Δ	IV	hph	Clumped yeasts, multiple buds, less chitin, and hyperpigmented at 37°C
wdchs5Δ	V	hph	No morphological change at 25°C, but conditionally lethal and hyperpigmented at 37°C
wdchs1Δwdchs2Δ	I, II	ble and hph	Abnormal hyperpigmented pseudohyphal-like growth with defective septa at 25°C, conditional lethal at 37°C
wdchs1Δwdchs3Δ	II, III	ble and hph/hph and ben	Like wdchs1Δ
wdchs1Δwdchs4Δ	II, IV	hph and sur	Combined phenotype of wdchs1Δ and wdchs4Δ
wdchs1Δwdchs5Δ	II, V	hph and sur	Like wdchs1Δ at 25°C; like wdchs5Δ at 37°C
wdchs2Δwdchs3Δ	I, III	sur and ble	Normal morphology except for greatly reduced chs activity at 37°C; less virulent in mouse models
wdchs2Δwdchs4Δ	I, IV	hph and sur	Combined phenotype of wdchs2Δ and wdchs4Δ
wdchs2Δwdchs5Δ	I, V	hph and sur	Like wdchs2Δ at 25°C; like wdchs5Δ at 37°C
wdchs3Δwdchs4Δ	III, IV	hph and sur	Like wdchs4Δ
wdchs3Δwdchs5Δ	III, V	hph and sur	No morphological change at 25°C; like wdchs5Δ at 37°C
wdchs4Δwdchs5Δ	IV, V	sur and hph	Like wdchs4Δ at 25°C; like wdchs5Δ at 37°C
wdchs1Δwdchs2Δwdchs3Δ	I, II, III	ble and sur and hph	Like wdchs1Δwdchs2Δ, except morphology even more abnormal at 25°C
wdchs1Δwdchs3Δwdchs4Δ	II, III, IV	ble and sur and hph	Combined phenotype of wdchs1Δ and wdchs4Δ

[a]Adapted from Table 2, reference 120.
[b]*hph*, hygromycin B resistance; *ble*, phleomycin resistance; *WdURA5*, orodidine 5'-monophosphate gene used for disruption in the *wdura5Δ* background; *ben*, benomyl resistance; *sur*, sulfonyl urea resistance. Markers separated by slashes (/) indicate the use of more than one marker for disruption, whereas those separated by "and" indicate markers used respectively for double and triple disruptions.

responsible for endowing any mutant with these properties completely understood. Also, two cases involve the disruption of two *WdCHS* genes in the same background (e.g., *WdCHS1* and *WdCHS2* or *WdCHS2* and *WdCHS3*) whereas only one case involves a single gene (*WdCHS5*). Finally, of the three types of mutations, only those that lead to defects in both WdChs2p and WdChs3p together or in WdChs5p alone are known to be solely responsible for the loss of virulence in the *wdchs2Δwdchs3Δ* and *wdchs5Δ* mutants, respectively, because virulence has been reestablished in them by complementation with the corresponding *WdCHS2*, *WdCHS3*, or *WdCHS5* wild-type genes. Similar complementation of a *wdchs1Δwdchs2Δ* mutant has not been accomplished, possibly because of its fragility even at 25°C.

It can be argued that a factor required for the survival of a fungus should not be considered a virulence factor, because a cell must be viable to be virulent (144). Under this scenario, only a nonessential property that lowers virulence, but does not affect the growth rate or viability of a fungus on standard laboratory culture media, would qualify. By these standards, melanin is clearly a virulence factor in

W. dermatitidis. However, the situation with respect to chitin is blurred, because some WdChsp are dispensable for growth at 25°C but not at 37°C. For example, all single *wdchsΔ* strains and most double *wdchsΔ* strains grow well at 25°C and exhibit no or only minor morphological differences compared to the wild type. The two exceptions that blur the classification of some WdChsp as bona fide virulence factors are the requirements of WdChs5p alone or either WdChs1p or WdChs2p for growth at 37°C. Of the two types of *wdchsΔ* mutants, those without a functional WdChs5p are the most interesting, because while they grow at wild-type rates at 25°C, yeast at 37°C slowly lose shape and die in late log or early stationary phase if not protected by osmotic stabilizers (73). Similarly, *wdchs1Δwdchs2Δ* mutants also do not survive at 37°C without osmotic protection (H. Liu and P. J. Szaniszlo, unpublished data). Only in the case of mutants with both *WdCHS2* and *WdCHS3* disrupted in the same background is virulence lost without growth being affected at 25 or 37°C. Nonetheless, and in spite of one's view of how to define a fungal virulence factor, a picture is emerging about how each WdChsp contributes to the viability, growth, morphology, and virulence of *W. dermatitidis*. The additional findings about each WdChsp reviewed below suggest that the regulatory circuits controlling the expression and function of the chitin synthases with no orthologs in *S. cerevisiae* and *Candida albicans*, and particularly WdChs5p, may be exceptionally vulnerable targets for antifungal drugs with efficacy against the black molds. They also may suggest ways in which others might determine if the discoveries made about chitin and the chitin synthases of *W. dermatitidis* have relevance to other pathogenic black molds, at least seven of which are already known to have multiple chitin synthase genes (8, 66, 107). Deduced amino acid sequences of DNA fragments amplified by PCR indicate that these genes encode isozymes of classes I, II, III, and IV, although to date none is known to have more than three. Also, no evidence as yet suggests whether any has a class V-type chitin synthase. Nonetheless, it is suspected that additional *CHS* genes, including ones that encode class V isozymes, would be found in these and additional black pathogenic molds should they be investigated to the extent of the model.

CELL BIOLOGY OF THE WdChsp ISOZYMES

Determining whether any WdChsp contributes to virulence required the development of a variety of molecular genetic protocols, because prior to 1992 virtually none existed for *W. dermatitidis* (17, 42, 70, 89, 106, 120, 135, 136, 148, 150, 151). As mentioned already, methods for molecular transformations and targeted gene disruptions are now at hand and have been used to produce many *wdchs*Δ disruption strains. Tests show that some are less virulent in acute mouse models than are their parents. Confirmation that the loss of virulence results solely from the disruption of a single gene and that the virulence can be restored by the wild-type version of that gene similarly required the development of complementation methods. These too are now available and have confirmed that WdPks1p and possibly as many as four WdChsp are virulence factors. Many other molecular biology-based tools were also required to begin to determine how each WdChsp might mechanistically contribute to the virulence and biology of *W. dermatitidis*. Most were developed by modifying the original basic transformation, gene replacement, integrative gene disruption, and complementation protocols. With these methods, new mutants are routinely created by random and targeted genomic vector integrations and then the disrupted genes are cloned by selection marker rescue strategies. In addition, gene sequences are frequently expressed and overexpressed after site-specific or random genomic integrations and fusion proteins monitored by the β-galactosidase assay, Western analysis, or fluorescence after gene sequences are added for *lacZ*, *myc*, and HA or for green fluorescent protein, respectively. The characterizations reviewed below of the *WdCHS* genes, WdChsp proteins, and *wdchs*Δ mutants, which have been derived by using these protocols, are beginning to reveal how the five *WdCHS* gene products of *W. dermatitidis* coordinately contribute to the normal biology of the model and to suggest which classes of chitin synthase isozymes of other black molds might most warrant similar investigations.

WdCHS1 and WdChs1p

W. dermatitidis was among the first molds of any type shown to have multiple chitin synthase structural genes (8). The initial identification of *WdCHS1*, as well as of *WdCHS2* and *WdCHS3*, was made by sequencing cloned PCR products amplified from the genomic DNA from 14 fungi with degenerate primers designed on the basis of conserved regions of Chs1p and Chs2p of *S. cerevisiae* and CaChs1p of *C. albicans*. Alignments of the 32 predicted amino acid sequences of the PCR products obtained suggested that they fell into three classes and that the *WdCHS1* fragment encoded a class II member, because it aligned in a group containing Chs2p of *S. cerevisiae*. Similarly amplified PCR fragments of *WdCHS1* were then used as a probe to clone the first full-length gene of *W. dermatitidis* and to localize it in chromosome 2 of the four CHEF-gel chromosomes of strain 8656 resolved to date (89). The *WdCHS1* open reading frame consists of 2,967 bp, which is interrupted by a 47-bp intron and encodes an isozyme of 988 amino acids with a calculated molecular mass of about 111 kDa and a putative pI of 8.46 (89, 120, 151). Transcription of *WdCHS1* is only slightly elevated by culture at 37°C instead of 25°C (134). At both temperatures, strains with *WdCHS1* disrupted also grow at wild-type rates, which documents that WdChs1p is not encoded by an essential gene (120, 151). However, *wdchs1*Δ mutants are hyperpigmented, frequently form short pseudohypha-like chains of yeast cells (some of which may be abnormally enlarged and contain more than one nucleus), and have chitin enrichments in septal regions. They also show little reduction in chitin synthase activity. These characteristics suggest that WdChs1p is involved primarily in septum formation and yeast cell separation, in the manner of its class II ortholog, Chs2p, in *S. cerevisiae* (11). The swellings exhibited by some *wdchs1*Δ cells and the hyperpigmentation of *wdchs1*Δ colonies further indicate that minor cell wall integrity defects, which are distinct from those associated with *CHS2* mutants of *S. cerevisiae*, also result from *WdCHS1* disruption. It is hypothesized that the product of another *WdCHS* gene compensates for the loss of WdChs1p function and allows good, but not perfect, yeast budding growth. As yet, the effects of *WdCHS1* disruption on hyphal or isotropic growth are not known.

WdCHS2 and WdChs2p

The amino acid alignments that identified the encoded polypeptide of the *WdCHS1* PCR product as a part of a class II chitin synthase similarly revealed that the *WdCHS2* PCR product encoded a portion of a class I chitin synthase (8). The product was then used as a probe to help isolate the full-length gene and to localize it also in chromosome 2 (89, 139, 151). *WdCHS2* is generally similar to *WdCHS1* because it too has only one intron and is transcribed only slightly more extensively at 37°C than at 25°C (120, 134, 139, 151). However, *WdCHS2* is shorter than *WdCHS1* and consists of a 2,787-bp open reading frame with a slightly larger 59-bp intron. The predicted WdChs2p has 928 amino acids, a calculated mass of 106 kDa, and a pI of 7.28. Phylogenetic analyses of the putative protein further suggest that of the five WdChsp isozymes, WdChs2p is most similar to WdChs1p, which may suggest it originated by gene duplication (72).

Strains with *WdCHS2* disrupted have drastically reduced chitin synthase activity (120, 139, 151). Nonetheless, they grow at wild-type rates and show no morphological defects in yeast growth or in the ability to carry out transitions to alternate phenotypes. However, disruption of *WdCHS2* in strains having either *WdCHS1* or *WdCHS3* disrupted, but not *WdCHS4* or *WdCHS5* disrupted, produces two unique double-mutant strains. As mentioned above, the *wdchs2Δwdchs3Δ* mutants show no morphological abnormalities but for unknown reasons have significantly lowered virulence in mouse models of acute infection. In contrast, *wdchs1Δwdchs2Δ* mutants have dramatically altered phenotypes. At 25°C they grow at lower rates than the wild type or each single-disruption mutant, are hyperpigmented in the manner of *wdchs1Δ* mutants, and form considerably longer, branching chains of elongated and enlarged cells similar to those in microcolonies of the wild type treated with polyoxin (21, 120, 140, 151). Most cells of the *wdchs1Δwdchs2Δ* double mutant are also multinucleate and exhibit considerably more intense general and septal region staining than those of the *wdchs1Δ* strain. Furthermore, electron microscopic images substantiate that the *wdchs1Δwdchs2Δ* mutants are more aberrant at septal regions than are mutants with only *WdCHS1* disrupted (L. Mendoza

and P. J. Szaniszlo, unpublished data). Taken together, the phenotypes resulting from the disruption of *WdCHS1* and *WdCHS2* singly and together suggest that their products have overlapping functions required for normal septation and cell separation at 25°C. They also suggest that the functions of WdChs1p and WdChs2p are not completely redundant, because mutants with only *WdCHS1* disrupted have minor but clearly evident septal defects whereas those with only *WdCHS2* disrupted do not. Possibly WdChs2p plays more of a repair role and is a backup of WdChs1p when the latter is defective. Support for this possibility is by analogy to Chs1p and Chs2p in *S. cerevisiae*, which are also class I and class II chitin synthases, respectively. In yeast, Chs2p is responsible for primary septum formation whereas Chs1p, which like WdChs2p is responsible for most of the latent enzymatic activity, acts primarily in an auxiliary capacity to repair defects that might occur during cell separations (11, 12). However, unlike the situation in *W. dermatitidis*, mutants with both Chs1p and Chs2p disrupted in the same background are not dramatically affected morphologically. Finally, even though *wdchs1Δwdchs2Δ* mutants grow at 25°C, albeit abnormally, they have no detectable virulence when tested in mice (Kauffman et al., unpublished). This lack of virulence is most probably because *wdchs1Δwdchs2Δ* mutants do not grow at 37°C unless protected by an osmotic stabilizer (Liu and Szaniszlo, unpublished). Nonetheless, the mechanistic basis for the loss of cell wall integrity of this double mutant at 37°C is not clear.

WdCHS3 and WdChs3p

As reviewed above, a fragment of *WdCHS3* was detected among the same PCR products that were used to identify *WdCHS1* and *WdCHS2* (8). Alignment of the predicted amino acid sequence of the *WdCHS3* fragment with those derived from the PCR products of the other 13 species revealed that it was a portion of a third group of chitin synthases, but not one that included Chs3p of *S. cerevisiae*. This led to the assignment of the WdChs3p-derived polypeptide and five others to the class III chitin synthase group. It also led to the realization that some fungi have at least one chitin synthase gene with no homolog in *S. cerevisiae*. Because of this, *WdCHS3* and WdChs3p received considerable early attention after *WdCHS3* was cloned from a cDNA library (137). The hypothesis driving this attention was that WdChs3p might be a mold-specific virulence factor, because of the absence of class III-type isozymes in nonpathogenic *S. cerevisiae* or pathogenic *C. albicans*. The potential for WdChs3p to act as a virulence factor was heightened by finding that *WdCHS3* transcription is significantly enhanced by shifting of cells from 25°C to temperatures associated with infection, as well as by a number of stress conditions that bring about transitions of yeast cells to hyphae or sclerotic forms (134, 137).

Unlike WdCHS1 and WdCHS2, WdCHS3 resides in chromosome 3 of W. dermatitidis and has two introns instead of one (89, 137). The WdCHS3 open reading frame consists of 2,658 bp, which encodes an isozyme of 885 amino acids with a calculated mass of 99.4 kDa and a pI of 8.47. Fusions of the lacZ gene with truncated portions of the WdCHS3 5′ upstream regulatory sequence (URS) suggest that at least one negative regulatory element exists between bp −780 and −1,600, which is most active at 37°C. Western blots of a myc-tagged WdChs3p fusion also show that WdChs3p itself is produced in increasing amounts commensurate with increases in temperature and also temporally with increases in time after a shift from 25 to 37°C. While the increased WdCHS3 transcript levels and WdChs3p protein levels produced at temperatures of infection strongly suggest that WdChs3p should be a virulence factor, such is apparently not the case. In fact, although disruption of WdCHS3 results in significantly reduced chitin synthase activities in wdchs3Δ mutants cultured at 37°C compared to those of the wild type, no loss of virulence or changes in morphology, growth rates, and chitin content result from this particular mutation.

The early attention directed toward WdCHS3 and WdChs3p was not without merit because, as mentioned above, mutants with WdCHS3 disrupted together with WdCHS2 in the same background are less virulent (139). This finding is surprising, because in the manner of wdchs2Δ and wdchs3Δ single disruption mutants, wdchs2Δwdchs3Δ double mutants exhibit no growth defects and have normal chitin contents, even though they have significantly reduced WdChsp enzymatic activity (137, 139). The loss of virulence associated with the disruption of WdCHS3 in the wdchs2Δ background is even more puzzling in light of the fact that the same disruption in other wdchsΔ mutant backgrounds does not significantly exaggerate the phenotype of the resulting mutant phenotype beyond that of its parent. Even when WdCHS3 is disrupted together with WdCHS1 and WdCHS2 or with WdCHS1 and WdCHS4, the phenotypes of the resulting triple mutants are essentially the same as those of the double mutants, wdchs1Δwdchs2Δ and wdchs1Δwdchs4Δ, respectively (i.e., abnormal at 25°C, death at 37°C, and clumped chains).

Of the five WdChsp proteins of W. dermatitidis, only the activity of WdChs3p is characterized in much detail (138). This is because, as with all other chitin synthases, none has been purified to homogeneity and no quadruple wdchsΔ mutant with only a single WdChsp activity has been derived, possibly because such mutants are not viable. To circumvent these two problems and begin for the first time to characterize a class III chitin synthase, a cDNA of WdCHS3, with and without a hemagglutinin (HA) epitope coding sequence, was heterologously expressed in S. cerevisiae under the regulation of the inducible GAL1 promoter. Under inducing conditions, total chitin synthase activity was increased about 120-fold. However, unexpectedly, this activity does not require trypsin activation. Nonetheless, and even though the WdChs3p activity was undoubtedly contaminated with low levels of S. cerevisiae Chsp activities, the enrichment was considered to be so great that WdChs3p still could be accurately characterized.

In agreement with the molecular mass calculation of WdChs3p, Western blots of the HA-tagged isozyme produced in S. cerevisiae also suggest a size of about 100 kDa (137, 138). Probably because it is already active when isolated, the WdChs3p activity is not increased when treated with trypsin and is, in fact, inactivated by trypsin in a time- and concentration-dependent fashion. In most other respects, the WdChs3p activity profiles are similar to those reported for other Chsp isozymes. For example, the activity is stimulated by Mg^{2+}, Mn^{2+}, and Co^{2+} and interestingly also by Ca^{2+}. It is also inhibited by Cu^{2+} and the divalent-cation chelator EDTA and is sensitive to the competitive inhibitors polyoxin and nikkomycin. Finally, the characterization of the activity produced in S. cerevisiae suggests that WdChs3p may be particularly amenable to further purification because it is stable over a broad range of pH and temperatures.

WdCHS4 and WdChs4p

The survey of PCR products that originally identified fragments of WdCHS1, WdCHS2, and WdCHS3 revealed none having a derived polypeptide orthologous to any portion of Chs3p of S. cerevisiae (8, 9). This is because the degenerate primers used were designed on the basis of conserved regions identified only in the three class I and class II isozymes of S. cerevisiae and C. albicans known at the time. Therefore, new primers that also took into account conserved regions in the class IV isozyme Chs3p of S. cerevisiae were designed and then used successfully to amplify a PCR fragment from genomic DNA of W. dermatitidis, which encoded a polypeptide more similar to portions of Chs3p than to portions of Chs1p or Chs2p (9, 66, 107, 129).

After the full-length gene was cloned by marker rescue, it was named *WdCHS4*, localized in chromosome 4, confirmed like *CHS3* to encode a class IV chitin synthase, and disrupted (140). Unlike the other genes of *W. dermatitidis*, *WdCHS4* has no introns but instead consists of 3,717 bp that encode a product of 1,238 amino acids with a calculated mass of 138.8 kDa and pI of 9.07. Although *WdCHS4* is transcriptionally up-regulated in yeast cells shifted to higher temperatures, the response is slower than that of *WdCHS3* (134).

Disruption of *WdCHS4* produces mutants with reduced chitin contents and yeast cells that tend to clump, are multiply budded, form hyperpigmented colonies, and grow more slowly than the wild-type strain, particularly at 37°C (140). However, chitin synthase activity is not reduced in *wdchs4Δ* mutants, a paradox suggesting that one or more of the other WdChsp proteins compensate when WdChs4p function is defective or absent. The significant increases in *WdCHS3* mRNA and WdChs3p associated with a shift of cells from 25 to 37°C indicate that WdChs3p is likely to be responsible for the compensation (134, 137). However, the increased levels of WdChs3p are not solely responsible for the retention of virulence by *wdchs4Δ* strains, because wild-type virulence is maintained even by a strain with *WdCHS1*, *WdCHS3*, and *WdCHS4* disrupted in the same background (139). Thus, even though this triple mutant has only WdChs2p and WdChs5p, its virulence is normal, in spite of having the combined phenotype of the *wdchs1Δ* and *wdchs4Δ* single mutants (i.e., clumps and chains, respectively [120]). Conversely, WdChs4p does not effectively compensate for the loss of WdChs1p and WdChs2p functions because, as noted above, *wdchs1Δwdchs2Δ* double mutants and *wdchs1Δwdchs2Δwdchs3Δ* triple mutants are both very aberrant at 25°C and not viable without cell wall stabilization at 37°C (120). This finding indicates that the class IV chitin synthase of *W. dermatitidis* is not as critical to its biology as is its class IV ortholog Chs3p in *S. cerevisiae*. Possibly this comparative lack of importance is because *W. dermatitidis* does not produce a detectable chitin ring prior to bud emergence, which in *S. cerevisiae* is formed by the chitin synthesized by Chs3p. The absence of chitin ring formation in *W. dermatitidis* yeast means that the functions of class IV chitin synthases have changed during evolution, even though some molds retain or have gained the ability to grow in a yeast morphotype. In this respect, it is interesting that the disruption of *CHS2* and *CHS3* together in *S. cerevisiae* is lethal whereas the disruption of their homologs (*WdCHS1* and *WdCHS4*) together in *W. dermatitidis* is not. This further supports the notion that the class IV chitin synthase WdChs4p is not a functional equivalent of the Chs3p isozyme of *S. cerevisiae*.

WdCHS5 and WdChs5p

New degenerate primers also had to be designed before a gene sequence encoding a portion of a class V chitin synthase in *W. dermatitidis* was found (74). After this sequence was identified, the whole gene was isolated from a cosmid library by using the PCR product as a probe, named *WdCHS5*, characterized, and disrupted (73). As might be expected of a gene encoding a class V chitin synthase, the *WdCHS5* open reading frame is the longest of the *WdCHS* genes. It consists of 5,658 bp, has two small introns, and encodes a derived protein, WdChs5p, of 1,885 amino acids with a calculated mass of 208.9 kDa and a pI of 7.76. In the manner of most other class V chitin synthases, WdChs5p appears to be a fusion protein having a myosin motor-like domain fused to a chitin synthase domain, which may be their most distinguishing feature. While the function of the myosin motor-like domain of WdChs5p and all other class V isozymes identified to date remains unknown, it is tempting to suggest that the putative motor localizes the isozyme to unique sites distinct from those found in *S. cerevisiae* or *C. albicans*. Supporting this possibility is the identification of a characteristic ATP- or GTP-binding site motif (i.e., P-loop; GESGSGKT) and switch I (TASKAG) and switch II (DFPGF) motifs in the myosin motor-like domains of WdChs5p.

Regulation of *WdCHS5* transcription is similar to that of *WdCHS3* (73, 134, 137). For example, every condition that induces increased *WdCHS3* transcription also induces increased *WdCHS5* transcription. Analyses of fusions of *lacZ* with truncated versions of the 5′ URS of *WdCHS5* also indicate that negative regulatory binding sites are present between bp −880 and −450. This region has putative *cis*-acting elements for such transcription factors as those for the repressor of *CAR1* (RepCar1p) in *S. cerevisiae*, StuAp, a transcriptional repressor in *Aspergilus nidulans* that also regulates development, and PacCp, a zinc finger transcription factor that activates alkali-expressed genes and represses acid-expressed genes also in *A. nidulans* (38, 41, 76, 91). Similar sequences are also found in the 5′ URS of *WdCHS3* (113). In spite of these similarities, it is clear that WdChs5p plays a more critical role in the biology of *W. dermatitidis* than does WdChs3p, because cells devoid of its activity eventually die at 37°C but appear normal at 25°C whereas cells devoid of WdChs3p activity appear normal at both temperatures (73, 137).

The death that occurs when *wdchs5Δ* mutants are cultured at 37°C is due to the loss of cell wall integrity, which in turn is responsible for the mutants' loss of virulence in mouse models, as discussed above (73). At the higher temperature, *wdchs5Δ* yeast are hyperpigmented, often become isotropically enlarged, have wrinkled instead of smooth wall surfaces, and often blow out at discrete loci in their more chitinous cell walls from which cytoplasm leaks. To a large extent, osmotic stabilizers reverse these defects and allow the mutant to grow at 37°C, more like the wild type. Nonetheless, why the disruption of *WdCHS5* affects cells only at 37°C is unclear. Among the possibilities, one is that another WdChsp, which is required for normal growth at 25°C, is not produced, is degraded, or does not function at 37°C. Under this scenario, it would be expected that *wdchs5Δ* strains with another *WdCHS* gene disrupted would not be viable at 25°C. However, this is not the case because strains with *WdCHS5* disrupted and with each of the other *WdCHS* genes disrupted singly in the same background have been recovered (Liu and Szaniszlo, unpublished). This result suggests that the products of more than one of the other *WdCHS* genes are responsible. It is suspected that those products are WdChs1p and WdChs2p, at least one of which must be present with WdChs5p to ensure viability of the *wdchs5Δ* mutants at 25°C. Supporting this idea is the fact that a *wdchs1Δwdchs3Δwdchs4Δ* triple mutant is viable and grows well at both 25 and 37°C with only the two functional isozymes WdChs2p and WdChs5p. This suggests that WdChs2p, possibly with the help of WdChs5p, allows good growth at 25°C and that WdChs5p, possibly with the help of either WdChs2p or WdChs1p, allows good growth at 37°C. Additional support for this hypothesis could result from the production of a *wdchs2Δwdchs3Δwdchs4Δ* triple mutant, which has only functional WdChs1p and WdChs5p isozymes. However, this particular mutant has not been derived, although numerous attempts have been made (Liu and Szaniszlo, unpublished). The reason for this failure is probably that this triple mutant is simply not viable at any temperature, but because *W. dermatitidis* is asexual and thus not amenable to tetrad analysis, absolute proof for such a complex synthetic lethal condition is elusive. Furthermore, in the absence of a complete genome sequence it is impossible to know with certainty whether this species has even another *WdCHS* gene that is responsible for the viability of the *wdchs5Δ* mutant at 25°C. Regardless of the ultimate truth, the class V chitin synthase of *W. dermatitidis* clearly represents a particularly vulnerable target for the design of an antifungal drug. Based on the evidence at hand, this drug would be effective in the treatment of infections caused by *W. dermatitidis* even if it was not an effective inhibitor of any other WdChsp. Should this possibility be true for other black mold agents of disease, then the development of such a drug might be warranted. Unfortunately, no data exist that support this appealing prospect.

OTHER GENE PRODUCTS THAT MAY CONTRIBUTE TO VIRULENCE

A number of other genes of *W. dermatitidis* have been identified and sometimes cloned, sequenced, and disrupted (120). Because it is beyond the scope of this chapter to describe them all, the four discussed in this short closing section were chosen as examples only to illustrate why certain genes receive in-depth study and others do not. In general, most attention is given to genes encoding proteins that affect chitin or melanin biosynthesis, because in theory any gene product that affects the incorporation of these virulence factors into the cell walls of *W. dermatitidis* should also affect its virulence. Thus, the extent to which each gene is investigated is dictated by whether its gene product is linked directly to enrichments with or depletions of cell wall melanin or chitin or indirectly by inducing or inhibiting such outcomes during morphotype transitions. In some cases, the genes selected for study were cloned by marker rescue strategies after vectors with and without gene-targeting sequences disrupted random genes and produced interesting mutants with abnormal pigmentation, morphology, or both. In other cases, specific genes were cloned in the manner of the *WdPKS1* and the *WdCHS* structural genes, because they were suspected to affect directly enrichments of cell walls with chitin.

Among the genes cloned by marker rescue strategies, two, *WdYG1* and *WdMOT1*, have already been described above. Therefore, they are mentioned again here only because the first affected 1,8-DHN melanin biosynthesis alone whereas the second affected both 1,8-DHN melanin biosynthesis and morphology. The former is also an example of a gene identified because of a fortuitous ectopic integration of a plasmid designed to disrupt *WdURA5* (152). This particular disruption, which was successfully accomplished in another strain, was being attempted to provide a stable *wdura5Δ* recipient strain for identifying transformants after complementation with the cloned wild-type gene and selection with 5-fluoroorotic acid (5FOA). In contrast, the latter was identified in a random disruption screen for pigment mutants (42, 43). Although neither mutant has been tested in mice, it is assumed that at least *wdyg1Δ* would be less virulent, because it cannot synthesize 1,8-DHN melanin. The virulence status of *wdmot1Δ* is more problematic, because while it grows more slowly than the wild type, it is also more filamentous and is hyperpigmented.

The two examples of genes that were cloned specifically because they might affect virulence by enriching the cell walls of *W. dermatitidis* with chitin are *WdCDC42* and *WdSTUA*. The study of *WdCDC42* was prompted by the knowledge that yeast of *S. cerevisiae* ts *cdc42* mutants, when shifted to the restrictive temperature, lose polarity and then produce unbudded terminal phenotypes that are isotropically enlarged and have cell walls enriched with chitin in all wall areas (62, 120). As noted above, similar phenotypes are also produced by certain ts mutants of *W. dermatitidis* (e.g., Mc2 [*wdcdc1*] and Mc3 [*wdcdc2*]), as well as by the wild type exposed to extreme acidity (pH 2.5) or to calcium limitation at neutral pH (22, 67, 109, 125). However, in the case of *W. dermatitidis*, the resulting isotropic forms are often considerably more enlarged, have even thicker and more chitin-enriched cell walls, and may become multicellular because mitosis and cytokinesis often continue (21, 22, 53, 109, 110, 120, 121, 124, 125). If cytokinesis in the isotropic forms is not executed, then the morphotype is a sclerotic cell, whereas if cytokinesis is completed, it is a sclerotic body. As with the isotropic forms of *S. cerevisiae*, if a sclerotic cell of the ts mutant Mc3 (*wdcdc2*) is returned relatively quickly to the permissive temperature, so that it forms a bud cell instead of a hypha, then the new bud has little chitin compared to its chitin-enriched mother cell (59, 121). These striking similarities between the terminal phenotypes of the ts mutants of the two species cultured at 37°C suggested that studies of the *CDC42* homolog of *S. cerevisiae* in *W. dermatitidis* would provide insights into the mechanisms regulating the conversion of yeast to the potentially more resistant sclerotic cells and sclerotic body morphotypes of the model.

To address this hypothesis, *WdCDC42* was cloned from a cDNA library, confirmed like *CDC42* of *S. cerevisiae* to encode a Rho-type GTPase, disrupted, and then molecularly mutated site specifically to produce constitutively active and dominant negative alleles for complementation and overexpression studies with *S. cerevisiae* and *W. dermatitidis*, respectively (149). In contrast to its *CDC42* homolog in *S. cerevisiae*, *WdCDC42* is not essential for viability. The WdCdc42p-derived protein is, however, highly similar (81% identity) to its *S. cerevisiae* ortholog and therefore not surprisingly cross-complements the ts *cdc42* mutant DJTD2-16A at the restrictive temperature. Whereas none of the *wdcdc42* mutant

alleles complement the DJTD2-16A strain at 37°C, their overexpression in *S. cerevisiae* produces phenotypes at the restrictive temperature that are virtually identical to those produced by the equivalent *cdc42* alleles. Interestingly, while *wdcdc42Δ* null mutants are viable and grow at near-wild-type rates at both 25 and 37°C, they still exhibit some morphological abnormalities. In particular, *wdcdc42Δ* yeast cultured in rich media are slimmer than wild-type yeast, and a small percentage (6%) have two nuclei, often separated from each other by a single transverse septum. The production of these cells, which have the characteristics of the so-called "planate" morphotype sometimes produced in vivo by *W. dermatitidis*, is strong evidence that WdCdc42p plays a role in maintaining normal yeast budding growth. Even stronger evidence is provided by the finding that overexpression of the constitutively active *wdcdc42^{G14V}* allele, but not the dominant negative *wdcdc42^{D120A}* and *wdcdc42^{T19N}* alleles, in the wild type and *wdcdc42Δ* mutant also produced largely unbudded forms at 37°C. In this case, however, most are typical, isotropically enlarged sclerotic cells and sclerotic bodies, which have thickened cell walls enriched with chitin and melanin and may have multiple nuclei and one or more internal septa. Similarly, overexpression of the constitutively active allele, but not the dominant negative alleles, inhibits hyphal apical extension in the ts Hf1 strain and delays isotropic development of the ts Mc3 (*wdcdc2*) strain at the restrictive temperature. Taken together, these data further implicate WdCdc42p in the morphotype switching displayed by *W. dermatitidis*. However, sequencing shows that *WdCDC42* is not the gene mutated in either of the ts Mc strains (*wdcdc1* and *wdcdc2*). This finding implies that some other protein that affects polarization and yeast bud emergence, such as those encoded by *CDC24*, *BEM1*, or *BEM2* in *S. cerevisiae*, may be responsible. Interestingly, no loss of virulence is associated with the disruption of the *WdCDC42* gene. Very likely, this is because in molds, unlike in fungi more closely related to *S. cerevisiae*, *WdCDC42* is not always essential and the loss of WdCdc42p functions is compensated for by another protein. Possibly the compensating protein is a Rac-type GTPase, which is also encoded by nonessential genes in molds and has no homologs in yeast (54). However, no experimental evidence supports such a functional redundancy between Cdc42p and a RacGTPase in any fungus, including *W. dermatitidis*.

The cloning of *WdSTUA*, a gene that encodes a putative transcription factor having a highly conserved APSES (for "Asm1p, Phd1p, StuAp, Efg1p, Sok2p") DNA binding domain, was prompted by the identification of two potential binding sites for this protein in the most active putative repression region (bp −880 to −450) in the 5′ URS of *WdCHS5* (73). Thus, *WdSTUA* was cloned, characterized, and disrupted, because WdStuAp might act directly as a transcriptional activator or repressor of *WdCHS5* (Q. Wang, H. Liu, and P. J. Szaniszlo, *Abstr. 104th Gen. Meet. Am. Soc. Microbiol. 2004*, abstr. F039, p. 269, 2004). Two other compelling reasons to clone *WdSTUA* were that StuAp orthologs are well-studied regulators of morphological transitions in several fungi and APSES family members are known to be involved in the determination of fungal virulence (7, 37, 46, 75, 116, 141). Thus, chitin enrichments could be affected by WdStuAp via its induction or inhibition of yeast-to-hypha or yeast-to-sclerotic-form transitions. After amplification of a *WdSTUA* fragment by PCR with primers targeting the conserved APSES DNA binding domain, the full-length gene was cloned from a cosmid library using the PCR fragment as a probe. The gene has three introns in an open reading frame of 1,881 bp that encodes a derived protein of 627 amino acids. The WdStuAp calculated molecular mass is 66.9 kDa, and its predicted pI is 8.47. As might be expected, the WdStuAp-derived protein has highest identity to orthologs in other known or suspected filamentous ascomycetes and has only relatively few regions of amino acid identity outside the APSES domain in those of *S. cerevisiae* and related yeasts. Nonetheless, heterologous overexpression of *WdSTUA* induces filamentation in *S. cerevisiae* in much the same manner as does the overexpression of *PHD1*, which encodes an APSES family member that influences pseudohyphal development in that fungus (46; Q. Wang and P. J. Szaniszlo, unpublished data).

Mutants with *WdSTUA* deleted retain wild-type growth rates in most rich liquid media at 37°C. However, on agar-solidified rich media, *wdstuA*Δ mutants produce rougher colonies and more pseudohyphae/moniliform hyphae. In contrast, on mycelium-inducing solid medium, the opposite occurs: the wild type produces considerably more filamentous growth than the *wdstuA*Δ mutant. In addition, the *wdstuA*Δ mutant is inhibited in conidiation. Whether these differences between the *wdstuA*Δ mutant and wild type result in a loss of virulence is unclear, because the mutant has not been evaluated in mice. However, it is clear that the absence of WdStuAp function does not significantly affect *WdCHS5* transcription (Wang and Szaniszlo, unpublished). This is prompting similar evaluations of other genes that encode transcription factors that might bind one of the other putative binding sites identified in the 5′ URS of *WdCHS5*, such as that for WdPacC. The genes encoding this particular transcription factor and others in W. *dermatitidis* are currently being cloned and disrupted in order to determine the effects of their products on chitin enrichment, and particularly that contributed by WdChs5p. It is suspected that the ability of *W. dermatitidis* to strengthen its cell walls with the chitin product of WdChs5p is directly related to one of these factors. Should this be the case, then the same mechanism of wall strengthening may eventually be identified as a virulence factor in other black mold pathogens of humans and in turn be deemed a suitable secondary target for the design of an antifungal drug with efficacy beyond the *W. dermatitidis* model.

Acknowledgments. Many thanks are due to all my former and present students, postdoctoral associates, and colleagues for their numerous contributions, which made this review possible. I also thank Susan Szaniszlo for her help with manuscript preparation, Gwen Gage for her graphics help, and the National Institute of Allergy and Infectious Diseases for the generous financial assistance required for the molecular aspects of my research.

REFERENCES

1. **Adamkewewicz, J. I., K. H. Hansen, W. A. Prud'homme, J. L. Davis, and J. Thorner.** 2001. High affinity interaction of yeast transcriptional regulator, Mot1, with TATA box-binding protein (TBP). *J. Biol. Chem.* **276:** 1883–1894.

2. **Ajello, L., L. K. George, R. T. Steigbigel, and C. J. K. Wang.** 1974. A case of phaeohyphomycosis caused by a new species of *Phialophora. Mycologia* **66:**490–498.

3. **Alviano, C. S., S. R. Farbiarz, W. deSouza, J. Angluster, and L. R. Travassos.** 1991. Characterization of *Fonsecaea pedrosoi* melanin. *J. Gen Microbiol.* **137:**837–844.

4. **Barenfanger, J., F. Ramierz, R. P. Tewari, and L. Eagleton.** 1989. Pulmonary phaeohyphomycosis in a patient with hemoptysis. *Chest* **95:**1158–1160.

5. **Bell, A. A., J. E. Puhalla, W. J. Tolmsoff, and R. D. Stipanovic.** 1976. Use of mutants to establish (+)-scytalone as an intermediate in melanin biosynthesis by *Verticillium dahliae. Can. J. Microbiol.* **22:**787–799.

6. **Borelli, D.** 1955. *Sporotichum gougerottii, Hormiscium dermatitidis, Phialophora jeanselmei, Phialophora gougerottii* (Matruchot, 1910) comb. n. *Mem. Congr. Venez. Cienc. Med.* **5:**2945–2971.

7. **Borneman, A. R., M. J. Hynes, and A. Andrianopoulos.** 2000. The *abaA* homologue of *Penicillium marneffei* participates in two developmental programs: conidiation and dimorphic growth. *Mol. Microbiol.* **38:**1034–1047.

8. **Bowen, A. R., J. L. Chen-Wu, M. Momany, R. Young, P. J. Szaniszlo, and P. W. Robbins.** 1992. Classification of fungal chitin synthases. *Proc. Natl. Acad. Sci. USA* **89:**519–523.

9. **Bulawa, C. E.** 1992. *CSD2, CSD3,* and *CSD4,* genes required for chitin synthesis in *Saccharomyces cerevisiae*: the *CSD2* gene product is related to chitin synthases and to developmentally regulated proteins in *Rhizobium* species and *Xenopus laevis. Mol. Cell. Biol.* **4:** 1764–1776.

10. **Butler, M. J., G. Lazarovits, V. J. Higgins, and M.-A. Lachance.** 1987. *Phaeococcomyces* sp. Analysis of

melanin production by the yeast. *Can. J. Microbiol.* **35**:728–734.

11. **Cabib, E., S. J. Silverman, A. Sburlati, and M. L. Slater.** 1990. Chitin synthesis in yeast (*Saccharomyces cerevisiae*), p. 31–41. *In* P. J. Kuhn, A. P. J. Trinici, M. J. Jung, M. W. Goosey, and L. G. Copping (ed.), *Biochemistry of Cell Walls and Membranes in Fungi.* Springer-Verlag, New York, N.Y.

12. **Cabib, E., S. J. Silverman, and J. A. Shaw.** 1992. Chitinase and chitin synthase 1: counterbalancing activities in cell separation of *Saccharomyces cerevisiae. J. Gen. Microbiol.* **138**:97–102.

13. **Campbell, C. K., and S. S. A. Al-Hedaithy.** 1993. Phaeohyphomycosis of the brain caused by *Rhamichloridium mackenziei* sp. nov. in Middle Eastern countries. *J. Med. Vet. Mycol.* **31**:325–332.

14. **Carrion, A. L.** 1950. Yeast-like dematiaceous fungi infecting the human skin. *Arch. Dermatol. Syphilol.* **61**:996–1009.

15. **Centers for Disease Control and Prevention.** 2002. *Exophiala* infection from contaminated injectable steroids prepared by a compounding pharmacy—United States, July–November 2002. *Morb. Mortal. Wkly. Rep.* **51**:1109–1112.

16. **Chang, C. L., D.-S. Kim, D. J. Park, H. J. Kim, C. H. Lee, and J. H. Shin.** 2000. Acute cerebral phaeohyphomycosis due to *Wangiella dermatitidis* accompanied by cerebrospinal fluid eosinophilia. *J. Clin. Microbiol.* **38**:1965–1966.

17. **Chen, W.** 1996. *Molecular Cloning and Characterization of the Actin Gene of Wangiella dermatitidis.* M.A. thesis. The University of Texas at Austin.

18. **Chua, J. D., S. M. Gordon, J. Banbury, G. S. Hall, and G. W. Procop.** 2001. Relapsing *Exophiala jeanselmei* phaeohyphomycosis in a lung-transplant patient. *Transpl. Infect. Dis.* **3**:235–238.

19. **Conant, N. F., D. T. Smith, R. D. Baker, and J. R. Callaway.** 1971. Chromoblastomycosis, p. 503–540. *In Manual of Clinical Mycology,* 3rd ed. The W. B. Saunders Co., Philadelphia, Pa.

20. **Cooper, B. H.** 1985. *Phialophora verrucosa* and other chromoblastomycotic fungi, p. 263–280. *In* P. J. Szaniszlo (ed.), *Fungal Dimorphism, with Emphasis on Fungi Pathogenic for Humans.* Plenum Press, New York, N.Y.

21. **Cooper, C. R., Jr., J. L. Harris, C. W. Jacobs, and P. J. Szaniszlo.** 1984. Effect of polyoxin AL on cellular development in *Wangiella dermatitidis. Exp. Mycol.* **8**:349–363.

22. **Cooper, C. R., Jr., and P. J. Szaniszlo.** 1993. Evidence for two cell division cycle (*CDC*) genes that govern yeast bud emergence in the pathogenic fungus *Wangiella dermatitidis. Infect. Immun.* **61**:2069–2081.

23. **Cooper, C. R., Jr., and P. J. Szaniszlo.** 1997. Melanin as a virulence factor in dematiaceous pathogenic fungi, p. 81–93. *In* H. V. Bossche, D. A. Stevens, and F. C. Odds (ed.), *Proceedings of the Host-Fungus Interplay Conference.* National Foundation for Infectious Diseases, Bethesda, Md.

24. **Crosby, J. H., M. H. O' Quinn, J. C. H. Steele, Jr., and R. N. Rao.** 1989. Fine-needle aspiration of subcutaneous phaeohyphomycosis caused by *Wangiella dermatitidis. Diagn. Cytopathol.* **5**:293–297.

25. **de Hoog, G. S.** 1983. On the potentially pathogenic dematiaceous Hyphomycetes, p. 149–216. *In* D. Howard (ed.), *Fungi Pathogenic for Humans and Animals,* part A. *Biology.* Marcel Dekker, Inc., New York, N.Y.

26. **de Hoog, G. S.** 1999. Ecology and evolution of black yeasts and their relatives. *Stud. Mycol.* **43**:3–4.

27. **de Hoog, G. S.** 1977. *Rhinocladiella* and allied genera. *Stud. Mycol.* **15**:1–177.

28. **de Hoog, G. S., K. Takeo, S. Yoshida, E. Gottlich, K. Nishimura, and M. Miyaji.** 1994. Pleoanamorphic life cycle of *Exophiala dermatitidis. Antonie Leeuwenhoek* **65**:143–153.

29. **de Monbrison, F., M. A. Piens, B. Ample, S. Euvard, P. Cochat, and S. Picot.** 2004. Two cases of subcutaneous phaeohyphomycosis due to *Exophiala jeanselmei,* in cardiac transplant and renal transplant patents. *Br. J. Dermatol.* **150**:596–624.

30. **Din, A. B., C. A. Spect, P. W. Robbins, and O. Yarden.** 1996. *chs4,* a class IV chitin synthase gene from *Neurospora crassa. Mol. Gen. Genet.* **250**:214–222.

31. **Dixon, D. M., and A. Polak-Wyss.** 1991. The medically important dematiaceous fungi and their identifications. *Mycoses* **34**:1–18.

32. **Dixon, D. M., A. Polak, and G. W. Connor.** 1989. Mel- mutants of *Wangiella dermatitidis* in mice: evaluation of multiple mouse and fungal strains. *J. Med. Vet. Mycol.* **27**:335–341.

33. **Dixon, D. M., A. Polak, and P. J. Szaniszlo.** 1987. Pathogenicity and virulence of wild-type and melanin-deficient *Wangiella dermatitidis. J. Med. Vet. Mycol.* **25**:97–106.

34. **Dixon, D. M., J. Migliozzi, C. R. Cooper, Jr., O. Solis, B. Breslin, and P. J. Szaniszlo.** 1992. Melanized and nonmelanized multicellular form mutants of *Wangiella dermatitidis* in mice: mortality and histopathology studies. *Mycoses* **35**:17–21.

35. **Dixon, D. M., P. J. Szaniszlo, and A. Polak.** 1991. Dihydroxynaphthalene (DHN) melanin and its relationship with virulence in the early stages of phaeohyphomycosis, p. 297–318. *In* G. T. Cole and H. C. Hoch (ed.), *The Fungal Spore and Disease Initiation in Plants and Animals.* Plenum Publishing Corp., New York, N.Y.

36. **Dixon, D. M., T. J. Walsh, W. G. Merz, and M. R. McGinnis.** 1989. Infections due to *Xylohypha bantiana* (*Cladosporium trichoides*). *Rev. Infect. Dis.* **11**:515–525.

37. **Doedt, T., S. Korishnamurthy, D. P. Bockmuhl, B. Tebarth, C. Stempel, C. L. Russell, A. J. Brown, and J. F. Ernst.** 2004. APSES proteins regulate morphogenesis and metabolism in *Candida albicans. Mol. Biol. Cell* **15**:3167–3180.

38. **Dutton, J. R., S. Johns, and B. L. Miller.** 1997. StuAp is a sequence specific transcription factor that regulates developmental complexity in *A. nidulans. EMBO J.* **16**:5710–5721.

39. **Ellis, M. B.** 1971. *Dematiaceous Hyphomycetes.* Commonwealth Mycological Institute, Kew, England.

40. **Emmons, C. W.** 1966. Pathogenic dematiaceous fungi. *Jpn. J. Med. Mycol.* **7**:233–245.

41. **Espeso, E. A., and M. A. Penala.** 1996. Three binding sites for the *Aspergillus nidulans* Pac C zinc-finger transcription factor are necessary and sufficient for regulation by ambient pH of the isopenicillin N synthase gene promoter. *J. Biol. Chem.* **46**:28825–28830.

42. **Feng, B.** 2000. *The Cloning and Study of Two Genes, WdPKS1 and WdMOT1, That Affect DHN-Melanin Biosynthesis in Wangiella dermatitidis.* Ph.D. thesis. The University of Texas at Austin.

43. **Feng, B., X. Wang, M. Hauser, S. Kaufmann, S. Jentsch, G. Haase, J. M. Becker, and P. J. Szaniszlo.** 2001.

Molecular cloning and characterization of *WdPKS1*, a gene involved in dihydroxynaphthalene melanin biosynthesis and virulence in *Wangiella* (*Exophiala*) *dermatitidis*. *Infect. Immun.* 69:1781–1794.

44. Geis, P. A., and P. J. Szaniszlo. 1984. Carotenoid pigments of the dematiaceous fungus *Wangiella dermatitidis*. *Mycologia* 76:268–273.

45. Geis, P. A., M. H. Wheeler, and P. J. Szaniszlo. 1984. Pentaketide metabolites of melanin synthesis in the dematiaceous fungus *Wangiella dermatitidis*. *Arch. Microbiol.* 137:324–328.

46. Gimeno, C. J., and G. R. Fink. 1994. Induction of pseudohyphal growth by overexpression of *PHD1*, a *Saccharomyces cerevisiae* gene related to transcriptional regulators of fungal development. *Mol. Cell. Biol.* 14:2100–2112.

47. Greig, J., M. Harkness, P. Taylor, C. Hasmi, S. Liang, and J. Kwan. 2003. Peritonitis due to *Exophiala dermatitidis* complicating continuous ambulatory peritoneal analysis. *Clin. Microbiol. Infect.* 9:713–715.

48. Grove, S. M., K. B. Oujezdsky, and P. J. Szaniszlo. 1973. Budding in the dimorphic fungus *Phialophora dermatitidis*. *J. Bacteriol.* 115:323–329.

49. Grur, K. E., G. P. Gross, P. H. Cooper, and S. A. Harding. 1979. Cystic chromomycosis due to *Wangiella dermatitidis*. *Arch. Dermatol.* 115:1433–1434.

50. Haase, G., H. Skopnik, T. Groten, H. Kwsenbach, and H. G. Posselt. 1991. Long-term fungal cultures from patients with cystic fibrosis. *Mycoses* 34:373–376.

51. Haase, G., L. Sonntag, B. Metzer-Kirk, and G. S. deHoog. 1999. Phylogenetic inferences by SSU-gene analysis of members of the Herpotrichiellaceae with special reference to human pathogenic species. *Stud. Mycol.* 43:80–97.

52. Hamza, S. H., P. J. Mercado, H. G. Skelton, and K. J. Smith. 2003. An unusual dematiaceous fungal infection of the skin caused by *Fonsecaea pedrosoi*: a case report and review of the literature. *J. Cutan. Pathol.* 30:340–343.

53. Harris, J. L., and P. J. Szaniszlo. 1986. Localization of chitin in walls of *Wangiella dermatitidis* using colloidal gold-labeled chitinase. *Mycologia* 78:853–857.

54. Harris, S. D., and M. Momany. 2004. Polarity in filamentous fungi: moving beyond the yeast paradigm. *Fungal Genet. Biol.* 41:391–400.

55. Hiruma, M., A. Kawada, T. Ohata, Y. Ohnishi, H. Takahashi, M. Mazaki, A. Ishibashi, K. Hatasuse, M. Kakihara, and M. Yoshida. 1993. Systemic phaeohyphomycosis caused by *Exophiala dermatitidis*. *Antonie Leeuwenhoek* 65:143–152.

56. Hohl, P. E., H. P. Holley, Jr., E. Prevost, L. Ajello, and A. A. Padye. 1983. Infections due to *Wangiella dermatitidis* in humans: report of the first documented case from the United States and a review of the literature. *Rev. Infect. Dis.* 5:854–864.

57. Horre, R., and G. S. de Hoog. 1999. Primary cerebral infections by melanized fungi: a review. *Stud. Mycol.* 43:176–193.

58. Jacobs, C. W., and P. J. Szaniszlo. 1982. Microtubule function and its relation to cellular development and the yeast cell cycle in *Wangiella dermatitidis*. *Arch. Microbiol.* 133:155–161.

59. Jacobs, C. W., R. L. Roberts, and P. J. Szaniszlo. 1985. Reversal of multicellular-form development in a conditional morphological mutant of the fungus *Wangiella dermatitidis*. *J. Gen. Microbiol.* 131:1719–1728.

60. Jacobson, E. S. 2000. Pathogenic roles for fungal melanins. *Clin. Microbiol. Rev.* 13:708–717.

61. Jacobson, E. S., E. Hove, and H. S. Emery. 1995. Antioxidant function of melanin in black fungi. *Infect. Immun.* 63:4944–4945.

62. Johnson, D. L., and J. R. Pringle. 1990. Molecular characterization of *CDC42*, a *Saccharomyces cerevisiae* gene involved in the development of cell polarity. *J. Cell Biol.* 111:143–153.

63. Jotisankasa, V., H. S. Neilson, Jr., and N. F. Conant. 1970. *Phialophora dermatitidis*: its morphology and biology. *Sabouraudia* 8:98–107.

64. Kanj, S. S., S. S. Amr, and G. D. Roberts. 2001. *Ramichloridium mackenziei* brain abscess: report of two cases and review of literature. *Med. Mycol.* 39:97–102.

65. Kano, K. 1938. Uber die chomoblastomykase durch einen noch nicht als pathogen beschriebene p. lz: *Hormiscium dermatitidis* n. sp. *Arch. Dermatol.* 176:282–294.

66. Karuppayil, S. M., M. Peng, L. Mendoza, T. A. Levins, and P. J. Szaniszlo. 1996. Identification of the conserved coding sequences of three chitin synthase genes in *Fonsecaea pedrosoi*. *J. Vet. Med. Mycol.* 34:117–125.

67. Karuppayil, S. M., and P. J. Szaniszlo. 1997. Importance of calcium to the regulation of polymorphism in *Wangiella* (*Exophiala*) *dermatitidis*. *J. Med. Vet. Mycol.* 35:379–388.

68. Kusenbach, G., H. Skopnik, G. Haase, F. Friedrichs, and H. Doehmen. 1992. *Exophiala dermatitidis* in cystic fibrosis. *Eur. J. Pediatr.* 151:344–346.

69. Kwon-Chung, K. J., and J. E. Bennett. 1992. Phaeohyphomycosis, p. 621–677. *In Medical Mycology*, Lea and Febiger, Philadelphia, Pa.

70. Kwon-Chung, K. J., W. E. Goldman, B. Klein, and P. J. Szaniszlo. 1998. Fate of transforming DNA in pathogenic fungi. *Med. Mycol. Suppl.* 36:38–44.

71. Levy, I., J. Stein, S. Ashkenazi, Z. Samra, G. Livni, and I. Yaniv. 2003. Ecthyma gangrenosum caused by *Exoserohilum* in a child with leukemia: a case report and review of the literature. *Pediatr. Dermatol.* 20:495–497.

72. Liu, H. 2003. WdChs5p of *Wangiella* (*Exophiala*) *dermatitidis*, a Class V Chitin Synthase, Is Essential for Sustained Growth at Temperature of Infection. Ph.D. thesis. The University of Texas at Austin.

73. Liu, H., S. Kauffman, J. M. Becker, and P. J. Szaniszlo. 2004. *Wangiella* (*Exophiala*) *dermatitidis* WdChs5p, a class V chitin synthase, is essential for sustained cell growth at temperature of infection. *Eukaryot. Cell* 3:40–51.

74. Liu, H., Z. Wang, L. Zheng, M. Hauser, S. Kauffman, J. M. Becker and P. J. Szaniszlo. 2001. Relevance of chitin and chitin synthases to virulence in *Wangiella* (*Exophiala*) *dermatitidis*, a model melanized pathogen of humans, p. 463–472. *In* R. A. A. Muzzarelli (ed.), *Chitin Enzymology* 2001. Atec Edizioni, Grottammare, Italy.

75. Lo, H. J., J. R. Kohler, B. DiDomenco, D. Loebenburg, A. Cacciapuoti, and G. R. Fink. 1997. Nonfilamentous *Candida albicans* mutants are avirulent. *Cell* 90:939–949.

76. Luche, R. M., R. Sumrada, and T. G. Cooper. 1990. A *cis*-acting element present in multiple genes serves as a repressor protein binding site for the yeast *CAR1* gene. *Mol. Cell. Biol.* 10:3884–3895.

77. Matos, T., G. S. de Hoog, A. G. de Boer, I. de Crom, and G. Haase. 2002. High prevalence of the neurotrope *Exophiala dermatitidis* and related oligotrophic black yeasts in sauna facilities. *Mycoses* 45:373–377.

78. Matsumoto, T., A. A. Padhye, L. Ajello, and P. G. Standard. 1984. Critical review of human isolates of *Wangiella dermatitidis*. *Mycologia* 76:232–249.

79. Matsumoto, T., A. A. Padhye, and L. Ajello. 1991. Medical significance of the so-called black yeast. *Eur. J. Epidemiol.* 3:91–95.

80. Matsumoto, T., and L. Ajello. 1991. Dematiaceous fungi potentially pathogenic to humans and lower animals, p. 117–162. *In* D. K. Arora, L. Ajello and K. G. Mukerji (ed.), *Handbook of Applied Mycology*, vol. 2. *Humans, Animals and Insects*. Marcel Dekker, Inc., New York, N.Y.

81. Matsumoto, T., L. Ajello, T. Matsuda, P. J. Szaniszlo, and T. J. Walsh. 1994. Developments in hyalohyphomycosis and phaeohyphomycosis. *J. Med. Vet. Mycol. Suppl.* 32: 329–349.

82. McGinnis, M. R. 1977. *Wangiella*, a new genus to accommodate *Hormiscium dermatitidis*. *Mycotaxon* 13:127–136.

83. McGinnis, M. R. 1980. *Laboratory Handbook of Medical Mycology*. Academic Press, Inc., New York, N.Y.

84. McGinnis, M. R. 1983. Chromoblastomycosis and phaeohyphomycosis: new concepts, diagnosis and mycology. *J. Am. Acad. Dermatol.* 8:1–16.

85. McGinnis, M. R., I. F. Salkin, W. A. Schell, and L. Pararell. 1991. Dematiaceous fungi, p. 644–658. *In* A. Balows, W. J. Hausler, Jr., K. L. Herrmann, H. D. Isenberg, and H. J. Shadomy (ed.), *Manual of Clinical Microbiology*, 5th ed. American Society for Microbiology, Washington, D.C.

86. McGinnis, M. R., S. M. Lemon, D. H. Walker, G. S. de Hoog, and G. Haase. 1999. Fatal cerebritis caused by a new species of *Cladophialophora*. *Stud. Mycol.* 43: 166–171.

87. McIntosh, N. D. P. 1996. *Yeast-to-Hypha Transition in Wangiella dermatitidis*. M.A. thesis. The University of Texas at Austin.

88. Meis, J. F., G. M. P. Wesseling, H. P. H. Kremer, and P. E. Verweij. 1999. Unsuspected cerebral phaeohyphomycosis presenting as chronic basilar meningitis. *Stud. Mycol.* 43:163–165.

89. Mendoza, A. L. 1995. *Cloning and Molecular Characterization of the Chitin Synthase 1 (WdCHS1) Gene of Wangiella dermatitidis*. Ph.D. thesis. The University of Texas at Austin.

90. Mendoza, L., S. M. Karruppayil, and P. J. Szaniszlo. 1993. Calcium regulates in vitro dimorphism in chromoblastomycotic fungi. *Mycoses* 36:157–164.

91. Miller, K. Y., J. Wu, and B. L. Miller. 1992. *stuA* is required for cell pattern determination in *Aspergillus*. *Genes Dev.* 6:1770–1782.

92. Montijn, R. C., P. V. Wolven, S. de Hoog, and F. M. Klis. 1997. β-Glycosylated proteins in the cell wall of the black yeast *Exophiala* (*Wangiella*) *dermatitidis*. *Microbiology* 143:1673–1680.

93. Munro, C. A., and N. A. R. Gow. 2001. Chitin synthesis in human pathogenic fungi. *Med. Mycol.* (Suppl. 1) 39:41–53.

94. Myoken, Y., T. Sugata, Y. Fujita, T. Kyo, M. Fujihara, M. Katsu, and Y. Mikami. 2003. Successful treatment of invasive stomatitis due to *Exophiala dermatitidis* in a patient with acute myeloid leukemia. *J. Oral Pathol. Med.* 32:51–54.

95. Nishimura, K., and M. Miyaji. 1983. Defense mechanisms of mice against *Exophiala dermatitidis* infection. *Mycopathologia* 81:9–21.

96. Nucci, M., T. Akiti, G. Barreiros, F. Silverra, S. G. Revankar, D. Sutton, and T. F. Patterson. 2001. Nosocomial fungemia due to *Exophiala jeanselmei* and *Rhinocladiella* species: newly described causes of bloodstream infection. *J. Clin. Microbiol.* 39:514–518.

97. Ohira, S., K. Isoda, H. Hamanaka, K. Takahashi, K. Nishimoto, and H. Mizutani. 2002. Phaeohyphomycosis caused by *Phialophora verrucosa* developed in a patient with non-HIV acquired immunodeficiency syndrome. *Mycoses* 45:50–54.

98. Ohkusu, M., M. Yamaguchi, K. Hata, S. Yoshida, R. Tanaka, K. Nishimura, G. S. de Hoog, and K. Takeo. 1999. Cellular and nuclear characteristics of *Exophiala dermatitidis*. *Stud. Mycol.* 43:143–150.

99. Okan, G., and M. Rendon. 2001. Chromoblastomycoses in a patient with chronic obstructive pulmonary disease. *Eur. Acad. Dermatol. Venereol.* 15:181–189.

100. Oujezdsky, K. B., and P. J. Szaniszlo. 1974. Conidial ontogeny in *Phialophora dermatitidis*. *Mycologia* 66:537–542.

101. Oujezdsky, K. B., S. N. Grove, and P. J. Szaniszlo. 1973. Morphological and structural changes during the yeast-to-mold conversion of *Phialophora dermatitidis*. *J. Bacteriol.* 113:468–477.

102. Padhye, A. A., J. D. Dunkel, R. M. Winn, S. Weber, E. P. Ewing, and G. S. de Hoog. 1999. Subcutaneous phaeohyphomycosis caused by an undescribed *Cladophialophora* sp. *Stud. Mycol.* 43:173–175.

103. Padhye, A. A., L. Ajello, A. W. Marion, and K. K. Steinbronn. 1986. Phaeohyphomycosis of the nasal sinuses caused by a new species of *Exoserhilum*. *J. Clin. Microbiol.* 24:245–249.

104. Padhye, A. A., M. S. Davis, D. Baer, A. Reddick, K. K. Sinha, and J. Ott. 1998. Phaeohyphomycosis caused by *Phaeoacremonium inflatipis*. *J. Clin. Microbiol.* 36: 2763–2765.

105. Pappagianis, D., and L. Ajello. 1994. Dematiaceous—a mycological misnomer? *J. Med. Vet. Mycol.* 32: 319–321.

106. Peng, M., C. R. Cooper, Jr., and P. J. Szaniszlo. 1995. Genetic transformation of the pathogenic fungus *Wangiella dermatitidis*. *Appl. Microbiol. Biotechnol.* 44:444–450.

107. Peng, M., S. M. Karuppayil, L. Mendoza, T. A. Levin, and P. J. Szaniszlo. 1995. Use of the polymerase chain reaction to identify coding sequences for chitin synthase isozymes in *Phialophora verrucosa*. *Curr. Genet.* 27:517–523.

108. Randhawa, H. S., V. Budimulja, G. Bazaz-Malik, K. Bramono, M. Hiramia, P. Kullaranjaya, and V. Rojanavanich. 1994. Recent developments in the diagnosis and treatment of subcutaneous mycoses. *J. Med. Vet. Mycol.* 32:299–307.

109. Roberts, R. L., and P. J. Szaniszlo. 1978. Temperature-sensitive multicellular mutants of *Wangiella dermatitidis*. *J. Bacteriol.* 135:622–632.

110. Roberts, R. L., R. J. Lo, and P. J. Szaniszlo. 1979. Nuclear division in temperature-sensitive multicellular mutants of *Wangiella dermatitidis*. *J. Bacteriol.* 137: 1456–1458.

111. Roberts, R. L., R. J. Lo, and P. J. Szaniszlo. 1980. Yeast-phase cell cycle of the polymorphic fungus *Wangiella dermatitidis*. *J. Bacteriol.* 144:721–731.

112. Roberts, R. L., R. J. Lo, and P. J. Szaniszlo. 1980. Induction of synchronous growth in the yeast phase of *Wangiella dermatitidis*. *J. Bacteriol.* 141:981–984.

113. Ruiz-Herrera, J., J. M. Gonzalez-Prieto, and R. Ruiz-Medrano. 2001. Significance of the multigenic control of chitin synthase in fungi, p. 451–462. *In* R. A. A. Muzzarelli (ed.), *Chitin Enzymology 2001.* Atec Edizioni, Grottammare, Italy.

114. Schnitzler, N., H. Peltroche-Llacsahuanga, N. Bestier, J. Zundorf, R. Lutticken, and G. Haase. 1999. Effect of melanin and carotenoids of *Exophiala (Wangiella) dermatitidis* on phagocytosis, oxidative burst, and killing of human neutrophils. *Infect. Immun.* **67:**94–101.

115. Schol-Schwarz, M. B. 1968. *Rhinocladiella,* synonym *Fonsecaea,* and its relation to *Phialophora. Antonie Leeuwenhoek* **34:**199–152.

116. Sohn, K., C. Urban, H. Brunner, and S. Rupp. 2003. EFG1 is a major regulator of cell wall dynamics in *Candida albicans* as revealed by DNA microarrays. *Mol. Microbiol.* **47:**89–102.

117. Spect, C. A., Y. Liu, P. W. Robbins, C. E. Bulawa, N. Iartchouk, K. R. Winter, P. J. Riggle, J. C. Rhodes, C. L. Dodge, D. W. Culp, and P. T. Borgia. 1996. The *chsD* and *chsE* genes of *Aspergillus nidulans* and their role in chitin synthesis. *Fungal Genet. Biol.* **20:**153–167.

118. Sterflinger, K., G. S. de Hoog, and G. Haase. 1999. Phylogeny and ecology of meristematic ascomycetes. *Stud. Mycol.* **43:**5–22.

119. Sutton, D. A., M. Slifkin, R. Yakulis, and M. G. Renaldi. 1998. U.S. case report of cerebral phaeohyphomycosis caused by *Ramichoridium obovoideum* (*R. mackenziei*): criteria for identification, therapy and review of the known dematiaceous neurotropic taxa. *J. Clin. Microbiol.* **36:**708–715.

120. Szaniszlo, P. J. 2002. Molecular genetic studies of the model dematiaceous pathogen *Wangiella dermatitidis. Int. J. Med. Microbiol.* **292:**381–390.

121. Szaniszlo, P. J., and M. Momany. 1993. Chitin, chitin synthase and chitin synthase conserved region homologues in *Wangiella dermatitidis,* p. 229–242. *In* B. Maresca, G. Kobayashi, and H. Yamaguchi (ed.), *NATO Workshop Proceedings on Molecular Biology and Its Application to Medical Mycology.* Springer-Verlag, New York, N.Y.

122. Szaniszlo, P. J., C. W. Jacobs, and P. A. Geis. 1983. Dimorphism in pathogenic fungi, p. 373–436. *In* D. Howard (ed.), *Fungi Pathogenic for Humans and Animals,* part A. *Biology.* Marcel Dekker, Inc., New York, N.Y.

123. Szaniszlo, P. J., L. Mendoza, and S. M. Karuppayil. 1993. Clues about chromoblastomycotic and other dematiaceous pathogens based on *Wangiella* as a model, p. 241–255. *In* H. Vanden Bossche, F. C. Odds, and D. Kerridge (ed.), *Dimorphic Fungi in Biology and Medicine.* Plenum Press, New York, N.Y.

124. Szaniszlo, P. J., P. A. Geis, C. W. Jacobs, C. R. Cooper, Jr., and J. L. Harris. 1983. Cell wall changes associated with yeast-to-multicellular form conversion in *Wangiella dermatitidis.* p. 239–244. *In* D. Schlessinger (ed.), *Microbiology—1983.* American Society for Microbiology, Washington, D.C.

125. Szaniszlo, P. J., P. H. Hsieh, and J. D. Marlowe. 1976. Induction and ultrastructure of the multiseptate (sclerotic) morphology in *Phialophora dermatitidis. Mycologia* **68:**117–130.

126. Szaniszlo, P. J., S. M. Karuppayil, L. Mendoza, and R. Rennard. 1993. Cell-cycle regulation of polymorphism in *Wangiella dermatitidis. Arch. Med. Res.* **54:**251–261.

127. Taylor, B. E., M. H. Wheeler, and P. J. Szaniszlo. 1991. Evidence for pentaketide melanin biosynthesis in

dematiaceous human pathogenic fungi. *Mycologia* **79:**320–322.

128. Tsai, H.-F., I. Fuji, A. Watanabe, M. H. Wheeler, Y. C. Chung, Y. Yasuoka, Y. Ebizaka, and K. J. Kwon Chung. 2001. Pentaketide melanin biosynthesis in *Aspergillus fumigatus* requires chain-length shortening of a heptaketide precursor. *J. Biol. Chem.* **276:**29292–29298.

129. Valdiveso, M. H., P. C. Mol, J. A. Shaw, E. Cabib, and A. Duran. 1991. *CAL1,* a gene required for activity of the chitin synthase 3 in *Saccharomyces cerevisiae. J. Cell Biol.* **114:**101–109.

130. Vieira, M. R., A. Milheiro, and F. A. Pacheco. 2001. Phaeohyphomycosis due to *Cladosporium cladosporioides. Med. Mycol.* **39:**135–137.

131. Walsh, T. J. 1998. Emerging fungal pathogens: evolving challenges to immunocompromised patients, p. 221–232. *In* W. M. Scheld, D. Armstrong, and J. M. Hughes (ed.), *Emerging Infections.* ASM Press, Washington, D.C.

132. Walsh, T. J., A. Groll, J. Hiemenz, R. Fleming, E. Roilides, and E. Anaissie. 2004. Infections due to emerging and uncommon medically important pathogens. *Clin. Microbiol. Infect.* **10**(Suppl. 1)**:**48–66.

133. Walsh, T. J., B. DePauw, E. Anaissie, and P. Martino. 1994. Recent advances in the epidemiology, prevention and treatment of invasive fungal infections in neutropenic patients. *J. Med. Vet. Mycol.* **32:**33–51.

134. Wang, Q., H. Liu, and P. J. Szaniszlo. 2002. Compensatory expression of five chitin synthase genes, a response to stress stimuli in *Wangiella (Exophiala) dermatitidis,* a melanized fungal pathogen of humans. *Microbiology* **148:**2811–2817.

135. Wang, X. 1998. *Construction of a Cosmid Library, and Its Use in the Cloning of the WdPKS1 gene of Wangiella dermatitidis by Mutant Complementation.* M.A. thesis. The University of Texas at Austin.

136. Wang, Z. 1998. *Cloning and Characterization of the Chitin Synthase 3 (WdCHS3) and Chitin Synthase 4 (WdCHS4) Genes of Wangiella dermatitidis and Expression Studies of WdCHS3.* Ph.D. thesis. The University of Texas at Austin.

137. Wang, Z., and P. J. Szaniszlo. 2000. *WdCHS3,* a gene that encodes a class III chitin synthase in *Wangiella (Exophiala) dermatitidis,* is expressed differentially under stress conditions. *J. Bacteriol.* **182:**874–881.

138. Wang, Z., and P. J. Szaniszlo. 2002. Characterization of WdChs3p, a class III chitin synthase of *Wangiella (Exophiala) dermatitidis,* overexpressed in *Saccharomyces cerevisiae. Med. Mycol.* **40:**283–289.

139. Wang, Z., L. Zheng, H. Liu, Q. Wang, M. Hauser, S. Kauffman, J. M. Becker, and P. J. Szaniszlo. 2001. WdChs2p, a class I chitin synthase, together with WdChs3p (class III), contributes to virulence in *Wangiella (Exophiala) dermatitidis. Infect. Immun.* **69:**7517–7526.

140. Wang, Z., L. Zheng, M. Hauser, J. M. Becker, and P. J. Szaniszlo. WdChs4p, a homolog of chitin synthase 3 in *Saccharomyces cerevisiae,* alone cannot support the growth of *Wangiella (Exophiala) dermatitidis* at the temperature of infection. *Infect. Immun.* **12:**6619–6630.

141. Ward, M., C. J. Gimeno, G. R. Fink, and S. Garret. 1996. *SOK2* may regulate cyclic AMP-dependent protein kinase-stimulated growth and pseudohyphal development by repressing transcription. *Mol. Cell. Biol.* **15:**6854–6863.

142. Wheeler, M. H., and A. A. Bell. 1987. Melanins and their importance to pathogenic fungi. *Curr. Top. Med. Mycol.* **2**:338–387.

143. Wheeler, M. H., and R. D. Stipanovic. 1985. Melanin biosynthesis and metabolism of flaviolin and 2-hydroxyjuglene in *Wangiella dermatitidis*. *Arch. Microbiol.* **142**:234–241.

144. Wickes, B. L., and K. J. Kwon-Chung. 2002. Genetic basis of pathogenicity in *Cryptococcus neoformans*, p. 25–49. *In* R. A. Calderone and R. L. Cihlar (ed.), *Fungal Pathogenesis: Principles and Clinical Applications.* Marcel Dekker, Inc., New York, N.Y.

145. Yamaguchi, M., S. K. Biswas, Y. Suzuki, H. Furukawa, M. Sameshima, and K. Takeo. 2002. The spindle pole body duplicates in early G1 phase in the pathogenic yeast *Exophiala dermatitidis*: an ultrastructural study. *Exp. Cell Res.* **279**:71–79.

146. Yamaguchi, M., Y. Kubabrua, M. Shimizu, H. Furukawa, H. Nishioka, and K. Takeo. 2003. The spindle pole body of the pathogenic yeast *Exophiala dermatitidis*: variation in morphology and positional relationship to the nucleolus and the bud in interphase cells. *Eur. J. Cell Biol.* **82**:531–538.

147. Yamaguchi, M., S. K. Biswas, Y. Suzuki, H. Furukawa, and K. Takeo. 2003. Three dimensional reconstruction of a pathogenic yeast *Exophiala dermatitidis* by freeze-substitution and serial sectioning electron mycoscopy. *FEMS Microbiol. Lett.* **219**:17–21.

148. Ye, X. 1998. Role of a *CDC42 Homologous Gene in the Regulation of Cell Polarity and Morphogenetic Transitions in Wangiella dermatitidis*. Ph.D. thesis. The University of Texas at Austin.

149. Ye, X., and P. J. Szaniszlo. 2000. Expression of a constitutively active Cdc42 homolog promotes development of sclerotic bodies but represses hyphal growth in the pathogenic fungus *Wangiella* (*Exophiala*) *dermatitidis*. *J. Bacteriol.* **182**:4941–4950.

150. Ye, X., B. Feng, and P. J. Szaniszlo. 1999. A color-selectable and site-specific integrative transformation system for gene expression studies in the dermatiaceous fungus *Wangiella* (*Exophiala*) *dermatitidis*. *Curr. Genet.* **36**:241–247.

151. Zheng, L. 1997. *Establishment of Genetic Transformation Systems in and Molecular Cloning of the Chitin Synthase 2 (WdCHS2) Gene, and Characterization of the WdCHS1 and WdCHS2 Genes of Wangiella dermatitidis*. Ph.D. thesis. The University of Texas at Austin.

152. Zheng, L., and P. J. Szaniszlo. 1999. Cloning and use of the *WdURA5* gene as a *hisG* cassette selection marker for potentially disrupting multiple genes in *Wangiella dermatitidis*. *Med. Mycol.* **37**:85–96.

Molecular Principles of Fungal Pathogenesis
Edited by Joseph Heitman et al.
©2006 ASM Press, Washington, D.C.

Chapter 29

Zygomycetes as Agents of Infectious Disease in Humans

Ashraf S. Ibrahim and Brad Spellberg

The zygomycoses are a group of infections caused by a variety of fungal organisms belonging to the class Zygomycetes. This class is composed of two orders, Mucorales and Entomophthorales. Fungi belonging to the order Mucorales are distributed into six families, all of which can cause cutaneous and deep infections, mainly in immunocompromised patients (84) (Fig. 1). In contrast, the order Entomophthorales contains two families of organisms that cause subcutaneous and mucocutaneous infections primarily in immunocompetent hosts (45, 92) (Fig. 1). Since organisms belonging to both orders cause distinct infections, we refer to disease due to the Mucorales as mucormycosis and disease due to the Entomophthorales as entomophthoramycosis. The focus of this chapter is to discuss the molecular principles of mucormycosis pathogenesis.

ETIOLOGY

The agents of mucormycosis are classified into six different families based on morphologic analysis of the fungus, carbohydrate assimilation, and the maximal growth temperature. The organisms are distinguished morphologically by the presence and location of rhizoids, the presence of apophyses, and the morphology of the columellae (92). Because the diseases caused by the different families of Mucorales are clinically indistinguishable, laboratory confirmation of the identity of the causative agent is the only way to differentiate among these fungi. Members belonging to the family Mucoraceae are isolated more frequently from patients with mucormycosis than are the members of any other family. Among the members of the Mucoraceae, *Rhizopus oryzae* (*Rhizopus arrhizus*) is by far the most common cause of infection (84, 92). Other less frequently isolated species of the Mucoraceae that cause a similar spectrum of

infections include *Rhizopus microsporus* var. *rhizopodiformis*, *Absidia corymbifera*, *Apophysomyces elegans*, *Mucor* species, and *Rhizomucor pusillus* (45, 58, 84). Increasing numbers of cases of mucormycosis due to infection with *Cunninghamella* spp. (in the Cunninghamellaceae family) have also been reported (23, 56, 59, 101). Rare case reports have demonstrated the ability of species belonging to the remaining four families to cause mucormycosis (11, 50, 52, 64, 84).

ECOLOGY

The agents of mucormycosis are ubiquitous, thermotolerant organisms that usually grow in decaying matter including bread, vegetables, fruits, and seeds. They can also be recovered from soil, compost piles, and animal excreta (92). Abundant growth with airborne spores is usually seen in any carbohydrate-containing media within a few days. These spores are readily recovered as contaminants in laboratory cultures and can lead to misdiagnosis (102).

CLINICAL MANIFESTATIONS

A clinical hallmark of mucormycosis is vascular invasion resulting in thrombosis and tissue infarction or necrosis. Mucormycosis virtually always occurs in patients with defects in host defense and/or with increased available iron in their serum, although rare cases have been found in apparently normal hosts (60, 61, 79). In most cases, the infection is relentlessly progressive and results in death unless treatment with a combination of surgical debridement and antifungal therapy is promptly initiated.

EPIDEMIOLOGY

Based on clinical presentation and the involvement of a particular anatomic site, mucormycosis

Ashraf S. Ibrahim and Brad Spellberg • Division of Infectious Diseases, David Geffen School of Medicine at the University of California Los Angeles, Los Angeles Biomedical Research Institute at Harbor-UCLA Medical Center, 1124 W. Carson St., Torrance, CA 90502.

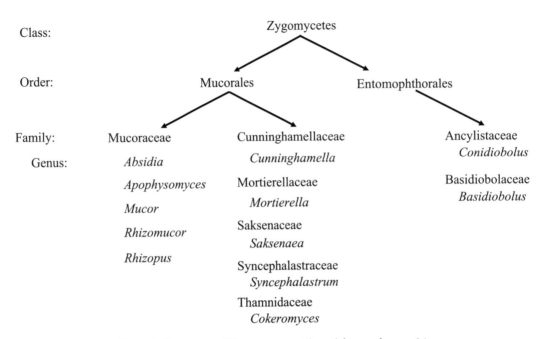

Figure 1. Taxonomy of Zygomycetes. Adapted from reference 84.

can be divided into at least six clinical categories: rhinocerebral, pulmonary, cutaneous, gastrointestinal, disseminated, and miscellaneous. Of note, these categories of invasive mucormycosis tend to occur in patients with specific defects in host defense. For example, diabetics in ketoacidosis typically develop rhinocerebral disease and more rarely develop pulmonary or disseminated disease (54, 70, 75, 78).

The mechanism by which ketoacidosis enhances susceptibility to rhinocerebral disease remains unclear. As discussed below, patients in ketoacidosis, or indeed those with any systemic acidosis, have increased available serum iron due to proton-mediated dissociation of iron from sequestering proteins (9). However, the disseminated form of the disease is most common in the context of deferoxamine therapy (32, 81, 97; J. J. Goodill and J. G. Abuelo, Letter, *N. Engl. J. Med.* **317**:54, 1987; J. H. Veis, R. Contiguglia, M. Klein, J. Mishell, A. C. Alfrey, and J. I. Shapiro, Letter, *Ann. Intern. Med.* **107**:258, 1987), a second setting in which increased iron is available to the fungus (see below), indicating that increased available iron cannot, by itself, explain the preferential occurrence of rhinocerebral disease in patients with ketoacidosis. Furthermore, while hyperglycemia and acidosis negatively impact neutrophil chemotaxis and phagocytic activity (22), this also cannot explain the preferential occurrence of rhinocerebral disease in patients with diabetic ketoacidosis because neutropenic patients tend to develop pulmonary rather than rhinocerebral disease (69, 96).

Much more readily understood are the risk factors for invasive skin and soft tissue infections caused by the agents of mucormycosis. These infections occur in patients with disrupted cutaneous barriers, either as a result of traumatic implantation of soil, direct access via intravenous catheters or subcutaneous injections (4, 53, 83), or even maceration beneath a contaminated, moist surface (5, 79). Indeed, a nationwide epidemic of cutaneous mucormycosis occurred in patients who had contaminated surgical dressings applied to their skin (37, 71). These latter cases illustrate a recent, alarming trend in the epidemiology of mucormycosis. Mucormycosis, formerly virtually always community acquired, is rapidly becoming a nosocomial infection in patients with malignancy or undergoing organ transplantation.

At transplant centers there has been an alarming rise in the incidence of mucormycosis (51, 89). For example, at the Fred Hutchinson Cancer Center, Marr et al. described a doubling in cases from 1985–1989 to 1995–1999 (67). Similarly, other investigators have described a greater than doubling in the incidence of mucormycosis in transplant patients over a similar time-span (57). In patients undergoing hematological stem cell transplantation, mucormycosis develops at least as commonly in non-neutropenic as in neutropenic periods. For example, two major transplant centers have reported that more than half the cases of mucormycosis occurred more than 90 days after transplantation (66, 67).

Major risk factors for mucormycosis in the transplant setting include underlying myelodysplastic syndrome (possibly due to iron overload from repeat blood transfusions) and graft-versus-host disease treated with steroids (66, 67, 77, 89). Administration of antithymocyte globulin may also be a risk factor for mucormcyosis (89). Although less than half of these patients are neutropenic at the time of disease onset, prolonged neutropenia is a risk factor for mucormycosis in this setting (85), as are diabetes mellitus and steroid use (85). The role of antifungal prophylaxis in predisposing patients to developing mucormycosis is increasingly being described. Prophylaxis with either itraconazole (85) or voriconazole (47, 48, 51, 68) has been implicated in predisposing to mucormycosis.

TREATMENT

General Principles

Four factors are critical for eradicating mucormycosis: rapidity of diagnosis, reversal of the underlying predisposing factors (if possible), appropriate surgical debridement of infected tissue, and appropriate antifungal therapy. Early diagnosis is important because small, focal lesions can often be surgically excised before they progress to involve critical structures or disseminate (75). Correcting or controlling predisposing problems is also essential for improving the treatment outcome. Hyperglycemia and acidemia in diabetic ketoacidotic patients should be corrected. Discontinuation of deferoxamine or immunosuppressive therapy, particularly steroids, should be strongly considered when the diagnosis of mucormycosis is made.

There have been no prospective randomized trials to define the optimal antifungal therapy for mucormycosis. Amphotericin B remains the only antifungal agent approved for the treatment of this infection (58, 84). Because many Mucorales isolates are relatively or highly resistant to amphotericin B, high doses of this drug are recommended (1.0 to 1.5 mg/kg/day). Unfortunately, at these high doses, nephrotoxicity is common. The lipid formulations of amphotericin B are significantly less nephrotoxic than amphotericin B deoxycholate and can be administered at higher doses for longer periods. Several case reports of patients with mucormycosis document successful treatment with a lipid formulation of amphotericin B at up to 15 mg/kg/day (19, 110; M. Ericsson, M. Anniko, M. Gustafsson, C. A. Hjalt, R. Stenling, and A. Tärnvik, Letter, *Clin. Infect. Dis.* **16**:585–586, 1993).

In our murine model, high-dose liposomal amphotericin B was considerably more effective than amphotericin B deoxycholate against disseminated infection caused by *R. oryzae* (43). In further support of the first-line role of lipid-based amphotericin B are the results of a recent retrospective review of 120 cases of mucormycosis in patients with hematological malignancies (38). Treatment with a lipid formulation of amphotericin B was associated with a 66% survival rate, compared to 40% survival when patients were treated with amphotericin B deoxycholate. Based in part on the historically poor success rates with amphotericin B deoxycholate and on the limited available animal and clinical data, there is a developing consensus that high doses of lipid formulation amphotericin B should be the initial antifungal therapy of choice for all patients with mucormycosis.

There is little to guide the choice of liposomal amphotericin B versus amphotericin B lipid complex for mucormycosis. On the basis of a single study with rabbits (39), it is thought that the brain penetration of liposomal amphotericin B is significantly greater than that of amphotericin B lipid complex. Therefore, in cases of central nervous system disease, liposomal amphotericin B may be preferred.

Itraconazole is the only marketed azole drug that has in vitro activity against members of the Mucorales (93). There are case reports of successful therapy with itraconazole alone (31, 83). However, itraconazole prophylaxis has been described as a risk factor for breakthrough mucormycosis (85), and the activity of itraconazole against the agents of mucormycosis in vitro and in animal models is somewhat limited (25, 26, 100). Voriconazole, a recently approved second-generation broad-spectrum triazole, is not active against the Mucorales in vitro (93). Conversely, posaconazole and ravuconazole, investigational triazoles, have promising in vitro activity against agents of mucormycosis (80, 93). In experimental animal models of disseminated mucormycosis, posaconazole is more efficacious than itraconazole but less efficacious than amphotericin B (25, 94). There are increasing reports of the use of salvage posaconazole therapy for refractory mucormycosis. Successful outcomes have been seen in patients with rhinocerebral mucormycosis in conjunction with amphotericin B (47), and in a heart and kidney transplant patient who failed to respond to amphotericin B therapy (98). Further data are needed to determine whether posaconazole, alone or in combination with amphotericin B, may be useful for the treatment of mucormycosis.

Caspofungin, the first member of the novel echinocandin class of antifungal drugs to be marketed in the United States, has minimal activity against the agents of mucormycosis when tested in vitro by standard techniques (29, 33). However, the accuracy of current in vitro testing of caspofungin activity against molds remains unclear. It is now known that *R. oryzae* expresses the target enzyme for caspofungin (44), and in the murine model of disseminated mucormycosis, caspofungin did have limited activity against *R. oryzae* (44). Furthermore, the combination of caspofungin plus amphotericin B lipid complex was synergistic, significantly improving survival compared to that of mice treated with either drug alone (91). Clinical experience with caspofungin in the setting of mucormycosis is extremely limited. In a report of a patient with necrotizing pancreatitis and abdominal mucormycosis, the addition of caspofungin to liposomal amphotericin B did not improve the outcome, but the patient had already clinically progressed while receiving liposomal amphotericin B prior to initiation of caspofungin therapy (103). There have also been reports of breakthrough mucormycosis in patients receiving either caspofungin alone (88) or caspofungin plus voriconazole (13). Conversely, the related experimental echinocandin, micafungin, has been added on a salvage basis to a patient failing to respond to amphotericin B therapy for craniofacial mucormycosis (49). The patient began responding to therapy shortly after the addition of micafungin and was ultimately cured. These data suggest that echinocandins may have a role as a second agent, in combination with amphotericin B or lipid-based amphotericin B, in serious cases of mucormycosis. Further study of the utility of this class of agents is needed in this setting.

Other Adjunctive Therapies

Case reports have suggested that hyperbaric oxygen may be a beneficial adjunct to the standard surgical and medical antifungal therapy of mucormycosis, particularly for patients with rhinocerebral disease (20, 36). It is hypothesized that hyperbaric oxygen might be useful for treating mucormycosis in conjunction with standard therapy because higher oxygen pressure improves the ability of neutrophils to kill the organism (24). Additionally, high oxygen pressure inhibits the germination of fungal spores and growth of mycelia in vitro (86).

The role of adjunctive cytokine therapy for mucormycosis has been insufficiently studied. A recent case report suggested a favorable outcome in a leukemic child with rhinocerebral mucormycosis following the addition of gamma interferon and granulocyte-macrophage colony-stimulating factor to the therapy (3). Further study of cytokines that activate host phagocyte function are warranted for this disease.

Prognosis

The overall survival rate of patients with mucormycosis is approximately 50%, although survival rates of up to 85% have been reported more recently. Much of the variability in outcome is due to the mix of patients in these studies. Rhinocerebral mucormycosis has a higher survival rate than does pulmonary or disseminated mucormycosis because the rhinocerebral disease can frequently be diagnosed earlier and the most common underlying predisposing factor, diabetic ketoacidosis, can be rapidly reversed (76). In contrast, pulmonary mucormycosis has a high mortality (65% at 1 year [67]) because it is difficult to diagnose and frequently occurs in myeloablated patients. For example, in one large study, only 44% of patients with pulmonary mucormycosis were diagnosed premortem and the overall survival rate was only 20% (96). In a separate study in which 93% of the infections were diagnosed premortem, the survival rate was 73% (76). Mortality in patients with disseminated disease approaches 100%, in large part because surgical removal of infected tissues is not feasible and in part because these patients tend to be the most highly immunocompromised (e.g., those undergoing allogeneic stem cell transplantation).

As these grim statistics emphasize, it is imperative to understand the pathogenesis of mucormycosis in order to devise better therapeutic strategies to prevent and/or treat these life-threatening infections.

PATHOGENESIS

Acquisition and Host Defense Strategies

Mucormycosis infections can be acquired through inhalation (e.g., rhinocerebral and pulmonary mucormycosis), ingestion of contaminated food (e.g., gastrointestinal mucormycosis), or direct penetration through abraded skin (cutaneous mucormycosis).

Both mononuclear and polymorphonuclear phagocytes of normal hosts kill members of the Mucorales by the generation of oxidative metabolites and the cationic peptides, defensins (30, 105, 109). Clinical evidence demonstrates that these phagocytes are the major host defense mechanism against mucormycosis. Neutropenic patients and those with dysfunctional phagocytes are at increased risk of developing mucormycosis. For example, hyperglycemia and acidosis impair the ability of phagocytes to use chemotaxis to reach the organisms and kill them by both oxidative and nonoxidative mechanisms (22). Additionally, corticosteroid treatment affects the ability of mouse bronchoalveolar macrophages to prevent spore germination in vitro or after in vivo infection induced by intranasal inoculation (109). The precise mechanisms by which ketoacidosis, diabetes, or steroids impair phagocyte function are unknown.

Role of Iron

As mentioned, another important clinical feature is the increased susceptibility of patients with elevated available iron levels in serum to mucormycosis. For example, multiple lines of evidence support the conclusion that patients with systemic acidosis have elevated levels of available iron in serum, probably due to release of iron from binding proteins in the presence of acidosis (9). In one study, sera collected from patients with diabetic ketoacidosis supported the growth of R. oryzae at acidic (7.3 to 6.88) but not alkaline (7.78 to 8.38) pH. Acidic sera that supported the growth of R. oryzae were found to contain increased available iron levels in serum (69 µg/dl versus 13 µg/dl for sera which did not support R. oryzae growth). Finally, simulated acidotic conditions decreased the iron-binding capacity of sera collected from normal volunteers (9). Therefore, the increased susceptibility of patients with diabetic ketoacidosis to mucormycosis is likely to be due at least in part to elevated available iron levels in serum during diabetic ketoacidosis.

Another context associated with elevated levels of available iron in serum is mucormycosis occurring in patients treated with the iron chelator deferoxamine (15). While deferoxamine is an iron chelator from the perspective of the human host, Rhizopus can utilize it as a siderophore to scavenge previously unavailable iron (14, 28). Rhizopus can accumulate 8- and 40-fold greater amounts of iron supplied by deferoxamine than can Aspergillus fumigatus and Candida albicans, respectively, and increased iron uptake by Rhizopus is linearly correlated with its growth in serum (14). Additionally, animal model studies emphasize an exceptional iron requirement for Rhizopus pathogenicity, since administration of deferoxamine or free iron impairs the survival of animals infected with Rhizopus but not C. albicans (1, 15, 28, 99). In contrast, other iron chelators, such as those belonging to the hydroxypyridinone family (e.g., L1 and CP94), cannot be used by the fungus to take up iron and hence do not support the growth of Rhizopus spp. in vitro (15). Moreover, the iron chelator L1 actually prolongs the survival time of guinea pigs infected with R. oryzae compared to that of placebo-treated animals (15). These data suggest that appropriate iron chelators may be useful as adjunctive therapies for mucormycosis.

Given the importance of available iron in mucormycosis infection, fungal iron transporters probably play a major role in disease pathogenesis. Two iron transportation systems have been identified in Rhizopus species. A high-affinity iron permease gene (rFTR1) has been isolated in R. oryzae and shown to complement the iron uptake defect of a Saccharomyces cerevisiae ftr1 mutant in iron-limiting medium (35). Importantly, rFTR1 gene expression is induced when R. oryzae is iron limited, indicating that this gene might operate in vivo where the iron supply is limited (35). The other iron transportation system that has been identified in R. oryzae includes a secreted siderophore from the hydroxamate family (40). However, the Rhizopus siderophore is inefficient in obtaining iron from serum (14, 28) and therefore probably contributes only minimally to virulence.

Role of Fungus-Endothelial Cell Interactions

As mentioned above, a hallmark of mucormycosis infections is extensive angioinvasion with vessel thrombosis and tissue necrosis. Angioinvasion is associated with hematogenous dissemination from the original site of infection to other target organs. Hence, damage and penetration of endothelial cells lining blood vessels is likely to be a critical step in the pathogenetic strategy of the organism. An earlier study demonstrated that R. oryzae spores and not pregerminated spores (germlings) adhere to subendothelial matrix proteins including laminin and type IV collagen in vitro (16). Similarly, we found that R. oryzae spores adhere to subendothelial matrix proteins significantly better than do R. oryzae hyphae but that spores and hyphae adhere equivalently to human umbilical vein endothelial cells (46). The disparity between spore and germ tube adherence to subendothelial matrix proteins, in the setting of equivalent adherence to endothelial cells, indicates that R. oryzae adhesins to endothelial cells are probably distinct from adhesins that bind subendothelial matrix proteins.

We also found that germlings of R. oryzae damage endothelial cells in vitro and that this damage is independent of serum factors and requires phagocytosis of R. oryzae by endothelial cells (46). Surprisingly, R. oryzae viability was not required for endothelial cell damage, but phagocytosis was required for dead R. oryzae cells to cause damage (46). The fact that dead R. oryzae cells damage human endothelial cells may, in part, explain the lack of efficacy of cidal antifungal agents during clinical disease. These results elucidate the nature of R. oryzae-endothelial cell interactions central to angioinvasion and tissue necrosis during mucormycosis infections.

Other putative virulence factors include the ability to secrete aspartic proteinases (34). Additionally,

Rhizopus species have an active ketone reductase system which may be an additional virulence factor by enhancing growth in the acidic, glucose-rich, ketoacidotic state (6).

Despite the availability of clinically relevant animal models of mucormycosis (43, 106–108), none of these putative virulence factors have been definitively proven to be essential for the development of mucormycosis, because necessary molecular biology tools to construct virulence factor mutants are in their infancy, as discussed further below.

GENE MANIPULATION

Selection Markers

Agents of the Mucorales are normally present as haploids. These organisms have a defined sexual cycle involving mating between the positive (+) and the negative (−) strains. However, classical genetic approaches have rarely been used for studying these organisms because germination of zygospores (resulting from fusing the nuclei of opposite mating types) takes months and the rate of germination is irregular and unpredictable. Most of the molecular biology techniques used to study the Mucorales have relied heavily on transforming the organisms with auxotrophic markers such as *leuA* (62), *pyrG* (*URA3*) (90), or *pyrF* (*URA5*) (unpublished data). These markers are often used to transform mutants that have been generated via chemical or UV mutagenesis and which therefore might harbor other undetected mutations that could complicate further investigations related to pathogenesis.

The use of dominant selection markers has the advantage of circumventing problems inherent to the use of auxotrophic markers. These advantages mean that there is no need to transform a mutated strain, the gene used for auxotrophic selection must be introduced in its native locus, there is no need to screen isolates on potentially mutagenic substances for recycling the auxotrophic selection marker (e.g., 5-fluoroorotic acid when selecting for *pyrG* or *pyrF* mutants), and there is less likelihood of isolating false transformants due to uracil carryover from rich medium resulting in growth of the transformed mutant on minimal medium without being transformed by the auxotrophic selection marker. Unfortunately, the agents of mucormycosis are generally resistant to commonly utilized dominant selection markers such as hygromycin B, phleomycin, nourseothericin, and zeocin. Additionally, although *R. oryzae* spores are sensitive to mycophenolic acid, this sensitivity is lost when the spores are germinated (unpublished data). To date, genes conferring resistance to neomycin or blasticidin S have been used in the transformation of *Absidia glauca* (112) and *R. niveus* (114), respectively. Recently, the *amdS* gene from *Aspergillus nidulans* driven by the pgk2 promoter of *R. niveus* was successfully used in transforming *R. oryzae* (72). The *amdS* gene allows the organism to grow on media containing acetamide as a sole source of nitrogen. Finally, the exogenous *aphI* marker, which confers resistance to kanamycin, was used to transform *Mucor miehei* by using the *Agrobacterium tumefaciens*-mediated method (73).

Transformation and Fate of Transformed DNA

Transformation of Mucorales has been achieved mainly by protoplasting. More recently, biolistic delivery and *A. tumefaciens*-mediated transformation have been successfully employed in gene transfer of *R. oryzae* (72, 90). In general, stable transformation of members of the Mucorales has been severely hampered by the inherent difficulty in generating integrative events, regardless of the method used (90). In most of the transformation systems described thus far, including the transformation of *R. oryzae* (72, 90), *R. delemar* (42), *R. niveus* (62, 95, 114), *Mucor circinelloides* (7, 12, 87), *A. glauca* (17), *Phycomyces blakesleeanus* (8), and *Rhizomucor pussilis* (104, 113), the transformed DNA remains extrachromosomal, replicating autonomously in a concatenated high-molecular-weight form that can be mistaken for integration in the genomic DNA. These transformants usually have very low mitotic stability when cultured on non-selection media. Of note, it has been shown for *Fusarium* (82), *Histoplasma capsulatum* (82, 111), and *Pestalotiopsis microspora* (63) that the fungus can introduce telomeric sequences onto exogenous plasmids that allow the plasmids to autonomously replicate. It is currently not known if the Mucorales introduce such sequences (e.g., autonomous replicating sequences or telomeric sequences) that would permit the transformed DNA to replicate autonomously. Given the autonomous replication of exogenous plasmids witnessed to date in Mucorales, it is possible that these fungi similarly introduce these sequences into plasmids.

Several investigators attempted to enhance the mitotic stability of the transformed DNA by using different methods to facilitate chromosomal integration, without a great deal of success. For example, although chromosomal integration of transformed DNA was achieved in *A. glauca* through the introduction of a repeated DNA sequence into an autonomously replicating plasmid, rearrangements associated with phenotypic changes were detected in the introduced DNA (18). Additionally, other attempts such as the inclusion of a SEG1 element in

A. glauca (17) or the introduction of a homologous rDNA region in *Marsupella alpina* (65) either failed to integrate or resulted in rearrangement of the introduced plasmid, respectively.

Recently it has been demonstrated that site directed integration can be achieved in *R. oryzae* by using a biolistic delivery system or *A. tumefaciens*-mediated transformation. In both techniques the *pyrG* gene was used as a selection marker (72, 90). Transformation with a biolistic delivery system, using a circular plasmid (transformation efficiency of 10 to 50 transformants per μg), resulted predominantly in the classic mitotically unstable transformants. These transformants contained autonomously replicated high-molecular-weight plasmid rather than integrations into the genome. When a plasmid was linearized within the *pyrG* gene and then transformed into an *R. oryzae pyrG* null mutant, 20% of the transformants underwent type I integration at the *pyrG* locus and 40% still occurred via extrachromosomal autonomous replication (90). A type III gene replacement of the mutated *pyrG* gene was also observed at a lower frequency and might represent homologous chromosomal integration, which could potentially be used for disruption of genes in *R. oryzae* (90).

Recently, transformation of fungi has been described by utilizing *A. tumefaciens* to insert the desired DNA into the fungal genome (2, 21, 27). However, unlike with other fungi, only protoplasts (not spores or germlings) of *R. oryzae* have been shown to be susceptible to transformation by this method (72). The transformation of protoplasts of *R. oryzae* could be problematic and inconsistent depending on the quality of protoplasts produced. However, this method yielded transformants that showed 100% mitotic stability resulting from integration of the transformed DNA (72). Because of the vigorous growth of the transformed *R. oryzae* on selection plates, no attempt was made to determine the transformation frequency. Nevertheless, Southern blot analysis of eight purified transformants revealed that six transformants had an extra copy of *pyrG*, indicating a type I integration at the same spot in the chromosome in all transformants. The integration at a common site in the chromosome suggested the presence of a hot spot for integration that has yet to be identified (72). In contrast, the other two transformants had undergone *pyrG* gene conversion replacement through type II integration since only one copy of the *pyrG* could be detected (72). It is important to mention that another member of the Mucorales, *M. miehei*, has also been transformed via *A. tumefaciens*-mediated transformation using the exogenous *aphI* marker. However, the resultant transformants were mitotically unstable, indicating that the transformed DNA had not stably integrated into the genome (73). It is possible that the Mucorales have a defense mechanism that detects foreign DNA and prevents its integration into the genome. This would explain why the use of an endogenous marker (e.g., *pyrG*) resulted in the high frequency of integration seen with *R. oryzae* transformed by the *Agrobacterium*-mediated method. These recent advances in gene transfer of members of the Mucorales have improved the stability of the resulting transformants. Such transformants can potentially be used for disruptions of specific targets to delineate the role of these targets in the pathogenesis of the disease. Of note, a single study has already described gene disruption in *M. circinelloides* by transforming a disruption cassette, utilizing *pyrG* as a marker flanked by 0.8 to 1.3 kb sequence corresponding to the upstream and downstream regions of the target gene (74).

Additionally, recently RNA interference to silence the lactate dehydrogenase gene (*LDH*) was applied to *R. oryzae*. By using a translational fusion construct driven by the *pgk* promoter of *R. oryzae*, it was feasible to significantly lower lactate dehydrogenase enzymatic activity and thereby reduce lactic acid production by up to 95% (C. D. Skory, *Abstr. 104th Gen. Meet. Am. Soc. Microbiol. 2004*, abstr. O-052, 2004). This method was shown to be stable (due to chromosomal integration of the construct) and gene specific. These molecular approaches will aid in understanding mucormycosis pathogenesis and may therefore contribute to establishing improved therapeutic strategies to treat and/or prevent mucormycosis infections.

Heterologous expression can also be used to delineate the function of a particular gene. This technique is widely used to study the function of *C. albicans* genes in isolation by expressing them in the surrogate organism *S. cerevisiae*. Not only can Mucorales genes be expressed in *S. cerevisiae*, but also they can complement *S. cerevisiae* null mutants. For example, we found that the high-affinity iron permease gene of *R. oryzae* (*rFTR1*), which is expressed in iron-limited medium, complements to restore the ability of an *S. cerevisiae ftr1* null mutant to grow on iron-limited medium (35). Of note, this expression in *S. cerevisiae* is dependant on the lack of introns in the transformed Mucorales gene (55). If the gene has an intron, it must be removed (41) or modified by the addition of *S. cerevisiae* splicing sequences (10). However, introns do not pose a barrier to gene expression in different species of a Mucorales genus, since the *pyrG* gene of *R. niveus* containing two putative introns was successfully expressed in *R. delemar* (42) and *R. oryzae* (72). Finally, heterologous expression of genes among different genera of the Mucorales has also been achieved. For example, the *LeuA* gene of *M. circinelloides* complemented an *R. niveus leu1⁻* mutant (95).

Genome Sequencing

Genome sequence data have the potential to revolutionize the study of mucormycosis, as has been done for other pathogenic fungi, including *Candida* and *Aspergillus*. A genome-sequencing project for *R. oryzae* was initiated as part of the Fungal Genome Initiative at the Broad Institute of Harvard and Massachusetts Institute of Technology. The project sequenced the genome of *R. oryzae* RA 99-880 (Fungal Testing Laboratory, University of Texas Health Science Center, San Antonio), a clinical isolate obtained from the brain of a patient with a fatal case of rhinocerebral mucormycosis. The genome sequence was recently completed, and a 10× assembly was released to the public and can be accessed through http://www.broad.mit.edu/annotation/fungi/rhizopus_oryzae/index.html. This sequence project will undoubtedly provide data invaluable in advancing our understanding of the pathogenesis of mucormycosis. Specifically, the sequence of individual genes, the presence of gene families, and the provision of raw data for development of large-scale functional genomics, such as microarrays, are likely to result from the sequence database.

SUMMARY

Mucormycoses are life-threatening infections that afflict the immunocompromised host. Despite current treatment options, mortality and morbidity of mucormycosis remain unacceptably high. An understanding of the pathogenesis of disease has the potential to lead to development of better prophylactic and/or therapeutic strategies. Therefore, advancing the molecular biology of the Mucorales, which is crucial to genetic manipulation of the organism to identify virulence factors, is of major importance.

REFERENCES

1. Abe, F., H. Inaba, T. Katoh, and M. Hotchi. 1990. Effects of iron and desferrioxamine on *Rhizopus* infection. *Mycopathologia* **110**:87–91.
2. Abuodeh, R. O., M. J. Orbach, M. A. Mandel, A. Das, and J. N. Galgiani. 2000. Genetic transformation of *Coccidioides immitis* facilitated by *Agrobacterium tumefaciens*. *J. Infect. Dis.* **181**:2106–2110.
3. Abzug, M. J., and T. J. Walsh. 2004. Interferon-gamma and colony-stimulating factors as adjuvant therapy for refractory fungal infections in children. *Pediatr. Infect. Dis. J.* **23**:769–773.
4. Adam, R. D., G. Hunter, J. DiTomasso, and G. Comerci, Jr. 1994. Mucormycosis: emerging prominence of cutaneous infections. *Clin. Infect. Dis.* **19**:67–76.
5. Alsuwaida, K. 2002. Primary cutaneous mucormycosis complicating the use of adhesive tape to secure the endotracheal tube. *Can. J. Anaesth.* **49**:880–882.
6. Anand, V. K., G. Alemar, and J. A. Griswold, Jr. 1992. Intracranial complications of mucormycosis: an experimental model and clinical review. *Laryngoscope* **102**:656–662.
7. Anaya, N., and M. I. Roncero. 1991. Transformation of a methionine auxotrophic mutant of *Mucor circinelloides* by direct cloning of the corresponding wild type gene. *Mol. Gen. Genet.* **230**:449–455.
8. Arnau, J., F. J. Murillo, and S. Torres-Martinez. 1988. Expression of Tn5-derived kanamycin resistance in the fungus *Phycomyces blakesleeanus*. *Mol. Gen. Genet.* **212**:375–377.
9. Artis, W. M., J. A. Fountain, H. K. Delcher, and H. E. Jones. 1982. A mechanism of susceptibility to mucormycosis in diabetic ketoacidosis: transferrin and iron availability. *Diabetes* **31**:1109–1114.
10. Ashikari, T., T. Amachi, H. Yoshizumi, H. Horiuchi, M. Takagi, and K. Yano. 1990. Correct splicing of modified introns of a *Rhizopus* proteinase gene in *Saccharomyces cerevisiae*. *Mol. Gen. Genet.* **223**:11–16.
11. Bearer, E. A., P. R. Nelson, M. Y. Chowers, and C. E. Davis. 1994. Cutaneous zygomycosis caused by *Saksenaea vasiformis* in a diabetic patient. *J. Clin. Microbiol.* **32**:1823–1824.
12. Benito, E. P., J. M. Diaz-Minguez, E. A. Iturriaga, V. Campuzano, and A. P. Eslava. 1992. Cloning and sequence analysis of the *Mucor circinelloides pyrG* gene encoding orotidine-5′-monophosphate decarboxylase: use of *pyrG* for homologous transformation. *Gene* **116**:59–67.
13. Blin, N., N. Morineau, F. Gaillard, O. Morin, N. Milpied, J. L. Harousseau, and P. Moreau. 2004. Disseminated mucormycosis associated with invasive pulmonary aspergillosis in a patient treated for post-transplant high-grade non-Hodgkin's lymphoma. *Leuk. Lymphoma* **45**:2161–2163.
14. Boelaert, J. R., M. de Locht, J. Van Cutsem, V. Kerrels, B. Cantinieaux, A. Verdonck, H. W. Van Landuyt, and Y. J. Schneider. 1993. Mucormycosis during deferoxamine therapy is a siderophore-mediated infection. In vitro and in vivo animal studies. *J. Clin. Investig.* **91**:1979–1986.
15. Boelaert, J. R., J. Van Cutsem, M. de Locht, Y. J. Schneider, and R. R. Crichton. 1994. Deferoxamine augments growth and pathogenicity of *Rhizopus*, while hydroxypyridinone chelators have no effect. *Kidney Int.* **45**:667–671.
16. Bouchara, J. P., N. A. Oumeziane, J. C. Lissitzky, G. Larcher, G. Tronchin, and D. Chabasse. 1996. Attachment of spores of the human pathogenic fungus *Rhizopus oryzae* to extracellular matrix components. *Eur. J. Cell Biol.* **70**:76–83.
17. Burmester, A., A. Wostemeyer, J. Arnau, and J. Wostemeyer. 1992. The SEG1 element: a new DNA region promoting stable mitotic segregation of plasmids in the zygomycete *Absidia glauca*. *Mol. Gen. Genet.* **235**:166–172.
18. Burmester, A., A. Wostemeyer, and J. Wostemeyer. 1990. Integrative transformation of a zygomycete, *Absidia glauca*, with vectors containing repetitive DNA. *Curr. Genet.* **17**:155–161.
19. Cagatay, A. A., S. S. Oncu, S. S. Calangu, T. T. Yildirmak, H. H. Ozsut, and H. H. Eraksoy. 2001.

Rhinocerebral mucormycosis treated with 32 gram liposomal amphotericin B and incomplete surgery: a case report. *BMC Infect. Dis.* **1**:22.

20. **Chassaing, N., L. Valton, M. Kany, E. Bonnet, E. Uro-Coste, M. B. Delisle, P. Bousquet, and G. Geraud.** 2003. Rhino-cerebral fungal infection successfully treated with supplementary hyperbaric oxygen therapy. *Rev. Neurol.* (Paris) **159**:1178–1180. (In French.)

21. **Chen, X., M. Stone, C. Schlagnhaufer, and C. P. Romaine.** 2000. A fruiting body tissue method for efficient *Agrobacterium*-mediated transformation of *Agaricus bisporus*. *Appl. Environ. Microbiol.* **66**:4510–4513.

22. **Chinn, R. Y., and R. D. Diamond.** 1982. Generation of chemotactic factors by *Rhizopus oryzae* in the presence and absence of serum: relationship to hyphal damage mediated by human neutrophils and effects of hyperglycemia and ketoacidosis. *Infect. Immun.* **38**:1123–1129.

23. **Cohen-Abbo, A., P. M. Bozeman, and C. C. Patrick.** 1993. *Cunninghamella* infections: review and report of two cases of *Cunninghamella* pneumonia in immunocompromised children. *Clin. Infect. Dis.* **17**:173–177.

24. **Couch, L., F. Theilen, and J. T. Mader.** 1988. Rhinocerebral mucormycosis with cerebral extension successfully treated with adjunctive hyperbaric oxygen therapy. *Arch. Otolaryngol. Head Neck Surg.* **114**:791–794.

25. **Dannaoui, E., J. F. Meis, D. Loebenberg, and P. E. Verweij.** 2003. Activity of posaconazole in treatment of experimental disseminated zygomycosis. *Antimicrob. Agents Chemother.* **47**:3647–3650.

26. **Dannaoui, E., J. Meletiadis, J. W. Mouton, J. F. Meis, and P. E. Verweij.** 2003. In vitro susceptibilities of zygomycetes to conventional and new antifungals. *J. Antimicrob. Chemother.* **51**:45–52.

27. **de Groot, M. J., P. Bundock, P. J. Hooykaas, and A. G. Beijersbergen.** 1998. *Agrobacterium tumefaciens*-mediated transformation of filamentous fungi. *Nat. Biotechnol.* **16**:839–842.

28. **de Locht, M., J. R. Boelaert, and Y. J. Schneider.** 1994. Iron uptake from ferrioxamine and from ferrirhizoferrin by germinating spores of *Rhizopus microsporus*. *Biochem. Pharmacol.* **47**:1843–1850.

29. **Del Poeta, M., W. A. Schell, and J. R. Perfect.** 1997. In vitro antifungal activity of pneumocandin L-743,872 against a variety of clinically important molds. *Antimicrob. Agents Chemother.* **41**:1835–1836.

30. **Diamond, R. D., C. C. Haudenschild, and N. F. Erickson III.** 1982. Monocyte-mediated damage to *Rhizopus oryzae* hyphae in vitro. *Infect. Immun.* **38**:292–297.

31. **Eisen, D. P., and J. Robson.** 2004. Complete resolution of pulmonary *Rhizopus oryzae* infection with itraconazole treatment: more evidence of the utility of azoles for zygomycosis. *Mycoses* **47**:159–162.

32. **Eiser, A. R., R. F. Slifkin, and M. S. Neff.** 1987. Intestinal mucormycosis in hemodialysis patients following deferoxamine. *Am. J. Kidney Dis.* **10**:71–73.

33. **Espinel-Ingroff, A.** 1998. Comparison of in vitro activities of the new triazole SCH56592 and the echinocandins MK-0991 (L-743,872) and LY303366 against opportunistic filamentous and dimorphic fungi and yeasts. *J. Clin. Microbiol.* **36**:2950–2956.

34. **Farley, P. C., and P. A. Sullivan.** 1998. The *Rhizopus oryzae* secreted aspartic proteinase gene family: an analysis of gene expression. *Microbiology* **144**:2355–2366.

35. **Fu, Y., H. Lee, M. Collins, H. F. Tsai, B. Spellberg, J. E. Edwards, Jr., K. J. Kwon-Chung, and A. S. Ibrahim.** 2004. Cloning and functional characterization of the *Rhizopus oryzae* high affinity iron permease (rFTR1) gene. *FEMS Microbiol. Lett.* **235**:169–176.

36. **Garcia-Covarrubias, L., D. M. Barratt, R. Bartlett, and K. Van Meter.** 2004. Treatment of mucormycosis with adjunctive hyperbaric oxygen: five cases treated at the same institution and review of the literature. *Rev. Investig. Clin.* **56**:51–55. (In Spanish.)

37. **Gartenberg, G., E. J. Bottone, G. T. Keusch, and I. Weitzman.** 1978. Hospital-acquired mucormycosis (*Rhizopus rhizopodiformis*) of skin and subcutaneous tissue: epidemiology, mycology and treatment. *N. Engl. J. Med.* **299**:1115–1118.

38. **Gleissner, B., A. Schilling, I. Anagnostopolous, I. Siehl, and E. Thiel.** 2004. Improved outcome of zygomycosis in patients with hematological diseases? *Leuk. Lymphoma* **45**:1351–1360.

39. **Groll, A. H., N. Giri, V. Petraitis, R. Petraitiene, M. Candelario, J. S. Bacher, S. C. Piscitelli, and T. J. Walsh.** 2000. Comparative efficacy and distribution of lipid formulations of amphotericin B in experimental *Candida albicans* infection of the central nervous system. *J. Infect. Dis.* **182**:274–282.

40. **Holzberg, M., and W. M. Artis.** 1983. Hydroxamate siderophore production by opportunistic and systemic fungal pathogens. *Infect. Immun.* **40**:1134–1139.

41. **Horiuchi, H., T. Ashikari, T. Amachi, H. Yoshizumi, M. Takagi, and K. Yano.** 1990. High-level secretion of a *Rhizopus niveus* aspartic proteinase in *Saccharomyces cerevisiae*. *Agric. Biol. Chem.* **54**:1771–1779.

42. **Horiuchi, H., N. Takaya, K. Yanai, M. Nakamura, A. Ohta, and M. Takagi.** 1995. Cloning of the *Rhizopus niveus pyr4* gene and its use for the transformation of *Rhizopus delemar*. *Curr. Genet.* **27**:472–478.

43. **Ibrahim, A. S., V. Avanessian, B. Spellberg, and J. E. Edwards, Jr.** 2003. Liposomal amphotericin B, and not amphotericin B deoxycholate, improves survival of diabetic mice infected with *Rhizopus oryzae*. *Antimicrob. Agents Chemother.* **47**:3343–3344.

44. **Ibrahim, A. S., J. C. Bowman, V. Avanessian, K. Brown, B. Spellberg, J. J. Edwards, and C. M. Douglas.** 2005. Caspofungin inhibits *Rhizopus oryzae* 1,3-β-D glucan synthase, lowers burden in brain measured by quantitative PCR, and improves survival at a low but not a high dose during murine disseminated zygomycosis. *Antimicrob. Agents Chemother.* **49**:721–727.

45. **Ibrahim, A. S., J. E. J. Edwards, and S. G. Filler.** 2003. Zygomycosis, p. 241–251. *In* W. E. Dismukes, P. G. Pappas, and J. D. Sobel (ed.), *Clinical Mycology.* Oxford University Press, New York, N.Y.

46. **Ibrahim, A. S., B. J. Spellberg, V. Avanessian, Y. Fu, and J. Edwards, Jr.** 2005. *Rhizopus oryzae* adheres to, is phagocytosed by, and damages endothelial cells in vitro. *Infect. Immun.* **73**:778–783.

47. **Ide, L., I. Buysschaert, H. Demuynck, R. De Man, A. Verlinde, E. De Laere, and I. Surmont.** 2004. Zygomycosis in neutropenic patients with past *Aspergillus* infection: a role for posaconazole? *Clin. Microbiol. Infect.* **10**:862–863.

48. **Imhof, A., S. A. Balajee, D. N. Fredricks, J. A. Englund, and K. A. Marr.** 2004. Breakthrough fungal infections in stem cell transplant recipients receiving voriconazole. *Clin. Infect. Dis.* **39**:743–746.

49. Jacobs, P., L. Wood, A. Du Toit, and K. Esterhuizen. 2003. Eradication of invasive mucormycosis—effectiveness of the echinocandin FK463. *Hematology* **8:**119–123.

50. Kamalam, A., and A. S. Thambiah. 1980. Cutaneous infection by *Syncephalastrum*. *Sabouraudia* **18:**19–20.

51. Kauffman, C. A. 2004. Zygomycosis: reemergence of an old pathogen. *Clin. Infect. Dis.* **39:**588–590.

52. Kemna, M. E., R. C. Neri, R. Ali, and I. F. Salkin. 1994. *Cokeromyces recurvatus*, a mucoraceous zygomycete rarely isolated in clinical laboratories. *J. Clin. Microbiol.* **32:**843–845.

53. Kerr, O. A., C. Bong, C. Wallis, and M. J. Tidman. 2004. Primary cutaneous mucormycosis masquerading as pyoderma gangrenosum. *Br. J. Dermatol.* **150:** 1212–1213.

54. Khor, B. S., M. H. Lee, H. S. Leu, and J. W. Liu. 2003. Rhinocerebral mucormycosis in Taiwan. *J. Microbiol. Immunol. Infect.* **36:**266–269.

55. Kohno, M., M. Enatsu, and W. Kugimiya. 1998. Cloning of genomic DNA of *Rhizopus niveus* lipase and expression in the yeast *Saccharomyces cerevisiae*. *Biosci. Biotechnol. Biochem.* **62:**2425–2427.

56. Kontoyiannis, D. P., S. Vartivarian, E. J. Anaissie, G. Samonis, G. P. Bodey, and M. Rinaldi. 1994. Infections due to *Cunninghamella bertholletiae* in patients with cancer: report of three cases and review. *Clin. Infect. Dis.* **18:**925–928.

57. Kontoyiannis, D. P., V. C. Wessel, G. P. Bodey, and K. V. Rolston. 2000. Zygomycosis in the 1990s in a tertiary-care cancer center. *Clin. Infect. Dis.* **30:**851–856.

58. Kwon-Chung, K. J., and J. E. Bennett. 1992. Mucormycosis, p. 524–559. *In Medical Mycology*. Lea & Febiger, Philadelphia, Pa.

59. Kwon-Chung, K. J., R. C. Young, and M. Orlando. 1975. Pulmonary mucormycosis caused by *Cunninghamella elegans* in a patient with chronic myelogenous leukemia. *Am. J. Clin. Pathol.* **64:**544–548.

60. Larsen, K., C. von Buchwald, B. Ellefsen, and D. Francis. 2003. Unexpected expansive paranasal sinus mucormycosis. *J. Otorhinolaryngol. Relat. Spec.* **65:** 57–60.

61. Lee, F. Y., S. B. Mossad, and K. A. Adal. 1999. Pulmonary mucormycosis: the last 30 years. *Arch. Intern. Med.* **159:**1301–1309.

62. Liou, C. M., K. Yanai, H. Horiuchi, and M. Takagi. 1992. Transformation of a Leu⁻ mutant of *Rhizopus niveus* with the *leuA* gene of *Mucor circinelloides*. *Biosci. Biotechnol. Biochem.* **56:**1503–1504.

63. Long, D. M., E. D. Smidansky, A. J. Archer, and G. A. Strobel. 1998. In vivo addition of telomeric repeats to foreign DNA generates extrachromosomal DNAs in the taxol-producing fungus *Pestalotiopsis microspora*. *Fungal Genet. Biol.* **24:**335–344.

64. Lye, G. R., G. Wood, and G. Nimmo. 1996. Subcutaneous zygomycosis due to *Saksenaea vasiformis*: rapid isolate identification using a modified sporulation technique. *Pathology* **28:**364–365.

65. Mackenzie, D. A., P. Wongwathanarat, A. T. Carter, and D. B. Archer. 2000. Isolation and use of a homologous histone H4 promoter and a ribosomal DNA region in a transformation vector for the oil-producing fungus *Mortierella alpina*. *Appl. Environ. Microbiol.* **66:** 4655–4661.

66. Maertens, J., H. Demuynck, E. K. Verbeken, P. Zachée, G. E. Verhoef, P. Vandenberghe, and M. A. Boogaerts. 1999. Mucormycosis in allogeneic bone marrow transplant recipients: report of five cases and review of the role of iron overload in the pathogenesis. *Bone Marrow Transplant.* **24:**307–312.

67. Marr, K. A., R. A. Carter, F. Crippa, A. Wald, and L. Corey. 2002. Epidemiology and outcome of mould infections in hematopoietic stem cell transplant recipients. *Clin. Infect. Dis.* **34:**909–917.

68. Marty, F. M., L. A. Cosimi, and L. R. Baden. 2004. Breakthrough zygomycosis after voriconazole treatment in recipients of hematopoietic stem-cell transplants. *N. Engl. J. Med.* **350:**950–952.

69. McAdams, H. P., M. Rosado de Christenson, D. C. Strollo, and E. F. Patz, Jr. 1997. Pulmonary mucormycosis: radiologic findings in 32 cases. *Am. J. Roentgenol.* **168:**1541–1548.

70. McNulty, J. S. 1982. Rhinocerebral mucormycosis: predisposing factors. *Laryngoscope* **92:**1140–1143.

71. Mead, J. H., G. P. Lupton, C. L. Dillavou, and R. B. Odom. 1979. Cutaneous *Rhizopus* infection. Occurrence as a postoperative complication associated with an elasticized adhesive dressing. *JAMA* **242:**272–274.

72. Michielse, C. B., K. Salim, P. Ragas, A. F. Ram, B. Kudla, B. Jarry, P. J. Punt, and C. A. van den Hondel. 2004. Development of a system for integrative and stable transformation of the zygomycete *Rhizopus oryzae* by *Agrobacterium*-mediated DNA transfer. *Mol. Genet. Genomics* **271:**499–510.

73. Monfort, A., L. Cordero, S. Maicas, and J. Polaina. 2003. Transformation of *Mucor miehei* results in plasmid deletion and phenotypic instability. *FEMS Microbiol. Lett.* **224:**101–106.

74. Navarro, E., J. M. Lorca-Pascual, M. D. Quiles-Rosillo, F. E. Nicolas, V. Garre, S. Torres-Martinez, and R. M. Ruiz-Vazquez. 2001. A negative regulator of light-inducible carotenogenesis in *Mucor circinelloides*. *Mol. Genet. Genomics* **266:**463–470.

75. Nithyanandam, S., M. S. Jacob, R. R. Battu, R. K. Thomas, M. A. Correa, and O. D'Souza. 2003. Rhino-orbito-cerebral mucormycosis. A retrospective analysis of clinical features and treatment outcomes. *Indian. J. Ophthalmol.* **51:**231–236.

76. Parfrey, N. A. 1986. Improved diagnosis and prognosis of mucormycosis. A clinicopathologic study of 33 cases. *Medicine* **65:**113–123.

77. Pavie, J., M. Lafaurie, C. Lacroix, A. Marie Zagdanski, D. Debrosse, G. Socie, F. Derouin, E. Gluckman, and J. Michel Molina. 2004. Successful treatment of pulmonary mucormycosis in an allogenic bone-marrow transplant recipient with combined medical and surgical therapy. *Scand. J. Infect. Dis.* **36:**767–769.

78. Peterson, K. L., M. Wang, R. F. Canalis, and E. Abemayor. 1997. Rhinocerebral mucormycosis: evolution of the disease and treatment options. *Laryngoscope* **107:**855–862.

79. Petrikkos, G., A. Skiada, H. Sambatakou, A. Toskas, G. Vaiopoulos, M. Giannopoulou, and N. Katsilambros. 2003. Mucormycosis: ten-year experience at a tertiary-care center in Greece. *Eur. J. Clin. Microbiol. Infect. Dis.* **22:**753–756.

80. Pfaller, M. A., S. A. Messer, R. J. Hollis, and R. N. Jones. 2002. Antifungal activities of posaconazole, ravuconazole, and voriconazole compared to those of itraconazole and amphotericin B against 239 clinical isolates of *Aspergillus* spp. and other filamentous fungi: report

from SENTRY Antimicrobial Surveillance Program, 2000. *Antimicrob. Agents Chemother.* **46:**1032–1037.

81. Ponz, E., J. M. Campistol, T. Ribalta, J. Montoliú, J. Ramírez, J. Almirall, and L. Revert. 1991. Disseminated mucormycosis in a hemodialyzed female patient treated with deferoxamine. *Rev. Clin. Esp.* **188:**85–87. (In Spanish).

82. Powell, W. A., and H. C. Kistler. 1990. In vivo rearrangement of foreign DNA by *Fusarium oxysporum* produces linear self-replicating plasmids. *J. Bacteriol.* **172:**3163–3171.

83. Quinio, D., A. Karam, J. P. Leroy, M. C. Moal, B. Bourbigot, O. Masure, B. Sassolas, and A. M. Le Flohic. 2004. Zygomycosis caused by *Cunninghamella bertholletiae* in a kidney transplant recipient. *Med. Mycol.* **42:**177–180.

84. Ribes, J. A., C. L. Vanover-Sams, and D. J. Baker. 2000. Zygomycetes in human disease. *Clin. Microbiol. Rev.* **13:**236–301.

85. Rickerts, V., A. Bohme, and G. Just-Nubling. 2002. Risk factor for invasive zygomycosis in patients with hematologic malignancies. *Mycoses* **45**(Suppl. 1):27–30.

86. Robb, S. M. 1966. Reactions of fungi to exposure to 10 atmospheres pressure of oxygen. *J. Gen. Microbiol.* **45:**17–29.

87. Ruiz-Hidalgo, M. J., A. P. Eslava, M. I. Alvarez, and E. P. Benito. 1999. Heterologous expression of the *Phycomyces blakesleeanus* phytoene dehydrogenase gene (*carB*) in *Mucor circinelloides*. *Curr. Microbiol.* **39:**259–264.

88. Safdar, A., S. O'Brien, and I. F. Kouri. 2004. Efficacy and feasibility of aerosolized amphotericin B lipid complex therapy in caspofungin breakthrough pulmonary zygomycosis. *Bone Marrow Transplant.* **34:**467–468.

89. Siwek, G. T., K. J. Dodgson, M. de Magalhaes-Silverman, L. A. Bartelt, S. B. Kilborn, P. L. Hoth, D. J. Diekema, and M. A. Pfaller. 2004. Invasive zygomycosis in hematopoietic stem cell transplant recipients receiving voriconazole prophylaxis. *Clin. Infect. Dis.* **39:**584–587.

90. Skory, C. D. 2002. Homologous recombination and double-strand break repair in the transformation of *Rhizopus oryzae*. *Mol. Genet. Genomics* **268:**397–406.

91. Spellberg, B., Y. Fu, J. Edwards, Jr., and A. S. Ibrahim. 2005. Combination therapy with amphotericin B lipid complex and caspofungin acetate of disseminated zygomycosis in diabetic ketoacidotic mice. *Antimicrob. Agents Chemother.* **49:**830–832.

92. Sugar, A. M. 1995. Agent of mucormycosis and related species, p. 2311–2321. *In* G. Mandell, J. Bennett, and R. Dolin (ed.), *Principles and Practices of Infectious Diseases*, 4th ed. Churchill Livingstone, New York, N.Y.

93. Sun, Q. N., A. W. Fothergill, D. I. McCarthy, M. G. Rinaldi, and J. R. Graybill. 2002. In vitro activities of posaconazole, itraconazole, voriconazole, amphotericin B, and fluconazole against 37 clinical isolates of zygomycetes. *Antimicrob. Agents Chemother.* **46:**1581–1582.

94. Sun, Q. N., L. K. Najvar, R. Bocanegra, D. Loebenberg, and J. R. Graybill. 2002. In vivo activity of posaconazole against *Mucor* spp. in an immunosuppressed-mouse model. *Antimicrob. Agents Chemother.* **46:**2310–2312.

95. Takaya, N., K. Yanai, H. Horiuchi, A. Ohta, and M. Takagi. 1996. Cloning and characterization of the *Rhizopus niveus leu1* gene and its use for homologous transformation. *Biosci. Biotechnol. Biochem.* **60:**448–452.

96. Tedder, M., J. A. Spratt, M. P. Anstadt, S. S. Hegde, S. D. Tedder, and J. E. Lowe. 1994. Pulmonary mucormycosis: results of medical and surgical therapy. *Ann. Thorac. Surg.* **57:**1044–1050.

97. Teggi, A., A. Fedele, R. Bagnato, F. Traditi, C. M. Lanzalone, and F. De Rosa. 1989. A case of disseminated mucormycosis. *Rec. Prog. Med.* **80:**308–313.

98. Tobon, A. M., M. Arango, D. Fernandez, and A. Restrepo. 2003. Mucormycosis (zygomycosis) in a heart-kidney transplant recipient: recovery after posaconazole therapy. *Clin. Infect. Dis.* **36:**1488–1491.

99. Van Cutsem, J., and J. R. Boelaert. 1989. Effects of deferoxamine, feroxamine and iron on experimental mucormycosis (zygomycosis). *Kidney Int.* **36:**1061–1068.

100. Van Cutsem, J., F. Van Gerven, J. Fransen, and P. A. Janssen. 1989. Treatment of experimental zygomycosis in guinea pigs with azoles and with amphotericin B. *Chemotherapy* **35:**267–272.

101. Ventura, G. J., H. M. Kantarjian, E. Anaissie, R. L. Hopfer, and V. Fainstein. 1986. Pneumonia with *Cunninghamella* species in patients with hematologic malignancies. A case report and review of the literature. *Cancer* **58:**1534–1536.

102. Verweij, P. E., A. Voss, J. P. Donnelly, B. E. de Pauw, and J. F. Meis. 1997. Wooden sticks as the source of a pseudoepidemic of infection with *Rhizopus microsporus* var. *rhizopodiformis* among immunocompromised patients. *J. Clin. Microbiol.* **35:**2422–2423.

103. Voitl, P., C. Scheibenpflug, T. Weber, O. Janata, and A. M. Rokitansky. 2002. Combined antifungal treatment of visceral mucormycosis with caspofungin and liposomal amphotericin B. *Eur. J. Clin. Microbiol. Infect. Dis.* **21:**632–634.

104. Wada, M., T. Beppu, and S. Horinouchi. 1996. Integrative transformation of the zygomycete *Rhizomucor pusillus* by homologous recombination. *Appl. Microbiol. Biotechnol.* **45:**652–657.

105. Waldorf, A. R. 1989. Pulmonary defense mechanisms against opportunistic fungal pathogens. *Immunol. Ser.* **47:**243–271.

106. Waldorf, A. R., and R. D. Diamond. 1984. Cerebral mucormycosis in diabetic mice after intrasinus challenge. *Infect. Immun.* **44:**194–195.

107. Waldorf, A. R., C. Halde, and N. A. Vedros. 1982. Murine model of pulmonary mucormycosis in cortisone-treated mice. *Sabouraudia* **20:**217–224.

108. Waldorf, A. R., S. M. Levitz, and R. D. Diamond. 1984. In vivo bronchoalveolar macrophage defense against *Rhizopus oryzae* and *Aspergillus fumigatus*. *J. Infect. Dis.* **150:**752–760.

109. Waldorf, A. R., N. Ruderman, and R. D. Diamond. 1984. Specific susceptibility to mucormycosis in murine diabetes and bronchoalveolar macrophage defense against *Rhizopus*. *J. Clin. Investig.* **74:**150–160.

110. Weng, D. E., W. H. Wilson, R. Little, and T. J. Walsh. 1998. Successful medical management of isolated renal zygomycosis: case report and review. *Clin. Infect. Dis.* **26:**601–605.

111. Woods, J. P., and W. E. Goldman. 1993. Autonomous replication of foreign DNA in *Histoplasma capsulatum*: role of native telomeric sequences. *J. Bacteriol.* **175:**636–641.

112. Wostemeyer, J., A. Burmester, and C. Weigel. 1987. Neomycin resistance as a dominantly selectable marker

for transformation of the zygomycete *Absidia glauca*. *Curr. Genet.* **12:**625–627.

113. **Yamazaki, H., Y. Ohnishi, K. Takeuchi, N. Mori, N. Shiraishi, Y. Sakata, H. Suzuki, and S. Horinouchi.** 1999. Genetic transformation of a *Rhizomucor pusillus* mutant defective in asparagine-linked glycosylation: production of a milk-clotting enzyme in a less-glycosylated form. *Appl. Microbiol. Biotechnol.* **52:**401–409.

114. **Yanai, K., H. Horiuchi, M. Takagi, and K. Yano.** 1991. Transformation of *Rhizopus niveus* using a bacterial blasticidin S resistance gene as a dominant selectable marker. *Curr. Genet.* **19:**221–226.

Molecular Principles of Fungal Pathogenesis
Edited by Joseph Heitman et al.
©2006 ASM Press, Washington, D.C.

Chapter 30

Signaling Pathways in the Dimorphic Human Fungal Pathogen *Penicillium marneffei*

ALEX ANDRIANOPOULOS AND SOPHIE ZUBER

SIGNALING FROM THE ENVIRONMENT

All organisms, whether free-living or intracellular in a host, must be able to sense their environment with respect to nutrients, light, temperature, and other signals and to respond appropriately. These signals can be considered either extrinsic or intrinsic depending on the exact point of interaction between the signal or stimulus and the receptor in the organism. For example, mating pheromones in fungi are detected by binding to transmembrane receptors (4). Many nutrients are also "sensed" by membrane-bound or associated receptors (25). In some instances these signals, although extrinsically derived, may not trigger detection until they are internalized and sometimes modified. In other cases, physical signals such as light or temperature (22, 45) are sensed internally.

Plasma membrane-localized receptors and their associated signal transduction pathways detect the presence or absence of a signal and effect the appropriate cellular response. Heterotrimeric guanine nucleotide-binding proteins (G proteins) link signals from receptors to targets by coupling cell surface receptors to cytoplasmic effector proteins. They are important components of signal transduction pathways in all eukaryotes. This chapter focuses on the heterotrimeric G proteins and their role in growth and morphogenesis in the thermally dimorphic human pathogen *Penicillium marneffei*.

P. marneffei is the only known dimorphic species in the genus or other closely related genera (1, 39, 56). Other dimorphic pathogenic fungi such as *Histoplasma capsulatum*, *Blastomyces dermatitidis*, *Coccidioides immitis*, and *C. posadasii* are distantly related to *P. marneffei* but closely related to each other (2). There are a number of species in closely related genera that are also important pathogens of animals or plants but are not dimorphic, such as the human pathogen *Aspergillus fumigatus*.

MORPHOGENESIS AND DEVELOPMENT IN *P. MARNEFFEI*

At either 25 or 37°C, and in the presence of a carbon source, *P. marneffei* conidia (spores) initiate germination by isotropic growth followed by a switch to highly polarized growth to produce a germ tube (Fig. 1). Persistent apical growth of the germ tube produces a hypha, and cellular compartments are established behind the growing tip by septation. Subapical cells are capable of repolarizing to produce branched cells with a new apical growth point, and this occurs at a higher frequency at 37°C than at 25°C. The compartments of actively growing cells are multinucleate, while most older hyphal compartments are often uninucleate.

At 25°C, *P. marneffei* grows by apical extension and branching to produce hyphae with multinucleate compartments. When grown in submerged liquid culture, *P. marneffei* continues to grow as vegetative hyphae. However, under the appropriate inductive conditions, which include an air interface, *P. marneffei* undergoes asexual development to produce complex, multicellular structures (conidiophores) bearing uninucleate spores (conidia) (Fig. 1). Asexual development begins with the emergence of a multinucleate aerial stalk from a specialized vegetative hyphal cell known as a foot cell. The tips of these stalk cells differentiate to produce the uninucleate metula and phialide cells, also collectively known as sterigmata cells, which are produced by a budding division rather than by apical growth and septation. Metula cells bud from the stalk tip, and these subsequently bud one or more of the sporogenous phialide cell type. Unlike the acropetal mode of division of all

Alex Andrianopoulos • Department of Genetics, University of Melbourne, Victoria 3010, Australia. **Sophie Zuber** • Quality and Safety Assurance Department, Nestle Research Center, Vers-chez-les-Blanc, CH-1000 Lausanne 26, Switzerland.

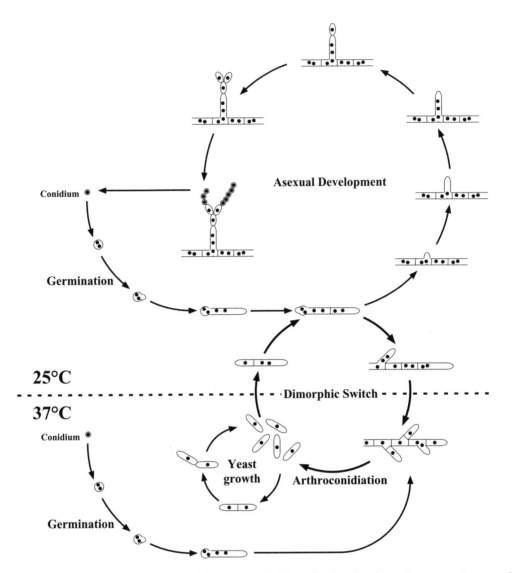

Figure 1. *P. marneffei* life cycle. A diagram of the *P. marneffei* life cycle, showing the various vegetative growth and developmental stages, is presented. The solid circles represent nuclei in cells. The filamentous phase exhibited at 25°C is shown above the dashed line and is characterized by hyphal growth and the asexual development program which produces conidia. The yeast phase expressed at 37°C is shown below the dashed line and is characterized by the arthroconidiation program, yeast growth, and division by fission. The dimorphic switch which links the yeast and hyphal growth phases in response to temperature is shown. (From reference 2 with permission.)

other cells produced at 25°C, phialides undergo repeated basipetal cell division to produce a chain of conidia such that older conidia are displaced by younger ones budded from the phialide. The mature conidia are pale green. Under conditions of carbon limitation (0.1% glucose), *P. marneffei* undergoes an accelerated and rudimentary form of asexual development where stalks are either very short and unbranched or absent such that sterigmata bud directly from foot cells. The temporal pattern of asexual development under carbon-limiting conditions (e.g., 0.1% glucose) is such that conidiophores appear several days before mature conidiophores are detected in carbon-rich medium (1% glucose) (2).

At 37°C, *P. marneffei* grows as uninucleate elongate yeast cells that divide by fission (13, 26, 56, 60) (Fig. 1). Yeast cells are produced from hyphal cells that have undergone a process called arthroconidiation. Arthroconidiation begins with the coupling of nuclear division and cell division such that uninucleate pre-arthroconidiating hyphal cells are formed. These cells are shorter (20 μm) along their long axis than are vegetative hyphal cells (40 μm) grown at 25°C and are separated by double septa. The pre-arthroconidial hyphal cells separate as the cell wall material between the double septa is degraded to liberate uninucleate, single cells called arthroconidia. Arthroconidia initiate polarized growth, maintaining the elongated cell shape, and divide by fission after nuclear division to produce the true yeast cells (2).

PATHOGENICITY IN *P. MARNEFFEI*

P. marneffei is endemic to Southeast Asia and causes the third most common opportunistic infection in AIDS patients (see references 18 and 70 and references therein). Confirmed cases of *P. marneffei* infection have been found in Europe, North America, Africa, and Australia although most of these incidents have been associated with travel in Southeast Asia (see references 18 and 72, and references therein). Untreated *P. marneffei* infections in immunocompromised individuals are fatal. Although infections respond to treatment with antimycotic agents, resistance to these drugs has developed and continued treatment is required to avoid relapse, especially in immunocompromised individuals (64–68).

Infection by *P. marneffei* is likely to occur through the inhalation of conidia produced by the hyphal form at 25°C. While germination of conidia at 37°C in vitro produces hyphal cells that subsequently undergo arthroconidiation, hyphal cells are not evident in infected individuals. Instead, *P. marneffei* is present as yeast cells in pulmonary alveolar macrophages and peripheral blood mononuclear cells (21). Although pulmonary alveolar macrophages appear to be the primary defense mechanism, *P. marneffei* yeast cells are capable of growing and dividing in these macrophages in immunocompromised hosts and lead to macrophage cell death.

HETEROTRIMERIC G PROTEINS

Extensive characterization of G proteins and their biological and biochemical activities has uncovered many key concepts in signal transduction (27, 28, 48, 49). G proteins are trimers whose activity is regulated by their association and dissociation from a heterotrimeric state to a Gα (α) monomer and a Gβγ (βγ) dimer. The α subunit consists of two well-characterized domains, a GTPase domain that contains the guanine nucleotide-binding pocket and a unique helical domain that buries the GTP in the core of the protein. The GTPase domain is structurally similar to the superfamily of GTPases including small G proteins and elongation factors. Although the βγ subunit is made up of two polypeptides, β and γ, it functions as a monomer and the two subunits cannot be dissociated without strong denaturants. The classical β subunit consists of seven WD-40 repeats which are arranged in a ring, forming a propeller structure. The WD-40 repeat motif is not unique to β subunits and signal transduction but is also found in proteins involved in pre-mRNA splicing, transcriptional regulation, assembly of the cytoskeleton,

and vesicular traffic (17). A new class of β subunits composed of kelch repeats, which are predicted to fold into structures strikingly similar to the WD-40-based β subunits, have been shown to interact with *Saccharomyces cerevisiae* α subunit Gpa2 and appear to play a similar role (30). The C-terminal region of the γ subunit is the site for prenylation that allows membrane attachment of the βγ complex (12).

G proteins are known to transmit signals from receptors that span the plasma membrane seven times and are therefore referred to as seven-transmembrane receptors or G-protein-coupled receptors. When the α subunit is bound to GDP, it associates with the βγ subunit to form an inactive heterotrimer that binds to the receptor. Although monomeric GDP-bound α subunits can interact with receptors, the association is greatly enhanced by βγ subunits. The C-terminal region of the α subunit, and in particular the last five residues, interacts directly with the receptor and is important for receptor specificity (9). Activation of the receptor by ligand binding leads to a conformational change in the receptor that in turn changes the conformation of the α subunit and leads to an exchange of GDP for GTP. This triggers the dissociation of both the receptor and the βγ complex from the α subunit (Fig. 2). The α subunit was thought to be the sole active component in signaling to downstream effectors, while the role of the βγ complex was simply to anchor the trimeric complex

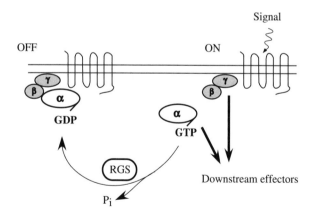

Figure 2. Heterotrimeric G protein signaling. Heterotrimeric G proteins are trimers whose function depends on the ability to dissociate into an α monomer and a βγ dimer. G proteins are generally coupled to 7-TM receptors. When GDP is bound, the α subunit associates with the βγ subunit to form an inactive complex that binds to the receptor. On binding of the signal to the receptor, the α subunit becomes activated and exchanges GDP for GTP, thus dissociating from the receptor and the βγ complex. Both the α and βγ subunits can trigger downstream signaling pathways. RGS proteins accelerate the GTP hydrolysis rate of the Gα subunit, thus driving the GTP-bound α subunit back to the inactive GDP-bound form.

and negatively regulate the signaling activity of the α subunit. This paradigm was fundamentally changed when biochemical and genetic experiments showed that the free βγ subunit and not the α subunit activates ion channels in mammals and the pheromone response pathway in *S. cerevisiae* (43, 71). It is now clear that both α and βγ subunits can interact with appropriate targets such as phosphodiesterases, protein kinases, adenylate cyclases, phospholipases, and ion channels to trigger downstream signaling pathways. Signaling is terminated when the intrinsic GTPase activity of the α subunit leads to the hydrolysis of GTP to GDP, thus leading to the reassociation of the α subunit and the βγ complex (Fig. 2).

Heterotrimeric G proteins transmit a broad range of biological signals. To maintain precise control over a particular signal and its effects, cells have a number of mechanisms that limit the duration of signaling beyond the slow intrinsic GTPase activity of the α subunit. Several molecular mechanisms that inhibit G protein signaling have been described. G protein signaling can be inhibited by negatively acting proteins called RGS proteins (for "regulator of G protein signaling"). These proteins accelerate the rate of GTP hydrolysis of α proteins, thus driving the GTP-bound α subunit back to the inactive GDP-bound form (Fig. 2) (37). Other mechanisms of signal attenuation operate at the level of the signal and its receptor. Ligands can be removed from the extracellular compartment by uptake or degradation. Receptors can be desensitized by phosphorylation due to the action of G protein coupled receptor kinases and second-messenger kinases such as protein kinase C (PKC) and PKA. Receptor phosphorylation leads to interaction with arrestins and the uncoupling from G proteins, effectively terminating the signaling. The plasma membrane can also be depleted of receptors by receptor endocytosis, which contributes to desensitization, resensitization, and polarization. A prolonged exposure of cells to a certain signal leads to increased degradation and diminished synthesis of receptors (5, 55).

CLASSIFICATION OF FUNGAL G PROTEIN SUBUNITS

At present at least 41 fungal genes encoding α subunits have been cloned, sequenced, and characterized, often by analysis of gene deletion mutants or mutants carrying dominant activating or dominant negative alleles. Many more fungal genes encoding α subunits are present in the genome sequence databases and are available for sequence comparison but have not been characterized and assigned a biological function. At least 14 genes encoding β and

4 genes encoding γ subunits have been cloned from fungi.

Phylogenetic analysis reveals three major classes within the family of fungal α subunits (46, 57). Some fungal species such as the yeasts *S. cerevisiae*, *Schizosaccharomyces pombe*, *Candida albicans*, and *Cryptococcus neoformans* possess two α subunits, while most filamentous fungi such as *Aspergillus nidulans*, *Neurospora crassa*, *Magnaporthe grisea*, and *Cryphonectria parasitica* possess three, with one α subunit in each class of fungal α subunits (78). Most fungal species, however, seem to have only one β and one γ subunit. The only α subunit that cannot be placed in any of the three classes is the *Ustilago maydis* Gpa4, which is only very distantly related to other fungal α proteins (57). Members of the class I fungal α subunits show high sequence homology to the mammalian αi class, containing a number of conserved functional motifs (69). However, *S. cerevisiae* Gpa1 and *C. albicans* Cag1 are exceptions. Class II and III fungal α subunits cannot be assigned clear homology to any of the mammalian α classes; however, members of class III fungal α subunits have been proposed to be related to the mammalian αs class (6).

There are three α subunits encoding genes identified in *P. marneffei*: *gasA*, *gasB*, and *gasC*. The *gasA* and *gasC* genes play distinct roles in *P. marneffei* morphogenesis yet show some overlap. The *gasA* gene encodes a major negative regulator of asexual development, while the *gasC* gene encodes a regulator of conidial germination. No clear role in morphogenesis or growth has been identified for the *gasB* gene. It is significant that none of these α subunits play a role in the hypha-yeast or yeast-hypha dimorphic switch in *P. marneffei*.

GasA SIGNALING REGULATES HYPHAL PROLIFERATION AND ASEXUAL DEVELOPMENT

GasA is a member of the class I fungal α subunits, which regulate various cellular processes such as growth, asexual development, virulence, and mating. The *P. marneffei gasA* gene encodes a key regulator of asexual development at 25°C. *P. marneffei* strains carrying mutations in *gasA* show defects in the decision between hyphal proliferation and asexual development. Mutants carrying a dominant activating *gasA*^{G42R} allele are locked in the vegetative growth mode and cannot initiate asexual development (Fig. 3) (76). In contrast, *P. marneffei* strains carrying a dominant negative *gasA*^{G203R} allele conidiate abundantly under conditions in which asexual development is inhibited such as growth in liquid culture. These data suggest that GasA signaling

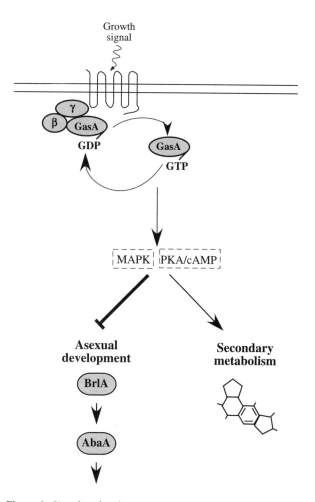

Figure 3. Signaling by the *gasA*-encoded α subunit. GasA signaling at 25°C is likely to involve both a cAMP-PKA pathway and a MAPK pathway in the negative regulation of asexual development and the expression of the core regulatory factors BrlA and AbaA. GasA also plays a minor positive role in the regulation of secondary metabolic pathways. However, it has no apparent role during yeast cell morphogenesis at 37°C. The signal/ligand for this process is not defined. The degree of regulation of each process is indicated by the thickness of the line.

antagonizes conidiophore development. Unlike the dominant negative *gasA*^{G203R} strain, the Δ*gasA* mutant does not produce conidiophores under noninducing conditions; instead, hyphal cells exhibit conidiophore-like characteristics. The difference between Δ*gasA* and *gasA*^{G203R} mutants is that in the former any proteins normally associated with the GasA α subunit are free to act in its absence. Therefore the lack of asexual development in the Δ*gasA* strain in liquid culture suggests that GasA acts with a βγ subunit and that both the α and βγ subunits antagonize conidiophore development. In the dominant negative *gasA*^{G203R} strain, where signaling of both the α and the βγ subunit is impaired, asexual development is evident (76).

The *brlA* gene encodes a transcriptional activator with a C$_2$H$_2$ zinc finger DNA-binding domain that is the central regulator of asexual development (A. R. Borneman, M. J. Hynes, and A. Andrianopoulos, unpublished data). Expression of the *brlA* gene is strictly limited to cells undergoing asexual development and is absent from all vegetative cell types. Loss of *brlA* function blocks asexual development under all conditions, while forced ectopic expression of *brlA* can drive asexual development under inappropriate conditions such as growth in liquid culture. The *brlA* gene is a downstream target for GasA-mediated signaling, since expression studies show that *brlA* expression is strongly induced in the dominant-negative *gasA*^{G203R} strain, moderately induced in the Δ*gasA* mutant, and absent in the wild type under conditions which normally block asexual development (76). Therefore, the levels of *brlA* expression are consistent with the phenotypic observations evident in *gasA* mutant strains.

Signaling by GasA negatively regulates asexual development and must be relieved before asexual development can proceed. One mechanism by which these negative effects are abrogated is through the downregulation of *gasA* transcription during asexual development, as evidenced by lower levels of *gasA* mRNA in conidiating cultures compared with vegetative hyphal cultures. This downregulation of *gasA* transcription may be more pronounced than observed if it is cell type restricted, since asexual development cultures consist of both vegetative and conidiophore cell types (7). Given that regulation of the GTPase activity of α subunits is the main mechanism by which signaling is regulated, in conjunction with specific regulators of this activity by RGS proteins, the transcriptional mechanism is likely to be a late or terminal event.

The role of *gasA* in regulation of growth and asexual development is analogous to that reported for *fadA* in *A. nidulans* (75). Strains carrying the dominant negative *fadA*^{G203R} allele conidiate in liquid culture, while the Δ*fadA* mutant does not (75). Introduction of the *P. marneffei gasA*^{G203R} allele into *A. nidulans* also leads to asexual development in liquid culture (76). Interestingly, the *A. nidulans fadA* gene does not show significant transcriptional control, unlike *P. marneffei gasA*. The *sfaD* and *gpgA* genes of *A. nidulans* encode Gβ (β) and Gγ (γ) subunits, respectively (59, 61). Deletion of either *sfaD* or both *sfaD* and *gpgA* leads to hyperactive asexual development, while deletion of *gpgA* alone does not (59, 61). This suggests that localization of the G protein complex, provided primarily by the γ subunit, is required for correct signaling by the complex.

Inactivation of *fadA*, *sfaD*, or *gpgA* all result in reduced vegetative growth (61).

The βγ subunit of the plant pathogen *M. grisea* also appears to be actively involved in signaling. Strains carrying either a dominant activating *magB*^{G42R} allele or a Δ*magB* allele show a severe reduction in asexual development, whereas no significant effect on asexual development is evident in strains carrying the dominant negative *magB*^{G203R} allele. This suggests that βγ in *M. grisea* actively downregulates asexual development in both null and constitutively active *magB*^{G42R} mutants (23).

In addition to controlling asexual development, the *gasA* gene is involved in regulating the expression of a carotinoid-based pigment produced by *P. marneffei* at 25°C but not at 37°C (76). Strains carrying the dominant activating *gasA*^{G42R} allele produce excessive pigment compared to the wild-type strain, but this does not extend to production of this pigment at 37°C. Strains carrying either the Δ*gasA* or the dominant negative alleles produce substantially less pigment at 25°C. Although the effects of these *gasA* alleles are clear, they are much weaker than those for *gasC* mutants (see below). In contrast, the *A. nidulans fadA* gene plays a major role in the control of secondary-metabolite production (33) (see below).

The *P. marneffei gasA* gene does not play a significant role in hyphal proliferation, since neither deletion of *gasA* nor expression of dominant activating or negative alleles affects hyphal growth (76). This shows that GasA functions by activating a signaling pathway that inhibits asexual development, as has been noted in *M. grisea* (23). In contrast, *A. nidulans fadA* both antagonizes conidiophore development and plays a significant role in the regulation of hyphal proliferation. A dominant activating *fadA*^{G42R} mutation promotes growth and eventually results in colony autolysis, while the Δ*fadA* and the dominant negative *fadA*^{G203R} mutants exhibit moderate and severe growth defects, respectively. This suggests that FadA acts positively to regulate growth (75). Hyphal proliferation is also regulated by the βγ subunit, since deletion of either the *sfaD* or *gpgA* genes results in restricted vegetative growth (59, 61). Therefore, the regulation of vegetative growth in *P. marneffei*, either hyphal at 25°C or yeast at 37°C, is likely to be regulated by one or more different, independent signal transduction pathway(s). Hyphal proliferation and asexual development are independently regulated in *N. crassa* by the α subunits GNA1 and GNA3 (36, 73).

Although the downstream signaling cascade activated by GasA has not been conclusively demonstrated by genetic and biochemical studies, toxicological studies suggest that a cyclic AMP (cAMP)-PKA pathway propagates the *gasA* signal (76, 78). A marked inhibition of growth is evident in the presence of the PKA inhibitor H-89 for both mutants negatively affected in *gasA* signaling (the dominant negative *gasA*^{G203R} and Δ*gasA* strains), which was not observed in the wild-type and dominant activating

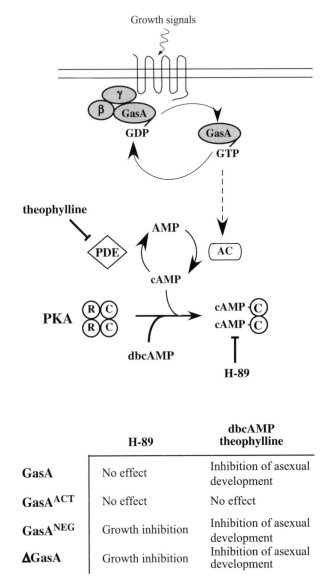

	H-89	dbcAMP theophylline
GasA	No effect	Inhibition of asexual development
GasA^{ACT}	No effect	No effect
GasA^{NEG}	Growth inhibition	Inhibition of asexual development
ΔGasA	Growth inhibition	Inhibition of asexual development

Figure 4. Toxicological studies of GasA signaling. The predicted signaling pathway through GasA to the cAMP-PKA pathway, based on genetic and physiological studies using chemical inhibitors and analogues as well as known pathways in related fungi, is shown. Adenylate cyclase (AC), phosphodiesterase (PDE), and PKA with its catalytic (C) and regulatory (R) subunits are illustrated, as is the action of H-89, theophylline, and dbcAMP on the cAMP-PKA pathway. The effect of these toxicological agents on growth and development for the wild type (GasA) and the strains carrying either a *gasA* dominant activating (GasA^{ACT}), dominant negative (GasA^{NEG}), or deletion (ΔGasA) allele is presented. The studies support the involvement of a cAMP-PKA pathway in GasA signaling.

gasA[G42R] strains (Fig. 4). Furthermore, use of nonmetabolizable dibutyryl cAMP and inhibition of phosphodiesterase activity with theophylline, which lead to an elevation in the intracellular cAMP level, results in the inhibition of asexual development in the wild-type, dominant negative *gasA*[G203R] and Δ*gasA* strains (Fig. 4). These results suggest that GasA activates a cAMP-PKA cascade and imply that GasA may function as an αs subunit stimulating adenylate cyclase despite its sequence similarity to mammalian αi subunits, which are known to inhibit adenylate cyclase. Therefore, GasA represents another example of fungal α proteins with similarity to the αi class but which are more likely to function as activators of adenylyl cyclase, such as FadA of *A. nidulans* (75) and GNA1 of *N. crassa* (35), while CPG1 of *C. parasitica* seems to work as a bona fide αi protein (15). It has been shown that FadA signaling in *A. nidulans* is mediated through PkaA, the catalytic subunit of a PKA. Interestingly, FadA also appears to play an additional PkaA-independent role in the regulation of asexual development, possibly through another PKA or mitogen-activated protein kinase (MAPK) cascade (62).

The growth inhibition for the dominant negative *gasA*[G203R] and the Δ*gasA* strains when grown in the presence of the PKA inhibitor H-89, which is not observed in the absence of H-89, indicates that a second signal transduction pathway may be signaling through the same cAMP-PKA cascade as GasA. This implies that inhibiting PKA with H-89 in a *gasA*[G203R] or Δ*gasA* background enhances an underlying effect on growth which was undetectable when the dominant negative *gasA*[G203R] and Δ*gasA* strains were grown in the absence of H-89, suggesting that PKA was sufficiently active due to the putative second signaling pathway. In *N. crassa*, two different α subunits, GNA1 and GNA3, regulate cAMP levels, leading to a dual regulation of adenylate cyclase. GNA1 is a direct positive regulator of adenylyl cyclase, and GNA3 influences cAMP levels by posttranscriptionally controlling the amount of adenylyl cyclase protein (35, 36).

The PKA inhibitor H-89 is unable to rescue the aconidial phenotype of the dominant activating *gasA*[G42R] strain. This may be due to incomplete inhibition of the *P. marneffei* PKA by H-89, such that GasA signaling in the dominant activating *gasA*[G42R] strain remains sufficient to suppress asexual development. Alternatively, GasA may signal not only through a cAMP-PKA cascade but also via another signal transduction cascade such as a MAPK pathway. In the presence of caffeine, an accentuated phenotype is evident in both mutants negatively affected in their *gasA* signaling (the dominant negative

gasA[G203R] and Δ*gasA* strains). Caffeine sensitivity is a characteristic phenotype in *S. cerevisiae* mutants affected in MAPK signaling, and its effect has best been described for the PKC1-mediated MAPK pathway essential for the maintenance of cell wall integrity (29, 47). In contrast to *S. cerevisiae*, where sensitivity to caffeine leads to growth inhibition, in *P. marneffei* a strong effect was observed on the production of aerial hyphae in the sensitive strains and no reduction in colony size. Since caffeine sensitivity in *S. cerevisiae* has been linked to a defect in the PKC1-MPK1 cell integrity pathway (47), the caffeine phenotype observed in *P. marneffei* may indicate that GasA not only signals to regulate asexual development but also signals through a MAPK to coordinate a cell integrity pathway. A caveat to these studies is that caffeine is also known to act as an inhibitor of mammalian phosphodiesterase (54), and a similar effect on phosphodiesterases of lower eukaryotes has not been ruled out at this stage. In *A. nidulans*, a dominant activating *fadA*[G42R] allele leads to colonies which autolyze, and in *N. crassa*, Δ*gna1* mutants are more sensitive to hyperosmotic medium than is the wild type (34, 74). This suggests that G protein signaling may also play a role in the regulation of a cell integrity pathway in filamentous fungi. However, no cell lysis defect on normal medium or sensitivity to hyperosmotic medium was observed in the *P. marneffei gasA* mutants compared to the wild type. The MAPK pathway proposed to be involved in GasA signaling is therefore more likely to be directly involved in the regulation of asexual development, as probably is the cAMP-PKA pathway.

GasA is a key regulator of asexual development but does not play a role in dimorphic switching or yeast growth. This is despite the fact that in *P. marneffei*, asexual development and dimorphic switching share transcriptional regulatory elements, such as the *abaA* gene, which is required for both asexual development and correct yeast cell morphogenesis (7). At 25°C, GasA regulates the expression of *brlA*, the asexual development-specific regulatory gene (A. R. Broneman, M. J. Hynes, and A. Andrianopoulos, unpublished data), and a dominant negative *gasA*[G203R] strain shows inappropriate *brlA* expression at this temperature (76).

However, at 37°C, the *brlA* transcript is not present in any of the *gasA* mutants, suggesting the presence of a *brlA* repressor and an *abaA* activator that are independent of GasA at this temperature. The α subunit GasA is therefore probably linked to a receptor and a signal which are specific for asexual development, indicating that signaling components regulating asexual development and dimorphic switching are not shared in *P. marneffei*.

P. MARNEFFEI gasC REGULATES CONIDIAL GERMINATION

The *P. marneffei gasC* gene encodes a G protein α subunit that shows high homology to members of class III fungal α subunits. Deletion of *gasC* severely delays the germination of conidia (Fig. 5) (77). A similar but slightly less severe delay is evident in strains carrying the dominant negative *gasC*^{G207R} allele, while strains carrying a dominant activating *gasC*^{G45R} allele show a significantly accelerated germination rate relative to the wild type. Once germination is effected, hyphal growth of *gasC* strains does not differ from that of the wild type. In many fungi, spores require water and a fermentable carbon source such as glucose in order to activate germination (19, 32, 52). GasC does not appear to be part of the carbon-sensing mechanism during germination, as strains carrying the dominant activating *gasC*^{G45R} allele remain dependent on a carbon source for germination. The nature of the carbon source used does not appear to selectively affect either the deletion strain or the dominant mutant strains; good carbon sources result in faster germination for all strains, and poor carbon sources produce slower germination rates. Therefore, it is unlikely that GasC is linked to a glucose sensor required for conidial germination. This contrasts with the action of *gasC* orthologues in *A. nidulans* and *S. pombe*. Dominant activating *A. nidulans ganB*^{Q208L} alleles, and dominant activating *P. marneffei gasC*^{G45R} alleles, in *A. nidulans* lead to accelerated conidial germination (14, 78). More importantly, *A. nidulans* strains carrying the dominant activating *ganB*^{Q208L}

allele are able to germinate in the absence of a carbon source (14). In contrast, only *A. nidulans* dominant negative *ganB*^{Q208L} alleles delay conidial germination in *A. nidulans* (14). The GasC homologue of *S. pombe*, Gpa2, mediates glucose sensing and is required for ascospore germination (31).

Ungerminated spores of the Δ*gasC* mutant show no morphological signs of germination long after wild-type spores have germinated. However, these Δ*gasC* mutant spores are only severely delayed, not absolutely blocked, in germination (77). This shows that GasC is linked to a very early sensor triggering germination, which can be overcome by activation of downstream factors or alternate pathways. Two other genes are known to be involved in germination in *P. marneffei*: the RAS orthologue *rasA* and the CDC42 orthologue *cflA* (10, 11). Strains carrying the dominant negative *rasA*^{D125A} or dominant activating *rasA*^{G19V} allele show significantly reduced rates of conidial germination compared to the wild type, while the number of germlings with abnormal morphology is substantially increased (Fig. 5) (11). These data suggest that cycling of RasA activity is necessary for correct temporal and spatial execution of the germination program. Mirroring the results for *gasC*, strains carrying a dominant negative *cflA*^{D120A} allele are severely delayed in germination while strains carrying a dominant activating *cflA*^{G14V} allele show precocious activation of germination (Fig. 5) (10). No apparent defect in morphology of the germlings was evident in either mutant strain. The *rasA* and *cflA* genes appear to be in the same genetic pathway, since the dominant activating *cflA*^{G14V} allele is able to suppress the conidial germination defect of the dominant negative *rasA*^{D125A} allele in double-mutant strains (11). It remains to be determined whether *gasC* is the upstream component of the *rasA*/*cflA* signaling pathway or part of an independent pathway. Examination of early markers of germination, such as trehalose mobilization, may help to delineate the role of GasC in germination (20, 24, 53). In *A. nidulans*, RasA plays an important role in the progression from isotropic to polarized growth during conidial germination (63) but its role during initiation of germination is unclear (24, 51). The dominant activating *rasA*^{G17V} allele produces giant, swollen, multinucleate spores in either the presence or absence of glucose (51). However, similar studies have been unable to reproduce these *rasA*^{G17V} phenotypes (24). The *P. marneffei rasA* strains are similarly unable to germinate in the absence of glucose (K. J. Boyce, M. J. Hynes, and A. Andrianopoulos, unpublished data), suggesting that the presence of a carbon source is crucial for germination.

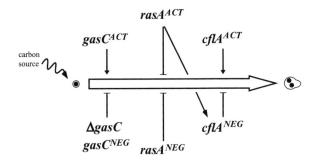

Figure 5. Control of conidial germination. Conidial germination in *P. marneffei* is controlled by the action of the α-subunit-encoding *gasC*, the Ras GTPase-encoding *rasA*, and the small Rho-type (CDC42) GTPase-encoding *cflA* genes. Dominant activated *gasC*^{ACT} or dominant activated *cflA*^{ACT} alleles result in precocious germination under inducing conditions (i.e., presence of a carbon source). Conidial germination is delayed in a Δ*gasC* strain or in strains carrying the dominant negative *cflA*^{NEG} allele or dominant negative or dominant activated *rasA*^{ACT}/*rasA*^{NEG} alleles. The *rasA* gene is upstream of *cflA* based on genetic interaction studies.

Like GasA, GasC also plays a role in the regulation of secondary-metabolite production and asexual development (Fig. 6). Strains carrying the dominant activating *gasC*^{G45R} allele show increased production of the *P. marneffei* red pigment at 25°C, while strains carrying Δ*gasC* or dominant negative *gasC*^{G207R} alleles fail to produce the red pigment (77). Conversely, strains carrying the dominant activating *gasC*^{G45R} allele show delayed onset of asexual development and reduced conidial yields while strains carrying the Δ*gasC*, but not the dominant negative *gasC*^{G207R}, allele undergo more extensive asexual development to produce denser asexual development (77). Neither of these processes implicates carbon source sensing with GasC. Wild-type *P. marneffei* asexual development is delayed by high levels of glucose. If *gasC* mediated a carbon signal, the Δ*gasC* mutant would have been expected to conidiate precociously regardless of the carbon source.

As with *gasA*, mutation of *gasC* has no detectable effect on the dimorphic switch or the maintenance of yeast growth at 37°C. This contrasts with the situation in *S. cerevisiae* and *U. maydis*, where the GasC orthologues Gpa2 and Gpa3 are required for pseudohyphal growth and dimorphic switching, respectively (38, 44, 57). Two major differences may explain this difference in regulation in *P. marneffei* compared to *S. cerevisiae* and *U. maydis*. First, the latter organisms are both predominantly yeasts that undergo a yeast-to-hypha transition while *P. marneffei* is predominantly hyphal and switches between a hyphal and yeast form. Second, the inducing signal(s) for dimorphic switching for the latter organisms is chemical in nature while the signal for *P. marneffei* is temperature, a physical signal. Given these differences, it is not surprising that the signaling mechanisms may have evolved independently.

The downstream signaling pathway activated by GasC is likely to be a PKA-cAMP cascade (Fig. 6). Although there is no direct evidence to support this hypothesis at this stage, the phenotype of the Δ*gasC* mutant, delayed germination and a highly asynchronous germination pattern, is similar to that seen in the *A. nidulans* adenylyl cyclase Δ*cyaA* mutant (24). The *A. nidulans* Δ*pkaA* mutant also demonstrates a delay in germination, and this mutation affects the production of secondary metabolites and asexual development (24, 62). These three processes, germination, asexual development, and secondary metabolism, are all regulated by GasC in *P. marneffei*. Recently, it has been shown that *A. nidulans* GanB mediates changes in cAMP levels early in germination and that it is likely to form a heterotrimeric G protein complex with SfaD (β) and GpgA (γ) (40). Similarly, the *gasC* orthologue in *A. fumigatus*, *gpaB*,

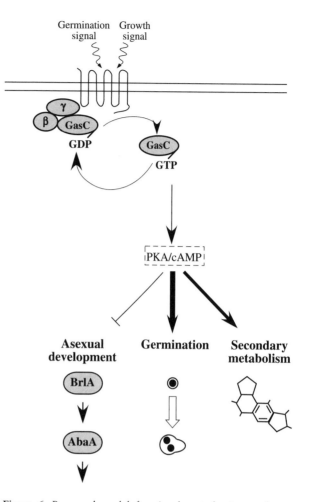

Figure 6. Proposed model for signal transduction pathways regulating germination, asexual development, and secondary metabolism in *P. marneffei*. The α subunit GasC plays a major role in the regulation of germination. GasC also regulates the production of secondary metabolites and, to a lesser extent, asexual development. The action of the βγ subunit in GasC signaling is unclear, and it is proposed that GasC acts through the cAMP-PKA pathway. The signals and receptors linked to this G protein are unknown. The degree of regulation of each process is indicated by the thickness of the line.

is required for PKA activity, which can be restored by the addition of exogenous cAMP. *A. fumigatus* Δ*gpaB* mutants show drastically reduced virulence in a mouse inhalation infection model (41).

With respect to asexual development and secondary metabolism, *P. marneffei* *gasA* and *gasC* have overlapping roles. GasA-mediated signaling strongly inhibits asexual development but only weakly regulates red-pigment production, while GasC-mediated signaling weakly regulates asexual development and strongly regulates pigment production (Fig. 7). The overlapping roles between these two α subunits suggests that they share signaling components or effectors.

Many fungi have three α subunit encoding genes but only one β and one γ, and so it had been suggested that either (i) these α subunits might share the β and γ subunits, (ii) some α subunits might function independently of a β and γ subunit, or (iii) a previously unrecognized βγ subunit may be involved. The last case was shown to be the case for the *S. cerevisiae* Gpa2 α subunit, which interacts with a new type of β subunit containing kelch repeats rather than WD-40 repeats (30). In addition, genetic experiments have shown that the *A. nidulans* SfaD (β) and GpgA (γ) subunits function with both the FadA and GanB α subunits (40, 59, 61). Therefore, it is also possible that the weak regulation of asexual development by GasC and pigment production by GasA could be indirect effects due to cross-activation of the main

programs regulated by each of the α subunits. If this is the case, it is interesting that this cross-reactivity does not extend to the control of germination, which is strictly a *gasC*-regulated process.

P. MARNEFFEI gasB IS A GENE WITHOUT A FUNCTION

P. marneffei GasB is a member of class II fungal α subunits. With the exceptions of *S. pombe* Gpa1 (50) and *M. grisea* MagC (42), most deletion mutants of members of this class have produced either very minor or no visible phenotype, making it difficult to assign a biological function to these proteins. For example, deletion of *U. maydis* gpa2 (57) or *N. crassa* gna2 (3) does not cause any abnormalities during

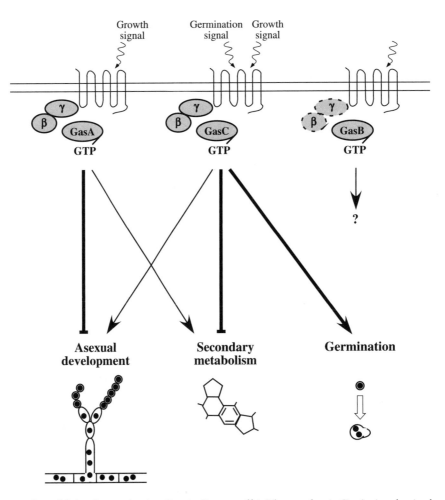

Figure 7. Proposed model for G protein signaling in *P. marneffei*. The α subunit GasA signals via the cAMP-PKA pathway and a MAPK pathway to regulate asexual development and, to a lesser extent, secondary metabolism. GasC plays a major role in the regulation of germination. GasC also regulates secondary metabolites and, to a lesser extent, asexual development and is proposed to signal through the cAMP-PKA pathway. The overlap between GasA and GasC function may reflect the fact that they may share a βγ subunit. The action of the α subunit GasB is unknown. The degree of regulation of each process is indicated by the thickness of the line.

mating or any other essential steps in the life cycle and development of these fungi.

Analysis of this third α subunit-encoding gene, *gasB*, in *P. marneffei* reveals that GasB does not affect essential steps in the growth and development of *P. marneffei* such as germination, asexual development, and dimorphic switching (S. Zuber, M. J. Hynes, and A. Andrianopoulos, unpublished data) (Fig. 7). Overexpression of *gasB* leads to reduced growth at low pH but not at high pH and indicates that the effects of overexpressing *gasB* are pH enhanced (Zuber et al., unpublished data).

Some pertinent questions arise from these studies. It is intriguing that reduced growth is not observed in the Δ*gasB* mutant or strains carrying the dominant activating *gasB^{G45R}* or dominant negative *gasB^{G208R}* allele and is seen only when *gasB* is overexpressed. The absence of phenotype in both the Δ*gasB* mutant and strains carrying the dominant activating *gasB^{G45R}* allele suggest that the βγ complex is the major signaling component and that the function of the α-subunit GasB is to sequester the βγ complex. Following this model, the βγ complex positively regulates growth while the α subunit inhibits growth by associating with the βγ complex and blocking signaling. However, this does not explain why reduced growth is not observed in the strains carrying the dominant negative *gasB^{G208R}* allele and suggests that this mutation blocks βγ signaling only partially. It is only when the α subunit is overexpressed and present in vast amounts that the βγ signaling is sufficiently masked to lead to a visible phenotype.

Some members of class II fungal α subunits have been implicated in sexual development; for example, *S. pombe* Gpa1 is essential for mating (50). The α-subunit MagC in *M. grisea* is highly homologous to GasB and appears to function in ascospore maturation and/or germination (42). The GasB homologue Gna2 in *N. crassa* also seems to play a role in sexual development, since Δ*gna1* Δ*gna2* double mutants have more pronounced defects in female sterility than do Δ*gna1* strains (3). The fact that *gasB* alleles do not interfere with the *A. nidulans* sexual cycle does not exclude the possibility of GasB playing a role in the cryptic sexual cycle presumed to be present in *P. marneffei* (8). Characterization of the GasB homologue GanA in *A. nidulans* would help to further answer this question on the function of this class of fungal α proteins. If a Δ*ganA* mutant is defective in the *A. nidulans* sexual cycle, the ability of GasB to complement this defect could be tested. Since the direct effectors for members of class II fungal α subunits remain completely unknown, future experiments should aim at identifying downstream

components linked to the GasB signaling cascade, as this could help to clarify the biological function of GasB in *P. marneffei*.

CONCLUSION

A growing number of genes encoding α subunits have been cloned from various fungal species, and their biological functions have been investigated. These studies have demonstrated the importance and the complexity of G protein signaling in the regulation of key processes of fungal life cycles, such as dimorphic switching, mating, asexual development, and pathogenicity (6).

Examination of the role of G protein signaling and its cellular roles in *P. marneffei* has contributed to our understanding of *P. marneffei* biology and the role of α subunits in fungi. The isolation and characterization of three genes encoding α subunits of heterotrimeric G proteins in the dimorphic fungus *P. marneffei* has made it possible to compare the contribution of G protein signaling on the various developmental programs displayed by this fungus, such as germination, asexual development, and dimorphic switching. GasC plays a major role in germination, and this represents the first report linking an α subunit to the regulation of germination in filamentous fungi. More importantly, studies of conidial germination, which is a key process in the early steps of fungal infections, and identification of a regulator of this process may provide novel insights into targets for preventing infection. Identifying the specific signal and receptor linked to GasC could be an important step toward a better understanding of the pathogenicity of *P. marneffei* at its outset. In addition, GasC is involved in the regulation of the production of secondary metabolites and asexual development.

Furthermore, GasC shows overlapping functions with GasA, which is a key regulator of asexual development and is highly similar to *A. nidulans* FadA. The basis of this cross-regulation remains to be discovered. The biological function of GasB is unclear at this stage. Answers to this intriguing question could come from future studies involving the identification of upstream and downstream components linked to this class of α subunits that, at present, remain completely unknown. Another central question concerns the receptors and their signals. G proteins are coupled to 7-TM receptors, but only very few receptors of fungal G proteins have so far been characterized.

Of particular significance is the fact that none of these α subunits appear to regulate dimorphic switching. The question of how temperature is sensed by *P. marneffei* and what signaling pathway is invoked to execute the dimorphic switching program remains to be answered.

Acknowledgments. This research was supported by grants from the Australian Research Council, the National Health and Medical Research Council of Australia, the University of Melbourne, and Novo Nordisk A/S. The contributions of past laboratory members are gratefully acknowledged.

REFERENCES

1. **Alexopoulos, C. J., C. W. Mims, and M. Blackwell.** 1996. *Introductory Mycology.* John Wiley & Sons, Inc., New York, N.Y.

2. **Andrianopoulos, A.** 2002. Control of morphogenesis in the human fungal pathogen *Penicillium marneffei. Int. J. Med. Microbiol.* **292:**331–347.

3. **Baasiri, R. A., X. Lu, P. S. Rowley, G. E. Turner, and K. A. Borkovich.** 1997. Overlapping functions for two G protein α subunits in *Neurospora crassa. Genetics* **147:**137–145.

4. **Bardwell, L.** 2005. A walk-through of the yeast mating pheromone response pathway. *Peptides* **26:**339–350.

5. **Böhm, S. K., E. F. Grady, and N. W. Bunnett.** 1997. Regulatory mechanisms that modulate signalling by G-protein-coupled receptors. *Biochem. J.* **322:**1–18.

6. **Bölker, M.** 1998. Sex and crime: heterotrimeric G proteins in fungal mating and pathogenesis. *Fungal Genet. Biol.* **25:**143–156.

7. **Borneman, A. R., M. J. Hynes, and A. Andrianopoulos.** 2000. The *abaA* homologue of *Penicillium marneffei* participates in two developmental programmes: conidiation and dimorphic growth. *Mol. Microbiol.* **38:**1034–1047.

8. **Borneman, A. R., M. J. Hynes, and A. Andrianopoulos.** 2001. An STE12 homolog from the asexual, dimorphic fungus *Penicillium marneffei* complements the defect in sexual development of an *Aspergillus nidulans steA* mutant. *Genetics* **157:**1003–1014.

9. **Bourne, H. R.** 1997. How receptors talk to trimeric G proteins. *Curr. Opin. Cell Biol.* **9:**134–142.

10. **Boyce, K. J., M. J. Hynes, and A. Andrianopoulos.** 2001. The CDC42 homolog of the dimorphic fungus *Penicillium marneffei* is required for correct cell polarization during growth but not development. *J. Bacteriol.* **183:**3447–3457.

11. **Boyce, K. J., M. J. Hynes, and A. Andrianopoulos.** 2005. The Ras and Rho GTPases genetically interact to coordinately regulate cell polarity during development in *Penicillium marneffei. Mol. Microbiol.* **55:**1487–1501.

12. **Casey, P. J.** 1994. Lipid modifications of G proteins. *Curr. Opin. Cell Biol.* **6:**219–225.

13. **Chan, Y., and T. C. Chow.** 1990. Ultrastructural observations on *Penicillium marneffei* in natural human infection. *Ultrastruct. Pathol.* **14:**439–452.

14. **Chang, M. H., K. S. Chae, D. M. Han, and K. Y. Jahng.** 2004. The GanB Gα-protein negatively regulates a sexual sporulation and plays a positive role in conidial germination in *Aspergillus nidulans. Genetics* **167:**1305–1315.

15. **Choi, G. H., B. Chen, and D. L. Nuss.** 1995. Virus-mediated or transgenic suppression of a G-protein α subunit and attenuation of fungal virulence. *Proc. Natl. Acad. Sci. USA* **92:**305–309.

16. **Choi, W., and R. A. Dean.** 1997. The adenylate cyclase gene *mac1* of *Magnaporthe grisea* controls appressorium formation and other aspects of growth and development. *Plant Cell* **9:**1973–1983.

17. **Clapham, D. E., and E. J. Neer.** 1997. G protein βγ subunits. *Annu. Rev. Pharmacol. Toxicol.* **37:**167–203.

18. **Cooper, C. R. J.** 1998. From bamboo rats to humans: the odyssey of *Penicillium marneffei. ASM News* **64:**390–397.

19. **d'Enfert, C.** 1997. Fungal spore germination: insights from the molecular genetics of *Aspergillus nidulans* and *Neurospora crassa. Fungal Genet. Biol.* **21:**163–172.

20. **d'Enfert, C., and T. Fontaine.** 1997. Molecular characterization of the *Aspergillus nidulans treA* gene encoding an acid trehalase required for growth on trehalose. *Mol. Microbiol.* **24:**203–216.

21. **Deng, Z. L., and D. H. Conner.** 1985. Progressive disseminated penicilliosis caused by *Penicillium marneffei:* report of eight cases and differentiation of the causative organism from *Histoplasma capsulatum. Am. J. Clin. Pathol.* **84:**323–327.

22. **Eriksson, S., R. Hurme, and M. Rhen.** 2002. Low-temperature sensors in bacteria. *Philos. Trans. R. Soc. Lond. Ser. B* **357:**887–893.

23. **Fang, E. G., and R. A. Dean.** 2000. Site-directed mutagenesis of the *magB* gene affects growth and development in *Magnaporthe grisea. Mol. Plant-Microbe Interact.* **13:**1214–1227.

24. **Fillinger, S., M. K. Chaveroche, K. Shimizu, N. Keller, and C. d'Enfert.** 2002. cAMP and RAS signalling independently control spore germination in the filamentous fungus *Aspergillus nidulans. Mol. Microbiol.* **44:**1001–1016.

25. **Forsberg, H., and P. O. Ljungdahl.** 2001. Sensors of extracellular nutrients in *Saccharomyces cerevisiae. Curr. Genet.* **40:**91–109.

26. **Garrison, R. G., and K. S. Boyd.** 1973. Dimorphism of *P. marneffei* as observed by electron microscopy. *Can. J. Microbiol.* **19:**1305–1309.

27. **Gilman, A. G.** 1987. G proteins: transducers of receptor-generated signals. *Annu. Rev. Biochem.* **56:**615–649.

28. **Hamm, H. E.** 1998. The many faces of G protein signaling. *J. Biol. Chem.* **273:**669–672.

29. **Hampsey, M.** 1997. A review of phenotypes in *Saccharomyces cerevisiae. Yeast* **13:**1099–1133.

30. **Harashima, T., and J. Heitman.** 2002. The Gα protein Gpa2 controls yeast differentiation by interacting with kelch repeat proteins that mimic Gβ subunits. *Mol. Cell* **10:**163–173.

31. **Hatanaka, M., and C. Shimoda.** 2001. The cyclic AMP/PKA signal pathway is required for initiation of spore germination in *Schizosaccharomyces pombe. Yeast* **18:**207–217.

32. **Herman, P. K., and J. Rine.** 1997. Yeast spore germination: a requirement for Ras protein activity during reentry into the cell cycle. *EMBO J.* **16:**6171–6181.

33. **Hicks, J. K., J.-H. Yu, N. P. Keller, and T. H. Adams.** 1997. *Aspergillus* sporulation and mycotoxin production both require inactivation of the FadA Gα protein-dependent signaling pathway. *EMBO J.* **16:**4916–4923.

34. **Ivey, F. D., P. N. Hodge, G. E. Turner, and K. A. Borkovich.** 1996. The Gαi homologue *gna-1* controls multiple differentiation pathways in *Neurospora crassa. Mol. Biol. Cell* **7:**1283–1297.

35. **Ivey, F. D., Q. Yang, and K. A. Borkovich.** 1999. Positive regulation of adenylyl cyclase activity by a Gαi homolog in *Neurospora crassa. Fungal Genet. Biol.* **26:**48–61.

36. **Kays, A. M., P. S. Rowley, R. A. Baasiri, and K. A. Borkovich.** 2000. Regulation of conidiation and adenylyl cyclase levels by the Gα protein GNA-3 in *Neurospora crassa. Mol. Cell. Biol.* **20:**7693–7705.

37. Kölle, M. R. 1997. A new family of G-protein regulators —the RGS proteins. *Curr. Opin. Cell Biol.* **9:**143–147.

38. Kübler, E., H. U. Mosch, S. Rupp, and M. P. Lisanti. 1997. Gpa2p, a G-protein α-subunit, regulates growth and pseudohyphal development in *Saccaromyces cerevisiae* via a cAMP-dependent mechanism. *J. Biol. Chem.* **272:**20321–20323.

39. Kwon-Chung, K. J., and J. E. Bennett. 1992. *Medical Mycology.* Lea & Febiger, Philadelphia, Pa.

40. Lafon, A., J. A. Seo, K. H. Han, J. H. Yu, and C. d'Enfert. 2005. The heterotrimeric G-protein GanB(α)-SfaD(β)-GpgA(γ) is a carbon source sensor involved in early cAMP-dependent germination in *Aspergillus nidulans.* *Genetics* **171:**71–80.

41. Liebmann B., M. Muller, A. Braun, and A. A. Brakhage. 2004. The cyclic AMP-dependent protein kinase a network regulates development and virulence in *Aspergillus fumigatus.* *Infect. Immun.* **72:**5193–5203.

42. Liu, S. H., and R. A. Dean. 1997. G protein α-subunit genes control growth, development and pathogenicity of *Magnaporthe grisea.* *Mol. Plant-Microbe Interact.* **10:**1075–1086.

43. Logothetis, D. E., Y. Kurachi, J. Galper, E. J. Neer, and D. E. Clapman. 1987. The βγ subunits of GTP-binding proteins activate the muscarinic K⁺ channel in heart. *Nature* **325:**321–326.

44. Lorenz, M. C., and J. Heitman. 1997. Yeast pseudohyphal growth is regulated by Gpa2, a G protein α homolog. *EMBO J.* **16:**7008–7018.

45. Loros, J. 2005. A kinase for light and time. *Mol. Microbiol.* **56:**299–302.

46. Loubradou, G., J. Begueret, and B. Turcq. 1999. MOD-D, a Gα subunit of the fungus *Podospora anserina*, is involved in both regulation of development and vegetative incompatibility. *Genetics* **152:**519–528.

47. Martin, H., J. M. Rodriguez-Pachon, C. Ruiz, C. Nombela, and M. Molina. 2000. Regulatory mechanisms for modulation of signaling through the cell integrity Slt2-mediated pathway in *Saccaromyces cerevisiae.* *J. Biol. Chem.* **275:**1511–1519.

48. Neer, E. J. 1995. Heterotrimeric G proteins: organizers of transmembrane signals. *Cell* **80:**249–257.

49. Neves, S. R., P. T. Ram, and R. Iyengar. 2002. G protein pathways. *Science* **296:**1636–1639.

50. Obara, T., M. Nakafuku, M. Yamamoto, and Y. Kaziro. 1991. Isolation and characterization of a gene encoding a G-protein α subunit from *Schizosaccharomyces pombe*: involvement in mating and sporulation pathways. *Proc. Natl. Acad. Sci. USA* **88:**5877–5881.

51. Osherov, N., and G. May. 2000. Conidial germination in *Aspergillus nidulans* requires RAS signaling and protein synthesis. *Genetics* **155:**647–656.

52. Osherov, N., and G. S. May. 2001. The molecular mechanisms of conidial germination. *FEMS Microbiol. Lett.* **199:**153–160.

53. Osherov, N., J. Mathew, A. Romans, and G. S. May. 2002. Identification of conidial-enriched transcripts in *Aspergillus nidulans* using suppression subtractive hybridization. *Fungal Genet. Biol.* **37:**197–204.

54. Parsons, W. J., V. Ramkumar, and G. L. Stiles. 1988. Isobutylmethylxanthine stimulates adenylate cyclase by blocking the inhibitory regulatory protein, Gi. *Mol. Pharmacol.* **34:**37–41.

55. Pitcher, J. A., N. J. Freedman, and R. J. Lefkowitz. 1998. G protein-coupled receptor kinases. *Annu. Rev. Biochem.* **67:**653–692.

56. Pitt, J. I. 1979. *The Genus Penicillium and Its Teleomorphic States Eupenicillium and Talaromyces.* Academic Press, Inc., New York, N.Y.

57. Regenfelder, E., T. Spellig, A. Hartmann, S. Lauenstein, M. Bölker, and R. Kahmann. 1997. G-proteins in *Ustilago maydis*: transmission of multiple signals? *EMBO J.* **16:**1934–1942.

58. Rens-Domiano, S., and H. E. Hamm. 1995. Structural and functional relationships of heterotrimeric G-proteins. *FASEB J.* **9:**1059–1066.

59. Rosen, S., J. H. Yu, and T. H. Adams. 1999. The *Aspergillus nidulans sfaD* gene encodes a G protein β subunit that is required for normal growth and repression of sporulation. *EMBO J.* **18:**5592–5600.

60. Segretain, G. 1959. *Penicillium marneffei* n. sp., agent d'une mycose du système réticuloendothélial. *Mycopathol. Mycol. Appl.* **11:**327–353.

61. Seo, J. A., K. H. Han, and J. H. Yu. 2005. Multiple roles of a heterotrimeric G protein γ subunit in governing growth and development of *Aspergillus nidulans.* *Genetics* **171:**81–89.

62. Shimizu, K., and N. P. Keller. 2001. Genetic involvement of a cAMP-dependent protein kinase in a G protein signaling pathway regulating morphological and chemical transitions in *Aspergillus nidulans.* *Genetics* **157:**591–600.

63. Som, T., and V. S. Kolaparthi. 1994. Developmental decisions in *Aspergillus nidulans* are modulated by Ras activity. *Mol. Cell. Biol.* **14:**5333–5348.

64. Sirisanthana, T., and K. Supparatpinyo. 1998. Epidemiology and management of penicilliosis in human immunodeficiency virus-infected patients. *Int. J. Infect. Dis.* **3:**48–53.

65. Sirisanthana, T., K. Supparatpinyo, J. Perriens, and K. E. Nelson. 1998. Amphotericin B and itraconazole for treatment of disseminated *Penicillium marneffei* infection in human immunodeficiency virus-infected patients. *Clin. Infect. Dis.* **26:**1107–1110.

66. Supparatpinyo, K., S. Chiewchanvit, P. Hirunsri, C. Uthammachai, K. E. Nelson, and T. Sirisanthana. 1992. *Penicillium marneffei* infection in patients infected with human immunodeficiency virus. *Clin. Infect. Dis.* **14:**871–874.

67. Supparatpinyo, K., K. E. Nelson, W. G. Merz, B. J. Breslin, C. R. Cooper, C. Kamwan, and T. Sirisanthana. 1993. Response to antifungal therapy by human immunodeficiency virus-infected patients with disseminated *Penicillium marneffei* infections and in vitro susceptibilities of isolates from clinical specimens. *Antimicrob. Agents Chemother.* **37:**2407–2411.

68. Supparatpinyo, K., C. Khamwan, V. Baosoung, K. E. Nelson, and T. Sirisanthana. 1994. Disseminated *Penicillium marneffei* infection in Southeast Asia. *Lancet* **344:**110–113.

69. Turner, G. E., and K. A. Borkovich. 1993. Identification of a G protein α subunit from *Neurospora crassa* that is a member of the Gi family. *J. Biol. Chem.* **268:**14805–14811.

70. Vossler, J. 2001. *Penicillium marneffei*: an emerging fungal pathogen. *Clin. Microbiol. News.* **23:**25–29.

71. Whiteway, M., L. Hougan, D. Dignard, D. Y. Thomas, L. Bell, G. C. Saari, F. J. Grant, P. O'Hara, and V. L. MacKay. 1989. The *STE4* and *STE18* genes of yeast encode potential β and γ subunits of the mating factor receptor-coupled G protein. *Cell* **56:**467–477.

72. Wong, S. S., H. Siau, and K. Y. Yuen. 1999. Penicilliosis marneffei—West meets East. *J. Med. Microbiol.* **48:**973–975.

73. **Yang, Q., and K. A. Borkovich.** 1999. Mutational activation of a Gαi causes uncontrolled proliferation of aerial hyphae and increased sensitivity to heat and oxidative stress in *Neurospora crassa. Genetics* **151:**107–117.

74. **Yu, J. H., R. A. Butchko, M. Fernandes, N. P. Keller, T. J. Leonard, and T. H. Adams.** 1996. Conservation of structure and function of the aflatoxin regulatory gene *aflR* from *Aspergillus nidulans* and *A. flavus. Curr. Genet.* **29:**549–555.

75. **Yu, J. H., J. Wieser, and T. H. Adams.** 1996. The *Aspergillus* FlbA RGS domain protein antagonizes G-protein signalling to block proliferation and allow development. *EMBO J.* **15:**5184–5190.

76. **Zuber, S., M. J. Hynes, and A. Andrianopoulos.** 2002. G-protein signaling mediates asexual development at 25°C but has no effect on yeast-like growth at 37°C in the dimorphic fungus *Penicillium marneffei. Eukaryot. Cell* **1:**440–447.

77. **Zuber, S., M. J. Hynes, and A. Andrianopoulos.** 2003. The G-protein α-subunit GasC plays a major role in germination in the dimorphic fungus *Penicillium marneffei. Genetics* **164:**487–499.

78. **Zuber, S.** 2003. Regulatory mechanisms controlling development in the dimorphic fungus *Penicillium marneffei.* PhD thesis. University of Melbourne, Melbourne, Australia.

IV. THE HOST

Molecular Principles of Fungal Pathogenesis
Edited by Joseph Heitman et al.
©2006 ASM Press, Washington, D.C.

Chapter 31

Innate Humoral Immunity to Fungi

THOMAS R. KOZEL

Normal adults have a high level of resistance to disseminated infection by most fungi. As a consequence, fungi such as *Aspergillus* spp., *Candida* spp., and *Cryptococcus neoformans* are generally regarded as opportunistic pathogens. Humoral components of the innate immune system contribute to this high level of natural host resistance. Innate immunity is of particular importance to fungal infection because it provides an immediate response to invasive fungal disease and it recognizes a broad spectrum of fungal pathogens.

The humoral innate immune system has three key elements: the complement system; pattern recognition by the collectin protein, mannan-binding lectin (MBL); and naturally occurring antibodies that are reactive with carbohydrate antigens found in the fungal cell wall. The recognition capabilities of these three systems are fundamentally different; however, these systems intersect in key places as their activation leads to augmentation of host resistance. All three systems rely on cellular effectors for their antifungal activity, most notably phagocytosis and killing of fungal cells by professional phagocytes such as monocytes, macrophages, and particularly neutrophils. Thus, it is not surprising that neutropenia is a key predisposing factor for disseminated infection by opportunistic pathogens such as *Candida* spp.

This chapter considers each of the three systems of innate humoral immunity in turn, with an emphasis on recognition, amplification cascades, interaction between the systems, and the similarities and differences between fungi in their interactions with these systems. Finally, the chapter concludes with a discussion of mechanisms used by fungi to avoid innate humoral immunity.

THE COMPLEMENT SYSTEM

The complement system is an effector of both acquired and innate immunity. It has an ancient origin and has been found in all major invertebrate deuterostome groups that have been studied to date (29). As a consequence, the complement system predates the appearance of antibodies, T-cell receptors, and major histocompatibility complex molecules of adaptive immunity.

Fungi are of considerable historic importance for our understanding of the complement system. Von Dungern reported in 1900 that incubation of serum with yeast cells led to inactivation of then unknown heat-sensitive components of serum that are normally destructive to bacteria and other cells (111). This observation was confirmed 2 years later by Ehrlich and Sachs (25). In the last 50 years, zymosan, a cell wall product derived from *Saccharomyces cerevisiae*, has been studied as a prototype activator of the complement system (28). Zymosan played a particularly important role in the discovery of the alternative pathway of complement activation (86–88).

The complement system is a group of proteins that are activated in a cascade-like fashion to facilitate humoral defense against invading microbes. Activation of the complement system may occur by one of three pathways. First, binding of immunoglobulin M (IgM) or several subclasses of IgG to the fungal surface can initiate the classical complement system. Second, binding of MBL to the fungal surface can initiate the lectin pathway. Finally, the alternative pathway is a complex system of proteins that provides for discrimination between self and nonself in a manner that does not involve the specific recognition elements of antibody or MBL. Regardless of the manner of activation, the classical, lectin, and alternative pathways converge and continue in a cascade to form the membrane attack complex. Fungi are not normally regarded as susceptible to the membrane attack complex, but several proteins that are bound to the fungal surface as a result of activation of the complement system

---

Thomas R. Kozel • Department of Microbiology and Immunology, University of Nevada School of Medicine, Reno, NV 59557.

are highly opsonic for phagocytosis by macrophages and neutrophils, and soluble peptides released as by-products of the cascade contribute to the inflammatory response to fungi in tissues. The reader is referred to any standard immunology text for a discussion of the classical, lectin, and alternative pathways.

Classical and Lectin Complement Pathways

The classical and lectin pathways are initiated when antibody or MBL, respectively, binds to the fungal cell surface. Antibodies may be generated as a consequence of fungal infection. Perhaps more important are natural antibodies found in the serum of normal individuals. The lectin pathway bypasses C1 of the classical pathway but utilizes the remaining classical-pathway proteins. The epitopes recognized by naturally occurring antibodies and MBL are considered in later sections of this chapter. The hallmark of activation of the classical pathway is rapid activation kinetics and a synchronous deposition of opsonic fragments of C3 on the fungal surface. Presumably, the rapid activation kinetics and uniform distribution of C3 fragments follow binding of antibody to sites that are distributed over the fungal surface. Such rapid activation kinetics are illustrated by the deposition of C3 onto *C. albicans* yeast cells following incubation in normal human serum (56). Normal human serum contains antibodies that are reactive with the *C. albicans* cell wall. Binding of these antibodies is followed by activation of the classical pathway with subsequent covalent binding of C3 fragments to the cell surface. If the antibody is removed by absorption of the serum with *C. albicans* yeast cells or immunoabsorption with purified mannan, the kinetics for C3 deposition are dramatically retarded. In contrast to the rather extensive study of antibody initiation of the classical pathway, the effects of MBL and the lectin pathway on complement activation by pathogenic fungi have received little study. The interaction of MBL with pathogenic fungi is discussed in a later section.

Alternative Complement Pathway

The alternative complement pathway is a central component of innate immunity to fungi. Activation of the alternative pathway occurs when metastable C3b that is formed in serum binds covalently through an ester or amide bond to the fungal surface. On nonactivating particles, such as cells of "self," complement factor H binds to the bound C3b and acts as a cofactor that enables the protease factor I to cleave the C3b to form iC3b. iC3b is unable to form a C3 convertase complex, and so the process stops (42). In contrast, binding of complement factor B to surface C3b is favored on activating particles. Factor B is cleaved by factor D to produce Bb, which, in association with C3b, forms the alternative-pathway C3 convertase that enables amplification of the amount of bound C3b.

Encapsulated *C. neoformans* cells are potent activators of the complement system. Incubation of encapsulated cryptococci in human serum leads to the binding of 10^7 to 10^8 C3 molecules per yeast cell (116). The amount of C3 fragments bound per cell greatly exceeds the numbers observed with other fungal cells such as *C. albicans*, zymosan, or acapsular cryptococci, most probably reflecting the three-dimensional matrix presented by the capsule for binding of C3. Encapsulated cryptococci are such powerful activators of the complement system that incubation of even small numbers of cryptococci in human serum can lead to depletion of the complement system (55). As a consequence, it is not surprising that some patients with disseminated cryptococcosis exhibit depletion of the complement system (70).

Activation and binding of C3 fragments to encapsulated cryptococci incubated in normal human serum proceeds entirely via the action of the alternative pathway. Activation and binding of C3 is not blocked or influenced by treatment of serum with EGTA, which chelates the Ca^{2+} that is needed for activation of the classical pathway (58). Conversely, heating of serum at 50°C to inactivate factor B of the alternative pathway blocks complement activation by encapsulated cryptococci (21). The most compelling evidence that encapsulated cryptococci activate the complement cascade solely via the alternative pathway is provided by experiments in which yeast cells were incubated in an alternative pathway that was reconstructed from the six purified proteins of the alternative pathway (59). The kinetics for C3 deposition, the molecular form of bound C3, and the sites of C3 deposition in the capsule of yeast cells incubated in a reconstituted pathway were indistinguishable from what was observed following incubation in normal human serum.

The kinetics and sites of C3 deposition on encapsulated cryptococci incubated in human serum are a paradigm for activation of the alternative pathway by fungal cells (58). The kinetics for C3 accumulation via the alternative pathway are much slower than the kinetics for C3 deposition by the classical pathway. Little C3 binding is apparent after 2 to 4 min of incubation in serum. After 4 min, there is a phase of rapid accumulation, which ends after 15 to 20 min incubation.

If the sites of C3 deposition on cryptococcal cells are assessed by immunofluorescence microscopy, a characteristic pattern is seen in which a limited number of minute focal sites of C3 binding are found after approximately 2 min of incubation (58). With increased incubation times, larger focal areas of C3 deposition are observed, accompanied by the appearance of new smaller focal areas of C3 binding. Finally, after 8 to 10 min, the patches of bound C3 are much larger and eventually coalesce such that most of the capsule appears to be filled with C3. This pattern of binding identifies several features of alternative-pathway activation by encapsulated cryptococci. First, activation of the alternative pathway is focal. Sites of C3 deposition appear to occur randomly within or on the capsule. Presumably, these foci are sites for deposition of metastable C3 that has escaped the surveillance activity of factors H and I. Second, the formation of initiation foci is asynchronous. New initiation sites occur as established sites expand. Finally, there appears to be a nuclear, circumferential expansion of the initiation foci. Indeed, studies have found that accumulation of C3 on encapsulated cryptococci follows the kinetics that would be predicted for accumulation on the surface of an expanding sphere (60).

The C3 fragments that are bound to the cryptococcal capsule are almost entirely in the form of iC3b. Given the apparently nuclear expansion of spheres of C3 deposition and the fact that C3b, not iC3b, is the active component in the alternative-pathway C3 convertase, the presence of large amounts of iC3b, even during the rapid-expansion phase, suggests that C3b within the expanding core of the sphere is rapidly converted by factors H and I to iC3b whereas the C3b needed to support sphere expansion would be rapidly formed at the surface of the sphere. The fact that almost all of the bound C3 is in the form of iC3b further suggests that phagocyte receptors for iC3b will be important in phagocytosis of the yeast.

Early studies using standard immunofluorescence and transmission electron microscopy found that the capsule itself is the site for binding of C3 to encapsulated cryptococci. Little or no binding to the cell wall was observed. With the advent of confocal microscopy, there has been an increased appreciation of the complexity of binding of C3 fragments within the capsular matrix. A study of capsule architecture found that the cryptococcal capsule is a matrix of variable porosity (34). The capsule is most dense near the cell wall and decreases in density as the capsule edge is approached. As a consequence of this variable matrix density, the capsule functions as a molecular sieve; small molecules are able to penetrate through the capsule to the cell wall, whereas large molecules may penetrate the periphery of the capsule but are excluded from the capsule interior.

One consequence of this variable molecular sieving is the apparent exclusion of complement proteins from the deep capsule interior.

Further studies have found that the location of C3 binding in the capsule is a function of capsule diameter (117). Incubation of small-capsule cryptococci in mouse serum led to C3 binding to the capsule edge. In contrast, incubation of large-capsule cryptococci in mouse serum produced binding of C3 beneath the capsule surface but not at the extreme capsule periphery. These results suggest that large-capsule cryptococci may avoid surveillance by the innate humoral immune system by directing C3 deposition to a site where potentially opsonic C3 fragments would not be available for interaction with phagocyte complement receptors.

C. albicans, acapsular *C. neoformans,* and *Blastomyces dermatitidis* have the potential for accumulation of C3 on the fungal surface via the alternative pathway. However, studies of alternative-pathway activation are characteristically obscured by the presence in normal sera of antibodies or MBL that are reactive with the cell walls of these fungi. As a consequence, studies of complement activation by these fungi typically measure activation initiated by the classical and/or lectin pathways. Alternative-pathway activation can be studied if activation of the classical and lectin pathways is blocked. As noted above, one means of blockade is the use of EGTA to chelate the Ca^{2+} that is required for both classical-pathway and lectin-pathway initiation but not for alternative-pathway activity. An alternative approach is absorption of serum to remove antibodies or MBL that would normally initiate the classical or lectin pathways. Results of studies of absorbed serum should be interpreted with caution, however, because *C. albicans* initiates the alternative pathway via antibody-independent and antibody-dependent mechanisms (see below) (120).

The pattern and kinetics for deposition of C3 onto *C. albicans* in the presence of EGTA closely resembles deposition of C3 onto encapsulated cryptococci that are incubated in normal human serum (56). C3 deposition is characterized by a delay of 2 to 4 min before readily measurable deposition of C3 occurs. Moreover, the pattern of binding seen by immunofluorescence microscopy is the focal, asynchronous pattern that is found with encapsulated cryptococci. As with encapsulated cryptococci, the sites of early C3 focus formation appear to be randomly distributed over the yeast surface and expand with continued incubation to coalesce and cover the yeast surface.

Anti-*Candida* antibody found in normal human serum plays a key role in facilitating alternative-pathway activation (120). If *C. albicans* yeast cells are incubated in serum that has been both treated with EGTA and absorbed to remove anti-*Candida* antibody, the pattern of C3 deposition resembles the focal initiation and expansion that is observed when the yeast cells have been incubated in serum that was treated only with EGTA. However, the kinetics for deposition using serum that has been absorbed to remove antibody and/or MBL is dramatically delayed relative to the kinetics found with EGTA-treated serum that was not absorbed. Focal binding sites take 15 min or more to appear, and incubation times of longer than 30 min are required for the cells to be coated with C3. Even with extended incubation times, the amount of C3 bound via antibody-independent activation (both EGTA treated and absorbed) is markedly smaller than the amounts that accumulate via either the classical pathway (untreated serum) or the antibody-dependent alternative pathway (EGTA-treated alone).

Acapsular cryptococci and *B. dermatitidis* behave in a manner quite similar to that of *C. albicans* yeast cells. Antibodies to both acapsular cryptococci and *B. dermatitidis* found in normal serum produce rapid activation and binding of C3 via the classical pathway. If the serum is treated with EGTA to block initiation of the classical pathway, the characteristic pattern of alternative-pathway activation emerges. C3 deposition is relatively slow, requiring several minutes before readily measurable amounts of C3 accumulate on the cells; C3 deposition is focal and asynchronous and appears to expand from focal initiation sites to cover the surface of the cells.

Biological Consequences of Complement Activation

A key feature of the complement system as a component of innate humoral immunity is the amplification that occurs at several points in the complement cascade. Amplification leads to coating of the yeast cells with potentially opsonic fragments of C3. The primary opsonic fragments are C3b and iC3b. The relative amounts of C3b versus iC3b that may coat the fungal surface varies somewhat between different fungi. The presence of a large proportion of C3b on a cell would suggest limited action by the regulatory proteins factors H and I. A large proportion of iC3b would suggest a cellular microenvironment that supported the binding of factor H to C3b and/or an environment that facilitated the action of factor I. Zymosan was the first fungal particle to be studied where it was found that the bound C3 fragments were primarily C3b (62). C3 fragments bound to *C. albicans* yeast cells are also almost entirely C3b (52). Conidia and hyphae of *Aspergillus fumigatus* present an interesting situation because the molecular form

of bound C3 varies with the cellular form of the fungus (57). C3 fragments bound to resting conidia are primarily iC3b. In contrast, C3 fragments bound to swollen conidia or hyphae are a mixture of C3b and iC3b. Finally, C3 bound to encapsulated cryptococci are almost entirely in the form of iC3b.

C3b and iC3b have different receptors on phagocytic cells. The primary receptor for C3b is complement receptor type 1 (CR1 or CD35). CR1 is present on erythrocytes, macrophages, monocytes, neutrophils, and B cells. The primary receptors for iC3b are complement receptor type 3 (CR3; CD11b/18) and complement receptor type 4 (CR4; CD11c/18). CR3 and CR4 are found on phagocytic cells such as macrophages, monocytes, and neutrophils. Given the variability between different fungi in expression of different proportions of surface C3b and iC3b, it is possible that different complement receptors may vary in importance for control of different fungal infections.

One of the most important consequences of complement activation is release of the soluble peptides, C4a, C3a, and C5a, that are mediators of inflammation. These peptides have activities that include chemotaxis and chemokinesis for neutrophils, stimulation of the respiratory burst, and targeting of vascular endothelial cells to increase vascular permeability. Importantly, C5a contributes to the killing of encapsulated cryptococci by neutrophils (67). The relative potency of these peptides is C5a > C3a > C4a (11), suggesting that C5a plays a significant role in resistance to fungal infection. Thus, it is not surprising that mice that are deficient in C5 have decreased resistance to disseminated cryptococcosis (67, 91), disseminated candidiasis (1, 40, 69, 77), and aspergillosis (40). Even *S. cerevisiae* can produce a lethal infection in C5-deficient mice (8).

Complement Activation by Pathogenic Fungi

Complement activation and the biological consequences of complement activation have been examined for most of the pathogenic fungi. Such studies demonstrate not only the marked propensity of a broad spectrum of fungi to activate the complement system but also the potential contribution of complement activation to innate resistance. Evidence for complement activation includes (i) production of biological activities such as opsonization or release of inflammatory proteins, (ii) depletion of complement activity as a consequence of complement activation, (iii) generation of cleavage products of proteins such as C3 or factor B that are discernible by immunochemical assay, and (iv) deposition of C3 onto the fungal surface that is measurable by use of radiolabeled C3 or

Table 1. Biological consequences of activation of the complement system by pathogenic fungi

| Microorganism | Published paper describing experimental system used to demonstrate complement activation | | | |
	Binding of C3 by immunofluorescence or use of radiolabeled C3	Opsonization	Generation of chemotactic peptides	Generation of cleavage fragments of C3 or factor B
C. neoformans	Goren and Warren (37), Diamond et al. (21), Kozel and Pfrommer (54), Truelsen et al. (107)	Mitchell and Friedman (76), Diamond et al. (21), Davies et al. (17)	Laxalt and Kozel (63), Diamond and Erickson (20)	Kozel and Pfrommer (54)
C. albicans	Kozel et al. (52)	Ferrante and Thong (27), Morrison and Cutler (78), Pereira and Hosking (84)	Ray and Wuepper (90), Thong and Ferrante (105)	Ray and Wuepper (90)
A. fumigatus	Kozel et al. (57)	Sturtevant and Latgé (94)	Budzko and Negroni (7)	Budzko and Negroni (7), De Bracco et al. (18)
Trichophyton spp.		Dahl and Carpenter (16), Dahl and Carpenter (16)	Tagami et al. (101)	Swan et al. (99)
P. brasiliensis	Munk and Da Silva (80)	Calich et al. (9)		Munk et al. (81)
C. immitis			Galgiani et al. (31)	Galgiani et al. (32)

immunofluorescence. Results of the application of such measures of complement activation to various fungi are summarized in Table 1.

MANNAN-BINDING LECTIN

The first description of what was eventually found to be MBL deficiency was a report in 1968 of an infant with severe recurrent infections, diarrhea, and failure to thrive (75). The deficiency was initially characterized as the lack of an inherited plasma factor that was necessary for optimal phagocytosis of yeast cells by peripheral blood leukocytes. This plasma factor was subsequently found to be MBL. MBL is a plasma glycoprotein of the collectin family that is composed of collagenous structures and C-type carbohydrate recognition domains. MBL was first identified on the basis of its strong binding to yeast mannan (49). It is a pattern recognition molecule that is synthesized in the liver, functions as an acute-phase protein, and appears in serum predominantly as a six-subunit oligomer (89, 102). The level of MBL in serum is genetically determined, and deficiency may be associated with frequent infection by a variety of microorganisms (96).

MBL has the ability to distinguish between self and nonself by recognizing certain patterns of carbohydrate structures. It is an oligomer of monomeric subunits, which consist of an N-terminal cysteine-rich domain, a collagen domain, a coiled-coil neck domain, and a C-type lectin domain known as the carbohydrate recognition domain (23). The monomeric units of MBL characteristically form octadecamers of six

trimeric subunits (68). The affinity of a single carbohydrate recognition domain is relatively low (10^{-3} to 10^{-4} M) (43, 46); however, through multivalent binding, the avidity of MBL binding is relatively high (10^{-8} to 10^{-10} M) (46, 48). As a consequence, the repeating array of ligands presented by carbohydrates on the surface of fungi that allow for multivalent binding is a critical factor in the recognition of fungi as nonself. The molecular basis for recognition by MBL lies in binding to sugars that contain equatorial hydroxyl groups such as those found at the 3 and 4 positions of mannose, N-acetylglucosamine, and fucose (64, 114).

Mannan is a prominent component of fungal cell walls, and there is clear evidence that MBL binds to the surface of several fungal pathogens. Various fungi have been treated with purified MBL, and MBL binding was assessed by flow cytometry using monoclonal antibodies specific for MBL (45). MBL binds to C. albicans, A. fumigatus, and acapsular C. neoformans (15, 44, 45, 83). In contrast, MBL does not bind to encapsulated cryptococci (15). The cryptococcal capsule is composed of an $\alpha(1\rightarrow3)$-linked mannose backbone (12) that would not have the requisite 3-hydroxyl available for recognition by MBL. The mannose backbone is decorated with single xylosyl and glucuronosyl side chains that would likewise fail to provide binding sites for MBL.

Once bound to a fungal surface, MBL is able to deploy a variety of potential antifungal activities. First, as noted in a previous section, MBL can serve as an initiator of the lectin complement pathway. MBL has a structure that closely resembles the structure of C1q. In serum, MBL interacts preferentially with homologues of C1r and C1s termed MBL-associated serine proteases (MASP), MASP-1, MASP-2, and MASP-3 (89). Binding of MBL to microbial carbohydrates such as yeast mannan activates the MASPs (103). In an activity similar to C1s, activated MASP, particularly MASP-2 (104), cleaves complement components C4 and C2 to form the C3 convertase C4b2a. In this manner, MBL can contribute to complement activation by fungal surfaces with the potential for subsequent opsonization and generation of inflammatory proteins of the complement cascade. To this end, treatment of *C. albicans* with purified MBL was found to enhance the deposition of C4 and C3 from MBL-deficient serum onto yeast cells (44).

A second mechanism by which MBL might contribute to host defense is direct opsonization for phagocytosis (61). Putative phagocyte receptors for MBL include C1qR (71), CR1 (35), and CD14 (13). Agglutination of target microbes by MBL has also been suggested as a means of altering the course of infection (110). MBL-induced agglutination was not seen with *C. albicans* yeast cells, but significant aggregation was observed when *C. albicans* hyphae were treated with MBL (44). The extent to which such aggregation might influence cutaneous or disseminated candidiasis remains to be determined.

A novel mechanism by which collectins might contribute to host resistance is via direct inhibition of fungal growth. MBL was found to inhibit the growth of *C. albicans* independently of complement activation; however, with complement activation, further growth inhibition was observed (44). Similarly, exposure of *Histoplasma capsulatum* yeast cells to the pulmonary collectins, surfactant proteins A and D, led to killing of the collectin-treated yeast cells through a mechanism that likely involved collectin-induced increased permeability of the yeast (72).

Whether MBL is absolutely necessary for optimal innate immunity to fungal infection requires more study. Most subjects who are MBL deficient appear to be healthy. However, some studies have noted an association between MBL deficiency and fungal infection, suggesting a role for MBL. For example, a study of patients with chronic necrotizing pulmonary aspergillosis found a significantly increased carriage of the codon 52 mutation compared with healthy controls (14). It is not known whether the mutant allele is directly associated with chronic necrotizing pulmonary aspergillosis or whether the allele is a marker for associated disease susceptibility factors. Yeast infections have also been associated with carriage of mutant MBL alleles (33,

98). Further evidence for a potential role for MBL in fungal infection is suggested by studies of MBL levels in mice injected with *C. albicans* yeast cells (100). Intravenous injection of *C. albicans* yeast cells produced a rapid decline in serum MBL levels, indicating binding of MBL to the yeast. Such declines in serum MBL levels were accompanied by increases in hepatic MBL mRNA concentrations, suggesting an acute-phase MBL response to fungal infection. Taken together, these clinical and experimental data suggest that MBL may be important in the initial response to disseminated fungal infection.

Although not an example of innate humoral immunity, recent studies of recurrent vulvovaginal candidiasis (RVVC) suggest a role for MBL in the control of mucosal candidal infection. A study of women with RVVC found that RVVC was highly associated with carriage of the mutant MBL codon 54 allele (2). This allele was also associated with a reduction in the vaginal concentration of MBL, suggesting that this MBL polymorphism is a determinant of susceptibility to RVVC.

Results from several in vitro and in vivo studies appear to argue against a role for MBL in host resistance. MBL did not facilitate opsonophagocytosis of *C. albicans* by monocyte-derived dendritic cells and proved to be inhibitory to phagocytosis at higher MBL concentrations (44). An earlier study similarly found that high concentrations of MBL were inhibitory for phagocytosis of *C. albicans* by bone marrow-derived murine macrophages (51). An absence of a role for MBL in resistance to disseminated candidiasis was suggested in a study that showed no difference in susceptibility between MBL-A$^{-/-}$ mice and wild-type mice (65). However, this absence of an apparent phenotype might be because MBL-C$^{-/-}$ mice or mice deficient in both MBL-A and MBL-C were not studied.

It is likely that MBL is one of several redundant systems that provide host resistance to pathogenic fungi. MBL may play a much greater role in host resistance in the immunocompromised patient who has lost one or more of these redundant systems. For example, the presence of variant MBL alleles has been associated with more frequent and more severe infections in patients with common variable immunodeficiency (79) or recipients of chemotherapy (82, 85). As a consequence, MBL replacement or adjunctive therapy may be useful in patients at greatest risk for disseminated fungal infection. Importantly, many of the conditions that predispose patients to opportunistic fungal infections are the same conditions that, when coupled with MBL deficiency, render patients at higher risk for infection.

NATURAL ANTIBODY

Normal adults have appreciable levels of "natural" antibodies to a wide variety of carbohydrate antigens. The list of antigens recognized by natural antibodies includes several polysaccharide antigens found in fungal cell walls. The origins and specificities of such antibodies have been the source of debate for many years. In some cases, antibodies are present for antigens to which there is no known exposure. As a consequence, the presence of natural antibodies is often attributed to exposure to cross-reactive antigens expressed by other microbes or antigens found in the environment (92). In the case of bacterial capsular polysaccharides, such natural antibodies may confer a considerable level of immunity on a large portion of the population (92). One might argue that natural antibodies that arise from ubiquitous exposure to fungal antigens or cross-reactive antigens are components of acquired immunity, but the near universal presence of such antibodies suggests a role for natural antibodies in resistance to disseminated fungal infection prior to the onset of an adaptive immune response.

Antibodies to cell wall mannan of *C. albicans* are common in sera from normal adults. Estimates of the occurrence of antimannan antibodies among adults range from 80% (19, 26) to 100% (47, 53, 66). There is much variability among individual levels of antibodies. Antibody levels are normally distributed (38, 47, 53, 66) and may vary as much as 600-fold or more between different donors (53, 66). The frequency and levels of antimannan antibodies in normal subjects varies with age; readily detectable antibodies are found in 18% of subjects aged 3 to 10 years, 48% aged 11 to 19 years, and 76% aged 20 to 80 years (26). Notably, antimannan antibodies are primarily IgG (38, 53, 66).

Natural antifungal antibodies probably contribute to host resistance by two mechanisms: direct opsonization for phagocytosis and activation of the classical complement cascade for rapid deposition of C3 fragments on the yeast. Activation of the classical pathway by natural anti-*Candida* antibodies differs in several key ways from activation of the alternative pathway described above. First, deposition of C3 is very rapid. C3 is readily detectable on the yeast by either immunofluorescence or by use of radiolabeled C3 after 2 min of incubation in normal serum. The process is largely complete after 4 min of incubation. This result contrasts with the much slower deposition of C3 via the alternative pathway, which requires many minutes to coat the cells with C3 fragments. Anti-*Candida* antibodies in normal sera also mediate deposition of C3 on germ tubes via the classical pathway; however, the kinetics are somewhat

slower than are observed with yeast cells (56). As noted above, an unusual aspect of anti-*Candida* antibodies in activation of the complement system is the involvement of antibody in initiation of the alternative pathway (120).

The antibody that is responsible for activation of the classical pathway by *C. albicans* is specific for mannan (56, 121). Absorption of serum with whole yeast cells or immunoabsorption with purified mannan introduces a marked delay into the kinetics for early deposition of C3 fragments that would be expected with the classical pathway. Early activation kinetics are restored by addition of antimannan IgG that is affinity purified from normal human serum.

Antimannan IgG is also directly opsonic for phagocytosis and killing of *C. albicans* yeast cells by neutrophils (53). Antimannan IgG that is affinity purified from normal human plasma contributes to killing by neutrophils in a dose-dependent fashion that does not require a functional complement system. However, affinity-purified antimannan antibody in very high concentrations is inhibitory to optimal opsonophagocytic killing, suggesting a prozone-like effect. In contrast, antimannan IgG is not needed for optimal opsonophagocytic killing if the complement system is functional, suggesting that naturally occurring mannan antibodies and the complement system are functionally redundant for opsonophagocytic killing by neutrophils.

Naturally occurring antibodies also influence complement activation by acapsular *C. neoformans* (58). Incubation of acapsular cryptococci in serum from normal adults leads to rapid, synchronous accumulation of C3 on the cells in a manner similar to that observed with *C. albicans*. Treatment of serum with EGTA produces the delayed, asynchronous, focal pattern of C3 accumulation that is characteristic of the alternative pathway, indicating that the rapid accumulation of C3 observed with untreated normal serum is due to the action of the classical pathway. Absorption of serum to remove antibodies produces a result identical to that found with EGTA-treated serum. Optimal C3 deposition kinetics are restored to absorbed serum by addition of purified human IgG (115). Unlike *C. albicans*, there is no evidence of a role for antibody in activation of the alternative pathway by acapsular cryptococci.

In contrast to the role of naturally occurring antimannan antibodies in complement activation by *C. albicans*, the antibody in normal serum that is reactive with acapsular cryptococci is specific for glucan (50). Antibodies reactive with acapsular cryptococci are blocked by treatment of serum with purified yeast glucan but not with mannan or chitin. Antibodies are also removed by absorption with zymosan but not *C. albicans* or *S. cerevisiae* yeast cells. Notably, zymosan consists of ghosts of *S. cerevisiae* that have been treated in a manner which appears to expose glucan that is not displayed on the surface of intact *S. cerevisiae* yeast cells (22).

B. dermatitidis provides a third example of the influence of natural antibodies on activation of the complement system (118, 119). Incubation of *B. dermatitidis* yeast cells in normal adult serum leads to rapid accumulation of C3 on the yeast in a manner that resembles results found with *C. albicans* or acapsular cryptococci, indicating activation of the classical pathway. If the classical pathway is blocked by treatment of serum with EGTA or by removal of antibody via absorption, delayed and asynchronous deposition of C3 occurs via the alternative pathway. The antibody in normal serum that mediates classical pathway activation by *B. dermatitidis* is specific for β-glucan.

The critical question with regard to naturally occurring antibodies to fungal antigens is the extent to which these antibodies contribute to host resistance. There is some evidence from clinical studies suggesting that natural antibodies are protective against fungal infection. Stratta et al. found that prophylaxis using intravenous immune globulin produced a significant reduction in the occurrence of disseminated fungal infection in liver transplant patients receiving OKT3 therapy (93); *Candida* spp. are the most common cause of fungal infection in this patient group (10). Further evidence of a role for naturally occurring antibodies is provided by an analysis of the contribution of sera from donors having variable levels of antimannan IgG to opsonophagocytic killing of *C. albicans* by neutrophils (53). In the absence of a functional complement system, there was a strong correlation between opsonophagocytic killing and antimannan antibody titer. In contrast, there was no apparent association between antimannan IgG levels and killing if the complement system was intact, once again emphasizing the probably redundant roles of antimannan IgG and the complement system in resistance to candidiasis.

EVASION OF INNATE HUMORAL IMMUNITY

There are several potential mechanisms by which pathogenic fungi might evade surveillance by innate humoral immunity. These include (i) use of molecular mimicry to avoid phagocytosis, (ii) binding of regulatory proteins or other means of disrupting activation of the complement system, and (iii) alteration of surface expression of ligands recognized by components of innate immunity.

The potential for molecular mimicry was first recognized by two independent discoveries that

C. albicans displays a surface receptor for C3 fragments (24, 41). In these studies, sheep erythrocytes coated with iC3b and C3d were found to bind to fungal cells. Further study found that erythrocytes coated with C3d and iC3b bound to *C. albicans* and its closely related species *C. stellatoidea* but not to other yeast species. Monoclonal antibodies that recognize CD11b and CD11c cross-react with surface proteins of *C. albicans*, indicating that the candidal iC3b receptor has structural homology to the α subunit of the mammalian leukocyte adhesion glycoproteins CD11b/CD18 and CD11c/CD18 (3, 4, 24). The candidal integrin-like protein probably has several properties that might contribute to virulence including molecular mimicry to avoid phagocytosis (36) and adhesion to host tissues (3, 30).

Several microbes bind the complement regulatory proteins factor H, FHL-1, and/or C4BP. C4BP binds to C4b, displacing C2a from the classical-pathway C3 convertase, and serves as a cofactor for cleavage of C4b by factor I. Factor H and FHL-1 bind to C3b, displacing Bb from the alternative pathway C3 convertase, and serve as cofactors for factor I in cleavage of C3b. As a consequence, binding of these negative regulatory proteins may disrupt amplification of the complement cascade. Bacteria that bind factor H, FHL-1, and/or C4BP include *Streptococcus pyogenes*, *Streptococcus pneumoniae*, *Neisseria gonorrhoeae*, *Borrelia burgdorferi*, and *Bordetella pertussis*. Recent studies have added *C. albicans* to the list of microbes that bind factor H, FHL-1, and C4BP (73, 74). Importantly, factor H, FHL-1, and C4BP that bind to the yeast surface retain their regulatory activities. C4BP and factor H display noncompetitive binding to *C. albicans*, suggesting distinct binding sites. However, FHL-1 competes for binding of C4BP, indicating that these two related complement regulators bind to the same structures on the yeast surface (73). The putative *C. albicans* surface receptors that serve as binding sites for complement regulatory proteins have yet to be identified.

A. fumigatus has several properties that block complement activation, including alterations in the capacity for accumulation of C3 fragments and regulation of the alternative pathway. Tsai et al. described a developmentaly regulated *alb1* gene that is required for conidial pigmentation (109). Disruption of *alb1* produced a significant increase in C3 binding to conidial surfaces, and the conidia of the *alb1* disruptant were ingested by neutrophils at a higher rate than were those of the wild type. Importantly, the *alb1* disruptant showed a significant loss of virulence compared to the wild-type and *alb1*-complemented strains in a murine model of aspergillosis. Additional studies demonstrated that a reddish-pink conidial pigmentation mutant exhibited enhanced binding of C3 relative to the wild type and that this mutation was complemented by a gene designated *arp1* (108). *A. fumigatus* conidia also express a surface protein that binds C3 (95). Finally, *A. fumigatus* secretes a soluble factor that inhibits complement activation and opsonization of fungal cells (112, 113). The inhibitor decreased the binding of C3b to fungal surfaces by selectively interfering with activation of the alternative pathway.

A strategy that has been used by a number of microbes for immune evasion is variation in expression of ligands or epitopes that are recognized by elements of acquired or innate immunity. There is clear evidence that *C. albicans* shows variability in the expression of epitopes between yeast and hyphal forms (5, 6). A preliminary report also suggests variability in the expression of *C. albicans* ligands that are recognized by MBL (J. Lillegard, P. Thorkildson, and T. Kozel, *Abstr. 103rd Gen. Meet. Am. Soc. Microbiol. 2003*, abstr. F-109, 2003). The MBL ligand is poorly expressed on yeast cells grown at room temperature but is abundant on yeast cells grown at 37°C. The specific ligand on *C. albicans* mannan that is recognized by MBL is not known. Similarly, the molecular basis for variation in surface expression of the ligand is not known. Phase variation in the ability of some bacteria to bind MBL has also been reported (106), suggesting that variability in ligand expression as a means for avoidance of innate immunity may be common to microbes that are targeted by MBL.

CONCLUSIONS

Resistance to many disseminated fungal infections is characteristically high in normal healthy adults. As a consequence, many fungi are regarded as opportunistic pathogens. The innate humoral immune system is an important contributor to this high natural resistance. Components of the innate humoral immune system include the alternative complement system, MBL, and naturally occurring antibodies that are reactive with glucan and mannan found in fungal cell walls. These elements of innate immunity function directly as opsonins or indirectly by activation of the complement system for cleavage and binding of opsonic fragments of C3 to the cell surface. The alternative pathway, MBL, and natural antibodies probably have redundant functions. As a consequence, it is possible that the host could lose one of these activities without loss of innate resistance.

Future studies should focus on identification of the mechanisms for recognition of fungi by the alternative pathway and MBL. What are the specific ligands or components of the cell wall that allow for recognition of fungi as nonself? Future studies should also take advantage of the increasing availability of transgenic mice to identify the components of innate humoral immunity that are essential for optimal resistance to disseminated fungal infection. Such information would aid in identification of individuals at greatest risk for fungal infection and point the way for therapies that might enhance innate resistance. To this end, MBL replacement therapy has been proposed for patients who are MBL deficient (97), and passive immunization with antibodies to *C. albicans* mannan has been shown to be protective in a murine model of hematogenously disseminated candidiasis (39).

REFERENCES

1. **Ashman, R. B., J. M. Papadimitriou, A. Fulurija, K. E. Drysdale, C. S. Farah, O. Naidoo, and T. Gotjamanos.** 2003. Role of complement C5 and T lymphocytes in pathogenesis of disseminated and mucosal candidiasis in susceptible DBA/2 mice. *Microb. Pathog.* **34:**103–113.

2. **Babula, O., G. Lazdane, J. Kroica, W. J. Ledger, and S. S. Witkin.** 2003. Relation between recurrent vulvovaginal candidiasis, vaginal concentrations of mannose-binding lectin, and a mannose-binding lectin gene polymorphism in Latvian women. *Clin. Infect. Dis.* **37:**733–737.

3. **Bendel, C. M., and M. K. Hostetter.** 1993. Distinct mechanisms of epithelial adhesion for *Candida albicans* and *Candida tropicalis*. Identification of the participating ligands and development of inhibitory peptides. *J. Clin. Investig.* **92:**1840–1849.

4. **Bendel, C. M., J. St. Sauver, S. Carlson, and M. K. Hostetter.** 1995. Epithelial adhesion in yeast species: correlation with surface expression of the integrin analog. *J. Infect. Dis.* **171:**1660–1663.

5. **Brawner, D. L., and J. E. Cutler.** 1986. Ultrastructural and biochemical studies of two dynamically expressed cell surface determinants on *Candida albicans*. *Infect. Immun.* **51:**327–336.

6. **Brawner, D. L., and J. E. Cutler.** 1985. Changes in surface topography of *Candida albicans* during morphogenesis. *J. Med. Vet. Mycol.* **23:**389–393.

7. **Budzko, D. B., and R. Negroni.** 1975. Hemolytic, cytotoxic and complement inactivating properties of extracts of different species of aspergillus. *Mycopathologia* **57:**23–26.

8. **Byron, J. K., K. V. Clemons, J. H. McCusker, R. W. Davis, and D. A. Stevens.** 1995. Pathogenicity of *Saccharomyces cerevisiae* in complement factor five-deficient mice. *Infect. Immun.* **63:**478–485.

9. **Calich, V. L. G., T. L. Kipnis, M. Mariano, C. F. Neto, and W. D. da Silva.** 1979. The activation of the complement system by *Paracoccidioides brasiliensis in vitro*: its opsonic effect and possible significance for an *in vivo* model of infection. *Clin. Immunol. Immunopathol.* **12:**20–30.

10. **Castaldo, P., R. J. Stratta, R. P. Wood, R. S. Markin, K. D. Patil, M. S. Shaefer, A. N. Langnas, E. C. Reed, S. Li, T. J. Pillen, and B. W. Shaw.** 1998. Clinical spectrum of fungal infections after orthotopic liver transplantation. *Arch. Surg.* **126:**149–156.

11. Chenoweth, D. E. 1986. Complement mediators of inflammation, p. 63–86. *In* G. D. Ross (ed.), *Immunobiology of the Complement System: an Introduction for Research and Clinical Medicine.* Academic Press, Inc., Orlando, Fla.

12. Cherniak, R., H. Valafar, L. C. Morris, and F. Valafar. 1998. *Cryptococcus neoformans* chemotyping by quantitative analysis of ^{1}H nuclear magnetic resonance spectra of glucuronoxylomannans with a computer-simulated artificial neural network. *Clin. Diagn. Lab. Immunol.* 5:146–159.

13. Chiba, H., H. Sano, D. Iwaki, S. Murakami, T. Takahashi, M. Konishi, H. Takahashi, and Y. Kuroki. 2004. Rat mannose-binding protein A binds CD14. *Infect. Immun.* 69:1587–1592.

14. Crosdale, D. J., K. V. Poulton, W. E. Ollier, W. Thomson, and D. W. Denning. 2001. Mannose-binding lectin gene polymorphisms as a susceptibility factor for chronic necrotizing pulmonary aspergillosis. *J. Infect. Dis.* 184:653–656.

15. Cross, C. E., and G. J. Bancroft. 1995. Ingestion of acapsular *Cryptococcus neoformans* occurs via mannose and β-glucan receptors, resulting in cytokine production and increased phagocytosis of the encapsulated form. *Infect. Immun.* 63:2604–2611.

16. Dahl, M. V., and R. Carpenter. 1986. Polymorphonuclear leukocytes, complement, and *Trichophyton rubrum.* *J. Investig. Dermatol.* 86:138–141.

17. Davies, S. F., D. P. Clifford, J. R. Hoidal, and J. E. Repine. 1982. Opsonic requirements for the uptake of *Cryptococcus neoformans* by human polymorphonuclear leukocytes and monocytes. *J. Infect. Dis.* 145:870–874.

18. De Bracco, M. M., D. B. Budzko, and R. Negroni. 1976. Mechanisms of activation of complement by extracts of *Aspergillus fumigatus.* *Clin. Immunol. Immunopathol.* 5:333–339.

19. de Saint Basile, G., A. Durandy, G. Somme, and C. Griscelli. 1987. Idiotypy of human anti-*Candida albicans* antibodies: recurrence, presence of a cross-reactive autoanti-idiotypic-like activity, and role in the induction of specific in vitro antibody response. *J. Immunol.* 138:417–422.

20. Diamond, R. D., and N. F. Erickson III. 1982. Chemotaxis of human neutrophils and monocytes induced by *Cryptococcus neoformans.* *Infect. Immun.* 38:380–382.

21. Diamond, R. D., J. E. May, M. A. Kane, M. M. Frank, and J. E. Bennett. 1974. The role of the classical and alternate complement pathways in host defenses against *Cryptococcus neoformans* infection. *J. Immunol.* 112: 2260–2270.

22. DiCarlo, F. J., and J. V. Fiore. 1958. On the composition of zymosan. *Science* 127:756–757.

23. Drickamer, K., M. S. Dordal, and L. Reynolds. 1986. Mannose-binding proteins isolated from rat liver contain carbohydrate-recognition domains linked to collagenous tails. *J. Biol. Chem.* 261:6878–6887.

24. Edwards, J. E., Jr., T. A. Gaither, J. J. O'Shea, D. Rotrosen, T. J. Lawley, S. A. Wright, M. M. Frank, and I. Green. 1986. Expression of specific binding sites on *Candida* with functional and antigenic characteristics of human complement receptors. *J. Immunol.* 137:3577–3583.

25. Ehrlich, P., and H. Sachs. 1902. Ueber die vielheit der complemente des serums. *Berl. Klin. Wochenschr.* 39:297–299.

26. Faux, J. A., A. E. Agbarakwe, S. A. Misbah, and H. M. Chapel. 1992. A comparison of specific IgG antibody levels to the cell wall mannan of *Candida albicans* in normal individuals and in patients with primary antibody deficiency. *J. Immunol. Methods* 153:167–172.

27. Ferrante, A., and Y. H. Thong. 1979. Requirement of heat-labile opsonins for maximal phagocytosis of *Candida albicans. Sabouraudia* 17:293–297.

28. Fizpatrick, F. W., and F. J. DiCarlo. 1964. Zymosan. *Ann. N. Y. Acad. Sci.* 118:235–261.

29. Fujita, T., Y. Endo, and M. Nonaka. 2004. Primitive complement system—recognition and activation. *Mol. Immunol.* 41:103–111.

30. Gale, C. A., C. M. Bendel, M. McClellan, M. Hauser, J. M. Becker, J. Berman, and M. K. Hostetter. 1998. Linkage of adhesion, filamentous growth, and virulence in *Candida albicans* to a single gene, *INT1. Science* 279:1355–1358.

31. Galgiani, J. N., R. A. Isenberg, and D. A. Stevens. 1978. Chemotaxigenic activity of extracts from the mycelial and spherule phases of *Coccidioides immitis* for human polymorphonuclear leukocytes. *Infect. Immun.* 21: 862–865.

32. Galgiani, J. N., P. Yam, L. D. Petz, P. L. Willams, and D. A. Stevens. 1980. Complement activation by *Coccidioides immitis:* in vitro and clinical studies. *Infect. Immun.* 28:944–949.

33. Garred, P., H. O. Madsen, B. Hofmann, and A. Svejgaard. 1995. Increased frequency of homozygosity of abnormal mannan-binding-protein alleles in patients with suspected immunodeficiency. *Lancet* 346:941–943.

34. Gates, M. A., P. Thorkildson, and T. R. Kozel. 2004. Molecular architecture of the *Cryptococcus neoformans* capsule. *Mol. Microbiol.* 52:13–24.

35. Ghiran, I., S. F. Barbashov, L. B. Klickstein, S. W. Tas, J. C. Jensenius, and A. Nicholson-Weller. 2000. Complement receptor 1/CD35 is a receptor for mannan-binding lectin. *J. Exp. Med.* 192:1797–1808.

36. Gilmore, B. J., E. M. Retsinas, J. S. Lorenz, and M. K. Hostetter. 1988. An iC3b receptor on *Candida albicans:* structure, function, and correlates for pathogenicity. *J. Infect. Dis.* 157:38–46.

37. Goren, M. B., and J. Warren. 1968. Immunofluorescence studies of reactions at the cryptococcal capsule. *J. Infect. Dis.* 118:215–229.

38. Greenfield, R. A., J. L. Stephens, M. J. Bussey, and J. M. Jones. 1983. Quantitation of antibody to *Candida* mannan by enzyme-linked immunosorbent assay. *J. Lab. Clin. Med.* 101:758–771.

39. Han, Y., and J. E. Cutler. 1995. Antibody response that protects against disseminated candidiasis. *Infect. Immun.* 63:2714–2719.

40. Hector, R. F., E. Yee, and M. S. Collins. 1990. Use of DBA/2N mice in models of systemic candidiasis and pulmonary and systemic aspergillosis. *Infect. Immun.* 58:1476–1478.

41. Heidenreich, F., and M. P. Dierich. 1985. *Candida albicans* and *Candida stellatoidea,* in contrast to other *Candida* species, bind iC3b and C3d but not C3b. *Infect. Immun.* 50:598–600.

42. Horstmann, R. D., M. K. Pangburn, and H. J. Muller-Eberhard. 1985. Species specificity of recognition by the alternative pathway of complement. *J. Immunol.* 134:1101–1104.

43. Iobst, S. T., M. R. Wormald, W. I. Weis, R. A. Dwek, and K. Drickamer. 1994. Binding of sugar ligands to

Ca^{2+}-dependent animal lectins. I. Analysis of mannose binding by site-directed mutagenesis and NMR. *J. Biol. Chem.* **269:**15505–15511.

44. Ip, W. K., and Y. L. Lau. 2004. Role of mannose-binding lectin in the innate defense against *Candida albicans*: enhancement of complement activation, but lack of opsonic function, in phagocytosis by human dendritic cells. *J. Infect. Dis.* **190:**632–640.

45. Jack, D. L., N. J. Klein, and M. W. Turner. 2001. Mannose-binding lectin: targeting the microbial world for complement attack and opsonophagocytosis. *Immunol. Rev.* **180:**86–99.

46. Jensenius, J. C., R. H. Jensen, K. McGwire, J. L. Larsen, and S. Thiel. 2003. Recombinant mannan-binding lectin (MBL) for therapy. *Biochem. Soc. Trans.* **31:**763–767.

47. Jones, J. M. 1980. Quantitation of antibody against cell wall mannan and a major cytoplasmic antigen of *Candida* in rabbits, mice, and humans. *Infect. Immun.* **30:**78–89.

48. Kawasaki, N., T. Kawasaki, and I. Yamashina. 1983. Isolation and characterization of a mannan-binding protein from human serum. *J. Biochem.* (Tokyo) **94:**937–947.

49. Kawasaki, T., R. Etoh, and I. Yamashina. 1978. Isolation and characterization of a mannan-binding protein from rabbit liver. *Biochem. Biophys. Res. Commun.* **81:**1018.

50. Keller, R. G., G. S. T. Pfrommer, and T. R. Kozel. 1994. Occurrences, specificities, and functions of ubiquitous antibodies in human serum that are reactive with the *Cryptococcus neoformans* cell wall. *Infect. Immun.* **62:**215–220.

51. Kitz, D. J., P. D. Stahl, and J. R. Little. 1992. The effect of a mannose binding protein on macrophage interactions with *Candida albicans*. *Cell. Mol. Biol.* **38:**407–412.

52. Kozel, T. R., R. R. Brown, and G. S. T. Pfrommer. 1987. Activation and binding of C3 by *Candida albicans*. *Infect. Immun.* **55:**1890–1894.

53. Kozel, T. R., R. S. MacGill, A. Percival, and Q. Zhou. 2004. Biological activities of naturally occurring antibodies reactive with *Candida albicans* mannan. *Infect. Immun.* **72:**209–218.

54. Kozel, T. R., and G. S. T. Pfrommer. 1986. Activation of the complement system by *Cryptococcus neoformans* leads to binding of iC3b to the yeast. *Infect. Immun.* **52:**1–5.

55. Kozel, T. R., A. Tabuni, B. J. Young, and S. M. Levitz. 1996. Influence of opsonization conditions on C3 deposition and phagocyte binding of large- and small-capsule *Cryptococcus neoformans* cells. *Infect. Immun.* **64:**2336–2338.

56. Kozel, T. R., L. C. Weinhold, and D. M. Lupan. 1996. Distinct characteristics of initiation of the classical and alternative complement pathways by *Candida albicans*. *Infect. Immun.* **64:**3360–3368.

57. Kozel, T. R., M. A. Wilson, T. P. Farrell, and S. M. Levitz. 1989. Activation of C3 and binding to *Aspergillus fumigatus* conidia and hyphae. *Infect. Immun.* **57:**3412–3417.

58. Kozel, T. R., M. A. Wilson, and J. W. Murphy. 1991. Early events in initiation of alternative complement pathway activation by the capsule of *Cryptococcus neoformans*. *Infect. Immun.* **59:**3101–3110.

59. Kozel, T. R., M. A. Wilson, G. S. T. Pfrommer, and A. M. Schlageter. 1989. Activation and binding of opsonic fragments of C3 on encapsulated *Cryptococcus*

neoformans by using an alternative complement pathway reconstituted from six isolated proteins. *Infect. Immun.* **57:**1922–1927.

60. Kozel, T. R., M. A. Wilson, and W. H. Welch. 1992. Kinetic analysis of the amplification phase for activation and binding of C3 to encapsulated and nonencapsulated *Cryptococcus neoformans*. *Infect. Immun.* **60:**3122–3127.

61. Kuhlman, M., K. Joiner, and R. A. B. Ezekowitz. 1989. The human mannose-binding protein functions as an opsonin. *J. Exp. Med.* **169:**1733–1745.

62. Law, S. K., and R. P. Levine. 1977. Interaction between the third complement protein and cell surface macromolecules. *Proc. Natl. Acad. Sci. USA* **74:**2701–2705.

63. Laxalt, K. A., and T. R. Kozel. 1979. Chemotaxigenesis and activation of the alternative complement pathway by encapsulated and non-encapsulated *Cryptococcus neoformans*. *Infect. Immun.* **26:**435–440.

64. Lee, R. T., Y. Ichikawa, M. Fay, K. Drickamer, M. C. Shao, and Y. C. Lee. 1991. Ligand-binding characteristics of rat serum-type mannose-binding protein (MBP-A). Homology of binding site architecture with mammalian and chicken hepatic lectins. *J. Biol. Chem.* **266:**4810–4815.

65. Lee, S. J., G. Gonzalez-Aseguinolaza, and M. C. Nussenzweig. 2002. Disseminated candidiasis and hepatic malarial infection in mannose-binding-lectin-A-deficient mice. *Mol. Cell. Biol.* **22:**8199–8203.

66. Lehmann, P. F., and E. Reiss. 1980. Comparison by ELISA of serum anti-*Candida albicans* mannan IgG levels of a normal population and in diseased patients. *Mycopathologia* **70:**89–93.

67. Lovchik, J. A., and M. F. Lipscomb. 1993. Role of C5 and neutrophils in the pulmonary intravascular clearance of circulating *Cryptococcus neoformans*. *J. Respir. Cell Mol. Biol.* **9:**617–627.

68. Lu, J., S. Thiel, H. Wiedemann, R. Timpl, and K. B. M. Reid. 1990. Binding of the pentamer/hexamer forms of mannan-binding protein to zymosan activates the proenzyme C1r$_2$C1s$_2$ complex, of the classical pathway of complement, without involvement of C1q. *J. Immunol.* **144:**2287–2294.

69. Lyon, F. L., R. F. Hector, and J. E. Domer. 1986. Innate and acquired immune responses against *Candida albicans* in congenic B10.D2 mice with deficiency of the C5 complement component. *J. Med. Vet. Mycol.* **24:**359–367.

70. Macher, A. M., J. E. Bennett, J. E. Gadek, and M. M. Frank. 1978. Complement depletion in cryptococcal sepsis. *J. Immunol.* **120:**1686–1690.

71. Malhotra, R., S. Thiel, K. B. M. Reid, and R. B. Sim. 1990. Human leukocyte C1q receptor binds other soluble proteins with collagen domains. *J. Exp. Med.* **172:**955–959.

72. McCormack, F. X., R. Gibbons, S. R. Ward, A. Kuzmenko, and G. S. Deepe. 2003. Macrophage-independent fungicidal action of the pulmonary collectins. *J. Biol. Chem.* **278:**36250–36256.

73. Meri, T., A. M. Blom, A. Hartmann, D. Lenk, S. Meri, and P. F. Zipfel. 2004. The hyphal and yeast forms of *Candida albicans* bind the complement regulator C3b-binding protein. *Infect. Immun.* **72:**6633–6641.

74. Meri, T., A. Hartmann, D. Lenk, R. Eck, R. Wurzner, J. Hellwage, S. Meri, and P. F. Zipfel. 2002. The yeast *Candida albicans* binds complement regulators factor H and FHL-1. *Infect. Immun.* **70:**5185–5192.

75. Miller, M. E., J. Seals, R. Kaye, and L. C. Levitsky. 1968. A familial, plasma associated defect of phagocytosis: a new cause of recurrent bacterial infections. *Lancet* ii:60–63.

76. Mitchell, T. G., and L. Friedman. 1972. In vitro phagocytosis and intracellular fate of variously encapsulated strains of *Cryptococcus neoformans*. *Infect. Immun.* 5:491–498.

77. Morelli, R., and L. T. Rosenberg. 1971. Role of complement during experimental *Candida* infection in mice. *Infect. Immun.* 3:521–523.

78. Morrison, R. P., and J. E. Cutler. 1981. In vitro studies of the interaction of murine phagocytic cells with *Candida albicans*. *J. Reticuloendothel. Soc.* 29:23–34.

79. Mullighan, C. G., S. E. Marshall, and K. I. Welsh. 2000. Mannose binding lectin polymorphisms are associated with early age of disease onset and autoimmunity in common variable immunodeficiency. *Scand. J. Immunol.* 51:111–122.

80. Munk, M. E., and W. Dias Da Silva. 1992. Activation of human complement system *Paracoccidioides brasiliensis* and its deposition on the yeast form cell surface. *J. Med. Vet. Mycol.* 30:481–484.

81. Munk, M. E., A. Kajdacsy-Balla, G. Del Negro, L. C. Cuce, and W. Dias Da Silva. 1992. Activation of human complement system in paracoccidioidomycosis. *J. Med. Vet. Mycol.* 30:317–321.

82. Neth, O., I. Hann, M. W. Turner, and N. J. Klein. 2001. Deficiency of mannose-binding lectin and burden of infection in children with malignancy: a prospective study. *Lancet* 358:614–618.

83. Neth, O., D. L. Jack, A. W. Dodds, H. Holzel, N. J. Klein, and M. W. Turner. 2000. Mannose-binding lectin binds to a range of clinically relevant microorganisms and promotes complement deposition. *Infect. Immun.* 68:688–693.

84. Pereira, H. A., and C. S. Hosking. 1984. The role of complement and antibody in opsonization and intracellular killing of *Candida albicans*. *Clin. Exp. Immunol.* 57:307–314.

85. Peterslund, N. A., C. Koch, and J. C. Jensenius. 2001. Association between deficiency of mannose-binding lectin and severe infections after chemotherapy. *Lancet* 358:637–638.

86. Pillemer, L., L. Blum, I. H. Lepow, O. A. Ross, E. W. Todd, and A. C. Wardlaw. 1954. Properdin system and immunity: demonstration and isolation of new serum protein, properdin, and its role in immune phenomena. *Science* 120:279–285.

87. Pillemer, L., L. Blum, I. H. Lepow, L. Wurz, and E. W. Todd. 1956. The properdin system and immunity. III. The zymosan assay of properdin. *J. Exp. Med.* 103:1–13.

88. Pillemer, L., and E. E. Ecker. 1941. Anticomplementary factor in fresh yeast. *J. Biol. Chem.* 137:139–142.

89. Presanius, J. S., M. Kojima, and R. B. Sim. 2003. Biochemistry and genetics of mannan-binding lectin (MBL). *Biochem. Soc. Trans.* 31:748–752.

90. Ray, T. L., and K. D. Wuepper. 1976. Activation of the alternative (properdin) pathway of complement by *Candida albicans* and related species. *J. Investig. Dermatol.* 67:700–703.

91. Rhodes, J. C. 1985. Contribution of complement component C5 to the pathogenesis of experimental murine cryptococcosis. *Sabouraudia* 23:225–234.

92. Robbins, J. B., R. Schneerson, M. P. Glode, W. Vann, M. S. Schiffer, T.-Y. Liu, J. C. Parke, Jr., and C. Huntley. 1975. Cross-reactive antigens and immunity to diseases caused by encapsulated bacteria. *J. Allergy Clin. Immunol.* 56:141–151.

93. Stratta, R. J., M. S. Shaefer, K. A. Cushing, R. S. Markin, E. C. Reed, A. N. Langnas, T. J. Pillen, and B. W. Shaw. 1992. A randomized prospective trial of acyclovar and immune globulin prophylaxis in liver transplant recipients receiving OKT3 therapy. *Arch. Surg.* 127:55–64.

94. Sturtevant, J., and J. P. Latgé. 1992. Participation of complement in the phagocytosis of the conidia of *Aspergillus fumigatus* by human polymorphonuclear cells. *J. Infect. Dis.* 166:580–586.

95. Sturtevant, J. E., and J. P. Latgé. 1992. Interactions between conidia of *Aspergillus fumigatus* and human complement component C3. *Infect. Immun.* 60:1913–1918.

96. Sumiya, M., and J. A. Summerfield. 1997. Mannose-binding protein, genetic variants and the risk of infection. *Q. J. Med.* 89:723–726.

97. Summerfield, J. A. 2003. Clinical potential of mannose-binding lectin-replacement therapy. *Biochem. Soc. Trans.* 31:770–773.

98. Summerfield, J. A., S. Ryder, M. Sumiya, M. Thursz, A. Gorchein, M. A. Monteil, and M. W. Turner. 1995. Mannose binding protein gene mutations associated with unusual and severe infections in adults. *Lancet* 345:886–889.

99. Swan, J. W., M. V. Dahl, P. A. Coppo, and D. E. Hammerschmidt. 1983. Complement activation by *Trichophyton rubrum*. *J. Investig. Dermatol.* 80:156–158.

100. Tabona, P., A. Mellor, and J. A. Summerfield. 1995. Mannose binding protein is involved in the first-line host defense: evidence from transgenic mice. *Immunology* 85:153–159.

101. Tagami, H., N. Natsume, T. Aoshima, F. Inuoe, S. Suehisa, and M. Yamada. 1982. Analysis of transepidermal leukocyte chemotaxis in experimental dermatophytosis in guinea pigs. *Arch. Dermatol. Res.* 273:205–217.

102. Thiel, S., U. Holmskov, L. Hviid, S. B. Laursen, and J. C. Jensenius. 1992. The concentration of the C-type lectin, mannan-binding protein, in human plasma increases during an acute phase response. *Clin. Exp. Immunol.* 90:31–35.

103. Thiel, S., S. V. Petersen, T. Vorup-Jensen, M. Matsushita, T. Fujita, C. M. Stover, W. J. Schwaeble, and J. C. Jensenius. 2000. Interaction of C1q and mannan-binding lectin (MBL) with C1r, C1s, MBL-associated serine proteases 1 and 2, and the MBL-associated protein MAp19. *J. Immunol.* 165:878–887.

104. Thiel, S., T. Vorup-Jensen, C. M. Stover, W. Schwaeble, S. B. Laursen, K. Poulsen, A. C. Willis, P. Eggleton, S. Hansen, U. Holmskov, K. B. Reid, and J. C. Jensenius. 1997. A second serine protease associated with mannan-binding lectin that activates complement. *Nature* 386:506–510.

105. Thong, Y. H., and A. Ferrante. 1978. Alternative pathway of complement activation by *Candida albicans*. *Aust. N. Z. J. Med.* 8:620–622.

106. Townsend, R., R. C. Read, M. W. Turner, N. J. Klein, and D. L. Jack. 2001. Differential recognition of obligate anaerobic bacteria by human mannose-binding lectin. *Clin. Exp. Immunol.* 124:223–228.

107. Truelsen, K., T. Young, and T. R. Kozel. 1992. In vivo complement activation and binding of C3 to encapsulated *Cryptococcus neoformans*. *Infect. Immun.* 60:1–4.

108. Tsai, H. F., R. G. Washburn, Y. C. Chang, and K. J. Kwon-Chung. 1997. Aspergillus fumigatus arp1 modulates conidial pigmentation and complement deposition. *Mol. Microbiol.* **26:**175–183.

109. Tsai, H.-F., Y. C. Chang, R. G. Washburn, M. H. Wheeler, and K. J. Kwon-Chung. 1998. The developmentally regulated *alb1* gene of *Aspergillus fumigatus*: its role in modulation of conidial morphology and virulence. *J. Bacteriol.* **180:**3031–3038.

110. van de Wetering, J. K., L. M. G. van Golde, and J. J. Batenburg. 2004. Collectins. Players of the innate immune system. *Eur. J. Biochem.* **271:**1229–1249.

111. von Dungern, E. 1900. Beitrage zur Immunitatslehre. *Munch. Med. Wochenschr.* **47:**677–680.

112. Washburn, R. G., D. J. DeHart, D. E. Agwu, B. J. Bryant-Varela, and N. C. Julian. 1990. *Aspergillus fumigatus* complement inhibitor: production, characterization, and purification by hydrophobic interaction and thin-layer chromatography. *Infect. Immun.* **58:**3508–3515.

113. Washburn, R. G., C. H. Hammer, and J. E. Bennett. 1986. Inhibition of complement by culture supernatants of *Aspergillus fumigatus*. *J. Infect. Dis.* **154:**944–951.

114. Weis, W. I., K. Drickamer, and W. A. Hendrickson. 1992. Structure of a C-type mannose-binding protein complexed with an oligosaccharide. *Nature* **360:**127–134.

115. Wilson, M. A., and T. R. Kozel. 1992. Contribution of antibody in normal human serum to early deposition of C3 onto encapsulated and nonencapsulated *Cryptococcus neoformans*. *Infect. Immun.* **60:**754–761.

116. Young, B. J., and T. R. Kozel. 1993. Effects of strain variation, serotype and structural modification on the kinetics for activation and binding of C3 to *Cryptococcus neoformans*. *Infect. Immun.* **61:**2966–2972.

117. Zaragoza, O., C. P. Taborda, and A. Casadevall. 2004. The efficacy of complement-mediated phagocytosis of *Cryptococcus neoformans* is dependent on the location of C3 in the polysaccharide capsule and involves both direct and indirect C3-mediated interactions. *Eur. J. Immunol.* **33:**1957–1967.

118. Zhang, M. X., T. T. Brandhorst, T. R. Kozel, and B. S. Klein. 2001. Role of glucan and surface protein BAD1 in complement activation by *Blastomyces dermatitidis* yeast. *Infect. Immun.* **69:**7559–7564.

119. Zhang, M. X., and B. Klein. 1997. Activation, binding, and processing of complement component 3 (C3) by *Blastomyces dermatitidis*. *Infect. Immun.* **65:**1849–1855.

120. Zhang, M. X., and T. R. Kozel. 1998. Mannan-specific immunoglobulin G antibodies in normal human serum accelerate binding of C3 to *Candida albicans* via the alternative complement pathway. *Infect. Immun.* **66:**4845–4850.

121. Zhang, M. X., D. M. Lupan, and T. R. Kozel. 1997. Mannan-specific IgG antibodies in normal human serum mediate classical pathway initiation of C3 binding to *Candida albicans*. *Infect. Immun.* **65:**3822–3827.

8

Molecular Principles of Fungal Pathogenesis
Edited by Joseph Heitman et al.
©2006 ASM Press, Washington, D.C.

Chapter 32

Innate and Acquired Cellular Immunity to Fungi

LUIGINA ROMANI

Human beings are constantly exposed to fungi. Most fungi (*Histoplasma capsulatum, Paracoccidioides brasiliensis, Coccidioides immitis, Blastomyces dermatitidis, Cryptococcus neoformans, Aspergillus fumigatus,* and *Pneumocystis carinii*) are ubiquitous in the environment. Some, including *Candida albicans,* establish lifelong commensalism on human body surfaces. Besides the frequent occurrence of allergic hypersensitivity, only a very limited number of fungi cause severe infections (107). Therefore, although some virulence factors are of obvious importance (109), pathogenicity is neither an invariant nor a stable characteristic of fungi. Although not unique among infectious agents, fungi possess complex and unusual relationships with the vertebrate immune system, partly due to some prominent features including their ability to exist in different forms and to reversibly switch from one to the other in infection. Examples are the dimorphic fungi (*H. capsulatum, P. brasiliensis, C. immitis,* and *B. dermatitidis*), which transform from saprobic filamentous molds to unicellular yeasts in the host; the filamentous fungi (for example, *Aspergillus* spp., *Fusarium* spp., and the Zygomycota) that, inhaled as unicellular conidia, may transform into branching hyphae in the lungs; some species of *Candida,* capable of growing in different forms such as yeasts, blastospores, pseudohyphae, and hyphae depending on infection sites; and *C. neoformans* yeasts, which become coated with a capsule. Because cycling between different morphotypes is not obligatory for fungi as it is for other organisms, morphological transition is a mechanism that fungi have evolved to adapt to different environments. Although associations between morphogenesis and virulence have long been presumed for fungi that are human pathogens (115), no molecular data unambiguously establish a role for fungal morphogenesis as a virulence factor (45). What fungal morphogenesis implicates, through antigenic variability, phenotypic switching, and dimorphic transition, is the existence of a multitude of recognition and effector mechanisms to oppose fungal infectivity at the different body sites.

Most fungi need a stable host-parasite interaction characterized by an immune response strong enough to allow host survival without pathogen elimination, thereby establishing commensalisms. Therefore, the balance of proinflammatory and anti-inflammatory signaling is a prerequisite for successful host-fungus interaction. In light of these considerations, although developments in fungal genomics may provide new insights into mechanisms of pathogenicity (78), the responsibility for virulence, regardless of the mode of its generation and maintenance, is shared by the host and the fungus at the pathogen-host interface. Studies with *C. albicans* have provided a paradigm that incorporates contributions from both the fungus and the host to explain the theme of the origin and maintenance of virulence for commensals (112). Through a high degree of flexibility, the model accommodates the concept of virulence as an important component of fungus fitness in vivo within the plasticity of the host immune system.

IMMUNITY TO FUNGI

Protective immunity against fungal pathogens is achieved by the integration of two distinct arms of the immune system, the innate and adaptive (or antigen-specific) responses (109). The majority of fungi are detected and destroyed within hours by innate defense mechanisms (53, 106, 112). The innate mechanisms appeared early in the evolution of multicellular organisms and act early after the infection. Innate defense strategies are designed to detect broad and conserved patterns which differ between pathogenic organisms and their multicellular hosts. Most of the innate mechanisms are inducible on infection, and their activation requires specific recognition of

Luigina Romani • Department of Experimental Medicine and Biochemical Science, University of Perugia, 06122 Perugia, Italy.

invariant evolutionarily conserved molecular structures shared by large groups of pathogens (also known as pathogen-associated molecular patterns [PAMPs]) by a set of pattern recognition receptors (PRRs), including Toll-like receptors (TLRs) (44, 70, 85, 94, 105). In vertebrates, however, if the infectious organism can breach these early lines of defense, an adaptive immune response will ensue, with generation of antigen-specific T-helper (Th) effector and B cells that specifically target the pathogen and memory cells that prevent subsequent infection with the same microorganism. Cytokines and other mediators play an essential role in the process and, indeed, may ultimately determine the type of inflammatory response that is generated toward the pathogens. The dichotomous Th-cell model has proven to be a useful construct that has shed light on the general principle that diverse effector functions are required for eradication of different fungal infections (109). To limit the pathologic consequences of an excessive inflammatory cell-mediated immune reaction, the immune system resorts to a number of protective mechanisms,

including the reciprocal cross-regulatory effects of Th1- and Th2-type effector cytokines, such as gamma interferon (IFN-γ) and interleukin-4 (IL-4), and the generation of regulatory T cells (Treg) (109). Thus, innate and adaptive immune responses are intimately linked and controlled by sets of molecules and receptors that act to generate the most effective form of immunity for protection against fungal pathogens (Fig. 1).

INNATE IMMUNITY

Antigen-independent recognition of fungi by the innate immune system leads to the immediate mobilization of immune effector and regulatory mechanisms that provide the host with three crucial survival advantages: (i) rapid initiation of the immune response (both innate and adaptive) and creation of the inflammatory and costimulatory context for antigen recognition; (ii) establishment of a first line of defense, which holds the pathogen in check during the maturation of the adaptive response; and (iii) steering of the adaptive response toward the

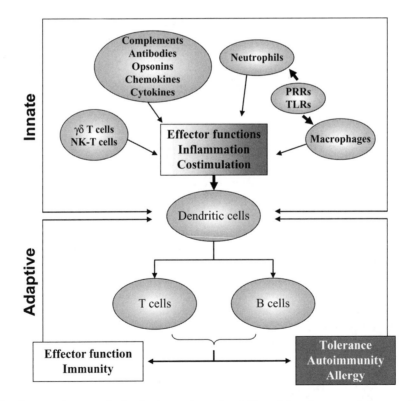

Figure 1. Interface between innate and adaptive immunity to fungi. Essential to the successful removal of pathogens is the early recognition of fungi by components of the innate immune systems. These involve the complement systems, opsonins, antibodies, and specialized receptors such as TLRs expressed on NK cells and phagocytes that recognize specific fungus-derived molecular structures. Successful engagement of some of these pathways leads to an inflammatory response with destruction of the pathogen alongside the establishment of DC and T-cell and/or B-cell interactions. A well-orchestrated innate and adaptive immune response will lead to pathogen eradication and host immunity (white box). Failure to efficiently discriminate self from nonself in innate as well as adaptive immunity can lead to pathogen proliferation and ultimately to dysregulated immunity such as autoimmunity, allergy, and sepsis (dark gray box).

cellular or humoral elements that are most appropriate for protection against the specific pathogen. Therefore, the goal to achieve the optimal activation of the antigen-specific immunity cannot be achieved without effectively activating the pathogen detection mechanisms of the innate immune response. Traditionally, the innate defense mechanisms used by the host against fungi have been considered of two types, constitutive and inducible, although the recent finding that the expression of constitutive mechanisms may be regulated by TLR signaling (130) highlights the interdependency of the two systems.

The constitutive defense mechanisms have two features in common: they are present at the sites of continuous interactions with fungi, and their destructive capacity is directed strictly at fungi but not at the host cells and tissues. This latter property is due to either the compartmentalization of these mechanisms (in which the route of exposure may condition host resistance to fungi) or the nature of their activities. The constitutive mechanisms include the barrier functions of the body surfaces and mucosal epithelia of the respiratory, gastrointestinal, and genitourinary tracts. Epithelial cells themselves may be endowed with antifungal effector activities (39), although epithelial cells and extracellular matrix may also promote fungal growth (65, 133). When hematogenous dissemination to visceral organs is an important step in the pathogenesis of the infection, the endothelial lining of blood vessels actively contributes to the process (7, 77, 100). Additional mechanisms include microbial antagonism, secretory immunoglobulin A (IgA), and locally produced enzymes and mediators with direct antifungal activity, such as antimicrobial peptides (27, 31), defensins (118), and collectins (41, 84). These molecules activate the innate immune system for microbe opsonization and the activation and recruitment of phagocytic cells. The mannose-binding lectin (MBL), a serum lectin, binds different fungi and promotes complement deposition (96). Recognition of diverse fungal and microbial species by MBL is mediated by the C-type lectin domain, which selectively binds to carbohydrates such as mannose and fucose. These carbohydrates are prevalent on glycolipids and glycoproteins that decorate the surface of fungi. MBL does not bind to sialic acid and other carbohydrates that commonly occur at terminal sugars on mammalian cell surface molecules. It is not surprising, therefore, that patients with defective MBL have increased susceptibility to fungal infections (5). In retaliation, fungi secrete a variety of enzymes such as phospholipases and proteases. These enzymes are considered to be major virulence factors, since they cause host cell damage and lysis and impair antifungal host defenses (42).

Humoral factors, such as complement (66) and antibodies (21), pivotally contribute to the innate defense mechanisms. They are discussed in other chapters of this book. The specific biological activities of the complement system and antibodies which contribute to host resistance are multifaceted and interdependent. For example, antibodies greatly contribute to the activation of the complement system by fungi, and complement is essential for antibody-mediated protection. Complement, antibodies, and collectins not only fulfill the requirement of a first line of defense against fungi but also have an impact on the inflammatory and adaptive immune responses through several mechanisms, including regulation of cytokine secretion by and costimulatory molecule expression on phagocytes (106).

Innate Immune Recognition Molecules

Host defense mechanisms inducible on infection require specific recognition of the infectious microorganisms by PRRs. They are expressed on professional and nonprofessional phagocytes such as polymorphonuclear leukocytes (PMNs, also known as neutrophils), mononuclear leukocytes (monocytes and macrophages), and dendritic cells (DCs). PRRs for fungi include TLRs, receptors for a variety of complement components (CRs), receptors for the Fc portion of immunoglobulins (FcRs), receptors for mannosyl/fucosyl glycoconjugate ligands (MRs), and receptors for β-glucan (dectin-1) (44, 70, 85, 94, 101, 122). Peptidoglycan recognition proteins are largely dispensable for mammalian immunity to fungi (139). Each receptor on phagocytes not only mediates distinct downstream intracellular events related to clearance but also participates in complex and disparate functions related to immunomodulation and activation of immunity, depending on cell types. The different receptors are called on to serve early-warning systems.

Not surprisingly, their ability to activate, in isolation, various effector functions is limited. With few exceptions (36), internalization via constitutively competent MRs does not represent an effective way of clearing fungi in the absence of opsonins. However, MRs on DCs activate specific programs that are relevant for the development of antifungal immune responses (see below). CR3 (also known as CD11b/CD18) engagement is one most efficient uptake mechanisms of opsonized fungi, but it has the remarkable characteristic of a broad capacity for recognition of diverse fungal ligands. The multiplicity of binding sites and the existence of different activation states enables CR3 of disparate (both positive and negative) effector activities against fungi (85). Thus, because signaling through CR3 may not lead to phagocyte activation without the concomitant engagement of FcR, the use of this receptor may contribute to intracellular fungal parasitism. It is of interest, indeed, that *H. capsulatum* uses this receptor for entry into macrophages, where it survives (76), but not into DCs, where it is rapidly degraded (43). Likewise, *Candida* exploits entry through CR3 to survive inside DCs (112). In contrast, ligation of FcR is usually sufficient to trigger phagocytosis, a vigorous oxidative burst, and generation of proinflammatory signals. Ultimately, recognition of antibody-opsonized particles represents a high-level threat. It is of interest that FcR-mediated phagocytosis may rescue the suppression of the respiratory burst, a finding explaining one possible mechanism through which T and B cells enhance the antifungal activity of macrophages (106).

The Toll protein of *Drosophila* is a transmembrane receptor involved in dorsoventral polarization during embryonic development and the induction of antifungal peptide genes (54). Mutation of the Toll gene led to impaired defense against *A. fumigatus* infection. Further evidence comes from studies with *Galleria mellonella*, an insect capable of sensing the pathogenicity of fungi in a manner similar to mammals (32). The sequence similarity of the cytoplasmic portion of *Drosophila* Toll and mammalian IL-1 receptor (IL-1R) intracellular domains suggested similarities in Toll and IL-1R signaling and illustrates the evolutionary conservation of both cellular signaling systems (98). TLRs are type I transmembrane proteins that are grouped into the same gene family based on their sequence similarity. Eleven mammalian TLRs have been described so far, and TLR ligands include PAMPs and additional ligands (98). The ability of PAMPs to induce costimulatory molecule expression on DCs suggests a permissive role of the PAMP/TLR system in the activation of T lymphocytes during antigen presentation.

All TLRs activate a core set of stereotyped responses, such as inflammation. However, individual TLRs can also induce specific programs in cells of the innate immune system that are tailored for the particular pathogen (98). TLRs and IL-1R have a similar signaling cascade culminating in activation of nuclear factor κB and mitogen-activated protein kinases that activate the transcription of the inflammatory and adaptive immune responses. The common signal pathways utilized by IL-1R and TLRs involve the recruitment of the adapter protein MyD88 (*Drosophila* myeloid differentiation primary response gene 88) through the homophilic interaction of the TIR domain. MyD88 in turn activates a series of IL-1R-associated kinases that are crucially involved in innate immunity, and the signal is propagated via a specific member of the tumor necrosis factor (TNF) receptor-associated factor family. However, in the case of TLR3- and TLR4-dependent lipopolysaccharide (LPS) signaling, other proteins may also serve as adapter molecules with or in place of MyD88 (140). It is recognized that the intricacies of how TLRs signal will ultimately provide an explanation for the molecular basis for how cells involved in innate immunity dictate the processes of host defense specific to the provoking pathogen.

The different impact of TLRs on the occurrence of the innate and adaptive Th immunity to fungi is consistent with the ability of each individual TLR to activate specialized antifungal effector functions on phagocytes and DCs (8, 9). Although not affecting the phagocytosis, TLRs affect specific antifungal programs of phagocytes, such as the respiratory burst, degranulation, and production of chemokines and cytokines (see below). Since the quantity and specificity of delivery of toxic neutrophil products ultimately determine the relative efficiency of fungicidal activity versus inflammatory cytotoxicity to host cells, this implies that TLRs may contribute to protection and immunopathology against fungi (8, 9). The emerging picture calls for (i) the essential requirement for the IL-1RI/MyD88-dependent pathway in the innate and Th1-mediated resistance to *C. albicans* (8); (ii) the essential requirement of the MyD88-dependent pathway in recognition of and response to *C. neoformans* (70, 141); (iii) the crucial involvement, although not essential, of the TLR4/MyD88 pathway in recognition of and resistance to *A. fumigatus* (8); (iv) the beneficial effect of TLR9 stimulation on immune-mediated resistance to pulmonary aspergillosis and cryptococcosis (11, 33, 111); (v) the dependency of TLR-dependent pathways from the site of the infection (8), also illustrated in flies infected with *C. neoformans* (1); and (vi) the occurrence of TLR signaling in a morphotype-specific manner (95), although the simultaneous engagement of multiple TLRs as well as TLR cooperativity in vivo makes it difficult to gauge the relative contribution of each single fungal morphotype in TLR activation and functioning. For instance, TLR4 and CD14 mediate the recognition of *C. albicans*-derived mannan (124). However, TNF-α and IL-1β production in response to *Candida* may also occur in a TLR4-independent manner, a finding consistent with the observation that resistance to infection is decreased in TLR4-deficient mice along with the release of chemokines (93). Therefore, TLR2 and TLR4 are both implicated in the elicitation of host defense against the fungus, a finding exemplifying recruitment of different TLRs by one microbial species.

Phagocytes

PMNs are the most heavily represented phagocyte population and generally the earliest to be recruited at sites of infection. Not surprisingly, therefore, quantitative or qualitative defects of neutrophils are major predisposing factors to disseminated candidiasis and aspergillosis (113). However, the finding that both human and murine PMNs also have fungistatic and fungicidal activities against other fungi (113) suggests a more general role of PMNs in fungal infections. Their functions may well go beyond microbicidal activity to include an immunoregulatory action on Th lymphocytes. Accumulating evidence indicates that cells of the myeloid lineage are capable of positive and negative regulation of T-cell function (17). Myeloid suppressor cells are responsible for the immunosuppression observed in conditions as dissimilar as tumor growth, immunosuppression, overwhelming infections, graft-versus-host disease, and pregnancy. A population of neutrophils suppressing Th1-cell activation is present in bone marrow-transplanted mice with candidiasis (86), a finding suggesting that myeloid suppressor cells may prevent functional immunoreconstitution in transplantation. The reciprocal influence between PMNs and T lymphocytes further implies that the immune resistance to fungi is a highly coordinate and unitary process, in the context of which the traditional dichotomy between neutropenia and T-cell-specific defects, as major predisposing factors to fungal infections, should be accommodated.

Studies with mice have confirmed the important role of PMNs as one first line of defense, as shown by the increased susceptibility to fungal infections of mice with defects of neutrophils, the potent antifungal activity of neutrophils, and the ability of fungal cells or their products to subvert neutrophil programs (35, 113, 116, 125). Binding and internalization of fungi by PMNs may occur through different receptors, of which CR3 (51) and FcγR, presumably FcγRI or FcαRI (129), are the most effective in terms of phagocytosis and fungicidal activity. As already mentioned, TLRs do not affect the phagocytosis of fungi, such as *C. albicans* or *A. fumigatus*, by PMNs but, by affecting the balance between fungicidal oxidative and nonoxidative mechanisms and the production of pro- and anti-inflammatory cytokines, TLR2, TLR4, and TLR9 ultimately affect the quality of microbicidal activity and inflammatory response of murine and human PMNs (8, 9).

Macrophages are a heterogeneous population of tissue-resident cells whose functionality varies according to anatomic location, state of activation, and fungal morphotype (81). Although macrophages possess the machinery for antigen presentation, their main contribution to antifungal defense is through phagocytosis and killing of fungi (81). Not surprisingly, therefore, fungi have exploited a variety of mechanisms or putative virulence factors to evade phagocytosis, escape destruction, and survive inside macrophages (106). Alveolar macrophages, known to represent the first line of defense against *Aspergillus* conidia (69, 117), ingest inhaled conidia very rapidly, destroy them intracellularly through oxidative mechanisms, and prevent germination to hyphae, the invasive form of the fungus. A member of the collectin family, pentraxin 3, is essential for prompt handling of *Aspergillus* conidia by alveolar macrophages, such that its deficiency is linked to the susceptibility to infection of otherwise immunocompetent mice (41). Murine macrophages recognize and internalize *A. fumigatus* conidia through lectin-like attachment sites in the presence of collectins. Various enzymes such as elastases and proteases produced by the fungus might play an important role in the ability of conidia to evade phagocytosis by alveolar macrophages and to resist hydrolysis by endogenous peptides and in the ability of germinating conidia to cross anatomical barriers (69). It is of interest that *Aspergillus* hyphae, at variance with conidia, appeared to be sensed by human monocytes through TLR4 and CD14 (131), a finding indicating the ability of TLRs to discriminate among fungal morphotypes. However, loss of TLR4-mediated signal transduction could also represent an evasion strategy of the fungus (95). Macrophages serve as a protected environment in which fungi multiply and disseminate. *H. capsulatum* is an example of a successful intracellular pathogen of mammalian macrophages (136). Subversion mechanisms that account for survival include defective phagosome-lysosome fusion, regulation of phagosomal pH, iron restriction, suppression of the respiratory burst, and lack of inhibition of conidia to yeast transition. *B. dermatitidis* utilizes a different mechanism of subversion that relies on the capacity of its adhesin BAD-1 to dysregulate the phagocyte production of proinflammatory cytokines by binding to CR3 and CD14 (16). *C. neoformans* utilizes a novel strategy for intracellular parasitism that includes the accumulation of intracellular polysaccharide in cytoplasmic vesicles (37). *C. albicans* promotes its survival by inducing host cell apoptosis (58).

As already emphasized with PMNs, the engagement of each receptor on mononuclear phagocytes may result in profoundly different downstream intracellular events. This differential downstream regulation has been observed in response to the different forms of *C. albicans* (59, 126) or to *Pneumocystis carinii*, whose internalization through the macrophage MR (36) accounts for the ability of alveolar macrophages to restrict the growth of the fungus, an activity dependent on the activation of the transcription factor GATA-2 (68). However, as already mentioned, internalization via constitutively competent MR does not represent an effective way of clearing yeasts such as *C. neoformans* and *C. albicans* (106), but opsonins are required for uptake of the organism and fungal growth inhibition in vivo. Unopsonized *C. albicans* yeasts can be internalized through the macrophage MRs; however, the highest killing activity is observed against opsonized yeasts. Actually, the enhancement of phagocytosis and killing of *C. albicans* by macrophages correlated with a decreased number of MRs. The MR-mediated uptake of unopsonized *Candida* yeasts may lead to phagocyte abuse if not accompanied by the coordinate activation of the cell cytotoxic machinery (82). Increased expression of the receptor with no induced cytotoxicity is induced on exposure of macrophages to IL-4. In contrast, IFN-γ downregulates the expression of macrophage MR but nevertheless results in effective killing, presumably via increased coupling of the receptor to cytotoxic functions (82). Thus, cooperation between Th1 and Th2 cytokines may be required for optimal stimulation of MR-mediated phagocytosis (106). TLR4 and CD14 on phagocytes mediate the recognition of fungal mannan and glucoronoxylomannan, a major component of the capsule of *C. neoformans* (121, 124). The finding that glucoronoxylomannan only partially activates TLR-dependent signal transduction pathways may account for its immunosuppressive and immunodysregulatory effects on the host.

The antifungal effector functions of effector phagocytes include mechanisms of killing and growth inhibition of fungi as well as pathways to oppose fungal infectivity, including effects on dimorphism and phenotypic switching (81, 106). The restriction of fungal growth occurs by both oxygen-dependent and -independent mechanisms, the latter consisting of intracellular or extracellular release of effector molecules, defensins, (neutrophil-cationic) peptides, and iron sequestration. Although phagocytes express intrinsic antifungal activity, in general this activity could be increased by opsonins and T-cell-derived cytokines, a finding suggesting that the innate and adaptive immune systems do not work independently but are reciprocally regulated. Enzymes such as the NADPH-oxidase and inducible nitric oxide (NO) synthase initiate the oxidative pathways known as respiratory burst. The respiratory burst produces toxic reactive oxygen intermediates and nitrogen oxidases whose nature varies depending on the nature of pathogens and type of phagocytic cells. In retaliation, fungi have evolved strategies to selectively inhibit the respiratory burst (50) or to adapt to oxidative stress (22). The activation of NADPH oxidase can be elicited by microbial products (such as LPS), by IFN-γ or IL-8, or by IgG binding to FcR. Myeloperoxidase, a lysosomal hemoprotein found in azurophilic granules of neutrophils and monocytes, but not macrophages, is also a mediator in the oxygen-dependent killing of fungi. Myeloperoxidase deficiency predisposes to pulmonary candidiasis and aspergillosis, although it is unable to play a role in host defense in the absence of the NADPH oxidase (2). Patients with inherited X-linked chronic granulomatous disease, resulting from a deficiency in oxidant formation due to defective NADPH-oxidase complex, have increased susceptibility to aspergillosis (40). However, transplantation of bone marrow cells transfected with the NADPH-oxidase gene restored the fungicidal activity of mice with chronic granulomatous disease, a finding opening the possibility of gene therapy in fungal infections (30).

Additional toxic molecules produced by phagocytes are the reactive nitrogen intermediates. Different NO synthases convert the amino acid L-arginine and molecular oxygen to L-citrulline and NO (106). It is remarkable that the NO synthase of macrophages is stimulated by LPS and oppositely regulated by Th1 and Th2 cytokines (106). Production of NO occurs in rat and murine macrophages in response to fungi; in retaliation, fungi also inhibit NO production (119).

Iron is an essential nutrient for fungi. Iron sequestration in response to infection is a demonstrated host defense mechanism, and thus iron acquisition is an important pathogenetic determinant (56). Iron restriction by activated macrophages represents a mechanism to control not only fungal viability, but also intracellular transition from conidia to yeasts and synthesis of cryptococcal polysaccharide capsule. Tight regulation of phagolysosomal pH is important to limit iron acquisition by fungi, since the element is not accessible at a pH exceeding 6.5. Thus, raising the phagolysosome pH, as by treatment with chloroquine, increases the antifungal activity of mononuclear phagocytes, although chloroquine may have additional pH-independent effector activities (134).

A number of studies have suggested that, in addition to phagocytes, natural killer and $\gamma\delta$ T cells participate in the early innate defense against fungi; however, their relative contribution is largely dependent on the type and site of infection (53, 106). Natural killer T cells, which recognize glycolipid antigens, have been involved in the early host protection against cryptococcal infection (62). In general, these cells serve two main functions, the direct inhibition or killing of the fungus and an assisting role for phagocytes, by the release of cytokines activating the phagocytic process. T lymphocytes bearing the γδ T-cell receptors are found in most epithelia and may contribute to immunosurveillance at the body surface (61).

Dendritic Cells

Since DCs are equipped with several TLRs, they are the main connectors of the innate and adaptive immune systems. DCs are bone marrow-derived cells of both lymphoid and myeloid stem cell origin that are found in all lymphoid organs as well as nearly all nonlymphoid tissues and organs. DCs acquire antigens in peripheral tissues in their immature form. As they mature, DCs migrate to the T-cell areas of lymphoid organs, where they translate the tissue-derived information into the language of Th cells, providing them with the appropriate signals (103). The dual activation and tolerization function of DCs is mediated by their capacity to change the context of antigen presentation and to communicate to T cells the nature of the antigens they are presenting (103). This process exemplifies the importance of TLRs not only in direct early immune responses but also in activation of adaptive immunity.

DCs are uniquely able to decode the fungus-associated information and translate it in qualitatively different adaptive Th immune responses, in vitro and in vivo (18, 38, 87, 112). PRRs and TLRs determine the functional plasticity of DCs in response to fungi and contribute to the discriminative recognition of the different fungal morphotypes. DCs (both human and murine) are now known to recognize and internalize a number of fungi, including *A. fumigatus*, *C. albicans*, *C. neoformans*, *H. capsulatum*, *C. immitis*, and *Malassezia furfur*. *Candida* and *Aspergillus* proved to be useful pathogen models to dissect events occurring at the fungus/DC interface. DCs internalize *Candida* yeasts, *Aspergillus* conidia, and hyphae of both. Phagocytosis occurs through distinct morphologies and involves different recognition receptors but not TLRs. Recognition and internalization of yeasts and conidia occurred predominantly by coiling phagocytosis, through the engagement of MRs of different sugar specificity, DC-SIGN, and, partly, CR3 (20, 43, 97, 114, 120, 123). In contrast, entry of hyphae occurred by a more conventional, zipper-type phagocytosis and involved the cooperative action of FcγR and CR3 (114). Phagocytosis does not require TLR2, TLR4, TLR9, or MyD88 (8).

The engagement of distinct receptors by different fungal morphotypes translates into downstream signaling events, ultimately regulating cytokine production and costimulation, an event greatly influenced by fungal opsonins (63, 89, 114). Entry of *Candida* yeasts or *Aspergillus* conidia through MRs results in the production of proinflammatory cytokines including IL-12, upregulation of costimulatory molecules and major histocompatibility complex class II antigens, and activation of protective Th1-cell responses. IL-12 production by DCs also occurs through the MyD88 pathway with the implication of distinct TLRs (IL-1RI and TLR9 for *Candida* and TLR4 and TLR9 for *Aspergillus*) (8, 15). Actually, these events are all suppressed on entry through CR3. In contrast, coligation of CR3 with FcγR, as in the phagocytosis of hyphae, results in the production of IL-4/IL-10, upregulation of costimulatory molecules and major histocompatibility complex class II antigens, and activation of Th2/Treg cells. Opsonization with MBL, C3, and/or IgG greatly modifies the receptor exploitation on DCs by the different fungal morphotypes and qualitatively affects DC activation (114). Thus, collectins appear to favor the phagocytosis of the fungus without implicating the production of cytokine messengers to the immune system, an activity compatible with a primitive mechanism of host defense and in line with their ability to downregulate the inflammatory response to fungi (127). Altogether, opsonins, by subverting the morphotype-specific program of activation of DCs, may qualitatively affect DC and Th-cell functioning.

A remarkable and important feature of Peyer's patches DCs is the production of IL-10 in response to *Candida*, an event occurring by signaling through CR3 in the presence of opsonizing antibodies. These IL-10-producing DCs activate CD4$^+$CD25$^+$ Treg that negatively affect antifungal Th1 reactivity (88, 89). Conceivably, tissue-dependent factors and opsonins may modulate receptor usage by DCs at different body sites and may contribute to the functional plasticity of DCs at the pathogen/immune-system interface (Table 1). The functional plasticity of DCs

Table 1. Functional plasticity of DCs at the fungus/immune system interface

1. DCs are unique phagocytic cells because they can phagocytose different fungal morphotypes through different forms of phagocytosis.
2. DCs are equipped with PRRs such as TLRs, though which they recognize fungi and decode the fungus-associated information.
3. The engagement of distinct receptors by fungal morphotypes translates into downstream signaling events, ultimately regulating costimulation, cytokine production, and development of qualitatively different Th/Treg responses, an event greatly influenced by fungal opsonins.

at the pathogen/immune-system interface may offer new interpretative clues to fungal virulence. In fact, the qualitative development of the Th response to a fungus may not depend primarily on the nature of the fungal form being phagocytosed and presented. Rather, the nature of the cell response is strongly affected by the type of cell signaling initiated by the ligand-receptor interaction in DCs. In this scenario, for instance, *Candida* dimorphism per se can no longer be considered the single most important factor in determining commensalism versus infection, nor can specific forms of the fungus be regarded as absolutely indicative of saprophytism or infection at a given site. The selective exploitation of receptor-mediated entry into DCs could explain the full range of host immune relationships with the fungus, including saprophytism and infection. Averting cellular activation through the interaction with complement receptors may ultimately represent an important evasive strategy for the fungus, a condition common to different pathogens (34).

Studies in vivo confirm that DCs sample fungi at sites of infection, transport them to the draining lymph nodes, and initiate disparate Th responses to the different fungal morphotypes (12, 89). Furthermore, adoptive transfer of ex vivo DCs transfected with fungal RNA restores protective antifungal immunity in a murine model of allogeneic bone marrow transplantation (6, 14). These results, along with the finding that fungus-pulsed DCs could reverse T-cell anergy of patients with fungal diseases (46, 104), may suggest the utility of DCs for fungal vaccines (13).

INFLAMMATORY RESPONSE

On contact with a pathogen, cells of the innate immune system release a battery of chemokines and cytokines. The local release of these effector molecules serves to regulate cell trafficking of various types of leukocytes, thus initiating the inflammatory response, to activate phagocytic cells to a microbicidal state, and to direct Th-cell development. The inflammatory response to fungi may serve to limit the infection but may also contribute to pathogenicity, as exemplified by the occurrence of severe fungal infection in patients with immunorestitution disease (23). These patients may experience intractable fungal infections despite the presence of innate and adaptive immune responses. Recovery from infection may depend not only on fungal growth restriction but also on resolution of inflammatory pathology. This resolution imposes a new function on the immune system. In addition to efficient control of pathogens, tight regulatory mechanisms are required to balance protective immunity and immunopathology.

In experimental candidiasis, the course and outcome of the infection in different strains of mice correlate with fungal load but also with immunopathology (3). To limit the pathologic consequences of excessive inflammatory cell-mediated immune reactions, the immune system resorts to a number of protective mechanisms, including the reciprocal cross-regulatory effects of Th1- and Th2-type effector cytokines, such as IFN-γ and IL-4, the generation of Treg, and the secretion of anti-inflammatory cytokines that are key for maintaining a healthy balance between protection and immunopathology (109).

Recent evidence suggests that the inflammatory response to fungi is strictly controlled by the metabolic pathway involved in tryptophan catabolism and mediated by the enzyme indoleamine 2,3-dioxygenase (IDO) (10). IDO plays a complex role in immunoregulation in infection, pregnancy, autoimmunity, transplantation, and neoplasia (47). In candidiasis, IDO activity was induced at sites of infection as well as in DCs and in effector PMNs via IFN-γ- and cytotoxic T-lymphocyte-associated antigen 4-dependent mechanisms. IDO inhibition greatly exacerbated infection and associated inflammatory pathology as a result of deregulated innate and adaptive/regulatory immune responses. Interestingly, a role for tryptophan catabolism was also demonstrated in a fungus-autonomous fashion, its blockade in vitro promoting yeast-to-hypha transition. These results provide novel mechanistic insights into complex events that, occurring at the fungus/pathogen interface, relate to the dynamics of host adaptation to the fungus. The production of IFN-γ may be squarely placed at this interface, where IDO activation probably exerts fine control over fungal morphology as well as inflammatory and adaptive antifungal responses (Fig. 2).

Chemokines and Cytokines

Chemokines are small polypeptides that are synthesized by many types of cells and promote the recruitment of PMNs, monocytes, and lymphocytes at the infectious foci (53). The long-standing observation that fungi have the ability to attract leukocytes at the site of infection is now supported by the findings that fungi stimulate the release of chemokines from many cells. Local production of chemokines has been described in several experimental models of fungal infections. Chemokine production in concert with upregulation of adhesin molecule expression mediates leukocyte extravasation and migration into sites of fungal infections. Evidence indicates that the production of selected chemokines is critical for the development of protective Th1 responses to fungi (53). The production of

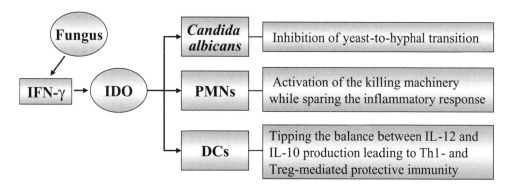

Figure 2. The crucial role for the IFN-γ/IDO-dependent metabolic pathway in *C. albicans* infection. The production of IFN-γ is placed squarely at the host/pathogen interface, where IDO activation exerts a fine control over fungal morphology as well as the inductive and effector pathways of immune resistance to the fungus.

chemokines and cytokines occurs mainly through the PAMP-TLR recognition system. In this regard, a number of cell wall components of fungi may act as PAMPs (106). TLR2 signaling by zymosan, β-glucan, *Candida* phospholipomannan, and *Aspergillus* conidia leads to the prevalent production of inflammatory (TNF-α and IL-1β) cytokines, although IL-10 is also produced occasionally. Signaling through TLR2 by zymosan occurs in collaboration with the β-glucan receptor dectin-1, a finding suggesting the occurrence of collaborative recognition of distinct microbial components by different classes of innate immune receptors (106). Dectin-1 also mediates the macrophage inflammatory responses to *P. carinii* (122). TLR4 signaling on PMNs mediates the production of TNF-α concomitantly with IL-10, a finding suggesting that the balance between proinflammatory and anti-inflammatory cytokine production can be differently affected by the different TLRs (9).

Immunoregulation by Cytokines

Transforming growth factor β (TGF-β) and IL-10 are potent immunosuppressive cytokines with both beneficial and detrimental effects on host responses to fungi (106, 113). TGF-β is a key component of the host response to fungi in mice and humans. It acts by inhibiting antifungal effector functions of phagocytes but also by dampening excessive, Th1-dependent inflammatory pathology. Thus, like IL-10, TGF-β is a double-edged sword in the fight against fungi. IL-10 is readily produced by PMNs, macrophages, DCs, and Treg in response to fungi and plays a crucial role in determining susceptibility to fungal infections in mice (106, 113) and humans (52, 73). High levels of IL-10 are detected in patients with chronic mucocutaneous candidiasis (CMC) or hepatosplenic candidiasis and in neutropenic or CGD patients with

aspergillosis (106). IL-10 acts by impairing the antifungal effector functions of phagocytes, including secretion of inflammatory cytokines and IL-12, and by inhibiting the development of protective cell-mediated immunity. Later in the course of the infection, high-level production of IL-10 may be beneficial by contributing to resolution of the inflammatory response (88). Circumstantial evidence suggests that IL-10 produced by cells of the innate immune system is responsible for the prevention of excessive activation of innate effector functions whereas IL-10 secreted by Treg is mainly responsible for the establishment of commensalism and, perhaps, fungal latency and persistence.

Proinflammatory cytokines that are inhibited by IL-10 and are important for the control of fungi are TNF-α, IL-1β, IL-6, and IL-12, all of which are produced in response to fungi (106, 113). In addition to regulating the recruitment of inflammatory cells and triggering the respiratory burst, these cytokines are implicated, directly or indirectly, in the activation of protective Th1 responses to fungi and in the occurrence of fungus-associated diseases. Blocking the biological activities of proinflammatory cytokines in patients may result in increased susceptibility to fungal infections (49, 92, 132, 135).

One of the more consistent dysregulations observed in response to fungi is their ability to subvert the production of IL-12, which, as the main physiological inducer of IFN-γ and Th1-cell differentiation, is an essential cytokine for the development of acquired resistance to fungi (106, 113). IL-12 production by PMNs, monocytes, and, importantly, DCs occurs in infections caused by all fungal species. IL-18 may play a supporting role in the IL-12-dependent Th1 resistance to fungi (106). IL-12 is produced in a morphotype-dependent manner, probably through the exploitation of different recognition receptors.

For instance, it is inhibited on CR3 engagement by *H. capsulatum* yeasts in macrophages and by *C. albicans* hyphae in DCs and is produced on MR engagement of *Candida* yeasts (106, 113). The production of IL-12 by human monocytes also is inversely correlated with fungal virulence (24, 75).

IFN-γ is a key cytokine in the innate control of fungal infections in mice and humans (106). Defective IFN-γ production is associated with reduced resistance to fungal infections in humans. It is produced by T and NK cells in response to both IL-12 and IL-18 released by cells of the innate immune system exposed to fungi. IFN-γ, in turn, stimulates migration, phagocytosis, and oxidative killing of PMNs and macrophages and favors the occurrence of Th1 reactivity. Experimental evidence indicates that IFN-γ may not work in a Th2 setting, a finding suggesting a degree of caution in evaluating patterns of IFN-γ production in human settings (106).

In addition to IL-10, IL-4 may act as one major discriminative factor of susceptibility and resistance in most fungal infections (110). Ablation of IL-4 renders susceptible mice resistant to fungal infections. IL-4 may both deactivate and activate phagocytes and DCs for certain specialized functions. For instance, it may inhibit the antifungal effector activities of phagocytes yet may promote IL-12 production by DCs (106). Thus, the most important mechanism underlying the inhibitory activity of IL-4 in infections relies on its ability to promote Th2 reactivity, thus dampening protective Th1 responses. However, susceptibility to fungal infections may not always be associated with overt production of IL-4. For instance, although an association between chronic disseminated candidiasis and genetic variants of IL-4 has been recently described (25), IL-4 or IL-5 levels are not always increased in patients with chronic mucocutaneous candidiasis, despite having defective type 1 cytokine production (72, 73).

ADAPTIVE IMMUNITY

Th1, Th2, and Treg Cells

Serological and skin reactivity surveys indicate that fungal infections are common but clinical disease is rare, consistent with the development of acquired immunity (90). For many fungal pathogens, the effective tissue response to invasion is granulomatous inflammation, a hallmark of cell-mediated immunity. For dimorphic fungi, the initial exposure either is asymptomatic or results in mild infection that confers protective immunity. For *C. neoformans*, the high prevalence of antibodies specific for cryptococcal antigens in normal individuals indicates that primary infection is followed by fungal growth restriction and lifelong immunity. Underlying acquired immunity to *C. albicans*, such as the expression of a positive delayed type hypersensitivity (DTH) response, is demonstrable in adult immunocompetent individuals and is presumed to prevent progression from mucosal colonization to symptomatic infection. Lymphocytes from healthy individuals show proliferative responses after stimulation with fungal antigens and produce several different cytokines (52, 92, 110).

There is extensive plasticity in the T-cell response to fungi. The heterogeneity of the CD4+ and CD8+ T-cell repertoire may account for the multiplicity and redundancy of effector mechanisms through which T cells participate in the control of fungal infections. These mechanisms include direct antifungal activity (71), release of antimicrobial peptides from CD8+ T cells (79), lysis of fungus-laden phagocytes (57), and effector functions resulting from dynamic interactions among T cells bearing selected members of the Vβ families of the T-cell receptor (28). This functional plasticity indicates the potential of vaccines for conditions of immunodeficiency, as highlighted by the ability of CD8+ T cells to compensate for CD4+ T-cell deficiency in experimental models of vaccine-induced resistance to endemic fungi (137, 138). The flexible program of T cells leads to the production of many mediators, including cytokines. Due to their action on circulating leukocytes, the cytokines produced by fungus-specific T cells are instrumental in mobilizing and activating antifungal effectors, thus providing prompt and effective control of infectivity once the fungus has established itself in tissues or spread to internal organs. Therefore, host resistance to fungi seems to depend on the induction of cellular immunity, mediated by T cells, cytokines, and effector phagocytes.

The clinical circumstances in which fungal infections occur are associated with impaired cell-mediated immunity. AIDS and severe hematological malignancies are examples of acquired defects in T-cell function that predispose to severe fungal infections. Furthermore, the occurrence of severe disseminated infections by filamentous fungi in nongranulocytopenic patients (48), as well as during the onset of graft-versus-host disease in bone marrow transplant recipients (83), provides compelling evidence of the pathogenic role of T-cell dysreactivity in infection. In endemic mycosis, the severity of the disease correlates with the degree of impairment of cell-mediated immunity and is associated with elevated levels of antibodies (80). Generation of a dominant Th1 response driven by IL-12 is required for the expression of protective immunity to fungi.

Experimental data have demonstrated the deleterious effects of IL-12 or IFN-γ ablation on the course and outcome of fungal infections (113). Through production of the signature cytokine IFN-γ and by providing help for opsonizing antibodies, the activation of Th1 cells is instrumental in the optimal activation of phagocytes at sites of infection. Therefore, the failure to deliver activating signals to effector phagocytes may predispose patients to overwhelming infections, limit the therapeutic efficacy of antifungals and antibodies, and favor persistency and/or commensalism. Immunological studies of patients with polar forms of paracoccidioidomycosis demonstrate an association between Th1-biased reactivity and asymptomatic and mild forms of the infection, in contrast to the correlation between Th2 responses and severe disease (19). Not surprisingly, therefore, patients with disseminated infection have defective production of IFN-γ and DTH anergy, associated with elevated levels of type 2 cytokines (IL-4 and IL-5), IgE, IgG4, and IgA and eosinophilia, which is a marker of poor prognosis in endemic mycoses (26). In patients with a defective IL-12/IFN-γ pathway, such as those with hyperimmunoglobulinemia E syndrome, fungal infections and allergy are both observed (91). However, as already mentioned, susceptibility to fungal infections may not always be associated with overt production of IL-4 (72).

Several clinical observations suggest an inverse relationship between IFN-γ and IL-10 production in patients with fungal infections. High levels of IL-10, negatively affecting IFN-γ production, are detected in chronic candidal diseases, in the severe form of endemic mycoses, and in neutropenic patients with aspergillosis. Fungal polysaccharides are known to negatively modulate cell-mediated immunity through the production of IL-10, which indicates that IL-10 production may be a consequence of infection. However, tolerance to fungi can also be achieved through the induction of Treg cells, which can finely tune antifungal Th reactivity. Therefore, suppressor T cells have recently regained their reputation as key controllers of antifungal immunity (90). Naturally occurring downregulatory mechanisms operating in the respiratory mucosa may account for the lack of pathology in *P. carinii*-infected mice (55). Administration of CD4$^+$ CD25$^+$ Treg cells prevents the inflammatory pathology associated with pathogen clearance. In mice with candidiasis, CD4$^+$ CD25$^+$ Treg cells, producing IL-10 and TGF-β, prevent complete elimination of the fungus from the gastrointestinal tract; fungal persistence allows the development of memory immunity (88). γδ T cells with a down-modulatory activity of Th1 responses

have recently been found to accumulate in the lungs of mice with cryptococcosis (128).

It has long been presumed that the ability of *C. albicans* to persist in host tissues involves mainly the immunosuppressive property of cell wall glycoproteins. Mannan and its oligosaccharide fragments could be potent inhibitors of cell-mediated immunity and seem to reproduce the immune deficit of patients with CMC (4). CMC, although encompassing various clinical entities, has been associated with autoimmune polyendocrinopathy-candidiasis-ectodermal dystrophy, a condition in which the mutated gene is involved in the ontogeny of CD25$^+$ Treg cells (74). In CMC, the defective type 1 cytokine production does not occur concomitantly with the increase in production of type 2 cytokines (namely IL-4 or IL-5) but, more often, with the increase in production of IL-10 (72). This finding has led to speculation that an inherent alteration in receptor-mediated signaling in response to fungal polysaccharide may predispose patients with CMC to a dysfunctional induction of Treg-cell activity, negatively affecting Th1-cell-dependent clearance of the fungus without implicating the activation of Th2 cells. Because both the recovery of the fungus from the gastrointestinal tract and the detection of underlying Th1 reactivity, such as DTH and lymphoproliferation, can fluctuate in healthy subjects, it is tempting to speculate that Treg cells mediate tolerance to the fungus at the site of colonization. The engagement of distinct TLRs on DCs and the ensuing production of IL-6 are crucial events mediating the inhibition of Treg-cell function (99). This is in agreement with the failure of IL-6-deficient mice to activate antifungal Th1 responses concomitantly with increased IL-10 production (113).

CONCLUSIONS: EXPLOITATION OF IMMUNITY FOR FUNGAL VACCINES

The therapeutic efficacy of antifungals is limited without the help of host immune reactivity. Various cytokines, including chemokines and growth factors, have proved to be beneficial in experimental and human refractory fungal infections (60, 67). The Th1-Th2 balance itself can be the target of immunotherapy (64, 102). The inhibition of Th2 cytokines or the addition of Th1 cytokines can increase the efficacy of antifungals, such as polyenes and azoles, in experimental mycoses (108). In the past decade, a dramatic shift has occurred in our mechanistic understanding of innate immunity. Precisely, the appreciation that activation of the innate immune system initiates, amplifies, and drives antigen-specific immune responses, together with the identification of discrete cell types, specific receptors, and the signaling pathways involved in the activation of innate immunity, has provided a multitude of new targets for exploitation by the development of adjuvants for vaccines (29). The model has brought DCs to center stage as promising targets for intervention for immunotherapy and vaccine development and has shifted the emphasis from the "antigen" toward the "adjuvant." The ultimate challenge will be to design fungal vaccines capable of inducing optimal immune responses by targeting specific receptors on DCs. This will require, however, further studies aimed at elucidating the convergence and divergence of pathways of immune protection elicited in infections or on vaccination.

Acknowledgments. I acknowledge financial support by the National Institute of Health (project 20D.27). I thank L. Bellocchio for editorial assistance.

REFERENCES

1. **Apidianakis, Y., L. G. Rahme, J. Heitman, F. M. Ausubel, S. B. Calderwood, and E. Mylonakis.** 2004. Challenge of *Drosophila melanogaster* with *Cryptococcus neoformans* and role of the innate immune response. *Eukaryot. Cell* **3**:413–419.
2. **Aratani, Y., F. Kura, H. Watanabe, H. Akagawa, Y. Takano, K. Suzuki, M. C. Dinauer, N. Maeda, and H. Koyama.** 2002. Relative contributions of myeloperoxidase and NADPH-oxidase to the early host defense against pulmonary infections with *Candida albicans* and *Aspergillus fumigatus*. *Med. Mycol.* **40**:557–563.
3. **Ashman, R. B.** 1998. *Candida albicans*: pathogenesis, immunity and host defence. *Res. Immunol.* **149**:281–288.
4. **Ashman, R. B., and J. M. Papadimitriou.** 1995. Production and function of cytokines in natural and acquired immunity to *Candida albicans* infection. *Microbiol. Rev.* **59**:646–672.
5. **Babula, O., G. Lazdane, J. Kroica, W. J. Ledger, and S. S. Witkin.** 2003. Relation between recurrent vulvovaginal candidiasis, vaginal concentrations of mannose-binding lectin, and a mannose-binding lectin gene polymorphism in Latvian women. *Clin. Infect. Dis.* **37**:733–737.
6. **Bacci, A., C. Montagnoli, K. Perruccio, S. Bozza, R. Gaziano, L. Pitzurra, A. Velardi, C. F. d'Ostiani, J. E. Cutler, and L. Romani.** 2002. Dendritic cells pulsed with fungal RNA induce protective immunity to *Candida albicans* in hematopoietic transplantation. *J. Immunol.* **168**:2904–2913.
7. **Belanger, P. H., D. A. Johnston, R. A. Fratti, M. Zhang, and S. G. Filler.** 2002. Endocytosis of *Candida albicans* by vascular endothelial cells is associated with tyrosine phosphorylation of specific host cell proteins. *Cell Microbiol.* **4**:805–812.
8. **Bellocchio, S., C. Montagnoli, S. Bozza, R. Gaziano, G. Rossi, S. S. Mambula, A. Vecchi, A. Mantovani, S. M. Levitz, and L. Romani.** 2004. The contribution of the Toll-like/IL-1 receptor superfamily to innate and adaptive immunity to fungal pathogens in vivo. *J. Immunol.* **172**:3059–3069.
9. **Bellocchio, S., S. Moretti, K. Perruccio, F. Fallarino, S. Bozza, C. Montagnoli, P. Mosci, G. B. Lipford, L. Pitzurra, and L. Romani.** 2004. TLRs govern neutrophil activity in aspergillosis. *J. Immunol.* **173**:7406–7415.
10. **Bozza, S., F. Fallarino, L. Pitzurra, T. Zelante, C. Montagnoli, S. Bellocchio, P. Mosci, C. Vacca, P. Puccetti, and L. Romani.** 2004. A crucial role for tryptophan catabolism at the host/*Candida albicans* interface. *J. Immunol.* **174**:2910–2918.
11. **Bozza, S., R. Gaziano, G. B. Lipford, C. Montagnoli, A. Bacci, P. Di Francesco, V. P. Kurup, H. Wagner, and L. Romani.** 2002. Vaccination of mice against invasive aspergillosis with recombinant *Aspergillus* proteins and CpG oligodeoxynucleotides as adjuvants. *Microbes Infect.* **4**:1281–1290.
12. **Bozza, S., R. Gaziano, A. Spreca, A. Bacci, C. Montagnoli, P. di Francesco, and L. Romani.** 2002. Dendritic cells transport conidia and hyphae of *Aspergillus fumigatus* from the airways to the draining lymph nodes and initiate disparate Th responses to the fungus. *J. Immunol.* **168**:1362–1371.
13. **Bozza, S., C. Montagnoli, R. Gaziano, G. Rossi, G. Nkwanyuo, S. Bellocchio, and L. Romani.** 2004. Dendritic cell-based vaccination against opportunistic fungi. *Vaccine* **2**:857–864.
14. **Bozza, S., K. Perruccio, C. Montagnoli, R. Gaziano, S. Bellocchio, E. Burchielli, G. Nkwanyuo, L. Pitzurra, A. Velardi, and L. Romani.** 2003. A dendritic cell vaccine against invasive aspergillosis in allogeneic hematopoietic transplantation. *Blood* **102**:3807–3814.
15. **Braedel, S., M. Radsak, H. Einsele, J. P. Latge, A. Michan, J. Loeffler, Z. Haddad, U. Grigoleit, H. Schild, and H. Hebart.** 2004. *Aspergillus fumigatus* antigens activate innate immune cells via Toll-like receptors 2 and 4. *Br. J. Haematol.* **125**:392–399.
16. **Brandhorst, T. T., M. Wuthrich, T. Warner, and B. Klein.** 1999. Targeted gene disruption reveals an adhesin indispensable for pathogenicity of *Blastomyces dermatitidis*. *J. Exp. Med.* **189**:1207–1216.
17. **Bronte, V., P. Serafini, A. Mazzoni, D. M. Segal, and P. Zanovello.** 2003. L-Arginine metabolism in myeloid cells controls T-lymphocyte functions. *Trends Immunol.* **24**:302–306.
18. **Buentke, E., and A. Scheynius.** 2003. Dendritic cells and fungi. *Apmis* **111**:789–796.
19. **Calich, V. L., C. A. Vaz, and E. Burger.** 1998. Immunity to *Paracoccidioides brasiliensis* infection. *Res. Immunol.* **149**:407–417.
20. **Cambi, A., K. Gijzen, J. M. de Vries, R. Torensma, B. Joosten, G. J. Adema, M. G. Netea, B. J. Kullberg, L. Romani, and C. G. Figdor.** 2003. The C-type lectin DC-SIGN (CD209) is an antigen-uptake receptor for *Candida albicans* on dendritic cells. *Eur. J. Immunol.* **33**:532–538.
21. **Casadevall, A.** 1995. Antibody immunity and invasive fungal infections. *Infect. Immun.* **63**:4211–4218.
22. **Chauhan, N., and R. Calderone.** 2004. Adaptation to oxidative, nutrient and pH stress by human pathogenic fungi, p. 129–166. *In* G. San-Blas and R. A. Calderone (ed.), *Pathogenic Fungi. Host Interactions and Emerging Strategies for Control.* Caister Academic Press, Wymondham, Norfolk, England.
23. **Cheng, V. C., K. Y. Yuen, W. M. Chan, S. S. Wong, E. S. Ma, and R. M. Chan.** 2000. Immunorestitution disease involving the innate and adaptive response. *Clin. Infect. Dis.* **30**:882–892.
24. **Chiani, P., C. Bromuro, and A. Torosantucci.** 2000. Defective induction of interleukin-12 in human monocytes by germ-tube forms of *Candida albicans*. *Infect. Immun.* **68**:5628–5634.

25. Choi, E. H., C. B. Foster, J. G. Taylor, H. C. Erichsen, R. A. Chen, T. J. Walsh, V. J. Anttila, T. Ruutu, A. Palotie, and S. J. Chanock. 2003. Association between chronic disseminated candidiasis in adult acute leukemia and common IL4 promoter haplotypes. *J. Infect. Dis.* **187:**1153–1156.

26. Clemons, K. V., and D. A. Stevens. 2001. Overview of host defense mechanisms in systemic mycoses and the basis for immunotherapy. *Semin. Respir. Infect.* **16:**60–66.

27. Cutuli, M., S. Cristiani, J. M. Lipton, and A. Catania. 2000. Antimicrobial effects of alpha-MSH peptides. *J. Leukoc. Biol.* **67:**233–239.

28. Deepe, G. S., Jr., and R. S. Gibbons. 2002. Functional properties of the T cell receptor repertoire in responding to the protective domain of heat-shock protein 60 from *Histoplasma capsulatum. J. Infect. Dis.* **186:**815–822.

29. Deepe, G. S., Jr. 2004. Preventative and therapeutic vaccines for fungal infections: from concept to implementation. *Expert Rev. Vaccines* **3:**701–709.

30. Dinauer, M. C., M. A. Gifford, N. Pech, L. L. Li, and P. Emshwiller. 2001. Variable correction of host defense following gene transfer and bone marrow transplantation in murine X-linked chronic granulomatous disease. *Blood* **97:**3738–3745.

31. Dorschner, R. A., B. Lopez-Garcia, J. Massie, C. Kim, and R. L. Gallo. 2004. Innate immune defense of the nail unit by antimicrobial peptides. *J. Am. Acad. Dermatol.* **50:**343–348.

32. Dunphy, G. B., U. Oberholzer, M. Whiteway, R. J. Zakarian, and I. Boomer. 2003. Virulence of *Candida albicans* mutants toward larval *Galleria mellonella* (Insecta, Lepidoptera, Galleridae). *Can. J. Microbiol.* **49:**514–524.

33. Edwards, L., A. E. Williams, A. M. Krieg, A. J. Rae, R. J. Snelgrove, and T. Hussell. 2004. Stimulation via Toll-like receptor 9 reduces *Cryptococcus neoformans*-induced pulmonary inflammation in an IL-12-dependent manner. *Eur. J. Immunol.* **35:**273–281.

34. Ehlers, M. R. 2000. CR3: a general purpose adhesion-recognition receptor essential for innate immunity. *Microbes Infect.* **2:**289–294.

35. Ellerbroek, P. M., D. J. Lefeber, R. van Veghel, J. Scharringa, E. Brouwer, G. J. Gerwig, A. Janbon, A. I. Hoepelman, and F. E. Coenjaerts. 2004. O-acetylation of cryptococcal capsular glucuronoxylomannan is essential for interference with neutrophil migration. *J. Immunol.* **173:**7513–7520.

36. Ezekowitz, R. A., D. J. Williams, H. Koziel, M. Y. Armstrong, A. Warner, F. F. Richards, and R. M. Rose. 1991. Uptake of *Pneumocystis carinii* mediated by the macrophage mannose receptor. *Nature* **351:**155–158.

37. Feldmesser, M., S. Tucker, and A. Casadevall. 2001. Intracellular parasitism of macrophages by *Cryptococcus neoformans. Trends Microbiol.* **9:**273–278.

38. Ferreira, K. S., J. D. Lopes, and S. R. Almeida. 2003. Regulation of T helper cell differentiation in vivo by GP43 from *Paracoccidioides brasiliensis* provided by different antigen-presenting cells. *Scand. J. Immunol.* **58:**290–297.

39. Fidel, P. L., Jr. 2002. The protective immune response against vaginal candidiasis: lessons learned from clinical studies and animal models. *Int. Rev. Immunol.* **21:**515–548.

40. Franzblau, M. J. 2003. Preventing fungal infections in chronic granulomatous disease. *N. Engl. J. Med.* **349:**1190–1191.

41. Garlanda, C., E. Hirsch, S. Bozza, A. Salustri, M. De Acetis, R. Nota, A. Maccagno, F. Riva, B. Bottazzi, G. Peri, A. Doni, L. Vago, M. Botto, R. De Santis, P. Carminati, G. Siracusa, F. Altruda, A. Vecchi, L. Romani, and A. Mantovani. 2002. Non-redundant role of the long pentraxin PTX3 in anti-fungal innate immune response. *Nature* **420:**182–186.

42. Ghannoum, M. A. 2000. Potential role of phospholipases in virulence and fungal pathogenesis. *Clin. Microbiol. Rev.* **13:**122–143.

43. Gildea, L. A., R. E. Morris, and S. L. Newman. 2001. *Histoplasma capsulatum* yeasts are phagocytosed via very late antigen-5, killed, and processed for antigen presentation by human dendritic cells. *J. Immunol.* **166:**1049–1056.

44. Gordon, S. 2002. Pattern recognition receptors: doubling up for the innate immune response. *Cell* **111:**927–930.

45. Gow, N. A., A. J. Brown, and F. C. Odds. 2002. Fungal morphogenesis and host invasion. *Curr. Opin. Microbiol.* **5:**366–371.

46. Grazziutti, M., D. Przepiorka, J. H. Rex, I. Braunschweig, S. Vadhan-Raj, and C. A. Savary. 2001. Dendritic cell-mediated stimulation of the in vitro lymphocyte response to *Aspergillus. Bone Marrow Transplant.* **27:**647–652.

47. Grohmann, U., F. Fallarino, and P. Puccetti. 2003. Tolerance, DCs and tryptophan: much ado about IDO. *Trends Immunol.* **24:**242–248.

48. Hage, C. A., M. Goldman, and L. J. Wheat. 2002. Mucosal and invasive fungal infections in HIV/AIDS. *Eur. J. Med. Res.* **7:**236–241.

49. Hage, C. A., K. L. Wood, H. T. Winer-Muram, S. J. Wilson, G. Sarosi, and K. S. Knox. 2003. Pulmonary cryptococcosis after initiation of anti-tumor necrosis factor-alpha therapy. *Chest* **124:**2395–2397.

50. Hamilton, A. J., and M. D. Holdom. 1999. Antioxidant systems in the pathogenic fungi of man and their role in virulence. *Med. Mycol.* **37:**375–389.

51. Han, Y., T. R. Kozel, M. X. Zhang, R. S. MacGill, M. C. Carroll, and J. E. Cutler. 2001. Complement is essential for protection by an IgM and an IgG3 monoclonal antibody against experimental, hematogenously disseminated candidiasis. *J. Immunol.* **167:**1550–1557.

52. Hebart, H., C. Bollinger, P. Fisch, J. Sarfati, C. Meisner, M. Baur, J. Loeffler, M. Monod, J. P. Latge, and H. Einsele. 2002. Analysis of T-cell responses to Aspergillus fumigatus antigens in healthy individuals and patients with hematologic malignancies. *Blood* **100:**4521–4528.

53. Herring, A. C., and G. B. Huffnagle. 2001. Innate immunity to fungi, p. 127–137. *In* S. H. E Kaufmann, A. Sher, and R. Ahmed (ed.), *Immunology of Infectious Diseases.* ASM Press, Washington, D.C.

54. Hoffmann, J. A., and J. M. Reichhart. 2002. *Drosophila* innate immunity: an evolutionary perspective. *Nat. Immunol.* **3:**121–126.

55. Hori, S., T. L. Carvalho, and J. Demengeot. 2002. CD25+ CD4+ regulatory T cells suppress CD4+ T cell-mediated pulmonary hyperinflammation driven by *Pneumocystis carinii* in immunodeficient mice. *Eur. J. Immunol.* **32:**1282–1291.

56. Howard, D. H. 1999. Acquisition, transport, and storage of iron by pathogenic fungi. *Clin. Microbiol. Rev.* **12:**394–404.

57. Huffnagle, G. B., J. L. Yates, and M. F. Lipscomb. 1991. Immunity to a pulmonary *Cryptococcus neoformans* infection requires both CD4[+] and CD8[+] T cells. *J. Exp. Med.* 173:793–800.

58. Ibata-Ombetta, S., T. Idziorek, P. A. Trinel, D. Poulain, and T. Jouault. 2003. *Candida albicans* phospholipomannan promotes survival of phagocytosed yeasts through modulation of bad phosphorylation and macrophage apoptosis. *J. Biol. Chem.* 278:13086–13093.

59. Kaposzta, R., L. Marodi, M. Hollinshead, S. Gordon, and R. P. da Silva. 1999. Rapid recruitment of late endosomes and lysosomes in mouse macrophages ingesting *Candida albicans*. *J. Cell Sci.* 112:3237–3248.

60. Kawakami, K. 2003. Promising immunotherapies with Th1-related cytokines against infectious diseases. *J. Infect. Chemother.* 9:201–209.

61. Kawakami, K. 2004. Regulation by innate immune T lymphocytes in the host defense against pulmonary infection with *Cryptococcus neoformans*. *Jpn. J. Infect. Dis.* 57:137–145.

62. Kawakami, K. 2002. Role of natural killer T cells in host defence against cryptococcal infection. *Rev. Iberoam. Micol.* 19:149–154.

63. Kelly, R. M., J. Chen, L. E. Yauch, and S. M. Levitz. 2005. Opsonic requirements for dendritic cell-mediated responses to *Cryptococcus neoformans*. *Infect. Immun.* 73:592–598.

64. Koguchi, Y., and K. Kawakami. 2002. Cryptococcal infection and Th1-Th2 cytokine balance. *Int. Rev. Immunol.* 21:423–438.

65. Kottom, T. J., J. R. Kohler, C. F. Thomas, Jr., G. R. Fink, and A. H. Limper. 2003. Lung epithelial cells and extracellular matrix components induce expression of *Pneumocystis carinii STE20*, a gene complementing the mating and pseudohyphal growth defects of STE20 mutant yeast. *Infect. Immun.* 71:6463–6471.

66. Kozel, T. R. 1998. Complement activation by pathogenic fungi. *Res. Immunol.* 149:309–320.

67. Kullberg, B. J., A. M. Oude Lashof, and M. G. Netea. 2004. Design of efficacy trials of cytokines in combination with antifungal drugs. *Clin. Infect. Dis.* 39:218–223.

68. Lasbury, M. E., X. Tang, P. J. Durant, and C. H. Lee. 2003. Effect of transcription factor GATA-2 on phagocytic activity of alveolar macrophages from *Pneumocystis carinii*-infected hosts. *Infect. Immun.* 71:4943–4952.

69. Latge, J. P. 2001. The pathobiology of *Aspergillus fumigatus*. *Trends Microbiol.* 9:382–389.

70. Levitz, S. M. 2004. Interactions of Toll-like receptors with fungi. *Microbes Infect.* 6:1351–1355.

71. Levitz, S. M., H. L. Mathews, and J. W. Murphy. 1995. Direct antimicrobial activity of T cells. *Immunol. Today* 16:387–391.

72. Lilic, D. 2002. New perspectives on the immunology of chronic mucocutaneous candidiasis. *Curr. Opin. Infect. Dis.* 15:143–147.

73. Lilic, D., I. Gravenor, N. Robson, D. A. Lammas, P. Drysdale, J. E. Calvert, A. J. Cant, and M. Abinun. 2003. Deregulated production of protective cytokines in response to *Candida albicans* infection in patients with chronic mucocutaneous candidiasis. *Infect. Immun.* 71:5690–5699.

74. Liston, A., S. Lesage, J. Wilson, L. Peltonen, and C. C. Goodnow. 2003. Aire regulates negative selection of organ-specific T cells. *Nat. Immunol.* 4:350–354.

75. Liu, L., K. Kang, M. Takahara, K. D. Cooper, and M. A. Ghannoum. 2001. Hyphae and yeasts of *Candida albicans* differentially regulate interleukin-12 production by human blood monocytes: inhibitory role of *C. albicans* germination. *Infect. Immun.* 69:4695–4697.

76. Long, K. H., F. J. Gomez, R. E. Morris, and S. L. Newman. 2003. Identification of heat shock protein 60 as the ligand on *Histoplasma capsulatum* that mediates binding to CD18 receptors on human macrophages. *J. Immunol.* 170:487–494.

77. Lopes Bezerra, L. M., and S. G. Filler. 2004. Interactions of *Aspergillus fumigatus* with endothelial cells: internalization, injury, and stimulation of tissue factor activity. *Blood* 103:2143–2149.

78. Lorenz, M. C., and G. R. Fink. 2001. The glyoxylate cycle is required for fungal virulence. *Nature* 412:83–86.

79. Ma, L. L., J. C. Spurrell, J. F. Wang, G. G. Neely, S. Epelman, A. M. Krensky, and C. H. Mody. 2002. CD8 T cell-mediated killing of *Cryptococcus neoformans* requires granulysin and is dependent on CD4 T cells and IL-15. *J. Immunol.* 169:5787–5795.

80. Magee, D. M., and R. A. Cox. 2002. Cell-mediated immunity and endemic mycoses, p. 279–292. *In* R. A. Calderone and R. L. Cihlar (ed.), *Fungal Pathogenesis: Principles and Clinical Applications*. Marcel Dekker, Inc., New York, N.Y.

81. Mansour, M. K., and S. M. Levitz. 2002. Interactions of fungi with phagocytes. *Curr. Opin. Microbiol.* 5:359–365.

82. Marodi, L., S. Schreiber, D. C. Anderson, R. P. MacDermott, H. M. Korchak, and R. B. Johnston, Jr. 1993. Enhancement of macrophage candidacidal activity by interferon-gamma. Increased phagocytosis, killing, and calcium signal mediated by a decreased number of mannose receptors. *J. Clin. Investig.* 91:2596–2601.

83. Marr, K. A., R. A. Carter, M. Boeckh, P. Martin, and L. Corey. 2002. Invasive aspergillosis in allogeneic stem cell transplant recipients: changes in epidemiology and risk factors. *Blood* 100:4358–4366.

84. McCormack, F. X., and J. A. Whitsett. 2002. The pulmonary collectins, SP-A and SP-D, orchestrate innate immunity in the lung. *J. Clin. Investig.* 109:707–712.

85. McKnight, A. J., and S. Gordon. 2000. Forum in immunology: innate recognition systems. *Microbes Infect.* 2:239–336.

86. Mencacci, A., C. Montagnoli, A. Bacci, E. Cenci, L. Pitzurra, A. Spreca, M. Kopf, A. H. Sharpe, and L. Romani. 2002. CD80[+] Gr-1[+] myeloid cells inhibit development of antifungal Th1 immunity in mice with candidiasis. *J. Immunol.* 169:3180–3190.

87. Montagnoli, C., A. Bacci, S. Bozza, R. Gaziano, S. Fiorucci, A. Spreca, and L. Romani. 2001. The plasticity of dendritic cells at the host/fungal interface. *Immunobiology* 204:582–589.

88. Montagnoli, C., A. Bacci, S. Bozza, R. Gaziano, P. Mosci, A. H. Sharpe, and L. Romani. 2002. B7/CD28-dependent CD4[+] CD25[+] regulatory T cells are essential components of the memory-protective immunity to *Candida albicans*. *J. Immunol.* 169:6298–6308.

89. Montagnoli, C., S. Bozza, A. Bacci, R. Gaziano, P. Mosci, J. Morschhauser, L. Pitzurra, M. Kopf, J. Cutler, and L. Romani. 2003. A role for antibodies in the generation of memory antifungal immunity. *Eur. J. Immunol.* 33:1193–1204.

90. Murphy, J. W. 1989. Immunity to fungi. *Curr. Opin. Immunol.* 2:360–367.

91. Netea, M. G., P. M. Schneeberger, E. de Vries, B. J. Kullberg, J. W. van der Meer, and M. I. Koolen. 2002. Th1/Th2 cytokine imbalance in a family with hyper-IgE syndrome. *Neth. J. Med.* **60:**349–353.

92. Netea, M. G., R. J. Stuyt, S. H. Kim, J. W. Van der Meer, B. J. Kullberg, and C. A. Dinarello. 2002. The role of endogenous interleukin (IL)-18, IL-12, IL-1beta, and tumor necrosis factor-alpha in the production of interferon-gamma induced by *Candida albicans* in human whole-blood cultures. *J. Infect. Dis.* **185:**963–970.

93. Netea, M. G., C. A. Van Der Graaf, A. G. Vonk, I. Verschueren, J. W. Van Der Meer, and B. J. Kullberg. 2002. The role of toll-like receptor (TLR) 2 and TLR4 in the host defense against disseminated candidiasis. *J. Infect. Dis.* **185:**1483–1489.

94. Netea, M. G., J. W. Van der Meer, and B. J. Kullberg. 2004. Toll-like receptors as an escape mechanism from the host defense. *Trends Microbiol.* **12:**484–488.

95. Netea, M. G., A. Warris, J. W. Van der Meer, M. J. Fenton, T. J. Verver-Janssen, L. E. Jacobs, T. Andresen, P. E. Verweij, and B. J. Kullberg. 2003. Aspergillus fumigatus evades immune recognition during germination through loss of Toll-like receptor-4-mediated signal transduction. *J. Infect. Dis.* **188:**320–326.

96. Neth, O., D. L. Jack, A. W. Dodds, H. Holzel, N. J. Klein, and M. W. Turner. 2000. Mannose-binding lectin binds to a range of clinically relevant microorganisms and promotes complement deposition. *Infect. Immun.* **68:**688–693.

97. Newman, S. L., and A. Holly. 2001. *Candida albicans* is phagocytosed, killed, and processed for antigen presentation by human dendritic cells. *Infect. Immun.* **69:**6813–6822.

98. O'Neill, L. A., K. A. Fitzgerald, and A. G. Bowie. 2003. The Toll-IL-1 receptor adaptor family grows to five members. *Trends Immunol.* **24:**286–290.

99. Pasare, C., and R. Medzhitov. 2003. Toll pathway-dependent blockade of CD4$^+$ CD25$^+$ T cell-mediated suppression by dendritic cells. *Science* **299:**1033–1036.

100. Phan, Q. T., R. A. Fratti, N. V. Prasadarao, J. E. Edwards, Jr., and S. G. Filler. 2005. N-cadherin mediates endocytosis of *Candida albicans* by endothelial cells. *J. Biol. Chem.* **280:**10455–10461.

101. Poulain, D., and T. Jouault. 2004. Candida albicans cell wall glycans, host receptors and responses: elements for a decisive crosstalk. *Curr. Opin. Microbiol.* **7:**342–349.

102. Puccetti, P., L. Romani, and F. Bistoni. 1995. A TH1-TH2-like switch in candidiasis: new perspectives for therapy. *Trends Microbiol.* **3:**237–240.

103. Reis e Sousa, C., A. Sher, and P. Kaye. 1999. The role of dendritic cells in the induction and regulation of immunity to microbial infection. *Curr. Opin. Immunol.* **11:**392–399.

104. Richards, J. O., N. M. Ampel, and D. F. Lake. 2002. Reversal of coccidioidal anergy in vitro by dendritic cells from patients with disseminated coccidioidomycosis. *J. Immunol.* **169:**2020–2025.

105. Roeder, A., C. J. Kirschning, R. A. Rupec, M. Schaller, and H. C. Korting. 2004. Toll-like receptors and innate antifungal responses. *Trends Microbiol.* **12:**44–49.

106. Romani, L. 2004. Innate immunity to fungi: the art of speed and specificity, p. 167–214. *In* G. San-Blas and R. A. Calderone (ed.), *Pathogenic Fungi. Host Interactions and Emerging Strategies for Control.* Caister Academic Press, Wymondham, Norfolk, England.

107. Romani, L. 2001. Overview of the fungal pathogens, p. 25–37. *In* S. H. E Kaufmann, A. Sher, and R. Ahmed

(ed.), *Immunology of Infectious Diseases.* ASM Press, Washington, D.C.

108. Romani, L. 2001. Host immune reactivity and antifungal chemotherapy: the power of being together. *J. Chemother.* **13:**347–353.

109. Romani, L. 2004. Immunity to fungal infections. *Nat. Rev. Immunol.* **4:**1–23.

110. Romani, L. 1997. The T cell response against fungal infections. *Curr. Opin. Immunol.* **9:**484–490.

111. Romani, L., F. Bistoni, R. Gaziano, S. Bozza, C. Montagnoli, K. Perruccio, L. Pitzurra, S. Bellocchio, A. Velardi, G. Rasi, P. Di Francesco, and E. Garaci. 2004. Thymosin alpha 1 activates dendritic cells for antifungal Th1 resistance through Toll-like receptor signaling. *Blood* **103:**4232–4239.

112. Romani, L., F. Bistoni, and P. Puccetti. 2002. Fungi, dendritic cells and receptors: a host perspective of fungal virulence. *Trends Microbiol.* **10:**508–514.

113. Romani, L., and S. H. Kaufmann. 1998. Immunity to fungi: editorial overview. *Res. Immunol.* **149:**277–281.

114. Romani, L., C. Montagnoli, S. Bozza, K. Perruccio, A. Spreca, P. Allavena, S. Verbeek, R. A. Calderone, F. Bistoni, and P. Puccetti. 2004. The exploitation of distinct recognition receptors in dendritic cells determines the full range of host immune relationships with *Candida albicans*. *Int. Immunol.* **16:**149–161.

115. Rooney, P. J., and B. S. Klein. 2002. Linking fungal morphogenesis with virulence. *Cell. Microbiol.* **4:**127–137.

116. Rubin-Bejerano, I., I. Fraser, P. Grisafi, and G. R. Fink. 2003. Phagocytosis by neutrophils induces an amino acid deprivation response in *Saccharomyces cerevisiae* and *Candida albicans*. *Proc. Natl. Acad. Sci. USA* **100:**11007–11012.

117. Schaffner, A., H. Douglas, and A. Braude. 1982. Selective protection against conidia by mononuclear and against mycelia by polymorphonuclear phagocytes in resistance to *Aspergillus*. Observations on these two lines of defense in vivo and in vitro with human and mouse phagocytes. *J. Clin. Investig.* **69:**617–631.

118. Schofield, D. A., C. Westwater, and E. Balish. 2004. Beta-defensin expression in immunocompetent and immunodeficient germ-free and *Candida albicans*-monoassociated mice. *J. Infect. Dis.* **190:**1327–1334.

119. Schroppel, K., M. Kryk, M. Herrmann, E. Leberer, M. Rollinghoff, and C. Bogdan. 2001. Suppression of type 2 NO-synthase activity in macrophages by *Candida albicans*. *Int. J. Med. Microbiol.* **290:**659–668.

120. Serrano-Gomez, D., A. Dominguez-Soto, J. Ancochea, J. A. Jimenez-Heffernan, J. A. Leal, and A. L. Corbi. 2004. Dendritic cell-specific intercellular adhesion molecule 3-grabbing nonintegrin mediates binding and internalization of *Aspergillus fumigatus* conidia by dendritic cells and macrophages. *J. Immunol.* **173:**5635–5643.

121. Shoham, S., C. Huang, J. M. Chen, D. T. Golenbock, and S. M. Levitz. 2001. Toll-like receptor 4 mediates intracellular signaling without TNF-alpha release in response to *Cryptococcus neoformans* polysaccharide capsule. *J. Immunol.* **166:**4620–4626.

122. Steele, C., L. Marrero, S. Swain, A. G. Harmsen, M. Zheng, G. D. Brown, S. Gordon, J. E. Shellito, and J. K. Kolls. 2003. Alveolar macrophage-mediated killing of *Pneumocystis carinii* f. sp. muris involves molecular recognition by the Dectin-1 beta-glucan receptor. *J. Exp. Med.* **198:**1677–1688.

123. Syme, R. M., J. C. Spurrell, E. K. Amankwah, F. H. Green, and C. H. Mody. 2002. Primary dendritic cells

phagocytose *Cryptococcus neoformans* via mannose receptors and Fcγ receptor II for presentation to T lymphocytes. *Infect. Immun.* 70:5972–5981.

124. Tada, H., E. Nemoto, H. Shimauchi, T. Watanabe, T. Mikami, T. Matsumoto, N. Ohno, H. Tamura, K. Shibata, S. Akashi, K. Miyake, S. Sugawara, and H. Takada. 2002. *Saccharomyces cerevisiae-* and *Candida albicans*-derived mannan induced production of tumor necrosis factor alpha by human monocytes in a CD14- and Toll-like receptor 4-dependent manner. *Microbiol. Immunol.* 46:503–512.

125. Torosantucci, A., P. Chiani, F. De Bernardis, A. Cassone, J. A. Calera, and R. Calderone. 2002. Deletion of the two-component histidine kinase gene (*CHK1*) of *Candida albicans* contributes to enhanced growth inhibition and killing by human neutrophils in vitro. *Infect. Immun.* 70:985–987.

126. Torosantucci, A., G. Romagnoli, P. Chiani, A. Stringaro, P. Crateri, S. Mariotti, R. Teloni, G. Arancia, A. Cassone, and R. Nisini. 2004. *Candida albicans* yeast and germ tube forms interfere differently with human monocyte differentiation into dendritic cells: a novel dimorphism-dependent mechanism to escape the host's immune response. *Infect. Immun.* 72:833–843.

127. Turner, M. W. 1996. Mannose-binding lectin: the pluripotent molecule of the innate immune system. *Immunol. Today* 17:532–540.

128. Uezu, K., K. Kawakami, K. Miyagi, Y. Kinjo, T. Kinjo, H. Ishikawa, and A. Saito. 2004. Accumulation of gammadelta T cells in the lungs and their regulatory roles in Th1 response and host defense against pulmonary infection with *Cryptococcus neoformans*. *J. Immunol.* 172:7629–7634.

129. van Spriel, A. B., I. E. van den Herik-Oudijk, N. M. van Sorge, H. A. Vile, J. A. van Strijp, and J. G. van de Winkel. 1999. Effective phagocytosis and killing of *Candida albicans* via targeting FcγRI (CD64) or FcαRI (CD89) on neutrophils. *J. Infect. Dis.* 179:661–669.

130. Vora, P., A. Youdim, L. S. Thomas, M. Fukata, S. Y. Tesfay, K. Lukasek, K. S. Michelsen, A. Wada, T. Hirayama, M. Arditi, and M. T. Abreu. 2004. Beta-defensin-2 expression is regulated by TLR signaling in intestinal epithelial cells. *J. Immunol.* 173:5398–5405.

131. Wang, J. E., A. Warris, E. A. Ellingsen, P. F. Jorgensen, T. H. Flo, T. Espevik, R. Solberg, P. E. Verweij, and A. O. Aasen. 2001. Involvement of CD14 and Toll-like receptors in activation of human monocytes by *Aspergillus fumigatus* hyphae. *Infect. Immun.* 69:2402–2406.

132. Warris, A., A. Bjorneklett, and P. Gaustad. 2001. Invasive pulmonary aspergillosis associated with infliximab therapy. *N. Engl. J. Med.* 344:1099–1100.

133. Wasylnka, J. A., and M. M. Moore. 2003. *Aspergillus fumigatus* conidia survive and germinate in acidic organelles of A549 epithelial cells. *J. Cell Sci.* 116:1579–1587.

134. Weber, S. M., S. M. Levitz, and T. S. Harrison. 2000. Chloroquine and the fungal phagosome. *Curr. Opin. Microbiol.* 3:349–353.

135. Wood, K. L., C. A. Hage, K. S. Knox, M. B. Kleiman, A. Sannuti, R. B. Day, L. J. Wheat, and H. L. Twigg, 3rd. 2003. Histoplasmosis after treatment with anti-tumor necrosis factor-alpha therapy. *Am. J. Respir. Crit. Care Med.* 167:1279–1282.

136. Woods, J. P. 2003. Knocking on the right door and making a comfortable home: *Histoplasma capsulatum* intracellular pathogenesis. *Curr. Opin. Microbiol.* 6:327–331.

137. Wuthrich, M., H. I. Filutowicz, T. Warner, G. S. Deepe, Jr., and B. S. Klein. 2003. Vaccine immunity to pathogenic fungi overcomes the requirement for CD4 help in exogenous antigen presentation to CD8[+] T cells: implications for vaccine development in immune-deficient hosts. *J. Exp. Med.* 197:1405–1416.

138. Wuthrich, M., H. I. Filutowicz, T. Warner, and B. S. Klein. 2002. Requisite elements in vaccine immunity to *Blastomyces dermatitidis*: plasticity uncovers vaccine potential in immune-deficient hosts. *J. Immunol.* 169:6969–6976.

139. Xu, M., Z. Wang, and R. M. Locksley. 2004. Innate immune responses in peptidoglycan recognition protein L-deficient mice. *Mol. Cell. Biol.* 24:7949–7957.

140. Yamamoto, M., S. Sato, H. Hemmi, K. Hoshino, T. Kaisho, H. Sanjo, O. Takeuchi, M. Sugiyama, M. Okabe, K. Takeda, and S. Akira. 2003. Role of adaptor TRIF in the MyD88-independent toll-like receptor signaling pathway. *Science* 301:640–643.

141. Yauch, L. E., M. K. Mansour, S. Shoham, J. B. Rottman, and S. M. Levitz. 2004. Involvement of CD14, Toll-like receptors 2 and 4, and MyD88 in the host response to the fungal pathogen *Cryptococcus neoformans* in vivo. *Infect. Immun.* 72:5373–5382.

Molecular Principles of Fungal Pathogenesis
Edited by Joseph Heitman et al.
©2006 ASM Press, Washington, D.C.

Chapter 33

Acquired Antibody-Mediated Immunity to Fungi

LIISE-ANNE PIROFSKI AND ARTURO CASADEVALL

The role that antibody-mediated immunity (AMI) plays in resistance to medically important fungi remains uncertain, largely because this question was not rigorously investigated until the last quarter of the 20th century, when the incidence of clinically relevant mycoses rose sharply. Factors that contributed to the increased incidence of mycotic diseases included the use of intravascular access devices and broad-spectrum antibiotics in the hospital setting; the use of chemotherapeutic and anti-inflammatory agents; advances in organ, bone, and stem cell transplantation; and the human immunodeficiency virus (HIV) pandemic. Each of these factors has the capacity to induce immune deficits that increase the risk for mycotic disease. However, the relatively low incidence of fungal disease was only partly responsible for the paucity of research on AMI against fungi. Another factor was the belief that immunity to fungi was conferred by cell-mediated immunity (CMI), and AMI had little or no role in protection. This view was part of a nearly century old dualism whereby CMI was thought to be responsible for immunity to fungi and intracellular microbes and AMI was thought to be responsible for immunity to bacteria and extracellular microbes. By the end of the 20th century, the discovery of new mechanisms of antibody (Ab) action and interplay between CMI and AMI rendered this view incorrect. Unraveling mechanisms of AMI against fungi was also slowed by the lack of suitable animal models and defined Ab reagents. The development of mice with targeted deficiencies of key components of the immune response, in combination with the advent of the hybridoma technology and advances in Ab engineering, provided a facile toolbox with which to address the importance of AMI against fungi (21). Studies with these reagents fueled new perspectives on AMI against fungi and other microbes for which the role

of AMI in natural protection is uncertain. This chapter provides an overview of available information on AMI against fungi.

CRITERIA FOR ANTIBODY-MEDIATED IMMUNITY

Historically, the role of AMI in immunity against a microbe was established by three criteria: (i) the observation that individuals with Ab or B-cell disorders were more susceptible to an infectious disease, (ii) the experimental demonstration that transfer of specific Ab from an immune host to a naive host protected against challenge with the relevant microbe, and (iii) the observation that those who are resistant to an infectious disease have specific Ab to the relevant microbe in their serum (19). While a role for AMI in protection against certain microbes, such as the pneumococcus, the menigococcus, and the agents of tetanus and diphtheria, was established with some or all of these criteria, they fail to establish a role for AMI in protection against other microbes, including fungi. For the most part, patients with B-cell and immunoglobulin (Ig) deficiency are not more susceptible to fungal diseases, with the caveat that combined B- and T-cell defects are often found in patients who develop cryptococcosis, candidiasis, and HIV-associated mycoses (145). Factors that could have undermined the ability of the above criteria to establish the efficacy of Ab-mediated protection against fungi include the following: they can be difficult to apply to microbes that elicit strong CMI responses (19); that fungal diseases are often chronic, making it difficult to predict when Ab administration might be effective; and heterogeneous Ab reagents may not contain a sufficient amount of the type of Ab necessary for protection (28).

The advent of hybridoma technology and subsequent advances in Ab engineering led to the generation

---

Liise-anne Pirofski and Arturo Casadevall • Departments of Medicine, Microbiology, and Immunology, Albert Einstein College of Medicine, Bronx, NY 10461.

of scores of defined, homogenous, monoclonal Ab (MAb) reagents. Studies with MAbs led to a paradigm shift which recognized that AMI could enhance host resistance to microbes, including fungi, for which standard approaches did not establish a role for AMI. MAbs have now been shown to be protective against experimental cryptococcosis (51, 60, 96, 126, 129, 146), candidiasis (15, 74, 77, 113), histoplasmosis (140), and pneumocystosis (64).

NATURAL ANTIBODY-MEDIATED IMMUNITY TO FUNGI

A common view in the mycology field is that AMI is not essential for natural immunity to fungal disease, but data for this conclusion are limited. The importance of AMI for resistance to fungal disease has not been rigorously examined for most clinically relevant fungi. Information from the fungi that have been studied suggests that AMI can play an immunoregulatory role.

C. neoformans

Abs to the *Cryptococcus neoformans* capsular polysaccharide glucuronoxylomannan (GXM) and/or cryptococcal proteins are nearly ubiquitous in sera from adults and children, including those with HIV infection (1, 34, 46, 49, 59, 70, 81, 173). Studies of the reactivity of serum with cryptococcal proteins in normal children indicate that cryptococcal infection is likely to be a benign childhood infection that usually does not cause clinically recognized disease (70). Despite serological evidence for infection and perhaps latency (1, 34, 46, 49, 59, 70, 81, 173), the importance of AMI in resistance to *C. neoformans* remains unresolved. However, populations at increased risk for cryptococcosis often have impaired B-cell or immunoglobulin (Ig) repertoires. In the pre-AIDS era, the highest prevalence of cryptococcal disease was among immunosuppressed patients with combined B- and T-cell defects resulting from hematologic malignancies and/or antilympholytic therapies (reviewed in reference 145). Since the beginning of the HIV pandemic, the highest prevalence of cryptococcosis has been in HIV-infected individuals (41). Although the prevalence of HIV-associated cryptococcosis has fallen among patients with access to highly active antiretroviral therapy, cryptococcosis occurs in 10 to 30% of patients with AIDS in the developing world (5, 61). Several lines of evidence suggest that the risk for HIV-associated cryptococcosis could be enhanced by dysregulated AMI. The Ab response to the cryptococcal capsular polysaccharide GXM is highly restricted (22), and VH3, the predominant Ig variable (V)-region gene family used in human Abs to GXM (60, 96, 145), is dysregulated and depleted in HIV-infected individuals (10, 33, 59, 166). This implies the existence of a "hole in the Ig repertoire" that could translate into a failure of AMI against *C. neoformans*. In support of this concept, a case-controlled study of HIV-infected individuals who did and did not develop cryptococcosis revealed that those who developed disease had lower levels of certain VH3-positive Abs than those who did not (59). The importance of establishing whether Ig repertoire defects contribute to impaired AMI to cryptococcosis lies in the possibility they could also impair the response to vaccines designed to elicit Abs to GXM (98).

C. neoformans infections also occur in patients with X-linked immunodeficiency (XID) (82, 177, 199), which is a disorder associated with defective AMI. Mice with an XID-like defect, CBA/N mice, were more susceptible to experimental cryptococcal infection (101). These associations are notable, because a key defect in human XID is an absence of CD27$^+$ B cells (2, 99). CD27$^+$ memory IgM is critical in the response to encapsulated pathogens (88, 195). Since reduced expression of CD27 is an HIV-associated B-cell defect (35, 134), it could contribute to the risk for cryptococcosis in certain HIV-infected patients. In support of this hypothesis, HIV-infected individuals had lower levels of IgM to GXM than did HIV-uninfected individuals in one cohort (81), and in another, HIV-infected individuals who developed cryptococcosis had lower levels of total IgM and other Ig repertoire defects than those who did not (59).

In experimental systems, mice rendered B-cell deficient by neonatal Ab depletion were no more susceptible to *C. neoformans* or *Candida albicans* than were undepleted mice (120, 172). Similarly, compared to normal mice, uMT knockout (KO) mice, which lack B cells and Ab (12), had no differences in the susceptibility to primary or secondary infection with *Histoplasma capsulatum* (4) but were more susceptible to infection with *Blastomyces dermatitidis* (200). However, a finding of no difference in susceptibility between B-cell-deficient and control mice is not conclusive for the absence of a role for AMI if control mice made predominantly nonprotective Abs (200), because the experiment would be comparing a condition with no Ab to one with nonprotective Ab. On the other hand, uMT KO mice were more susceptible to reinfection with *C. albicans* (121), and adoptively transferred lymphocytes from *C. neoformans*-immunized normal mice promoted reduced growth of *C. neoformans* in SCID mice but lymphocytes from uMT KO mice did not (3). These observations are consistent with the concept that

B cells can enhance cellular immune responses by promoting the development of Ag-specific T-cell responses and immunologic memory (11, 103, 157, 170). A role for B cells and/or Ab against *C. neoformans* is implied by the observation that uMT KO mice were more susceptible to experimental cryptococcosis (160).

P. jiroveci (Formerly P. carinii)

Infection with *Pneumocystis jiroveci* is thought to be ubiquitous based on serological evidence that specific Ab is acquired in childhood (144). However, *Pneumocystis* pneumonia (PCP) was uncommon until the HIV pandemic. In the pre-AIDS era, PCP occurred predominantly as a complication of lympholytic therapy for organ transplantation or hematologic malignancy (78, 80). The incidence of PCP rose markedly with the onset of the HIV epidemic, as PCP was among the first recognized AIDS-defining illnesses. Serological studies showed that the Ab response to *P. jiroveci* antigens (Ags) was reduced in some HIV-infected patients (55, 90), although an Ab response to a 95-kDa carbohydrate Ag was noted in some patients with PCP (93). Serology was not developed further to diagnose HIV-associated PCP, and the role of AMI in resistance to HIV-associated PCP is unknown. However, the importance of AMI in resistance to PCP has been established in animal models. PCP was found to develop spontaneously in uMT KO mice (100), and adoptive transfer of T cells without B cells was insufficient to protect SCID mice against PCP (79). The mechanism by which AMI facilitates resistance to PCP is complex. Delayed clearance of *P. carinii* from the lungs of B-cell-deficient CD40 and Fcγ R KO mice was associated with an absence of specific opsonic IgG and reduced numbers of CD4$^+$ T cells (92). These observations suggest that AMI against PCP involves enhancement of host defense by specific IgG and induction of the development of T-cell mediators of protection.

INDUCED ANTIBODY-MEDIATED IMMUNITY TO FUNGI

Despite uncertainty about the importance of AMI in natural resistance to fungi, the efficacy of Ab against *C. neoformans* (51, 60, 96, 126, 129, 146), *C. albicans* (15, 74, 77, 113), *H. capsulatum* (140), and *P. jiroveci* (64) is now established in experimental models. The demonstration that Ab can mediate protection against these fungi provides proof of the principle that AMI can be effective against medically relevant fungi. Studies with Abs to the above fungi led to the discovery of new mechanisms of Ab action and a new view of AMI, which is focused on the ability of Ab to affect the host response to the relevant

fungal species. This chapter is focused primarily on *C. neoformans*, *P. jiroveci*, and *C. albicans* because there is an extensive literature for each. Our goal is to be comprehensive without being exhaustive, and, recognizing that some significant literature may not have been cited, we apologize to investigators whose work may not have been cited.

Antibody-Mediated Immunity to *C. neoformans*
Studies of induced AMI to *C. neoformans* have provided a wealth of new information on mechanisms of Ab efficacy, including paradigms to explain Ab efficacy in systems where a role for AMI in natural resistance has not been discovered (25).

Abs to cryptococcal polysaccharide

The main antigenic determinant of *C. neoformans* is the GXM moiety of the polysaccharide capsule, and the majority of Abs to *C. neoformans* that have been studied bind GXM. Such Abs have been elicited in normal mice, human Ig transgenic mice, or humans by investigational conjugate vaccines consisting of serotype A GXM (strain NIH 371 [47]) and tetanus toxoid (GXM-TT [23, 124]), serotype D GXM (strain 24067) and diphtheria toxoid (GXM-DT [96]), or GXM peptide mimetics (P13-TT [58], P13-DT [98], P206.1-protein conjugates [7], or P315-KLH [114]). Like other microbial capsular polysaccharides, GXM is a T-independent type 2 Ag (174), which induces an Ab response characterized by variable (V)- and constant (C)-region restriction (146). The molecular genetic structure of Abs to GXM was revealed by studies of the gene use of MAb reagents. Mouse MAbs to GXM display marked restriction, as evidenced by their oligoclonal derivation (124), use of limited V-region gene segments irrespective of the eliciting Ag (22, 126), and the expression of a shared idiotype by Abs with similar specificity (22). The 7183 V region that is used in mouse Abs to GXM is a homolog of the VH3 genes (42), which are used in human Abs to GXM (59, 60, 96, 147). The 7183 and VH3 gene elements belong to the V-region clan III (85), which is composed of the predominant V regions used in Abs to encapsulated pathogens. The importance of V-region restriction for immunity to encapsulated pathogens has not been established. However, a biological explanation for such restriction may reflect structural limitations on Ag-Ab binding, whereby Ab receptors with a certain structure preferentially bind GXM epitopes. This concept is consistent with the existence of V-region clans with shared Ag specificity (85), the use of canonical V-region sequences in Abs to carbohydrate Ags (188), and evidence that CDR3 diversity with a single non-clan III V region was insufficient to generate carbohydrate Abs (201).

Abs to GXM are also restricted with respect to C-region (isotype) use. The IgG subclass profiles of Abs to GXM resemble those to other capsular polysaccharides, with IgG3 and IgG2 predominating in mice and humans, respectively (1, 23, 46, 146, 208). The importance of C-region restriction for immunity to encapsulated microbes has not been established. However, C-region structure was shown to influence the specificity of V-region-identical Abs to capsular polysaccharides, including Abs to GXM (38, 39, 116), and human Ig transgenic mice expressing IgG2 but not IgG4 were able to mount an Ab response to GXM (98). These observations are consistent with the concept that certain C-region/V-region structures generate a unique or "best-fit" Ab conformation, which translates into a greater likelihood of Ag selection. However, naturally selected C regions may not manifest the best efficacy as individual reagents. For example, IgG3 was the predominant IgG subclass in the mouse response to GXM (124), but IgG3 MAbs were not protective against experimental C. neoformans intravenous infection (129) (see below). At present, our understanding of relationships between affinity, avidity, and function for serum Abs to GXM, even if they are oligoclonal, is insufficient to predict the importance of C-region restriction for immunity to C. neoformans.

MAbs to GXM produced from mice vaccinated with GXM-TT (which contains strain NIH 371 serotype A GXM [47]) and serotype A capsular polysaccharide also bind to GXM serotypes D and AD and less to serotypes B and C (23, 24, 53). A human MAb to GXM produced from lymphocytes of a volunteer GXM-TT recipient reacted with all four GXM serotypes, albeit better with serotypes D and A (147). MAbs produced from human Ig transgenic mice were largely specific for serotype D GXM but were reactive with serotype A C. neoformans cells, suggesting differential expression of the epitope(s) they recognize (96). The specificity of mouse Abs to GXM is influenced by both V- and C-region characteristics. Mutational analysis of protective and nonprotective mouse MAbs with identical light chains revealed that residues located in the VH CDR1 and CDR2 conferred binding to defined epitopes, which translated into unique immunofluorescent binding patterns and MAb efficacy (135, 136). An annular pattern of binding to C. neoformans serotype D (strain 24067) was observed with MAbs that protected against experimental serotype D infection, and a punctate binding pattern was observed with MAbs that were not protective (37). The GXM specificity of MAbs that produce annular and punctate binding patterns is likely to be similar to those that result in rim and puffy capsular (quelling) reaction patterns, respectively (54). GXM specificity can also be influenced by isotype. V-region-identical MAbs with different isotypes manifested different immunofluorescence binding patterns and specificity, based on reactivity with peptide mimetics of GXM (116, 182).

Biological activity of Abs to C. neoformans

The opsonic efficacy of naturally occurring human Ig for C. neoformans is minimal and is probably limited to amplification of complement-mediated opsonization (198). Normal human serum IgG binds to and promotes effector cell opsonization of unencapsulated but not encapsulated C. neoformans (87, 115), whereas immune sera from human volunteers vaccinated with GXM-TT promoted human effector cell opsonization and phagocytosis of C. neoformans, with the predominant opsonic activity mediated by IgG2 Abs (208).

Classical concepts of Ab activity include direct and indirect antimicrobial effects, which can be a function of V- or C-region domains, or both (26). Direct mechanisms do not require effector cells or molecules and were classically associated with viral and toxin neutralization, whereas indirect mechanisms depend on effector cells and/or mediators and were classically associated with phagocytosis and antibody-dependent cellular cytotoxicity. Both types of Ab activity, mediated by classical and novel mechanisms, have been demonstrated for MAbs to C. neoformans. Direct effects include the capacity of certain MAbs to GXM, C. neoformans melanin, and the cell wall glucosylceramide to reduce C. neoformans growth and/or inhibit fungal budding in culture (104, 162, 163). Another direct effect is that protective GXM-binding Abs, but not nonprotective Abs, prevent GXM shedding in solution; this phenomenon may enhance host immunity by reducing the immunosuppressive effects of soluble polysaccharide (104).

Indirect mechanisms of Ab action include the ability of defined MAbs to GXM to promote phagocytosis and enhance effector cell antifungal activity against C. neoformans (52, 133, 209). IgG1 MAbs to GXM promote phagocytosis, enhance Ag presentation, increase costimulatory molecule expression, and mediate changes in cytokine expression by host effector cells in vitro (158, 189–192). Dendritic cell antigen presentation to lymphocytes is most efficient when C. neoformans is opsonized by Fc receptors (176). V-region-identical IgG switch variants of mouse MAbs to GXM promoted human and mouse macrophage-mediated opsonization of C. neoformans, with the activity of IgG2a being greater than that of IgG1 and IgG3 (167). Isotype was a critical determinant of the opsonic activity of V-region-identical MAbs to GXM. IgGs had more activity than IgM or IgA, with IgG1 having the greatest activity (133). For IgM MAbs, opsonic activity was a function of GXM specificity (37). One such MAb promoted complement-independent phagocytosis by enabling *C. neoformans* capsular polysaccharide to bind complement receptor directly, presumably via a lectin-like interaction that was exposed by MAb binding (180). Similarly, F(ab′)₂ fragments of certain MAbs that bound *C. neoformans* in a rim pattern also promoted complement-independent phagocytosis (137). These observations underscore the relationship between Ab structure, specificity, and biological function while providing new paradigms to explain mechanisms of Ab action. Certain opsonic Abs whose biological activity depends on effector cells enhanced the release of effector cell and T-cell proinflammatory cytokines (119, 123, 190). Hence, MAbs to *C. neoformans* can exert either direct or indirect antifungal effects by preventing growth and/or enhancing host immune mechanisms that can contain and kill *C. neoformans*, respectively.

Efficacy of Abs to GXM

Specific immune serum was first shown to enhance the efficacy of antifungal therapy against *C. neoformans* in the early 1960s (71), but it was not until nearly 25 years later that the ability of Abs to protect against cryptococcosis was rigorously established. This interval reflected the prevailing view that AMI was not effective against fungi. This view was reinforced by negative data obtained with serum reagents, which failed to demonstrate Ab efficacy against experimental cryptococcosis (18). Results obtained with serum reagents were difficult to interpret because of the heterogeneous nature of serum Abs and the possibility that the activity of a protective Ab could be undermined or counteracted by a nonprotective Ab (18, 28). The advent of the MAb technology and development of MAbs to GXM provided defined, homogenous reagents that ushered in a new era in the investigation of Ab efficacy in fungal infections based on the study of homogenous, monospecific reagents.

The MAb E1, a mouse IgG1 to serotype A GXM, was the first MAb to GXM shown to be protective in mice. E1 prolonged survival after intraperitoneal (i.p.) challenge with serotype A *C. neoformans* and enhanced the antifungal activity of amphotericin B (50, 51). E1 protected against pneumonia in C5-deficient mice but did not prevent pneumonia or meningitis in normal BALB/c mice (52). Similar to findings from the serum therapy era (28), these observations suggested that the benefit of Ab was

against early manifestations of disease (cryptococcosis) and anticipated later studies that linked Ab efficacy to the host inflammatory milieu (see below) (56, 117, 159). A human MAb to GXM produced from lymphocytes of a volunteer GXM-TT recipient was also protective against systemic *C. neoformans* challenge, as were mouse MAbs produced with different GXM immunogens (126).

A rich body of information about AMI to *C. neoformans* emerged from studies with mouse MAbs elicited by the investigational GXM-TT conjugate vaccine. These studies established that defined MAbs to GXM prolonged the survival of mice with lethal systemic and intracranial challenge with *C. neoformans* (128–130, 132). MAb-mediated protection in mice paralleled in vitro functional activity in that both V- and C-region characteristics influenced MAb efficacy. For V-region-identical mouse MAbs, IgG1s were protective but IgG3s were nonprotective or deleterious (127, 141). Also, an IgG2a antibody had diminished efficacy in normal or C5-deficient mouse strains (142). However, IgG3 MAbs were protective in mouse strains deficient in certain aspects of cellular immunity (see below).

In studies with IgM MAbs that used identical light chains, efficacy was a function of certain CDR1 and CDR2 somatic mutations which conferred a unique specificity; MAbs with different residues in the same positions had different specificity and were nonprotective (37, 135, 136). Protective human IgM MAbs produced from a volunteer recipient of GXM-TT or human Ig transgenic mice vaccinated with serotype D (strain 24067) GXM-DT shared a similar motif including CDR1 and CDR2 amino acid residues (96). This motif includes the germ line residues in 6 of 22 functional gene segments from the VH3 gene family, 4 of which were used in five human MAbs, to GXM (96, 147). Human MAbs that were derived from GXM-DT-vaccinated mice and that used the identical light chain gene to the VH3 MAbs, but the single VH6 gene segment, were not protective (96). GXM consists of multiple epitopes that elicit Abs that are nonprotective and/or deleterious in a defined model system. Competition experiments with nonprotective mouse and human MAbs revealed a different specificity (96). Taken together, available data show that certain light-chain structures are likely to confer GXM reactivity but heavy-chain structure determines specificity and efficacy, with the caveat that the C region also contributes to fine specificity (116). The fact that VH gene segments that are overrepresented in the naturally occurring Ab repertoire (43) were used in protective and nonprotective human Abs to GXM suggests that the inefficacy of naturally occurring AMI in HIV-infected individuals could in part be due to dysregulated VH3 gene expression in HIV-infected individuals (33).

Mechanisms of MAb-mediated protection

Studies of the efficacy of MAbs to GXM in normal and immunodeficient mouse models have provided new paradigms to explain AMI against cryptococcosis. Evidence that certain MAbs to GXM enhanced the antifungal activity of effector cells and stimulated CMI mechanisms in vitro (see above) led to the investigation of similar questions in vivo. The ability of an IgG1 to GXM to prolong survival after lethal intravenous challenge with serotype D *C. neoformans* required CD4$^+$ T cells and/or gamma interferon, whereas an IgG3 that enhanced infection in the setting of CD4$^+$-T-cell deficiency was protective in CD8$^+$-T-cell deficient mice (203). Consistent with the concept that MAb efficacy depends on the enhancement of CMI mechanisms, interleukin-12 (IL-12), IL-6, and IL-4 were each required for IgG1-, IgG2a-, or IgG2b-mediated protection (8). However, MAbs to GXM also required IL-10 and inducible nitric oxide synthase to mediate protection against intravenous and pulmonary infection, respectively (8, 159). The generation of a granulomatous inflammatory response is critical for immunity to *C. neoformans* (62). Therefore, the fact that both Th1- and Th2-type cytokines are required for Ab efficacy suggests that MAb-*C. neoformans* GXM immune complexes that produce Th1-type responses require a counterregulatory mechanism, and vice versa. In support of this concept, immunized mice that lacked the inhibitory FcRII receptor were more susceptible to pneumococcal infection than were unimmunized mice despite having more serum opsonophagocytic activity (36).

The efficacy of a human IgM MAb to GXM against experimental cryptococcosis required complement component 3 (C3) (60), but the efficacy of mouse IgGs did not (169). Mouse IgG3, which was nonprotective or enhancing in normal mice (142), was protective in C3-deficient mice (169). Protective mouse IgM and IgG1 and a human MAb to GXM each displayed a prozone-like phenomenon, whereby MAb efficacy decreased as a function of the MAb amount administered and/or the inoculum (96, 178, 179). IgM-mediated prozone phenomena may reflect interference with oxidative killing mechanisms as a consequence of Ig coating of the polysaccharide capsule (178). IgG-mediated prozone-like effects were associated with the administration of large amounts of MAb and immunomodulatory effects resulting in increased cytokine production in the brain (179).

These observations provide additional evidence for a relationship between detrimental MAbs and the host inflammatory milieu (see above) (8, 203) and illustrate that AMI against cryptococcosis is not a singular function of Ab isotype or amount but, rather, a complex function of these factors and the cellular immune response of the host.

Proof of the principle that AMI can prolong survival and enhance host immune mechanisms against experimental cryptococcosis fueled a paradigm shift that has resulted in reconsideration of Ab-based therapies for many infectious diseases. A phase I clinical trial of the mouse IgG1 to GXM, 18B7 (20), in HIV-infected individuals with a history of cryptococcosis has been completed, and a reduction in serum polysaccharide levels was observed in patients receiving larger doses (89). Since MAbs to GXM can enhance the efficacy of antifungal therapy (50, 125, 131), the promise of MAbs for cryptococcosis includes their use for primary or adjunctive therapy or both.

Abs to other *C. neoformans* determinants

Abs to *C. neoformans* determinants other than GXM have been described. The best-studied Abs are to *C. neoformans* melanin. Melanin is a biological pigment found in *C. neoformans*, which contributes to its pathogenicity as an independent virulence factor (27). Like GXM, melanin has antigenic and anti-inflammatory properties and elicits a T-cell-independent type 2 response in mice (139). Mouse MAbs to *C. neoformans* melanin prolonged survival in mice and inhibited *C. neoformans* growth in culture. The *C. neoformans* cell wall glucosylceramide (162) and mannoproteins (148) elicit human Ab responses. MAbs to galactoxylomannan (186) have been described, but their protective efficacy is unknown. At present, the contribution of such Abs to natural AMI or their ability to confer protection upon passive transfer is unknown.

Vaccines for *C. neoformans*

A major focus of vaccine strategies for *C. neoformans* has been on the use of GXM as a vaccine antigen. Similar to the rationale for the development of capsular polysaccharide-based vaccines for the pneumococcus and *Haemophilus influenzae* type B (161), this approach endeavors to elicit opsonic Abs to GXM that would prevent disease. The immunogenicity of early cryptococcal capsular polysaccharide-protein conjugate vaccines was inconsistent (72). Advances in carbohydrate chemistry and conjugation led to the production of a GXM-TT conjugate from immunochemically defined GXM from serotype A (NIH 371) (see above) (47). GXM-TT was immunogenic in mice and human volunteers (23, 147, 208) and prolonged the survival of mice after a lethal intravenous challenge with *C. neoformans* (48). GXM-TT efficacy was associated with a reduction in the fungal burden in organs and the GXM levels in serum, and GXM-TT-elicited immune serum was protective in naive mice (48). GXM-TT did not undergo further evaluation, for reasons unrelated to its efficacy or promise (102).

Another approach to the development of a vaccine for *C. neoformans* was to search for peptide mimics that serve as antigenic surrogates for GXM (145). Antigenic peptide mimetics of other capsular polysaccharides have been identified (40, 73, 91, 171). The rationale for investigating the feasibility of a peptide mimetic-based vaccine for *C. neoformans* was twofold. First, since the response to GXM requires part of the Ig repertoire that is dysregulated in HIV-infected patients, it might not elicit the same type of restricted response (145). Second, since certain GXM epitopes elicited nonprotective or detrimental Abs, it could focus the response on epitopes that would elicit protective Abs (9, 145). Protective mouse (IgG1) and human (IgM) MAbs to GXM were used to select peptide mimetics of GXM (184, 185, 204). The mouse IgG1 selected a variety of peptide mimetics, which proved to be invaluable reagents for mapping the GXM specificity of mouse MAbs to GXM and their switch variants (135, 184, 185, 202). Studies with mouse IgG1-selected peptide mimetic-protein conjugate vaccines revealed that higher-affinity mimetics were more antigenic, that priming with GXM-TT increased the level of Abs reactive with GXM determinants that elicit protective rather than nonprotective Abs, but that certain protein carriers induced Abs with the same specificity as nonprotective Abs (7, 114). The relationship between Ab amount and vaccine efficacy for peptide mimetic-based vaccines is not understood (reviewed in references 14 and 98).

Studies with a peptide mimetic selected by a human IgM MAb (P13) showed that it was an immunologic mimic of GXM, based on its ability to induce an Ab response to GXM and inhibit the GXM binding of naturally occurring and GXM-TT-elicited Abs (58, 204). P13-TT and P13-bovine serum albumin conjugates induced Ab responses in mice that prolonged survival after lethal *C. neoformans* challenge (58). P13-diphtheria toxoid (P13-DT) induced a protective response in human Ig transgenic mice expressing IgG2 but not IgG4 (98). The fact that serum IgG4 from these mice bound P13 but not GXM suggests that the C region influences GXM binding and specificity, as reported for MAbs to GXM (see above) (116). Similarly, the idiotype expression of P13-induced Abs to GXM resembled that of GXM-induced Abs (98), suggesting that

restriction in the human Ab response to GXM is likely to reflect constraints on GXM binding, which result in the selection of Abs with the "best fit." Immunization with P13-TT did not prolong survival after *C. neoformans* challenge in human Ig transgenic mice (98). These mice had the highest levels of IgM to GXM, suggesting that a prozone-like phenomenon could have undermined the activity of potentially protective Abs, as described for a mouse IgM MAb (178). Passive transfer of immune sera from P13-DT/CpG-, P13-TT/CpG-, and CpG-vaccinated to naive mice prolonged survival as a function of the level of IgM to GXM in early sera and both IgM and IgG in late sera (97). The ability of CpG, an adjuvant that binds TLR9, to induce a specific Ab response to GXM is consistent with the structure of human MAbs to GXM, which demonstrates that the natural human Ab repertoire has the capacity to react with GXM (96). The ability of GXM- and P13-protein conjugates to prolong the survival of *C. neoformans*-infected mice provides proof of the principle that it is feasible to consider vaccines that elicit Abs to GXM for *C. neoformans*.

Antibody-Mediated Immunity to *Candida* spp.

Early studies with polyclonal sera to *C. albicans* produced inconclusive results regarding the efficacy of acquired AMI to *C. albicans* (18). However, in the past decade, studies with MAbs have significantly clarified the potential of AMI against this microbe.

Abs to *Candida* hsp90

Studies of AMI to *C. albicans* in patients with disseminated candidiasis led to the discovery of disease-specific fungal Ags that induce protective Abs, which are currently under development as therapeutic reagents. A clinical study in the 1980s revealed that sera from patients who recovered from disseminated candidiasis had unique reactivity with a 47-kDa *C. albicans* Ag (111). The Ag, which was found to be cytoplasmic and cell wall associated, was identified as a carboxy-terminal determinant of *C. albicans* hsp90 (105, 109, 175). The association between an Ab response to hsp90 and recovery from disseminated candidiasis in patients taking amphotericin B and the discovery of Abs to hsp90 in individuals with AIDS, an immunocompromised population that does not have an increased incidence of invasive candidiasis (106), supported the concept that Ab to hsp90 could serve as a therapeutic reagent (112, 156). Investigation of the efficacy of Abs to hsp90 against *C. albicans* revealed that the survival of mice with disseminated murine candidiasis was significantly enhanced by passive administration of (i) sera from patients with Ab to hsp90 (110), (ii) a mouse MAb (110), and (iii) a recombinant human Ab (108) to the hsp90 epitope associated with protection in patients (107, 113). Based on evidence that hsp90 is required for yeast viability, a recombinant Ab without an Fc was developed as an inhibitor of hsp90 (15, 113). The resulting Ab, Mycograb, had antifungal activity that was synergistic with amphotericin B in vitro against *C. albicans*, *C. krusei*, *C. tropicalis*, *C. glabrata*, and *C. parapsilosis* (113). In vivo studies revealed that Mycograb alone did not induce the clearance of *Candida* from infected mouse tissues but that Mycograb in combination with amphotericin B induced the clearance of *C. albicans*, *C. krusei,* and *C. glabrata* (113). Recently, Mycograb was also found to be synergistic with amphotericin B against *C. neoformans* in vitro (138). Mycograb was first tested in an open-label pharmacokinetic study in the United Kingdom in patients with disseminated candidiasis who were receiving liposomal amphotericin B; it is now being tested in a double-blinded clinical trial in 10 countries (15).

Abs to *Candida* mannan

Mannoproteins function as adhesins that promote the adherence of *Candida* to host tissues (84). The Ab response to *Candida* mannans was protective in naive animals (45) and was associated with resistance to reinfection in a rat model of *Candida* vaginitis (30). MAbs to *Candida* mannans display diverse specificities for different *Candida* species (31, 83, 118). *Candida* mannan-elicited murine immune serum was protective against disseminated candidiasis in naive mice. An agglutinating IgM MAb and an IgG3 to *Candida* mannan that binds an acid-labile phosphomannan determinant (β-1,2-mannotriose) were protective against disseminated murine candidaisis (74, 77), but another agglutinating MAb that recognized a different, acid-stable epitope was not (74). The epitope of the protective MAbs was homogenously distributed on the yeast cell surface, whereas the epitope of the nonprotective MAb had a patchy distribution (77), suggesting a possible parallel to the paradigm of protective annular or rim and nonprotective punctuate or puffy binding patterns discovered for protective and nonprotective MAbs to *C. neoformans* (37, 54).

A protective MAb to *Candida* mannan (B6.1) enhanced the antifungal activity of mouse neutrophils, whereas a nonprotective MAb (B6) did not (17). However, both MAbs were protective in a mouse model of vaginal candidiasis (76), and both MAbs activated the classical complement pathway (205). Mannan-reactive human serum IgG promotes complement activation by *C. albicans* and C3 deposition on the yeast cell surface (206, 207). Although human serum IgG is not required for opsonophagocytic

killing of *C. albicans* in the presence of an intact complement pathway (86), the efficacy of protective IgM and IgG3 mouse MAbs against disseminated disease depended on complement (75). Vaccination of mice with mannoprotein complexes prior to intravaginal infection with *C. albicans* was associated with a significant reduction in the vaginal fungal burden (44, 76), and passive transfer of immune serum reduced the vaginal fungal burden in naive mice (76). The efficacy of Abs to *Candida* mannans is likely to involve more than one mechanism of action, including blocking adherence and/or inhibiting growth (122). A single-chain variable Ab fragment to *C. albicans* mannan was able to promote phagocytosis in the absence of complement through an unknown mechanism (196). One MAb to *Candida* mannan displayed fungicidal activity in vitro against a variety of *Candida* species, in addition to *C. neoformans* and *Aspergillus fumigatus* (122).

Abs to other *Candida* Ags

The finding that rabbit antiserum to a mouse MAb produced to the yeast killer toxin of *Pichia anomala* (154) inhibited the growth of *C. albicans* (155) led to further work which established that such anti-idiotypic Abs had antimicrobial activity similar to that of the native toxin. The use of a MAb to yeast killer toxin as an anti-idiotypic vaccine improved the survival of mice and rats that had received a lethal systemic or intravaginal *C. albicans* challenge, respectively (151, 152). Previously uninfected MAb-immunized rats manifested a recall response on intravaginal inoculation with *C. albicans* (150). Naturally occurring human anti-killer toxin-like Abs have been described previously (94). The activity of anti-idiotypic Abs to a MAb or single-chain Fv fragment to yeast killer toxin against a variety of microbes that express killer toxin receptors (149), including fungi for which current anti-fungal therapy is dismal, such as *Aspergillus* (32), has fueled efforts to develop anti-idiotypic vaccines or reagents with killer toxin activity. A peptide produced from a recombinant anti-idiotypic Fv to an anti-killer toxin MAb had activity against *C. albicans* in vitro and against *Candida* vaginitis in rats (153) and had in vitro and in vivo activity against *Paracoccidioides brasiliensis* (183).

Abs to *C. albicans* cell wall Ags could conceivably mediate protection by preventing adherence or interfering with fungal dimorphism. A MAb to a C-terminal epitope of a 58-kDa surface mannoprotein of *C. albicans* was found to be protective in mouse models of candidiasis, and patients who survived candidiasis had elevated Abs to this antigen (194). A Fab derived from a MAb to cell wall mannoprotein was demonstrated to prevent germ tube formation in *C. albicans* (29). Similar effects were reported with a panel of MAbs to cell wall antigens (165). This effect could contribute to host defense, since germ tube formation has been associated with virulence and adhesion to host tissues. Specific IgA to *C. albicans* surface determinants has been demonstrated to reduce adhesion to dental prosthetic materials (164). Since adhesion is the initial step in infection, Abs that prevent this phenomenon could contribute to host defense.

Ab responses to *C. albicans* proteinases have been associated with protection against vaginitis in a rat model (45). Direct evidence for the potential utility of Abs to proteinases comes from the observation that passive transfer of Abs to *C. albicans* secreted proteinase 2 reduced the fungal burden in a mouse model of infection (193). Similarly, some protection was observed with passive transfer of rabbit Abs to enolase in a murine model of candidiasis (187). Yet another potential target of AMI is a *C. albicans* immunosuppresive mitogenic protein, since Abs to this antigen mediate protection against experimental murine candidiasis in mice (181).

Vaccines for *Candida*

Candida mannan-conjugate vaccines are under development (16). Anti-idiotypic vaccination with MAbs and Ab fragments that induce Abs to killer toxin are also under development (see above and references 94, 95, 152, and 153). By functioning as both anti-idiotypic vaccines and antimicrobials, reagents that bear the internal image of killer toxin hold promise of activity against a diverse array of microbes that express killer toxin (see above). Another vaccine strategy for *Candida* based on a proteomic approach revealed that protective and nonprotective antisera had different specificities, and that levels of specific IgG2 were higher in protective antisera (57). The ability of Abs to certain *Candida* Ags to block the efficacy of others was previously noted (13). Protective and nonprotective Abs to *Candida* represent another system in which antifungal Abs can be distinguished by their structure and specificity and further clarifies the basis of historical difficulties in establishing the efficacy of sera against fungi.

Antibody-Mediated Immunity to *P. jiroveci*

Immune sera from infected animals and specific Abs can mediate protection against experimental PCP (6, 64, 66) despite evidence that secreted Ab is dispensable in the setting of intact CD4$^+$ T cells and T-cell memory (92). Anti-idiotypic Abs with killer toxin-like activity are active against *Pneumocystis*. Such antibodies inhibit attachment in vitro and

reduce infectivity and limit dissemination in the lungs in immunosuppressed mice (168). Specific Ab can alter the course of rodent PCP in active and passive immunization studies, with the caveat that Ab reactivity and efficacy are species specific (63, 65, 66, 68). Nasal administration of MAbs to *Pneumocystis* glycoprotein A (gpA) and the kexin-like protease Kex1 were protective as IgMs or IgG1 switch variants in SCID mice (64); however, immunization with gpA was not protective in an immunization-immunosuppression-infection model (69). The mechanism of protection for Abs to *Pneumocystis* is unknown but is proposed to involve inhibition of adherence or enhancement of opsonization (64). Protective MAbs that bind the *Pneumocystis* Kex1 Ag (64) react with *Pneumocystis* organisms from different species, including humans, by binding an epitope with positively charged proline-rich regions (197). These regions were found in *Streptococcus pneumoniae* PspA by database search, and the MAbs were also found to react with PspA (197). Intranasal immunization of mice with soluble *Pneumocystis* Ags protected against *Pneumocystis* challenge in CD4$^+$-T-cell-depleted mice (143). Since protection was associated with Ab to a 55- to 60-kDa Ag, AMI did not require T cells in this model.

Antibody-Mediated Immunity to Other Fungi

Although Ab is dispensable for natural resistance to *H. capsulatum* (4), MAbs that mediate protection against experimental histoplasmosis have been identified (140). IgM MAbs to a histone-2B-like protein on the *H. capsulatum* surface prolonged survival and enhanced the antifungal activity of amphotericin B in a murine intranasal challenge model of histoplasmosis (140). MAb efficacy was associated with increased levels of proinflammatory mediators in the lung in vivo and with complement-receptor-mediated opsonophagocytic killing in vitro (140). These observations suggest that AMI against histoplasmosis depends on the presence of Abs that can enhance the host cellular response to infection. MAbs to a surface determinant of *B. dermatitidis* (Bad1, formerly W1-1) promoted macrophage uptake of *B. dermatitidis* in vitro, but did not enhance killing and were not protective against murine experimental blastomycosis (200). In light of evidence that B cells enhanced experimental *B. dermatitidis* infection (200), it could be hypothesized that a role for AMI against blastomycosis might depend on Abs that downregulate the Ab response or prevent fungal growth. Such functions have been described for other Abs to fungi (104, 179). A role for natural AMI in protection against aspergillosis has not been defined. However, anti-idiotypic MAbs to killer toxin prolonged the survival of T-cell-depleted mice with invasive pulmonary aspergillosis (32), and a MAb to *Candida* mannan inhibited *A. fumigatus* growth in vitro (122), suggesting that acquired AMI could hold promise against invasive aspergillosis.

SUMMARY AND CONCLUSIONS

There is overwhelming evidence that some Ab responses to medically important fungi can make a decisive contribution to host defense. The success of passive Ab in preventing and treating fungal diseases in animal models and certain vaccines that elicit AMI is beginning to be translated from the bench to the bedside. Two Ab-based therapeutics are currently undergoing clinical evaluation (15, 20, 89). However, it is apparent that the antigenic complexity of fungal cells, which has been unraveled in part by the study of AMI, leads to highly diverse and heterogeneous Ab responses that differ markedly with respect to their protective quality. Ab efficacy against fungi is a a complex function of Ab specificity, isotype, amount, and host CMI function. This complexity has almost certainly contributed to the historical uncertainty regarding the role of AMI against mycotic diseases. First-generation studies of AMI against fungi lacked the tools to dissect the factors that can influence AMI. Second-generation studies were fueled by the development of tools to investigate the contributions of host and Ab characteristics to AMI against fungi. In recent years, these studies gathered the momentum to become a major engine driving new immunological concepts of AMI (26) and establish new paradigms of Ab structure-function relationships. The next generation of studies of AMI against fungi promises to assemble a mechanistic understanding of Ab action that incorporates structure-function relationships and the complex interplay of host and Ab characteristics in successful AMI against fungi.

REFERENCES

1. **Abadi, J., and L. Pirofski.** 1999. Antibodies reactive with the cryptococcal capsular polysaccharide glucuronoxylomannan are present in sera from children with and without HIV infection. *J. Infect. Dis.* **180:**915–919.
2. **Agematsu, K., H. Nagumo, K. Shinozaki, S. Hokibara, K. Yasui, K. Terada, N. Kawamura, T. Toba, S. Nonoyama, H. D. Ochs, and A. Komiyama.** 1998. Absence of IgD-CD27(+) memory B cell population in X-linked hyper-IgM syndrome. *J. Clin. Investig.* **102:**853–860.
3. **Aguirre, K. M., and L. L. Johnson.** 1997. A role for B cells in resistance to *Cryptococcus neoformans* in mice. *Infect. Immun.* **65:**525–530.
4. **Allendorfer, R., G. D. Brunner, and G. S. Deepe, Jr.** 1999. Complex requirements for nascent and memory immunity in pulmonary histoplasmosis. *J. Immunol.* **162:**7389–7396.

5. **Banerjee, U., K. Datta, T. Majumdar, and K. Gupta.** 2001. Cryptococcosis in India: the awakening of a giant? *Med. Mycol.* **39:**51–67.
6. **Bartlett, M. S., W. C. Angus, M. M. Shaw, P. J. Durant, C. H. Lee, J. M. Pascale, and J. W. Smith.** 1998. Antibody to *Pneumocystis carinii* protects rats and mice from developing pneumonia. *Clin. Diagn. Lab. Immunol.* **5:**74–77.
7. **Beenhouwer, D. O., R. J. May, P. Valadon, and M. D. Scharff.** 2002. High affinity mimotope of the polysaccharide capsule of *Cryptococcus neoformans* identified from an evolutionary phage peptide library. *J. Immunol.* **169:**6992–6999.
8. **Beenhouwer, D. O., S. Shapiro, M. Feldmesser, A. Casadevall, and M. D. Scharff.** 2001. Both Th1 and Th2 cytokines affect the ability of monoclonal antibodies to protect mice against *Cryptococcus neoformans*. *Infect. Immun.* **69:**6445–6455.
9. **Beenhouwer, D. O., P. Valadon, R. May, and M. D. Scharff.** 2000. Peptide mimicry of the polysaccharide capsule of *Cryptococcus neoformans*, p. 143–160. *In* M. W. Cunningham and R. S. Fujinami (ed.), *Molecular Mimicry: Microbes and Autoimmunity*. ASM Press, Washington, D.C.
10. **Berberian, L., L. Goodglick, T. J. Kipps, and J. Braun.** 1993. Immunoglobulin V$_H$3 gene products: natural ligands for HIV gp120. *Science* **261:**1588–1591.
11. **Bergmann, C. C., C. Ramakrishna, M. Kornacki, and S. A. Stohlman.** 2001. Impaired T cell immunity in B cell-deficient mice following viral central nervous system infection. *J. Immunol.* **167:**1575–1583.
12. **Bot, A.** 1996. Immunoglobulin deficient mice generated by gene targeting as models for studying the immune response. *Int. Rev. Immunol.* **13:**327–340.
13. **Bromuro, C., A. Torosantucci, P. Chiani, S. Conti, L. Polonelli, and A. Cassone.** 2002. Interplay between protective and inhibitory antibodies dictates the outcome of experimentally disseminated candidiasis in recipients of a *Candida albicans* vaccine. *Infect. Immun.* **70:**5462–5470.
14. **Buchwald, U. K., A. Lees, M. Steinitz, and L. Pirofski.** 2005. A peptide mimotope of type 8 pneumococcal capsular polysaccharide induces a protective immune response in mice. *Infect. Immun.* **73:**325–333.
15. **Burnie, J., and R. Matthews.** 2004. Genetically recombinant antibodies: new therapeutics against candidiasis. *Expert Opin. Biol. Ther.* **4:**233–241.
16. **Bystricky, S., E. Paulovicova, and E. Machova.** 2003. *Candida albicans* mannan-protein conjugate as vaccine candidate. *Immunol. Lett.* **85:**251–255.
17. **Caesar-TonThat, T. C., and J. E. Cutler.** 1997. A monoclonal antibody to *Candida albicans* enhances mouse neutrophil candidacidal activity. *Infect. Immun.* **65:**5354–5357.
18. **Casadevall, A.** 1995. Antibody immunity to invasive fungal infections. *Infect. Immun.* **63:**4211–4218.
19. **Casadevall, A.** 2004. The methodology for determining the efficacy of antibody-mediated immunity. *J. Immunol. Methods* **291:**1–10.
20. **Casadevall, A., W. Cleare, M. Feldmesser, R. Glatman-Freedman, T. R. Kozel, N. Lendvai, J. Mukherjee, L. Pirofski, J. Rivera, A. L. Rosas, M. D. Scharff, P. Valadon, K. Westin, and Z. Zhong.** 1998. Characterization of a murine monoclonal antibody to *Cryptococcus neoformans* polysaccharide which is a candidate for human therapeutic studies. *Antimicrob. Agents Chemother.* **42:**1437–1446.

21. Casadevall, A., E. Dadachova, and L. A. Pirofski. 2004. Passive antibody therapy for infectious diseases. *Nat. Rev. Microbiol.* 2:695–703.

22. Casadevall, A., M. DeShaw, M. Fan, F. Dromer, T. R. Kozel, and L. Pirofski. 1994. Molecular and idiotypic analysis of antibodies to *Cryptococcus neoformans* glucuronoxylomannan. *Infect. Immun.* 62:3864–3872.

23. Casadevall, A., J. Mukherjee, S. J. Devi, R. Schneerson, J. B. Robbins, and M. D. Scharff. 1992. Antibodies elicited by a *Cryptococcus neoformans*-tetanus toxoid conjugate vaccine have the same specificity as those elicited in infection. *J. Infect. Dis.* 165:1086–1093.

24. Casadevall, A., J. Mukherjee, and M. D. Scharff. 1992. Monoclonal antibody based ELISAs for cryptococcal polysaccharide. *J. Immunol. Methods* 154:27–35.

25. Casadevall, A., and L. Pirofski. 2005. Insights into mechanisms of antibody-mediated immunity from studies with *Cryptococcus neoformans. Curr. Mol. Med.* 5:421-433.

26. Casadevall, A., and L. A. Pirofski. 2004. New concepts in antibody-mediated immunity. *Infect. Immun.* 72:6191–6196.

27. Casadevall, A., A. L. Rosas, and J. D. Nosanchuk. 2000. Melanin and virulence in *Cryptococcus neoformans. Curr. Opin. Microbiol.* 3:354–358.

28. Casadevall, A., and M. D. Scharff. 1994. Serum therapy revisited: animal models of infection and development of passive antibody therapy. *Antimicrob. Agents Chemother.* 38:1695–1702.

29. Casanova, M., J. P. Martinez, and W. L. Chaffin. 1990. Fab fragments from a monoclonal antibody against a germ tube mannoprotein block the yeast-to-mycelium transition in *Candida albicans. Infect. Immun.* 58:3810–3812.

30. Cassone, A., M. Boccanera, D. Adriani, G. Santoni, and F. De Bernardis. 1995. Rats clearing a vaginal infection by *Candida albicans* acquire specific, antibody-mediated resistance to vaginal reinfection. *Infect. Immun.* 63:2619–2624.

31. Cassone, A., A. Torosantucci, M. Boccanera, G. Pellegrini, C. Palma, and F. Malavasi. 1988. Production and characterisation of a monoclonal antibody to a cell-surface, glucomannoprotein constituent of *Candida albicans* and other pathogenic *Candida* species. *J. Med. Microbiol.* 27:233–238.

32. Cenci, E., A. Mencacci, A. Spreca, C. Montagnoli, A. Bacci, K. Perruccio, A. Velardi, W. Magliani, S. Conti, L. Polonelli, and L. Romani. 2002. Protection of killer anti-idiotypic antibodies against early invasive aspergillosis in a murine model of allogeneic T-cell-depleted bone marrow transplantation. *Infect. Immun.* 70:2375–2382.

33. Chang, Q., P. Alpert, J. Abadi, and L. Pirofski. 2000. A pneumococcal capsular polysaccharide vaccine induces a repertoire shift with increased V$_H$3 expression in peripheral B cells from HIV-uninfected, but not HIV-infected individuals. *J. Infect. Dis.* 181:1313–1321.

34. Chen, L. C., D. L. Goldman, T. L. Doering, L. Pirofski, and A. Casadevall. 1999. Antibody response to *Cryptococcus neoformans* proteins in rodents and humans. *Infect. Immun.* 67:2218–2224.

35. Chong, Y., H. Ikematsu, K. Kikuchi, M. Yamamoto, M. Murata, M. Nishimura, S. Nabeshima, S. Kashiwagi, and J. Hayashi. 2004. Selective CD27$^+$ (memory) B cell reduction and characteristic B cell alteration in drug-naive and HAART-treated HIV type 1-infected patients. *AIDS Res. Hum. Retroviruses* 20:219–226.

36. Clatworthy, M. R., and K. G. Smith. 2004. FcgammaRIIb balances efficient pathogen clearance and the cytokine-mediated consequences of sepsis. *J. Exp. Med.* 199:717–723.

37. Cleare, W., and A. Casadevall. 1998. The different binding patterns of two immunoglobulin M monoclonal antibodies to *Cryptococcus neoformans* serotype A and D strains correlate with serotype classification and differences in functional assays. *Clin. Diagn. Lab. Immunol.* 5:125–129.

38. Cooper, L. J., D. Robertson, R. Granzow, and N. S. Greenspan. 1994. Variable domain-identical antibodies exhibit IgG subclass-related differences in affinity and kinetic constants as determined by surface plasmon resonance. *Mol. Immunol.* 31:577–584.

39. Cooper, L. J., A. R. Shikhman, D. D. Glass, D. Kangisser, M. W. Cunningham, and N. S. Greenspan. 1993. Role of heavy chain constant domains in antibody-antigen interaction. Apparent specificity differences among streptococcal IgG antibodies expressing identical variable domains. *J. Immunol.* 150:2231–2242.

40. Cunto-Amesty, G., P. Luo, B. Monzavi-Karbassi, A. Lees, J. Alexander, M. F. del Guercio, M. H. Nahm, C. Artaud, J. Stanley, and T. Kieber-Emmons. 2003. Peptide mimotopes as prototypic templates of broad-spectrum surrogates of carbohydrate antigens. *Cell Mol. Biol.* 49:245–254.

41. Currie, B. P., and A. Casadevall. 1994. Estimation of the prevalence of cryptococcal infection among patients infected with the human immunodeficiency virus in New York City. *Clin. Infect. Dis.* 19:1029–1033.

42. Dammers, P. M., and F. G. Kroese. 2001. Evolutionary relationship between rat and mouse immunoglobulin IGHV5 subgroup genes (PC7183) and human IGHV3 subgroup genes. *Immunogenetics* 53:511–517.

43. Davidkova, G., S. Pettersson, D. Holmberg, and I. Lundkvist. 1997. Selective usage of VH genes in adult human B lymphocyte repertoires. *Scand. J. Immunol.* 45:62–73.

44. De Bernardis, F., M. Boccanera, D. Adriani, A. Girolamo, and A. Cassone. 2002. Intravaginal and intranasal immunizations are equally effective in inducing vaginal antibodies and conferring protection against vaginal candidiasis. *Infect. Immun.* 70:2725–2729.

45. De Bernardis, F., M. Boccanera, D. Adriani, E. Spreghini, G. Santoni, and A. Cassone. 1997. Protective role of antimannan and anti-aspartyl proteinase antibodies in an experimental model of *Candida albicans* vaginitis in rats. *Infect. Immun.* 65:3399–3405.

46. DeShaw, M., and L. Pirofski. 1995. Antibodies to *Cryptococcus neoformans* capsular polysaccharide glucuronoxylomannan are ubiquitous in the serum of HIV$^+$ and HIV$^-$ individuals. *Clin. Exp. Immunol.* 99:425–432.

47. Devi, S. G. N., R. Schneerson, W. Egan, T. J. Ulrich, D. Bryla, J. B. Robbins, and J. E. Bennett. 1991. *Cryptococcus neoformans* serotype A glucuronoxylomannan-protein conjugate vaccines: synthesis, characterization, and immunogenicity. *Infect. Immun.* 59:3700–3707.

48. Devi, S. J. 1996. Preclinical efficacy of a glucuronoxylomannan-tetanus toxoid conjugate vaccine of *Cryptococcus neoformans* in a murine model. *Vaccine* 14:841–844.

49. Dromer, F., P. Aucouturier, J.-P. Clauvel, G. Saimot, and P. Yeni. 1988. *Cryptococcus neoformans* antibody levels in patients with AIDS. *Scand. J. Infect. Dis.* 20:283–285.

50. Dromer, F., and J. Charreire. 1991. Improved amphotericin B activity by a monoclonal anti-*Cryptococcus neoformans* antibody: study during murine cryptococcosis and mechanisms of action. *J. Infect. Dis.* 163:1114–1120.

51. Dromer, F., J. Charreire, A. Contrepois, C. Carbon, and P. Yeni. 1987. Protection of mice against experimental cryptococcosis by an anti-*Cryptococcus neoformans* monoclonal antibody. *Infect. Immun.* 55:749–752.

52. Dromer, F., C. Perronne, J. Barge, J. L. Vilde, and P. Yeni. 1989. Role of IgG and complement component C5 in the initial course of experimental cryptococcosis. *Clin. Exp. Immunol.* 78:412–417.

53. Dromer, F., J. Salamero, A. Contrepois, C. Carbon, and P. Yeni. 1987. Production, characterization, and antibody specificity of a mouse monoclonal antibody reactive with *Cryptococcus neoformans* capsular polysaccharide. *Infect. Immun.* 55:742–748.

54. Duro, R. M., D. Netski, P. Thorkildson, and T. R. Kozel. 2003. Contribution of epitope specificity to the binding of monoclonal antibodies to the capsule of *Cryptococcus neoformans* and the soluble form of its major polysaccharide, glucuronoxylomannan. *Clin. Diagn. Lab. Immunol.* 10:252–258.

55. Elvin, K., A. Bjorkman, N. Heurlin, B. M. Eriksson, L. Barkholt, and E. Linder. 1994. Seroreactivity to *Pneumocystis carinii* in patients with AIDS versus other immunosuppressed patients. *Scand. J. Infect. Dis.* 26:33–40.

56. Feldmesser, M., A. Mednick, and A. Casadevall. 2002. Antibody-mediated protection in murine *Cryptococcus neoformans* infection is associated with pleotrophic effects on cytokine and leukocyte responses. *Infect. Immun.* 70:1571–1580.

57. Fernandez-Arenas, E., G. Molero, C. Nombela, R. Diez-Orejas, and C. Gil. 2004. Contribution of the antibodies response induced by a low virulent *Candida albicans* strain in protection against systemic candidiasis. *Proteomics* 4:1204–1215.

58. Fleuridor, R., A. Lees, and L. Pirofski. 2001. A cryptococcal capsular polysaccharide mimotope prolongs the survival of mice with *Cryptococcus neoformans* infection. *J. Immunol.* 166:1087–1096.

59. Fleuridor, R., R. H. Lyles, and L. Pirofski. 1999. Quantitative and qualitative differences in the serum antibody profiles of HIV-infected persons with and without *Cryptococcus neoformans* meningitis. *J. Infect. Dis.* 180:1526–1536.

60. Fleuridor, R., Z. Zhong, and L. Pirofski. 1998. A human IgM monoclonal antibody prolongs survival of mice with lethal cryptococcosis. *J. Infect. Dis.* 178:1213–1216.

61. French, N., K. Gray, C. Watera, J. Nakiyingi, E. Lugada, M. Moore, D. Lalloo, J. A. Whitworth, and C. F. Gilks. 2002. Cryptococcal infection in a cohort of HIV-1-infected Ugandan adults. *AIDS* 16:1031–1038.

62. Fries, B. C., C. P. Taborda, E. Serfass, and A. Casadevall. 2001. Phenotypic switching of *Cryptococcus neoformans* occurs in vivo and influences the outcome of infection. *J. Clin. Investig.* 108:1639–1648.

63. Gigliotti, F., and C. G. Haidaris. 1998. Antigenic characterization of *Pneumocystis carinii*. *Semin. Respir. Infect.* 13:313–322.

64. Gigliotti, F., C. G. Haidaris, T. W. Wright, and A. G. Harmsen. 2002. Passive intranasal monoclonal antibody prophylaxis against murine *Pneumocystis carinii* pneumonia. *Infect. Immun.* 70:1069–1074.

65. Gigliotti, F., and A. G. Harmsen. 1997. *Pneumocystis carinii* host origin defines the antibody specificity and protective response induced by immunization. *J. Infect. Dis.* 176:1322–1326.

66. Gigliotti, F., and W. T. Hughes. 1988. Passive immunoprophylaxis with specific monoclonal antibody confers partial protection against *Pneumocystis carinii* pneumonitis in animal models. *J. Clin. Investig.* 81:1666–1668.

67. (Reference deleted.)

68. Gigliotti, F., D. C. Stokes, A. B. Cheatham, D. S. Davis, and W. T. Hughes. 1986. Development of murine monoclonal antibodies to *Pneumocystis carinii*. *J. Infect. Dis.* 154:315–322.

69. Gigliotti, F., J. A. Wiley, and A. G. Harmsen. 1998. Immunization with *Pneumocystis carinii* gpA is immunogenic but not protective in a mouse model of *P. carinii* pneumonia. *Infect. Immun.* 66:3179–3182.

70. Goldman, D. L., H. Khine, J. Abadi, D. L. Lindenberg, L. Pirofski, R. Niang, and A. Casadevall. 2001. Serologic evidence for *Cryptococcus neoformans* infection in early childhood. *Pediatrics* 107:e66.

71. Gordon, M. A., and E. Lapa. 1964. Serum protein enhancement of antibiotic therapy in cryptococcosis. *J. Infect. Dis.* 114:373–377.

72. Goren, M. B., and G. M. Middlebrook. 1967. Protein conjugates of polysaccharide from *Cryptococcus neoformans*. *J. Immunol.* 98:901.

73. Grothaus, M. C., N. Srivastava, S. L. Smithson, T. Kieber-Emmons, D. B. Williams, G. M. Carlone, and M. A. J. Westerink. 2000. Selection of an immunogenic peptide mimic of the capsular polysaccharide of *Neisseria meningitidis* serogroup A using a peptide display library. *Vaccine* 18:1253–1263.

74. Han, Y., and J. E. Cutler. 1995. Antibody response that protects against disseminated candidiasis. *Infect. Immun.* 63:2714–2719.

75. Han, Y., T. R. Kozel, M. X. Zhang, R. S. MacGill, M. C. Carroll, and J. E. Cutler. 2001. Complement is essential for protection by an IgM and an IgG3 monoclonal antibody against experimental, hematogenously disseminated candidiasis. *J. Immunol.* 167:1550–1557.

76. Han, Y., R. P. Morrison, and J. E. Cutler. 1998. A vaccine and monoclonal antibodies that enhance mouse resistance to *Candida albicans* vaginal infection. *Infect. Immun.* 66:5771–5776.

77. Han, Y., M. H. Riesselman, and J. E. Cutler. 2000. Protection against candidiasis by an immunoglobulin G3 (IgG3) monoclonal antibody specific for the same mannotriose as an IgM protective antibody. *Infect. Immun.* 68:1649–1654.

78. Hardy, A. M., C. P. Wajszczuk, A. F. Suffredini, T. R. Hakala, and M. Ho. 1984. *Pneumocystis carinii* pneumonia in renal-transplant recipients treated with cyclosporine and steroids. *J. Infect. Dis.* 149:143–147.

79. Harmsen, A. G., and M. Stankiewicz. 1991. T cells are not sufficient for resistance to *Pneumocystis carinii* pneumonia in mice. *J. Protozool.* 38:44S–45S.

80. Harris, R. E., J. A. McCallister, S. A. Allen, A. S. Barton, and R. L. Baehner. 1980. Prevention of pneumocystis pneumonia. Use of continuous sulfamethoxazole-trimethroprim therapy. *Am. J. Dis. Child.* 134:35–38.

81. Houpt, D. C., G. S. Pfrommer, B. J. Young, T. A. Larson, and T. R. Kozel. 1994. Occurrences, immunoglobulin classes, and biological activities of antibodies in normal human serum that are reactive with *Cryptococcus*

neoformans glucuronoxylomannan. *Infect. Immun.* **62:** 2857–2864.

82. Jo, E. K., H. S. Kim, M. Y. Lee, M. Iseki, J. H. Lee, C. H. Song, J. K. Park, T. J. Hwang, and H. Kook. 2002. X-linked hyper-IgM syndrome associated with *Cryptosporidium parvum* and *Cryptococcus neoformans* infections: the first case with molecular diagnosis in Korea. *J. Korean Med. Sci.* **17:**116–120.

83. Kagaya, K., Y. Miyakawa, H. Fujihara, M. Suzuki, G. Soe, and Y. Fukazawa. 1989. Immunologic significance of diverse specificity of monoclonal antibodies against mannans of *Candida albicans. J. Immunol.* **143:**3353–3358.

84. Kanbe, T., Y. Han, B. Redgrave, M. H. Riesselman, and J. E. Cutler. 1993. Evidence that mannans of *Candida albicans* are responsible for adherence of yeast forms to spleen and lymph node tissue. *Infect. Immun.* **61:** 2578–2584.

85. Kirkham, P. M., R. F. Mortari, J. A. Newton, and H. W. Schroeder. 1992. Immunoglobulin V_H clan and family identity predicts variable domain structure and may influence antigen binding. *EMBO J.* **11:**603–609.

86. Kozel, T. R., R. S. MacGill, A. Percival, and Q. Zhou. 2004. Biological activities of naturally occurring antibodies reactive with *Candida albicans* mannan. *Infect. Immun.* **72:**209–218.

87. Kozel, T. R. and T. G. McGaw. 1979. Opsonization of *Cryptococcus neoformans* by human immunoglobulin G: role of immunoglobulin G in phagocytosis by macrophages. *Infect. Immun.* **18:**701–707.

88. Kruetzmann, S., M. M. Rosado, H. Weber, U. Germing, O. Tournilhac, H. H. Peter, R. Berner, A. Peters, T. Boehm, A. Plebani, I. Quinti, and R. Carsetti. 2003. Human immunoglobulin M memory B cells controlling *Streptococcus pneumoniae* infections are generated in the spleen. *J. Exp. Med.* **197:**939–945.

89. Larsen, R. A., P. G. Pappas, J. R. Perfect, J. A. Aberg, A. Casadevall, G. A. Cloud, R. James, S. Filler, W. E. Dismukes, and the National Institute of Allergy and Infectious Diseases Mycoses Study Group. 2005. Phase I evaluation of the safety and pharmacodynamic activity of a murine derived anti-cryptococcal antibody 18B7 in subjects with treated cryptococcal meningitis. *Antmicrob. Agents Chemother.* **49:**952–958.

90. Laursen, A. L., and P. L. Andersen. 1998. Low levels of IgG antibodies against *Pneumocystis carinii* among HIV-infected patients. *Scand. J. Infect. Dis.* **30:**495–499.

91. Lesinski, G. B., S. L. Smithson, N. Srivastava, D. Chen, G. Widera, and M. A. Westerink. 2001. A DNA vaccine encoding a peptide mimic of *Streptococcus pneumoniae* serotype 4 capsular polysaccharide induces specific anticarbohydrate antibodies in Balb/c mice. *Vaccine* **19:** 1717–1726.

92. Lund, F. E., K. Schuer, M. Hollifield, T. D. Randall, and B. A. Garvy. 2003. Clearance of *Pneumocystis carinii* in mice is dependent on B cells but not on *P. carinii*-specific antibody. *J. Immunol.* **171:**1423–1430.

93. Lundgren, B., J. D. Lundgren, T. Nielsen, L. Mathiesen, J. O. Nielsen, and J. A. Kovacs. 1992. Antibody responses to a major *Pneumocystis carinii* antigen in human immunodeficiency virus-infected patients with and without *P. carinii* pneumonia. *J. Infect. Dis.* **165:** 1151–1155.

94. Magliani, W., S. Conti, S. Arseni, R. Frazzi, A. Salati, and L. Polonelli. 2001. Killer anti-idiotypes in the control of fungal infections. *Curr. Opin. Investig. Drugs* **2:** 477–479.

95. Magliani, W., S. Conti, A. Salati, S. Arseni, L. Ravanetti, R. Frazzi, and L. Polonelli. 2003. Biotechnological approaches to the production of idiotypic vaccines and antiidiotypic antibiotics. *Curr. Pharm. Biotechnol.* **4:**91–97.

96. Maitta, R., K. Datta, Q. Chang, R. Luo, K. Subramaniam, B. Witover, and L. Pirofski. 2004. Protective and non-protective human IgM monoclonal antibodies to *Cryptococcus neoformans* glucuronoxylomannan manifest different specificity and gene use. *Infect. Immun.* **72:**4810–4818.

97. Maitta, R., K. Datta, and L. Pirofski. 2004. Efficacy of immune sera from human immunoglobulin transgenic mice immunized with a peptide mimotope of *Cryptococcus neoformans* glucuronoxylomannan. *Vaccine* **22/28–29:**4062–4068.

98. Maitta, R. W., K. Datta, A. Lees, S. S. Belouski, and L. A. Pirofski. 2004. Immunogenicity and efficacy of *Cryptococcus neoformans* capsular polysaccharide glucuronoxylomannan peptide mimotope-protein conjugates in human immunoglobulin transgenic mice. *Infect. Immun.* **72:**196–208.

99. Malbran, A., L. Belmonte, B. Ruibal-Ares, P. Bare, I. Massud, C. Parodi, M. Felippo, R. Hodinka, K. Haines, K. E. Nichols, and M. M. de Bracco. 2004. Loss of circulating $CD27^+$ memory B cells and $CCR4^+$ T cells occurring in association with elevated EBV loads in XLP patients surviving primary EBV infection. *Blood* **103:**1625–1631.

100. Marcotte, H., D. Levesque, K. Delanay, A. Bourgeault, R. de la Durantaye, S. Brochu, and M. C. Lavoie. 1996. *Pneumocystis carinii* infection in transgenic B cell-deficient mice. *J. Infect. Dis.* **173:**1034–1037.

101. Marquis, G., S. Montplaisir, M. Pelletier, S. Mousseau, and P. Auger. 1985. Genetic resistance to murine cryptococcosis: increased susceptibility in the CBA/N XID mutant strain of mice. *Infect. Immun.* **47:**282–287.

102. Marshall, E. 1995. Dispute slows paper on "remarkable" vaccine. *Science* **268:**1712–1715.

103. Marten, N. W., S. A. Stohlman, and C. C. Bergmann. 2001. MHV infection of the CNS: mechanisms of immune-mediated control. *Viral Immunol.* **14:**1–18.

104. Martinez, L. R., D. Moussai, and A. Casadevall. 2004. Antibody to *Cryptococcus neoformans* glucuronoxylomannan inhibits the release of capsular antigen. *Infect. Immun.* **72:**3674–3679.

105. Matthews, R., and J. Burnie. 1989. Cloning of a DNA sequence encoding a major fragment of the 47 kilodalton stress protein homologue of *Candida albicans. FEMS Microbiol. Lett.* **51:**25–30.

106. Matthews, R., J. Burnie, D. Smith, I. Clark, J. Midgley, M. Conolly, and B. Gazzard. 1988. *Candida* and AIDS: evidence for protective antibody. *Lancet* **ii:**263–266.

107. Matthews, R., J. P. Burnie, and W. Lee. 1991. The application of epitope mapping in the development of a new serological test for systemic candidosis. *J. Immunol. Methods* **143:**73–79.

108. Matthews, R., S. Hodgetts, and J. Burnie. 1995. Preliminary assessment of a human recombinant antibody fragment to hsp90 in murine invasive candidiasis. *J. Infect. Dis.* **171:**1668–1671.

109. Matthews, R., C. Wells, and J. P. Burnie. 1988. Characterisation and cellular localisation of the immunodominant 47-kda antigen of *Candida albicans. J. Med. Microbiol.* **27:**227–232.

110. Matthews, R. C., J. P. Burnie, D. Howat, T. Rowland, and F. Walton. 1991. Autoantibody to heat-shock

protein 90 can mediate protection against systemic candidosis. *Immunology* **74**:20–24.

111. **Matthews, R. C., J. P. Burnie, and S. Tabaqchali.** 1984. Immunoblot analysis of the serological response in systemic candidosis. *Lancet* **ii**:1415–1418.

112. **Matthews, R. C., J. P. Burnie, and S. Tabaqchali.** 1987. Isolation of immunodominant antigens from sera of patients with systemic candidiasis and characterization of serological response to *Candida albicans*. *J. Clin. Microbiol.* **25**:230–237.

113. **Matthews, R. C., G. Rigg, S. Hodgetts, T. Carter, C. Chapman, C. Gregory, C. Illidge, and J. Burnie.** 2003. Preclinical assessment of the efficacy of mycograb, a human recombinant antibody against fungal HSP90. *Antimicrob. Agents Chemother.* **47**:2208–2216.

114. **May, R. J., D. O. Beenhouwer, and M. D. Scharff.** 2003. Antibodies to keyhole limpet hemocyanin cross-react with an epitope on the polysaccharide capsule of *Cryptococcus neoformans* and other carbohydrates: implications for vaccine development. *J. Immunol.* **171**:4905–4912.

115. **McGaw, T. G., and T. R. Kozel.** 1979. Opsonization of *Cryptococcus neoformans* by human immunoglobulin G: masking of immunoglobulin G by cryptococcal polysaccharide. *Infect. Immun.* **25**:262–267.

116. **McLean, G. R., M. Torres, N. Elguezabal, A. Nakouzi, and A. Casadevall.** 2002. Isotype can affect the fine specificity of an antibody for a polysaccharide antigen. *J. Immunol.* **169**:1379–1386.

117. **Mednick, A. J., M. Feldmesser, J. Rivera, and A. Casadevall.** 2003. Neutropenia alters lung cytokine production in mice and reduces their susceptibility to pulmonary cryptococcosis. *Eur. J. Immunol.* **33**:1744–1753.

118. **Miyakawa, Y., K. Kagaya, Y. Fukazawa, and G. Soe.** 1986. Production and characterization of agglutinating monoclonal antibodies against predominant antigenic factors for *Candida albicans*. *J. Clin. Microbiol.* **23**:881–886.

119. **Monari, C., A. Casadevall, C. Retini, F. Baldelli, F. Bistoni, and A. Vecchiarelli.** 1999. Antibody to capsular polysaccharide enhances the function of neutrophils from patients with AIDS against *Cryptococcus neoformans*. *AIDS* **13**:653–660.

120. **Monga, D. P., R. Kumar, L. N. Mohapatra, and A. N. Malaviya.** 1979. Experimental cryptococcosis in normal and B-cell-deficient mice. *Infect. Immun.* **26**:1–3.

121. **Montagnoli, C., S. Bozza, A. Bacci, R. Gaziano, P. Mosci, J. Morschhaauser, L. Pitzurra, M. Kopf, J. Cutler, and L. Romani.** 2003. A role for antibodies in the generation of memory antifungal immunity. *Eur. J. Immunol.* **33**:1183–1192.

122. **Moragues, M. D., M. J. Omaetxebarria, N. Elguezabal, M. J. Sevilla, S. Conti, L. Polonelli, and J. Ponton.** 2003. A monoclonal antibody directed against a *Candida albicans* cell wall mannoprotein exerts three anti-*C. albicans* activities. *Infect. Immun.* **71**:5273–5279.

123. **Mozaffarian, N., J. W. Berman, and A. Casadevall.** 1995. Immune complexes increase nitric oxide production by interferon-gamma-stimulated murine macrophage-like J774.16 cells. *J. Leukos. Biol.* **57**:657–662.

124. **Mukherjee, J., A. Casadevall, and M. D. Scharff.** 1993. Molecular characterization of the antibody responses to *Cryptococcus neoformans* infection and glucuronoxylomannan-tetanus toxoid conjugate immunization. *J. Exp. Med.* **177**:1105–1116.

125. **Mukherjee, J., M. Feldmesser, M. D. Scharff, and A. Casadevall.** 2000. Monoclonal antibodies to *Cryptococcus neoformans* glucuronoxylomannan enhance fluconazole efficacy. *Antimicrob. Agents Chemother.* **39**:1398–1405.

126. **Mukherjee, J., T. R. Kozel, and A. Casadevall.** 1998. Monoclonal antibodies reveal additional epitopes of serotype D *Cryptococcus neoformans* capsular glucuronoxylomannan that elicit protective antibodies. *J. Immunol.* **161**:3557–3567.

127. **Mukherjee, J., G. Nussbaum, M. D. Scharff, and A. Casadevall.** 1995. Protective and nonprotective monoclonal antibodies to *Cryptococcus neoformans* originating from one B cell. *J. Exp. Med.* **181**:405–409.

128. **Mukherjee, J., L. Pirofski, M. D. Scharff, and A. Casadevall.** 1993. Antibody-mediated protection in mice with lethal intracerebral *Cryptococcus neoformans* infection. *Proc. Natl. Acad. Sci. USA* **90**:3636–3640.

129. **Mukherjee, J., M. D. Scharff, and A. Casadevall.** 1992. Protective murine monoclonal antibodies to *Cryptococcus neoformans*. *Infect. Immun.* **60**:4534–4541.

130. **Mukherjee, J., M. D. Scharff, and A. Casadevall.** 1995. Variable efficacy of passive antibody administration against diverse *Cryptococcus neoformans* strains. *Infect. Immun.* **63**:3353–3359.

131. **Mukherjee, J., L. S. Zuckier, M. D. Scharff, and A. Casadevall.** 1994. Therapeutic efficacy of monoclonal antibodies to *Cryptococcus neoformans* glucuronoxylomannan alone and in combination with amphotericin B. *Antimicrob. Agents Chemother.* **38**:580–587.

132. **Mukherjee, S., S. Lee, J. Mukherjee, M. D. Scharff, and A. Casadevall.** 1994. Monoclonal antibodies to *Cryptococcus neoformans* capsular polysaccharide modify the course of intravenous infection in mice. *Infect. Immun.* **62**:1079–1088.

133. **Mukherjee, S., S. C. Lee, and A. Casadevall.** 1995. Antibodies to *Cryptococcus neoformans* glucuronoxylomannan enhance antifungal activity of murine macrophages. *Infect. Immun.* **63**:573–579.

134. **Nagase, H., K. Agematsu, K. Kitano, M. Takamoto, Y. Okubo, A. Komiyama, and K. Sugane.** 2001. Mechanism of hypergammaglobulinemia by HIV infection: circulating memory B-cell reduction with plasmacytosis. *Clin. Immunol.* **100**:250–259.

135. **Nakouzi, A., and A. Casadevall.** 2003. The function of conserved amino acids in or near the complementarity determining regions for related antibodies to *Cryptococcus neoformans* glucuronoxylomannan. *Mol. Immunol.* **40**:351–361.

136. **Nakouzi, A., P. Valadon, J. D. Nosanchuk, N. Green, and A. Casadevall.** 2001. Molecular basis for immunoglobulin M specificity to epitopes in *Cryptococcus neoformans* polysaccharide that elicit protective and non-protective antibodies. *Infect. Immun.* **69**:3398–3409.

137. **Netski, D., and T. R. Kozel.** 2002. Fc-dependent and Fc-independent opsonization of *Cryptococcus neoformans* by anticapsular monoclonal antibodies: importance of epitope specificity. *Infect. Immun.* **70**:2812–2819.

138. **Nooney, L., R. C. Matthews, and J. P. Burnie.** 2005. Evaluation of Mycograb(R), amphotericin B, caspofungin, and fluconazole in combination against *Cryptococcus neoformans* by checkerboard and time-kill methodologies. *Diagn. Microbiol. Infect. Dis.* **51**:19–29.

139. Nosanchuk, J. D., A. L. Rosas, and A. Casadevall. 1998. The antibody response to fungal melanin in mice. *J. Immunol.* **160:**6026–6031.

140. Nosanchuk, J. D., J. N. Steenbergen, L. Shi, G. S. Deepe, Jr., and A. Casadevall. 2003. Antibodies to a cell surface histone-like protein protect against *Histoplasma capsulatum*. *J. Clin. Investig.* **112:**1164–1175.

141. Nussbaum, G., W. Cleare, A. Casadevall, M. D. Scharff, and P. Valadon. 1997. Epitope location in the *Cryptococcus neoformans* capsule is a determinant of antibody efficacy. *J. Exp. Med.* **185:**685–694.

142. Nussbaum, G., R. Yuan, A. Casadevall, and M. D. Scharff. 1996. Immunoglobulin G3 blocking antibodies to the fungal pathogen *Cryptococcus neoformans*. *J. Exp. Med.* **183:**1905–1909.

143. Pascale, J. M., M. M. Shaw, P. J. Durant, A. A. Amador, M. S. Bartlett, J. W. Smith, R. L. Gregory, and G. L. McLaughlin. 1999. Intranasal immunization confers protection against murine *Pneumocystis carinii* lung infection. *Infect. Immun.* **67:**805–809.

144. Pifer, L. L., W. T. Hughes, S. Stagno, and D. Woods. 1978. *Pneumocystis carinii* infection: evidence for high prevalence in normal and immunosuppressed children. *Pediatrics* **61:**35–41.

145. Pirofski, L. 2001. Polysaccharides, mimotopes and vaccines for encapsulated pathogens. *Trends Microbiol.* **9:**445–452.

146. Pirofski, L., and A. Casadevall. 1996. *Cryptococcus neoformans*: paradigm for the role of antibody in immunity. *Zentbl. Bakteriol.* **284:**475–495.

147. Pirofski, L., R. Lui, M. DeShaw, A. B. Kressel, and Z. Zhong. 1995. Analysis of human monoclonal antibodies elicited by vaccination with a *Cryptococcus neoformans* glucuronoxylomannan capsular polysaccharide vaccine. *Infect. Immun.* **63:**3005–3014.

148. Pitzurra, L., S. Perito, F. Baldelli, F. Bistoni, and A. Vecchiarelli. 2003. Humoral response against *Cryptococcus neoformans* mannoprotein antigens in HIV-infected patients. *Clin. Exp. Immunol.* **133:**91–96.

149. Polonelli, L. 2000. Human, monoclonal and recombinant candidacidal, pneumocysticidal and mycobactericidal antibodies. *Indian J. Chest Dis. Allied Sci.* **42:**259–263.

150. Polonelli, L., F. De Bernadis, S. Conti, M. Boccanera, W. Magliani, M. Gerloni, C. Cantelli, and A. Cassone. 1996. Human natural yeast killer toxin-like candidacidal antibodies. *J. Immunol.* **156:**1880–1885.

151. Polonelli, L., F. De Bernardis, S. Conti, M. Boccanera, M. Gerloni, G. Morace, W. Magliani, C. Chezzi, and A. Cassone. 1994. Idiotypic intravaginal vaccination to protect against candidal vaginitis by secretory, yeast killer toxin-like anti-idiotypic antibodies. *J. Immunol.* **152:**3175–3182.

152. Polonelli, L., R. Lorenzini, F. De Bernardis, M. Gerloni, S. Conti, G. Morace, W. Magliani, and C. Chezzi. 1993. Idiotypic vaccination: immunoprotection mediated by anti-idiotypic antibodies with antibiotic activity. *Scand. J. Immunol.* **37:**105–110.

153. Polonelli, L., W. Magliani, S. Conti, L. Bracci, L. Lozzi, P. Neri, D. Adriani, F. De Bernardis, and A. Cassone. 2003. Therapeutic activity of an engineered synthetic killer antiidiotypic antibody fragment against experimental mucosal and systemic candidiasis. *Infect. Immun.* **71:**6205–6212.

154. Polonelli, L., and G. Morace. 1987. Production and characterization of yeast killer toxin monoclonal antibodies. *J. Clin. Microbiol.* **25:**460–462.

155. Polonelli, L., and G. Morace. 1988. Yeast killer toxin-like anti-idiotypic antibodies. *J. Clin. Microbiol.* **26:**602–604.

156. Porsius, J. C., H. J. van Vliet, J. H. van Zeijl, W. H. Goessens, and M. F. Michel. 1990. Detection of an antibody response in immunocompetent patients with systemic candidiasis or *Candida albicans* colonisation. *Eur. J. Clin. Microbiol. Infect. Dis.* **9:**352–355.

157. Ramakrishna, C., S. A. Stohlman, R. D. Atkinson, M. J. Shlomchik, and C. C. Bergmann. 2002. Mechanisms of central nervous system viral persistence: the critical role of antibody and B cells. *J. Immunol.* **168:**1204–1211.

158. Retini, C., A. Casadevall, D. Pietrella, C. Monari, B. Palazzetti, and A. Vecchiarelli. 1999. Specific activated T cells regulate IL-12 production by human monocytes stimulated with *Cryptococcus neoformans*. *J. Immunol.* **162:**1618–1623.

159. Rivera, J., J. Mukherjee, L. M. Weiss, and A. Casadevall. 2002. Antibody efficacy in murine pulmonary *Cryptococcus neoformans* infection: a role for nitric oxide. *J. Immunol.* **168:**3419–3427.

160. Rivera, J., O. Zaragoza, and A. Casadevall. 2005. Antibody-mediated protection against *Cryptococcus neoformans* is dependent on B cells. *Infect. Immun.* **73:**1141–1150.

161. Robbins, J. B., R. Schneerson, and S. C. Szu. 1995. Perspective: hypothesis: serum IgG antibody is sufficient to confer protection against infectious diseases by inactivating the inoculum. *J. Infect. Dis.* **171:**1387–1398.

162. Rodrigues, M. L., L. R. Travassos, K. R. Miranda, A. J. Franzen, S. Rozental, W. de Souza, C. S. Alviano, and E. Barreto-Bergter. 2000. Human antibodies against a purified glucosylceramide from *Cryptococcus neoformans* inhibit cell budding and fungal growth. *Infect. Immun.* **68:**7049–7060.

163. Rosas, A. L., J. D. Nosanchuk, and A. Casadevall. 2001. Passive immunization with melanin-binding monoclonal antibodies prolongs survival of mice with lethal *Cryptococcus neoformans* infection. *Infect. Immun.* **69:**3410–3412.

164. San Millan, R., N. Elguezabal, P. Regulez, M. D. Moragues, G. Quindos, and J. Ponton. 2000. Effect of salivary secretory IgA on the adhesion of *Candida albicans* to polystyrene. *Microbiology* **146:**2105–2112.

165. San Millan, R., P. A. Ezkurra, G. Quindos, R. Robert, J. M. Senet, and J. Ponton. 1996. Effect of monoclonal antibodies directed against *Candida albicans* cell wall antigens on the adhesion of the fungus to polystyrene. *Microbiology* **142:**2271–2277.

166. Scamurra, R. W., D. J. Miller, M. Abrahamsen, V. Kapur, S. M. Wahl, E. C. Milner, and E. N. Janoff. 2000. Impact of HIV-1 infection on VH3 gene repertoire of naive human B cells. *J. Immunol.* **164:**5482–5491.

167. Schlageter, A. M., and T. R. Kozel. 1990. Opsonization of *Cryptococcus neoformans* by a family of isotype-switch variant antibodies specific for the capsular polysaccharide. *Infect. Immun.* **58:**1914–1918.

168. Seguy, N., L. Polonelli, E. Dei-Cas, and J. C. Cailliez. 1998. Effect of a killer toxin of *Pichia anomala* to *Pneumocystis*. Perspectives in the control of pneumocystosis. *FEMS Immunol. Med. Microbiol.* **22:**145–149.

169. Shapiro, S., D. O. Beenhouwer, M. Feldmesser, C. Taborda, M. C. Carroll, A. Casadevall, and M. D. Scharff. 2002. Immunoglobulin G monoclonal antibodies to *Cryptococcus neoformans* protect mice deficient in

complement component C3. *Infect. Immun.* 70: 2598–2604.

170. Shen, H., J. K. Whitmire, X. Fan, D. J. Shedlock, S. M. Kaech, and R. Ahmed. 2003. A specific role for B cells in the generation of CD8 T cell memory by recombinant *Listeria monocytogenes. J. Immunol.* 170:1443–1451.

171. Shin, J. S., J. Yu, J. Lin, L. Zhong, K. L. Bren, and M. H. Nahm. 2002. Peptide mimotopes of pneumococcal capsular polysaccharide of 6B serotype: a peptide mimotope can bind to two unrelated antibodies. *J. Immunol.* 168: 6273–6278.

172. Sinha, B. K., S. Prasad, and D. P. Monga. 1987. Studies of the role of B-cells in the resistance of mice to experimental candidiasis. *Zentbl. Bakteriol. Mikrobiol. Hyg. Ser. A* 266:316–322.

173. Speed, B. R., J. Kaldor, B. Cairns, and M. Pegorer. 1996. Serum antibody response to active infection with *Cryptococcus neoformans* and its varieties in immunocompetent subjects. *J. Med. Vet. Mycol.* 34:187–193.

174. Sundstrom, J. B., and R. Cherniak. 1992. The glucuronoxylomannan of *Cryptococcus neoformans* serotype A is a type 2 T-cell-independent antigen. *Infect. Immun.* 60:4080–4087.

175. Swoboda, R. K., G. Bertram, S. Budge, G. W. Gooday, N. A. Gow, and A. J. Brown. 1995. Structure and regulation of the HSP90 gene from the pathogenic fungus *Candida albicans. Infect. Immun.* 63:4506–4514.

176. Syme, R. M., J. C. Spurrell, E. K. Amankwah, F. H. Green, and C. H. Mody. 2002. Primary dendritic cells phagocytose *Cryptococcus neoformans* via mannose receptors and Fcγ receptor II for presentation to T lymphocytes. *Infect. Immun.* 70:5972–5981.

177. Tabone, M. D., G. Leverger, J. Landman, C. Aznar, L. Boccon-Gibon, and G. Lasfargues. 1994. Disseminated lymphonodular cryptococcosis in a child with X-linked hyper-IgM immunodeficiency. *Pediatr. Infect. Dis. J.* 13:77–79.

178. Taborda, C., and A. Casadevall. 2001. Immunoglobulin M efficacy against *Cryptococcus neoformans*: mechanism, dose dependence, and prozone-like effects in passive protection experiments. *J. Immunol.* 166: 2100–2107.

179. Taborda, C., J. Rivera, O. Zaragoza, and A. Casadevall. 2003. More is not necessarily better: prozone-like effects in passive immunization with IgG. *J. Immunol.* 170:3621–3620.

180. Taborda, C. P., and A. Casadevall. 2002. CR3 (CD11b/CD18) and CR4 (CD11c/CD18) are involved in complement-independent antibody-mediated phagocytosis of *Cryptococcus neoformans. Immunity* 16:791–802.

181. Tavares, D., P. Ferreira, M. Vilanova, A. Videira, and M. Arala-Chaves. 1995. Immunoprotection against systemic candidiasis in mice. *Int. Immunol.* 7:785–796.

182. Torres, M., R. May, M. D. Scharff, and A. Casadevall. 2005. Variable-region identical antibodies differing in isotype demonstrate differences in fine specificity and idiotype. *J. Immunol.* 174:2132–2142.

183. Travassos, L. R., L. S. Silva, E. G. Rodrigues, S. Conti, A. Salati, W. Magliani, and L. Polonelli. 2004. Therapeutic activity of a killer peptide against experimental paracoccidioidomycosis. *J. Antimicrob. Chemother.* 54:956–958.

184. Valadon, P., G. Nussbaum, L. F. Boyd, D. H. Margulies, and M. D. Scharff. 1996. Peptide libraries define the fine specificity of anti-polysaccharide antibodies to *Cryptococcus neoformans. J. Mol. Biol.* 261:11–22.

185. Valadon, P., G. Nussbaum, J. Oh, and M. D. Scharff. 1998. Aspects of antigen mimicry revealed by immunization with a peptide mimetic of *Cryptococcus neoformans* polysaccharide. *J. Immunol.* 161:1829–1836.

186. van de Moer, A., S. L. Salhi, R. Cherniak, B. Pau, M. L. Garrigues, and J. M. Bastide. 1990. An anti-*Cryptococcus neoformans* monoclonal antibody directed against galactoxylomannan. *Res. Immunol.* 141:33–42.

187. van Deventer, H. J., W. H. Goessens, A. J. van Vliet, and H. A. Verbrugh. 1996. Anti-enolase antibodies partially protective against systemic candidiasis in mice. *Clin. Microbiol. Infect.* 2:36–43.

188. Vargas-Madrazo, E., F. Lara-Ochoa, and J. C. Almagro. 1995. Canonical structure repertoire of the antigen-binding site of immunoglobulins suggests strong geometrical restrictions associated to the mechanism of immune recognition. *J. Mol. Biol.* 254:497–504.

189. Vecchiarelli, A., C. Monari, C. Retini, D. Pietrella, B. Palazzetti, L. Pitzurra, and A. Casadevall. 1998. *Cryptococcus neoformans* differently regulates B7-1 (CD80) and B7-2 (CD86) expression on human monocytes. *Eur. J. Immunol.* 28:114–121.

190. Vecchiarelli, A., D. Pietrella, F. Bistoni, T. R. Kozel, and A. Casadevall. 2002. Antibody to *Cryptococcus neoformans* capsular glucuronoxylomannan promotes expression of interleukin-12Rbeta2 subunit on human T cells in vitro through effects mediated by antigen-presenting cells. *Immunology* 106:267–272.

191. Vecchiarelli, A., C. Retini, A. Casadevall, C. Monari, D. Pietrella, and T. R. Kozel. 1998. Involvement of C3a and C5a in interleukin-8 secretion by human polymorphonuclear cells in response to capsular material of *Cryptococcus neoformans. Infect. Immun.* 66: 4324–4330.

192. Vecchiarelli, A., C. Retini, C. Monari, and A. Casadevall. 1998. Specific antibody to *Cryptococcus neoformans* alters human leukocyte cytokine synthesis and promotes T-cell proliferation. *Infect. Immun.* 66:1244–1247.

193. Vilanova, M., L. Teixeira, I. Caramalho, E. Torrado, A. Marques, P. Madureira, A. Ribeiro, P. Ferreira, M. Gama, and J. Demengeot. 2004. Protection against systemic candidiasis in mice immunized with secreted aspartic proteinase 2. *Immunology* 111:334–342.

194. Viudes, A., A. Lazzell, S. Perea, W. R. Kirkpatrick, J. Peman, T. F. Patterson, J. P. Martinez, and J. L. Lopez-Ribot. 2004. The C-terminal antibody binding domain of *Candida albicans* mp58 represents a protective epitope during candidiasis. *FEMS Microbiol. Lett.* 232:133–138.

195. Weller, S., M. C. Braun, B. K. Tan, A. Rosenwald, C. Cordier, M. E. Conley, A. Plebani, D. S. Kumararatne, D. Bonnet, O. Tournilhac, G. Tchernia, B. Steiniger, L. M. Staudt, J. L. Casanova, C. A. Reynaud, and J. C. Weill. 2004. Human blood IgM "memory" B cells are circulating splenic marginal zone B cells harboring a prediversified immunoglobulin repertoire. *Blood.* 104: 3647–3654.

196. Wellington, M., J. M. Bliss, and C. G. Haidaris. 2003. Enhanced phagocytosis of *Candida* species mediated by opsonization with a recombinant human antibody single-chain variable fragment. *Infect. Immun.* 71:7228–7231.

197. Wells, J., F. Gigliotti, P. J. Simpson-Haidaris, and C. G. Haidaris. 2004. Epitope mapping of a protective monoclonal antibody against *Pneumocystis carinii* with

shared reactivity to *Streptococcus pneumoniae* surface antigen PspA. *Infect. Immun.* **72:**1548–1556.

198. **Wilson, M. A., and T. R. Kozel.** 1992. Contribution of antibody in normal human serum to early deposition of C3 onto encapsulated and nonencapsulated *Cryptococcus neoformans*. *Infect. Immun.* **60:**754–761.

199. **Winkelstein, J. A., M. C. Marino, H. Ochs, R. Fuleihan, P. R. Scholl, R. Geha, E. R. Stiehm, and M. E. Conley.** 2003. The X-linked hyper-IgM syndrome: clinical and immunologic features of 79 patients. *Medicine* **82:**373–384.

200. **Wuthrich, M., and B. S. Klein.** 2000. Investigation of anti-WI-1 adhesin antibody-mediated protection in experimental pulmonary blastomycosis. *J. Infect. Dis.* **181:**1720–1728.

201. **Xu, J. L., and M. M. Davis.** 2000. Diversity in the CDR3 region of V(H) is sufficient for most antibody specificities. *Immunity* **13:**37–45.

202. **Young, A. C., P. Valadon, A. Casadevall, M. D. Scharff, and J. C. Sacchettini.** 1997. The three dimensional structures of a polysaccharide binding antibody to *Cryptococcus neoformans* and its complex with a peptide from a phage library: implications for the identification of peptide mimotopes. *J. Mol. Biol.* **74:**22–34.

203. **Yuan, R., A. Casadevall, and M. D. Scharff.** 1997. T cells cooperate with passive antibody to modify the course of *Cryptococcus neoformans* in mice. *Proc. Natl. Acad. Sci. USA* **94:**2483–2488.

204. **Zhang, H., Z. Zhong, and L. Pirofski.** 1997. Peptide epitopes recognized by a human anti-cryptococcal glucuronoxylomannan antibody. *Infect. Immun.* **65:**1158–1164.

205. **Zhang, M. X., J. E. Cutler, Y. Han, and T. R. Kozel.** 1998. Contrasting roles of mannan-specific monoclonal immunoglobulin M antibodies in the activation of classical and alternative pathways by *Candida albicans*. *Infect. Immun.* **66:**6027–6029.

206. **Zhang, M. X., and T. R. Kozel.** 1998. Mannan-specific immunoglobulin G antibodies in normal human serum accelerate binding of C3 to *Candida albicans* via the alternative complement pathway. *Infect. Immun.* **66:**4845–4850.

207. **Zhang, M. X., D. M. Lupan, and T. R. Kozel.** 1997. Mannan-specific immunoglobulin G antibodies in normal human serum mediate classical pathway initiation of C3 binding to *Candida albicans*. *Infect. Immun.* **65:**3822–3827.

208. **Zhong, Z., and L. Pirofski.** 1996. Opsonization of *Cryptococcus neoformans* by human anti-glucuronoxylomannan antibodies. *Infect. Immun.* **64:**3446–3450.

209. **Zhong, Z., and L. Pirofski.** 1998. Antifungal activity of a human antiglucuronoxylomannan antibody. *Clin. Diagn. Lab. Immunol.* **5:**58–64.

Molecular Principles of Fungal Pathogenesis
Edited by Joseph Heitman et al.
©2006 ASM Press, Washington, D.C.

Chapter 34

Site-Specific Mucosal Immunity to Fungi: Lessons Learned from *Candida albicans* Applied to Other Fungi

PAUL L. FIDEL, JR., AND BEATRIZ FINKEL-JIMENEZ

Candida albicans is a commensal organism of the gastrointestinal (GI) tract of most human adults as well as the genital tract of adult females. Point prevalence studies show asymptomatic colonization by *C. albicans* in the vaginas of 15 to 25% of healthy women and in the mouths and GI tracts of up to 60 and 90% of healthy adults, respectively (reviewed in reference 69). However, *C. albicans* can also be a pathogen of these same mucosal tissues and produce infections in both immunocompetent and immunocompromised individuals. Prior to the human immunodeficiency virus (HIV) epidemic, host defense against *C. albicans* at these mucosal sites was largely considered one-dimensional. Although general anatomical distinctions were always understood relative to immunity and host defense, a concept that "all mucosal sites were equal" was often implied for mucosal candidiasis whereby discussions of host defense at one mucosal site were generally applied to other mucosal sites as well. This was primarily due to the lack of large populations of individuals with a type of mucosal candidiasis at any one geographical site that could be studied in-depth. Oral candidiasis was relatively uncommon until transplantation and other forms of therapeutic immunosuppression became more evident. However, the numbers of patients available to study at any one place remained very small. Vaginal candidiasis has always been common but has never been taken seriously by the research community as a mucosal fungal infection because of anecdotal causes and treatment regimens. GI tract infections are difficult to diagnose and usually not studied in humans.

Prior to in-depth analyses of specific mucosal sites, the majority of mucosal *Candida* infections were thought to be associated with some form of immunoglobulin A (IgA) antibody or humoral immune deficiency. However, when antibody deficiency was difficult to demonstrate (110, 135, 149), it was the syndrome of chronic mucocutaneous candidiasis (CMC) that stimulated the next series of explanations. Since CMC was considered to be due to a deficiency in peripheral (blood) T-cell-mediated immunity (108–111), most other mucosal *Candida* infections were also considered to be caused by a similar T-cell deficiency despite their tendency to be site specific and not all-encompassing.

The HIV epidemic created large numbers of individuals with severe immunosuppression and a significant increase in mucosal candidiasis. Accordingly, mucosal candidiasis became a means to refer predominantly to oropharyngeal candidiasis (OPC) or esophageal candidiasis since they were being seen most often. In fact, esophageal candidiasis is considered an AIDS-defining illness (114, 132, 153). And as much as vulvovaginal candidiasis (VVC) was considered common in HIV-infected women and a possible AIDS-defining illness at one point (29, 62, 179), it soon became clear that vaginal candidiasis, while common, was really no more common than in HIV-negative women (47, 98, 124, 162, 174, 191). This finding, together with basic immunology research that was coming of age for vaginal candidiasis (reviewed in reference 79), began to suggest that factors associated with susceptibility to oral and vaginal candidiasis were different. Following some additional studies in HIV-infected persons with OPC, immunological data too began to follow the physiological anatomical distinctions and revealed that host defenses against *Candida* at the oral and vaginal mucosas were unique, distinct, and independent (124). As a result, it is now considered that all mucosas are not equal for mucosal *Candida* infections and that host defense at the various mucosal

Paul L. Fidel, Jr., and Beatriz Finkel-Jimenez • Department of Microbiology, Immunology, and Parasitology, Louisiana State University Health Sciences Center, New Orleans, LA 70112.

sites needs to be studied independently and exclusively at the local level with respect to cells and soluble immune factors.

The vast amount of data accumulated over the past 15 years showing distinct T-cell percentages or composition at different mucosal sites (e.g., vagina and GI tract), especially in the intraepithelial layers, antigen-presenting cells (e.g., dendritic cells, macrophages, Langerhans' cells), or B cells (81, 95, 97, 99, 100, 102, 145), have wholly supported the concept that mucosal sites are indeed independent and unique relative to host immune reactivity. Thus, a clear divergence of immunity at mucosal sites has taken shape. There is perhaps more dogma with reference to host defense against OPC cell-mediated immunity (CMI) by CD4$^+$ Th1-type responses, yet other nonconventional host defenses are also critical. On the other hand, natural host defense against VVC does not appear to include T cells or any type of adaptive immunity but relies on innate immunity for both protection and susceptibility to infection.

This chapter uses *Candida* as the example to emphasize the uniqueness of mucosal immunity to fungi. Following a general discussion of the dogmatic host defense mechanisms against *Candida albicans*, a fairly detailed discussion of host defense against oral, vaginal, and GI candidiasis is reviewed to stress the uniqueness of the responses at each anatomical site and what this implies for immunity to other fungal organisms that infect multiple sites. The chapter concludes with a discussion of other fungi that infect multiple sites and on which studies on host defense have been focused to date, together with a road map for research on immunity to fungi that affect different mucosal tissues or anatomical sites.

HOST RESPONSE TO *C. ALBICANS*: A DOGMATIC VIEW

Innate Resistance

The polymorphonuclear leukocyte (PMN) is the first line of defense against *C. albicans*. The role of macrophages in defense is less clear, although they readily phagocytize the yeast form of *Candida*. Anti-*Candida* activities by PMN or macrophages can be oxidative (e.g., hydrogen peroxide, nitric oxide, and superoxide dismutase) or nonoxidative (i.e., lactoferrin, lysozyme, calprotectin, and defensins) (reviewed in reference 68). Phagocytosis is usually required for oxidative mechanisms, although there is some evidence that PMN can secrete oxidative products that affect nearby fungal hyphal forms too large to be phagocytized (126).

PMN are relatively effective as a first line of defense, providing that organisms gain access to the vasculature or that the PMN attach to hyphae. However, PMN function can be greatly enhanced by cytokines such as tumor necrosis factor alpha (TNF-α), gamma interferon (IFN-γ), interleukin-2 (IL-2), IL-8, and colony-stimulating factors (59). Oxidative mechanisms are enhanced in the presence of serum as well. There are also both classical and alternate complement pathways that become activated by *C. albicans* which promote increased adherence to erythrocytes (63, 94, 115).

The functional mechanisms by which monocytes and macrophages affect *C. albicans* are less well defined (reviewed in reference 68). In general, once phagocytized in vitro, killing is at least as effective as that by PMNs and involves oxidative mechanisms and candidacidal molecules. Activation of macrophages with cytokines appears to increase the killing activity, although the effects appear short-lived at best. Killing of both blastospores and hyphae has been demonstrated, but the rate and efficiency depend on the source of the cells. For example, pulmonary macrophages appear to be particularly efficient against *C. albicans*.

A role for dendritic cells against fungal infections has also been suggested. Immature myeloid dendritic cells can phagocytose both yeast and hyphae of *C. albicans* (166). Yeast forms are recognized through the mannose-fucose and complement receptor 3 (CR3) receptors, while hyphae are recognized through the FcγR and CR3 receptors (166). Mature dendritic cells are less phagocytic than immature cells, but they have increased expression of major histocompatibility complex (MHC) class II, which leads to more efficient antigen presentation (10, 157, 161). Dendritic cells are also capable of directing the acquired immune response to *Candida*, depending on the form phagocytosed and the receptors used, and the type of response induced can affect whether a protective or nonprotective response is generated (166).

The activity of NK cells against *C. albicans* has been controversial. While a limited number of studies showed inhibition of *C. albicans* growth by murine leukocyte suspensions containing NK cells (24, 140, 183), the general consensus is that NK cells are not effective against *Candida* (6, 60). Suffice it to say that although inhibition of *C. albicans* can be mediated by cells from lymphoid tissue, the primary and secondary effector cells are innate, not specific, and consist of PMNs and macrophages, respectively.

A final type of cellular innate resistance mechanism has only recently been discovered; it involves epithelial cells lining the surface of mucosal tissues. Oral and vaginal epithelial cells from mice, macaques, and humans have the ability to inhibit the growth of *Candida* in vitro (12, 183, 184). The inhibition is static rather than cidal (148) and requires contact between viable epithelial cells and *Candida*, with no role for soluble factors or phagocytosis (181, 183).

Humoral Immunity

C. albicans antigens readily induce the production of immunoglobulins, as evidenced by the presence of *Candida*-specific IgG in serum and IgA in mucosal secretions of healthy individuals. Thus, presumably humoral immune mechanisms are important host defenses in both the systemic circulation and at mucosal sites. However, a role for antibody-mediated immunity at mucosal sites has been controversial. A major argument against a role for humoral immunity comes from clinical observations wherein individuals with congenital or acquired B-cell abnormalities are not more susceptible to mucosal (or systemic) candidiasis than are immunocompetent individuals (164). Other observations suggestive of a lack of a role for antibody include the following: (i) levels of *Candida*-specific IgA were higher in the saliva of HIV-positive patients than in that of HIV-negative controls (46, 51, 137); (ii) levels of *Candida*-specific IgA, secretory IgA (sIgA), IgG, or subclasses did not differ in saliva of HIV-positive OPC-positive patients compared to HIV-negative OPC-negative controls (197); (iii) women with recurrent VVC (RVVC) had normal levels of *Candida*-specific IgG in serum and IgA in vaginal secretions (73, 110, 135); (iv) mice protected from a secondary vaginal challenge did not have detectable levels of *Candida*-specific antibodies in vaginal secretions (198); and (v) in vitro studies have provided little evidence that serum antibody and complement can kill *C. albicans* (164) despite increased phagocytosis of the organism in the presence of complement.

Despite the vast amount of information showing little role for *Candida*-specific antibodies or at least little evidence of any deficiency in such naturally induced antibodies, there are considerable data derived from experiments with animals that support a role for *Candida*-specific IgA and IgG antibodies in protection against systemic or mucosal infection (43, 54, 90, 91, 156). These contradictory findings may be reconciled by taking into consideration a theory proposed by Casadevall (40). He suggested that a pool of natural antibodies may consist of a mixture of immunoprotective, nonprotective, and deleterious antibodies. The type that dominates the pool of antibodies dictates the outcome (protection versus infection). If this theory is correct, then "protective" antibodies should be present under conditions of "immunity" and should be identifiable and able to be isolated. Unfortunately, the lack of protection observed in the presence of normal levels of *Candida*-specific antibodies suggests that "protective" antibodies, if they exist, are in the minority of most pools examined. On the other hand, the existence of protection in animal models using "protective" antibodies with known specificities and of a specific class and subclass provides considerable hope for the use of these antibodies for vaccine development.

Cell-Mediated Immunity

The mechanisms associated with CMI against *C. albicans* appear to be dominated by CD4$^+$ T cells. Protection often correlates with delayed-type hypersensitivity (DTH), but a direct role for DTH per se has been difficult to assess. There have been reports where the presence of DTH did not correlate with protection (77, 85). Nevertheless, *Candida*-specific DTH develops in vivo in immunocompetent individuals. DTH is demonstrable at an early age and probably results from early exposure to *C. albicans*. Thus, a pool of *Candida*-specific memory T cells continues to circulate in the peripheral blood. The response is presumably initiated at the mucosal site where antigen-presenting cells such as Langerhans' cells or dendritic cells process and present the antigen to locally associated T cells or transport it to the draining lymph nodes where T cells can be primed. On migration of the *Candida*-specific T cells to the site of antigen deposition and contact with additional processed antigen, cytokine secretion by the T cells activates tissue macrophages and signals other phagocytic cells to infiltrate the site of the insult (mononuclear infiltrate), in an attempt to eliminate or reduce the organism. The current dogma is that Th1-type responses are associated with resistance to intracellular pathogens while Th2-type responses are associated with resistance to extracellular pathogens. However, for *C. albicans*, and in fact most medically important fungi that are predominantly extracellular, Th1-type responses have been associated with resistance to candidal infection, with Th2-type responses showing an association with susceptibility to infection (31, 44, 133, 167, 185, 201).

In contrast to the role played by CD4$^+$ cells in protection against *C. albicans* infections at mucosal tissues, a role for CD8$^+$ cells as an acquired host defense is unknown. To date, there has been no evidence that CD8$^+$ cytotoxic T lymphocytes (CTL) are directly involved in host defense against *C. albicans*. On the other hand, CD8$^+$ T cells have a role in suppression of CMI in candidiasis. There are numerous reports showing that CD8$^+$ T cells specific for *C. albicans* cell wall antigens, both glycoprotein and mannan, suppress DTH and lymphocyte proliferation in infected mice (61, 84, 85). Suppressor or regulatory cells also appear to hinder the ability to clear the infection (85). It appears that the downregulation of DTH includes an antagonistic function of IL-12 p40 and that the production of CD8$^+$ downregulatory T cells is dependent on the presence of IL-10, IL-4, and IFN-γ in the absence of biologically active IL-12 (128, 190). Additionally, naive CD8$^+$ T cells stimulated with IL-2 inhibit the growth of *C. albicans* (15). This was first shown in cells from uninfected mice but more recently has also been suggested clinically to occur in persons who have had a recent episode of OPC (50). The mechanism appears to be non-MHC restricted through the action of excreted granzymes (15, 134) and may in fact be a form of innate immunity.

HOST DEFENSE AGAINST OROPHARYNGEAL CANDIDIASIS

OPC involves infections of the hard and soft palate, tongue, buccal mucosa, and floor of the mouth. It presents as reddened patches (erythematous) or white curd-like lesions (pseudomembranous). Chewing and swallowing can be difficult under these conditions. Infections can be acute or recurrent and are common in immunocompromised patients, especially those infected with HIV. OPC is also a common manifestation of CMC (reviewed in reference 149) and occurs in patients with lymphoma, patients undergoing steroid therapy, patients receiving cytotoxic chemotherapy for cancer, and transplant recipients. Although OPC occurs under several immunocompromising conditions, it appears to be much more common in HIV-infected persons than under any other condition (88, 147, 171, 173). In fact, OPC is often one of the first clinical signs of underlying HIV infection and occurs in 50 to 95% of all HIV-positive persons sometime during their progression to full-blown AIDS (159). Thus, it is possible that a link between HIV and OPC is present that enhances susceptibility to OPC. Interestingly, highly active antiretroviral therapy has reduced the incidence of OPC (150). This reduction is postulated to be due to increased immune responsiveness as well as the action of the protease inhibitors in highly active antiretroviral therapy on the secretory aspartyl proteases (Saps), an important virulence factor of *C. albicans* (32, 42, 89).

Host defense against OPC has only recently begun to be studied in greater detail as a result of the HIV epidemic. Animal models have not been of great value in understanding the natural host defense, since the disease resolves too quickly in most cases with variable and inconsistent fungal titers (reviewed in reference 169). On the other hand, considerable information has accumulated from studies of humans who are immunocompromised. In general, most data to date suggest that CD4$^+$ Th1-type cells are critical for host defense against infection. Clinically, OPC is most common in HIV-positive persons when CD4$^+$ cell numbers drop below 200 cells/μl (88, 147, 159, 173). In vitro immune analyses using peripheral blood mononuclear cells (PBMC) show that cells from most individuals respond to *Candida* antigens with Th1-type cytokines. Thus, consistent with dogma, susceptibility to OPC is enhanced under reduced numbers of CD4$^+$ T cells, due to a lack of Th1-type responses and/or a shift to Th2-type responses. Some early studies of HIV-positive individuals suggested that reduced Th1-type cytokine responses were responsible for OPC (158). However, a study evaluating PBMC reactivity in HIV-negative and -positive persons with or without symptomatic OPC stratified by CD4$^+$ T-cell number showed little to no appreciable differences in *Candida*-specific proliferation or cytokine production between the two groups (124). In another study, no demonstrable deficiencies in clonal responses of the PBMC from such patients to a variety of antigenic peptides were shown (116). These results suggested that the *Candida*-specific T cells themselves were not becoming defective with immunosuppression but that a threshold number of CD4$^+$ T cells was required to protect the oral cavity against infection by this commensal organism. Below this threshold number of cells, local immune mechanisms must function exclusively. The prevalence of OPC may then depend on the status of the local immune mechanisms. Indeed, some individuals with <200 CD4 cells/μl never have OPC while others have recurrent bouts of OPC.

Local immunity first focused on saliva. In support of the Th1/Th2 dichotomy concept, it was recently reported that HIV-negative individuals had Th1/Th0 cytokines in their saliva whereas HIV-positive individuals had primarily Th2-type cytokines, which were present at higher concentrations in patients with OPC (125). Interestingly, the Th2-type profile was the result of reduced amounts of Th1-type cytokines rather than increased amounts of Th2-type cytokines. Subsequent studies investigated tissue lymphocytes, where both CD4$^+$ and CD8$^+$ cells have been identified in OPC lesions (165). Other investigators suggested that only CD8$^+$ cells were present (144, 165, 192). Data from our laboratory support the latter observation, and, in fact, we have shown an accumulation of CD8$^+$ T cells at a considerable distance from *Candida* located superficially at the outer epithelium (143). This cellular accumulation is not seen in individuals without OPC. This finding suggested a role for CD8$^+$ T cells against infection, with a potential problem in cell trafficking leading to OPC.

The latest data continue to support a role for CD8[+] T cells. Immunohistochemical analyses showed that the majority of the cells possess the αβ T-cell receptor (TCR), although considerable numbers of γδ TCR[+] cells are also present. In those with OPC, the numbers of αβ TCR[+] cells are significantly increased whereas the numbers of γδ TCR[+] cells increased only slightly. Thus, the primary leukocytes acting in response to the infection when CD4[+] T-cell numbers are reduced and unavailable are thought to be αβ TCR[+] CD8[+] T cells. Additional studies showed that the CD8 antigen was composed of the standard αβ heterodimer that is thymically derived rather than the αα homodimer often found on CD8 cells in GI tissue and thought to be extrathymically derived (163, 202, 135a). Also examined were homing receptors on the CD8[+] T cells. Cellular migration is controlled by heterodimer homing receptors made up of single-chain integrins and reciprocal adhesion molecules present on tissue. While some homing receptors govern the migration of cells out of blood and into mucosal tissues, others govern migration through mucosal tissues. For example, α4β7 on T cells binds to MAdCAM on tissue for migration into mucosal tissue, α4β1 on T cells binds to VCAM-1 on tissue to migrate into lymphoid tissue, and αeβ7 on T cells binds to E-cadherin on tissue to migrate through mucosal tissue. The CD8[+] T cells present in both OPC-negative and -positive tissue have positive integrin expression with different combinations of α and β chains resulting in different degrees of individual homing receptors (e.g., α4β7, α4β1, and αeβ7). Although difficult to evaluate with several possible combinations, there was no discernable difference in OPC-negative versus OPC-positive tissue except in the increased number of cells in OPC-positive tissue with homing receptors. Thus, there appears to be no deficiency in homing receptors on the CD8[+] T cells. In contrast, there were differences in adhesion molecule expression. MAdCAM expression is significantly increased in tissue of OPC patients, in support of the increased presence of T cells. The level of E-cadherin, on the other hand, is significantly decreased in tissue from OPC patients (135a). Interestingly, the expression of E-cadherin is localized to the epithelium, which is critical for migration of cells to the outer epithelium, where Candida infection is primarily located. Therefore, the decrease in the amount of E-cadherin may limit the ability of the CD8[+] T cells to migrate to the outer epithelium and as such represents a possible dysfunction associated in those with OPC. This E-cadherin deficiency would provide a reasonable explanation for the accumulation of cells at the lamina propria/epithelium interface and play a role in the susceptibility to OPC, in support of our underlying hypothesis. This general hypothesis is illustrated in Fig. 1.

Studies using experimental models support a role for a combination of T cells and innate cells against OPC. Both CD4[+] and CD8[+] T cells were shown to be recruited along with macrophages into the mucosal tissue, and intraepithelial CD4[+] T cells persisted after resolution of the infection (1, 45). Moreover, there was a time-dependent recruitment of γ/δ TCR[+] cells that correlated with the resolution of the disease (45, 65). Interestingly, a murine AIDS model showed a rate of 30% recurrent OPC in inoculated mice, with a predominance of CD8[+] T cells recruited into the tissues (58). Analysis of cytokines showed a role for both Th1- and Th2-type cytokines in resistance to infection, depending on the strain of mice used (65). A recent study examined cytokines in oral tissues from infected mice and found increases in IL-6, IFN-γ, and TNF-α levels in mice recovering from oral infection (67). However, no IL-2, IL-4, or IL-10 was detected during infection (67). In a recent study, in vivo cellular depletions were used to create immunocompromised mice. CD4[+] T cells, PMN, and macrophages all appeared to be important in resistance to disease (66).

Humoral immunity, on the other hand, again does not appear to play a role in protection against or susceptibility to OPC. Early studies of Candida-specific antibodies in the saliva of HIV-positive persons with or without OPC yielded similar or elevated levels of IgA or IgG, although in one study a reduced affinity of Candida-specific IgA antibodies was found in AIDS patients (51). There is no evidence to date, however, that a deficiency in Candida-specific antibodies is present in HIV-positive persons that could account for the increased prevalence of OPC (137, 199). A recent comprehensive analysis of Candida-specific IgA and IgG in the saliva of a large number of HIV-negative and HIV-positive persons stratified by OPC status as well as CD4[+] T-cell numbers, which included subclass analysis, supports the earlier findings (197).

Other anti-OPC mechanisms involve innate cellular defenses. PMN appear to play a role, since neutropenic patients are susceptible to OPC (5). However, in-depth studies with PMN in the oral cavity have not been conducted. Epithelial cells represent another cell type with potential innate function. As stated above, epithelial cells inhibit the growth of Candida species in vitro by a static mechanism (148, 181, 183). While both oral and vaginal epithelial

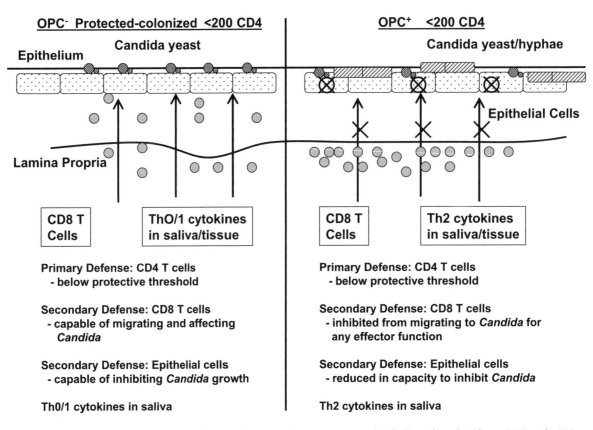

Figure 1. Proposed immune function of the oral mucosa in HIV-positive individuals with and without OPC and <200 CD4 cells/μl. (Left) Protective mechanisms associated with the HIV-positive OPC-negative individual colonized with *C. albicans* when the primary defense of CD4⁺ T cells is below the protective threshold (200 cells/μl). The secondary defense consists of migration of CD8⁺ T cells to the outer epithelium and functional oral epithelial cell anti-*Candida* activity, which together keep *Candida* in check and prevent *Candida* infection. Th1 cytokines in saliva may also act in protection against infection, along with cytokines in tissue. (Right) Condition of OPC. In this scenario, where CD4⁺ T-cell numbers are below the protective threshold, CD8⁺ T cells are inhibited from migrating to *Candida* for effector function and the epithelial cells have a reduced capacity to inhibit *Candida*, resulting in increased susceptibility to infection. Reduced levels of Th1 cytokines in saliva may also contribute to the susceptibility. The presence of CD8⁺ T cells in the tissue is supported by the presence of CD8⁺ T-cell-associated cytokines and chemokines. Reprinted from reference 124a with permission.

cells have anti-*Candida* activity via cell contact through a putative carbohydrate moiety (182), oral epithelial cells have a much stronger capacity to inhibit the growth of *Candida* (up to 80% versus 40% for vaginal cells) (181). Analysis of oral epithelial cells in HIV-positive persons showed significantly reduced activity when the cells were derived from patients with OPC than when they were derived from patients without OPC, potentially contributing to the susceptibility to infection (181). Additionally, epithelial cells produce both cytokines and chemokines in response to *Candida*, which may contribute to the innate and/or adaptive immune response(s) (180). The latest mechanism studies show that while a putative carbohydrate was suggested by the sensitivity of the anti-*Candida* activity following treatment of the epithelial cells with periodic acid, several attempts to

identify the carbohydrate by enzyme inhibitors were unsuccessful (148, 182). In fact, the most recent data challenge a role for carbohydrates. First, neutralized supernatants from periodic acid-treated oral epithelial cells containing the putative carbohydrate could not compete with fresh epithelial cells for inhibition of *Candida*. Second, the abrogation of anti-*Candida* activity by periodic acid did not depend on the amount of carbohydrate released, which at times did not show any release over buffer-treated cells. Together, these results suggested that the epithelial cell antifungal activity was potentially acid labile rather than occurring via an effector carbohydrate. This in fact was confirmed by additional studies showing abrogation of anti-*Candida* activity by treatment of epithelial cells with several other acids that release little to no carbohydrate from the cells (203).

In yet other studies with the epithelial cells, the interesting observation that the antifungal activity was partially resistant to fixation (182) was revisited. Recognizing that fixation has permanent effects on cells, effectively killing them, it was interesting that the antifungal activity remained fairly intact despite the high viability (or impermeability) by standard trypan blue exclusion following fixation. In fact, viability assessed by trypan blue exclusion averaged 75% in fixed or unfixed cells whereas viability assessed by propidium iodide and fluorescein diacetate (PI/FDA) was 16% for fixed cells and 75% for unfixed cells. Thus, it appeared that antifungal activity was dependent on intact epithelial cells impermeable to trypan blue but was not dependent on live epithelial cells. This hypothesis was further supported by experiments evaluating viability or permeability by trypan blue, PI/FDA, and antifungal activity in freshly fixed and unfixed epithelial cells compared to those held in culture for 4 days prior to the coculture with *Candida*, where viability by trypan blue was reduced to <25%. In each case, antifungal activity directly correlated with trypan blue permeability. Taken together, these latest data suggest that antifungal activity is fungistatic and dependent on cell contact by intact but not necessarily live epithelial cells through an acid-labile mechanism (203).

Thus, several lines of defense may be important for protection against OPC, many of which do not become evident until CD4$^+$ cell numbers are reduced below the protective threshold. Overall, we postulate that protection against OPC involves both primary and secondary host defense mechanisms. Primary defense involves CD4$^+$ T cells and represents the main protective mechanism for all those who are immunocompetent. This CD4$^+$ T-cell-mediated defense becomes insignificant, though, in patients who become more and more immunocompromised (e.g., those with AIDS). What is critical, however, to the patients with significant progression to AIDS is the secondary host defenses. These include local immune entities such as cytokines in saliva that can promote protection (Th1 type or proinflammatory), the epithelial cell anti-*Candida* activity, and tissue-associated CD8$^+$ T cells with the appropriate homing receptors and adhesion molecules that allow them to maximize their presence.

HOST DEFENSE AGAINST VAGINAL CANDIDIASIS

VVC affects a significant number of women, predominantly in their reproductive years (107, 175, 176). An estimated 75% of all women experience an episode of acute VVC in their lifetime, with another 5 to 10% developing RVVC (175, 176). Vulvovaginitis involves infections of the vaginal lumen as well as the vulva. Symptoms include burning, itching, soreness, an abnormal discharge, and dyspareunia. Signs include vaginal and vulvar erythema and edema. Acute VVC has several known predisposing factors including antibiotic and oral contraceptive usage, hormone replacement therapy, pregnancy, uncontrolled diabetes mellitus, and immunosuppressive therapy (107, 175, 176). RVVC is multifactorial in etiology but is usually defined as idiopathic with no known predisposing factors in the majority of those affected (175, 176). For women with RVVC, antifungal therapy is highly effective for individual symptomatic attacks but does not prevent subsequent recurrence. There is little evidence that resistance to antifungal drugs plays a role in the pathogenesis of RVVC (131). Instead, susceptibility to RVVC is postulated to be immune based in that these otherwise healthy women experience repeated symptomatic episodes as a result of some immunological dysfunction or deficiency. Furthermore, presumably recurrences are a result of relapse rather than reinfection since strain types of *C. albicans* tend to remain the same in women for multiple recurrences over several years (187). This observation is consistent with the fact that most antifungal drugs are static rather than cidal and, as such, do not completely eliminate the organisms. Overall, the differences in the populations susceptible to OPC and VCC, together with the differences in how the two infections present, is an indicator that host defenses are different. The following discussion reveals these differences.

In contrast to the fairly well documented role for CMI by CD4$^+$ T cells against oral candidiasis, the role for CMI against vaginal candidiasis has not been clear and more recently has been all but discounted. Thus, despite intense efforts to date, the natural protective host defenses against VVC remain poorly understood.

Studies of host defenses against VVC began with clinical observations in which women with RVVC were examined for systemic *Candida*-specific responsiveness by culture of their PBMC in vitro or through skin testing in search of DTH responses. The results over time conflicted with the results of some studies wherein it was suggested that there was a deficiency in *Candida*-specific CMI, or a non-antigen-specific deficiency, or no deficiency at all (reviewed in reference 79). Thus, these studies did little to identify an immunological deficiency, if any, in women with RVVC.

In subsequent studies, animal models have been employed to attempt to understand the natural host defense mechanisms against vaginal *C. albicans* infection. Such studies have included rat, mouse, and macaque models of VVC (reviewed in reference 80). All are dependent on a state of pseudoestrus for the infection to become established. The rat model has been particularly helpful in suggesting a potential role of antibodies against *C. albicans* at the vaginal mucosa. The mouse model, on the other hand, has been used primarily to study the role of CMI (reviewed in reference 177). The macaque model is relatively new and to date has been described with only limited immunological data (184). Mice and rats, in contrast to macaques and humans, are not normally colonized with *C. albicans* and do not have any preexisting *Candida*-specific immunity.

Mice inoculated under pseudoestrus conditions acquire a persistent vaginal infection together with systemic Th1-type responses, as evidenced by Th1-type cytokine production in response to *Candida* antigen by lymph node cells draining the vaginal tissue, and *Candida*-specific DTH. However, this systemic Th1-type CMI or that preinduced by systemic immunization with *Candida* antigen in adjuvant did not protect mice against VVC (77). Partial protection against vaginitis was achieved, however, in animals given a second inoculation following the spontaneous resolution of a primary infection in the absence of estrogen (75). Interestingly, suppression of *Candida*-specific systemic CMI or systemic T-cell depletion had no effect on the protection, suggesting that systemic CMI played a limited role at the vaginal mucosa in protection against vaginal candidiasis. In support of this suggestion as well, mice either resistant or susceptible to systemic *C. albicans* infection (93) were shown to be equally susceptible to vaginitis and could all be partially protected against a second vaginal infection (72). These data were supported by data gathered in another study, in which six different strains of mice representing four haplotypes were equally susceptible to infection (16). Finally, adoptive transfer of *Candida*-sensitized T cells into T-cell-deficient (nude) mice had no effect on the course of vaginitis (18). Based on these results, it was postulated that some form of locally acquired mucosal immunity, T-cell and/or antibody mediated, was responsible for protecting mice against vaginal *C. albicans* infection and that the vaginal mucosa had a significant level of immunological independence with little influence from systemic sources.

The independent compartmentalization concept correlates well with a clinical study showing that RVVC patients had normal levels of *Candida*-specific Th1-type CMI in the peripheral circulation (76). It should be mentioned, however, that other researchers have obtained data to the contrary (39, 52).

Nevertheless, the lack of identifiable deficiencies in systemic CMI in women with RVVC correlates with two relevant clinical observations: women with RVVC are not susceptible to CMC or other forms of cutaneous candidiasis (175), and women with CMC are generally not susceptible to RVVC (149). From these data, the putative immune deficiency in RVVC patients is presumed to be localized to the vagina and does not occur at the systemic level. These data were the first evidence to suggest some level of immunological independence regarding CMI host defense mechanisms at the vaginal mucosa against *C. albicans*. Further support for these observations comes from studies of HIV-infected women, who, despite the higher incidence of OPC under conditions of immunosuppression, do not contract VVC any more frequently than do HIV-negative women (47, 124, 174, 191). Furthermore, peripheral blood lymphocyte proliferation and Th1/Th2 cytokine production in response to *Candida* antigens were relatively similar in HIV-positive women with or without symptomatic VVC (124), similar to RVVC patients.

Subsequent studies focused on the presence of T-cell subpopulations in the vaginal mucosa. The vaginal mucosa, while not composed of organized lymphoid tissues like the Peyer's patches in the GI tract, is immunologically competent for both CMI and T-cell-dependent antibody production with the presence of both T cells and MHC class II$^+$ cells. To this end, we and others reported that vaginal lymphocytes are phenotypically distinct from those in the peripheral circulation (81, 145). Although CD4$^+$ α/β TCR$^+$ cells predominate in the vaginal tissue, a higher percentage of γ/δ TCR$^+$ cells are present with few if any CD8$^+$ cells than in the systemic circulation. Interestingly, human vaginal lymphocytes are also phenotypically distinct in comparison to those in the peripheral circulation, but humans and mice differ in the types and relative proportions of cells; the human vagina has increased numbers of CD4$^+$ and γ/δ T cells but also considerable numbers of CD8$^+$ T cells (82, 95).

Studies to examine changes in vaginal T cells during experimental vaginal *Candida* infection have all shown little evidence for modulation of α/β or γ/δ vaginal T cells during either primary or secondary experimental vaginal *C. albicans* infections and no evidence for systemic T-cell infiltration during a vaginal *Candida* infection (74, 196). These results are consistent with those of studies of immunodeficient or knockout mice (18, 38) and correlate with the systemic immune studies of immunocompetent mice (72, 75, 77, 78). This lack of responsiveness to *Candida* infection in mice is in direct contrast, however, to a murine model of a genital *Chlamydia trachomatis* or herpes simplex virus type 2 infection, where CD4$^+$ T cells infiltrate readily into the vaginal mucosa in response to infection (106, 152). Taken as a whole, these data emphasize the uniqueness of the host vaginal response(s) to *C. albicans*. Interestingly, a study of a dual infection with *Candida* and *Chlamydia* showed a complete independence of "responsiveness" in the upper genital tract (*Chlamydia*) and "lack of responsiveness" in the lower tract (*Candida*) (105), illustrating further the complexity of immunity in the genital tract. Together, these data suggested that instead of a simple lack of a role for systemic or local CMI against *C. albicans* at the vaginal mucosa, there may be some form of immunoregulation or tolerance that prohibits a more profound CMI response.

There are several pieces of evidence, both experimentally and clinically, that support immunoregulation against CMI at the vaginal mucosa. First, experimental vaginitis in δ-chain TCR knockout mice has been found to be less severe, suggestive of a tolerance role for vaginal γ/δ T cells (195). Evaluation of Th cytokines in the vaginal tissue homogenates from mice showed high constitutive concentrations of transforming growth factor β (TGF-β) in naive mice, which were increased further in estrogen-treated and/or vaginally infected mice; the same mice had low levels of other Th1- or Th2-type cytokines (186). Clinically, TGF-β was found in fairly high concentrations in human vaginal secretions, with modulations upward over the course of the menstrual cycle (71). Additionally, although Th1/Th2 cytokines are detectable in vaginal secretions of women with RVVC, there are no discernible patterns that would account for susceptibility to RVVC when comparisons to control women were made (73). In the mouse model of infection, T cells with appropriate homing receptors for infiltration into the vaginal mucosa are reduced in number or lost during a vaginal *Candida* infection despite the upregulation of reciprocal adhesion molecules on the vaginal endothelium (196). Further supporting immunoregulation, a T-cell-derived antigen-binding molecule has been identified in serum of RVVC patients and to a lesser extent in control women; it can become associated with TGF-β and inhibit *Candida*-specific PBL proliferation and cytokine production (130). Finally, despite some preliminary evidence that the intravaginal challenge of normal healthy women with *Candida* antigen stimulated Th1-type cytokine secretion in vaginal lavages 16 to 18 h postchallenge (73), the results of a more recent study in which responsiveness to the antigen in a large number of women was tested during different stages of the menstrual cycle showed no evidence of

immune responsiveness (71). Thus, there is convincing evidence that immunoregulation or tolerance is functioning at the vaginal mucosa and that it inhibits the ability of *Candida*-specific systemic CMI to reach the vaginal mucosa.

The rat model, used primarily to study humoral immunity, was recently used to evaluate CMI (56). The progression of infection, however, is quite different in the rat from the mouse, in that spontaneous resolution occurs within a 3-week period in the rat. Consistent with this property is the fact that T cells could be demonstrated infiltrating the vaginal mucosa following infection and Th1-type cytokines were produced and secreted into vaginal secretions. Additionally, in contrast to data obtained from the mouse model of vaginitis, a recent study of the rat model showed that passive transfer of both CD4$^+$ and CD8$^+$ T cells from previously infected rats accelerated the clearance of *Candida* from the vaginas of naive animals (172). Also, the CD4$^+$ T cells were more effective at clearance than were the CD8$^+$ T cells, suggesting that, in the rat model, vaginal lymphocytes, especially CD4$^+$ T cells, are important in clearance of *Candida* from the vagina (172). Thus, the rat, in contrast to the mouse and human, may exemplify what would occur in response to infection in the absence of immunoregulation.

The role of humoral immunity in recovery from vaginitis is equally controversial. Immunoglobulins are clearly present in vaginal secretions, including *Candida*-specific antibodies (IgG and IgA) (160, 193, 194). Interestingly, though, B cells are a minor constituent of the mouse, rat, or human vaginal mucosa (151). In contrast to other mucosal sites, IgG antibodies predominate over IgA antibodies, at least on the basis of total immunoglobulin (73, 120). Despite that, clinical studies show that serum IgG and vaginal IgG and IgA levels are similar in women with RVVC (symptomatic or in remission) and healthy control women with no history of RVVC (73, 110, 135). Experimentally, similar results were observed in infected and uninfected rhesus macaques, including the predominance of vaginal IgA over IgG immunoglobulins (184). Additionally, in a mouse model of vaginal infection, *Candida*-specific IgA and IgG were not detectable in lavage fluids of animals protected against secondary challenge, suggesting a minor or no protective role for *Candida*-specific antibodies against the murine vaginal infection (198). While these experimental and clinical observations largely fail to support a role for antibody against candidiasis, there are studies with rodents that indicate a protective role for humoral immunity in experimental vaginal candidiasis. Pollonelli et al. showed that rats given an intravaginal immunization using antibodies that neutralized the anti-*Candida* activity of a yeast killer toxin were protected against a vaginal infection (155). The protection was found to be associated with rising vaginal titers of anti-idiotypic IgA antibodies that could transfer protection passively to unimmunized rats. Interestingly, antibodies similar to the killer toxin anti-idiotypic antibodies were detected in women with symptomatic vaginitis (156). Similarly, Cassone and colleagues (43, 54) have shown that a primary *C. albicans* vaginal infection in estrogen-treated rats induced antimannan and antiaspartyl IgA antibodies in vaginal secretions and that those antibodies appeared to protect the same rats against a secondary vaginal infection. Moreover, recipient rats treated intravaginally with vaginal fluids from rats with primary infection were protected. In a follow-up study, vaginal lymphocyte analysis during primary or secondary vaginal *C. albicans* infections in rats revealed an increase in the number of B cells in the tissue (56). Successful protection against vaginitis in the presence of these IgA antibodies could reflect "protective" antibodies based on the Casadevall concept of nonprotective, protective, and indifferent antibodies (40). In support of this, Han et al. have shown that two different "protective" antibodies specific for mannan (IgM and IgG3) protect mice against *Candida* vaginitis when given either locally or systemically (90, 91). Taking this into account, one might speculate that when antibodies are detected clinically that do not appear to have a protective effect, indifferent or unprotective antibodies may predominate in the antibody pool. If indeed there is some precedent for this, it appears to be the norm rather than the exception.

Innate immunity appears to play a significant role in protection against vaginitis as well. This was initially predicted based on a leukocytic infiltrate of predominantly PMN that is often observed in the vaginal lavage fluid of infected animals. These leukocytic cells are usually associated with or attached to the hyphae and/or sheets of epithelial cells. On the other hand, when this infiltrate is present during infection, it does not correlate with a lower vaginal fungal burden (74, 168). Additionally, this leukocytic infiltrate is generally not observed in clinical cases of *Candida* vaginitis on the basis of normal clinical diagnostic procedures. Although PMN and macrophages are potential candidates for anti-*Candida* innate resistance and are present at or near the vaginal mucosa, Cantorna et al. (38) showed that animals with the beige mutation, which are immunodeficient in phagocytic cells (*bg/bg*), were not more susceptible to a natural *C. albicans* infection under nonestrogenized conditions. Two more recent studies addressed the specific role of PMN during experimental vaginitis.

The results indicated that depletion of PMN under estrogen-treated or non-estrogen-treated conditions had no effect on the vaginal fungal burden (17, 74), suggesting that PMN do not play a significant role against *C. albicans* in the vagina despite their presence in the vaginal lumen during some experimental infections. Perhaps this reflects their inability to function against *C. albicans* in the vaginal microenvironment compared to their ability to function well in blood or tissue culture. It might reflect local immunoregulatory activity as well. Alternatively, the presence of PMN during an infection might not represent a response to the yeast but might simply reflect their normal presence during the diestrus stage of the menstrual cycle (every 2 days in mice), when they are deployed to phagocytose the apoptotic squamous epithelial cells from the estrus (2-day cycle) stage.

As mentioned above, vaginal epithelial cells represent a potential innate anti-*Candida* host defense mechanism. The mechanism of the vaginal epithelial cell anti-*Candida* activity is identical to that of oral epithelial cells (148, 203), although the level of activity is weaker than that for oral cells (181). If these cells are important as an innate mechanism, weaker activity by vaginal cells may reflect a higher prevalence of vaginal versus oral candidiasis when both tissues are colonized (181). Interestingly, a study of humans showed that while the vaginal epithelial cell anti-*Candida* activity was not different at the various stages of the menstrual cycle, it was significantly reduced in women with a history of RVVC (12). Thus, reduced epithelial cell anti-*Candida* activity may represent, in part, a local immune deficiency associated with RVVC.

Although these studies conducted over the past two decades suggested that adaptive immunity is not likely to be involved in either resistance or susceptibility to infection, they had failed to reveal immune events that protect women from vaginitis and which events occur or are associated with susceptibility to acute or recurrent symptomatic infection. Thus, either there are few protective immune factors or the models and clinical studies have not been able to reveal them. Support for the latter came from a recent study evaluating adolescents for vaginal yeast colonization and local and systemic immune sensitization. Interestingly, adolescents were found to be asymptomatically colonized with yeast in the vagina at a rate of 26%, which is at the high end of the normal range for adults. These adolescents had the same species distribution of *Candida* in the vagina (90% *C. albicans*) and were sensitized to *Candida* systemically (peripheral blood responses and antibodies) (11). However, what set the adolescents apart from the adults was an extremely high vaginal fungal burden in those asymptomatically colonized (50% had between 200 and 1,000 CFU compared to 95% of adults having <200 CFUs). Probably most intriguing was that the attack rate of acute VVC in this population (in a longitudinal study over 3 years) was <3%. Thus, the adolescents could maintain large numbers of *Candida* organisms vaginally without contracting a symptomatic infection, suggesting strong protective activity/responsiveness.

In light of all the findings to date, we reasoned that the design that was likely to provide valuable information on putative vaginal protective host defenses against *C. albicans* was a live challenge that offers a controlled yet natural condition, whereby immunoreactivity can be monitored. Live-challenge models in humans are not new. Currently there are live-challenge models for *Haemophilus ducreyi*, *Neisseria gonorrhoeae*, and Norwalk virus (4, 49, 129). The primary rationale for these models is the inability of the organism to colonize or infect animals. There is considerable risk with these models since the organisms have pathogenic potential. Nevertheless, the ability to treat the infections successfully provided confidence that a well-constructed experimental design would have minimal risk to the subject. Vaginal *C. albicans* infections are similarly treated with great success, providing a high level of confidence in the use of *Candida* organisms in a live challenge. Furthermore, the fact that *C. albicans* is a commensal organism of the vaginal mucosa with no known migration to extravaginal sites make *C. albicans* an excellent organism for a live-challenge model with minimal risk.

The results of studies to date evaluating the natural history following intravaginal inoculation of live *C. albicans* organisms into healthy women have completely reshaped our hypotheses relative to host defense against vaginitis and susceptibility to infection. In a very conservative initial approach involving testing women with no history of vaginitis and inoculating small numbers (yet bolus) of *Candida* organisms, few women became symptomatic with vaginitis (~10%). Broadening the inoculating condition to include using larger numbers of *Candida* organisms, inoculating them at different stages of the menstrual cycle, and adding *Candida* growth-promoting supplements to the inoculum vehicle (growth media, glucose, and estrogen) also had very little effect. In fact, a high percentage of women did not even become asymptomatically colonized. In contrast, inclusion of women with documented infrequent episodes of VVC (antibiotic usage, oral contraceptive usage, pregnancy, etc.) resulted in a higher rate of symptomatic infection (~50%), with the majority of the remaining women becoming asymptomatically

colonized (70). Interestingly, in either design, protection occurred in the absence of any evidence of inflammation or an inflammatory response, whereas those with symptomatic infection had a heavy vaginal cellular infiltrate consisting almost entirely of PMNs. Furthermore, a high PMN infiltration score correlated with a higher fungal burden. Although the more pathogenic hyphal form of *Candida* was present along with blastoconidia in those with symptomatic infection, hyphae were also detected in some of those asymptomatically colonized.

Based on these data, we hypothesize that VVC is associated with signals following *Candida*-vaginal epithelial cell interactions that promote a nonprotective inflammatory leukocytic response and concomitant clinical symptoms while resistance to VVC is associated with a lack of signals and/or appropriate local milieu of innate mediators that is noninflammatory. We also postulate that a threshold number of organisms in the vagina following inoculation (not the inoculum itself) is crucial to the signal(s) in the face of T-cell immunoregulation and that this threshold is different for different groups of women. For example, women with primary RVVC inevitably acquire an infection shortly after completing a regimen of antifungal therapy. In these women, the threshold of *Candida* required to signal the PMN infiltration would be extremely low. Women with an infrequent history of VVC or secondary RVVC have a higher threshold of *Candida* required to signal the PMNs. In these women, the threshold is usually met when the population numbers of *Candida* increase following antibiotic therapy or hormone replacement therapy, following the use of high-estrogen oral contraceptives, during pregnancy, or due to diabetes mellitus. In women with no history of VVC, the threshold of *Candida* is extremely high (levels rarely attainable). Thus, although the population numbers of *Candida* can increase, with reduced numbers of the bacterial flora, during antibiotic therapy or increased estrogen levels from oral contraceptives, they never reach the threshold number to signal the PMN migration, and hence symptoms of vaginitis rarely occur. We can also extend this concept to adolescents, whose threshold of *Candida* is expected to be high (similar to adults with no history of vaginitis). This hypothesis regarding the high threshold in adolescence would explain the extremely large population numbers of *Candida* in the vaginas of adolescents with few symptomatic infections observed. A schematic diagram of this hypothesis is shown in Fig. 2.

Taken together, instead of susceptibility to VVC being associated with a deficient or missing immune component and protection associated with a T-cell inflammatory response, it is now suggested

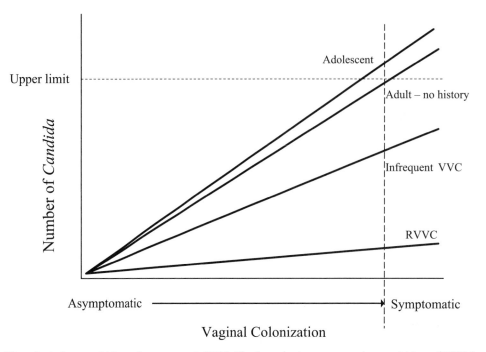

Figure 2. Hypothesis for acquisition of symptomatic VVC. The hypothesis centers on the acquisition of VVC by achieving a threshold of *Candida* organisms in the vagina. Below the threshold, the vaginal presence of *Candida* is considered commensal and asymptomatic. Once the threshold is crossed (horizontal dotted line), the signals are initiated that allow for migration of PMNs into the vagina that cause the symptoms associated with vaginitis. Accordingly, there are different thresholds for different groups of women. For women with RVVC, the threshold is very low; it increases incrementally in women with infrequent episodes of VVC, women with no history of VVC, and adolescents. The upper limit is an arbitrary level of organism number that would be considered unattainable. Reprinted from reference 70 with permission.

that susceptibility to infection is associated with an aggressive inflammatory response that causes the symptoms while resistance is noninflammatory. Thus, both resistance and susceptibility appear to be associated with innate immunity, with no demonstrable role for adaptive immunity.

HOST DEFENSE AGAINST GASTROINTESTINAL CANDIDIASIS

For the purposes of the discussion at hand, GI candidiasis is defined as infections of the stomach and small and large intestines. GI candidiasis is not easily recognized clinically but does occur in cancer patients who are heavily immunosuppressed and patients receiving antibiotic prophylaxis (5, 170). It can be difficult to diagnose due to the lack of specific symptoms and the lack of the ability to differentiate between infectious and commensal *C. albicans*. However, compared to OPC and VVC, GI candidiasis is relatively rare. This comparative rarity appears to be due to mechanisms involving gut flora, innate immunity, and immune regulation. Thus, the GI tract represents yet another mucosal site of infection with a seemingly different or at least modified set of rules for host defense.

Host defense against GI candidiasis until recently has been studied primarily through the use of immunodeficient mice. Studies using both congenitally deficient and severe combined immunodeficient (*scid*) mice have shown that T cells are critical for effective protection against GI candidiasis (37). *scid* mice were shown to be more susceptible to GI tract colonization but were not susceptible to sublethal intravenous challenge or dissemination from the GI tract. Mice with congenital T-cell (*nude/nude*) and phagocytic cell (*beige/beige*) deficiencies (*nu/nu*, *bg/bg*) developed persistent GI candidiasis, while those with only phagocytic cell deficiencies (*nu/+*, *bg/bg*) cleared *C. albicans* from the GI tract effectively. The *nu/+* mice (wild type) responded to *Candida* antigens with IL-2 production and DTH, while *nu/nu* mice did not, suggesting the correlation between a Th1-type response and clearance of the organism. There is also evidence that gnotobiotic mice deficient in CD8+ T cells are susceptible to GI candidiasis (9, 103) providing yet another example of the potential role for CD8+ T cells against *Candida*. An important question, however, is the nature of the role played by circulating versus mucosal T cells in the protection. The mucosal tissue of the GI tract has intraepithelial lymphocytes, Peyer's patches (organized lymphoid tissue), T cells, and antigen-presenting cells that function in many types of GI infections. Much less is known, however, about the function of any of these cell populations against *C. albicans*, although more recent studies have begun to address this issue. Peyer's patch lymphocytes from colonized mice which went on to clear the fungus from the GI tract produced Th1- but not Th2-type cytokines in response to *Candida* antigen (14). Thus, GI candidiasis appears to conform to the Th1/Th2-type immune response pattern seen in oral candidiasis, namely, that Th1-type responses are associated with resistance to infection while Th2-type responses promote susceptibility (31, 44, 133, 167, 185, 201).

The role of antibodies against GI candidiasis has not been evaluated experimentally except through early studies, which showed no protection from *Candida* infection by antibody-containing sera (8, 87). There are some more recent studies, though, which suggest a possible role for antibodies, but more so as an immune exclusion entity (neutralizing) than an immune elimination entity; in other words, antibodies bind to *Candida* organisms and inhibit their attachment to epithelial cells. This phenomenon was shown by enhanced binding of *Candida* to epithelial cells in B-cell knockout mice compared to wild-type mice (189). Thus, antibodies may play a role in resistance to invasion of gut epithelium by *Candida*.

The most recent data on host defense against gastrointestinal candidiasis involves innate mechanisms. These include dendritic cells (DC), Toll-like receptor (TLR) activation, and defensins (101, 146). While formal studies of cells isolated specifically from the gut mucosa are yet limiting, data continue to be collected. DC act both as a phagocytic cell against *Candida* and as an antigen-presenting cell to stimulate T-cell responses. This dual role is true in the gut as well. Three populations of DC are recognized in the murine gut Peyer's patches: myeloid (CD11b+ CD8-), lymphoid (CD11b- CD8+), and double negative (CD11b- CD8-) (101). The myeloid DC secrete Th2-type cytokines, while the lymphoid and double-negative DC secrete Th1-type cytokines (101). Moreover, *C. albicans* blastoconidia and hyphae activate DC differently. Whereas blastoconidia stimulate DC in a manner that stimulates T-cell responses, hyphae induce the DC to tolerance (166). DC and macrophages also have TLR that are activated and direct immune responses when stimulated by *Candida* (146). TLR4 appears to be the major TLR on DC with which *Candida* interacts to stimulate immune responses (146). A similar mechanism is true of macrophages (127).

The gut flora may also modulate host responses to *Candida*. Wagner et al. have shown the induction of candidiasis of endogenous origin in immunocompetent gnotobiotic mice that lack an intestinal microflora and have also shown that prior colonization of the mice with a pure culture of probiotic bacteria reduces the incidence of infection (188). The possible mechanisms that contribute to this activity include direct anti-*Candida* activity by bacteria and/or immunomodulation.

Immunoregulation in more general terms refers to the concept of tolerance in the gut. There are now substantial roles for TGF-β and regulatory T cells (Tr1) (117), Th3-type T cells that produce TGF-β (104), and CD4+ CD25+ T regulatory cells (7) that may also play a role in tolerance in the gut. While studies of the interactions and/or functions of these cell populations isolated from the gut have not been conducted with *Candida*, the likelihood that they are involved in responses or lack thereof to *Candida* is fairly high and probably involves DC. The reason why GI candidiasis is fairly rare may be due to a balance of innate and adaptive immune reactivity that favors commensalism.

Taken together, the GI tract response to *Candida* is a mix between what is seen in OPC and VVC. While T cells may be involved, there is also a significant level of tolerance against such responses. However, in the GI tract, there does not seem to be the hyper-innate response that is seen in the vaginal cavity, and hence *Candida* is maintained more in a commensal state with less incidence of infection.

IMPLICATIONS FOR MUCOSAL IMMUNITY TO FUNGI

Recognizing the distinct differences in host responses to *Candida* in the oral cavity, gut, and vaginal cavity, one can begin to appreciate the uniqueness of the mucosa at these sites and how they handle a commensal organism, together with the limitations in host responses that precipitate a pathogenic state and infection. One cannot speak of host defense against "mucosal candidiasis" in an attempt to address host defense against *C. albicans* at any one mucosal site. Host defense at each site must be thought of differently and spoken of independently. Table 1 provides a summary of the distinctions between host defenses against oral, GI, and vaginal candidiasis as well as what is still questionable. Overall, the major distinctions are that CD4+ Th1-type responses and CD8+ T cells are involved in GI and oral candidiasis but not VVC. Humoral immunity by naturally induced antibodies is possibly involved in GI candidiasis but not OPC or VVC. Innate immunity is involved at all three sites, but at

Table 1. Host defenses against mucosal *Candida* infections

Host defense mechanism	Mechanism operating against:		
	GI infection	Oral infection	Vaginal infection
Innate			
PMN	+/−	+/?	+/+[a]
Macrophages	+/−	−	−
NK cells	−	−	−
Epithelial cells	?	++	++
Humoral			
IgM	−/?	−	+/−?
IgG	−/?	−	+/−?
IgA	−/?	−/?	+/−?
Cell-mediated			
CD4+ T cells (Th1)	++	+++	−
CD8+ T cells[b]	+	++	−

[a]Activity is associated with susceptibility to infection rather than protection.
[b]May be an innate rather than adaptive response.

different levels (DC, PMN, and epithelial cells), including promoting susceptibility to VVC. Finally, immunoregulation is involved in GI candidiasis and VVC but not OPC. Immunoregulation is interesting in and of itself as an issue for systemic and local immunity. While systemic immunity can be induced from the local infection, immunoregulation can prohibit its function at the local site of infection (i.e., VVC). Thus, evaluating systemic immune responses that may exist during a local infection may be limiting if the systemic immune reactivity is not measured or evaluated at the local site as well.

These distinctive host responses to *Candida* at various anatomic sites serve as a prime example of other medically important fungi that can infect multiple sites, both mucosal and nonmucosal. This includes fungi that infect the lungs and disseminate, those that infect the skin and mucosa, those that infect the lungs and central nervous system, etc. Historically, most studies of host defense against other medically important fungi focused on cells in the lymph nodes, spleen, or peripheral blood. However, inasmuch as systemic populations of cells might play a role in host defense, the microenvironment is critical to the function of those cell populations and should be considered in evaluations of host responses along with locally derived immune reactivity. The following is a short review of what is known regarding evaluations of host defense against other fungi that infect multiple sites.

HOST DEFENSE AGAINST OTHER FUNGI THAT INFECT MULTIPLE SITES

Histoplasmosis is a mycosis of the reticuloendothelial system involving the lungs, spleen, and kidneys. *Histoplasma capsulatum* conidia produced in the environment are inhaled into the lungs, where transformation to the pathogenic yeast form occurs (57). Prior to the advent of animal models, studies of the host responses in humans were performed with lymphocytes isolated from peripheral blood. Subsequent studies turned to an animal model of disseminated disease by inoculating mice intravenously and examining the fungal burden, cellular infiltration, or cytokines produced in the spleen after infection (13, 23, 200). From these studies, it was established that both natural and acquired immunity were important in histoplasmosis. Disseminated disease usually observed in those with an impaired immune system such as HIV-infected individuals has changed the global studies of host defenses against histoplasmosis. To understand the perturbations of immunity that led to disseminated disease, it was vital to define the evolution of the immune response at the primary site of infection, the lungs. Studies of the lungs show a role for DC, neutrophils, and macrophages at various levels of activity (30, 118). Regarding adaptive immunity, both CD4+ and CD8+ T cells play a role in protection, with CD4+ T cells as the primary effector cells. Furthermore, both proinflammatory and Th1-type cytokines play a role in primary and secondary infection (2, 3, 48, 204).

Paracoccidioidomycosis is acquired by inhaling airborne propagules produced by the fungal mycelium. The disease localized to the lungs forms an inflammatory granulomatous reaction or can disseminate to other organs, including mucosal surfaces and skin (139). As in histoplasmosis, early experimental and clinical studies focused on cells isolated from human blood and on cells from the spleen and lymph nodes of infected animals with disseminated disease (21, 178). The results showed that phagocytic neutrophils and macrophages from infected hosts had reduced anti-*Paracoccidioides* activity, indicating the important role of these innate immune cells against paracoccidioidomycosis (19, 21). Studies of adaptive immunity demonstrated that benign forms of the disease were associated with low levels of antibodies and positive T-cell function measured by DTH reactions whereas disseminated forms were associated with high levels of antibodies and T-cell anergy (33, 141). Establishment of a pulmonary infection in mice by intratracheal inoculation has been instrumental in understanding host defenses against *P. brasiliensis* at the site of infection (19, 36). Site-specific immune analyses show a strong role for both macrophages and neutrophils in controlling *P. brasiliensis*. Several chemokines and proinflamatory cytokines are important as well (34, 83, 86). A Th1-type response by CD4$^+$ T cells in the lungs is also a critical adaptive response (35, 154). Studies of other sites of infection have not been performed.

Besides the lungs, *Coccidioides* can infect skin, bones and joints, meninges, and the genitourinary tract (64). Host immune defense against *C. immitis* infection has been studied in humans and animal models. Throughout the years, peripheral blood lymphocytes from infected patients as well as spleen and lymph node cells from infected mice have been used to evaluate the host responses to *Coccidioides* (53). Despite differences in the course of the disease in mice and humans, the mouse model has been an important tool in evaluating host responses to the fungus in the lungs (53, 178). These studies have shown that alveolar macrophages phagocytose *C. immitis* arthroconidia and endospores, with endospores being the most vulnerable to growth inhibition and killing by activated phagocytes (53). Protection against pulmonary *C. immitis* infection is dominated by Th1-associated immune responses. Macrophages and/or DC acting as antigen-presenting cells lead to the induction of IFN-γ and other Th1 cytokines (53). Studies of the skin, meninges, and genital tract have not been conducted.

Blastomyces produces a wide spectrum of illness in humans and canines. As in histoplasmosis, paracocidioidomycoses, and coccidioidomycoses, the primary site of blastomycosis infection is the lung, and the infection can resolve spontaneously. Disseminated disease compromises the skin, central nervous system, genitourinary tract, bones, and other organs. Early experimental studies to identify the host defense mechanisms in disseminated disease were performed with cells isolated from the spleen and lymph nodes (25, 92, 178). These studies showed that neutrophils and macrophages did not kill *Blastomyces* efficiently even when these cells were activated. Although CMI appeared critical,

the process was not well understood (22, 26). Subsequent studies have focused on the lungs as the primary site of infection. The role of neutrophils and macrophages in pulmonary *B. dermatitidis* infection has been evaluated by using virulent and nonvirulent strains of *B. dermatitidis* (20, 113). Expression of BAD1, which is a virulence factor expressed during transition, impairs the activity of phagocytes, neutrophils, and macrophages against the yeast. In healthy individuals, adaptive immunity depends on T cells rather than B cells and antibody, with Th1-type CD4$^+$ T cells and cytokines playing crucial roles in the response to the fungus (112). No studies have been conducted at any other sites of infection.

As with the other fungi, the lung has also been established as the primary port of entry for *Cryptococcus neoformans*. The infection may be asymptomatic but has the potential to disseminate or reactivate in immunocompromised patients (41). Other sites of infection include the central nervous system, skin, bones, and prostate gland (41). Early studies conducted to investigate the innate and acquired host defense mechanisms that participate in restricting cryptococcal infections were performed using in vitro assays with peripheral blood lymphocytes and spleen and lymph node effector cells (142). With the increase in the incidence of cryptococcal infections observed in HIV-infected patients, efforts were concentrated on cells in the lungs active against *C. neoformans* (138). In the alveolar spaces, the yeast cells are confronted by alveolar macrophages. Following this interaction, experimental data have demonstrated the importance of both CD4$^+$ and CD8$^+$ T cells in controlling the infection. Th1 and Th2 cytokines are produced in the lungs during the first weeks of infection, but the ability to maintain Th1 cytokine production is required for clearance of infection (96). Increased susceptibility of HIV patients to cryptococcosis indicates that the loss of CD4 cells predisposes individuals to infection with *C. neoformans* (138). Although much has been learned about CMI against *C. neoformans* in the lungs, less is known about host resistance in the brain. To date, studies have shown that the same cells and mechanisms involved in host defense in the lungs are also important in the brain, albeit delayed (27, 28).

Aspergillosis refers to a broad range of disease states: allergic aspergillosis, aspergilloma, and invasive aspergillosis. *Aspergillus fumigatus* is the most common disease-causing species (122, 123). All individuals are constantly in contact with airborne conidia. In the lungs of immunocompromised patients, the conidia germinate and invade the tissue, causing severe disease that ranges from an allergic reaction to the most severe invasive aspergillosis (123). Mouse models and clinical observations have shown the importance of epithelial cells, phagocytic cells, and intact CMI in the lungs (121). Alveolar macrophages and neutrophils play an essential role in the killing of *A. fumigatus* conidia and further prevent their germination to hyphae (119, 136). Protective immunity is associated with the production of TNF-α and Th1 cytokines (123). In contrast, invasive aspergillosis is associated with a CD4 Th2 response, with increased levels of IL-4 and IL-10 (123).

Studies of host defense against these medically important fungi are certainly following the correct path and moving to evaluations at sites of infection. In most instances, early studies of disseminated disease that evaluated systemic immune mechanisms have given way to site-specific infections evaluating systemic and local immune responses. Clearly, this trend must continue and ultimately move to secondary sites of infection utilizing appropriate models.

CONCLUSION

We would like this review to serve as an example of what the future should be relative to host defense against the medically important fungi. The days of using the systemic immune system to study host defenses against fungal infections at specific sites without strict evaluations of those responses or other local responses at the site of infection should be past. Also past should be any assumption that host defenses against fungi at one mucosal site are similar to that at another mucosal site. While it is recognized that evaluating site-specific immune responses will not always be easy, such an evaluation remains critical to a fuller understanding of the true host response for each infection. Such concepts should serve as a paradigm shift of sorts or road map for how host defenses should be envisioned henceforth. This paradigm shift will also be important for vaccine research. Development of vaccines must consider the site-specific host responses and target the site of infection with the vaccine. It is already well known that immunization at systemic sites does not affect mucosal sites whereas immunization of mucosal sites usually affects both mucosal and systemic sites. However, the problem is still how best to stimulate strong responses at mucosal sites. Solving the problem remains a challenge for vaccine research targeting protection against infections by the medically important fungi. Yet suffice to say that studies that address host responses at the sites of infection will do much to advance our understanding of immunity against fungal infections, the pathogenesis of fungal infections, and the development of effective vaccines to protect against them.

REFERENCES

1. **Allen, C. M., A. Saffer, R. K. Meister, F. M. Beck, and S. Bradway.** 1994. Comparison of a lesion-inducing isolate and a non-lesional isolate of *Candida albicans* in an immunosuppressed rat model of oral candidiasis. *J. Oral Pathol. Med.* **23:**133–139.
2. **Allendoerfer, R., G. P. Biovin, and G. S. Deepe.** 1997. Modulation of immune responses in murine pulmonary histoplasmosis. *J. Infect. Dis.* **175:**905–914.
3. **Allendoerfer, R., and G. S. Deepe.** 1997. Intrapulmonary response to *Histoplasma capsulatum* in gamma interferon knockout mice. *Infect. Immun.* **65:**2564–2569.
4. **Al-Tawfig, J. A., K. L. Palmer, C. Chen, J. C. Haley, B. P. Katz, A. F. Hood, and S. M. Spinola.** 1999. Experimental infection of human volunteers with *Haemophilus ducreyi* does not confer protection against subsequent challenge. *J. Infect. Dis.* **179:**1283–1287.
5. **Anaissie, E. J., and G. P. Bodey.** 1990. Fungal infections in patients with cancer. *Pharmacotherapy* **10:**164S–169S.
6. **Arancia, G., A. Molinari, P. Crateri, A. Stringaro, C. Ramoni, M. L. Dupuis, M. J. Gomez, A. Torosantucci, and A. Cassone.** 1995. Noninhibitory binding of human interleukin-2-activated natural killer cells to the germ tube forms of *Candida albicans*. *Infect. Immun.* **63:**280–288.
7. **Baecher-Allan, C., J. A. Brown, G. J. Freeman, and D. A. Hafler.** 2001. CD4$^+$ CD25^{high} regulatory cells in human peripheral blood. *J. Immunol.* **167:**1245–1253.
8. **Balish, E., and H. Filutowicz.** 1991. Serum antibody response of gnotobiotic athymic and euthymic mice following alimentary tract colonization and infection with *Candida albicans*. *Can. J. Microbiol.* **37:**204–210.
9. **Balish, E., A. Vazquez-Torres, J. Jones-Carson, R. D. Wagner, and T. Warner.** 1996. Importance of β$_2$-microglobulin in murine resistance to mucosal and systemic candidiasis. *Infect. Immun.* **64:**5092–5097.
10. **Banchereau, J., and R. M. Steinman.** 1998. Dendritic cells and the control of immunity. *Nature* **392:**245–252.
11. **Barousse, M., B. J. Van Der Pol, D. Fortenberry, D. Orr, and P. L. Fidel, Jr.** 2004. Vaginal yeast colonization, prevalence of vaginitis, and associated local immunity in adolescents. *Sex. Transm. Infect.* **80:**48–53.
12. **Barousse, M. M., C. Steele, K. Dunlap, T. Espinosa, D. Boikov, J. D. Sobel, and P. L. Fidel, Jr.** 2001. Growth inhibition of *Candida albicans* by human vaginal epithelial cells. *J. Infect. Dis.* **184:**1489–1493.
13. **Baugman, R. P., C. K. Kim, A. Vinegar, D. E. Hendricks, D. J. Schmidt, and W. E. Bullock.** 1986. The pathogenesis of experimental pulmonary histoplasmosis. Correlative studies of histopathology, bronchoalveolar lavage, and respiratory function. *Am. Rev. Respir. Dis.* **134:**771–776.
14. **Bennett, J. E.** 1990. Antimicrobial agents: antifungal agents, p. 1165–1181. *In* A. G. Gilman, T. W. Rall, A. S. Nies, and P. Taylor (ed.), *Goodman and Gilman's the Pharmacological Basis of Therapeutics*. Pergamon Press, Inc., Elmsford, N.Y.
15. **Beno, D. W. A., A. G. Stover, and H. L. Mathews.** 1995. Growth inhibition of *Candida albicans* hyphae by CD8$^+$ lymphocytes. *J. Immunol.* **154:**5273–5281.

16. Black, C. A., F. M. Eyers, M. L. Dunkley, R. L. Clancy, and K. W. Beagley. 1999. Major histocompatibility haplotype does not impact the course of experimentally induced murine vaginal candidiasis. *Lab. Anim. Sci.* **49:**668–672.

17. Black, C. A., F. M. Eyers, A. Russell, M. L. Dunkley, R. L. Clancy, and K. W. Beagley. 1998. Acute neutropenia decreases inflammation associated with murine vaginal candidiasis but has no effect on the course of infection. *Infect. Immun.* **66:**1273–1275.

18. Black, C. A., F. M. Eyers, A. Russell, M. L. Dunkley, R. L. Clancy, and K. W. Beagley. 1999. Increased severity of *Candida* vaginitis in BALB/c *nu/nu* mice versus the parent strain is not abrogated by adoptive transfer of T cell enriched lymphocytes. *J. Reprod. Immunol.* **45:**1–18.

19. Borges-Walmsley, M. I., D. Chen, X. Shu, and A. R. Walmsley. 2002. The pathobiology of *Paracoccidioides brasiliensis. Trends Microbiol.* **10:**80–87.

20. Brandhorst, T., M. Wüthrich, T. Warner, and B. S. Klein. 1999. Targeted gene disruption reveals an adhesin indispensable for pathogenicity of *Blastomyces dermatitidis. J. Exp. Med.* **189:**1207–1216.

21. Brummer, E. 1994. Interaction of *Paracoccidioides brasiliensis* with host defense cells, p. 213–223. *In* M. Franco, C. S. Lacaz, A. M. Restrepo, and G. Del Negro (ed.), *Paracoccidioidomycoses.* CRC Press, Inc., Boca Raton, Fla.

22. Brummer, E., L. H. Hanson, A. Restrepo, and D. A. Stevens. 1988. In vivo and in vitro activation of pulmonary macrophages by IFN-gamma for enhanced killing of *Paracoccidioides brasiliensis* or *Blastomyces dermatitidis. J. Immunol.* **140:**2786–2789.

23. Brummer, E., N. Kurita, S. Yosihida, K. Nishimura, and M. Miyahi. 1991. Fungistatic activity of human neutrophils against *Histoplasma capsulatum:* correlation with phagocytosis. *J. Infect. Dis.* **164:**158–162.

24. Brummer, E., J. G. McEwen, and D. A. Stevens. 1986. Fungicidal activity of murine inflammatory polymorphonuclear neutrophils: comparison with murine peripheral blood PMN. *Clin. Exp. Immunol.* **66:**681–690.

25. Brummer, E., P. A. Morozumi, P. T. Vo, and D. A. Stevens. 1982. Protection against pulmonary blastomycosis: adoptive transfer with T lymphocytes, but not serum, from resistant mice. *Cell. Immunol.* **73:**349–359.

26. Brummer, E., and D. A. Stevens. 1987. Activation of pulmonary macrophages for fungicidal activity by gamma-interferon or lymphokines. *Clin. Exp. Immunol.* **70:**520–528.

27. Buchanan, K. L. 2005. Innate and acquired immunity in *Cryptococcus neoformans* infection of the central nervous system, *In* P. L. Fidel and G. Huffnagle (ed.), *Fungal Immunology: from an Organ Perspective,* in press. Springer-Verlag, London, United Kingdom.

28. Buchanan, K. L., and H. A. Doyle. 2000. Requirement for CD4+ T lymphocytes in host resistance against *Cryptococcus neoformans* in the central nervous system of immunized mice. *Infect. Immun.* **68:**456–462.

29. Burns, D. N., R. Tuomala, B. H. Chang, R. Hershow, H. Minkoff, E. Rodriquez, C. Zorrilla, H. Hammill, and J. Regan. 1997. Vaginal colonization or infection with *Candida albicans* in human immunodeficiency virus-infected women during pregnancy and during postpartum period. *Clin. Infect. Dis.* **24:**201–210.

30. Cain, J. A., and G. S. Deepe. 1998. Evolution of the primary immune response to *Histoplasma capsulatum* in murine lung. *Infect. Immun.* **66:**1473–1481.

31. Cain, J. A., and G. S. Deepe, Jr. 2000. Interleukin-12 neutralization alters lung inflammation and leukocyte expression of CD80, CD86, and major histocompatibility complex class II in mice infected with *Histoplasma capsulatum. Infect. Immun.* **68:**2069–2076.

32. Calderone, R. A., and W. A. Fonzi. 2001. Virulence factors of *Candida albicans. Trends Microbiol.* **9:**327–335.

33. Camargo, Z. P., and L. E. Cano. 1994. Humoral immunity, p. 31–38. *In* M. Franco, C. S. Lacaz, A. M. Restrepo, and G. Del Negro (ed.), *Paracoccidioidomycosis.* CRC Press, Inc., Boca Raton, Fla.

34. Cano, L. E., S. S. Kashino, C. Arruda, D. André, C. F. Xidieh, L. M. Singer-Vermes, C. A. C. Vaz, E. Burger, and V. L. G. Calich. 1998. Protective role of gamma interferon in experimental pulmonary paracoccidioidomycosis. *Infect. Immun.* **66:**800–806.

35. Cano, L. E., L. M. Singer-Vermes, T. A. Costa, J. O. Mengel, C. F. Xidieh, C. Arruda, D. André, C. A. C. Vaz, E. Burger, and V. L. G. Calich. 2000. Deletion of CD8+ T cells in vivo impairs host defense of mice resistant and susceptible to pulmonary paracoccidioidomycosis. *Infect. Immun.* **68:**352–359.

36. Cano, L. E., L. M. Singer-Vermes, C. A. C. Vaz, M. Russo, and V. L. G. Calich. 1995. Pulmonary paracoccidioidomycosis in resistant and susceptible mice: relationship among progression of infection, bronchoalveolar cell activation, cellular immune response, and specific isotype patterns. *Infect. Immun.* **63:**1777–1783.

37. Cantorna, M. T., and E. Balish. 1990. Mucosal and systemic candidiasis in congenitally immunodeficient mice. *Infect. Immun.* **58:**1093–1100.

38. Cantorna, M. T., D. Mook, and E. Balish. 1990. Resistance of congenitally immunodeficient gnotobiotic mice to vaginal candidiasis. *Infect. Immun.* **58:**3813–3815.

39. Carvalho, L. P., O. Bacellar, N. Neves, A. R. de Jesus, and E. M. Carvalho. 2002. Downregulation of IFN-gamma production in patients with recurrent vaginal candidiasis. *J. Allergy Clin. Immmunol.* **109:**102–105.

40. Casadevall, A. 1995. Antibody immunity and invasive fungal infections. *Infect. Immun.* **63:**4211–4218.

41. Casadevall, A., and J. R. Perfect. 1998. *Cryptococcus neoformans.* ASM Press, Washington, D.C.

42. Cassone, A., F. De Bernardis, A. Torosantucci, E. Tacconelli, M. Tumbarello, and R. Cauda. 1999. *In vitro* and *in vivo* anticandidal activity of human immunodeficiency virus protease inhibitors. *J. Infect. Dis.* **180:**448–453.

43. Cassone, A., M. Boccanera, D. A. Adriani, G. Santoni, and F. De Bernardis. 1995. Rats clearing a vaginal infection by *Candida albicans* acquire specific, antibody-mediated resistance to vaginal infection. *Infect. Immun.* **63:**2619–2624.

44. Cenci, E., A. Mencacci, G. Del Sero, A. Bacci, C. Montagnoli, C. F. D'Ostiani, P. Mosci, M. Bachmann, F. Bistoni, M. Kopf, and L. Romani. 1999. Interleukin-4 causes susceptibility to invasive pulmonary aspergillosis through suppression of protective type I responses. *J. Infect. Dis.* **180:**1957–1968.

45. Chakir, J., L. Cote, C. Coulombe, and N. Deslauriers. 1994. Differential pattern of infection and immune response during experimental oral candidiasis in BALB/c and DBA/2 (H-2d) mice. *Oral Microbiol. Immunol.* **9:**88–94.

46. Challacombe, S. J., and S. P. Sweet. 1997. Salivary and mucosal immune responses to HIV and its co-pathogens. *Oral Dis.* 3(Suppl. 1):S79–S84.

47. Clark, R. A., S. A. Blakley, J. Rice, and W. Brandon. 1995. Predictors of HIV progression in women. *J. Acquir. Immune Defic. Syndr. Hum. Retrovirol.* 9:43–50.

48. Clemons, K. V., W. C. Darbonne, J. T. Curnutte, R. A. Sobel, and D. A. Stevens. 2000. Experimental histoplasmosis in mice treated with anti-murine interferon-γ antibody and in interferon-γ knockout mice. *Microb. Infect.* 2:997–1001.

49. Cohen, M. S., J. G. Cannon, A. E. Jerse, L. M. Charniga, S. F. Isbey, and L. G. Whicker. 1994. Human experimentation with *Neisseria gonorrhoeae*: rationale, methods, and implications for the biology of infection and vaccine development. *J. Infect. Dis.* 169:532–537.

50. Colon, M., N. Toledo, C. L. Valiente, N. Rodriguez, N. Yano, H. L. Mathews, and Y. Yamamura. 1998. Antifungal and cytokine producing activities of CD8⁺ T lymphocytes from HIV-1 infected individuals. *Bol. Asoc. Med. Puerto Rico* 90:21–26.

51. Coogan, M. M., S. P. Sweet, and S. J. Challacombe. 1994. Immunoglobulin A (IgA), IgA1, and IgA2 antibodies to *Candida albicans* in whole and parotid saliva in human immunodeficiency virus infection and AIDS. *Infect. Immun.* 62:892–896.

52. Corrigan, E. M., R. L. Clancy, M. L. Dunkley, F. M. Eyers, and K. W. Beagley. 1998. Cellular immunity in recurrent vulvovaginal candidiasis. *Clin. Exp. Immunol.* 111:574–578.

53. Cox, R. A., and D. M. Magee. 2004. Coccidioidomycosis: host response and vaccine development. *Clin. Microbiol. Rev.* 17:804–839.

54. De Bernardis, F., M. Boccanera, D. Adriana, E. Spreghini, G. Santoni, and A. Cassone. 1997. Protective role of antimannan and anti-aspartyl proteinase antibodies in an experimental model of *Candida albicans* vaginitis in rats. *Infect. Immun.* 65:3399–3405.

55. (Reference deleted.)

56. De Bernardis, F., G. Santoni, M. Boccanera, E. Spreghini, D. Adriani, L. Morelli, and A. Cassone. 2000. Local anticandidal immune responses in a rat model of vaginal infection by and protection against *Candida albicans*. *Infect. Immun.* 68:3297–3304.

57. Deepe, G. S. 2000. *Histoplasma capsulatum*, p. 2718–2732. *In* G. L. Mandell, R. Dolin, and J. E. Bennett (ed.), *Principles and Practices in Infectious Diseases*. Churchill Livingstone, Inc., Philadelphia, Pa.

58. Deslauriers, N., L. Cote, S. Montplaisir, and L. De Repentigny. 1997. Oral carriage of *Candida albicans* in murine AIDS. *Infect. Immun.* 65:661–667.

59. Djeu, J. Y. 1993. Modulators of immune response to fungi, p. 521–532. *In* J. W. Murphy, H. Friedman, and M. Bendinelli (ed.), *Fungal Infections and Immune Responses*. Plenum Press, Inc., New York, N.Y.

60. Djeu, J. Y., and D. K. Blanchard. 1987. Regulation of human polymorphonuclear neutrophil (PMN) activity against *Candida albicans* by large granular lymphocytes via release of a PMN-activating factor. *J. Immunol.* 139:2761–2767.

61. Domer, J. E., R. E. Garner, and R. N. Befidi-Mengue. 1989. Mannan as an antigen in cell-mediated immunity (CMI) assays and as a modulator of mannan-specific CMI. *Infect. Immun.* 57:693–700.

62. Duerr, A., M. F. Sierra, J. Feldman, L. M. Clarke, I. Ehlich, and J. Dehovitz. 1997. Immune compromise and prevalence of *Candida* vulvovaginitis in human immunodeficiency virus-infected women. *Obstet. Gynecol.* 90:252–256.

63. Edwards, J. E., Jr., T. A. Gaither, J. J. O'Shea, D. Rotrosen, T. L. Lawley, S. A. Wright, M. M. Frank, and I. Green. 1986. Expression of specific binding sites on *Candida* with functional and antigenic characteristics of human complement receptors. *J. Immunol.* 137:3577–3583.

64. Einstein, H., and C. Catanzaro. 1996. *Coccidioidomycosis*. National Foundation for Infectious Disease, Washington, D.C.

65. Elahi, S., G. Pang, R. Clancy, and R. B. Ashman. 2000. Cellular and cytokine correlates of mucosal protection in murine model of oral candidiasis. *Infect. Immun.* 68:5771–5777.

66. Farah, C. S., S. Elahi, G. Pang, T. Gotjamanos, G. J. Seymour, R. L. Clancy, and R. B. Ashman. 2001. T cells augment monocyte and neutrophil function in host resistance against oropharyngeal candidiasis. *Infect. Immun.* 69:6110–6118.

67. Farah, C. S., T. Gotjamanos, G. J. Seymour, and R. B. Ashman. 2002. Cytokines in the oral mucosa of mice infected with *Candida albicans*. *Oral Microbiol. Immunol.* 17:375–378.

68. Fidel, P. L., Jr. 1999. *Candida albicans*: from commensal to pathogen, p. 441–476. *In* G. W. Tannock (ed.), *Medical Importance of the Normal Microflora*. Chapman & Hall, London, United Kingdom.

69. Fidel, P. L., Jr. 2002. Distinct protective host defenses against oral and vaginal candidiasis. *Med. Mycol.* 40:359–375.

70. Fidel, P. L., Jr., M. Barousse, T. Espinosa, C. Camaratti, M. Ficarra, D. H. Martin, A. J. Quayle, and K. Dunlap. 2004. A live intravaginal *Candida* challenge in humans reveals new hypotheses for the immunopathogenesis of vulvovaginal candidiasis. *Infect. Immun.* 72:2939–2946.

71. Fidel, P. L., Jr., M. Barousse, T. Espinosa, R. R. Chesson, and K. Dunlap. 2003. Local immune responsiveness following intravaginal challenge with *Candida* antigen in adult women at different stages of the menstrual cycle. *Med. Mycol.* 41:97–109.

72. Fidel, P. L., Jr., J. L. Cutright, and J. D. Sobel. 1995. Effects of systemic cell-mediated immunity on vaginal candidiasis in mice resistant and susceptible to *Candida albicans* infections. *Infect. Immun.* 63:4191–4194.

73. Fidel, P. L., Jr., K. A. Ginsburg, J. L. Cutright, N. A. Wolf, D. Leaman, K. Dunlap, and J. D. Sobel. 1997. Vaginal-associated immunity in women with recurrent vulvovaginal candidiasis: evidence for vaginal Th1-type responses following intravaginal challenge with *Candida* antigen. *J. Infect. Dis.* 176:728–739.

74. Fidel, P. L., Jr., W. Luo, C. Steele, J. Chabain, M. Baker, and F. L. Wormley. 1999. Analysis of vaginal cell populations during experimental vaginal candidiasis. *Infect. Immun.* 67:3135–3140.

75. Fidel, P. L., Jr., M. E. Lynch, D. H. Conaway, L. Tait, and J. D. Sobel. 1995. Mice immunized by primary vaginal *C. albicans* infection develop acquired vaginal mucosal immunity. *Infect. Immun.* 63:547–553.

76. Fidel, P. L., Jr., M. E. Lynch, V. Redondo-Lopez, J. D. Sobel, and R. Robinson. 1993. Systemic cell-mediated immune reactivity in women with recurrent vulvovaginal candidiasis (RVVC). *J. Infect. Dis.* 168:1458–1465.

77. Fidel, P. L., Jr., M. E. Lynch, and J. D. Sobel. 1994. Effects of preinduced *Candida*-specific systemic cell-mediated

immunity on experimental vaginal candidiasis. *Infect. Immun.* **62:**1032–1038.

78. **Fidel, P. L., Jr., M. E. Lynch, and J. D. Sobel.** 1995. Circulating CD4 and CD8 T cells have little impact on host defense against experimental vaginal candidiasis. *Infect. Immun.* **63:**2403–2408.

79. **Fidel, P. L., Jr., and J. D. Sobel.** 1996. Immunopathogenesis of recurrent vulvovaginal candidiasis. *Clin. Microbiol. Rev.* **9:**335–348.

80. **Fidel, P. L., Jr., and J. D. Sobel.** 1999. Murine models of *Candida* vaginal infections, p. 741–748. *In* O. Zak and M. Sande (ed.), *Experimental Models in Antimicrobial Chemotherapy.* Academic Press, Ltd., London, United Kingdom.

81. **Fidel, P. L., Jr., N. A. Wolf, and M. A. KuKuruga.** 1996. T lymphocytes in the murine vaginal mucosa are phenotypically distinct from those in the periphery. *Infect. Immun.* **64:**3793–3799.

82. **Fidel, P. L., Jr., F. L. Wormley, Jr., J. Chaiban, R. R. Chesson, and V. Lounev.** 2001. Analysis of the CD4 protein on human vaginal CD4$^+$ T cells. *Am. J. Reprod. Immunol.* **45:**200–204.

83. **Fornazim, M. C., A. Balthazar, R. Quagliato, Jr., R. L. Mamoni, C. Garcia, and M. H. Blotta.** 2003. Evaluation of bronchoalveolar cells in pulmonary paracoccidioidomycosis. *Eur. Respir. J.* **22:**895–899.

84. **Garner, R. E., A. M. Childress, L. G. Human, and J. E. Domer.** 1990. Characterization of *Candida albicans* mannan-induced, mannan-specific delayed-hypersensitivity suppressor cells. *Infect. Immun.* **58:**2613–2620.

85. **Garner, R. E., and J. E. Domer.** 1994. Lack of effect of *Candida albicans* mannan on development of protective immune responses in experimental murine candidiasis. *Infect. Immun.* **62:**738–741.

86. **Gonzalez, A., J. H. Sahaza, B. L. Ortiz, A. Restrepo, and L. E. Cano.** 2003. Production of pro-inflammatory cytokines during early stages of experimental *Paracoccidioides brasiliensis* infection. *Med. Mycol.* **41:**391–399.

87. **Greenfield, R. A.** 1992. Host defense system interactions with *Candida.* *J. Med. Vet. Mycol.* **30:**89–104.

88. **Greenspan, D., E. Komaroff, M. Redford, J. A. Phelan, M. Navazesh, M. E. Alves, H. Kamrath, R. Mulligan, C. E. Barr, and J. S. Greenspan.** 2000. Oral mucosal lesions and HIV viral load in the Women's Interagency HIV study (WIHS). *J. Acquir. Immune Defic. Syndr.* **25:**44–50.

89. **Gruber, A., C. Speth, E. Lukasser-Vogl, R. Zangerle, M. Borg-von Zepelin, M. P. Dierich, and R. Wurzner.** 1999. Human immunodeficiency virus type 1 protease inhibitor attenuates *Candida albicans* virulence properties in vitro. *Immunopharmacology* **41:**227–234.

90. **Han, Y., R. P. Morrison, and J. E. Cutler.** 1998. A vaccine and monoclonal antibodies that enhance mouse resistance to *Candida albicans* vaginal infection. *Infect. Immun.* **66:**5771–5776.

91. **Han, Y., M. H. Riesselman, and J. E. Cutler.** 2000. Protection against candidiasis by an immunoglobulin G3 (IgG3) monoclonal antibody specific for the same mannotriose as an IgM protective antibody. *Infect. Immun.* **68:**1649–1654.

92. **Harvey, R. P., E. S. Schmid, C. C. Carrington, and D. A. Stevens.** 1978. Mouse model of pulmonary blastomycosis: utility, simplicity, and quantitative parameters. *Am. Rev. Respir. Dis.* **117:**695–703.

93. **Hector, R. F., J. E. Domer, and E. W. Carrow.** 1982. Immune responses to *Candida albicans* in genetically distinct mice. *Infect. Immun.* **38:**1020–1028.

94. **Heidenreich, F., and M. P. Dierich.** 1985. *Candida albicans* and *Candida stellatoidea*, in contrast to other *Candida* species, bind iC3b and C3d but not C3b. *Infect. Immun.* **50:**598–600.

95. **Hladik, F., G. Lentz, E. Delpit, A. McElroy, and M. J. McElrath.** 1999. Coexpression of CCR5 and IL-2 in human genital but not blood cells: implications for the ontogeny of the CCR5$^+$ Th1 phenotype. *J. Immunol.* **163:**2306–2313.

96. **Huffnagle, G. B., and M. F. Lipscomb.** 1998. Cells and cytokines in pulmonary cryptococcosis. *Res. Immunol.* **149:**387–396.

97. **Ibraghimov, A. R., R. E. Sacco, M. Sandor, Z. Iakoubov, and R. G. Lynch.** 1995. Resident CD4$^+$ αβ T cells of the murine female genital tract: a phenotypically distinct T cell lineage that rapidly proliferates in response to systemic T cell activation stimuli. *Int. Immunol.* **7:**1763–1769.

98. **Imam, N., C. C. J. Carpenter, K. H. Mayer, A. Fisher, M. Stein, and S. B. Danforth.** 1990. Hierarchical pattern of mucosal *Candida* infections in HIV-seropositive women. *Am. J. Med.* **89:**142–146.

99. **Inghirami, G., B. Y. Zhu, L. Chess, and D. M. Knowles.** 1990. Flow cytometric and immunohistochemical characterization of the gamma/delta T-lymphocyte population in normal human lymphoid tissue and peripheral blood. *Am. J. Pathol.* **136:**357–367.

100. **Itohara, S., A. G. Farr, J. J. Lafaille, M. Bonneville, Y. Takagaki, W. Haas, and S. Tonegawa.** 1990. Homing of a gamma/delta thymocyte subset with homogenous T-cell receptors to mucosal epithelia. *Nature* **343:**754–757.

101. **Iwasaki, A., and B. Kelsall.** 2001. Unique functions of CD11b$^+$, CD8α$^+$, and double-negative Peyer's patch dendritic cells. *J. Immunol.* **166:**4884–4890.

102. **Johansson, E.-L., A. Rudin, L. Wassen, and J. Holmgren.** 2000. Distribution of lymphocytes and adhesion molecules in human cervix and vagina. *Immunology* **96:**272–277.

103. **Jones-Carson, J., A. Vazquez-Torres, T. Warner, and E. Balish.** 2000. Disparate requirement for T cells in resistance to mucosal and acute systemic candidiasis. *Infect. Immun.* **68:**2363–2365.

104. **Jonuleit, H., and E. Schmitt.** 2003. The regulatory T cell family: distinct subsets and the interrelations. *J. Immunol.* **171:**6323–6327.

105. **Kelly, K. A., H. L. Gray, J. C. Walker, R. G. Rank, F. L. Wormley, Jr., and P. L. Fidel, Jr.** 2001. *Chlamydia trachomatis* infection does not enhance local cellular immunity against concurrent *Candida albicans* vaginal infection. *Infect. Immun.* **69:**3451–3454.

106. **Kelly, K. A., and R. G. Rank.** 1997. Identification of homing receptors that mediate the recruitment of CD4 T cells to the genital tract following intravaginal infection with *Chlamydia trachomatis.* *Infect. Immun.* **65:**5198–5208.

107. **Kent, H. L.** 1991. Epidemiology of vaginitis. *Am. J. Obstet. Gynecol.* **165:**1168–1175.

108. **Kirkpatrick, C. H.** 1984. Host factors in defense against fungal infections. *Am. J. Med.* **77:**1–12.

109. **Kirkpatrick, C. H., J. W. Chandler, and R. N. Schimke.** 1970. Chronic mucocutaneous moniliasis with impaired delayed hypersensitivity. *Clin. Exp. Immunol.* **6:**375–385.

110. **Kirkpatrick, C. H., R. R. Rich, and J. E. Bennett.** 1971. Chronic mucocutaneous candidiasis: model building in cellular immunity. *Ann. Intern. Med.* **74:**955–978.

111. Kirkpatrick, C. H., and P. G. Sohnle. 1981. Chronic mucocutaneous candidiasis, p. 495–514. *In* B. Safai and R. A. Good (ed.), *Immunodermatology*. Plenum Press, New York, N.Y.

112. Klein, B. S. 2000. Molecular basis of pathogenicity in *Blastomyces dermatitidis*: the importance of adhesion. *Curr. Opin. Microbiol.* 3:339–343.

113. Klein, B. S., S. Chaturvedi, L. H. Hogan, J. M. Jones, and S. L. Newman. 1994. Altered expression of the surface protein WI-1 in genetically related strains of *Blastomyces dermatitidis* that differ in virulence regulates recognition of yeasts in human macrophages. *Infect. Immun.* 62:3536–3542.

114. Klein, R. S., C. A. Harris, C. B. Small, B. Moll, M. Lesser, and G. H. Friedland. 1984. Oral candidiasis in high-risk patients as the initial manifestation of the acquired immunodeficiency syndrome. *N. Engl. J. Med.* 311:354–357.

115. Kozel, T. R., R. R. Brown, and G. S. T. Pfrommer. 1987. Activation and binding of C3 by *Candida albicans*. *Infect. Immun.* 55:1890–1894.

116. Kunkl, A., L. Mortara, M. T. Valle, D. Fenoglio, M. Paola, A. M. Megiovanni, A. Alessandrini, G. Li Pira, G. Mazzarello, V. Del Bono, A. Canessa, D. Bassetti, and F. Manca. 1998. Recognition of antigenic clusters of *Candida albicans* by T lymphocytes from human immunodeficiency virus-infected persons. *J. Infect. Dis.* 178:488–496.

117. Kunzmann, S., J. G. Wohlfahrt, S. Itoh, H. Asao, M. Komada, C. A. Akdis, K. Blaser, and C. B. Schmidt-Weber. 2003. SARA and Hgs attentuate susceptibility to TGF-β1-mediated T cell suppression. *FASEB J.* 17:194–202.

118. Kurita, N., E. Brummer, S. Yoshida, K. Nishimura, and M. Miyahi. 1991. Antifungal activity of murine polymorphonuclear neutrophils against *Histoplasma capsulatum*. *J. Med. Vet. Mycol.* 29:133–143.

119. Kurup, V. P., and G. Grunig. 2002. Animal models of allergic bronchopulmonary aspergillosis. *Mycopathologia* 153:165–177.

120. Kutteh, W. H., Z. Moldoveanu, and J. Mestecky. 1998. Mucosal immunity in the female reproductive tract: correlation of immunoglobulins, cytokines, and reproductive hormones in human cervical mucus around the time of ovulation. *AIDS Res. Hum. Retroviruses* 14:51–55.

121. Latgé, J. P. 1999. *Aspergillus fumigatus* and aspergillosis. *Clin. Infect. Dis.* 26:781–805.

122. Latgé, J. P. 2001. The pathobiology of *Aspergillus fumigatus*. *Trends Microbiol.* 9:382–389.

123. Latgé, J. P., and R. Calderon. 2002. Host-microbe interactions: fungi invasive human fungal opportunistic infections. *Curr. Opin. Microbiol.* 5:355–358.

124. Leigh, J. E., M. Barousse, R. K. Swoboda, T. Myers, S. Hager, N. A. Wolf, J. L. Cutright, J. Thompson, J. D. Sobel, and P. L. Fidel, Jr. 2001. *Candida*-specific systemic cell-mediated immune reactivities in HIV-infected persons with and without mucosal candidiaisis. *J. Infect. Dis.* 183:277–285.

124a. Leigh, J. E., K. Shetty, and P. L. Fidel, Jr. 2004. Oral opportunistic infections in the HIV-positive individual: review and role of mucosal immunity. *AIDS Pat. Care STDs* 18:13–26.

125. Leigh, J. E., C. Steele, F. L. Wormley, Jr., W. Luo, R. A. Clark, W. R. Gallaher, and P. L. Fidel, Jr. 1998. Th1/Th2 cytokine expression in saliva of HIV-positive and HIV-negative individuals: a pilot study in HIV-positive individuals with oropharyngeal candidiasis. *J. Acquir. Immun. Defic. Syndr. Hum. Retrovirol.* 19:373–380.

126. Levitz, S. M., and T. P. Farrell. 1990. Human neutrophil degranulation stimulated by *Aspergillus fumigatus*. *J. Leukoc. Biol.* 47:170–175.

127. Li, Q., and B. J. Cheravil. 2003. Role of Toll-like receptor 4 in macrophage activation and tolerance during *Salmonella enterica* serovar Typhimurium infection. *Infect. Immun.* 71:4873–4882.

128. Li, S. P., S. I. Lee, and J. E. Domer. 1998. Alterations in frequency of interleukin-2 (IL-2)-, gamma interferon-, or IL-4-secreting splenocytes induced by *Candida albicans* mannan and/or monophosphoryl lipid A. *Infect. Immun.* 66:1392–1399.

129. Lindesmith, L., C. Moe, S. Marionneau, N. Ruvoen, X. Jiang, L. Lindblad, P. Stewart, J. LePendu, and R. Baric. 2003. Human susceptibility and resistance to Norwalk virus infection. *Nat. Med.* 9:548–553.

130. Little, C. H., G. M. Georgiou, A. Marceglia, H. Ogedgebe, R. E. Cone, and D. Mazza. 2000. Measurement of T-cell-derived antigen binding molecules and immunoglobulin G specific to *Candida albicans* mannan in sera of patients with recurrent vulvovaginal candidiasis. *Infect. Immun.* 68:3840–3847.

131. Lynch, M. E., J. D. Sobel, and P. L. Fidel, Jr. 1996. Role of antifingal drug resistance in the pathogenesis of recurrent vulvovaginal candidasis. *J. Med. Vet. Mycol.* 34:337–339.

132. Macher, A. M. 1988. The pathology of AIDS. *Public Health Rep.* 103:246–254.

133. Magee, D. M., and R. A. Cox. 1995. Roles of gamma interferon and interleukin-4 in genetically determined resistance to *Coccidioides immitis*. *Infect. Immun.* 63:3514–3519.

134. Mathews, H. L., and D. W. A. Beno. 1993. Quantitative measurement of lymphocyte mediated growth inhibition of *Candida albicans*. *J. Immunol. Methods* 164:155–157.

135. Mathur, S., G. Virella, J. Koistinen, E. O. Horger, T. A. Mahvi, and H. H. Fudenberg. 1977. Humoral immunity in vaginal candidiasis. *Infect. Immun.* 15:287–294.

135a. McNulty, K. M., J. Plianrungsi, J. E. Leigh, D. Mercante, and P. L. Fidel, Jr. 2005. Characterization of CD8+ T cells and microenvironment in oral lesions of HIV-infected persons with oropharyngeal candidiasis. *Infect. Immun.* 73:3659–3667.

136. Mehrad, B., T. A. Moore, and T. J. Standiford. 2000. Macrophage inflammatory protein-1 alpha is critical mediator of host defense against invasive pulmonary aspergillosis in neutropenic hosts. *J. Immunol.* 165:962–968.

137. Millon, L., C. Drobacheff, R. Piarroux, M. Monod, G. Reboux, R. Laurent, and D. Meillet. 2001. Longitudinal study of anti-*Candida albicans* mucosal immunity against aspartic proteinases in HIV-infected patients. *J. Acquir. Immune Defic. Syndr.* 26:137–144.

138. Mitchell, T. G., and J. R. Perfect. 1995. Cryptococcosis in the era of AIDS—100 years after the discovery of *Cryptococcus neoformans*. *Clin. Microbiol. Rev.* 8:515–548.

139. Montenegro, M. R., and M. Franco. 1994. Pathology, p. 19–26. *In* M. Franco, C. S. Lacaz, A. M. Restrepo, and G. Del Negro (ed.), *Paracoccidioidomycoses*. CRC Press, Inc., Boca Raton, Fla.

140. Morrison, C. J., E. Brummer, and D. A. Stevens. 1987. Effect of a local immune reaction on peripheral blood

polymorphonuclear neutrophil microbicidal function: studies with fungal targets. *Cell. Immunol.* **110:**176–182.

141. **Mota, N. G. S., M. T. Rezhallah-Iwasso, M. T. S. Peraçoli, R. C. Audi, R. P. Mendes, J. Marcondes, S. A. Marques, N. L. Dillon, and M. F. Franco.** 1985. Correlation between cell-mediated immunity and clinical forms of paracoccidioidomycosis. *Trans. R. Soc. Trop. Med. Hyg.* **79:**765–772.

142. **Murphy, J. W.** 1998. Protective cell-mediated immunity against *Cryptococcus neoformans. Res. Immunol.* **149:**373–386.

143. **Myers, T. A., J. E. Leigh, A. Arribas, S. Hager, R. A. Clark, E. Lilly, and P. L. Fidel, Jr.** 2003. Immunohistochemical evaluation of T cells in oral lesions from human immunodeficiency virus-positive persons with oropharyngeal candidiasis. *Infect. Immun.* **71:**956–963.

144. **Nagai, Y., N. Takeshita, and T. Saku.** 1992. Histopathologic and ultrastructural studies of oral mucosa with *Candida* infection. *J. Oral Pathol. Med.* **21:**171–175.

145. **Nandi, D., and J. P. Allison.** 1991. Phenotypic analysis and gamma/delta-T cell receptor repertoire of murine T cells associated with the vaginal epithelium. *J. Immunol.* **147:**1773–1778.

146. **Netea, M. G., C. A. A. Van der Graaf, A. G. Vonk, I. Verschueren, J. W. M. van der Meer, and B. J. Kullberg.** 2002. The role of Toll-like receptor (TLR) 2 and 4 in the host defense against disseminated candidiasis. *J. Infect. Dis.* **185:**1483–1489.

147. **Nielsen, H., K. D. Bentsen, L. Hojtved, E. H. Willemoes, F. Scheutz, M. Schiodt, K. Stoltze, and J. J. Pindborg.** 1994. Oral candidiasis and immune status of HIV-infected patients. *J. Oral Pathol. Med.* **23:**140–143.

148. **Nomanbhoy, F., C. Steele, J. Yano, and P. L. Fidel, Jr.** 2002. Vaginal and oral epithelial cell anti-*Candida* activity. *Infect. Immun.* **70:**7081–7088.

149. **Odds, F. C.** 1988. Chronic mucocutaneous candidiosis, p. 104–110. *In* F. C. Odds (ed.), *Candida and Candidosis.* University Park Press, Baltimore, Md.

150. **Palella, F. J., Jr., K. M. Delaney, A. C. Moorman, M. O. Loveless, J. Fuhrer, G. A. Satten, D. J. Aschman, S. D. Holmberg, and HIV Outpatient Study Investigators.** 1998. Declining morbidity and mortality among patients with advanced human immunodeficiency virus infection. *N. Engl. J. Med.* **338:**853–860.

151. **Parr, M. B., and E. L. Parr.** 1991. Langerhans cells and T lymphocyte subsets in the murine vagina and cervix. *Biol. Reprod.* **44:**491–498.

152. **Parr, M. B., and E. L. Parr.** 1997. Protective immunity against HSV-2 in the mouse vagina. *J. Reprod. Immunol.* **36:**77–92.

153. **Phelan, J. A., B. R. Salzman, G. H. Friendland, and R. S. Klein.** 1987. Oral findings in patients with acquired immune deficiency syndrome. *Oral Surg.* **64:**50–56.

154. **Pina, A., R. C. Valente-Ferreira, E. E. W. Molinari-Madlum, C. A. C. Vaz, A. C. Keller, and V. L. G. Calich.** 2004. Absence of interleukin-4 determines less severe pulmonary paracoccidioidomycosis associated with impaired Th2 response. *Infect. Immun.* **72:**2369–2378.

155. **Polonelli, L., F. De Bernardis, S. Conti, M. Boccanera, M. Gerloni, G. Morace, W. Magliani, C. Chezzi, and A. Cassone.** 1994. Idiotypic intravaginal vaccination to protect against candidal vaginitis by secretory, yeast killer toxin-like anti-idiotypic antibodies. *J. Immunol.* **152:**3175–3182.

156. **Polonelli, L., F. De Bernardis, S. Conti, M. Boccanera, W. Magliani, C. Gerloni, C. Cantella, and A. Cassone.** 1996. Human natural yeast killer toxin-like candidicidal antibodies. *J. Immunol.* **156:**1880–1885.

157. **Pulendran, B., K. Palucka, and J. Banchereau.** 2001. Sensing pathogens and tuning immune responses. *Science* **293:**253–256.

158. **Quinti, I., C. Palma, E. C. Guerra, M. J. Gomez, I. Mezzaroma, F. Aiuti, and A. Cassone.** 1991. Proliferative and cytotxic responses to mannoproteins of *Candida albicans* by peripheral blood lymphocyes of HIV-infected subjects. *Clin. Exp. Immunol.* **85:**1–8.

159. **Rabeneck, L., M. M. Crane, J. M. Risser, C. E. Lacke, and N. P. Wray.** 1993. A simple clinical staging system that predicts progression to AIDS using CD4 count, oral thrush, and night sweats. *J. Gen. Intern. Med.* **8:**5–9.

160. **Regulez, P., J. F. Garcia Fernandez, M. D. Moragues, J. Schneider, G. Quindos, and J. Ponton.** 1994. Detection of anti-*Candida albicans* IgE antibodies in vaginal washes from patients with acute vulvovaginal candidiasis. *Gynecol. Obstet. Investig.* **37:**110–114.

161. **Reis e Sousa, C.** 2001. Dendritic cells as sensors of infection. *Cell* **14:**495–498.

162. **Rhoads, J. L., D. C. Wright, R. R. Redfield, and D. S. Burke.** 1987. Chronic vaginal candidiasis in women with human immunodeficiency virus infection. *JAMA* **257:**3105–3107.

163. **Rocha, B., P. Vassalli, and D. Guy-Grand.** 1994. Thymic and extrathymic origins of gut intraepithelial lymphocyte populations in mice. *J. Exp. Med.* **180:**681–686.

164. **Rogers, T. J., and E. Balish.** 1980. Immunity to *Candida albicans. Microbiol. Rev.* **44:**660–682.

165. **Romagnoli, P., N. Pimpinelli, M. Mori, P. A. Reichart, L. R. Eversole, and G. Ficarra.** 1997. Immunocompetent cells in oral candidiasis of HIV-infected patients: an immunohistochemical and electron microscopical study. *Oral Dis.* **3:**99–105.

166. **Romani, L., F. Bistoni, and P. Puccetti.** 2002. Fungi, dendritic cells and receptors: a host perspective of fungal virulence. *Trends Microbiol.* **10:**508–514.

167. **Romani, L., P. Puccetti, and F. Bistoni.** 1996. Biological role of Th cell subsets in candidiasis, p. 114–137. *In* S. Romagnani (ed.), *Th1 and Th2 Cells in Health and Disease.* Karger, Farmington, Conn.

168. **Saavedra, M., B. Taylor, N. W. Lukacs, and P. L. Fidel, Jr.** 1999. Local production of chemokines during experimental vaginal candidiasis. *Infect. Immun.* **67:**5820–5829.

169. **Samaranayake, Y. H., and L. P. Samaranayake.** 2001. Experimental oral candidiasis in animal models. *Clin. Microbiol. Rev.* **14:**398–429.

170. **Samonis, G., A. Gikas, P. Toloudis, S. Maraki, G. Vrentzos, Y. Tselentis, N. Tsaparas, and G. Bodey.** 1994. Prospective study of the impact of broad-spectrum antibiotics on the yeast flora of the human gut. *Eur. J. Clin. Microbiol. Infect. Dis.* **13:**665–667.

171. **Sangeorzan, J. A., S. F. Bradley, X. He, L. T. Zarins, G. L. Ridenour, R. N. Tiballi, and C. A. Kauffman.** 1994. Epidemiology of oral candidiasis in HIV infected patients: colonization, infection, treatment, and emergence of fluconazole resistance. *Am. J. Med.* **97:**339–346.

172. **Santoni, G., M. Boccanera, D. Adriani, R. Lucciarini, C. Amantini, S. Morrone, A. Cassone, and F. De Bernardis.** 2002. Immune cell-mediated protection against vaginal candidiasis: evidence for a major role of vaginal CD4$^+$

T cells and possible participation of other local lymphocyte effectors. *Infect. Immun.* **70:**4791–4797.

173. Schuman, P., S. Ohmit, J. D. Sobel, K. H. Mayer, V. Greene, A. Rompalo, and R. S. Klein. 1998. Oral lesions among women living with or at risk for HIV infection. *Am. J. Med.* **104:**559–563.

174. Schuman, P., J. D. Sobel, S. Ohmit, K. H. Mayer, A. Rompalo, A. Duerr, D. K. Smith, D. Warren, and R. S. Klein. 1998. Mucosal candidal colonization and candidiasis in women with or at risk for human immunodeficiency virus infection. *Clin. Infect. Dis.* **27:**1161–1167.

175. Sobel, J. D. 1988. Pathogenesis and epidemiology of vulvovaginal candidiasis. *Ann. N. Y. Acad. Sci.* **544:**547–557.

176. Sobel, J. D. 1992. Pathogenesis and treatment of recurrent vulvovaginal candidiasis. *Clin. Infect. Dis.* **14:**S148–S153.

177. Sobel, J. D., A. Hasegawa, F. De Bernardis, D. A. Adriani, G. Pellegrini, A. Cassone, P. L. Fidel, Jr., C. G. Haidaris, F. Gigliotti, A. G. Harmsen, S. Fujita, K. Yamamoto, K. Shibuya, K. Uchida, and H. Yamaguchi. 1998. Selected animal models: vaginal candidiosis, *Pneumocystis* pnemonia, dermatophytosis and trichosporonosis. *Med. Mycol.* **36:**129–136.

178. Sorensen, K. N., K. V. Clemons, and D. A. Stevens. 1999. Murine models of blastomycosis, coccidioidomycosis, and histoplasmosis. *Mycopathologia* **146:**53–65.

179. Spinillo, A., G. Michelone, C. Cavanna, L. Colonna, E. Capuzzo, and S. Nicola. 1994. Clinical and microbiological characteristics of symptomatic vulvovaginal candidiasis in HIV-seropositive women. *Genitourin. Med.* **70:**268–272.

180. Steele, C., and P. L. Fidel, Jr. 2002. Cytokine and chemokine production by human oral and vaginal epithelial cells in response to *Candida albicans. Infect. Immun.* **70:**577–583.

181. Steele, C., J. E. Leigh, R. K. Swoboda, and P. L. Fidel, Jr. 2000. Growth inhibition of *Candida* by human oral epithelial cells. *J. Infect. Dis.* **182:**1479–1485.

182. Steele, C., J. E. Leigh, R. K. Swoboda, H. Ozenci, and P. L. Fidel, Jr. 2001. Potential role for a carbohydrate moiety in anti-*Candida* activity of human oral epithelial cells. *Infect. Immun.* **69:**7091–7099.

183. Steele, C., H. Ozenci, W. Luo, M. Scott, and P. L. Fidel, Jr. 1999. Growth inhibition of *Candida albicans* by vaginal cells from naive mice. *Med. Mycol.* **37:**251–260.

184. Steele, C., M. Ratterree, and P. L. Fidel, Jr. 1999. Differential susceptibility to experimental vaginal candidiasis in macaques. *J. Infect. Dis.* **180:**802–810.

185. Tachibana, T., T. Matsuyama, and M. Mitsuyama. 1999. Involvement of CD4+ T cells and macrophages in acquired protection against infection with *Sporothrix schenckii* in mice. *Med. Mycol.* **37:**397–404.

186. Taylor, B. N., M. Saavedra, and P. L. Fidel, Jr. 2000. Local Th1/Th2 cytokine production during experimental vaginal candidiasis. *Med. Mycol.* **38:**419–431.

187. Vazquez, J. A., J. D. Sobel, R. Demitriou, J. Vaishampayan, M. Lynch, and M. J. Zervos. 1994. Karyotyping of *Candida albicans* isolates obtained longitudinally in women with recurrent vulvovaginal candidiasis. *J. Infect. Dis.* **170:**1566–1569.

188. Wagner, R. D., C. L. Pierson, T. Warner, M. Dohnalek, J. Farmer, L. Roberts, M. Hilty, and E. Balish. 1997. Biotherapeutic effects of probiotic bacteria on candidiasis in immunodeficient mice. *Infect. Immun.* **65:**4165–4172.

189. Wagner, R. D., A. Vazquez-Torres, J. Jones-Carson, and E. Balish. 1996. B cell knockout mice are resistant to

mucosal and systemic candidiasis of endogenous origin but susceptible to experimental systemic candidasis. *J. Infect. Dis.* **174:**589–597.

190. Wang, Y. Y., S. K. P. Li, S. A. Moser, K. L. Bost, and J. E. Domer. 1998. Cytokine involvement in immunomodulatory activity affected by *Candida albicans* mannan. *Infect. Immun.* **66:**1384–1391.

191. White, M. H. 1996. Is vulvovaginal candidiasis an AIDS-related illness? *Clin. Infect. Dis.* **22**(Suppl. 2):S124–S127.

192. Williams, D. W., A. J. C. Potts, M. J. Wilson, J. B. Matthews, and M. A. O. Lewis. 1997. Characterisation of the inflammatory cell infiltrate in chronic hyperplastic candidosis of the oral mucosa. *J. Oral Pathol. Med.* **26:**83–89.

193. Witkin, S. S., J. Jeremias, and W. J. Ledger. 1988. A localized vaginal allergic response in women with recurrent vaginitis. *J. Allergy Clin. Immmunol.* **81:**412–416.

194. Witkin, S. S., J. Jeremias, and W. J. Ledger. 1989. Vaginal eosinophils and IgE antibodies to *Candida albicans* in women with recurrent vaginitis. *J. Med. Vet. Mycol.* **27:**57–58.

195. Wormley, F. L., Jr., C. Steele, K. Wozniak, K. Fujihashi, J. R. McGhee, and P. L. Fidel, Jr. 2001. Resistance of TCR δ-chain knock-out mice to experimental *Candida* vaginitis. *Infect. Immun.* **69:**7162–7164.

196. Wormley, F. L., Jr., J. Chaiban, and P. L. Fidel, Jr. 2001. Cell adhesion molecule and lymphocyte activation marker expression during experimental vaginal candidiasis. *Infect. Immun.* **69:**5072–5079.

197. Wozniak, K. L., J. E. Leigh, S. Hager, R. K. Swoboda, and P. L. Fidel, Jr. 2002. A comprehensive study of *Candida*-specific antibodies in saliva of HIV-infected persons with oropharyngeal candidiasis. *J. Infect. Dis.* **185:**1269–1276.

198. Wozniak, K. L., F. L. Wormley, Jr., and P. L. Fidel, Jr. 2002. *Candida*-specific antibodies during experimental vaginal candidiasis in mice. *Infect. Immun.* **70:**5790–5799.

199. Wray, D., D. H. Felix, and C. G. Cumming. 1990. Alteration of humoral responses to *Candida* in HIV infection. *Br. Dent. J.* **168:**326–329.

200. Wu-Hsieh, B., and D. H. Howard. 1989. Histoplasmosis, p. 199–225. *In* R. Cox (ed.), *Immunology of the Fungal Diseases.* CRC Press, Inc., Boca Raton, Fla.

201. Wuthrich, M., B. E. Finkel-Jiminez, and B. S. Klein. 2000. Interleukin 12 as an adjuvant to WI-1 adhesin immunization augments delayed-type hypersensitivity, shifts the subclass distribution of immunoglobulin G antibodies, and enhances protective immunity to *Blastomyces dermatitidis* infection. *Infect. Immun.* **68:**7172–7174.

202. Yamada, K., Y. Kimura, H. Nishimura, Y. Namii, M. Murase, and Y. Yoshikai. 1999. Characterization of CD4+ CD8αα+ and CD4− CD8αα+ intestinal intraepithelial lymphocytes in rats. *Int. Immunol.* **11:**21–28.

203. Yano, J., E. Lilly, C. Steele, D. Fortenberry, and P. L. Fidel, Jr. 2005. Oral and vaginal epithelial cell anti-*Candida* activity is acid labile and does not require live epithelial cells. *Oral Microbiol. Immunol.* **20:**199–205.

204. Zhou, P., G. G. Miller, and R. A. Seder. 1998. Factors involved in regulating primary and secondary immunity to infection with *Histoplasma capsulatum*: TNFα plays a critical role in maintaining seconday immunity in absence of IFNγ. *J. Immunol.* **160:**1359–1368.

Molecular Principles of Fungal Pathogenesis
Edited by Joseph Heitman et al.
©2006 ASM Press, Washington, D.C.

Chapter 35

Intracellular Pathogenesis of *Histoplasma capsulatum*

SIMON L. NEWMAN

Histoplasma capsulatum is a dimorphic fungal pathogen that causes a broad spectrum of disease activity. In most instances, immunocompetent individuals infected with this organism experience a mild flu-like illness, and only a small number of people actually exhibit acute clinical symptoms that require medical attention. However, depending on the size of the inoculum, some individuals may develop a chronic cavitary pneumonia or a progressive disseminated infection that is life-threatening (16). This latter disease is most often observed in individuals whose immune system is compromised either by immunosupressive agents or by infection with human immunodeficiency virus (HIV) (82). Although *H. capsulatum* is distributed worldwide, it is particularly endemic in the Ohio and Mississippi river valleys in the United States, where it has been estimated that as many as 500,000 new infections occur each year (1).

In nature, *H. capsulatum* lives in soil rich in bird and bat guano as a mold made up of filamentous mycelia or hyphae that bear both macro- and microconidia (20, 40). Infection with *H. capsulatum* occurs via inhalation of aerosolized microconidia or small mycelial fragments, which are deposited in the terminal bronchioles and alveoli of the lungs. The inhaled conidia then convert into the yeast form that is responsible for the pathogenesis of histoplasmosis. It is unclear whether conidia convert to the yeast form within alveolar macrophages (AM) or extracelluarly, since a temperature of 37°C is sufficient for transformation of conidia into yeasts (51). However, *H. capsulatum* conidia are phagocytized as rapidly as yeasts by AM (56), and in vitro the conidia convert into yeasts more rapidly within murine macrophages (Mφ) than in tissue culture medium (41). Thus, in the lungs, conidia may convert into yeasts both intracellularly and extracelluarly.

Once the conidia have converted into yeasts, the yeasts are phagocytized by AM, within which they multiply (9, 10, 16, 68). The replicating yeasts destroy the AM and are subsequently ingested by both other resident AM and inflammatory phagocytes that are recruited into the lungs. Repetition of this cycle leads to lymphohematogenous spread of the organisms to lymph nodes and other organs rich in mononuclear phagocytes. The hallmark of host defense against *H. capsulatum* is the development of specific cell-mediated immunity, which leads to activation of Mφ, killing of the yeast, and resolution of the infection (19, 55).

The interaction of *H. capsulatum* with Mφ is the key event in the pathogenesis of histoplasmosis. Mφ first provide an environment for fungal replication and dissemination and subsequently act as the final effector cells to remove the organism from the host. In addition, recent studies have suggested that dendritic cells (DC), the major antigen-presenting cells of the immune system, may also play a key role in host defense against *H. capsulatum*. This chapter reviews what is known about the strategies used by *H. capsulatum* yeasts to survive within human and animal Mφ, how Mφ become activated, and the mechanisms by which activated Mφ kill *H. capsulatum* yeasts. Since there are major differences in the interaction of *H. capsulatum* with human versus murine Mφ, these differences are discussed where applicable (Fig. 1). The interaction of *H. capsulatum* with DC is also discussed, as are the different roles that Mφ and DC may play in the development of cell-mediated immunity to *H. capsulatum*.

RECOGNITION OF *H. CAPSULATUM* YEASTS AND CONIDIA BY MACROPHAGES AND DENDRITIC CELLS

During the early phase of infection, Mφ recognize and phagocytose *H. capsulatum*, thereby providing

Simon L. Newman • Department of Medicine, Division of Infectious Diseases, University of Cincinnati College of Medicine, Cincinnati, OH 45267.

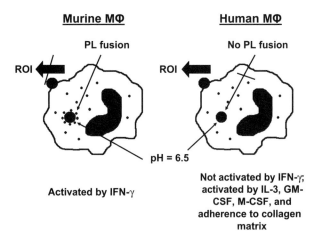

Murine MΦ

PL fusion

ROI

Human MΦ

No PL fusion

ROI

pH = 6.5

Activated by IFN-γ

Not activated by IFN-γ;
activated by IL-3, GM-
CSF, M-CSF, and
adherence to collagen
matrix

Figure 1. Differential response of human and murine MΦ to *H. capsulatum* yeasts. Upon ingestion of *H. capsulatum* yeasts, the respiratory burst of murine MΦ is not stimulated, but PL fusion does occur. In contrast, there is production of toxic oxygen radicals by human MΦ, but minimal PL fusion. Further, murine MΦ are activated by IFN-γ to inhibit the intracellular replication of yeasts, whereas IFN-γ does not activate human MΦ. In both human and murine MΦ, the intraphagosomal pH is maintained at 6.5.

them access to a permissive intracellular environment for replication. This initial interaction between *H. capsulatum* and MΦ is crucial for the pathogenesis of histoplasmosis. MΦ recognize *H. capsulatum* yeasts and microconidia via the CD18 family of adhesion-promoting β$_2$-integrins (LFA-1 [CD11a], CR3 [CD11b], and p150,95 [CD11c]) (12, 56). Based on experiments using monoclonal antibodies to the individual α-chains and the β$_2$ chain, *H. capsulatum* yeasts and conidia appear to bind independently to each of the three receptors on human monocyte-derived MΦ and human AM. For efficient attachment of *H. capsulatum*, the receptors must be mobile within the MΦ membrane, and mobility requires intact actin microfilaments of the cellular cytoskeleton (56). Other MΦ receptors do not appear to be involved in recognition of *H. capsulatum* yeasts, and there is no requirement for MΦ-derived complement components (12). To date, similar studies have not been performed with murine MΦ. However, murine MΦ also possess CD18 receptors (71), and it is therefore likely that they are used in recognition of *H. capsulatum*.

Recent studies demonstrate that the ligand on the surface of *H. capsulatum* yeasts and conidia that is recognized by MΦ CD18 receptors is heat shock protein 60 (hsp60) (50). Several lines of evidence confirm that hsp60 is on the cell wall of *H. capsulatum* yeasts in discrete clusters and that there are about 250 hsp60 antigenic sites per yeast cell. This small number of sites explains why MΦ CD18 receptors

must be mobile to efficiently bind *H. capsulatum* yeasts. Further, when hsp60 is depleted from a freeze-thaw extract of yeasts that contains *H. capsulatum* surface proteins, the freeze-thaw extract no longer inhibits the binding of *H. capsulatum* yeasts to MΦ, suggesting that hsp60 is the major, if not the sole, ligand by which MΦ recognize *H. capsulatum*.

hsp60 is a 60-kDa heat shock protein homologous to GroEL of *Escherichia coli* (31). In the cytoplasm, heat shock proteins chaperone nascent or aberrantly folded proteins, prevent protein aggregation, promote proper folding of nonnative proteins, assist in the assembly of multisubunit protein complexes, and target proteins for membrane translocation (27, 44). Heat shock proteins of a remarkable number of microbial pathogens including *H. capsulatum* are immunogenic. Indeed, vaccination of mice with *H. capsulatum* rhsp60 is protective against pulmonary histoplasmosis (30). Thus, *H. capsulatum* hsp60 serves a dual function in the immunopathology of histoplasmosis.

The mechanism by which *H. capsulatum* hsp60 reaches the surface of the cell wall is unclear since the gene does not contain a classical secretory leader sequence (GenBank accession number P50142). In fact, the *H. capsulatum hsp60* gene possesses a unique leader sequence upstream from the sequence encoding the amino terminus of the mature polypeptide, which has high homology to the *hsp60* leader sequence in two other pathogenic fungi, *Coccidioides immitis* (GenBank accession number O660008) and *Paracoccidioides brasiliensis* (GenBank accession number U81786). This leader sequence has not been found in the *hsp60* sequence of higher organisms such as mammals, in nonpathogenic yeasts such as *Saccharomyces cerevisiae* (GenBank accession number M33301), or in commensal opportunists such as *Candida albicans* (GenBank accession number AF085694). Interestingly, the hsp60 associated with the cell wall and cell membrane of *H. capsulatum* is N glycosylated while the cytosolic form is not, suggesting that the surface form of hsp60 traffics through the Golgi apparatus. Rapid amplification of cDNA ends analysis of the hsp60 mRNA suggests that there is an alternative exon with a canonical leader sequence, which may direct the export of an alternatively spliced form of hsp60 to the surface of the yeasts (F. Gomez, personal communication). Further experiments to confirm this hypothesis are under way.

Human monocyte-derived DC (28) and mouse lung DC (S. L. Newman, unpublished observations) phagocytose *H. capsulatum* yeasts and conidia as well as do MΦ. In contrast to MΦ, human DC recognize *H. capsulatum* yeasts via the fibronectin receptor,

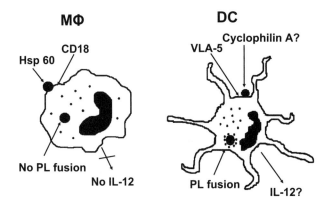

Figure 2. Differential recognition and response of human Mφ and DC to *H. capsulatum* yeasts. Mφ utilize the CD18 receptor family to recognize the ligand Hsp60 on the surface of *H. capsulatum* yeasts. Upon ingestion of the yeasts, PL fusion is blocked and the yeasts replicate. In contrast, DC use the VLA-5 receptor to recognize cyclophilin A on the surface of *H. capsulatum* yeasts. Upon ingestion of the yeasts, PL fusion takes place and the yeasts are killed and degraded.

very late antigen 5 (VLA-5; integrin-$\alpha_5\beta_1$), and not via β_2-integrins (Fig. 2). The reason for the different receptor usage by Mφ and DC is unknown, but we hypothesize that the difference may be due to differential mobility of these receptors. On the surface of DC, VLA-5 is mobile whereas β_2-integrins are not, and the converse is true with these molecules on the surface of Mφ.

Preliminary evidence from our laboratory suggests that the ligand on *H. capsulatum* yeasts recognized by DC VLA-5 is cyclophilin A (CycA). Like hsp60, cyclophilin is normally an intracellular molecule. However, flow cytometric analysis using an antibody to recombinant *H. capsulatum* cyclophilin (rCycA) clearly shows that this molecule is present on the surface of *H. capsulatum* yeasts. In addition, rCycA inhibits the binding of *H. capsulatum* yeasts to DC but not to Mφ. Small interfering RNA technology has been used to silence the *cycA* gene, and these yeasts lack CycA on their surface (F. Gomez and S. L. Newman, unpublished observations).

PHAGOCYTOSIS OF *H. CAPSULATUM* YEASTS AND CONIDIA BY MACROPHAGES AND DENDRITIC CELLS

On attachment, phagocytosis of unopsonized *H. capsulatum* yeasts and conidia by human cultured Mφ and freshly adherent AM is rapid, and yeasts and conidia are phagocytosed with equal efficiency (56). Peritoneal Mφ (PM) and AM from outbred Swiss albino mice (41) and mouse lung DC also phagocytose *H. capsulatum* yeasts and conidia with similar efficiency (Newman, unpublished). Based on electron

micrographs, it appears that each yeast resides within its own phagosome.

INTRACELLULAR FATE OF *H. CAPSULATUM* YEASTS IN MACROPHAGES AND DENDRITIC CELLS

In vitro, *H. capsulatum* yeasts multiply readily within rabbit, mouse, and human AM (21, 26, 41, 59), mouse, guinea pig, and human PM (26, 33, 35–37), and human monocytes and cultured Mφ (26, 59). The intracellular generation time of yeasts in mouse and guinea pig PM is 10 to 12 h (35, 36), whereas *H. capsulatum* yeasts multiply within human monocyte/Mφ with intracellular generation times of 14 to 20 h (59). Serum is not required for phagocytosis of *H. capsulatum* yeasts by Mφ. Furthermore, opsonization of the yeasts with normal or *H. capsulatum*-immune serum does not affect the intracellular growth of *H. capsulatum* yeasts in either mouse or guinea pig PM (36) or in human Mφ (59).

In contrast to their ability replicate in Mφ, *H. capsulatum* yeasts phagocytosed by human DC (28) are killed, and partially degraded yeasts can be seen within DC as early as 2 h. The mechanism of killing and yeast digestion appears to be through massive phagolysosomal fusion (27a). Compared to mouse PM, mouse lung DC also potently inhibit the growth of *H. capsulatum* yeasts (Newman, unpublished). It is not known if the yeasts are actually killed by lung DC. Thus, Mφ and DC utilize different receptors to recognize *H. capsulatum* and to recognize different ligands on the surface of *H. capsulatum*; once ingested, the yeasts have different intracellular fates (Fig. 2).

STRATEGIES FOR SURVIVAL AND INTRACELLULAR REPLICATION OF *H. CAPSULATUM* YEASTS IN MACROPHAGES

Once inside the phagosome, *H. capsulatum* must avoid the microbicidal armamentarium of the Mφ that consists of the generation of toxic oxygen molecules (respiratory burst) and lysosomal hydrolases (phagolysosomal fusion [PL fusion]). Intracellular pathogens have evolved numerous strategies to avoid these Mφ defense mechanisms, including inhibition of PL fusion (e.g., *Mycobacterium tuberculosis*), escaping from the phagocytic vacuole into the cytoplasm (e.g., *Listeria monocytogenes*), and initiating phagocytosis without stimulating the respiratory burst (e.g., *Toxoplasma gondii*). Knowledge of the strategies used by *H. capsulatum* yeasts to adapt to the intracellular environment of Mφ is critical to the development of novel immunotherapies for histoplasmosis. Over the past several years, considerable progress has been made in understanding the mechanisms by which *H. capsulatum* yeasts avoid being killed in a normally hostile environment. Provocatively, these studies demonstrate that there are several significant differences in the manner in which *H. capsulatum* yeasts interact with human versus murine Mφ, and these differences are highlighted where appropriate (Fig. 1).

First, it is apparent that the thick cell wall of the yeasts does not make *H. capsulatum* impervious to destruction by Mφ. Both human (63) and mouse (37) Mφ readily digest heat-killed or fixed *H. capsulatum* yeasts. Second, *H. capsulatum* yeasts must rapidly synthesize new proteins to survive within Mφ. Thus, the addition of cycloheximide to human Mφ infected with viable *H. capsulatum* yeasts leads to killing and digestion of the fungus (63). Since cycloheximide inhibits protein synthesis in both yeasts (76) and mammalian cells (25), the data indicate that Mφ do not need to synthesize new proteins to kill and digest the yeasts. Presumably the yeasts are killed and degraded by the fusion of preformed lysosomal hydrolases with yeast-containing phagocytic vacuoles.

Third, although *H. capsulatum* yeasts must synthesize new proteins to survive within Mφ, the yeasts do not secrete a product(s) that globally paralyzes Mφ fungicidal mechanisms. In human Mφ that phagocytose both viable yeasts and fluorescein isothiocyanate-labeled heat-killed yeasts, viable yeasts survive and multiply whereas heat-killed yeasts are digested (63). Furthermore, in P388D1 mouse Mφ that contain both *H. capsulatum* yeasts and fluorescence-labeled zymosan particles, phagosomes containing zymosan acidify normally whereas phagosomes containing *H. capsulatum* maintain a relatively neutral pH of 6.5 (23). Clearly, the survival strategy of *H. capsulatum* yeasts focuses on regulating its phagosomal environment.

STIMULATION OF THE RESPIRATORY BURST

In vitro *H. capsulatum* yeasts can be killed by a combination of H_2O_2, Fe^{2+}, and iodide (73) or H_2O_2, horseradish peroxidase, and iodide (11, 32). However, in these experiments, millimolar quantities of toxic oxygen radicals were generated, in contrast to the nanomolar quantities of oxygen radicals produced by phagocytes. Further, a considerable amount of evidence suggests that *H. capsulatum* yeasts are impervious to reactive oxygen intermediates produced by human and murine phagocytes. *H. capsulatum* yeasts stimulate the respiratory burst of human monocyte/Mφ and neutrophils upon phagocytosis (12, 73, 74) but are not killed (26, 59, 73). Further, the intracellular replication of *H. capsulatum* yeasts in human monocytes and Mφ occurs at a similar rate

(59) even though Mφ have lost their myeloperoxidase and are unable to make toxic oxygen metabolites distal to H_2O_2 (53, 65, 72). Finally, phagocytosis of serum-opsonized *H. capsulatum* yeasts by human neutrophils stimulates a strong respiratory burst, and superoxide anion (O_2^-) remains trapped within the phagocytic vacuole (74). Thus, yeasts should be exposed directly to a lethal concentration of oxygen radicals. Nevertheless, all of the fungistatic activity of human neutrophils is mediated by the contents of the azurophil granules (60, 61).

In contrast to human Mφ, phagocytosis of unopsonized *H. capsulatum* yeasts by murine PM that are either resident (22, 85) or gamma interferon (IFN-γ) activated (84) does not stimulate the respiratory burst. Phagocytosis of yeasts opsonized in fresh or heat-inactivated normal mouse serum or *H. capsulatum*-immune serum does stimulate the respiratory burst of mouse PM (85), but the intracellular growth of yeasts still is not impaired (36).

PHAGOLYSOSOMAL FUSION

Phagocytosis of unopsonized *H. capsulatum* yeasts by the mouse P388D1 Mφ cell line (24) and resident PM (79) leads to normal PL fusion. In contrast, there is minimal PL fusion after ingestion of *H. capsulatum* yeasts by human Mφ (62) or the RAW 264.7 mouse Mφ cell line (78). As *H. capsulatum* yeasts multiply readily within all of these Mφ populations, the yeasts apparently utilize different strategies to avoid the destructive effects of lysosomal hydrolases in different types of Mφ. In some Mφ, *H. capsulatum* avoids lysosomal hydrolases altogether, and in other Mφ, *H. capsulatum* neutralizes the activity of the hydrolases. The key to this unique behavior lies in the regulation of intraphagosomal pH and the absolute necessity of *H. capsulatum* to acquire intracellular iron to survive and multiply.

INTRAPHAGOSOMAL pH
AND IRON ACQUISITION

In mouse P388D1 Mφ (23) and RAW Mφ (78) and in human Mφ, *H. capsulatum* yeasts maintain an intraphagosomal pH of ~6.5 (61a). At this pH, lysosomal hydrolases that may enter the phagosome presumably are inactive, and therefore do not kill and degrade *H. capsulatum* yeasts. At a pH of 6.5, *H. capsulatum* yeasts still can acquire iron from transferrin, the major source of iron within the Mφ. At this pH, transferrin is half-saturated (67), leaving some free iron available for yeast survival and growth. Indeed, disruption of the capacity of *H. capsulatum* yeasts to acquire iron within the Mφ leads to its demise. Chloroquine, which prevents the release of iron from transferrin by raising the endocytic and lysosomal pH (43), induces human Mφ to kill *H. capsulatum* yeasts (58). No killing occurs when iron nitriloacetate, an iron compound that is soluble at neutral to alkaline pH (8), is included with the chloroquine.

Chloroquine given intraperitoneally for 6 days to *H. capsulatum*-infected C57BL/6 mice significantly reduces the growth of *H. capsulatum* in the spleen and liver. Furthermore, treatment of mice with chloroquine for 10 days following a lethal inoculum of *H. capsulatum* protects 67% of the mice whereas all control mice are dead by day 11 (58). Thus, both in vitro and in vivo, the ability of *H. capsulatum* yeasts to survive in phagosomes depends on regulating the intraphagosomal pH.

So how does *H. capsulatum* acquire iron from Mφ? At present, there is no definitive answer to this question. However, it is likely that *H. capsulatum* uses more than one strategy to acquire iron from the host. One possible mechanism is the production of low-molecular-weight siderophores that bind ferric iron. Indeed, it has long been known that *H. capsulatum* yeasts produce hydroxamic acid siderophores under low-iron conditions (13, 14). Interestingly, these siderophores are detected only after 4 days of culture in low-iron media (39). Since *H. capsulatum* yeasts must respond rapidly on ingestion by Mφ, it seems unlikely that *H. capsulatum* utilizes siderophores to obtain iron in Mφ phagosomes. However, the environment inside a Mφ phagosome is considerably different from that in tissue culture medium, and the yeasts still might utilize siderophores to acquire iron under these conditions.

A second possibility is the production of a ferric reductase. Under iron-limiting conditions, *H. capsulatum* yeasts produce three different types of iron-reducing compounds: a reduced glutathione-dependent enzymatic reductase, low-molecular-weight nonenzymatic reductants, and a cell surface ferric reductase (80, 81). We have found that although a variety of ferric reductase inhibitors restrain the replication of *H. capsulatum* yeasts in tissue culture media, they do not block the growth of yeasts within Mφ (61a).

These results prompted us to search for yet a third mechanism by which *H. capsulatum* yeasts might acquire iron. *S. cerevisiae* utilizes a complex of a multicopper oxidase (Fet3) and a ferric permease (Ftr1) for high-affinity transport of ferric iron. A BLAST search of the *H. capsulatum* genome revealed sequence homologies of 61% for Ftr1 and 65% for Fet3. Tetrathiomolybdate, an inhibitor of multicopper oxidases, inhibits the replication of *H. capsulatum* yeasts in tissue culture medium and in human Mφ, suggesting that FET3 is required for growth.

Furthermore, reverse transcription-PCR shows that the expression of mRNA for FET3 and FTR1 is increased under iron-limiting conditions (J. Hilty and S. L. Newman, unpublished observations). The requirement for FET3 and FTR1 for *H. capsulatum* growth in Mφ currently is being evaluated by silencing the genes with siRNA constructs. Until the results of these experiments are known, the requirement for FET3 and FTR1 in iron acquisition within Mφ remains speculative.

This is not the end of the pH story, however. As stated above, we have determined that in yeast containing phagosomes in human Mφ the pH is ~6.5, the same as for P388D1 Mφ. However, when human Mφ ingest heat-killed or fixed yeasts, the pH remains 6.5, and no acidification of the phagosome is observed for up to 24 h. This result has been obtained using two different techniques to quantify intraphagosomal pH and is in contrast to the results obtained with P388D1 Mφ (23) and RAW Mφ (78). In addition, the intraphagosomal pH is ~6.5 when Mφ phagocytose viable *S. cerevisiae* cells, which are easily killed and digested. However, when monocytes or Mφ ingest zymosan particles, acidification occurs, indicating that phagosomal acidification depends on the target particle (61a).

These data suggest that human Mφ may not require phagosomal acidification to kill *H. capsulatum* yeasts. To test this hypothesis, bafilomycin A was used to block the activity of the vacuolar ATPase (V-ATPase), under conditions in which Mφ are induced to kill *H. capsulatum* yeasts. Stimulation of human Mφ fungicidal activity by coculture with chloroquine (58) or by adherence to type 1 collagen matrices (62) is not reversed by bafilomycin. Furthermore, bafilomycin does not affect the capacity of human Mφ to completely degrade heat-killed *H. capsulatum* yeasts. In contrast, bafilomycin completely blocks the ability of murine PM to digest heat-killed yeasts (61a). Thus, mouse PM, but not human Mφ, require phagosomal acidification for digestion of *H. capsulatum* yeasts. These data demonstrate that the lysosomal enzymes in mouse PM and human monocyte-derived Mφ are considerably different and require different pHs for optimum enzyme activity.

Interestingly, bafilomycin also does not inhibit mouse PM fungistatic activity induced by IFN-γ (61a). Based on this observation, we hypothesize that phagosomal acidification still does not occur in IFN-γ-activated murine PM. If acidification did occur, the lysosomal enzymes that enter the phagosome should be able to kill and digest the yeasts. This hypothesis currently is being tested.

The different data obtained on intraphagosomal pH and PL fusion in murine versus human Mϕ can now be explained as follows. In murine Mϕ that have ingested *H. capsulatum* yeasts, there is PL fusion but there is modulation of the intraphagosomal pH to 6.5, presumably through alkalinization. In this scenario, a pH of 6.5 serves a dual purpose: lysosomal hydrolases are not active at the relatively neutral pH, and transferrin is half-saturated, leaving free iron for use by the yeasts (Fig. 1). In human Mϕ that have ingested *H. capsulatum* yeasts, PL fusion has to be inhibited because the lysosomal hydrolases of these Mϕ are fully active at pH 6.5. However, a pH of 6.5 still is necessary for the yeasts to acquire iron. In murine RAW Mϕ, PL fusion also is inhibited, as is the insertion of the V-ATPase into the phagosomal membrane (78). If the V-ATPase also is excluded from the phagosomal membrane in human Mϕ, it means that the yeasts probably have to slightly acidify the phagosome to obtain a pH of 6.5.

CALCIUM-BINDING PROTEIN

In addition to the necessity to acquire iron, the expression of the calcium-binding protein, Cbp1, appears to be a prerequisite for the intracellular survival of *H. capsulatum* yeasts in Mϕ (75). Cbp1 is a low-molecular-weight protein produced in copious quantities by *H. capsulatum* yeasts but not mycelia during exponential growth in liquid culture media (6, 7, 66). The *CBP1* gene is yeast phase specific, and its expression is transcriptionally regulated (45, 66). *CBP1* is expressed when *H. capsulatum* yeasts replicate within P388D1 Mϕ and hamster tracheal epithelial cells, and Cpb1 is secreted in vivo in a murine model of histoplasmosis (6, 45). Most importantly, *CBP1* is a virulence gene for *H. capsulatum*. A *cpb1* null strain grows slowly in liquid culture when calcium is limited, fails to replicate in and kill P388D1 Mϕ in vitro, and is unable to infect the lungs of mice. Complementation of the null mutant with a full-length copy of *CBP1* restores the wild-type phenotype, including virulence, thereby fulfilling Koch's molecular postulates (75).

At first glance, it would appear that Cbp1 acts in a nutritional capacity as a scavenger of calcium. *H. capsulatum* mycelia, which do not produce Cbp1, do not grow well under calcium-limiting conditions, whereas the yeasts replicate relatively well (7). Thus, resistance to calcium limitation directly correlates with the expression of *CBP1*. However, in yeasts, the expression of *CBP1* is not regulated by the availability of calcium, and this fact is inconsistent with a simple nutritional role. Thus, although Cbp1 is clearly a virulence factor for *H. capsulatum* yeasts, its role in virulence probably is more complex than the mere scavenging of calcium.

ACTIVATION OF Mϕ FUNGICIDAL/FUNGISTATIC ACTIVITY

Murine Peritoneal Macrophages

In addition to differences in stimulation of the respiratory burst and PL fusion, human and murine Mϕ have differential requirements for activation of anti-*Histoplasma* activity, which makes sense because the Mϕ have to overcome different intracellular survival strategies (Fig. 1).

Since the original experiments of Salvin (70), it has been known that the induction of cell-mediated immunity is required for host defense against *H. capsulatum*. These early experiments demonstrated that resistance to a lethal inoculum of *H. capsulatum* yeasts in mice and guinea pigs could be induced by prior injection of dead yeasts or by a sublethal infection with viable yeasts. This resistance to a lethal inoculum of yeasts did not correlate with the titer of complement-fixing antibodies and could not be transferred by homologous immune serum. Over the years, these studies have been confirmed and extended by numerous investigators (4, 17, 29, 34, 38, 83, 86). Our current understanding of murine host defense against *H. capsulatum* is that a successful immune response requires the induction of Th1-type helper T cells and the production of the cytokines interleukin-12 (IL-12), IFN-γ, and tumor necrosis factor alpha (TNF-α) (2, 3, 77, 87, 89, 90). Further, the secretion of IL-12 is required for the production of IFN-γ (2, 89), which presumably is required for activation of Mϕ to destroy the yeasts. Although TNF-α production is required for a successful host response in both naive and *H. capsulatum*-immune animals (3, 77, 87, 89, 90), the mechanism of action of TNF-α is unknown. However, TNF-α does not appear to directly activate Mϕ (3, 48). Interestingly, the production of granulocyte-Mϕ colony-stimulating factor (GM-CSF) also is necessary for survival in naive but not *H. capsulatum*-immune mice and appears to act via suppression of the Th2 cytokines IL-4 and IL-10 (18).

The data obtained from murine models of histoplasmosis are reflected by the results of in vitro experiments. In vitro, IFN-γ activates resident mouse PM to inhibit the intracellular growth of yeasts (86). Although the replication of yeasts is virtually nil, they remain viable. Interestingly, in vitro, splenic Mϕ are not activated by IFN-γ alone but require the presence of lipopolysaccharide (LPS) as a second signal (48). Again, the replication of yeasts is inhibited but the yeasts remain viable. These data suggest that, both in vitro and in vivo, another cytokine(s) in addition to IFN-γ is required to induce fungicidal activity in murine Mϕ. We predict that this additional cytokine(s) will stimulate phagosomal acidification.

Activated murine Mφ inhibit the growth of *H. capsulatum* by two mechanisms. First, activation of resident PM by IFN-γ (49) or of splenic Mφ by IFN-γ and LPS (48) can be reversed by the addition of holotransferrin or ferrous sulfate, demonstrating that Mφ activation by IFN-γ leads to growth inhibition by restriction of intracellular iron. Second, only Mφ that produce NO have anti-*Histoplasma* activity. Thus, incubation of the RAW 264.7 mouse Mφ cell line with IFN-γ leads to inhibition of the intracellular replication of *H. capsulatum* yeasts and the production of NO. In contrast, IFN-γ-treated P388D1 mouse Mφ do not secrete NO and are incapable of inhibiting the growth of *H. capsulatum* (46). In addition, the anti-*Histoplasma* activity of murine splenic Mφ that have been stimulated with IFN-γ and LPS is completely blocked by *N*-monomethyl-L-arginine, a competitive inhibitor of NO synthase (47). Collectively, these data suggest that IFN-γ plus LPS activate Mφ to synthesize NO and that this NO participates in the nitrosylation of iron transferrin so that iron is no longer available to the yeasts. In vivo studies of mice also demonstrate that the production of NO is required for successful clearance of *H. capsulatum* yeasts in the primary immune response. Remarkably, however, NO does not appear to be essential for clearance of fungi in the secondary immune response to *H. capsulatum* (88).

Human Macrophages

In contrast to our knowledge of murine immunology, the cytokines required to activate human Mφ fungistatic or fungicidal activity against *H. capsulatum* remain obscure. Unlike mouse PM, human Mφ are not activated to an anti-*Histoplasma* state by IFN-γ or TNF-α (26, 57). This observation is puzzling because human Mφ are activated by IFN-γ to kill and/or inhibit the intracellular growth of other intracellular pathogens such as *Trypanosoma cruzi* (64), *Leishmania donovani* (53), *Toxoplasma gondii* (54), and *Chlamydia psittaci* (69). Perhaps part of the explanation for a lack of effectiveness of IFN-γ is that the mechanism(s) of killing and growth inhibition of these other pathogens is mediated, at least in part, through the production of toxic oxygen metabolites and, as discussed above, *H. capsulatum* yeasts appear to be impervious to the toxic oxygen radicals of phagocytes.

However, culture of human Mφ with IFN-γ also downregulates transferrin receptors (15), which should result in decreased uptake of iron by Mφ. Since *H. capsulatum* yeasts are extremely sensitive to the iron concentration within Mφ, one might expect that downregulation of transferrin receptors would be sufficient to restrict the intracellular growth of *H. capsulatum* yeasts. The probable explanation for

this apparent contradiction is that after 5 to 7 days of culture in serum-containing medium, human Mφ have more than sufficient iron stores to support the intracellular replication of *H. capsulatum* yeasts and therefore downregulation of transferrin receptors by IFN-γ does not significantly reduce the intraphagosomal iron concentration.

Although IFN-γ is without effect, culture of monocytes into Mφ in the presence of IL-3, GM-CSF, or monocyte CSF (M-CSF) does activate Mφ fungistatic (but not fungicidal) activity against *H. capsulatum* yeasts. Optimal activation of Mφ by CSFs requires their presence during the differentiation process. Addition of IFN-γ or TNF-α to CSF-activated Mφ during the last 24 h prior to infection with *H. capsulatum* yeasts does not further enhance Mφ fungistatic activity (57). The mechanism(s) by which CSFs stimulate human Mφ fungistatic activity against *H. capsulatum* yeasts is unknown.

Although more potent Mφ-activating cytokines have not been identified, human Mφ fungicidal activity can be activated by adherence to type 1 collagen matrices (62). Other extracellular matrix proteins, such as fibronectin, laminin, and vitronectin, or non-gelled collagen do not induce Mφ anti-*Histoplasma* activity. Mφ are activated within 1 h of adherence to the collagen matrices, and fungicidal activity is maintained for up to 4 days postinfection. Culture of collagen-adherent Mφ with IFN-γ or TNF-α, or IL-3, GM-CSF, or M-CSF does not augment Mφ fungicidal activity against *H. capsulatum* yeasts (62).

The mechanism(s) of Mφ-mediated fungistasis directly counteracts the yeast survival strategy of inhibiting PL fusion. In *H. capsulatum*-infected collagen-adherent Mφ, massive PL fusion occurs and the yeasts are destroyed, presumably by preformed lysosomal hydrolases. There is no evidence that toxic oxygen radicals, NO, or the restriction of intracellular iron participates in Mφ fungicidal activity (62).

ROLES OF MACROPHAGES AND DENDRITIC CELLS IN IMMUNITY TO *H. CAPSULATUM*: A HYPOTHESIS

As discussed above, Mφ provide a permissive intracellular environment for the replication of *H. capsulatum* yeasts whereas DC are able to kill and degrade *H. capsulatum*. Further, compared to Mφ, DC are infinitely superior at processing and presenting protein antigens to T cells (5). Indeed, immature human DC that have ingested viable *H. capsulatum* yeasts stimulate lymphocyte proliferation, even when the DC and lymphocytes are obtained from individuals who have never been infected with *H. capsulatum* (as in a primary immune response) (28). In contrast, mature human monocyte-derived Mφ stimulate very little lymphocyte proliferation even when incubated with heat-killed yeasts (Newman, unpublished).

We propose that in the lungs, both Mφ and DC take up *H. capsulatum* yeasts. Yeasts that enter Mφ are able to replicate in and disseminate from the lungs to other organs. Yeasts that are phagocytosed by DC are killed, and fungal antigens are processed for presentation to T cells in nearby lymph nodes. Stimulated T cells then produce IFN-γ and other cytokines to activate Mφ and resolve the infection. This differential process presumably also would occur in organs such as the liver and spleen.

Evidence that supports this hypothesis comes from the fact that ingestion of *H. capsulatum* yeasts by human monocytes causes downregulation of IL-12 secretion stimulated by heat-killed *Staphylococcus aureus* and IFN-γ. This downregulation is induced via signal transduction through CR3 (CD11b/CD18), and antibodies to CR3 also downregulate IL-12 production (52). Since IL-12 production is critical for the induction of IFN-γ in a murine model of histoplasmosis (2, 89), it must be produced by another cell, presumably the DC. Since DC recognize *H. capsulatum* via VLA-5 and not CR3, phagocytosis of yeasts should not block the production of IL-12. Indeed, this scenario occurs with *Leishmania major*. DC that have ingested *L. major* promastigotes produce IL-12, whereas IL-12 is not detected in murine Mφ that have phagocytosed these cells (42). Thus, DC and Mφ probably play distinct but complementary roles in host defense against *H. capsulatum*.

REFERENCES

1. Ajello, L. 1971. Distribution of *Histoplasma capsulatum* in the United States, p. 103–122. *In* L. Ajello, E. W. Chick, and M. F. Furcolow (ed.), *Histoplasmosis*. Charles C Thomas, Springfield, Ill.

2. Allendoerfer, R., G. P. Biovin, and G. S. Deepe, Jr. 1997. Modulation of immune responses in murine pulmonary histoplasmosis. *J. Infect. Dis.* **175**:905–914.

3. Allendoerfer, R., and G. S. Deepe, Jr. 1998. Blockade of endogenous TNF-alpha exacerbates primary and secondary pulmonary histoplasmosis by differential mechanisms. *J. Immunol.* **160**:6072–6082.

4. Allendoerfer, R., D. M. Magee, G. S. Deepe, Jr., and J. R. Graybill. 1993. Transfer of protective immunity in murine histoplasmosis by a CD4⁺ T-cell clone. *Infect. Immun.* **61**:714–718.

5. Banchereau, J., and R. M. Steinman. 1998. Dendritic cells and the control of immunity. *Nature* **392**:245–252.

6. Batanghari, J. W., G. S. Deepe, Jr., E. Di Cera, and W. E. Goldman. 1998. Histoplasma acquisition of calcium and expression of CBP1 during intracellular parasitism. *Mol. Microbiol.* **27**:531–539.

7. Batanghari, J. W., and W. E. Goldman. 1997. Calcium dependence and binding in cultures of *Histoplasma capsulatum*. *Infect. Immun.* **65**:5257–5261.

8. Bates, G. W., and J. Wernicke. 1971. The kinetics and mechanism of iron (III) exchange between chelates and transferrin. *J. Biol. Chem.* **246**:3679–3686.

9. Baughman, R. P., C. K. Kim, A. Vinegar, D. E. Hendricks, D. J. Schmidt, and W. E. Bullock. 1986. The pathogenesis of experimental pulmonary histoplasmosis. Correlative studies of histopathology, bronchoalveolar lavage, and respiratory function. *Am. Rev. Respir. Dis.* **134**:771–776.

10. Berry, C. L. 1969. The production of disseminated histoplasmosis in the mouse: the effects of changes in reticuloendothelial function. *J. Pathol.* **97**:441–457.

11. Brummer, E., N. Kurita, S. Yosihida, K. Nishimura, and M. Miyaji. 1991. Fungistatic activity of human neutrophils against *Histoplasma capsulatum*: correlation with phagocytosis. *J. Infect. Dis.* **164**:158–162.

12. Bullock, W. E., and S. D. Wright. 1987. Role of the adherence-promoting receptors, CR3, LFA-1, and p150,95, in binding of *Histoplasma capsulatum* by human macrophages. *J. Exp. Med.* **165**:195–210.

13. Burt, W. R. 1982. Identification of coprogen B and its breakdown products from *Histoplasma capsulatum*. *Infect. Immun.* **35**:990–996.

14. Burt, W. R., A. L. Underwood, and G. L. Appleton. 1981. Hydroxamic acid from *Histoplasma capsulatum* that displays growth factor activity. *Appl. Environ. Microbiol.* **42**:560–563.

15. Byrd, T. F., and M. A. Horwitz. 1989. Interferon gamma-activated human monocytes downregulate transferrin receptors and inhibit the intracellular multiplication of *Legionella pneumophila* by limiting the availability of iron. *J. Clin. Investig.* **83**:1457–1465.

16. Deepe, G. S., and W. E. Bullock. 1992. Histoplasmosis: a granulomatous inflammatory response, p. 943–958. *In* J. I. Gallin, I. M. Goldstein, and R. Snyderman (ed.), *Inflammation: Basic Principles and Clinical Correlates*. Raven Press, New York, N.Y.

17. Deepe, G. S., Jr. 1988. Protective immunity in murine histoplasmosis: functional comparison of adoptively transferred T-cell clones and splenic T cells. *Infect. Immun.* **56**:2350–2355.

18. Deepe, G. S., Jr., R. Gibbons, and E. Woodward. 1999. Neutralization of endogenous granulocyte-macrophage colony-stimulating factor subverts the protective immune response to *Histoplasma capsulatum*. *J. Immunol.* **163**:4985–4993.

19. Deepe, G. S., Jr., and R. A. Seder. 1998. Molecular and cellular determinants of immunity to *Histoplasma capsulatum*. *Res. Immunol.* **149**:397–406; discussion 509–510.

20. DeMonbruen, W. A. 1934. The cultivation and characteristics of Darling's *Histoplasma capsulatum*. *Am. J. Trop. Med.* **14**:93–125.

21. De Sanchez, S. B., and L. M. Carbonell. 1975. Immunological studies on *Histoplasma capsulatum*. *Infect. Immun.* **11**:387–394.

22. Eissenberg, L. G., and W. E. Goldman. 1987. *Histoplasma capsulatum* fails to trigger release of superoxide from macrophages. *Infect. Immun.* **55**:29–34.

23. Eissenberg, L. G., W. E. Goldman, and P. H. Schlesinger. 1993. *Histoplasma capsulatum* modulates the acidification of phagolysosomes. *J. Exp. Med.* **177**:1605–1611.

24. Eissenberg, L. G., P. H. Schlesinger, and W. E. Goldman. 1988. Phagosome-lysosome fusion in P388D1 macrophages infected with *Histoplasma capsulatum*. *J. Leukoc. Biol.* **43**:483–491.

25. Ennis, H. L., and M. Lubin. 1964. Cycloheximide: aspects of inhibition of protein synthesis in mammalian cells. *Science* **146**:1474–1476.

26. Fleischmann, J., B. Wu-Hsieh, and D. H. Howard. 1990. The intracellular fate of *Histoplasma capsulatum* in

human macrophages is unaffected by recombinant human interferon-gamma. *J. Infect. Dis.* **161:**143–145.

27. Frydman, J., and F. Hartl. 1994. Molecular chaperone of Hsp70 and Hsp60 in protein folding, p. 251–283. *In* R. I. Morimoto, A. Tissieres, and C. Georgopoulos (ed.), *The Biology of Heat Shock Protein and Molecular Chaperones.* Cold Spring Harbor Laboratory Press, Plain View, N.Y.

27a. Gildea, L. A., G. M. Ciraolo, R. E. Morris, and S. L. Newman. 2005. Human dendritic cell activity against *Histoplasma capsulatum* is mediated by phagolysosomal fusion. *Infect. Immun.* **73:**6803–6811.

28. Gildea, L. A., R. E. Morris, and S. L. Newman. 2001. *Histoplasma capsulatum* yeasts are phagocytosed via very late antigen-5, killed, and processed for antigen presentation by human dendritic cells. *J. Immunol.* **166:**1049–1056.

29. Gomez, A. M., W. E. Bullock, C. L. Taylor, and G. S. Deepe, Jr. 1988. Role of L3T4$^+$ T cells in host defense against *Histoplasma capsulatum. Infect. Immun.* **56:**1685–1691.

30. Gomez, F. J., R. Allendoerfer, and G. S. Deepe, Jr. 1995. Vaccination with recombinant heat shock protein 60 from *Histoplasma capsulatum* protects mice against pulmonary histoplasmosis. *Infect. Immun.* **63:**2587–2595.

31. Hallberg, R. L. 1990. A mitochondrial chaperonin: genetic, biochemical, and molecular characteristics. *Semin. Cell Biol.* **1:**37–45.

32. Howard, D. H. 1981. Comparative sensitivity of *Histoplasma capsulatum* conidiospores and blastospores to oxidative antifungal systems. *Infect. Immun.* **32:**381–387.

33. Howard, D. H. 1960. Effect of Mycostatin and Fungizone on the growth of *Histoplasma capsulatum* in tissue culture. *J. Bacteriol.* **79:**442–449.

34. Howard, D. H. 1973. Further studies on the inhibition of *Histoplasma capsulatum* within macrophages from immunized animals. *Infect. Immun.* **8:**577–581.

35. Howard, D. H. 1964. Intracellular behavior of *Histoplasma capsulatum. J. Bacteriol.* **87:**33–38.

36. Howard, D. H. 1965. Intracellular growth of *Histoplasma capsulatum. J. Bacteriol.* **89:**518–523.

37. Howard, D. H. 1959. Observation on tissue cultures of mouse peritoneal exudates inoculated with *Histoplasma capsulatum. J. Bacteriol.* **78:**69–78.

38. Howard, D. H., V. Otto, and R. K. Gupta. 1971. Lymphocyte-mediated cellular immunity in histoplasmosis. *Infect. Immun.* **4:**605–610.

39. Howard, D. H., R. Rafie, A. Tiwari, and K. F. Faull. 2000. The hydroxamate siderophores of *Histoplasma capsulatum. Infect. Immun.* **68:**2338–2343.

40. Howell, A. 1939. Studies on *Histoplasma capsulatum* and similar form species. 1. Morphology and development. *Mycologia* **31:**191–216.

41. Kimberlin, C. L., A. R. Hariri, H. O. Hempel, and N. L. Goodman. 1981. Interactions between *Histoplasma capsulatum* and macrophages from normal and treated mice: comparison of the mycelial and yeast phases in alveolar and peritoneal macrophages. *Infect. Immun.* **34:**6–10.

42. Konecny, P., A. J. Stagg, H. Jebbari, N. English, R. N. Davidson, and S. C. Knight. 1999. Murine dendritic cells internalize *Leishmania major* promastigotes, produce IL-12 p40 and stimulate primary T cell proliferation in vitro. *Eur. J. Immunol.* **29:**1803–1811.

43. Krogstad, D. J., and P. H. Schlesinger. 1987. Acid-vesicle function, intracellular pathogens, and the action of chloroquine against Plasmodium falciparum. *N. Engl. J. Med.* **317:**542–549.

44. Kubota, H., G. Hynes, and K. Willison. 1995. The chaperonin containing t-complex polypeptide 1 (TCP-1).

Multisubunit machinery assisting in protein folding and assembly in the eukaryotic cytosol. *Eur. J. Biochem.* **230:**3–16.

45. Kugler, S., B. Young, V. L. Miller, and W. E. Goldman. 2000. Monitoring phase-specific gene expression in Histoplasma capsulatum with telomeric GFP fusion plasmids. *Cell. Microbiol.* **2:**537–547.

46. Lane, T. E., G. C. Otero, B. A. Wu-Hsieh, and D. H. Howard. 1994. Expression of inducible nitric oxide synthase by stimulated macrophages correlates with their antihistoplasma activity. *Infect. Immun.* **62:**1478–1479.

47. Lane, T. E., B. A. Wu-Hsieh, and D. H. Howard. 1994. Antihistoplasma effect of activated mouse splenic macrophages involves production of reactive nitrogen intermediates. *Infect. Immun.* **62:**1940–1945.

48. Lane, T. E., B. A. Wu-Hsieh, and D. H. Howard. 1993. Gamma interferon cooperates with lipopolysaccharide to activate mouse splenic macrophages to an antihistoplasma state. *Infect. Immun.* **61:**1468–1473.

49. Lane, T. E., B. A. Wu-Hsieh, and D. H. Howard. 1991. Iron limitation and the gamma interferon-mediated antihistoplasma state of murine macrophages. *Infect. Immun.* **59:**2274–2278.

50. Long, K. H., F. J. Gomez, R. E. Morris, and S. L. Newman. 2003. Identification of heat shock protein 60 as the ligand on *Histoplasma capsulatum* that mediates binding to CD18 receptors on human macrophages. *J. Immunol.* **170:**487–494.

51. Maresca, B., and G. S. Kobayashi. 1989. Dimorphism in Histoplasma capsulatum: a model for the study of cell differentiation in pathogenic fungi. *Microbiol. Rev.* **53:**186–209.

52. Marth, T., and B. L. Kelsall. 1997. Regulation of interleukin-12 by complement receptor 3 signaling. *J. Exp. Med.* **185:**1987–1995.

53. Murray, H. W., and D. M. Cartelli. 1983. Killing of intracellular Leishmania donovani by human mononuclear phagocytes. Evidence for oxygen-dependent and -independent leishmanicidal activity. *J. Clin. Investig.* **72:**32–44.

54. Murray, H. W., B. Y. Rubin, S. M. Carriero, A. M. Harris, and E. A. Jaffee. 1985. Human mononuclear phagocyte antiprotozoal mechanisms: oxygen-dependent vs oxygen-independent activity against intracellular *Toxoplasma gondii. J. Immunol.* **134:**1982–1988.

55. Newman, S. L. 2001. Cell-mediated immunity to *Histoplasma capsulatum. Semin. Respir. Infect.* **16:**102–108.

56. Newman, S. L., C. Bucher, J. Rhodes, and W. E. Bullock. 1990. Phagocytosis of *Histoplasma capsulatum* yeasts and microconidia by human cultured macrophages and alveolar macrophages. Cellular cytoskeleton requirement for attachment and ingestion. *J. Clin. Investig.* **85:**223–230.

57. Newman, S. L., and L. Gootee. 1992. Colony-stimulating factors activate human macrophages to inhibit intracellular growth of *Histoplasma capsulatum* yeasts. *Infect. Immun.* **60:**4593–4597.

58. Newman, S. L., L. Gootee, G. Brunner, and G. S. Deepe, Jr. 1994. Chloroquine induces human macrophage killing of *Histoplasma capsulatum* by limiting the availability of intracellular iron and is therapeutic in a murine model of histoplasmosis. *J. Clin. Investig.* **93:**1422–1429.

59. Newman, S. L., L. Gootee, C. Bucher, and W. E. Bullock. 1991. Inhibition of intracellular growth of *Histoplasma capsulatum* yeast cells by cytokine-activated human monocytes and macrophages. *Infect. Immun.* **59:**737–741.

60. Newman, S. L., L. Gootee, and J. E. Gabay. 1993. Human neutrophil-mediated fungistasis against *Histoplasma*

capsulatum. Localization of fungistatic activity to the azurophil granules. *J. Clin. Investig.* **92**:624–631.

61. Newman, S. L., L. Gootee, J. E. Gabay, and M. E. Selsted. 2000. Identification of constituents of human neutrophil azurophil granules that mediate fungistasis against *Histoplasma capsulatum*. *Infect. Immun.* **68**:5668–5672.

61a. Newman, S. L., L. Gootee, J. Hilty, and R. E. Morris. 2006. Human macrophages do not require phagosome acidification to mediate fungistatic/fungicidal activity against *Histoplasma capsulatum*. *J. Immunol.* **176**:1806–1813.

62. Newman, S. L., L. Gootee, C. Kidd, G. M. Ciraolo, and R. Morris. 1997. Activation of human macrophage fungistatic activity against *Histoplasma capsulatum* upon adherence to type 1 collagen matrices. *J. Immunol.* **158**:1779–1786.

63. Newman, S. L., L. Gootee, R. Morris, and W. E. Bullock. 1992. Digestion of *Histoplasma capsulatum* yeasts by human macrophages. *J. Immunol.* **149**:574–580. (Erratum, **149**:3127.)

64. Nogueira, N., S. Chaplan, M. Reesink, J. Tydings, and Z. A. Cohn. 1982. *Trypanosoma cruzi*: induction of microbicidal activity in human mononuclear phagocytes. *J. Immunol.* **128**:2142–2146.

65. Pabst, M. J., H. B. Hedegaard, and R. B. Johnston, Jr. 1982. Cultured human monocytes require exposure to bacterial products to maintain an optimal oxygen radical response. *J. Immunol.* **128**:123–128.

66. Patel, J. B., J. W. Batanghari, and W. E. Goldman. 1998. Probing the yeast phase-specific expression of the *CBP1* gene in *Histoplasma capsulatum*. *J. Bacteriol.* **180**: 1786–1792.

67. Princiotto, J. V., and E. J. Zapolski. 1975. Difference between the two iron-binding sites of transferrin. *Nature* **255**:87–88.

68. Procknow, J. J., M. I. Page, and C. G. Loosli. 1960. Early pathogenesis of experimental histoplasmosis. *Arch. Pathol.* **69**:413–426.

69. Rothermel, C. D., B. Y. Rubin, E. A. Jaffe, and H. W. Murray. 1986. Oxygen-independent inhibition of intracellular *Chlamydia psittaci* growth by human monocytes and interferon-gamma-activated macrophages. *J. Immunol.* **137**:689–692.

70. Salvin, S. B. 1955/56. Acquired resistance in experimental histoplasmosis. *Trans. N. Y. Acad. Sci.* **18**:462–468.

71. Sanchez-Madrid, F., J. A. Nagy, E. Robbins, P. Simon, and T. A. Springer. 1983. A human leukocyte differentiation antigen family with distinct alpha-subunits and a common beta-subunit: the lymphocyte function-associated antigen (LFA-1), the C3bi complement receptor (OKM1/Mac-1), and the p150,95 molecule. *J. Exp. Med.* **158**:1785–1803.

72. Sasada, M., A. Kubo, T. Nishimura, T. Kakita, T. Moriguchi, K. Yamamoto, and H. Uchino. 1987. Candidacidal activity of monocyte-derived human macrophages: relationship between *Candida* killing and oxygen radical generation by human macrophages. *J. Leukoc. Biol.* **41**:289–294.

73. Schaffner, A., C. E. Davis, T. Schaffner, M. Markert, H. Douglas, and A. I. Braude. 1986. In vitro susceptibility of fungi to killing by neutrophil granulocytes discriminates between primary pathogenicity and opportunism. *J. Clin. Investig.* **78**:511–524.

74. Schnur, R. A., and S. L. Newman. 1990. The respiratory burst response to *Histoplasma capsulatum* by human neutrophils. Evidence for intracellular trapping of superoxide anion. *J. Immunol.* **144**:4765–4772.

75. Sebghati, T. S., J. T. Engle, and W. E. Goldman. 2000. Intracellular parasitism by *Histoplasma capsulatum*: fungal

virulence and calcium dependence. *Science* **290**:1368–1372.

76. Siegel, M. R., and H. D. Sisler. 1964. Site of action of cycloheximide in cells of *Saccharomyces pastorianus*. II. The nature of inhibition of protein synthesis in a cell free system. *Biochim. Biophys. Acta* **87**:83–89.

77. Smith, J. G., D. M. Magee, D. M. Williams, and J. R. Graybill. 1990. Tumor necrosis factor-alpha plays a role in host defense against *Histoplasma capsulatum*. *J. Infect. Dis.* **162**:1349–1353.

78. Strasser, J. E., S. L. Newman, G. M. Ciraolo, R. E. Morris, M. L. Howell, and G. E. Dean. 1999. Regulation of the macrophage vacuolar ATPase and phagosome-lysosome fusion by *Histoplasma capsulatum*. *J. Immunol.* **162**:6148–6154.

79. Taylor, M. L., M. E. Espinosa-Schoelly, R. Iturbe, B. Rico, J. Casasola, and F. Goodsaid. 1989. Evaluation of phagolysosome fusion in acridine orange stained macrophages infected with *Histoplasma capsulatum*. *Clin. Exp. Immunol.* **75**:466–470.

80. Timmerman, M. M., and J. P. Woods. 1999. Ferric reduction is a potential iron acquisition mechanism for *Histoplasma capsulatum*. *Infect. Immun.* **67**:6403–6408.

81. Timmerman, M. M., and J. P. Woods. 2001. Potential role for extracellular glutathione-dependent ferric reductase in utilization of environmental and host ferric compounds by *Histoplasma capsulatum*. *Infect. Immun.* **69**:7671–7678.

82. Wheat, J. 1994. Histoplasmosis and coccidioidomycosis in individuals with AIDS. A clinical review. *Infect. Dis. Clin. North Am.* **8**:467–482.

83. Williams, D. M., J. R. Graybill, and D. J. Drutz. 1981. Adoptive transfer of immunity to *Histoplasma capsulatum* in athymic nude mice. *Sabouraudia* **19**:39–48.

84. Wolf, J. E., A. L. Abegg, S. J. Travis, G. S. Kobayashi, and J. R. Little. 1989. Effects of *Histoplasma capsulatum* on murine macrophage functions: inhibition of macrophage priming, oxidative burst, and antifungal activities. *Infect. Immun.* **57**:513–519.

85. Wolf, J. E., V. Kerchberger, G. S. Kobayashi, and J. R. Little. 1987. Modulation of the macrophage oxidative burst by *Histoplasma capsulatum*. *J. Immunol.* **138**:582–586.

86. Wu-Hsieh, B. A., and D. H. Howard. 1987. Inhibition of the intracellular growth of *Histoplasma capsulatum* by recombinant murine gamma interferon. *Infect. Immun.* **55**:1014–1016.

87. Wu-Hsieh, B. A., G. S. Lee, M. Franco, and F. M. Hofman. 1992. Early activation of splenic macrophages by tumor necrosis factor alpha is important in determining the outcome of experimental histoplasmosis in mice. *Infect. Immun.* **60**:4230–4238. (Erratum, **60**:5324.)

88. Zhou, P., G. Miller, and R. A. Seder. 1998. Factors involved in regulating primary and secondary immunity to infection with *Histoplasma capsulatum*: TNF-alpha plays a critical role in maintaining secondary immunity in the absence of IFN-gamma. *J. Immunol.* **160**: 1359–1368.

89. Zhou, P., M. C. Sieve, J. Bennett, K. J. Kwon-Chung, R. P. Tewari, R. T. Gazzinelli, A. Sher, and R. A. Seder. 1995. IL-12 prevents mortality in mice infected with *Histoplasma capsulatum* through induction of IFN-gamma. *J. Immunol.* **155**:785–795.

90. Zhou, P., M. C. Sieve, R. P. Tewari, and R. A. Seder. 1997. Interleukin-12 modulates the protective immune response in SCID mice infected with *Histoplasma capsulatum*. *Infect. Immun.* **65**:936–942.

Molecular Principles of Fungal Pathogenesis
Edited by Joseph Heitman et al.
©2006 ASM Press, Washington, D.C.

Chapter 36

Interactions of Immune Cells with the Facultative Intracellular Pathogen *Cryptococcus neoformans*

LAUREN E. YAUCH AND STUART M. LEVITZ

Cryptococcosis is an invasive fungal infection that is caused by species of *Cryptococcus*, most commonly *C. neoformans* or *C. gattii*. *C. neoformans* is composed of two varieties: *C. neoformans* var. *grubii* (serotype A isolates) and *C. neoformans* var. *neoformans* (serotype D isolates). *C. gattii* includes serotypes B and C. The majority of cryptococcal infections in immunocompromised individuals are due to var. *grubii* or var. *neoformans*, whereas *C. gattii* causes infections primarily in immunocompetent persons (172). *C. neoformans* is ubiquitous in the environment, and exposure is thought to most commonly occur via inhalation. Disease occurs mainly in those with impaired cell-mediated immunity, including AIDS patients and organ transplant recipients undergoing immunosuppressive therapy. Cryptococcosis is the leading cause of meningitis in AIDS patients (7, 63).

Cryptococcus has a polysaccharide capsule, which is the major virulence factor of the organism. Acapsular mutant strains of *C. neoformans* are essentially avirulent (22, 52). The major component of the capsule is glucuronoxylomannan (GXM), an approximately 1,000-kDa polysaccharide that consists of a (1→3)-linked linear α-D-mannopyranan backbone, with β-D-glucopyranosyluronic acid (Glc*p*A), β-D-xylopyranosyl (Xyl*p*), and O-acetyl substitutions (27). The pattern of Glc*p*A and Xyl*p* substitutions imparts serotype specificity. The capsule, which also contains galactoxylomannan (GalXM), has a number of immunomodulatory properties, including inhibition of phagocytosis.

Exposure to *C. neoformans* most commonly occurs following inhalation of thinly encapsulated basidiospores or yeast cells. Once inside the lungs, low iron and elevated bicarbonate concentrations stimulate the capsule size to increase (60, 176, 182). Serologic studies suggest that asymptomatic infection is common

(1, 25, 57), implying that "normal" human immune systems can control the infection. There is evidence the organism can remain latent and can reactivate later if the immune system becomes compromised (53).

Clinical and experimental data have established that T cells are required for resistance to *C. neoformans*. Depletion of CD4[+] or CD8[+] T cells in mice significantly reduces survival after infection with live *C. neoformans* (81, 137). The immune response to *C. neoformans* is initiated in the lungs by phagocytic cells including alveolar macrophages, polymorphonuclear leukocytes (PMN), and dendritic cells (DC). In immunocompetent hosts, the organism is contained in the lungs in granulomas, in a host response that requires macrophages and CD4[+] and CD8[+] T cells (70, 81, 137). However, if the infection is not contained in the lungs, *C. neoformans* can hematogenously disseminate. The organism has a predilection for the central nervous system (CNS), and cryptococcal meningoencephalitis is the most common clinical manifestation of cryptococcosis.

This chapter covers the interactions of *Cryptococcus* with immune cells, with a focus on macrophages. The mechanisms of uptake of the organism, including the opsonic requirements, are reviewed, as well as the fate of the organism inside the macrophage: growth inhibition or killing versus intracellular replication. The mechanisms of growth inhibition and killing are covered, as well as the effects of *C. neoformans* on cytokine production and T-cell activation. While much of what is known about *C. neoformans* interactions with macrophages comes from in vitro studies, in vivo mouse and rat studies have been critical in confirming or disproving in vitro observations. The interactions of the organism with other cell types, including neutrophils, T cells, and NK cells, are also covered. Finally, the

---

Lauren E. Yauch and Stuart M. Levitz • Departments of Medicine and Microbiology, Boston University, Boston, MA 02118.

interactions of soluble cryptococcal antigens with immune cells are reviewed.

MACROPHAGES: BINDING, PHAGOCYTOSIS, AND ANTIFUNGAL ACTIVITY

Macrophages are an important first line of defense against pathogens, as they can phagocytose and kill a number of microorganisms. They also present antigens to T cells, although not as efficiently as DC do. Macrophages produce cytokines and chemokines, thereby recruiting and activating other immune cells such as T cells and neutrophils. The antimicrobial activities of macrophages are enhanced by T-cell-derived gamma interferon (IFN-γ) and other cytokines, as well as by cell-cell contact with T cells.

Acapsular mutant strains of *C. neoformans* are readily internalized by murine macrophages in a process mediated by mannose and β-glucan receptors (33). The relevance of studies with acapsular *C. neoformans* to human infection is uncertain, though, as environmental isolates are thinly encapsulated and the capsule size rapidly increases in vivo. Thinly encapsulated *C. neoformans* cells are not phagocytosed in the absence of opsonins. The capsule inhibits phagocytosis by presenting an unrecognizable surface to phagocytic cells (100). However, opsonization with complement or anticapsular antibodies allows for phagocytosis via complement receptors (CR) or immunoglobulin Fc receptors (FcR), respectively. Following incubation with normal human serum, acapsular strains of *C. neoformans* activate both the alternative and classical complement pathways whereas encapsulated *C. neoformans* is an efficient activator of just the alternative complement pathway (98). Large amounts of the complement component C3, mostly in the form of iC3b, are deposited on the capsular surface. CR2 (CD21), CR3 (CD11b/CD18), and CR4 (CD11c/CD18) are all receptors for iC3b (although only the last two are generally found on phagocytes), while CR1 (CD35) primarily recognizes C3b. CR1, CR3, and/or CR4 mediates complement-dependent phagocytosis of encapsulated organisms by human monocyte-derived macrophages (MDM) (119), murine peritoneal macrophages (34), rat pulmonary macrophages (64), and murine bronchoalveolar macrophages (BAM) (111). The activation of complement by the organism is discussed in more detail in chapter 13 of this volume.

Numerous studies have revealed the antiphagocytic properties of the cryptococcal capsule. While acapsular organisms are avidly taken up by murine peritoneal macrophages, encapsulated yeasts are taken up at a low rate (102). Similarly, binding of serum-opsonized acapsular *C. neoformans* by human MDM leads to rapid internalization, whereas for encapsulated *C. neoformans*, phagocytosis proceeds slowly (Fig. 1) (119; unpublished observations). Phagocytosis of serum-opsonized *C. neoformans* by murine and rat alveolar macrophages, and rat peritoneal macrophages, is greater for thinly encapsulated organisms than for yeast cells possessing thick capsules (11, 136, 191), perhaps a consequence of complement depletion by the capsule (104). Likewise, phagocytosis of *C. neoformans* by human alveolar macrophages is greater for thinly encapsulated organisms than for organisms with a thick capsule (175). Soluble cryptococcal polysaccharide inhibits phagocytosis of acapsular *C. neoformans* by macrophages, apparently by binding to cell wall α-1,3-glucans (13, 99, 162).

The addition of a complement source, such as normal human serum, promotes the binding of encapsulated organisms to multiple cell types, including human PMN, monocytes, and MDM, as well as murine BAM and peritoneal macrophages (101, 111, 120). Complement-mediated phagocytosis can be increased by the activation of macrophages, for example with certain cytokines. Phagocytosis of complement-opsonized encapsulated *C. neoformans* by resident peritoneal macrophages is increased by tumor necrosis factor alpha (TNF-α) and/or granulocyte-macrophage colony-stimulating factor (GM-CSF) treatment (30). Anticapsular antibodies also mediate phagocytosis of the organism and can enhance macrophage anticryptococcal activity (17). Antibody-*Cryptococcus* interactions are discussed in chapter 33.

There are three possible outcomes for an organism following phagocytosis: it may replicate intracellularly, remain latent in the cell, or be killed by the phagocyte. The fate of *Cryptococcus* inside phagocytic cells depends on a number of factors, including the host species, cell type (including anatomical source), activation state of the cell, the mechanism of uptake (for instance CR or FcR), and the strain of *Cryptococcus*. In vitro and in vivo data have demonstrated that *C. neoformans* can grow both extracellularly and intracellularly in macrophages in vitro and in vivo but that under certain conditions, macrophages are able to inhibit the growth of the organism or even kill it. Murine macrophages are activated by microbial products and cytokines, such as IFN-γ. Activated macrophages are potent antimicrobial effectors via the induction of inducible nitric oxide synthase (iNOS), also known as NOS2, and the production of oxygen radicals. On activation, these cells also produce cytokines, including TNF-α, and upregulate costimulatory molecules, major histocompatibility complex (MHC) class I, and MHC class II, allowing for T-cell activation.

In vitro studies have revealed that growth inhibition of *C. neoformans* depends on the source of

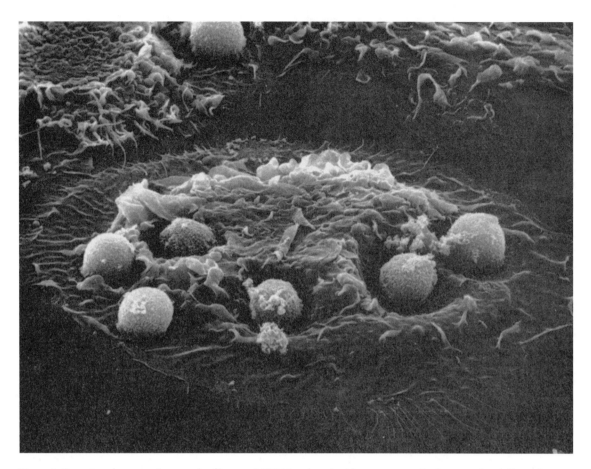

Figure 1. Scanning electron micrograph of human MDM incubated with serum-opsonized encapsulated *C. neoformans* for 30 min. Reprinted from reference 119a by permission of Routledge/Taylor Francis Group, LLC.

the macrophages and often requires macrophage activation. Several studies have demonstrated the importance of a source of complement for the anti-cryptococcal activity of murine peritoneal macrophages and BAM (12, 111). Killing of encapsulated *C. neoformans* by murine and rat BAM does not require macrophage activation (10, 111). However, the activation of macrophages (e.g., by IFN-γ or GM-CSF) can enhance their anticryptococcal activity (24, 51, 138, 154). The ability of macrophages to inhibit cryptococcal growth depends on the mouse strain as well. While resident peritoneal macrophages from one outbred (Swiss Webster) and three inbred (BALB/c, C57BL/6, and DBA/2J) mouse strains mediate fungistasis of serum-opsonized encapsulated *C. neoformans*, outbred CD-1 strain macrophages are unable to inhibit fungal growth unless activated by IFN-γ and lipopolysaccharide (LPS) (12).

Human monocytes/macrophages exhibit anticryptococcal activity in vitro, although intracellular replication in macrophages can also occur, depending on experimental conditions. Human MDM cultured

with lymphocytes internalize encapsulated *C. neoformans* in the presence of serum but are unable to inhibit growth during extended in vitro cultures (38). The organism does not inhibit phagolysosomal fusion in human MDM or human microglial cells (109, 116). Indeed, studies have shown the organism has adapted to live in the acidic environment of macrophage phagolysosomes. Chloroquine and ammonium chloride enhance the anticryptococcal activity of human MDM by alkalinizing the phagolysosome, and the growth of *C. neoformans* in a cell-free system decreases as the pH increases (115).

Human MDM inhibit the growth of serum-opsonized encapsulated *C. neoformans*, yet killing cannot be induced (114). In fact, IFN-γ treatment of the MDM or the addition of anticapsular antibody impairs the growth inhibition of encapsulated *C. neoformans*. In contrast to macrophages, human monocytes are efficient at killing serum-opsonized encapsulated *C. neoformans* (40, 135), and the anticryptococcal activity can be increased by culturing the monocytes on endothelial cells (167). The percent killing by monocytes varies among cryptococcal strains, with recently obtained clinical isolates being more resistant to killing (135). Human alveolar macrophages can also kill serum-opsonized *C. neoformans* but are less effective at killing the organism than are peripheral blood monocytes or PMN (183). In contrast to data obtained using murine BAM (111), human alveolar macrophages kill acapsular *C. neoformans* more efficiently than they kill encapsulated organisms (183). Data on the effect of IFN-γ on the anticryptococcal activity of human alveolar macrophages are conflicting (16, 160, 183).

Although many studies have shown that macrophages can inhibit or kill *C. neoformans*, the organism can also replicate in the cells. As discussed above, human MDM internalize serum-opsonized encapsulated *C. neoformans* but are unable to inhibit growth during extended in vitro cultures (38). In fact, intracellular organisms grow better than yeasts cultured in the absence of leukocytes, demonstrating that at least under some conditions, macrophages provide a favorable environment for the growth of *C. neoformans*.

Complement- and antibody-opsonized encapsulated *C. neoformans* cells are able to replicate in the murine J774 macrophage cell line and in murine BAM in vitro, and this is associated with cytotoxicity (50, 180). The organism begins dividing 2 h after being phagocytosed, with up to 40 organisms accumulating in the cell before macrophage lysis occurs around 6 to 18 h after infection (180). Replication of *C. neoformans* in phagolysosomes results in destruction of the phagolysosomal membrane, allowing the contents to leak out. Infected macrophages also demonstrate accumulation of capsular polysaccharide-containing vesicles throughout the cytoplasm and have been termed Hueco cells, after the Spanish word for "hole" (50, 180).

Thus, *C. neoformans* is taken up by macrophages in vitro but can survive and replicate, ultimately leading to the death of the macrophages. Presumably, then, the development of a CD4+ T-cell response contributes to host defenses against cryptococcosis by activating macrophages to become more potent anticryptococcal effector cells. Interestingly, the ability of the organism to replicate in phagocytic cells may have developed evolutionarily in response to interactions with environmental amoebae, which are able to internalize *C. neoformans* (14, 173). This subject is discussed in more detail in chapter 16.

ROLE OF MACROPHAGES IN VIVO

Reflecting the in vitro data, acapsular mutants are essentially avirulent in vivo, with virulence being restored by complementation of the defective gene (21, 52, 106). While intratracheal (i.t.) infection of mice with encapsulated *C. neoformans* results in a persistent infection, acapsular organisms are cleared from the lungs by day 14 (50). Monocytes/macrophages are recruited to the lungs of i.t.-infected mice (75, 76, 80). Mice infected i.t. with heat-killed encapsulated *C. neoformans* demonstrate an increase in the numbers of macrophages, T cells, and B cells in the lungs 7 days postinfection (88). The macrophages, many of which contain internalized yeast cells, display an activated phenotype, as demonstrated by upregulated MHC class II, FcR, and intercellular cell adhesion molecule 1 (ICAM-1) (87). Macrophage activation and phagocytosis are reduced if the mice are depleted of CD4+ and CD8+ T cells or treated with anti-IFN-γ.

After i.t. infection of mice, encapsulated *C. neoformans* is internalized by alveolar macrophages and neutrophils (50). By 7 days, most organisms are intracellular and can be detected in multinucleated giant cells up to 28 days later. Intracellular parasitism is associated with replication, accumulation of soluble capsular polysaccharide in vesicles, and cytotoxicity. In support of in vitro data, *C. neoformans* does not inhibit phagolysosomal fusion in murine alveolar macrophages in vivo.

Rats infected i.t. with a serotype D strain of *C. neoformans* develop a persistent pulmonary cryptococcosis with little extrapulmonary dissemination (56). Early in the infection, a predominance of extracellular organisms is seen, whereas after granuloma formation the majority of the yeasts are found within epithelioid cells or macrophages. The formation of granulomas correlates with decreased CFU in the lungs. *C. neoformans* persists in granulomas for up to 18 months, demonstrating the ability of this organism to live intracellularly in a latent state (59). When the infection is reactivated, the lung CFU increase and extracellular organisms predominate.

Histopathological studies of the lungs of humans with cryptococcosis demonstrate granulomatous responses, with minimal inflammation (3, 72). Lung sections from patients with cryptococcosis reveal both extracellular and intracellular organisms, with intracellular organisms seen in macrophages and more often in giant cells (3).

While macrophages are critical for the host response to *C. neoformans*, they may also play a detrimental role in cryptococcal infection by allowing intracellular replication and may serve as a "Trojan horse." The organism was shown to disseminate to the CNS via infected macrophages in a murine model of cryptococcosis (170).

Thus, *C. neoformans* is internalized by macrophages in vitro and in vivo. While activation of murine macrophages in vitro can result in killing, the organism can also live and replicate in these cells, ultimately lysing them. Human monocytes and alveolar macrophages can kill *C. neoformans* in vitro, but MDM cannot do so and may actually provide a favorable environment for replication. It is important to point out that even when macrophages have been shown to kill *C. neoformans* in vitro, only a fraction of the organisms are killed, and those surviving could replicate and overwhelm the macrophages. In vitro studies have shed important light on macrophage-*Cryptococcus* interactions, but the differences seen between human and murine cells and between macrophages from different anatomical sites make it difficult to draw conclusions about the requirements for macrophage anticryptococcal activity. However, under the in vitro conditions that most closely approximate conditions following inhalation of the fungus, human alveolar macrophages are able to kill complement-opsonized encapsulated *C. neoformans*. Moreover, because exposure is common but disease is rare in immunocompetent humans, it can be inferred that host defenses, including macrophages, are effective at keeping *C. neoformans* in check until a T-cell response is initiated and the organism is cleared or a granulomatous response contains the organism in the lungs.

INDUCTION OF CYTOKINES AND CHEMOKINES IN MACROPHAGES BY *C. NEOFORMANS*

An important function of macrophages is the production of cytokines and chemokines, which recruit cells to sites of infection, and can direct the activities of the recruited cells, for instance by skewing the T-cell response to Th1 or Th2. In general, a Th1 response, including the production of TNF-α, interleukin-12 (IL-12), and IFN-γ, is protective in murine models of cryptococcosis (97) whereas the Th2 cytokine IL-4 and the immunosuppressive IL-10 are deleterious (5, 84).

The cytokines and chemokines induced by *C. neoformans* depend on the cell type, *C. neoformans* strain, and opsonization. Acapsular organisms are more potent at inducing proinflammatory cytokine release from macrophages than are encapsulated organisms, which downregulate the production of some cytokines. Murine peritoneal macrophages stimulated with acapsular *C. neoformans* upregulate TNF-α, GM-CSF, macrophage inflammatory protein 1α (MIP-1α), and IL-1β mRNA (33). In contrast, encapsulated organisms do not induce TNF-α from murine thioglycolate-elicited peritoneal macrophages and actually inhibit the production of TNF-α induced by LPS plus IFN-γ (89). Similarly, encapsulated

C. neoformans cells downregulate the production of IL-12p40 and enhance the production of IL-10 by J774 macrophages stimulated with LPS plus IFN-γ (90).

Similar to murine data, encapsulation of the organism inhibits the production of proinflammatory cytokines by human macrophages and induces the release of IL-10. Encapsulated organisms induce less TNF-α and IL-1β from human monocytes and alveolar macrophages than do acapsular mutants (187). Human alveolar macrophages stimulate a greater autologous T-cell response, including proliferation and IL-2 and IFN-γ release, following stimulation with acapsular, compared with encapsulated, *C. neoformans* (183). Encapsulated *C. neoformans* cells also induce significantly less IL-12 from human monocytes than do acapsular organisms, due in part to IL-10 induced by the encapsulated organisms (163, 164). Opsonization with anticapsular antibody, or the addition of autologous T cells, increases the amount of IL-12 produced (163). Encapsulated *C. neoformans* cells induce more IL-10 release from human monocytes than do acapsular organisms, and blocking IL-10 increases the amount of TNF-α and IL-1β produced by monocytes in response to encapsulated organisms (186).

The opsonization conditions also affect the cytokines induced. For instance, rat alveolar macrophages induce the transcription of IL-6 following phagocytosis of *C. neoformans* opsonized with rabbit capsular antiserum but not with normal rabbit serum (121). Rat alveolar macrophages also produce more monocyte chemotactic protein 1 (MCP-1) when stimulated with anticapsular antibody-opsonized *C. neoformans* than do serum-opsonized organisms (64).

Susceptible mouse strains, including C57BL/6, do not develop a protective Th1 response to pulmonary *C. neoformans*; instead, they initiate a deleterious response that results in chronic eosinophilia and nonresolving pneumonia (74). Heavily encapsulated and highly virulent *C. neoformans* strains generally elicit poor inflammatory responses (5, 75, 82, 92). Mice infected i.t. with a highly virulent strain produce IL-4 and IL-10 in the lungs but little TNF-α or IFN-γ (93). Similarly, mice infected with another highly virulent strain induce less TNF-α, IFN-γ, and MCP-1 in the lungs than do mice infected with a less virulent strain (153). TNF-α treatment of mice infected i.t. with a virulent strain of *C. neoformans* increases survival; however, neutralizing TNF-α does not decrease survival, since TNF-α is not induced by the organism (89). In another model of pulmonary cryptococcosis involving a less virulent strain of *C. neoformans*, neutralizing TNF-α results in less IL-12 and IFN-γ production, demonstrating the importance of TNF-α in the initiation of a protective Th1 response (66). IL-12 treatment also has protective effects. BALB/c mice are susceptible to i.t. infection with a highly virulent strain of *C. neoformans*; however, IL-12 treatment increases the survival of these mice (94). Treatment with IL-12 induces the production of the chemokines MIP-1α, MCP-1, and RANTES in the lungs and increases the recruitment of macrophages and CD4$^+$ T cells (92). Chemokines are critical to the host response to *C. neoformans*. MIP-1α is made by numerous types of immune cells and is induced in the lungs of *C. neoformans*-infected mice (79). While MIP-1α$^{-/-}$ mice demonstrate no defect in leukocyte recruitment after i.t. infection, they have reduced survival and increased lung CFU compared with wild-type mice (152). The MIP-1α$^{-/-}$ mice also display increased levels of the Th2 cytokines IL-4 and IL-13 in the lungs, as well as eosinophilia, suggesting that MIP-1α is important in mediating a protective Th1 response during cryptococcosis. MIP-1α is important in the immune response to *C. neoformans* in the CNS as well, as MIP-1α$^{-/-}$ mice have impaired leukocyte recruitment to the brain and delayed clearance of the organism (77). CCR5, the receptor for MIP-1α, MIP-1β, and RANTES, is also involved in the host response to *C. neoformans* in vivo, as CCR5$^{-/-}$ mice are significantly more susceptible to i.t. infection than are wild-type mice (78). Pulmonary clearance of the organism and leukocyte recruitment to the lungs are similar in wild-type and CCR5$^{-/-}$ mice; however, the CCR5$^{-/-}$ mice demonstrate reduced recruitment of leukocytes to the brain.

Like MIP-1α, MCP-1, which is chemotactic for macrophages and T cells, and its receptor, CCR2, are important in the development of a protective Th1 response during pulmonary cryptococcosis in mice (80, 179). Human monocytes, but not BAM, produce MCP-1 when stimulated with serum-opsonized encapsulated *C. neoformans* (118).

POLYMORPHONUCLEAR LEUKOCYTES

PMN (neutrophils) are phagocytic cells that are quickly recruited to sites of infection, including the lungs of mice following i.t. infection with *C. neoformans* (50). Serum activated by *C. neoformans* generates the complement protein C5a, which is chemotactic for human neutrophils (39). In addition, encapsulated *C. neoformans* releases products with direct chemotactic activity for human PMN (45). Human neutrophils phagocytose encapsulated *C. neoformans* when the organism is opsonized with normal human serum or anticapsular antibodies (101). However, while it promotes binding to human neutrophils, complement deposition on encapsulated *C. neoformans* does not necessarily lead to phagocytosis (103).

PMN stimulated with serum-opsonized *C. neoformans* release IL-1β, IL-6, IL-8, and TNF-α (165). The amounts of cytokines and the chemokine IL-8 released correlate with the size of the capsule, with a heavily encapsulated strain inducing larger amounts than the thinly encapsulated or acapsular strains do. Human PMN are also potent anticryptococcal effector cells and can kill serum- or antibody-opsonized organisms with greater efficiency than mononuclear cells do (37, 40, 134, 135). Similar to data obtained using human alveolar macrophages, phagocytosis and killing by PMN are greater for thinly encapsulated *C. neoformans* than for organisms with a thicker capsule (175). Akin to human monocytes, PMN demonstrate increased anticryptococcal activity when cultured on endothelial cells compared with their activity when cultured on plastic (167).

In support of a role for neutrophils as anticryptococcal effector cells in vivo, some of the recruited neutrophils in the lungs of infected mice contain *C. neoformans* (50). Interestingly, despite their ability to kill *C. neoformans* in vitro, neutrophils were shown to be nonprotective in a mouse model of pulmonary cryptococcosis (133). Mice depleted of neutrophils prior to i.t. infection have lower CFU in the lungs 1 day postinfection and survive significantly longer than do control mice. Another study found that depletion of neutrophils in the first week following i.t. *C. neoformans* infection does not prevent the development of a protective immune response (65).

The role of neutrophils in human infection is not clear. Histopathological examination of lungs from patients with cryptococcosis reveals marked variation in the PMN response, ranging from numerous to absent (3). Unlike candidiasis and aspergillosis, cryptococcosis is not generally associated with neutropenia.

DENDRITIC CELLS

DC are powerful antigen-presenting cells capable of initiating primary T-cell responses. As with other phagocyte populations, human and murine DC do not recognize encapsulated *C. neoformans* unless the fungi are opsonized with complement and/or anticapsular antibody (96, 184). In contrast, human DC internalize unopsonized acapsular *C. neoformans* via a process that is partially dependent on mannose receptors (174, 184). Phagocytosis of acapsular *C. neoformans* induces the upregulation of MHC class I, MHC class II, and the DC maturation markers CD40 and CD83, as well as proliferation of autologous T cells and IFN-γ release (174, 184). In contrast, encapsulated organisms induce upregulation of CD40 but have no effect on the expression of MHC class I, MHC class II, or CD83 (184). However, opsonization of encapsulated *C. neoformans* with anticapsular antibody increases the expression of CD40, CD83, and CD86 on the DC, as well as T-cell proliferation and IFN-γ production.

Anticryptococcal activity of human DC is seen when the organisms are opsonized with serum or anticapsular antibody, while murine bone marrow-derived DC inhibit the growth of *C. neoformans* only following antibody opsonization (96). Human, but not murine, DC produce small amounts of TNF-α when stimulated with serum and/or antibody-opsonized *C. neoformans*, although neither population releases detectable amounts of IL-10 or IL-12.

In vivo data support a role for DC in the anticryptococcal response. Immunization of mice with cryptococcal culture filtrate antigens results in migration of DC to the draining lymph nodes, which is associated with a protective immune response (4). Neutralization of TNF-α significantly reduces the migration of DC to lung-associated lymph nodes following pulmonary *C. neoformans* challenge, and this is associated with an impaired antifungal response (65).

MECHANISMS OF PHAGOCYTE KILLING

Phagocytes possess a number of antimicrobial mediators, including reactive oxygen intermediates (ROIs), the reactive nitrogen intermediates (RNIs), and antimicrobial peptides. The α-defensins HNP1 to HNP3 are the major antimicrobial peptides found in human neutrophils (62). In contrast, murine neutrophils do not express α-defensins. ROIs and RNIs are generated by NADPH oxidase, myeloperoxidase, and/or iNOS (NOS2) (9). Superoxide (O_2^-), an ROI, can be converted in the phagosome into H_2O_2 by superoxide dismutase (SOD) or to hydroxyl radicals (·OH) or hydroxyl anions (OH^-). To evade oxidative killing, microbes produce a number of effectors, including SOD and catalase. SOD detoxifies oxygen radicals by converting superoxide anions to H_2O_2 and O_2, while catalase degrades H_2O_2.

Killing of serum-opsonized *C. neoformans* by resident and elicited rat peritoneal macrophages is inhibited by the addition of SOD or catalase, suggesting a role for ROI in anticryptococcal activity (169). Both acapsular and encapsulated organisms activate the respiratory burst (as measured by H_2O_2 production) in murine resident peritoneal cells (110). While acapsular *C. neoformans* cells activate the burst more efficiently when unopsonized than when serum opsonized, activation by encapsulated cryptococci requires opsonization with anticapsular immunoglobulin G (IgG) and serum. In this system, however, the respiratory burst is not the main mechanism of killing. Likewise, the anticryptococcal activity of murine J774 macrophages and BAM is observed in the presence of

ROI scavengers (111, 143). A role for ROI in growth inhibition of *C. neoformans* by human macrophages has not been demonstrated. However, human PMN can kill serum-opsonized *C. neoformans* via oxidative mechanisms, since inhibition of myeloperoxidase inhibits early killing by PMN (40). PMN from patients with chronic granulomatous disease, who have defective NADPH oxidase activity and therefore little to no respiratory burst, demonstrate defective antibody-dependent killing of *C. neoformans*, illustrating the importance of ROI in killing of the organism (36, 134). SOD, an O_2^- scavenger, and the $OH^·$ scavengers mannitol and dimethylsulfoxide, but not the H_2O_2 inhibitor, catalase, inhibit the killing of serum-opsonized *C. neoformans* by PMN (23). These data implicate ROI distal to H_2O_2 in PMN-mediated killing. In support of $OH^·$ as an important anticryptococcal effector, it was found that a strain of *C. neoformans* that is impaired in mannitol production is more susceptible to killing by PMN (23). Inhibition of the respiratory burst reduces, but does not abrogate, the antifungal activity of human DC against antibody-opsonized organisms (96).

Activated macrophages use iNOS to oxidize L-arginine to L-citrulline and the antimicrobial reactive nitrogen metabolite NO. The expression of iNOS in macrophages can be regulated in a number of ways, including cell-cell contact, cytokines, and microbial products (8). IFN-γ plus LPS is frequently used to induce NO in macrophages. Numerous in vitro studies have demonstrated the importance of RNI in the anticryptococcal activity of murine macrophages, yet in some systems antifungal activity is independent of RNI, and the organism can also inhibit NO production. In a cell-free system, chemically generated $NO^·$ is able to inhibit cryptococcal growth (2), and N^G-monomethyl-L-arginine (*N*-MMA), a competitive inhibitor of NO synthesis, inhibits the anticryptococcal activity of murine macrophages in vitro (12, 177). In addition, IFN-γ enhances the anticryptococcal activity of murine macrophages at least in part via the induction of NO (110, 154, 177). IFN-γ also stimulates TNF-α release by macrophages, which enhances the IFN-γ-induced NO production, thus potentiating the macrophage anticryptococcal activity (91).

Acapsular *C. neoformans* induces NO in unactivated murine resident peritoneal cells (RPC) and IFN-γ-primed J774 macrophages (110, 149). In contrast, NO production by RPC stimulated with encapsulated organisms requires opsonization with anticapsular IgG or activation of the RPC with IFN-γ (110). NO is not produced in IFN-γ-primed J774 macrophages stimulated with encapsulated *C. neoformans*, even when the organism is opsonized with anticapsular antibody, possibly due to a lack of

TNF-α production (149). However, J774 macrophages do demonstrate anticryptococcal activity which is increased by opsonization with an anticapsular antibody and is observed when NOS is inhibited (143).

In contrast to the studies demonstrating that *C. neoformans* induces NO, one study found that acapsular and encapsulated strains could inhibit LPS-plus IFN-γ-induced NO production by thioglycolate-elicited peritoneal macrophages (95). Thus, for murine cells in vitro, it appears that, depending on the cell type and activation state, strain of *C. neoformans*, and opsonins, NO may be induced, not induced, or inhibited by the organism and anticryptococcal activity can be NO dependent or independent.

In vivo animal studies support a role for NO in the host response to *C. neoformans*. Survival of rats infected intraperitoneally with *C. neoformans* is significantly impaired when NOS activity is inhibited in vivo (168). NOS2 is detected in macrophages, neutrophils, vascular endothelium, and respiratory epithelium in the lungs of rats with pulmonary cryptococcosis (55). The induction of NOS2 correlates with clearance of the organism from the lungs. In a rat model of chronic pulmonary cryptococcosis, NOS2 is detected in epithelioid cells and some macrophages, and the decrease in expression seen over the course of infection correlates with organism persistence (59). Interestingly, areas of the lung containing large amounts of GXM have little NOS2 reactivity, suggesting another potential inhibitory effect of GXM.

Mice infected with *C. neoformans* i.t. upregulate NOS2 mRNA in the lungs and have increased levels of systemic NO (126). Inflammatory cells from infected lungs mediate NO-dependent growth inhibition of *C. neoformans* ex vivo. Additionally, clearance of the organism from the lungs is inhibited by anti-IFN-γ or *N*-MMA treatment, both of which inhibit NO production. Levels of NO production have been associated with resistance in a mouse model of cryptococcosis (125). While C.B-17 and C57BL/6 mice demonstrate similar levels of iNOS protein in the lungs after i.t. *C. neoformans* infection, the C.B-17 mice have greater NO production, which correlates with their increased resistance to infection. Further supporting a role for NO, NOS2$^{-/-}$ mice have significantly reduced survival compared with wild-type mice following i.t. *C. neoformans* infection, although the ability of alveolar macrophages from NOS2$^{-/-}$ mice to phagocytose and kill antibody-opsonized *C. neoformans* in vitro is normal (166). Treatment of the NOS2$^{-/-}$ mice with anticapsular antibody prior to infection does not have the protective effect seen in wild-type mice, suggesting that NOS is required for antibody-mediated protection.

In contrast to the animal data, a role for NO in anticryptococcal activity has not been found for human monocytes/macrophages. The anticryptococcal activity of human alveolar macrophages is independent of NO production (16, 183). While studies have shown that human macrophages can make NO (48), the amount produced may not be fungicidal.

A cocktail of respiratory burst inhibitors decreases the anticryptococcal activity of PMN by approximately 50%, demonstrating the ability of these cells to kill *C. neoformans* by nonoxidative mechanisms as well (130). Anticryptococcal activity is found in the cytoplasm and in primary (azurophil) granules of PMN. Calprotectin is responsible for the antifungal activity in the cytoplasm, while the activity in the primary granules is mediated in part by defensins.

VIRULENCE FACTORS

While the polysaccharide capsule is the major virulence factor that helps *Cryptococcus* organisms elude phagocyte defenses, other virulence factors have been defined, including SOD, melanin, and phospholipases. SOD contributes to the survival of the yeast within macrophages, as *sod1* mutant *C. neoformans* strains are more susceptible to growth inhibition by unstimulated, but not stimulated, macrophages of a murine alveolar macrophage cell line (31). SOD plays a role in vivo as well, since a *sod1* mutant strain of *C. neoformans* is less virulent than the wild-type strain in a mouse model of cryptococcosis (31). Similarly, a *sod1* mutant of *C. gattii* is more susceptible to killing by PMN and is less virulent in a mouse model of cryptococcosis (148).

The phenoloxidase system is also an important virulence factor of the yeast. Phenoloxidase catalyzes the oxidative polymerization of phenolic and/or indolic compounds (such as L-DOPA) to melanin (150). Melanized organisms are more resistant to ROI and RNI than are nonmelanized organisms (190). In vitro, melanized *C. neoformans* organisms are also more resistant to phagocytosis by J774 macrophages and more resistant to the antifungal activity of microglial cells (6, 189). Moreover, melanin production by the organism is associated with increased virulence in vivo (105, 106).

Phospholipases are enzymes secreted by bacteria and pathogenic fungi that aid in tissue invasion by breaking down host cell membranes, resulting in cell lysis (54). *C. neoformans* possesses phospholipase activity, which could contribute to cryptococcal virulence by aiding in tissue invasion, as well as escape from the phagosome (26, 32, 188). A phospholipase, phospholipase B (PLB1), is responsible for the phospholipase B, lysophospholipase transacylase, and lysophospholipase activities of the organism (32). While wild-type and *plb1* mutant organisms are phagocytosed equivalently by J774 macrophages, the mutant strain demonstrates slower intracellular growth (32). Similarly, the *plb1* mutant is susceptible to killing by an alveolar macrophage cell line in vitro, whereas wild-type organisms are not killed (151). Growth of the phospholipase *plb1* mutant is impaired in alveolar macrophages in vivo, and the *plb1* mutant has reduced virulence in animal models of cryptococcosis (32, 151, 170).

In addition to the capsule, the organism has another antiphagocytic virulence factor, termed antiphagocytic protein (App1) (127). App1 is secreted by the organism into growth media and can be detected in the sera of patients with cryptococcal meningitis. Phagocytosis of Δ*app1* *C. neoformans* by murine alveolar macrophages is enhanced compared with that of wild-type organisms, and the addition of recombinant App1 inhibits the phagocytosis of a wild-type *C. neoformans* strain.

EFFECT OF HUMAN IMMUNODEFICIENCY VIRUS ON ANTICRYPTOCOCCAL ACTIVITY OF MACROPHAGES AND NEUTROPHILS

Owing to *C. neoformans* being one of the most common opportunistic infections in AIDS patients, a number of studies have examined the effects of human immunodeficiency virus (HIV) infection on *Cryptococcus*-phagocyte interactions. Human bronchoalveolar lavage cells from early-stage HIV-infected individuals have no defect in fungistasis, but patients with late-stage infection do demonstrate impaired anticryptococcal activity (161). Similarly, peripheral blood monocytes from AIDS patients have impaired anticryptococcal activity compared with monocytes from healthy donors (139). The monocytes are defective in phagocytosis and killing of serum-opsonized acapsular and encapsulated organisms and in superoxide anion production, and they are less capable of inducing an autologous T-cell response.

In vitro infection with an M-tropic strain of HIV inhibits the anticryptococcal activity of human blood monocytes (15) and human alveolar macrophages (83). PMN from late-stage HIV-infected patients are able to kill acapsular *C. neoformans*, but they demonstrate impaired killing of encapsulated organisms (29, 141). However, opsonization with an anticapsular antibody increases killing (141). PMN from late-stage HIV-infected patients also exhibit impaired IL-8 production when stimulated with serum-opsonized acapsular or encapsulated *C. neoformans* compared with PMN from healthy donors. The amounts of the HIV-inhibitory chemokines MIP-1α, MIP-1β, and RANTES produced by peripheral blood mononuclear cells stimulated with serum-opsonized encapsulated *C. neoformans* are similar for HIV-negative and HIV-infected donors (73).

EOSINOPHILS

Animal models of cryptococcosis have suggested a role for eosinophils, which are granulocytic leukocytes involved in allergic reactions and parasite elimination. In vitro, rat eosinophils are able to internalize IgG1- or IgE-opsonized *C. neoformans* cells (49). Eosinophils are recruited to the lungs of rats and some mouse strains following pulmonary *C. neoformans* infection (56, 75). In the lungs of infected C57BL/6 mice, eosinophils are found closely associated with yeast cells (49). While GXM can be detected in the cytoplasm of some eosinophils, phagocytosis of the organism is very rare and requires anticapsular antibody treatment prior to infection. Eosinophils appear to play a detrimental role in murine cryptococcosis, as the persistence of eosinophils in the lungs of infected mice correlates with a Th2 response and an inability to clear the infection (74). In spite of the recruitment of these cells to the lungs of infected nonhuman animals, eosinophils are rarely found in human cryptococcosis (3).

NK CELLS

NK cells are innate immune cells that mediate the cytotoxicity of infected cells and tumor cells. These cells possess a number of mechanisms for mediating cytotoxicity, including the release of perforin and granzymes, which are contained in secretory granules and cooperate to induce apoptosis in target cells (178). Granulysin is another component of the secretory granules of human cytotoxic lymphocytes, and it can lyse microbes and tumor cells (28). Murine splenic NK cells bind to, inhibit, and kill *C. neoformans* under some conditions in vitro (69, 86, 145, 147). Growth inhibition is abrogated when the secretory system is inhibited, and the secretory granule component perforin demonstrates anticryptococcal activity (68).

In support of a role for NK cells in vivo, C57BL/6 *bg/bg* mice which have impaired NK cell activity, have significantly reduced survival and increased organ fungal burdens compared with control mice following i.v. infection with *C. neoformans* (67, 132). However, *bg/bg* mice have other immune defects, and so a role for NK cells in this system is suggested but not proven. Indeed, mice depleted of NK cells prior to i.v. infection with *C. neoformans* demonstrate no defect in survival, and depletion of NK cells before i.t. infection has no effect on lung, spleen, or brain CFU (124).

Human NK cells exhibit direct anticryptococcal activity. IL-2-activated NK cells form conjugates with and inhibit the growth of *C. neoformans* (112). Freshly isolated NK cells also interact with *C. neoformans* and demonstrate direct anticryptococcal activity, independent of opsonization (113, 144). The direct anticryptococcal activity of human NK cells is mediated by constitutively expressed perforin (129).

A critical function of NK cells is the production of IFN-γ. Encapsulated *C. neoformans* stimulates the release of IFN-γ from human NK cells, even in the absence of complement activation (117). Stimulation of murine NK cells with IL-12 and IL-18 elicits the production of IFN-γ, resulting in enhanced fungicidal activity of thioglycolate-elicited peritoneal macrophages (86). In vivo, IL-12 plus IL-18 treatment results in increased levels of IFN-γ in the lungs of *C. neoformans*-infected mice; these levels are reduced when NK cells are depleted (159).

T CELLS

CD4$^+$ and CD8$^+$ T cells are vital to the anticryptococcal response in the lungs, and CD4$^+$ T cells are important in the response in the CNS as well (18, 70, 71, 81). Not only is a cell-mediated immune response critical in activating macrophages during cryptococcosis, but also T cells have direct anticryptococcal activity in vitro, although the in vivo relevance remains to be determined. Early studies showed that human mixed mononuclear cells and phagocyte-depleted lymphoid cells are able to kill antibody-opsonized *C. neoformans* (37). Human CD4$^+$ and CD8$^+$ T cells, cultured with IL-2 or freshly isolated, form conjugates with *C. neoformans* and are able to directly mediate fungistasis, independent of opsonization (Fig. 2) (112, 113, 144). The mechanism of growth inhibition by human CD8$^+$ T cells is dependent on granulysin and independent of perforin (128). Unlike NK cells, in which the anticryptococcal activity is constitutive, the induction of granulysin in the CD8$^+$ T cells stimulated with cryptococcal antigen requires IL-15 made by CD4$^+$ T-cell-activated accessory cells. Like NK cells, T cells produce IFN-γ, but not IL-4 or IL-10, when stimulated with encapsulated *C. neoformans*, and this does not require complement (117).

γδ T CELLS AND NKT CELLS

Like NK cells, NKT cells and γδ T cells are innate immune lymphocytes. NKT cells and γδ T cells are both involved in the host response to *C. neoformans* in mice. NKT cells are recruited to the lungs of i.t.-infected mice in an MCP-1-dependent manner (85). IFN-γ production by lung leukocytes is decreased in NKT-cell deficient mice, and clearance of the organism from the lungs is impaired as well. γδ

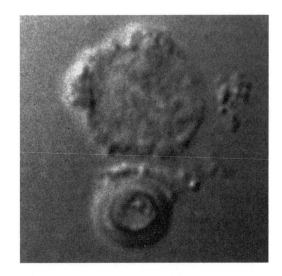

Figure 2. Nomarski differential interference contrast microscopy of the interaction between a human CD8$^+$ T cell (top) and encapsulated *C. neoformans* (bottom). Granules released by the T cell are visible on the yeast cell capsule. Reprinted from reference 113 with permission.

T cells are also recruited to the lungs of mice infected i.t. with *C. neoformans* (181). Compared with control mice, infected mice lacking γδ T cells have increased IFN-γ levels in the lungs and serum and reduced lung CFU, suggesting a downregulatory role for γδ T cells in pulmonary cryptococcosis.

IMMUNE CELL INTERACTIONS WITH SOLUBLE CRYPTOCOCCAL ANTIGENS

Many studies have investigated the role of the soluble cryptococcal antigens GXM and cryptococcal mannoproteins (MP) during cryptococcosis. Efforts to identify immunogenic antigens of *C. neoformans* led to the identification of MP. These highly mannosylated proteins induce TNF-α from peripheral blood mononuclear cells (PBMC), induce IL-12 from human monocytes, and elicit a stronger delayed-type hypersensitivity response than the capsule components GXM or GalXM (20, 146, 157). Moreover, MP induce the maturation of human DC and subsequent T-cell activation (156). Vaccination with MP has proven protective in mouse models of cryptococcosis (131, 155).

In contrast to MP, GXM has a number of immunoinhibitory properties. During cryptococcosis, GXM is shed into the blood and cerebrospinal fluid at up to microgram-per-milliliter concentrations and can be detected even after successful antifungal therapy (158). GXM in the brains of patients with cryptococcal meningitis is associated with microglial cells, which may serve as a reservoir for GXM once the organism is cleared (107, 108). GXM clearance studies have found a role for tissue macrophages in clearance of the polysaccharide from serum (58, 61). A number of receptors for GXM have been identified, including CD14, CD18, Toll-like receptor 2 (TLR2), TLR4, and FcγRII (41, 142, 171). While MDM accumulate GXM for up to 1 week in vitro, PMN rapidly internalize GXM and then expel or degrade it (142). In mice with pulmonary cryptococcosis, GXM is detected only in neutrophils that had internalized yeast, suggesting that these cells do not take up the soluble polysaccharide in vivo (50).

While two studies have found that GXM induces small amounts of TNF-α from PBMC in the presence of fresh serum (20, 35), other groups have found that GXM does not induce TNF-α from PBMC or MDM (140, 171) and in fact can downregulate TNF-α production (187). GXM, like encapsulated organisms, induces IL-10 release by human monocytes (186). GXM also induces the production of TNF-α, IL-1β, and IL-8 from PMN in the presence of serum, with induction of IL-8 release being mediated by the complement components C5 and C3a (165, 185). GXM also affects costimulatory molecule expression. Soluble GXM inhibits the upregulation of the costimulatory molecule CD80 on human monocytes induced by acapsular *C. neoformans* (164). While upregulation of CD86 and CD40 is seen on MDM that had internalized GXM, the polysaccharide decreases the expression of MHC class II on bystander MDM that did not take up GXM (140). However, GXM does not modulate the expression of CD40, CD86, or MHC class II on human monocyte-derived DC (184).

Cryptococcal antigens affect leukocyte migration. GXM is directly chemotactic for PMN and also activates a chemotactic heat-labile serum component (45). A culture filtrate from encapsulated *C. neoformans* is chemotactic for human neutrophils in vitro and murine neutrophils in vivo (43). Interestingly, the chemotactic activity of the culture filtrate is seen with *C. neoformans* but not *C. gattii* strains. Although GXM is directly chemotactic for PMN, it also inhibits PMN migration toward chemoattractants in vitro and in vivo (44, 122). In the presence and absence of serum, GXM induces L-selectin shedding from the surface of PMN (42). L-selectin binding to vascular endothelium is a critical first step in extravasation of neutrophils from the bloodstream to sites of inflammation. GXM interferes with the initial rolling of PMN on the endothelium and E-selectin-transfected cells in vitro as well (47). Firm adhesion of PMN to TNF-α-stimulated endothelium is also partially inhibited by GXM treatment of PMN or the endothelium (46). Additionally, binding of cryptococcal antigens to CD18 on PMN could interfere with the second step of extravasation, which is the tight binding of β₂-integrins to their ligands on endothelial cells.

Data from patients with cryptococcosis have supported a role for GXM in inhibition of leukocyte migration. Despite high levels of IL-8 in the cerebrospinal fluid of patients with cryptococcal meningitis, there is a paucity of leukocytes, possibly due to GXM in the cerebrospinal fluid (19). In fact, there is an inverse correlation between GXM titers and numbers of leukocytes in the cerebospinal fluid of patients with cryptococcal meningitis (123).

SUMMARY

C. neoformans is an important human pathogen in the population of patients with impaired T-cell function, particularly those with AIDS. Recent infections due to *C. gattii* in British Columbia have underscored the risk of cryptococcosis in the immunocompetent population as well. The pathogenesis of cryptococcosis has been the subject of much research. In vitro studies have demonstrated the capacity of macrophages and neutrophils to phagocytose *C. neoformans*, mediate growth inhibition, and produce cytokines and chemokines in response to the organism. Although in vitro studies illustrate the antiphagocytic property of the capsule, the organism is nevertheless internalized by phagocytes in vivo. The anticryptococcal activity of immune cells in vitro depends on numerous factors, including the species, cell type, and anatomical source of the host cells, the availability of opsonins, and the target *C. neoformans* strain. In vitro and in vivo studies have demonstrated that macrophages can provide a favorable environment for replication of *C. neoformans* but can also be activated by T-cell-derived cytokines to inhibit cryptococcal growth. Studies of *Cryptococcus*–immune-cell interactions have led to the identification of candidate vaccines and immunomodulators, some of which have been studied in clinical trials. The challenge for the future will be to continue to apply the research findings to the development of novel approaches to the prevention and treatment of cryptococcosis.

REFERENCES

1. **Abadi, J., and L. Pirofski.** 1999. Antibodies reactive with the cryptococcal capsular polysaccharide glucuronoxylomannan are present in sera from children with and without human immunodeficiency virus infection. *J. Infect. Dis.* **180:**915–919.

2. **Alspaugh, J. A., and D. L. Granger.** 1991. Inhibition of *Cryptococcus neoformans* replication by nitrogen oxides supports the role of these molecules as effectors of macrophage-mediated cytostasis. *Infect. Immun.* **59:**2291–2296.

3. **Baker, R. D., and R. K. Haugen.** 1955. Tissue changes and tissue diagnosis in cryptococcosis; a study of 26 cases. *Am. J. Clin. Pathol.* **25:**14–24.

4. **Bauman, S. K., K. L. Nichols, and J. W. Murphy.** 2000. Dendritic cells in the induction of protective and

nonprotective anticryptococcal cell-mediated immune responses. *J. Immunol.* **165:**158–167.

5. **Blackstock, R., K. L. Buchanan, A. M. Adesina, and J. W. Murphy.** 1999. Differential regulation of immune responses by highly and weakly virulent *Cryptococcus neoformans* isolates. *Infect. Immun.* **67:**3601–3609.

6. **Blasi, E., R. Barluzzi, R. Mazzolla, B. Tancini, S. Saleppico, M. Puliti, L. Pitzurra, and F. Bistoni.** 1995. Role of nitric oxide and melanogenesis in the accomplishment of anticryptococcal activity by the BV-2 microglial cell line. *J. Neuroimmunol.* **58:**111–116.

7. **Bogaerts, J., D. Rouvroy, H. Taelman, A. Kagame, M. A. Aziz, D. Swinne, and J. Verhaegen.** 1999. AIDS-associated cryptococcal meningitis in Rwanda (1983–1992): epidemiologic and diagnostic features. *J. Infect.* **39:**32–37.

8. **Bogdan, C.** 2001. Nitric oxide and the immune response. *Nat. Immunol.* **2:**907–916.

9. **Bogdan, C., M. Rollinghoff, and A. Diefenbach.** 2000. Reactive oxygen and reactive nitrogen intermediates in innate and specific immunity. *Curr. Opin. Immunol.* **12:**64–76.

10. **Bolanos, B., and T. G. Mitchell.** 1989. Killing of *Cryptococcus neoformans* by rat alveolar macrophages. *J. Med. Vet. Mycol.* **27:**219–228.

11. **Bolanos, B., and T. G. Mitchell.** 1989. Phagocytosis of *Cryptococcus neoformans* by rat alveolar macrophages. *J. Med. Vet. Mycol.* **27:**203–217.

12. **Brummer, E., and D. A. Stevens.** 1994. Anticryptococcal activity of macrophages: role of mouse strain, C5, contact, phagocytosis, and L-arginine. *Cell. Immunol.* **157:**1–10.

13. **Bulmer, G. S., and J. R. Tacker.** 1975. Phagocytosis of *Cryptococcus neoformans* by alveolar macrophages. *Infect. Immun.* **11:**73–79.

14. **Bunting, L. A., J. B. Neilson, and G. S. Bulmer.** 1979. *Cryptococcus neoformans*: gastronomic delight of a soil ameba. *Sabouraudia* **17:**225–232.

15. **Cameron, M. L., D. L. Granger, T. J. Matthews, and J. B. Weinberg.** 1994. Human immunodeficiency virus (HIV)-infected human blood monocytes and peritoneal macrophages have reduced anticryptococcal activity whereas HIV-infected alveolar macrophages retain normal activity. *J. Infect. Dis.* **170:**60–67.

16. **Cameron, M. L., D. L. Granger, J. B. Weinberg, W. J. Kozumbo, and H. S. Koren.** 1990. Human alveolar and peritoneal macrophages mediate fungistasis independently of L-arginine oxidation to nitrite or nitrate. *Am. Rev. Respir. Dis.* **142:**1313–1319.

17. **Casadevall, A.** 1995. Antibody immunity and invasive fungal infections. *Infect. Immun.* **63:**4211–4218.

18. **Cauley, L. K., and J. W. Murphy.** 1979. Response of congenitally athymic (nude) and phenotypically normal mice to *Cryptococcus neoformans* infection. *Infect. Immun.* **23:**644–651.

19. **Chaka, W., R. Heyderman, I. Gangaidzo, V. Robertson, P. Mason, J. Verhoef, A. Verheul, A. I. Hoepelman, and the University of Zimbabwe Meningitis Group.** 1997. Cytokine profiles in cerebrospinal fluid of human immunodeficiency virus-infected patients with cryptococcal meningitis: no leukocytosis despite high interleukin-8 levels. *J. Infect. Dis.* **176:**1633–1636.

20. **Chaka, W., A. F. Verheul, V. V. Vaishnav, R. Cherniak, J. Scharringa, J. Verhoef, H. Snippe, and I. M. Hoepelman.** 1997. *Cryptococcus neoformans* and cryptococcal glucuronoxylomannan, galactoxylomannan, and mannoprotein induce different levels of tumor necrosis factor alpha in human peripheral blood mononuclear cells. *Infect. Immun.* **65:**272–278.

21. **Chang, Y. C., R. Cherniak, T. R. Kozel, D. L. Granger, L. C. Morris, L. C. Weinhold, and K. J. Kwon-Chung.** 1997. Structure and biological activities of acapsular *Cryptococcus neoformans* 602 complemented with the *CAP64* gene. *Infect. Immun.* **65:**1584–1592.

22. **Chang, Y. C., and K. J. Kwon-Chung.** 1994. Complementation of a capsule-deficient mutation of *Cryptococcus neoformans* restores its virulence. *Mol. Cell. Biol.* **14:**4912–4919.

23. **Chaturvedi, V., B. Wong, and S. L. Newman.** 1996. Oxidative killing of *Cryptococcus neoformans* by human neutrophils. Evidence that fungal mannitol protects by scavenging reactive oxygen intermediates. *J. Immunol.* **156:**3836–3840.

24. **Chen, G. H., J. L. Curtis, C. H. Mody, P. J. Christensen, L. R. Armstrong, and G. B. Toews.** 1994. Effect of granulocyte-macrophage colony-stimulating factor on rat alveolar macrophage anticryptococcal activity in vitro. *J. Immunol.* **152:**724–734.

25. **Chen, L. C., D. L. Goldman, T. L. Doering, L. Pirofski, and A. Casadevall.** 1999. Antibody response to *Cryptococcus neoformans* proteins in rodents and humans. *Infect. Immun.* **67:**2218–2224.

26. **Chen, S. C., M. Muller, J. Z. Zhou, L. C. Wright, and T. C. Sorrell.** 1997. Phospholipase activity in *Cryptococcus neoformans*: a new virulence factor? *J. Infect. Dis.* **175:**414–420.

27. **Cherniak, R., and J. B. Sundstrom.** 1994. Polysaccharide antigens of the capsule of *Cryptococcus neoformans*. *Infect. Immun.* **62:**1507–1512.

28. **Clayberger, C., and A. M. Krensky.** 2003. Granulysin. *Curr. Opin. Immunol.* **15:**560–565.

29. **Coffey, M. J., S. M. Phare, S. George, M. Peters-Golden, and P. H. Kazanjian.** 1998. Granulocyte colony-stimulating factor administration to HIV-infected subjects augments reduced leukotriene synthesis and anticryptococcal activity in neutrophils. *J. Clin. Investig.* **102:**663–670.

30. **Collins, H. L., and G. J. Bancroft.** 1992. Cytokine enhancement of complement-dependent phagocytosis by macrophages: synergy of tumor necrosis factor-alpha and granulocyte-macrophage colony-stimulating factor for phagocytosis of *Cryptococcus neoformans*. *Eur. J. Immunol.* **22:**1447–1454.

31. **Cox, G. M., T. S. Harrison, H. C. McDade, C. P. Taborda, G. Heinrich, A. Casadevall, and J. R. Perfect.** 2003. Superoxide dismutase influences the virulence of *Cryptococcus neoformans* by affecting growth within macrophages. *Infect. Immun.* **71:**173–180.

32. **Cox, G. M., H. C. McDade, S. C. Chen, S. C. Tucker, M. Gottfredsson, L. C. Wright, T. C. Sorrell, S. D. Leidich, A. Casadevall, M. A. Ghannoum, and J. R. Perfect.** 2001. Extracellular phospholipase activity is a virulence factor for *Cryptococcus neoformans*. *Mol. Microbiol.* **39:**166–175.

33. **Cross, C. E., and G. J. Bancroft.** 1995. Ingestion of acapsular *Cryptococcus neoformans* occurs via mannose and β-glucan receptors, resulting in cytokine production and increased phagocytosis of the encapsulated form. *Infect. Immun.* **63:**2604–2611.

34. **Cross, C. E., H. L. Collins, and G. J. Bancroft.** 1997. CR3-dependent phagocytosis by murine macrophages: different cytokines regulate ingestion of a defined CR3

ligand and complement-opsonized *Cryptococcus neoformans*. *Immunology* **91**:289–296.

35. **Delfino, D., L. Cianci, M. Migliardo, G. Mancuso, V. Cusumano, C. Corradini, and G. Teti.** 1996. Tumor necrosis factor-inducing activities of *Cryptococcus neoformans* components. *Infect. Immun.* **64**:5199–5204.

36. **Diamond, R. D.** 1974. Antibody-dependent killing of *Cryptococcus neoformans* by human peripheral blood mononuclear cells. *Nature* **247**:148–150.

37. **Diamond, R. D., and A. C. Allison.** 1976. Nature of the effector cells responsible for antibody-dependent cell-mediated killing of *Cryptococcus neoformans*. *Infect. Immun.* **14**:716–720.

38. **Diamond, R. D., and J. E. Bennett.** 1973. Growth of *Cryptococcus neoformans* within human macrophages in vitro. *Infect. Immun.* **7**:231–236.

39. **Diamond, R. D., and N. F. Erickson III.** 1982. Chemotaxis of human neutrophils and monocytes induced by *Cryptococcus neoformans*. *Infect. Immun.* **38**:380–382.

40. **Diamond, R. D., R. K. Root, and J. E. Bennett.** 1972. Factors influencing killing of *Cryptococcus neoformans* by human leukocytes in vitro. *J. Infect. Dis.* **125**:367–376.

41. **Dong, Z. M., and J. W. Murphy.** 1997. Cryptococcal polysaccharides bind to CD18 on human neutrophils. *Infect. Immun.* **65**:557–563.

42. **Dong, Z. M., and J. W. Murphy.** 1996. Cryptococcal polysaccharides induce L-selectin shedding and tumor necrosis factor receptor loss from the surface of human neutrophils. *J. Clin. Investig.* **97**:689–698.

43. **Dong, Z. M., and J. W. Murphy.** 1995. Effects of the two varieties of *Cryptococcus neoformans* cells and culture filtrate antigens on neutrophil locomotion. *Infect. Immun.* **63**:2632–2644.

44. **Dong, Z. M., and J. W. Murphy.** 1995. Intravascular cryptococcal culture filtrate (CneF) and its major component, glucuronoxylomannan, are potent inhibitors of leukocyte accumulation. *Infect. Immun.* **63**:770–778.

45. **Dong, Z. M., and J. W. Murphy.** 1993. Mobility of human neutrophils in response to *Cryptococcus neoformans* cells, culture filtrate antigen, and individual components of the antigen. *Infect. Immun.* **61**:5067–5077.

46. **Ellerbroek, P. M., A. I. Hoepelman, F. Wolbers, J. J. Zwaginga, and F. E. Coenjaerts.** 2002. Cryptococcal glucuronoxylomannan inhibits adhesion of neutrophils to stimulated endothelium in vitro by affecting both neutrophils and endothelial cells. *Infect. Immun.* **70**:4762–4771.

47. **Ellerbroek, P. M., L. H. Ulfman, A. I. Hoepelman, and F. E. Coenjaerts.** 2004. Cryptococcal glucuronoxylomannan interferes with neutrophil rolling on the endothelium. *Cell. Microbiol.* **6**:581–592.

48. **Fang, F. C., and A. Vazquez-Torres.** 2002. Nitric oxide production by human macrophages: there's NO doubt about it. *Am. J. Physiol. Ser. L* **282**:L941–L943.

49. **Feldmesser, M., A. Casadevall, Y. Kress, G. Spira, and A. Orlofsky.** 1997. Eosinophil-*Cryptococcus neoformans* interactions in vivo and in vitro. *Infect. Immun.* **65**:1899–1907.

50. **Feldmesser, M., Y. Kress, P. Novikoff, and A. Casadevall.** 2000. *Cryptococcus neoformans* is a facultative intracellular pathogen in murine pulmonary infection. *Infect. Immun.* **68**:4225–4237.

51. **Flesch, I. E., G. Schwamberger, and S. H. Kaufmann.** 1989. Fungicidal activity of IFN-γ-activated macrophages.

52. **Fromtling, R. A., H. J. Shadomy, and E. S. Jacobson.** 1982. Decreased virulence in stable, acapsular mutants of *Cryptococcus neoformans*. *Mycopathologia* **79**:23–29.

53. **Garcia-Hermoso, D., G. Janbon, and F. Dromer.** 1999. Epidemiological evidence for dormant *Cryptococcus neoformans* infection. *J. Clin. Microbiol.* **37**:3204–3209.

54. **Ghannoum, M. A.** 2000. Potential role of phospholipases in virulence and fungal pathogenesis. *Clin. Microbiol. Rev.* **13**:122–143.

55. **Goldman, D., Y. Cho, M. Zhao, A. Casadevall, and S. C. Lee.** 1996. Expression of inducible nitric oxide synthase in rat pulmonary *Cryptococcus neoformans* granulomas. *Am. J. Pathol.* **148**:1275–1282.

56. **Goldman, D., S. C. Lee, and A. Casadevall.** 1994. Pathogenesis of pulmonary *Cryptococcus neoformans* infection in the rat. *Infect. Immun.* **62**:4755–4761.

57. **Goldman, D. L., H. Khine, J. Abadi, D. J. Lindenberg, L. Pirofski, R. Niang, and A. Casadevall.** 2001. Serologic evidence for *Cryptococcus neoformans* infection in early childhood. *Pediatrics* **107**:E66.

58. **Goldman, D. L., S. C. Lee, and A. Casadevall.** 1995. Tissue localization of *Cryptococcus neoformans* glucuronoxylomannan in the presence and absence of specific antibody. *Infect. Immun.* **63**:3448–3453.

59. **Goldman, D. L., S. C. Lee, A. J. Mednick, L. Montella, and A. Casadevall.** 2000. Persistent *Cryptococcus neoformans* pulmonary infection in the rat is associated with intracellular parasitism, decreased inducible nitric oxide synthase expression, and altered antibody responsiveness to cryptococcal polysaccharide. *Infect. Immun.* **68**:832–838.

60. **Granger, D. L., J. R. Perfect, and D. T. Durack.** 1985. Virulence of *Cryptococcus neoformans*. Regulation of capsule synthesis by carbon dioxide. *J. Clin. Investig.* **76**:508–516.

61. **Grinsell, M., L. C. Weinhold, J. E. Cutler, Y. Han, and T. R. Kozel.** 2001. In vivo clearance of glucuronoxylomannan, the major capsular polysaccharide of *Cryptococcus neoformans*: a critical role for tissue macrophages. *J. Infect. Dis.* **184**:479–487.

62. **Gudmundsson, G. H., and B. Agerberth.** 1999. Neutrophil antibacterial peptides, multifunctional effector molecules in the mammalian immune system. *J. Immunol. Methods* **232**:45–54.

63. **Hakim, J. G., I. T. Gangaidzo, R. S. Heyderman, J. Mielke, E. Mushangi, A. Taziwa, V. J. Robertson, P. Musvaire, and P. R. Mason.** 2000. Impact of HIV infection on meningitis in Harare, Zimbabwe: a prospective study of 406 predominantly adult patients. *AIDS* **14**:1401–1407.

64. **He, W., A. Casadevall, S. C. Lee, and D. L. Goldman.** 2003. Phagocytic activity and monocyte chemotactic protein expression by pulmonary macrophages in persistent pulmonary cryptococcosis. *Infect. Immun.* **71**:930–936.

65. **Herring, A. C., N. R. Falkowski, G. H. Chen, R. A. McDonald, G. B. Toews, and G. B. Huffnagle.** 2005. Transient neutralization of tumor necrosis factor alpha can produce a chronic fungal infection in an immunocompetent host: potential role of immature dendritic cells. *Infect. Immun.* **73**:39–49.

66. **Herring, A. C., J. Lee, R. A. McDonald, G. B. Toews, and G. B. Huffnagle.** 2002. Induction of interleukin-12 and gamma interferon requires tumor necrosis factor

Extracellular killing of *Cryptococcus neoformans*. *J. Immunol.* **142**:3219–3224.

67. Hidore, M. R., and J. W. Murphy. 1986. Natural cellular resistance of beige mice against *Cryptococcus neoformans*. *J. Immunol.* **137**:3624–3631.

68. Hidore, M. R., N. Nabavi, C. W. Reynolds, P. A. Henkart, and J. W. Murphy. 1990. Cytoplasmic components of natural killer cells limit the growth of *Cryptococcus neoformans*. *J. Leukoc. Biol.* **48**:15–26.

69. Hidore, M. R., N. Nabavi, F. Sonleitner, and J. W. Murphy. 1991. Murine natural killer cells are fungicidal to *Cryptococcus neoformans*. *Infect. Immun.* **59**:1747–1754.

70. Hill, J. O. 1992. CD4+ T cells cause multinucleated giant cells to form around *Cryptococcus neoformans* and confine the yeast within the primary site of infection in the respiratory tract. *J. Exp. Med.* **175**:1685–1695.

71. Hill, J. O., and K. M. Aguirre. 1994. CD4+ T cell-dependent acquired state of immunity that protects the brain against *Cryptococcus neoformans*. *J. Immunol.* **152**:2344–2350.

72. Hoang, L. M., J. A. Maguire, P. Doyle, M. Fyfe, and D. L. Roscoe. 2004. *Cryptococcus neoformans* infections at Vancouver Hospital and Health Sciences Centre (1997–2002): epidemiology, microbiology and histopathology. *J. Med. Microbiol.* **53**:935–940.

73. Huang, C., and S. M. Levitz. 2000. Stimulation of macrophage inflammatory protein-1α, macrophage inflammatory protein-1β, and RANTES by *Candida albicans* and *Cryptococcus neoformans* in peripheral blood mononuclear cells from persons with and without human immunodeficiency virus infection. *J. Infect. Dis.* **181**:791–794.

74. Huffnagle, G. B., M. B. Boyd, N. E. Street, and M. F. Lipscomb. 1998. IL-5 is required for eosinophil recruitment, crystal deposition, and mononuclear cell recruitment during a pulmonary *Cryptococcus neoformans* infection in genetically susceptible mice (C57BL/6). *J. Immunol.* **160**:2393–2400.

75. Huffnagle, G. B., G. H. Chen, J. L. Curtis, R. A. McDonald, R. M. Strieter, and G. B. Toews. 1995. Down-regulation of the afferent phase of T cell-mediated pulmonary inflammation and immunity by a high melanin-producing strain of *Cryptococcus neoformans*. *J. Immunol.* **155**:3507–3516.

76. Huffnagle, G. B., M. F. Lipscomb, J. A. Lovchik, K. A. Hoag, and N. E. Street. 1994. The role of CD4+ and CD8+ T cells in the protective inflammatory response to a pulmonary cryptococcal infection. *J. Leukoc. Biol.* **55**:35–42.

77. Huffnagle, G. B., and L. K. McNeil. 1999. Dissemination of *C. neoformans* to the central nervous system: role of chemokines, Th1 immunity and leukocyte recruitment. *J. Neurovirol.* **5**:76–81.

78. Huffnagle, G. B., L. K. McNeil, R. A. McDonald, J. W. Murphy, G. B. Toews, N. Maeda, and W. A. Kuziel. 1999. Cutting edge: role of C-C chemokine receptor 5 in organ-specific and innate immunity to *Cryptococcus neoformans*. *J. Immunol.* **163**:4642–4646.

79. Huffnagle, G. B., R. M. Strieter, L. K. McNeil, R. A. McDonald, M. D. Burdick, S. L. Kunkel, and G. B. Toews. 1997. Macrophage inflammatory protein-1α (MIP-1α) is required for the efferent phase of pulmonary cell-mediated immunity to a *Cryptococcus neoformans* infection. *J. Immunol.* **159**:318–327.

80. Huffnagle, G. B., R. M. Strieter, T. J. Standiford, R. A. McDonald, M. D. Burdick, S. L. Kunkel, and G. B. Toews. 1995. The role of monocyte chemotactic protein-1 (MCP-1) in the recruitment of monocytes and CD4+ T cells during a pulmonary *Cryptococcus neoformans* infection. *J. Immunol.* **155**:4790–4797.

81. Huffnagle, G. B., J. L. Yates, and M. F. Lipscomb. 1991. Immunity to a pulmonary *Cryptococcus neoformans* infection requires both CD4+ and CD8+ T cells. *J. Exp. Med.* **173**:793–800.

82. Huffnagle, G. B., J. L. Yates, and M. F. Lipscomb. 1991. T cell-mediated immunity in the lung: a *Cryptococcus neoformans* pulmonary infection model using SCID and athymic nude mice. *Infect. Immun.* **59**:1423–1433.

83. Ieong, M. H., C. C. Reardon, S. M. Levitz, and H. Kornfeld. 2000. Human immunodeficiency virus type 1 infection of alveolar macrophages impairs their innate fungicidal activity. *Am. J. Respir. Crit. Care Med.* **162**:966–970.

84. Kawakami, K., M. Hossain Qureshi, T. Zhang, Y. Koguchi, Q. Xie, M. Kurimoto, and A. Saito. 1999. Interleukin-4 weakens host resistance to pulmonary and disseminated cryptococcal infection caused by combined treatment with interferon-gamma-inducing cytokines. *Cell. Immunol.* **197**:55–61.

85. Kawakami, K., Y. Kinjo, K. Uezu, S. Yara, K. Miyagi, Y. Koguchi, T. Nakayama, M. Taniguchi, and A. Saito. 2001. Monocyte chemoattractant protein-1-dependent increase of V α 14 NKT cells in lungs and their roles in Th1 response and host defense in cryptococcal infection. *J. Immunol.* **167**:6525–6532.

86. Kawakami, K., Y. Koguchi, M. H. Qureshi, S. Yara, Y. Kinjo, K. Uezu, and A. Saito. 2000. NK cells eliminate *Cryptococcus neoformans* by potentiating the fungicidal activity of macrophages rather than by directly killing them upon stimulation with IL-12 and IL-18. *Microbiol. Immunol.* **44**:1043–1050.

87. Kawakami, K., S. Kohno, J. Kadota, M. Tohyama, K. Teruya, N. Kudeken, A. Saito, and K. Hara. 1995. T cell-dependent activation of macrophages and enhancement of their phagocytic activity in the lungs of mice inoculated with heat-killed *Cryptococcus neoformans*: involvement of IFN-gamma and its protective effect against cryptococcal infection. *Microbiol. Immunol.* **39**:135–143.

88. Kawakami, K., S. Kohno, N. Morikawa, J. Kadota, A. Saito, and K. Hara. 1994. Activation of macrophages and expansion of specific T lymphocytes in the lungs of mice intratracheally inoculated with *Cryptococcus neoformans*. *Clin. Exp. Immunol.* **96**:230–237.

89. Kawakami, K., X. Qifeng, M. Tohyama, M. H. Qureshi, and A. Saito. 1996. Contribution of tumour necrosis factor-alpha (TNF-α) in host defence mechanism against *Cryptococcus neoformans*. *Clin. Exp. Immunol.* **106**:468–474.

90. Kawakami, K., M. H. Qureshi, Y. Koguchi, K. Nakajima, and A. Saito. 1999. Differential effect of *Cryptococcus neoformans* on the production of IL-12p40 and IL-10 by murine macrophages stimulated with lipopolysaccharide and gamma interferon. *FEMS Microbiol. Lett.* **175**:87–94.

91. Kawakami, K., M. H. Qureshi, Y. Koguchi, T. Zhang, H. Okamura, M. Kurimoto, and A. Saito. 1999. Role of TNF-α in the induction of fungicidal activity of mouse peritoneal exudate cells against *Cryptococcus neoformans* by IL-12 and IL-18. *Cell. Immunol.* **193**:9–16.

92. Kawakami, K., K. Shibuya, M. H. Qureshi, T. Zhang, Y. Koguchi, M. Tohyama, Q. Xie, S. Naoe, and A. Saito. 1999. Chemokine responses and accumulation of inflammatory cells in the lungs of mice infected with highly virulent *Cryptococcus neoformans*: effects of interleukin-12. *FEMS Immunol. Med. Microbiol.* 25:391–402.

93. Kawakami, K., M. Tohyama, X. Qifeng, and A. Saito. 1997. Expression of cytokines and inducible nitric oxide synthase mRNA in the lungs of mice infected with *Cryptococcus neoformans*: effects of interleukin-12. *Infect. Immun.* 65:1307–1312.

94. Kawakami, K., M. Tohyama, Q. Xie, and A. Saito. 1996. IL-12 protects mice against pulmonary and disseminated infection caused by *Cryptococcus neoformans. Clin. Exp. Immunol.* 104:208–214.

95. Kawakami, K., T. Zhang, M. H. Qureshi, and A. Saito. 1997. *Cryptococcus neoformans* inhibits nitric oxide production by murine peritoneal macrophages stimulated with interferon-gamma and lipopolysaccharide. *Cell. Immunol.* 180:47–54.

96. Kelly, R. M., J. Chen, L. E. Yauch, and S. M. Levitz. 2005. Opsonic requirements for dendritic cell-mediated responses to *Cryptococcus neoformans. Infect. Immun.* 73:592–598.

97. Koguchi, Y., and K. Kawakami. 2002. Cryptococcal infection and Th1-Th2 cytokine balance. *Int. Rev. Immunol.* 21:423–438.

98. Kozel, T. R. 1996. Activation of the complement system by pathogenic fungi. *Clin. Microbiol. Rev.* 9:34–46.

99. Kozel, T. R. 1977. Non-encapsulated variant of *Cryptococcus neoformans*. II. Surface receptors for cryptococcal polysaccharide and their role in inhibition of phagocytosis by polysaccharide. *Infect. Immun.* 16:99–106.

100. Kozel, T. R., and E. C. Gotschlich. 1982. The capsule of *Cryptococcus neoformans* passively inhibits phagocytosis of the yeast by macrophages. *J. Immunol.* 129:1675–1680.

101. Kozel, T. R., B. Highison, and C. J. Stratton. 1984. Localization on encapsulated *Cryptococcus neoformans* of serum components opsonic for phagocytosis by macrophages and neutrophils. *Infect. Immun.* 43:574–579.

102. Kozel, T. R., and R. P. Mastroianni. 1976. Inhibition of phagocytosis by cryptococcal polysaccharide: dissociation of the attachment and ingestion phases of phagocytosis. *Infect. Immun.* 14:62–67.

103. Kozel, T. R., G. S. Pfrommer, A. S. Guerlain, B. A. Highison, and G. J. Highison. 1988. Strain variation in phagocytosis of *Cryptococcus neoformans*: dissociation of susceptibility to phagocytosis from activation and binding of opsonic fragments of C3. *Infect. Immun.* 56:2794–2800.

104. Kozel, T. R., A. Tabuni, B. J. Young, and S. M. Levitz. 1996. Influence of opsonization conditions on C3 deposition and phagocyte binding of large- and small-capsule *Cryptococcus neoformans* cells. *Infect. Immun.* 64:2336–2338.

105. Kwon-Chung, K. J., I. Polacheck, and T. J. Popkin. 1982. Melanin-lacking mutants of *Cryptococcus neoformans* and their virulence for mice. *J. Bacteriol.* 150:1414–1421.

106. Kwon-Chung, K. J., and J. C. Rhodes. 1986. Encapsulation and melanin formation as indicators of virulence in *Cryptococcus neoformans. Infect. Immun.* 51:218–223.

107. Lee, S. C., A. Casadevall, and D. W. Dickson. 1996. Immunohistochemical localization of capsular polysaccharide antigen in the central nervous system cells in cryptococcal meningoencephalitis. *Am. J. Pathol.* 148:1267–1274.

108. Lee, S. C., D. W. Dickson, and A. Casadevall. 1996. Pathology of cryptococcal meningoencephalitis: analysis of 27 patients with pathogenetic implications. *Hum. Pathol.* 27:839–847.

109. Lee, S. C., Y. Kress, M. L. Zhao, D. W. Dickson, and A. Casadevall. 1995. *Cryptococcus neoformans* survive and replicate in human microglia. *Lab. Investig.* 73:871–879.

110. Levitz, S. M., and D. J. DiBenedetto. 1988. Differential stimulation of murine resident peritoneal cells by selectively opsonized encapsulated and acapsular *Cryptococcus neoformans. Infect. Immun.* 56:2544–2551.

111. Levitz, S. M., and D. J. DiBenedetto. 1989. Paradoxical role of capsule in murine bronchoalveolar macrophage-mediated killing of *Cryptococcus neoformans. J. Immunol.* 142:659–665.

112. Levitz, S. M., and M. P. Dupont. 1993. Phenotypic and functional characterization of human lymphocytes activated by interleukin-2 to directly inhibit growth of *Cryptococcus neoformans* in vitro. *J. Clin. Investig.* 91:1490–1498.

113. Levitz, S. M., M. P. Dupont, and E. H. Smail. 1994. Direct activity of human T lymphocytes and natural killer cells against *Cryptococcus neoformans. Infect. Immun.* 62:194–202.

114. Levitz, S. M., and T. P. Farrell. 1990. Growth inhibition of *Cryptococcus neoformans* by cultured human monocytes: role of the capsule, opsonins, the culture surface, and cytokines. *Infect. Immun.* 58:1201–1209.

115. Levitz, S. M., T. S. Harrison, A. Tabuni, and X. Liu. 1997. Chloroquine induces human mononuclear phagocytes to inhibit and kill *Cryptococcus neoformans* by a mechanism independent of iron deprivation. *J. Clin. Investig.* 100:1640–1646.

116. Levitz, S. M., S. H. Nong, K. F. Seetoo, T. S. Harrison, R. A. Speizer, and E. R. Simons. 1999. *Cryptococcus neoformans* resides in an acidic phagolysosome of human macrophages. *Infect. Immun.* 67:885–890.

117. Levitz, S. M., and E. A. North. 1996. Gamma interferon gene expression and release in human lymphocytes directly activated by *Cryptococcus neoformans* and *Candida albicans. Infect. Immun.* 64:1595–1599.

118. Levitz, S. M., E. A. North, Y. Jiang, S. H. Nong, H. Kornfeld, and T. S. Harrison. 1997. Variables affecting production of monocyte chemotactic factor 1 from human leukocytes stimulated with *Cryptococcus neoformans. Infect. Immun.* 65:903–908.

119. Levitz, S. M., and A. Tabuni. 1991. Binding of *Cryptococcus neoformans* by human cultured macrophages. Requirements for multiple complement receptors and actin. *J. Clin. Investig.* 87:528–535.

119a. Levitz, S. M., A. Tabuni, and B. Liss. 1994. Macrophage-*Cryptococcus* interactions, p. 533–543. In B. S. Zwilling and T. K. Eisenstein(ed.), *Macrophage-Pathogen Interactions*. Marcel Dekker, New York, N.Y.

120. Levitz, S. M., A. Tabuni, R. Wagner, H. Kornfeld, and E. H. Smail. 1992. Binding of unopsonized *Cryptococcus neoformans* by human bronchoalveolar macrophages: inhibition by a large-molecular-size serum component. *J. Infect. Dis.* 166:866–873.

121. Li, R. K., and T. G. Mitchell. 1997. Induction of interleukin-6 mRNA in rat alveolar macrophages by in vitro exposure to both *Cryptococcus neoformans* and

anti-*C. neoformans* antiserum. *J. Med. Vet. Mycol.* **35:** 327–334.

122. **Lipovsky, M. M., G. Gekker, S. Hu, L. C. Ehrlich, A. I. Hoepelman, and P. K. Peterson.** 1998. Cryptococcal glucuronoxylomannan induces interleukin (IL)-8 production by human microglia but inhibits neutrophil migration toward IL-8. *J. Infect. Dis.* **177:**260–263.

123. **Lipovsky, M. M., L. J. van Elden, A. M. Walenkamp, J. Dankert, and A. I. Hoepelman.** 1998. Does the capsule component of the *Cryptococcus neoformans* glucuronoxylomannan impair transendothelial migration of leukocytes in patients with cryptococcal meningitis? *J. Infect. Dis.* **178:**1231–1232.

124. **Lipscomb, M. F., T. Alvarellos, G. B. Toews, R. Tompkins, Z. Evans, G. Koo, and V. Kumar.** 1987. Role of natural killer cells in resistance to *Cryptococcus neoformans* infections in mice. *Am. J. Pathol.* **128:** 354–361.

125. **Lovchik, J., M. Lipscomb, and C. R. Lyons.** 1997. Expression of lung inducible nitric oxide synthase protein does not correlate with nitric oxide production in vivo in a pulmonary immune response against *Cryptococcus neoformans*. *J. Immunol.* **158:**1772–1778.

126. **Lovchik, J. A., C. R. Lyons, and M. F. Lipscomb.** 1995. A role for gamma interferon-induced nitric oxide in pulmonary clearance of *Cryptococcus neoformans*. *Am. J. Respir. Cell Mol. Biol.* **13:**116–124.

127. **Luberto, C., B. Martinez-Marino, D. Taraskiewicz, B. Bolanos, P. Chitano, D. L. Toffaletti, G. M. Cox, J. R. Perfect, Y. A. Hannun, E. Balish, and M. Del Poeta.** 2003. Identification of App1 as a regulator of phagocytosis and virulence of *Cryptococcus neoformans*. *J. Clin. Investig.* **112:**1080–1094.

128. **Ma, L. L., J. C. Spurrell, J. F. Wang, G. G. Neely, S. Epelman, A. M. Krensky, and C. H. Mody.** 2002. CD8 T cell-mediated killing of *Cryptococcus neoformans* requires granulysin and is dependent on CD4 T cells and IL-15. *J. Immunol.* **169:**5787–5795.

129. **Ma, L. L., C. L. Wang, G. G. Neely, S. Epelman, A. M. Krensky, and C. H. Mody.** 2004. NK cells use perforin rather than granulysin for anticryptococcal activity. *J. Immunol.* **173:**3357–3365.

130. **Mambula, S. S., E. R. Simons, R. Hastey, M. E. Selsted, and S. M. Levitz.** 2000. Human neutrophil-mediated nonoxidative antifungal activity against *Cryptococcus neoformans*. *Infect. Immun.* **68:**6257–6264.

131. **Mansour, M. K., L. E. Yauch, J. B. Rottman, and S. M. Levitz.** 2004. Protective efficacy of antigenic fractions in mouse models of cryptococcosis. *Infect. Immun.* **72:**1746–1754.

132. **Marquis, G., S. Montplaisir, M. Pelletier, S. Mousseau, and P. Auger.** 1985. Genetic resistance to murine cryptococcosis: the beige mutation (Chediak-Higashi syndrome) in mice. *Infect. Immun.* **47:**288–293.

133. **Mednick, A. J., M. Feldmesser, J. Rivera, and A. Casadevall.** 2003. Neutropenia alters lung cytokine production in mice and reduces their susceptibility to pulmonary cryptococcosis. *Eur. J. Immunol.* **33:**1744–1753.

134. **Miller, G. P., and S. Kohl.** 1983. Antibody-dependent leukocyte killing of *Cryptococcus neoformans*. *J. Immunol.* **131:**1455–1459.

135. **Miller, M. F., and T. G. Mitchell.** 1991. Killing of *Cryptococcus neoformans* strains by human neutrophils and monocytes. *Infect. Immun.* **59:**24–28.

136. **Mitchell, T. G., and L. Friedman.** 1972. In vitro phagocytosis and intracellular fate of variously encapsulated strains of *Cryptococcus neoformans*. *Infect. Immun.* **5:**491–498.

137. **Mody, C. H., M. F. Lipscomb, N. E. Street, and G. B. Toews.** 1990. Depletion of CD4$^+$ (L3T4$^+$) lymphocytes in vivo impairs murine host defense to *Cryptococcus neoformans*. *J. Immunol.* **144:**1472–1477.

138. **Mody, C. H., C. L. Tyler, R. G. Sitrin, C. Jackson, and G. B. Toews.** 1991. Interferon-gamma activates rat alveolar macrophages for anticryptococcal activity. *Am. J. Respir. Cell Mol. Biol.* **5:**19–26.

139. **Monari, C., F. Baldelli, D. Pietrella, C. Retini, C. Tascini, D. Francisci, F. Bistoni, and A. Vecchiarelli.** 1997. Monocyte dysfunction in patients with acquired immunodeficiency syndrome (AIDS) versus *Cryptococcus neoformans*. *J. Infect.* **35:**257–263.

140. **Monari, C., F. Bistoni, A. Casadevall, E. Pericolini, D. Pietrella, T. R. Kozel, and A. Vecchiarelli.** 2005. Glucuronoxylomannan, a microbial compound, regulates expression of costimulatory molecules and production of cytokines in macrophages. *J. Infect. Dis.* **191:**127–137.

141. **Monari, C., A. Casadevall, D. Pietrella, F. Bistoni, and A. Vecchiarelli.** 1999. Neutrophils from patients with advanced human immunodeficiency virus infection have impaired complement receptor function and preserved Fcgamma receptor function. *J. Infect. Dis.* **180:**1542–1549.

142. **Monari, C., C. Retini, A. Casadevall, D. Netski, F. Bistoni, T. R. Kozel, and A. Vecchiarelli.** 2003. Differences in outcome of the interaction between *Cryptococcus neoformans* glucuronoxylomannan and human monocytes and neutrophils. *Eur. J. Immunol.* **33:**1041–1051.

143. **Mukherjee, S., M. Feldmesser, and A. Casadevall.** 1996. J774 murine macrophage-like cell interactions with *Cryptococcus neoformans* in the presence and absence of opsonins. *J. Infect. Dis.* **173:**1222–1231.

144. **Murphy, J. W., M. R. Hidore, and S. C. Wong.** 1993. Direct interactions of human lymphocytes with the yeast-like organism, *Cryptococcus neoformans*. *J. Clin. Investig.* **91:**1553–1566.

145. **Murphy, J. W., and D. O. McDaniel.** 1982. In vitro reactivity of natural killer (NK) cells against *Cryptococcus neoformans*. *J. Immunol.* **128:**1577–1583.

146. **Murphy, J. W., R. L. Mosley, R. Cherniak, G. H. Reyes, T. R. Kozel, and E. Reiss.** 1988. Serological, electrophoretic, and biological properties of *Cryptococcus neoformans* antigens. *Infect. Immun.* **56:**424–431.

147. **Nabavi, N., and J. W. Murphy.** 1985. In vitro binding of natural killer cells to *Cryptococcus neoformans* targets. *Infect. Immun.* **50:**50–57.

148. **Narasipura, S. D., J. G. Ault, M. J. Behr, V. Chaturvedi, and S. Chaturvedi.** 2003. Characterization of Cu,Zn superoxide dismutase (SOD1) gene knock-out mutant of *Cryptococcus neoformans* var. *gattii*: role in biology and virulence. *Mol. Microbiol.* **47:**1681–1694.

149. **Naslund, P. K., W. C. Miller, and D. L. Granger.** 1995. *Cryptococcus neoformans* fails to induce nitric oxide synthase in primed murine macrophage-like cells. *Infect. Immun.* **63:**1298–1304.

150. **Nosanchuk, J. D., and A. Casadevall.** 2003. The contribution of melanin to microbial pathogenesis. *Cell. Microbiol.* **5:**203–223.

151. **Noverr, M. C., G. M. Cox, J. R. Perfect, and G. B. Huffnagle.** 2003. Role of PLB1 in pulmonary inflammation and cryptococcal eicosanoid production. *Infect. Immun.* **71:**1538–1547.

152. Olszewski, M. A., G. B. Huffnagle, R. A. McDonald, D. M. Lindell, B. B. Moore, D. N. Cook, and G. B. Toews. 2000. The role of macrophage inflammatory protein-1 α/CCL3 in regulation of T cell-mediated immunity to *Cryptococcus neoformans* infection. *J. Immunol.* 165:6429–6436.

153. Olszewski, M. A., G. B. Huffnagle, T. R. Traynor, R. A. McDonald, D. N. Cook, and G. B. Toews. 2001. Regulatory effects of macrophage inflammatory protein 1α/CCL3 on the development of immunity to *Cryptococcus neoformans* depend on expression of early inflammatory cytokines. *Infect. Immun.* 69:6256–6263.

154. Perfect, J. R., D. L. Granger, and D. T. Durack. 1987. Effects of antifungal agents and gamma interferon on macrophage cytotoxicity for fungi and tumor cells. *J. Infect. Dis.* 156:316–323.

155. Pietrella, D., R. Cherniak, C. Strappini, S. Perito, P. Mosci, F. Bistoni, and A. Vecchiarelli. 2001. Role of mannoprotein in induction and regulation of immunity to *Cryptococcus neoformans*. *Infect. Immun.* 69:2808–2814.

156. Pietrella, D., C. Corbucci, S. Perito, G. Bistoni, and A. Vecchiarelli. 2005. Mannoproteins from *Cryptococcus neoformans* promote dendritic cell maturation and activation. *Infect. Immun.* 73:820–827.

157. Pitzurra, L., R. Cherniak, M. Giammarioli, S. Perito, F. Bistoni, and A. Vecchiarelli. 2000. Early induction of interleukin-12 by human monocytes exposed to *Cryptococcus neoformans* mannoproteins. *Infect. Immun.* 68:558–563.

158. Powderly, W. G. 1993. Cryptococcal meningitis and AIDS. *Clin. Infect. Dis.* 17:837–842.

159. Qureshi, M. H., T. Zhang, Y. Koguchi, K. Nakashima, H. Okamura, M. Kurimoto, and K. Kawakami. 1999. Combined effects of IL-12 and IL-18 on the clinical course and local cytokine production in murine pulmonary infection with *Cryptococcus neoformans*. *Eur. J. Immunol.* 29:643–649.

160. Reardon, C. C., S. J. Kim, R. P. Wagner, and H. Kornfeld. 1996. Interferon-gamma reduces the capacity of human alveolar macrophages to inhibit growth of *Cryptococcus neoformans* in vitro. *Am. J. Respir. Cell Mol. Biol.* 15:711–715.

161. Reardon, C. C., S. J. Kim, R. P. Wagner, H. Koziel, and H. Kornfeld. 1996. Phagocytosis and growth inhibition of *Cryptococcus neoformans* by human alveolar macrophages: effects of HIV-1 infection. *AIDS* 10:613–618.

162. Reese, A. J., and T. L. Doering. 2003. Cell wall α-1,3-glucan is required to anchor the *Cryptococcus neoformans* capsule. *Mol. Microbiol.* 50:1401–1409.

163. Retini, C., A. Casadevall, D. Pietrella, C. Monari, B. Palazzetti, and A. Vecchiarelli. 1999. Specific activated T cells regulate IL-12 production by human monocytes stimulated with *Cryptococcus neoformans*. *J. Immunol.* 162:1618–1623.

164. Retini, C., T. R. Kozel, D. Pietrella, C. Monari, F. Bistoni, and A. Vecchiarelli. 2001. Interdependency of interleukin-10 and interleukin-12 in regulation of T-cell differentiation and effector function of monocytes in response to stimulation with *Cryptococcus neoformans*. *Infect. Immun.* 69:6064–6073.

165. Retini, C., A. Vecchiarelli, C. Monari, C. Tascini, F. Bistoni, and T. R. Kozel. 1996. Capsular polysaccharide of *Cryptococcus neoformans* induces proinflammatory cytokine release by human neutrophils. *Infect. Immun.* 64:2897–2903.

166. Rivera, J., J. Mukherjee, L. M. Weiss, and A. Casadevall. 2002. Antibody efficacy in murine pulmonary

167. Roseff, S. A., and S. M. Levitz. 1993. Effect of endothelial cells on phagocyte-mediated anticryptococcal activity. *Infect. Immun.* 61:3818–3824.

168. Rossi, G. R., L. A. Cervi, M. M. Garcia, L. S. Chiapello, D. A. Sastre, and D. T. Masih. 1999. Involvement of nitric oxide in protecting mechanism during experimental cryptococcosis. *Clin. Immunol.* 90:256–265.

169. Rossi, G. R., D. A. Sastre, H. R. Rubinstein, and D. T. Masih. 1994. Biochemical basis for the killing of *Cryptococcus neoformans* by rat peritoneal cells. *J. Med. Vet. Mycol.* 32:405–414.

170. Santangelo, R., H. Zoellner, T. Sorrell, C. Wilson, C. Donald, J. Djordjevic, Y. Shounan, and L. Wright. 2004. Role of extracellular phospholipases and mononuclear phagocytes in dissemination of cryptococcosis in a murine model. *Infect. Immun.* 72:2229–2239.

171. Shoham, S., C. Huang, J. M. Chen, D. T. Golenbock, and S. M. Levitz. 2001. Toll-like receptor 4 mediates intracellular signaling without TNF-α release in response to *Cryptococcus neoformans* polysaccharide capsule. *J. Immunol.* 166:4620–4626.

172. Sorrell, T. C. 2001. *Cryptococcus neoformans* variety gattii. *Med. Mycol.* 39:155–168.

173. Steenbergen, J. N., H. A. Shuman, and A. Casadevall. 2001. *Cryptococcus neoformans* interactions with amoebae suggest an explanation for its virulence and intracellular pathogenic strategy in macrophages. *Proc. Natl. Acad. Sci. USA* 98:15245–15250.

174. Syme, R. M., J. C. Spurrell, E. K. Amankwah, F. H. Green, and C. H. Mody. 2002. Primary dendritic cells phagocytose *Cryptococcus neoformans* via mannose receptors and Fcγ receptor II for presentation to T lymphocytes. *Infect. Immun.* 70:5972–5981.

175. Tabeta, H., N. Kohno, K. Kamei, A. Honda, H. Unno, K. Nagao, T. Yamaguchi, and M. Miyaji. 1991. The defensive role of human pulmonary alveolar macrophages and polymorphonuclear leukocytes against strongly or weakly virulent strains of *Cryptococcus neoformans*. *Nihon Kyobu Shikkan Gakkai Zasshi* 29:1174–1179. (In Japanease.)

176. Takeo, K., I. Uesaka, K. Uehira, and M. Nishiura. 1973. Fine structure of *Cryptococcus neoformans* grown in vivo as observed by freeze-etching. *J. Bacteriol.* 113:1449–1454.

177. Tohyama, M., K. Kawakami, M. Futenma, and A. Saito. 1996. Enhancing effect of oxygen radical scavengers on murine macrophage anticryptococcal activity through production of nitric oxide. *Clin. Exp. Immunol.* 103:436–441.

178. Trapani, J. A., and M. J. Smyth. 2002. Functional significance of the perforin/granzyme cell death pathway. *Nat. Rev. Immunol.* 2:735–747.

179. Traynor, T. R., W. A. Kuziel, G. B. Toews, and G. B. Huffnagle. 2000. CCR2 expression determines T1 versus T2 polarization during pulmonary *Cryptococcus neoformans* infection. *J. Immunol.* 164:2021–2027.

180. Tucker, S. C., and A. Casadevall. 2002. Replication of *Cryptococcus neoformans* in macrophages is accompanied by phagosomal permeabilization and accumulation of vesicles containing polysaccharide in the cytoplasm. *Proc. Natl. Acad. Sci. USA* 99:3165–3170.

181. Uezu, K., K. Kawakami, K. Miyagi, Y. Kinjo, T. Kinjo, H. Ishikawa, and A. Saito. 2004. Accumulation of γδ T cells in the lungs and their regulatory roles in Th1

response and host defense against pulmonary infection with *Cryptococcus neoformans*. *J. Immunol.* **172:** 7629–7634.

182. **Vartivarian, S. E., E. J. Anaissie, R. E. Cowart, H. A. Sprigg, M. J. Tingler, and E. S. Jacobson.** 1993. Regulation of cryptococcal capsular polysaccharide by iron. *J. Infect. Dis.* **167:**186–190.

183. **Vecchiarelli, A., D. Pietrella, M. Dottorini, C. Monari, C. Retini, T. Todisco, and F. Bistoni.** 1994. Encapsulation of *Cryptococcus neoformans* regulates fungicidal activity and the antigen presentation process in human alveolar macrophages. *Clin. Exp. Immunol.* **98:**217–223.

184. **Vecchiarelli, A., D. Pietrella, P. Lupo, F. Bistoni, D. C. McFadden, and A. Casadevall.** 2003. The polysaccharide capsule of *Cryptococcus neoformans* interferes with human dendritic cell maturation and activation. *J. Leukoc. Biol.* **74:**370–378.

185. **Vecchiarelli, A., C. Retini, A. Casadevall, C. Monari, D. Pietrella, and T. R. Kozel.** 1998. Involvement of C3a and C5a in interleukin-8 secretion by human polymorphonuclear cells in response to capsular material of *Cryptococcus neoformans*. *Infect. Immun.* **66:**4324–4330.

186. **Vecchiarelli, A., C. Retini, C. Monari, C. Tascini, F. Bistoni, and T. R. Kozel.** 1996. Purified capsular polysaccharide of *Cryptococcus neoformans* induces interleukin-10 secretion by human monocytes. *Infect. Immun.* **64:**2846–2849.

187. **Vecchiarelli, A., C. Retini, D. Pietrella, C. Monari, C. Tascini, T. Beccari, and T. R. Kozel.** 1995. Downregulation by cryptococcal polysaccharide of tumor necrosis factor alpha and interleukin-1β secretion from human monocytes. *Infect. Immun.* **63:**2919–2923.

188. **Vidotto, V., A. Sinicco, D. Di Fraia, S. Cardaropoli, S. Aoki, and S. Ito-Kuwa.** 1996. Phospholipase activity in *Cryptococcus neoformans*. *Mycopathologia* **136:**119–123.

189. **Wang, Y., P. Aisen, and A. Casadevall.** 1995. *Cryptococcus neoformans* melanin and virulence: mechanism of action. *Infect. Immun.* **63:**3131–3136.

190. **Wang, Y., and A. Casadevall.** 1994. Susceptibility of melanized and nonmelanized *Cryptococcus neoformans* to nitrogen- and oxygen-derived oxidants. *Infect. Immun.* **62:**3004–3007.

191. **Yasuoka, A., S. Kohno, H. Yamada, M. Kaku, and H. Koga.** 1994. Influence of molecular sizes of *Cryptococcus neoformans* capsular polysaccharide on phagocytosis. *Microbiol. Immunol.* **38:**851–856.

Molecular Principles of Fungal Pathogenesis
Edited by Joseph Heitman et al.
©2006 ASM Press, Washington, D.C.

Chapter 37

Fungal Interactions with Leukocytes

Mairi C. Noverr, Dennis M. Lindell, Galen B. Toews, and Gary B. Huffnagle

The respiratory tract is the main portal of entry for numerous fungal pathogens including *Histoplasma capsulatum*, *Blastomyces dermatitidis*, *Cryptococcus neoformans*, *Pneumocystis jiroveci*, *Aspergillus fumigatus*, *Coccidioides immitis*, *Penicillium marneffei*, and *Paracoccidioides brasiliensis*. The respiratory tract is constantly exposed to mold spores, fungal elements, desiccated yeast, and numerous other nonfungal microorganisms that are inhaled from the environment. Many fungi are essentially environmental organisms that have developed mechanisms to survive and grow with the nonmammalian world of plants, bacteria, and amoeba. If these mechanisms allow an organism to survive inside a mammalian host or evade host defenses, they have been classically termed "virulence factors." This chapter discusses host defenses in the lungs and the role played by fungal virulence factors in allowing the organism to survive inside the host and either cause disease or remain latent.

Host defense against fungal infection in the lungs requires both innate and adaptive immune defenses. Some fungi can cause disease in healthy individuals and are considered primary pathogens (*H. capsulatum*, *B. dermatitidis*, *C. immitis*, and *P. brasiliensis*). Some are largely opportunistic (*A. fumigatus* and *P. jiroveci*), while others can be pathogenic in multiple settings (*C. neoformans*, *A. fumigatus*, and *H. capsulatum*). Many cause disease only in hypersensitive (allergic) individuals (*Penicillium* and *Alternaria*). *Candida albicans* is the only fungus that is exclusively found as a part of our normal microbiota (in small numbers in healthy hosts), and while it is a significant cause of opportunistic fungal infections, *C. albicans* is not a common cause of pulmonary fungal infection.

Overall, the relative requirement for innate versus adaptive immunity depends on the species of fungus. In general, the pulmonary host defense against fungi uses a common set of hierarchical immunologic mechanisms. These mechanisms include mechanical barriers, innate defenses, and adaptive responses. If an organism is not eliminated or controlled at any step, signals are generated that promote the development of the next phase of the response, which includes increasing antigen specificity and regulated vigorous inflammatory processes.

PULMONARY DEFENSES

On inhalation from the environment, fungi first encounter the barrier defenses of the lungs (cough reflexes, airway secretions, and mucociliary action of the upper airway epithelium). The single most effective defense mechanism of mammals against the thousands of airborne fungal spores that are inhaled from the environment (especially in individuals who are working in the garden or on a construction site where the number of inhaled fungal cells can be in the millions) is their elevated body temperature. While fungi are environmental in origin, the vast majority of soil fungi cannot grow at temperatures above 35°C.

However, if an inhaled organism can grow at elevated temperatures and bypasses or evades these barrier defenses, innate immunity is critical for the eradication of the fungal pathogen. A number of effector mechanisms are induced early to combat a fungal infection, and these effector mechanisms also generate subsequent activating/inflammatory signals. Alveolar macrophages, dendritic cells (DC), and recruited neutrophils and monocytes compose the phagocytic cells of the pulmonary innate immune system. In addition, structural cells such as epithelial cells, endothelial cells, smooth muscle cells, and mast cells, in combination with resident natural killer

Mairi C. Noverr, Dennis M. Lindell, Galen B. Toews, and Gary B. Huffnagle • Division of Pulmonary & Critical Care Medicine, Department of Internal Medicine, University of Michigan, Ann Arbor, MI 48109.

cells and T cells, are responsible for the production of inflammatory cytokines and chemokines that augment innate immunity and can modulate the development of an adaptive immune response. All of these cells possess various forms of recognition receptors for microbial compounds including those of fungal origin. Signaling through these receptors can influence the cytokine and chemokine profile generated by the innate cells, which, in turn, directs adaptive responses. Thus, the cells of the innate immune system possess many immunoregulatory functions that provide a link between innate and adaptive immunity.

Adaptive immunity is initiated by cells that possess antigen receptors that are capable of specific recognition of a diverse array of microbial structures. These antigen receptors, such as those present on T and B lymphocytes, are generated through germ line DNA rearrangement, mutation, and transcriptional regulation. T cells up- or down-regulate inflammatory responses and phagocyte activation, while antibodies augment antifungal T-cell immunity in the lungs and provide additional opsonins for uptake of fungi by phagocytes. In almost all fungal infections, CD4$^+$ T cells play a role in recruiting macrophages to the lungs and are the major producers of gamma interferon (IFN-γ). Surprisingly, CD8$^+$ T cells contribute to host defense against pulmonary fungal infection, probably because most fungal pathogens have an intracellular phase.

RECOGNITION OF FUNGI IN THE LUNGS

There are a number of recognition receptors and endogenous opsonins for fungal components in the airways. These include complement, complement receptors, mannose-binding proteins, mannose receptors, surfactant proteins, Fc receptors, CD18, VLA-5, pentraxins, and Toll-like receptors (TLR) (23, 24, 44, 46, 49, 66). TLR2 and TLR4 play a significant role in antifungal immunity. They signal through their common intracellular adaptor protein MyD88 (40, 64). Different strains of inbred mice vary in their basal TLR expression levels. This difference in TLR expression is one potential mechanism underlying differences among these mouse strains in susceptibility to various fungal pathogens (43). For example, C57BL/6 mice are susceptible to pulmonary infection with *C. neoformans* while CBA/J mice are relatively resistant (27). It remains to be determined whether differences in these mouse strains can be traced to differences in TLR interactions. The list of fungal ligands for TLRs is growing and currently includes HSP60, cell wall β-glucans, and polysaccharides (40, 64).

CELL TYPES ACTIVE IN DEFENSE AGAINST FUNGI

Macrophages

Macrophages can efficiently recognize and phagocytose fungal pathogens. The airspaces of the lungs contain a large number of resident macrophages (alveolar macrophages), while the parenchyma of the lungs contains a distinct population of macrophages (interstitial macrophages). Inflammation caused by fungal infection also results in the influx of monocytes from the circulation. These monocytes quickly differentiate into macrophages once in the lungs (recruited macrophages). *C. neoformans*, *A. fumigatus*, *P. jiroveci*, and *H. capsulatum* are recognized, bound, and ingested by alveolar macrophages. Furthermore, these organisms can survive and replicate within alveolar macrophages rather than being killed by these cells (38, 48, 53). Despite their high phagocytic and microbicidal ability, alveolar macrophages are poor antigen-presenting cells. In the lungs, antigen presentation is performed largely by DC. Alveolar macrophages also secrete numerous cytokines and lipid mediators that regulate both the innate and adaptive responses. Alveolar macrophages secrete cytokines such as tumor necrosis factor alpha (TNF-α), interleukin-12 (IL-12), IL-6, and chemokines such as macrophage inflammatory protein 1α (MIP-1α), IL-8, MIP-2, and KC. These cells are also important source of prostaglandins such as prostaglandin E$_2$, which is an important modulator of host responses (20). Macrophages, especially recruited macrophages, are the major antifungal effector cells and kill fungi through a variety of mechanisms including the release of reactive oxygen intermediates and nitric oxide (NO). Production of NO by macrophages is induced by a variety of inflammatory cytokines such as IFN-γ, TNF-α, IL-12, or IL-18 (19). For example, TNF-α, together with IL-12 and IL-18, potentiates murine macrophage fungicidal activities against *C. neoformans* by enhancing IFN-γ and NO production (35, 36). An important difference between murine and human macrophages is that NO and other reactive nitrogen intermediates constitute a key mechanism of the antimicrobial activities of murine macrophages. In contrast, human macrophages make relatively little NO, and oxidant-mediated killing is accomplished by the production of reactive oxygen intermediates.

Neutrophils

Neutrophil influx into the lungs accompanies almost all pulmonary fungal infections; however, the role of neutrophils in host defense is varied. For some fungi, such as *C. neoformans*, neutrophils are recruited to the site of infection but do not play a protective role (51). However, for pulmonary *Aspergillus* infections, neutrophils are the primary line of defense (16). The predominant chemotactic signals for recruiting neutrophils into lungs are CXC chemokines, leukotrienes, and complement components. Fungal virulence factors of *C. neoformans*, such as the capsular polysaccharide (glucuronoxylomannan [GXM]) and the mannoprotein fraction (MP-4), can diminish neutrophil recruitment to sites of infection (18). Neutrophils possess both oxidative and nonoxidative killing mechanisms. A number of fungus-neutrophil interactions induce an oxidative burst and degranulation; these interactions include contact between neutrophils and *A. fumigatus* or *C. neoformans* (12, 41). Nonoxidative mechanisms include the defensin family of antimicrobial peptides. Defensins from human neutrophils have potent fungicidal activity against *C. neoformans* and *C. albicans* and fungistatic activity against *H. capsulatum* (47, 54)

Dendritic Cells

The lungs are rich in resident DC, and the number of DC increase significantly during an inflammatory response. DC precursors migrate from the circulation into the lungs, where they differentiate to become "immature" DC. These DC are highly phagocytic and have low expression of major histocompatibility complex class II and costimulatory molecules. Immature DC become activated following phagocytosis of fungi and signaling from proinflammatory cytokines such as TNF-α. These DC migrate via the lymphatics to the draining lymph nodes, where they become mature DC. Mature DC are poorly phagocytic and express high levels of major histocompatibility complex class II and costimulatory molecules. They also stimulate T-cell responses (6, 8, 24, 66, 68, 69). DC can phagocytose and degrade *H. capsulatum*, *C. neoformans*, *C. albicans*, and *A. fumigatus* and subsequently process the appropriate antigens for stimulation of lymphocyte proliferation. Two major chemokine receptors involved in DC migration from the blood to the lungs to the lymph nodes are CCR2 and CCR7. CCR2 is probably involved in the blood-to-lung migration of precursor DC that subsequently differentiate into immature DC (68). CCR7 mediates DC migration from the lungs into T-cell areas of the secondary lymphoid tissues (22, 63). Critical to this whole process of DC migration is the proinflammatory cytokine TNF-α.

Secretion of TNF-α by innate immune cells is critical for the development of protective adaptive immunity. Neutralization of TNF-α diminishes mature DC accumulation in draining lymph nodes during *C. neoformans* infection (6, 26). Lung phagocytic cells produce TNF-α on infection with *P. jiroveci*,

B. dermatitidis, *H. capsulatum*, *A. fumigatus*, *C. immitis*, *C. neoformans*, *C. albicans*, and *P. brasiliensis* (1, 6, 9, 13, 14, 21, 52, 65). In general, treatment of mice with anti-TNF-α early in fungal infections diminishes pulmonary clearance, decreases leukocyte recruitment, lowers production of IFN-γ, and prevents the development of protective T-cell-mediated immunity (2, 21, 26, 31). Similarly, infection of TNF receptor 1 (TNFR1) or TNF receptor 2 (TNFR2) knockout mice with *H. capsulatum* results in high mortality compared to that of wild-type mice (3). TNF-α regulation of adaptive immunity occurs through the up-regulation of Th1 responses or suppression of Th2 responses, including elevated levels of IL-4 and IL-10 and significantly lower levels of IFN-γ during infection (2, 3, 21, 26, 31). Unfortunately, the increasing use of anti-TNF-α monoclonal antibody in human inflammatory diseases such as rheumatoid arthritis has confirmed the critical role of this inflammatory mediator in antifungal immunity. There are a number of reports of both reactivation and primary *Histoplasma*, *Aspergillus*, *Pneumocystis*, and *Cryptococcus* infections in patients being treated with anti-TNF-α therapy (39, 67, 70, 72, 73). Thus, even though TNF-α is a mediator of inflammatory diseases, it is also a critical mediator of protective immunity against fungal infections.

CD4 Th1 Cells

In pulmonary fungal infections that escape innate defense mechanisms, clearance relies on the production of Th1 cytokines and the suppression of Th2 responses. Th1 and Th2 CD4$^+$ T cells are defined by the cytokines they produce. Th1 cells produce IFN-γ, while Th2 cells produce IL-4, IL-5, and IL-13. Th1 cells drive inflammatory responses, augment intracellular killing of fungi by macrophages, drive delayed-type hypersensitivity responses to fungi, and drive immunoglobulin G isotype switching. Th2 cells also drive inflammatory responses, but the response is different from that mediated by Th1 cells. Th2 cells mediate what is classically referred to as the allergic response: eosinophilia, increased mucus production, immunoglobulin E production, mast cell activation, and induction of alternatively activated macrophages (which phagocytose fungi but cannot kill them). These latter leukocytes play a major role in granuloma formation, fibrosis, and tissue repair, which can be either beneficial or detrimental, depending on the timing of the response.

Dysregulation of the Th1/Th2 balance toward a Th2 profile results in increased susceptibility to fungal infection. IFN-γ production is critical for clearance of fungal infections in the lungs by activating macrophages and other phagocytes to ingest and kill fungi. Numerous studies have reported that IFN-γ is produced at high concentrations in mice resistant to *C. neoformans*, *P. brasiliensis*, *H. capsulatum*, *C. immitis*, and *B. dermatitidis*. This IFN-γ production occurs early in the course of infection, usually after the secretion of TNF-α and IL-12. The early production of IFN-γ drives the lung environment toward Th1. Murine models of *C. immitis* infection have demonstrated that susceptible mice can be rendered more resistant by treatment with recombinant IFN-γ (45). In addition, IFN-γ knockout mice or mice treated with monoclonal antibodies against IFN-γ are more susceptible to sublethal challenges of *H. capsulatum* (2, 75). Susceptibility to *P. brasiliensis* infection also correlates with an absence of IFN-γ production (33). Numerous studies have also demonstrated that defective IFN-γ production leads to decreased resistance to *C. neoformans* infection (42). Taken together, these observations demonstrate the importance of Th1-cell production of IFN-γ in pulmonary host defense against fungi.

CD4 Th2 Cells

Susceptibility against pulmonary fungal infections in otherwise healthy individuals is associated with the production of Th2 cytokines such as IL-4, IL-5, and IL-10 (although, as discussed below, IL-10 is more appropriately classified as a regulatory T cell [Treg] cytokine). In animal models, susceptibility to infection correlates with early production of IL-4, and elimination of IL-4 can provide protection against fungal infection (5, 11, 15, 29, 34, 45). Incubation of monocytes with IL-4 or IL-10 in vitro significantly suppresses their antifungal activity against *A. fumigatus* hyphae (61, 62). Consistent with this idea, patients suffering from *Aspergillus*-induced asthma and allergic bronchopulmonary aspergillosis exhibit overexuberant Th2 responses to the organism (27). Elevated levels of IL-5 correlate with pulmonary eosinophilia and susceptibility to *P. brasiliensis* and *C. neoformans* infection (30, 33). Thus, the balance between Th1 and Th2 cytokines is crucial in determining whether a pulmonary fungal infection induces an allergic response, is cleared, or becomes invasive.

CD8 T Cells

CD8⁺ T cells are critical for host defense against a number of fungal pathogens, including *C. neoformans*, *P. jiroveci*, *B. dermatitidis*, *H. capsulatum*, and *P. brasiliensis* infections (4, 7, 10, 42, 74). CD8⁺ T cells play a role in adaptive immunity to fungi because most of the pathogenic fungi can also survive intracellularly in macrophages (48). The mechanisms underlying the protective function of CD8⁺ T cells

remain to be determined, but it is likely that the production of IFN-γ by CD8⁺ T cells is a key role for these cells.

CD4 T Regulatory Cells

Regulatory T cells (Treg) are a subset of CD4⁺ T cells that possess potent anti-inflammatory capabilities and have cytokine profiles distinct from either Th1 or Th2 cells (50). These cells are present both in mucosal surfaces and systemically and are induced on infection. They comprise a diverse group of cells that, even in small numbers, can suppress antigen-specific responses (largely via the secretion of IL-10 and/or transforming growth factor β). IL-10 is an immunosuppressive cytokine produced by a variety of cell types including macrophages, monocytes, T cells, and B cells. IL-10 inhibits the expression of proinflammatory cytokines and enhances humoral immune responses while attenuating cell-mediated immune reactions (25). Transforming growth factor β is produced by monocytes and T cells and also exhibits immunosuppressive activity, inhibiting T-cell activation and cell proliferation (71). Several subsets of Treg have been identified, including Tr1 cells, Th3 cells, and CD4⁺ CD25⁺ Tr cells. Tr1 cells produce high levels of IL-10 and are induced during immune responses in the respiratory tract. IL-10 production by Tr1 cells results in local immunosuppression of Th1 responses. The CD25⁺ CD4⁺ Treg subtype appears to regulate immunopathology during *P. jiroveci* infection (28). Treg clearly play a role not only in the Th1/Th2 balance of adaptive immunity but also in controlling the severity and strength of overall adaptive T-cell responses. While these are the focus of a number of current investigations, little is known about their role in pulmonary fungal infections and fungal exposure. However, Treg are induced on exposure to or infection by a number of microbes, and so it is likely that they are involved in the pathology and/or hypersensitivity generated in response to fungal infection.

Polarization of Antifungal T-Cell Responses by Dendritic Cells

The maturation state or phenotype of DC is perhaps the most important regulatory step in the development of T-cell immunity in the lungs (and systemically). Mature DC activate naive T cells in the lymph nodes via antigen presentation and cytokine production and begin the process of T-cell polarization toward a Th1 or Th2 phenotype (which is probably completed at the site of infection). Recent studies have demonstrated that different forms of fungi induce different functional states (maturation) in DC. DC are able to discriminate between *A. fumigatus* conidia and hyphae in terms of cytokine production.

Phagocytosis of conidia by DC induces the secretion of IL-12, whereas phagocytosis of hyphae induces the secretion of IL-4 and IL-10 (8). Similarly, ingestion of *C. albicans* yeast by DC activates DC to produce IL-12 and prime Th1 cells whereas ingestion of hyphae inhibits IL-12 and Th1 priming and induces IL-4 production (17). Together, these observations indicate that DC have the ability to distinguish between different forms of a fungal species and to drive Th1 versus Th2 responses. On the other hand, because clearance of hyphae from the host generally requires a Th1 response, the Th2 response induced by DC may be deleterious to immunocompromised hosts.

DC are also responsible for influencing the development of Treg. While mature DC (activated by inflammatory signals) induce the differentiation of inflammatory T cells (Th1 and Th2), immature DC (absence of inflammatory signals) appear to induce the differentiation of Treg (32). As opposed to Th1 or Th2 cells, Treg are involved in limiting or down-regulating inflammatory responses. These cells are especially important in controlling inflammation in the lung environment, where inflammatory cells come into contact with a large variety of different airborne antigens. Under normal conditions, resident DC in the airways are exposed to antigens under noninflammatory conditions and are thought to induce Treg responses to the antigens, thereby limiting T-cell-mediated inflammatory responses. In relation to fungal exposure in the lungs, fungal aeroallergens induce minimal, if any, response in the lungs on repeated exposure, demonstrating that they are subject to tolerogenic mechanisms. However, intact mold spores and hyphal elements contain a number of immunogenic moieties. Many of these moieties can stimulate inflammatory cytokine production by lung leukocytes when adminstered in pure form (37).

Fungal Modulation of Dendritic Cell Biology

Fungal virulence factors and secreted or shed fungal products can interfere with DC maturation, which may influence the host response to infection. As described above, hyphal forms of *Candida* and *Aspergillus* stimulate IL-4 and/or IL-10 production by DC while yeast and conidia stimulate IL-12 production (8, 17). One set of factors that can modulate DC function is the eicosanoids, metabolic derivatives of arachidonic acid that include prostaglandins, leukotrienes, and lipoxins (55). Fungi also produce a wide array of bioactive lipids related to prostaglandins, leukotrienes, and lipoxins (55, 59, 60). It seems possible that fungal lipid metabolites will play a role in immune deviation and chronic fungal infection,

most probably by targeting DC function, since prostaglandins are potent modulators of DC biology. Currently, there is no evidence that traditional inhibitors of mammalian eicosanoid production (nonsteroidal anti-inflammatory drugs) affect the course of fungal infections in vivo. However, the enzymes involved in microbial (fungal, parasite, and protistan) eicosanoid synthesis are distinct and may not respond to such inhibitors. Definitive experiments to determine the role of fungal eicosanoids await the identification of the biochemical pathways in eicosanoid synthesis (in order to create null mutants).

B Cells and Antibody

Although B cells are discussed last in the list of cells that interact with fungi, their role (via production of antifungal antibodies) in modulating host defense against fungi cannot be understated. The role of antibodies in the host defense against fungi is described in depth in chapter 33.

CAN RESPONSES TO FUNGI IN THE LUNGS BE CONTROLLED BY EVENTS OUTSIDE THE LUNGS?

Fungal allergies are a major health concern in the field of fungus-human interactions. The lung environment is exposed to a variety of inhaled fungal elements and mold spores on a daily basis. This low-level exposure to fungal antigens normally elicits a tolerogenic immune response, which prevents the development of an allergic response. Such a response is compared to an active pulmonary fungal infection, which generates an inflammatory immune response. An active fungal infection can also be accompanied by an allergic response to the infection, known as bronchopulmonary mycosis.

The events leading to a breakdown of immunological tolerance and subsequent allergic responses to normal fungal exposure are mostly unknown. However, evidence suggests that tolerance to pulmonary mold spore exposure can be influenced and/or controlled by the events in the gut (56, 58). The gastrointestinal tract serves as the largest immune organ and comes in contact with a barrage of innocuous antigens from food and our symbiotic gut microbiota. As such, the gastrointestinal tract also exerts incredible tolerogenic capacity, known as oral tolerance (tolerance to swallowed antigens). However, the ability to induce oral tolerance requires the presence of the normal microbiota, suggesting that the microbiota may play an active role in promoting immune tolerance and/or preventing inflammation. Therefore, it has been proposed that tolerance mechanisms initiated in the gut also regulate tolerance at

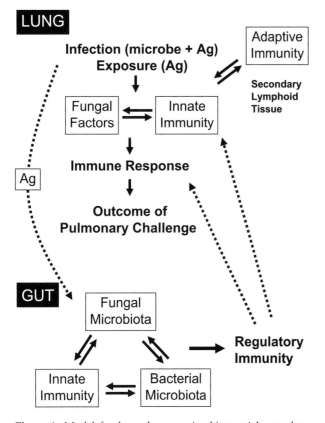

LUNG

Infection (microbe + Ag)
Exposure (Ag)

Adaptive
Immunity

Secondary
Lymphoid
Tissue

Fungal
Factors ⇌ Innate
Immunity

Ag

Immune Response

**Outcome of
Pulmonary Challenge**

GUT

Fungal
Microbiota

Regulatory
Immunity

Innate
Immunity ⇌ Bacterial
Microbiota

Figure 1. Model for how the gut microbiota might regulate pulmonary immune responses. Antigen (Ag) initially comes in contact with the lung during infection (microbe + Ag) or inhalation (Ag). This antigen is captured by cells of the innate immune system, which then stimulate adaptive immune responses in secondary lymphoid organs such as the lymph nodes or spleen. Antigen-specific T cells are recruited to the lungs on subsequent antigen exposure. To prevent the development of allergic inflammatory (Th2) responses to inhaled antigens that are noninfectious (pollens) or of extremely low infectivity (mold spores), regulatory T-cell networks produce anti-inflammatory mediators that down-regulate inflammatory T-cell activities. Inhaled antigens are also swallowed because the anatomy of the sinuses and upper airways is designed to trap environmental antigens (aerosols, microparticulates, and macroparticulates) in the mucus layer and then "sweep" them into the throat, where they are swallowed. Swallowed Ags are then acquired by DC in the gut. Under noninflammatory conditions, these DC promote the development of a regulatory T-cell response to the Ag, which can control reexposure responses in the lungs in the absence of infection. The precise mechanism by which the gastrointestinal microbiota balance helps to maintain the T-cell regulatory networks that mediate oral tolerance is unknown.

other mucosal sites, such as the lungs (reviewed in reference 57) (Fig. 1). Using an animal model of fungal allergy, it was found that microbiota disruption (antibiotic treatment combined with increased *C. albicans* colonization in the gut) resulted in increased Th2-dependent allergic responses to *A. fumigatus* conidia in the lungs. This suggests that

overgrowth of fungal microbiota (*C. albicans*) results in a break in mucosal tolerance, leading to overexuberant allergic responses to inhaled mold spores. The mechanism whereby the microbiota regulates tolerance mechanisms remains to be determined; however, it is likely that DC phenotype and regulatory T-cell development play a role (Fig. 1).

SUMMARY

The majority of human fungal pathogens are opportunistic ubiquitous organisms, causing disease only under permissive conditions. Prevention of growth, invasion, and disease after colonization by a fungal pathogen and effective clearance of infection from the lungs involve a series of coordinated host defense mechanisms including lung barrier function, resident host defenses, recruited host defenses, and adaptive immunity. Innate immunity provides the first line of defense in the form of nonspecific antifungal activities of phagocytic cells. These cells, along with other nonphagocytic cells, also direct subsequent adaptive responses by secretion of proinflammatory cytokines such as TNF-α, IL-12, IL-18, and granulocyte-macrophage colony-stimulating factor, which promote the development of protective Th1 responses. Th1 cells secrete IFN-γ and other proinflammatory cytokines and chemokines which drive macrophage activation, granuloma formation, and delayed-type hypersensitivity responses to the infection. Humoral immunity plays a role in coordinating an effective and controlled cell-mediated response; thus, low levels of Th2 cytokines are probably important. The pathogenesis of pulmonary fungal infections is also influenced by host genetics and fungal virulence factors. Pulmonary fungal infections are controlled by either completely eliminating the infection (e.g., *Aspergillus*) or reducing the fungal burden to a very low level whereby the remaining organisms are controlled by granuloma formation (e.g., *Histoplasma*) or latency (e.g., *C. neoformans*).

REFERENCES

1. **Allendoerfer, R., and G. S. Deepe, Jr.** 1998. Blockade of endogenous TNF-α exacerbates primary and secondary pulmonary histoplasmosis by differential mechanisms. *J. Immunol.* **160:**6072–6082.

2. **Allendoerfer, R., and G. S. Deepe, Jr.** 1997. Intrapulmonary response to *Histoplasma capsulatum* in gamma interferon knockout mice. *Infect. Immun.* **65:**2564–2569.

3. **Allendoerfer, R., and G. S. Deepe, Jr.** 2000. Regulation of infection with *Histoplasma capsulatum* by TNFR1 and -2. *J. Immunol.* **165:**2657–2664.

4. **Allendoerfer, R., G. D. Brunner, and G. S. Deepe, Jr.** 1999. Complex requirements for nascent and memory immunity in pulmonary histoplasmosis. *J. Immunol.* **162:**7389–7396.

5. **Arruda, C., R. C. Valente-Ferreira, A. Pina, S. S. Kashino, R. A. Fazioli, C. A. Vaz, M. F. Franco, A. C. Keller, and**

V. L. Calich. 2004. Dual role of interleukin-4 (IL-4) in pulmonary paracoccidioidomycosis: endogenous IL-4 can induce protection or exacerbation of disease depending on the host genetic pattern. *Infect. Immun.* **72:**3932–3940.

6. **Bauman, S. K., G. B. Huffnagle, and J. W. Murphy.** 2003. Effects of tumor necrosis factor alpha on dendritic cell accumulation in lymph nodes draining the immunization site and the impact on the anticryptococcal cell-mediated immune response. *Infect. Immun.* **71:**68–74.

7. **Beck, J. M., R. L. Newbury, B. E. Palmer, M. L. Warnock, P. K. Byrd, and H. B. Kaltreider.** 1996. Role of CD8+ lymphocytes in host defense against *Pneumocystis carinii* in mice. *J. Lab. Clin. Med.* **128:**477–487.

8. **Bozza, S., R. Gaziano, A. Spreca, A. Bacci, C. Montagnoli, P. di Francesco, and L. Romani.** 2002. Dendritic cells transport conidia and hyphae of *Aspergillus fumigatus* from the airways to the draining lymph nodes and initiate disparate Th responses to the fungus. *J. Immunol.* **168:**1362–1371.

9. **Brieland, J. K., C. Jackson, F. Menzel, D. Loebenberg, A. Cacciapuoti, J. Halpern, S. Hurst, T. Muchamuel, R. Debets, R. Kastelein, T. Churakova, J. Abrams, R. Hare, and A. O'Garra.** 2001. Cytokine networking in lungs of immunocompetent mice in response to inhaled *Aspergillus fumigatus. Infect. Immun.* **69:**1554–1560.

10. **Cano, L. E., L. M. Singer-Vermes, T. A. Costa, J. O. Mengel, C. F. Xidieh, C. Arruda, D. C. Andre, C. A. Vaz, E. Burger, and V. L. Calich.** 2000. Depletion of CD8+ T cells in vivo impairs host defense of mice resistant and susceptible to pulmonary paracoccidioidomycosis. *Infect. Immun.* **68:**352–359.

11. **Cenci, E., A. Mencacci, G. Del Sero, A. Bacci, C. Montagnoli, C. F. d'Ostiani, P. Mosci, M. Bachmann, F. Bistoni, M. Kopf, and L. Romani.** 1999. Interleukin-4 causes susceptibility to invasive pulmonary aspergillosis through suppression of protective type I responses. *J. Infect. Dis.* **180:**1957–1968.

12. **Chaturvedi, V., B. Wong, and S. L. Newman.** 1996. Oxidative killing of *Cryptococcus neoformans* by human neutrophils. Evidence that fungal mannitol protects by scavenging reactive oxygen intermediates. *J. Immunol.* **156:**3836–3840.

13. **Chen, W., E. A. Havell, and A. G. Harmsen.** 1992. Importance of endogenous tumor necrosis factor alpha and gamma interferon in host resistance against *Pneumocystis carinii* infection. *Infect. Immun.* **60:**1279–1284.

14. **Cox, R. A., and D. M. Magee.** 1995. Production of tumor necrosis factor alpha, interleukin-1 alpha, and interleukin-6 during murine coccidioidomycosis. *Infect. Immun.* **63:**4178–4180.

15. **Deepe, G. S., Jr.** 2000. Immune response to early and late *Histoplasma capsulatum* infections. *Curr. Opin. Microbiol.* **3:**359–362.

16. **Diamond, R. D.** 1993. Invasive aspergillosis: host defenses. *Recent Results Cancer Res.* **132:**109–115.

17. **d'Ostiani, C. F., G. Del Sero, A. Bacci, C. Montagnoli, A. Spreca, A. Mencacci, P. Ricciardi-Castagnoli, and L. Romani.** 2000. Dendritic cells discriminate between yeasts and hyphae of the fungus *Candida albicans.* Implications for initiation of T helper cell immunity in vitro and in vivo. *J. Exp. Med.* **191:**1661–1674.

18. **Ellerbroek, P. M., A. M. Walenkamp, A. I. Hoepelman, and F. E. Coenjaerts.** 2004. Effects of the capsular polysaccharides of *Cryptococcus neoformans* on phagocyte migration and inflammatory mediators. *Curr. Med. Chem.* **11:**253–266.

19. **Fang, F. C.** 2004. Antimicrobial reactive oxygen and nitrogen species: concepts and controversies. *Nat. Rev. Microbiol.* **2:**820–832.

20. **Fels, A. O., and Z. A. Cohn.** 1986. The alveolar macrophage. *J. Appl. Physiol.* **60:**353–369.

21. **Finkel-Jimenez, B., M. Wuthrich, T. Brandhorst, and B. S. Klein.** 2001. The WI-1 adhesin blocks phagocyte TNF-α production, imparting pathogenicity on *Blastomyces dermatitidis. J. Immunol.* **166:**2665–2673.

22. **Forster, R., A. Schubel, D. Breitfeld, E. Kremmer, I. Renner-Muller, E. Wolf, and M. Lipp.** 1999. CCR7 coordinates the primary immune response by establishing functional microenvironments in secondary lymphoid organs. *Cell* **99:**23–33.

23. **Garlanda, C., E. Hirsch, S. Bozza, A. Salustri, M. De Acetis, R. Nota, A. Maccagno, F. Riva, B. Bottazzi, G. Peri, A. Doni, L. Vago, M. Botto, R. De Santis, P. Carminati, G. Siracusa, F. Altruda, A. Vecchi, L. Romani, and A. Mantovani.** 2002. Non-redundant role of the long pentraxin PTX3 in anti-fungal innate immune response. *Nature* **420:**182–186.

24. **Gildea, L. A., R. E. Morris, and S. L. Newman.** 2001. *Histoplasma capsulatum* yeasts are phagocytosed via very late antigen-5, killed, and processed for antigen presentation by human dendritic cells. *J. Immunol.* **166:**1049–1056.

25. **Grutz, G.** 2005. New insights into the molecular mechanism of interleukin-10-mediated immunosuppression. *J. Leukoc. Biol.* **77:**3–15.

26. **Herring, A. C., J. Lee, R. A. McDonald, G. B. Toews, and G. B. Huffnagle.** 2002. Induction of interleukin-12 and gamma interferon requires tumor necrosis factor alpha for protective T1-cell-mediated immunity to pulmonary *Cryptococcus neoformans* infection. *Infect. Immun.* **70:**2959–2964.

27. **Hogaboam, C. M., K. Blease, and J. M. Schuh.** 2003. Cytokines and chemokines in allergic bronchopulmonary aspergillosis (ABPA) and experimental *Aspergillus*-induced allergic airway or asthmatic disease. *Front. Biosci.* **8:**e147–e156.

28. **Hori, S., T. L. Carvalho, and J. Demengeot.** 2002. CD25+ CD4+ regulatory T cells suppress CD4+ T cell-mediated pulmonary hyperinflammation driven by *Pneumocystis carinii* in immunodeficient mice. *Eur. J. Immunol.* **32:**1282–1291.

29. **Hostetler, J. S., E. Brummer, R. L. Coffman, and D. A. Stevens.** 1993. Effect of anti-IL-4, interferon-gamma and an antifungal triazole (SCH 42427) in paracoccidioidomycosis: correlation of IgE levels with outcome. *Clin. Exp. Immunol.* **94:**11–16.

30. **Huffnagle, G. B., M. B. Boyd, N. E. Street, and M. F. Lipscomb.** 1998. IL-5 is required for eosinophil recruitment, crystal deposition, and mononuclear cell recruitment during a pulmonary *Cryptococcus neoformans* infection in genetically susceptible mice (C57BL/6). *J. Immunol.* **160:**2393–2400.

31. **Huffnagle, G. B., G. B. Toews, M. D. Burdick, M. B. Boyd, K. S. McAllister, R. A. McDonald, S. L. Kunkel, and R. M. Strieter.** 1996. Afferent phase production of TNF-α is required for the development of protective T cell immunity to *Cryptococcus neoformans. J. Immunol.* **157:**4529–4536.

32. **Jonuleit, H., E. Schmitt, K. Steinbrink, and A. H. Enk.** 2001. Dendritic cells as a tool to induce anergic and regulatory T cells. *Trends Immunol.* **22:**394–400.

33. **Kashino, S. S., R. A. Fazioli, C. Cafalli-Favati, L. H. Meloni-Bruneri, C. A. Vaz, E. Burger, L. M. Singer, and**

V. L. Calich. 2000. Resistance to *Paracoccidioides brasiliensis* infection is linked to a preferential Th1 immune response, whereas susceptibility is associated with absence of IFN-γ production. *J. Interferon Cytokine Res.* 20:89–97.

34. Kawakami, K., M. Hossain Qureshi, T. Zhang, Y. Koguchi, Q. Xie, M. Kurimoto, and A. Saito. 1999. Interleukin-4 weakens host resistance to pulmonary and disseminated cryptococcal infection caused by combined treatment with interferon-gamma-inducing cytokines. *Cell. Immunol.* 197:55–61.

35. Kawakami, K., M. H. Qureshi, Y. Koguchi, T. Zhang, H. Okamura, M. Kurimoto, and A. Saito. 1999. Role of TNF-α in the induction of fungicidal activity of mouse peritoneal exudate cells against *Cryptococcus neoformans* by IL-12 and IL-18. *Cell. Immunol.* 193:9–16.

36. Kawakami, K., M. Tohyama, X. Qifeng, and A. Saito. 1997. Expression of cytokines and inducible nitric oxide synthase mRNA in the lungs of mice infected with *Cryptococcus neoformans*: effects of interleukin-12. *Infect. Immun.* 65:1307–1312.

37. Kheradmand, F., A. Kiss, J. Xu, S. H. Lee, P. E. Kolattukudy, and D. B. Corry. 2002. A protease-activated pathway underlying Th cell type 2 activation and allergic lung disease. *J. Immunol.* 169:5904–5911.

38. Latge, J. P. 2001. The pathobiology of *Aspergillus fumigatus*. *Trends Microbiol.* 9:382–389.

39. Lee, J. H., N. R. Slifman, S. K. Gershon, E. T. Edwards, W. D. Schwieterman, J. N. Siegel, R. P. Wise, S. L. Brown, J. N. Udall, Jr., and M. M. Braun. 2002. Life-threatening histoplasmosis complicating immunotherapy with tumor necrosis factor alpha antagonists infliximab and etanercept. *Arthritis Rheum.* 46:2565–2570.

40. Levitz, S. M. 2004. Interactions of Toll-like receptors with fungi. *Microbes Infect.* 6:1351–1355.

41. Levitz, S. M., and T. P. Farrell. 1990. Human neutrophil degranulation stimulated by *Aspergillus fumigatus*. *J. Leukoc. Biol.* 47:170–175.

42. Lindell, D. M., and G. B. Huffnagle. 2005. Cell-mediated immunity to *C. neoformans* in the lungs, p. 157–180. *In* P. L. Fidel and G. B. Huffnagle (ed.), *Fungal Immunology from an Organ Perspective*. Kluwer, New York, N.Y.

43. Liu, T., T. Matsuguchi, N. Tsuboi, T. Yajima, and Y. Yoshikai. 2002. Differences in expression of Toll-like receptors and their reactivities in dendritic cells in BALB/c and C57BL/6 mice. *Infect. Immun.* 70:6638–6645.

44. Long, K. H., F. J. Gomez, R. E. Morris, and S. L. Newman. 2003. Identification of heat shock protein 60 as the ligand on *Histoplasma capsulatum* that mediates binding to CD18 receptors on human macrophages. *J. Immunol.* 170:487–494.

45. Magee, D. M., and R. A. Cox. 1995. Roles of gamma interferon and interleukin-4 in genetically determined resistance to *Coccidioides immitis*. *Infect. Immun.* 63:3514–3519.

46. Mambula, S. S., K. Sau, P. Henneke, D. T. Golenbock, and S. M. Levitz. 2002. Toll-like receptor (TLR) signaling in response to *Aspergillus fumigatus*. *J. Biol. Chem.* 277:39320–39326.

47. Mambula, S. S., E. R. Simons, R. Hastey, M. E. Selsted, and S. M. Levitz. 2000. Human neutrophil-mediated nonoxidative antifungal activity against *Cryptococcus neoformans*. *Infect. Immun.* 68:6257–6264.

48. Mansour, M. K., and S. M. Levitz. 2002. Interactions of fungi with phagocytes. *Curr. Opin. Microbiol.* 5:359–365.

49. Mansour, M. K., L. S. Schlesinger, and S. M. Levitz. 2002. Optimal T cell responses to *Cryptococcus neoformans* mannoprotein are dependent on recognition of conjugated carbohydrates by mannose receptors. *J. Immunol.* 168:2872–2879.

50. McGuirk, P., and K. H. Mills. 2002. Pathogen-specific regulatory T cells provoke a shift in the Th1/Th2 paradigm in immunity to infectious diseases. *Trends Immunol.* 23:450–455.

51. Mednick, A. J., M. Feldmesser, J. Rivera, and A. Casadevall. 2003. Neutropenia alters lung cytokine production in mice and reduces their susceptibility to pulmonary cryptococcosis. *Eur. J. Immunol.* 33:1744–1753.

52. Mencacci, A., E. Cenci, G. Del Sero, C. Fe d'Ostiani, P. Mosci, C. Montagnoli, A. Bacci, F. Bistoni, V. F. Quesniaux, B. Ryffel, and L. Romani. 1998. Defective costimulation and impaired Th1 development in tumor necrosis factor/lymphotoxin-alpha double-deficient mice infected with *Candida albicans*. *Int. Immunol.* 10:37–48.

53. Newman, S. L. 2001. Cell-mediated immunity to *Histoplasma capsulatum*. *Semin. Respir. Infect.* 16:102–108.

54. Newman, S. L., L. Gootee, J. E. Gabay, and M. E. Selsted. 2000. Identification of constituents of human neutrophil azurophil granules that mediate fungistasis against *Histoplasma capsulatum*. *Infect. Immun.* 68:5668–5672.

55. Noverr, M. C., J. R. Erb-Downward, and G. B. Huffnagle. 2003. Production of eicosanoids and other oxylipins by pathogenic eukaryotic microbes. *Clin. Microbiol. Rev.* 16:517–533.

56. Noverr, M. C., N. R. Falkowski, R. A. McDonald, A. N. McKenzie, and G. B. Huffnagle. 2005. Development of allergic airway disease in mice following antibiotic therapy and fungal microbiota increase: role of host genetics, antigen, and interleukin-13. *Infect. Immun.* 73:30–38.

57. Noverr, M. C., and G. B. Huffnagle. 2004. Does the microbiota regulate immune responses outside the gut? *Trends Microbiol.* 12:562–568.

58. Noverr, M. C., R. M. Noggle, G. B. Toews, and G. B. Huffnagle. 2004. Role of antibiotics and fungal microbiota in driving pulmonary allergic responses. *Infect. Immun.* 72:4996–5003.

59. Noverr, M. C., S. M. Phare, G. B. Toews, M. J. Coffey, and G. B. Huffnagle. 2001. Pathogenic yeasts *Cryptococcus neoformans* and *Candida albicans* produce immunomodulatory prostaglandins. *Infect. Immun.* 69:2957–2963.

60. Noverr, M. C., G. B. Toews, and G. B. Huffnagle. 2002. Production of prostaglandins and leukotrienes by pathogenic fungi. *Infect. Immun.* 70:400–402.

61. Roilides, E., A. Dimitriadou, I. Kadiltsoglou, T. Sein, J. Karpouzas, P. A. Pizzo, and T. J. Walsh. 1997. IL-10 exerts suppressive and enhancing effects on antifungal activity of mononuclear phagocytes against *Aspergillus fumigatus*. *J. Immunol.* 158:322–329.

62. Roilides, E., I. Kadiltsoglou, A. Dimitriadou, M. Hatzistilianou, A. Manitsa, J. Karpouzas, P. A. Pizzo, and T. J. Walsh. 1997. Interleukin-4 suppresses antifungal activity of human mononuclear phagocytes against *Candida albicans* in association with decreased uptake of blastoconidia. *FEMS Immunol. Med. Microbiol.* 19:169–180.

63. Saeki, H., A. M. Moore, M. J. Brown, and S. T. Hwang. 1999. Cutting edge: secondary lymphoid-tissue chemokine (SLC) and CC chemokine receptor 7 (CCR7) participate in the emigration pathway of mature dendritic cells from the skin to regional lymph nodes. *J. Immunol.* 162:2472–2475.

64. Sieling, P. A., and R. L. Modlin. 2002. Toll-like receptors: mammalian "taste receptors" for a smorgasbord of microbial invaders. *Curr. Opin. Microbiol.* **5:**70–75.

65. Souto, J. T., F. Figueiredo, A. Furlanetto, K. Pfeffer, M. A. Rossi, and J. S. Silva. 2000. Interferon-gamma and tumor necrosis factor-alpha determine resistance to *Paracoccidioides brasiliensis* infection in mice. *Am. J. Pathol.* **156:**1811–1820.

66. Syme, R. M., J. C. Spurrell, E. K. Amankwah, F. H. Green, and C. H. Mody. 2002. Primary dendritic cells phagocytose *Cryptococcus neoformans* via mannose receptors and Fcγ receptor II for presentation to T lymphocytes. *Infect. Immun.* **70:**5972–5981.

67. Tai, T. L., K. P. O'Rourke, M. McWeeney, C. M. Burke, K. Sheehan, and M. Barry. 2002. *Pneumocystis carinii* pneumonia following a second infusion of infliximab. *Rheumatology* (Oxford) **41:**951–952.

68. Traynor, T. R., A. C. Herring, M. E. Dorf, W. A. Kuziel, G. B. Toews, and G. B. Huffnagle. 2002. Differential roles of CC chemokine ligand 2/monocyte chemotactic protein-1 and CCR2 in the development of T1 immunity. *J. Immunol.* **168:**4659–4666.

69. Traynor, T. R., W. A. Kuziel, G. B. Toews, and G. B. Huffnagle. 2000. CCR2 expression determines T1 versus T2 polarization during pulmonary *Cryptococcus neoformans* infection. *J. Immunol.* **164:**2021–2027.

70. True, D. G., M. Penmetcha, and S. J. Peckham. 2002. Disseminated cryptococcal infection in rheumatoid arthritis treated with methotrexate and infliximab. *J. Rheumatol.* **29:**1561–1563.

71. von Boehmer, H. 2005. Mechanisms of suppression by suppressor T cells. *Nat. Immunol.* **6:**338–344.

72. Warris, A., A. Bjorneklett, and P. Gaustad. 2001. Invasive pulmonary aspergillosis associated with infliximab therapy. *N. Engl. J. Med.* **344:**1099–1100.

73. Wood, K. L., C. A. Hage, K. S. Knox, M. B. Kleiman, A. Sannuti, R. B. Day, L. J. Wheat, and H. L. Twigg, III. 2003. Histoplasmosis after treatment with anti-tumor necrosis factor-alpha therapy. *Am. J. Respir. Crit. Care Med.* **167:**1279–1282.

74. Wuthrich, M., H. I. Filutowicz, T. Warner, G. S. Deepe, Jr., and B. S. Klein. 2003. Vaccine immunity to pathogenic fungi overcomes the requirement for CD4 help in exogenous antigen presentation to CD8$^+$ T cells: implications for vaccine development in immune-deficient hosts. *J. Exp. Med.* **197:**1405–1416.

75. Zhou, P., M. C. Sieve, J. Bennett, K. J. Kwon-Chung, R. P. Tewari, R. T. Gazzinelli, A. Sher, and R. A. Seder. 1995. IL-12 prevents mortality in mice infected with *Histoplasma capsulatum* through induction of IFN-γ. *J. Immunol.* **155:**785–795.

Molecular Principles of Fungal Pathogenesis
Edited by Joseph Heitman et al.
©2006 ASM Press, Washington, D.C.

Chapter 38

Fungal Vaccine Development

JOSHUA D. NOSANCHUK AND GEORGE S. DEEPE, JR.

INTRODUCTION

This chapter reviews the existing information regarding the advances made in developing fungal vaccines. The most important question that needs to be addressed upfront is whether a vaccine(s) against the pathogenic fungi is a necessity. An affirmative answer may seem obvious, but the topic itself has been controversial to say the least. The disagreement is a result of a serious shortage of data regarding the incidence and prevalence of these infections. Most of the medically important fungal diseases, with a few exceptions, are not reportable to state or federal health boards, and this fact limits the ability to perform a cost-benefit analysis. The only fungal infection that has been subjected to this type of analysis is vaccination for *Coccidioides immitis* or *C. posadasii*. In that analysis, the authors determined that if all adults at risk for exposure to this fungus are vaccinated, cost savings would approximate $3,000,000 annually. The authors suggested that the cost benefits would be even greater if vaccination of children were to be included (7). Although this amount of money is seemingly small, given the enormous U.S. budget for health care, one must consider that these data are for only one of many medically important fungal infections and certainly not the most common. Hence, the use of vaccination to prevent fungal infections is an exceptionally valuable means of limiting the scope of fungal diseases in humans.

Demography of Vaccination

A number of issues need to be confronted prior to implementation of any fungal vaccination strategy. They include identifying who should be vaccinated and when is it feasible and reasonable to vaccinate. Much of the planning for such a program requires good epidemiological data regarding the prevalence and incidence of disease, the age at onset of exposure, and those who are at highest risk. Unfortunately, the information regarding much of the preceding issues just does not exist. Since fungal diseases are not reportable in states within the United States, the Centers for Disease Control and Prevention has relied on hospital discharge data to assess the numbers of individuals afflicted with several of these pathogenic agents. Hence, only the most severe illnesses are documented. Also, since invasive fungal infections are often extremely difficult to diagnose, a large number of these infections are not identified. This difficulty is best demonstrated by data from studies of fungal endocarditis, an intravascular infection, where blood cultures are positive in only 53% of patients and a delayed or incorrect diagnosis occurs in 82% of patients (46). As for acquisition of infection, it is known that colonization by *Candida albicans* most probably occurs during birth and that exposure to *Cryptococcus neoformans* or *Pneumocystis jiroveci* is acquired during childhood (57, 61). Nevertheless, there is much to be learned regarding the epidemiology of fungal infections.

Much more information is known about the at-risk population. In general, infections with *Histoplasma capsulatum*, *Blastomyces dermatitidis*, *C. immitis*, and *Paracoccidioides brasiliensis* occur in otherwise normal hosts. Infections with *Candida* spp. arise in individuals with T-cell dysfunction or those who have other risk factors including administration of broad-spectrum antibiotics or hyperalimentation fluid and/or presence of an indwelling plastic catheter (44). Likewise, infections with *Aspergillus* spp., *C. neoformans*, or *P. jiroveci* develop in hosts who manifest immunological disturbances (139, 163). Thus, one of the major challenges becomes the ability to vaccinate hosts with underlying immune dysfunction. New cellular targets must be identified depending on the immune cells that are

Joshua D. Nosanchuk • Department of Medicine, Albert Einstein College of Medicine, Bronx, NY 10461.   **George S. Deepe, Jr.** • Department of Medicine, University of Cincinnati College of Medicine, 231 Bethesda Ave, Cincinnati, OH 45267.

disabled, or vaccines may have to be delivered with soluble factors that are necessary but absent in the immunocompromised host.

It is highly improbable that one antigen or even a small number of antigens will induce protection against a wide spectrum of fungal pathogens. Although this development would be ideal, the vastly different immunopathogenic properties employed by the pathogenic fungi most probably exclude the possibility of a universal vaccine. For example, *H. capsulatum* is an intracellular pathogen whereas *B. dermatitidis* and *Coccidioides* spp. are largely extracellular. The immune elements activated in response to intracellular and extracellular pathogens differ. Another distinction is that a few fungi such as *C. albicans* or *P. jiroveci* have adapted to survive within the human host. The former can be found in the gastrointestinal tract, and the latter is found the lungs. All of the other fungi mentioned above reside within the external environment and have evolved the ability to survive and cause disease in the human host while maintaining their environmental niche. For them, the human is an incidental host.

Approaches to Vaccination

This chapter encompasses two distinct approaches to vaccination. The first and most commonly understood and used is preventive vaccination. In this approach, immunization is delivered prior to exposure to the pathogen and consequently engenders an immune response that will obviate or ameliorate subsequent exposure to the microbe. Classically, the goal of this approach has been to induce an antibody response, which will neutralize the infectivity of viruses or the biological activity of toxins. The second method is therapeutic vaccination, in which hosts are vaccinated during the disease process. The principal diseases for which therapeutic vaccines are most useful are chronic infections. Most often, the focus is on chronic viral infections such as hepatitis B or C or human immunodeficiency virus infection. Since several of the pathogenic fungi produce chronic infections, therapeutic vaccines may mitigate the course of disease. This type of immunization may be passive, in which antibody or cells are transferred into hosts to modify the course of disease, or active, in which an immunogen is injected into the host to stimulate the immune system to modulate the disease process. One example is vaccination of herpes simplex virus-infected animals with glycoproteins B and D (147). This injection reduces the frequency of reactivation disease in the experimental animal model.

Vaccination and Innate Immunity

The innate immune system is the arm of immunity that does not undergo somatic recombination and does not express memory but stands as a "first responder" to foreign antigens. Although it was once considered to be of lesser importance than the adaptive immune response, the explosion in understanding of the innate immune system has elevated it to equal scientific footing. The knowledge that a functional innate immune system is vital to an optimal adaptive immune response has elevated the position of the former as an area that requires critical examination.

The mammalian innate immune system consists of numerous effector elements including but not limited to complement, collectins, cytokines, chemokines, mononuclear and polymorphonuclear phagocytes, dendritic cells, and natural killer (NK) cells. Specific receptors such as Toll receptors are extremely important in the recognition of microbes and their products by professional phagocytes and dendritic cells. A fundamental feature of innate immunity is to respond to microbial invasion promptly in order to limit the extent of the infectious process. On contact with a pathogen or foreign antigen, innate immunity transmits a danger signal. Another key property of innate immunity is to serve as a bridge to the adaptive response and promulgate the maturation of T- and B-cell responses. The outcome is the development of memory cells that will remain vigilant against subsequent invasion long after the initial insult. Stimulation of the innate immune system also directs the differentiation pathway of T cells into Th1 or Th2 cells (144). These qualities are critical for host defense, since in large measure Th1 cells enhance the protective immune response to fungi whereas Th2 cells dampen this response. As an example, production of the cytokine interleukin-12 (IL-12) by macrophages and/or dendritic cells is critical for subsequent synthesis of gamma interferon (IFN-γ) (144). The latter is a potent activator of cellular immunity, and its absence is associated with impaired host defenses against several fungi including *H. capsulatum*, *C. immitis*, *C. neoformans*, *P. brasiliensis*, and *C. albicans* (2, 16, 78, 92, 98, 111, 136, 137, 178, 179). Thus, the IL-12–IFN-γ axis is essential in the full expression of protective immunity to many medically important fungi.

Dendritic cells also play a critical role in antigen presentation and may be one of the few populations that can present fungal antigens (135). Many of the native antigens produced by fungi are glycoproteins, with the exception of pure polysaccharides such as cryptococcal glucuronoxylomannan (GXM). The glycoproteins are engulfed by antigen-presenting cells that bear the mannose receptor (100). In fact, the immunogenic filtrate from actively growing *C. neoformans* cells is not presented by cells that lack mannose receptors whereas it is in receptor-positive cells. These receptors are present on macrophages and dendritic cells but not B cells, another population that can process antigens for presentation to either CD4$^+$ or CD8$^+$ cells. Therefore, the ability of T cells to recognize antigenic epitopes in conjunction with the appropriate major histocompatibility complex molecules is limited to interaction with cells that bear mannose receptors.

Among the many innate elements engaged in enhancement of protective immunity to fungi are several cytokines including IL-12, tumor necrosis factor alpha (TNF-α), granulocyte-macrophage colony-stimulating factor (GM-CSF), and IFN-γ. The ability of lymphocytes and phagocytes to produce these cytokines constitutes a major effector mechanism of host resistance (134). The absence of these cytokines in mice often is associated with overwhelming infection and subsequent death. The findings from experiments with murine models have been extended to the human condition since recipients of monoclonal antibody (MAb) to TNF-α are more susceptible to various infectious diseases, especially tuberculosis and histoplasmosis (85, 93, 166).

The importance of these cytokines in regulating the efficacy of vaccines for fungi has been demonstrated in several murine models. Immunization of mice with recombinant heat shock protein 60 from *H. capsulatum* confers a vigorous protective immune response (63). Vaccination with this protective antigen is associated with greater production of IL-10, IL-12, and IFN-γ by splenocytes than that in mice injected with bovine serum albumin or *H. capsulatum* heat shock protein 70, an antigen that fails to protect mice from a challenge with this fungus (31). IL-10 synthesized in response to vaccination is not generated by T cells, whereas IFN-γ seems to be largely if not exclusively a product of this cell population and, in particular, the CD4$^+$ subpopulation. The protective efficacy of recombinant heat shock protein 60 is abolished if endogenous IL-10, IL-12, or IFN-γ is neutralized. The finding that IL-10 is required for vaccine efficacy is surprising since this cytokine is largely considered to be anti-inflammatory and an inhibitor of Th1 responsiveness as well as production of other cytokines including TNF-α. Nevertheless, there is precedent for this finding since other vaccines have been reported to be dependent on the presence of IL-10.

Injection of heat shock protein 70 from *H. capsulatum* elicits a T-cell-dependent immune response but fails to confer protection against this fungus (2). Thus, heat shock protein 70 is antigenic but not immunogenic. The aforementioned results provide one explanation for the lack of protective activity associated with heat shock protein 70. Its failure to elicit brisk production of IL-12 distinguishes it from heat shock protein 60. Although the data concern

only a single protein, it may be possible to extrapolate the findings to suggest that production of IL-12 by vaccines which consist of a protein(s) may be an indicator of vaccine efficacy.

The necessity of IL-12 for vaccine efficacy also has been demonstrated in a model of murine candidiasis. Succesful vaccination of mice with an attenuated *C. albicans* yeast is dependent on the presence of endogenous IL-12 (108). Likewise, this is true for vaccination against the parasite *Leishmania major* (67).

There are at least two principal reasons why a fungal vaccine should engage innate immunity. The first is that optimal activation of the innate system enhances the likelihood that a vigorous cellular immune response will develop. Abundant data exist to show that innate immunity is a crucial link to the adaptive immune response. This branch of immune responsiveness is necessary to combat most fungal infections long after the vaccine has been processed by the immune system. The second reason centers on the fact that many of the patients who develop serious fungal infections manifest some form of immune deficiency. Altered immunity may be manifest either as quantitative deficits in cells and/or soluble effector molecules or as dysfunctional cells that are a consequence of pharmacologic agents. A fungal vaccine that can stimulate innate immunity in hosts with defective adaptive immunity may be useful as a therapeutic vaccine candidate. This approach may augment protective immunity, albeit transiently perhaps, in the face of unremitting infection.

One strategy for therapeutic vaccination is to utilize the innate immune system as an alternative or salvage pathway. Natural killer (NK) cells are capable of producing several cytokines such as IFN-γ or TNF-α, key factors in the generation of protective immunity to fungal diseases (172). Although phagocytes also are capable of releasing these cytokines, their production may be dampened by the intracellular residence of the fungus or by fungal products. Since NK cells do not harbor fungi, it is possible that they could maintain their integrity during infection and thus be responsive to a therapeutic vaccine.

The approach would be to emulsify an antigen(s) with an adjuvant that would target the NK cells. As a consequence of this engagement, these cells would release cytokines that are essential for host resistance. That production of cytokines from NK cells can modify the course of infection has been demonstrated in mice with severe combined immunodeficiency that were infected with *H. capsulatum*. Treatment of these mice with recombinant IL-12 sharply reduces the fungal burden, and the success of treatment depends on IL-12 stimulation of NK cells to produce IFN-γ (178, 179).

Another approach to therapeutic vaccination using innate immunity is dendritic cell transfer. These cells are central to the development of adaptive immune responses and are the most potent antigen-presenting cells. Therefore, they can serve as powerful adjuvants for vaccine-induced immunity. To harness their immunological properties, dendritic cells can be exposed to an antigen, to a mixture of antigens, or to a microbe and then transferred into recipients. Transfer of murine dendritic cells exposed to *Aspergillus fumigatus* or *C. albicans* resulted in a significant reduction in fungal burden in normal recipients and in those that underwent hematopoietic stem cell transplantation. The beneficial effect of dendritic cell transfer could be mimicked by exposing dendritic cells to RNA from *C. albicans* yeast or *A. fumigatus* conidia (5, 10–12). These findings have been extended to *C. neoformans* (11). This exciting finding eliminates the possibility of viable microbes escaping from within the confines of dendritic cells and exacerbating infection.

Dendritic cells that were engineered to express CD40 ligand, a costimulatory molecule, and exposed to *P. carinii* could adoptively transfer protection against this fungus in mice that lacked CD4+ cells (177). The principal mediator of protection is antibody, since injection of serum from immunized healthy animals can transfer protection. This finding provides evidence for the complex interactions between innate and adaptive immunity. Moreover, the absence of this critical subset of T cells does not prohibit the utility of vaccines.

VACCINATION AND ADAPTIVE IMMUNITY

T-Cell-Mediated Responses

Much of the focus on vaccine development for pathogenic microbes has been on the humoral arm of the adaptive immune response. To date, the successful vaccines that are currently in use for humans generate antibodies that can mitigate the effects of viral or bacterial infections (e.g., those caused by poliovirus, rubella virus, pneumococcus, or *Haemophilus influenzae*) or neutralize toxins generated by bacterial pathogens (e.g., tetanus toxoid). Several prokaryotic and eukaryotic pathogens such as *Mycobacterium tuberculosis*, *Plasmodium* spp., *H. capsulatum*, and *B. dermatitidis* require effector T cells to mediate clearance (3, 51, 65, 116, 170). These properties are independent of the capacity of T cells to cooperate with B cells and induce antibody production. Vaccines that must prompt T cells to express effector function exclusive of antibody have yet to be developed completely. The only one in use for humans whose principal feature is to engage T cells is

M. bovis BCG. However, the efficacy of this vaccine that is used worldwide for vaccination to prevent tuberculosis appears to vary widely (91). One explanation is that there is no uniform preparation of this vaccine.

The prevailing dogma is that optimal vaccine efficacy requires the presence of CD4+ and CD8+ T cells. Immunization with either fungal molecules or attenuated organisms requires the presence of one or both subsets of T cells to express functional activity. The efficacy of vaccination of mice with heat shock protein 60 from *H. capsulatum* is abolished in the absence of CD4+ cells (31). On the other hand, vaccination of CD4+-deficient mice via the subcutaneous route with either a *Bad1* knockout strain of *B. dermatitidis* or live yeast from *H. capsulatum* does confer vigorous protection in mice subsequently challenged intranasally (169). This result, in conjunction with the findings stated above regarding dendritic cell vaccination of *P. carinii* with CD40 ligand-expressing cells, provides two very important points. First, CD4+ cells are thought to be essential for the function of CD8+ cells. It is unequivocal that this dictum is not true for these experimental approaches. Thus, the results indicate that CD8+ T cells can confer protection in the absence of CD4+ cells. Moreover, the findings also stress the fact that memory cells can be generated in the absence of CD4+ cells, and this finding appears to contradict data from other models. More importantly, the two studies that clearly show the efficacy of vaccination in animals lacking CD4+ cells indicate that advances can be made to vaccinate individuals with diseases such as AIDS, in which the CD4+ cell numbers are profoundly depleted. These discoveries concerning vaccination indicate that it may be possible to immunize effectively those with profoundly altered immunity.

The IL-12–IFN-γ axis is critical to the success of some of the vaccines for fungi. The protective efficacy of *H. capsulatum* heat shock protein 60 is abolished if mice are given MAb to IFN-γ during the vaccination process (31). Likewise, neutralization of this cytokine following vaccination also eliminates this biological property of heat shock protein 60. This result should be contrasted with that obtained for the *Bad1* mutant of *B. dermatitidis*. Vaccination with this replicating hypovirulent yeast protects CD4+ T-cell-deficient mice in the absence of IFN-γ, TNF-α, or GM-CSF (169). The efficacy of the vaccine is blunted if IFN-γ knockout mice that lack CD4+ T cells are given MAb to TNF-α or GM-CSF. On the other hand, protective immunity induced by the mutant yeast in CD4+ cell-depleted, TNF-α knockout mice is impaired if GM-CSF, but not IFN-γ, is neutralized. The data unequivocally demonstrate the utility of this mutant as a vaccine candidate.

The vaccine functioned quite well in mice lacking both a major cellular and soluble effector. Moreover, the requirements for the maintenance of protective immunity differed when endogenous IFN-γ or TNF-α was lacking. Thus, the adaptive immune response to this vaccine manifests powerful compensatory mechanisms when vital elements of the immune system are absent. Another conclusion that can be drawn is that vaccination with a recombinant mutant may be more efficacious than vaccination with a single molecule or even a group of molecules. The major difference is that the mutant can replicate whereas the latter are inert. An obvious risk is that a mutant may cause disease or that its virulence may be altered by the immune system and that it might become a more aggressive agent as a result.

Humoral Immunity

Historically, antibody-mediated immunity is associated with opsonization, toxin and viral neutralization, complement fixation, and antibody-dependent cellular cytotoxicity. These classical activities of antibody have been shown to occur against human-pathogenic fungi. Additionally, antibodies can interact with and modify the activities of the other arms of the immune system.

Extensive studies of C. neoformans have provided significant insights into the effects of a fungal infection on the humoral immune system. The polysaccharide capsule of C. neoformans is a major virulence factor and is associated with deleterious effects on the immune system, such as a reduction in phagocytosis, an alteration in proinflammatory cytokine production, and an inhibition of lymphoproliferation (reviewed in reference 20). Initial attempts in the 1960s to alter the host response with antibody utilized serum therapy for human cryptococcosis by combining rabbit anticryptococcal antibodies with the administration of amphotericin B (66). Subsequent studies utilized MAbs since the mechanism of antibody action is extremely complex and studies using sera to investigate the usefulness of antibody may be difficult to reproduce because (i) the concentration of protective antibodies may be too low to modify the course of infection, (ii) blocking antibodies may interfere with the activity of protective antibodies, and/or (iii) antigens that illicit protective antibodies may not be immunogenic (21). Interestingly, a recent report on C. albicans demonstrates how the existence of blocking antibodies in polyclonal responses abolishes the protective power of immune sera (15). Another important caveat to the use of antibody is that under certain conditions, administration of large doses of MAb can be less effective than that of smaller MAb (151, 154), a phenomenon which is consistent with a prozone-like effect.

In 1987, the administration of a MAb to C. neoformans polysaccharide (GXM) was shown to protect mice subsequently infected with the fungus (39). More recently, a phase I clinical trial using a murine MAb to C. neoformans GXM (19) for adjunctive therapy of cryptococcosis in patients with AIDS has been completed. Antibodies to the polysaccharide capsule of C. neoformans are potent opsonins that enhance the fungicidal activity of macrophages (114, 143). The antibodies also increase NK-cell (76, 77, 110, 115) and leukocyte (35, 36, 109) killing of C. neoformans. Antibody administration enhances the clearance of the polysaccharide antigen of C. neoformans from infected hosts (62, 94), which may benefit the host by limiting the adverse effects of the polysaccharide (73, 74, 90). Antibody can augment complement deposition on C. neoformans (40, 89, 164), which can affect phagocytosis and fungal killing. Classical antibody activities for C. albicans include opsonization and phagocytosis (26), neutralization of extracellular proteases (23), and direct fungicidal activity (126).

Recent advances in our understanding of antibody-mediated immunity have revealed that a major function of antibody is the regulation of cell-mediated activities and thus a modification of the inflammatory response to infection. In the classical associations described above, the interplay of antibody with cellular and innate immune responses is implicit in the interactive nature of antibody with phagocytic cells, complement, and NK cells. The acknowledgment of the interrelationship between the various arms of the immune system led to the hypothesis that antibody immunity may involve the regulation of the host inflammatory response that inadvertently damages host tissues. For example, antibody efficacy in cryptococcosis is dependent on cell-mediated immunity, as demonstrated by the findings that passive antibody administration is not protective in mice deficient in CD4+ T cells (173), IFN-γ (173), inducible nitric oxide synthase (132), or other Th1- and Th2-associated cytokines (8). Antibody may also act independently of cell-mediated immune responses by augmenting innate immunity by promoting opsonization and killing and/or enhancing complement-mediated fungicidal activity. However, phagocytosis involving the engagement of Fc receptors results in signal transduction events inducing the release of many cytokines (103, 158) and chemokines (104) and promotes NO synthesis (112). Cell-mediated responses can also be enhanced by the increased antigen presentation that may follow internalization and destruction of pathogens. Furthermore, antibody activation of complement produces C3a and C5a fragments that serve as chemoattractants which stimulate acquired immune responses.

Antibody efficacy in fungal infections may not be dependent on cellular or innate responses if the activity of the antibody results in direct killing (25, 125, 127) or in antigen clearance (23, 62, 94). For *C. neoformans*, MAbs to melanin (138) and to β-glucosylceramide (133) inhibit cellular replication in vitro. Since these compounds are components of the *C. neoformans* cell wall, binding of antibody to melanin or the ceramide interferes with the ability of the cell to grow and bud. In fact, administration of the MAb to melanin significantly prolongs the survival of lethally infected mice (138).

A recombinant MAb to candidal heat shock protein 90 (HSP90) is in clinical trials for treatment of invasive candidias (107). Since the antibody lacks an Fc component, its mechanism of action is postulated to occur via inhibition HSP90 chaperone functions (107). Also, the MAb may interact with human HSP90, which could result in an altered host cytokine or chemokine response or NO production (106). In addition to the MAb to HSP90, protective and nonprotective MAbs to surface components of *C. albicans* have been generated (68). MAbs to the *C. albicans* polysaccharide are protective in murine models (69–72). Additional putatively protective MAbs to cell surface components have also been described previously (reviewed in reference 105). Furthermore, Fab fragments from a MAb to a surface mannoprotein on *C. albicans* can inhibit germ tube formation (22).

Although less well developed, antibody against endemic dimorphic fungi has been evaluated. Recently, a MAb to *H. capsulatum* histone H2B expressed on the fungal cell surface was shown to modify the pathogenesis of murine infection (118). The MAbs resulted in increased phagocytosis of *H. capsulatum* yeast cells by macrophages, reduced cellular replication, and modest fungal cell death. The protective effect of the antibody was significantly augmented by the administration of suboptimal concentrations of amphotericin B. This effect could be due to the activation of Toll-like receptors by amphotericin B (141).

Antibodies in human serum can promote the activation of complement, and complement is required for killing of *B. dermatitidis* yeast by human neutrophils (41). Activation of the classical pathway occurs rapidly in the presence of pooled nonimmune human serum (176). Absorption of nonimmune serum with *B. dermatitidis* or glucan abolishes classical-pathway activity, but the alternative pathway is operative in a delayed and less efficient fashion (176). Hence, antibody in nonimmune individuals may facilitate the killing of *B. dermatitidis* during infection.

The BAD-1 (WI-1) adhesin of *B. dermatitidis* is an immunodominant antigen that elicits cellular and antibody-mediated responses to blastomycosis (87, 167). Interestingly, MAb to BAD-1 significantly enhances the C3 binding capacity of *B. dermatitidis* and incubation of yeast with serum from patients with blastomycosis increases the accumulation of C3 on the organism twofold compared to incubation with nonimmune serum (175). The MAbs to the adhesin enhance binding of *B. dermatitidis* yeast to macrophage cell lines and entry of the yeast into the cells in vitro, but they do not inhibit the growth or enhance the killing of the fungus (171). Passive administration of the MAbs generated to this antigen is not beneficial in infected mice and can even worsen infection (171). In their study, the investigators varied the inocula of *B. dermatitidis*, tested two strains of yeast, and used four different mouse strains, but only immunoglobulin G1 (IgG1) and IgG2a antibodies were tested, which may not be the appropriate isotypes for protection. For example, only one MAb isotype to listeriolysin O is protective against infection with *Listeria monocytogenes* (42). Additionally, the minimum amount of MAb to BAD-1 administered was 1 mg, which may have resulted in a prozone-like effect that abrogated any beneficial activity of the MAbs (151, 154). The finding that certain MAbs to BAD-1 worsened disease is consistent with the antibody enhancement seen with some MAbs against *C. neoformans* (113, 174). Furthermore, passive antibody administration may interfere with effective cell-mediated responses by biasing the immune response toward a Th2 response by directing antigen to macrophage Fcγ receptors (4). Immunization of mice with BAD-1 and Freund's adjuvant induces a mixed Th1 and Th2 immune response (167). Passive transfer of antibody from mice with this dichotomous response would not necessarily be expected to protect infected mice. For example, in *Plasmodium falciparum*, IgG2 antibodies can interfere with the protective activities of IgG1 and IgG3 isotypes (55, 56).

In paracoccidioidomycoses, Th2-biased responses are more common in individuals unable to control infection with *P. brasiliensis*. Patients with severe paracoccidioidomycosis, acute "juvenile" or multifocal "adult" forms, characteristically have increased levels of IgE, IgA, and IgG4 compared to individuals with mild, unifocal disease (6, 9, 83, 99). It is unclear whether the generation of this biased antibody response modifies the pathogenesis of the mycosis or is due to an inappropriate response by the host to infection or a combination of these processes.

In candidiasis, surface mannans function as adhesins during yeast cell attachment to macrophages (84). *C. albicans* mannan extract-protein conjugates induce a protective immune response to experimental candidiasis (72), and an IgM MAb specific for a β-1,2-mannotriose protects against the disease (68, 71). Similarly, the extracellular matrix protein laminin binds specifically to *P. brasiliensis* yeast and enhances adhesion of the fungus to the surface of diverse cells (58, 160). The gp43 glycoprotein, biochemically defined as a high-mannose concanavalin A-binding glycoprotein, is an immunodominant antigen on the cell surface of *P. brasiliensis* (149) and is responsible for cellular adhesion (160). MAbs to gp43 (12 different MAbs of IgG2b isotype with kappa light chains) modify the binding of the yeast to cells in vitro and alter the pathogenicity of the fungus in vivo (58). For example, administration of a MAb to gp43 that increased the adhesion of *P. brasiliensis* to epithelial cells in vitro caused more severe disease in a hamster testicle model of paracoccidioidomycoses whereas administration of a different MAb or anti-gp43 polyclonal antiserum was protective. The findings of protective and disease-enhancing MAbs to a defined antigen are again similar to those seen with protective and nonprotective MAbs to *C. neoformans* polysaccharide (113) and anti-glucan antibodies in candidiasis (15).

Candidate Fungal Vaccines

In recent years, there has been a plethora of vaccine candidates for the prevention of fungal diseases. Successful vaccines have been reported with each of the major medically important fungi: *A. fumigatus*, *C. albicans*, *C. immitis*, *C. posadasii*, *H. capsulatum*, *B. dermatitidis*, *P. jiroveci* and *C. neoformans*. Unfortunately, no universal immunogen for fungal diseases has been identified that confers protection, and, because of the widely divergent pathogenic mechanisms of each of the fungi, it is highly unlikely that one will be identified. In addition, the host requirements for effective clearance do differ substantially among these fungi. The consequence is that each fungus will require its own immunogen. A review of some of the vaccine candidates is given below.

Aspergillus spp.

Aspergillus species are ubiquitous in the environment and are thought to be transmitted to humans by inhalation or exposure to environmental sources including water (102). Infections are frequently complicated and often lethal, particularly in patients with advanced HIV infection or prolonged neutropenia or individuals receiving immunosuppressive therapies (33). Diagnosis of aspergillosis is frequently delayed, and successful clinical outcomes in different treatment trials, even with newer agents, are often no higher than 40 to 50% (75, 122). Results of animal studies have suggested that host defects in both mononuclear phagocytes and neutrophils are required to permit the

germination of *Aspergillus* conidia and infection of tissues.

Given the epidemiology of the individuals at risk for invasive disease, the lack or dysfunction of the cellular immune system is a primary obstacle to immunization. Nevertheless, a vaccine for aspergillosis would have tremendous impact (148). Vaccine studies have focused largely on examining the role of immunization with filtrates or sonicates from hyphae on disease (130, 131). Materials from both filtrates and sonicates protect glucocorticosteroid-treated mice (81). However, results have not been consistent. For example, in one study the filtrate mediated protective immunity and in another the sonicate of hyphae induced protection. Furthermore, the efficacy of the filtrate depends on the presence of CD4$^+$ T cells, IFN-γ, and IL-2. The effect of the culture filtrate could not be mimicked by one of the cloned antigens, Asp f 2, which elicits a Th2 response (24). Nevertheless, the results suggest that within complex extracts of this fungus there are antigens that can mediate protection. A candidate antigen, Asp f 16, when administered with unmethylated CpG oligodeoxynucleotides, induces a Th1 response that confers protection in mice (10). A significant limitation of the animal models used in these studies is that normal mice are vaccinated and then immunosuppressed to induce infection. Given the epidemiology of aspergillosis, a desirable goal is to be able to vaccinate therapeutically at the time of illness. Along these lines, work is under way to develop antibodies that could be administered to high-risk individuals to inhibit the germination of the fungus, which would abort infection.

An interesting approach to vaccine development for aspergillosis has targeted the activation of dendritic cells (11, 12). In a murine hematopoietic transplantation model, myeloid-lineage dendritic cells exposed to conidia or RNA from conidia transferred into the immunodeficient mice resulted in a reduction in the fungal burden and a prolongation of survival but did not result in sterilizing immunity. These studies have served to expand the potential avenues in the continued search for optimal vaccination against fungal pathogens.

In contrast to invasive aspergillosis, bronchopulmonary aspergillosis may be amenable to vaccination. This illness, which manifests as a form of asthma, can be mitigated by antifungal treatment of the fungus (161). Hence, a vaccine that modifies the immune response to *Aspergillus* antigens from a Th2 to a Th1 response would be exceptionally useful.

B. dermatitidis

B. dermatitidis is presumably a soil-based fungus that has been associated with riverbanks (88).

Infection with *B. dermatitidis* is thought to occur after inhalation of aerosolized conidia, and the common clinical manifestations include pneumonia, skin and bone lesions, and involvement of the genitourinary tract (14). Vaccine development has focused on a surface-expressed molecule, termed *BAD1* (170). This virulence determinant is a target of the humoral and cellular immune response to this fungus. *BAD1* contains a hydrophobic amino terminus, a large stretch of tandem repeats with homology to the invasion protein of *Yersinia*, and a carboxy terminus with homology to epidermal growth factor (86). Investigators have postulated that this region may be involved in binding of *B. dermatitidis* to the extracellular matrix.

Although immunization of mice with native *BAD1* induces only modest protection, vaccination of mice with a mutant strain of *B. dermatitidis* lacking the gene encoding *BAD1* results in a dramatic clearance of infection (168). Vaccination efficacy is independent of CD4$^+$ T cells, since CD8$^+$ T cells can confer protection because they are able to produce IFN-γ, TNF-α, and GM-CSF that are involved in the generation of the protective immune response (169). The protective effect of vaccination with the mutant *B. dermatitidis* strain is abolished in TNF-α-deficient mice following neutralization of GM-CSF. Similarly, IFN-γ-deficient mice are poorly protected by vaccination following neutralization of either TNF-α or GM-CSF. This work provides a new paradigm for understanding how protective immunity to fungal pathogens can be elicited, and it challenges the concept that CD4$^+$ T cells are required to generate the Th1-type milieu that is thought to mediate protection. The unexpected redundancy demonstrated by the ability of the mutant *B. dermatitidis* to induce and maintain protective immunity by CD8$^+$ T cells in the absence of CD4$^+$ lymphocytes is a significant result that provides support for the development of live attenuated fungal vaccines for hosts with defective CD4$^+$ T-cell immunity, such as individuals infected with HIV (169, 170).

Candida spp.

This section concentrates on vaccine development for *C. albicans*, which is the most common *Candida* species currently causing invasive disease (165). *C. albicans* is part of the normal flora of the human skin, gastrointestinal tract, and oral cavity, as well as the genital tract of women. Since it is a commensal, the majority of infections are endogenous. *C. albicans* is one of the leading causes of bloodstream infections in the United States (165) and is a major cause of morbidity and mortality in patients requiring intensive care (45). Important risk factors

for candidal disease include impaired cellular immunity, intravenous catheters, exposure to broad-spectrum antibiotics, dysfunctional or absent phagocytes, total parenteral nutrition, and implantation of prosthetics (reviewed in reference 117).

Since the organism is a commensal, it is difficult to create an effective preventative vaccine. Hence, efforts toward vaccine development have focused on disease prevention. A major emphasis has been the development of therapeutic antibodies. It is particularly notable that a recombinant human antibody, Mycograb, reactive with candidal HSP90 is currently being used in an international double-blinded, placebo-controlled efficacy and safety study for invasive candidiasis (106). Since Mycograb lacks an Fc component, its mechanism of action is postulated to occur via inhibition HSP90 chaperone functions (107). Also, Mycograb may interact with human HSP90, which could result in an altered host cytokine or chemokine response or NO production (106).

In addition to Mycograb, MAbs to the *C. albicans* polysaccharide are protective in murine models (69–72). Protective MAbs to additional cell surface components have also been described (reviewed in reference 105). Administration of a mannoprotein complex or a secreted aspartyl proteinase can induce protective immunity in a murine model of *Candida* vaginitis. Administration of either of these extracts results in an antibody response, but it is unclear if the induced antibodies are responsible for the protective effect. It has been shown that Fab fragments from a MAb to a surface mannoprotein on *C. albicans* can inhibit germ tube formation (22), which may affect pathogenesis. One of the secreted aspartyl proteinases, Sap2, appears to be responsible for the protective effect. Similarly, an antigen known as mannoprotein 65 also confers protection, principally in the model of vaginitis. The immunological requirements for the protective effect are not elucidated. A different approach has utilized the development of an antibody that binds to β(1-3)-glucan, and this "yeast killer antibody" is therapeutic in systemic candidiasis (127).

Whole-cell active vaccination can induce protective cellular responses to *C. albicans* (15, 18, 108). Recently, a promising vaccine using an adhesin to enhance cell-mediated responses to *C. albicans* has been reported (80). Using surrogate genetics to search for a dominant *C. albicans* adhesin, the investigators have identified an agglutination-like sequence (*ALS1*) gene that encodes a protein that mediates binding to epithelial and endothelial cells (52, 53). The cell binding region of Als1 is localized to the N terminus (97). Immunization of mice with the recombinant N-terminal domain of Als1 significantly prolongs survival after challenge with a lethal inoculum of *C. albicans* (80). The vaccine selectively enhances cell-mediated immune responses, increasing the stimulation of Th1 splenocytes by *C. albicans* and increasing delayed-type hypersensitivity in vivo. Titers of antibody to the recombinant protein do not correlate with survival. Also, the vaccine is not protective in T-cell-deficient mice whereas the efficacy is maintained in B-cell-deficient animals.

Coccidioides spp.

Two major species of *Coccidioides* exist. *C. posadasii* and *C. immitis* reside in the soil in arid environments and are endemic to the Lower Sonoran Valley and in sections of Central and South America. Disturbances of soil result in aerosolization of the arthroconidial form of the organism, and inhalation results in pulmonary infection (1). However, only a small fraction of exposed individuals develop disease, which includes pneumonia, cavitary pulmonary disease, and a disseminated form that can involve the skin, bone marrow, central nervous system, and visceral organs (27).

Despite the failure of the original vaccine to confer protection in human subjects (119, 120), a major effort has resulted in the identification of several immunogens for the prevention of coccidioidomycosis (reviewed in reference 28). The utility of several recombinant antigens as vaccine candidates has been analyzed. One of the most thoroughly studied is the antigen known as proline-rich antigen, a 19-kDa protein that, when used as recombinant protein or as a naked DNA, is associated with a reduction in fungal burden in two strains of mice (145). The protective activity of the proline-rich antigen has been localized to amino acids 1 to 106 (123). Since there is genetic diversity of *Coccidioides* spp., this antigen has been subjected to sequence analysis among different strains. At the amino acid level, the sequence is highly conserved, which suggests that immunization with this antigen may induce protection against most if not all invading strains of the fungus. Immunization with this antigen induces a Th1 response that is critical for fungal clearance.

Although the proline-rich antigen has been the major focus of vaccine research, additional vaccine candidates have been identified (28). For example, vaccination with a recombinant urease or the DNA encoding the urease gene protects a susceptible mouse strain, BALB/c, against *C. immitis* (96). Protection is associated with a brisk Th1 response. Vaccination studies with an expression library in which pools of cDNA are used have identified a vaccine candidate gene. Although the specific function of the gene is unknown, vaccination with DNA containing the gene protected mice against challenge with *C. immitis*

(82). In addition, immunization with a recombinant protein from a gene with homology to β(1-3)-glucanosyltransferase reduces the fungal burden in mice and extends survival (32).

C. neoformans

C. neoformans is an encapsulated, yeast-like fungus that is ubiquitous in nature and is the leading cause of fungal meningitis (20). In particular, cryptococcosis most often develops in individuals infected with HIV or patients receiving corticosteroids or other immunosuppressive agents, especially in post-transplantation patients. The major environmental reservoir for *C. neoformans* is bird guano, from either wild or pet (e.g., cockatiels) birds.

Several laboratories have generated MAbs that are protective in murine cryptococcosis (19, 38, 39, 50, 133, 138), and the majority are specific for the capsular polysaccharide GXM. Interestingly, the protective effect of MAbs to the capsule is not necessarily associated with a reduction in the fungal burden. A phase I clinical trial using a murine MAb to *C. neoformans* GXM (19) has recently been completed to evaluate adjunctive therapy of cryptococcosis in patients with AIDS. In addition to MAb to GXM, antibodies to melanin (138) or a glucosylceramide (133) of *C. neoformans* can modify infection with the fungus. The experience with *C. neoformans* demonstrates that the efficacy of antibody is complex and variable, depending on antibody dose, isotype, idiotype, and specificity as well as the yeast inoculum (47, 95, 151, 154, 159, 174). The efficacy of MAb is dependent on the presence of T cells and both Th1 and Th2 cytokines.

GXM conjugated to tetanus toxoid has been used in an experimental vaccine (34). The vaccine induced a protective immune response following challenge with the fungus by various routes. One concern of such an approach stems from observations that particular subsets of MAbs to the carbohydrate capsule could mediate protection while others caused exacerbation of cryptococcosis. Thus, it is imperative that any vaccine containing GXM generate only protective B-cell clones. Peptide mimetopes are being developed for use as immunogens that may limit the repertoire of elicited antibodies.

Aside from using the capsule as a potential immunogen, mannosylated proteins have shown promise as vaccine candidates (79, 100, 101). A 25-kDa polysaccharide deacetylase from *C. neoformans* protects mice and reduces fungal burden activity. The protective efficacy is correlated with the induction of a Th1 response. The filtrate from actively growing yeast can induce a protective immune response in mice. Both a mannoprotein containing fraction and one that is free of mannosylated moieties mediate protection against disseminated cryptococcosis in mice. The loss of the carbohydrate portion of the material abolishes protection. Immunization of B-cell- but not T-cell-deficient mice confers protection.

H. capsulatum

H. capsulatum is a dimorphic fungus that pro- duces a broad spectrum of disease ranging from a mild influenza-like illness to a disseminated form that may involve virtually any tissue (13, 17, 129). This fungus is the most common cause of fungal res- piratory infections, infecting approximately 200,000 to 500,000 individuals per year. It is endemic to the midwestern and southeastern United States but is found worldwide. H. capsulatum resides in the soil and thrives in decaying excreta of chickens, bats, starlings, and blackbirds.

More recent work has shown that both native and recombinant heat shock protein 60 (hsp60) con- fer protective immunity to this fungus in several mouse strains (63, 64). The immunogenic activity of this protein is dependent on the presence of $CD4^+$ T cells, IL-10, IL-12, and IFN-γ (31). Furthermore, immunization with hsp60 induces the expansion of Vβ 8.1/8.2$^+$ T cells, which are essential to the bio- logical function of the vaccine since the elimination of these cells abrogates protection (142). The protec- tive activity of the entire protein has been mapped to amino acids 171 to 443. Interestingly, only a fraction of Vβ 8.1/8.2$^+$ T cells recognize F3, and those that secrete IFN-γ and respond to this polypeptide are the cell population responsible for the efficacy of vacci- nation with hsp60. These results indicate that vacci- nation induces a small population of cells that confer protection.

Although the major focus on vaccine develop- ment for H. capsulatum has recently focused on hsp60, initial vaccination studies of H. capsulatum demonstrated the protective efficacy of a ribosome- protein complex and an ethylenediamine extract of the fungus (54). The H antigen of H. capsulatum, a secreted β-glucosidase (29, 48, 49), is a historically important immunodiagnostic antigen (162). Protection against lethal pulmonary infection can be achieved by immunization with recombinant H antigen (30). Antibody also modifies the pathogenesis of histoplas- mosis (118). Although the effect of antibody was mod- est, there is a remarkable improvement in the survival of mice when the antibody is administered with a sub- inhibitory amount of the antifungal agent amphoteri- cin B. The antibody binds to histone H2B located on the fungal cell surface, and work is under way to determine whether immunization with the histone protein is protective. These results suggest that there are several avenues to pursue in the search for a safe and effective vaccine for histoplasmosis.

P. brasiliensis

P. brasiliensis is a dimorphic, soil-based fungus that is the major causative agent of systemic mycosis in Latin America (140). The majority of infections are asymptomatic. Pneumonia is the most common form of disease. A minority of individuals develop extrapulmonary disease, which typically involves the skin, mucosa, and adrenal glands.

A 43-kDa glycoprotein of P. brasiliensis, gp43, is a laminin binding molecule (160) that has been iden- tified in a soluble form in the plasma, urine, and saliva of patients (128, 157). In addition to serving as the major diagnostic antigen of this fungus, gp43 is a virulence factor because it inhibits the lymphoprolif- erative responses and decreases the phagocytic and fungistatic activity of macrophages. Vaccination gp43 with protein or the DNA encoding gp43 can protect susceptible mice from infection (124). The protective activity of the entire protein has been mapped to 15 amino acids within the core of the protein (152, 153). The success of vaccination with either the entire pro- tein or the peptide is dependent on eliciting a Th1 response. Additional work has shown that vaccina- tion with anionic exchange fractions of a crude admixture of P. brasiliensis antigens can also confer protection (37). These results suggest that additional antigens may be employed in the development of a vaccine against P. brasiliensis.

Pneumocystis spp.

In 1988, Pneumocystis spp. were classified as fungi on the basis of their rRNA (43). Previously, they were thought to be protozoa. The species of Pneumocystis that causes diseases in humans was also recently renamed from P. carinii to P. jiroveci (150).

P. jiroveci is an airborne pathogen that can be transmitted from human to human, resulting in mild to severe respiratory disease. There is no known envi- ronmental reservoir. Prior to the AIDS epidemic, dis- ease occurred primarily in patients with lymphoma and leukemia, particularly in the setting of a reduc- tion in the daily amount of corticosteroid adminis- tered (156). In this setting, patients classically develop a sudden onset of fever and hypoxia, which is likely to occur due to the resumption of the host's ability to develop an intense inflammatory response to the organism. Prior to the advent of Pneumocystis prophylaxis in individuals with AIDS, Pneumocystis was the most frequent cause of pneumonia and was a leading cause of death in these patients. The need for a vaccine could not be more apparent.

The major challenge for a vaccine for this organ- ism, as it is for many of the fungi, is that the vaccine must be effective in the absence of key elements of the immune system. Much of the investigation of the protective efficacy of antigens or antibodies has been done with either rat or murine models of infection. This is important because the species of *Pneumocystis* that causes infection in these rodents differs from the one (*P. jiroveci*) that causes infection in humans (150). Nevertheless, these species share a number of genes, and the results are most likely to be applicable to the human condition.

Immunotherapy with MAb to this fungus can ameliorate the course of disease (59). Immunization of rats with recombinant p55 or with recombinant or native major surface glycoprotein from this fungus can confer protection in a rat model of pneumocystis induced by treatment with corticosteroids (60, 146, 155). In addition, immunization of mice with an extract of the fungus combined with cholera toxin prevents infection in CD4$^+$-depleted animals exposed to *Pneumocystis* (121).

The above results indicate that vaccination of immunocompetent hosts can confer protection when they become immunosuppressed. One of the major challenges would be to vaccinate an immunosuppressed individual. The induction of protection against experimental blastomycosis in mice lacking CD4$^+$ T cells by vaccination with a mutant *B. dermatitidis* strain (169) demonstrates that this challenge is not insurmountable. CD4$^+$ cell-deficient mice have been protected against *Pneumocystis* by the transfer of primed dendritic cells (177). The murine dendritic cells were genetically engineered to express CD40 and were exposed to *Pneumocystis* prior to adoptive transfer. The transfer of dendritic cells into these mice confers protection apparently through induction of a humoral response. Antibodies that are generated recognize a 55-kDa molecule from the organism that may be identical or similar to the 55-kDa antigen used to vaccinate rats. The import of this study is that an immunodeficient host could be vaccinated against a common infectious threat to this population.

SUMMARY

This chapter provides an overview of the accomplishments in the field of fungal vaccination. Clearly, there are numerous candidate vaccines and at least two have entered the early phases of clinical study. Although the usefulness of the MAb to *C. neoformans* polysaccharide or the antibody to hsp90 of *C. albicans* is not known, the mere fact that these compounds made it into the clinical arena is a sufficient impetus for further development. The area is an exciting one, and one that requires an interdisciplinary approach utilizing expertise in fungal biology, pathogenesis, and immunology.

REFERENCES

1. Ajello, L. 1971. Coccidioidomycosis and histoplasmosis. A review of their epidemiology and geographical distribution. *Mycopathol. Mycol. Appl.* **45:**221–230.
2. Allendoerfer, R., B. Maresca, and G. S. Deepe, Jr. 1996. Cellular immune responses to recombinant heat shock protein 70 from *Histoplasma capsulatum*. *Infect. Immun.* **64:**4123–4128.
3. Allendorfer, R., G. D. Brunner, and G. S. Deepe, Jr. 1999. Complex requirements for nascent and memory immunity in pulmonary histoplasmosis. *J. Immunol.* **162:**7389–7396.
4. Anderson, C. F., and D. M. Mosser. 2002. Cutting edge: biasing immune responses by directing antigen to macrophage Fc gamma receptors. *J. Immunol.* **168:**3697–3701.
5. Bacci, A., C. Montagnoli, K. Perruccio, S. Bozza, R. Gaziano, L. Pitzurra, A. Velardi, C. F. d'Ostiani, J. E. Cutler, and L. Romani. 2002. Dendritic cells pulsed with fungal RNA induce protective immunity to *Candida albicans* in hematopoietic transplantation. *J. Immunol.* **168:**2904–2913.
6. Baida, H., P. J. Biselli, M. Juvenale, G. M. Del Negro, M. J. Mendes-Giannini, A. J. Duarte, and G. Benard. 1999. Differential antibody isotype expression to the major *Paracoccidioides brasiliensis* antigen in juvenile and adult form paracoccidioidomycosis. *Microbes Infect.* **1:**273–278.
7. Barnato, A. E., G. D. Sanders, and D. K. Owens. 2001. Cost-effectiveness of a potential vaccine for *Coccidioides immitis*. *Emerg. Infect. Dis.* **7:**797–806.
8. Beenhouwer, D. O., S. Shapiro, M. Feldmesser, A. Casadevall, and M. D. Scharff. 2001. Both Th1 and Th2 cytokines affect the ability of monoclonal antibodies to protect mice against *Cryptococcus neoformans*. *Infect. Immun.* **69:**6445–6455.
9. Biselli, P. J., M. Juvenale, M. J. Mendes-Giannini, A. J. Duarte, and G. Benardi. 2001. IgE antibody response to the main antigenic component of *Paracoccidioides brasiliensis* in patients with paracoccidioidomycosis. *Med. Mycol.* **39:**475–478.
10. Bozza, S., R. Gaziano, G. B. Lipford, C. Montagnoli, A. Bacci, P. Di Francesco, V. P. Kurup, H. Wagner, and L. Romani. 2002. Vaccination of mice against invasive aspergillosis with recombinant *Aspergillus* proteins and CpG oligodeoxynucleotides as adjuvants. *Microbes Infect.* **4:**1281–1290.
11. Bozza, S., C. Montagnoli, R. Gaziano, G. Rossi, G. Nkwanyuo, S. Bellocchio, and L. Romani. 2004. Dendritic cell-based vaccination against opportunistic fungi. *Vaccine* **22:**857–864.
12. Bozza, S., K. Perruccio, C. Montagnoli, R. Gaziano, S. Bellocchio, E. Burchielli, G. Nkwanyuo, L. Pitzurra, A. Velardi, and L. Romani. 2003. A dendritic cell vaccine against invasive aspergillosis in allogeneic hematopoietic transplantation. *Blood* **102:**3807–3814.
13. Bradsher, R. W. 1996. Histoplasmosis and blastomycosis. *Clin. Infect. Dis.* **22:**S102–S111.
14. Bradsher, R. W., S. W. Chapman, and P. G. Pappas. 2003. Blastomycosis. *Infect. Dis. Clin. North Am.* **17:**21–40, vii.
15. Bromuro, C., A. Torosantucci, P. Chiani, S. Conti, L. Polonelli, and A. Cassone. 2002. Interplay between protective and inhibitory antibodies dictates the outcome of experimentally disseminated candidiasis in recipients of a *Candida albicans* vaccine. *Infect. Immun.* **70:**5462–5470.

16. Cano, L. E., S. S. Kashino, C. Arruda, D. Andre, C. F. Xidieh, L. M. Singer-Vermes, C. A. Vaz, E. Burger, and V. L. Calich. 1998. Protective role of gamma interferon in experimental pulmonary paracoccidioidomycosis. *Infect. Immun.* **66**:800–806.

17. Cano, M. V., and R. A. Hajjeh. 2001. The epidemiology of histoplasmosis: a review. *Semin. Respir. Infect.* **16**:109–118.

18. Cardenas-Freytag, L., E. Cheng, P. Mayeux, J. E. Domer, and J. D. Clements. 1999. Effectiveness of a vaccine composed of heat-killed *Candida albicans* and a novel mucosal adjuvant, LT(R192G), against systemic candidiasis. *Infect. Immun.* **67**:826–833.

19. Casadevall, A., W. Cleare, M. Feldmesser, A. Glatman-Freedman, D. L. Goldman, T. R. Kozel, N. Lendvai, J. Mukherjee, L.A. Pirofski, J. Rivera, A. L. Rosas, M. D. Scharff, P. Valadon, K. Westin, and Z. Zhong. 1998. Characterization of a murine monoclonal antibody to *Cryptococcus neoformans* polysaccharide that is a candidate for human therapeutic studies. *Antimicrob. Agents Chemother.* **42**:1437–1446.

20. Casadevall, A., and J. R. Perfect. 1998. *Cryptococcus neoformans*. American Society for Microbiology, Washington, D.C.

21. Casadevall, A., and M. D. Scharff. 1995. Return to the past: the case for antibody-based therapies in infectious diseases. *Clin. Infect. Dis.* **21**:150–161.

22. Casanova, M., J. P. Martinez, and W. L. Chaffin. 1990. Fab fragments from a monoclonal antibody against a germ tube mannoprotein block the yeast-to-mycelium transition in *Candida albicans*. *Infect. Immun.* **58**:3810–3812.

23. Cassone, A., M. Boccanera, D. Adriani, G. Santoni, and F. De Bernardis. 1995. Rats clearing a vaginal infection by *Candida albicans* acquire specific, antibody-mediated resistance to vaginal reinfection. *Infect. Immun.* **63**:2619–2624.

24. Cenci, E., A. Mencacci, A. Bacci, F. Bistoni, V. P. Kurup, and L. Romani. 2000. T cell vaccination in mice with invasive pulmonary aspergillosis. *J. Immunol.* **165**:381–388.

25. Cenci, E., A. Mencacci, A. Spreca, C. Montagnoli, A. Bacci, K. Perruccio, A. Velardi, W. Magliani, S. Conti, L. Polonelli, and L. Romani. 2002. Protection of killer antiidiotypic antibodies against early invasive aspergillosis in a murine model of allogeneic T-cell-depleted bone marrow transplantation. *Infect. Immun.* **70**:2375–2382.

26. Chilgren, R. A., R. Hong, and P. G. Quie. 1968. Human serum interactions with *Candida albicans*. *J. Immunol.* **101**:128-132.

27. Chiller, T. M., J. N. Galgiani, and D. A. Stevens. 2003. Coccidioidomycosis. *Infect. Dis. Clin. North Am.* **17**:41–57, viii.

28. Cole, G. T., J. M. Xue, C. N. Okeke, E. J. Tarcha, V. Basrur, R. A. Schaller, R. A. Herr, J. J. Yu, and C. Y. Hung. 2004. A vaccine against coccidioidomycosis is justified and attainable. *Med. Mycol.* **42**:189–216.

29. Deepe, G. S., Jr., and G. G. Durose. 1995. Immunobiological activity of recombinant H antigen from *Histoplasma capsulatum*. *Infect. Immun.* **63**:3151–3157.

30. Deepe, G. S., Jr., and R. Gibbons. 2001. Protective efficacy of H antigen from *Histoplasma capsulatum* in a murine model of pulmonary histoplasmosis. *Infect. Immun.* **69**:3128–3134.

31. Deepe, G. S., Jr., and R. S. Gibbons. 2002. Cellular and molecular regulation of vaccination with heat shock

32. protein 60 from *Histoplasma capsulatum*. *Infect. Immun.* **70**:3759–3767.

33. Delgado, N., J. Xue, J. J. Yu, C. Y. Hung, and G. T. Cole. 2003. A recombinant β-1,3-glucanosyltransferase homolog of *Coccidioides posadasii* protects mice against coccidioidomycosis. *Infect. Immun.* **71**:3010–3019.

33. Denning, D. W. 1998. Invasive aspergillosis. *Clin. Infect. Dis.* **26**:781–803.

34. Devi, S. J., R. Schneerson, W. Egan, T. J. Ulrich, D. Bryla, J. B. Robbins, and J. E. Bennett. 1991. *Cryptococcus neoformans* serotype A glucuronoxylomannan-protein conjugate vaccines: synthesis, characterization, and immunogenicity. *Infect. Immun.* **59**:3700–3707.

35. Diamond, R. D. 1974. Antibody-dependent killing of *Cryptococcus neoformans* by human peripheral blood mononuclear cells. *Nature* **247**:148–150.

36. Diamond, R. D., and A. C. Allison. 1976. Nature of the effector cells responsible for antibody-dependent cell-mediated killing of *Cryptococcus neoformans*. *Infect. Immun.* **14**:716–720.

37. Diniz, S. N., B. S. Reis, T. S. Goes, C. S. Zouain, M. F. Leite, and A. M. Goes. 2004. Protective immunity induced in mice by F0 and FII antigens purified from *Paracoccidioides brasiliensis*. *Vaccine* **22**:485–492.

38. Dromer, F., and J. Charreire. 1991. Improved amphotericin B activity by a monoclonal anti-*Cryptococcus neoformans* antibody: study during murine cryptococcosis and mechanisms of action. *J. Infect. Dis.* **163**:1114–1120.

39. Dromer, F., J. Charreire, A. Contrepois, C. Carbon, and P. Yeni. 1987. Protection of mice against experimental cryptococcosis by anti-*Cryptococcus neoformans* monoclonal antibody. *Infect. Immun.* **55**:749–752.

40. Dromer, F., C. Perronne, J. Barge, J. L. Vilde, and P. Yeni. 1989. Role of IgG and complement component C5 in the initial course of experimental cryptococcosis. *Clin. Exp. Immunol.* **78**:412–417.

41. Drutz, D. J., and C. L. Frey. 1985. Intracellular and extracellular defenses of human phagocytes against *Blastomyces dermatitidis* conidia and yeasts. *J. Lab. Clin. Med.* **105**:737–750.

42. Edelson, B. T., P. Cossart, and E. R. Unanue. 1999. Cutting edge: paradigm revisited: antibody provides resistance to *Listeria* infection. *J. Immunol.* **163**:4087–4090.

43. Edman, J. C., J. A. Kovacs, H. Masur, D. V. Santi, H. J. Elwood, and M. L. Sogin. 1988. Ribosomal RNA sequence shows *Pneumocystis carinii* to be a member of the fungi. *Nature* **334**:519–522.

44. Edwards, J. E. 2000. *Candida* species, p. 2656–2674. *In* G. L. Mandell, J. E. Bennett, and R. Dolin (ed.), *Principles and Practice of Infectious Diseases*, 5th ed, vol. 2. Churchill Livingstone, Inc., Philadelphia, Pa.

45. Eggimann, P., J. Garbino, and D. Pittet. 2003. Management of *Candida* species infections in critically ill patients. *Lancet Infect. Dis.* **3**:772–785.

46. Ellis, M. E., H. Al-Abdely, A. Sandridge, W. Greer, and W. Ventura. 2001. Fungal endocarditis: evidence in the world literature, 1965–1995. *Clin. Infect. Dis.* **32**:50–62.

47. Feldmesser, M., and A. Casadevall. 1998. Mechanism of action of antibody to capsular polysaccharide in *Cryptococcus neoformans* infection. *Front. Biosci.* **3**:D136–D151.

48. Fisher, K. L., G. S. Deepe, Jr., and J. P. Woods. 1999. *Histoplasma capsulatum* strain variation in both H antigen production and β-glucosidase activity and

overexpression of HAG1 from a telomeric linear plasmid. *Infect. Immun.* **67:**3312–3316.

49. **Fisher, K. L., and J. P. Woods.** 2000. Determination of β-glucosidase enzymatic function of the *Histoplasma capsulatum* H antigen using a native expression system. *Gene* **247:**191–197.

50. **Fleuridor, R., Z. Zhong, and L. Pirofski.** 1998. A human IgM monoclonal antibody prolongs survival of mice with lethal cryptococcosis. *J. Infect. Dis.* **178:**1213–1216.

51. **Flynn, J. L., and J. Chan.** 2001. Immunology of tuberculosis. *Annu. Rev. Immunol.* **19:**93–129.

52. **Fu, Y., A. S. Ibrahim, D. C. Sheppard, Y. C. Chen, S. W. French, J. E. Cutler, S. G. Filler, and J. E. Edwards, Jr.** 2002. *Candida albicans* Als1p: an adhesin that is a downstream effector of the EFG1 filamentation pathway. *Mol. Microbiol.* **44:**61–72.

53. **Fu, Y., G. Rieg, W. A. Fonzi, P. H. Belanger, J. E. Edwards, Jr., and S. G. Filler.** 1998. Expression of the *Candida albicans* gene ALS1 in *Saccharomyces cerevisiae* induces adherence to endothelial and epithelial cells. *Infect. Immun.* **66:**1783–1786.

54. **Garcia, J. P., and D. H. Howard.** 1971. Characterization of antigens from the yeast phase of *Histoplasma capsulatum. Infect. Immun.* **4:**116–125.

55. **Garraud, O., S. Mahanty, and R. Perraut.** 2003. Malaria-specific antibody subclasses in immune individuals: a key source of information for vaccine design. *Trends Immunol.* **24:**30–35.

56. **Garraud, O., R. Perraut, G. Riveau, and T. B. Nutman.** 2003. Class and subclass selection in parasite-specific antibody responses. *Trends Parasitol.* **19:**300–304.

57. **Gerrard, M. P., O. B. Eden, B. Jameson, and A. W. Craft.** 1987. Serological study of *Pneumocystis carinii* infection in the absence of immunosuppression. *Arch. Dis. Child.* **62:**177–179.

58. **Gesztesi, J. L., R. Puccia, L. R. Travassos, A. P. Vicentini, J. Z. de Moraes, M. F. Franco, and J. D. Lopes.** 1996. Monoclonal antibodies against the 43,000 Da glycoprotein from *Paracoccidioides brasiliensis* modulate laminin-mediated fungal adhesion to epithelial cells and pathogenesis. *Hybridoma* **15:**415–422.

59. **Gigliotti, F., C. G. Haidaris, T. W. Wright, and A. G. Harmsen.** 2002. Passive intranasal monoclonal antibody prophylaxis against murine *Pneumocystis carinii* pneumonia. *Infect. Immun.* **70:**1069–1074.

60. **Gigliotti, F., J. A. Wiley, and A. G. Harmsen.** 1998. Immunization with *Pneumocystis carinii* gpA is immunogenic but not protective in a mouse model of *P. carinii* pneumonia. *Infect. Immun.* **66:**3179–3182.

61. **Goldman, D. L., H. Khine, J. Abadi, D. J. Lindenberg, L. Pirofski, R. Niang, and A. Casadevall.** 2001. Serologic evidence for *Cryptococcus neoformans* infection in early childhood. *Pediatrics* **107:**E66.

62. **Goldman, D. L., S. C. Lee, and A. Casadevall.** 1995. Tissue localization of *Cryptococcus neoformans* glucuronoxylomannan in the presence and absence of specific antibody. *Infect. Immun.* **63:**3448–3453.

63. **Gomez, F. J., R. Allendoerfer, and G. S. Deepe, Jr.** 1995. Vaccination with recombinant heat shock protein 60 from *Histoplasma capsulatum* protects mice against pulmonary histoplasmosis. *Infect. Immun.* **63:**2587–2595.

64. **Gomez, F. J., A. M. Gomez, and G. S. Deepe, Jr.** 1991. Protective efficacy of a 62-kilodalton antigen, HIS-62, from the cell wall and cell membrane of *Histoplasma capsulatum* yeast cells. *Infect. Immun.* **59:**4459–4464.

65. **Good, M. F., J. A. Berzofsky, and L. H. Miller.** 1988. The T cell response to the malaria circumsporozoite protein: an immunological approach to vaccine development. *Annu. Rev. Immunol.* **6:**663–688.

66. **Gordon, M. A., and A. Casadevall.** 1995. Serum therapy for cryptococcal meningitis. *Clin. Infect. Dis.* **21:**1477–1479.

67. **Gurunathan, S., C. Prussin, D. L. Sacks, and R. A. Seder.** 1998. Vaccine requirements for sustained cellular immunity to an intracellular parasitic infection. *Nat. Med.* **4:**1409–1415.

68. **Han, Y., and J. E. Cutler.** 1995. Antibody response that protects against disseminated candidiasis. *Infect. Immun.* **63:**2714–2719.

69. **Han, Y., and J. E. Cutler.** 1997. Assessment of a mouse model of neutropenia and the effect of an anti-candidiasis monoclonal antibody in these animals. *J. Infect. Dis.* **175:**1169–1175.

70. **Han, Y., T. Kanbe, R. Cherniak, and J. E. Cutler.** 1997. Biochemical characterization of *Candida albicans* epitopes that can elicit protective and nonprotective antibodies. *Infect. Immun.* **65:**4100–4107.

71. **Han, Y., R. P. Morrison, and J. E. Cutler.** 1998. A vaccine and monoclonal antibodies that enhance mouse resistance to *Candida albicans* vaginal infection. *Infect. Immun.* **66:**5771–5776.

72. **Han, Y., M. A. Ulrich, and J. E. Cutler.** 1999. *Candida albicans* mannan extract-protein conjugates induce a protective immune response against experimental candidiasis. *J. Infect. Dis.* **179:**1477–1484.

73. **Henderson, D. K., J. E. Bennett, and M. A. Huber.** 1982. Long-lasting, specific immunologic unresponsiveness associated with cryptococcal meningitis. *J. Clin. Investig.* **69:**1185–1190.

74. **Henderson, D. K., V. L. Kan, and J. E. Bennett.** 1986. Tolerance to cryptococcal polysaccharide in cured cryptococcosis patients: failure of antibody secretion in vitro. *Clin. Exp. Immunol.* **65:**639–646.

75. **Herbrecht, R., D. W. Denning, T. F. Patterson, J. E. Bennett, R. E. Greene, J. W. Oestmann, W. V. Kern, K. A. Marr, P. Ribaud, O. Lortholary, R. Sylvester, R. H. Rubin, J. R. Wingard, P. Stark, C. Durand, D. Caillot, E. Thiel, P. H. Chandrasekar, M. R. Hodges, H. T. Schlamm, P. F. Troke, and B. de Pauw.** 2002. Voriconazole versus amphotericin B for primary therapy of invasive aspergillosis. *N. Engl. J. Med.* **347:**408–415.

76. **Hidore, M. R., and J. W. Murphy.** 1986. Correlation of natural killer cell activity and clearance of *Cryptococcus neoformans* from mice after adoptive transfer of splenic nylon wool-nonadherent cells. *Infect. Immun.* **51:**547–555.

77. **Hidore, M. R., and J. W. Murphy.** 1986. Natural cellular resistance of beige mice against *Cryptococcus neoformans. J. Immunol.* **137:**3624–3631.

78. **Hoag, K. A., M. F. Lipscomb, A. A. Izzo, and N. E. Street.** 1997. IL-12 and IFN-gamma are required for initiating the protective Th1 response to pulmonary cryptococcosis in resistant C.B-17 mice. *Am. J. Respir. Cell Mol. Biol.* **17:**733–739.

79. **Huang, C., S. H. Nong, M. K. Mansour, C. A. Specht, and S. M. Levitz.** 2002. Purification and characterization of a second immunoreactive mannoprotein from *Cryptococcus neoformans* that stimulates T-cell responses. *Infect. Immun.* **70:**5485–5493.

80. **Ibrahim, A. S., B. J. Spellberg, V. Avenissian, Y. Fu, S. G. Filler, and J. E. Edwards, Jr.** 2005. Vaccination with

recombinant N-terminal domain of Als1p improves survival during murine disseminated candidiasis by enhancing cell-mediated, not humoral, immunity. *Infect. Immun.* **73**:999–1005.

81. Ito, J. I., and J. M. Lyons. 2002. Vaccination of corticosteroid immunosuppressed mice against invasive pulmonary aspergillosis. *J. Infect. Dis.* **186**:869–871.

82. Ivey, F. D., D. M. Magee, M. D. Woitaske, S. A. Johnston, and R. A. Cox. 2003. Identification of a protective antigen of *Coccidioides immitis* by expression library immunization. *Vaccine* **21**:4359–4367.

83. Juvenale, M., G. M. Del Negro, A. J. Duarte, and G. Benard. 2001. Antibody isotypes to a *Paracoccidioides brasiliensis* somatic antigen in sub-acute and chronic form paracoccidioidomycosis. *J. Med. Microbiol.* **50**:127–134.

84. Kanbe, T., Y. Han, B. Redgrave, M. H. Riesselman, and J. E. Cutler. 1993. Evidence that mannans of *Candida albicans* are responsible for adherence of yeast forms to spleen and lymph node tissue. *Infect. Immun.* **61**:2578–2584.

85. Keane, J., S. Gershon, R. P. Wise, E. Mirabile-Levens, J. Kasznica, W. D. Schwieterman, J. N. Siegel, and M. M. Braun. 2001. Tuberculosis associated with infliximab, a tumor necrosis factor alpha-neutralizing agent. *N. Engl. J. Med.* **345**:1098–1104.

86. Klein, B. S., L. H. Hogan, and J. M. Jones. 1993. Immunologic recognition of a 25-amino acid repeat arrayed in tandem on a major antigen of *Blastomyces dermatitidis*. *J. Clin. Investig.* **92**:330–337.

87. Klein, B. S., and J. M. Jones. 1994. Purification and characterization of the major antigen WI-1 from *Blastomyces dermatitidis* yeasts and immunological comparison with A antigen. *Infect. Immun.* **62**:3890–3900.

88. Klein, B. S., J. M. Vergeront, R. J. Weeks, U. N. Kumar, G. Mathai, B. Varkey, L. Kaufman, R. W. Bradsher, J. F. Stoebig, and J. P. Davis. 1986. Isolation of *Blastomyces dermatitidis* in soil associated with a large outbreak of blastomycosis in Wisconsin. *N. Engl. J. Med.* **314**:529–534.

89. Kozel, T. R., B. C. deJong, M. M. Grinsell, R. S. MacGill, and K. K. Wall. 1998. Characterization of anticapsular monoclonal antibodies that regulate activation of the complement system by the *Cryptococcus neoformans* capsule. *Infect. Immun.* **66**:1538–1546.

90. Kozel, T. R., W. F. Gulley, and J. Cazin, Jr. 1977. Immune response to *Cryptococcus neoformans* soluble polysaccharide: immunological unresponsiveness. *Infect. Immun.* **18**:701–707.

91. Kumar, H., D. Malhotra, S. Goswami, and R. N. Bamezai. 2003. How far have we reached in tuberculosis vaccine development? *Crit. Rev. Microbiol.* **29**:297–312.

92. Lavigne, L. M., L. R. Schopf, C. L. Chung, R. Maylor, and J. P. Sypek. 1998. The role of recombinant murine IL-12 and IFN-γ in the pathogenesis of a murine systemic *Candida albicans* infection. *J. Immunol.* **160**:284–292.

93. Lee, J. H., N. R. Slifman, S. K. Gershon, E. T. Edwards, W. D. Schwieterman, J. N. Siegel, R. P. Wise, S. L. Brown, J. N. Udall, Jr., and M. M. Braun. 2002. Life-threatening histoplasmosis complicating immunotherapy with tumor necrosis factor alpha antagonists infliximab and etanercept. *Arthritis Rheum.* **46**:2565–2570.

94. Lendvai, N., A. Casadevall, Z. Liang, D. L. Goldman, J. Mukherjee, and L. Zuckier. 1998. Effect of immune mechanisms on the pharmacokinetics and organ distribution of cryptococcal polysaccharide. *J. Infect. Dis.* **177**:1647–1659.

95. Lendvai, N., X. W. Qu, W. Hsueh, and A. Casadevall. 2000. Mechanism for the isotype dependence of antibody-mediated toxicity in *Cryptococcus neoformans*-infected mice. *J. Immunol.* **164**:4367–4374.

96. Li, K., J. J. Yu, C. Y. Hung, P. F. Lehmann, and G. T. Cole. 2001. Recombinant urease and urease DNA of *Coccidioides immitis* elicit an immunoprotective response against coccidioidomycosis in mice. *Infect. Immun.* **69**:2878–2887.

97. Loza, L., Y. Fu, A. S. Ibrahim, D. C. Sheppard, S. G. Filler, and J. E. Edwards, Jr. 2004. Functional analysis of the *Candida albicans ALS1* gene product. *Yeast* **21**:473–482.

98. Magee, D. M., and R. A. Cox. 1996. Interleukin-12 regulation of host defenses against *Coccidioides immitis*. *Infect. Immun.* **64**:3609–3613.

99. Mamoni, R. L., S. A. Nouer, S. J. Oliveira, C. C. Musatti, C. L. Rossi, Z. P. Camargo, and M. H. Blotta. 2002. Enhanced production of specific IgG4, IgE, IgA and TGF-β in sera from patients with the juvenile form of paracoccidioidomycosis. *Med. Mycol.* **40**:153–159.

100. Mansour, M. K., L. S. Schlesinger, and S. M. Levitz. 2002. Optimal T cell responses to *Cryptococcus neoformans* mannoprotein are dependent on recognition of conjugated carbohydrates by mannose receptors. *J. Immunol.* **168**:2872–2879.

101. Mansour, M. K., L. E. Yauch, J. B. Rottman, and S. M. Levitz. 2004. Protective efficacy of antigenic fractions in mouse models of cryptococcosis. *Infect. Immun.* **72**:1746–1754.

102. Marr, K. A., T. Patterson, and D. Denning. 2002. Aspergillosis. Pathogenesis, clinical manifestations, and therapy. *Infect. Dis. Clin. North Am.* **16**:875–894, vi.

103. Marsh, C. B., J. E. Gadek, G. C. Kindt, S. A. Moore, and M. D. Wewers. 1995. Monocyte Fcγ receptor cross-linking induces IL-8 production. *J. Immunol.* **155**:3161–3167.

104. Marsh, C. B., M. D. Wewers, L. C. Tan, and B. H. Rovin. 1997. Fcγ receptor cross-linking induces peripheral blood mononuclear cell monocyte chemoattractant protein-1 expression: role of lymphocyte Fcγ RIII. *J. Immunol.* **158**:1078–1084.

105. Martinez, J. P., M. L. Gil, J. L. Lopez-Ribot, and W. L. Chaffin. 1998. Serologic response to cell wall mannoproteins and proteins of *Candida albicans*. *Clin. Microbiol. Rev.* **11**:121–141.

106. Matthews, R. C., and J. P. Burnie. 2004. Recombinant antibodies: a natural partner in combinatorial antifungal therapy. *Vaccine* **22**:865–871.

107. Matthews, R. C., G. Rigg, S. Hodgetts, T. Carter, C. Chapman, C. Gregory, C. Illidge, and J. Burnie. 2003. Preclinical assessment of the efficacy of mycograb, a human recombinant antibody against fungal HSP90. *Antimicrob. Agents Chemother.* **47**:2208–2216.

108. Mencacci, A., G. Del Sero, E. Cenci, C. F. d'Ostiani, A. Bacci, C. Montagnoli, M. Kopf, and L. Romani. 1998. Endogenous interleukin 4 is required for development of protective CD4+ T helper type 1 cell responses to *Candida albicans*. *J. Exp. Med.* **187**:307–317.

109. Miller, G. P., and S. Kohl. 1983. Antibody-dependent leukocyte killing of *Cryptococcus neoformans*. *J. Immunol.* **131**:1455–1459.

110. Miller, M. F., T. G. Mitchell, W. J. Storkus, and J. R. Dawson. 1990. Human natural killer cells do not inhibit growth of *Cryptococcus neoformans* in the absence of antibody. *Infect. Immun.* **58**:639–645.

111. Mody, C. H., C. L. Tyler, R. G. Sitrin, C. Jackson, and G. B. Toews. 1991. Interferon-gamma activates rat alveolar macrophages for anticryptococcal activity. *Am. J. Respir. Cell Mol. Biol.* **5:**19–26.

112. Mozaffarian, N., J. W. Berman, and A. Casadevall. 1995. Immune complexes increase nitric oxide production by interferon-gamma-stimulated murine macrophage-like J774.16 cells. *J. Leukoc. Biol.* **57:**657–663.

113. Mukherjee, J., G. Nussbaum, M. D. Scharff, and A. Casadevall. 1995. Protective and nonprotective monoclonal antibodies to *Cryptococcus neoformans* originating from one B cell. *J. Exp. Med.* **181:**405–409.

114. Mukherjee, S., S. C. Lee, and A. Casadevall. 1995. Antibodies to *Cryptococcus neoformans* glucuronoxylomannan enhance antifungal activity of murine macrophages. *Infect. Immun.* **63:**573–579.

115. Nabavi, N., and J. W. Murphy. 1986. Antibody-dependent natural killer cell-mediated growth inhibition of *Cryptococcus neoformans*. *Infect. Immun.* **51:**556–562.

116. Nardin, E. H., and R. S. Nussenzweig. 1993. T cell responses to pre-erythrocytic stages of malaria: role in protection and vaccine development against pre-erythrocytic stages. *Annu. Rev. Immunol.* **11:**687–727.

117. Nosanchuk, J. D. 2002. Fungal myocarditis. *Front. Biosci.* **7:**d1423–d1438.

118. Nosanchuk, J. D., J. N. Steenbergen, L. Shi, G. S. Deepe, Jr., and A. Casadevall. 2003. Antibodies to a cell surface histone-like protein protect against *Histoplasma capsulatum*. *J. Clin. Investig.* **112:**1164–1175.

119. Pappagianis, D., and The Valley Fever Vaccine Study Group. 1993. Evaluation of the protective efficacy of the killed *Coccidioides immitis* spherule vaccine in humans. *Am. Rev. Respir. Dis.* **148:**656–660.

120. Pappagianis, D., and H. B. Levine. 1975. The present status of vaccination against coccidioidomycosis in man. *Am. J. Epidemiol.* **102:**30–41.

121. Pascale, J. M., M. M. Shaw, P. J. Durant, A. A. Amador, M. S. Bartlett, J. W. Smith, R. L. Gregory, and G. L. McLaughlin. 1999. Intranasal immunization confers protection against murine *Pneumocystis carinii* lung infection. *Infect. Immun.* **67:**805–809.

122. Patterson, T. F., W. R. Kirkpatrick, M. White, J. W. Hiemenz, J. R. Wingard, B. Dupont, M. G. Rinaldi, D. A. Stevens, J. R. Graybill, and the I3 Aspergillus Study Group. 2000. Invasive aspergillosis. Disease spectrum, treatment practices, and outcomes. *Medicine* (Baltimore) **79:**250–260.

123. Peng, T., L. Shubitz, J. Simons, R. Perrill, K. I. Orsborn, and J. N. Galgiani. 2002. Localization within a proline-rich antigen (Ag2/PRA) of protective antigenicity against infection with *Coccidioides immitis* in mice. *Infect. Immun.* **70:**3330–3335.

124. Pinto, A. R., R. Puccia, S. N. Diniz, M. F. Franco, and L. R. Travassos. 2000. DNA-based vaccination against murine paracoccidioidomycosis using the gp43 gene from *Paracoccidioides brasiliensis*. *Vaccine* **18:**3050–3058.

125. Polonelli, L., F. De Bernadis, S. Conti, M. Boccanera, W. Magliani, M. Gerloni, C. Cantelli, and A. Cassone. 1996. Human natural yeast killer toxin-like candidacidal antibodies. *J. Immunol.* **156:**1880–1885.

126. Polonelli, L., F. De Bernadis, S. Conti, M. Boccanera, M. Gerloni, G. Morace, W. Magliani, C. Chezzi, and A. Cassone. 1994. Idiotypic intravaginal vaccination to protect against candidal vaginitis by secretory, yeast killer toxin-like anti-idiotypic antibodies. *J. Immunol.* **152:**3175–3182.

127. Polonelli, L., W. Magliani, S. Conti, L. Bracci, L. Lozzi, P. Neri, D. Adriani, F. De Bernardis, and A. Cassone. 2003. Therapeutic activity of an engineered synthetic killer antiidiotypic antibody fragment against experimental mucosal and systemic candidiasis. *Infect. Immun.* **71:**6205–6212.

128. Puccia, R., and L. R. Travassos. 1991. 43-kilodalton glycoprotein from *Paracoccidioides brasiliensis*: immunochemical reactions with sera from patients with paracoccidioidomycosis, histoplasmosis, or Jorge Lobo's disease. *J. Clin. Microbiol.* **29:**1610–1615.

129. Retallack, D. M., and J. P. Woods. 1999. Molecular epidemiology, pathogenesis, and genetics of the dimorphic fungus *Histoplasma capsulatum*. *Microbes Infect.* **1:**817–825.

130. Richard, J. L., W. M. Peden, and J. M. Sacks. 1991. Effects of adjuvant-augmented germling vaccines in turkey poults challenged with *Aspergillus fumigatus*. *Avian Dis.* **35:**93–99.

131. Richard, J. L., J. R. Thurston, R. C. Cutlip, and A. C. Pier. 1982. Vaccination studies of aspergillosis in turkeys: subcutaneous inoculation with several vaccine preparations followed by aerosol challenge exposure. *Am. J. Vet. Res.* **43:**488–492.

132. Rivera, J., J. Mukherjee, L. M. Weiss, and A. Casadevall. 2002. Antibody efficacy in murine pulmonary *Cryptococcus neoformans* infection: a role for nitric oxide. *J. Immunol.* **168:**3419–3427.

133. Rodrigues, M. L., L. R. Travassos, K. R. Miranda, A. J. Franzen, S. Rozental, W. de Souza, C. S. Alviano, and E. Barreto-Bergter. 2000. Human antibodies against a purified glucosylceramide from *Cryptococcus neoformans* inhibit cell budding and fungal growth. *Infect. Immun.* **68:**7049–7060.

134. Romani, L. 2004. Immunity to fungal infections. *Nat. Rev. Immunol.* **4:**1–23.

135. Romani, L., F. Bistoni, and P. Puccetti. 2002. Fungi, dendritic cells and receptors: a host perspective of fungal virulence. *Trends Microbiol.* **10:**508–514.

136. Romani, L., A. Mencacci, U. Grohmann, S. Mocci, P. Mosci, P. Puccetti, and F. Bistoni. 1992. Neutralizing antibody to interleukin 4 induces systemic protection and T helper type 1-associated immunity in murine candidiasis. *J. Exp. Med.* **176:**19–25.

137. Romani, L., S. Mocci, C. Bietta, L. Lanfaloni, P. Puccetti, and F. Bistoni. 1991. Th1 and Th2 cytokine secretion patterns in murine candidiasis: association of Th1 responses with acquired resistance. *Infect. Immun.* **59:**4647–4654.

138. Rosas, A. L., J. D. Nosanchuk, and A. Casadevall. 2001. Passive immunization with melanin-binding monoclonal antibodies prolongs survival of mice with lethal *Cryptococcus neoformans* infection. *Infect. Immun.* **69:**3410–3412.

139. Roths, J. B., and C. L. Sidman. 1992. Both immunity and hyperresponsiveness to *Pneumocystis carinii* result from transfer of CD4$^+$ but not CD8$^+$ T cells into severe combined immunodeficiency mice. *J. Clin. Investig.* **90:**673–678.

140. San-Blas, G., G. Nino-Vega, and T. Iturriaga. 2002. *Paracoccidioides brasiliensis* and paracoccidioidomycosis: molecular approaches to morphogenesis, diagnosis, epidemiology, taxonomy and genetics. *Med. Mycol.* **40:**225–242.

141. Sau, K., S. S. Mambula, E. Latz, P. Henneke, D. T. Golenbock, and S. M. Levitz. 2003. The antifungal drug amphotericin B promotes inflammatory cytokine release

by a Toll-like receptor- and CD14-dependent mechanism. *J. Biol. Chem.* **278:**37561–37568.

142. Scheckelhoff, M., and G. S. Deepe, Jr. 2002. The protective immune response to heat shock protein 60 of *Histoplasma capsulatum* is mediated by a subset of Vβ 8.1/8.2⁺ T cells. *J. Immunol.* **169:**5818–5826.

143. Schlageter, A. M., and T. R. Kozel. 1990. Opsonization of *Cryptococcus neoformans* by a family of isotype-switch variant antibodies specific for the capsular polysaccharide. *Infect. Immun.* **58:**1914–1918.

144. Seder, R., and W. Paul. 1994. Acquisition of lymphokine-producing phenotype CD4⁺ T cells. *Ann. Rev. Immunol.* **12:**635–674.

145. Shubitz, L., T. Peng, R. Perrill, J. Simons, K. Orsborn, and J. N. Galgiani. 2002. Protection of mice against *Coccidioides immitis* intranasal infection by vaccination with recombinant antigen 2/PRA. *Infect. Immun.* **70:**3287–3289.

146. Smulian, A. G., D. W. Sullivan, and S. A. Theus. 2000. Immunization with recombinant *Pneumocystis carinii* p55 antigen provides partial protection against infection: characterization of epitope recognition associated with immunization. *Microbes Infect.* **2:**127–136.

147. Stanberry, L. R., D. I. Bernstein, R. L. Burke, C. Pachl, and M. G. Myers. 1987. Vaccination with recombinant herpes simplex virus glycoproteins: protection against initial and recurrent genital herpes. *J. Infect. Dis.* **155:**914–920.

148. Stevens, D. A. 2004. Vaccinate against aspergillosis! A call to arms of the immune system. *Clin. Infect. Dis.* **38:**1131–1136.

149. Straus, A. H., E. Freymuller, L. R. Travassos, and H. K. Takahashi. 1996. Immunochemical and subcellular localization of the 43 kDa glycoprotein antigen of *Paracoccidioides brasiliensis* with monoclonal antibodies. *J. Med. Vet. Mycol.* **34:**181–186.

150. Stringer, J. R., C. B. Beard, R. F. Miller, and A. E. Wakefield. 2002. A new name (*Pneumocystis jiroveci*) for *Pneumocystis* from humans. *Emerg. Infect. Dis.* **8:**891–896.

151. Taborda, C. P., and A. Casadevall. 2001. Immunoglobulin M efficacy against *Cryptococcus neoformans*: mechanism, dose dependence, and prozone-like effects in passive protection experiments. *J. Immunol.* **166:**2100–2107.

152. Taborda, C. P., M. A. Juliano, R. Puccia, M. Franco, and L. R. Travassos. 1998. Mapping of the T-cell epitope in the major 43-kilodalton glycoprotein of *Paracoccidioides brasiliensis* which induces a Th-1 response protective against fungal infection in BALB/c mice. *Infect. Immun.* **66:**786–793.

153. Taborda, C. P., C. R. Nakaie, E. M. Cilli, E. G. Rodrigues, L. S. Silva, M. F. Franco, and L. R. Travassos. 2004. Synthesis and immunological activity of a branched peptide carrying the T-cell epitope of gp43, the major exocellular antigen of *Paracoccidioides brasiliensis*. *Scand. J. Immunol.* **59:**58–65.

154. Taborda, C. P., J. Rivera, O. Zaragoza, and A. Casadevall. 2003. More is not necessarily better: prozone-like effects in passive immunization with IgG. *J. Immunol.* **170:**3621–3630.

155. Theus, S. A., A. G. Smulian, P. Steele, M. J. Linke, and P. D. Walzer. 1998. Immunization with the major surface glycoprotein of *Pneumocystis carinii* elicits a protective response. *Vaccine* **16:**1149–1157.

156. Thomas, C. F., Jr., and A. H. Limper. 2004. Pneumocystis pneumonia. *N. Engl. J. Med.* **350:**2487–2498.

157. Travassos, L. R., R. Puccia, P. Cisalpino, C. Taborda, E. G. Rodrigues, M. Rodrigues, J. F. Silveira, and I. C. Almeida. 1995. Biochemistry and molecular biology of the main diagnostic antigen of *Paracoccidioides brasiliensis*. *Arch. Med. Res.* **26:**297–304.

158. Tripp, C. S., K. P. Beckerman, and E. R. Unanue. 1995. Immune complexes inhibit antimicrobial responses through interleukin-10 production. Effects in severe combined immunodeficient mice during *Listeria* infection. *J. Clin. Investig.* **95:**1628–1634.

159. Vecchiarelli, A., and A. Casadevall. 1998. Antibody-mediated effects against *Cryptococcus neoformans*: evidence for interdependency and collaboration between humoral and cellular immunity. *Res. Immunol.* **149:**321–333.

160. Vicentini, A. P., J. L. Gesztesi, M. F. Franco, W. de Souza, J. Z. de Moraes, L. R. Travassos, and J. D. Lopes. 1994. Binding of *Paracoccidioides brasiliensis* to laminin through surface glycoprotein gp43 leads to enhancement of fungal pathogenesis. *Infect. Immun.* **62:**1465–1469.

161. Wark, P., P. Gibson, and A. Wilson. 2004. Azoles for allergic bronchopulmonary aspergillosis associated with asthma. *Cochrane Database Syst. Rev.* **3:**CD001108.

162. Wheat, J., M. L. French, R. B. Kohler, S. E. Zimmerman, W. R. Smith, J. A. Norton, H. E. Eitzen, C. D. Smith, and T. G. Slama. 1982. The diagnostic laboratory tests for histoplasmosis: analysis of experience in a large urban outbreak. *Ann. Intern. Med.* **97:**680–685.

163. Wheat, L. J., M. Goldman, and G. Sarosi. 2002. State-of-the-art review of pulmonary fungal infections. *Semin. Respir. Infect.* **17:**158–181.

164. Wilson, M. A., and T. R. Kozel. 1992. Contribution of antibody in normal human serum to early deposition of C3 onto encapsulated and nonencapsulated *Cryptococcus neoformans*. *Infect. Immun.* **60:**754–761.

165. Wisplinghoff, H., T. Bischoff, S. M. Tallent, H. Seifert, R. P. Wenzel, and M. B. Edmond. 2004. Nosocomial bloodstream infections in US hospitals: analysis of 24,179 cases from a prospective nationwide surveillance study. *Clin. Infect. Dis.* **39:**309–317.

166. Wood, K. L., C. A. Hage, K. S. Knox, M. B. Kleiman, A. Sannuti, R. B. Day, L. J. Wheat, and H. L. Twigg, III. 2003. Histoplasmosis after treatment with anti-tumor necrosis factor-alpha therapy. *Am. J. Respir. Crit. Care Med.* **167:**1279–1282.

167. Wuthrich, M., W. L. Chang, and B. S. Klein. 1998. Immunogenicity and protective efficacy of the WI-1 adhesin of *Blastomyces dermatitidis*. *Infect. Immun.* **66:**5443–5449.

168. Wuthrich, M., H. I. Filutowicz, and B. S. Klein. 2000. Mutation of the WI-1 gene yields an attenuated *Blastomyces dermatitidis* strain that induces host resistance. *J. Clin. Investig.* **106:**1381–1389.

169. Wuthrich, M., H. I. Filutowicz, T. Warner, G. S. Deepe, Jr., and B. S. Klein. 2003. Vaccine immunity to pathogenic fungi overcomes the requirement for CD4 help in exogenous antigen presentation to CD8⁺ T cells: implications for vaccine development in immune-deficient hosts. *J. Exp. Med.* **197:**1405–1416.

170. Wuthrich, M., H. I. Filutowicz, T. Warner, and B. S. Klein. 2002. Requisite elements in vaccine immunity to *Blastomyces dermatitidis*: plasticity uncovers vaccine

potential in immune-deficient hosts. *J. Immunol.* **169:**6969–6976.

171. **Wuthrich, M., and B. S. Klein.** 2000. Investigation of anti-WI-1 adhesin antibody-mediated protection in experimental pulmonary blastomycosis. *J. Infect. Dis.* **181:**1720–1728.

172. **Yokoyama, W. M., S. Kim, and A. R. French.** 2004. The dynamic life of natural killer cells. *Annu. Rev. Immunol.* **22:**405–429.

173. **Yuan, R. R., A. Casadevall, J. Oh, and M. D. Scharff.** 1997. T cells cooperate with passive antibody to modify *Cryptococcus neoformans* infection in mice. *Proc. Natl. Acad. Sci. USA* **94:**2483–2488.

174. **Yuan, R. R., G. Spira, J. Oh, M. Paizi, A. Casadevall, and M. D. Scharff.** 1998. Isotype switching increases efficacy of antibody protection against *Cryptococcus neoformans* infection in mice. *Infect. Immun.* **66:**1057–1062.

175. **Zhang, M. X., T. T. Brandhorst, T. R. Kozel, and B. S. Klein.** 2001. Role of glucan and surface protein BAD1 in complement activation by *Blastomyces dermatitidis* yeast. *Infect. Immun.* **69:**7559–7564.

176. **Zhang, M. X., and B. Klein.** 1997. Activation, binding, and processing of complement component 3 (C3) by *Blastomyces dermatitidis*. *Infect. Immun.* **65:**1849–1855.

177. **Zheng, M., J. E. Shellito, L. Marrero, Q. Zhong, S. Julian, P. Ye, V. Wallace, P. Schwarzenberger, and J. K. Kolls.** 2001. CD4$^+$ T cell-independent vaccination against *Pneumocystis carinii* in mice. *J. Clin. Investig.* **108:**1469–1474.

178. **Zhou, P., M. C. Sieve, J. Bennett, K. J. Kwon-Chung, R. P. Tewari, R. T. Gazzinelli, A. Sher, and R. A. Seder.** 1995. IL-12 prevents mortality in mice infected with *Histoplasma capsulatum* through induction of IFN-γ. *J. Immunol.* **155:**785–795.

179. **Zhou, P., M. C. Sieve, R. P. Tewari, and R. A. Seder.** 1997. Interleukin-12 modulates the protective immune response in SCID mice infected with *Histoplasma capsulatum*. *Infect. Immun.* **65:**936–942.

Molecular Principles of Fungal Pathogenesis
Edited by Joseph Heitman et al.
©2006 ASM Press, Washington, D.C.

Chapter 39

Molecular Immunopathogenesis of Innate Host Defense against Chronic Disseminated (Hepatosplenic) Candidiasis

THOMAS J. WALSH, EMMANUEL ROILIDES, KAROLL CORTEZ, AND CARON LYMAN

Chronic disseminated candidiasis or hepatosplenic candidiasis is a persistent infection of the liver, spleen, and other tissues in neutropenic patients following recovery from neutropenia, during which time the patients have sustained hematogenous seeding of *Candida* spp. to the liver and other tissue sites (10, 27). This infection is characterized by a chronic, debilitating, progressively deteriorating course that may be refractory to conventional therapy or require an extensive course of antifungal therapy, often requiring 6 to 12 months of management (1, 8, 33, 34). Infection of the liver and spleen during candidemia in nonneutropenic immunocompetent patients is unusual. Neutropenia, which is a critical factor in establishment of this distinctive pattern of disseminated candidiasis, appears to have a permissive effect on the early initiation of this infection. The immunopathogenesis of this infection until recently has not been well understood.

CLINICAL, HISTOLOGICAL, AND MICROBIOLOGICAL CHARACTERISTICS

Principally involving the liver and spleen, chronic disseminated candidiasis may initially manifest only as an earlier episode of candidemia or even with no evidence of antecedent candidemia until recovery from neutropenia. In addition to the liver and spleen, other sites, such as the kidneys, lungs, brain, eyes, and other tissues, may be infected by antecedent candidemia. Marcus et al. found that multiple hepatic abscesses may be due to *Candida* spp. or bacteria while solitary hepatic abscesses were caused by bacterial infections but not by candidiasis (13).

Patients with chronic disseminated candidiasis typically present with persistent fever after recovery from an episode of neutropenia. Hepatic infection may manifest as right upper quadrant discomfort and increased serum alkaline phosphatase levels. Splenic involvement may present as left upper quadrant pain. Nausea and weight loss also accompany the infection. Blood cultures may be negative prior to the diagnosis of hepatic candidiasis and are almost invariably negative on recovery from neutropenia. Patients may be in remission of their hematological neoplasia at the time of these symptoms and signs. These findings should prompt the further diagnostic imaging for evaluation of patients with suspected chronic disseminated candidiasis. Chronic disseminated candidiasis is characterized by the presence of multiple "bull's eye" lesions in the liver or spleen on computed tomography, magnetic resonance imaging, or ultrasonography. These lesions are not apparent in patients who are neutropenic but, rather, become recognizable at the time of neutrophil recovery.

The lesions of chronic disseminated candidiasis are histologically characterized as pyogranulomas, consisting of an inner core of necrosis, blastoconidia, and pseudohyphae surrounded by an admixture of polymorphonuclear leukocytes (PMNs), monocytes (MNCs), macrophages, lymphocytes, and multinucleate giant cells and further by an outer ring of fibrosis (13). The inflammatory component of these lesions evolves over time. With treatment, some lesions develop central calcification. The presence of calcification is an important end point of therapy.

Chronic disseminated candidiasis may pose a formidable therapeutic challenge. As mentioned above, protracted courses of treatment are necessary, with antifungal therapy with deoxycholate amphotericin B administered over approximately 6 to 12 months (1, 8, 33, 34). Current experience with amphotericin B lipid complex indicates that it is effective in the treatment of hepatosplenic candidiasis, with curative total

Thomas J. Walsh, Emmanuel Roilides, Karoll Cortez, and Caron Lyman • Immunocompromised Host Section, Pediatric Oncology Branch, National Cancer Institute, Bethesda, MD 20892.

doses of drug being delivered in a much shorter time and with fewer side effects than have been seen with conventional deoxycholate amphotericin B (33). Experimental data and several encouraging reports indicate that fluconazole also has favorable efficacy in the treatment of hepatosplenic candidiasis in patients in whom amphotericin B has failed to control the infection or who have had serious amphotericin B-related toxicities (1, 8, 29, 32). Fluconazole has been increasingly used as initial therapy in stable patients in an outpatient setting. There is a paucity of data for the use of echinocandins in the treatment of chronic disseminated candidiasis.

Whether patients with chronic disseminated candidiasis should continue to receive antineoplastic therapy, which may cause neutropenia with the risk for progressive hepatosplenic involvement or breakthrough fungemia, can be a major dilemma. Further study found that chronic disseminated candidiasis in patients with cancer can be treated successfully under careful observation through repeated courses of chemotherapy-induced neutropenia without progression of deep tissue candidiasis or breakthrough fungemia (34). In the management of hepatosplenic candidiasis, the therapeutic end point is measured in terms of resolution or calcification of lesions. Children tend to have calcified lesions more than adults. Premature discontinuation of therapy may lead to recurrence of symptoms and progression of the lesions. Moreover, some patients remain refractory to all efforts of antifungal chemotherapy. Such patients may have a favorable response to the addition of recombinant cytokines, such as granulocyte-macrophage colony-stimulating factor (GM-CSF) or gamma interferon (6).

CHRONIC DISSEMINATED CANDIDIASIS IS A DISEASE OF Th1/Th2 DYSIMMUNOREGULATION

Until recently, the pathogenesis of chronic disseminated candidiasis has not been well understood. Chronic disseminated candidiasis presents a paradox of innate host defense; i.e., hematogenous candidiasis ensues during neutropenia, and despite recovery from neutropenia, deep tissue disease becomes clinically apparent and progressive in severity. Recovery from neutropenia should resolve the infection. Instead, the progression of infection despite recovery from neutropenia suggests a systemic dysimmunoregulation that has a permissive effect for disseminated candidiasis.

In endeavoring to understand the need for chronic therapy and the inability of the host to eradicate organisms in chronic disseminated candidiasis, we found that patients with this infection have an ongoing state of immunosuppression despite having recovered from neutropenia. This chronic immunosuppression appears to be related to a Th1/Th2 dysimmunoregulation, with increased expression of a Th2 phenotype that allows persistence of the pathogen in hepatic, splenic, and other tissues (Fig. 1–3).

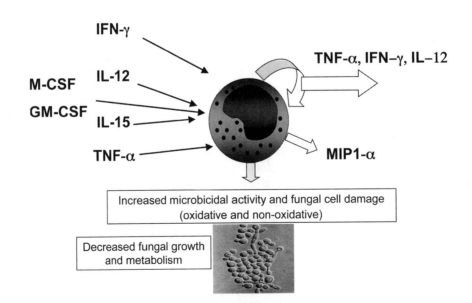

Figure 1. Normal patterns of the innate host response of human monocytes to *C. albicans*.

Th2-RELATED CYTOKINES IN THE IMMUNOPATHOGENESIS OF DISSEMINATED CANDIDIASIS

IL-4 and IL-10

Early studies by Cenci et al., Romani et al., and others demonstrated that interleukin-4 (IL-4) and IL-10 reduce the killing of *Candida albicans* in vitro and in vivo (3, 4, 14, 28). Hypothesizing that the pathogenesis of disseminated candidiasis in cancer patients may also involve the immunosuppressive effect of Th2-related cytokines, we found that IL-4 suppresses the mononuclear microbicidal activity and phagocytic host response to *C. albicans* (18). Consistent with this hypothesis were our observations that gamma interferon, GM-CSF, and M-CSF partially reversed IL-4-induced immunosuppression. We subsequently found that IL-10 also suppresses the host response of MNCs and PMNs to *C. albicans*, as evidenced by decreased fungicidal activity against serum-opsonized blastoconidia, reduced capacity to damage unopsonized hyphae, and suppressed O_2^- production (16, 19). These suppressive effects probably extend to other MNC-derived cells, including splenic macrophages and Kupffer cells in their response to *C. albicans* (20, 21) (Fig. 2).

Despite their recovery from neutropenia, patients with chronic disseminated candidiasis have protracted courses of infection, which often require several months of therapy. We therefore hypothesized that a switch from Th1 to a Th2 cytokine profile may contribute to this refractoriness. Consistent with this hypothesis, we found that circulating levels of IL-10 were significantly increased in patients with chronic disseminated

candidiasis in comparison to healthy donors, cancer patients without infections, and cancer patients with bacterial infections (21) (Fig. 3).

Molecular genetic studies from our laboratories further substantiate a role of Th2 dysimmunoregulation in patients with chronic disseminated candidiasis (5). We found that a common haplotype of the *IL4* promoter containing the −589 single-nucleotide polymorphism (previously reported to influence transcriptional activity) was significantly overrepresented in leukemic patients with chronic disseminated candidiasis ($P = 0.01$, odds ratio [OR] = 2.16). Another common haplotype of the *IL4* promoter not containing the −589C single-nucleotide polymorphism appears to significantly protect against chronic disseminated candidiasis ($P = 0.018$, OR = 0.47).

Transforming Growth Factor β

Transforming growth factor β (TGF-β) is known to suppress phagocyte function at multiple points, including its activity against *C. albicans* (12, 26, 30, 31). Further substantiating the dysimmunoregulation observed in chronic disseminated candidiasis, we documented the increased expression, cellular source, and role of TGF-β in the pathogenesis of chronic disseminated candidiasis in animal models and humans (11). We found that human peripheral blood MNCs incubated with *C. albicans* in vitro also released large amounts of biologically active TGF-β1. These results further correlated with the immunolocalization of TGF-β observed in livers from the rabbit model of chronic disseminated (hepatosplenic) candidiasis (11). Liver biopsy specimens from patients with chronic disseminated candidiasis demonstrated intense staining for extracellular

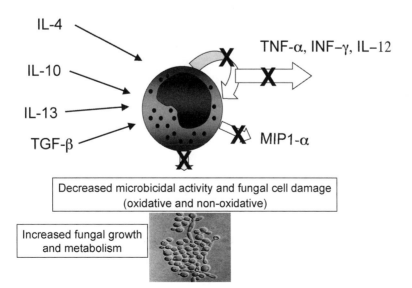

Figure 2. Cytokine-mediated suppression of the innate host response to *C. albicans*.

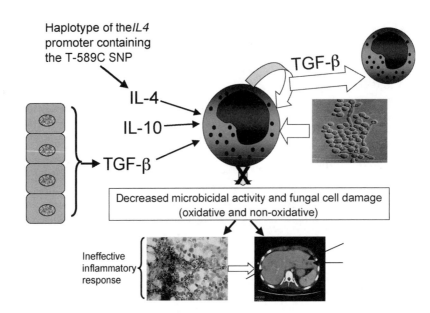

Figure 3. Immunopathogenesis of Th1/Th2 dysimmunoregulations of chronic disseminated (hepatosplenic) candidiasis.

matrix-associated TGF-β1 within inflammatory granulomas, as well as staining for TGF-β1 and TGF-β3 within adjacent hepatocytes. That localized expression of TGF-β in hepatic tissue infected with *C. albicans* may contribute to the immunopathogenesis is also observed in other eukaryotic infections, such as leishmaniasis, where Th1/Th2 dysimmunoregulation is also thought to play an important pathophysiological role (2).

Effect of Other Suppressive Cytokine Responses on Innate Immunity

IL-13 may exert suppressive and enhancing effects against *C. albicans*, depending on the predominant fungal form. IL-13 suppresses MNC-mediated killing of *C. albicans* blastoconidia but does not alter the damage caused by MNC to hyphae (7). Moreover, IL-13 enhances phagocytosis by MNCs of *C. albicans*. These IL-13-mediated effects are blocked by anti-IL-13 antibody.

Functional Genomic Responses of Monocyte-Derived Innate Host Defense Molecules to *C. albicans*

The early stages of infection during disseminated candidiasis appear to be a critical factor in establishment of infection in the liver and spleen. To better understand these early events, we investigated the functional response and kinetics of expression of innate host defense molecules of human monocytes against *C. albicans* by microarray analysis on cDNA human genomic chips (9). Expression of genes encoding proinflammatory cytokines, including tumor necrosis factor alpha, IL-1, IL-6, and leukemia

inhibitory factor, is markedly enhanced during the first 6 h and coincides with an increase in the phagocytosis of blastoconidia. Expression of these genes returns to near baseline by 18 h.

Genes encoding chemokines (including IL-8), chemokine receptors (including CCR1, CCR5, CCR7, and CXCR5), macrophage inflammatory proteins 1, 3, and 4, and MNC chemoattractant protein 1, also are strongly up-regulated, attaining peak expression at 4 to 6 h, in response to *C. albicans*. Expression of genes whose products may protect MNC viability (e.g., BCL2-related protein, metallothioneins, CD71, and SOCS3) is increased at 4 to 6 h and remains so throughout the 18 h. Thus, *C. albicans* induces an early dynamic cascade of expression of genes whose products are related to the recruitment, activation, and protection of neutrophils and MNC. Perturbations of these early events in neutropenic hosts may impair early clearance and allow the establishment of infection in the liver and spleen.

Th1 RESPONSES TO *C. ALBICANS* AND THE IMPLICATIONS FOR IMMUNOTHERAPEUTIC INTERVENTIONS

Early studies of animal models of *C. albicans* infection have illustrated the protective role of Th1 cytokines and the suppressive effect of Th2-type cytokines on the host response to infection (14, 15, 25) (see Fig. 1). For example, IL-12 is required for Th1 differentiation in murine candidiasis (24), and production of this cytokine by neutrophils correlates with a protective response. Exogenous IL-12 is effective in protecting neutropenic hosts who are susceptible to infection and can enhance the host phagocytic response to *C. albicans* (23).

IL-15 also is a Th1-related cytokine that has biological properties similar to those of IL-2 (22). We found that IL-15 directly up-regulates the oxidative and microbicidal activity of human MNCs against *C. albicans* (29). IL-15-treated human MNCs demonstrated enhanced fungicidal activity against *C. albicans*. IL-15 enhances O_2^- production by elutriated human MNCs in response to opsonized *C. albicans* blastoconidia, without inducing the release of detectable levels of tumor necrosis factor alpha, IL-1β, or IL-12. That we found no significant increase in the number of mRNA transcripts for components of the NADPH-oxidase complex, p40-phox, p47-phox, and gp91-phox suggests a posttranscriptional regulation of enhanced O_2^- production by IL-15.

Th1-based immunotherapeutic interventions against chronic disseminated candidiasis could include one or more of the following interventions: gamma interferon, IL-2, IL-12, IL-15, anti-IL-4 antibody, anti-IL-10 antibody, or soluble IL-4 receptor. Among these potential Th1-based interventions, only gamma interferon is readily available on a practical clinical basis. Available recombinant but non-Th1-based cytokines may augment the activity of monocyte or neutrophil against chronic disseminated candidiasis.

RECOMBINANT CYTOKINES

Granulocyte Colony-Stimulating Factor

The advent of recombinant cytokines offers the potential for augmenting quantitative and qualitative elements of host response to invasive fungal pathogens (22). G-CSF, GM-CSF, and gamma interferon reverse the corticosteroid-induced immunosuppression of PMNs and MNCs and MDMs against *C. albicans* and non-*albicans Candida* spp. (17).

Macrophage Colony-Stimulating Factor: Augmentation of Systemic Host Defenses

To better understand the immunopathogenesis of disseminated candidiasis, we compared the antifungal activities of macrophages residing in the spleen, liver, and lungs of rabbits against blastoconidia and hyphae of *C. albicans* (20). We found that splenic macrophages, Kupffer cells, and pulmonary alveolar macrophages were similar in their phagocytic capacity but differed in their capacity to damage *Candida* hyphae, in the order splenic macrophages > Kupffer cells > pulmonary alveolar macrophages. M-CSF significantly activated phagocytosis and growth inhibition of these cell populations to a similar degree while having a greater effect on the ability of splenic macrophages to damage unopsonized hyphae.

Cytokines and Gamma Interferon in Treatment of Refractory Chronic Disseminated Candidiasis

The laboratory findings described above collectively support the concept that chronic disseminated candidiasis is a disease of Th1/Th2 dysimmunoregulation. Thus, as logical extensions of our laboratory investigations, we recently reported the use of recombinant gamma interferon and GM-CSF in the successful treatment of chronic disseminated candidiasis refractory to available antifungal chemotherapy (6). Although larger clinical trials of the immunotherapeutic management of chronic disseminated candidiasis are desirable, such studies are unlikely due to the resources involved in conducting such a trial. Instead, carefully selected patients who are unresponsive to conventional antifungal therapy should be monitored carefully while receiving a rationally selected immunomodulatory agent for this often tenacious form of disseminated candidiasis.

REFERENCES

1. **Anaissie, E., G. P. Bodey, H. Kantarjian, C. David, K. Barnett, E. Bow, R. Defelice, N. Downs, T. File, G. Karam, D. Potts, M. Shelton, and A. Sugar.** 1991. Fluconazole therapy for chronic disseminated candidiasis in patients with leukemia and prior amphotericin B therapy. *Am. J. Med.* **91:**142–150.
2. **Barral, A., M. Barral-Netto, E. C. Yong, C. E. Brownell, D. R. Twardzik, and S. G. Reed.** 1993. Transforming growth factor beta as a virulence mechanism for *Leishmania braziliensis. Proc. Natl. Acad. Sci. USA* **90:**3442–3446.
3. **Cassatella, M. A.** 1998. The neutrophil: one of the cellular targets of interleukin-10. *Int. J. Clin. Lab. Res.* **28:**148–161.
4. **Cenci, E., L. Romani, A. Mencacci, R. Spaccapelo, E. Schiaffella, P. Puccetti, and F. Bistoni.** 1993. Interleukin-4 and interleukin-10 inhibit nitric oxide-dependent macrophage killing of *Candida albicans. Eur. J. Immunol.* **23:**1034–1038.
5. **Choi, E. H., C. B. Foster, J. G. Taylor, H. C. Erichsen, R. A. Chen, T. J. Walsh, V. J. Anttila, T. Ruutu, A. Palotie, and S. J. Chanock.** 2003. Association between chronic disseminated candidiasis in adult acute leukemia and common IL4 promoter haplotypes. *J. Infect. Dis.* **187:**1153–1156.
6. **Dignani, M. C., J. H. Rex, K. W. Chan, G. Dow, M. deMagalhaes-Silverman, A. Maddox, T. Walsh, and E. Anaissie.** 2005. Immunomodulation with interferon-gamma and colony-stimulating factors for refractory fungal infections in patients with leukemia. *Cancer* **104:**199–204.
7. **Katsifa, H., S. Tsaparidou, E. Diza, C. Gil-Lamaignere, T. J. Walsh, and E. Roilides.** 2001. Effects of interleukin-13 on antifungal activity of human monocytes against *Candida albicans. FEMS Immunol. Med. Microbiol.* **31:**211–217.
8. **Kauffman, C. A., S. F. Bradley, S. C. Ross, and D. R. Weber.** 1991. Hepatosplenic candidiasis: successful treatment with fluconazole. *Am. J. Med.* **91:**137–141.
9. **Kim, H. S., E. H. Choi, J. Khan, E. Roilides, A. Francesconi, M. Kasai, T. Sein, R. L. Schaufele,**

K. Sakurai, C. G. Son, B. T. Greer, S. Chanock, C. A. Lyman, and T. J. Walsh. 2005. Expression of genes encoding innate host defense molecules in normal human monocytes in response to *Candida albicans*. *Infect. Immun.* **73:**3714–3724.

10. Kontoyiannis, D. P., M. A. Luna, B. I. Samuels, and G. P. Bodey. 2000. Hepatosplenic candidiasis. A manifestation of chronic disseminated candidiasis. *Infect. Dis. Clin. North Am.* **14:**721–739.

11. Letterio, J. J., T. Lehrnbecher, G. Pollack, T. J. Walsh, S. J. Chanock, J. J. Letterio, and A. B. Roberts. 2001. Invasive candidiasis stimulates hepatocyte and monocyte production of active transforming growth factor beta. *Infect. Immun.* **69:**5115–5120.

12. Letterio, J. J., and A. B. Roberts. 1998. Regulation of immune responses by TGF-β. *Annu. Rev. Immunol.* **16:**137–161.

13. Marcus, S. G., T. J. Walsh, P. A. Pizzo, and D. N. Danforth, Jr. 1993. Hepatic abscess in cancer patients. Characterization and management. *Arch. Surg.* **128:**1358–1364; discussion, 1364.

14. Puccetti, P., A. Mencacci, E. Cenci, R. Spaccapelo, P. Mosci, K.-H. Enssle, L. Romani, and F. Bistoni. 1994. Cure of murine candidiasis by recombinant soluble interleukin-4 receptor. *J. Infect. Dis.* **169:**1325–1331.

15. Puccetti, P., L. Romani, and F. Bistoni. 1995. A Th1-Th2-like switch in candidiasis: new perspectives for therapy. *Trends Microbiol.* **3:**237–240.

16. Roilides, E., A. Anastasiou-Katsiardani, A. Dimitriadou-Georgiadou, I. Kadiltsoglou, S. Tsaparidou, C. Panteliadis, and T. J. Walsh. 1998. Suppressive effects of interleukin-10 on human mononuclear phagocyte function against *Candida albicans* and *Staphylococcus aureus*. *J. Infect. Dis.* **178:**1734–1742.

17. Roilides, E., A. Holmes, C. Blake, P. A. Pizzo, and T. J. Walsh. 1995. Effects of granulocyte colony-stimulating factor and interferon-gamma on antifungal activity of human polymorphonuclear neutrophils against pseudohyphae of different medically important *Candida* species. *J. Leukoc. Biol.* **57:**651–656.

18. Roilides, E., I. Kadiltsoglou, A. Dimitriadou, M. Hatzistilianou, A. Manitsa, J. Karpouzas, P. A. Pizzo, and T. J. Walsh. 1997. Interleukin-4 suppresses antifungal activity of human mononuclear phagocytes against *Candida albicans* in association with decreased uptake of blastoconidia. *FEMS Immunol. Med. Microbiol.* **19:**169–180.

19. Roilides, E., H. Katsifa, S. Tsaparidou, T. Stergiopoulou, C. Panteliadis, and T. J. Walsh. 2000. Interleukin-10 suppresses phagocytic and antihyphal activities of human neutrophils. *Cytokine* **12:**379–387.

20. Roilides, E., C. A. Lyman, T. Sein, C. Gonzalez, and T. J. Walsh. 2000. Antifungal activity of splenic, liver and pulmonary macrophages against *Candida albicans* and effects of macrophage colony-stimulating factor. *Med. Mycol.* **38:**161–168.

21. Roilides, E., T. Sein, R. Schaufele, S. J. Chanock, and T. J. Walsh. 1998. Increased serum concentrations of interleukin-10 in patients with hepatosplenic candidiasis. *J. Infect. Dis.* **178:**589–592.

22. Roilides, E., and T. J. Walsh. 2004. Recombinant cytokines in augmentation and immunomodulation of host defenses against *Candida* spp. *Med. Mycol.* **42:**1–13.

23. Romani, L., F. Bistoni, A. Mencacci, E. Cenci, R. Spaccapelo, and P. Puccetti. 1995. IL12 in *Candida albicans* infections. *Res. Immunol.* **146:**532–538.

24. Romani, L., F. Bistoni, and P. Puccetti. 1997. Initiation of T-helper cell immunity to *Candida albicans* by IL-12: the role of neutrophils. *Chem. Immunol.* **68:**110–135.

25. Romani, L., E. Cenci, A. Menacci, F. Bistoni, and P. Puccetti. 1995. T helper cell dichotomy to *Candida albicans*: implications for pathology, therapy, and vaccine design. *Immunol. Res.* **14:**148–162.

26. Spaccapelo, R., L. Romani, L. Tonnetti, E. Cenci, A. Mencacci, G. Del Sero, R. Tognellini, S. G. Reed, P. Puccetti, and F. Bistoni. 1995. TGF-β is important in determining the in vivo patterns of susceptibility or resistance in mice infected with *Candida albicans*. *J. Immunol.* **155:**1349–1360.

27. Thaler, M., B. Pastakia, T. H. Shawker, T. O'Leary, and P. A. Pizzo. 1988. Hepatic candidiasis in cancer patients: the evolving picture of the syndrome. *Ann. Intern. Med.* **108:**88–100.

28. Tonnetti, L., R. Spaccapelo, E. Cenci, A. Mencacci, P. Puccetti, R. L. Coffman, F. Bistoni, and L. Romani. 1995. Interleukin-4 and -10 exacerbate candidiasis in mice. *Eur. J. Immunol.* **25:**1559–1565.

29. Vazquez, N., T. J. Walsh, D. Friedman, S. J. Chanock, and C. A. Lyman. 1998. Interleukin-15 augments superoxide production and microbicidal activity of human monocytes against *Candida albicans*. *Infect. Immun.* **66:**145–150.

30. Vodovotz, Y., C. Bogdan, J. Paik, Q. W. Xie, and C. Nathan. 1993. Mechanisms of suppression of macrophage nitric oxide release by transforming growth factor beta. *J. Exp. Med.* **178:**605–613.

31. Wahl, S. M., J. B. Allen, B. S. Weeks, H. L. Wong, and P. E. Klotman. 1993. Transforming growth factor beta enhances integrin expression and type IV collagenase secretion in human monocytes. *Proc. Natl. Acad. Sci. USA* **90:**4577–4581.

32. Walsh, T. J., S. Aoki, F. Mechinaud, J. Bacher, J. Lee, M. Rubin, and P. A. Pizzo. 1990. Effects of preventive, early, and late antifungal chemotherapy with fluconazole in different granulocytopenic models of experimental disseminated candidiasis. *J. Infect. Dis.* **161:**755–760.

33. Walsh, T. J., N. L. Seibel, C. Arndt, R. E. Harris, M. J. Dinubile, and A. Reboli. 1999. Amphotericin B lipid complex in pediatric patients with invasive fungal infections. *Pediatr. Infect. Dis. J.* **18:**702–708.

34. Walsh, T. J., P. O. Whitcomb, S. G. Revankar, and P. A. Pizzo. 1995. Successful treatment of hepatosplenic candidiasis through repeated cycles of chemotherapy and neutropenia. *Cancer* **76:**2357–2362.

V. FUTURE DIRECTIONS

Molecular Principles of Fungal Pathogenesis
Edited by Joseph Heitman et al.
©2006 ASM Press, Washington, D.C.

Chapter 40

Studying Fungal Virulence by Using Genomics

MICHAEL C. LORENZ

The ongoing flood of fungal genome sequences, unimaginable only a few years ago, combined with the development of sensitive and high-throughput technologies to take advantage of these data, makes this a very propitious time for mycologists. Long used in studies of the model yeast *Saccharomyces cerevisiae*, genome-scale approaches are now available for many fungi, including pathogens of both plants and animals. Thus, genomics can now be used to probe fungal pathogenesis, adding a powerful new tool to the investigation of this difficult problem. Investigators using genomic approaches as part of their pathogenesis studies have the opportunity to make significant and rapid advances toward a more complete understanding of these remarkable organisms.

However, we are not quite at the point where genome-scale approaches can easily be applied in most species. Genome sequence is emerging at ever higher rates, but this is only the first of many long steps to making full use of genomic technologies. The smaller size of the fungal pathogenesis community, compared to those for *Saccharomyces* and *Schizosaccharomyces*, limits the number of hands available to build the tools and resources necessary. Thus, while we can see the potential inherent in genomics, there are many obvious challenges to overcome. In this and subsequent chapters the advantages and disadvantages of genomics, large-scale genetics, and proteomics are discussed, focusing on a small but diverse group of organisms.

As I hope this chapter will make evident, genomics, in particular genomic microarrays, hold great promise for mycological research. To this end, the chapter covers the current state of fungal genome sequencing and the advantages of and obstacles to using this emerging data for genomic projects, the "pregenomic" methods used for finding in vivo-expressed genes and how these efforts provide intellectual support for expression profiling studies, and a

discussion of options for microarray development. It reviews a few selected microarray-based studies directly relevant to fungal pathogenesis. Most importantly, it addresses a series of challenges that must be overcome to take full advantage of these technologies. Readers are cautioned that this is a fast-moving field and that information regarding commercial array suppliers and genome-sequencing status, in particular, is changing rapidly: during the final preparations of this chapter, two genome sequence papers went into press and the sequences of five additional species were released.

WHY GENOMICS?

Many investigators who work primarily with fungal pathogens are converts from the *S. cerevisiae* research community. To those used to the genetic tools and approaches available for this model yeast, working with fungal pathogens can be a frustrating experience. While substantial advances in technology are closing the gap, there remain significant limitations in the molecular genetics of these organisms. *Cryptococcus neoformans* is perhaps the best developed, with a defined sexual cycle, episomes, a homologous recombination/gene disruption system, and at least a partly complete genome sequence, although these tools remain inefficient in comparison to *S. cerevisiae*. The reasonably advanced molecular genetics of *Candida albicans* is still complicated by its diploidy and a lack of episomal vectors. At the other end of the spectrum, the life cycle of *Coccidioides* is still a bit mysterious and there is not a long-term in vitro culture system for *Pneumocystis carinii*. In general, the means for genetic manipulation in pathogenic fungi is still rudimentary, and this has greatly impacted our study of these organisms.

Regardless of the molecular genetic tools available for any given species, it has always been impractical to

Michael C. Lorenz • Department of Microbiology and Molecular Genetics, The University of Texas Health Science Center, 6431 Fannin, Houston, TX 77030.

do classical genetic screens for "virulence." The number of animals required for anything remotely saturating is prohibitive, both ethically and fiscally. Newer technologies, such as signature-tagged mutagenesis, allowing the testing of large pools of mutants, and the availability of simple model hosts (both described elsewhere in this volume) provide opportunities for in vivo screens but, to date, have been employed in limited fashion for only a few species. Instead, most studies focus on virulence-related phenotypes, such as adhesion, stress resistance, or morphogenetic processes. This, of course, requires that the investigator make assumptions about what processes are important in vivo and convert that to an appropriate in vitro test. The promise of genomics is the ability to screen for genes of interest in organisms for which genetic tools are limited—to let the organism tell us what processes are important. Since we understand so little about fungal pathogenesis, in particular, such (relatively) unbiased screening methods are critical to our further understanding of virulence.

FUNGAL GENOME SEQUENCING

In May 1992, Oliver et al. published the complete sequence of chromosome III of *S. cerevisiae* (41).

Reporting the first eukaryotic chromosome to be fully sequenced, this paper was a landmark that presaged the impending genomic sequencing revolution. Still, it is hard to believe that the *S. cerevisiae* genome project concluded less than a decade ago. The blistering pace of technology development has taken the work of sequencing a genome from a Herculean effort requiring years of effort by large consortia to one that can be done by a single center in, quite literally, a matter of days. In parallel, new technologies have emerged to make full use of these data, from genome databases to whole-genome deletion libraries and microarrays.

To the great benefit of those interested in fungal pathogens, this accelerated pace means that the genome sequences of most species of interest are either already available or are in progress. Tables 1 and 2 show a comparison of the genomes of a number of species for which at least draft sequences exist including both published and unpublished work. As can be seen, there are significant differences in the genome size and gene number, confirming the diversity of fungal species. The Genomes OnLine Database (GOLD; http://www.genomeonline.org) lists no fewer than 75 species for which there exist

Table 1. Published fungal genome sequences

Organism	Strain sequenced	Genome size[a] (Mb)	Estimated no. of genes	Completion date[b]	Primary website(s)	Reference(s)
S. cerevisiae	S288c	12.1	5,749	1996	www.yeastgenome.org	21
Other Saccharomyces spp.	—[c]			2003	genome.wustl.edu/projects/yeast www.broad.mit.edu/annotation/ fungi/comp_yeasts	10, 26
S. pombe	972	12.5	4,940	2002	www.genedb.org/genedb/pombe	62
N. crassa	OR74A	38.0	10,082	2003	www.broad.mit.edu/resources.html	20
Abshbya gossypii	ATCC 10895	9.2	4,178	2004	agd.unibas.ch	14
C. albicans	SC5314	14.9[d]	6,419	2004	www.stanford.edu/group/candida www.candidagenome.org candida.bri.nrc.ca/candida http://genolist.pasteur.fr/CandidaDB/	4, 24
C. glabrata	CBS138	12.3	5,283	2004	cbi.labri.fr/Genolevures/	16
Kluyveromyces lactis	CLIB210	10.6	5,329	2004	cbi.labri.fr/Genolevures/	16
Debaromyces hansenii	CBS767	12.2	6,906	2004	cbi.labri.fr/Genolevures/	16
Yarrowia lipolytica	CLIB99	20.5	6,703	2004	cbi.labri.fr/Genolevures/	16
Phanerochaete chrysosporium	RP-78	29.6	11,777	2004	genome.jgi-psf.org/whiterot1	37
M. grisea	70-15	38.8	11,108	2005	www.broad.mit.edu/resources.html	12
C. neoformans serotype D	JEC21 B-3501A	19.0 18.5	6,572	2005	www.tigr.org/tdb/tgi/fungi.shtml	32
A. nidulans	FGSC A4	31.0	9,457	2005	http://www.broad.mit.edu/resources.html	20a
A. fumigatus	Af293	29.4	9,926	2005	http://www.tigr.org/tdb/e2kl/aful/new.shtml http://www.sanger.ac.uk/Projects/ A_fumigatus	40a
A. oryzae	RIB40	37.0	12,074	2005		35a

[a]Genome size generally excludes the rDNA repeats.

[b]This is the date of publication unless otherwise indicated.

[c]The two groups partially sequenced the related species *S. castellii*, *S. kluyveri*, *S. mikatae*, *S. bayanus*, *S. kudriavzevii*, and *S. paradoxus* for the purpose of comparative analysis with *S. cerevisiae*.

[d]Estimated haploid genome size.

Table 2. Selected unpublished fungal sequences of medical or agricultural importance

Organism	Strain sequenced	Genome size[a] (Mb)	Estimated no. of genes	Primary website(s)
Fusarium graminearum	PH-1 (NRRL 31084)	36.1	11,640	www.broad.mit.edu/resources.html
U. maydis	521	19.7	6,522	www.broad.mit.edu/resources.html
C. neoformans serotype A	H99	24.0		www.broad.mit.edu/resources.html
C. immitis	RS	28.8		www.broad.mit.edu/resources.html
C. posadasii	C735			www.tigr.org/tdb/tgi/fungi.shtml
H. capsulatum	G217B	23–25		www.genome.wustl.edu/projects/hcapsulatum/
C. tropicalis	MYA-3404	14.5 (15)[b]		www.broad.mit.edu/resources.html
C. guilliermondii	ATCC6260	10.6 (12)[b]		www.broad.mit.edu/resources.html
C. lusitaniae	ATCC42720	12.1 (16)[b]		www.broad.mit.edu/resources.html
C. dubliniensis	CD36	16.1		http://www.sanger.ac.uk/Projects/C_dubliniensis/
C. neoformans serotype B	R265	17.2 (20)[b]		www.broad.mit.edu/resources.html
Chaetomium globosum	CBS 148.51	34.3 (36)[b]		www.broad.mit.edu/resources.html
Rhizopus oryzae	RA99–880	39.6 (40)[b]		www.broad.mit.edu/resources.html

[a]Genome size generally excludes the rDNA repeats.

[b]The first number is the amount of sequence currently available; in parentheses is the estimated total (haploid) genome size.

either a genome-sequencing project or a large-scale expressed sequence tag EST project (or, sometimes, both). For many species, notably *C. albicans*, *C. neoformans*, and *Aspergillus fumigatus*, multiple strains are being sequenced.

What Does "Complete Genome Sequence" Mean?

The process by which chromosome III and the rest of the *S. cerevisiae* genome was sequenced produced extremely high-quality sequence data, in terms of both accuracy and completeness. Starting from the construction of a detailed physical map, overlapping clones were sequenced to generate virtually complete chromosome sequence from telomere to telomere, with few gaps. From the perspective of current genome projects, the *S. cerevisiae* effort was almost absurdly laborious, and the physical mapping approach has given way to shotgun sequencing in which the genome, in its entirety, is fragmented into short pieces and cloned between universal primer sites in a standard vector. Libraries of sufficient diversity to substantially overrepresent the genome are constructed and sequenced.

Shotgun sequencing in this manner is extremely fast and relatively inexpensive. Famously, the U.S. Department of Energy Joint Genome Institute, in collaboration with the Baylor College of Medicine Genome Sequencing Center, completed the sequencing of the 2.8-Mb *Enterococcus faecium* genome in a single day in 2000. The 40-Mb *Neurospora crassa* genome was sequenced in 11 days by the Whitehead/Massachusetts Institute of Technology Center for Genome Research (now the Broad Institute). Raw sequence reads, however, do not

make a genome, and assembling large sequence contigs remains a slow and iterative process. When finished, the assembled contigs generally fall short of being complete chromosomes (although they can be hundreds of kilobases). The difficulty of assembly can be seen in the fact that the *E. faecium* genome has not been formally published although the data have been in the public domain for 4 years.

An example of the shotgun sequencing and assembly process is the *Magnaporthe grisea* project at the Broad Institute. Currently, the draft sequence (12) exists in 159 supercontigs averaging 243 kb. This accounts for 39 Mb of assembled sequence out of an estimated genome size of 40 Mb, for a 97% coverage rate. Unfortunately, however, there are no plans to continue sequencing or assembly, and so unless interested researchers pick up from this point, there will not be a "complete" genome in the sense of end-to-end sequence. This is true for most projects now under way or recently finished.

The above should not be construed as criticism of the sequencing efforts. On the contrary, given the limited resources likely to be expended on fungal genomes, most mycologists would prefer to have 20 species that are 95% complete rather than 5 "finished" genomes. There are probably some interesting genes in the 1 Mb of the *Magnaporthe* sequence that has been missed, but the existing sequence contains 11,108 (predicted) genes, which should occupy the *Magnaporthe* community for some time.

Genome Annotation

A second challenge is gene annotation. Most mycologists have used the Saccharomyces Genome Database (SGD; http://www.yeastgenome.org) at some point and will appreciate that the development of that very valuable resource involved a significant effort by many people and, importantly, requires continual maintenance to keep it up to date. A critical but sometimes overlooked part of the ongoing curation is as a clearinghouse for gene naming, open reading frame (ORF) calls, and functional classification, thus standardizing the annotation of the yeast genome.

For species with a smaller research community and less molecular information about gene function (including all fungal pathogens), it is even more important to develop a genome annotation in a cooperative manner. Most sequencing projects will include a semiautomated process of ORF prediction, leading to a list of putative genes that meet certain criteria (ORF length, codon bias, homology to known genes, etc.). They will not, in most cases, attempt to name or categorize these genes beyond a list of BLAST hits. It is thus tempting for individual groups to produce a more detailed annotation to support their genomic projects. Such was the case with the *C. albicans* sequencing project, in which no fewer than six groups developed independent annotations in conjunction with array construction. Each annotation predicted and named genes in different ways, and it was often not easy to compare genes among them. In the end, this unworkable scenario was evident to all and resulted in a community-supported effort to develop a single annotation, which has been published recently (4) (see http://candida.bri.nrc.ca for details), and a database based on SGD (http://www.candidagenome.org; several other useful databases exist as well [Table 1]). Although the outcome is positive, an earlier recognition of the existence of multiple independent projects would certainly have reduced the duplication of effort and produced an annotation sooner.

In summary, mycologists are being presented with a great resource in the plethora of fungal genome sequences now (or soon to be) available. Capitalizing on this resource requires a community-based effort to develop a standardized annotation, including a gene list, predicted start-stop sites, nomenclature, and functional categorization, when possible. Further, the experience with the *S. cerevisiae* genome, which is being continually updated as new data emerge, emphasizes the need for a process to determine how changes are adopted and a clearinghouse through which they are disseminated. Without these resources, genome sequence data will be underutilized.

EXPRESSION ANALYSIS AS A TOOL TO UNDERSTAND FUNGAL PATHOGENESIS

The underlying principle behind analyzing genes expressed in vivo to understand pathogenicity is that microorganisms generally express genes only when they are likely to be required for optimal growth or survival in a particular condition. Applied to virulence, this assumes that there is a unique set of genes preferentially expressed in the context of infection (e.g., genes that are not expressed during growth in vitro) and, further, that these genes contribute to some aspect of disease progression, just as genes induced by uracil depletion usually have functions in nucleotide biosynthesis. For the medical mycologist, these markers are then useful as molecular diagnostics or targets for therapeutic intervention. For the biologist, the genes tell us something about the nature of life within the host and about what is necessary for survival in this complex and rapidly changing environment.

Early Approaches

Several random genetic approaches have been developed to identify genes expressed preferentially in vivo. Although these are not "genomic," per se, because they do not require genome sequence information, they are important tools and their success in finding in vivo-expressed genes (many of which affect virulence) supports the development of microarray-based technologies for pathogenesis studies. A summary of these techniques is given in Table 3.

In vivo expression technology

In vivo expression technology (IVET) is a promoter-trap system that is used to find genes expressed only in vivo. It was developed for *Salmonella* out of the observation that *purA* mutants (deficient in purine metabolism) are avirulent. Random genetic fragments were cloned upstream of a promoterless artificial operon containing the *purA* and *lacZ* genes and integrated into a *purA Salmonella* mutant (36). Bacteria that persist in mice after injection must contain a construct in which *purA* is expressed; hence, the genomic fragment must have a promoter that is active in vivo. Plating recovered colonies in vitro on MacConkey indicator medium and looking for white colonies (in which *lacZ* is not expressed and, hence, the promoter is not active) can identify in vivo-specific promoters.

For fungal pathogens, IVET has been used for screening purposes only with *Histoplasma capsulatum* (48). Retallack et al. (48) adapted the bacterial system by using a construct including a promoterless *URA5* gene and a hygromycin resistance marker. Random genomic fragments were placed upstream of *URA5* either by integrating the linear construct through nonhomologous recombination or through shotgun cloning of short DNA pieces in a plasmid. Strains prototrophic for uracil in vitro (e.g., the genomic fragment inserted upstream of *URA5* contains a promoter active

Table 3. Comparison of expression-based approaches

Technique	Genome sequence?[a]	Means of detection	Comments
IVET	None	Promoter activity	Good reporters and integration/plasmid system not available in many species.
IVIAT	None	Antibody presence	Ideal for finding vaccine targets, but need source of sera and might miss many important but nonantigenic proteins.
dd-RTPCR	None	mRNA abundance	Well established, if complex, molecular biology. Not saturating, but has found many valid targets.
SAGE	Moderate (ESTs)	mRNA abundance	Need substantial sequence data to make sense of results. Does not require specialized equipment. Can be saturating.
Microarrays	High	mRNA abundance	Large upfront investment in equipment and reagents. Saturating and high-throughput. Data analysis can be challenging.

[a]Degree to which the system requires genome sequence data.

in vitro) were eliminated from the pool. These strains were then tested either in whole animals or in a cell culture model of *Histoplasma* infection. Uracil auxotrophs are inviable in both models, and so surviving strains have promoter fragments active in vivo. Several in vivo-specific promoters were identified, although the genes and functions have not yet been reported.

IVET has also been used with *C. albicans*, not as a screening tool but as a means of characterizing the in vivo expression of specific genes by using a recombination-based system (7) adapted for *C. albicans* by integrating a mycophenolic acid resistance (MPAr) gene surrounded by FLP recognition sites (54). The FLP recombinase is fused with the promoter of genes of interest; active promoters express FLP, and the cell loses the MPAr gene via FLP-mediated recombination. Several of the SAP secreted aspartyl proteases were tested in animal models, allowing characterization of the temporal and spatial regulation of these genes in vivo (54).

IVET has several drawbacks that limit its utility as a screening tool in many fungal pathogens. First, fungi have more genes, more noncoding DNA, and many more promoters than bacteria do, and so libraries must be much larger to be saturating. For many fungi (*C. albicans*, for instance), integrative transformation is not efficient enough to make a complete library. It also requires growth in vitro, which is problematic in other fungi (e.g., *Pneumocystis*). The reporter gene must be required for persistence in vivo, but if very high (or precise) expression is necessary, it may not be very sensitive, such as for *URA3* in *C. albicans*.

In vivo-induced antigen technology

Infected patients develop antibodies to exposed antigens on the pathogen. In vivo-induced antigen technology (IVIAT) uses these antibodies to identify proteins expressed in vivo. Sera from infected patients are pooled and then adsorbed to in vitro-grown

pathogen cells and cell lysates, removing antibodies to most cellular proteins. The sera are then used to probe an expression library of the organism's genes in *Escherichia coli*. Positive hits are presumably present in vivo, since the patients have made antibodies to them, but are not present in vitro, since these antibodies would have been removed during the adsorption step. IVIAT has the advantage of finding exposed antigens (since antibodies to the protein are present), which can be particularly useful if the goal is a vaccine candidate.

Since the system is based on antigenicity of the target proteins, proteins that are not expressed highly, not expressed on the cell surface, or masked by oligosaccharides may not be seen. It also requires the identification and enrollment of suitable patients in the study. Alternatively, sera can be obtained from experimental animals, but the spectrum of antigens recognized by human and animals is not always the same.

Sera from patients with oral thrush (due to *C. albicans*) were used to identify in vivo-expressed genes by this technology. Among the 10 genes found were those encoding known virulence determinants (including HWP1 and CST20) and several novel proteins, one of which, NOT5, was shown to reduce adhesion and virulence when mutated (9).

Differential display

In differential display (also called differential-display reverse transcription-PCR [dd-RTPCR]), cDNA is made from two populations of cells, labeled, and directly compared to identify differences in abundance. Since there are thousands of mRNA species in each cell, this is done in multiple pools with different combinations of oligo(dT)-anchored and random primers to reduce the complexity in any one pool. Differential display is a complicated technique and has been reviewed in detail elsewhere (57). Its primary disadvantage is the difficulty of treating the control and experimental populations exactly equally. Since there are several amplification steps, even small differences in starting quantities, mRNA quality, or primer abundance can skew the final results, and it is essentially impossible to do this in a saturating manner.

dd-RTPCR has been used successfully to find in vivo-expressed genes in fungal pathogens. Two examples from Maresca's group are illustrative. In the first, they identified *C. albicans* genes preferentially expressed after macrophage phagocytosis, finding a number of genes related to alternative carbon utilization and energy metabolism (44), a spectrum similar to that found in a similar model by microarray analysis (33, 34). The same laboratory also used dd-RTPCR for *H. capsulatum* in contact with macrophages (11); one gene identified encodes a homolog of the human p100 coactivator protein, which plays a role in macrophage activation (43).

Serial analysis of gene expression

Like differential display, serial analysis of gene expression (SAGE) is a simple idea backed by complicated molecular biology. It has been described elsewhere (60), but the basic idea is that cDNA is synthesized from the poly(A) RNA population, enzymatically fragmented into short (9- to 13-bp) pieces, and ligated into concatemers to be sequenced. The number of times a tag appears is proportional to its abundance. BLAST searches determine which gene the tag came from. Hence, at least some sequence information from the organism is needed to ascribe the tag to a longer, recognizable sequence, or the SAGE data will not be useful (see below). The sequence need not be genomic, however; an EST database is adequate. The details of the molecular biology of SAGE have been described elsewhere (43a).

The Kronstad laboratory has used SAGE to identify genes regulated by temperature or growth in vivo in *C. neoformans*. SAGE was performed with both H99 (serotype A) and B3501 (serotype D) strains at 25 and 37°C. In an initial report (published in 2002), 29 (58%) of the top 50 genes upregulated at 37°C did not match any sequence then available, and it was therefore impossible to predict how the genes represented by these tags were involved in temperature-regulated processes (55). A year—and a lot of sequencing—later, an expanded study that also included cells recovered from the cerebrospinal fluid of infected rabbits found hits for all but 22% of the SAGE tags present exclusively in vivo (56). Clearly, then, substantial sequence resources are necessary to make full use of the SAGE technology.

Microarray Technology

Although the technology is less than 10 years old, whole-genome DNA microarrays are now sufficiently commonplace that a brief description will suffice. The principles are the same as for Northern hybridizations—a nucleotide probe will bind to total cellular RNA in a sequence-specific manner. As long as the probe is in excess, binding is proportional to the concentration of the corresponding mRNA in the population. By labeling one of the nucleic acids, one can quantify the mRNA level in one population relative to another. It is therefore a means of determining whether gene expression is altered by environmental or genetic changes. At their core, both Northern analysis and microarray technologies rely on the assumption that altered expression is a sign of biological relevance, similar to the systems described in the previous section.

In a Northern blot, the investigator selects a single gene of interest as the probe. This highlights its biggest drawbacks: it can examine only one preselected gene at a time. Thus, Northern analysis is not useful as a means of finding unknown genes regulated by a particular manipulation. Genomic microarrays can determine the relative abundance of thousands of mRNA species at a time, potentially the entire genome. Because no assumptions are made about which genes are interesting, microarrays can be used for the initial identification of genes of interest on a scale that is close to saturating.

In a genomic microarray, gene fragments are immobilized on a substrate (types of arrays are discussed below) such that a large number of genes (thousands) are arrayed in a known pattern. An important distinction from Northern analysis is that microarray hybridization labels the cellular RNA (or, rather, cDNA made from total mRNA) with fluorescent dyes that emit at different wavelengths, rather than a single gene probe. Two labeled pools, prepared from different cell populations, are combined and hybridized. Fluorescence is detected via a scanner that excites the fluorophores and quantitates the resulting signal intensity of both dyes at each spot on the array (each spot corresponding to a single gene). The difference in signal between the two populations is proportional to the difference in abundance of the corresponding mRNA.

A schematic for microarray experiments is presented in Color Plate 8.

Commercial sources of arrays

DNA microarray projects present many hurdles, although many of these are being overcome as the technology spreads. First and foremost is microarray construction. The only fungal pathogen for which arrays are available "off the shelf" as of this writing is *C. albicans*. Eurogentec sells a ~6,000-gene microarray developed in collaboration with the Galar

Fungail consortium (www.eurogentec.be). A ~6,800-ORF microarray is available though the Microarray Laboratory of the Biotechnology Research Institute of the National Research Council of Canada (www.bri.nrc.gc.ca/business/microarraylab). Finally, Qiagen/Operon sells a ~6,400-ORF oligonucleotide set that can be used by investigators to print microarrays themselves (www.operon.com). This company also sells an array set for *M. grisea*. It is beyond the purview (and wisdom) of this chapter to assess the quality of each of these reagents, but it should be noted that all were constructed prior to the development of a standardized community-based annotation. Some of this can be seen in the difference in total gene number, and this adds a complication to comparison of data collected on different platforms.

Microarrays exist for other fungal pathogens (*Histoplasma*, *Cryptococcus*, and *C. glabrata*, among others), but they have been made by academic consortia and are not commercially available. Several companies are offering custom array services in which they will work with clients to build complete genome arrays. While utilizing the expertise and quality control of professional services often makes sense, this obviously comes at a high cost.

Types of microarrays

Because there is no commercial source for microarrays for most pathogenic fungi, investigators interested in pursuing this type of research will have to take a leading role in array construction, even if parts of the project are to be contracted out. For this reason, a description of the options and variables for array construction is warranted. There are multiple platforms for microarray analysis, all based on the principles

described above. These differ in the array substrate and the size and synthesis of the gene fragments spotted. This, of course, leads to differences in the cost, availability, sensitivity, and experimental designs supported. A comparison is shown in Table 4, and they are described below. Many readers will be familiar with the array format made by Affymetrix; these arrays are described at the end of this section for comparison, although their product line does not currently include any pathogenic fungi.

Substrate. Arrays have been printed on glass microscope slides and on nylon membranes. The former is by far the most common. Spotting on nylon is effective and has been used for *C. albicans* genomic projects (19, 39), but the technology for both spotting DNA and analyzing the results is less well developed. In addition, probes are labeled radioactively, which many investigators prefer to avoid. Scanning membrane arrays can be done using a phosphorimager, however, reducing the need for specialized equipment.

Glass slides are the substrate of choice for most applications. A 1- by 3-cm microscope slide has the theoretical capacity to print any fungal genome in duplicate. Initially, slides were treated with polylysine to facilitate DNA binding to the glass. Currently, homemade polylysine treatments are being superseded by commercially available slides with a variety of surface coatings that are generally more uniform and are associated with less background. Examples include Amersham's Codelink and Corning's GAPS slides. Detection is through fluorescent labeling. The equipment necessary to print and scan glass slide arrays is both specialized and expensive but is increasingly available in university core facilities.

Table 4. Relevant features of microarray types

Array Type	DNA probes	Array construction	Made where?	Labeling	Advantages	Disadvantages
Glass slides	PCR products; oligonucleotides	DNA spotted on substrate	In-house or contract	Fluorescent	Flexible, amenable to in-house projects. Equipment readily available.	Quality control (particularly in-house). Expense of array construction and equipment.
Filters	PCR products	DNA spotted on substrate	Mostly in-house	Radioactive	Inexpensive. Requires little specialized equipment.	Radioactive detection. Limited sensitivity and capacity on filters.
Affymetrix	Oligonucleotides	Direct synthesis on substrate	Proprietary/ contract	Fluorescent	Sensitive. Excellent statistical models.	High up-front cost. Limited control of probe sets.
Nimblegen	Oligonucleotides	Direct synthesis on substrate	Proprietary/ contract	Fluorescent	Custom arrays feasible. High feature density.	Higher cost than slide arrays.

Spotted DNA. There are two basic options for the DNA that is spotted on the array: PCR products and oligonucleotides. The first DNA microarrays used PCR products amplified with gene-specific primers. The primers can be short (~20-mers) and are therefore inexpensive. As anyone who has ever done large-scale PCR can attest, however, getting consistent success and yields is a rarely achieved goal. The setup, analysis, and cleanup of the PCR amplifications is labor-intensive, but the process can make virtually limitless amounts of DNA. Because of variations in yield, however, there can be significant differences in the concentration of DNA per spot, which affects both signal intensity and uniformity of spotting (all other things equal, solutions of different DNA concentrations will make different-sized spots on the slide).

Increasingly, the use of long oligonucleotides (~70-mers) is the preferred choice. Advantages include uniform concentration and optimized hybridization parameters (such as melting temperature and complexity). Further, oligonucleotides to divergent regions of gene families can be designed, to discriminate between members. Although PCR primers specific for single family members can be designed, they often amplify conserved stretches that cross-hybridize. Long oligonucleotide arrays are less sensitive because the DNA fragments are shorter, and so more starting material may be necessary to obtain a good signal. Oligonucleotides 70 bases long are substantially more expensive than a pair of 20-mers, but the cost difference is decreasing and the cost can be quite reasonable as part of a commericially available oligonucleotide set, although these are currently available only for *C. albicans* and *M. grisea* (from Qiagen/Operon).

Microarrays are accessible even for organisms without a completely known genome sequence. A project with *H. capsulatum*, which is described in greater detail in chapter 41, is illustrative. Hwang et al. constructed a plasmid-based genomic library with short inserts (1 to 2 kb). The inserts were PCR amplified directly from bacterial cultures grown in microtiter format, using common primer sites flanking the insert, and spotted on glass slides. Simultaneously, the plasmid inserts were end sequenced (23). In addition to the need for large-scale sequencing, this array, with 10,000 elements, contains 10 to 20 Mb of DNA, short of the 23- to 25-Mb estimate for the total genome length. There will be some redundancy in the library as well, so that this falls far short of being a complete genome array. Nevertheless, as their very nice paper attests, this is a useful tool in the absence of a complete sequence. A similar array has been constructed for *Cryptococcus* (27).

Affymetrix arrays. The proprietary Affymetrix platform synthesizes oligonucleotides directly on the surface of a silicon wafer by using photoactivatable nucleotides and light masks to control where activation occurs to build oligonucleotides of known sequence. This is a very precise process, and features can be small and closely packed. For the *S. cerevisiae* genome arrays, there are 32 25-mer oligonucleotides (16 exact matches and 16 single-base mismatches) for each gene, for a total of more than 200,000 features on a single wafer of about 1 cm². These are one-color experiments, and intensity is calculated based on the signal difference between the exact and mismatched oligonucleotides.

A limitation of Affymetrix-style arrays is the need to synthesize the complex light masks, a substantial upfront cost that must be recouped through sales. The reduced interest in antifungal development by pharmaceutical companies means that the market is limited for arrays for pathogenic fungi, and Affymetrix does not sell arrays for any pathogenic fungus, although a *C. albicans* array has been constructed and used in a few studies (28, 29).

A related technology developed by NimbleGen reduces this upfront cost by using a maskless process. This company now sells arrays for 117 microbial genomes; curiously, it does not seem to consider fungi to be "microbial," and so *S. cerevisiae* is the only fungus represented. Nevertheless, NimbleGen's technology offers the hope of economically feasible custom arrays for many organisms.

Prospects. Many of the barriers to entry into the microarray world are disappearing as commercial suppliers of arrays arise and institutions are developing genomics core facilities offering shared access to the expensive equipment necessary. Unfortunately, the day when arrays can be bought off the shelf for most fungal pathogens is still far off. Hopefully, with heightened interest in these organisms and continuing decreases in the cost of the technologies for array and oligonucleotide synthesis, this will change.

Experimental Uses of Microarrays for Fungal Pathogenesis

Microarrays are specific and saturating and can be used for gene identification. These features make them ideal for studying fungal pathogenesis using appropriate model systems. To date, there have been no reports of microarray-based studies of fungal cells isolated from infected animal tissue. There have, however, been a number of microarray-based studies that examine virulence-related phenotypes or interactions between pathogens and host cells or tissues in culture. The following sections review which experiments have been done and what has been learned thus far. Given the different levels of development of genomic technologies in fungal pathogens, it makes sense to organize this discussion by organism.

C. albicans

The research community for *C. albicans* is somewhat larger than for other fungal pathogens, and its genome sequencing project was the first started. As a result, forays into genomics for this organism are the most advanced. There have been three major foci of microarray projects in *C. albicans*: in vitro studies of morphogenesis, stress response, and the interaction of *C. albicans* with mammalian cells.

Morphogenesis. It has long been assumed that the polymorphic nature of *C. albicans* contributes to its virulence. Multiple morphologies can be seen in infected tissues, with hyphae often adjacent to yeast cells. This is in contrast to most other fungal pathogens, even dimorphic species such as *Histoplasma*, that are present exclusively in the yeast morphology in vivo. Experimental evidence that cells locked in one morphological state are less virulent is mounting to support this assumption (31, 52), and it is safe to say that polymorphism is a significant factor in pathogenesis in this species.

To say that morphogenesis in *C. albicans* is complicated is a decided understatement. Distinct environmental conditions, including temperature, pH, nutrient deprivation, and the presence of serum, govern morphogenesis through multiple signaling pathways. Specific conditions or combinations control whether the growth form is present as yeast, pseudo-hyphae, or true hyphae, and at least a dozen transcriptional regulators have been implicated in the control of morphogenesis. Readers are referred to chapter 22 in this volume and to earlier reviews (6) for a more complete description.

The broadest study of morphogenesis has been reported by Nantel et al.; the authors described 18 separate comparisons of hyphal induction by temperature, serum, pH (in Lee's medium), and regulation by the transcription factors EFG1 and/or CPH1 by using a >6,000-gene PCR/glass slide array (40). Reassuringly, several of the most highly induced genes, including *ECE1*, *HWP1*, *RBT1*, and *SAP4-6*, were previously identified as hypha-specific or hypha-induced genes. The largest category of genes reported as induced are those for cell surface or secreted proteins. Given the known differences between yeast and hyphal cells in adhesion and immunoreactivity, this is not too surprising. Related functions, including cytoskeletal structure and secretion, were also represented. Regulation of most of these genes required either CPH1 or EFG1.

Other array-based studies have focused more closely on specific genes, including *EFG1* and its APSES homolog *EFH1* (15); *NRG1* and *TUP1* (39); *EFG1*, *CPH1*, and its homolog *CPH2* (30); and the forkhead transcription factor-encoding gene, *FKH1* (2). Taken as a whole, the results confirm the regulatory complexity and implicate a system of redundancy whereby the downstream genes are under the control of multiple inputs, allowing the cell maximal flexibility to initiate hyphal growth under many conditions.

Stress responses. The in vivo environment presents specific stresses that pathogens must neutralize in order to persist. These include temperatures at minimum of 37°C and up to 44°C in feverish animals, well above the growth optima for most fungi. Contact with immune cells can lead to oxidative and cell wall damage. Some bodily environments do not have constant nutrient availability and can differ dramatically in pH. *S. cerevisiae* has a well-coordinated stress response, easily seen by microarray, that activates a common system on exposure to a number of insults. Curiously, although a few *C. albicans* heat shock proteins are modestly induced by heat shock, oxidative, and osmotic stresses, there does not appear to be a coordinated stress response (18). This is corroborated by the lack of an induced cross-tolerance phenotype (where, for example, exposure to mild oxidative stress increases survival following a subsequent severe heat shock). It is possible that the stresses used by these authors, similar to the *S. cerevisiae* experiments, were not sufficiently extreme to elicit the complete response: the "heat shock" was a shift from 23 to 37°C, the normal growth temperature for *C. albicans*.

Another well-known stress of the in vivo environment is the acquisition of iron, an essential cofactor for many cellular processes. Iron uptake systems are critical for virulence in many fungal (and bacterial) species, including *C. albicans* (47). Low iron availability is one inducer of capsule synthesis in *C. neoformans* (59) and induces secretion of hydroxamate siderophores in *H. capsulatum* (22). The Agabian laboratory has studied the transcriptional response to low iron availability in *C. albicans* by using a custom Affymetrix array (29). More than 1,000 genes were significantly regulated, including those involved in iron uptake, heme synthesis, oxidative stress responses, and mitochondrial function. From this work, they also identified a homolog (SFU1) of the *U. maydis* URBS1 transcription factor known to repress siderophore biosynthesis (61) and found 31 potential targets of SFU1, many of which had known roles in iron homeostasis.

Candida-**mammalian cell interactions.** Three studies reported to date have used microarrays to describe what happens in *C. albicans* cells on contact with human cells. Two of these are from the Fink laboratory and investigated interactions with primary human neutrophils or cultured murine macrophages. The third study, from Hube and colleagues, describes the response of *C. albicans* to incubation in human whole blood. The pathways in *C. albicans* regulated during these interactions are somewhat surprising.

Rubin-Bejerano et al. studied the responses of *C. albicans* and *S. cerevisiae* to phagocytosis by human neutrophils isolated from blood using Affymetrix (*S. cerevisiae*) or ~6,700-gene glass slide (*C. albicans*) microarrays (51). These neutrophils are actively fungicidal, killing both species effectively. The profiles of the two yeasts were similar: activation of genes involved in some amino acid biosynthetic pathways, particularly for sulfur-containing amino acids and for arginine. The amino acid response in *Candida* was somewhat broader than in *Saccharomyces*. Surprisingly, there was little evidence for a stress response, even to oxidative damage, which should occur in this environment.

Using the same approach, an almost completely different set of responses was seen following phagocytosis by cultured J774 murine macrophages (33, 34). These cells are mildly antifungal: they kill *S. cerevisiae* efficiently but have very little activity against *C. albicans*. *S. cerevisiae* cells, prior to death, have a limited response (assayed using Affymetrix arrays) involving the induction of the glyoxylate cycle and gluconeogenesis, suggesting, not surprisingly, that the phagolysosome is glucose-deficient. This response is seen in *C. albicans* as well (using the same 6,700-gene array, expanded to include genes involved in fatty acid degradation). In addition, there is a substantial repression of translation, including translation factors, tRNA synthases, and ribosomal proteins. There are stress responses to oxidative damage and DNA repair. The only similarity to the neutrophil profile is the induction of arginine biosynthesis. In all, the response of *C. albicans* to macrophages involves 10-fold more genes than in *S. cerevisiae* (~600 versus 53) (33).

The role of alternative carbon metabolism, identified in this array project, was later also seen in experiments using differential display of macrophage phagocytosed cells, including the glyoxylate enzyme ICL1 and several proteins involved in peroxisome function (44). ICL1 has been demonstrated to be required for full virulence in *C. albicans*, suggesting that alternative carbon metabolism is necessary in vivo as well (34). This has also been shown for the intracellular pathogen *Mycobacterium tuberculosis* (38).

When incubated in whole blood, *C. albicans* showed a partly conflicting set of responses to a 2,002-gene membrane array (19). In contrast to the results described above, there was a substantial upregulation of translation functions in blood. Furthermore, induction was seen for both the glyoxylate cycle/gluconeogenesis and glycolysis, which is counterintuitive. Some stress-responsive genes were activated, but there was relatively high expression of some of these genes in the preculture condition, complicating this analysis.

The fact that these three studies gave conflicting results emphasizes two important points about microarray analysis of host-pathogen interactions. First, the interaction is complex and dynamic, with both cells responding to the other. Contrary to initial expectations, different cell types provoke different responses so that the profiles of *C. albicans* cells incubated with neutrophils and macrophages are distinct. These profiles are superimposed in the blood study. Second, these studies used different comparison conditions; cells in blood were compared to $t = 0$ cells essentially in stationary phase (19), while the macrophage experiment used time-matched cells grown in the same media without macrophages (33). The issues of complexity and comparison conditions are discussed in more detail in a later section.

Other studies. This chapter is not intended to be a comprehensive review of the growing number of microarray studies completed to date. Recent papers have addressed genomic responses to azole exposure (1, 13, 50), phenotype switching (28), mating response (58), pH responses (3, 35), and two-component signaling (8). With the lifestyle of *C. albicans* as an organism predominantly associated with the host, it can certainly be argued that any study of *C. albicans* biology is relevant to virulence. However, the studies described above should certainly be an adequate representation of the kinds of experiments now possible in this species.

Summary. The lessons learned thus far from *C. albicans* microarray studies are that responses are complex and only sometimes do they bear a close resemblance to similar *S. cerevisiae* models, emphasizing the need for thorough studies of *C. albicans* itself. Furthermore, *C. albicans* can have very different responses to different cell types, as seen with the neutrophil-versus-macrophage experiments. It has been very tempting to consider an in vivo microarray experiment (such as profiling *C. albicans* from a mouse kidney) as an "ideal" experiment. The existing studies, however, cast doubt on how interpretable the results would be, since the cells would represent a mixed population and it is not clear what the appropriate control would be.

H. capsulatum

Histoplasma is a fascinating organism. Its dimorphism is controlled by temperature; it lives as a mycelium at lower temperatures and as a yeast at 37°C. Temperature-regulated or phase-specific genes have been identified in vitro using various methods; at least one of these, which encodes a secreted calcium binding protein (CBP1), is required in vivo (53). In the body, it is able to divide within macrophages partly by blocking acidification of the phagolysosome. Several phase-regulated genes are known, but recent transcriptional analysis has greatly added to our understanding of the mycelium-yeast transition (23). Using arrays made from random genomic fragments (see above), the investigators probed the transcriptional response to temperature and hence to genes expressed only in the mycelial or yeast phases.

A more detailed description of this work will appear in chapter 41, and so this chapter merely summarizes the results obtained in that study. This microarray was used to identify genes specifically expressed in the yeast or mycelial forms, and about 500 clones were differentially regulated. CBP1, a calcium binding protein, and YPS3 were both identified as yeast-specific genes, as had been shown previously (25, 42, 53). Through annotation of the induced genes, the authors were able to identify functions upregulated in yeast cells, including sulfur metabolism, accelerated growth, and some nutrient-related genes. Mycelial functions included genes thought to contribute to polarized growth and conidiation (based on homologs in other organisms) and a distinct set of nutrient-related genes. Interestingly, they also discovered phase-specific regulation of transcriptional start sites in several genes, adding another layer of complexity to the maintenance of the yeast-versus-filamentous states (23).

C. neoformans

Two recent reports have described a *C. neoformans* genomic microarray containing 6,274 elements, constructed in a manner similar to the *Histoplasma* project described above. In the first study, analysis of transcripts induced by growth at 37°C identified, among other genes, a homolog of the gene encoding the MGA2 transcription factor. Mutant strains lacking MGA2, constructed as a result of the microarray data, do not grow well at elevated temperatures, marking MGA2 as a regulator of high-temperature growth (27). In the second study, the array was used to find genes regulated by the cyclic AMP cascade by comparing strains with or without the GPA1 Gα protein that regulates cAMP production, which has previously been shown to control melanin production and capsule synthesis.

Included in the set of induced genes are several genes with known roles in capsule biosynthesis and two melanin-producing laccase genes (*LAC1* and *LAC2*), one of which (*LAC2*) was novel. The investigators demonstrated that LAC1 is the dominant laccase, but the newly identified LAC2 does participate in melanin production (45).

REMAINING CHALLENGES AND EXPERIMENTAL CONSIDERATIONS

The data reviewed above demonstrate both the power of microarray analysis for fungal pathogenesis studies and the difficulties inherent in such work. Several of the studies have identified hundreds of responsive genes, risking a data overload that complicates analysis. Further, seemingly minor differences in starting conditions can produce large disparities in results if not controlled carefully. Finally, the fungi appear to have very specific responses to various aspects of host contact: witness the limited overlap between neutrophil and macrophage profiles. Similarly, the *Histoplasma* study did not distinguish between regulation by temperature and by phase; although linked, these may have independent contributions to virulence.

At the same time, genomic efforts are finding new genes that alter growth in vivo. From the temperature-regulated *MGA2* gene of *Cryptococcus* to the machinery of alternative carbon utilization in *Candida*, we have gained new insights into pathogenicity based on this work. To continue this success, several challenges remain. In the sections below, I address some of these issues and the possible solutions.

Annotation

Microarrays with spots containing unknown fragments of the genome may be interesting as abstract art, but they are certainly not biologically useful. In each of the projects described above, a substantial effort was made by the group that constructed the arrays to annotate the genes or genomic fragments spotted. While not trivial, this is an essential step. Investigators interested in the newly emerging fungal genomes would be well served to proceed in a collaborative manner to both share the workload and standardize the annotation.

A second challenge with annotation is that few genes have been cloned for most pathogenic fungi. For *C. albicans*, the annotation effort found that only 355 genes (of ~6,300 total) had been named in publications prior to the release of the genome sequence (4). For other species, the number will be a fraction of even this meager total. Thus, for most genes, the best functional information will come from sequence homology to proteins in other species (mostly *S. cerevisiae*). While this is very useful, it is important to realize the limitations of "inferred by sequence homology," to use the gene ontology term. It is almost certain that the *Cryptococcus* homolog of ScENO1 is enolase; however, the same cannot be said for cell surface proteins and signaling pathways, which may have significant functional divergence from the *S. cerevisiae* model. Even in the case of well-conserved enzymes of metabolism, there may be interesting regulatory differences that are directly relevant to pathogenic processes not present in *S. cerevisiae*.

Comparison Conditions

Studies described above, assaying *C. albicans* in human blood or in cultured macrophages (19, 33, 34), demonstrate the importance of the selection of control conditions. Microarray studies are inherently comparative—an absolute expression level for a particular gene is essentially meaningless, whereas quantitating the change in expression from one condition to another is the goal. It is not surprising, then, that the choice of the control or comparison conditions significantly affects the spectrum of genes that appear to be regulated. This does not mean that the approach used in either study was "wrong," but it does mean that it is very important to understand the control conditions used. The blood study compared all data to the transcript profile of cells essentially in stationary phase, i.e., ones grown overnight in medium and then incubated in phosphate-buffered saline for 30 min prior to the coculture. These were, however, the direct precursors of the cells introduced to the blood culture. In contrast, the macrophage experiment compared cells incubated with macrophages for 1 h to cells grown in medium for 1 h (and the same for other time points). However, there was no "starting" ($t = 0$) population in this study. In an animal experiment, these challenges are compounded. What, exactly, is an appropriate control condition for *Candida* in a mouse kidney? Log-phase cells in YPD? Because the control conditions partially determine which genes are identified, an understanding of what those conditions are (something often glossed over in array papers) is critical.

Complexity

The dissimilarity seen between neutrophil- and macrophage-phagocytosed *C. albicans* cells demonstrates the distinct transcript profiles that can be seen in specific contexts. In blood, these profiles overlapped (since both neutrophils and macrophage/monocytes were present) and interpretation of the data was made much easier by the monoculture experiments. If one extrapolates this situation to organisms recovered in vivo, say from the kidneys, it is reasonable to suspect that there will be overlapping transcript profiles as well, with some bound to endothelial cells of the kidney plumbing, others penetrating the tissue itself, and some interacting with invading macrophages. The polymorphism seen commonly in histological preparations of *C. albicans* infections will further complicate the transcriptional picture. Imagining transcriptional data from fully formed conidiophores from pulmonary aspergillosis is even more daunting. Hence, the question: can we even hope to understand such data?

The best answer is, "Yes, but not yet." Two steps will greatly assist in the analysis. The first is to start, not by isolating cells from the whole kidney but from plaques on the outer cortex, from cells that have penetrated the medulla, or from renal blood vessels. Dissecting complex organs into more homogenous pieces will reduce the complexity, although there are clearly limitations to this approach. Two challenges exist for this. The first is how to work with the small number of cells collected from such an experiment; this is addressed below. The second is that an investigator must be a more accomplished physiologist than most fungal biologists (at least, far more accomplished than *this* fungal biologist!) to do this accurately.

The other precursor to understanding this experiment is to know the profiles of the pathogen exposed to isolated cell types. Having an endothelial cell program, a macrophage program, and an epithelial cell program will help to decipher the complex responses. Some of this is currently being done for *C. albicans*, and it will have to be repeated for other organisms as well.

Low Cell Numbers, Host Contamination, and Amplification

Working in vitro, microarray experiments typically start with hundreds of milliliters of dense cultures containing $>10^9$ cells to collect the large amounts of RNA necessary. In vivo experiments have a big (or, rather, a small) problem: typical tissue burdens with *C. albicans* are 10^6 to 10^7 cells per g of tissue in the kidney, an organ that may only weigh 150 mg. In the *Cryptococcus* SAGE study (56), rabbits were inoculated with 10^9 organisms; by day 5, only 5.2×10^7 cells were recovered in the pooled cerebrospinal fluid of 12 rabbits. By day 9, numbers had fallen another 10-fold. Isolating *H. capsulatum* or *A. fumigatus* from a lung poses even more severe problems of this sort. Because it may take 10^8 cells or more to collect enough RNA for a standard array experiment, this in vivo experiment will require either large numbers of animals, which is time-consuming, expensive, and ethically questionable, or in vitro amplification of recovered RNA. Anything that decreases the recovery of fungal cells, such as dissecting an organ to homogenize the population, will exacerbate this problem.

RNA amplification has been used in many studies but rarely in microorganisms. Amplification to allow transcript profiling from small biopsy samples has been successful and, with some variation, accurately reflects the starting RNA population. The amplification process involves first-strand cDNA synthesis from an oligo(dT)-primed reaction, then second strand synthesis, and in vitro transcription to regenerate RNA to be labeled. Additional rounds of cDNA synthesis and in vitro transcription can further amplify the initial RNA pool, but the average size of the RNA and the degree to which the amplified pool reflects the original population declines with each successive cycle. Although the molecular biology for amplification appears complex, several commercially available kits facilitate this, and amplification has been used for at least one of the studies described above (the *C. albicans*-macrophage study [33]).

In addition to recovery of only small numbers of fungal cells from in vivo sites, the cells are contaminated with host tissue. One advantage of the microarray system is its specificity: mammalian DNA should not cross-hybridize to a fungal genome array except maybe for a very few highly conserved genes. However, the presence of mammalian RNA competes with the fungal RNA during cDNA synthesis and labeling and would be particularly problematic during amplification procedures. In the macrophage study that used amplification, the ratio of *C. albicans* cells to macrophages was 2:1 and macrophages were rapidly lysed in cold water. However, there was still detectable macrophage RNA, and achieving an adequate signal on the arrays required twice as much input RNA as normally used (33). From an organ, the ratio of fungal to mammalian cells could be much more skewed than this, with the fungal cells as a small minority of the overall population. Furthermore, tissue homogenization and lysis is more difficult and time-consuming than is lysis of cultured cells; during this time, the fungal cells are exposed to environments different from that intended (the organ of a live animal). These issues are not insurmountable, but they reinforce the idea that some kind of dissection will be required to increase the proportion of fungal cells in the recovered material. To date, no in vivo studies have been reported, and so determination of how serious these concerns might be is a matter of conjecture.

Statistical Considerations

There are as many statistical models for analyzing microarray data as there are platforms on which to do the experiments (frankly, probably more). Other reviews present comprehensive descriptions of various methods and caveats (5, 17, 46), and so this discussion is confined to a few considerations that directly impact experimental design.

Amplification

As discussed above, accumulation of enough RNA for a microarray experiment may require an amplification step. Amplification distorts the profile of nucleic acids in the pool and thus will distort the array data. In general, very-low-abundance mRNAs are lost during amplification and very-high-abundance mRNAs are reduced (as a percentage of total RNA). Because of this, if one directly compares an amplified population to an unamplified population (imagine an animal experiment in which in vitro-grown cells are the comparison condition), there can be "expression" changes that actually result from the distortions inherent in amplification. If multiple rounds of amplification are required, this problem is exacerbated. The simple solution to this is to subject both RNA populations to the same amplification protocol; this is not a perfect fix, but it applies the same distortions to both conditions.

Dye-swapping

The fluorescent labels used for the majority of array experiments (at least for glass slide arrays), Cy3 and Cy5, have different stabilities and bleaching properties, with Cy5 being the less stable dye. If one labels two identical RNA populations with these two dyes, the overall signal from the Cy5 channel will be lower, but this is not uniformly consistent across the slide due to inconsistent photobleaching across the slide. This, again, can imply "expression" changes that actually result from the array processing. The typical solution to this is a dye-swap, in which the array hybridizations are done in duplicate, one in which population 1 is labeled with Cy3 and population 2 is labeled with Cy5 and the other in which this is reversed. Since most investigators perform all array hybridizations in duplicate, this does not require additional hybridizations.

Replicates

In addition to hybridization of a single sample in duplicate, biological replicates are necessary. The choice of how many replicates to do depends significantly on the difficulty of collecting the samples, the number of statisticians on the research team, and the depth of the research budget. Two to four replicates are typical, each of which is hybridized in duplicate.

Reference RNAs

Because microarray experiments are inherently comparative, the choice of the control condition is particularly important, as discussed above. It also, however, affects the questions that can be asked on the basis of the data. As an example, a schematic for two simple experiments is shown in Fig. 1. In

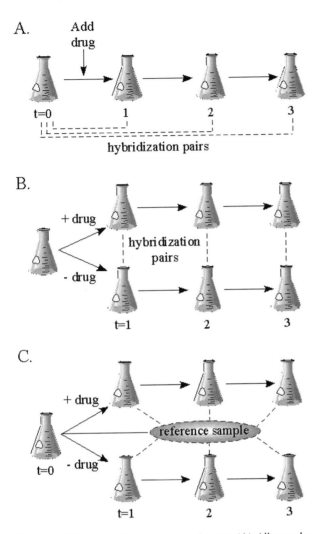

Figure 1. Schematics for array experiments. (A) All samples are compared to the precursor ($t = 0$) sample. (B) All samples are compared to time-matched controls. (C) Reference sample design in which all samples are compared to a constant RNA population. See the text for a description of the merits of each design.

conditions (as nutrients are exhausted, for instance). In the experiment in Fig. 1B, these differences are apparent but it is harder to directly compare the early and late time points. Which of these experimental designs is appropriate depends on the questions to be asked, but both are limiting. In addition, both of these designs make it more difficult to compare experiments done on different days (since they will have different $t = 0$ denominators).

One solution to this is to create a constant reference RNA sample that is used for all hybridizations throughout an entire project. Figure 1C demonstrates this. Because all conditions are hybridized against the same pool of RNA (so that all ratios have the same denominator), any data comparison desired can be made ($t = 3+/ t = 0, t = 3+/ t = 1+, t = 1+/ t = 1-$). Furthermore, this makes interexperiment comparisons, done on different days or by different people, more legitimate. The ideal reference RNA would be pooled from many conditions to represent as many mRNA species as possible.

The drawbacks to the reference RNA approach are the increased expenditure of time and money involved. This requires some effort in making the pool in the first place and requires twice as many arrays as would be needed for a simpler experimental design. Determination of whether the more dynamic statistical analysis is worth these expenses is dependent on the nature of the experiments involved and other practical considerations.

Spectrum of Genes Identified by Microarrays

Microarray experiments have some inherent biases that are important to understand. The most highly regulated genes in a particular condition are usually downstream effectors directly involved in responding to that change. The genomics of pheromone response in *Saccharomyces* is illustrative. An informaticist, unfamiliar with the reams of molecular genetics of the mating response, would identify genes like *FIG1* and *AGA1*; in other words, he would find the targets of STE12. This experiment has been done (49) and the top 20 induced genes are listed in Table 5 (for the sake of simplicity, this list is from one of the many conditions examined in reference 49). From these data, the most important genes for mating would appear to be those encoding proteins involved in cell-cell contact, polarization, and karyogamy. While these are clearly important aspects of mating, subsequent knockouts of the induced genes would have revealed fairly disappointing phenotypes, since few of the induced genes have clear sterile phenotypes.

The transcript profiling would not find, however, the bulk of the well-characterized mitogen-activated protein kinase pathway that governs these mating

Fig. 1A, three time points after addition of a drug are all compared to the $t = 0$ culture (right before adding the compound). In other words, the ratio generated is the expression in the presence of the drug versus that at $t = 0$. In Fig. 1B, the initial culture is split into drug-treated and mock-treated samples. Subsequent time points are compared for the plus-drug and minus-drug samples.

Why does this matter? In the experiment in Fig. 1A, it is easy to see how prolonged exposure to the drug affects gene expression (since the denominators are the same, one can directly compare the $t = 1$ and $t = 3$ samples to see the difference between the acute and steady-state responses). It does not, however, allow one to determine whether the changes at $t = 3$ are due to the drug or to alterations in the culture

Table 5. Top 20 *S. cerevisiae* genes induced by α-factor in microarray experiments[a]

ORF	Gene	Ratio[b]	Pheromone regulated?[c]	Description[d]
YCL075W		100.0	No	Pseudogene: encodes fragment of Ty Pol protein
YBR040W	FIG1	100.0	Yes	Membrane protein required for efficient mating, possibly through Ca^{2+} homeostasis
YML047C	PRM6	68.2	Yes	Predicted to have 2 transmembrane segments; regulated by Ste12p during mating
YLR445W		42.4	No	Hypothetical ORF
YIL037C	PRM2	32.8	Yes	Predicted transmembrane protein with a coiled-coil domain; regulated by Ste12p
YPL192C		29.2	Yes	Required for karyogamy; localizes to the inner membrane of the nuclear envelope
YMR232W	FUS2	28.8	Yes	Cytoplasmic protein localized to the shmoo tip; required for nuclear fusion
YNL279W	PRM1	25.2	Yes	Membrane protein at the shmoo tip involved in membrane fusion; regulated by Ste12p
YJL170C	ASG7	23.8	Yes	An **a**-specific gene that is induced to a higher expression level by α-factor
YDR085C	AFR1	17.6	Yes	Regulates α-factor receptor signaling and polarization; arrestin-like cytoskeletal protein
YML048W		16.6	No	Hypothetical ORF (dubious)
YDR124W		16.1	No	Hypothetical ORF
YCR089W	FIG2	15.5	Yes	Cell wall adhesin; may be involved in maintenance of cell wall integrity during mating
YIL082W		15.0	No	Hypothetical ORF (dubious)
YMR198W	CIK1	14.8	Yes	Spindle pole body protein essential for karyogamy; regulated by KAR4 and mating
YCL027W	FUS1	13.3	Yes	Membrane protein at the shmoo tip; may coordinate signaling, fusion, and polarization
YNR044W	AGA1	13.2	Yes	Glycosylphosphatidylinositol-anchored subunit of α-agglutinin of **a**-cells, highly O-glycosylated protein
YIL080W		12.4	No	TyB Gag-Pol protein; proteolytically processed to make Gag, reverse transcriptase, protease, and integrase proteins
YCL076W		11.4	No	Hypothetical ORF (dubious)
YDR125C	ECM18	10.7	No	Protein of unknown function, similar to Rlp24p
YCL055W	KAR4	8.7	Yes	Transcription factor required to induce KAR3 and CIK1 during mating and meiosis
YKL221W	MCH2	7.5	No	Putative monocarboxylate permeases; mutant not deficient in monocarboxylate transport
YMR065W	KAR5	7.4	Yes	Required for nuclear membrane fusion; may form a complex with Jem1p and Kar2p
YKL189W	HYM1	7.4	No	Hypothetical ORF (dubious)

[a]Data from reference 49.

[b]Mean ratio of expression 30 min after addition of 50 nM synthetic α-factor versus vehicle only.

[c]Whether the gene was known to be pheromone induced from previous work.

[d]Abbreviated description from the Saccharomyces Genome Database (www.yeastgenome.org).

responses. Only the α-factor receptor STE2 and the kinase FUS3 are significantly induced, with the Gα GPA1 being moderately induced (Fig. 2). The phenotypes of knockouts of these three proteins would have been frustrating indeed—*ste2* strains are **a**-specific sterile strains and have no phenotype in α cells, *fus3* mutants are fully fertile, and *gpa1* mutants are inviable!

The bias of genomics toward effectors and not transducers is well known and is certainly not a reason to avoid such experiments. It does, however, make combinatorial approaches important. A thorough understanding of fungal pathogenesis will require parallel efforts in genomics, proteomics, genetics, and immunology.

Unknown Genes

The final challenge to be discussed is actually a great opportunity to direct future studies and a sobering reminder of how little we have learned about fungal pathogens. It is well appreciated that each organism will have genes unique to that species; that is, one will not be able to find sequence homologs in current databases for some genes. This ranges from species to species; in both *S. cerevisiae* and *C. albicans*, these "orphan" genes account for 19% of the

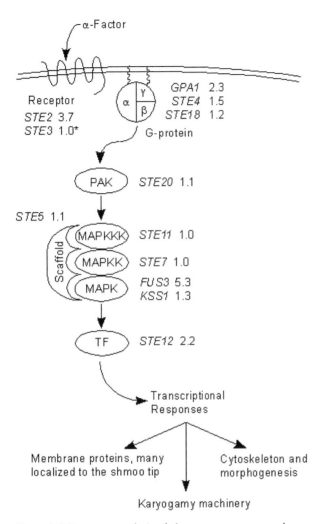

Figure 2. Microarray analysis of pheromone response pathway components. The induction of each gene is indicated after treatment with 50 nM α-factor for 30 min. Data from reference 48.

if the distribution was random (33). Although these are only two examples, they suggest that not only will there be unique genes in fungal pathogens but also that these genes will play a large role in the virulence of these organisms. Studying these novel genes can be frustrating without some kind of foothold into their function, but it should certainly keep medical mycologists busy for a long time to come.

CONCLUSIONS

Connecting specific genes to roles in virulence has always been a challenge, nowhere more so than for fungal pathogens. A paucity of tools has made it difficult at all stages of research, from identifying candidate virulence factors to determining how they contribute to disease progression. In this chapter, I have highlighted some of the technological platforms, some early successes, and some areas in which further development is needed. Hopefully, the chapter has both conveyed the promise of this field and emphasized the importance of combinatorial approaches to studies in medical mycology. Genomics, along with proteomics, large-scale and signature-tagged mutagenesis, and sequencing efforts, have added powerful new approaches to our repertoire. Although most of these have not been fully utilized to date, the incorporation of these technologies into mycological studies will allow unprecedented progress in understanding the biology and pathology of these fascinating and important organisms.

genome, and in *N. crassa* they make up 41% of the genes (4, 20). As additional species are sequenced, these numbers are reduced, but it is still to be expected that a substantial number of genes will be unique to each species. This presents problems, of course, because it is not possible to rely on studies of other species to shed light on their function. An understanding of these genes, though, can tell us much about the unique circumstances seen by these species during evolution.

For fungal pathogens, this clearly provides an expectation that there will be "orphan" genes critical for pathogenesis. Indeed, in the SAGE study of *Cryptococcus* from rabbit cerebrospinal fluid 35% of the tags found only in vivo are orphans (56). Microarray analysis of phagocytosed *C. albicans* cells identified 36% of upregulated genes as unique to this species, roughly twice what would have been expected

REFERENCES

1. **Barker, K. S., S. Crisp, N. Wiederhold, R. E. Lewis, B. Bareither, J. Eckstein, R. Barbuch, M. Bard, and P. D. Rogers.** 2004. Genome-wide expression profiling reveals genes associated with amphotericin B and fluconazole resistance in experimentally induced antifungal resistant isolates of *Candida albicans*. *J. Antimicrob. Chemother.* **54:**376–385.
2. **Bensen, E. S., S. G. Filler, and J. Berman.** 2002. A forkhead transcription factor is important for true hyphal as well as yeast morphogenesis in *Candida albicans*. *Eukaryot. Cell* **1:**787–798.
3. **Bensen, E. S., S. J. Martin, M. Li, J. Berman, and D. A. Davis.** 2004. Transcriptional profiling in *Candida albicans* reveals new adaptive responses to extracellular pH and functions for Rim101p. *Mol. Microbiol.* **54:** 1335–1351.
4. **Braun, B. R., M. van het Hoog, C. d'Enfert, M. Martchenko, J. Dungan, A. Kuo, D. O. Inglis, M. A. Uhl, H. Hogues, M. Berriman, M. C. Lorenz, A. Levitin, U. Oberholzer, C. Bachewich, D. Harcus, A. Marcil, D. Dignard, T. Iouk, R. Zito, L. Frangeul, F. Tekaia, K. Rutherford, E. Wang, N. A. Gow, L. L. Hoyer, G. Kohler, J. Morschhauser, G. Newport, S. Znaidi, M. Raymond, B. Turcotte, G. Sherlock, M. Costanzo, J. Ihmels, J. Berman, D. Sanglard, N. Agabian, A. P. Mitchell, A. D. Johnson, M. Whiteway, and A. Nantel.** 2005. A human-curated annotation of the *Candida albicans* genome. *PLoS Genet.* **1:**36–57.

5. Brazma, A., P. Hingamp, J. Quackenbush, G. Sherlock, P. Spellman, C. Stoeckert, J. Aach, W. Ansorge, C. A. Ball, H. C. Causton, T. Gaasterland, P. Glenisson, F. C. Holstege, I. F. Kim, V. Markowitz, J. C. Matese, H. Parkinson, A. Robinson, U. Sarkans, S. Schulze-Kremer, J. Stewart, R. Taylor, J. Vilo, and M. Vingron. 2001. Minimum information about a microarray experiment (MIAME)—toward standards for microarray data. *Nat. Genet.* **29:**365–371.

6. Brown, A. J. 2002. Morphogenetic signaling pathways in *Candida albicans*, p. 95–106. *In* R. Calderone (ed.), *Candida and Candidiasis.* ASM Press, Washington, D.C.

7. Camilli, A., D. T. Beattie, and J. J. Mekalanos. 1994. Use of genetic recombination as a reporter of gene expression. *Proc. Natl. Acad. Sci. USA* **91:**2634–2638.

8. Chauhan, N., D. Inglis, E. Roman, J. Pla, D. Li, J. A. Calera, and R. Calderone. 2003. *Candida albicans* response regulator gene *SSK1* regulates a subset of genes whose functions are associated with cell wall biosynthesis and adaptation to oxidative stress. *Eukaryot. Cell* **2:**1018–1024.

9. Cheng, S., C. J. Clancy, M. A. Checkley, M. Handfield, J. D. Hillman, A. Progulske-Fox, A. S. Lewin, P. L. Fidel, and M. H. Nguyen. 2003. Identification of *Candida albicans* genes induced during thrush offers insight into pathogenesis. *Mol. Microbiol.* **48:**1275–1288.

10. Cliften, P., P. Sudarsanam, A. Desikan, L. Fulton, B. Fulton, J. Majors, R. Waterston, B. A. Cohen, and M. Johnston. 2003. Finding functional features in *Saccharomyces* genomes by phylogenetic footprinting. *Science* **301:**71–76.

11. Colonna-Romano, S., A. Porta, A. Franco, G. S. Kobayashi, and B. Maresca. 1998. Identification and isolation by DDRT-PCR of genes differentially expressed by *Histoplasma capsulatum* during macrophages infection. *Microb. Pathog.* **25:**55–66.

12. Dean, R. A., N. J. Talbot, D. J. Ebbole, M. L. Farman, T. K. Mitchell, M. J. Orbach, M. Thon, R. Kulkarni, J. R. Xu, H. Pan, N. D. Read, Y. H. Lee, I. Carbone, D. Brown, Y. Y. Oh, N. Donofrio, J. S. Jeong, D. M. Soanes, S. Djonovic, E. Kolomiets, C. Rehmeyer, W. Li, M. Harding, S. Kim, M. H. Lebrun, H. Bohnert, S. Coughlan, J. Butler, S. Calvo, L. J. Ma, R. Nicol, S. Purcell, C. Nusbaum, J. E. Galagan, and B. W. Birren. 2005. The genome sequence of the rice blast fungus *Magnaporthe grisea. Nature* **434:**980–986.

13. De Backer, M. D., T. Ilyina, X. J. Ma, S. Vandoninck, W. H. Luyten, and H. Vanden Bossche. 2001. Genomic profiling of the response of *Candida albicans* to itraconazole treatment using a DNA microarray. *Antimicrob. Agents Chemother.* **45:**1660–1670.

14. Dietrich, F. S., S. Voegeli, S. Brachat, A. Lerch, K. Gates, S. Steiner, C. Mohr, R. Pohlmann, P. Luedi, S. Choi, R. A. Wing, A. Flavier, T. D. Gaffney, and P. Philippsen. 2004. The *Ashbya gossypii* genome as a tool for mapping the ancient *Saccharomyces cerevisiae* genome. *Science* **304:**304–307.

15. Doedt, T., S. Krishnamurthy, D. P. Bockmuhl, B. Tebarth, C. Stempel, C. L. Russell, A. J. Brown, and J. F. Ernst. 2004. APSES proteins regulate morphogenesis and metabolism in *Candida albicans. Mol. Biol. Cell* **15:**3167–3180.

16. Dujon, B., D. Sherman, G. Fischer, P. Durrens, S. Casaregola, I. Lafontaine, J. De Montigny, C. Marck, C. Neuveglise, E. Talla, N. Goffard, L. Frangeul, M. Aigle, V. Anthouard, A. Babour, V. Barbe, S. Barnay, S. Blanchin, J. M. Beckerich, E. Beyne, C. Bleykasten, A. Boisrame, J. Boyer, L. Cattolico, F. Confanioleri, A. De Daruvar, L. Despons, E. Fabre, C. Fairhead, H. Ferry-Dumazet, A. Groppi, F. Hantraye, C. Hennequin, N. Jauniaux, P. Joyet, R. Kachouri, A. Kerrest, R. Koszul, M. Lemaire, I. Lesur, L. Ma, H. Muller, J. M. Nicaud, M. Nikolski, S. Oztas, O. Ozier-Kalogeropoulos, S. Pellenz, S. Potier, G. F. Richard, M. L. Straub, A. Suleau, D. Swennen, F. Tekaia, M. Wesolowski-Louvel, E. Westhof, B. Wirth, M. Zeniou-Meyer, I. Zivanovic, M. Bolotin-Fukuhara, A. Thierry, C. Bouchier, B. Caudron, C. Scarpelli, C. Gaillardin, J. Weissenbach, P. Wincker, and J. L. Souciet. 2004. Genome evolution in yeasts. *Nature* **430:**35–44.

17. Eisen, M. B., P. T. Spellman, P. O. Brown, and D. Botstein. 1998. Cluster analysis and display of genome-wide expression patterns. *Proc. Natl. Acad. Sci. USA* **95:**14863–14868.

18. Enjalbert, B., A. Nantel, and M. Whiteway. 2003. Stress-induced gene expression in *Candida albicans*: absence of a general stress response. *Mol. Biol. Cell* **14:**1460–1467.

19. Fradin, C., M. Kretschmar, T. Nichterlein, C. Gaillardin, C. d'Enfert, and B. Hube. 2003. Stage-specific gene expression of *Candida albicans* in human blood. *Mol. Microbiol.* **47:**1523–1543.

20. Galagan, J. E., S. E. Calvo, K. A. Borkovich, E. U. Selker, N. D. Read, D. Jaffe, W. FitzHugh, L. J. Ma, S. Smirnov, S. Purcell, B. Rehman, T. Elkins, R. Engels, S. Wang, C. B. Nielsen, J. Butler, M. Endrizzi, D. Qui, P. Ianakiev, D. Bell-Pedersen, M. A. Nelson, M. Werner-Washburne, C. P. Selitrennikoff, J. A. Kinsey, E. L. Braun, A. Zelter, U. Schulte, G. O. Kothe, G. Jedd, W. Mewes, C. Staben, E. Marcotte, D. Greenberg, A. Roy, K. Foley, J. Naylor, N. Stange-Thomann, R. Barrett, S. Gnerre, M. Kamal, M. Kamvysselis, E. Mauceli, C. Bielke, S. Rudd, D. Frishman, S. Krystofova, C. Rasmussen, R. L. Metzenberg, D. D. Perkins, S. Kroken, C. Cogoni, G. Macino, D. Catcheside, W. Li, R. J. Pratt, S. A. Osmani, C. P. DeSouza, L. Glass, M. J. Orbach, J. A. Berglund, R. Voelker, O. Yarden, M. Plamann, S. Seiler, J. Dunlap, A. Radford, R. Aramayo, D. O. Natvig, L. A. Alex, G. Mannhaupt, D. J. Ebbole, M. Freitag, I. Paulsen, M. S. Sachs, E. S. Lander, C. Nusbaum, and B. Birren. 2003. The genome sequence of the filamentous fungus *Neurospora crassa. Nature* **422:**859–868.

20a. Galagan, J. E., et al. 2005. Sequencing of *Aspergillus nidulans* and comparative analysis with *A. fumigatus* and *A. oryzae. Nature* **438:**1105–1115.

21. Goffeau, A., B. G. Barrell, H. Bussey, R. W. Davis, B. Dujon, H. Feldmann, F. Galibert, J. D. Hoheisel, C. Jacq, M. Johnston, E. J. Louis, H. W. Mewes, Y. Murakami, P. Philippsen, H. Tettelin, and S. G. Oliver. 1996. Life with 6000 genes. *Science* **274:**546, 563–567.

22. Howard, D. H., R. Rafie, A. Tiwari, and K. F. Faull. 2000. Hydroxamate siderophores of *Histoplasma capsulatum. Infect. Immun.* **68:**2338–2343.

23. Hwang, L., D. Hocking-Murray, A. K. Bahrami, M. Andersson, J. Rine, and A. Sil. 2003. Identifying phase-specific genes in the fungal pathogen *Histoplasma capsulatum* using a genomic shotgun microarray. *Mol. Biol. Cell* **14:**2314–2326.

24. Jones, T., N. A. Federspiel, H. Chibana, J. Dungan, S. Kalman, B. B. Magee, G. Newport, Y. R. Thorstenson, N. Agabian, P. T. Magee, R. W. Davis, and S. Scherer. 2004. The diploid genome sequence of *Candida albicans. Proc. Natl. Acad. Sci. USA* **101:**7329–7334.

25. Keath, E. J., and F. E. Abidi. 1994. Molecular cloning and sequence analysis of *yps-3*, a yeast-phase-specific gene in the dimorphic fungal pathogen *Histoplasma capsulatum*. *Microbiology* **140:**759–767.

26. Kellis, M., N. Patterson, M. Endrizzi, B. Birren, and E. S. Lander. 2003. Sequencing and comparison of yeast species to identify genes and regulatory elements. *Nature* **423:**241–254.

27. Kraus, P. R., M. J. Boily, S. S. Giles, J. E. Stajich, A. Allen, G. M. Cox, F. S. Dietrich, J. R. Perfect, and J. Heitman. 2004. Identification of *Cryptococcus neoformans* temperature-regulated genes with a genomic-DNA microarray. *Eukaryot. Cell* **3:**1249–1260.

28. Lan, C. Y., G. Newport, L. A. Murillo, T. Jones, S. Scherer, R. W. Davis, and N. Agabian. 2002. Metabolic specialization associated with phenotypic switching in *Candida albicans*. *Proc. Natl. Acad. Sci. USA* **99:**14907–14912.

29. Lan, C. Y., G. Rodarte, L. A. Murillo, T. Jones, R. W. Davis, J. Dungan, G. Newport, and N. Agabian. 2004. Regulatory networks affected by iron availability in *Candida albicans*. *Mol. Microbiol.* **53:**1451–1469.

30. Lane, S., C. Birse, S. Zhou, R. Matson, and H. Liu. 2001. DNA array studies demonstrate convergent regulation of virulence factors by Cph1, Cph2, and Efg1 in *Candida albicans*. *J. Biol. Chem.* **276:**48988–48996.

31. Lo, H. J., J. R. Kohler, B. DiDomenico, D. Loebenberg, A. Cacciapuoti, and G. R. Fink. 1997. Nonfilamentous *C. albicans* mutants are avirulent. *Cell* **90:**939–949.

32. Loftus, B. J., E. Fung, P. Roncaglia, D. Rowley, P. Amedeo, D. Bruno, J. Vamathevan, M. Miranda, I. J. Anderson, J. A. Fraser, J. E. Allen, I. E. Bosdet, M. R. Brent, R. Chiu, T. L. Doering, M. J. Donlin, C. A. D'Souza, D. S. Fox, V. Grinberg, J. Fu, M. Fukushima, B. J. Haas, J. C. Huang, G. Janbon, S. J. Jones, H. L. Koo, M. I. Kryzwinski, K. J. Kwon-Chung, K. B. Lengeler, R. Maiti, M. Marra, R. E. Marra, C. A. Mathewson, T. G. Mitchell, M. Pertea, F. R. Riggs, S. L. Salzberg, J. E. Schein, A. Shvartsbeyn, H. Shin, M. Shumway, C. A. Specht, B. B. Suh, A. Tenney, T. R. Utterback, B. L. Wickes, J. R. Wortman, N. H. Wye, J. W. Kronstad, J. K. Lodge, J. Heitman, R. W. Davis, C. M. Fraser, and R. W. Hyman. 2005. The genome of the basidiomycetous yeast and human pathogen *Cryptococcus neoformans*. *Science* **307:**1321–1324.

33. Lorenz, M. C., J. A. Bender, and G. R. Fink. 2004. Transcriptional response of *Candida albicans* upon internalization by macrophages. *Eukaryot. Cell* **3:**1076–1087.

34. Lorenz, M. C., and G. R. Fink. 2001. The glyoxylate cycle is required for fungal virulence. *Nature* **412:**83–86.

35. Lotz, H., K. Sohn, H. Brunner, F. A. Muhlschlegel, and S. Rupp. 2004. *RBR1*, a novel pH-regulated cell wall gene of *Candida albicans*, is repressed by RIM101 and activated by NRG1. *Eukaryot. Cell* **3:**776–784.

35a. Machida, M., et al. 2005. Genome sequencing and analysis of *Aspergillus oryzae*. *Nature* **438:**1157–1161.

36. Mahan, M. J., J. M. Slauch, and J. J. Mekalanos. 1993. Selection of bacterial virulence genes that are specifically induced in host tissues. *Science* **259:**686–688.

37. Martinez, D., L. F. Larrondo, N. Putnam, M. D. Gelpke, K. Huang, J. Chapman, K. G. Helfenbein, P. Ramaiya, J. C. Detter, F. Larimer, P. M. Coutinho, B. Henrissat, R. Berka, D. Cullen, and D. Rokhsar. 2004. Genome sequence of the lignocellulose degrading fungus *Phanerochaete chrysosporium* strain RP78. *Nat. Biotechnol.* **22:**695–700.

38. McKinney, J. D., K. Honer zu Bentrup, E. J. Munoz-Elias, A. Miczak, B. Chen, W. T. Chan, D. Swenson, J. C. Sacchettini, W. R. Jacobs, Jr., and D. G. Russell. 2000. Persistence of *Mycobacterium tuberculosis* in macrophages and mice requires the glyoxylate shunt enzyme isocitrate lyase. *Nature* **406:**735–738.

39. Murad, A. M., C. d'Enfert, C. Gaillardin, H. Tournu, F. Tekaia, D. Talibi, D. Marechal, V. Marchais, J. Cottin, and A. J. Brown. 2001. Transcript profiling in *Candida albicans* reveals new cellular functions for the transcriptional repressors CaTup1, CaMig1 and CaNrg1. *Mol. Microbiol.* **42:**981–993.

40. Nantel, A., D. Dignard, C. Bachewich, D. Harcus, A. Marcil, A. P. Bouin, C. W. Sensen, H. Hogues, M. van het Hoog, P. Gordon, T. Rigby, F. Benoit, D. C. Tessier, D. Y. Thomas, and M. Whiteway. 2002. Transcription profiling of *Candida albicans* cells undergoing the yeast-to-hyphal transition. *Mol. Biol. Cell* **13:**3452–3465.

40a. Nierman, W. C., et al. 2005. Genomic sequence of the pathogenic and allergenic filamentous fungus *Aspergillus fumigatus*. *Nature* **438:**1151–1156.

41. Oliver, S. G., Q. J. van der Aart, M. L. Agostoni-Carbone, M. Aigle, L. Alberghina, D. Alexandraki, G. Antoine, R. Anwar, J. P. Ballesta, P. Benit, et al. 1992. The complete DNA sequence of yeast chromosome III. *Nature* **357:**38–46.

42. Patel, J. B., J. W. Batanghari, and W. E. Goldman. 1998. Probing the yeast phase-specific expression of the *CBP1* gene in *Histoplasma capsulatum*. *J. Bacteriol.* **180:**1786–1792.

43. Porta, A., S. Colonna-Romano, I. Callebaut, A. Franco, L. Marzullo, G. S. Kobayashi, and B. Maresca. 1999. An homologue of the human 100-kDa protein (p100) is differentially expressed by *Histoplasma capsulatum* during infection of murine macrophages. *Biochem. Biophys. Res. Commun.* **254:**605–613.

43a. Powell, J. 2000. SAGE: the serial analysis of gene expression. *Methods Mol. Biol.* **99:**297–319.

44. Prigneau, O., A. Porta, J. A. Poudrier, S. Colonna-Romano, T. Noel, and B. Maresca. 2003. Genes involved in beta-oxidation, energy metabolism and glyoxylate cycle are induced by *Candida albicans* during macrophage infection. *Yeast* **20:**723–730.

45. Pukkila-Worley, R., Q. D. Gerrald, P. R. Kraus, M. J. Boily, M. J. Davis, S. S. Giles, G. M. Cox, J. Heitman, and J. A. Alspaugh. 2005. Transcriptional network of multiple capsule and melanin genes governed by the *Cryptococcus neoformans* cyclic AMP cascade. *Eukaryot. Cell* **4:**190–201.

46. Quackenbush, J. 2001. Computational analysis of microarray data. *Nat. Rev. Genet.* **2:**418–427.

47. Ramanan, N., and Y. Wang. 2000. A high-affinity iron permease essential for *Candida albicans* virulence. *Science* **288:**1062–1064.

48. Retallack, D. M., G. S. Deepe, Jr., and J. P. Woods. 2000. Applying in vivo expression technology (IVET) to the fungal pathogen *Histoplasma capsulatum*. *Microb. Pathog.* **28:**169–182.

49. Roberts, C. J., B. Nelson, M. J. Marton, R. Stoughton, M. R. Meyer, H. A. Bennett, Y. D. He, H. Dai, W. L. Walker, T. R. Hughes, M. Tyers, C. Boone, and S. H. Friend. 2000. Signaling and circuitry of multiple MAPK pathways revealed by a matrix of global gene expression profiles. *Science* **287:**873–880.

50. Rogers, P. D., and K. S. Barker. 2002. Evaluation of differential gene expression in fluconazole-susceptible and -resistant isolates of *Candida albicans* by cDNA microarray analysis. *Antimicrob. Agents Chemother.* **46:**3412–3417.

51. **Rubin-Bejerano, I., I. Fraser, P. Grisafi, and G. R. Fink.** 2003. Phagocytosis by neutrophils induces an amino acid deprivation response in *Saccharomyces cerevisiae* and *Candida albicans*. *Proc. Natl. Acad. Sci. USA* **100:** 11007–11012.

52. **Saville, S. P., A. L. Lazzell, C. Monteagudo, and J. L. Lopez-Ribot.** 2003. Engineered control of cell morphology in vivo reveals distinct roles for yeast and filamentous forms of *Candida albicans* during infection. *Eukaryot. Cell* **2:**1053–1060.

53. **Sebghati, T. S., J. T. Engle, and W. E. Goldman.** 2000. Intracellular parasitism by *Histoplasma capsulatum*: fungal virulence and calcium dependence. *Science* **290:** 1368–1372.

54. **Staib, P., M. Kretschmar, T. Nichterlein, H. Hof, and J. Morschhauser.** 2000. Differential activation of a *Candida albicans* virulence gene family during infection. *Proc. Natl. Acad. Sci. USA* **97:**6102–6107.

55. **Steen, B. R., T. Lian, S. Zuyderduyn, W. K. MacDonald, M. Marra, S. J. Jones, and J. W. Kronstad.** 2002. Temperature-regulated transcription in the pathogenic fungus *Cryptococcus neoformans*. *Genome Res.* **12:** 1386–1400.

56. **Steen, B. R., S. Zuyderduyn, D. L. Toffaletti, M. Marra, S. J. Jones, J. R. Perfect, and J. Kronstad.** 2003. *Cryptococcus neoformans* gene expression during experimental cryptococcal meningitis. *Eukaryot. Cell* **2:** 1336–1349.

57. **Sturtevant, J.** 2000. Applications of differential-display reverse transcription-PCR to molecular pathogenesis and medical mycology. *Clin. Microbiol. Rev.* **13:**408–427.

58. **Tsong, A. E., M. G. Miller, R. M. Raisner, and A. D. Johnson.** 2003. Evolution of a combinatorial transcriptional circuit: a case study in yeasts. *Cell* **115:**389–399.

59. **Vartivarian, S. E., E. J. Anaissie, R. E. Cowart, H. A. Sprigg, M. J. Tingler, and E. S. Jacobson.** 1993. Regulation of cryptococcal capsular polysaccharide by iron. *J. Infect. Dis.* **167:**186–190.

60. **Velculescu, V. E., L. Zhang, B. Vogelstein, and K. W. Kinzler.** 1995. Serial analysis of gene expression. *Science* **270:**484–487.

61. **Voisard, C., J. Wang, J. L. McEvoy, P. Xu, and S. A. Leong.** 1993. *urbs1*, a gene regulating siderophore biosynthesis in *Ustilago maydis*, encodes a protein similar to the erythroid transcription factor GATA-1. *Mol. Cell. Biol.* **13:**7091–7100.

62. **Wood, V., R. Gwilliam, M. A. Rajandream, M. Lyne, R. Lyne, A. Stewart, J. Sgouros, N. Peat, J. Hayles, S. Baker, D. Basham, S. Bowman, K. Brooks, D. Brown, S. Brown, T. Chillingworth, C. Churcher, M. Collins, R. Connor, A. Cronin, P. Davis, T. Feltwell, A. Fraser, S. Gentles, A. Goble, N. Hamlin, D. Harris, J. Hidalgo, G. Hodgson, S. Holroyd, T. Hornsby, S. Howarth, E. J. Huckle, S. Hunt, K. Jagels, K. James, L. Jones, M. Jones, S. Leather, S. McDonald, J. McLean, P. Mooney, S. Moule, K. Mungall, L. Murphy, D. Niblett, C. Odell, K. Oliver, S. O'Neil, D. Pearson, M. A. Quail, E. Rabbinowitsch, K. Rutherford, S. Rutter, D. Saunders, K. Seeger, S. Sharp, J. Skelton, M. Simmonds, R. Squares, S. Squares, K. Stevens, K. Taylor, R. G. Taylor, A. Tivey, S. Walsh, T. Warren, S. Whitehead, J. Woodward, G. Volckaert, R. Aert, J. Robben, B. Grymonprez, I. Weltjens, E. Vanstreels, M. Rieger, M. Schafer, S. Muller-Auer, C. Gabel, M. Fuchs, C. Fritzc, E. Holzer, D. Moestl, H. Hilbert, K. Borzym, I. Langer, A. Beck, H. Lehrach, R. Reinhardt, T. M. Pohl, P. Eger, W. Zimmermann, H. Wedler, R. Wambutt, B. Purnelle, A. Goffeau, E. Cadieu, S. Dreano, S. Gloux, V. Lelaure, S. Mottier, F. Galibert, S. J. Aves, Z. Xiang, C. Hunt, K. Moore, S. M. Hurst, M. Lucas, M. Rochet, C. Gaillardin, V. A. Tallada, A. Garzon, G. Thode, R. R. Daga, L. Cruzado, J. Jimenez, M. Sanchez, F. del Rey, J. Benito, A. Dominguez, J. L. Revuelta, S. Moreno, J. Armstrong, S. L. Forsburg, L. Cerrutti, T. Lowe, W. R. McCombie, I. Paulsen, J. Potashkin, G. V. Shpakovski, D. Ussery, B. G. Barrell, and P. Nurse.** 2002. The genome sequence of *Schizosaccharomyces pombe*. *Nature* **415:**871–880.

Molecular Principles of Fungal Pathogenesis
Edited by Joseph Heitman et al.
©2006 ASM Press, Washington, D.C.

Chapter 41

Future of Functional Genomics of *Histoplasma capsulatum*

ANITA SIL AND LENA HWANG

Histoplasma capsulatum is the causative agent of the respiratory and systemic disease histoplasmosis. It is a primary pathogen that infects healthy as well as immunocompromised people and is thought to be the most common cause of fungal respiratory infections in the world. It is thought to cause approximately 500,000 infections a year in the United States (20, 55, 93, 96). Most of these infections cause limited disease in healthy hosts. However, immunocompromised individuals often develop progressive, disseminated disease that can be fatal. This is especially true for patients undergoing immunosuppressant chemotherapy and those with lymphoreticular neoplasms (8–10, 39, 76). *H. capsulatum* is endemic in the Mississippi and Ohio River valleys and is a leading pathogen affecting AIDS patients in the Midwestern United States (84).

H. capsulatum is a haploid dimorphic fungus, existing in either a filamentous mycelial form or a budding yeast form as shown in Fig. 1 (20, 51, 95, 96). The mycelial form is found primarily in the soil. This form of the organism is specialized both for mating (and subsequent production of meiotic spores) and for the production of vegetative spores, or conidia. *H. capsulatum* enters the host via inhalation of conidia and mycelial fragments that aerosolize when the soil is disrupted. Within the lungs of the host, *H. capsulatum* is taken up by macrophages and other phagocytic cells and the conversion from the conidial or mycelial form to the budding yeast form is initiated (11, 20, 96). The yeast cells somehow subvert the antimicrobial environment of the alveolar macrophage and survive and multiply within the phagosome or phagolysosome of these cells. *H. capsulatum* then spreads to multiple organs of the reticuloendothelial system such as the spleen, liver, lymph nodes, and bone marrow. The infection in most hosts is limited by the development of cell-mediated immunity and the activation of macrophages (20, 24, 34, 66). However, even in these healthy hosts, the organism may cause a latent infection that can be reactivated later if the host becomes immunocompromised (24, 30).

For *H. capsulatum*, the ability to grow as mycelia and as yeast is key to establishing and maintaining disease in the host. The mycelial form initiates the infection through its ability to aerosolize and produce infectious conidiospores. The yeast form is thought to be essential for *H. capsulatum* to survive and proliferate within the host, since mycelial cells that are artificially blocked from converting to the yeast form are unable to cause disease (54, 60). This change in morphology in response to environmental signals and the role of the morphologic switch in disease are particularly interesting features of the biology of *H. capsulatum*.

All of the systemic dimorphic fungal pathogens, including *H. capsulatum*, *Coccidioides immitis*, *Blastomyces dermatitidis*, and *Paracoccidioides brasiliensis*, use temperature as a key signal determining morphology and cell specialization (52). The conversion of *H. capsulatum* from mycelial cells to budding yeast cells, and vice versa, can be induced in culture simply by shifting the growth temperature (51, 52). At 25°C, *H. capsulatum* grows in the mycelial form. When the cells are shifted to 37°C, they convert to the budding yeast form. Little is known about how any of the systemic dimorphic fungi sense temperature and respond appropriately. This type of dimorphic switch is also a key virulence determinant for many pathogenic fungi (7), including *H. capsulatum*, as described above. Again, little is known about how morphology plays a role in virulence. *H. capsulatum* is poised to become a model system for studying temperature-regulated morphology and survival as an intracellular eukaryotic pathogen.

Anita Sil and Lena Hwang • Department of Microbiology and Immunology, University of California San Francisco, 513 Parnassus, Box 0414, San Francisco, CA 94143-0414.

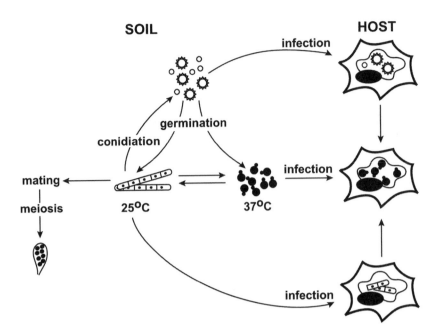

Figure 1. Life cycle of *H. capsulatum*. This figure illustrates the various stages of the *H. capsulatum* life cycle in the soil and in the host. The mycelial form grows in the soil (or at 25°C in the laboratory). It can mate and undergo meiosis, forming ascospores. The mycelia can also produce vegetative conidiospores through the process of conidiation, generating at least two types of conidiospores, macroconidia and microconidia. The conidia can germinate into either the mycelial form or the yeast form. Infection occurs when the soil is disrupted and the host inhales aerosolized conidia or hyphal fragments. These cells convert to the yeast form inside the host. The yeast form is infectious if introduced into an animal in the laboratory, but there is normally no host-to-host transmission of the yeast form. Schematic courtesy of Davina Hocking Murray.

Functional genomics will play a large role in the study of this organism. Classical genetic studies have been limited, and with the genome sequence of three strains of *H. capsulatum* soon to be completed, genomic studies will be a powerful means of identifying candidate genes by virtue of regulated gene expression, homology, and comparative genomics. The power of genomics is dependent on the accurate annotation of the genome. Microarray technologies not only are instrumental in determining regulated gene expression but also can potentially be used to identify genes and their exon and intron structures. This chapter describes (i) the previous use of functional genomics in *H. capsulatum*, (ii) the current status of genome-sequencing projects for this organism, (iii) potential annotation of the genome by using tiling microarrays, and (iv) future uses of functional genomics to dissect *H. capsulatum* biology.

FUNCTIONAL GENOMICS OF *H. CAPSULATUM* WITHOUT A GENOME SEQUENCE

Generation of a Shotgun Genomic DNA Microarray

Functional genomics usually begins with the genome sequence of an organism. However, for *H. capsulatum*, functional genomic analysis preceded a genome sequence, demonstrating that genomics can be applied to organisms that have not been sequenced. Hwang et al. (35) generated a shotgun genomic DNA microarray for *H. capsulatum* even though very little sequence information was available for this organism at the time. They used this array to compare the gene expression profiles of yeast and mycelial cells. To construct the array, 1- to 2-kb genomic DNA fragments were generated by partial restriction enzyme digestion of genomic DNA from the virulent strain G217B (Fig. 2). These fragments were then ligated into the *Escherichia coli* vector pBLUESCRIPT, generating a library of genomic fragments. Using a common pair of primers from the pBLUESCRIPT vector that flanked the genomic insert, 9,600 independent library clones were amplified by PCR. The PCR products were then spotted onto glass slides to generate the microarray. Each array element, or "spot," corresponds to a PCR product from a bacterial clone. The bacterial clones were also subjected to DNA sequencing, resulting in limited sequence information for the majority of array elements. Based on the estimated genome size at the time, it was thought that approximately one-third of the genome was represented on the array.

To monitor gene expression changes, array elements must include coding sequence. Using random

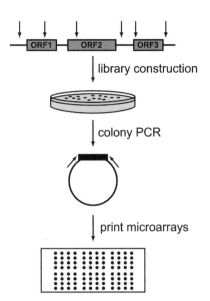

Figure 2. Construction of a shotgun genomic microarray. A schematic of *H. capsulatum* genomic DNA is shown at the top of the figure. The genomic DNA was partially digested with Sau3AI, size selected, cloned into pBluescript, and transformed into *E. coli*. Approximately 10,000 independent colonies were inoculated into 96-well plates, and these cultures were subjected to colony PCR with common primers in the vector. The resultant PCR products were spotted on glass slides to generate shotgun genomic microarrays. Reprinted from reference 35 with permission.

genomic DNA as a template for a microarray could result in many nonproductive array elements if the genome contains large intergenic regions or large introns that will interfere with hybridization to a cDNA probe. The structure of approximately 30 sequenced *H. capsulatum* genes was known at the time of construction of the shotgun genomic array. Available information predicted that the average intron was fairly small (~100 bp), with up to six introns per gene. To determine if the majority of array elements contained coding sequence, a trial shotgun genomic microarray was used to competitively hybridize genomic DNA versus a cDNA probe generated from *H. capsulatum* yeast cells. The genomic DNA probe generates a signal for all of the array elements on a genomic microarray, whereas the cDNA probe gives a signal only for elements that contain coding sequence. Hwang et al. (35) observed that 75% of their spots hybridized to the cDNA probe, indicating that a minimum of 75% of the array elements corresponded to coding sequence. Since the cDNA probe was generated from a single growth condition, it is likely that certain transcripts were not represented in the probe. Therefore, it is probable that even more than 75% of the microarray contained coding sequence.

Comparing the Gene Expression of Morphologic Forms

One of the fundamental features of *H. capsulatum* biology is its ability to change morphology between a mycelial form and a yeast form; however, little is known about the molecular differences between each form and the regulation of the transition between them. A number of yeast- and mycelial-specific genes of *H. capsulatum* had been identified in a variety of laboratories (17, 25, 31, 32, 36, 40, 42, 69, 87, 88), but the shotgun microarray approach was the first large-scale analysis of gene expression differences in this organism. To develop a more complete description of the two morphologic phases, Hwang et al. (35) used their random genomic microarray to compare the gene expression profiles of yeast and mycelial cells. They hoped to identify morphology-related transcriptional regulators, as well as yeast- and mycelial-specific genes that would help to explain the phenotypic differences between the two morphological phases.

To compare the gene expression profiles of mycelial and yeast cells, Hwang et al. (35) isolated RNA from *H. capsulatum* growing in the mycelial form at 25°C and in the yeast form at 37°C. Differentially labeled cDNA probes were generated and competitively hybridized to the shotgun genomic microarray. Approximately 500 spots on the microarray gave at least fivefold higher expression in one phase than in the other (Fig. 3). Previously identified yeast-phase-specific genes *CBP1* and *yps-3* (40, 69) were present as control spots on the array and, as expected, showed yeast-specific expression on the microarray. To validate whether the array elements that appeared to be differentially expressed on the microarray were truly phase regulated, a small subset of the phase-specific clones were analyzed by Northern blotting (Fig. 3). The bacterial clones that corresponded to the array elements were retrieved from the bacterial archives, and the genomic DNA insert was used as a probe. The Northern analysis confirmed that the clones were indeed phase regulated, and the authors went on to annotate the most highly phase-regulated array clones.

The clones were annotated as follows: (i) the ends of each array clone were sequenced; (ii) the sequence information from each clone was used to map the location of each array clone sequence onto contigs from an ongoing genome-sequencing project (described below); (iii) each contig was subjected to Basic Local Alignment Search Tool (BLAST) analysis against the nonredundant (nr) database of genes at the National Center for Bioinformatics to identify the location of putative genes based on homologs in other organisms; and (iv) spots that overlapped with a single

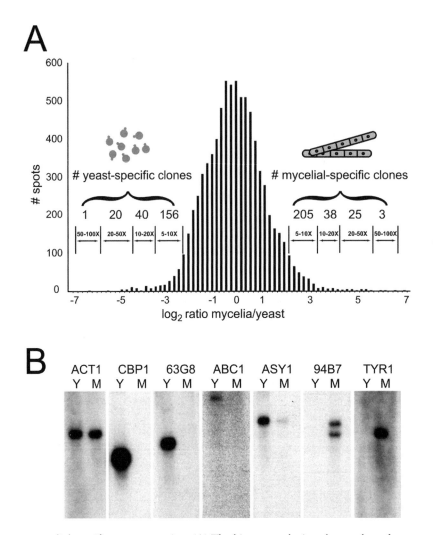

Figure 3. Yeast- or mycelial-specific gene expression. (A) The histogram depicts the number of spots on the microarray (y axis) versus the $\log_2$ of the ratio of the mycelial signal to the yeast signal (x axis). A $\log_2$ ratio of zero indicates spots that are equivalently expressed. The numbers of clones that show differential expression from 5- to 100-fold in one morphologic form compared to the other are labeled in the figure. (B) Northern blot analysis of gene expression in yeast and mycelia. Total RNA from yeast (Y) is in the left lane, and total RNA from mycelia (M) is in the right lane. *ACT1* (actin) is equivalently expressed between the two RNA samples. *CBP1* (calcium binding protein), *63G8* (unknown microarray clone), *ABC1*, and *ASY1* are yeast specific by microarray and Northern analysis. *94B7* (unknown microarray clone) and *TYR1* are mycelium specific by microarray and Northern analysis. Reprinted from reference 35 with permission.

putative gene were annotated. Approximately half of the differentially expressed genes were annotated by this method.

A number of the genes that were identified to be phase regulated may shed some light on the phenotypic differences of each morphologic form (Fig. 4). Mycelial and yeast cells differ not just in morphology but also in growth rate, growth environment (host versus soil), sulfur metabolism, mating, melanin production, and conidiation. Differentially expressed genes can be categorized based on the function of their BLAST homologs in other organisms. These categories can be used to generate hypotheses about

the function of putative *H. capsulatum* orthologs. Yeast-specific genes involved in nutrient acquisition (such as amino acid permeases) may play a role in the higher growth rate of yeast than of mycelia. Furthermore, these genes may play an important role in acquisition of nutrients within the phagosome of the host cell. In addition to temperature, sulfhydryl groups appear to play a role in the regulation of morphology (53, 60, 75). The mechanism of how the sulfhydryl oxidation state influences morphology is unclear and may be elucidated by further study of yeast-specific genes implicated in sulfur metabolism, such as cysteine dioxygenase and ATP sulfurylase.

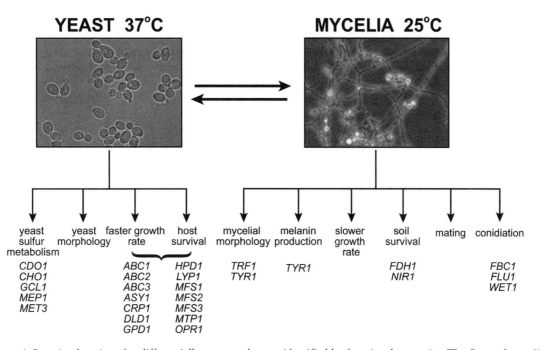

Figure 4. Putative functions for differentially expressed genes identified by functional genomics. The figure shows *H. capsulatum* cells growing in either the yeast form at 37°C or the mycelial form at 25°C. Genes were annotated based on homology and categorized based on the function of their ortholog in other organisms. A number of yeast-specific genes were annotated as potentially being involved in sulfur metabolism and growth rate/host survival. Mycelial-specific genes were implicated in polarized cell growth, melanin production, soil survival, and conidiation. Adapted from reference 35 with permission.

Since the mycelial sample used for gene expression studies also contained conidia (35), the mycelium-specific genes could include genes involved in spore formation. Mycelium-specific genes included three putative orthologs of *Aspergillus* regulatory factors involved in conidiation (56, 94). These genes, which have homology to the *Aspergillus fluG, wetA,* and *flbC* genes, are the first candidates implicated in conidiation in *H. capsulatum*. In *Aspergillus nidulans, fluG* initiates conidiophore development (49) and *flbC* and *wetA* function downstream of *fluG*. Although Hwang et al. (35) did not determine the order of expression of the *H. capsulatum* homologs during conidiation, they did determine that the *H. capsulatum fluG* ortholog *FLU1* is expressed during the stationary phase of yeast-form cells whereas the *wetA* and *flbC* orthologs are not. Since yeast phase cells transform into mycelia if grown for extended periods in stationary phase (52), a subset of genes transcribed late in stationary phase, such as *FLU1*, may function to establish mycelial growth. In contrast, the *wetA* and *flbC* orthologs may be expressed downstream of *FLU1* once robust mycelial growth occurs. Further experimentation is required to determine the relationship between these genes.

Although temperature is a key regulator of morphology in *H. capsulatum*, it is possible that some of the genes identified in this study are temperature regulated rather than morphology regulated. Future gene expression profiling experiments using mutants that are trapped in a particular morphology independent of temperature may distinguish morphology-regulated genes from temperature-regulated genes (V. Nguyen and A. Sil, unpublished data).

Determination of the function of each of the genes identified in this study requires further experimentation. However, the identification of a large set of genes with interesting expression patterns in an organism where few genes had previously been identified emphasizes the power of genomics. A whole-genome microarray for *H. capsulatum* has recently been constructed (see below). Coverage of the entire genome will allow a comprehensive analysis of differential gene expression in this organism.

H. CAPSULATUM GENOME-SEQUENCING PROJECTS

Choice of Strains To Be Sequenced

Genome sequencing has stimulated the development of a vast array of genomic tools. The results of current genome-sequencing projects for *H. capsulatum* will allow researchers to exploit these tools to probe the biology of this pathogen. Two separate projects are under way to sequence three distinct, unrelated strains of *H. capsulatum*. The choice of strains was determined based on a number of criteria, including phylogeny and previous use of the strain for laboratory analysis. Here we describe the characteristics of the three strains chosen for sequencing.

Most of the molecular studies of *H. capsulatum* have taken place with G217B, G186A, and the Downs strain. G217B and the Downs strains are North American clinical isolates, and G186A is a clinical isolate from Panama. Originally, *H. capsulatum* strains were classified into two "chemotypes" based on the polysaccharide composition of their cell walls (14, 73). G217B, a member of chemotype I, lacks α-(1,3)-glucan in its cell wall. G186A, a member of chemotype II, contains large amounts of α-(1,3)-glucan. Interestingly, G186A variants that lack α-(1,3)-glucan are avirulent (43, 44) whereas G217B contains no α-(1,3)-glucan but is fully virulent.

The Downs strain is a temperature-sensitive strain that was isolated from a patient who was probably immunocompromised (59). In fact, a number of *H. capsulatum* isolates from AIDS patients have been found to be closely related to the Downs strain (82). Several studies of dimorphism have taken place with this strain; however, unlike G217B and G186A, it is avirulent in standard animal models (47).

A number of experiments support the idea that the genomes of the G217B, G186A, and Downs strains are fundamentally different. Restriction fragment length polymorphism analysis of genomic and mitochondrial DNA was used to study G217B, G186B (a strain very closely related to G186A), and the Downs strain (41). This analysis indicated that the strains fall into distinct classes. Additionally, although the resolution is limited, contour-clamped homogenous electric field gel electrophoresis and field-inversion gel electrophoresis suggested that G217B, G186B, and the Downs strain may have chromosomes whose size and number differ (83). Furthermore, DNA renaturation kinetics suggested that the genome of the Downs strain is considerably larger than the genome of G186A (12).

Recent studies using minimal genome sequence to analyze the phylogenetic relationships between 137 individual isolates of *H. capsulatum* also indicated major differences between G217B, G186A, and the Downs strain. The 137 isolates segregate into at least eight clades (a group of organisms which includes the most recent common ancestor of all of its members and all of the descendants of that most recent common ancestor) (37, 38). The clades were identified as North American class 1 (NAm 1), North American class 2 (NAm 2), Latin American group A (LAm A), Latin American group B (LAm B), Australian, Netherlands, Eurasian, and African. These eight clades form seven

phylogenetic species, with the Eurasian clade originating from within Latin American group A. In addition, there were seven lone lineages identified in Latin America. These lineages are not members of any clades and are distinct from each other.

This phylogenetic analysis supports the idea that G186A, G217B, and the Downs strain are not close relatives. G186A represents a lone lineage from Panama. It is distantly related to the genotypes in North America and Latin America. G217B falls into the NAm 2 clade, and Downs is in the NAm 1 clade. NAm 1 and NAm 2 are two distinct phylogenetic species and are as distant from each other as from any other *Histoplasma* clade. As described below, G217B, G186A, and a NAm1 strain that is closely related to the Downs strain are currently being sequenced. Genome information from these three diverse strains will be used to probe their phenotypic differences.

Current Genome-Sequencing Projects

At present, there are two genome-sequencing projects for three strains of *H. capsulatum*: G217B, G186A, and WU24. The Fungal Genome Initiative at the Broad Institute (http://www.broad.mit.edu/annotation/fungi/fgi/) has sequenced WU24, a NAm 1 strain that is closely related to the Downs strain. Sequence information from this strain is now available.

The Genome Sequencing Center (GSC) at Washington University in St. Louis, Mo., funded by the National Institute for Allergy and Infectious Disease, has sequenced strains G217B and G186A (http://www.genome.wustl.edu/projects/hcapsulatum/). The investigators have completed whole-genome shotgun sequencing of paired-end reads from plasmids and fosmids, developed a fosmid-based physical map (50), and assembled both genomes. Most previously identified *H. capsulatum* genes contain small introns that can hinder the identification of putative coding sequences. To address this, the GSC sequenced cDNAs from yeast and mycelia of both G217B and G186A to aid gene finding and annotation of the genome. Current annotation of the genome using a combination of evidence-based and ab initio predictions (EAnnot, Fgenesh-plus, Genewise, Fgenesh, and SNAP) identified 11,223 predicted genes for G217B and 10,381 for G186A.

The genomes of G217B and G186A appear to be quite different. The genome size of G217B is approximately 41 Mb, compared to a much smaller genome of approximately 30 Mb for G186A. This may be due in part to a much larger repeat content for G217B. Of the G217B genome, 34% is composed of repeat sequence (including transposon sequence), whereas G186A contains only 7.5% repeat sequence. Analysis of the repeat sequence is under way. Annotation of both genomes is also in progress. Comparison of all three genomes to identify sequences that are well conserved will aid the annotation. In addition, it will be interesting to identify genes that are unique to each of these strains and determine the role they have in the biology of *H. capsulatum*.

As mentioned above, phylogenetic studies indicate the existence of a wide variety of *Histoplasma* isolates; perhaps future genome projects will shed light on divergent *Histoplasma* strains with different disease manifestations. G217B, G186A, and WU24 are all designated *H. capsulatum* var. *capsulatum*, and they cause respiratory and systemic disease in humans. In contrast, two other varieties of *Histoplasma* exist that cause distinct diseases. In Africa, *H. capsulatum* var. *duboisii* causes a disease (African histoplasmosis) that is characterized by cutaneous and subcutaneous lesions (74). The recent phylogenetic analysis of *H. capsulatum* found an isolate of *H. capsulatum* var. *capsulatum* in the *H. capsulatum* var. *duboisii* clade, indicating that a monophyletic group of *H. capsulatum* can cause both systemic histoplasmosis and African histoplasmosis (38). Characterizing the genome from individuals in this clade could shed light on microbial differences that influence disease manifestation in the host. Additionally, it was originally thought that equine histoplasmosis was caused by a distinct species, *H. capsulatum* var. *farciminosum*. However, Kasuga et al. (38) showed that *H. capsulatum* var. *farciminosum* isolates were found in three different clades, indicating that the ability to infect horses has evolved independently at least three times. Sequence analysis of *H. capsulatum* var. *farciminosum* strains could yield interesting information on the evolution of this pathogen and determinants of host tropism.

Functional Genomics of *H. capsulatum*

With the completion of the genome sequence for G217B and G186A, a whole-genome oligonucleotide microarray has been designed for each strain. This microarray will represent each putative gene from both genomes as a unique 70-mer oligonucleotide. In addition to its standard use as a tool for gene expression profiling, the *H. capsulatum* microarray will be used to aid in the annotation of the genome. Essentially, identification of an expressed transcript by the microarray will confirm the validity of a predicted gene. As a part of the genome project, the G217B microarray is being used to determine the global gene expression of *H. capsulatum* cells grown in the yeast form compared with the mycelial form (V. Nguyen and A. Sil, unpublished data). One reason why these samples were selected to confirm the annotation of the genome is that previous experiments indicate that they are significantly different in their gene expression profile (35), thus allowing identification of a broad set of transcripts.

This examination of yeast versus mycelial gene expression on the whole-genome microarray will significantly extend the identification of phase-regulated genes. In addition, since the mycelial form is specialized for conidiation, mating, and growth in soil at 25°C and the yeast form is specialized for growth in the host at 37°C, the gene expression profile of these two phases can begin to shed light on many aspects of the biology of *H. capsulatum*. Future experiments examining each morphologic phase in more detail will tease out genes that are specific to various functions. For example, the expression pattern of cells undergoing the transition from yeast to mycelia and vice versa will be determined. The gene expression profiles will be subjected to Cluster analysis (19), which will identify sets of genes with similar expression patterns. A temporal analysis of the data set may be informative: genes that are expressed early in the transition may be important for the establishment of the yeast or mycelial phase, and genes that are expressed later may be important for the maintenance of each phase. In addition, some genes with similar expression patterns may be regulated by common transcription factors. Pattern-finding algorithms such as MEME (http://meme.sdsc.edu/meme/website/intro.html) can be used to compare the promoter regions of coregulated genes to identify putative regulatory sequences that are bound by the same transcription factor(s). Sequence comparisons between G217B, G186A, and WU24 can also be used to facilitate the identification of putative regulatory elements, which, unlike most noncoding sequence, are often well conserved. Comparison of gene expression profiles from G217B and G186A will assist in the identification of orthologous genes in the two strains and extend the analysis of the similarities and differences in the regulation of morphology between these two strains.

Experimental Annotation of the Genome

Although technologic advances have increased the rate at which genomes can be sequenced, genome annotation remains nontrivial. This is especially true for *H. capsulatum*, for the following reasons. First, few experimental data exist for use in verification of gene predictions. To compensate for the lack of functional data, the GSC has sequenced sufficient numbers of cDNAs to design rules for gene finding in *H. capsulatum*. However, there are multiple regions of the genome that do not contain cDNA coverage, and it is still difficult to predict all gene structures accurately. For example, in *Cryptococcus neoformans*, another fungal pathogen with complex gene structures, only 60% of known genes were predicted by a gene-finding algorithm that was trained on a relatively large data set (86). Additionally, because the closest relatives of *H. capsulatum* are not highly studied model systems, many of the top BLAST hits to regions of the *H. capsulatum* genome are unverified "hypothetical" or "putative" proteins in these other organisms. These hits may or may not reflect real coding sequences. Furthermore, genes that are novel to *H. capsulatum* may not have any BLAST hits in other organisms. Finally, the work of Hwang et al. indicates that transcripts can initiate far upstream of the relevant open reading frame (35). Hence, experimental confirmation of the gene structure will be quite valuable.

To facilitate the acquisition of experimental data to confirm gene structure, *Histoplasma* researchers are taking advantage of tiling microarrays, which have been a powerful means of obtaining experimental data for annotation purposes in other organisms (77, 79, 86, 97). Tiling arrays are microarrays where contiguous genomic regions are most commonly represented as end-to-end or overlapping oligonucleotides (for example, a 10-kb region could be represented as a moving window of 50-mer oligonucleotides with an overlap of 10 bp). These microarrays are subjected to hybridizations with fluorescently labeled cDNA probes derived from RNA samples. Tiles that bind fluorescent probe represent expressed genomic sequence. Intron sequences do not bind the fluorescent probe and therefore can be distinguished from exon sequences. These types of experiments also allow the boundaries of the 5' and 3' ends of transcripts to be defined. Furthermore, tiling is a potent gene discovery tool: even in an extremely well-studied organism such as *E. coli*, 25% of the transcripts identified by tiling studies were not previously characterized (90).

Tiling arrays were used to analyze expression data generated from the shotgun genomic microarray described at the beginning of this chapter (68a). Although tiling is not a new idea, recent technologic advances make it a reasonably cost-effective strategy. These researchers used a 12,000-element semiconductor array (www.combimatrix.com) where individual oligonucleotides are synthesized at each of the 12,000 microelectrodes (hereafter referred to as Combimatrix arrays). Specifically, the goal of the experiments was to examine the gene expression profile of *H. capsulatum* to nitrosative stress (caused by exposure to nitric oxide [•NO] and other reactive nitrogen species). These experiments are of interest because •NO produced by infected macrophages is thought to be an important component of the host's

ability to contain *H. capsulatum* infection. Mutant mice that are unable to produce •NO, for example, are more susceptible to *H. capsulatum* infection than are their wild-type counterparts (G. Deepe, personal communication). When studied in culture, the ability of infected macrophages to halt the growth of intracellular yeasts is dependent on the production of •NO by the macrophages, but interestingly, the inhibitory effect of •NO on *H. capsulatum* is fungistatic rather than fungicidal (48, 62, 66). This result suggests that *H. capsulatum* is able to resist killing by •NO. In addition, although the production of •NO is crucial to controlling infection, the immune system is not able to completely eliminate *H. capsulatum*; instead, yeast cells remain latent in the host for many years (11, 20, 22). Therefore, one hypothesis is that, on exposure to reactive nitrogen species, *H. capsulatum* might induce the transcription of genes required for detoxification of these reactive intermediates.

To identify such nitrosative stress response genes, these researchers used the 10,000-element shotgun genomic array described above (35) to detect genes that are transcriptionally upregulated at least fourfold when *H. capsulatum* is exposed to an •NO generator in culture (68a). Array elements that showed transcriptional induction in multiple nitrosative stress experiments were flagged for further analysis.

Since the spots on the microarray represent shotgun genomic fragments rather than known coding sequences, the next hurdle was to identify the genes that correspond to each microarray spot. Each array clone was mapped onto contigs from the genome-sequencing project. Each contig was subjected to BLAST analysis against the nr database of genes at the National Center for Bioinformatics to identify the locations of putative genes based on homologs in other organisms. Spots that clearly overlapped with putative genes were annotated as corresponding to those genes. However, in approximately 300 cases, a particular microarray spot did not unambiguously correspond to a single gene, either because the spot sequence overlapped with two adjacent genes or because the region of the contig that overlapped the spot did not contain a BLAST hit (Fig. 5). To determine the location of the gene for these induced spots, a region of the genome surrounding each spot was represented as end-to-end 50-mers on Combimatrix tiling arrays. Both the plus and minus strands were tiled for each region, since the coding sequence of interest could fall on either strand.

These tiling arrays were subjected to competitive hybridizations with fluorescently labeled cDNA generated from either control cells or cells treated

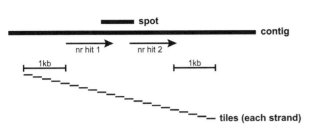

Figure 5. Schematic of tiling array. Spots from a shotgun genomic microarray that were induced in nitrosative stress and did not have a simple correspondence to a putative gene were selected for tiling. In the example shown, the spot overlaps two putative hits from the National Center for Biotechnology Information nonredundant database (nr hit 1 and nr hit 2). A region that extends 1 kb beyond the boundaries of the nr hit homologies was tiled as end-to-end 50-mer oligonucleotides.

with •NO donors. By generating a graph of the intensity of the fluorescent signal for each tile on the array, the location of expressed genes became obvious (Fig. 6). Contiguous regions of increased intensity over baseline indicated the location of an induced gene, and short dips in the intensity suggested the locations of potential introns. This technology was an expeditious way to identify nitrosative stress response genes.

This type of experiment is currently being used to confirm gene predictions generated by the GSC (C. Foo, M. Voorhies, and A. Sil, unpublished data), and it will probably be extended in order to obtain experimental evidence for the annotation of the entire genome. To maximize the ability to identify expressed sequences, it will be important to generate

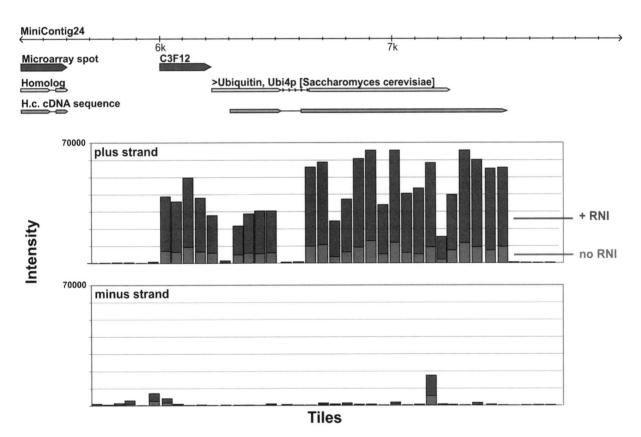

Figure 6. Gene expression analysis using tiling arrays allows definition of gene boundaries. Microarray spot C3F12 is a spot on the shotgun genomic array that is upregulated consistently in nitrosative stress. It does not overlap with any BLAST hits or sequenced cDNAs. The locations of (i) C3F12, (ii) an nr hit from *S. cerevisiae* (ubiquitin [Ubi4p]), and (iii) an unknown cDNA from the genome project are all mapped onto MiniContig 24. The relevant region of MiniContig 24 was tiled on a Combimatrix array as end-to-end 50-mers. This array was subjected to a competitive hybridization with differentially labeled probes (Cy3 [gray] was used to label cDNA generated from control cells, whereas Cy5 [black] was used to label cDNA generated from wild-type cells treated with reactive nitrogen intermediates [RNI]). Each vertical bar represents signal intensity obtained for the tile at that position. Both the plus and minus strands were tiled. The data indicate that the induced gene of interest is present on the plus strand and that the 5′ gene boundary extends significantly upstream of the homology with *S. cerevisiae* Ubi4p. The location of putative introns is revealed by transient interruptions in the signal intensity.

RNA from multiple conditions (both morphologic forms of *H. capsulatum*, conidiospores, cells growing within macrophages, and cells subjected to a variety of growth conditions including stationary phase, reactive oxygen and nitrogen species, oxidizing and reducing conditions, and other perturbations). In addition to defining gene boundaries, this technology will also assist in the identification of introns, alternative spliced products, antisense RNAs, and microRNAs.

FUTURE OF FUNCTIONAL GENOMICS IN *H. CAPSULATUM*

Functional genomics, through the use of whole-genome microarrays and related technologies, will open up genome-wide experimental approaches for the study of many aspects of the biology of *H. capsulatum*.

Macrophage Infection

Macrophages are the primary host cell for *H. capsulatum*, and functional genomics is an excellent tool to probe the interaction between *H. capsulatum* and the host cell.

H. capsulatum is phagocytosed by macrophages. Once inside the macrophage, the microbe continues to divide (Fig. 7); however, its fate seems to differ depending on the type of macrophage infected. The intracellular fate of *H. capsulatum* in a variety of macrophages, including human monocyte-derived macrophages, human alveolar macrophages, murine peritoneal or alveolar macrophages, and murine macrophage-like cell lines such as RAW264.7, and P388D1, has been studied (20, 66, 85, 96). In some but not all of these host cells, a respiratory burst is triggered after phagocytosis of *H. capsulatum* but the viability of *H. capsulatum* seems to be unaffected by reactive oxygen intermediates.

Phagocytosis of a nonpathogenic microbe results in maturation of the phagosome into a phagolysosome and destruction of the microbe (18, 91). The phagolysosome displays several degradative properties including acidic pH, the presence of acid-activated lysosomal hydrolases, antimicrobial peptides, and toxic oxidative compounds. During infection of human macrophages and the murine RAW264.7 cell line, *H. capsulatum* resides in a phagosome that does not fuse with the lysosome; it displays reduced levels of the host vacuolar membrane ATPase/proton pump that is responsible for acidification of the phagosome during maturation (67, 85). Thus, *H. capsulatum* interferes with normal phagosome maturation in these cells. In other types of macrophage host cells such as the murine P388D1 cell line, *H. capsulatum* resides in the phagolysosome (23). In either case, *H. capsulatum* prevents acidification of either the phagosome or the phagolysosome, instead maintaining

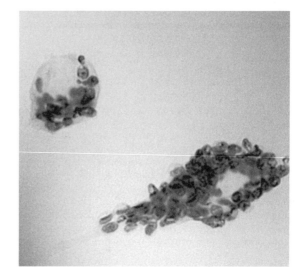

Figure 7. *H. capsulatum* infection of RAW264.7 cells. Monolayers of macrophages were infected with *H. capsulatum*. After 34 h, the monolayer was fixed and fungal cells were stained with periodic acid–Schiff base. The two macrophages shown are filled with *H. capsulatum* cells. Image courtesy of Dervla Isaac.

the pH at approximately 6.0 to 6.5 (21, 66, 85). Inhibition of acidification is thought to interfere with the function of lysosomal hydrolases. Furthermore, *H. capsulatum* is thought to maintain the pH at 6 to 6.5 to maximize acquisition of iron from host transferrin (66). Thus, *H. capsulatum* is thought to survive and grow in macrophages by virtue of its ability to prevent acidification of the phagosome while maintaining an optimum pH for iron acquisition.

Functional genomics could be used in a number of different ways to probe the interaction between the host cell and *H. capsulatum*. One obvious experiment is to examine the gene expression of *H. capsulatum* over a time course of macrophage infection. At each time point, host cells are lysed in a guanidinium thiocyanate buffer, the *H. capsulatum* cells are pelleted by centrifugation, RNA is prepared from these cells, and microarray analysis is used to determine how the gene expression profile evolves with infection (M. P. Nittler, M. Andersson, C. Berkes, and A. Sil, unpublished data). These experiments, coupled with an analysis of the cell biology of infection, will allow correlation of the timing of gene expression with trafficking through and replication within the host cell. The goal is to identify putative virulence factors by identifying genes whose expression changes during infection. For example, genes that are induced when *H. capsulatum* is present in early phagosomes might be important for blocking phagosome maturation. Because the fate of *H. capsulatum* seems to differ depending on the type of macrophage studied (as described above), gene expression profiling of *H. capsulatum* in a variety of host macrophages may provide a sensitive assay to determine whether the microbe is actually responding differently in each case.

This experimental setup also allows the purification of host cell RNA from the lysate supernatant after the *H. capsulatum* has been pelleted. Transcriptional profiling of host cells (using mouse or human microarrays) has been employed to understand how macrophages are manipulated by pathogens (6, 15, 57, 63, 64). By detecting changes in the expression profiles of both the host and *Histoplasma,* it will be possible to monitor the communication between host and pathogen.

Functional genomics can also be used to identify potential virulence factors by subjecting *H. capsulatum* to environmental conditions that mimic those experienced during infection. This type of strategy has been used successfully to identify virulence factors in *Salmonella* by performing gene expression profiling after exposing the microbe to conditions that mimic the environment of the macrophage phagosome (16). For *H. capsulatum,* it could be informative to determine the response to exogenous antimicrobial stimuli that mimic the antimicrobial activities of the macrophage, such as reactive oxygen and nitrogen species. Additionally, stimuli such as low iron levels or low pH could trigger gene expression programs that are important for survival in host cells. These experiments may identify candidate virulence genes, whose roles in infection should be explored by further experimentation.

Development and Germination of Conidia

Asexual sporulation during growth of the mycelial form of *Histoplasma* gives rise to at least two types of conidia, macroconidia and microconidia (70). Microconidia range in size from 2 to 6 μm, whereas macroconidia (Fig. 8) have been reported to range in size from 8 to 14 μm or 10 to 25 μm, depending on the strain and growth conditions. Microconidia are thought to be the most common infectious *Histoplasma* particle due to their small size, which permits them to access the alveoli of the lungs. Interestingly, even a single viable spore is thought to be capable of causing disease in mice (1), and infection of mice with spores results in high mortality (71). A number of studies published prior to 1973 discuss conditions that promote conidiation, as well as the morphology of the resultant spores (2, 3, 28, 29, 65, 70, 80, 81). Additionally, there have been cell biological studies monitoring the germination of microconidia and macroconidia (26, 27, 29, 33, 71). However, none of the molecules that are required for conidial development or germination have been identified.

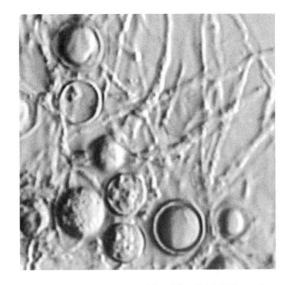

Figure 8. G186AR macroconidia. The G186AR strain was grown under conidiating conditions. The image shows hyphae and tuberculate macroconidia. Image courtesy of Diane Inglis.

Microarray analysis of gene expression would be an ideal method to identify genes involved in spore formation and germination in *H. capsulatum.* Conidiation in the fungus *A. nidulans* has been studied extensively, and several steps are known to be regulated at the level of transcription (89). In *H. capsulatum,* the comparison of yeast and conidiating mycelia using a shotgun genomic microarray as described above identified several orthologs of *Aspergillus* conidiation genes that are specifically expressed in *H. capsulatum* mycelia but not in yeast (35). Examining the gene expression profile of conidiating mycelia by using the whole *H. capsulatum* genome microarray will be instrumental in identifying genes required for conidial development. Gene expression profiling of pure populations of microconidia and macroconidia (compared to yeast and mycelia) will allow the identification of conidium-specific transcripts. These data will shed light on the molecular properties of conidia. Of particular interest is a greater molecular understanding of the pathogenicity of microconidia, which could be revealed by studying genes that are specifically expressed in these cells.

Study of Mating Type

H. capsulatum has two mating types: − and +. Although the two mating types are thought to be equivalently represented in the soil, the − mating type is more prevalent in immunocompetent hosts (46). Both mating types are equally represented in immunocompromised hosts, however, suggesting that each has equivalent access to the host (45). Taken as a whole, these data suggest that the − mating type is more virulent than the + mating type in immunocompetent hosts. However, next to nothing is known about how these mating types differ, and the mating-type locus itself has not been identified. Additionally, since strains seem to lose the ability to mate through serial passaging in the laboratory, many of the laboratory strains (such as G217B and G186A) are of unknown mating type. However, + and − mating-type strains are available at the American Type Culture Collection; these strains are a wonderful starting point for genomic analysis of mating type.

By comparing the gene expression profile of + and − strains by microarray analysis, it will now be possible to judge which genes are differentially expressed between the two mating types. One caveat is that the mating-competent strains are likely to have sequence differences from the reference strain represented on the microarray. Therefore, these studies are better accomplished with a microarray where each spot contains the entire coding sequence per gene (rather than a 70-mer oligonucleotide) so that sequence mismatches between the mating strains and the reference strain will be better tolerated. Once mating-type-specific genes are identified, these genes are candidates for determinants of the mating-type locus or for targets of the mating-type locus. One might expect to identify genes such as pheromones, pheromone receptors, transcription factors, and signaling components. If putative pheromone-encoding genes are identified, it could be very informative to synthesize pheromone in vitro, apply it to the opposite mating type, and determine the gene expression profile. Such experiments would make it possible to identify pheromone response genes (5).

Additionally, since the − mating type seems to cause a higher frequency of clinical cases than the + mating type does, any differentially expressed genes are candidates that might influence virulence in the host. It may be informative to examine the gene expression profile of each mating type in macrophages to see if − strains induce a different set of genes from those induced by + strains. This information could initiate an understanding of how mating type might influence virulence.

REVERSE GENETICS OF *H. CAPSULATUM*

Functional genomics identifies candidate genes and suggests function. To ascertain the true function of a particular gene, further examination and experimentation is necessary. A number of molecular genetic techniques are available for *H. capsulatum* in order to explore the role of genes identified through functional genomic screens.

Gene Disruption

Gene disruption is a vital step in the analysis of gene function. Unfortunately, in *H. capsulatum* only a handful of genes have been disrupted due to a low rate of homologous recombination and a high rate of illegitimate recombination. The current strategy for disruption of genes in *H. capsulatum* requires a positive and negative selection strategy whereby the disruption construct is maintained extrachromosomally for several generations to allow time for the rare homologous recombination event to occur (78).

Increasing the frequency of homologous versus nonhomologous recombination in *H. capsulatum* could make targeted gene disruption a more practical strategy. There are two routes for recombination in eukaryotes: homologous recombination, which recombines homologous sequences, and nonhomologous end joining (NHEJ), which recombines sequences with little or no homology (4). The core components of NHEJ are the DNA-dependent protein kinase catalytic subunit, the Ku70-Ku80 complex, and the DNA ligase IV-Xrcc4 complex (13, 92). Recently, work with *Neurospora crassa* revealed that mutating components of the NHEJ process resulted in increased homologous recombination. When the *N. crassa* Ku70 or Ku80 homologs were disrupted, the frequency of homologous recombination increased dramatically (68). It is possible that disruption of the Ku70 or Ku80 homologs in *H. capsulatum* would have the same effect, allowing for more straightforward disruption of target genes. If so, a transient disruption of Ku70 or Ku80 by RNA interference (see the following section) would be the ideal method of facilitating gene disruption.

RNA Interference

To exploit the large amount of information that will be generated by the genome project and functional genomics screens, a rapid method for disrupting gene function is essential. RNA interference (RNAi) technology can be an efficient and powerful alternative to gene disruption (61). The laboratory of William Goldman recently developed RNAi technology for *H. capsulatum* (72). They expressed long RNA stem-loop hairpins consisting of relatively large pieces of coding sequence and its reverse complement separated by a short spacer region. These constructs were maintained extrachromosomally on telomeric plasmids. The expression of both exogenous (green fluorescent protein) and endogenous genes was effectively reduced when these constructs were used. In other systems, short interfering RNAs have been used to induce silencing (58). These constructs can be generated quickly, and if they are effective in *H. capsulatum*, they would make going from gene sequence to phenotype even more expeditious. It may even be possible to perform genome-scale RNAi to screen for mutant phenotypes.

LOOKING FORWARD

These are exciting times in the *H. capsulatum* research field. The era of genomics, coupled with increasingly robust molecular biology-based tools, makes it a propitious time to study this dimorphic fungus. It is likely that genomics will contribute to the identification of genes and regulatory circuits that allow *H. capsulatum* to sense, respond to, and manipulate its environment, whether that be the soil or a mammalian host. In addition, comparative genomic analysis of *H. capsulatum* and other fungi will contribute to our understanding of the diversity of the fungal kingdom.

REFERENCES

1. **Ajello, L., and L. C. Runyon.** 1953. Infection of mice with single spores of *Histoplasma capsulatum*. *J. Bacteriol.* **66:**34–40.

2. **Anderson, K. L., and S. Marcus.** 1968. Sporulation characteristics of *Histoplasma capsulatum*. *Mycopathol. Mycol. Appl.* **36:**179–187.

3. **Artis, D., and G. L. Baum.** 1963. Tuberculate spore formation by thirty-two strains of *Histoplasma capsulatum*. *Mycopathol. Mycol. Appl.* **21:**29–35.

4. **Aylon, Y., and M. Kupiec.** 2004. DSB repair: the yeast paradigm. *DNA Repair* **3:**797–815.

5. **Bennett, R. J., M. A. Uhl, M. G. Miller, and A. D. Johnson.** 2003. Identification and characterization of a *Candida albicans* mating pheromone. *Mol. Cell. Biol.* **23:**8189–8201.

6. **Boldrick, J. C., A. A. Alizadeh, M. Diehn, S. Dudoit, C. L. Liu, C. E. Belcher, D. Botstein, L. M. Staudt, P. O. Brown, and D. A. Relman.** 2002. Stereotyped and specific gene expression programs in human innate immune responses to bacteria. *Proc. Natl. Acad. Sci. USA* **99:**972–977.

7. **Borges-Walmsley, M. I., and A. R. Walmsley.** 2000. cAMP signalling in pathogenic fungi: control of dimorphic switching and pathogenicity. *Trends Microbiol.* **8:**133–141.

8. **Bradsher, R. W.** 1996. Histoplasmosis and blastomycosis. *Clin. Infect. Dis.* **22**(Suppl. 2):S102–S111.

9. **Brown, A. E.** 1990. Overview of fungal infections in cancer patients. *Semin. Oncol.* **17:**2–5.

10. **Bryan, C. S., and A. F. DiSalvo.** 1979. Overwhelming opportunistic histoplasmosis. *Sabouraudia* **17:**209–212.

11. **Bullock, W. E.** 1993. Interactions between human phagocytic cells and *Histoplasma capsulatum*. *Arch. Med. Res.* **24:**219–223.

12. **Carr, J., and G. Shearer, Jr.** 1998. Genome size, complexity, and ploidy of the pathogenic fungus *Histoplasma capsulatum*. *J. Bacteriol.* **180:**6697–6703.

13. **Critchlow, S. E., and S. P. Jackson.** 1998. DNA end-joining: from yeast to man. *Trends Biochem. Sci.* **23:**394–398.

14. **Davis, T. E., Jr., J. E. Domer, and Y. T. Li.** 1977. Cell wall studies of *Histoplasma capsulatum* and *Blastomyces dermatitidis* using autologous and heterologous enzymes. *Infect. Immun.* **15:**978–987.

15. **Detweiler, C. S., D. B. Cunanan, and S. Falkow.** 2001. Host microarray analysis reveals a role for the *Salmonella* response regulator *phoP* in human macrophage cell death. *Proc. Natl. Acad. Sci. USA* **98:**5850–5855.

16. **Detweiler, C. S., D. M. Monack, I. E. Brodsky, H. Mathew, and S. Falkow.** 2003. *virK*, *somA* and *rcsC* are important for systemic *Salmonella enterica* serovar Typhimurium infection and cationic peptide resistance. *Mol. Microbiol.* **48:**385–400.

17. **Di Lallo, G., S. Gargano, and B. Maresca.** 1994. The *Histoplasma capsulatum cdc2* gene is transcriptionally regulated during the morphologic transition. *Gene* **140:**51–57.

18. **Duclos, S., and M. Desjardins.** 2000. Subversion of a young phagosome: the survival strategies of intracellular pathogens. *Cell. Microbiol.* **2:**365–377.

19. **Eisen, M. B., P. T. Spellman, P. O. Brown, and D. Botstein.** 1998. Cluster analysis and display of genome-wide expression patterns. *Proc. Natl. Acad. Sci. USA* **95:**14863–14868.

20. **Eissenberg, L. G., and W. E. Goldman.** 1991. *Histoplasma* variation and adaptive strategies for parasitism: new perspectives on histoplasmosis. *Clin. Microbiol. Rev.* **4:**411–421.

21. **Eissenberg, L. G., W. E. Goldman, and P. H. Schlesinger.** 1993. *Histoplasma capsulatum* modulates the acidification of phagolysosomes. *J. Exp. Med.* **177:**1605–1611.

22. **Eissenberg, L. G., Goldman, W. E.** 1994. The interplay between *Histoplasma capsulatum* and its host cells. *Balliere's Clin. Infect. Dis.* **1:**265–283.

23. **Eissenberg, L. G., P. H. Schlesinger, and W. E. Goldman.** 1988. Phagosome-lysosome fusion in P388D1 macrophages infected with *Histoplasma capsulatum*. *J. Leukoc. Biol.* **43:**483–491.

24. **Eissenberg, L. G., J. L. West, J. P. Woods, and W. E. Goldman.** 1991. Infection of P388D1 macrophages and respiratory epithelial cells by *Histoplasma capsulatum*: selection of avirulent variants and their potential role in persistent histoplasmosis. *Infect. Immun.* **59:**1639–1646.

25. **Gargano, S., G. Di Lallo, G. S. Kobayashi, and B. Maresca.** 1995. A temperature-sensitive strain of *Histoplasma capsulatum* has an altered delta 9-fatty acid desaturase gene. *Lipids* **30:**899–906.

26. **Garrison, R. G., and K. S. Boyd.** 1978. Electron microscopy of yeastlike cell development from the microconidium of *Histoplasma capsulatum*. *J. Bacteriol.* **133:**345–353.

27. **Garrison, R. G., and K. S. Boyd.** 1977. The fine structure of microconidial germination and vegetative cells of *Histoplasma capsulatum*. *Ann. Microbiol.* (Paris) **128:**135–149.

28. **Garrison, R. G., and J. W. Lane.** 1973. Scanning-beam electron microscopy of the conidia of the brown and albino filamentous varieties of *Histoplasma capsulatum*. *Mycopathol. Mycol. Appl.* **49:**185–191.

29. **Goos, R. D.** 1964. Germination of the macroconidia of *Histoplasma capsulatum*. *Mycologia* **56:**662–671.

30. **Hajjeh, R. A.** 1995. Disseminated histoplasmosis in persons infected with human immunodeficiency virus. *Clin. Infect. Dis.* **21**(Suppl. 1):S108–S110.

31. **Harris, G. S., E. J. Keath, and J. Medoff.** 1989. Characterization of alpha and beta tubulin genes in the dimorphic fungus *Histoplasma capsulatum*. *J. Gen. Microbiol.* **135:**1817–1832.

32. **Harris, G. S., E. J. Keath, and J. Medoff.** 1989. Expression of alpha- and beta-tubulin genes during dimorphic-phase transitions of *Histoplasma capsulatum*. *Mol. Cell. Biol.* **9:**2042–2049.

33. **Howard, D. H.** 1959. Observations on tissue cultures of mouse peritoneal exudates inoculated with *Histoplasma capsulatum*. *J. Bacteriol.* **78:**69–78.

34. **Huffnagle, G. B., and G. S. Deepe.** 2003. Innate and adaptive determinants of host susceptibility to medically important fungi. *Curr. Opin. Microbiol.* **6:**344–350.

35. **Hwang, L., D. Hocking-Murray, A. K. Bahrami, M. Andersson, J. Rine, and A. Sil.** 2003. Identifying phase-specific genes in the fungal pathogen *Histoplasma capsulatum* using a genomic shotgun microarray. *Mol. Biol. Cell* **14:**2314–2326.

36. Johnson, C. H., M. G. Klotz, J. L. York, V. Kruft, and J. E. McEwen. 2002. Redundancy, phylogeny and differential expression of *Histoplasma capsulatum* catalases. *Microbiology* **148**:1129–1142.

37. Kasuga, T., J. W. Taylor, and T. J. White. 1999. Phylogenetic relationships of varieties and geographical groups of the human pathogenic fungus *Histoplasma capsulatum* Darling. *J. Clin. Microbiol.* **37**:653–663.

38. Kasuga, T., T. J. White, G. Koenig, J. McEwen, A. Restrepo, E. Castaneda, C. Da Silva Lacaz, E. M. Heins-Vaccari, R. S. De Freitas, R. M. Zancope-Oliveira, Z. Qin, R. Negroni, D. A. Carter, Y. Mikami, M. Tamura, M. L. Taylor, G. F. Miller, N. Poonwan, and J. W. Taylor. 2003. Phylogeography of the fungal pathogen *Histoplasma capsulatum*. *Mol. Ecol.* **12**:3383–3401.

39. Kauffman, C. A., K. S. Israel, J. W. Smith, A. C. White, J. Schwarz, and G. F. Brooks. 1978. Histoplasmosis in immunosuppressed patients. *Am. J. Med.* **64**:923–932.

40. Keath, E. J., and F. E. Abidi. 1994. Molecular cloning and sequence analysis of *yps-3*, a yeast-phase-specific gene in the dimorphic fungal pathogen *Histoplasma capsulatum*. *Microbiology* **140**:759–767.

41. Keath, E. J., G. S. Kobayashi, and G. Medoff. 1992. Typing of *Histoplasma capsulatum* by restriction fragment length polymorphisms in a nuclear gene. *J. Clin. Microbiol.* **30**:2104–2107.

42. Keath, E. J., A. A. Painter, G. S. Kobayashi, and G. Medoff. 1989. Variable expression of a yeast-phase-specific gene in *Histoplasma capsulatum* strains differing in thermotolerance and virulence. *Infect. Immun.* **57**:1384–1390.

43. Klimpel, K. R., and W. E. Goldman. 1988. Cell walls from avirulent variants of *Histoplasma capsulatum* lack α-(1,3)-glucan. *Infect. Immun.* **56**:2997–3000.

44. Klimpel, K. R., and W. E. Goldman. 1987. Isolation and characterization of spontaneous avirulent variants of *Histoplasma capsulatum*. *Infect. Immun.* **55**:528–533.

45. Kwon-Chung, K. J., M. S. Bartlett, and L. J. Wheat. 1984. Distribution of the two mating types among *Histoplasma capsulatum* isolates obtained from an urban histoplasmosis outbreak. *Sabouraudia* **22**:155–157.

46. Kwon-Chung, K. J., R. J. Weeks, and H. W. Larsh. 1974. Studies on *Emmonsiella capsulata* (*Histoplasma capsulatum*). II. Distribution of the two mating types in 13 endemic states of the United States. *Am. J. Epidemiol.* **99**:44–49.

47. Lambowitz, A. M., G. S. Kobayashi, A. Painter, and G. Medoff. 1983. Possible relationship of morphogenesis in pathogenic fungus, *Histoplasma capsulatum*, to heat shock response. *Nature* **303**:806–808.

48. Lane, T. E., B. A. Wu-Hsieh, and D. H. Howard. 1994. Antihistoplasma effect of activated mouse splenic macrophages involves production of reactive nitrogen intermediates. *Infect. Immun.* **62**:1940–1945.

49. Lee, B. N., and T. H. Adams. 1996. *fluG* and *flbA* function interdependently to initiate conidiophore development in *Aspergillus nidulans* through *brlA* beta activation. *EMBO J.* **15**:299–309.

50. Magrini, V., W. C. Warren, J. Wallis, W. E. Goldman, J. Xu, E. R. Mardis, and J. D. McPherson. 2004. Fosmid-based physical mapping of the *Histoplasma capsulatum* genome. *Genome Res.* **14**:1603–1609.

51. Maresca, B., L. Carratu, and G. S. Kobayashi. 1994. Morphological transition in the human fungal pathogen *Histoplasma capsulatum*. *Trends Microbiol.* **2**:110–114.

52. Maresca, B., and G. S. Kobayashi. 1989. Dimorphism in *Histoplasma capsulatum*: a model for the study of cell differentiation in pathogenic fungi. *Microbiol. Rev.* **53**:186–209.

53. Maresca, B., A. M. Lambowitz, V. B. Kumar, G. A. Grant, G. S. Kobayashi, and G. Medoff. 1981. Role of cysteine in regulating morphogenesis and mitochondrial activity in the dimorphic fungus *Histoplasma capsulatum*. *Proc. Natl. Acad. Sci. USA* **78**:4596–4600.

54. Maresca, B., G. Medoff, D. Schlessinger, and G. S. Kobayashi. 1977. Regulation of dimorphism in the pathogenic fungus *Histoplasma capsulatum*. *Nature* **266**:447–448.

55. Marques, S. A., A. M. Robles, A. M. Tortorano, M. A. Tuculet, R. Negroni, and R. P. Mendes. 2000. Mycoses associated with AIDS in the Third World. *Med. Mycol.* **38**(Suppl 1):269–279.

56. Marshall, M. A., and W. E. Timberlake. 1991. *Aspergillus nidulans wetA* activates spore-specific gene expression. *Mol. Cell. Biol.* **11**:55–62.

57. McCaffrey, R. L., P. Fawcett, M. O'Riordan, K. D. Lee, E. A. Havell, P. O. Brown, and D. A. Portnoy. 2004. A specific gene expression program triggered by Gram-positive bacteria in the cytosol. *Proc. Natl. Acad. Sci. USA* **101**:11386–11391.

58. McManus, M. T., C. P. Petersen, B. B. Haines, J. Chen, and P. A. Sharp. 2002. Gene silencing using micro-RNA designed hairpins. *RNA* **8**:842–850.

59. Medoff, G., B. Maresca, A. M. Lambowitz, G. Kobayashi, A. Painter, M. Sacco, and L. Carratu. 1986. Correlation between pathogenicity and temperature sensitivity in different strains of *Histoplasma capsulatum*. *J. Clin. Investig.* **78**:1638–1647.

60. Medoff, G., M. Sacco, B. Maresca, D. Schlessinger, A. Painter, G. S. Kobayashi, and L. Carratu. 1986. Irreversible block of the mycelial-to-yeast phase transition of *Histoplasma capsulatum*. *Science* **231**:476–479.

61. Meister, G., and T. Tuschl. 2004. Mechanisms of gene silencing by double-stranded RNA. *Nature* **431**:343–349.

62. Nakamura, L. T., B. A. Wu-Hsieh, and D. H. Howard. 1994. Recombinant murine gamma interferon stimulates macrophages of the RAW cell line to inhibit intracellular growth of *Histoplasma capsulatum*. *Infect. Immun.* **62**:680–684.

63. Nau, G. J., J. F. Richmond, A. Schlesinger, E. G. Jennings, E. S. Lander, and R. A. Young. 2002. Human macrophage activation programs induced by bacterial pathogens. *Proc. Natl. Acad. Sci. USA* **99**:1503–1508.

64. Nau, G. J., A. Schlesinger, J. F. Richmond, and R. A. Young. 2003. Cumulative Toll-like receptor activation in human macrophages treated with whole bacteria. *J. Immunol.* **170**:5203–5209.

65. Neilsen, G. E., and R. E. Evans. 1964. A study of the sporulation of *Histoplasma capsulatum*. *J. Bacteriol.* **68**:261–264.

66. Newman, S. L. 1999. Macrophages in host defense against *Histoplasma capsulatum*. *Trends Microbiol.* **7**:67–71.

67. Newman, S. L., L. Gootee, C. Kidd, G. M. Ciraolo, and R. Morris. 1997. Activation of human macrophage fungistatic activity against *Histoplasma capsulatum* upon adherence to type 1 collagen matrices. *J. Immunol.* **158**:1779–1786.

68. Ninomiya, Y., K. Suzuki, C. Ishii, and H. Inoue. 2004. Highly efficient gene replacements in *Neurospora* strains deficient for nonhomologous end-joining. *Proc. Natl. Acad. Sci. USA* **101**:12248–12253.

68a. Nittler, M. P., D. Hocking-Murray, C. K. Foo, and A. Sil. 2005. Identification of *Histoplasma capsulatum* transcripts induced in response to reactive nitrogen species. *Mol. Biol. Cell* **16**:4792–4813.

69. Patel, J. B., J. W. Batanghari, and W. E. Goldman. 1998. Probing the yeast phase-specific expression of the *CBP1* gene in *Histoplasma capsulatum*. *J. Bacteriol.* **180:** 1786–1792.

70. Pine, L. (ed.). 1960. *Morphological and Physiological Characteristics of Histoplasma capsulatum*. Charles C Thomas, Springfield, Ill.

71. Procknow, J. J., M. I. Page, and C. G. Loosli. 1960. Early pathogenesis of experimental histoplasmosis. *Arch. Pathol.* **69:**413–426.

72. Rappleye, C. A., J. T. Engle, and W. E. Goldman. 2004. RNA interference in *Histoplasma capsulatum* demonstrates a role for α-(1,3)-glucan in virulence. *Mol. Microbiol.* **53:**153–165.

73. Reiss, E. 1977. Serial enzymatic hydrolysis of cell walls of two serotypes of yeast-form *Histoplasma capsulatum* with α(1→3)-glucanase, β(1→3)-glucanase, pronase, and chitinase. *Infect. Immun.* **16:**181–188.

74. Rippon, J. W. 1988. *Medical Mycology*. The W. B. Saunders Co., Philadelphia, Pa.

75. Sacco, M., G. Medoff, A. M. Lambowitz, B. V. Kumar, G. S. Kobayashi, and A. Painter. 1983. Sulfhydryl induced respiratory "shunt" pathways and their role in morphogenesis in the fungus, *Histoplasma capsulatum*. *J. Biol. Chem.* **258:**8223–8230.

76. Samonis, G., and D. Bafaloukos. 1992. Fungal infections in cancer patients: an escalating problem. *In Vivo* **6:**183–193.

77. Schadt, E. E., S. W. Edwards, D. GuhaThakurta, D. Holder, L. Ying, V. Svetnik, A. Leonardson, K. W. Hart, A. Russell, G. Li, G. Cavet, J. Castle, P. McDonagh, Z. Kan, R. Chen, A. Kasarskis, M. Margarint, R. M. Caceres, J. M. Johnson, C. D. Armour, P. W. Garrett-Engele, N. F. Tsinoremas, and D. D. Shoemaker. 2004. A comprehensive transcript index of the human genome generated using microarrays and computational approaches. *Genome Biol.* **5:**R73.

78. Sebghati, T. S., J. T. Engle, and W. E. Goldman. 2000. Intracellular parasitism by *Histoplasma capsulatum*: fungal virulence and calcium dependence. *Science* **290:** 1368–1372.

79. Shoemaker, D. D., E. E. Schadt, C. D. Armour, Y. D. He, P. Garrett-Engele, P. D. McDonagh, P. M. Loerch, A. Leonardson, P. Y. Lum, G. Cavet, L. F. Wu, S. J. Altschuler, S. Edwards, J. King, J. S. Tsang, G. Schimmack, J. M. Schelter, J. Koch, M. Ziman, M. J. Marton, B. Li, P. Cundiff, T. Ward, J. Castle, M. Krolewski, M. R. Meyer, M. Mao, J. Burchard, M. J. Kidd, H. Dai, J. W. Phillips, P. S. Linsley, R. Stoughton, S. Scherer, and M. S. Boguski. 2001. Experimental annotation of the human genome using microarray technology. *Nature* **409:**922–927.

80. Smith, C. D. 1964. Evidence of the presence in yeast extract of substances which stimulate the growth of *Histoplasma capsulatum* and *Blastomyces dermatitidis* similarly to that found in starling manure extract. *Mycopathol. Mycol. Appl.* **22:**99–105.

81. Smith, C. D., and M. L. Furcolow. 1964. The demonstration of growth stimulating substances for *Histoplasma capsulatum* and *Blastomyces dermatitidis* in infusions of starling (*Sturnis vulgaris*) manure. *Mycopathol. Mycol. Appl.* **22:**73–80.

82. Spitzer, E. D., E. J. Keath, S. J. Travis, A. A. Painter, G. S. Kobayashi, and G. Medoff. 1990. Temperature-sensitive variants of *Histoplasma capsulatum* isolated from patients with acquired immunodeficiency syndrome. *J. Infect. Dis.* **162:**258–261.

83. Steele, P. E., G. F. Carle, G. S. Kobayashi, and G. Medoff. 1989. Electrophoretic analysis of *Histoplasma capsulatum* chromosomal DNA. *Mol. Cell. Biol.* **9:** 983–987.

84. Sternberg, S. 1994. The emerging fungal threat. *Science* **266:**1632–1634.

85. Strasser, J. E., S. L. Newman, G. M. Ciraolo, R. E. Morris, M. L. Howell, and G. E. Dean. 1999. Regulation of the macrophage vacuolar ATPase and phagosome-lysosome fusion by *Histoplasma capsulatum*. *J. Immunol.* **162:**6148–6154.

86. Tenney, A. E., R. H. Brown, C. Vaske, J. K. Lodge, T. L. Doering, and M. R. Brent. 2004. Gene prediction and verification in a compact genome with numerous small introns. *Genome Res.* **14:**2330–2335.

87. Tian, X., and G. Shearer, Jr. 2001. Cloning and analysis of mold-specific genes in the dimorphic fungus *Histoplasma capsulatum*. *Gene* **275:**107–114.

88. Tian, X., and G. Shearer, Jr. 2002. The mold-specific *MS8* gene is required for normal hypha formation in the dimorphic pathogenic fungus *Histoplasma capsulatum*. *Eukaryot. Cell* **1:**249–256.

89. Timberlake, W. E. 1991. Temporal and spatial controls of *Aspergillus* development. *Curr. Opin. Genet. Dev.* **1:** 351–357.

90. Tjaden, B., R. M. Saxena, S. Stolyar, D. R. Haynor, E. Kolker, and C. Rosenow. 2002. Transcriptome analysis of *Escherichia coli* using high-density oligonucleotide probe arrays. *Nucleic Acids Res.* **30:**3732–3738.

91. Vieira, O. V., R. J. Botelho, and S. Grinstein. 2002. Phagosome maturation: aging gracefully. *Biochem. J.* **366:** 689–704.

92. Walker, J. R., R. A. Corpina, and J. Goldberg. 2001. Structure of the Ku heterodimer bound to DNA and its implications for double-strand break repair. *Nature* **412:**607–614.

93. Wheat, L. J., and C. A. Kauffman. 2003. Histoplasmosis. *Infect. Dis. Clin. North Am.* **17:**1–19, vii.

94. Wieser, J., B. N. Lee, J. Fondon III, and T. H. Adams. 1994. Genetic requirements for initiating asexual development in *Aspergillus nidulans*. *Curr. Genet.* **27:**62–69.

95. Woods, J. P. 2002. *Histoplasma capsulatum* molecular genetics, pathogenesis, and responsiveness to its environment. *Fungal Genet. Biol.* **35:**81–97.

96. Woods, J. P. 2003. Knocking on the right door and making a comfortable home: *Histoplasma capsulatum* intracellular pathogenesis. *Curr. Opin. Microbiol.* **6:** 327–331.

97. Yamada, K., J. Lim, J. M. Dale, H. Chen, P. Shinn, C. J. Palm, A. M. Southwick, H. C. Wu, C. Kim, M. Nguyen, P. Pham, R. Cheuk, G. Karlin-Newmann, S. X. Liu, B. Lam, H. Sakano, T. Wu, G. Yu, M. Miranda, H. L. Quach, M. Tripp, C. H. Chang, J. M. Lee, M. Toriumi, M. M. Chan, C. C. Tang, C. S. Onodera, J. M. Deng, K. Akiyama, Y. Ansari, T. Arakawa, J. Banh, F. Banno, L. Bowser, S. Brooks, P. Carninci, Q. Chao, N. Choy, A. Enju, A. D. Goldsmith, M. Gurjal, N. F. Hansen, Y. Hayashizaki, C. Johnson-Hopson, V. W. Hsuan, K. Iida, M. Karnes, S. Khan, E. Koesema, A. Ishida, P. X. Jiang, T. Jones, J. Kawai, A. Kamiya, C. Meyers, M. Nakajima, M. Narusaka, M. Seki, T. Sakurai, M. Satou, R. Tamse, M. Vaysberg, E. K. Wallender, C. Wong, Y. Yamamura, S. Yuan, K. Shinozaki, R. W. Davis, A. Theologis, and J. R. Ecker. 2003. Empirical analysis of transcriptional activity in the *Arabidopsis* genome. *Science* **302:**842–846.

Molecular Principles of Fungal Pathogenesis
Edited by Joseph Heitman et al.
©2006 ASM Press, Washington, D.C.

Chapter 42

Functional Genomic Approaches to Fungal Pathogenesis, Drug Target Validation, and Antifungal Drug Discovery

Ainslie B. Parsons, Howard Bussey, and Charles Boone

As the number of reported fungal infections increases worldwide and new pathogenic fungi continue to emerge, there is an ongoing need for novel antifungal therapies coupled with a thorough understanding of fungal pathogenesis. Fortunately, the post-genome-sequencing era of functional genomics is providing new insights, technologies, and methodologies for tackling these classical problems. In particular, genomic studies of the nonpathogenic budding yeast *Saccharomyces cerevisiae* are proving to be highly applicable to antifungal drug discovery. Functional genomics is also playing an increasing role in the identification of novel drug targets in pathogenic fungi such as *Candida albicans*. Among other applications, expanding knowledge of genetic interactions in fungi on a global scale is revealing new ideas and paradigms for combinatorial drug therapy.

FUNGAL PATHOGENS AND ANTIFUNGAL DRUGS

As indicated throughout this text, human fungal pathogens are numerous and diverse and pose a serious public health problem. Life-threatening infections can be caused by the well-known opportunist fungi such as *C. albicans*, *Cryptococcus neoformans*, and *Aspergillus fumigatus*. Fungal infections are a serious concern in immunocompromised patients (for example, organ transplant recipients and patients infected with human immunodeficiency virus), a population which continues to grow (13). In addition, new and emerging fungal pathogens, including species of *Candida* and *Aspergillus* other than *C. albicans* and *A. fumigatus* (60), as well as drug-resistant strains of common pathogens, continue to present new challenges (4).

There are currently four major classes of antifungal drugs licensed for therapy of invasive fungal infections. These include the polyenes (e.g., amphotericin B), which alter membrane function; nucleoside analogs (e.g., flucytosine), which inhibit DNA or RNA synthesis; the azoles (e.g., fluconazole, intraconazole, and voriconazole), which inhibit ergosterol biosynthesis; and the newest class, the echinocandins (e.g., caspofungin), which inhibit the synthesis of glucan, a necessary component of the yeast cell wall.

The polyenes, produced by *Streptomyces* species and introduced as drugs approximately 50 years ago, are the original antifungal drugs. The only polyene currently in clinical use is amphotericin B deoxycholate. Amphotericin B binds to ergosterol in the fungal plasma membrane, resulting in the leakage of intracellular univalent and divalent cations and leading to cell death (7). Amphotericin B also binds to the mammalian equivalent of ergosterol, cholesterol, but with less affinity and as such is more toxic to fungal cell membranes. Amphotericin B has the widest spectrum of activity against fungi, including most *Candida* and *Aspergillus* species (9); however, treatment often leads to acute and chronic side effects, particularly nephrotoxicity. More recently, lipid formulations of the drug have been developed that are less toxic (56). Resistance to amphotericin B is rare but can occur through alterations in the ergosterol content of the cell membrane (59, 67).

Flucytosine is a pyrimidine analog that is transported into susceptible fungi by cytosine permease and then deaminated into the active form (5-fluorouracil), which interferes with nucleic acid biosynthesis. Flucytosine has a relatively limited activity spectrum; *Candida* and *Cryptococcus* are normally susceptible to flucytosine, whereas most molds, including *Aspergillus*, are resistant. In addition, when it is used as a single agent, resistance quickly develops, normally though mutations in cytosine

Ainslie B. Parsons and Charles Boone • Banting and Best Department of Medical Research, University of Toronto, Toronto, ON M5G 1L6, Canada. **Howard Bussey** • Department of Biology, McGill University, Montreal, QC H2A 1B1, Canada.

permease; as such, it is most commonly used in combination with other antifungals such as amphotericin B (73).

The azole antifungals, synthetic compounds discovered in the late 1960s, are a continually expanding group of antifungal agents (20). Systemic azoles are fungistatic and have broad-spectrum activity against most yeasts and filamentous fungi. Azoles inhibit a key enzyme required for ergosterol biosynthesis, fungal cytochrome P450 14-α-sterol demethylase, which leads to a reduction in the amount of ergosterol on the fungal cell membrane and an inhibition of fungal cell growth. Several azoles are licensed for clinical use whose molecular structures differ as a result of specific side chains; these structural differences result in differences in pharmacological properties, toxicity profiles, and spectra of activity. These azoles include fluconazole, itraconazole, and voriconazole. Resistance to the azole drugs can occur through several mechanisms including alteration of the target enzyme and upregulation of efflux transporters; *C. albicans* resistance to fluconazole has emerged after long-term suppressive therapy (61). The azoles have a favorable safety profile and are generally well tolerated.

The newest antifungal agents with a novel mechanism of action are the echinocandins. Echinocandins are natural products discovered in the 1970s that inhibit β-1,3-glucan synthase (14, 16) which synthesizes β-1,3-glucan, a major essential component of the fungal cell wall and a target absent from humans. The sole member of the echinocandin family currently approved for clinical use is caspofungin, which has excellent activity against *Candida* (including azole-resistant species) and *Aspergillus* (12). Few toxicities have been associated with caspofungin; however, since it can be administered only intravenously, its use is limited to a hospital setting.

Despite the availability of the antifungal drugs described above, emerging resistance to the classic antifungal drugs and the increasing prevalence of fungal infections worldwide necessitates the development of new antifungals. Pertinently, there is an acute unmet need for a safe and effective new antifungal chemical entity that can be taken orally.

S. CEREVISIAE AS A MODEL AND TOOL FOR ANTIFUNGAL DRUG DISCOVERY

Budding Yeast as a Model Organism

Most mycologists view both *S. cerevisiae* and its scientists as being decidedly odd. For one thing, baker's yeast lacks much of the fungal repertoire that we expect of the molds and the mushrooms. Even odder to mycologists is that this reduced fungus, one

barely worthy of serious attention, should occupy such a central role in eukaryotic cell biology and have a large and often condescending research community working on it with little knowledge or interest in the fungi.

However, as a model organism, *S. cerevisiae*, also known as budding yeast, has many attributes. It is a unicellular organism with a compact genome of approximately 6,000 genes (29) and a life cycle well suited to classical genetic studies. Additionally, the high conservation of many cellular processes between yeast and higher organisms, especially with regard to basic cellular metabolism and cell division, makes yeast a fundamental model eukaryote (8). Indeed, recent sequencing of the yeast (29) and human (44, 74) genomes has revealed thousands of yeast proteins with amino acid sequence similarity to at least one human protein (~2,700 at BLAST E-value of $<10^{-10}$, ~1,100 at BLAST E-value of $<10^{-50}$; batch BLAST with default setting; each of 6,357 known or predicted *S. cerevisiae* open reading frames) (36).

A new era of yeast research is under way: functional genomics and proteomics. Because the *S. cerevisiae* genome was the first eukaryotic genome to be completely sequenced, yeast researchers have had a head start in the application of functional genomic technologies and systems biology. To date, gene expression (35), protein-protein interactions (22, 33, 39, 72), protein abundance (23) and localization (38), and genetic interactions (70, 71) have all been studied on a genome-wide scale in yeast. Numerous Web-based databases exist cataloguing yeast genes, proteins, and interactions; these include the Saccharomyces Genome Database (SGD) (17) and the Comprehensive Yeast Genome Database (CYGD) at the Munich Information Center for Protein Sequences (MIPS) (52) and the Yeast Proteome Database (YPD), operated by Incyte (10), allowing yeast researchers relatively easy access to this wealth of information.

Major Fungal Pathogens and their Taxonomic Relationships with S. cerevisiae

Putting aside our disdain for the humble baker's yeast and its acolytes, can knowledge of the fungal taxonomy of this well understood "model cell" help us in the fight to combat fungal disease? The Kingdom Fungi has many phyla containing human pathogens, with many of them discussed in this book (see chapter 44). The phylum containing *S. cerevisiae* and most of the major human pathogens is the Ascomycota. Here, *S. cerevisiae* is a member of the small class called the Hemiascomycetes. Two human pathogens also belong to this class: *Candida glabrata*, the human pathogen most closely related to *S. cerevisiae*, and *C. albicans*, which is more divergent. The largest ascomycete class is the Euascomycetes, containing the filamentous fungi, including the human pathogens *A. fumigatus*, *Penicillium marneffeii*, *Histoplasma capsulatum*, and *Coccidioides immitis*. Thus, despite its reduced form, knowledge of *S. cerevisiae* will be of great value in understanding the basic cell biology and hence in defining antifungal targets in these highly related pathogenic fungi. As for the other fungal phyla, including the Basidiomycota (containing *C. neoformans*), and the Zygomycota (see chapter 29), these are more divergent from *S. cerevisiae*. However, even here if we reflect on how useful *S. cerevisiae* continues to be in unraveling basic processes relevant to the animal kingdom (including humans), then, comparatively, baker's yeast remains closely related to all fungi and experiments with this organism will be highly informative in our study of all fungal pathogens.

S. cerevisiae as a Model for Fungal Pathogenesis and Drug Discovery

C. albicans is the most prevalent and well-studied pathogenic fungus. Despite the availability of the complete genome sequence of *C. albicans* (40) and the continued development of new molecular biology-based tools (11), the genetic manipulation of *C. albicans* remains difficult. While there are *C. albicans*-specific genes and functions, the molecular basis of many cellular processes appears to be conserved between *C. albicans* and *S. cerevisiae*. Thus, paradigms derived from *S. cerevisiae* studies will continue to direct *C. albicans* research and contribute to our understanding of fungal pathogenesis and the identification of antifungal drug targets. For example, the rediscovery of an *S. cerevisiae* dimorphic switch from budding to pseudohyphal growth (27) and a dissection of the *S. cerevisiae* pathways involved in this process (46, 64) prompted similar analyses of *C. albicans* (42, 65), which demonstrated that *C. albicans* mutants defective for the dimorphic switch are avirulent (47). In addition, the discovery that *S. cerevisiae* adheres to polystyrene and forms floral mats (62) has identified the first genetically tractable system for the study of fungal biofilms, which play a major role in infections because they are often associated with implanted prosthetic devices, such as catheters, and resistance to antimicrobial therapies. It has also become apparent that *S. cerevisiae* can infect immunocompromised patients (50). Clinically isolated yeast grow at higher temperatures (41°C) than do laboratory strains (50), and this characteristic has been correlated with their survival in a mouse model system (51). The first genetic studies of

clinically isolated yeast have shown that metabolic genes required for *C. albicans* virulence are also required for the survival of *S. cerevisiae* clinical isolates in mice (30) and that disruption of the *SSD1* gene causes clinical isolates to become hypervirulent (75).

S. cerevisiae also provides a good model system for antifungal drug discovery. The most commonly used antifungal compounds, including the azoles (43), polyenes (2), and echinocandins (16), are all active against *S. cerevisiae*. In many cases, the genetics underlying the gene products targeted by these drugs and mechanisms of resistance have been elucidated in *S. cerevisiae*. For example, *S. cerevisiae* has been used to characterize new azoles (6) and determine the mechanisms of resistance to fluconazole in both clinical (41) and theoretical (1) applications. *S. cerevisiae* was instrumental for the identification of β-1,3-glucan synthase, encoded by *FKS1* and *FKS2*, as the caspofungin target through the characterization of caspofungin-resistant mutants (14–16).

S. cerevisiae Gene Deletion Mutant Collection

There are a number of new genomic resources under development and in use with *S. cerevisiae*, including genome-wide microarrays (35), two-hybrid clone sets (39, 72), protein chips (78), epitope-tagged yeast strains (38), and a complete set of deletion mutant strains (24). Many of these resources and their corresponding technologies have direct applications to antifungal drug discovery, but perhaps the most relevant resource is the yeast deletion mutant set (24). This collection consists of a comprehensive set of over 20,000 *S. cerevisiae* gene deletion strains, created by an international consortium of laboratories. Included in the collection are heterozygous diploid strains corresponding to deletions in each of the ~6,000 yeast genes, including the ~1,000 essential genes, and a homozygous diploid strain, a *MATa* strain, and a *MATα* strain for each of the ~5,000 nonessential genes. Each gene deletion mutation is marked by two unique 20-mer nucleotide sequences, which are flanked by common PCR primers (69, 77). These "molecular barcodes" enable the identification of each strain within a mixed population of mutants through PCR-based amplification of the barcodes followed by a microarray-based quantitative detection (24). The barcode system is particularly useful for a chemical-genetic analysis of the antifungal drug response, as discussed below.

So far, the *S. cerevisiae* deletion mutant collection has been constructed in only one genetic background, S288c, the common laboratory strain background that was sequenced (29). For a systematic genetic analysis of fungal phenotypes relevant to pathogenicity, there is the potential to construct similar deletion mutant sets in distantly related *S. cerevisiae* strains (76) such as Σ1278b, which is competent for the dimorphic switch and biofilm formation, or even pathogenic *S. cerevisiae* strains. Indeed, the broad utility of the deletion mutant collection for the systematic genetic analysis of specific phenotypes has clearly demonstrated the efficiency of this approach over random-mutagenesis-based strategies (68). Given the incredible utility of the yeast deletion mutant collection, the construction of systematic deletion mutant collections for the major fungal pathogens will probably be one of the most important functional genomics resources for antifungal target and drug discovery.

FUNCTIONAL GENOMICS APPROACHES FOR ANTIFUNGAL DRUG DISCOVERY

A fundamental property of living systems is that they are robust to perturbation (32); therefore, it is no longer a surprise (28) that single deletion of most genes in an organism is not lethal. In *S. cerevisiae* the proportion of essential genes is ~20% of the total. While these genes are the major candidates for target screening, an exclusive focus on them means excluding the bulk of the genome from use in the drug discovery process. A discussion of how to exploit the knowledge of both essential genes and biological networks containing nonessential genes (the remaining ~80% of genes) to identify drug targets is offered below.

Nonessential Yeast Genome: Chemical-Genetic Profiling of the Set of ~5,000 Viable Deletion Mutants

An immediately applicable result of the *S. cerevisiae* gene deletion project was the finding that most yeast genes (~80%) are not required for viability and cell growth and the creation of a set of ~5,000 viable deletion mutants, which can be screened for particular phenotypes systematically. Comprehensive chemical-genetic profiles of drug-sensitive deletion mutant strains can provide clues to the mechanism of drug action and can establish an index of potential genetic and pharmacological interactions. For example, gene deletions that render cells hypersensitive to established antifungal drugs identify all the pathways that buffer the cell against the toxic effects of the drug. Genome-wide screens provide a sensitivity profile of all mutants regardless of the severity of the drug-induced phenotype. Consequently, even moderate sensitivities can be identified, many of which may be relevant to the mode of action of a drug but can easily be missed in traditional screens. This is powerful information, but it also brings with it a number of limitations both theoretical and practical. Happily, many of these limitations can now be at least partially addressed.

Because many compounds are in limited supply, a major challenge is to screen the largest number of mutants in the most efficient manner possible while using the smallest amount of growth medium. Parallel analysis of large numbers of pooled deletion strains in a minimal amount of medium is possible due to the unique molecular barcodes that tag and identify each deletion strain (69, 77) (Fig. 1). In this strategy, strains

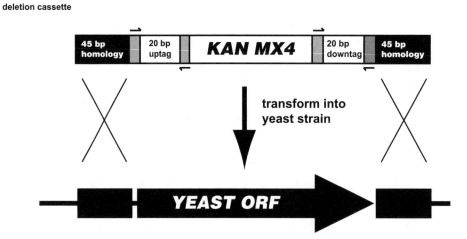

Figure 1. Construction strategy for the yeast deletion mutant collection. Each yeast open reading frame (ORF) is replaced with a deletion cassette consisting of an antibiotic resistance marker, KAN MX4, conferring resistance to kanamycin, and two unique 20-mer molecular barcodes (uptag and downtag). Each barcode is flanked by common primer sites (indicated by half arrows). Incorporation of the cassette is accomplished through homologous recombination of 45-bp regions of homology up- and downstream of the yeast open reading frame.

are pooled and grown in parallel in liquid culture under selective conditions (for example, in the presence of a compound) (Fig. 2). Genomic DNA is extracted from the pool, and the barcodes from each strain in the pool are simultaneously amplified using common flanking primers fluorescently labeled with Cy3 and Cy5. This pool of PCR amplimers is then hybridized to high-density DNA microarrays containing oligonucleotides corresponding to the barcodes, and the relative abundance of each strain in the pool is assessed by the strength of the resulting signals from the microarray readout compared to a mock control. Using this approach, the entire yeast deletion set can be pooled, grown competitively, and quantitatively assessed in small volumes of growth medium (24).

A number of groups have used this strategy to screen the yeast deletion mutant collection for drug hypersensitivity in order to uncover the pathways and cellular functions affected by drug treatment on a global scale. Giaever et al. profiled the nearly complete set of *S. cerevisiae* deletion strains under a number of different conditions including high salt, sorbitol, galactose, pH 8, minimal medium, and nystatin treatment (24). A similar method was used to

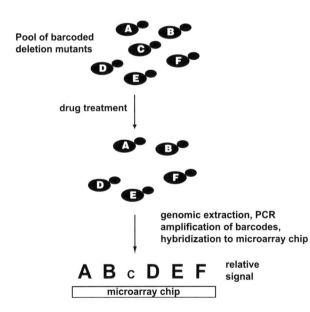

Figure 2. Parallel analysis of large pools of deletion mutants. Populations of pooled mutant cells, each marked with unique molecular barcodes, are grown in the presence or absence of a growth-inhibitory drug. Genomic DNA is extracted from the pool of mutants, and barcodes representing each strain are amplified by PCR with common primers labeled with the fluorescent marker Cy3 or Cy5. Drug-sensitive mutants are identified by competitive hybridization of the barcoded PCR products to a microarray containing oligonucleotides corresponding to each barcode.

screen the yeast genome for deletion mutants affected by two proteosome inhibitors; here, 52 strains were identified as hypersensitive to the drugs (19). Likewise, Hanway et al. (31) analyzed a collection of 2,827 yeast deletion strains for UV and methyl methanesulfonate sensitivity and implicated six genes not previously known to be involved in DNA damage repair pathways in the DNA damage response. Giaever et al. (25) also screened the deletion mutant collection for sensitivity to cisplatin, an antineoplastic agent that inhibits cell proliferation by modifying purines, and identified a number of genes involved in DNA repair.

Chemical-genetic profiling offers a rapid and easy method to obtain a "first glance" at the mode of action of a drug by using only a minimal volume of drug-containing medium. As outlined conceptually for drug-induced changes in gene expression patterns (37), compounds with similar biological effects will lead to similar chemical sensitivity profiles, and thus a compendium of chemical sensitivity profiles should provide new insights into the structural and functional relationships of the compounds. A detailed compendium of chemical-genetic interaction profiles will also organize the yeast genes into functional pathways because mutants that are sensitive to similar sets of compounds have highly similar phenotypes and thus are likely to occur within the same cellular pathway (Fig. 3). Moreover, the analysis of numerous structurally diverse compounds also provides a means of identifying deletion mutations that lead to hypersensitivity to a number of different compounds. Indeed, Parsons et al. (58) identified 65 different genes that appear to confer a multidrug-resistant phenotype on yeast cells, a set that includes genes encoding drug efflux pumps, ergosterol biosynthetic enzymes, and the vacuolar ATPase, which highlights the roles of pumps, membrane permeability, and the vacuole in cellular detoxification. Because chemical-genetic profiling focuses on compounds that impair cell growth, it can also be applied to natural-product extracts, which often contain only one growth-inhibitory active compound. Consequently, chemical-genetic profiling allows investigators to classify natural-product extracts before embarking on the time-consuming purification of the active component (A. B. Parsons et al., unpublished data).

SGA Analysis: Large-Scale Genetic Interaction Mapping in Yeast

The yeast haploid deletion mutant set has allowed the development of a high-throughput and genome-wide method to identify and study genetic interactions

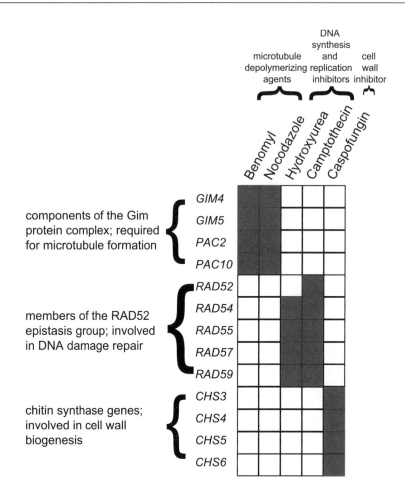

Figure 3. A compendium of chemical-genetic interaction profiles organizes yeast genes into functional pathways. Here, deletion mutants that are sensitive to a particular compound are indicated by the black boxes and an absence of a chemical-genetic interaction is indicated by a white box. Benomyl and nocodazole are microtubule-depolymerizing agents, hydroxyurea inhibits DNA synthesis, camptothecin inhibits topoisomerase function, and caspofungin is a cell wall inhibitor.

in *S. cerevisiae* (70, 71) (Fig. 4). This system, referred to as synthetic genetic array (SGA) analysis, allows the automated construction of double mutants, thereby enabling large-scale genetic interaction screens for synthetic lethal analysis and phenotypic enhancement. Synthetic lethal interactions are defined by two genes that are nonessential when individually deleted but essential when combined as a double mutant. In SGA, a series of robotic pinning steps are used to cross the ordered array of *MAT*a deletion mutants to a *MAT*α starting strain containing a query mutation of interest. The resultant diploids are selected for and pinned onto sporulation medium, and then double-mutant meiotic progeny are isolated by selective growth and scored for fitness defects. The key to this system was the development of an SGA reporter, *MFA1pr-HIS3*, which places the *HIS3* selectable marker under the control of a *MAT*a-specific reporter and enables the germination of *MAT*a meiotic progeny from a population containing both *MAT*a and *MAT*α spores.

Because fitness defects of double mutants can also be examined using the barcode approach with a microarray-based readout, SGA analysis can also be carried out using a pooled population of deletion mutants. Moreover, rather than crossing a mutation of interest into the set of ~5,000 viable deletion mutants, the query mutation can be introduced by transformation into heterozygous diploids containing the SGA reporter, a method referred to as dSLAM (57).

The *FKS1* gene, encoding a target for the antifungal caspofungin, provides a useful example of the type of genetic data generated by an SGA screen (45). In this case, an *FKS1* gene knockout was used as a query and crossed into the rest of the deletion mutant set such that the resultant double mutants were scored for a synthetic lethal or "sick" phenotype. *FKS1* is a nonessential gene that encodes a plasma membrane protein, Fks1p, thought to be the catalytic subunit of β-1,3-glucan synthase responsible for synthesis of the

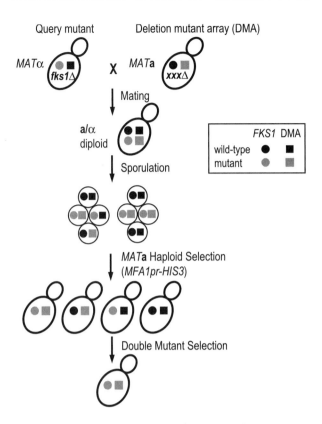

Figure 4. Synthetic genetic array analysis. A *MATα* strain carrying a query mutation (for example, *fks1Δ*) linked to a dominant selectable marker, such as the nourseothricin resistance marker *natMX*, and an *MFA1pr-HIS3* reporter is mated to ~5,000 *MATa* deletion mutants, each linked to a kanamycin resistance marker, *kanMX*. Growth of the resultant diploids is selected for on medium containing nourseothricin and kanamycin. The diploids are pinned onto medium designed to induce spore formation. *MATa* haploids are recovered using the *MFA1pr-HIS3* reporter gene, and double mutants are selected for medium containing nourseothricin and kanamycin. Inviable double mutants are scored as synthetic lethal, and slow-growing double mutants are scored as synthetic sick.

major cell wall polymer, β-1,3-glucan (15, 49). Fig. 5 shows the results of the *FKS1* SGA screen: 76 different genes showed a synthetic lethal or sick relationship with *FKS1*, defining a set we refer to as the *FKS1* genetic interaction neighborhood. Many of the genes within the *FKS1* neighborhood have been annotated with a general role in cell wall synthesis and assembly (e.g., genes encoding a second β-1,3-glucan synthase [*GSC2/FKS2*] and its stress-induced regulation, chitin and mannoprotein synthesis, secretion, and cell polarity). Genes with related biological functions are connected by synthetic genetic interactions more often than expected by chance (71), indicating that synthetic genetic interactions tend to occur among functionally related genes.

Large-scale SGA analysis of 132 different query genes has mapped a network containing ~4,000

interactions and ~1,000 genes (71). The number of interactions per query gene ranged from 1 to 146, with an average of 34 interactions per screen. This analysis predicts that there are on the order of ~100,000 digenic combinations associated with a synthetic lethal or sick phenotype, 100-fold more than the total number of essential genes. Genetic interactions are generally orthogonal (nonoverlapping) with protein-protein interactions, but two-dimensional hierarchical clustering of the interactions revealed that genes within the same pathway or complex often show similar patterns of genetic interactions. Clustering of previously uncharacterized genes with well studied pathways enables the prediction of precise molecular roles. A complete map of synthetic genetic interactions for yeast will identify sets of complexes and pathways that buffer each other.

A Compendium of Synthetic Lethal Interactions Provides a Key for Interpreting Chemical-Genetic Interaction Profiles

Chemical-genetic drug screens against the haploid deletion set can be interpreted by comparison with genome-wide genetic interaction screens generated by SGA or dSLAM. In theory, if an inhibitory molecule targets one gene product specifically, then a strain deleted for the drug target should provide a good model for drug activity (Fig. 6). In particular, mutants that are synthetically lethal with the deletion mutant of a drug target should also be hypersensitive to the drug. The pattern of growth defects among the ~5,000 deletion strains exposed to a drug should therefore mimic the pattern of synthetic lethality obtained from a mutation in the gene encoding the drug target. Thus, by comparing the chemical-genetic profiles of unknown compounds with a compendium of synthetic genetic profiles of characterized genes, it should be possible to identify the pathways and targets affected by drug treatment. Proof-of-concept experiments by Parsons et al. (58) demonstrated that the chemical-genetic interaction profiles of several compounds clustered with the corresponding SGA genetic interaction profile associated with the target gene. In theory, the chemical-genetic interactions for a specific drug that differ from those associated with the genetic interaction profiles of the target gene may provide an insight into the off-target drug effects; however, the two profiles are not expected to be exactly the same because mutations usually do not mimic the effect of a compound precisely. For example, some genes that show a chemical-genetic interaction with a drug but fail to show a genetic interaction with the drug target may be involved in cellular import or export of the compound or may provide insight into off-target drug effects.

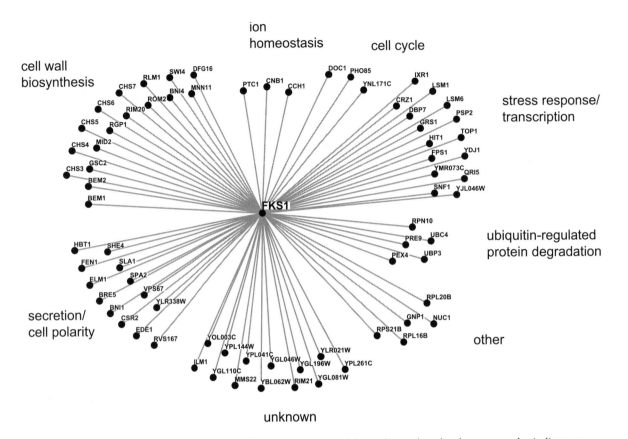

Figure 5. *FKS1* genetic interaction network. Genes are represented by nodes, and spokes between nodes indicate synthetic lethal or synthetic sick relationships. Genes are grouped according to their biological function.

Essential Yeast Genome: Essential Genes Are Hard To Define Precisely

The concept of gene essentiality is a complex and long-standing problem. Essential for what, growth or viability, and under what conditions? Blocking a target that is required for cell growth can lead to a static outcome in which the cells remain alive but fail to divide, whereas inhibiting a target required for viability leads to a cidal (lethal) terminal phenotype. Despite the apparent advantage of the latter, in practice they both serve as targets for effective clinical antibacterials. A gene in a fungal pathogen that is essential for viability in an infected immunocompromised human host would meet the gold-standard metric, but in most cases we can only tangentially deduce the likelihood of this being the case. The underlying genetic assumption here is that a drug that inhibits an essential gene will have the same effect as a null mutation in the gene.

Working with Essential Genes: the Tet System

At a practical level, how does one work with essential genes? Shutting them off blocks cell growth and/or viability; therefore, it is important to devise an efficient system to identify essential genes and maintain mutants with mutations in these genes.

How to stably propagate lethal mutants is not a trivial problem. The establishment of gene essentiality in *S. cerevisiae* was done by merely demonstrating that the null mutant, initially generated as a heterozygote in a diploid, failed to grow after germination as a haploid spore following meiosis and tetrad dissection. Thus, the set of essential genes was first defined by tetrad analysis and maintained as heterozygous deletion mutants. Identification of the set of essential yeast genes is one of the most important results to immediately fall out of the deletion project because it immediately defined ~1,000 candidate antifungal targets.

Another solution to stably maintain mutants defective for essential genes comes through the use of conditional alleles. For example, each essential gene could be placed under the control of a conditional promoter, such as the widely used tetracycline regulation system. Here, the native gene promoter is replaced by a tetracycline-regulatable promoter cassette, whose function is modulated by the addition of tetracycline, which either activates or inactivates a tetracycline-responsive transcription factor. In the "Tet-off" configuration, addition of tetracycline, or the related doxycycline, will shut off transcription of the essential gene, leading to growth inhibition and in some cases loss of viability (21).

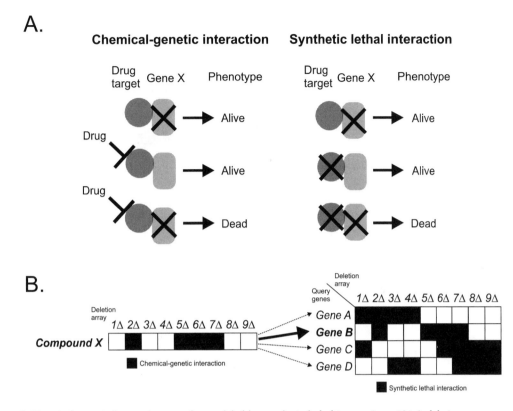

Figure 6. Chemical-genetic interactions can be modeled by synthetic lethal interactions. (A) A deletion mutant sensitive to a particular drug should also be synthetically lethal with the drug target. (B) Genome-wide chemical-genetic profiles can be compared to genome-wide synthetic lethal profiles to link compounds with their target protein or pathway.

Although there are a number of caveats with the use of such systems, it provides an effective system for demonstrating gene essentiality and for enabling phenotypic characterization of the mutant. In *S. cerevisiae*, tetracycline-regulated alleles of most of the essential genes have been constructed (53), allowing the terminal phenotype of hundreds of mutants to be examined for genome-wide changes in gene expression pattern, cell size, or morphology (53). The tetracycline-regulated alleles can also be used for SGA analysis, examining genetic interactions under semipermissive conditions (i.e., low levels of doxycycline that only partially inactivate the essential gene and lead to a slow growth phenotype), and proof-of-principle experiments have revealed that essential gene function is highly buffered by numerous nonessential and essential pathways (53).

The Haploinsufficiency of Essential Genes Allows Drug Screening

A more subtle use of genetics permits the gene dosage phenotypes of essential genes to be used in drug screening. For the most part, such heterozygotes, which retain half the gene dosage of the wild type for each essential gene, grow quite well. However, heterozygote mutants can be less robust than wild-type cells to perturbations that affect essential functions, a phenotype termed haploinsufficiency.

Although initially exploited in classical *Drosophila* genetics, the general use of gene dosage as a tool has found utility in experiments with fungi, providing a selection in cloning (63) and more recently using the heterozygous null yeast set of mutants in compound screening (25, 26, 48). The idea is simple but powerful. Having half the normal amount of an essential protein will render yeast cells more sensitive to an inhibitor of that protein than if they had the full amount (Fig. 7). This is because the MIC of an inhibitor is dependent on the protein concentration of its target. Viewed in another way, cellular overproduction of a protein will raise the MIC of an inhibitor, leading to cellular resistance (63). Thus, the protein concentration of a drug target is a variable that shifts the drug inhibition curve along the drug concentration axis. An important corollary is that mutants heterozygous at one essential locus will remain diploid at all others, such that the hypersensitivity shown by a given heterozygote is specific to compounds that inhibit that gene product. Consequently, heterozygotes display selective hypersensitivity to compounds that inhibit their protein products, allowing a screen that links compounds directly to their target.

Functional genomics allows the full potential of haploinsufficiency to be applied to drug screening, enabling the simultaneous screening of a compound

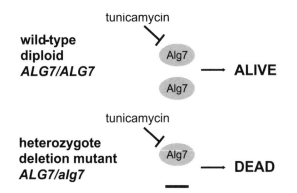

wild-type diploid *ALG7/ALG7*

tunicamycin

Alg7
Alg7

→ **ALIVE**

heterozygote deletion mutant *ALG7/alg7*

tunicamycin

Alg7

→ **DEAD**

Figure 7. Drug-induced haploinsufficiency. Lowering the gene dosage of a drug target from two copies to one in a diploid cell results in increased sensitivity to drugs acting on the gene product. For example, Alg7 is the target of tunicamycin. At a semi-inhibitory concentration of tunicamycin, wild-type cells are viable. However, lowering the gene dosage of *ALG7* to one copy in the heterozygote deletion mutant results in hypersensitivity to tunicamycin.

against all cellular targets by using a pooled set of barcoded heterozygous deletion mutants. Such a culture can be grown in the presence of a subinhibitory concentration of a bioactive compound; the strains heterozygous for the target gene will be selectively inhibited, haploinsufficient in the presence of the drug, and diluted out by the growth of the remainder of the population. Barcoded deletion mutants lost or remaining in the mutant pool following such a drug challenge regimen can be identified following PCR amplification and hybridization to a chip containing sequences specific to all the barcodes. In this way it is possible in principle to take any bioactive compound that inhibits a yeast protein and identify its target (5, 25, 26, 48). In practice, genes biologically related to the target are also found, but these can often be resolved by additional experimentation.

Application of *S. cerevisiae* Functional Genomics to Fungal Pathogens

Technologies and information including new compounds active against *S. cerevisiae* have at some stage to be validated in experiments with actual fungal pathogens, and efforts have begun to move the power of yeast genomics directly to such pathogens. Initially there were technical lags in acquiring genomic information, but this technique is becoming increasingly available for a broad range of fungi. Indeed, all that is required is an annotated genome sequence for the pathogen and some basic recombinant methodology. *C. albicans* is presently the best developed example of this trend, but other pathogenic fungi are likely to quickly follow in its footsteps.

1. One copy of the target gene is replaced by a barcoded *HIS3* marker

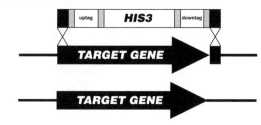

2. The second copy of the target gene is put under the control of a regulated promoter

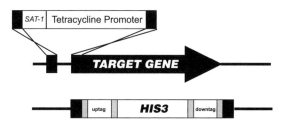

Figure 8. Construction of an essential *C. albicans* gene set. (Step 1) Heterozygote strains are constructed by transforming a wild-type *C. albicans* strain with a PCR-generated disruption cassette containing a *HIS3* selectable marker flanked with appropriate homologous sequence to replace one allele of the target gene. Two unique barcodes (uptag and downtag) flanked by primer sites common to all strains are introduced into the disruption cassette during PCR amplification. (Step 2) The endogenous promoter of the remaining wild-type allele is replaced with a PCR-generated tetracycline promoter replacement cassette containing the *SAT-1* dominant selectable marker engineered for expression in *C. albicans*.

C. albicans (discussed in more detail elsewhere in this volume) is a diploid with no classical genetics. Its genome is sequenced (40), and a predicted set of gene annotations is available. A genome comparison of *S. cerevisiae* and *C. albicans* is provided elsewhere in this book (see chapter 3). Recombinant technologies are well advanced for this organism, and Roemer et al. (66) have made a good start in compiling an essential *C. albicans* gene set. This group addressed the problem by disrupting one copy of a given gene, which generated a heterozygous diploid mutant collection for haploinsufficiency drug profiling, and building a Tet-regulatable promoter for the second copy in the diploid (Fig. 8).

Essential Gene List Comparison

Taking the *C. albicans* example as a case history in fungal pathogen comparative genomics directed to drug target identification: what have we learned? Of the 1,105 essential genes in *S. cerevisiae*, 864 orthologs of these were clearly identifiable in *C. albicans*, and of some 823 orthologs disrupted, 506 (61%) were essential. Some additional sampling of null phenotypes in *C. albicans*, using genes essential in *Schizosaccharomyces pombe* but not essential in *S. cerevisiae*, and of random nonessential genes in *S. cerevisiae* allowed an estimate of ~1,400 genes that are essential in *C. albicans* (66). Thus, the two key conclusions to be drawn are that the estimates of the number of singly essential genes are similar in *S. cerevisiae* and *C. albicans* (18 to 20%) and that the essential genes overlap substantially but not completely (~60% overlap between the two species). This species-specific variability raises issues of genome buffering and genetic interaction networks, which are discussed further in following sections.

"Universal Fungal Essential Gene Set"

Iteration of the comparative gene disruption process into other taxonomically more diverse pathogens such as *A. fumigatus* and *C. neoformans* will ultimately result in a smaller common set of conserved fungal genes that are singly essential, the "universal fungal essential gene set." This list is the Holy Grail of classic Big Pharma "blockbuster" broad-spectrum antifungal drug targets. We can expect that this list will be fairly short, further reducing the number of exploitable targets. The list can be expected to be further reduced by the application of more stringent criteria for essentiality in the pathogen. Having shown gene essentiality on nutrient agar, one can ask if the gene is essential for growth or viability on infection in a host. For example, given a *C. albicans* mutant carrying a Tet-regulated gene, the requirement for the gene during infection can be tested in a mouse model simply by examining the outcome of the infection in the presence and absence of tetracycline (66). The *C. albicans RHO1* gene was shown to be a therapeutic target for treating a systemic infection following such experimentation. Functional and comparative genomics used in this way offers a way to globally prioritize druggable targets for broad-spectrum antifungals. Parenthetically, it should come as no surprise that classic approaches had homed in on and used some of these key targets long before genomics did. For example, caspofungin, the most recent new antifungal to enter the clinic, targets the *FKS* genes, encoding the β-1,3-glucan synthase subunit. These *FKS* genes are essential, are common to all fungi, and are absent from humans.

SYNTHETIC LETHALITY AND COMBINATORIAL DRUG EFFECTS

Developments in functional genomics of fungi allow extensions of drug discovery beyond the current paradigms. We should not let commercial pharmaceutical priorities limit our thinking about drug targets to conserved, essential, fungus-specific genes. Indeed, the incredible developments in fungal functional genomics allow the research community to be creative about considering as many targets as possible for use in antifungal therapy. In particular, given that most genes are nonessential and that cells work through interconnected robustness, it makes sense to design drug combinations that inhibit such cellular systems, as outlined below.

A systematic way to identify drugs that may have synergistic effects is by selecting pairs of drugs that act on targets that themselves are synthetically lethal. For example, if two drugs target different essential gene products and a double mutant carrying conditional alleles of the essential genes dies at what is normally a permissive condition for each of the single mutants, then such synthetic drug combinations should act synergistically by working together at lower MICs than if used singly. Most importantly, this type of combination therapy is not limited to essential gene products but, rather, encompasses the entire synthetic lethal genetic interaction network. Amazingly, while there are only 1,000 essential genes in *S. cerevisiae*, there are estimated to be on the order of 100,000 synthetic digenic combinations (71). Thus, by using combinations of drugs that cut at the Achilles' heel of cell function, we can find a 100-fold-wider repertoire of drugs that act in a way that exploits a fundamental weakness of the cellular networks.

In an antifungal example, the nonessential genes *CHS3*, encoding chitin synthase 3, and *FKS1*, encoding a glucan synthase subunit, show a synthetic genetic interaction (55). Antifungal inhibitors are known that work on each of these products: the echinocandins for Fks1 and nikkomycin for Chs3. Thus, a combination of an echinocandin and nikkomycin may act synergistically since their respective targets normally buffer each other. Indeed, the pneumocandin L-735,560 and nikkomycin K have empirically been found to act in this way in *S. cerevisiae* (18) (Fig. 9). To further illustrate the power of this approach, a synthetic analysis of *FKS1* in *S. cerevisiae* (45) indicates that there are 11 more genes in addition to *CHS3* that show synthetic interactions with *FKS1* (Fig. 5) and which, when mutated singly, show an enhanced sensitivity to the echinocandin caspofungin. Compounds that inhibit such gene products specifically would be candidates for combination therapies with caspofungin.

Synthetic pairs are widespread among nonessential genes, and their use as synthetic targets opens a new avenue of drug discovery by permitting the use

Synthetic lethal interaction between *FKS1* and *CHS3*

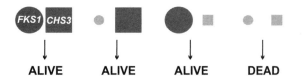

Synergistic effect of drugs acting on *FKS1* and *CHS3*

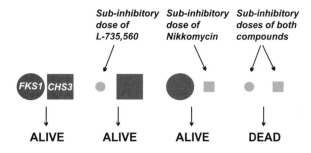

Figure 9. Synthetic drug combinations as predicted by synthetic lethal relationships can lead to synergistic effects. (Top) Synthetic lethal interaction between hypomorphic alleles of *FKS1* and *CHS3*. (Bottom) The synthetic lethal interaction between *FKS1* and *CHS3* is mimicked by treatment with specific drugs inhibiting the gene products of both genes. While this example describes an interaction between two essential genes, it is equally applicable to interactions between two nonessential genes or one essential gene and one nonessential gene.

of the majority of fungal genes as targets in combination therapies. While one can use existing inhibitors and even Food and Drug Administration approved drugs for the synthetic pairs, the exercise highlights the urgent need to have biological inhibitors for every gene product, a goal that is far from attained at present.

MAJOR FUTURE CHALLENGES

Functional Genomics Must Extend Drug Discovery beyond the Conventional

The generation of basic research tools is one compelling reason for the research community to move the implementation of "a small molecule inhibitor for every gene product" idea forward, and it forms the rationale for the National Institutes of Health Molecular Libraries Initiative (3). This initiative seeks to obtain small-molecule inhibitors of all human gene products as research tools and as "proof of principle" drug hits for targets, and its implementation would revolutionize the initial stages of drug discovery. Bringing the initial stages of drug discovery back to its basic research roots in this way is an

important part of the solution. To do so will require the assembly of high-quality chemical libraries and natural-product collections. But what of the expensive and time-consuming later stages of drug discovery, where chemical biology hits go to leads and medicinal chemistry turns some leads into candidates for clinical trails? Currently, this part of drug discovery is dominated by pharmaceutical and biotechnology companies. Here, commercial constraints limit drug discovery to druggable targets (34) that meet additional "expected-profit" criteria. Arguably, current drug discovery for "small-market" diseases such as antifungals cannot be effectively pursued commercially despite the acute need for treatment of these life-threatening infections. Genomic advances question the status quo assumption that profit should drive drug discovery priorities. Genomics allied to chemistry offers enormous hope for wide-ranging therapeutic intervention. Such considerations have prompted a thoughtful essay by Nathan on antibiotics (54), in which he suggests that one solution is to set up a not-for-profit pharmaceutical company. Such a company could be supported by government and private foundation funds and could contract out much of the drug discovery process. Several models exist for this far from utopian idea, for example, the "Medicines for Malaria Venture" with support from The Bill and Melinda Gates Foundation and the Rockefeller Foundation.

Summary

In summary, this chapter highlights the incredible power of combining systematic genetics with antifungal drug target identification. The basic principle underlying the chemical-genomic methodologies presented here is the concept that inhibiting a gene product through genetic or through chemical means is fundamentally equivalent. This model greatly expands the number of potential drug targets by looking beyond single protein targets to combinations of two (or more) targets that are both required for viability. While the majority of this pioneering work has been undertaken with a nonpathogenic strain of the fungus *S. cerevisiae*, clearly the next push must be to adopt the methodologies and approaches that have been applied to the budding yeast and establish similar systems for pathogenic fungi. In particular, the availability of a complete barcoded set of deletion mutants, including conditional alleles of essential genes, for the major fungal pathogens would permit identification of the "universal fungal essential gene set," allow a broad application of chemical genomics in these organisms, and enable a systematic analysis of phenotypes associated with pathogenicity. These types of large-scale projects should not only lead to the development of new antifungal therapies but also contribute to the larger goal of identifying a small-molecule inhibitor for every protein.

REFERENCES

1. Anderson, J. B., C. Sirjusingh, A. B. Parsons, C. Boone, C. Wickens, L. E. Cowen, and L. M. Kohn. 2003. Mode of selection and experimental evolution of antifungal drug resistance in *Saccharomyces cerevisiae*. *Genetics* **163:** 1287–1298.

2. Aucott, J. N., J. Fayen, H. Grossnicklas, A. Morrissey, M. M. Lederman, and R. A. Salata. 1990. Invasive infection with *Saccharomyces cerevisiae*: report of three cases and review. *Rev. Infect. Dis.* **12:**406–411.

3. Austin, C. P., L. S. Brady, T. R. Insel, and F. S. Collins. 2004. NIH Molecular Libraries Initiative. *Science* **306:**1138–1139.

4. Baddley, J. W., and S. A. Moser. 2004. Emerging fungal resistance. *Clin. Lab. Med.* **24:**721–735, vii.

5. Baetz, K., L. McHardy, K. Gable, T. Tarling, D. Reberioux, J. Bryan, R. J. Andersen, T. Dunn, P. Hieter, and M. Roberge. 2004. Yeast genome-wide drug-induced haploinsufficiency screen to determine drug mode of action. *Proc. Natl. Acad. Sci. USA* **101:**4525–4530.

6. Bammert, G. F., and J. M. Fostel. 2000. Genome-wide expression patterns in Saccharomyces cerevisiae: comparison of drug treatments and genetic alterations affecting biosynthesis of ergosterol. *Antimicrob. Agents Chemother.* **44:**1255–1265.

7. Bolard, J. 1986. How do the polyene macrolide antibiotics affect the cellular membrane properties? *Biochim. Biophys. Acta* **864:**257–304.

8. Botstein, D., S. A. Chervitz, and J. M. Cherry. 1997. Yeast as a model organism. *Science* **277:**1259–1260.

9. Brajtburg, J., W. G. Powderly, G. S. Kobayashi, and G. Medoff. 1990. Amphotericin B: current understanding of mechanisms of action. *Antimicrob. Agents Chemother.* **34:**183–188.

10. Csank, C., M. C. Costanzo, J. Hirschman, P. Hodges, J. E. Kranz, M. Mangan, K. O'Neill, L. S. Robertson, M. S. Skrzypek, J. Brooks, and J. I. Garrels. 2002. Three yeast proteome databases: YPD, PombePD, and CalPD (MycoPathPD). *Methods Enzymol.* **350:**347–373.

11. De Backer, M. D., P. T. Magee, and J. Pla. 2000. Recent developments in molecular genetics of *Candida albicans*. *Annu. Rev. Microbiol.* **54:**463–498.

12. Deresinski, S. C., and D. A. Stevens. 2003. Caspofungin. *Clin. Infect. Dis.* **36:**1445–1457.

13. Dixon, D. M., M. M. McNeil, M. L. Cohen, B. G. Gellin, and J. R. La Montagne. 1996. Fungal infections: a growing threat. *Public Health Rep.* **111:**226–235.

14. Douglas, C. M., J. A. D'Ippolito, G. J. Shei, M. Meinz, J. Onishi, J. A. Marrinan, W. Li, G. K. Abruzzo, A. Flattery, K. Bartizal, A. Mitchell, and M. B. Kurtz. 1997. Identification of the FKS1 gene of *Candida albicans* as the essential target of 1,3-β-D-glucan synthase inhibitors. *Antimicrob. Agents Chemother.* **41:**2471–2479.

15. Douglas, C. M., F. Foor, J. A. Marrinan, N. Morin, J. B. Nielsen, A. M. Dahl, P. Mazur, W. Baginsky, W. Li, M. el-Sherbeini, et al. 1994. The *Saccharomyces cerevisiae* FKS1 (ETG1) gene encodes an integral membrane protein which is a subunit of 1,3-beta-D-glucan synthase. *Proc. Natl. Acad. Sci. USA* **91:**12907–12911.

16. Douglas, C. M., J. A. Marrinan, W. Li, and M. B. Kurtz. 1994. A *Saccharomyces cerevisiae* mutant with echinocandin-resistant 1,3-beta-D-glucan synthase. *J. Bacteriol.* **176:**5686–5696.

17. Dwight, S. S., R. Balakrishnan, K. R. Christie, M. C. Costanzo, K. Dolinski, S. R. Engel, B. Feierbach, D. G. Fisk, J. Hirschman, E. L. Hong, L. Issel-Tarver, R. S. Nash, A. Sethuraman, B. Starr, C. L. Theesfeld, R. Andrada, G. Binkley, Q. Dong, C. Lane, M. Schroeder, S. Weng, D. Botstein, and J. M. Cherry. 2004. Saccharomyces genome database: underlying principles and organisation. *Brief Bioinform.* **5:**9–22.

18. el-Sherbeini, M., and J. A. Clemas. 1995. Nikkomycin Z supersensitivity of an echinocandin-resistant mutant of *Saccharomyces cerevisiae*. *Antimicrob. Agents Chemother.* **39:**200–207.

19. Fleming, J. A., E. S. Lightcap, S. Sadis, V. Thoroddsen, C. E. Bulawa, and R. K. Blackman. 2002. Complementary whole-genome technologies reveal the cellular response to proteasome inhibition by PS-341. *Proc. Natl. Acad. Sci. USA* **99:**1461–1466.

20. Fromtling, R. A. 1988. Overview of medically important antifungal azole derivatives. *Clin. Microbiol. Rev.* **1:**187–217.

21. Gari, E., L. Piedrafita, M. Aldea, and E. Herrero. 1997. A set of vectors with a tetracycline-regulatable promoter system for modulated gene expression in *Saccharomyces cerevisiae*. *Yeast* **13:**837–848.

22. Gavin, A. C., M. Bosche, R. Krause, P. Grandi, M. Marzioch, A. Bauer, J. Schultz, J. M. Rick, A. M. Michon, C. M. Cruciat, M. Remor, C. Hofert, M. Schelder, M. Brajenovic, H. Ruffner, A. Merino, K. Klein, M. Hudak, D. Dickson, T. Rudi, V. Gnau, A. Bauch, S. Bastuck, B. Huhse, C. Leutwein, M. A. Heurtier, R. R. Copley, A. Edelmann, E. Querfurth, V. Rybin, G. Drewes, M. Raida, T. Bouwmeester, P. Bork, B. Seraphin, B. Kuster, G. Neubauer, and G. Superti-Furga. 2002. Functional organization of the yeast proteome by systematic analysis of protein complexes. *Nature* **415:**141–147.

23. Ghaemmaghami, S., W. K. Huh, K. Bower, R. W. Howson, A. Belle, N. Dephoure, E. K. O'Shea, and J. S. Weissman. 2003. Global analysis of protein expression in yeast. *Nature* **425:**737–741.

24. Giaever, G., A. M. Chu, L. Ni, C. Connelly, L. Riles, S. Veronneau, S. Dow, A. Lucau-Danila, K. Anderson, B. Andre, A. P. Arkin, A. Astromoff, M. El-Bakkoury, R. Bangham, R. Benito, S. Brachat, S. Campanaro, M. Curtiss, K. Davis, A. Deutschbauer, K. D. Entian, P. Flaherty, F. Foury, D. J. Garfinkel, M. Gerstein, D. Gotte, U. Guldener, J. H. Hegemann, S. Hempel, Z. Herman, D. F. Jaramillo, D. E. Kelly, S. L. Kelly, P. Kotter, D. LaBonte, D. C. Lamb, N. Lan, H. Liang, H. Liao, L. Liu, C. Luo, M. Lussier, R. Mao, P. Menard, S. L. Ooi, J. L. Revuelta, C. J. Roberts, M. Rose, P. Ross-Macdonald, B. Scherens, G. Schimmack, B. Shafer, D. D. Shoemaker, S. Sookhai-Mahadeo, R. K. Storms, J. N. Strathern, G. Valle, M. Voet, G. Volckaert, C. Y. Wang, T. R. Ward, J. Wilhelmy, E. A. Winzeler, Y. Yang, G. Yen, E. Youngman, K. Yu, H. Bussey, J. D. Boeke, M. Snyder, P. Philippsen, R. W. Davis, and M. Johnston. 2002. Functional profiling of the *Saccharomyces cerevisiae* genome. *Nature* **418:**387–391.

25. Giaever, G., P. Flaherty, J. Kumm, M. Proctor, C. Nislow, D. F. Jaramillo, A. M. Chu, M. I. Jordan, A. P. Arkin, and R. W. Davis. 2004. Chemogenomic profiling: identifying the functional interactions of small molecules in yeast. *Proc. Natl. Acad. Sci. USA* **101:**793–798.

26. Giaever, G., D. D. Shoemaker, T. W. Jones, H. Liang, E. A. Winzeler, A. Astromoff, and R. W. Davis. 1999. Genomic profiling of drug sensitivities via induced haploinsufficiency. *Nat. Genet.* **21**:278–283.

27. Gimeno, C. J., P. O. Ljungdahl, C. A. Styles, and G. R. Fink. 1992. Unipolar cell divisions in the yeast *S. cerevisiae* lead to filamentous growth: regulation by starvation and RAS. *Cell* **68**:1077–1090.

28. Goebl, M. G., and T. D. Petes. 1986. Most of the yeast genomic sequences are not essential for cell growth and division. *Cell* **46**:983–992.

29. Goffeau, A., B. G. Barrell, H. Bussey, R. W. Davis, B. Dujon, H. Feldmann, F. Galibert, J. D. Hoheisel, C. Jacq, M. Johnston, E. J. Louis, H. W. Mewes, Y. Murakami, P. Philippsen, H. Tettelin, and S. G. Oliver. 1996. Life with 6000 genes. *Science* **274**:546, 563–567.

30. Goldstein, A. L., and J. H. McCusker. 2001. Development of *Saccharomyces cerevisiae* as a model pathogen. A system for the genetic identification of gene products required for survival in the mammalian host environment. *Genetics* **159**:499–513.

31. Hanway, D., J. K. Chin, G. Xia, G. Oshiro, E. A. Winzeler, and F. E. Romesberg. 2002. Previously uncharacterized genes in the UV- and MMS-induced DNA damage response in yeast. *Proc. Natl. Acad. Sci. USA* **99**:10605–10610.

32. Hartman, J. L. t., B. Garvik, and L. Hartwell. 2001. Principles for the buffering of genetic variation. *Science* **291**:1001–1004.

33. Ho, Y., A. Gruhler, A. Heilbut, G. D. Bader, L. Moore, S. L. Adams, A. Millar, P. Taylor, K. Bennett, K. Boutilier, L. Yang, C. Wolting, I. Donaldson, S. Schandorff, J. Shewnarane, M. Vo, J. Taggart, M. Goudreault, B. Muskat, C. Alfarano, D. Dewar, Z. Lin, K. Michalickova, A. R. Willems, H. Sassi, P. A. Nielsen, K. J. Rasmussen, J. R. Andersen, L. E. Johansen, L. H. Hansen, H. Jespersen, A. Podtelejnikov, E. Nielsen, J. Crawford, V. Poulsen, B. D. Sorensen, J. Matthiesen, R. C. Hendrickson, F. Gleeson, T. Pawson, M. F. Moran, D. Durocher, M. Mann, C. W. Hogue, D. Figeys, and M. Tyers. 2002. Systematic identification of protein complexes in *Saccharomyces cerevisiae* by mass spectrometry. *Nature* **415**:180–183.

34. Hopkins, A. L., and C. R. Groom. 2002. The druggable genome. *Nat. Rev. Drug Discov.* **1**:727–730.

35. Horak, C. E., and M. Snyder. 2002. Global analysis of gene expression in yeast. *Funct. Integr. Genomics* **2**:171–180.

36. Hughes, T. R. 2002. Yeast and drug discovery. *Funct. Integr. Genomics* **2**:199–211.

37. Hughes, T. R., M. J. Marton, A. R. Jones, C. J. Roberts, R. Stoughton, C. D. Armour, H. A. Bennett, E. Coffey, H. Dai, Y. D. He, M. J. Kidd, A. M. King, M. R. Meyer, D. Slade, P. Y. Lum, S. B. Stepaniants, D. D. Shoemaker, D. Gachotte, K. Chakraburtty, J. Simon, M. Bard, and S. H. Friend. 2000. Functional discovery via a compendium of expression profiles. *Cell* **102**:109–126.

38. Huh, W. K., J. V. Falvo, L. C. Gerke, A. S. Carroll, R. W. Howson, J. S. Weissman, and E. K. O'Shea. 2003. Global analysis of protein localization in budding yeast. *Nature* **425**:686–691.

39. Ito, T., T. Chiba, R. Ozawa, M. Yoshida, M. Hattori, and Y. Sakaki. 2001. A comprehensive two-hybrid analysis to explore the yeast protein interactome. *Proc. Natl. Acad. Sci. USA* **98**:4569–4574.

40. Jones, T., N. A. Federspiel, H. Chibana, J. Dungan, S. Kalman, B. B. Magee, G. Newport, Y. R. Thorstenson, N. Agabian, P. T. Magee, R. W. Davis, and S. Scherer. 2004. The diploid genome sequence of *Candida albicans*. *Proc. Natl. Acad. Sci. USA* **101**:7329–7334.

41. Kakeya, H., Y. Miyazaki, H. Miyazaki, K. Nyswaner, B. Grimberg, and J. E. Bennett. 2000. Genetic analysis of azole resistance in the Darlington strain of *Candida albicans*. *Antimicrob. Agents Chemother.* **44**:2985–2990.

42. Kohler, J. R., and G. R. Fink. 1996. Candida albicans strains heterozygous and homozygous for mutations in mitogen-activated protein kinase signaling components have defects in hyphal development. *Proc. Natl. Acad. Sci. USA* **93**:13223–13228.

43. Lamb, D., D. Kelly, and S. Kelly. 1999. Molecular aspects of azole antifungal action and resistance. *Drug Resist. Updat.* **2**:390–402.

44. Lander, E. S., L. M. Linton, B. Birren, C. Nusbaum, M. C. Zody, J. Baldwin, K. Devon, K. Dewar, M. Doyle, W. FitzHugh, R. Funke, D. Gage, K. Harris, A. Heaford, J. Howland, L. Kann, J. Lehoczky, R. LeVine, P. McEwan, K. McKernan, J. Meldrim, J. P. Mesirov, C. Miranda, W. Morris, J. Naylor, C. Raymond, M. Rosetti, R. Santos, A. Sheridan, C. Sougnez, N. Stange-Thomann, N. Stojanovic, A. Subramanian, D. Wyman, J. Rogers, J. Sulston, R. Ainscough, S. Beck, D. Bentley, J. Burton, C. Clee, N. Carter, A. Coulson, R. Deadman, P. Deloukas, A. Dunham, I. Dunham, R. Durbin, L. French, D. Grafham, S. Gregory, T. Hubbard, S. Humphray, A. Hunt, M. Jones, C. Lloyd, A. McMurray, L. Matthews, S. Mercer, S. Milne, J. C. Mullikin, A. Mungall, R. Plumb, M. Ross, R. Shownkeen, S. Sims, R. H. Waterston, R. K. Wilson, L. W. Hillier, J. D. McPherson, M. A. Marra, E. R. Mardis, L. A. Fulton, A. T. Chinwalla, K. H. Pepin, W. R. Gish, S. L. Chissoe, M. C. Wendl, K. D. Delehaunty, T. L. Miner, A. Delehaunty, J. B. Kramer, L. L. Cook, R. S. Fulton, D. L. Johnson, P. J. Minx, S. W. Clifton, T. Hawkins, E. Branscomb, P. Predki, P. Richardson, S. Wenning, T. Slezak, N. Doggett, J. F. Cheng, A. Olsen, S. Lucas, C. Elkin, E. Uberbacher, M. Frazier, et al. 2001. Initial sequencing and analysis of the human genome. *Nature* **409**:860–921.

45. Lesage, G., A. M. Sdicu, P. Menard, J. Shapiro, S. Hussein, and H. Bussey. 2004. Analysis of β-1,3-glucan assembly in *Saccharomyces cerevisiae* using a synthetic interaction network and altered sensitivity to caspofungin. *Genetics* **167**:35–49.

46. Liu, H., C. A. Styles, and G. R. Fink. 1993. Elements of the yeast pheromone response pathway required for filamentous growth of diploids. *Science* **262**:1741–1744.

47. Lo, H. J., J. R. Kohler, B. DiDomenico, D. Loebenberg, A. Cacciapuoti, and G. R. Fink. 1997. Nonfilamentous C. albicans mutants are avirulent. *Cell* **90**:939–949.

48. Lum, P. Y., C. D. Armour, S. B. Stepaniants, G. Cavet, M. K. Wolf, J. S. Butler, J. C. Hinshaw, P. Garnier, G. D. Prestwich, A. Leonardson, P. Garrett-Engele, C. M. Rush, M. Bard, G. Schimmack, J. W. Phillips, C. J. Roberts, and D. D. Shoemaker. 2004. Discovering modes of action for therapeutic compounds using a genome-wide screen of yeast heterozygotes. *Cell* **116**:121–137.

49. Mazur, P., N. Morin, W. Baginsky, M. el-Sherbeini, J. A. Clemas, J. B. Nielsen, and F. Foor. 1995. Differential expression and function of two homologous subunits of yeast 1,3-β-D-glucan synthase. *Mol. Cell. Biol.* **15**:5671–5681.

50. McCusker, J. H., K. V. Clemons, D. A. Stevens, and R. W. Davis. 1994. Genetic characterization of pathogenic

Saccharomyces cerevisiae isolates. *Genetics* **136**:1261–1269.

51. McCusker, J. H., K. V. Clemons, D. A. Stevens, and R. W. Davis. 1994. *Saccharomyces cerevisiae* virulence phenotype as determined with CD-1 mice is associated with the ability to grow at 42°C and form pseudohyphae. *Infect. Immun.* **62**:5447–5455.

52. Mewes, H. W., K. Albermann, K. Heumann, S. Liebl, and F. Pfeiffer. 1997. MIPS: a database for protein sequences, homology data and yeast genome information. *Nucleic Acids Res.* **25**:28–30.

53. Mnaimneh, S., A. P. Davierwala, J. Haynes, J. Moffat, W. T. Peng, W. Zhang, X. Yang, J. Pootoolal, G. Chua, A. Lopez, M. Trochesset, D. Morse, N. J. Krogan, S. L. Hiley, Z. Li, Q. Morris, J. Grigull, N. Mitsakakis, C. J. Roberts, J. F. Greenblatt, C. Boone, C. A. Kaiser, B. J. Andrews, and T. R. Hughes. 2004. Exploration of essential gene functions via titratable promoter alleles. *Cell* **118**:31–44.

54. Nathan, C. 2004. Antibiotics at the crossroads. *Nature* **431**:899–902.

55. Osmond, B. C., C. A. Specht, and P. W. Robbins. 1999. Chitin synthase III: synthetic lethal mutants and "stress related" chitin synthesis that bypasses the CSD3/CHS6 localization pathway. *Proc. Natl. Acad. Sci. USA* **96**:11206–11210.

56. Ostrosky-Zeichner, L., K. A. Marr, J. H. Rex, and S. H. Cohen. 2003. Amphotericin B: time for a new "gold standard". *Clin. Infect. Dis.* **37**:415–425.

57. Pan, X., D. S. Yuan, D. Xiang, X. Wang, S. Sookhai-Mahadeo, J. S. Bader, P. Hieter, F. Spencer, and J. D. Boeke. 2004. A robust toolkit for functional profiling of the yeast genome. *Mol. Cell* **16**:487–496.

58. Parsons, A. B., R. L. Brost, H. Ding, Z. Li, C. Zhang, B. Sheikh, G. W. Brown, P. M. Kane, T. R. Hughes, and C. Boone. 2004. Integration of chemical-genetic and genetic interaction data links bioactive compounds to cellular target pathways. *Nat. Biotechnol.* **22**:62–69.

59. Perea, S., and T. F. Patterson. 2002. Antifungal resistance in pathogenic fungi. *Clin. Infect. Dis.* **35**:1073–1080.

60. Pfaller, M. A., and D. J. Diekema. 2004. Rare and emerging opportunistic fungal pathogens: concern for resistance beyond *Candida albicans* and *Aspergillus fumigatus*. *J. Clin. Microbiol.* **42**:4419–4431.

61. Rex, J. H., M. G. Rinaldi, and M. A. Pfaller. 1995. Resistance of *Candida* species to fluconazole. *Antimicrob. Agents Chemother.* **39**:1–8.

62. Reynolds, T. B., and G. R. Fink. 2001. Bakers' yeast, a model for fungal biofilm formation. *Science* **291**:878–881.

63. Rine, J., W. Hansen, E. Hardeman, and R. W. Davis. 1983. Targeted selection of recombinant clones through gene dosage effects. *Proc. Natl. Acad. Sci. USA* **80**:6750–6754.

64. Roberts, R. L., and G. R. Fink. 1994. Elements of a single MAP kinase cascade in *Saccharomyces cerevisiae* mediate two developmental programs in the same cell type: mating and invasive growth. *Genes Dev.* **8**:2974–2985.

65. Rocha, C. R., K. Schroppel, D. Harcus, A. Marcil, D. Dignard, B. N. Taylor, D. Y. Thomas, M. Whiteway, and E. Leberer. 2001. Signaling through adenylyl cyclase is essential for hyphal growth and virulence in the pathogenic fungus *Candida albicans*. *Mol. Biol. Cell* **12**:3631–3643.

66. Roemer, T., B. Jiang, J. Davison, T. Ketela, K. Veillette, A. Breton, F. Tandia, A. Linteau, S. Sillaots, C. Marta, N. Martel, S. Veronneau, S. Lemieux, S. Kauffman, J. Becker, R. Storms, C. Boone, and H. Bussey. 2003. Large-scale essential gene identification in *Candida albicans* and applications to antifungal drug discovery. *Mol. Microbiol.* **50**:167–181.

67. Safe, L. M., S. H. Safe, R. E. Subden, and D. C. Morris. 1977. Sterol content and polyene antibiotic resistance in isolates of *Candida krusei*, *Candida parakrusei*, and *Candida tropicalis*. *Can. J. Microbiol.* **23**:398–401.

68. Scherens, B., and A. Goffeau. 2004. The uses of genome-wide yeast mutant collections. *Genome Biol.* **5**:229.

69. Shoemaker, D. D., D. A. Lashkari, D. Morris, M. Mittmann, and R. W. Davis. 1996. Quantitative phenotypic analysis of yeast deletion mutants using a highly parallel molecular bar-coding strategy. *Nat. Genet.* **14**:450–456.

70. Tong, A. H., M. Evangelista, A. B. Parsons, H. Xu, G. D. Bader, N. Page, M. Robinson, S. Raghibizadeh, C. W. Hogue, H. Bussey, B. Andrews, M. Tyers, and C. Boone. 2001. Systematic genetic analysis with ordered arrays of yeast deletion mutants. *Science* **294**:2364–2368.

71. Tong, A. H., G. Lesage, G. D. Bader, H. Ding, H. Xu, X. Xin, J. Young, G. F. Berriz, R. L. Brost, M. Chang, Y. Chen, X. Cheng, G. Chua, H. Friesen, D. S. Goldberg, J. Haynes, C. Humphries, G. He, S. Hussein, L. Ke, N. Krogan, Z. Li, J. N. Levinson, H. Lu, P. Menard, C. Munyana, A. B. Parsons, O. Ryan, R. Tonikian, T. Roberts, A. M. Sdicu, J. Shapiro, B. Sheikh, B. Suter, S. L. Wong, L. V. Zhang, H. Zhu, C. G. Burd, S. Munro, C. Sander, J. Rine, J. Greenblatt, M. Peter, A. Bretscher, G. Bell, F. P. Roth, G. W. Brown, B. Andrews, H. Bussey, and C. Boone. 2004. Global mapping of the yeast genetic interaction network. *Science* **303**:808–813.

72. Uetz, P., L. Giot, G. Cagney, T. A. Mansfield, R. S. Judson, J. R. Knight, D. Lockshon, V. Narayan, M. Srinivasan, P. Pochart, A. Qureshi-Emili, Y. Li, B. Godwin, D. Conover, T. Kalbfleisch, G. Vijayadamodar, M. Yang, M. Johnston, S. Fields, and J. M. Rothberg. 2000. A comprehensive analysis of protein-protein interactions in *Saccharomyces cerevisiae*. *Nature* **403**:623–627.

73. Vanden Bossche, H., P. Marichal, and F. C. Odds. 1994. Molecular mechanisms of drug resistance in fungi. *Trends Microbiol.* **2**:393–400.

74. Venter, J. C., M. D. Adams, E. W. Myers, P. W. Li, R. J. Mural, G. G. Sutton, H. O. Smith, M. Yandell, C. A. Evans, R. A. Holt, J. D. Gocayne, P. Amanatides, R. M. Ballew, D. H. Huson, J. R. Wortman, Q. Zhang, C. D. Kodira, X. H. Zheng, L. Chen, M. Skupski, G. Subramanian, P. D. Thomas, J. Zhang, G. L. Gabor Miklos, C. Nelson, S. Broder, A. G. Clark, J. Nadeau, V. A. McKusick, N. Zinder, A. J. Levine, R. J. Roberts, M. Simon, C. Slayman, M. Hunkapiller, R. Bolanos, A. Delcher, I. Dew, D. Fasulo, M. Flanigan, L. Florea, A. Halpern, S. Hannenhalli, S. Kravitz, S. Levy, C. Mobarry, K. Reinert, K. Remington, J. Abu-Threideh, E. Beasley, K. Biddick, V. Bonazzi, R. Brandon, M. Cargill, I. Chandramouliswaran, R. Charlab, K. Chaturvedi, Z. Deng, V. Di Francesco, P. Dunn, K. Eilbeck, C. Evangelista, A. E. Gabrielian, W. Gan, W. Ge, F. Gong, Z. Gu, P. Guan, T. J. Heiman, M. E. Higgins, R. R. Ji, Z. Ke, K. A. Ketchum, Z. Lai, Y. Lei, Z. Li, J. Li, Y. Liang, X. Lin, F. Lu, G. V. Merkulov, N. Milshina, H. M. Moore, A. K. Naik, V. A. Narayan, B. Neelam, D. Nusskern, D. B. Rusch, S. Salzberg, W. Shao, B. Shue, J. Sun, Z. Wang, A. Wang, X. Wang, J. Wang, M. Wei, R. Wides, C.

Xiao, C. Yan, et al. 2001. The sequence of the human genome. *Science* **291:**1304–1351.

75. Wheeler, R. T., M. Kupiec, P. Magnelli, C. Abeijon, and G. R. Fink. 2003. A *Saccharomyces cerevisiae* mutant with increased virulence. *Proc. Natl. Acad. Sci. USA* **100:**2766–2770.

76. Winzeler, E. A., and R. W. Davis. 1997. Functional analysis of the yeast genome. *Curr. Opin. Genet. Dev.* **7:**771–776.

77. Winzeler, E. A., D. D. Shoemaker, A. Astromoff, H. Liang, K. Anderson, B. Andre, R. Bangham, R. Benito, J. D. Boeke, H. Bussey, A. M. Chu, C. Connelly, K. Davis, F. Dietrich, S. W. Dow, M. El Bakkoury, F. Foury, S. H. Friend, E. Gentalen, G. Giaever, J. H. Hegemann, T. Jones, M. Laub, H. Liao, R. W. Davis, et al. 1999. Functional characterization of the *S. cerevisiae* genome by gene deletion and parallel analysis. *Science* **285:**901–906.

78. Zhu, H., J. F. Klemic, S. Chang, P. Bertone, A. Casamayor, K. G. Klemic, D. Smith, M. Gerstein, M. A. Reed, and M. Snyder. 2000. Analysis of yeast protein kinases using protein chips. *Nat. Genet.* **26:**283–289.

Molecular Principles of Fungal Pathogenesis
Edited by Joseph Heitman et al.
©2006 ASM Press, Washington, D.C.

Chapter 43

Genetic and Proteomic Analysis of Fungal Virulence

JENNIFER K. LODGE AND MICHAEL C. LORENZ

The advent of genome sequencing of pathogenic fungi has significantly changed the kinds of questions that can be tested by using these systems. This chapter explores the genetic and proteomic approaches that are now feasible for many fungal systems. Genome sequencing has been covered in chapter 40. Useful genetic approaches to the analysis of gene function include systematic gene function "knockout" techniques to determine the contribution of a gene to a particular phenotype. This approach also includes saturation random mutagenesis, targeted gene deletion on a genome-wide scale, and modulation of gene expression. Post-genomic approaches to the analysis of biological function, networks, and processes often include techniques that permit global analysis of gene expression at the protein level. Analysis of protein expression during modulation of external conditions or during particular developmental states can provide useful clues about which genes might be important for a particular function. These proteomic approaches are complementary to techniques, such as serial analysis of gene expression (SAGE) and microarrays, that measure gene expression at the RNA level. These techniques are discussed in chapter 40.

GENETIC APPROACHES

The final proof that a gene contributes to virulence is the demonstration that a mutant that lacks a functional copy of that gene has reduced ability to cause disease in an animal model. As a result, the generation of mutants will always be a necessary component of pathogenesis studies. However, the classical mutational approaches that have been tremendously important in molecular biology are not practical in studies of many fungal pathogens because of technical obstacles, including diploidy and difficulties in cloning a gene by complementation. Even in species where

random mutagenesis is possible, there are many reasons why one could not test the ~40,000 random mutants necessary for a saturating screen in animals, including both ethical and practical considerations. One solution to this problem has been to screen in vitro phenotypes thought to be related to virulence (such as filamentation or adherence to host substrates).

For mutation-based schemes to be practical for testing in animals, the number of mutants to be screened for a comprehensive study must be reduced to the fewest possible and there must be a simple means of identifying the mutation site. In addition, the ability to screen multiple mutants in pools, thus reducing the number of animals used, is also very valuable. Fortunately, technologies that meet some or all of these criteria are now available.

Insertional Mutagenesis

A mutation is of limited use if the affected gene cannot be identified. However, for pathogenicity studies neither the standard microbial cloning-by-complementation approaches nor the positional cloning strategies used for animals are feasible. Insertional mutagenesis, on the other hand, uses heterologous DNA as a mutagen. There are numerous insertional mutagenesis schemes that have been used successfully. The basic idea is that mutations are caused by the insertion of a DNA construct into random sites of the genome (which will, of course, sometimes interrupt genes or control sequences). The inserted DNA must contain, at a minimum, a selectable marker to identify mutants. This is not unlike a transposon hopping into the genome, and, in fact, transposon-based systems are commonly used for insertional mutagenesis. Importantly, the sequence of the inserted heterologous DNA serves as a tag that

Jennifer K. Lodge • Edward A. Doisy Department of Biochemistry and Molecular Biology, Saint Louis University School of Medicine, 1402 S. Grand Blvd., St. Louis, MO 63104. **Michael C. Lorenz** • Department of Microbiology and Molecular Genetics, The University of Texas Health Science Center, 6431 Fannin, Houston, TX 77030.

allows identification of the insertion site. Again, multiple techniques are available to identify the insertion site, but all eventually use a primer directed to the inserted DNA near the junction point to sequence into the adjacent genomic DNA. With the genome sequence readily available, only a few dozen base pairs of sequence information is required to determine the insertion point.

Insertional mutagenesis usually leads to complete loss of function of the targeted gene. In rare cases, insertion into control sequences can result in phenotypes from either increased or decreased expression levels. Even more rarely, an insertion into coding sequences can alter protein function independently of expression level, for instance by removing a negative regulatory sequence. Thus, the spectrum of mutants created by this technology is limited, but, to some degree, this is true of any mutagen. Because insertional mutagenesis can sometimes lead to unexpected phenotypes, the results obtained with this technique are usually confirmed by complementation and by construction of mutants in which the entire protein coding sequence of the gene of interest is deleted.

Finally, insertional mutagenesis is not, by itself, useful for pathogenicity studies. It can certainly speed in vitro screens of pathogenic organisms in which cloning by complementation is difficult or impossible, but it does not meet all the criteria for animal studies. It is mentioned here, however, because it is the basis for two approaches that are very useful for studying fungal pathogenesis, signature-tagged mutagenesis (STM) and targeted gene deletion.

Signature-Tagged Mutagenesis

The principle behind STM is the same as for insertional mutagenesis: creating mutants through the insertion of DNA of known sequence. STM, developed by David Holden for bacterial pathogens (28), incorporates a unique sequence tag into the insertional DNA element that allows the mutant to be identified within a pool of other mutants (Fig. 1). These "signature tags" are usually short (~20 to 40 bp) and consist of defined sequences with optimized hybridization profiles. In most STM schemes there are 96 distinct tags and thus 96 unique insertional cassettes. This means that a 96-well plate can be filled with insertional mutants, each of which has a different tag. At the extreme end, the *Saccharomyces cerevisiae* knockout library, containing deletions of all nonessential genes, uses a different tag for each of the at least 5,000 deletions (62).

The purpose of the tag is to allow one to distinguish one mutant from another in a pool of strains. Competition experiments are a common way of determining the relative fitness of two mutant strains

or of a mutant strain and a wild-type strain. In these experiments, each strain is uniquely marked in a way that does not affect fitness, such as with a dominant drug-resistant marker. The number of conventional markers that both are neutral and can be assessed by growth on indicator media is limited, though. With STM, however, the tag serves as the unique marker, so that the number of strains that can be compared is much larger. There is probably a limit to the number of strains that can be tested together, but all the viable haploid mutants in the *S. cerevisiae* deletion library (~4,700 strains) have been tested in single experiments using batch fermentors, and all strains can be detected. For pathogenesis studies, the animal model is the primary limiting factor for the number of strains that can be tested in parallel.

Tags are detected using specific gene chips or filter sets that contain each tag. Genomic DNA is prepared from the target population, and the tags are PCR amplified via common primer sites that flank the tag sequence. The amplicons are labeled and hybridized to the chip. The intensity of the signal is representative of the abundance of the tag in the population.

With this technology, it is easy to quantify fitness under a given set of conditions based on the relative proportion of a specific tag (representing a specific mutation) in the starting versus subsequent populations, after 20 generations, for example. A strain whose tag decreases in abundance is less fit than the average, whereas one whose tag increases is more fit. This technique is particularly useful for analysis of mutants in vivo. Mutants, each with a unique tag, can be tested in the animal in pools of up to 96 strains and compared to the growth in the same pool of mutants in vitro or to the starting inoculum. Mutations disadvantageous in vivo will be underrepresented in the recovered population. The result is the ability to use the animal model itself as a virulence screen.

Holden applied this system to *Aspergillus fumigatus*, testing 4,648 STM strains in a mouse model of invasive aspergillosis (6). Two insertions were reproducibly defective, one of which was upstream of the *para*-aminobenzoic acid (PABA) synthetase gene (*pabaA*). This mutant had no in vitro phenotypes except for a growth defect in the absence of PABA, but it was profoundly attenuated in virulence unless the diet of the mice was supplemented with PABA. It is important to note that availability of PABA had not previously been associated with virulence in fungi.

Cormack et al. (14) constructed 96 *Candida glabrata* strains by integrating a signature tag into the disrupted *URA3* locus and then used insertional mutagenesis to create the STM pool. A total of 4,800 mutants (in pools of 96) were screened in vitro for

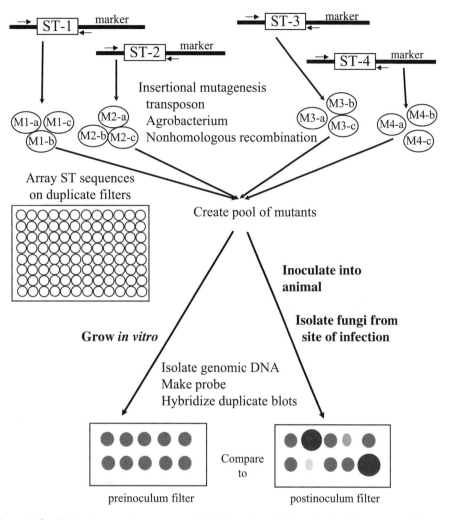

Figure 1. Schematic for STM. A unique signature tag (ST) oligonucleotide is synthesized, incorporated into a plasmid vector, and flanked by primer binding sites. Each "tagged" vector is used to transform cells producing many uniquely tagged mutants. One mutant representing each uniquely tagged vector (up to 96 in total) is assembled into an input pool. Tags present in the input pool are PCR amplified, labeled, and used to probe a filter containing dot blots of each tag forming the preinoculum filter. The pooled organisms are inoculated into an animal, and the infection is allowed to proceed for a determined period. Organisms are recovered from tissue at the site of infection, and the tags are amplified using common primers, labeled, and used to probe a duplicate filter (postinoculum filter). Tags missing from the pool are identified by a loss of signal on the blot. Mutants that overproliferate can also be identified by having a much stronger signal compared to the input blot. Mutants that did not have a change in their virulence have a similar signal to that of the input blot.

adherence to epithelial cells; adherence was increased in 5 strains, reduced in 10, and eliminated in 16. Of the nonadherent mutants, 14 were integrations into the *EPA1* gene, a member of a large family of fungal cell surface proteins. Heterologous expression of *EPA1* in *S. cerevisiae* cells allows this nonadherent species to bind to epithelial cells. A complete deletion of *C. glabrata EPA1* is nonadherent but still fully virulent; thus, there may be additional adhesins redundant with *EPA1* in vivo.

STM was also used to identify *Cryptococcus neoformans* var. *grubii* mutants with altered virulence in a mouse model of disseminated disease (43). In this study, circular plasmids containing signature tag sequences were biolistically transformed into *C. neoformans* to produce random insertion mutants. Pools of mutants were inoculated into mice and subsequently recovered from the brain. Several mutants with reduced virulence and one hypervirulent mutant were identified. The disadvantage in this study was that the method of transformation produced tandem arrays of plasmid, making identification of the flanking sequence difficult.

However, newer methods for transformation and the availability of genome sequence are improving the recovery of *C. neoformans* inserted sequences and the rapidity with which the insertion sites are identified, which is important for successful STM.

Mutagenesis using *Agrobacterium tumefaciens*-mediated transformation has been highly successful for generating and recovering insertional mutants (34). Major advantages of *A. tumefaciens*-mediated transformation are that (i) all transformants are stable and have DNA integrated into the genome rather than being present as unstable, extrachromosomal copies of the transformation vector; (ii) the ends of the insertion are defined by T-DNA borders; and (iii) single copies of the transforming DNA are integrated into the genome. These advantages reduce screening and allow for easy cloning of the insertion. The limited data currently available suggest that insertion of the T-DNA may not be completely random. There may be a slight sequence bias and a predilection for insertion into promoter regions. Because of this bias, the phenotype was linked to the inserted marker in only two-thirds of the mutants tested. For all insertion mutants, it is critical to link the insertion site to the phenotype, such as by complementation of the phenotype with a wild-type gene, demonstration that multiple independent insertions into the same gene result in the same phenotype, recapitulation of the phenotype by targeted gene deletion, and/or genetic segregation.

A major improvement in analysis of random insertion in *C. neoformans* is the recent availability of mating pairs for serotype A. This has permitted rapid analysis of whether the marker is linked to the phenotype, allowing confirmation that the phenotype is caused by the insertion. Comparison of a limited set of mutants generated by biolistic transformation of plasmid DNA and *A. tumefaciens*-mediated transformation suggested that the rate of generation of unlinked phenotypes was similar (34).

Screening 590 biolistically transformed mutants for defects in melanin biosynthesis, sensitivity to NO, and temperature sensitivity resulted in 4 mutants with defects in melanin formation, 2 that were hypersensitive to NO, and 2 that were unable to grow at 37°C (34). Of these eight, five had phenotypes that were linked to the selectable marker. The flanking sequences were cloned, and the mutants with melanin defects had an insertion into a predicted gene of unknown function or into a mitogen-activated kinase gene (*MPK2*); the NO-hypersensitive mutants had insertions into the *FHB1* gene, which encodes the flavohemoglobin shown to be important for resistance to NO stress; and the temperature-sensitive mutant had an insertion between a predicted protein and a helicase. Screening of 576 mutants generated by *A. tumefaciens*-mediated transformation resulted in the isolation of three melanin-defective strains. Two of these mutants had phenotypes that cosegregated with the selectable marker, and cloning of the flanking sequence showed that one mutant had an insertion in the promoter region of the laccase gene and the other had an insertion in a voltage-gated chloride channel.

Restriction enzyme-mediated insertion has also been used with fungi to promote relatively random insertions in fungal genomes. It has been commonly used in phytopathogens, but the only published studies in a human pathogen are those done with *C. albicans* (4) and *A. fumigatus* (5). In this method, the transforming DNA is cleaved with a restriction enzyme and is electroporated into a fungal cell along with a restriction enzyme that generates ends compatible with the enzyme that cut the transforming DNA. The transformed enzyme cleaves the genomic DNA at low frequency, and the transforming DNA is inserted into the restriction site by virtue of the compatible "sticky" ends of the DNA.

Targeted Gene Deletion

A more systematic mutagenesis approach is to construct a library of strains in which every (nonessential) gene is mutated. This approach also permits a saturating screen with a limited number of strains, an important consideration in virulence studies. An *S. cerevisiae* deletion set, incorporating signature tags, was created through a multilaboratory project and is available from the American Type Culture Collection, Research Genetics, and EUROSCARF. The creation of similar libraries in other fungi will depend on genome sequence, of course, and is facilitated by efficient PCR-based disruption protocols. It usually requires a community project to share the labor and costs of primer design, amplification, transformation, and genotyping, but this resource is well worth the effort.

As demonstrated for *S. cerevisiae*, systematic gene deletions have been a valuable tool for analysis of gene function (23, 62). Useful gene disruptions rely on accurate prediction of protein-coding sequences. Gene deletion by homologous recombination generally involves transformation of the organism with a construct that contains DNA corresponding to the 3'- and 5'-flanking sequences joined to a selectable marker. A double-crossover event integrates the marker specifically into the coding sequence. Gene deletion experiments are most informative when the boundaries of the protein-coding sequence of the gene in question are known, so that the entire protein can be eliminated without deleting genetic information in the surrounding genes. Deletions also require the development of efficient protocols for transformation, homologous recombination, and screening.

A fungal pathogen for which a significant effort toward a systematic gene deletion set has been expended is *C. neoformans*. In comparison with other common fungal pathogens, *C. neoformans* has several advantages for generation of a set of gene-specific deletions. First, unlike the diploid *C. albicans*, *C. neoformans* is haploid, which facilitates targeted gene disruptions because there is only one allele of each gene. Second, targeted gene deletions require homologous recombination, and although homologous recombination does not occur in *C. neoformans* at as high a rate as in *S. cerevisiae* or *C. albicans*, it is high enough to recover deletions with a reasonable number of screens. Third, gene deletion techniques have become relatively routine in *C. neoformans* and can be scaled up to a high-throughput method.

The reported frequency of homologous recombination in *C. neoformans* varies considerably due to some of the same factors listed previously for variation in transformation efficiencies, i.e., strain of the recipient cell, origin of the transforming DNA, construction of transforming vector, and procedure used to transform the cells. The published homologous recombination frequencies of various experiments range from 0.008 to 50% (21, 51), but because of the variety of methods used in published gene replacement experiments, direct comparison of the homologous recombination frequencies may lead to incorrect inferences. It is clear that higher rates of homologous recombination are obtained when biolistic transformation is used. HR rates in *C. neoformans* var. *neoformans,* obtained by electroporation, vary from 0.008 to 0.1%, but rates obtained by biolistic transformation in *C. neoformans* var. *neoformans* vary from 1 to 87% (15, 16).

The effects of varying the parameters of a homologous-recombination experiment are largely unknown. The minimum flanking sequence requirement for efficient homologous recombination in serotype A has been investigated (44). A systematic homologous recombination experiment using the selectable marker for hygromycin B resistance was performed. In these experiments, two independent loci, *LAC1* and *CAP59*, were chosen for disruption. The *CAP59* gene is involved with the production of the polysaccharide capsule, and *cap59* mutants have a rough and dry colony phenotype, whereas *lac1* mutants are unable to produce melanin when grown on L-dopa medium. Five constructs of each disruption vector were created with 400, 300, 200, 100, and 50 bp of flanking sequence on each side of the selectable marker. In addition, two asymmetric disruption vectors were created, one with 400 bp on one side of the marker and 50 bp on the other side and the other in which the length of the flanking sequences was reversed. Homologous recombination frequencies of the constructs varied. Constructs with 50 bp or less of flanking sequence

had a homologous recombination frequency of <1% of the total transformants, whereas constructs with flanking sequences longer than 200 bp had homologous recombination frequencies of >8%. The asymmetric constructs had intermediate rates of homologous recombination. In general, they were greater than the rates for the constructs with 50 bp on both sides but were lower than those for the constructs with 500 bp on each side. If unstable transformants were removed from the total transformants, the homologous recombination frequency of the longer constructs varied between 10 and 75%. Taken together, these data indicate that *C. neoformans* has a relatively robust homologous recombination system and can be relied on for the efficient creation of targeted gene disruptions.

A system for rapidly generating constructs for use in targeted gene disruption using homologous recombination has been devised (15). This system relies on PCR amplification to efficiently construct disruption vectors. In this system, homologous sequences from *C. neoformans* var. *neoformans* and *C. neoformans* var. *grubii* were PCR amplified from genomic DNA by using primers with short (35- to 42-bp) overlaps to a selectable marker cassette at their 5′ ends. These PCR products were then "linked" together with the selectable marker cassette separating the two flanking sequences in a third PCR amplification. This product was then used to disrupt the genomic copy of the target gene. Because this method is not dependent on sequential cloning of the target locus and knockout vector construction, it provides a starting point for the formulation of a method which is being used in the high-throughput generation of gene deletions.

Several laboratories have begun generating gene deletions with the long-term goal of deleting every nonessential gene in the serotype A strain H99. Each of the deletions is being tagged with a unique DNA sequence so that pools of mutants can be analyzed in vivo, similar to STM strategies. In vitro and in vivo analyses of these gene deletions have already revealed that the *PKC1* pathway and regulators of this pathway are important for cell integrity and for virulence and that deletion of a homolog of *SSD1* makes *C. neoformans* more virulent (K. J. Gerik and J. K. Lodge, unpublished data). Gene deletions cannot be used to analyze the function of genes essential for vegetative growth in a haploid organism. However, a stable diploid strain has been made for *C. neoformans* var. *neoformans* (54) and has proved useful for analysis of essential genes.

C. albicans is a diploid and so presents unusual problems for gene deletion because both alleles need to be sequentially disrupted (reviewed in reference 9).

However, there was a lack of selectable markers available for *C. albicans*. This paucity of markers is partly due to a lack of auxotrophic strains but is also due to the inherent resistance of *C. albicans* to many of the drugs, including G418 and hygromycin, that are commonly used for dominant selection in other fungi. Because of this problem, a system was developed so that a single *URA3* marker could be regenerated. The Ura Blaster (2) technique has been used extensively. This technique uses a *URA3* marker flanked by direct repeats. Important features of this technique are that the *URA3* marker can be subjected to both positive and negative selection and the direct repeats allow homologous recombination to delete the *URA3* marker. Typically, the first allele is disrupted and then the *URA3* selectable marker is looped out using direct repeats that flank the *URA3* gene. Ura⁻ transformants are selected by plating on medium containing 5-fluoroorotic acid. Next, the same disruption construct can be used to target the second allele. Although this method works very well, and has been improved upon (61), it is time-consuming and has not been easy to scale up to delete all the genes in *C. albicans*. There are two additional disadvantages of this approach. First, 5-fluoroorotic acid by itself is mutagenic in *C. albicans* (60). Second, the chromosomal locus at which *URA3* is integrated can significantly affect filamentation and virulence (11, 53, 55). Therefore, it is important that *URA3* be integrated at the same locus in all strains that are used in the same experiment.

A system to eliminate the need for two sequential gene disruption transformations has been developed by the Mitchell laboratory and promises a much more facile route to systematic gene deletions (17). This method uses a "UAU1" cassette, so named for its constituent parts: the *URA3* gene split by a complete *ARG4* gene such that the marker has the 5′ half of *URA3*, full-length *ARG4*, and the 3′ half of *URA3* (UAU1). A small region of *URA3* is duplicated on both sides of *ARG4* to allow recombination. Homologous integration is targeted by gene-specific flanking sequences and selected by arginine prototrophy. Accurate integration is confirmed, and correct heterozygotes are grown to select for strains prototrophic for uracil and arginine. The most likely way this prototrophy occurs is through duplication of the genomic region surrounding the marker (by mitotic gene conversion) followed by a recombination event that restores a functional *URA3* gene on one allele only. Thus, a homozygous mutation can be made via a single transformation and selection. In a minority of cases, the Ura⁺ Arg⁺ phenotype results from triploidization of the chromosomal region, but this can be easily detected. The

Mitchell laboratory has constructed a pilot library of 217 mutants and has screened this library for phenotypes in pH sensitivity (17, 35). A larger deletion project is in progress, and UAU1 deletion constructs for several thousand *C. albicans* genes have been constructed by The Institute for Genomic Research. A list of these constructs is available at http://www.tigr.org/tigr-scripts/e2k1/qzhao/complete_list.pl.

Another approach has been to screen for heterozygous phenotypes, which are readily apparent in *C. albicans*, presumably due to widespread haploinsufficiency in this obligate diploid. This observation has been taken advantage of to screen a transposon mutagenesis library in *C. albicans* (56) for genes that affect the switch between the yeast form and the filamentous form of the organism. A modified Tn7 was transposed in vitro into random genomic fragments from *C. albicans* and the resulting mutagenized fragments were transformed in *C. albicans*. A total of 146 genes that altered filamentous growth were identified in this screen; of these genes, 30% had no homolog in *S. cerevisiae*. Since approximately 30% of all *C. albicans* genes have no homolog in *S. cerevisiae*, it suggests that there was no skewing of the distribution of genes that showed haploinsufficiency for filamentous growth. Studies of genome-wide libraries of haploinsufficient *C. albicans* could be useful for the identification of novel antifungal targets and virulence.

For other pathogenic fungi, such as *Coccidioides immitis*, *Histoplasma capsulatum*, or *A. fumigatus*, gene deletion techniques are still relatively laborious and inefficient and there are no current efforts to generate a genome-wide set of gene deletions. The experience with *C. neoformans*, where technology development has turned the process of targeted gene disruption from heroic to routine, gives hope that future advances will improve the genetic tractability of these organisms.

Manipulation of mRNA Levels

In addition to generating mutants with mutations that disrupt the coding region, gene expression can be suppressed by modulating the mRNA that is produced. This can be done by selective inhibition or destruction of the mRNA using antisense RNA or RNA interference (RNAi) or by altering the synthesis of mRNA by replacing the promoter with a regulatable promoter. Lastly, there is tremendous interest in identifying essential genes in fungal systems since they are potential antifungal targets. For most fungal systems, this is a difficult problem that cannot be addressed by straightforward gene deletion. Manipulation of RNA levels to test for essentiality of genes is a feasible approach that can be scaled up to test the entire complement of genes.

RNAi and Antisense RNA

Gene expression can be posttranscriptionally modulated by two methods, antisense repression and RNAi. RNA antisense technology relies on the ability of RNA which is complementary to the mRNA of a target gene to bind to that mRNA, render it incompetent for translation, and target it for degradation. Antisense RNA has been used in several systems to modulate RNA levels. De Backer et al. (18) used a large-scale antisense RNA approach to identify essential *C. albicans* genes. A directional cDNA library was constructed so that antisense RNA was produced in the presence of galactose. Over 200 random clones were assayed in parallel for growth characteristics in inducing and noninducing media. A total of 86 different genes were identified with reduced growth in the inducing medium, and approximately half of these were hypothetical or unknown genes. Tests of heterozygous gene deletions of six of the putative essential genes indicated that four had clear growth defects when one copy of the gene was removed, supporting the idea that these genes are essential.

This kind of approach could easily be adapted to other fungal systems. Along with *C. albicans*, antisense RNA is effective in *C. neoformans* and is likely to work in other fungal systems as well. The expression of the calcineurin A gene (*CNA1*) from *C. neoformans* var. *neoformans* and the laccase gene (*LAC1*) from *C. neoformans* var. *grubii* has been repressed by using antisense RNA to each mRNA (24). In this system, cDNA for each gene was cloned in an antisense orientation in a vector containing the inducible promoter *GAL7*. This construct was used to transform the respective variety of *C. neoformans*. When the transformed cells were grown on medium containing galactose, mRNA levels of both genes were significantly reduced, producing the expected phenotypes of reduced melanin production (*LAC1*) and growth defect at 37°C (*CNA1*).

Gene expression can also be modified by RNAi. RNAi relies on the ability of an organism to degrade double-stranded RNA (dsRNA). In RNAi-competent organisms, dsRNA is specifically degraded into short (21- to 25-bp) fragments. Any mRNA that is homologous to these fragments is targeted for degradation (reviewed in reference 32). RNAi has been used as a genome-wide screen in *Caenorhabditis elegans* (reviewed in reference 10) and has been used to modulate the expression of genes in several fungal systems, including *C. neoformans*, *A. fumigatus*, and *H. capsulatum*. Liu et al. (36) used inverted repeats of portions of the *CAP59* or *ADE2* genes from *C. neoformans* cloned into a vector and separated from each other by an unrelated sequence. Expression of

this sequence was driven by the promoter of the constitutively expressed cryptococcal actin (*ACT*) gene. Cells with the appropriate mutant phenotypes were recovered after transformation with the RNAi constructs. Analysis of these cells indicated that mRNA expression for each targeted gene was significantly reduced compared to that in the wild-type strain. RNAi is also effective in *A. fumigatus* (41). dsRNAs corresponding to the *FKS1* or *ALB1/PKSP* genes were shown to have phenotypic consequences. In *H. capsulatum*, a hairpin RNA molecule that formed a dsRNA was used to modulate the expression of several genes, including the gene encoding an introduced green fluorescent protein, *ADE2*, *URA5*, and *AGS1*, the gene responsible for synthesis of the α-glucan in the cell wall (49). An important aspect of this work was that the expression of the RNAi was shown to be stable through macrophage infection, allowing the investigators to assess the impact of inhibition of *AGS1* on the virulence of the organism. The ability to specifically and temporally modulate gene expression in fungi will allow the identification of genes that are required for viability of the organisms as well as enabling the observation of the role of individual genes in the development of these organisms. The development of promoter systems that will allow the introduction of a single cDNA or genomic fragment into a vector with inducible expression of both strands of the dsRNA, rather than the construction of more complicated hairpin structures, will facilitate large-scale screening of genes by RNAi. Unfortunately, RNAi is not effective in all fungal pathogens, particularly *Candida* (and *Saccharomyces*) species, since they lack a homolog of the Dicer RNase necessary for RNAi-mediated down-regulation.

Promoters

A third method of analyzing the contribution of specific genes to virulence or vegetative growth is to modulate their expression by placing an inducible promoter in front of the protein-coding sequence. Use of an inducible promoter requires knowledge about the start site for translation of the protein and the locations of genes adjacent to the gene of interest so that the normal regulation of surrounding genes is unaffected. Additionally, if the expression of the promoter in the "off" situation is not low enough, the gene may still be expressed at levels high enough to provide adequate function. Tightly regulated inducible promoters have been used in *S. cerevisiae* and other systems but were not readily available in fungal pathogens. However, a synthetic promoter was derived that uses the Tet operator/repressor system (42) for use in *C. albicans*. In the absence of tetracycline the promoter is expressed, and in the presence of tetracycline the promoter is repressed. This promoter system is tightly regulated, and genes in the "off" mode are expressed at very low levels. A major advantage is that this promoter system can be regulated in vivo, since animals can be fed doxycycline to repress the promoter. This allows conditional mutants to be tested in vivo and adds the flexibility of temporally controlling expression. For example, Saville et al. (52) used the Tet operator/repressor system to regulate the expression of *NRG1* in *C. albicans* and investigate the role of the yeast-to-hypha transition in the mouse model of disseminated candidiasis. This technology has been used on a genomewide scale with *C. albicans* (50) to identify essential genes in a technique called GRACE. One copy of the gene is deleted, and the second copy is placed under the control of a synthetic Tet promoter. This work is described in detail in chapter 42. A variety of in vitro inducible promoters, responsive to maltose, galactose, and methionine, among others, are also available for *C. albicans*, but these promoters are less useful for virulence studies.

A synthetic Tet promoter has recently been constructed for *A. fumigatus* (57), but there is not yet a promoter for other medically important fungi. However, inducible promoters have been used, particularly for *C. neoformans*. The gene encoding topoisomerase was placed under the control of the *GAL7* promoter, and although the expression was reduced, the cells were still viable (20). However, a more tightly regulated Cu promoter, *CTR4*, has been described previously (45), and it was used to demonstrate that the thioredoxin reductase gene is essential in *C. neoformans* (40).

PROTEOMICS

Proteomics encompasses many different technologies used to analyze protein interactions and complexes, localization, posttranslational modification, and differential expression. Although many of these analyses have been done with the model system, *S. cerevisiae*, proteomics of pathogenic fungi have largely been limited to analysis of the protein components of specific organelles or differential expression of proteins under specific conditions.

Although transcriptome studies are relatively easy to do and provide a wealth of data, the proteome is the final product of the several complex layers of regulation, including transcription initiation, transcript processing and turnover, translational regulation, and posttranslational modifications. Many of these regulatory elements would not be detected by transcriptional analysis. Although for the majority of proteins and genes there is a concordance between the transcript level and the protein level, it is

clear that for many genes and proteins, there are discrepancies. Gygi et al. (27) compared protein levels in *S. cerevisiae* based on metabolic labeling and RNA levels based on frequency in SAGE libraries. They found that one could not predict the protein level from the RNA level, nor was there a correlation between codon bias and protein abundance. MacKay et al. (37) did a careful analysis of translational efficiency in *S. cerevisiae*. They isolated and separated mono- and polyribosomes and calculated that the more ribosomes on a transcript, the higher the translational efficiency. They analyzed which transcripts were part of various fractions of polyribosomes (translation state array analysis) and found that the codon adaptation index correlated with ribosome density. Sequestration of mRNAs into monosomes and mRNA protein particles can result from translational regulation. They then exposed yeast to mating pheromone and tested the translational efficiency for specific transcripts. Of the 816 proteins that changed expression levels, 617 were driven by changes in transcription while the rest were driven by changes in translational efficiency. These studies emphasize the importance of proteomic analysis and demonstrate how measurement of protein expression is complementary to measurement of mRNA expression. Finally, in order to develop a gene regulatory network, both functional genomics and proteomics information must be integrated (1, 3, 33).

The basics of proteomics require proteins to be isolated, separated, perhaps quantitated, and then identified by peptide mass fingerprinting or internal sequencing. Techniques such as two-dimensional (2D) gel electrophoresis or liquid chromatography that can separate and quantify proteins can be used to determine which proteins are regulated under specific conditions (Fig. 2). There are advantages and disadvantages to each system. The 2D gels are relatively simple to set up and run and have been the mainstay of proteomic analysis. However, only limited types of proteins can be analyzed on 2D gels. For example, proteins with high or low molecular weights or proteins with pIs of less than 3 or greater than 10 can be difficult to resolve. Different solubilization procedures have helped with separation of membrane or cell wall proteins. Detection of proteins by staining is based on abundance, so typically only the more abundant cellular proteins are analyzed.

Peptide mass fingerprinting or internal sequencing can be used to identify the protein. For identification of fungal proteins, both of these methods require a high-quality database of predicted proteins. Peptide mass fingerprinting uses the molecular mass of tryptic peptides from the protein in question and matches them to a database of proteins. With highly conserved

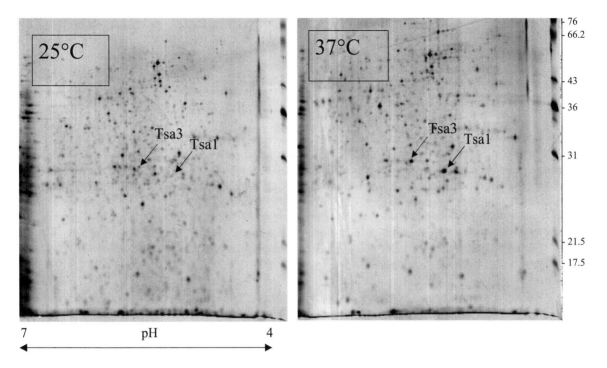

Figure 2. Comparison of lysates from *C. neoformans* grown at 25 and 37°C by 2D gel electrophoresis. The two thiol peroxidases, Tsa1 and Tsa3, that are up-regulated at higher temperature are indicated by arrows. Reprinted from reference 39.

proteins, a heterologous database could provide a good match; however, with divergent or unique proteins, other databases would not be useful. For fungi with few introns, a database generated from translating all of the open reading frames in the genome could be useful, but for those with many introns per gene, a well-annotated set of predicted proteins is optimal.

Many proteomic techniques that have been applied to the model system *S. cerevisiae* have not yet been used with pathogenic fungi. These techniques include genome-wide identification of protein complexes (30), protein-protein interactions (67), protein interaction networks (64, 65), analysis of posttranslational modifications (25, 46), genome-wide localization of proteins (31), and measurements of the absolute level of protein expression on a genome-wide scale (22). If applied to pathogenic fungi, these technologies have tremendous potential to advance the study of pathogenicity and identification of novel antifungal drug targets.

In addition, advanced techniques that have improved the range of protein analyzed or the sensitivity and accuracy of protein quantitation have been developed and applied to *S. cerevisiae*. Multidimensional protein identification technology (MudPIT) is an automated system for shotgun proteomics that eliminates many of the limitations of 2D gel electrophoresis described above (58, 63). Essentially, total protein lysates are digested into

peptides using a protease, the peptides are separated using 2D liquid chromatography, and the separated peptides are automatically identified by tandem mass spectrometry. The advantages of MudPIT are that it can be automated and is largely unbiased. Peptides from low-abundance proteins, membrane-bound proteins, and proteins with extremes of molecular weight or pI are represented similarly to other proteins. Finally, MudPIT is quantitative and reproducible, so that it could be used for analysis of differential protein expression (59). Another option is to develop a database of accurate mass tags for an organism (13). If measurement of peptide mass is done accurately, once a database of peptide masses and identifications by tandem mass spectrometry has been generated, proteins can be identified by accurate measurement of a single peptide. This requires significant effort to generate the initial database, but subsequent experiments are rapid since peptides can be separated and their molecular masses can be determined within a few hours. Other methods hold promise for analyzing differential expression of proteins. The isotope-coded affinity tag method covalently attaches tags to specific amino acids (e.g., cysteines) (26). Two different tags are used so that two different protein lysates can be labeled. The proteins are cleaved into peptides by a protease, and the labeled peptides are purified. The two samples are mixed and then separated by liquid chromatography.

Peptides that are present in a different ratio from most of the other peptides are chosen for sequencing.

Generation of a complete 2D protein map of *C. albicans* proteins has been initiated (29, 48, 66). 2D gel electrophoresis of protein lysates from *C. albicans* yeast cells resulted in a map containing 942 spots. Of those, 391 spots representing 316 different proteins were identified (66). Analysis of 106 spots from 2D gel electrophoresis of lysates from hyphal cells resulted in the identification of 66 proteins (29). The availability of this information on the Web is a valuable tool for other laboratories attempting to identify proteins. Additionally, significant efforts have been made to identify proteins in the cell wall of fungi. One study isolated cell walls from *C. albicans* and identified proteins that were covalently bound to the wall (19). Proteins covalently bound to the wall through a glycosylphosphatidylinositol (GPI) anchor were released from the wall preparations by incubation with HF-pyridine, and Pir proteins were released from the wall by incubation with NaOH. Proteins were fractionated on ion-exchange columns and identified by tandem mass spectrometry. This method identified 14 proteins that were covalently bound to the cell wall of *C. albicans*. Five of these proteins were involved in cell wall biosynthesis, three were adhesions, one was a superoxide dismutase, and five were of unknown function but had homologs in *S. cerevisiae* that were cell wall associated. Another study of cell wall proteins utilized sequential fractionation of *C. albicans* cell walls from both yeast and hyphal forms (47). The walls were extracted with dithiothreitol and hot sodium dodecyl sulfate to remove loosely associated proteins, alkali, glucanase, and exochitinase, and the resulting proteins were separated by 2D gel electrophoresis. In this analysis, over 30 proteins were identified from the loosely associated fraction, and although a few of these were known or predicted to be cell wall or cell surface associated, many were involved in translation elongation or glycolysis. In the subsequent fractions, fewer spots were isolated and fewer corresponded to enzymes that were predicted to be cytoplasmic.

Proteomics was used to identify GPI-anchored proteins that could possibly be important for cell wall remodeling from *A. fumigatus* (7). Membrane-bound proteins were isolated, and the GPI-anchored proteins were released by incubation with a GPI-specific phospholipase. This enzymatic cleavage leaves a terminal anchor that can be recognized by antibodies. 2D separation of the cleaved proteins followed by Western blotting with the antibody to the remnant of the GPI anchor revealed the presence of at least 19 spots from *A. fumigatus*. From these, nine proteins were identified, and six of these were predicted to

have enzymatic functions, including two glucanases, a phosphatase, a phospholipase, and two glucano-transferases. Other proteins included a homolog of ECM33, a protein associated with cell wall maintenance in *S. cerevisiae* and *C. albicans* (38).

Differential Expression Analysis

As discussed above, direct analysis of protein expression levels is complementary to analysis of mRNA expression levels because these two values do not always correlate with one another. Therefore, it is important to investigate both protein and mRNA expression levels. Mechanisms of antifungal resistance can be elucidated by identification of proteins induced by treatment with antifungal agents. To determine mechanisms of fungal virulence, it is important to study the expression of proteins during the interaction of a pathogen with host cells, either in vitro or in vivo. However, the use of proteomics to determine changes in fungal protein expression under such conditions can be problematic due to contamination by host proteins and the limitation of obtaining sufficient material for analysis. As more sensitive proteomic methods are adopted for the pathogenic fungi, it may be possible to identify proteins expressed in vivo.

Most of the proteomic studies have focused on *C. albicans*. One area of intense interest is the difference between yeast and hyphal cells. Proteomic maps of *C. albicans* yeast cells and hyphal cells were analyzed by 2D gel electrophoresis for comparison, and proteins were identified by peptide mass fingerprinting (12). Three proteins were highly induced in hyphal cells: Phr1, Pra1, and Tsa1. *PHR1* and *PRA1* had been previously identified as being important for the formation of hyphae. Analysis of *phr1Δ* and *pra1Δ* strains did not reveal any striking deviation from wild-type protein expression, but the putative Phr1 and Pra1 spots were missing, confirming their identification. Multiple spots of the thiol peroxidase, Tsa1, were found, and one that was increased in protein content potentially represents the oxidized form of the enzyme. Comparison of the protein expression data with Northern analysis suggested that *PRH1* is transcriptionally regulated in hyphae whereas *TSA1* and *PRA1* are not.

An analysis of the proteomic changes that are triggered by amino acid starvation was used to study the GCN4 response in *C. albicans* (66). Comparisons of the protein maps of wild-type cells and *gcn4Δ* mutant cells were performed using organisms grown in the presence of histidine and the absence of 3-aminotriazole as the amino acid-rich condition or absence of histidine and the presence of 3-aminotriazole as the amino acid-starved condition. This analysis revealed sets of proteins that were induced or repressed by amino acid starvation in a *GCN4*-dependent manner. Identification of 24 proteins that were repressed by amino acid starvation and 31 proteins that were induced by amino acid starvation suggested that the *GCN4* response of *C. albicans* is very similar to what has been observed in *S. cerevisiae*. One exception was that proteins involved in purine biosynthesis were up-regulated during amino acid starvation in *S. cerevisiae* but not in *C. albicans*.

Proteomic analysis of the *C. albicans* response to several antifungal drugs showed that inhibition of specific targets resulted in a unique profile (8). The profile of proteins of *C. albicans* cells that were treated with two different β(1,3)-glucan inhibitors and two different azoles (which inhibit ergosterol biosynthesis) were compared with the protein profile of cells grown in the absence of these inhibitors. The protein profiles from cells treated with the β(1,3)-glucan inhibitors were 69% similar, and the profiles from cells treated with the two azoles were 62% similar. However, comparison of the β(1,3)-glucan inhibitors to the azoles showed that only 30 to 40% of the changes were similar. These results suggest that proteomics could be used to cluster the response to antifungals with unknown mechanisms of action and potentially identify new targets for antifungal activity.

To date, only one study of proteomics in *C. neoformans* has been published, and none have been reported for the other pathogenic fungi. However, with genome sequencing and the creation of a well-annotated set of predicted proteins, proteomic studies are likely to become more common. A proteomic comparison of *C. neoformans* strain H99 grown at 25 versus 37°C using 2D gel analysis revealed that two proteins were highly up-regulated at 37°C (39). Both of these proteins were identified as thiol peroxidases, proteins that have been shown to be important for resistance to oxidative stress in other organisms. Deletion of both of these genes, as well as a third thiol peroxidase that was identified through a genomic search, demonstrated that only one thiol peroxidase, Tsa1, played a role in protection from exogenous oxidative stress and virulence. The *tsa1*Δ strain also was sensitive to nitrosative stress. In addition, 2D gels of the lysates from the deletion strains demonstrated conclusively that the correct identification of the protein spots had been made—the spots that had been identified as Tsa1 and Tsa3 were absent in the lysates from the appropriate deletion strain.

CONCLUSIONS AND FUTURE IMPACT

Fungal genome-sequencing projects have already had a major impact on research into the virulence mechanisms of these important pathogens,

but we have only seen the tip of the iceberg. The model system, *S. cerevisiae*, has provided examples of the kinds of experiments and analyses that are possible. However, this organism is rarely pathogenic and so is not the optimal model for studying mechanisms of fungal pathogenecity. To date, much of the genomic and proteomic studies have focused on *C. albicans* because the genome sequence has been available for several years, many molecular biology-based tools are in place, and the number of researchers working on *C. albicans* is relatively large. However, genetic analysis of *C. albicans* is limited by the diploid nature of the organism and the lack of a functional sexual cycle. Other fungi, notably *C. neoformans*, are catching up and have been useful model systems for studying mechanisms of fungal pathogenecity. With genome sequences available for many medically important fungal pathogens and the development of facile molecular biology-based tools, there are opportunities for analysis of pathogenecity mechanisms and antifungal drug targets on a genome-wide and proteome-wide scale. It will be interesting to see the comparison between different fungal pathogens and to determine which mechanisms they have in common and which are unique to each pathogen.

Acknowledgments. We are grateful to Drew Walton and Kim Gerik for reading the manuscript. J.K.L. acknowledges the support of NIH-NIAID (RO1-AI051209 and RO1-AI50184) for work on functional genomics and proteomics in her laboratory.

REFERENCES

1. **Aggarwal, K., and K. H. Lee.** 2003. Functional genomics and proteomics as a foundation for systems biology. *Brief Funct. Genomic Proteomic* 2:175–184.
2. **Alani, E., L. Cao, and N. Kleckner.** 1987. A method for gene disruption that allows repeated use of URA3 selection in the construction of multiply disrupted yeast strains. *Genetics* 116:541–545.
3. **Bader, G. D., A. Heilbut, B. Andrews, M. Tyers, T. Hughes, and C. Boone.** 2003. Functional genomics and proteomics: charting a multidimensional map of the yeast cell. *Trends Cell Biol.* 13:344–356.
4. **Brown, D. H., I. V. Slobodkin, and C. A. Kumamoto.** 1996. Stable transformation and regulated expression of an inducible reporter construct in *Candida albicans* using restriction enzyme-mediated integration. *Mol. Gen. Genet.* 251:75–80.
5. **Brown, J. S., A. Aufauvre-Brown, and D. W. Holden.** 1998. Insertional mutagenesis of *Aspergillus fumigatus. Mol. Gen. Genet.* 259:327–335.
6. **Brown, J. S., A. Aufauvre-Brown, J. Brown, J. M. Jennings, H. Arst, Jr., and D. W. Holden.** 2000. Signature-tagged and directed mutagenesis identify PABA synthetase as essential for *Aspergillus fumigatus* pathogenicity. *Mol. Microbiol.* 36:1371–1380.
7. **Bruneau, J. M., T. Magnin, E. Tagat, R. Legrand, M. Bernard, M. Diaquin, C. Fudali, and J. P. Latge.** 2001. Proteome analysis of *Aspergillus fumigatus* identifies glycosylphosphatidylinositol-anchored proteins associated to the cell wall biosynthesis. *Electrophoresis* 22:2812–2823.

8. Bruneau, J. M., I. Maillet, E. Tagat, R. Legrand, F. Supatto, C. Fudali, J. P. Caer, V. Labas, D. Lecaque, and J. Hodgson. 2003. Drug induced proteome changes in *Candida albicans*: comparison of the effect of β(1,3)-glucan synthase inhibitors and two triazoles, fluconazole and itraconazole. *Proteomics* 3:325–336.

9. Bruno, V. M., and A. P. Mitchell. 2004. Large-scale gene function analysis in *Candida albicans*. *Trends Microbiol.* 12:157–161.

10. Carpenter, A. E., and D. M. Sabatini. 2004. Systematic genome-wide screens of gene function. *Nat. Rev. Genet.* 5:11–22.

11. Cheng, S., M. H. Nguyen, Z. Zhang, H. Jia, M. Handfield, and C. J. Clancy. 2003. Evaluation of the roles of four *Candida albicans* genes in virulence by using gene disruption strains that express *URA3* from the native locus. *Infect. Immun.* 71:6101–6103.

12. Choi, W., Y. J. Yoo, M. Kim, D. Shin, H. B. Jeon, and W. Choi. 2003. Identification of proteins highly expressed in the hyphae of *Candida albicans* by two-dimensional electrophoresis. *Yeast* 20:1053–1060.

13. Conrads, T. P., G. A. Anderson, T. D. Veenstra, L. Pasa-Tolic, and R. D. Smith. 2000. Utility of accurate mass tags for proteome-wide protein identification. *Anal. Chem.* 72:3349–3354.

14. Cormack, B. P., N. Ghori, and S. Falkow. 1999. An adhesin of the yeast pathogen *Candida glabrata* mediating adherence to human epithelial cells. *Science* 285:578–582.

15. Davidson, R. C., J. R. Blankenship, P. R. Kraus, M. de Jesus Berrios, C. M. Hull, C. D'Souza, P. Wang, and J. Heitman. 2002. A PCR-based strategy to generate integrative targeting alleles with large regions of homology. *Microbiology* 148:2607–2615.

16. Davidson, R. C., T. D. Moore, A. R. Odom, and J. Heitman. 2000. Characterization of the MFα pheromone of the human fungal pathogen *Cryptococcus neoformans*. *Mol. Microbiol.* 38:1017–1026.

17. Davis, D. A., V. M. Bruno, L. Loza, S. G. Filler, and A. P. Mitchell. 2002. *Candida albicans* Mds3p, a conserved regulator of pH responses and virulence identified through insertional mutagenesis. *Genetics* 162:1573–1581.

18. De Backer, M. D., B. Nelissen, M. Logghe, J. Viaene, I. Loonen, S. Vandoninck, R. de Hoogt, S. Dewaele, F. A. Simons, P. Verhasselt, G. Vanhoof, R. Contreras, and W. H. Luyten. 2001. An antisense-based functional genomics approach for identification of genes critical for growth of *Candida albicans*. *Nat. Biotechnol.* 19:235–241.

19. de Groot, P. W., A. D. de Boer, J. Cunningham, H. L. Dekker, L. de Jong, K. J. Hellingwerf, C. de Koster, and F. M. Klis. 2004. Proteomic analysis of *Candida albicans* cell walls reveals covalently bound carbohydrate-active enzymes and adhesins. *Eukaryot. Cell* 3:955–965.

20. Del Poeta, M., D. L. Toffaletti, T. H. Rude, C. C. Dykstra, J. Heitman, and J. R. Perfect. 1999. Topoisomerase I is essential in *Cryptococcus neoformans*: role in pathobiology and as an antifungal target. *Genetics* 152:167–178.

21. Fox, D. S., M. C. Cruz, R. A. Sia, H. Ke, G. M. Cox, M. E. Cardenas, and J. Heitman. 2001. Calcineurin regulatory subunit is essential for virulence and mediates interactions with FKBP12-FK506 in *Cryptococcus neoformans*. *Mol. Microbiol.* 39:835–849.

22. Ghaemmaghami, S., W. K. Huh, K. Bower, R. W. Howson, A. Belle, N. Dephoure, E. K. O'Shea, and J. S. Weissman. 2003. Global analysis of protein expression in yeast. *Nature* 425:737–741.

23. Giaever, G., A. M. Chu, L. Ni, C. Connelly, L. Riles, et al. 2002. Functional profiling of the *Saccharomyces cerevisiae* genome. *Nature* 418:387–391.

24. Gorlach, J. M., H. C. McDade, J. R. Perfect, and G. M. Cox. 2002. Antisense repression in *Cryptococcus neoformans* as a laboratory tool and potential antifungal strategy. *Microbiology* 148:213–219.

25. Graves, P. R., and T. A. Haystead. 2003. A functional proteomics approach to signal transduction. *Recent Prog. Horm. Res.* 58:1–24.

26. Gygi, S. P., B. Rist, S. A. Gerber, F. Turecek, M. H. Gelb, and R. Aebersold. 1999. Quantitative analysis of complex protein mixtures using isotope-coded affinity tags. *Nat. Biotechnol.* 17:994–999.

27. Gygi, S. P., Y. Rochon, B. R. Franza, and R. Aebersold. 1999. Correlation between protein and mRNA abundance in yeast. *Mol. Cell. Biol.* 19:1720–1730.

28. Hensel, M., J. E. Shea, C. Gleeson, M. D. Jones, E. Dalton, and D. W. Holden. 1995. Simultaneous identification of bacterial virulence genes by negative selection. *Science* 269:400–403.

29. Hernandez, R., C. Nombela, R. Diez-Orejas, and C. Gil. 2004. Two-dimensional reference map of *Candida albicans* hyphal forms. *Proteomics* 4:374–382.

30. Ho, Y., A. Gruhler, A. Heilbut, G. D. Bader, L. Moore, S. L. Adams, A. Millar, P. Taylor, K. Bennett, K. Boutilier, L. Yang, C. Wolting, I. Donaldson, S. Schandorff, J. Shewnarane, M. Vo, J. Taggart, M. Goudreault, B. Muskat, C. Alfarano, D. Dewar, Z. Lin, K. Michalickova, A. R. Willems, H. Sassi, P. A. Nielsen, K. J. Rasmussen, J. R. Andersen, L. E. Johansen, L. H. Hansen, H. Jespersen, A. Podtelejnikov, E. Nielsen, J. Crawford, V. Poulsen, B. D. Sorensen, J. Matthiesen, R. C. Hendrickson, F. Gleeson, T. Pawson, M. F. Moran, D. Durocher, M. Mann, C. W. Hogue, D. Figeys, and M. Tyers. 2002. Systematic identification of protein complexes in *Saccharomyces cerevisiae* by mass spectrometry. *Nature* 415:180–183.

31. Huh, W. K., J. V. Falvo, L. C. Gerke, A. S. Carroll, R. W. Howson, J. S. Weissman, and E. K. O'Shea. 2003. Global analysis of protein localization in budding yeast. *Nature* 425:686–691.

32. Hutvagner, G., and P. D. Zamore. 2002. RNAi: nature abhors a double-strand. *Curr. Opin. Genet. Dev.* 12:225–232.

33. Ideker, T., T. Galitski, and L. Hood. 2001. A new approach to decoding life: systems biology. *Annu. Rev. Genomics Hum. Genet.* 2:343–372.

34. Idnurm, A., J. L. Reedy, J. C. Nussbaum, and J. Heitman. 2004. *Cryptococcus neoformans* virulence gene discovery through insertional mutagenesis. *Eukaryot. Cell* 3:420–429.

35. Li, M., S. J. Martin, V. M. Bruno, A. P. Mitchell, and D. A. Davis. 2004. *Candida albicans* Rim13p, a protease required for Rim101p processing at acidic and alkaline pHs. *Eukaryot. Cell* 3:741–751.

36. Liu, H., T. R. Cottrell, L. M. Pierini, W. E. Goldman, and T. L. Doering. 2002. RNA interference in the pathogenic fungus *Cryptococcus neoformans*. *Genetics* 160:463–470.

37. MacKay, V. L., X. Li, M. R. Flory, E. Turcott, G. L. Law, K. A. Serikawa, X. L. Xu, H. Lee, D. R. Goodlett, R. Aebersold, L. P. Zhao, and D. R. Morris. 2004. Gene expression analyzed by high-resolution state array analysis and quantitative proteomics: response of yeast to mating pheromone. *Mol. Cell. Proteomics* 3:478–489.

38. Martinez-Lopez, R., L. Monteoliva, R. Diez-Orejas, C. Nombela, and C. Gil. 2004. The GPI-anchored protein

CaEcm33p is required for cell wall integrity, morphogenesis and virulence in *Candida albicans*. *Microbiology* 150:3341–3354.

39. Missall, T. A., M. E. Pusateri, and J. K. Lodge. 2004. Thiol peroxidase is critical for virulence and resistance to nitric oxide and peroxide in the fungal pathogen, *Cryptococcus neoformans. Mol. Microbiol.* 51:1447–1458.

40. Missall, T. A., and J. K. Lodge. 2005. Thioredoxin reductase is essential for viability in the fungal pathogen, *Cryptococcus neoformans. Eukaryot. Cell* 4:487–489.

41. Mouyna, I., C. Henry, T. L. Doering, and J. P. Latge. 2004. Gene silencing with RNA interference in the human pathogenic fungus *Aspergillus fumigatus. FEMS Microbiol. Lett.* 237:317–324.

42. Nakayama, H., T. Mio, S. Nagahashi, M. Kokado, M. Arisawa, and Y. Aoki. 2000. Tetracycline-regulatable system to tightly control gene expression in the pathogenic fungus *Candida albicans. Infect. Immun.* 68:6712–6719.

43. Nelson, R. T., J. Hua, B. Pryor, and J. K. Lodge. 2001. Identification of virulence mutants of the fungal pathogen *Cryptococcus neoformans* using signature-tagged mutagenesis. *Genetics* 157:935–947.

44. Nelson, R. T., B. A. Pryor, and J. K. Lodge. 2003. Sequence length required for homologous recombination in *Cryptococcus neoformans. Fungal Genet. Biol.* 38:1–9.

45. Ory, J. J., C. L. Griffith, and T. L. Doering. 2004. An efficiently regulated promoter system for *Cryptococcus neoformans* utilizing the CTR4 promoter. *Yeast* 21:919–926.

46. Panse, V. G., U. Hardeland, T. Werner, B. Kuster, and E. Hurt. 2004. A proteome-wide approach identifies sumoylated substrate proteins in yeast. *J. Biol. Chem.* 279:41346–41351.

47. Pitarch, A., M. Sanchez, C. Nombela, and C. Gil. 2002. Sequential fractionation and two-dimensional gel analysis unravels the complexity of the dimorphic fungus *Candida albicans* cell wall proteome. *Mol. Cell. Proteomics* 1:967–982.

48. Pitarch, A., M. Sanchez, C. Nombela, and C. Gil. 2003. Analysis of the *Candida albicans* proteome. I. Strategies and applications. *J. Chromatogr. Ser. B.* 787:101–128.

49. Rappleye, C. A., J. T. Engle, and W. E. Goldman. 2004. RNA interference in *Histoplasma capsulatum* demonstrates a role for α-(1,3)-glucan in virulence. *Mol. Microbiol.* 53:153–165.

50. Roemer, T., B. Jiang, J. Davison, T. Ketela, K. Veillette, A. Breton, F. Tandia, A. Linteau, S. Sillaots, C. Marta, N. Martel, S. Veronneau, S. Lemieux, S. Kauffman, J. Becker, R. Storms, C. Boone, and H. Bussey. 2003. Large-scale essential gene identification in *Candida albicans* and applications to antifungal drug discovery. *Mol. Microbiol.* 50:167–181.

51. Salas, S. D., J. E. Bennett, K. J. Kwon-Chung, J. R. Perfect, and P. R. Williamson. 1996. Effect of the laccase gene *CNLAC1*, on virulence of *Cryptococcus neoformans. J. Exp. Med.* 184:377–386.

52. Saville, S. P., A. L. Lazzell, C. Monteagudo, and J. L. Lopez-Ribot. 2003. Engineered control of cell morphology

in vivo reveals distinct roles for yeast and filamentous forms of *Candida albicans* during infection. *Eukaryot. Cell* 2:1053–1060.

53. Sharkey, L. L., W. L. Liao, A. K. Ghosh, and W. A. Fonz. 2005. Flanking direct repeats of HisG alter URA3 marker expression at the *HWP1* locus of *Candida albicans. Microbiology* 151:1061–1071.

54. Sia, R. A., K. B. Lengeler, and J. Heitman. 2000. Diploid strains of the pathogenic basidiomycete *Cryptococcus neoformans* are thermally dimorphic. *Fungal Genet. Biol.* 29:153–163.

55. Staab, J. F., and P. Sundstrom. 2003. *URA3* as a selectable marker for disruption and virulence assessment of *Candida albicans* genes. *Trends Microbiol.* 11:69–73.

56. Uhl, M. A., M. Biery, N. Craig, and A. D. Johnson. 2003. Haploinsufficiency-based large-scale forward genetic analysis of filamentous growth in the diploid human fungal pathogen *C. albicans. EMBO J.* 22:2668–2678.

57. Vogt, K., R. Bhabhra, J. C. Rhodes, and D. S. Askew. 2005. Doxycycline-regulated gene expression in the opportunistic fungal pathogen *Aspergillus fumigatus. BMC Microbiol.* 5:1.

58. Washburn, M. P., D. Wolters, and J. R. Yates III. 2001. Large-scale analysis of the yeast proteome by multidimensional protein identification technology. *Nat. Biotechnol.* 19:242–247.

59. Washburn, M. P., R. R. Ulaszek, and J. R. Yates III. 2003. Reproducibility of quantitative proteomic analyses of complex biological mixtures by multidimensional protein identification technology. *Anal. Chem.* 75:5054–5061.

60. Wellington, M., and E. Rustchenko. 2005. 5-Fluoro-orotic acid induces chromosome alterations in *Candida albicans. Yeast* 22:57–70.

61. Wilson, R. B., D. Davis, B. M. Enloe, and A. P. Mitchell. 2000. A recyclable *Candida albicans* URA3 cassette for PCR product-directed gene disruptions. *Yeast* 16:65–70.

62. Winzeler, E. A., D. D. Shoemaker, A. Astromoff, H. Liang, K. Anderson, et al. 1999. Functional characterization of the *Saccharomyces cerevisiae* genome by gene deletion and parallel analysis. *Science* 285:901–906.

63. Wolter, D. A., M. P. Washburn, and J. R. Yates. 2001. An automated multidimensional protein identification technology for shotgun proteomics. *Anal. Chem.* 73:5683–5690.

64. Wuchty, S. 2002. Interaction and domain networks of yeast. *Proteomics* 2:1715–1723.

65. Wuchty, S., and E. Almaas. 2005. Peeling the yeast protein network. *Proteomics* 5:444–449.

66. Yin, Z., D. Stead, L. Selway, J. Walker, I. Riba-Garcia, T. McLnerney, S. Gaskell, S. G. Oliver, P. Cash, and A. J. Brown. 2004. Proteomic response to amino acid starvation in *Candida albicans* and *Saccharomyces cerevisiae. Proteomics* 4:2425–2436.

67. Zhu, H., M. Bilgin, R. Bangham, D. Hall, A. Casamayor, P. Bertone, N. Lan, R. Jansen, S. Bidlingmaier, T. Houfek, T. Mitchell, P. Miller, R. A. Dean, M. Gerstein, and M. Snyder. 2001. Global analysis of protein activities using proteome chips. *Science* 293:2101–2105.

Molecular Principles of Fungal Pathogenesis
Edited by Joseph Heitman et al.
©2006 ASM Press, Washington, D.C.

Chapter 44

Genomic Perspectives on the Fungal Kingdom

JASON E. STAJICH AND FRED S. DIETRICH

The fungal kingdom has been the test bed of eukaryotic genomics. The workhorse of the kingdom has been *Saccharomyces cerevisiae,* the first eukaryote whose genome was sequenced (14), the first eukaryote whose entire transcriptome has been expression profiled (41), and the first eukaryote for which systematic gene knockouts of nearly all nonessential genes (13) have been constructed. Building on this, whole-genome sets of tagged proteins have been constructed (11) and localized (19) and whole-genome protein-protein interaction maps have been constructed by two-hybrid analysis (43) and by coimmunoprecipitation (17). *S. cerevisiae* is also a test bed for proteomics (35), metabolomics (37), and systematic protein structure determination (36). At the same time, systematic genome-sequencing projects for a range of species spanning the fungal kingdom have been undertaken. These genome sequences provide an opportunity to pose comparative questions about fungal evolution; discover differences in genome structure, content, and organization; and identify novel genes.

The *S. cerevisiae* genome-sequencing project took some 10 years to complete, with over 1,000 individuals involved in the project. Robert Mortimer established a key part of the foundation of this project with his development and maintenance of the yeast genetic map (33). This genomic structure, particularly once the number of chromosomes was known, was instrumental in developing a physical map of the yeast genome. Much of the work on the *S. cerevisiae* physical map was carried out in the laboratory of Maynard Olson, including chromosome separation by pulsed-field gel electrophoresis (4) and whole-genome restriction mapping (27). More important for the *S. cerevisiae* genome-sequencing effort was the development by Maynard Olson's group of a set of overlapping lambda and later cosmid clones spanning the yeast genome

(34, 38). This allowed for the clone-by-clone sequencing of the yeast genome by a consortium of laboratories around the world. This method of sequencing, involving intensive library construction, clone mapping, and, finally, sequencing, was also used to complete the genome of *Schizosaccharomyces pombe* by a group led by Bart Barrell at the Sanger Centre. The clone-by-clone sequencing method has now been supplanted by whole-genome shotgun sequencing as the method of choice for fungal genomic sequencing.

Genome sequencing has provided a means for describing the complete genetic makeup of an organism. The application of sequencing technology to fungi has provided a wealth of data of interest to medical mycologists and to evolutionary and cellular biologists alike. Following the success of the *S. cerevisiae* genome project, more than 40 fungal genomes have been sequenced since 1995. These fungi were selected because of their proven interest in evolutionary, medical, industrial, and agricultural research. Several groups have contributed to or are now contributing to these ongoing sequencing efforts. These include the Fungal Genome Initiative (http://www.broad.mit.edu/annotations/fungi/fgi/), launched in 2000 as a multi-investigator effort centered at the Broad Institute at the Massachusetts Institute of Technology to sequence, assemble, and annotate fungal genomes. The U.S. Department of Energy Joint Genome Institute (http://www.jgi.doe.gov/) has sequenced environmentally and evolutionarily important fungi as part of the Genomes to Life initiative. The Génolevures (http://cbi.labri.fr/Genolevures/) consortium has also contributed many partial and finished Hemiascomycetes genomes as part of an evolutionary genomics approach to understanding yeast evolution. Several other groups including The Institute for Genomic Research (TIGR), Sanger Centre, and the

Jason E. Stajich • University Program in Genetics & Genomics, Department of Molecular Genetics and Microbiology, Institute for Genome Science & Policy, Duke University, Durham, NC 27710.   **Fred S. Dietrich** • 287 CARL Bldg., Box 3568, Duke University Medical Center, Durham, NC 27710.

groups of Ron Davis at Stanford University and Peter Philippsen at the University of Basel have also contributed whole-genome fungal sequences. Table 1 lists the completed and in-progress fungal genome projects as of April 2005.

The genome projects have provided a valuable tool for researchers studying individual fungal species. Genomic sequencing also gives us an idea of how far we still have to go to understand the biology of eukaryotic systems when, as we approach the 10th anniversary of the completion of the genome of the extensively studied fungus *S. cerevisiae*, the function of one-third of the genes is still completely unknown. The sequencing of multiple fungal genomes is giving us an initial view of the degree of conservation and diversity within the fungal kingdom. Furthermore, the

Table 1. List of currently funded and ongoing fungal genome sequencing projects

Species	Clade	Genome status	Citation or sequencing group[a]
Schizosaccharomyces pombe	Archaeascomycota	Finished	45
Coprinus cinereus	Basidiomycota	WGS contigs	Broad-FGI
Cryptococcus neoformans var. *grubii* (serotype A) strain H99	Basidiomycota	WGS contigs	Broad-FGI Duke University Medical Center
Cryptococcus neoformans var. *neoformans* (serotype D) strain B3501	Basidiomycota	Finished	28
Cryptococcus neoformans var. *neoformans* (serotype D) strain JEC21	Basidiomycota	Finished	28
Cryptococcus gattii (serotype B) strain WM276	Basidiomycota	WGS contigs	University of British Columbia Genome Sequencing Center
Cryptococcus gattii (serotype B) strain R265	Basidiomycota	WGS contigs	Broad-FGI
Laccaria bicolor	Basidiomycota	In progress	JGI-DOE
Phakopsora pachyrhizi	Basidiomycota	In progress	JGI-DOE
Phanerochaete chrysosporium	Basidiomycota	WGS contigs	31
Sporobolomyces roseus	Basidiomycota	In progress	JGI-DOE
Ustilago maydis	Basidiomycota	WGS contigs	Broad-FGI
Aspergillus fumigatus	Euascomycota	WGS contigs	TIGR and Sanger Centre
Aspergillus nidulans	Euascomycota	WGS contigs	Broad-FGI
Aspergillus oryzae	Euascomycota	WGS contigs	NITE (Japan)
Aspergillus terreus	Euascomycota	WGS contigs	Microbia
Coccidioides immitis	Euascomycota	WGS contigs	Broad-FGI
Chaetomium globosum	Euascomycota	WGS contigs	Broad-FGI
Gibberella zeae (anamorph *Fusarium graminearum*)	Euascomycota	WGS contigs	Broad-FGI
Histoplasma capsulatum (*Ajellomyces capsulatus*) strain NAmI	Euascomycota	In progress	Broad-FGI
Histoplasma capsulatum strain 186AR	Euascomycota	WGS contigs	Washington University, St. Louis, Mo.
Histoplasma capsulatum strain 217B	Euascomycota	WGS contigs	Washington University, St. Louis, Mo.
Magnaporthe grisea	Euascomycota	WGS contigs	7
Mycosphaerella fijiensis	Euascomycota	In progress	JGI-DOE
Mycosphaerella graminicola	Euascomycota	In progress	JGI-DOE
Neurospora crassa	Euascomycota	WGS contigs	10
Podosopora anserina	Euascomycota	In progress	Broad-FGI
Podosopora anserina strain S mat+	Euascomycota	WGS contigs	Genoscope and CNRS
Trichoderma reesei	Euascomycota	In progress	JGI-DOE
Unicinocarpus reesei	Euascomycota	In progress	Broad-FGI
Glomus intraradices	Glomeromycota	In progress	JGI-DOE
Ashbya gossypii	Hemiascomycota	Finished	8
Candida albicans	Hemiascomycota	WGS contigs	20
Candida albicans strain MTLα/α (WO1)	Hemiascomycota	In progress	Broad-FGI
Candida dubliniensis	Hemiascomycota	WGS contigs	Sanger Centre
Candida glabrata	Hemiascomycota	Finished	9
Candida guilliermondii strain ATCC 6260	Hemiascomycota	WGS contigs	Broad-FGI

Continued on following page

Table 1. *Continued*

Species	Clade	Genome status	Citation or sequencing group[a]
Candida lusitaniae strain ATCC 42720	Hemiascomycota	WGS contigs	Broad-FGI
Candida tropicalis strain MYA-3404	Hemiascomycota	WGS contigs	Broad-FGI
Debaryomyces hansenii	Hemiascomycota	Finished	9
Kluyveromyces waltii	Hemiascomycota	WGS contigs	22
Kluyveromyces lactis	Hemiascomycota	Finished	9
Lodderomyces elongisporus	Hemiascomycota	WGS contigs	Broad-FGI
Pichia stipitis	Hemiascomycota	In progress	JGI-DOE
Saccharomyces bayanus	Hemiascomycota	WGS contigs	6, 23
Saccharomyces castellii	Hemiascomycota	WGS contigs	6
Saccharomyces cerevisiae strain S288C	Hemiascomycota	Finished	14
Saccharomyces cerevisiae strain RM11–1A	Hemiascomycota	WGS contigs	Broad-FGI
Saccharomyces cerevisiae strain YJM789	Hemiascomycota	WGS contigs	Stanford Genome Technology Center
Saccharomyces kluyveri	Hemiascomycota	WGS contigs	6
Saccharomyces kudriavzevii	Hemiascomycota	WGS contigs	6
Saccharomyces mikatae	Hemiascomycota	WGS contigs	6, 23
Saccharomyces paradoxus	Hemiascomycota	WGS contigs	23
Yarrowia lipolytica	Hemiascomycota	Finished	9
Encephalitozoon cuniculi	Microsporidia	Finished	21
Nosema locustae	Microsporidia	WGS contigs	Marine Biological Laboratory, Woods Hole, Mass.
Pneumocystis carinii	Pneumocystidomycetes	In progress	Cincinnati Children's Hospital
Rhizopus oryzae	Zygomycota	WGS contigs	Broad-FGI

[a]Broad-FGI, MIT/Harvard Broad Institute, funded through the Fungal Genome Initiative; JGI-DOE, Department of Energy Joint Genome Institute, Walnut Creek, Calif.; Microbia, Cambridge, Mass.; Sanger Centre, Wellcome Trust Sanger Centre, Hinxton, Cambridge, United Kingdom; Genoscope, Evry, France; CNRS, Centre National de la Recherche Scientifique, Paris, France.

comparison of these fungal genomes with the genomes of other eukaryotes gives us a more precise view both of the scale of conservation of eukaryotic genes and of novel genes restricted to the fungi. While much research on fungi has focused on understanding conserved eukaryotic functions, the genomic sequence also will be important in characterizing fungus-specific pathways such as those involved in secondary metabolism, including antibiotics, and specialized degradation pathways, such as for cellulose.

Improvements in genome-sequencing technology, particularly 96-capillary fluorescent sequencers, automation of template preparation, and base calling and assembly software, mean that for large genome centers the time now required to go from DNA sample to assembled sequence has decreased from several years to less than 6 months. This increased speed and decreased cost mean that it is no longer possible to sequence the genomes of only the most important research organisms and pathogens but that a much wider range of species of clinical, environmental, or evolutionary interest can be sequenced. Future improvements in sequencing technology will probably drive the cost of sequencing lower and make it increasingly feasible to sequence previously uncharacterized fungal genomes. This capability will enhance and fundamentally change the approaches used to find and map genes of interest and provide a way to

characterize strain variation and to study evolutionary mechanisms in depth. For example, numerous experiments have shown that when *S. cerevisiae* is grown in continuous culture, the fitness of the strain will improve for growth under those specific conditions. In the past it has in general not been possible to identify the specific mutations giving rise to the improved fitness. As the cost of sequencing drops, it will soon be possible to sequence a fungal strain, grow it in a chemostat, and then sequence the derived strain and identify the changes that have occurred.

SEQUENCE FINISHING

Once whole-genome shotgun sequencing is completed, the initial product is a set of contigs (regions of continuous sequence) and scaffolds (contigs that are linked by clone pair information, i.e., sequences from opposite ends of clones are in different contigs, implying that those contigs must be adjacent in the genome). At this point, assuming that the shotgun sequence was determined to sufficient depth, the data can be searched and genes can be identified. For some genomes, such as *Kluyveromyces waltii* (22), the data are available in this form. For other genomes, such as *Kluyveromyces lactis* (9), the sequencing has been taken to completion by sequencing across the gaps in the assembly. This process, generally referred to as sequence finishing, results in the sequence being available in the form of chromosomes instead of contigs and scaffolds. While for most biologists the notion of chromosomes is more satisfying than contigs and scaffolds, completing the last 0.1% of the genome tends to be labor-intensive and expensive. Therefore, until more automated ways are developed to sequence genomes to completion, it is likely that many of the sequenced genomes will continue to be available at the level of 99.9% sequence completion and in hundreds to thousands of contigs instead of being available as 100% sequenced chromosomes.

GENOME ANNOTATION

Examination of the genomic sequence of an organism, whether finished or not, reveals little to the human eye but an incomprehensible string of nucleotides. Annotation, the process of identifying biological features within the sequence, is thus a crucial step in making the sequence useful to most biologists. The process of genome annotation typically involves both computational analysis and human curation.

Genome annotation typically starts with attempts to identify protein-coding genes encoded in the genome. At the simplest level, computational gene prediction consists of searching for open reading frames (ORFs) within the genome. ORFs are defined as a stretch of DNA, beginning with a start codon (ATG) and terminating with a stop codon, that must be a multiple of 3 bases long. One can identify all the ORFs which are at least a certain size, say 100 codons or greater in length, as the potential coding region of the genome. For members of the Hemiascomycetes like *S. cerevisiae* and *Ashbya gossypii*, which have few introns, this is a reasonable approach to annotating the coding region; to a first approximation, it was the way in which the initial annotation of the *S. cerevisiae* genome was generated. This method, however, resulted in the annotation of approximately 800 "genes" in *S. cerevisiae* that are spurious and do not encode proteins and which are now referred to in the Saccharomyces Genome Database (SGD) (5) as "dubious." It also resulted in many small protein-coding genes being overlooked.

Sequencing of multiple fungal genomes has revealed that certain characteristics of *S. cerevisiae* are not typical of fungi in general. One example of this is that most fungal genomes contain far more introns than those found in the Hemiascomycetes. These introns mean that the ORF approach is too simplistic for gene identification. More sophisticated algorithms score segments of DNA against a model of gene structure of exon and intron states and properties such as no stop codons in exons and the GT^AG splice site rule for starting and ending an intron. Programs that predict genes in a genome based solely on a model of genes with no external information are referred to as ab initio gene predictors. These include Genscan (3), GlimmerHMM (30), and SNAP (24) and are very fast but typically have a limited ability to predict the full gene structure accurately for most genes (15). Another approach is to predict genes only when there is additional experimental evidence for them. This evidence includes short bits of transcript called expressed sequence tags or full-length transcript sequences. Additionally, protein sequence from annotations of related species can be aligned to the genome sequence to predict potential coding regions. Several algorithms exist to predict the existence of genes based on this experimental evidence, including Genewise and Genomewise (2), Exonerate (40), and Procrustes (12). These evidence-based methods are limited in the available evidence, and so poorly expressed genes are unlikely to have cDNA evidence in an expressed sequence tag library and will not be represented in the predicted gene set. Similarly, the protein sequence for rapidly evolving genes will not be alignable between species. An alternative approach to gene prediction is to use paired alignments of genome sequence of two or more species to refine the ab initio predictions. Implementations of this strategy include Twinscan (25) and DoubleScan (32).

There are therefore several approaches to predict genes in DNA sequence, each with its inherent errors. To make more accurate gene calls, predictions from several of these programs can be combined with a program that is aware of coding sequence and splice-site rules. Several implementations of this approach are available, including GAZE (18), Genomewise (2), Combiner (1), and GLEAN (A. J. Mackey, F. Pereira, and D. S. Roos, unpublished data), and results suggest that multiple lines of evidence produce a better overall gene set than do individual gene prediction programs. Once the structure of the transcript is identified, including the transcription start sites and polyadenylation sites, the regulatory elements remain to be identified. In genes containing multiple introns, possible alternative splice forms remain to be identified. Proteomic data can allow the identification of sites of posttranslational modification.

Identification of tRNA- and rRNA-encoding genes is fairly straightforward due to the high level of conservation of these genes and to programs such as tRNAscan-SE (29). Identifying other less highly conserved RNA coding genes can be difficult and computationally demanding. Efforts such as Rfam have begun cataloging alignments of these molecules, and these alignments can be searched against the genome to find the homologous genes in unannotated genomes. A thorough annotation of a fungal genome also includes the identification of noncoding chromosomal features such as centromeres, telomeres, and origins of replication.

Once genes are identified, one would like to have a guess at their function. Usually this is inferred by finding similar genes in other species that have an annotated function. Tools including BLAST and FASTA are used to compare sequences to databases of sequence with the assumption that sequences which can be aligned with high confidence are likely to be homologous. Usually at least 30% similarity is necessary to infer homology. More sophisticated tools, such as PSI-BLAST, a special version of BLAST, can identify more distant homologs by constructing a position-specific scoring matrix (PSSM), which, instead of searching a single protein sequence against the database, builds an alignment of similar sequences and assigns a probability for each column of the alignment of seeing each of the 20 amino acids. This matrix is used to search a database of sequences and is more sensitive than searching with a single amino acid sequence. Other specialized tools to identify protein similarities use the mathematical techniques of hidden Markov models, similar to a PSSM but with probabilities associated with insertions and deletions throughout the alignment.

Identification of genes that are probably homologous by one of the above methods suggests that the genes have evolved from a common ancestral gene and thus may have the same or similar function. Thus, at the simplest level a gene can be annotated by transferring the annotation from its best BLAST hit in a search against previously annotated genomes. This approach, however, has pitfalls such as misidentifying members of gene families and giving erroneous gene assignments for genes that are present in a species being annotated. Another pitfall to consider is that even though genes may be homologs, this does not guarantee that they carry out the same function in different species. Another potential problem is propagation of error, where an incorrectly annotated gene in one species can result in the error being inadvertently propagated throughout the fungal kingdom. A better approach, often termed phylogenomics, evaluates the functional annotation of the entire set of homologous genes for a given query gene and makes a prediction about the function for the overall set of genes. The overall prediction is assigned to the query sequence and can have an associated confidence value in this prediction based on the consistency of all the functional annotation of the set of homologs.

In addition to whole-protein comparison, protein domains can be searched. Curated databases of aligned protein domains called Pfam, SMART, and TIGRFAM have been created and store models for protein domains such as zinc fingers and kinase. These models are used by tools such as HMMER to identify known domains in proteins.

An important recent development in gene identification is gene ontology (GO). GO attempts to classify our knowledge of gene function by describing three aspects of each gene: molecular function, biological process, and cellular component. GO describes the set of known gene functions, processes, and cellular localalization components and organizes them in a network of increasingly specific terms. For example, a general category of "DNA binding" has the more specific terms beneath it "DNA replication" and "transcription factor." Many of these functions, processes, and components are found in common among all eukaryotes, thus allowing comparisons among species. There are also fungus-specific terms and species-specific terms. An important aspect of GO is that an integral part of the description of each gene by function, process, and component is the evidence code. Examples of evidence codes are IDA (inferred from direct assay) and TAS (traceable author statement). While this may seem like a minor point, in many databases such as GenBank genes are annotated without necessarily including a link to the evidence supporting the annotation. If the annotation is in doubt, it can often be difficult to trace back to the origin of the annotation.

Annotation of the function of a gene depends ultimately on someone having done an experiment to determine gene function. While genome sequencers are quick to point out the uses of genome sequence to biologists, in a large sense it is actually the bench biologists who are providing the data to the genome-sequencing community. A genome sequence, without function attached to it, is of fairly limited use. One of the bigger challenges to genomics remains identification of the role of genes whose function is currently unknown.

Computational biology tools currently cannot perfectly predict the gene structure of all genes in a genome even when using the available EST and protein evidence. Human curation is needed to verify cases not programmed into the gene prediction models or to interpret what may be artifacts in the data or new biological discoveries. Alternative splicing and RNA-editing events are difficult to identify accurately with computational tools alone and require a human intellect to synthesize the different lines of evidence properly into a gene model. Human curation can identify cases where sequence assembly errors or low sequence quality may have introduced errors by identification of inconsistencies during comparisons of multiple data sets. Curators evaluate the combined data from gene prediction, alignments of the genomic locus to similar loci from different species, and alignment of EST and cDNA evidence in order to annotate a final gene model. This gene model is updated when researchers add additional experimental evidence, which includes corrected exons and introns, untranslated regions, promoters, polyadenylation sites, or additional isoforms.

The combined human curation and automated analysis amasses information about the likely set of proteins and RNAs encoded in the genome. However, this genome annotation is a hypothesis that must be tested to confirm and correct the annotated transcriptome and proteome of an organism. One approach to this verification involves using reverse transcription-PCR to test predictions about specific exons in order to test the accuracy of computationally derived gene models (42). Proteomic data, genetic data, and serial analysis of gene expression data can also be used to confirm annotations. Together, high-throughput data collection, automated analyses, and manual detailed gene annotation are needed to properly assess and improve the annotation of a genome.

GENE CONTENT

Fungi inhabit a variety of habitats and ecosystems, yet their genomes have roughly the same core set of genes necessary for life as a fungus. Fifty percent of the genes in *A. gossypii*, the species currently with the smallest known gene set of any free-living fungus, have identifiable homologs in the Basidiomycota. Within the Hemiascomycetes, over 90% of the *A. gossypii* genes have homologs in *S. cerevisiae*, and at least 50% of the *A. gossypii* genes have an identifiable homolog in mammalian genomes.

Protein domains correspond to evolutionarily conserved portions of proteins and typically have specific folding patterns and identified functions. The most abundant domains in the fungi are the kinase domains and the sugar transporters. Table 2 is a list of the top 12 most abundant Pfam domains. Approximately 40% of the fungal genes have no recognizable homologs in nonfungal metazoans. These are probably genes that have evolved in the fungal lineage.

GENOME EVOLUTION

The genome sequences of related organisms provide more than just catalogs of species gene content, they allow comparisons between genomes to study elements that are highly diverged and those that are highly conserved. Alignment of genes from multiple organisms, using sequence similarity search tools like BLAST or FASTA, typically identifies the most highly conserved and slowly evolving genes since they share the greatest sequence conservation. These genes also

Table 2. Top 12 most abundant fungal protein domains[a]

Pfam domain	Description
WD40	WD40 repeat
HEAT	HEAT repeat
Extensin_2	Extensin-like (cell wall proteins)
TPR_2	Tetratricopeptide repeat
MFS_1	Major facilitator superfamily
Ank	Ankaryin repeat
TPR_1	Tetratricopeptide repeat
sugar_tr	Sugar transporter
zf-C2H2	C2H2-type zinc finger
LRR_1	Leucine rich repeat
MMR_HSR1	GTPase of unknown function
Pkinase	Protein kinase

[a]Based on nine fungal proteomes: *C. glabrata*, *D. hansenii*, *K. lactis*, *M. grisea*, *N. crassa*, *S. cerevisiae*, *S. pombe*, *U. maydis*, and *Y. lipolytica*.

tend to be the most functionally important for the core processes of the organism, such as genes that function in the cell cycle, cytoskeleton, DNA replication, and packaging, metabolism, transcription, and translation (39). A family of genes in opposition to this trend include those involved in meiosis, whose members tend to have lower than average sequence conservation when compared between species.

There is a major caveat when identifying genes shared among a set of species. For highly conserved genes, the presence or absence of genes in two species is often fairly clear. However, for genes that are less well conserved, the methodology breaks down with increasing evolutionary distance. For example, the *KAR1* genes of *S. cerevisiae* and *A. gossypii* are less than 24% identical, and thus the homology is near the limit of detection by BLAST; it is therefore not surprising that no homolog of this gene can be found in the much more distantly related fungus *Neurospora crassa*. This does not mean that *N. crassa* does not have an orthologue of KAR1, only that these tools cannot detect it, if it is present.

Examples of this rapid evolution of gene sequences beyond the detectable limits of BLAST have been uncovered in pairs of duplicate genes created by the whole-genome duplication in *S. cerevisiae*. In this case, the genes had evolved rapidly enough to not be identified as significantly similar by BLAST, but the duplicate pair can be identified when gene order is taken into account in comparison with the unduplicated genomes of the fungi *A. gossypii* and *K. waltii* (8, 22). Examples of this method of detection include the spindle pole body protein-encoding genes *CNM67* and *ADY3*, which are less than 21% identical yet are a duplicate pair of genes resulting from the whole-genome duplication in the *S. cerevisiae* lineage (44). The gene order in the regions of *CNM67* and *ADY3* in *S. cerevisiae* relative to the gene order in *A. gossypii* unambiguously shows the relationship between the two *S. cerevisiae* genes. This is thus a case where synteny, in general terms the conservation of gene order and orientation, is a more powerful tool than BLAST at identifying paralogues.

Synteny is becoming an increasingly important tool in identification of orthologues between closely related species. While there is very little synteny between species as distantly related as *S. cerevisiae* and *S. pombe*, within groups of species such as the Hemiascomycetes there is extensive synteny. Synteny is an example of an annotation tool that grows more powerful as more genome sequences become available.

Sequencing of additional Hemiascomycetes genomes has thus allowed reannotation of *S. cerevisiae*, so that the current annotation available from SGD (http://www.yeastgenome.org/) is considerably

refined from the one initially released, with hundreds of corrections listed in the update table.

While genome sequencing indicated that the *S. cerevisiae* lineage underwent an ancient genome duplication resulting in hundreds of duplicate genes, comparison of fungal genomes identifies other duplicate gene patterns as well. Genomes of the Hemiascomycetes contain two to five copies of *ADH1*-related alcohol dehydrogenase genes. These genes, while all paralogues, are scattered around the genome. Comparison of the positions of these genes suggests that the common ancestor of the Hemiascomycetes had multiple copies of this gene. Other genes, such as *CUP1* of *S. cerevisiae*, are tandemly duplicated and appear to have duplicated much more recently. Even the very small genome of *A. gossypii* contains more than 20 pairs of tandemly duplicated genes. Thus tandem duplication followed by genome rearrangement to

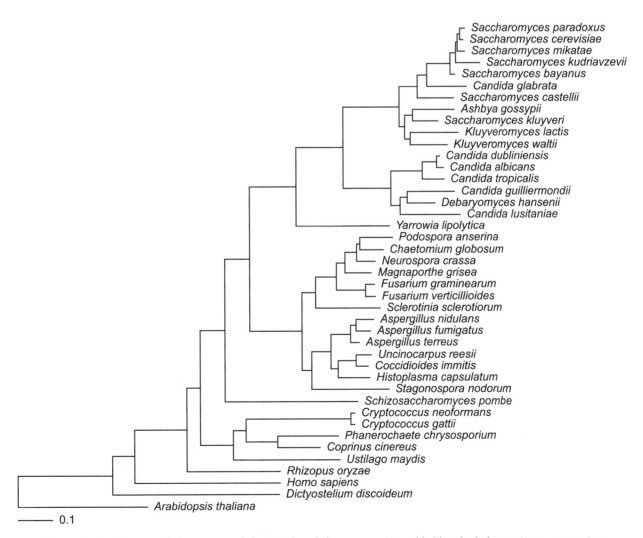

Figure 1. Species tree of the sequenced fungi inferred from a maximum-likelihood phylogenetic reconstruction algorithm on proteins from 25 orthologous genes gleaned from whole-genome comparisons.

separate the duplicate genes appears to be one mechanism by which new genes arise.

Classic genetic analysis with *S. cerevisiae* indicated that most *S. cerevisiae* mutant phenotypes show a Mendelian pattern of inheritance; hence, most genes are single-copy genes. Some genes, such as those encoding the hexose transporters, were clearly present in larger numbers and showed considerable variation between strains. Genomic sequencing has confirmed and extended these observations. Many genes are present only in single copy among all the studied fungi. Essential genes tend to be less duplicated, and duplicated genes tend to have evolved at increased rates (26). A pattern has also emerged of a relationship between a gene's "duplicability" and its connectedness within a gene network (16, 46). While many fungal genomes contain evidence that gene duplication may be an important mechanism by which new genes evolve, in some fungi there is a paucity of duplicates. For *N. crassa* this has been attributed to the repeat-induced point mutations. Repeat-induced point mutation has shaped the *N. crassa* genome by limiting both the number of transposable elements in the genome and the number of gene families with closely related paralogues. This suggests that the recent evolution of *N. crassa* may have proceeded differently from that of most other fungal species, where novel genes originate from duplicated ones. It is also clear that, unlike what is thought to have occurred in the evolution of the bacterial genome, horizontal gene transfer appears to have played at most a minor role in the evolution of the fungi. Only a very small number of genes of recent bacterial origin are found in *S. cerevisiae* or other sequenced fungal genomes. For most fungal genes, the protein phylogeny is similar to the rDNA phylogeny, suggesting inheritance of these genes from a common ancestor. A phylogenetic tree of the sequenced fungi is shown in Fig. 1.

SUMMARY

Genomics is now being brought to bear on the fungal species, providing a "parts list" of gene products, repetitive elements, and conserved elements. The availability of multiple genomes from a range of species with different evolutionary distances provides perspective on the evolution of gene order, content, and substitution rates. This exciting pot of data opens up new challenges and new opportunities in fungal research.

REFERENCES

1. Allen, J. E., M. Pertea, and S. L. Salzberg. 2004. Computational gene prediction using multiple sources of evidence. *Genome Res.* **14:**142–148.
2. Birney, E., M. Clamp, and R. Durbin. 2004. GeneWise and Genomewise. *Genome Res.* **14:**988–995.
3. Burge, C., and S. Karlin. 1997. Prediction of complete gene structures in human genomic DNA. *J. Mol. Biol.* **268:**78–94.
4. Carle, G. F., and M. V. Olson. 1984. Separation of chromosomal DNA molecules from yeast by orthogonal-field-alternation gel electrophoresis. *Nucleic Acids Res.* **12:**5647–5664.
5. Christie, K. R., S. Weng, R. Balakrishnan, M. C. Costanzo, K. Dolinski, S. S. Dwight, S. R. Engel, B. Feierbach, D. G. Fisk, J. E. Hirschman, E. L. Hong, L. Issel-Tarver, R. Nash, A. Sethuraman, B. Starr, C. L. Theesfeld, R. Andrada, G. Binkley, Q. Dong, C. Lane, M. Schroeder, D. Botstein, and J. M. Cherry. 2004. Saccharomyces Genome Database (SGD) provides tools to identify and analyze sequences from *Saccharomyces cerevisiae* and related sequences from other organisms. *Nucleic Acids Res.* **32**(Database issue):D311-D314.
6. Cliften, P., P. Sudarsanam, A. Desikan, L. Fulton, B. Fulton, J. Majors, R. Waterston, B. A. Cohen, and M. Johnston. 2003. Finding functional features in *Saccharomyces* genomes by phylogenetic footprinting. *Science* **301:**71–76.
7. Dean, R. A., N. J. Talbot, D. J. Ebbole, M. L. Farman, T. K. Mitchell, M. J. Orbach, M. Thon, R. Kulkarni, J. R. Xu, H. Pan, N. D. Read, Y. H. Lee, I. Carbone, D. Brown, Y. Y. Oh, N. Donofrio, J. S. Jeong, D. M. Soanes, S. Djonovic, E. Kolomiets, C. Rehmeyer, W. Li, M. Harding, S. Kim, M. H. Lebrun, H. Bohnert, S. Coughlan, J. Butler, S. Calvo, L. J. Ma, R. Nicol, S. Purcell, C. Nusbaum, J. E. Galagan, and B. W. Birren. 2005. The genome sequence of the rice blast fungus *Magnaporthe grisea*. *Nature* **434:**980–986.
8. Dietrich, F. S., S. Voegeli, S. Brachat, A. Lerch, K. Gates, S. Steiner, C. Mohr, R. Pohlmann, P. Luedi, S. Choi, R. A. Wing, A. Flavier, T. D. Gaffney, and P. Philippsen. (2004). The *Ashbya gossypii* genome as a tool for mapping the ancient *Saccharomyces cerevisiae* genome. *Science* **304:**304–307.
9. Dujon, B., D. Sherman, G. Fischer, P. Durrens, S. Casaregola, I. Lafontaine, J. De Montigny, C. Marck, C. Neuveglise, E. Talla, N. Goffard, L. Frangeul, M. Aigle, V. Anthouard, A. Babour, V. Barbe, S. Barnay, S. Blanchin, J. M. Beckerich, E. Beyne, C. Bleykasten, A. Boisrame, J. Boyer, L. Cattolico, F. Confanioleri, A. De Daruvar, L. Despons, E. Fabre, C. Fairhead, H. Ferry-Dumazet, A. Groppi, F. Hantraye, C. Hennequin, N. Jauniaux, P. Joyet, R. Kachouri, A. Kerrest, R. Koszul, M. Lemaire, I. Lesur, L. Ma, H. Muller, J. M. Nicaud, M. Nikolski, S. Oztas, O. Ozier-Kalogeropoulos, S. Pellenz, S. Potier, G. F. Richard, M. L. Straub, A. Suleau, D. Swennen, F. Tekaia, M. Wesolowski-Louvel, E. Westhof, B. Wirth, M. Zeniou-Meyer, I. Zivanovic, M. Bolotin-Fukuhara, A. Thierry, C. Bouchier, B. Caudron, C. Scarpelli, C. Gaillardin, J. Weissenbach, P. Wincker and J. L. Souciet. 2004. Genome evolution in yeasts. *Nature* **430:**35–44.
10. Galagan, J. E., S. E. Calvo, K. A. Borkovich, E. U. Selker, N. D. Read, D. Jaffe, W. FitzHugh, L. J. Ma, S. Smirnov, S. Purcell, B. Rehman, T. Elkins, R. Engels, S. Wang, C. B. Nielsen, J. Butler, M. Endrizzi, D. Qui, P. Ianakiev, D. Bell-Pedersen, M. A. Nelson, M. Werner-Washburne, C. P. Selitrennikoff, J. A. Kinsey, E. L. Braun, A. Zelter, U. Schulte, G. O. Kothe, G. Jedd, W. Mewes, C. Staben, E. Marcotte, D. Greenberg, A. Roy, K. Foley, J. Naylor, N. Stange-Thomann, R. Barrett, S. Gnerre, M. Kamal, M. Kamvysselis, E. Mauceli, C. Bielke, S. Rudd, D. Frishman, S. Krystofova, C. Rasmussen, R. L. Metzenberg, D. D. Perkins, S. Kroken, C. Cogoni, G. Macino, D. Catcheside, W. Li, R. J. Pratt,

S. A. Osmani, C. P. DeSouza, L. Glass, M. J. Orbach, J. A. Berglund, R. Voelker, O. Yarden, M. Plamann, S. Seiler, J. Dunlap, A. Radford, R. Aramayo, D. O. Natvig, L. A. Alex, G. Mannhaupt, D. J. Ebbole, M. Freitag, I. Paulsen, M. S. Sachs, E. S. Lander, C. Nusbaum, and B. Birren. 2003. The genome sequence of the filamentous fungus *Neurospora crassa. Nature* 422:859–868.

11. Gavin, A. C., M. Bosche, R. Krause, P. Grandi, M. Marzioch, A. Bauer, J. Schultz, J. M. Rick, A. M. Michon, C. M. Cruciat, M. Remor, C. Hofert, M. Schelder, M. Brajenovic, H. Ruffner, A. Merino, K. Klein, M. Hudak, D. Dickson, T. Rudi, V. Gnau, A. Bauch, S. Bastuck, B. Huhse, C. Leutwein, M. A. Heurtier, R. R. Copley, A. Edelmann, E. Querfurth, V. Rybin, G. Drewes, M. Raida, T. Bouwmeester, P. Bork, B. Seraphin, B. Kuster, G. Neubauer and G. Superti-Furga. 2002. Functional organization of the yeast proteome by systematic analysis of protein complexes. *Nature* 415:141–147.

12. Gelfand, M. S., A. A. Mironov, and P. A. Pevzner. 1996. Gene recognition via spliced sequence alignment. *Proc. Natl. Acad. Sci. USA* 93:9061–9066.

13. Giaever, G., A. M. Chu, L. Ni, C. Connelly, L. Riles, S. Veronneau, S. Dow, A. Lucau-Danila, K. Anderson, B. Andre, A. P. Arkin, A. Astromoff, M. El-Bakkoury, R. Bangham, R. Benito, S. Brachat, S. Campanaro, M. Curtiss, K. Davis, A. Deutschbauer, K. D. Entian, P. Flaherty, F. Foury, D. J. Garfinkel, M. Gerstein, D. Gotte, U. Guldener, J. H. Hegemann, S. Hempel, Z. Herman, D. F. Jaramillo, D. E. Kelly, S. L. Kelly, P. Kotter, D. LaBonte, D. C. Lamb, N. Lan, H. Liang, H. Liao, L. Liu, C. Luo, M. Lussier, R. Mao, P. Menard, S. L. Ooi, J. L. Revuelta, C. J. Roberts, M. Rose, P. Ross-Macdonald, B. Scherens, G. Schimmack, B. Shafer, D. D. Shoemaker, S. Sookhai-Mahadeo, R. K. Storms, J. N. Strathern, G. Valle, M. Voet, G. Volckaert, C. Y. Wang, T. R. Ward, J. Wilhelmy, E. A. Winzeler, Y. Yang, G. Yen, E. Youngman, K. Yu, H. Bussey, J. D. Boeke, M. Snyder, P. Philippsen, R. W. Davis, and M. Johnston. 2002. Functional profiling of the *Saccharomyces cerevisiae* genome. *Nature* 418:387–391.

14. Goffeau, A., B. G. Barrell, H. Bussey, R. W. Davis, B. Dujon, H. Feldmann, F. Galibert, J. D. Hoheisel, C. Jacq, M. Johnston, E. J. Louis, H. W. Mewes, Y. Murakami, P. Philippsen, H. Tettelin, and S. G. Oliver. 1996. Life with 6000 genes. *Science* 274:546, 563–567.

15. Guigo, R., P. Agarwal, J. F. Abril, M. Burset, and J. W. Fickett. 2000. An assessment of gene prediction accuracy in large DNA sequences. *Genome Res.* 10:1631–1642.

16. Hahn, M. W., and A. D. Kern. 2005. Comparative genomics of centrality and essentiality in three eukaryotic protein-interaction networks. *Mol. Biol. Evol.* 22:803–806.

17. Ho, Y., A. Gruhler, A. Heilbut, G. D. Bader, L. Moore, S. L. Adams, A. Millar, P. Taylor, K. Bennett, K. Boutilier, L. Yang, C. Wolting, I. Donaldson, S. Schandorff, J. Shewnarane, M. Vo, J. Taggart, M. Goudreault, B. Muskat, C. Alfarano, D. Dewar, Z. Lin, K. Michalickova, A. R. Willems, H. Sassi, P. A. Nielsen, K. J. Rasmussen, J. R. Andersen, L. E. Johansen, L. H. Hansen, H. Jespersen, A. Podtelejnikov, E. Nielsen, J. Crawford, V. Poulsen, B. D. Sorensen, J. Matthiesen, R. C. Hendrickson, F. Gleeson, T. Pawson, M. F. Moran, D. Durocher, M. Mann, C. W. Hogue, D. Figeys, and M. Tyers. 2002. Systematic identification of protein complexes in *Saccharomyces cerevisiae* by mass spectrometry. *Nature* 415:180–183.

18. Howe, K. L., T. Chothia, and R. Durbin. 2002. GAZE: a generic framework for the integration of gene-prediction data by dynamic programming. *Genome Res.* 12: 1418–1427.

19. Huh, W. K., J. V. Falvo, L. C. Gerke, A. S. Carroll, R. W. Howson, J. S. Weissman, and E. K. O'Shea. 2003. Global analysis of protein localization in budding yeast. *Nature* 425:686–691.

20. Jones, T., N. A. Federspiel, H. Chibana, J. Dungan, S. Kalman, B. B. Magee, G. Newport, Y. R. Thorstenson, N. Agabian, P. T. Magee, R. W. Davis, and S. Scherer. 2004. The diploid genome sequence of *Candida albicans. Proc. Natl. Acad. Sci. USA* 101:7329–7334.

21. Katinka, M. D., S. Duprat, E. Cornillot, G. Metenier, F. Thomarat, G. Prensier, V. Barbe, E. Peyretaillade, P. Brottier, P. Wincker, F. Delbac, H. El Alaoui, P. Peyret, W. Saurin, M. Gouy, J. Weissenbach, and C. P. Vivares. 2001. Genome sequence and gene compaction of the eukaryote parasite *Encephalitozoon cuniculi. Nature* 414:450–453.

22. Kellis, M., B. W. Birren, and E. S. Lander. 2004. Proof and evolutionary analysis of ancient genome duplication in the yeast *Saccharomyces cerevisiae. Nature* 428:617–624.

23. Kellis, M., N. Patterson, M. Endrizzi, B. Birren, and E. S. Lander. 2003. Sequencing and comparison of yeast species to identify genes and regulatory elements. *Nature* 423:241–254.

24. Korf, I. 2004. Gene finding in novel genomes. *BMC Bioinformatics* 5:59.

25. Korf, I., P. Flicek, D. Duan, and M. R. Brent. 2001. Integrating genomic homology into gene structure prediction. *Bioinformatics* 17(Suppl. 1):S140–S148.

26. Langkjaer, R. B., P. F. Cliften, M. Johnston, and J. Piskur. 2003. Yeast genome duplication was followed by asynchronous differentiation of duplicated genes. *Nature* 421:848–852.

27. Link, A. J., and M. V. Olson. 1991. Physical map of the *Saccharomyces cerevisiae* genome at 110-kilobase resolution. *Genetics* 127:681–698.

28. Loftus, B. J., E. Fung, P. Roncaglia, D. Rowley, P. Amedeo, D. Bruno, J. Vamathevan, M. Miranda, I. J. Anderson, J. A. Fraser, J. E. Allen, I. E. Bosdet, M. R. Brent, R. Chiu, T. L. Doering, M. J. Donlin, C. A. D'Souza, D. S. Fox, V. Grinberg, J. Fu, M. Fukushima, B. J. Haas, J. C. Huang, G. Janbon, S. J. Jones, H. L. Koo, M. I. Krzywinski, J. K. Kwon-Chung, K. B. Lengeler, R. Maiti, M. A. Marra, R. E. Marra, C. A. Mathewson, T. G. Mitchell, M. Pertea, F. R. Riggs, S. L. Salzberg, J. E. Schein, A. Shvartsbeyn, H. Shin, M. Shumway, C. A. Specht, B. B. Suh, A. Tenney, T. R. Utterback, B. L. Wickes, J. R. Wortman, N. H. Wye, J. W. Kronstad, J. K. Lodge, J. Heitman, R. W. Davis, C. M. Fraser, and R. W. Hyman. 2005. The genome of the basidiomycetous yeast and human pathogen *Cryptococcus neoformans. Science* 307:1321–1324.

29. Lowe, T. M., and S. R. Eddy. 1997. tRNAscan-SE: a program for improved detection of transfer RNA genes in genomic sequence. *Nucleic Acids Res.* 25:955–964.

30. Majoros, W. H., M. Pertea, and S. L. Salzberg. 2004. TigrScan and GlimmerHMM: two open source ab initio eukaryotic gene-finders. *Bioinformatics* 20: 2878–2879.

31. Martinez, D., L. F. Larrondo, N. Putnam, M. D. Gelpke, K. Huang, J. Chapman, K. G. Helfenbein, P. Ramaiya, J. C. Detter, F. Larimer, P. M. Coutinho, B. Henrissat, R. Berka, D. Cullen, and D. Rokhsar. 2004. Genome sequence of the lignocellulose degrading fungus *Phanerochaete chrysosporium* strain RP78. *Nat. Biotechnol.* 22:695–700.

32. **Meyer, I. M., and R. Durbin.** 2002. Comparative ab initio prediction of gene structures using pair HMMs. *Bioinformatics* **18:**1309–1318.

33. **Mortimer, R. K., D. Schild, C. R. Contopoulou, and J. A. Kans.** 1989. Genetic map of *Saccharomyces cerevisiae,* edition 10. *Yeast* **5:**321–403.

34. **Olson, M. V., J. E. Dutchik, M. Y. Graham, G. M. Brodeur, C. Helms, M. Frank, M. MacCollin, R. Scheinman, and T. Frank.** 1986. Random-clone strategy for genomic restriction mapping in yeast. *Proc. Natl. Acad. Sci. USA* **83:**7826–7830.

35. **Perrot, M., F. Sagliocco, T. Mini, C. Monribot, U. Schneider, A. Shevchenko, M. Mann, P. Jeno, and H. Boucherie.** 1999. Two-dimensional gel protein database of *Saccharomyces cerevisiae* (update 1999). *Electrophoresis* **20:**2280–2298.

36. **Quevillon-Cheruel, S., D. Liger, N. Leulliot, M. Graille, A. Poupon, I. L. de La Sierra-Gallay, C. Z. Zhou, B. Collinet, J. Janin, and H. Van Tilbeurgh.** 2004. The Paris-Sud yeast structural genomics pilot-project: from structure to function. *Biochimie* **86:**617–623.

37. **Raamsdonk, L. M., B. Teusink, D. Broadhurst, N. Zhang, A. Hayes, M. C. Walsh, J. A. Berden, K. M. Brindle, D. B. Kell, J. J. Rowland, H. V. Westerhoff, K. van Dam, and S. G. Oliver.** 2001. A functional genomics strategy that uses metabolome data to reveal the phenotype of silent mutations. *Nat. Biotechnol.* **19:**45–50.

38. **Riles, L., J. E. Dutchik, A. Baktha, B. K. McCauley, E. C. Thayer, M. P. Leckie, V. V. Braden, J. E. Depke, and M. V. Olson.** 1993. Physical maps of the six smallest chromosomes of *Saccharomyces cerevisiae* at a resolution of 2.6 kilobase pairs. *Genetics* **134:**81–150.

39. **Rubin, G. M., M. D. Yandell, J. R. Wortman, G. L. Gabor Miklos, C. R. Nelson, I. K. Hariharan, M. E. Fortini, P. W. Li, R. Apweiler, W. Fleischmann, J. M. Cherry, S. Henikoff, M. P. Skupski, S. Misra, M. Ashburner, E. Birney, M. S. Boguski, T. Brody, P. Brokstein, S. E. Celniker, S. A. Chervitz, D. Coates, A. Cravchik, A. Gabrielian, R. F. Galle, W. M. Gelbart, R. A. George, L. S. Goldstein, F. Gong, P. Guan, N. L. Harris, B. A. Hay, R. A. Hoskins, J. Li, Z. Li, R. O. Hynes, S. J. Jones, P. M. Kuehl, B. Lemaitre, J. T. Littleton, D. K. Morrison, C. Mungall, P. H. O'Farrell, O. K. Pickeral, C. Shue, L. B. Vosshall, J. Zhang, Q. Zhao, X. H. Zheng, and S. Lewis.** 2000. Comparative genomics of the eukaryotes. *Science* **287:**2204–2215.

40. **Slater, G. S., and E. Birney.** 2005. Automated generation of heuristics for biological sequence comparison. *BMC Bioinformatics* **6:**31.

41. **Spellman, P. T., G. Sherlock, M. Q. Zhang, V. R. Iyer, K. Anders, M. B. Eisen, P. O. Brown, D. Botstein, and B. Futcher.** 1998. Comprehensive identification of cell cycle-regulated genes of the yeast *Saccharomyces cerevisiae* by microarray hybridization. *Mol. Biol. Cell* **9:**3273–3297.

42. **Tenney, A. E., R. H. Brown, C. Vaske, J. K. Lodge, T. L. Doering, and M. R. Brent.** 2004. Gene prediction and verification in a compact genome with numerous small introns. *Genome Res.* **14:**2330–2335.

43. **Uetz, P., L. Giot, G. Cagney, T. A. Mansfield, R. S. Judson, J. R. Knight, D. Lockshon, V. Narayan, M. Srinivasan, P. Pochart, A. Qureshi-Emili, Y. Li, B. Godwin, D. Conover, T. Kalbfleisch, G. Vijayadamodar, M. Yang, M. Johnston, S. Fields, and J. M. Rothberg.** 2000. A comprehensive analysis of protein-protein interactions in *Saccharomyces cerevisiae. Nature* **403:**623–627.

44. **Wolfe, K.** 2004. Evolutionary genomics: yeasts accelerate beyond BLAST. *Curr. Biol.* **14:**R392–R394.

45. **Wood, V., R. Gwilliam, M. A. Rajandream, M. Lyne, R. Lyne, A. Stewart, J. Sgouros, N. Peat, J. Hayles, S. Baker, D. Basham, S. Bowman, K. Brooks, D. Brown, S. Brown, T. Chillingworth, C. Churcher, M. Collins, R. Connor, A. Cronin, P. Davis, T. Feltwell, A. Fraser, S. Gentles, A. Goble, N. Hamlin, D. Harris, J. Hidalgo, G. Hodgson, S. Holroyd, T. Hornsby, S. Howarth, E. J. Huckle, S. Hunt, K. Jagels, K. James, L. Jones, M. Jones, S. Leather, S. McDonald, J. McLean, P. Mooney, S. Moule, K. Mungall, L. Murphy, D. Niblett, C. Odell, K. Oliver, S. O'Neil, D. Pearson, M. A. Quail, E. Rabbinowitsch, K. Rutherford, S. Rutter, K. Saunders, K. Seeger, S. Sharp, J. Skelton, M. Simmonds, R. Squares, S. Squares, K. Stevens, K. Taylor, R. G. Taylor, A. Tivey, S. Walsh, T. Warren, S. Whitehead, J. Woodward, G. Volckaert, R. Aert, J. Robben, B. Grymonprez, I. Weltjens, E. Vanstreels, M. Rieger, M. Schafer, S. Muller-Auer, C. Gabel, M. Fuchs, A. Dusterhoft, C. Fritzc, E. Holzer, D. Moestl, H. Hilbert, K. Borzym, I. Langer, A. Beck, H. Lehrach, R. Reinhardt, T. M. Pohl, P. Eger, W. Zimmermann, H. Wedler, R. Wambutt, B. Purnelle, A. Goffeau, E. Cadieu, S. Dreano, S. Gloux, V. Lelaure, S. Mottier, F. Galibert, S. J. Aves, Z. Xiang, C. Hunt, K. Moore, S. M. Hurst, M. Lucas, M. Rochet, C. Gaillardin, V. A. Tallada, A. Garzon, G. Thode, R. R. Daga, L. Cruzado, J. Jimenez, M. Sanchez, F. del Rey, J. Benito, A. Dominguez, J. L. Revuelta, S. Moreno, J. Armstrong, S. L. Forsburg, L. Cerutti, T. Lowe, W. R. McCombie, I. Paulsen, J. Potashkin, G. V. Shpakovski, D. Ussery, B. G. Barrell, and P. Nurse.** 2002. The genome sequence of *Schizosaccharomyces pombe. Nature* **415:**871–880.

46. **Yang, J., R. Lusk, and W. H. Li.** 2003. Organismal complexity, protein complexity, and gene duplicability. *Proc. Natl. Acad. Sci. USA* **100:**15661–15665.

INDEX